Table of Problem-Solving Strategies

Note for users of the five-volume edition:
Volume 1 (pp. 1–443) includes chapters 1–15.
Volume 2 (pp. 444–559) includes chapters 16–19.
Volume 3 (pp. 560–719) includes chapters 20–24.
Volume 4 (pp. 720–1101) includes chapters 25–36.
Volume 5 (pp. 1102–1279) includes chapters 36–42.

Chapters 37–42 are not in the Standard Edition.

Brief Contents

PEARSON ALWAYS LEARNING

Randall D. Knight

Physics for Scientists and Engineers

A Strategic Approach with Modern Physics

Physics 211
Third Custom Edition for Pennsylvania State University

Taken from:
Physics for Scientists and Engineers:
A Strategic Approach with Modern Physics, Third Edition
by Randall D. Knight

Cover Art: Courtesy of Pearson Learning Solutions.

Taken from:

Physics for Scientists and Engineers: A Strategic Approach with Modern Physics, Third Edition
by Randall D. Knight

This special edition published in cooperation with Pearson Learning Solutions.

Pearson Learning Solutions, 501 Boylston Street, Suite 900, Boston, MA 02116
A Pearson Education Company
www.pearsoned.com

Printed in the United States of America

5 6 7 8 9 10 VOUD 19 18 17 16 15

000200010271898163

MS

PEARSON

ISBN 10: 1-269-91935-0
ISBN 13: 978-1-269-91935-7

About the Author

Randy Knight has taught introductory physics for over 30 years at Ohio State University and California Polytechnic University, where he is currently Professor of Physics. Professor Knight received a bachelor's degree in physics from Washington University in St. Louis and a Ph.D. in physics from the University of California, Berkeley. He was a post-doctoral fellow at the Harvard-Smithsonian Center for Astrophysics before joining the faculty at Ohio State University. It was at Ohio State that he began to learn about the research in physics education that, many years later, led to this book.

Professor Knight's research interests are in the field of lasers and spectroscopy, and he has published over 25 research papers. He also directs the environmental studies program at Cal Poly, where, in addition to introductory physics, he teaches classes on energy, oceanography, and environmental issues. When he's not in the classroom or in front of a computer, you can find Randy hiking, sea kayaking, playing the piano, or spending time with his wife Sally and their seven cats.

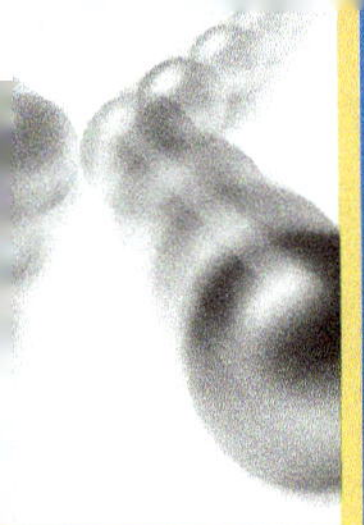

Builds problem-solving skills and confidence…

… through a carefully structured and research-proven program of problem-solving techniques and practice materials.

At the heart of the problem-solving instruction is the consistent 4-step MODEL/ VISUALIZE/ SOLVE/ ASSESS approach, used throughout the book and all supplements. ***Problem-Solving Strategies*** provide detailed guidance for particular topics and categories of problems, often drawing on key skills outlined in the step-by-step procedures of ***Tactics Boxes.*** Problem-Solving Strategies and Tactics Boxes are also illustrated in dedicated MasteringPhysics ***Skill-Builder Tutorials***.

PROBLEM-SOLVING STRATEGY 10.1 Conservation of mechanical energy

1 MODEL Choose a system that is isolated and has no friction or other losses of mechanical energy.

2 VISUALIZE Draw a before-and-after pictorial representation. Define symbols, list known values, and identify what you're trying to find.

3 SOLVE The mathematical representation is based on the law of conservation of mechanical energy:

$$K_f + U_f = K_i + U_i$$

4 ASSESS Check that your result has the correct units, is reasonable, and answers the question.

Exercise 8

TACTICS BOX 9.1 Drawing a before-and-after pictorial representation

EXAMPLE 4.15 Analyzing rotational data

You've been assigned the task of measuring the start-up characteristics of a large industrial motor. After several seconds, when the motor has reached full speed, you know that the angular acceleration will be zero, but you hypothesize that the angular acceleration may be constant during the first couple of seconds as the motor speed increases. To find out, you attach a shaft encoder to the 3.0-cm-diameter axle. A shaft encoder is a device that converts the angular position of a shaft or axle to a signal that can be read by a computer. After setting the computer program to read four values a second, you start the motor and acquire the following data:

Time (s)	Angle(°)
0.00	0
0.25	16
0.50	69
0.75	161
1.00	267
1.25	428
1.50	620

a. Do the data support your hypothesis of a constant angular acceleration? If so, what is the angular acceleration? If not, is the angular acceleration increasing or decreasing with time?

b. A 76-cm-diameter blade is attached to the motor shaft. At what time does the acceleration of the tip of the blade reach 10 m/s^2?

1 MODEL The axle is rotating with nonuniform circular motion. Model the tip of the blade as a particle.

2 VISUALIZE FIGURE 4.38 shows that the blade tip has both a tangential and a radial acceleration.

$\alpha = 2m$. If the graph is not a straight line, our observation of whether it curves upward or downward will tell us whether the angular acceleration us increasing or decreasing.

FIGURE 4.39 is the graph of θ versus t^2, and it confirms our hypothesis that the motor starts up with constant angular acceleration. The best-fit line, found using a spreadsheet, gives a slope of 274.6°/s^2. The units come not from the spreadsheet but by looking at the units of rise (°) over run (s^2 because we're graphing t^2 on the x-axis). Thus the angular acceleration is

$$\alpha = 2m = 549.2°/s^2 \times \frac{\pi \text{ rad}}{180°} = 9.6 \text{ rad/s}^2$$

where we used $180° = \pi$ rad to convert to SI units of rad/s^2.

FIGURE 4.39 Graph of θ versus t^2 for the motor shaft.

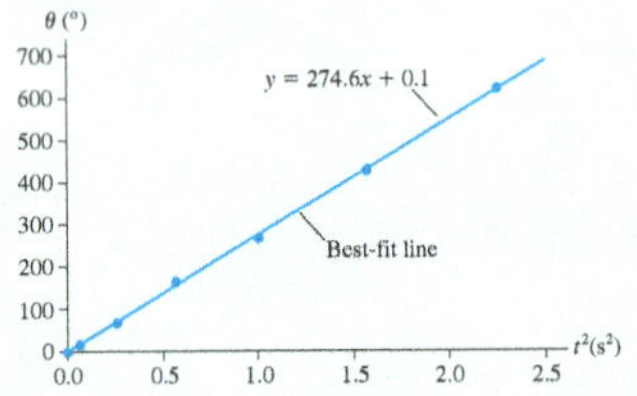

b. The magnitude of the linear acceleration is

$$a = \sqrt{a_r^2 + a_t^2}$$

Worked Examples walk the student carefully through detailed solutions, focusing on underlying reasoning and common pitfalls to avoid.

NEW! ***Data-based Examples*** (shown here) help students with the skill of drawing conclusions from laboratory data.

CHALLENGE EXAMPLE 10.10 A rebounding pendulum

A 200 g steel ball hangs on a 1.0-m-long string. The ball is pulled sideways so that the string is at a 45° angle, then released. At the very bottom of its swing the ball strikes a 500 g steel paperweight that is resting on a frictionless table. To what angle does the ball rebound?

NEW! ***Challenge Examples*** illustrate how to integrate multiple concepts and use more sophisticated reasoning.

NEW! The Mastering Study Area also has ***Video Tutor Solutions***, created by Randy Knight's College Physics co-author Brian Jones. These engaging and helpful videos walk students through a representative problem for each main topic, often starting with a qualitative overview in the context of a lab- or real-world demo.

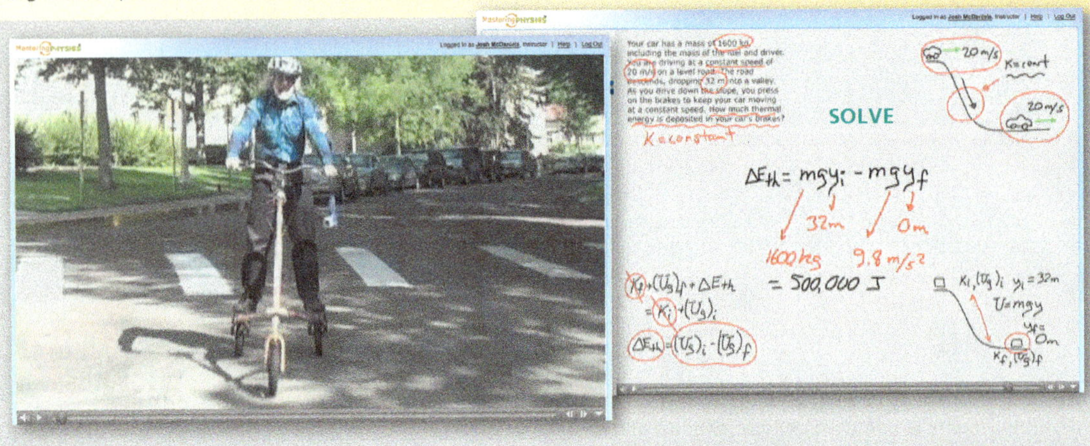

Promotes deeper understanding…

… using powerful techniques from multimedia learning theory that focus and structure student learning, and improve engagement and retention.

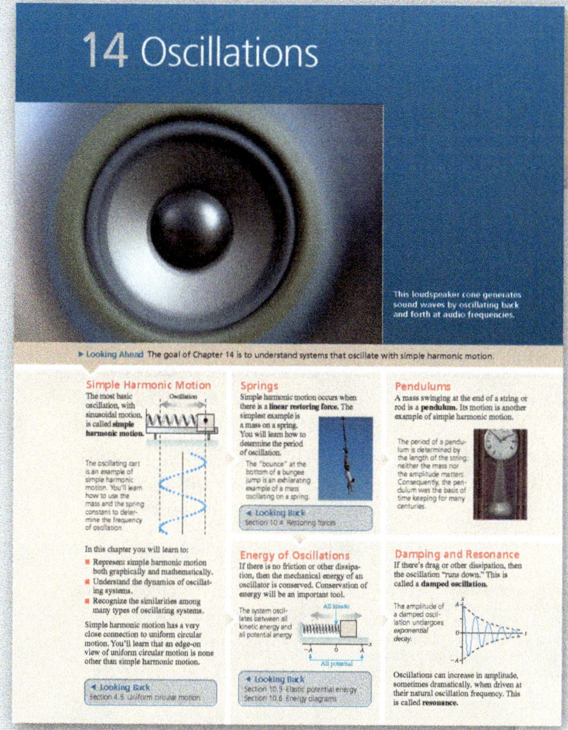

NEW! ***Illustrated Chapter Previews*** give an overview of the upcoming ideas for each chapter, setting them in context, explaining their utility, and tying them to existing knowledge (through ***Looking Back*** references).

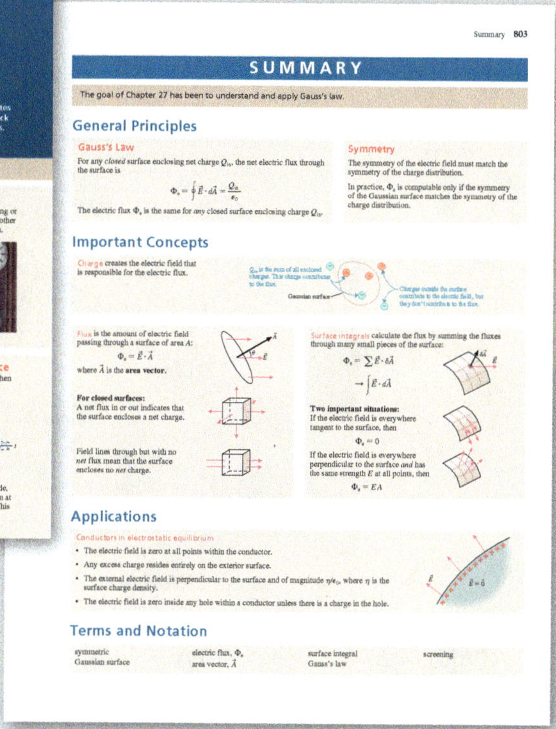

Critically acclaimed ***Visual Chapter Summaries*** and ***Part Knowledge Structures*** consolidate understanding by providing key concepts and principles in words, math, and figures and organizing these into a hierarchy.

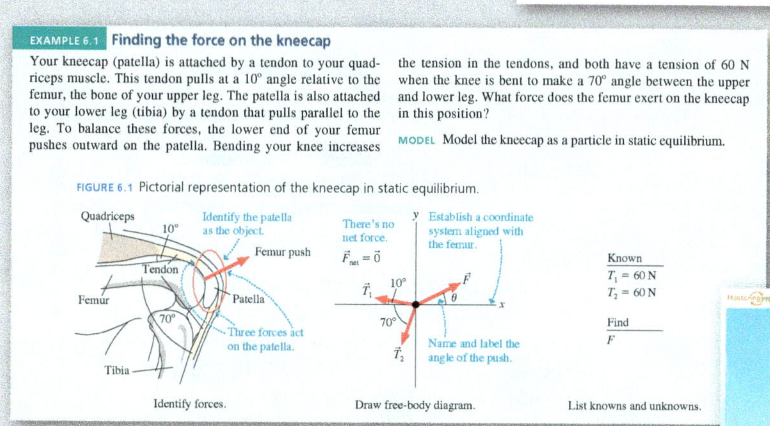

NEW! ***Life-science and bioengineering examples*** provide general interest, and specific context for biosciences students.

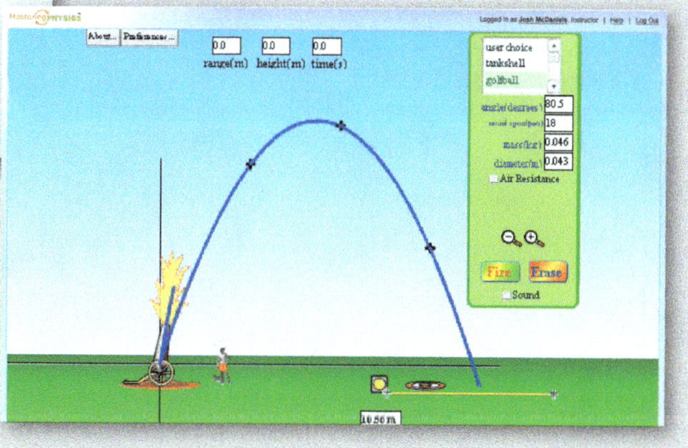

NEW! ***PhET Simulations and Tutorials*** allow students to explore real-life phenomena and discover the underlying physics. Sixteen tutorials are provided in the MasteringPhysics item library, and 76 PhET simulations are available in the Study Area and Pearson eText, along with the comprehensive library of ActivPhysics applets and applet-based tutorials.

NEW! ***Video Tutor Demonstrations*** feature "pause-and-predict" demonstrations of key physics concepts and incorporate assessment as the student progresses to actively engage them in understanding the key conceptual ideas underlying the physics principles.

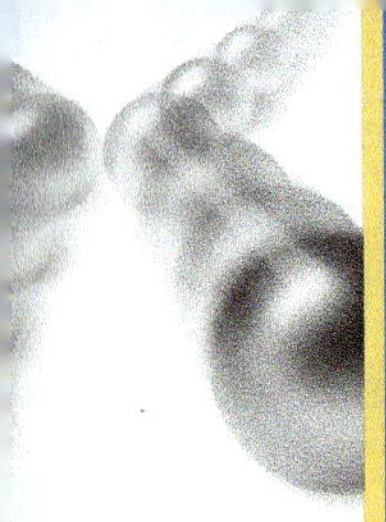

Provides research-enhanced problems...

... extensively class-tested and calibrated using MasteringPhysics data.

Data captured by MasteringPhysics® has been thoroughly analyzed by the author to ensure an optimal range of difficulty (indicated in the textbook using a three-bar rating), problem types, and topic coverage are being met.

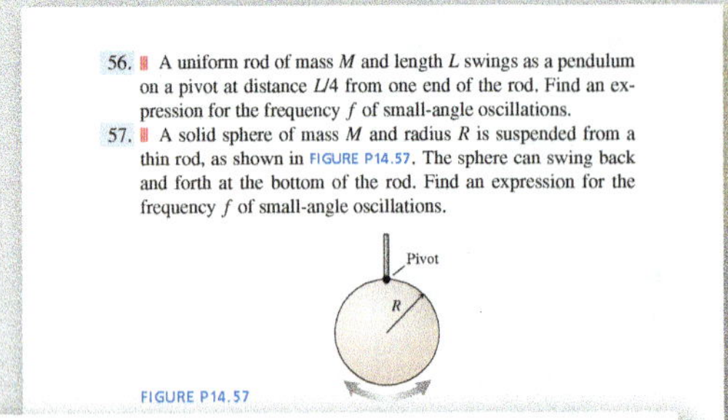

56. A uniform rod of mass M and length L swings as a pendulum on a pivot at distance $L/4$ from one end of the rod. Find an expression for the frequency f of small-angle oscillations.

57. A solid sphere of mass M and radius R is suspended from a thin rod, as shown in FIGURE P14.57. The sphere can swing back and forth at the bottom of the rod. Find an expression for the frequency f of small-angle oscillations.

FIGURE P14.57

An ***increased emphasis on symbolic answers*** encourages students to work algebraically.

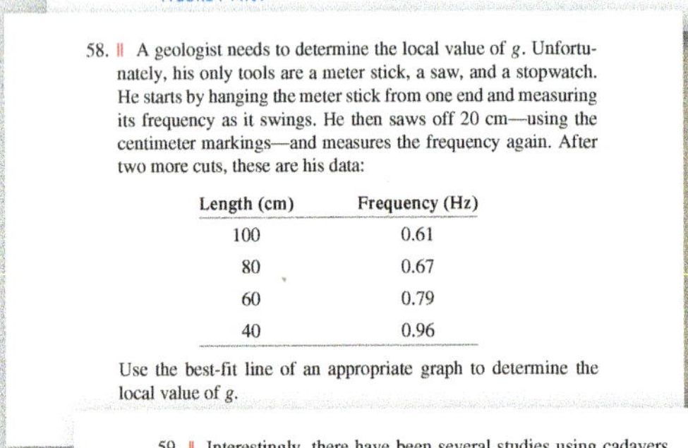

58. A geologist needs to determine the local value of g. Unfortunately, his only tools are a meter stick, a saw, and a stopwatch. He starts by hanging the meter stick from one end and measuring its frequency as it swings. He then saws off 20 cm—using the centimeter markings—and measures the frequency again. After two more cuts, these are his data:

Length (cm)	Frequency (Hz)
100	0.61
80	0.67
60	0.79
40	0.96

Use the best-fit line of an appropriate graph to determine the local value of g.

NEW! ***Data-based end-of-chapter problems*** allow students to practice drawing conclusions from data (as demonstrated in the new data-based examples in the text).

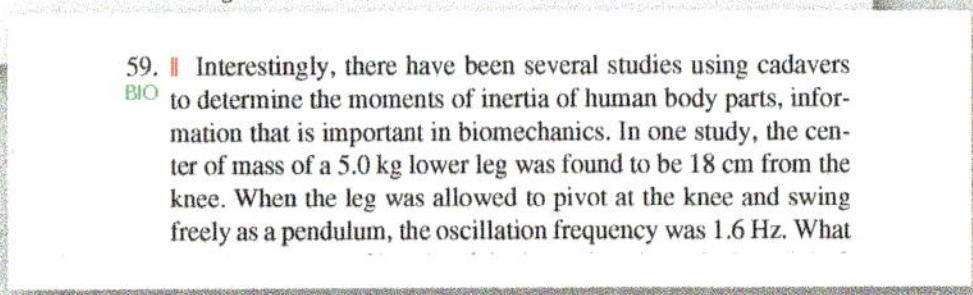

59. BIO Interestingly, there have been several studies using cadavers to determine the moments of inertia of human body parts, information that is important in biomechanics. In one study, the center of mass of a 5.0 kg lower leg was found to be 18 cm from the knee. When the leg was allowed to pivot at the knee and swing freely as a pendulum, the oscillation frequency was 1.6 Hz. What

NEW! ***BIO problems*** are set in life-science, bioengineering, or biomedical contexts.

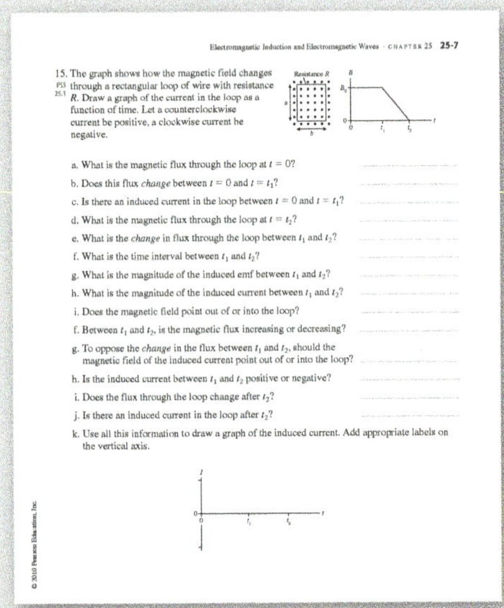

NEW! ***Student Workbook exercises*** help students work through a full solution symbolically, structured around the relevant textbook Problem-Solving Strategy.

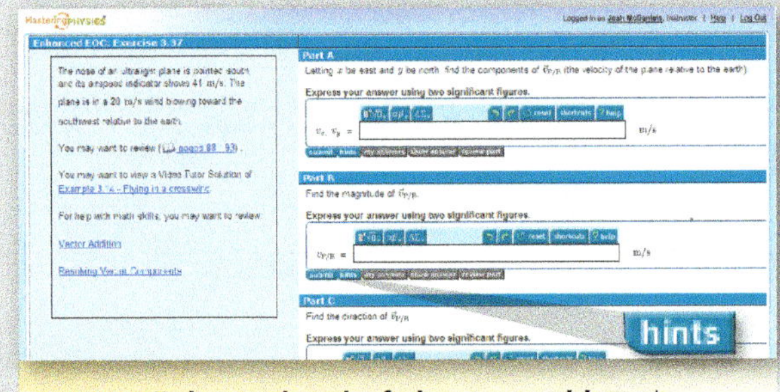

NEW! ***Enhanced end-of-chapter problems*** in MasteringPhysics now offer additional support such as problem-solving strategy hints, relevant math review and practice, links to the eText, and links to the related ***Video Tutor Solution.***

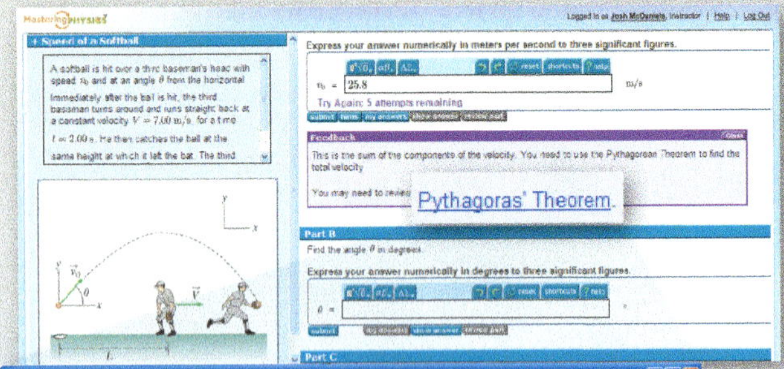

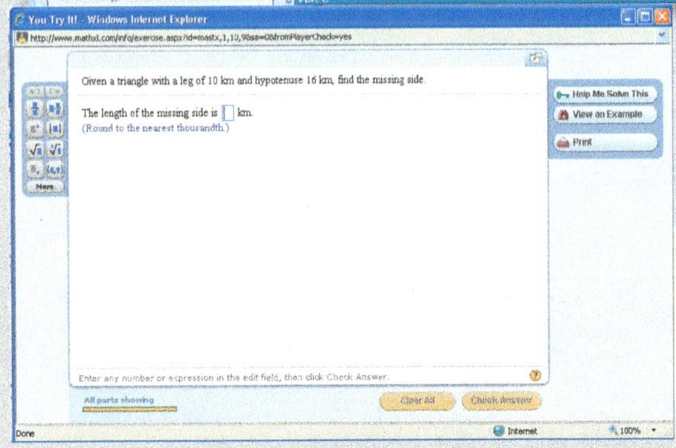

NEW! ***Math Remediation*** found within selected tutorials provide just-in-time math help and allow students to brush up on the most important mathematical concepts needed to successfully complete assignments. This new feature links students directly to math review and practice helping students make the connection between math and physics.

Make a difference with MasteringPhysics...

... the most effective and widely used online science tutorial, homework, and assessment system available.

MasteringPHYSICS www.masteringphysics.com

Pre-Built Assignments. For every chapter in the book, MasteringPhysics provides pre-built assignments that cover the material with a tested mix of tutorials and end-of-chapter problems of graded difficulty. Professors may use these assignments as-is or take them as a starting point for modification.

NEW! ***Quizzing and Testing Enhancements.***
These include options to:

- Hide item titles.
- Add password protection.
- Limit access to completed assignments.
- Randomize question order in an assignment.

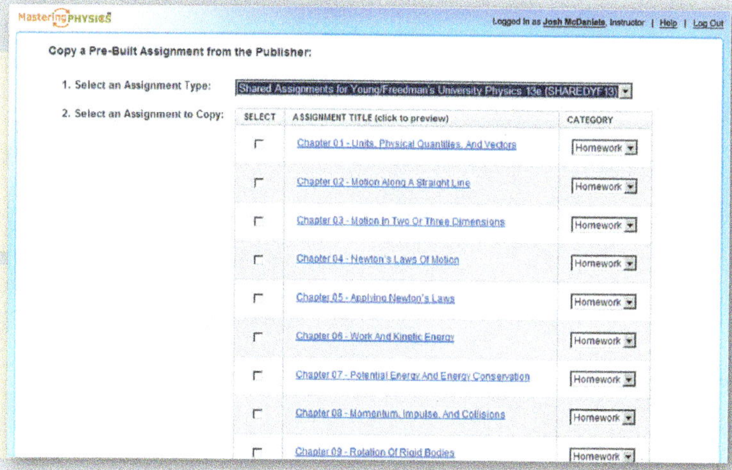

Gradebook

- Every assignment is graded automatically.
- Shades of red highlight vulnerable students and challenging assignments.
- The ***Gradebook Diagnostics*** screen provides your favorite weekly diagnostics, summarizing grade distribution, improvement in scores over the course, and much more.

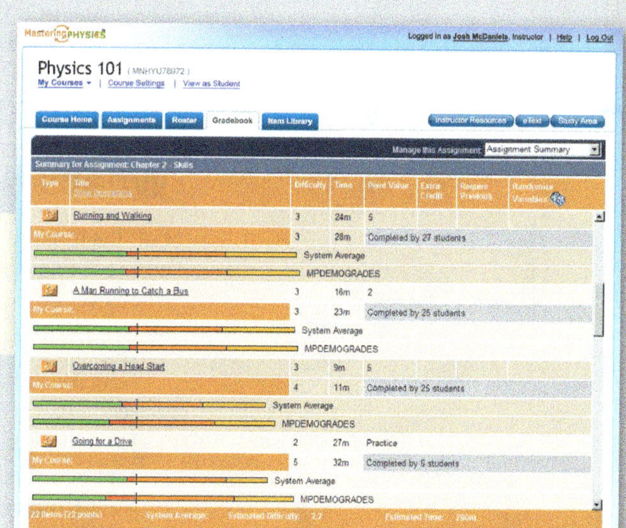

Class Performance on Assignment. Click on a problem to see which step your students struggled with most, and even their most common wrong answers. Compare results at every stage with the national average or with your previous class.

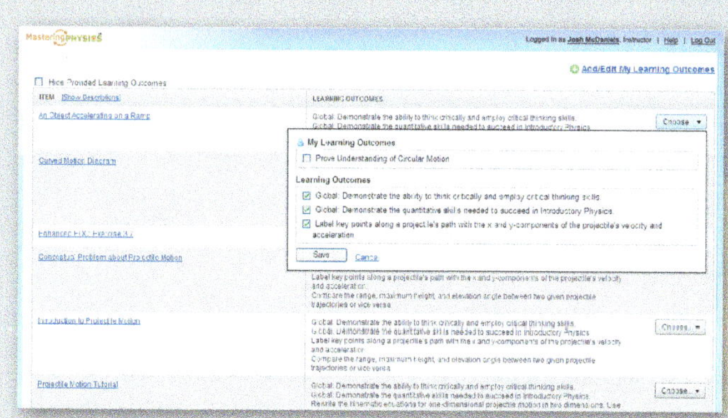

NEW! ***Learning Outcomes.*** In addition to being able to create your own learning outcomes to associate with questions in an assignment, you can now select content that is tagged to a large number of publisher-provided learning outcomes. You can also print or export student results based on learning outcomes for your own use or to incorporate into reports for your administration.

Preface to the Instructor

In 2003 we published *Physics for Scientists and Engineers: A Strategic Approach*. This was the first comprehensive introductory textbook built from the ground up on research into how students can more effectively learn physics. The development and testing that led to this book had been partially funded by the National Science Foundation. This first edition quickly became the most widely adopted new physics textbook in more than 30 years, meeting widespread critical acclaim from professors and students. For the second edition, and now the third, we have built on the research-proven instructional techniques introduced in the first edition and the extensive feedback from thousands of users to take student learning even further.

Objectives

My primary goals in writing *Physics for Scientists and Engineers: A Strategic Approach* have been:

- To produce a textbook that is more focused and coherent, less encyclopedic.
- To move key results from physics education research into the classroom in a way that allows instructors to use a range of teaching styles.
- To provide a balance of quantitative reasoning and conceptual understanding, with special attention to concepts known to cause student difficulties.
- To develop students' problem-solving skills in a systematic manner.
- To support an active-learning environment.

These goals and the rationale behind them are discussed at length in the *Instructor Guide* and in my small paperback book, *Five Easy Lessons: Strategies for Successful Physics Teaching*. Please request a copy from your local Pearson sales representative if it is of interest to you (ISBN 978-0-8053-8702-5).

What's New to This Edition

For this third edition, we continue to apply the best results from educational research, and to refine and tailor them for this course and its students. At the same time, the extensive feedback we've received has led to many changes and improvements to the text, the figures, and the end-of-chapter problems. These include:

- New illustrated **Chapter Previews** give a visual overview of the upcoming ideas, set them in context, explain their utility, and tie them to existing knowledge (through **Looking Back** references). These previews build on the cognitive psychology concept of an "advance organizer."
- New **Challenge Examples** illustrate how to integrate multiple concepts and use more sophisticated reasoning in problem-solving, ensuring an optimal range of worked examples for students to study in preparation for homework problems.
- New **Data-based Examples** help students with the skill of drawing conclusions from laboratory data. Designed to supplement lab-based instruction, these examples also help students in general with mathematical reasoning, graphical interpretation, and assessment of results.

End-of-chapter problem enhancements include the following:

- **Data from Mastering Physics® have been thoroughly analyzed** to ensure an optimal range of difficulty, problem types, and topic coverage. In addition, the wording of every problem has been reviewed for clarity. Roughly 20% of the end-of-chapter problems are new or significantly revised.

- **Data-based problems** allow students to practice drawing conclusions from data (as demonstrated in the new data-based examples in the text).
- **An increased emphasis on symbolic answers** encourages students to work algebraically. The *Student Workbook* also contains new exercises to help students work through symbolic solutions.
- **Bio problems** are set in life-science, bioengineering, or biomedical contexts.

Targeted content changes have been carefully implemented throughout the book. These include:

- **Life-science and bioengineering worked examples and applications** focus on the physics of life-science situations in order to serve the needs of life-science students taking a calculus-based physics class.
- **Descriptive text throughout has been streamlined** to focus the presentation and generate a shorter text.
- The chapter on ***Modern Optics and Matter Waves*** has been re-worked into Chapters 38 and 39 to streamline the coverage of this material.

At the front of the book, you'll find an illustrated walkthrough of the new pedagogical features in this third edition. The *Preface to the Student* demonstrates how all the book's features are designed to help your students.

Textbook Organization

The 42-chapter extended edition (ISBN 978-0-321-73608-6/0-321-73608-7) of *Physics for Scientists and Engineers* is intended for a three-semester course. Most of the 36-chapter standard edition (ISBN 978-0-321-75294-9/0-321-75294-5), ending with relativity, can be covered in two semesters, although the judicious omission of a few chapters will avoid rushing through the material and give students more time to develop their knowledge and skills.

There's a growing sentiment that quantum physics is quickly becoming the province of engineers, not just scientists, and that even a two-semester course should include a reasonable introduction to quantum ideas. The *Instructor Guide* outlines a couple of routes through the book that allow most of the quantum physics chapters to be included in a two-semester course. I've written the book with the hope that an increasing number of instructors will choose one of these routes.

The full textbook is divided into seven parts: Part I: *Newton's Laws*, Part II: *Conservation Laws*, Part III: *Applications of Newtonian Mechanics*, Part IV: *Thermodynamics*, Part V: *Waves and Optics*, Part VI: *Electricity and Magnetism*, and Part VII: *Relativity and Quantum Physics*. Although I recommend covering the parts in this order (see below), doing so is by no means essential. Each topic is self-contained, and Parts III–VI can be rearranged to suit an instructor's needs. To facilitate a reordering of topics, the full text is available in the five individual volumes listed in the margin.

Organization Rationale: Thermodynamics is placed before waves because it is a continuation of ideas from mechanics. The key idea in thermodynamics is energy, and moving from mechanics into thermodynamics allows the uninterrupted development of this important idea. Further, waves introduce students to functions of two variables, and the mathematics of waves is more akin to electricity and magnetism than to mechanics. Thus moving from waves to fields to quantum physics provides a gradual transition of ideas and skills.

The purpose of placing optics with waves is to provide a coherent presentation of wave physics, one of the two pillars of classical physics. Optics as it is presented in introductory physics makes no use of the properties of electromagnetic fields. There's little reason other than historical tradition to delay optics until after E&M.

- **Extended edition,** with modern physics (ISBN 978-0-321-73608-6 / 0-321-73608-7): Chapters 1–42.
- **Standard edition** (ISBN 978-0-321-75294-9 / 0-321-75294-5): Chapters 1–36.
- **Volume 1** (ISBN 978-0-321-75291-8 / 0-321-75291-0) covers mechanics: Chapters 1–15.
- **Volume 2** (ISBN 978-0-321-75318-2 / 0-321-75318-6) covers thermodynamics: Chapters 16–19.
- **Volume 3** (ISBN 978-0-321-75317-5 / 0-321-75317-8) covers waves and optics: Chapters 20–24.
- **Volume 4** (ISBN 978-0-321-75316-8 / 0-321-75316-X) covers electricity and magnetism, plus relativity: Chapters 25–36.
- **Volume 5** (ISBN 978-0-321-75315-1 / 0-321-75315-1) covers relativity and quantum physics: Chapters 36–42.
- **Volumes 1–5** boxed set (ISBN 978-0-321-77265-7 / 0-321-77265-2).

The documented difficulties that students have with optics are difficulties with waves, not difficulties with electricity and magnetism. However, the optics chapters are easily deferred until the end of Part VI for instructors who prefer that ordering of topics.

The Student Workbook

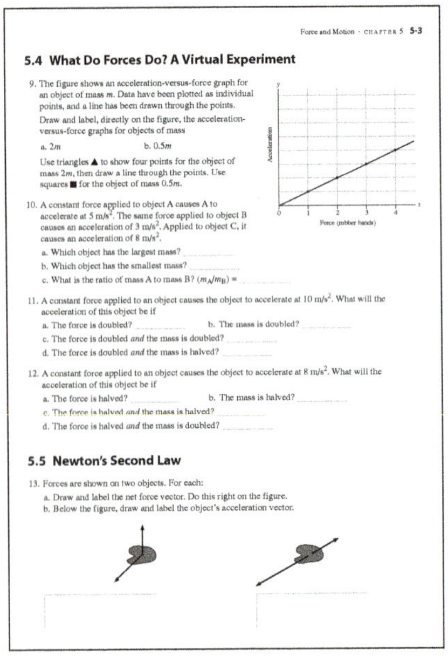
Force and Motion · CHAPTER 5 5-3

5.4 What Do Forces Do? A Virtual Experiment

9. The figure shows an acceleration-versus-force graph for an object of mass m. Data have been plotted as individual points, and a line has been drawn through the points.
Draw and label, directly on the figure, the acceleration-versus-force graphs for objects of mass
a. $2m$ b. $0.5m$
Use triangles ▲ to show four points for the object of mass $2m$, then draw a line through the points. Use squares ■ for the object of mass $0.5m$.

10. A constant force applied to object A causes A to accelerate at 5 m/s^2. The same force applied to object B causes an acceleration of 3 m/s^2. Applied to object C, it causes an acceleration of 8 m/s^2.
a. Which object has the largest mass? ________
b. Which object has the smallest mass? ________
c. What is the ratio of mass A to mass B? (m_A/m_B) = ________

11. A constant force applied to an object causes the object to accelerate at 10 m/s^2. What will the acceleration of this object be if
a. The force is doubled? ________ b. The mass is doubled? ________
c. The force is doubled *and* the mass is doubled? ________
d. The force is doubled *and* the mass is halved? ________

12. A constant force applied to an object causes the object to accelerate at 8 m/s^2. What will the acceleration of this object be if
a. The force is halved? ________ b. The mass is halved? ________
c. The force is halved *and* the mass is halved? ________
d. The force is halved *and* the mass is doubled? ________

5.5 Newton's Second Law

13. Forces are shown on two objects. For each:
a. Draw and label the net force vector. Do this right on the figure.
b. Below the figure, draw and label the object's acceleration vector.

A key component of *Physics for Scientists and Engineers: A Strategic Approach* is the accompanying *Student Workbook*. The workbook bridges the gap between textbook and homework problems by providing students the opportunity to learn and practice skills prior to using those skills in quantitative end-of-chapter problems, much as a musician practices technique separately from performance pieces. The workbook exercises, which are keyed to each section of the textbook, focus on developing specific skills, ranging from identifying forces and drawing free-body diagrams to interpreting wave functions.

The workbook exercises, which are generally qualitative and/or graphical, draw heavily upon the physics education research literature. The exercises deal with issues known to cause student difficulties and employ techniques that have proven to be effective at overcoming those difficulties. The workbook exercises can be used in class as part of an active-learning teaching strategy, in recitation sections, or as assigned homework. More information about effective use of the *Student Workbook* can be found in the *Instructor Guide*.

Available versions: Extended (ISBN 978-0-321-75308-3/0-321-75308-9), Standard (ISBN 978-0-321-75309-0/0-321-75309-7), Volume 1 (ISBN 978-0-321-75314-4/0-321-75314-3), Volume 2 (ISBN 978-0-321-75313-7/0-321-75313-5), Volume 3 (ISBN 978-0-321-75312-0/0-321-75310-0), Volume 4 (ISBN 978-0-321-75311-3/0-321-75311-9), and Volume 5 (ISBN 978-0-321-75310-6/0-321-75310-0).

Instructor Supplements

- The **Instructor Guide for *Physics for Scientists and Engineers*** (ISBN 978-0-321-74765-5/0-321-74765-8) offers detailed comments and suggested teaching ideas for every chapter, an extensive review of what has been learned from physics education research, and guidelines for using active-learning techniques in your classroom. This invaluable guide is available on the Instructor Resource DVD, and via download, either from the MasteringPhysics Instructor Area or from the Instructor Resource Center (www.pearsonhighered.com/educator).
- The **Instructor Solutions** (ISBN 978-0-321-76940-4/0-321-76940-6), written by the author, Professor Larry Smith (Snow College), and Brett Kraabel (Ph.D., University of California, Santa Barbara), provide *complete* solutions to all the end-of-chapter problems. The solutions follow the four-step Model/Visualize/Solve/Assess procedure used in the Problem-Solving Strategies and in all worked examples. The solutions are available by chapter as editable Word® documents and as PDFs for your own use or for posting on your password-protected course website. Also provided are PDFs of handwritten solutions to all of the exercises in the *Student Workbook*, written by Professor James Andrews and Brian Garcar (Youngstown State University). All solutions are available only via download, either from the MasteringPhysics Instructor Area or from the Instructor Resource Center (www.pearsonhighered.com/educator).
- The cross-platform **Instructor Resource DVD** (ISBN 978-0-321-75456-1/0-321-75456-5) provides a comprehensive library of more than 220 applets from **ActivPhysics OnLine** and 76 **PhET simulations**, as well as all figures, photos, tables, summaries, and key equations from the textbook in JPEG format. In addition, all the Problem-Solving Strategies, Tactics Boxes, and Key Equations are provided in editable Word format. PowerPoint® **Lecture Outlines** with embedded **Classroom Response System "Clicker" Questions** (including reading quizzes) are also provided.
- **MasteringPhysics®** (www.masteringphysics.com) is the most advanced, educationally effective, and widely used physics homework and tutorial system in the world. Eight years in development, it provides instructors with a library of extensively pre-tested end-of-chapter problems and rich, multipart, multistep tutorials that incorporate a wide variety of answer types, wrong answer feedback, individualized help (comprising hints or simpler sub-problems upon request), all driven by the largest metadatabase of student problem-solving in the world. NSF-sponsored published research (and subsequent

studies) show that MasteringPhysics has dramatic educational results. MasteringPhysics allows instructors to build wide-ranging homework assignments of just the right difficulty and length and provides them with efficient tools to analyze in unprecedented detail both class trends and the work of any student.

MasteringPhysics routinely provides instant and individualized feedback and guidance to more than 100,000 students every day. A wide range of tools and support make MasteringPhysics fast and easy for instructors and students to learn to use. Extensive class tests show that by the end of their course, an unprecedented nine of ten students recommend MasteringPhysics as their preferred way to study physics and do homework.

For the third edition of *Physics for Scientists and Engineers,* MasteringPhysics now has the following functionalities:

- **Learning Outcomes:** In addition to being able to create their own learning outcomes to associate with questions in an assignment, professors can now select content that is tagged to a large number of publisher-provided learning outcomes. They can also print or export student results based on learning outcomes for their own use or to incorporate into reports for their administration.
- **Quizzing and Testing Enhancements:** These include options to hide item titles, add password protection, limit access to completed assignments, and to randomize question order in an assignment.
- **Math Remediation:** Found within selected tutorials, special links provide just-in-time math help and allow students to brush up on the most important mathematical concepts needed to successfully complete assignments. This new feature links students directly to math review and practice helping students make the connection between math and physics.
- **Enhanced End-of-Chapter Problems:** A subset of homework problems now offer additional support such as problem-solving strategy hints, relevant math review and practice, links to the eText, and links to the related Video Tutor Solution.

- **ActivPhysics OnLine™** (accessed through the Self Study area within www.masteringphysics.com) provides a comprehensive library of more than 220 tried and tested ActivPhysics core applets updated for web delivery using the latest online technologies. In addition, it provides a suite of highly regarded applet-based tutorials developed by education pioneers Alan Van Heuvelen and Paul D'Alessandris.

  The online exercises are designed to encourage students to confront misconceptions, reason qualitatively about physical processes, experiment quantitatively, and learn to think critically. The highly acclaimed ActivPhysics OnLine companion workbooks help students work through complex concepts and understand them more clearly. The applets from the ActivPhysics OnLine library are also available on the Instructor Resource DVD for this text.
- The **Test Bank** (ISBN 978-0-321-74766-2/0-321-74766-6) contains more than 2,000 high-quality problems, with a range of multiple-choice, true/false, short-answer, and regular homework-type questions. Test files are provided both in TestGen (an easy-to-use, fully networkable program for creating and editing quizzes and exams) and Word format. They are available only via download, either from the MasteringPhysics Instructor Area or from the Instructor Resource Center (www.pearsonhighered.com/educator).

Student Supplements

- The **Student Solutions Manuals Chapters 1–19** (ISBN 978-0-321-74767-9/0-321-74767-4) and **Chapters 20–42** (ISBN 978-0-321-77269-5/0-321-77269-5), written by the author, Professor Larry Smith (Snow College), and Brett Kraabel (Ph.D., University of California, Santa Barbara), provide *detailed* solutions to more than half of the odd-numbered end-of-chapter problems. The solutions follow the four-step Model/Visualize/Solve/Assess procedure used in the Problem-Solving Strategies and in all worked examples.
- **MasteringPhysics®** (www.masteringphysics.com) is a homework, tutorial, and assessment system based on years of research into how students work physics problems and precisely where they need help. Studies show that students who use MasteringPhysics significantly increase their scores compared to handwritten homework. MasteringPhysics achieves this improvement by providing students with instantaneous feedback specific to their wrong answers, simpler sub-problems upon request when they get stuck, and partial credit for their method(s). This individualized, 24/7 Socratic tutoring is recommended by 9 out of 10 students to their peers as the most effective and time-efficient way to study.
- **Pearson eText** is available through MasteringPhysics, either automatically when MasteringPhysics is packaged with new books, or available as a purchased upgrade online. Allowing students access to the text wherever they have access to the Internet, Pearson eText comprises the full text, including figures that can be enlarged for better viewing. With eText, students are also able to pop up definitions and terms to help with vocabulary and the reading of the material. Students can also take notes in eText using the annotation feature at the top of each page.

- **Pearson Tutor Services** (www.pearsontutorservices.com) Each student's subscription to MasteringPhysics also contains complimentary access to Pearson Tutor Services, powered by Smarthinking, Inc. By logging in with their MasteringPhysics ID and password, they will be connected to highly qualified e-instructors who provide additional interactive online tutoring on the major concepts of physics. Some restrictions apply; offer subject to change.
- **ActivPhysics OnLine™** (accessed through the Self Study area within www.masteringphysics.com) provides students with a suite of highly regarded applet-based tutorials (see above). The following workbooks help students work through complex concepts and understand them more clearly:
- **ActivPhysics OnLine Workbook, Volume 1: Mechanics • Thermal Physics • Oscillations & Waves** (ISBN 978-0-8053-9060-5/0-8053-9060-X)
- **ActivPhysics OnLine Workbook, Volume 2: Electricity & Magnetism • Optics • Modern Physics** (ISBN 978-0-8053-9061-2/0-8053-9061-8)

Acknowledgments

I have relied upon conversations with and, especially, the written publications of many members of the physics education research community. Those who may recognize their influence include Arnold Arons, Uri Ganiel, Ibrahim Halloun, Richard Hake, Ken Heller, Paula Heron, David Hestenes, Leonard Jossem, Jill Larkin, Priscilla Laws, John Mallinckrodt, Kandiah Manivannan, Lillian McDermott and members of the Physics Education Research Group at the University of Washington, David Meltzer, Edward "Joe" Redish, Fred Reif, Jeffery Saul, Rachel Scherr, Bruce Sherwood, Josip Slisko, David Sokoloff, Richard Steinberg, Ronald Thornton, Sheila Tobias, Alan Van Heuleven, and Michael Wittmann. John Rigden, founder and director of the Introductory University Physics Project, provided the impetus that got me started down this path. Early development of the materials was supported by the National Science Foundation as the *Physics for the Year 2000* project; their support is gratefully acknowledged.

I especially want to thank my editor Jim Smith, development editor Alice Houston, project editor Martha Steele, and all the other staff at Pearson for their enthusiasm and hard work on this project. Production project manager Beth Collins, Rose Kernan and the team at Nesbitt Graphics, Inc., and photo researcher Eric Schrader get a good deal of the credit for making this complex project all come together. Larry Smith and Brett Kraabel have done an outstanding job of checking the solutions to every end-of-chapter problem and updating the *Instructor Solutions Manual*. Jim Andrews and Brian Garcar must be thanked for so carefully writing out the solutions to *The Student Workbook* exercises, and Jason Harlow for putting together the Lecture Outlines. In addition to the reviewers and classroom testers listed below, who gave invaluable feedback, I am particularly grateful to Charlie Hibbard for his close scrutiny of every word and figure.

Finally, I am endlessly grateful to my wife Sally for her love, encouragement, and patience, and to our many cats, past and present, who understand clearly that their priority is not deadlines but "Pet me, pet me, pet me."

Randy Knight, September 2011
rknight@calpoly.edu

Reviewers and Classroom Testers

Special thanks go to our third edition review panel: Kyle Altman, Taner Edis, Kent Fisher, Marty Gelfand, Elizabeth George, Jason Harlow, Bob Jacobsen, David Lee, Gary Morris, Eric Murray, and Bruce Schumm.

Gary B. Adams, *Arizona State University*
Ed Adelson, *Ohio State University*
Kyle Altmann, *Elon University*
Wayne R. Anderson, *Sacramento City College*
James H. Andrews, *Youngstown State University*
Kevin Ankoviak, *Las Positas College*
David Balogh, *Fresno City College*
Dewayne Beery, *Buffalo State College*
Joseph Bellina, *Saint Mary's College*
James R. Benbrook, *University of Houston*
David Besson, *University of Kansas*
Randy Bohn, *University of Toledo*
Richard A. Bone, *Florida International University*
Gregory Boutis, *York College*
Art Braundmeier, *University of Southern Illinois, Edwardsville*
Carl Bromberg, *Michigan State University*
Meade Brooks, *Collin College*
Douglas Brown, *Cabrillo College*
Ronald Brown, *California Polytechnic State University, San Luis Obispo*
Mike Broyles, *Collin County Community College*
Debra Burris, *University of Central Arkansas*
James Carolan, *University of British Columbia*
Michael Chapman, *Georgia Tech University*
Norbert Chencinski, *College of Staten Island*
Kristi Concannon, *King's College*

Sean Cordry, *Northwestern College of Iowa*
Robert L. Corey, *South Dakota School of Mines*
Michael Crescimanno, *Youngstown State University*
Dennis Crossley, *University of Wisconsin–Sheboygan*
Wei Cui, *Purdue University*
Robert J. Culbertson, *Arizona State University*
Danielle Dalafave, *The College of New Jersey*
Purna C. Das, *Purdue University North Central*
Chad Davies, *Gordon College*
William DeGraffenreid, *California State University–Sacramento*
Dwain Desbien, *Estrella Mountain Community College*
John F. Devlin, *University of Michigan, Dearborn*
John DiBartolo, *Polytechnic University*
Alex Dickison, *Seminole Community College*
Chaden Djalali, *University of South Carolina*
Margaret Dobrowolska, *University of Notre Dame*
Sandra Doty, *Denison University*
Miles J. Dresser, *Washington State University*
Charlotte Elster, *Ohio University*
Robert J. Endorf, *University of Cincinnati*
Tilahun Eneyew, *Embry-Riddle Aeronautical University*
F. Paul Esposito, *University of Cincinnati*
John Evans, *Lee University*
Harold T. Evensen, *University of Wisconsin–Platteville*
Michael R. Falvo, *University of North Carolina*
Abbas Faridi, *Orange Coast College*
Nail Fazleev, *University of Texas–Arlington*
Stuart Field, *Colorado State University*
Daniel Finley, *University of New Mexico*
Jane D. Flood, *Muhlenberg College*
Michael Franklin, *Northwestern Michigan College*
Jonathan Friedman, *Amherst College*
Thomas Furtak, *Colorado School of Mines*
Alina Gabryszewska-Kukawa, *Delta State University*
Lev Gasparov, *University of North Florida*
Richard Gass, *University of Cincinnati*
J. David Gavenda, *University of Texas, Austin*
Stuart Gazes, *University of Chicago*
Katherine M. Gietzen, *Southwest Missouri State University*
Robert Glosser, *University of Texas, Dallas*
William Golightly, *University of California, Berkeley*
Paul Gresser, *University of Maryland*
C. Frank Griffin, *University of Akron*
John B. Gruber, *San Jose State University*
Stephen Haas, *University of Southern California*
John Hamilton, *University of Hawaii at Hilo*
Jason Harlow, *University of Toronto*
Randy Harris, *University of California, Davis*
Nathan Harshman, *American University*
J. E. Hasbun, *University of West Georgia*
Nicole Herbots, *Arizona State University*
Jim Hetrick, *University of Michigan–Dearborn*
Scott Hildreth, *Chabot College*
David Hobbs, *South Plains College*
Laurent Hodges, *Iowa State University*
Mark Hollabaugh, *Normandale Community College*
John L. Hubisz, *North Carolina State University*
Shane Hutson, *Vanderbilt University*
George Igo, *University of California, Los Angeles*
David C. Ingram, *Ohio University*
Bob Jacobsen, *University of California, Berkeley*
Rong-Sheng Jin, *Florida Institute of Technology*
Marty Johnston, *University of St. Thomas*
Stanley T. Jones, *University of Alabama*
Darrell Judge, *University of Southern California*
Pawan Kahol, *Missouri State University*
Teruki Kamon, *Texas A&M University*
Richard Karas, *California State University, San Marcos*
Deborah Katz, *U.S. Naval Academy*
Miron Kaufman, *Cleveland State University*
Katherine Keilty, *Kingwood College*
Roman Kezerashvili, *New York City College of Technology*
Peter Kjeer, *Bethany Lutheran College*
M. Kotlarchyk, *Rochester Institute of Technology*
Fred Krauss, *Delta College*
Cagliyan Kurdak, *University of Michigan*
Fred Kuttner, *University of California, Santa Cruz*
H. Sarma Lakkaraju, *San Jose State University*
Darrell R. Lamm, *Georgia Institute of Technology*
Robert LaMontagne, *Providence College*
Eric T. Lane, *University of Tennessee–Chattanooga*
Alessandra Lanzara, *University of California, Berkeley*
Lee H. LaRue, *Paris Junior College*
Sen-Ben Liao, *Massachusetts Institute of Technology*
Dean Livelybrooks, *University of Oregon*
Chun-Min Lo, *University of South Florida*
Olga Lobban, *Saint Mary's University*
Ramon Lopez, *Florida Institute of Technology*
Vaman M. Naik, *University of Michigan, Dearborn*
Kevin Mackay, *Grove City College*
Carl Maes, *University of Arizona*
Rizwan Mahmood, *Slippery Rock University*
Mani Manivannan, *Missouri State University*
Richard McCorkle, *University of Rhode Island*
James McDonald, *University of Hartford*
James McGuire, *Tulane University*
Stephen R. McNeil, *Brigham Young University–Idaho*
Theresa Moreau, *Amherst College*
Gary Morris, *Rice University*
Michael A. Morrison, *University of Oklahoma*
Richard Mowat, *North Carolina State University*
Eric Murray, *Georgia Institute of Technology*
Taha Mzoughi, *Mississippi State University*
Scott Nutter, *Northern Kentucky University*
Craig Ogilvie, *Iowa State University*
Benedict Y. Oh, *University of Wisconsin*
Martin Okafor, *Georgia Perimeter College*
Halina Opyrchal, *New Jersey Institute of Technology*
Yibin Pan, *University of Wisconsin–Madison*
Georgia Papaefthymiou, *Villanova University*
Peggy Perozzo, *Mary Baldwin College*

Brian K. Pickett, *Purdue University, Calumet*
Joe Pifer, *Rutgers University*
Dale Pleticha, *Gordon College*
Marie Plumb, *Jamestown Community College*
Robert Pompi, *SUNY-Binghamton*
David Potter, *Austin Community College–Rio Grande Campus*
Chandra Prayaga, *University of West Florida*
Didarul Qadir, *Central Michigan University*
Steve Quon, *Ventura College*
Michael Read, *College of the Siskiyous*
Lawrence Rees, *Brigham Young University*
Richard J. Reimann, *Boise State University*
Michael Rodman, *Spokane Falls Community College*
Sharon Rosell, *Central Washington University*
Anthony Russo, *Okaloosa-Walton Community College*
Freddie Salsbury, *Wake Forest University*
Otto F. Sankey, *Arizona State University*
Jeff Sanny, *Loyola Marymount University*
Rachel E. Scherr, *University of Maryland*
Carl Schneider, *U. S. Naval Academy*
Bruce Schumm, *University of California, Santa Cruz*
Bartlett M. Sheinberg, *Houston Community College*
Douglas Sherman, *San Jose State University*
Elizabeth H. Simmons, *Boston University*
Marlina Slamet, *Sacred Heart University*
Alan Slavin, *Trent College*
Larry Smith, *Snow College*
William S. Smith, *Boise State University*
Paul Sokol, *Pennsylvania State University*
LTC Bryndol Sones, *United States Military Academy*
Chris Sorensen, *Kansas State University*
Anna and Ivan Stern, *AW Tutor Center*
Gay B. Stewart, *University of Arkansas*
Michael Strauss, *University of Oklahoma*
Chin-Che Tin, *Auburn University*
Christos Valiotis, *Antelope Valley College*
Andrew Vanture, *Everett Community College*
Arthur Viescas, *Pennsylvania State University*
Ernst D. Von Meerwall, *University of Akron*
Chris Vuille, *Embry-Riddle Aeronautical University*
Jerry Wagner, *Rochester Institute of Technology*
Robert Webb, *Texas A&M University*
Zodiac Webster, *California State University, San Bernardino*
Robert Weidman, *Michigan Technical University*
Fred Weitfeldt, *Tulane University*
Jeff Allen Winger, *Mississippi State University*
Carey Witkov, *Broward Community College*
Ronald Zammit, *California Polytechnic State University, San Luis Obispo*
Darin T. Zimmerman, *Pennsylvania State University, Altoona*
Fredy Zypman, *Yeshiva University*

Preface to the Student

From Me to You

The most incomprehensible thing about the universe is that it is comprehensible.
—Albert Einstein

The day I went into physics class it was death.
—Sylvia Plath, *The Bell Jar*

Let's have a little chat before we start. A rather one-sided chat, admittedly, because you can't respond, but that's OK. I've talked with many of your fellow students over the years, so I have a pretty good idea of what's on your mind.

What's your reaction to taking physics? Fear and loathing? Uncertainty? Excitement? All of the above? Let's face it, physics has a bit of an image problem on campus. You've probably heard that it's difficult, maybe downright impossible unless you're an Einstein. Things that you've heard, your experiences in other science courses, and many other factors all color your *expectations* about what this course is going to be like.

It's true that there are many new ideas to be learned in physics and that the course, like college courses in general, is going to be much faster paced than science courses you had in high school. I think it's fair to say that it will be an *intense* course. But we can avoid many potential problems and difficulties if we can establish, here at the beginning, what this course is about and what is expected of you—and of me!

Just what is physics, anyway? Physics is a way of thinking about the physical aspects of nature. Physics is not better than art or biology or poetry or religion, which are also ways to think about nature; it's simply different. One of the things this course will emphasize is that physics is a human endeavor. The ideas presented in this book were not found in a cave or conveyed to us by aliens; they were discovered and developed by real people engaged in a struggle with real issues. I hope to convey to you something of the history and the process by which we have come to accept the principles that form the foundation of today's science and engineering.

You might be surprised to hear that physics is not about "facts." Oh, not that facts are unimportant, but physics is far more focused on discovering *relationships* that exist between facts and *patterns* that exist in nature than on learning facts for their own sake. As a consequence, there's not a lot of memorization when you study physics. Some—there are still definitions and equations to learn—but less than in many other courses. Our emphasis, instead, will be on thinking and reasoning. This is important to factor into your expectations for the course.

Perhaps most important of all, *physics is not math!* Physics is much broader. We're going to look for patterns and relationships in nature, develop the logic that relates different ideas, and search for the reasons *why* things happen as they do. In doing so, we're going to stress qualitative reasoning, pictorial and graphical reasoning, and reasoning by analogy. And yes, we will use math, but it's just one tool among many.

It will save you much frustration if you're aware of this physics–math distinction up front. Many of you, I know, want to find a formula and plug numbers into it—that is,

(a) X-ray diffraction pattern

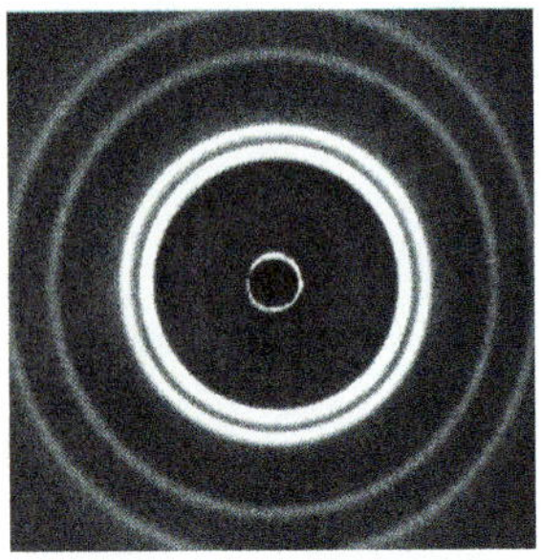

(b) Electron diffraction pattern

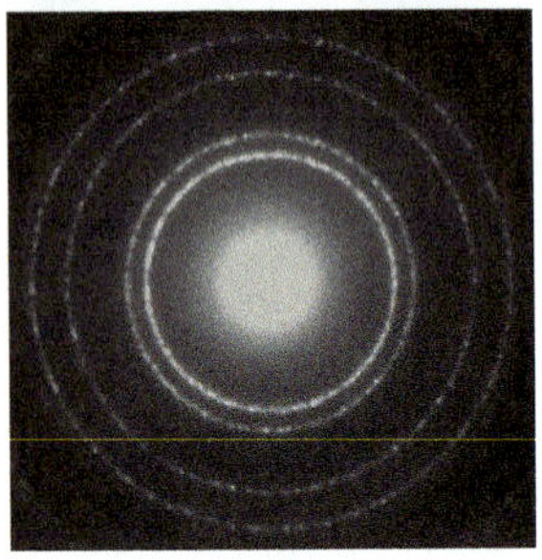

to do a math problem. Maybe that worked in high school science courses, but it is *not* what this course expects of you. We'll certainly do many calculations, but the specific numbers are usually the last and least important step in the analysis.

Physics is about recognizing patterns. For example, the top photograph is an x-ray diffraction pattern showing how a focused beam of x rays spreads out after passing through a crystal. The bottom photograph shows what happens when a focused beam of electrons is shot through the same crystal. What does the obvious similarity in these two photographs tell us about the nature of light and the nature of matter?

As you study, you'll sometimes be baffled, puzzled, and confused. That's perfectly normal and to be expected. Making mistakes is OK too *if* you're willing to learn from the experience. No one is born knowing how to do physics any more than he or she is born knowing how to play the piano or shoot basketballs. The ability to do physics comes from practice, repetition, and struggling with the ideas until you "own" them and can apply them yourself in new situations. There's no way to make learning effortless, at least for anything worth learning, so expect to have some difficult moments ahead. But also expect to have some moments of excitement at the joy of discovery. There will be instants at which the pieces suddenly click into place and you *know* that you understand a powerful idea. There will be times when you'll surprise yourself by successfully working a difficult problem that you didn't think you could solve. My hope, as an author, is that the excitement and sense of adventure will far outweigh the difficulties and frustrations.

Getting the Most Out of Your Course

Many of you, I suspect, would like to know the "best" way to study for this course. There is no best way. People are different, and what works for one student is less effective for another. But I do want to stress that *reading the text* is vitally important. Class time will be used to clarify difficulties and to develop tools for using the knowledge, but your instructor will *not* use class time simply to repeat information in the text. The basic knowledge for this course is written down on these pages, and the *number-one expectation* is that you will read carefully and thoroughly to find and learn that knowledge.

Despite there being no best way to study, I will suggest *one* way that is successful for many students. It consists of the following four steps:

1. **Read each chapter *before* it is discussed in class.** I cannot stress too strongly how important this step is. Class attendance is much more effective if you are prepared. When you first read a chapter, focus on learning new vocabulary, definitions, and notation. There's a list of terms and notations at the end of each chapter. Learn them! You won't understand what's being discussed or how the ideas are being used if you don't know what the terms and symbols mean.
2. **Participate actively in class.** Take notes, ask and answer questions, and participate in discussion groups. There is ample scientific evidence that *active participation* is much more effective for learning science than passive listening.
3. **After class, go back for a careful re-reading of the chapter.** In your second reading, pay closer attention to the details and the worked examples. Look for the *logic* behind each example (I've highlighted this to make it clear), not just at what formula is being used. Do the *Student Workbook* exercises for each section as you finish your reading of it.
4. **Finally, apply what you have learned to the homework problems at the end of each chapter.** I strongly encourage you to form a study group with two or three classmates. There's good evidence that students who study regularly with a group do better than the rugged individualists who try to go it alone.

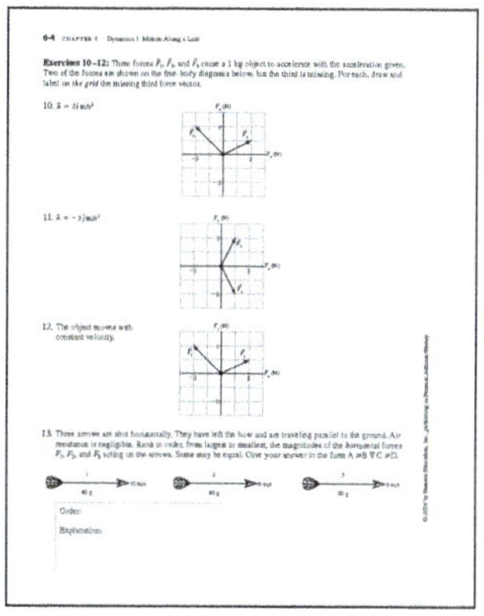

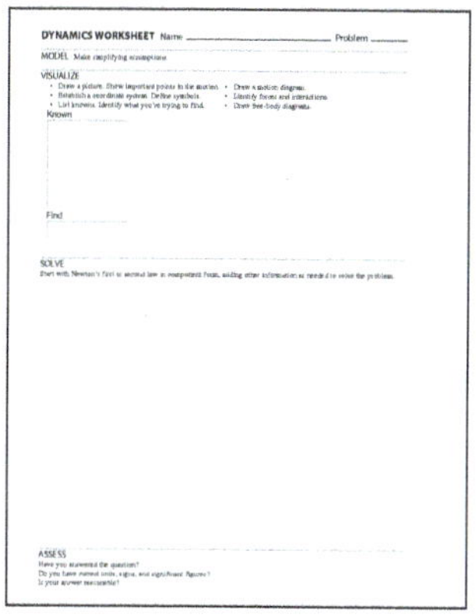

Did someone mention a workbook? The companion *Student Workbook* is a vital part of the course. Its questions and exercises ask you to reason *qualitatively,* to use graphical information, and to give explanations. It is through these exercises that you will learn what the concepts mean and will practice the reasoning skills appropriate to the chapter. You will then have acquired the baseline knowledge and confidence you need *before* turning to the end-of-chapter homework problems. In sports or in music, you would never think of performing before you practice, so why would you want to do so in physics? The workbook is where you practice and work on basic skills.

Many of you, I know, will be tempted to go straight to the homework problems and then thumb through the text looking for a formula that seems like it will work. That approach will not succeed in this course, and it's guaranteed to make you frustrated and discouraged. Very few homework problems are of the "plug and chug" variety where you simply put numbers into a formula. To work the homework problems successfully, you need a better study strategy—either the one outlined above or your own—that helps you learn the concepts and the relationships between the ideas.

A traditional guideline in college is to study two hours outside of class for every hour spent in class, and this text is designed with that expectation. Of course, two hours is an average. Some chapters are fairly straightforward and will go quickly. Others likely will require much more than two study hours per class hour.

Getting the Most Out of Your Textbook

Your textbook provides many features designed to help you learn the concepts of physics and solve problems more effectively.

- TACTICS BOXES give step-by-step procedures for particular skills, such as interpreting graphs or drawing special diagrams. Tactics Box steps are explicitly illustrated in subsequent worked examples, and these are often the starting point of a full *Problem-Solving Strategy*.

TACTICS BOX 5.3 Drawing a free-body diagram (MP)

1. **Identify all forces acting on the object.** This step was described in Tactics Box 5.2.
2. **Draw a coordinate system.** Use the axes defined in your pictorial representation.
3. **Represent the object as a dot at the origin of the coordinate axes.** This is the particle model.
4. **Draw vectors representing each of the identified forces.** This was described in Tactics Box 5.1. Be sure to label each force vector.
5. **Draw and label the *net force* vector $\vec{F}_{net}$.** Draw this vector beside the diagram, not on the particle. Or, if appropriate, write $\vec{F}_{net} = \vec{0}$. Then check that $\vec{F}_{net}$ points in the same direction as the acceleration vector $\vec{a}$ on your motion diagram.

Exercises 24–29

TACTICS BOX 32.3 Evaluating line integrals (MP)

1. If $\vec{B}$ is everywhere perpendicular to a line, the line integral of $\vec{B}$ is

$$\int_i^f \vec{B}\cdot d\vec{s} = 0$$

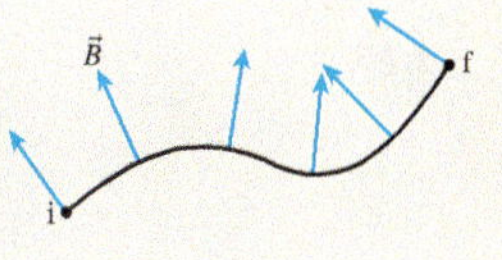

2. If $\vec{B}$ is everywhere tangent to a line of length l *and* has the same magnitude B at every point, then

$$\int_i^f \vec{B}\cdot d\vec{s} = Bl$$

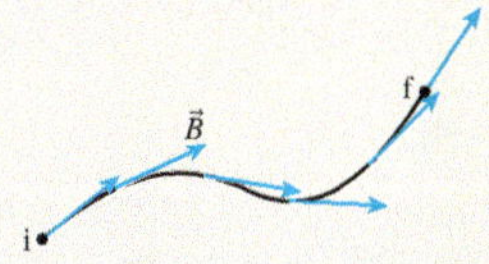

Exercises 23–24

- **PROBLEM-SOLVING STRATEGIES** are provided for each broad class of problems—problems characteristic of a chapter or group of chapters. The strategies follow a consistent four-step approach to help you develop confidence and proficient problem-solving skills: **MODEL, VISUALIZE, SOLVE, ASSESS.**

PROBLEM-SOLVING STRATEGY 6.2 Dynamics problems

MODEL Make simplifying assumptions.

VISUALIZE Draw a **pictorial representation.**

- Show important points in the motion with a sketch, establish a coordinate system, define symbols, and identify what the problem is trying to find.
- Use a motion diagram to determine the object's acceleration vector $\vec{a}$.
- Identify all forces acting on the object *at this instant* and show them on a free-body diagram.
- It's OK to go back and forth between these steps as you visualize the situation.

SOLVE The mathematical representation is based on Newton's second law:

$$\vec{F}_{\text{net}} = \sum_i \vec{F}_i = m\vec{a}$$

The vector sum of the forces is found directly from the free-body diagram. Depending on the problem, either

- Solve for the acceleration, then use kinematics to find velocities and positions; or
- Use kinematics to determine the acceleration, then solve for unknown forces.

ASSESS Check that your result has the correct units, is reasonable, and answers the question.

Exercise 22

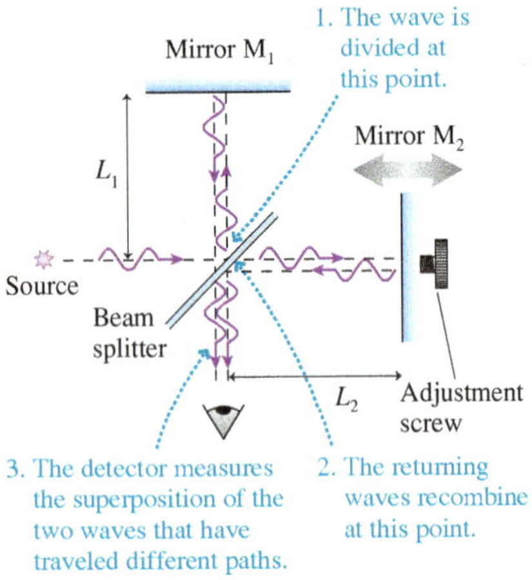

Annotated FIGURE showing the operation of the Michelson interferometer.

- Worked **EXAMPLES** illustrate good problem-solving practices through the consistent use of the four-step problem-solving approach and, where appropriate, the Tactics Box steps. The worked examples are often very detailed and carefully lead you through the *reasoning* behind the solution as well as the numerical calculations. A careful study of the reasoning will help you apply the concepts and techniques to the new and novel problems you will encounter in homework assignments and on exams.
- **NOTE ▶** paragraphs alert you to common mistakes and point out useful tips for tackling problems.
- **STOP TO THINK** questions embedded in the chapter allow you to quickly assess whether you've understood the main idea of a section. A correct answer will give you confidence to move on to the next section. An incorrect answer will alert you to re-read the previous section.
- Blue annotations on figures help you better understand what the figure is showing. They will help you to interpret graphs; translate between graphs, math, and pictures; grasp difficult concepts through a visual analogy; and develop many other important skills.
- *Pencil sketches* provide practical examples of the figures you should draw yourself when solving a problem.

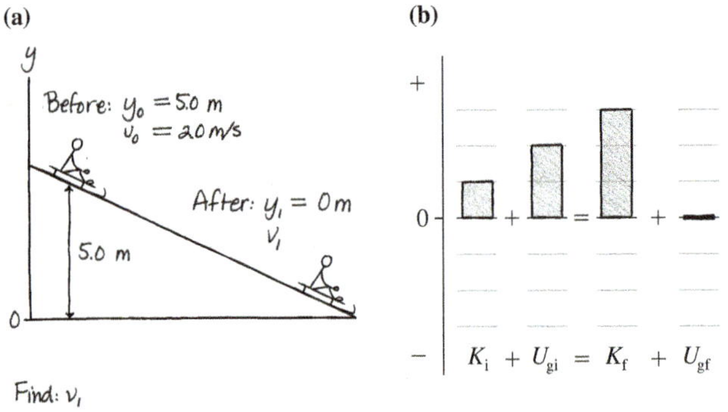

Pencil-sketch FIGURE showing a toboggan going down a hill and its energy bar chart.

- Each chapter begins with a *Chapter Preview*, a visual outline of the chapter ahead with recommendations of important topics you should review from previous chapters. A few minutes spent with the Preview will help you organize your thoughts so as to get the most out of reading the chapter.
- Schematic *Chapter Summaries* help you organize what you have learned into a hierarchy, from general principles (top) to applications (bottom). Side-by-side pictorial, graphical, textual, and mathematical representations are used to help you translate between these key representations.
- *Part Overviews* and *Summaries* provide a global framework for what you are learning. Each part begins with an overview of the chapters ahead and concludes with a broad summary to help you to connect the concepts presented in that set of chapters. KNOWLEDGE STRUCTURE tables in the Part Summaries, similar to the Chapter Summaries, help you to see the forest rather than just the trees.

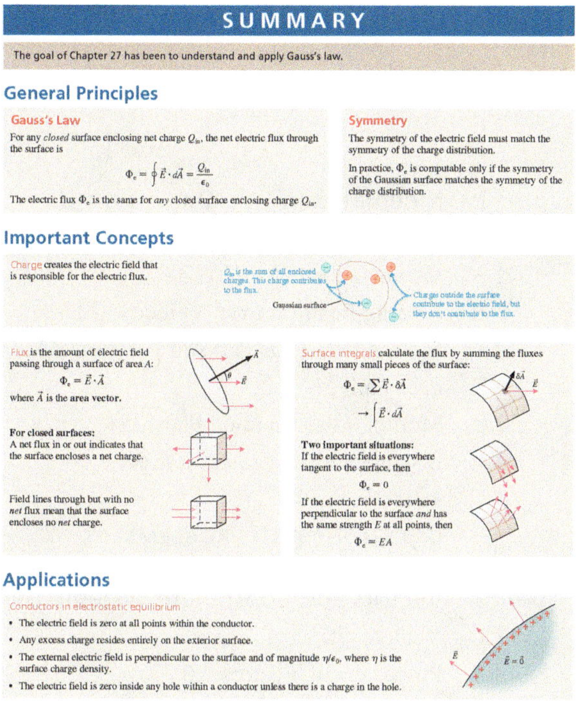

SUMMARY

The goal of Chapter 27 has been to understand and apply Gauss's law.

General Principles

Gauss's Law

For any *closed* surface enclosing net charge Q_{in}, the net electric flux through the surface is

$$\Phi_e = \oint \vec{E} \cdot d\vec{A} = \frac{Q_{in}}{\epsilon_0}$$

The electric flux Φ_e is the same for *any* closed surface enclosing charge Q_{in}.

Symmetry

The symmetry of the electric field must match the symmetry of the charge distribution.

In practice, Φ_e is computable only if the symmetry of the Gaussian surface matches the symmetry of the charge distribution.

Important Concepts

Charge creates the electric field that is responsible for the electric flux.

Flux is the amount of electric field passing through a surface of area A:

$$\Phi_e = \vec{E} \cdot \vec{A}$$

where $\vec{A}$ is the area vector.

For closed surfaces:
A net flux in or out indicates that the surface encloses a net charge.

Field lines through but with no *net* flux mean that the surface encloses no *net* charge.

Surface integrals calculate the flux by summing the fluxes through many small pieces of the surface:

$$\Phi_e = \sum \vec{E} \cdot \delta\vec{A} \rightarrow \int \vec{E} \cdot d\vec{A}$$

Two important situations:
If the electric field is everywhere tangent to the surface, then

$$\Phi_e = 0$$

If the electric field is everywhere perpendicular to the surface *and* has the same strength E at all points, then

$$\Phi_e = EA$$

Applications

Conductors in electrostatic equilibrium

- The electric field is zero at all points within the conductor.
- Any excess charge resides entirely on the exterior surface.
- The external electric field is perpendicular to the surface and of magnitude η/ϵ_0, where η is the surface charge density.
- The electric field is zero inside any hole within a conductor unless there is a charge in the hole.

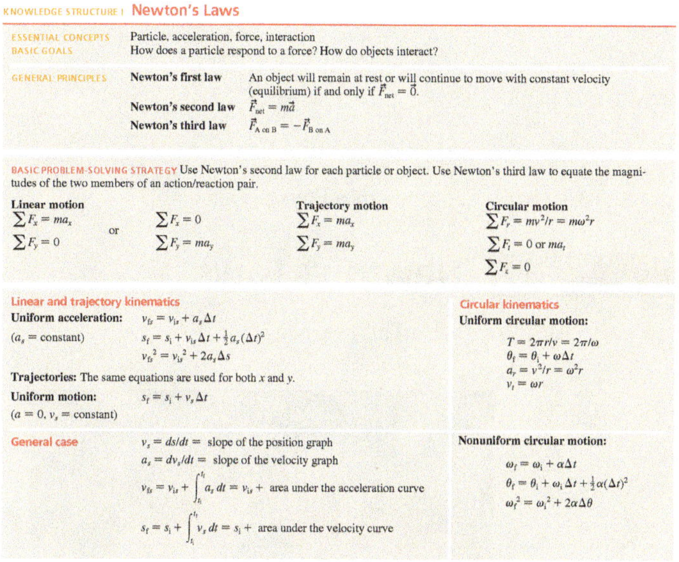

KNOWLEDGE STRUCTURE I Newton's Laws

ESSENTIAL CONCEPTS Particle, acceleration, force, interaction
BASIC GOALS How does a particle respond to a force? How do objects interact?

GENERAL PRINCIPLES
Newton's first law: An object will remain at rest or will continue to move with constant velocity (equilibrium) if and only if $\vec{F}_{net} = \vec{0}$.
Newton's second law: $\vec{F}_{net} = m\vec{a}$
Newton's third law: $\vec{F}_{A\ on\ B} = -\vec{F}_{B\ on\ A}$

BASIC PROBLEM-SOLVING STRATEGY Use Newton's second law for each particle or object. Use Newton's third law to equate the magnitudes of the two members of an action/reaction pair.

Linear motion: $\sum F_x = ma_x$, $\sum F_y = 0$ or $\sum F_x = 0$, $\sum F_y = ma_y$

Trajectory motion: $\sum F_x = ma_x$, $\sum F_y = ma_y$

Circular motion: $\sum F_r = mv^2/r = m\omega^2 r$, $\sum F_t = 0$ or ma_t, $\sum F_z = 0$

Linear and trajectory kinematics

Uniform acceleration: (a_s = constant)
$v_{fs} = v_{is} + a_s \Delta t$
$s_f = s_i + v_{is}\Delta t + \frac{1}{2}a_s(\Delta t)^2$
$v_{fs}^2 = v_{is}^2 + 2a_s \Delta s$

Trajectories: The same equations are used for both x and y.

Uniform motion: ($a = 0$, v_s = constant)
$s_f = s_i + v_s \Delta t$

General case
$v_s = ds/dt =$ slope of the position graph
$a_s = dv_s/dt =$ slope of the velocity graph
$v_{fs} = v_{is} + \int_{t_i}^{t_f} a_s\, dt = v_{is} +$ area under the acceleration curve
$s_f = s_i + \int_{t_i}^{t_f} v_s\, dt = s_i +$ area under the velocity curve

Circular kinematics

Uniform circular motion:
$T = 2\pi r/v = 2\pi/\omega$
$\theta_f = \theta_i + \omega \Delta t$
$a_r = v^2/r = \omega^2 r$
$v_t = \omega r$

Nonuniform circular motion:
$\omega_f = \omega_i + \alpha \Delta t$
$\theta_f = \theta_i + \omega_i \Delta t + \frac{1}{2}\alpha(\Delta t)^2$
$\omega_f^2 = \omega_i^2 + 2\alpha\Delta\theta$

Now that you know more about what is expected of you, what can you expect of me? That's a little trickier because the book is already written! Nonetheless, the book was prepared on the basis of what I think my students throughout the years have expected—and wanted—from their physics textbook. Further, I've listened to the extensive feedback I have received from thousands of students like you, and their instructors, who used the first and second editions of this book.

You should know that these course materials—the text and the workbook—are based on extensive research about how students learn physics and the challenges they face. The effectiveness of many of the exercises has been demonstrated through extensive class testing. I've written the book in an informal style that I hope you will find appealing and that will encourage you to do the reading. And, finally, I have endeavored to make clear not only that physics, as a technical body of knowledge, is relevant to your profession but also that physics is an exciting adventure of the human mind.

I hope you'll enjoy the time we're going to spend together.

Detailed Contents

Part III Applications of Newtonian Mechanics

Part V Waves and Optics

Introduction

Journey into Physics

Said Alice to the Cheshire cat,
"Cheshire-Puss, would you tell me, please, which way I ought to go from here?"
"That depends a good deal on where you want to go," said the Cat.
"I don't much care where—" said Alice.
"Then it doesn't matter which way you go," said the Cat.
—Lewis Carroll, *Alice in Wonderland*

Have you ever wondered about questions such as

Why is the sky blue?

Why is glass an insulator but metal a conductor?

What, really, is an atom?

These are the questions of which physics is made. Physicists try to understand the universe in which we live by observing the phenomena of nature—such as the sky being blue—and by looking for patterns and principles to explain these phenomena. Many of the discoveries made by physicists, from electromagnetic waves to nuclear energy, have forever altered the ways in which we live and think.

You are about to embark on a journey into the realm of physics. It is a journey in which you will learn about many physical phenomena and find the answers to questions such as the ones posed above. Along the way, you will also learn how to use physics to analyze and solve many practical problems.

As you proceed, you are going to see the methods by which physicists have come to understand the laws of nature. The ideas and theories of physics are not arbitrary; they are firmly grounded in experiments and measurements. By the time you finish this text, you will be able to recognize the *evidence* upon which our present knowledge of the universe is based.

Which Way Should We Go?

We are rather like Alice in Wonderland, here at the start of the journey, in that we must decide which way to go. Physics is an immense body of knowledge, and without specific goals it would not much matter which topics we study. But unlike Alice, we *do* have some particular destinations that we would like to visit.

The physics that provides the foundation for all of modern science and engineering can be divided into three broad categories:

- Particles and energy.
- Fields and waves.
- The atomic structure of matter.

A particle, in the sense that we'll use the term, is an idealization of a physical object. We will use particles to understand how objects move and how they interact with each other. One of the most important properties of a particle or a collection of particles is *energy*. We will study energy both for its value in understanding physical processes and because of its practical importance in a technological society.

A scanning tunneling microscope allows us to "see" the individual atoms on a surface. One of our goals is to understand how an image such as this is made.

Particles are discrete, localized objects. Although many phenomena can be understood in terms of particles and their interactions, the long-range interactions of gravity, electricity, and magnetism are best understood in terms of *fields,* such as the gravitational field and the electric field. Rather than being discrete, fields spread continuously through space. Much of the second half of this book will be focused on understanding fields and the interactions between fields and particles.

Certainly one of the most significant discoveries of the past 500 years is that matter consists of atoms. Atoms and their properties are described by quantum physics, but we cannot leap directly into that subject and expect that it would make any sense. To reach our destination, we are going to have to study many other topics along the way—rather like having to visit the Rocky Mountains if you want to drive from New York to San Francisco. All our knowledge of particles and fields will come into play as we end our journey by studying the atomic structure of matter.

The Route Ahead

Here at the beginning, we can survey the route ahead. Where will our journey take us? What scenic vistas will we view along the way?

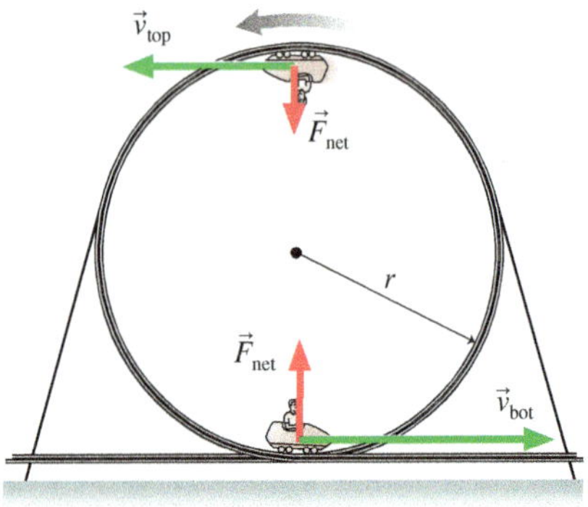

Parts I and II, *Newton's Laws* and *Conservation Laws,* form the basis of what is called *classical mechanics.* Classical mechanics is the study of motion. (It is called *classical* to distinguish it from the modern theory of motion at the atomic level, which is called *quantum mechanics.*) The first two parts of this textbook establish the basic language and concepts of motion. Part I will look at motion in terms of *particles* and *forces.* We will use these concepts to study the motion of everything from accelerating sprinters to orbiting satellites. Then, in Part II, we will introduce the ideas of *momentum* and *energy.* These concepts—especially energy—will give us a new perspective on motion and extend our ability to analyze motion.

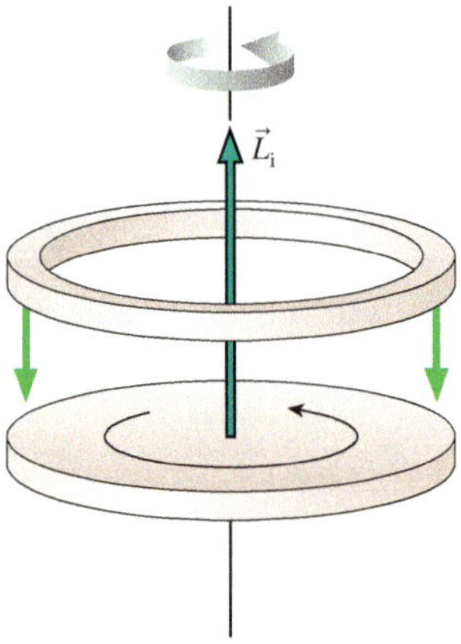

Part III, *Applications of Newtonian Mechanics,* will pause to look at four important applications of classical mechanics: Newton's theory of gravity, rotational motion, oscillatory motion, and the motion of fluids. Only oscillatory motion is a prerequisite for later chapters. Your instructor may choose to cover some or all of the other chapters, depending upon the time available, but your study of Parts IV–VII will not be hampered if these chapters are omitted.

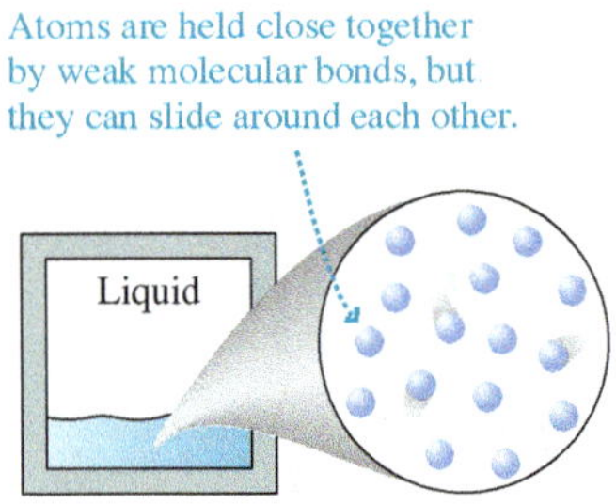

Part IV, *Thermodynamics,* extends the ideas of particles and energy to systems such as liquids and gases that contain vast numbers of particles. Here we will look for connections between the *microscopic* behavior of large numbers of atoms and the *macroscopic* properties of bulk matter. You will find that some of the properties of gases that you know from chemistry, such as the ideal gas law, turn out to be direct consequences of the underlying atomic structure of the gas. We will also expand the concept of energy and study how energy is transferred and utilized.

Waves are ubiquitous in nature, whether they be large-scale oscillations like ocean waves, the less obvious motions of sound waves, or the subtle undulations of light waves and matter waves that go to the heart of the atomic structure of matter. In **Part V,** *Waves and Optics,* we will emphasize the unity of wave physics and find that many diverse wave phenomena can be analyzed with the same concepts and mathematical language. Light waves are of special interest, and we will end this portion of our journey with an exploration of optical instruments, ranging from microscopes and telescopes to that most important of all optical instruments—your eye.

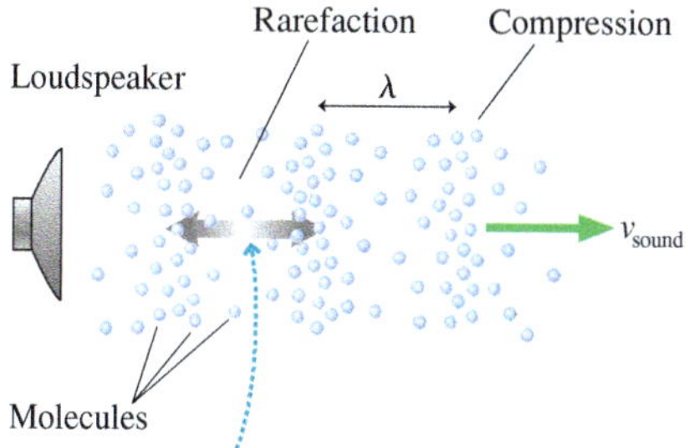

Individual molecules oscillate back and forth with displacement D. As they do so, the compressions propagate forward at speed v_{sound}. Because compressions are regions of higher pressure, a sound wave can be thought of as a pressure wave.

Part VI, *Electricity and Magnetism,* is devoted to the *electromagnetic force,* one of the most important forces in nature. In essence, the electromagnetic force is the "glue" that holds atoms together. It is also the force that makes this the "electronic age." We'll begin this part of the journey with simple observations of static electricity. Bit by bit, we'll be led to the basic ideas behind electrical circuits, to magnetism, and eventually to the discovery of electromagnetic waves.

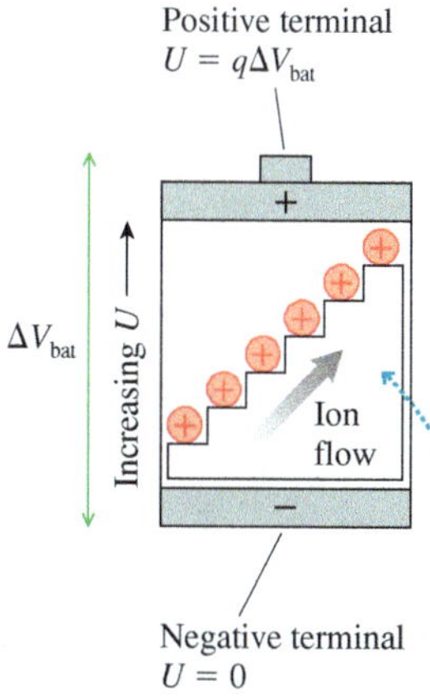

The charge escalator "lifts" charge from the negative side to the positive side. Charge q gains energy $\Delta U = q\Delta V_{bat}$.

Part VII is *Relativity and Quantum Physics.* We'll start by exploring the strange world of Einstein's theory of *relativity,* a world in which space and time aren't quite what they appear to be. Then we will enter the microscopic domain of *atoms,* where the behaviors of light and matter are at complete odds with what our common sense tells us is possible. Although the mathematics of quantum theory quickly gets beyond the level of this text, and time will be running out, you will see that the quantum theory of atoms and nuclei explains many of the things that you learned simply as rules in chemistry.

This picture of an atom would need to be 10 m in diameter if it were drawn to the same scale as the dot representing the nucleus.

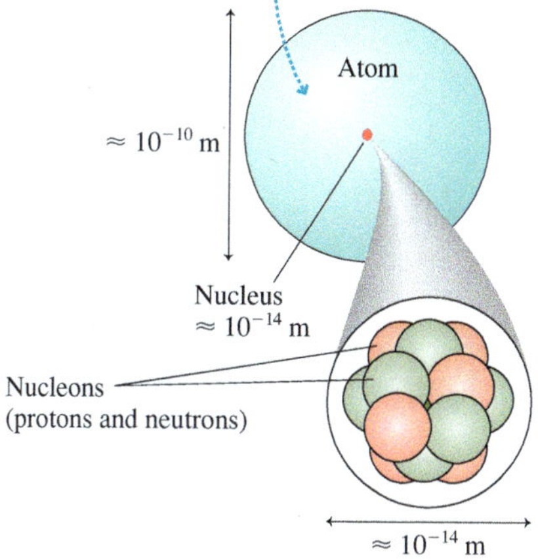

We will not have visited all of physics on our travels. There just isn't time. Many exciting topics, ranging from quarks to black holes, will have to remain unexplored. But this particular journey need not be the last. As you finish this text, you will have the background and the experience to explore new topics further in more advanced courses or for yourself.

With that said, let us take the first step.

PART I

Newton's Laws

Motion can be exhilarating and beautiful. These sailboats are responding to forces of wind, water, and the weight of the crew as they balance precariously on the edge.

OVERVIEW

Why Things Change

Each of the seven parts of this book opens with an overview to give you a look ahead, a glimpse at where your journey will take you in the next few chapters. It's easy to lose sight of the big picture while you're busy negotiating the terrain of each chapter. In Part I, the big picture, in a word, is *change.*

Simple observations of the world around you show that most things change, few things remain the same. Some changes, such as aging, are biological. Others, such as sugar dissolving in your coffee, are chemical. We're going to study change that involves *motion* of one form or another—the motion of balls, cars, and rockets.

There are two big questions we must tackle:

- **How do we describe motion?** It is easy to say that an object moves, but it's not obvious how we should measure or characterize the motion if we want to analyze it mathematically. The mathematical description of motion is called *kinematics,* and it is the subject matter of Chapters 1 through 4.
- **How do we explain motion?** Why do objects have the particular motion they do? Why, when you toss a ball upward, does it go up and then come back down rather than keep going up? Are there "laws of nature" that allow us to predict an object's motion? The explanation of motion in terms of its causes is called *dynamics,* and it is the topic of Chapters 5 through 8.

Two key ideas for answering these questions are *force* (the "cause") and *acceleration* (the "effect"). A variety of pictorial and graphical tools will be developed in Chapters 1 through 5 to help you develop an *intuition* for the connection between force and acceleration. You'll then put this knowledge to use in Chapters 5 through 8 as you analyze motion of increasing complexity.

Another important tool will be the use of *models.* Reality is extremely complicated. We would never be able to develop a science if we had to keep track of every little detail of every situation. A model is a simplified description of reality—much as a model airplane is a simplified version of a real airplane—used to reduce the complexity of a problem to the point where it can be analyzed and understood. We will introduce several important models of motion, paying close attention, especially in these earlier chapters, to where simplifying assumptions are being made, and why.

The "laws of motion" were discovered by Isaac Newton roughly 350 years ago, so the study of motion is hardly cutting-edge science. Nonetheless, it is still extremely important. Mechanics—the science of motion—is the basis for much of engineering and applied science, and many of the ideas introduced here will be needed later to understand things like the motion of waves and the motion of electrons through circuits. Newton's mechanics is the foundation of much of contemporary science, thus we will start at the beginning.

1 Concepts of Motion

Motion takes many forms. The snowboarder seen here is an example of translational motion.

▶ **Looking Ahead** The goal of Chapter 1 is to introduce the fundamental concepts of motion.

The Chapter Preview

Each chapter will start with an overview of the material to come. You should read these chapter previews carefully to get a sense of the road ahead.

Arrows show the flow of ideas in the chapter.

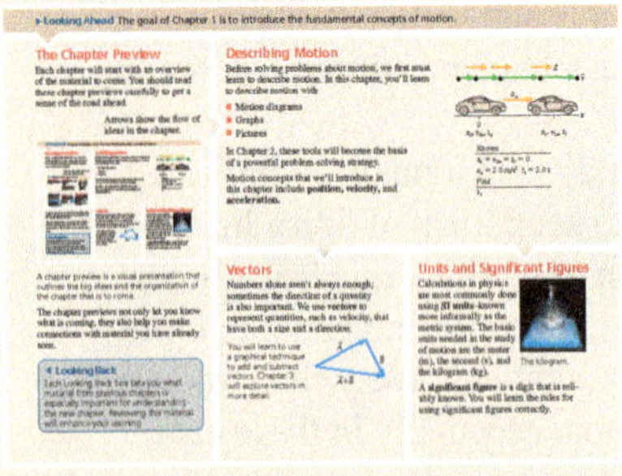

A chapter preview is a visual presentation that outlines the big ideas and the organization of the chapter that is to come.

The chapter previews not only let you know what is coming, they also help you make connections with material you have already seen.

◀ Looking Back

Each Looking Back box tells you what material from previous chapters is especially important for understanding the new chapter. Reviewing this material will enhance your learning.

Describing Motion

Before solving problems about motion, we first must learn to describe motion. In this chapter, you'll learn to describe motion with

- Motion diagrams
- Graphs
- Pictures

In Chapter 2, these tools will become the basis of a powerful problem-solving strategy.

Motion concepts that we'll introduce in this chapter include **position, velocity,** and **acceleration.**

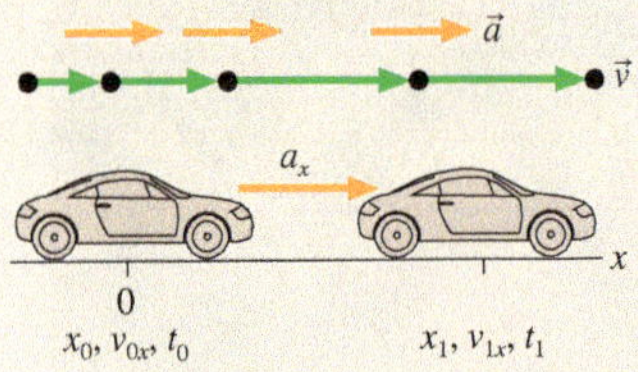

Known
$x_0 = v_{0x} = t_0 = 0$
$a_x = 2.0\ \text{m/s}^2 \quad t_1 = 2.0\ \text{s}$
Find
x_1

Vectors

Numbers alone aren't always enough; sometimes the direction of a quantity is also important. We use **vectors** to represent quantities, such as velocity, that have both a size and a direction.

You will learn to use a graphical technique to add and subtract vectors. Chapter 3 will explore vectors in more detail.

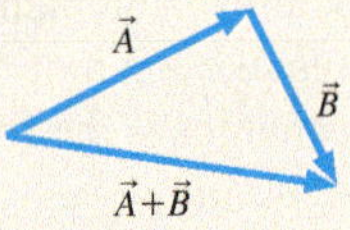

Units and Significant Figures

Calculations in physics are most commonly done using **SI units**–known more informally as the metric system. The basic units needed in the study of motion are the meter (m), the second (s), and the kilogram (kg).

The kilogram.

A **significant figure** is a digit that is reliably known. You will learn the rules for using significant figures correctly.

1.1 Motion Diagrams

Motion is a theme that will appear in one form or another throughout this entire book. Although we all have intuition about motion, based on our experiences, some of the important aspects of motion turn out to be rather subtle. So rather than jumping immediately into a lot of mathematics and calculations, this first chapter focuses on *visualizing* motion and becoming familiar with the *concepts* needed to describe a moving object. Our goal is to lay the foundations for understanding motion.

FIGURE 1.1 Four basic types of motion.

As a starting point, let's define **motion** as the change of an object's position with time. FIGURE 1.1 shows four basic types of motion that we will study in this book. The first three—linear, circular, and projectile motion—in which the object moves through space are called **translational motion.** The path along which the object moves, whether straight or curved, is called the object's **trajectory.** Rotational motion is somewhat different in that rotation is a change of the object's *angular* position. We'll defer rotational motion until later and, for now, focus on translational motion.

FIGURE 1.2 Four frames from the movie of a car.

Making a Motion Diagram

An easy way to study motion is to make a movie of a moving object. A movie camera, as you probably know, takes photographs at a fixed rate, typically 30 photographs every second. Each separate photo is called a *frame,* and the frames are all lined up one after the other in a *filmstrip.* As an example, FIGURE 1.2 shows four frames from the movie of a car going past. Not surprisingly, the car is in a somewhat different position in each frame.

Suppose we cut the individual frames of the filmstrip apart, stack them on top of each other, and project the entire stack at once onto a screen for viewing. The result is shown in FIGURE 1.3. This composite photo, showing an object's position at several *equally spaced instants of time,* is called a **motion diagram.** As the example below shows, we can define concepts such as at rest, constant speed, speeding up, and slowing down in terms of how an object appears in a motion diagram.

NOTE ▶ It's important to keep the camera in a *fixed position* as the object moves by. Don't "pan" it to track the moving object. ◀

FIGURE 1.3 A motion diagram of the car shows all the frames simultaneously.

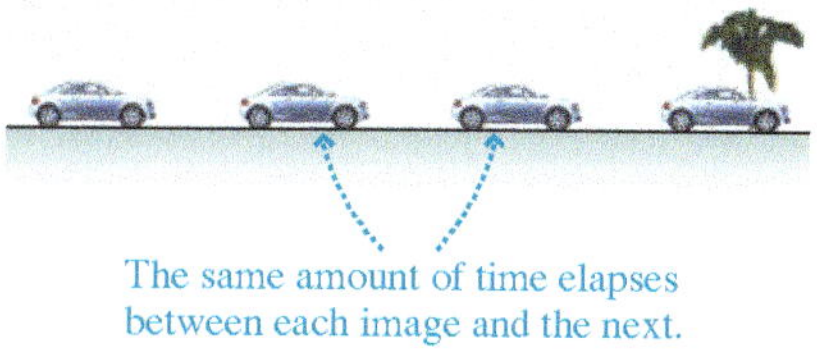

Examples of motion diagrams

An object that occupies only a *single position* in a motion diagram is *at rest.*

A stationary ball on the ground.

Images that are *equally spaced* indicate an object moving with *constant speed.*

A skateboarder rolling down the sidewalk.

An *increasing distance* between the images shows that the object is *speeding up.*

A sprinter starting the 100 meter dash.

A *decreasing distance* between the images shows that the object is *slowing down.*

A car stopping for a red light.

A more complex motion shows aspects of both slowing down (as the ball rises) and speeding up (as the ball falls).

A jump shot from center court.

STOP TO THINK 1.1 Which car is going faster, A or B? Assume there are equal intervals of time between the frames of both movies.

NOTE ▶ Each chapter will have several *Stop to Think* questions. These questions are designed to see if you've understood the basic ideas that have been presented. The answers are given at the end of the chapter, but you should make a serious effort to think about these questions before turning to the answers. If you answer correctly, and are sure of your answer rather than just guessing, you can proceed to the next section with confidence. But if you answer incorrectly, it would be wise to reread the preceding sections before proceeding onward. ◀

1.2 The Particle Model

For many types of motion, such as that of balls, cars, and rockets, the motion of the object *as a whole* is not influenced by the details of the object's size and shape. All we really need to keep track of is the motion of a single point on the object, so we can treat the object *as if* all its mass were concentrated into this single point. An object that can be represented as a mass at a single point in space is called a **particle.** A particle has no size, no shape, and no distinction between top and bottom or between front and back.

If we treat an object as a particle, we can represent the object in each frame of a motion diagram as a simple dot rather than having to draw a full picture. FIGURE 1.4 shows how much simpler motion diagrams appear when the object is represented as a particle. Note that the dots have been numbered 0, 1, 2, . . . to tell the sequence in which the frames were exposed.

FIGURE 1.4 Motion diagrams in which the object is represented as a particle.

(a) Motion diagram of a rocket launch

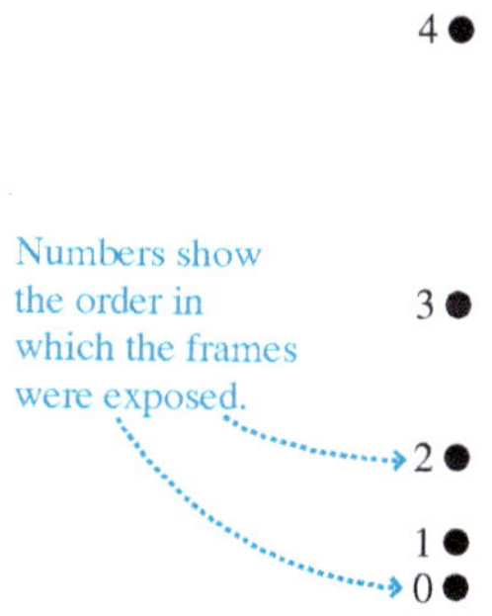

(b) Motion diagram of a car stopping

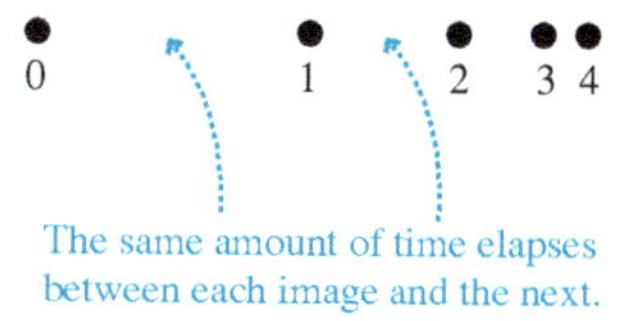

Using the Particle Model

Treating an object as a particle is, of course, a simplification of reality. As we noted in the Part I Overview, such a simplification is called a *model.* Models allow us to focus on the important aspects of a phenomenon by excluding those aspects that play only a minor role. The **particle model** of motion is a simplification in which we treat a moving object as if all of its mass were concentrated at a single point. The particle model is an excellent approximation of reality for the translational motion of cars, planes, rockets, and similar objects. In later chapters, we'll find that the motion of more complex objects, which cannot be treated as a single particle, can often be analyzed as if the object were a collection of particles.

Not all motions can be reduced to the motion of a single point. Consider a rotating gear. The center of the gear doesn't move at all, and each tooth on the gear is moving in a different direction. Rotational motion is qualitatively different than translational motion, and we'll need to go beyond the particle model later when we study rotational motion.

STOP TO THINK 1.2 Three motion diagrams are shown. Which is a dust particle settling to the floor at constant speed, which is a ball dropped from the roof of a building, and which is a descending rocket slowing to make a soft landing on Mars?

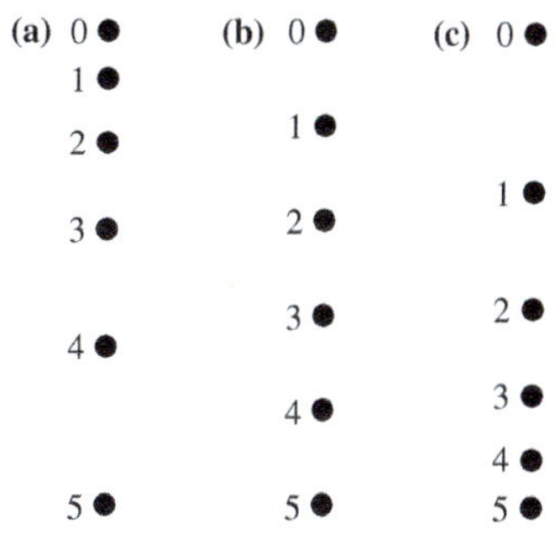

1.3 Position and Time

As we look at a motion diagram, it would be useful to know *where* the object is (i.e., its *position*) and *when* the object was at that position (i.e., the *time*). Position measurements can be made by laying a coordinate system grid over a motion diagram. You can then measure the (x, y) coordinates of each point in the motion diagram. Of course, the world does not come with a coordinate system attached. A coordinate system is an artificial grid that *you* place over a problem in order to analyze the motion. You place the origin of your coordinate system wherever you wish, and different observers of a moving object might all choose to use different origins. Likewise, you can choose the orientation of the x-axis and y-axis to be helpful for that particular problem. The conventional choice is for the x-axis to point to the right and the y-axis to point upward, but there is nothing sacred about this choice. We will soon have many occasions to tilt the axes at an angle.

Time, in a sense, is also a coordinate system, although you may never have thought of time this way. You can pick an arbitrary point in the motion and label it "$t = 0$ seconds."

This is simply the instant you decide to start your clock or stopwatch, so it is the origin of your time coordinate. Different observers might choose to start their clocks at different moments. A movie frame labeled "$t = 4$ seconds" was taken 4 seconds after you started your clock.

We typically choose $t = 0$ to represent the "beginning" of a problem, but the object may have been moving before then. Those earlier instants would be measured as negative times, just as objects on the x-axis to the left of the origin have negative values of position. Negative numbers are not to be avoided; they simply locate an event in space or time *relative to an origin.*

To illustrate, FIGURE 1.5a shows an xy-coordinate system and time information superimposed over the motion diagram of a basketball. You can see that the ball's position is $(x_4, y_4) = (12\text{ m}, 9\text{ m})$ at time $t_4 = 2.0$ s. Notice how we've used subscripts to indicate the time and the object's position in a specific frame of the motion diagram.

FIGURE 1.5 Position and time measurements made on the motion diagram of a basketball.

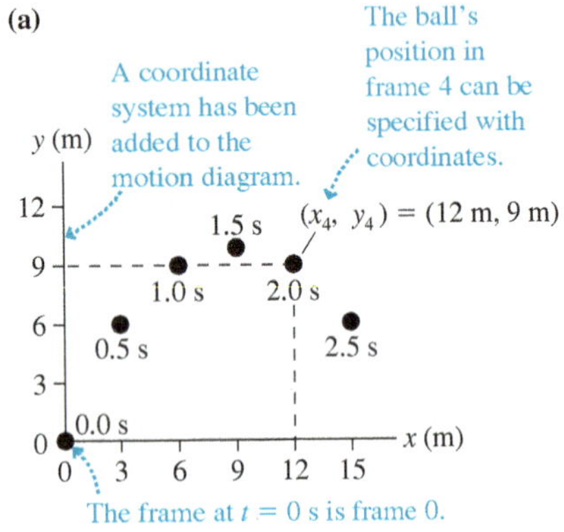

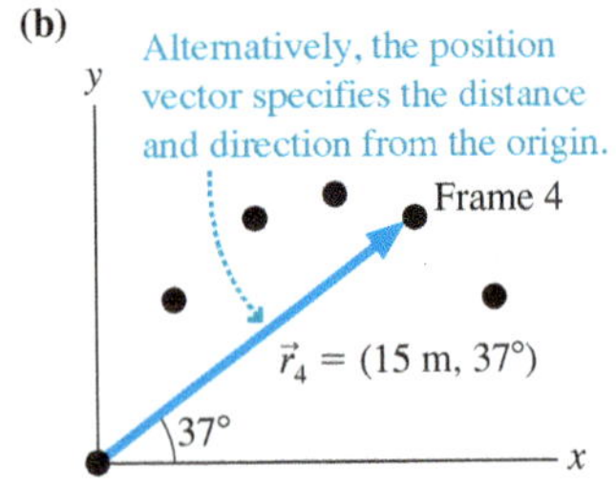

NOTE ▶ The frame at $t = 0$ is frame 0. That is why the fifth frame is labeled 4. ◀

Another way to locate the ball is to draw an arrow from the origin to the point representing the ball. You can then specify the length and direction of the arrow. An arrow drawn from the origin to an object's position is called the **position vector** of the object, and it is given the symbol $\vec{r}$. FIGURE 1.5B shows the position vector $\vec{r}_4 = (15\text{ m}, 37°)$.

The position vector $\vec{r}$ does not tell us anything different than the coordinates (x, y). It simply provides the information in an alternative form. Although you're more familiar with coordinates than with vectors, you will find that vectors are a useful way to describe many concepts in physics.

A Word About Vectors and Notation

Some physical quantities, such as time, mass, and temperature, can be described completely by a single number with a unit. For example, the mass of an object is 6 kg and its temperature is 30°C. A physical quantity described by a single number (with a unit) is called a **scalar quantity.** A scalar can be positive, negative, or zero.

Many other quantities, however, have a directional quality and cannot be described by a single number. To describe the motion of a car, for example, you must specify not only how fast it is moving, but also the *direction* in which it is moving. A **vector quantity** is a quantity having both a *size* (the "How far?" or "How fast?") and a *direction* (the "Which way?"). The size or length of a vector is called its *magnitude.* The magnitude of a vector can be positive or zero, but it cannot be negative. Vectors will be studied thoroughly in Chapter 3, so all we need for now is a little basic information.

We indicate a vector by drawing an arrow over the letter that represents the quantity. Thus $\vec{r}$ and $\vec{A}$ are symbols for vectors, whereas r and A, without the arrows, are symbols for scalars. In handwritten work you must draw arrows over all symbols that represent vectors. This may seem strange until you get used to it, but it is very important because we will often use both r and $\vec{r}$, or both A and $\vec{A}$, in the same problem, and they mean different things! Without the arrow, you will be using the same symbol with two different meanings and will likely end up making a mistake. Note that the arrow over the symbol always points to the right, regardless of which direction the actual vector points. Thus we write $\vec{r}$ or $\vec{A}$, never $\overleftarrow{r}$ or $\overleftarrow{A}$.

Displacement

Consider the following:

> Sam is standing 50 feet (ft) east of the corner of 12th Street and Vine. He then walks northeast for 100 ft to a second point. What is Sam's change of position?

FIGURE 1.6 shows Sam's motion in terms of position vectors. Sam's initial position is the vector $\vec{r}_0$ drawn from the origin to the point where he starts walking. Vector $\vec{r}_1$ is his position after he finishes walking. You can see that Sam has changed position, and a *change* of position is called a **displacement.** His displacement is the vector labeled $\Delta\vec{r}$. The Greek letter delta (Δ) is used in math and science to indi cate the *change* in a quantity. Here it indicates a change in the position $\vec{r}$.

NOTE ▶ $\Delta\vec{r}$ is a *single* symbol. You cannot cancel out or remove the Δ in algebraic operations. ◀

FIGURE 1.6 Sam undergoes a displacement $\Delta\vec{r}$ from position $\vec{r}_0$ to position $\vec{r}_1$.

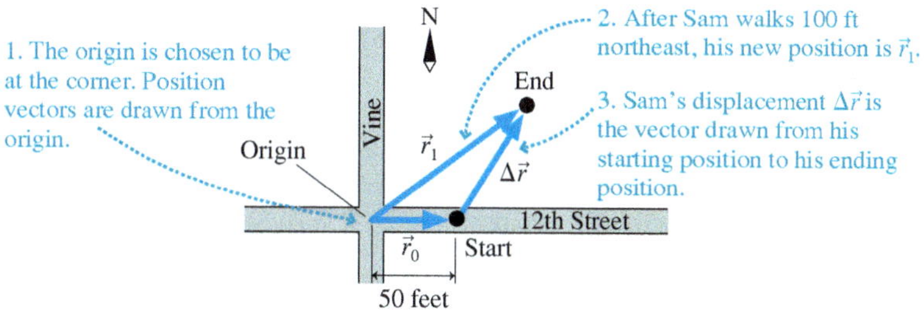

Displacement is a vector quantity; it requires both a length and a direction to describe it. Specifically, the displacement $\Delta\vec{r}$ is a vector drawn *from* a starting position *to* an ending position. Sam's displacement is written

$$\Delta\vec{r} = (100 \text{ ft, northeast})$$

The length, or magnitude, of a displacement vector is simply the straight-line distance between the starting and ending positions.

Sam's final position in Figure 1.6, vector $\vec{r}_1$, can be seen as a combination of where he started, vector $\vec{r}_0$, plus the vector $\Delta\vec{r}$ representing his change of position. In fact, $\vec{r}_1$ is the *vector sum* of vectors $\vec{r}_0$ and $\Delta\vec{r}$. This is written

$$\vec{r}_1 = \vec{r}_0 + \Delta\vec{r} \tag{1.1}$$

Notice, however, that we are adding vector quantities, not numbers. Vector addition is a different process from "regular" addition. We'll explore vector addition more thoroughly in Chapter 3, but for now you can add two vectors $\vec{A}$ and $\vec{B}$ with the three-step procedure shown in Tactics Box 1.1.

TACTICS BOX 1.1 Vector addition

To add $\vec{B}$ to $\vec{A}$:

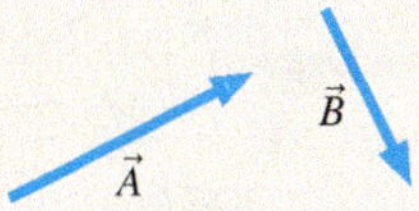

❶ Draw $\vec{A}$.

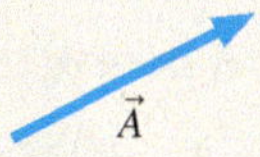

❷ Place the tail of $\vec{B}$ at the tip of $\vec{A}$.

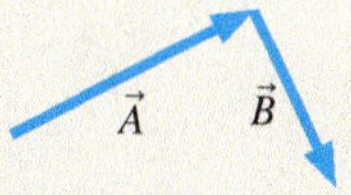

❸ Draw an arrow from the tail of $\vec{A}$ to the tip of $\vec{B}$. This is vector $\vec{A} + \vec{B}$.

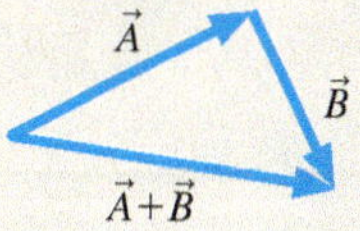

If you examine Figure 1.6, you'll see that the steps of Tactics Box 1.1 are exactly how $\vec{r}_0$ and $\Delta\vec{r}$ are added to give $\vec{r}_1$.

NOTE ▶ A vector is not tied to a particular location on the page. You can move a vector around as long as you don't change its length or the direction it points. Vector $\vec{B}$ is not changed by sliding it to where its tail is at the tip of $\vec{A}$. ◀

FIGURE 1.7 Sam's displacement $\Delta\vec{r}$ is unchanged by using a different coordinate system.

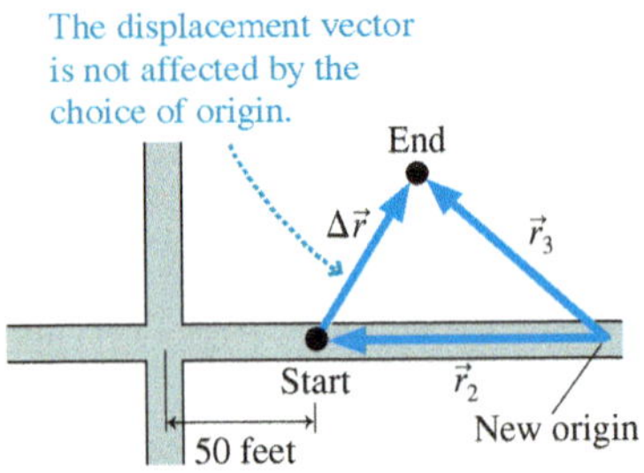

In Figure 1.6, we chose *arbitrarily* to put the origin of the coordinate system at the corner. While this might be convenient, it certainly is not mandatory. FIGURE 1.7 shows a different choice of where to place the origin. Notice something interesting. The initial and final position vectors $\vec{r}_0$ and $\vec{r}_1$ have become new vectors $\vec{r}_2$ and $\vec{r}_3$, but the displacement vector $\Delta\vec{r}$ has not changed! **The displacement is a quantity that is independent of the coordinate system.** In other words, the arrow drawn from one position of an object to the next is the same no matter what coordinate system you choose.

This observation suggests that the displacement, rather than the actual position, is what we want to focus on as we analyze the motion of an object. Equation 1.1 told us that $\vec{r}_1 = \vec{r}_0 + \Delta\vec{r}$. This is easily rearranged to give a more precise definition of displacement: **The displacement $\Delta\vec{r}$ of an object as it moves from an initial position $\vec{r}_i$ to a final position $\vec{r}_f$ is**

$$\Delta\vec{r} = \vec{r}_f - \vec{r}_i \tag{1.2}$$

Graphically, $\Delta\vec{r}$ is a vector arrow drawn from position $\vec{r}_i$ to position $\vec{r}_f$. The displacement vector is independent of the coordinate system.

NOTE ▶ To be more general, we've written Equation 1.2 in terms of an *initial position* and a *final position,* indicated by subscripts i and f. We'll frequently use i and f when writing general equations, then use specific numbers or values, such as 0 and 1, when working a problem. ◀

FIGURE 1.8 The negative of a vector.

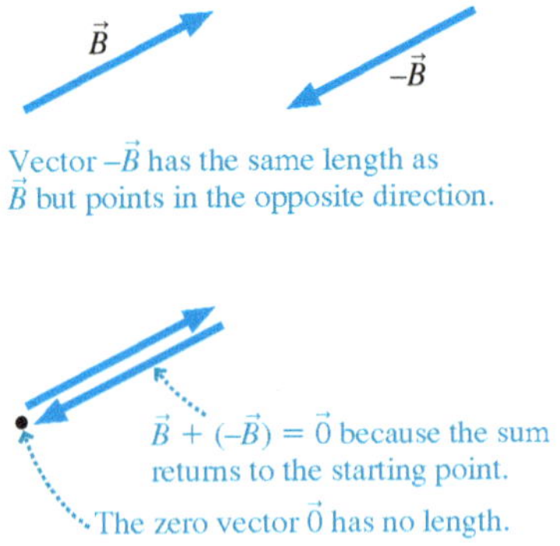

This definition of $\Delta\vec{r}$ involves *vector subtraction.* With numbers, subtraction is the same as the addition of a negative number. That is, $5 - 3$ is the same as $5 + (-3)$. Similarly, we can use the rules for vector addition to find $\vec{A} - \vec{B} = \vec{A} + (-\vec{B})$ if we first define what we mean by $-\vec{B}$. As FIGURE 1.8 shows, the negative of vector $\vec{B}$ is a vector with the same length but pointing in the opposite direction. This makes sense because $\vec{B} - \vec{B} = \vec{B} + (-\vec{B}) = \vec{0}$, where $\vec{0}$, a vector with zero length, is called the **zero vector.**

TACTICS BOX 1.2 Vector subtraction

To subtract $\vec{B}$ from $\vec{A}$:

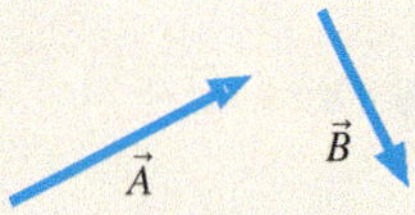

❶ Draw $\vec{A}$.

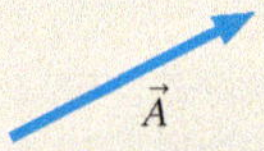

❷ Place the tail of $-\vec{B}$ at the tip of $\vec{A}$.

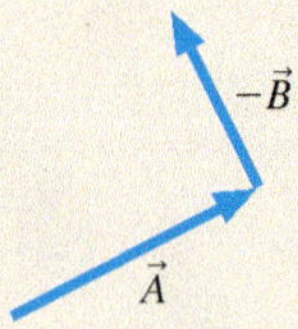

❸ Draw an arrow from the tail of $\vec{A}$ to the tip of $-\vec{B}$. This is vector $\vec{A} - \vec{B}$.

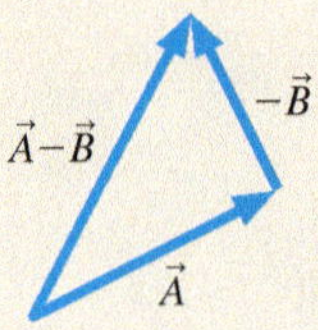

FIGURE 1.9 uses the vector subtraction rules of Tactics Box 1.2 to prove that the displacement $\Delta\vec{r}$ is simply the vector connecting the dots of a motion diagram.

FIGURE 1.9 Using vector subtraction to find $\Delta\vec{r} = \vec{r}_f - \vec{r}_i$.

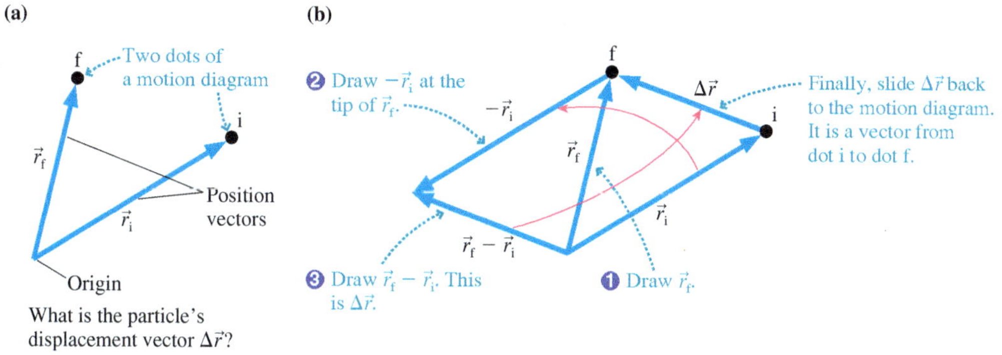

Application to Motion Diagrams

The first step in analyzing a motion diagram is to determine all of the displacement vectors. As Figure 1.9 shows, the displacement vectors are simply the arrows connecting each dot to the next. Label each arrow with a *vector* symbol $\Delta\vec{r}_n$, starting with $n = 0$. FIGURE 1.10 shows the motion diagrams of Figure 1.4 redrawn to include the displacement vectors. You do not need to show the position vectors.

FIGURE 1.10 Motion diagrams with the displacement vectors.

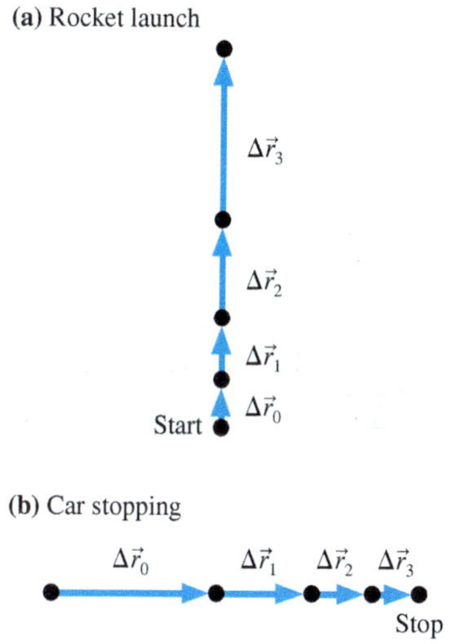

NOTE ▶ When an object either starts from rest or ends at rest, the initial or final dots are *as close together* as you can draw the displacement vector arrow connecting them. In addition, just to be clear, you should write "Start" or "Stop" beside the initial or final dot. It is important to distinguish stopping from merely slowing down. ◀

Now we can conclude, more precisely than before, that, as time proceeds:

- An object is speeding up if its displacement vectors are increasing in length.
- An object is slowing down if its displacement vectors are decreasing in length.

EXAMPLE 1.1 Headfirst into the snow

Alice is sliding along a smooth, icy road on her sled when she suddenly runs headfirst into a large, very soft snowbank that gradually brings her to a halt. Draw a motion diagram for Alice. Show and label all displacement vectors.

MODEL Use the particle model to represent Alice as a dot.

VISUALIZE FIGURE 1.11 shows Alice's motion diagram. The problem statement suggests that Alice's speed is very nearly constant until she hits the snowbank. Thus her displacement vectors are of equal length as she slides along the icy road. She begins slowing when she hits the snowbank, so the displacement vectors then get shorter until she stops. We're told that her stop is gradual, so we want the vector lengths to get shorter gradually rather than suddenly.

FIGURE 1.11 Alice's motion diagram.

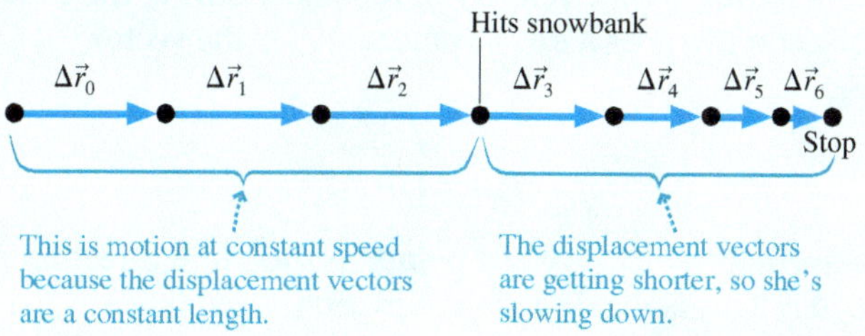

A stopwatch is used to measure a time interval.

Time Interval

It's also useful to consider a *change* in time. For example, the clock readings of two frames of film might be t_1 and t_2. The specific values are arbitrary because they are timed relative to an arbitrary instant that you chose to call $t = 0$. But the **time interval** $\Delta t = t_2 - t_1$ is *not* arbitrary. It represents the elapsed time for the object to move from one position to the next. All observers will measure the same value for Δt, regardless of when they choose to start their clocks.

The time interval $\Delta t = t_f - t_i$ measures the elapsed time as an object moves from an initial position $\vec{r}_i$ at time t_i to a final position $\vec{r}_f$ at time t_f. The value of Δt is independent of the specific clock used to measure the times.

To summarize the main idea of this section, we have added coordinate systems and clocks to our motion diagrams in order to measure *when* each frame was exposed and *where* the object was located at that time. Different observers of the motion may choose different coordinate systems and different clocks. However, all observers find the *same* values for the displacements $\Delta\vec{r}$ and the time intervals Δt because these are independent of the specific coordinate system used to measure them.

1.4 Velocity

The victory goes to the runner with the highest average speed.

It's no surprise that, during a given time interval, a speeding bullet travels farther than a speeding snail. To extend our study of motion so that we can compare the bullet to the snail, we need a way to measure how fast or how slowly an object moves.

One quantity that measures an object's fastness or slowness is its **average speed,** defined as the ratio

$$\text{average speed} = \frac{\text{distance traveled}}{\text{time interval spent traveling}} = \frac{d}{\Delta t} \tag{1.3}$$

If you drive 15 miles (mi) in 30 minutes ($\frac{1}{2}$ h), your average speed is

$$\text{average speed} = \frac{15 \text{ mi}}{\frac{1}{2} \text{ h}} = 30 \text{ mph} \tag{1.4}$$

Although the concept of speed is widely used in our day-to-day lives, it is not a sufficient basis for a science of motion. To see why, imagine you're trying to land a jet plane on an aircraft carrier. It matters a great deal to you whether the aircraft carrier is moving at 20 mph (miles per hour) to the north or 20 mph to the east. Simply knowing that the boat's speed is 20 mph is not enough information!

It's the displacement $\Delta\vec{r}$, a vector quantity, that tells us not only the distance traveled by a moving object, but also the *direction* of motion. Consequently, a more useful ratio than $d/\Delta t$ is the ratio $\Delta\vec{r}/\Delta t$. This ratio is a vector because $\Delta\vec{r}$ is a vector, so it has both a magnitude and a direction. The size, or magnitude, of this ratio will be larger for a fast object than for a slow object. But in addition to measuring how fast an object moves, this ratio is a vector that points in the direction of motion.

It is convenient to give this ratio a name. We call it the **average velocity,** and it has the symbol $\vec{v}_{avg}$. **The average velocity of an object during the time interval Δt, in which the object undergoes a displacement $\Delta\vec{r}$, is the vector**

$$\vec{v}_{avg} = \frac{\Delta\vec{r}}{\Delta t} \tag{1.5}$$

An object's average velocity vector points in the same direction as the displacement vector $\Delta\vec{r}$. This is the direction of motion.

NOTE ▶ In everyday language we do not make a distinction between speed and velocity, but in physics *the distinction is very important.* In particular, speed is simply "How fast?" whereas velocity is "How fast, and in which direction?" As we go along we will be giving other words more precise meanings in physics than they have in everyday language. ◀

As an example, FIGURE 1.12a shows two ships that move 5 miles in 15 minutes. Using Equation 1.5 with $\Delta t = 0.25$ h, we find

$$\vec{v}_{\text{avg A}} = (20 \text{ mph, north})$$
$$\vec{v}_{\text{avg B}} = (20 \text{ mph, east}) \tag{1.6}$$

Both ships have a speed of 20 mph, but their velocities are different. Notice how the velocity *vectors* in FIGURE 1.12b point in the direction of motion.

NOTE ▶ Our goal in this chapter is to *visualize* motion with motion diagrams. Strictly speaking, the vector we have defined in Equation 1.5, and the vector we will show on motion diagrams, is the *average* velocity $\vec{v}_{\text{avg}}$. But to allow the motion diagram to be a useful tool, we will drop the subscript and refer to the average velocity as simply $\vec{v}$. Our definitions and symbols, which somewhat blur the distinction between average and instantaneous quantities, are adequate for visualization purposes, but they're not the final word. We will refine these definitions in Chapter 2, where our goal will be to develop the mathematics of motion. ◀

FIGURE 1.12 The displacement vectors and velocities of ships A and B.

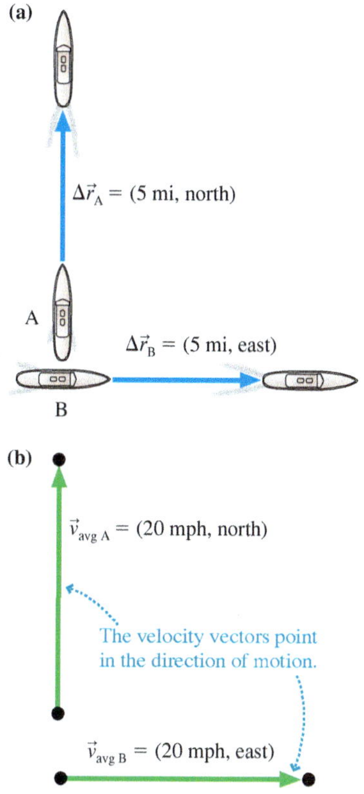

Motion Diagrams with Velocity Vectors

The velocity vector points in the same direction as the displacement $\Delta\vec{r}$, and the length of $\vec{v}$ is directly proportional to the length of $\Delta\vec{r}$. Consequently, the vectors connecting each dot of a motion diagram to the next, which we previously labeled as displacements, could equally well be identified as velocity vectors.

This idea is illustrated in FIGURE 1.13, which shows four frames from the motion diagram of a tortoise racing a hare. The vectors connecting the dots are now labeled as velocity vectors $\vec{v}$. **The length of a velocity vector represents the average speed with which the object moves between the two points.** Longer velocity vectors indicate faster motion. You can see that the hare moves faster than the tortoise.

Notice that the hare's velocity vectors do not change; each has the same length and direction. We say the hare is moving with *constant velocity.* The tortoise is also moving with its own constant velocity.

FIGURE 1.13 Motion diagram of the tortoise racing the hare.

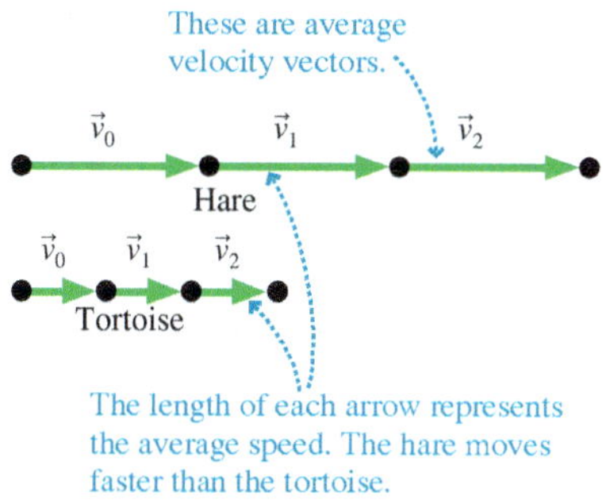

EXAMPLE 1.2 **Accelerating up a hill**

The light turns green and a car accelerates, starting from rest, up a 20° hill. Draw a motion diagram showing the car's velocity.

MODEL Use the particle model to represent the car as a dot.

VISUALIZE The car's motion takes place along a straight line, but the line is neither horizontal nor vertical. Because a motion diagram is made from frames of a movie, it will show the object moving with the correct orientation—in this case, at an angle of 20°. FIGURE 1.14 shows several frames of the motion diagram, where we see the car speeding up. The car starts from rest, so the first arrow is drawn as short as possible and the first dot is labeled "Start." The displacement vectors have been drawn from each dot to the next, but then they are identified and labeled as average velocity vectors $\vec{v}$.

FIGURE 1.14 Motion diagram of a car accelerating up a hill.

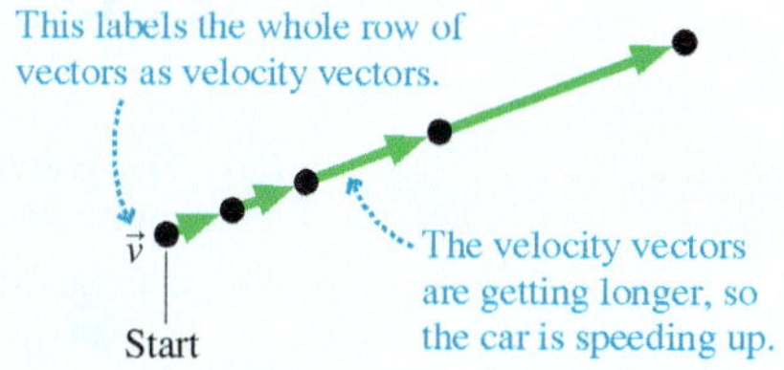

NOTE ▶ Rather than label every single vector, it's easier to give one label to the entire row of velocity vectors. You can see this in Figure 1.14. ◀

EXAMPLE 1.3 It's a hit!

Jake hits a ball at a 60° angle above horizontal. It is caught by Jose. Draw a motion diagram of the ball.

MODEL This example is typical of how many problems in science and engineering are worded. The problem does not give a clear statement of where the motion begins or ends. Are we interested in the motion of the ball just during the time it is in the air between Jake and Jose? What about the motion *as* Jake hits it (ball rapidly speeding up) or *as* Jose catches it (ball rapidly slowing down)? The point is that *you* will often be called on to make a *reasonable interpretation* of a problem statement. In this problem, the details of hitting and catching the ball are complex. The motion of the ball through the air is easier to describe, and it's a motion you might expect to learn about in a physics class. So our *interpretation* is that the motion diagram should start as the ball leaves Jake's bat (ball already moving) and should end the instant it touches Jose's hand (ball still moving). We will model the ball as a particle.

VISUALIZE With this interpretation in mind, **FIGURE 1.15** shows the motion diagram of the ball. Notice how, in contrast to the car of Figure 1.14, the ball is already moving as the motion diagram movie begins. As before, the average velocity vectors are found by connecting the dots with *straight* arrows. You can see that the average velocity vectors get shorter (ball slowing down), get longer (ball speeding up), and change direction. Each $\vec{v}$ is different, so this is *not* constant-velocity motion.

FIGURE 1.15 Motion diagram of a ball traveling from Jake to Jose.

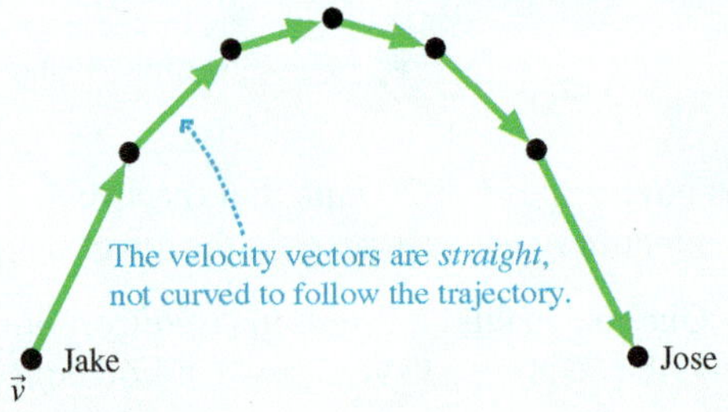

STOP TO THINK 1.3 A particle moves from position 1 to position 2 during the interval Δt. Which vector shows the particle's average velocity?

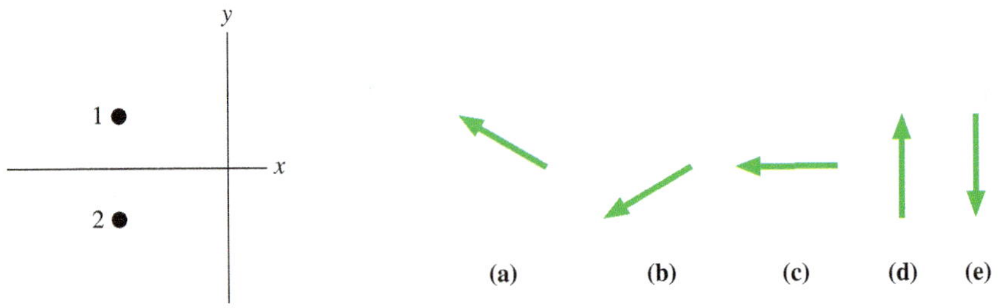

1.5 Linear Acceleration

The goal of this chapter is to find a set of concepts with which to describe motion. Position, time, and velocity are important concepts, and at first glance they might appear to be sufficient. But that is not the case. Sometimes an object's velocity is constant, as it was in Figure 1.13. More often, an object's velocity changes as it moves, as in Figure 1.14 and 1.15. We need one more motion concept, one that will describe a *change* in the velocity.

Because velocity is a vector, it can change in two possible ways:

1. The magnitude can change, indicating a change in speed; or
2. The direction can change, indicating that the object has changed direction.

We will concentrate for now on the first case, a change in speed. The car accelerating up a hill in Figure 1.14 was an example in which the magnitude of the velocity vector changed but not the direction. We'll return to the second case in Chapter 4.

When we wanted to measure changes in position, the ratio $\Delta\vec{r}/\Delta t$ was useful. This ratio is the *rate of change of position.* By analogy, consider an object whose velocity changes from $\vec{v}_1$ to $\vec{v}_2$ during the time interval Δt. Just as $\Delta\vec{r} = \vec{r}_2 - \vec{r}_1$ is the change of position, the quantity $\Delta\vec{v} = \vec{v}_2 - \vec{v}_1$ is the change of velocity. The ratio $\Delta\vec{v}/\Delta t$ is then the *rate of change of velocity.* It has a large magnitude for objects that speed up quickly and a small magnitude for objects that speed up slowly.

The ratio $\Delta\vec{v}/\Delta t$ is called the **average acceleration**, and its symbol is $\vec{a}_{avg}$. **The average acceleration of an object during the time interval Δt, in which the object's velocity changes by $\Delta\vec{v}$, is the vector**

$$\vec{a}_{avg} = \frac{\Delta\vec{v}}{\Delta t} \qquad (1.7)$$

The average acceleration vector points in the same direction as the vector $\Delta\vec{v}$.

Acceleration is a fairly abstract concept. Yet it is essential to develop a good intuition about acceleration because it will be a key concept for understanding why objects move as they do. Motion diagrams will be an important tool for developing that intuition.

NOTE ▶ As we did with velocity, we will drop the subscript and refer to the average acceleration as simply $\vec{a}$. This is adequate for visualization purposes, but not the final word. We will refine the definition of acceleration in Chapter 2. ◀

The Audi TT accelerates from 0 to 60 mph in 6 s.

Finding the Acceleration Vectors on a Motion Diagram

Let's look at how we can determine the average acceleration vector $\vec{a}$ from a motion diagram. From its definition, Equation 1.7, we see that $\vec{a}$ points in the same direction as $\Delta\vec{v}$, the change of velocity. This critical idea is the basis for a technique to find $\vec{a}$.

TACTICS BOX 1.3 Finding the acceleration vector

MP

To find the acceleration as the velocity changes from $\vec{v}_n$ to $\vec{v}_{n+1}$, we must determine the *change* of velocity $\Delta\vec{v} = \vec{v}_{n+1} - \vec{v}_n$.

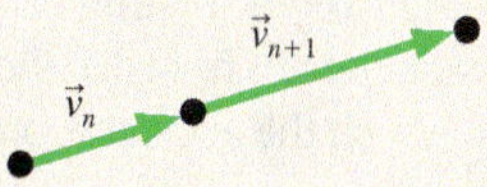

❶ Draw the velocity vector $\vec{v}_{n+1}$.

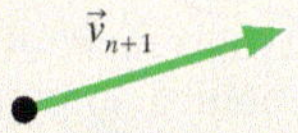

❷ Draw $-\vec{v}_n$ at the tip of $\vec{v}_{n+1}$.

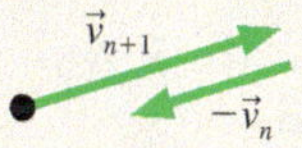

❸ Draw $\Delta\vec{v} = \vec{v}_{n+1} - \vec{v}_n$
$= \vec{v}_{n+1} + (-\vec{v}_n)$
This is the direction of $\vec{a}$.

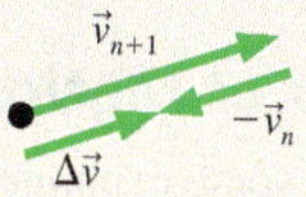

❹ Return to the original motion diagram. Draw a vector at the middle point in the direction of $\Delta\vec{v}$; label it $\vec{a}$. This is the average acceleration at the midpoint between $\vec{v}_n$ and $\vec{v}_{n+1}$.

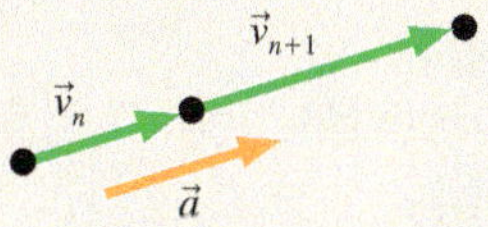

Exercises 21–24

Many Tactics Boxes will refer you to exercises in the *Student Workbook* where you can practice the new skill.

Notice that the acceleration vector goes beside the middle dot, not beside the velocity vectors. This is because each acceleration vector is determined as the *difference* between the *two* velocity vectors on either side of a dot. The length of $\vec{a}$ does not have to be the exact length of $\Delta\vec{v}$; it is the direction of $\vec{a}$ that is most important.

The procedure of Tactics Box 1.3 can be repeated to find $\vec{a}$ at each point in the motion diagram. Note that we cannot determine $\vec{a}$ at the first and last points because we have only one velocity vector and can't find $\Delta\vec{v}$.

The Complete Motion Diagram

You've now seen several *Tactics Boxes* that help you accomplish specific tasks. Tactics Boxes will appear in nearly every chapter in this book. We'll also, where appropriate, provide *Problem-Solving Strategies.*

PROBLEM-SOLVING STRATEGY 1.1 Motion diagrams

MP

MODEL Represent the moving object as a particle. Make simplifying assumptions when interpreting the problem statement.

VISUALIZE A complete motion diagram consists of:

- The position of the object in each frame of the film, shown as a dot. Use five or six dots to make the motion clear but without overcrowding the picture. More complex motions may need more dots.
- The average velocity vectors, found by connecting each dot in the motion diagram to the next with a vector arrow. There is *one* velocity vector linking each *two* position dots. Label the row of velocity vectors $\vec{v}$.
- The average acceleration vectors, found using Tactics Box 1.3. There is *one* acceleration vector linking each *two* velocity vectors. Each acceleration vector is drawn at the dot between the two velocity vectors it links. Use $\vec{0}$ to indicate a point at which the acceleration is zero. Label the row of acceleration vectors $\vec{a}$.

STOP TO THINK 1.4 A particle undergoes acceleration $\vec{a}$ while moving from point 1 to point 2. Which of the choices shows the velocity vector $\vec{v}_2$ as the particle moves away from point 2?

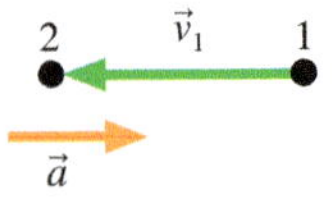

Examples of Motion Diagrams

Let's look at some examples of the full strategy for drawing motion diagrams.

EXAMPLE 1.4 The first astronauts land on Mars

A spaceship carrying the first astronauts to Mars descends safely to the surface. Draw a motion diagram for the last few seconds of the descent.

MODEL Represent the spaceship as a particle. It's reasonable to assume that its motion in the last few seconds is straight down. The problem ends as the spacecraft touches the surface.

VISUALIZE FIGURE 1.16 shows a complete motion diagram as the spaceship descends and slows, using its rockets, until it comes to rest on the surface. Notice how the dots get closer together as it slows. The inset shows how the acceleration vector $\vec{a}$ is determined at one point. All the other acceleration vectors will be similar, because for each pair of velocity vectors the earlier one is longer than the later one.

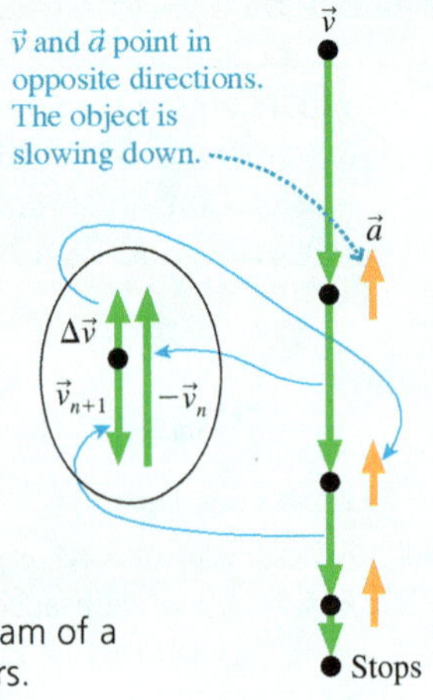

FIGURE 1.16 Motion diagram of a spaceship landing on Mars.

EXAMPLE 1.5 Skiing through the woods

A skier glides along smooth, horizontal snow at constant speed, then speeds up going down a hill. Draw the skier's motion diagram.

MODEL Represent the skier as a particle. It's reasonable to assume that the downhill slope is a straight line. Although the motion as a whole is not linear, we can treat the skier's motion as two separate linear motions.

VISUALIZE FIGURE 1.17 shows a complete motion diagram of the skier. The dots are equally spaced for the horizontal motion, indicating constant speed; then the dots get farther apart as the skier speeds up going down the hill. The insets show how the average acceleration vector $\vec{a}$ is determined for the horizontal motion and along the slope. All the other acceleration vectors along the slope will be similar to the one shown because each velocity vector is longer than the preceding one. Notice that we've explicitly written $\vec{0}$ for the acceleration beside the dots where the velocity is constant. The acceleration at the point where the direction changes will be considered in Chapter 4.

FIGURE 1.17 Motion diagram of a skier.

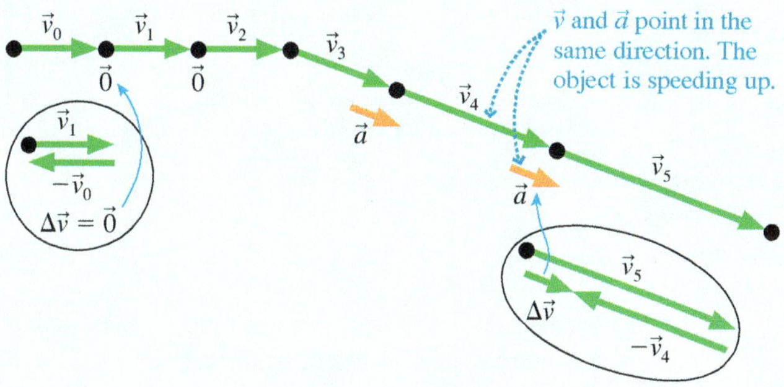

Notice something interesting in Figure 1.16 and 1.17. Where the object is speeding up, the acceleration and velocity vectors point in the *same direction.* Where the object is slowing down, the acceleration and velocity vectors point in *opposite directions.* These results are always true for motion in a straight line. **For motion along a line:**

- **An object is speeding up if and only if $\vec{v}$ and $\vec{a}$ point in the same direction.**
- **An object is slowing down if and only if $\vec{v}$ and $\vec{a}$ point in opposite directions.**
- **An object's velocity is constant if and only if $\vec{a} = \vec{0}$.**

NOTE ▶ In everyday language, we use the word *accelerate* to mean "speed up" and the word *decelerate* to mean "slow down." But speeding up and slowing down are both changes in the velocity and consequently, by our definition, *both* are accelerations. In physics, *acceleration* refers to changing the velocity, no matter what the change is, and not just to speeding up. ◀

EXAMPLE 1.6 Tossing a ball

Draw the motion diagram of a ball tossed straight up in the air.

MODEL This problem calls for some interpretation. Should we include the toss itself, or only the motion after the ball is released? Should we include the ball hitting the ground? It appears that this problem is really concerned with the ball's motion through the air. Consequently, we begin the motion diagram at the moment that the tosser releases the ball and end the diagram at the moment the ball hits the ground. We will consider neither the toss nor the impact. And, of course, we will represent the ball as a particle.

VISUALIZE We have a slight difficulty here because the ball retraces its route as it falls. A literal motion diagram would show

Continued

the upward motion and downward motion on top of each other, leading to confusion. We can avoid this difficulty by horizontally separating the upward motion and downward motion diagrams. This will not affect our conclusions because it does not change any of the vectors. FIGURE 1.18 shows the motion diagram drawn this way. Notice that the very top dot is shown twice—as the end point of the upward motion and the beginning point of the downward motion.

The ball slows down as it rises. You've learned that the acceleration vectors point opposite the velocity vectors for an object that is slowing down along a line, and they are shown accordingly. Similarly, $\vec{a}$ and $\vec{v}$ point in the same direction as the falling ball speeds up. Notice something interesting: The acceleration vectors point downward both while the ball is rising *and* while it is falling. Both "speeding up" and "slowing down" occur with the *same* acceleration vector. This is an important conclusion, one worth pausing to think about.

Now let's look at the top point on the ball's trajectory. The velocity vectors are pointing upward but getting shorter as the ball approaches the top. As the ball starts to fall, the velocity vectors are pointing downward and getting longer. There must be a moment—just an instant as $\vec{v}$ switches from pointing up to pointing down—when the velocity is zero. Indeed, the ball's velocity *is* zero for an instant at the precise top of the motion!

But what about the acceleration at the top? The inset shows how the average acceleration is determined from the last upward velocity before the top point and the first downward velocity. We find that the acceleration at the top is pointing downward, just as it does elsewhere in the motion.

Many people expect the acceleration to be zero at the highest point. But recall that the velocity at the top point *is* changing—from up to down. If the velocity is changing, there *must* be an acceleration. A downward-pointing acceleration vector is needed to turn the velocity vector from up to down. Another way to think about this is to note that zero acceleration would mean no change of velocity. When the ball reached zero velocity at the top, it would hang there and not fall if the acceleration were also zero!

FIGURE 1.18 Motion diagram of a ball tossed straight up in the air.

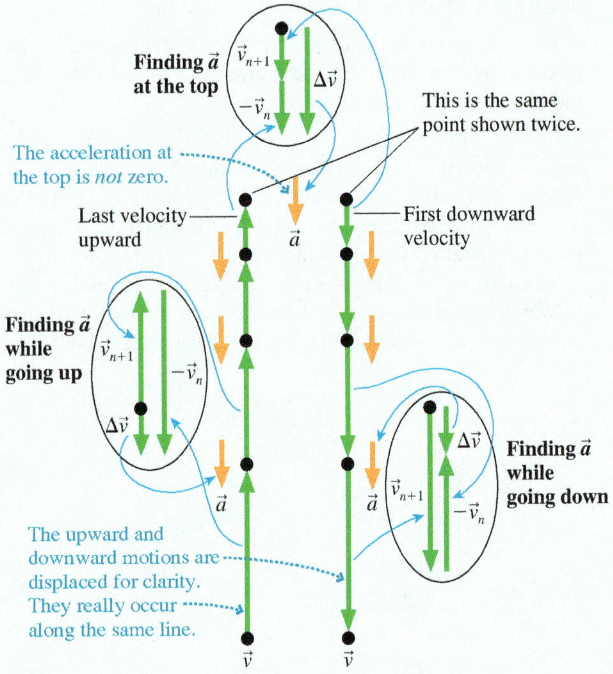

1.6 Motion in One Dimension

As you've seen, an object's motion can be described in terms of three fundamental quantities: its position $\vec{r}$, velocity $\vec{v}$, and acceleration $\vec{a}$. These quantities are vectors, having a direction as well as a magnitude. But for motion in one dimension, the vectors are restricted to point only "forward" or "backward." Consequently, we can describe one-dimensional motion with the simpler quantities x, v_x, and a_x (or y, v_y, and a_y). However, we need to give each of these quantities an explicit *sign,* positive or negative, to indicate whether the position, velocity, or acceleration vector points forward or backward.

Determining the Signs of Position, Velocity, and Acceleration

Position, velocity, and acceleration are measured with respect to a coordinate system, a grid or axis that *you* impose on a problem to analyze the motion. We will find it convenient to use an x-axis to describe both horizontal motion and motion along an inclined plane. A y-axis will be used for vertical motion. A coordinate axis has two essential features:

1. An origin to define zero; and
2. An x or y label to indicate the positive end of the axis.

We will adopt the convention that **the positive end of an x-axis is to the right and the positive end of a y-axis is up.** The signs of position, velocity, and acceleration are based on this convention.

TACTICS BOX 1.4 **Determining the sign of the position, velocity, and acceleration**

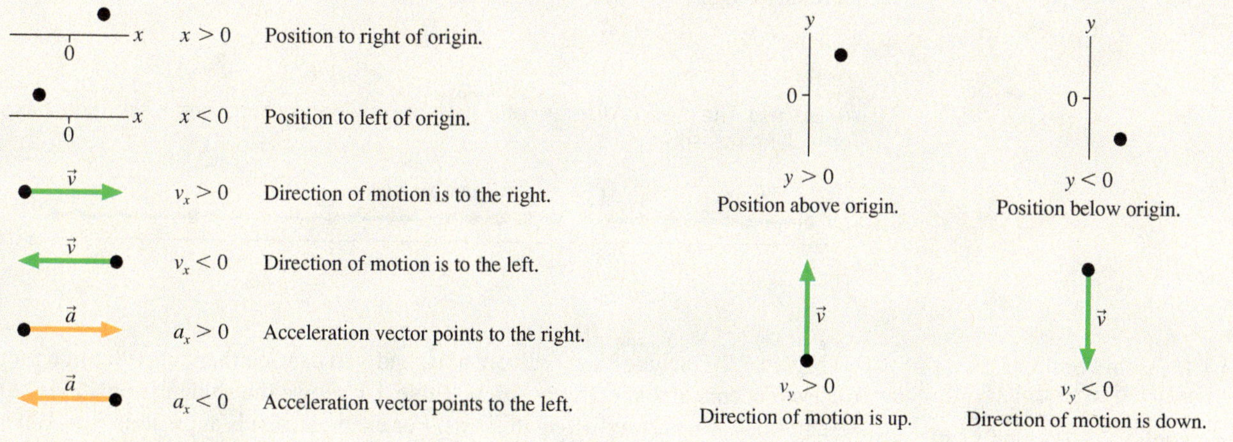

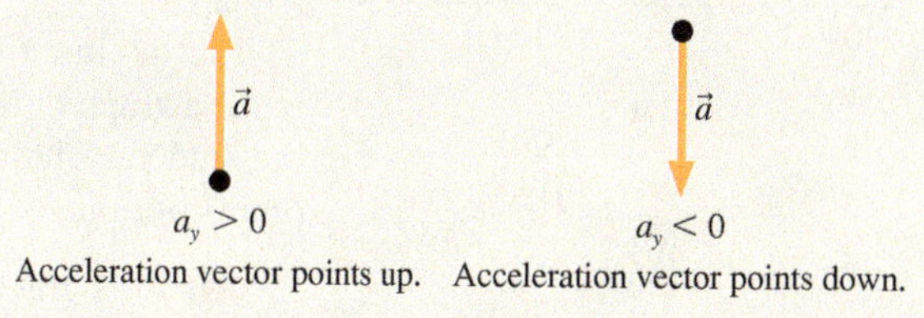

- The sign of position (x or y) tells us *where* an object is.
- The sign of velocity (v_x or v_y) tells us *which direction* the object is moving.
- The sign of acceleration (a_x or a_y) tells us which way the acceleration vector points, *not* whether the object is speeding up or slowing down.

Exercises 30–31

Acceleration is where things get a bit tricky. A natural tendency is to think that a positive value of a_x or a_y describes an object that is speeding up while a negative value describes an object that is slowing down (decelerating). However, this interpretation *does not work.*

Acceleration was defined as $\vec{a}_{avg} = \Delta\vec{v}/\Delta t$. The direction of $\vec{a}$ can be determined by using a motion diagram to find the direction of $\Delta\vec{v}$. The one-dimensional acceleration a_x (or a_y) is then positive if the vector $\vec{a}$ points to the right (or up), negative if $\vec{a}$ points to the left (or down).

FIGURE 1.19 shows that this method for determining the sign of a does not conform to the simple idea of speeding up and slowing down. The object in Figure 1.19a has a positive acceleration ($a_x > 0$) not because it is speeding up but because the vector $\vec{a}$ points in the positive direction. Compare this with the motion diagram of Figure 1.19b. Here the object is slowing down, but it still has a positive acceleration ($a_x > 0$) because $\vec{a}$ points to the right.

FIGURE 1.19 One of these objects is speeding up, the other slowing down, but they both have a positive acceleration a_x.

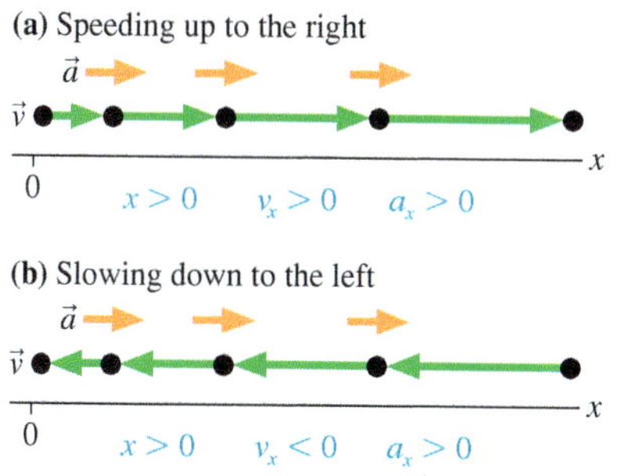

We found that an object is speeding up if $\vec{v}$ and $\vec{a}$ point in the same direction, slowing down if they point in opposite directions. For one-dimensional motion this rule becomes:

- An object is speeding up if and only if v_x and a_x have the same sign.
- An object is slowing down if and only if v_x and a_x have opposite signs.
- An object's velocity is constant if and only if $a_x = 0$.

Notice how the first two of these rules are at work in Figure 1.19.

Position-versus-Time Graphs

FIGURE 1.20 is a motion diagram, made at 1 frame per minute, of a student walking to school. You can see that she leaves home at a time we choose to call $t = 0$ min and makes steady progress for a while. Beginning at $t = 3$ min there is a period where the distance traveled during each time interval becomes less—perhaps she slowed down to speak with a friend. Then she picks up the pace, and the distances within each interval are longer.

FIGURE 1.20 The motion diagram of a student walking to school and a coordinate axis for making measurements.

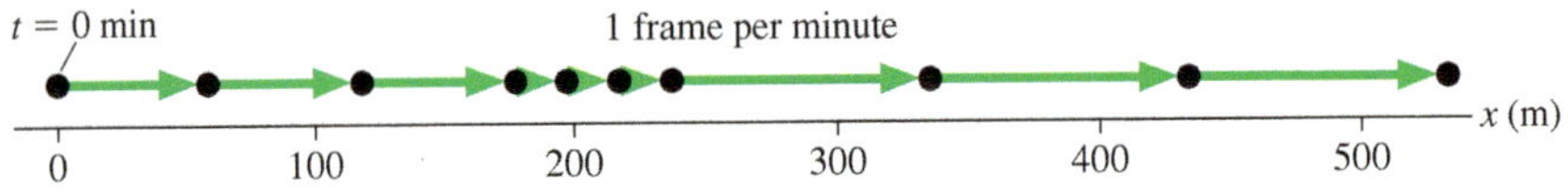

TABLE 1.1 Measured positions of a student walking to school

Time t (min)	Position x (m)
0	0
1	60
2	120
3	180
4	200
5	220
6	240
7	340
8	440
9	540

Figure 1.20 includes a coordinate axis, and you can see that every dot in a motion diagram occurs at a specific position. Table 1.1 shows the student's positions at different times as measured along this axis. For example, she is at position $x = 120$ m at $t = 2$ min.

The motion diagram is one way to represent the student's motion. Another is to make a graph of the measurements in Table 1.1. FIGURE 1.21a is a graph of x versus t for the student. The motion diagram tells us only where the student is at a few discrete points of time, so this graph of the data shows only points, no lines.

NOTE ▶ A graph of "a versus b" means that a is graphed on the vertical axis and b on the horizontal axis. Saying "graph a versus b" is really a shorthand way of saying "graph a as a function of b." ◀

FIGURE 1.21 Position graphs of the student's motion.

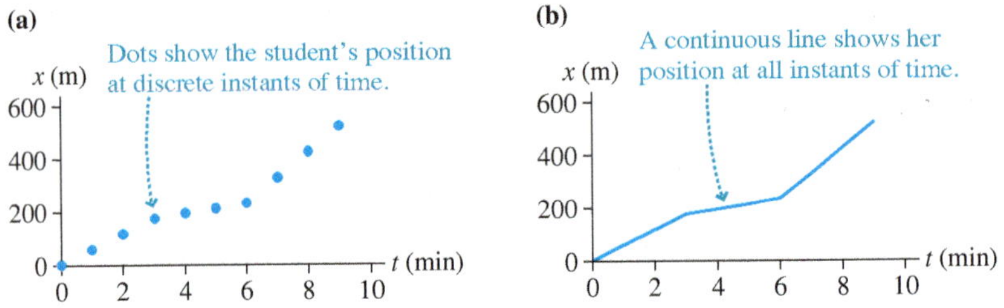

However, common sense tells us the following. First, the student was *somewhere specific* at all times. That is, there was never a time when she failed to have a well-defined position, nor could she occupy two positions at one time. (As reasonable as this belief appears to be, it will be severely questioned and found not entirely accurate when we get to quantum physics!) Second, the student moved *continuously* through all intervening points of space. She could not go from $x = 100$ m to $x = 200$ m without passing through every point in between. It is thus quite reasonable to believe that her motion can be shown as a continuous line passing through the measured points, as shown in FIGURE 1.21b. A continuous line or curve showing an object's position as a function of time is called a **position-versus-time graph** or, sometimes, just a *position graph*.

NOTE ▶ A graph is *not* a "picture" of the motion. The student is walking along a straight line, but the graph itself is not a straight line. Further, we've graphed her position on the vertical axis even though her motion is horizontal. Graphs are *abstract representations* of motion. We will place significant emphasis on the process of interpreting graphs, and many of the exercises and problems will give you a chance to practice these skills. ◀

EXAMPLE 1.7 Interpreting a position graph

The graph in FIGURE 1.22a represents the motion of a car along a straight road. Describe the motion of the car.

MODEL Represent the car as a particle.

VISUALIZE As FIGURE 1.22b shows, the graph represents a car that travels to the left for 30 minutes, stops for 10 minutes, then travels back to the right for 40 minutes.

FIGURE 1.22 Position-versus-time graph of a car.

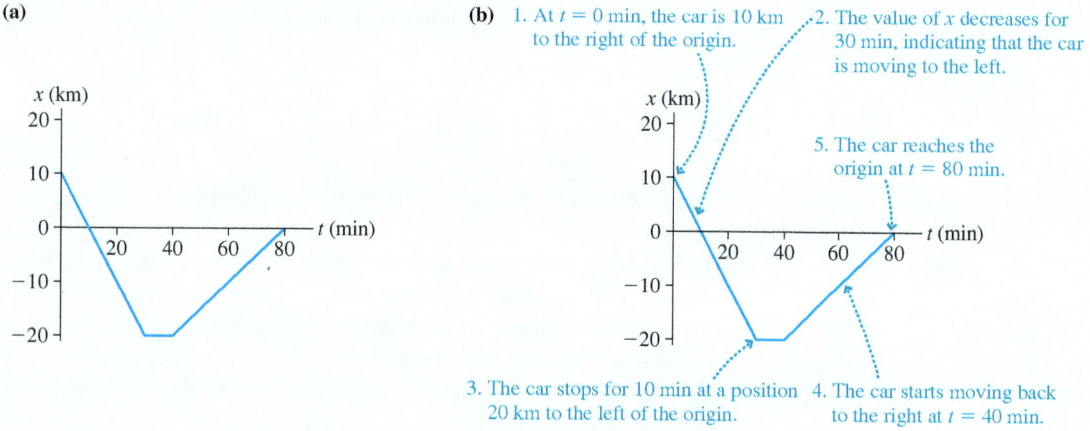

1.7 Solving Problems in Physics

Physics is not mathematics. Math problems are clearly stated, such as "What is 2 + 2?" Physics is about the world around us, and to describe that world we must use language. Now, language is wonderful—we couldn't communicate without it—but language can sometimes be imprecise or ambiguous.

The challenge when reading a physics problem is to translate the words into symbols that can be manipulated, calculated, and graphed. **The translation from words to symbols is the heart of problem solving in physics.** This is the point where ambiguous words and phrases must be clarified, where the imprecise must be made precise, and where you arrive at an understanding of exactly what the question is asking.

Using Symbols

Symbols are a language that allows us to talk with precision about the relationships in a problem. As with any language, we all need to agree to use words or symbols in the same way if we want to communicate with each other. Many of the ways we use symbols in science and engineering are somewhat arbitrary, often reflecting historical roots. Nonetheless, practicing scientists and engineers have come to agree on how to use the language of symbols. Learning this language is part of learning physics.

We will use subscripts on symbols, such as x_3, to designate a particular point in the problem. Scientists usually label the starting point of the problem with the subscript "0," not the subscript "1" that you might expect. When using subscripts, make sure that all symbols referring to the same point in the problem have the *same numerical subscript.* To have the same point in a problem characterized by position x_1 but velocity v_{2x} is guaranteed to lead to confusion!

Drawing Pictures

You may have been told that the first step in solving a physics problem is to "draw a picture," but perhaps you didn't know why, or what to draw. The purpose of drawing a picture is to aid you in the words-to-symbols translation. Complex problems have far more information than you can keep in your head at one time. Think of a picture as a "memory extension," helping you organize and keep track of vital information.

Although any picture is better than none, there really is a *method* for drawing pictures that will help you be a better problem solver. It is called the **pictorial representation** of the problem. We'll add other pictorial representations as we go along, but the following procedure is appropriate for motion problems.

TACTICS BOX 1.5 Drawing a pictorial representation

MP

1. **Draw a motion diagram.** The motion diagram develops your intuition for the motion.
2. **Establish a coordinate system.** Select your axes and origin to match the motion. For one-dimensional motion, you want either the x-axis or the y-axis parallel to the motion. The coordinate system determines whether the signs of v and a are positive or negative.
3. **Sketch the situation.** Not just any sketch. Show the object at the *beginning* of the motion, at the *end,* and at any point where the character of the motion changes. Show the object, not just a dot, but very simple drawings are adequate.
4. **Define symbols.** Use the sketch to define symbols representing quantities such as position, velocity, acceleration, and time. *Every* variable used later in the mathematical solution should be defined on the sketch. Some will have known values, others are initially unknown, but all should be given symbolic names.
5. **List known information.** Make a table of the quantities whose values you can determine from the problem statement or that can be found quickly with simple geometry or unit conversions. Some quantities are implied by the problem, rather than explicitly given. Others are determined by your choice of coordinate system.
6. **Identify the desired unknowns.** What quantity or quantities will allow you to answer the question? These should have been defined as symbols in step 4. Don't list every unknown, only the one or two needed to answer the question.

It's not an overstatement to say that a well-done pictorial representation of the problem will take you halfway to the solution. The following example illustrates how to construct a pictorial representation for a problem that is typical of problems you will see in the next few chapters.

EXAMPLE 1.8 Drawing a pictorial representation

Draw a pictorial representation for the following problem: A rocket sled accelerates horizontally at 50 m/s^2 for 5.0 s, then coasts for 3.0 s. What is the total distance traveled?

VISUALIZE The motion diagram shows an acceleration phase followed by a coasting phase. Because the motion is horizontal, the appropriate coordinate system is an x-axis. We've chosen to place the origin at the starting point. The motion has a beginning, an end, and a point where the nature of the motion changes from accelerating to coasting. These are the three sled positions sketched in FIGURE 1.23. The quantities x, v_x, and t are needed at each of three *points,* so these have been defined on the sketch and distinguished by subscripts. Accelerations are associated with *intervals* between the points, so only two accelerations are defined. Values for three quantities are given in the problem statement, although we need to use the motion diagram, where $\vec{a}$ points to the right, and our choice of coordinate system to know that $a_{0x} = +50$ m/s^2 rather than -50 m/s^2.

The values $x_0 = 0$ m and $t_0 = 0$ s are choices we made when setting up the coordinate system. The value $v_{0x} = 0$ m/s is part of our *interpretation* of the problem. Finally, we identify x_2 as the quantity that will answer the question. We now understand quite a bit about the problem and would be ready to start a quantitative analysis.

FIGURE 1.23 A pictorial representation.

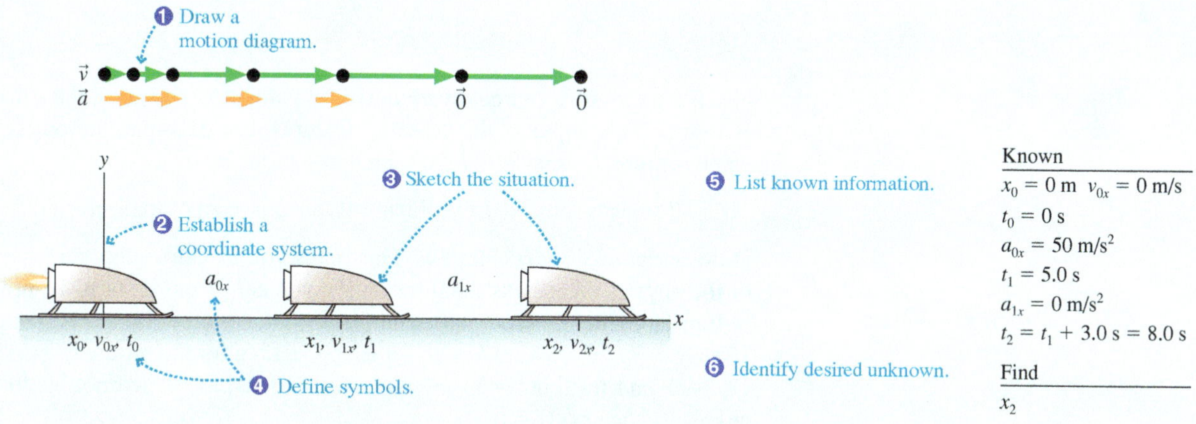

We didn't *solve* the problem; that is not the purpose of the pictorial representation. The pictorial representation is a systematic way to go about interpreting a problem and getting ready for a mathematical solution. Although this is a simple problem, and you probably know how to solve it if you've taken physics before, you will soon be faced with much more challenging problems. Learning good problem-solving skills at the beginning, while the problems are easy, will make them second nature later when you really need them.

Representations

A picture is one way to *represent* your knowledge of a situation. You could also represent your knowledge using words, graphs, or equations. Each **representation of knowledge** gives us a different perspective on the problem. The more tools you have for thinking about a complex problem, the more likely you are to solve it.

There are four representations of knowledge that we will use over and over:

1. The *verbal* representation. A problem statement, in words, is a verbal representation of knowledge. So is an explanation that you write.
2. The *pictorial* representation. The pictorial representation, which we've just presented, is the most literal depiction of the situation.
3. The *graphical* representation. We will make extensive use of graphs.
4. The *mathematical* representation. Equations that can be used to find the numerical values of specific quantities are the mathematical representation.

NOTE ▶ The mathematical representation is only one of many. Much of physics is more about thinking and reasoning than it is about solving equations. ◀

A new building requires careful planning. The architect's visualization and drawings have to be complete before the detailed procedures of construction get under way. The same is true for solving problems in physics.

A Problem-Solving Strategy

One of the goals of this textbook is to help you learn a *strategy* for solving physics problems. The purpose of a strategy is to guide you in the right direction with minimal wasted effort. The four-part problem-solving strategy shown below—**Model, Visualize, Solve, Assess**—is based on using different representations of knowledge. You will see this problem-solving strategy used consistently in the worked examples throughout this textbook, and you should endeavor to apply it to your own problem solving.

Throughout this textbook we will emphasize the first two steps. They are the *physics* of the problem, as opposed to the mathematics of solving the resulting equations. This is not to say that those mathematical operations are always easy—in many cases they are not. But our primary goal is to understand the physics.

General Problem-Solving Strategy

MP

MODEL It's impossible to treat every detail of a situation. Simplify the situation with a model that captures the essential features. For example, the object in a mechanics problem is usually represented as a particle.

VISUALIZE This is where expert problem solvers put most of their effort.

- Draw a *pictorial representation.* This helps you visualize important aspects of the physics and assess the information you are given. It starts the process of translating the problem into symbols.
- Use a *graphical representation* if it is appropriate for the problem.
- Go back and forth between these representations; they need not be done in any particular order.

SOLVE Only after modeling and visualizing are complete is it time to develop a *mathematical representation* with specific equations that must be solved. All symbols used here should have been defined in the pictorial representation.

ASSESS Is your result believable? Does it have proper units? Does it make sense?

Textbook illustrations are obviously more sophisticated than what you would draw on your own paper. To show you a figure very much like what you *should* draw, the final example of this section is in a "pencil sketch" style. We will include one or more pencil-sketch examples in nearly every chapter to illustrate exactly what a good problem solver would draw.

EXAMPLE 1.9 Launching a weather rocket

Use the first two steps of the problem-solving strategy to analyze the following problem: A small rocket, such as those used for meteorological measurements of the atmosphere, is launched vertically with an acceleration of 30 m/s^2. It runs out of fuel after 30 s. What is its maximum altitude?

MODEL We need to do some interpretation. Common sense tells us that the rocket does not stop the instant it runs out of fuel. Instead, it continues upward, while slowing, until it reaches its maximum altitude. This second half of the motion, after running out of fuel, is like the ball that was tossed upward in the first half of Example 1.6. Because the problem does not ask about the rocket's descent, we conclude that the problem ends at the point of maximum altitude. We'll represent the rocket as a particle.

VISUALIZE FIGURE 1.24 shows the pictorial representation in pencil-sketch style. The rocket is speeding up during the first half of the motion, so $\vec{a}_0$ points upward, in the positive y-direction. Thus the initial acceleration is $a_{0y} = 30$ m/s^2. During the second half, as the rocket slows, $\vec{a}_1$ points downward. Thus a_{1y} is a negative number.

FIGURE 1.24 Pictorial representation for the rocket.

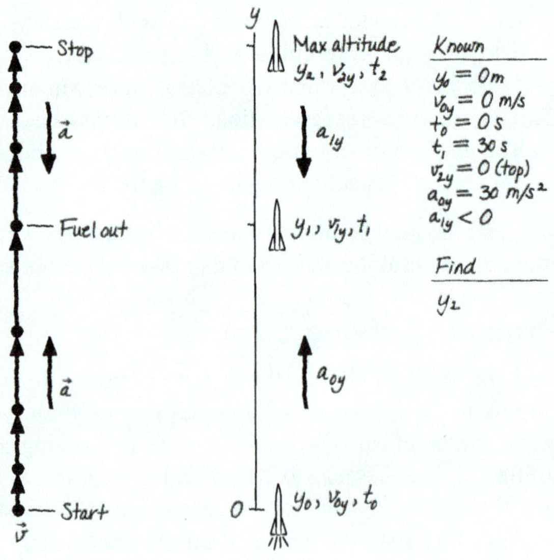

This information is included with the known information. Although the velocity v_{2y} wasn't given in the problem statement, we know it must be zero at the very top of the trajectory. Last, we have identified y_2 as the desired unknown. This, of course, is not the only unknown in the problem, but it is the one we are specifically asked to find.

ASSESS If you've had a previous physics class, you may be tempted to assign a_{1y} the value $-9.8\ \text{m/s}^2$, the free-fall acceleration. However, that would be true only if there is no air resistance on the rocket. We will need to consider the *forces* acting on the rocket during the second half of its motion before we can determine a value for a_{1y}. For now, all that we can safely conclude is that a_{1y} is negative.

Our task in this section is not to *solve* problems—all that in due time—but to focus on what is happening in a problem. In other words, to make the translation from words to symbols in preparation for subsequent mathematical analysis. Modeling and the pictorial representation will be our most important tools.

1.8 Units and Significant Figures

Science is based upon experimental measurements, and measurements require *units*. The system of units used in science is called *le Système Internationale d'Unités*. These are commonly referred to as **SI units**. Older books often referred to *mks units,* which stands for "meter-kilogram-second," or *cgs units,* which is "centimeter-gram-second." For practical purposes, SI units are the same as mks units. In casual speaking we often refer to *metric units,* although this could mean either mks or cgs units.

All of the quantities needed to understand motion can be expressed in terms of the three basic SI units shown in Table 1.2. Other quantities can be expressed as a combination of these basic units. Velocity, expressed in meters per second or m/s, is a ratio of the length unit to the time unit.

TABLE 1.2 The basic SI units

Quantity	Unit	Abbreviation
time	second	s
length	meter	m
mass	kilogram	kg

Time

The standard of time prior to 1960 was based on the *mean solar day*. As time-keeping accuracy and astronomical observations improved, it became apparent that the earth's rotation is not perfectly steady. Meanwhile, physicists had been developing a device called an *atomic clock*. This instrument is able to measure, with incredibly high precision, the frequency of radio waves absorbed by atoms as they move between two closely spaced energy levels. This frequency can be reproduced with great accuracy at many laboratories around the world. Consequently, the SI unit of time—the second—was redefined in 1967 as follows:

One *second* is the time required for 9,192,631,770 oscillations of the radio wave absorbed by the cesium-133 atom. The abbreviation for second is the letter s.

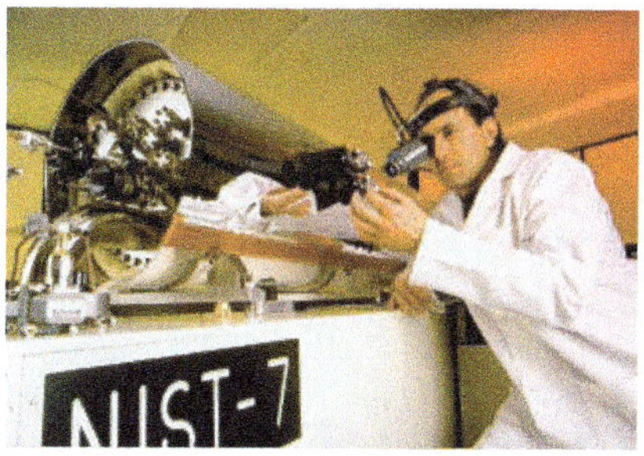

An atomic clock at the National Institute of Standards and Technology is the primary standard of time.

Several radio stations around the world broadcast a signal whose frequency is linked directly to the atomic clocks. This signal is the time standard, and any time-measuring equipment you use was calibrated from this time standard.

Length

The SI unit of length—the meter—was originally defined as one ten-millionth of the distance from the North Pole to the equator along a line passing through Paris. There are obvious practical difficulties with implementing this definition, and it was later abandoned in favor of the distance between two scratches on a platinum-iridium bar stored in a special vault in Paris. The present definition, agreed to in 1983, is as follows:

One *meter* is the distance traveled by light in vacuum during 1/299,792,458 of a second. The abbreviation for meter is the letter m.

This is equivalent to defining the speed of light to be exactly 299,792,458 m/s. Laser technology is used in various national laboratories to implement this definition and to calibrate secondary standards that are easier to use. These standards

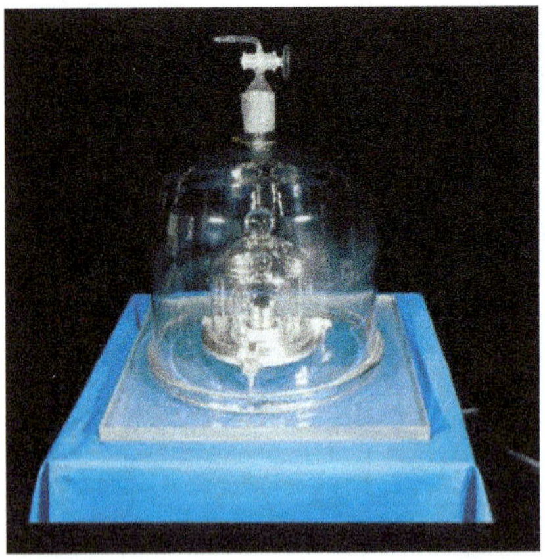

By international agreement, this metal cylinder, stored in Paris, is the definition of the kilogram.

TABLE 1.3 Common prefixes

Prefix	Power of 10	Abbreviation
giga-	10^9	G
mega-	10^6	M
kilo-	10^3	k
centi-	10^{-2}	c
milli-	10^{-3}	m
micro-	10^{-6}	μ
nano-	10^{-9}	n

TABLE 1.4 Useful unit conversions

1 in = 2.54 cm
1 mi = 1.609 km
1 mph = 0.447 m/s
1 m = 39.37 in
1 km = 0.621 mi
1 m/s = 2.24 mph

ultimately make their way to your ruler or to a meter stick. It is worth keeping in mind that any measuring device you use is only as accurate as the care with which it was calibrated.

Mass

The original unit of mass, the gram, was defined as the mass of 1 cubic centimeter of water. That is why you know the density of water as 1 g/cm^3. This definition proved to be impractical when scientists needed to make very accurate measurements. The SI unit of mass—the kilogram—was redefined in 1889 as:

> One *kilogram* is the mass of the international standard kilogram, a polished platinum-iridium cylinder stored in Paris. The abbreviation for kilogram is kg.

The kilogram is the only SI unit still defined by a manufactured object. Despite the prefix *kilo,* it is the kilogram, not the gram, that is the SI unit.

Using Prefixes

We will have many occasions to use lengths, times, and masses that are either much less or much greater than the standards of 1 meter, 1 second, and 1 kilogram. We will do so by using *prefixes* to denote various powers of 10. Table 1.3 lists the common prefixes that will be used frequently throughout this book. Memorize it! Few things in science are learned by rote memory, but this list is one of them. A more extensive list of prefixes is shown inside the cover of the book.

Although prefixes make it easier to talk about quantities, the SI units are meters, seconds, and kilograms. Quantities given with prefixed units must be converted to SI units before any calculations are done. Unit conversions are best done at the very beginning of a problem, as part of the pictorial representation.

Unit Conversions

Although SI units are our standard, we cannot entirely forget that the United States still uses English units. Thus it remains important to be able to convert back and forth between SI units and English units. Table 1.4 shows several frequently used conversions, and these are worth memorizing if you do not already know them. While the English system was originally based on the length of the king's foot, it is interesting to note that today the conversion 1 in = 2.54 cm is the *definition* of the inch. In other words, the English system for lengths is now based on the meter!

There are various techniques for doing unit conversions. One effective method is to write the conversion factor as a ratio equal to one. For example, using information in Table 1.3 and 1.4, we have

$$\frac{10^{-6}\text{ m}}{1\ \mu\text{m}} = 1 \qquad \text{and} \qquad \frac{2.54\text{ cm}}{1\text{ in}} = 1$$

Because multiplying any expression by 1 does not change its value, these ratios are easily used for conversions. To convert 3.5 μm to meters we compute

$$3.5\ \cancel{\mu\text{m}} \times \frac{10^{-6}\text{ m}}{1\ \cancel{\mu\text{m}}} = 3.5 \times 10^{-6}\text{ m}$$

Similarly, the conversion of 2 feet to meters is

$$2.00\ \cancel{\text{ft}} \times \frac{12\ \cancel{\text{in}}}{1\ \cancel{\text{ft}}} \times \frac{2.54\ \cancel{\text{cm}}}{1\ \cancel{\text{in}}} \times \frac{10^{-2}\text{ m}}{1\ \cancel{\text{cm}}} = 0.610\text{ m}$$

Notice how units in the numerator and in the denominator cancel until only the desired units remain at the end. You can continue this process of multiplying by 1 as many times as necessary to complete all the conversions.

Assessment

As we get further into problem solving, we will need to decide whether or not the answer to a problem "makes sense." To determine this, at least until you have more experience with SI units, you may need to convert from SI units back to the English units in which you think. But this conversion does not need to be very accurate. For example, if you are working a problem about automobile speeds and reach an answer of 35 m/s, all you really want to know is whether or not this is a realistic speed for a car. That requires a "quick and dirty" conversion, not a conversion of great accuracy.

Table 1.5 shows several approximate conversion factors that can be used to assess the answer to a problem. Using 1 m/s ≈ 2 mph, you find that 35 m/s is roughly 70 mph, a reasonable speed for a car. But an answer of 350 m/s, which you might get after making a calculation error, would be an unreasonable 700 mph. Practice with these will allow you to develop intuition for metric units.

TABLE 1.5 Approximate conversion factors. Use these only for assessment, not in problem solving.

1 cm ≈ $\frac{1}{2}$ in
10 cm ≈ 4 in
1 m ≈ 1 yard
1 m ≈ 3 feet
1 km ≈ 0.6 mile
1 m/s ≈ 2 mph

NOTE ▶ These approximate conversion factors are accurate to only one significant figure. This is sufficient to assess the answer to a problem, but do *not* use the conversion factors from Table 1.5 for converting English units to SI units at the start of a problem. Use Table 1.4. ◀

Significant Figures

It is necessary to say a few words about a perennial source of difficulty: significant figures. Mathematics is a subject where numbers and relationships can be as precise as desired, but physics deals with a real world of ambiguity. It is important in science and engineering to state clearly what you know about a situation—no less and, especially, no more. Numbers provide one way to specify your knowledge.

If you report that a length has a value of 6.2 m, the implication is that the actual value falls between 6.15 m and 6.25 m and thus rounds to 6.2 m. If that is the case, then reporting a value of simply 6 m is saying less than you know; you are withholding information. On the other hand, to report the number as 6.213 m is wrong. Any person reviewing your work—perhaps a client who hired you—would interpret the number 6.213 m as meaning that the actual length falls between 6.2125 m and 6.2135 m, thus rounding to 6.213 m. In this case, you are claiming to have knowledge and information that you do not really possess.

The way to state your knowledge precisely is through the proper use of **significant figures.** You can think of a significant figure as being a digit that is reliably known. A number such as 6.2 m has *two* significant figures because the next decimal place—the one-hundredths—is not reliably known. As FIGURE 1.25 shows, the best way to determine how many significant figures a number has is to write it in scientific notation.

FIGURE 1.25 Determining significant figures.

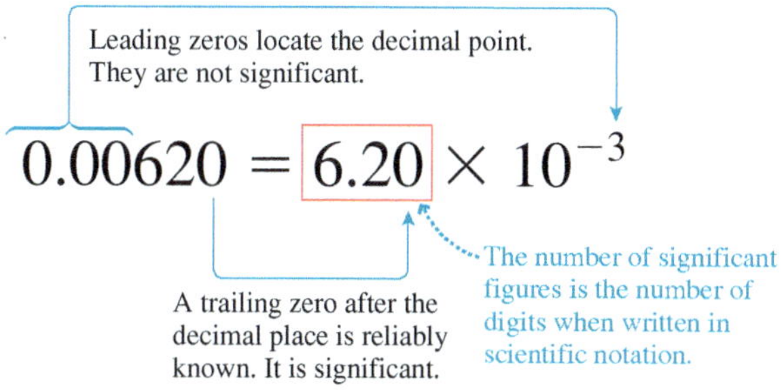

- The number of significant figures ≠ the number of decimal places.
- In whole numbers, trailing zeros are not significant. 320 is 3.2×10^2 and has 2 significant figures, not 3.
- Changing units shifts the decimal point but does not change the number of significant figures.

Calculations with numbers follow the "weakest link" rule. The saying, which you probably know, is that "a chain is only as strong as its weakest link." If nine out of ten links in a chain can support a 1000 pound weight, that strength is meaningless if the tenth link can support only 200 pounds. Nine out of the ten numbers used in a calculation might be known with a precision of 0.01%; but if the tenth number is poorly known, with a precision of only 10%, then the result of the calculation cannot possibly be more precise than 10%.

TACTICS BOX 1.6 Using significant figures

1. When multiplying or dividing several numbers, or taking roots, the number of significant figures in the answer should match the number of significant figures of the *least* precisely known number used in the calculation.
2. When adding or subtracting several numbers, the number of decimal places in the answer should match the *smallest* number of decimal places of any number used in the calculation.
3. It is acceptable to keep one or two extra digits during intermediate steps of a calculation, as long as the final answer is reported with the proper number of significant figures. The goal is to minimize round-off errors in the calculation. But only one or two extra digits, not the seven or eight shown in your calculator display.

Exercises 38–39

EXAMPLE 1.10 Using significant figures

An object consists of two pieces. The mass of one piece has been measured to be 6.47 kg. The volume of the second piece, which is made of aluminum, has been measured to be $4.44 \times 10^{-4}\ m^3$. A handbook lists the density of aluminum as $2.7 \times 10^3\ kg/m^3$. What is the total mass of the object?

SOLVE First, calculate the mass of the second piece:

$$m = (4.44 \times 10^{-4}\ m^3)(2.7 \times 10^3\ kg/m^3)$$
$$= 1.199\ kg = 1.2\ kg$$

The number of significant figures of a product must match that of the *least* precisely known number, which is the two-significant-figure density of aluminum. Now add the two masses:

$$\begin{array}{r} 6.47\ \text{kg} \\ +1.2\ \ \text{kg} \\ \hline 7.7\ \ \text{kg} \end{array}$$

The sum is 7.67 kg, but the hundredths place is not reliable because the second mass has no reliable information about this digit. Thus we must round to the one decimal place of the 1.2 kg. The best we can say, with reliability, is that the total mass is 7.7 kg.

Some quantities can be measured very precisely—three or more significant figures. Others are inherently much less precise—only two significant figures. Examples and problems in this textbook will normally provide data to either two or three significant figures, as is appropriate to the situation. **The appropriate number of significant figures for the answer is determined by the data provided.**

NOTE ▶ Be careful! Many calculators have a default setting that shows two decimal places, such as 5.23. This is dangerous. If you need to calculate 5.23/58.5, your calculator will show 0.09 and it is all too easy to write that down as an answer. By doing so, you have reduced a calculation of two numbers having three significant figures to an answer with only one significant figure. The proper result of this division is 0.0894 or 8.94×10^{-2}. You will avoid this error if you keep your calculator set to display numbers in *scientific notation* with two decimal places. ◀

Proper use of significant figures is part of the "culture" of science and engineering. We will frequently emphasize these "cultural issues" because you must learn to speak the same language as the natives if you wish to communicate effectively. Most students know the rules of significant figures, having learned them in high school, but many fail to apply them. It is important to understand the reasons for significant figures and to get in the habit of using them properly.

TABLE 1.6 Some approximate lengths

	Length (m)
Circumference of the earth	4×10^7
New York to Los Angeles	5×10^6
Distance you can drive in 1 hour	1×10^5
Altitude of jet planes	1×10^4
Distance across a college campus	1000
Length of a football field	100
Length of a classroom	10
Length of your arm	1
Width of a textbook	0.1
Length of your little fingernail	0.01
Diameter of a pencil lead	1×10^{-3}
Thickness of a sheet of paper	1×10^{-4}
Diameter of a dust particle	1×10^{-5}

TABLE 1.7 Some approximate masses

	Mass (kg)
Large airliner	1×10^5
Small car	1000
Large human	100
Medium-size dog	10
Science textbook	1
Apple	0.1
Pencil	0.01
Raisin	1×10^{-3}
Fly	1×10^{-4}

Orders of Magnitude and Estimating

Precise calculations are appropriate when we have precise data, but there are many times when a very rough estimate is sufficient. Suppose you see a rock fall off a cliff and would like to know how fast it was going when it hit the ground. By doing a mental comparison with the speeds of familiar objects, such as cars and bicycles, you might judge that the rock was traveling at "about" 20 mph.

This is a one-significant-figure estimate. With some luck, you can distinguish 20 mph from either 10 mph or 30 mph, but you certainly cannot distinguish 20 mph from 21 mph. A one-significant-figure estimate or calculation, such as this, is called an **order-of-magnitude estimate.** An order-of-magnitude estimate is indicated by the symbol $\sim$, which indicates even less precision than the "approximately equal" symbol $\approx$. You would say that the speed of the rock is $v \sim 20$ mph.

A useful skill is to make reliable estimates on the basis of known information, simple reasoning, and common sense. This is a skill that is acquired by practice. Many chapters in this book will have homework problems that ask you to make order-of-magnitude estimates. The following example is a typical estimation problem.

Table 1.6 and 1.7 have information that will be useful for doing estimates.

EXAMPLE 1.11 Estimating a sprinter's speed

Estimate the speed with which an Olympic sprinter crosses the finish line of the 100 m dash.

SOLVE We do need one piece of information, but it is a widely known piece of sports trivia. That is, world-class sprinters run the 100 m dash in about 10 s. Their *average* speed is $v_{avg} \approx (100 \text{ m})/(10 \text{ s}) \approx 10$ m/s. But that's only average. They go slower than average at the beginning, and they cross the finish line at a speed faster than average. How much faster? Twice as fast, 20 m/s, would be ≈ 40 mph. Sprinters don't seem like they're running as fast as a 40 mph car, so this probably is too fast. Let's *estimate* that their final speed is 50% faster than the average. Thus they cross the finish line at $v \sim 15$ m/s.

STOP TO THINK 1.5 Rank in order, from the most to the least, the number of significant figures in the following numbers. For example, if b has more than c, c has the same number as a, and a has more than d, you could give your answer as b > c = a > d.

a. 82 b. 0.0052 c. 0.430 d. 4.321×10^{-10}

SUMMARY

The goal of Chapter 1 has been to introduce the fundamental concepts of motion.

General Strategy

Motion Diagrams

- Help visualize motion.
- Provide a tool for finding acceleration vectors.

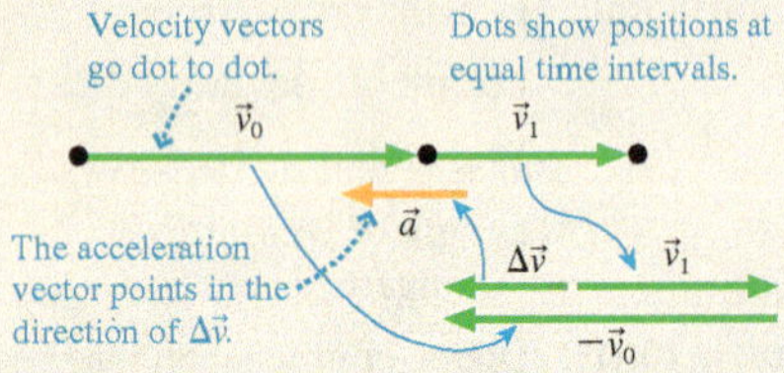

▶ These are the average velocity and the average acceleration vectors.

Problem Solving

MODEL Make simplifying assumptions.

VISUALIZE Use:

- **Pictorial representation**
- **Graphical representation**

SOLVE Use a **mathematical representation** to find numerical answers.

ASSESS Does the answer have the proper units? Does it make sense?

Important Concepts

The particle model represents a moving object as if all its mass were concentrated at a single point.

Position locates an object with respect to a chosen coordinate system. Change in position is called displacement.

Velocity is the rate of change of the position vector $\vec{r}$.

Acceleration is the rate of change of the velocity vector $\vec{v}$.

An object has an acceleration if it

- Changes speed and/or
- Changes direction.

Pictorial Representation

❶ Draw a motion diagram.

❷ Establish coordinates.

❸ Sketch the situation.

❹ Define symbols.

❺ List knowns.

❻ Identify desired unknown.

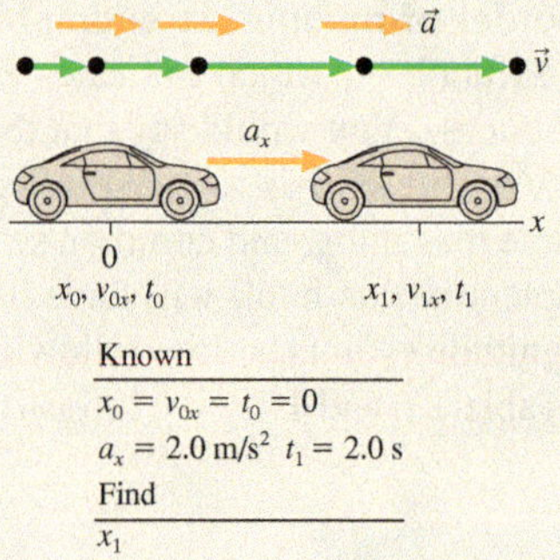

Applications

For **motion along a line:**

- Speeding up: $\vec{v}$ and $\vec{a}$ point in the same direction, v_x and a_x have the same sign.
- Slowing down: $\vec{v}$ and $\vec{a}$ point in opposite directions, v_x and a_x have opposite signs.
- Constant speed: $\vec{a} = \vec{0}$, $a_x = 0$.

Acceleration a_x is positive if $\vec{a}$ points right, negative if $\vec{a}$ points left. The sign of a_x does *not* imply speeding up or slowing down.

Significant figures are reliably known digits. The number of significant figures for:

- **Multiplication, division, powers** is set by the value with the fewest significant figures.
- **Addition, subtraction** is set by the value with the smallest number of decimal places.

The appropriate number of significant figures in a calculation is determined by the data provided.

Terms and Notation

motion
translational motion
trajectory
motion diagram
particle
particle model
position vector, $\vec{r}$
scalar quantity
vector quantity
displacement, $\Delta\vec{r}$
zero vector, $\vec{0}$
time interval, Δt
average speed
average velocity, $\vec{v}$
average acceleration, $\vec{a}$
position-versus-time graph
pictorial representation
representation of knowledge
SI units
significant figures
order-of-magnitude estimate

CONCEPTUAL QUESTIONS

1. How many significant figures does each of the following numbers have?
   a. 53.2 b. 0.53 c. 5.320 d. 0.0532
2. How many significant figures does each of the following numbers have?
   a. 310 b. 0.00310 c. 1.031 d. 3.10×10^5
3. Is the particle in FIGURE Q1.3 speeding up? Slowing down? Or can you tell? Explain.

FIGURE Q1.3

4. Does the object represented in FIGURE Q1.4 have a positive or negative value of a_x? Explain.
5. Does the object represented in FIGURE Q1.5 have a positive or negative value of a_y? Explain.

FIGURE Q1.4

FIGURE Q1.5

6. Determine the signs (positive or negative) of the position, velocity, and acceleration for the particle in FIGURE Q1.6.

FIGURE Q1.6
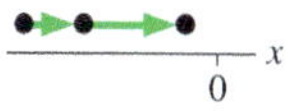

7. Determine the signs (positive or negative) of the position, velocity, and acceleration for the particle in FIGURE Q1.7.

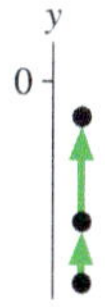
FIGURE Q1.7

FIGURE Q1.8

8. Determine the signs (positive or negative) of the position, velocity, and acceleration for the particle in FIGURE Q1.8.

EXERCISES AND PROBLEMS

Exercises

Section 1.1 Motion Diagrams

1. | A car skids to a halt to avoid hitting an object in the road. Draw a basic motion diagram, using the images from the movie, from the time the skid begins until the car is stopped.
2. | A rocket is launched straight up. Draw a basic motion diagram, using the images from the movie, from the moment of liftoff until the rocket is at an altitude of 500 m.
3. | You're driving along the highway at 60 mph until you enter a town where the speed limit is 30 mph. You slow quickly, but not instantly, to 30 mph. Draw a basic motion diagram of your car, using images from the movie, from 30 s before reaching the city limit until 30 s afterward.

Section 1.2 The Particle Model

4. | a. Write a paragraph describing the particle model. What is it, and why is it important?
   b. Give two examples of situations, different from those described in the text, for which the particle model is appropriate.
   c. Give an example of a situation, different from those described in the text, for which it would be inappropriate.

Section 1.3 Position and Time

Section 1.4 Velocity

5. | You drop a soccer ball from your third-story balcony. Use the particle model to draw a motion diagram showing the ball's position and average velocity vectors from the time you release the ball until the instant it touches the ground.
6. | A softball player hits the ball and starts running toward first base. Use the particle model to draw a motion diagram showing her position and her average velocity vectors during the first few seconds of her run.
7. | A softball player slides into second base. Use the particle model to draw a motion diagram showing his position and his average velocity vectors from the time he begins to slide until he reaches the base.

Section 1.5 Linear Acceleration

8. | a. FIGURE EX1.8 shows the first three points of a motion diagram. Is the object's average speed between points 1 and 2 greater than, less than, or equal to its average speed between points 0 and 1? Explain how you can tell.
   b. Use Tactics Box 1.3 to find the average acceleration vector at point 1. Draw the completed motion diagram, showing the velocity vectors and acceleration vector.

FIGURE EX1.8

FIGURE EX1.9

9. | a. FIGURE EX1.9 shows the first three points of a motion diagram. Is the object's average speed between points 1 and 2 greater than, less than, or equal to its average speed between points 0 and 1? Explain how you can tell.
   b. Use Tactics Box 1.3 to find the average acceleration vector at point 1. Draw the completed motion diagram, showing the velocity vectors and acceleration vector.

10. ‖ FIGURE EX1.10 shows two dots of a motion diagram and vector $\vec{v}_1$. Copy this figure and add vector $\vec{v}_2$ and dot 3 if the acceleration vector $\vec{a}$ at dot 2 (a) points up and (b) points down.

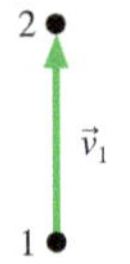

FIGURE EX1.10

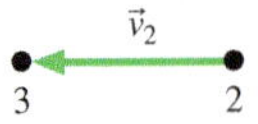

FIGURE EX 1.11

11. | FIGURE EX1.11 shows two dots of a motion diagram and vector $\vec{v}_2$. Copy this figure and add vector $\vec{v}_1$ and dot 1 if the acceleration vector $\vec{a}$ at dot 2 (a) points to the right and (b) points to the left.

12. | A car travels to the left at a steady speed for a few seconds, then brakes for a stop sign. Draw a complete motion diagram of the car.

13. | A child is sledding on a smooth, level patch of snow. She encounters a rocky patch and slows to a stop. Draw a complete motion diagram of the child and her sled.

14. | You use a long rubber band to launch a paper wad straight up. Draw a complete motion diagram of the paper wad from the moment you release the stretched rubber band until the paper wad reaches its highest point.

15. | A roof tile falls straight down from a two-story building. It lands in a swimming pool and settles gently to the bottom. Draw a complete motion diagram of the tile.

16. | Your roommate drops a tennis ball from a third-story balcony. It hits the sidewalk and bounces as high as the second story. Draw a complete motion diagram of the tennis ball from the time it is released until it reaches the maximum height on its bounce. Be sure to determine and show the acceleration at the lowest point.

17. | A toy car rolls down a ramp, then across a smooth, horizontal floor. Draw a complete motion diagram of the toy car.

Section 1.6 Motion in One Dimension

18. ‖ FIGURE EX1.18 shows the motion diagram of a drag racer. The camera took one frame every 2 s.

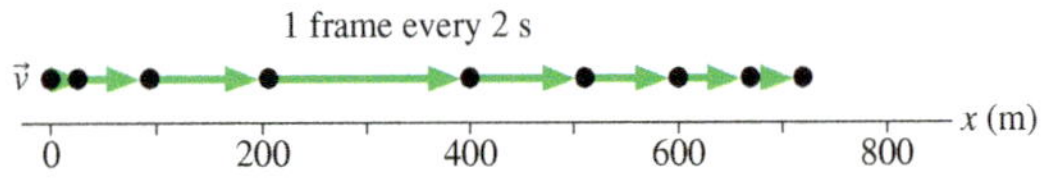

FIGURE EX1.18

a. Measure the x-value of the racer at each dot. List your data in a table similar to Table 1.1, showing each position and the time at which it occurred.

b. Make a position-versus-time graph for the drag racer. Because you have data only at certain instants, your graph should consist of dots that are not connected together.

19. | Write a short description of the motion of a real object for which FIGURE EX1.19 would be a realistic position-versus-time graph.

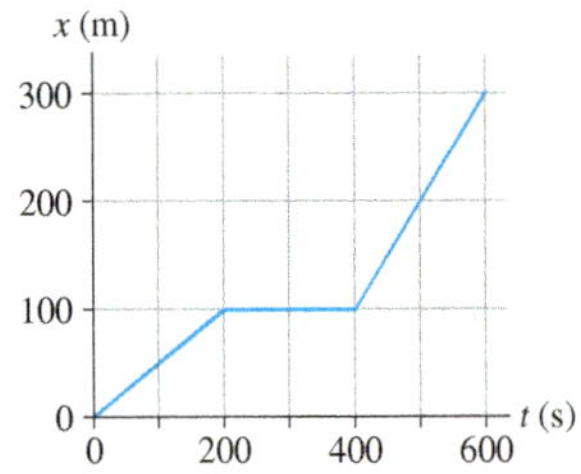

FIGURE EX1.19

20. | Write a short description of the motion of a real object for which FIGURE EX1.20 would be a realistic position-versus-time graph.

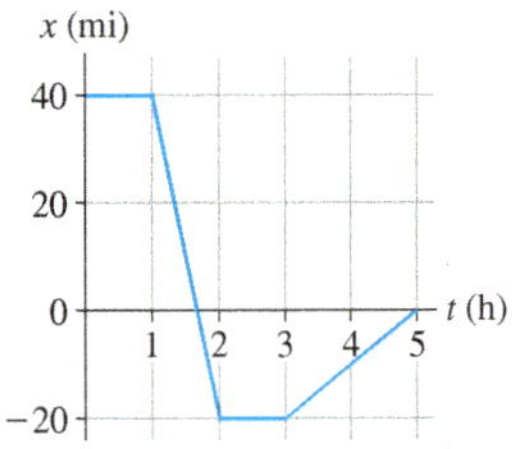

FIGURE EX1.20

Section 1.7 Solving Problems in Physics

21. ‖ Draw a pictorial representation for the following problem. Do *not* solve the problem. The light turns green, and a bicyclist starts forward with an acceleration of 1.5 m/s^2. How far must she travel to reach a speed of 7.5 m/s?

22. ‖ Draw a pictorial representation for the following problem. Do *not* solve the problem. What acceleration does a rocket need to reach a speed of 200 m/s at a height of 1.0 km?

Section 1.8 Units and Significant Figures

23. | Convert the following to SI units:
 a. 6.15 ms b. 27.2 km
 c. 112 km/h d. 72 μm/ms

24. | Convert the following to SI units:
 a. 8.0 in b. 66 ft/s
 c. 60 mph d. 14 in^2

25. | Convert the following to SI units:
 a. 3 hours b. 2 days
 c. 1 year d. 215 ft/s

26. | Using the approximate conversion factors in Table 1.5, convert the following to SI units *without* using your calculator.
 a. 20 ft b. 60 mi
 c. 60 mph d. 8 in

27. ‖ Using the approximate conversion factors in Table 1.5, convert the following SI units to English units *without* using your calculator.
 a. 30 cm b. 25 m/s
 c. 5 km d. 0.5 cm

28. | Compute the following numbers, applying the significant figure rule adopted in this textbook.
 a. 33.3×25.4 b. $33.3 - 25.4$
 c. $\sqrt{33.3}$ d. $333.3 \div 25.4$

29. | Compute the following numbers, applying the significant figure rule adopted in this textbook.
 a. 12.5^3 b. 12.5×5.21
 c. $\sqrt{12.5} - 1.2$ d. 12.5^{-1}

30. | Estimate (don't measure!) the length of a typical car. Give your answer in both feet and meters. Briefly describe how you arrived at this estimate.

31. | Estimate the height of a telephone pole. Give your answer in both feet and meters. Briefly describe how you arrived at this estimate.

32. | Estimate the average speed with which you go from home to campus via whatever mode of transportation you use most commonly. Give your answer in both mph and m/s. Briefly describe how you arrived at this estimate.

33. | Estimate the average speed with which the hair on your head grows. Give your answer in both m/s and μm/hour. Briefly describe how you arrived at this estimate.

Problems

For Problems 34 through 43, draw a complete pictorial representation. Do *not* solve these problems or do any mathematics.

34. | A Porsche accelerates from a stoplight at 5.0 m/s^2 for five seconds, then coasts for three more seconds. How far has it traveled?
35. | A jet plane is cruising at 300 m/s when suddenly the pilot turns the engines up to full throttle. After traveling 4.0 km, the jet is moving with a speed of 400 m/s. What is the jet's acceleration as it speeds up?
36. | Sam is recklessly driving 60 mph in a 30 mph speed zone when he suddenly sees the police. He steps on the brakes and slows to 30 mph in three seconds, looking nonchalant as he passes the officer. How far does he travel while braking?
37. | You would like to stick a wet spit wad on the ceiling, so you toss it straight up with a speed of 10 m/s. How long does it take to reach the ceiling, 3.0 m above?
38. | A speed skater moving across frictionless ice at 8.0 m/s hits a 5.0-m-wide patch of rough ice. She slows steadily, then continues on at 6.0 m/s. What is her acceleration on the rough ice?
39. | Santa loses his footing and slides down a frictionless, snowy roof that is tilted at an angle of 30°. If Santa slides 10 m before reaching the edge, what is his speed as he leaves the roof?
40. | A motorist is traveling at 20 m/s. He is 60 m from a stoplight when he sees it turn yellow. His reaction time, before stepping on the brake, is 0.50 s. What steady deceleration while braking will bring him to a stop right at the light?
41. | A car traveling at 30 m/s runs out of gas while traveling up a 10° slope. How far up the hill will the car coast before starting to roll back down?
42. || Ice hockey star Bruce Blades is 5.0 m from the blue line and gliding toward it at a speed of 4.0 m/s. You are 20 m from the blue line, directly behind Bruce. You want to pass the puck to Bruce. With what speed should you shoot the puck down the ice so that it reaches Bruce exactly as he crosses the blue line?
43. || David is driving a steady 30 m/s when he passes Tina, who is sitting in her car at rest. Tina begins to accelerate at a steady 2.0 m/s^2 at the instant when David passes. How far does Tina drive before passing David?

Problems 44 through 48 show a motion diagram. For each of these problems, write a one or two sentence "story" about a *real object* that has this motion diagram. Your stories should talk about people or objects by name and say what they are doing. Problems 34 through 43 are examples of motion short stories.

44. |

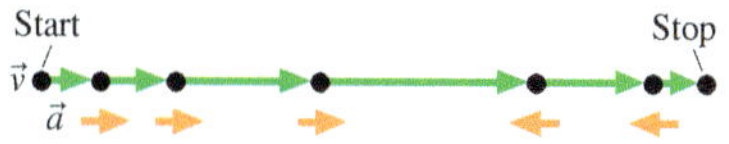

FIGURE P1.44

45. |

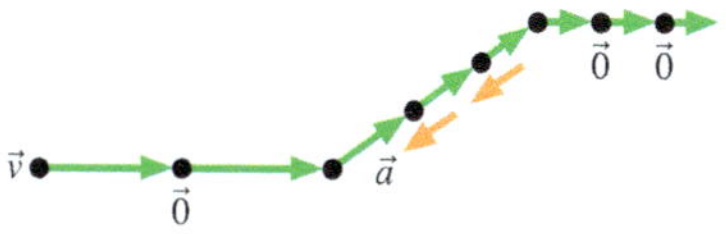

FIGURE P1.45

46. |

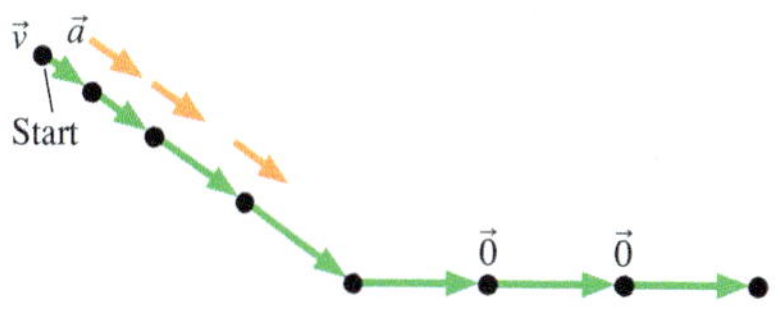

FIGURE P1.46

47. |

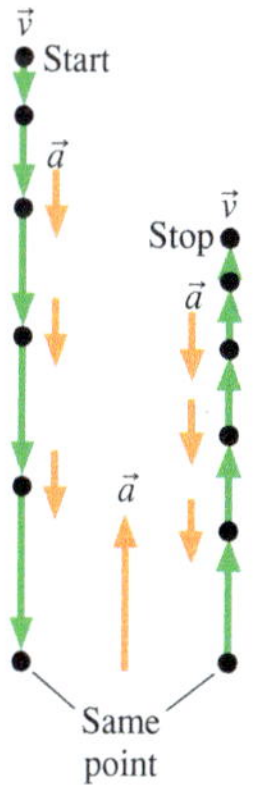

FIGURE P1.47

48. |

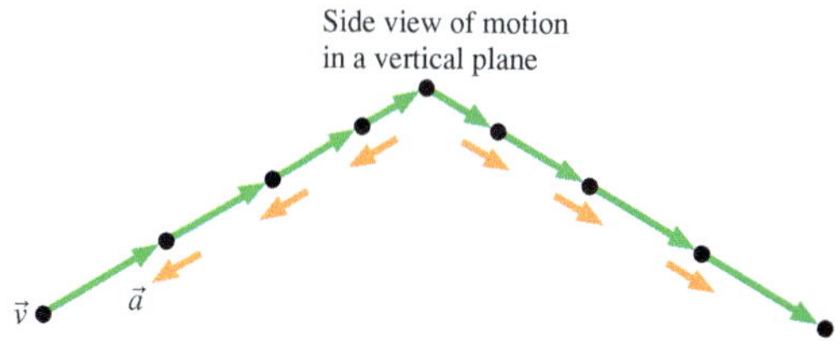

FIGURE P1.48

Problems 49 through 52 show a partial motion diagram. For each:

a. Complete the motion diagram by adding acceleration vectors.
b. Write a physics *problem* for which this is the correct motion diagram. Be imaginative! Don't forget to include enough information to make the problem complete and to state clearly what is to be found.
c. Draw a pictorial representation for your problem.

49. |

FIGURE P1.49

50. |

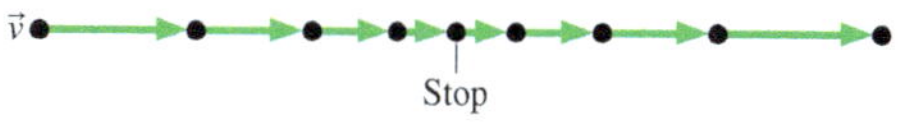

FIGURE P1.50

51. |

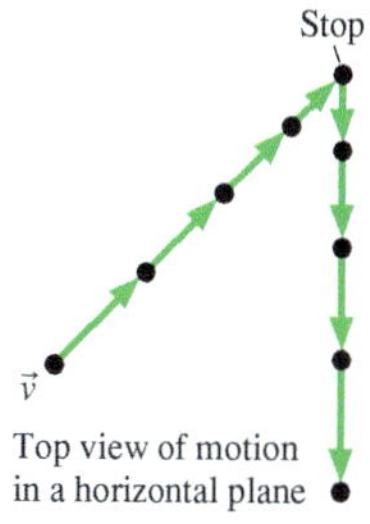

FIGURE P1.51

52. | 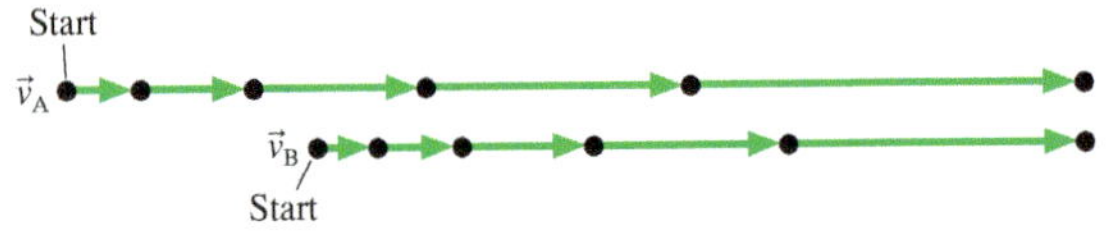

FIGURE P1.52

53. | A regulation soccer field for international play is a rectangle with a length between 100 m and 110 m and a width between 64 m and 75 m. What are the smallest and largest areas that the field could be?

54. || The quantity called *mass density* is the mass per unit volume of a substance. Express the following mass densities in SI units.
   a. Aluminum, 2.7×10^{-3} kg/cm^3
   b. Alcohol, 0.81 g/cm^3

55. || FIGURE P1.55 shows a motion diagram of a car traveling down a street. The camera took one frame every 10 s. A distance scale is provided.

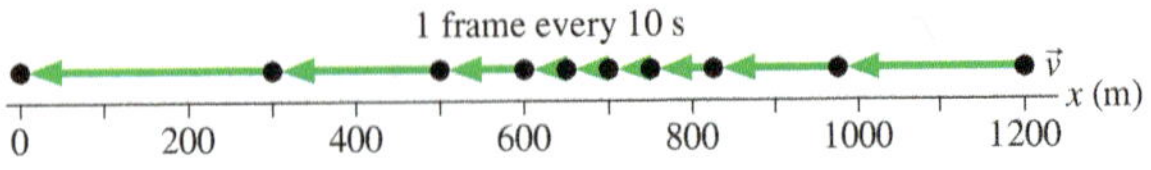

FIGURE P1.55

   a. Measure the x-value of the car at each dot. Place your data in a table, similar to Table 1.1, showing each position and the instant of time at which it occurred.
   b. Make a position-versus-time graph for the car. Because you have data only at certain instants of time, your graph should consist of dots that are not connected together.

56. | Write a short description of a real object for which FIGURE P1.56 would be a realistic position-versus-time graph.

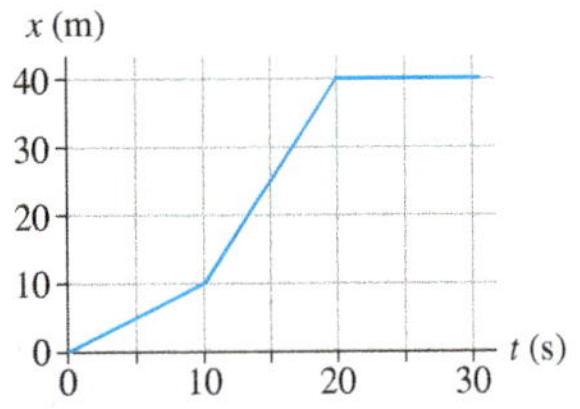

FIGURE P1.56

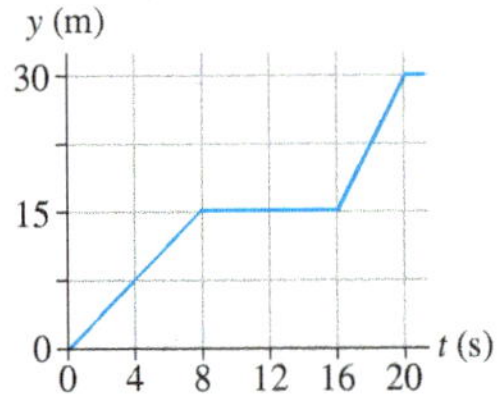

FIGURE P1.57

57. | Write a short description of a real object for which FIGURE P1.57 would be a realistic position-versus-time graph.

STOP TO THINK ANSWERS

Stop to Think 1.1: B. The images of B are farther apart, so it travels a larger distance than does A during the same intervals of time.

Stop to Think 1.2: a. Dropped ball. **b.** Dust particle. **c.** Descending rocket.

Stop to Think 1.3: e. The average velocity vector is found by connecting one dot in the motion diagram to the next.

Stop to Think 1.4: b. $\vec{v}_2 = \vec{v}_1 + \Delta\vec{v}$, and $\Delta\vec{v}$ points in the direction of $\vec{a}$.

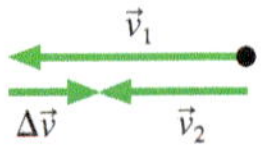

Stop to Think 1.5: d > c > b = a.

2 Kinematics in One Dimension

This Japanese "bullet train" accelerates slowly but steadily until reaching a speed of 300 km/h.

▶ **Looking Ahead** The goal of Chapter 2 is to learn how to solve problems about motion in a straight line.

Kinematics

Kinematics is the name for the mathematical description of motion. We begin with motion along a straight line; for example, runners, rockets, and skiers. Kinematics in two dimensions—projectile motion and circular motion—will be taken up in Chapter 4.

The motion of an object is described by its *position*, *velocity*, and *acceleration*. In one dimension, these quantities are represented by x, v_x, and a_x. You learned to show these on motion diagrams in Chapter 1.

Now we will use calculus to give precise meaning to velocity and acceleration.

◀ **Looking Back**
Sections 1.4–1.5 Velocity and acceleration

It is very important to know when velocity and acceleration are positive and when they are negative.

◀ **Looking Back**
Tactics Box 1.4 The signs of position, velocity, and acceleration

Graphical Representations of Motion

Position, velocity, and acceleration are related graphically.

- You will learn how to draw the position-versus-time, velocity-versus-time, and acceleration-versus-time graphs that describe various types of motion.
- You will learn that the slope of the position-versus-time graph is the instantaneous value of velocity.
- Similarly, the slope of the velocity-versus-time graph is the instantaneous value of acceleration.

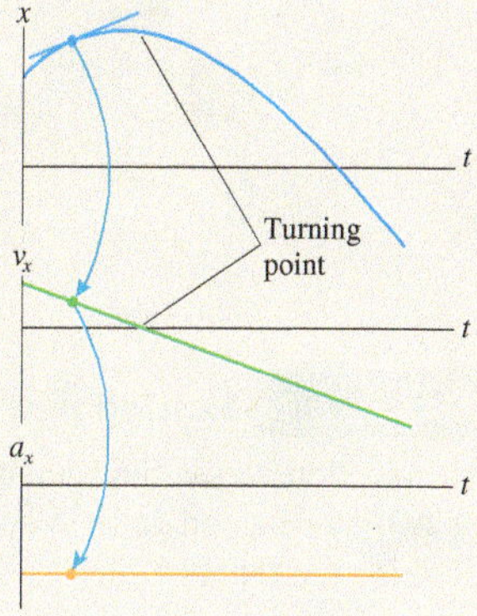

Solving Problems in Kinematics

You will begin learning to solve problems using a four-part *problem-solving strategy*.

MODEL Use the particle model.

VISUALIZE Draw a pictorial representation. Use a graphical representation.

SOLVE Use three kinematic equations that we'll develop in this chapter.

ASSESS Check whether the result makes sense.

◀ **Looking Back**
Tactics Box 1.5 Drawing a pictorial representation

Problems you will learn to solve in this chapter include free fall and motion on an inclined plane.

2.1 Uniform Motion

Riding steadily over level ground is a good example of uniform motion.

If you drive your car at a perfectly steady 60 miles per hour (mph), you will cover 60 mi during the first hour, another 60 mi during the second hour, yet another 60 mi during the third hour, and so on. In this case, 60 mi is not your position, but rather the *change* in your position during each hour; that is, your displacement Δx. Similarly, 1 hour is a time interval Δt rather than a specific instant of time. **Straight-line motion in which equal displacements occur during any successive equal-time intervals is called uniform motion.**

FIGURE 2.1 shows how uniform motion appears in motion diagrams and position-versus-time graphs. Because all equal-time intervals have equal displacements, the position graph is a straight line. In fact, an alternative definition of uniform motion is: **An object's motion is uniform if and only if its position-versus-time graph is a straight line.**

The slope of a straight-line graph is defined as "rise over run." Because position is graphed on the vertical axis, the "rise" of a position-versus-time graph is the object's displacement Δx. The "run" is the time interval Δt. Consequently, the slope is $\Delta x/\Delta t$.

Chapter 1 defined the **average velocity** as $\Delta\vec{r}/\Delta t$. For one-dimensional motion this is simply

$$v_{avg} \equiv \frac{\Delta x}{\Delta t} \text{ or } \frac{\Delta y}{\Delta t} = \text{slope of the position-versus-time graph} \qquad (2.1)$$

That is, **the average velocity is the slope of the position-versus-time graph.** Velocity has units of "length per time," such as "miles per hour." The SI units of velocity are meters per second, abbreviated m/s.

NOTE ▶ The symbol $\equiv$ in Equation 2.1 stands for "is defined as" or "is equivalent to." This is a stronger statement than the two sides simply being equal. ◀

In the case of uniform motion, where the slope $\Delta x/\Delta t$ is the same at all times, it appears that the average velocity is constant and unchanging. Consequently, a final definition of uniform motion is: **An object's motion is uniform if and only if its velocity v_x or v_y does not change.** There's no real need to specify "average" for a velocity that doesn't change, so we will drop the subscript and refer to the average velocity as v_x or v_y.

FIGURE 2.1 Motion diagram and position graph for uniform motion.

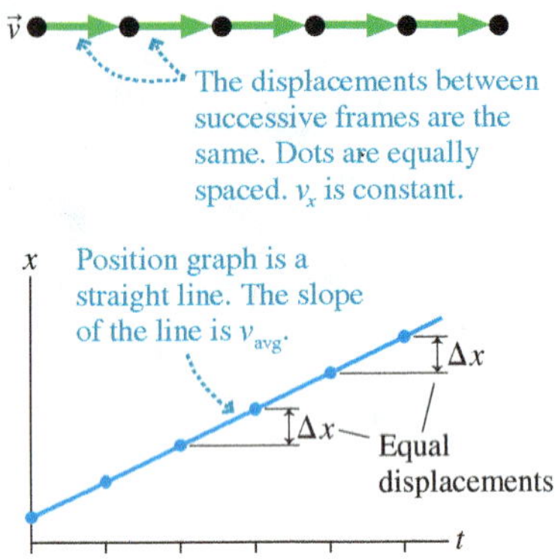

EXAMPLE 2.1 Skating with constant velocity

The position-versus-time graph of FIGURE 2.2 represents the motion of two students on roller blades. Determine their velocities and describe their motion.

MODEL Represent the two students as particles.

VISUALIZE Figure 2.2 is a graphical representation of the students' motion. Both graphs are straight lines, telling us that both skaters are moving uniformly with constant velocities.

SOLVE We can determine the students' velocities by measuring the slopes of the graphs. Skater A undergoes a displacement $\Delta x_A = 2.0$ m during the time interval $\Delta t_A = 0.40$ s Thus his velocity is

$$(v_x)_A = \frac{\Delta x_A}{\Delta t_A} = \frac{2.0 \text{ m}}{0.40 \text{ s}}$$
$$= 5.0 \text{ m/s}$$

FIGURE 2.2 Graphical representations of two students on roller blades.

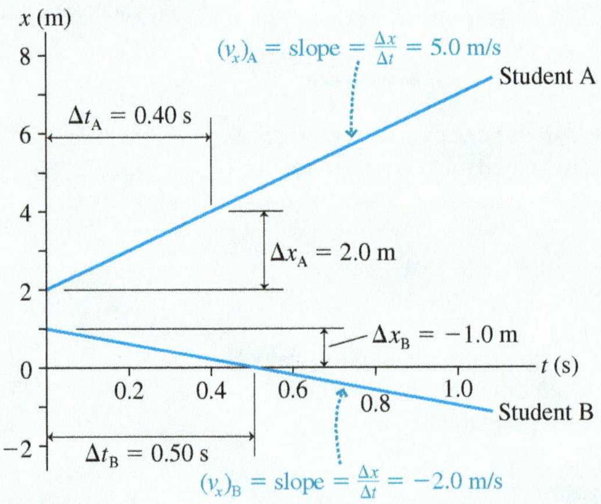

We need to be more careful with skater B. Although he moves a distance of 1.0 m in 0.50 s, his *displacement* is

$$\Delta x_B = x_{\text{at } 0.5\text{ s}} - x_{\text{at } 0.0\text{ s}} = 0.0\text{ m} - 1.0\text{ m} = -1.0\text{ m}$$

Careful attention to the signs is very important! This leads to

$$(v_x)_B = \frac{\Delta x_B}{\Delta t_B} = \frac{-1.0\text{ m}}{0.50\text{ s}} = -2.0\text{ m/s}$$

ASSESS The minus sign indicates that skater B is moving to the left. Our interpretation of this graph is that two students on roller blades are moving with constant velocities in opposite directions. Skater A starts at $x = 2.0$ m and moves to the right with a velocity of $x = 5.0$ m/s. Skater B starts at $x = 1.0$ m and moves to the left with a velocity of -2.0 m/s. Their speeds, of ≈ 10 mph and ≈ 4 mph, are reasonable for skaters on roller blades.

Example 2.1 brought out several points that are worth emphasizing. These are summarized in Tactics Box 2.1.

TACTICS BOX 2.1 Interpreting position-versus-time graphs

1. Steeper slopes correspond to faster speeds.
2. Negative slopes correspond to negative velocities and, hence, to motion to the left (or down).
3. The slope is a ratio of intervals, $\Delta x/\Delta t$, not a ratio of coordinates. That is, the slope is *not* simply x/t.
4. We are distinguishing between the *actual* slope and the *physically meaningful* slope. If you were to use a ruler to measure the rise and the run of the graph, you could compute the actual slope of the line as drawn on the page. That is not the slope to which we are referring when we equate the velocity with the slope of the line. Instead, we find the *physically meaningful* slope by measuring the rise and run using the scales along the axes. The "rise" Δx is some number of meters; the "run" Δt is some number of seconds. The physically meaningful rise and run include units, and the ratio of these units gives the units of the slope.

Exercises 1–3

An object's **speed** v is how fast it's going, independent of direction. This is simply $v = |v_x|$ or $v = |v_y|$ the magnitude or absolute value of the object's velocity. In Example 2.1, for example, skater B's *velocity* is -2.0 m/s but his *speed* is 2.0 m/s. Speed is a scalar quantity, not a vector.

NOTE ▶ Our mathematical analysis of motion is based on velocity, not speed. The subscript in v_x or v_y is an essential part of the notation, reminding us that, even in one dimension, the velocity is a vector. ◀

The Mathematics of Uniform Motion

We need a mathematical analysis of motion that will be valid regardless of whether an object moves along the x-axis, the y-axis, or any other straight line. Consequently, it will be convenient to write equations for a "generic axis" that we will call the s-axis. The position of an object will be represented by the symbol s and its velocity by v_s.

NOTE ▶ In a specific problem you should use either x or y, whichever is appropriate, rather than s. ◀

FIGURE 2.3 The velocity is found from the slope of the position-versus-time graph.

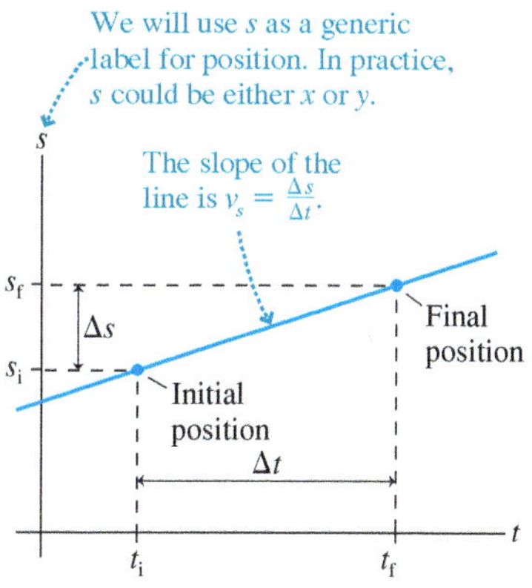

Consider an object in uniform motion along the s-axis with the linear position-versus-time graph shown in FIGURE 2.3. The object's **initial position** is s_i at time t_i. The term *initial position* refers to the starting point of our analysis or the starting point in a problem; the object may or may not have been in motion prior to t_i. At a later time t_f, the ending point of our analysis, the object's **final position** is s_f.

The object's velocity v_s along the s-axis can be determined by finding the slope of the graph:

$$v_s = \frac{\text{rise}}{\text{run}} = \frac{\Delta s}{\Delta t} = \frac{s_f - s_i}{t_f - t_i} \tag{2.2}$$

Equation 2.2 is easily rearranged to give

$$s_f = s_i + v_s \, \Delta t \quad \text{(uniform motion)} \tag{2.3}$$

The velocity of a uniformly moving object tells us the amount by which its position changes during each second. A particle with a velocity of 20 m/s *changes* its position by 20 m during every second of motion: by 20 m during the first second of its motion, by another 20 m during the next second, and so on. If the object starts at $s_i = 10$ m, it will be at $s = 30$ m after 1 second of motion and at $s = 50$ m after 2 seconds of motion. Thinking of velocity like this will help you develop an intuitive understanding of the connection between velocity and position.

EXAMPLE 2.2 Relating a velocity graph to a position graph

FIGURE 2.4 is the position-versus-time graph of a car.

a. Draw the car's velocity-versus-time graph.
b. Describe the car's motion.

MODEL Represent the car as a particle, with a well-defined position at each instant of time.

VISUALIZE Figure 2.4 is the graphical representation.

SOLVE

a. The car's position-versus-time graph is a sequence of three straight lines. Each of these straight lines represents uniform motion at a constant velocity. We can determine the car's velocity during each interval of time by measuring the slope of the line. From $t = 0$ s to $t = 2$ s ($\Delta t = 2.0$ s) the car's displacement is $\Delta x = -4.0 \text{ m} - 0.0 \text{ m} = -4.0$ m. The velocity during this interval is

$$v_x = \frac{\Delta x}{\Delta t} = \frac{-4.0 \text{ m}}{2.0 \text{ s}} = -2.0 \text{ m/s}$$

The car's position does not change from $t = 2$ s to $t = 4$ s ($\Delta x = 0$), so $v_x = 0$. Finally, the displacement between $t = 4$ s and $t = 6$ s is $\Delta x = 10.0$ m. Thus the velocity during this interval is

$$v_x = \frac{10.0 \text{ m}}{2.0 \text{ s}} = 5.0 \text{ m/s}$$

These velocities are shown on the velocity-versus-time graph of FIGURE 2.5.

b. The car backs up for 2 s at 2.0 m/s, sits at rest for 2 s, then drives forward at 5.0 m/s for at least 2 s. We can't tell from the graph what happens for $t > 6$ s.

ASSESS The velocity graph and the position graph look completely different. The *value* of the velocity graph at any instant of time equals the *slope* of the position graph.

FIGURE 2.4 Position-versus-time graph.

FIGURE 2.5 The corresponding velocity-versus-time graph.

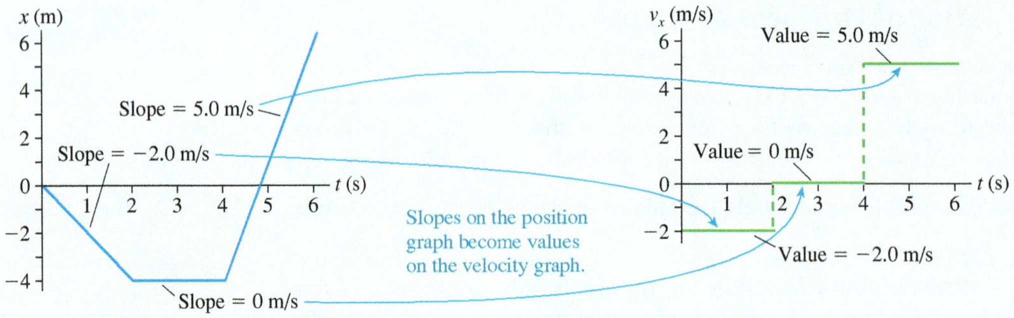

EXAMPLE 2.3 Lunch in Cleveland?

Bob leaves home in Chicago at 9:00 A.M. and travels east at a steady 60 mph. Susan, 400 miles to the east in Pittsburgh, leaves at the same time and travels west at a steady 40 mph. Where will they meet for lunch?

MODEL Here is a problem where, for the first time, we can really put all four aspects of our problem-solving strategy into play. To begin, represent Bob and Susan as particles.

VISUALIZE FIGURE 2.6 shows the pictorial representation. The equal spacings of the dots in the motion diagram indicate that the motion is uniform. In evaluating the given information, we recognize that the starting time of 9:00 A.M. is not relevant to the problem. Consequently, the initial time is chosen as simply $t_0 = 0$ h. Bob and Susan are traveling in opposite directions, hence one of the velocities must be a negative number. We have chosen a coordinate system in which Bob starts at the origin and moves to the right (east) while Susan is moving to the left (west). Thus Susan has the negative velocity. Notice how we've assigned position, velocity, and time symbols to each point in the motion. Pay special attention to how subscripts are used to distinguish different points in the problem and to distinguish Bob's symbols from Susan's.

One purpose of the pictorial representation is to establish what we need to find. Bob and Susan meet when they have the same position at the same time t_1. Thus we want to find $(x_1)_B$ at the time when $(x_1)_B = (x_1)_S$. Notice that $(x_1)_B$ and $(x_1)_S$ are Bob's and Susan's *positions,* which are equal when they meet, not the distances they have traveled.

SOLVE The goal of the mathematical representation is to proceed from the pictorial representation to a mathematical solution of the problem. We can begin by using Equation 2.3 to find Bob's and Susan's positions at time t_1 when they meet:

$$(x_1)_B = (x_0)_B + (v_x)_B(t_1 - t_0) = (v_x)_B t_1$$

$$(x_1)_S = (x_0)_S + (v_x)_S(t_1 - t_0) = (x_0)_S + (v_x)_S t_1$$

Notice two things. First, we started by writing the *full* statement of Equation 2.3. Only then did we simplify by dropping those terms known to be zero. You're less likely to make accidental errors if you follow this procedure. Second, we replaced the generic symbol s with the specific horizontal-position symbol x, and we replaced the generic subscripts i and f with the specific symbols 0 and 1 that we defined in the pictorial representation. This is also good problem-solving technique.

The condition that Bob and Susan meet is

$$(x_1)_B = (x_1)_S$$

By equating the right-hand sides of the above equations, we get

$$(v_x)_B t_1 = (x_0)_S + (v_x)_S t_1$$

Solving for t_1 we find that they meet at time

$$t_1 = \frac{(x_0)_S}{(v_x)_B - (v_x)_S} = \frac{400 \text{ miles}}{60 \text{ mph} - (-40) \text{ mph}} = 4.0 \text{ hours}$$

Finally, inserting this time back into the equation for $(x_1)_B$ gives

$$(x_1)_B = \left(60 \, \frac{\text{miles}}{\text{hour}}\right) \times (4.0 \text{ hours}) = 240 \text{ miles}$$

While this is a number, it is not yet the answer to the question. The phrase "240 miles" by itself does not say anything meaningful. Because this is the value of Bob's *position,* and Bob was driving east, the answer to the question is, "They meet 240 miles east of Chicago."

ASSESS Before stopping, we should check whether or not this answer seems reasonable. We certainly expected an answer between 0 miles and 400 miles. We also know that Bob is driving faster than Susan, so we expect that their meeting point will be *more* than halfway from Chicago to Pittsburgh. Our assessment tells us that 240 miles is a reasonable answer.

FIGURE 2.6 Pictorial representation for Example 2.3.

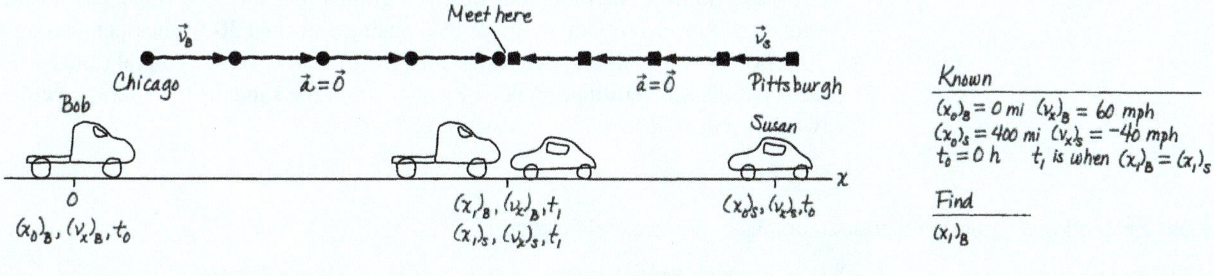

FIGURE 2.7 Position-versus-time graphs for Bob and Susan.

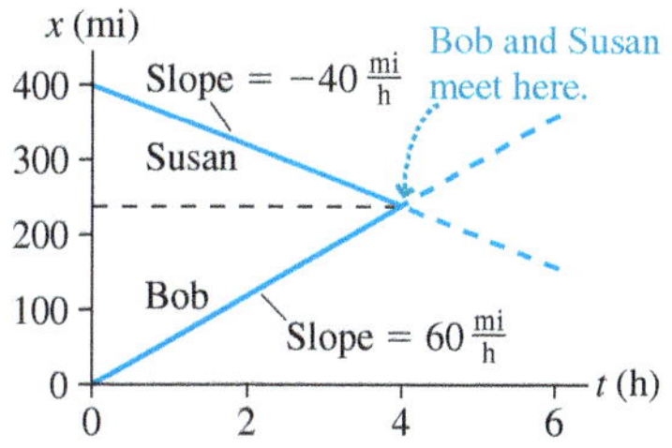

It is instructive to look at this example from a graphical perspective. FIGURE 2.7 shows position-versus-time graphs for Bob and Susan. Notice the negative slope for Susan's graph, indicating her negative velocity. The point of interest is the intersection of the two lines; this is where Bob and Susan have the same position at the same time. Our method of solution, in which we equated $(x_1)_B$ and $(x_1)_S$, is really just solving the mathematical problem of finding the intersection of two lines.

STOP TO THINK 2.1 Which position-versus-time graph represents the motion shown in the motion diagram?

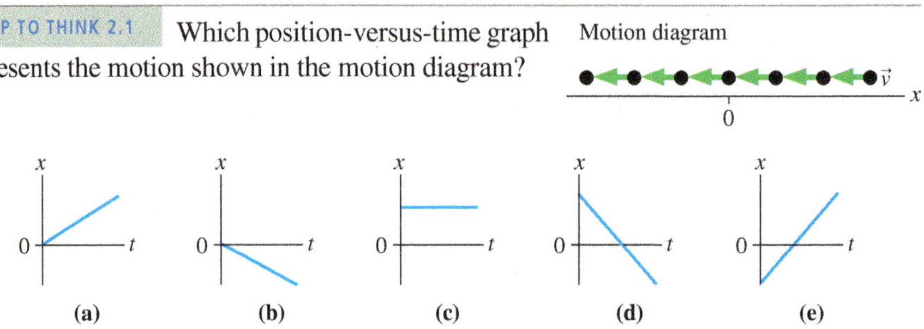

2.2 Instantaneous Velocity

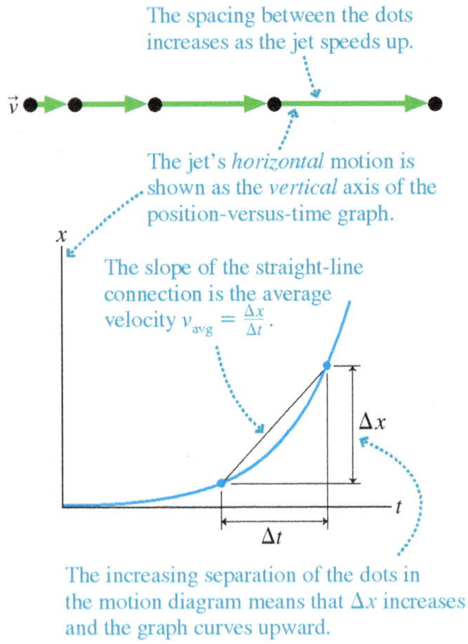

FIGURE 2.8 Motion diagram and position graph of a jet during takeoff.

FIGURE 2.8 shows the motion diagram of a jet as it takes off. The increasing length of the velocity vectors tells us that the jet is speeding up, so this is *not* uniform motion. Consequently, the position-versus-time graph is *not* a straight line.

We can determine the jet's average speed v_{avg} between any two times t_i and t_f by finding the slope of the straight-line connection between the two points. However, average velocity has only limited usefulness for an object whose velocity isn't constant. The jet's average velocity during takeoff might be 30 m/s, but the speedometer in the cockpit would show the jet traveling at less than 30 m/s during the first few seconds. Similarly, the speedometer would read more than 30 m/s just before the wheels leave the ground.

In contrast to a velocity averaged over some interval of time, the speedometer reading tells you how fast you're going *at that instant.* We define an object's **instantaneous velocity** to be its velocity—a speed *and* a direction—at a single *instant* of time *t*.

As we've seen, velocity is the *rate* at which an object changes its position. Rates tell us how quickly or how slowly things change, and that idea is conveyed by the word "per." An instantaneous velocity of 80 miles *per* hour means that the rate at which your car's position is changing—at that exact instant—is such that it would travel 80 miles in 1 hour *if* it continued at that rate without change. Whether or not it actually does travel at that velocity for another hour, or even for another millisecond, is not relevant.

Using Motion Diagrams and Graphs

Let's use motion diagrams and position graphs to analyze a rocket as it takes off. **FIGURE 2.9a** shows a motion diagram made using a normal 30-frames-per-second movie camera. We would like to determine the *instantaneous* velocity v_{2y} at time t_2. Because the rocket is accelerating, its velocity at t_2 is not the same as the average velocity between t_1 and t_3. How can we measure v_{2y}?

FIGURE 2.9 Motion diagrams and position graphs of an accelerating rocket.

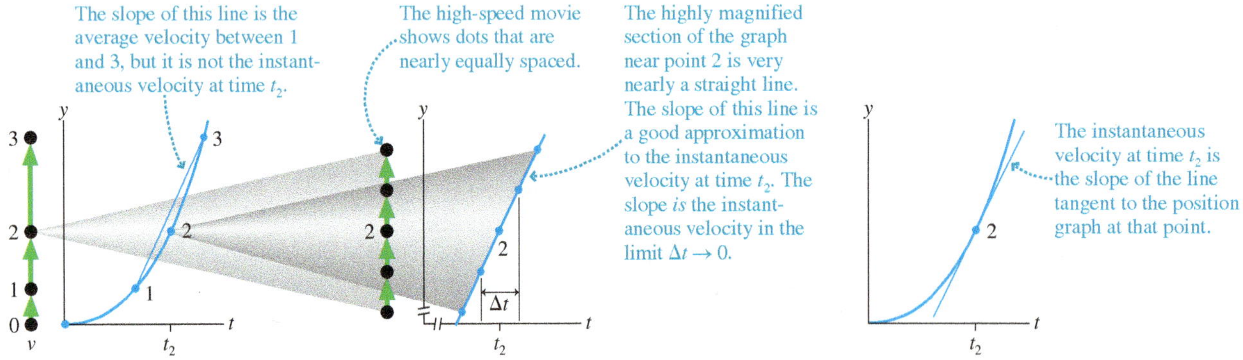

Suppose we use a high-speed camera, one that takes 3000 frames per second, to film just the segment of motion right around time t_2. This "magnified" motion diagram is shown in FIGURE 2.9b. At this level of magnification, each velocity vector is *almost* the same length. Further, the greatly magnified section of the curved position graph is *almost* a straight line. That is, the motion appears very nearly uniform on this time scale. If the rocket suddenly changed to *constant*-velocity motion at time t_2, it would continue to move with a velocity given by the slope of the graph in Figure 2.9b.

In other words, the average velocity $v_{avg} = \Delta s/\Delta t$ becomes a better and better approximation to the instantaneous velocity v_s as the time interval Δt over which the average is taken gets smaller and smaller. By magnifying the motion diagram, we are using smaller and smaller time intervals Δt.

We can state this idea mathematically in terms of the limit $\Delta t \to 0$:

$$v_s \equiv \lim_{\Delta t \to 0} \frac{\Delta s}{\Delta t} = \frac{ds}{dt} \qquad \text{(instantaneous velocity)} \qquad (2.4)$$

As Δt continues to get smaller, the average velocity $v_{avg} = \Delta s/\Delta t$ reaches a constant or *limiting* value. That is, **the instantaneous velocity at time *t* is the average velocity during a time interval Δt, centered on *t*, as Δt approaches zero.** In calculus, this limit is called *the derivative of s with respect to t*, and it is denoted *ds/dt*.

Graphically, $\Delta s/\Delta t$ is the slope of a straight line. As Δt gets smaller (i.e., more and more magnification), the straight line becomes a better and better approximation of the curve *at that one point*. In the limit $\Delta t \to 0$, the straight line is tangent to the curve. As FIGURE 2.9c shows, **the instantaneous velocity at time *t* is the slope of the line that is tangent to the position-versus-time graph at time *t*.** That is,

$$v_s = \text{slope of the position-versus-time graph at time } t \qquad (2.5)$$

EXAMPLE 2.4 Finding velocity from position graphically

FIGURE 2.10 shows the position-versus-time graph of an elevator.

a. At which labeled point or points does the elevator have the least speed?
b. At which point or points does the elevator have maximum velocity?
c. Sketch an approximate velocity-versus-time graph for the elevator.

FIGURE 2.10 Position-versus-time graph.

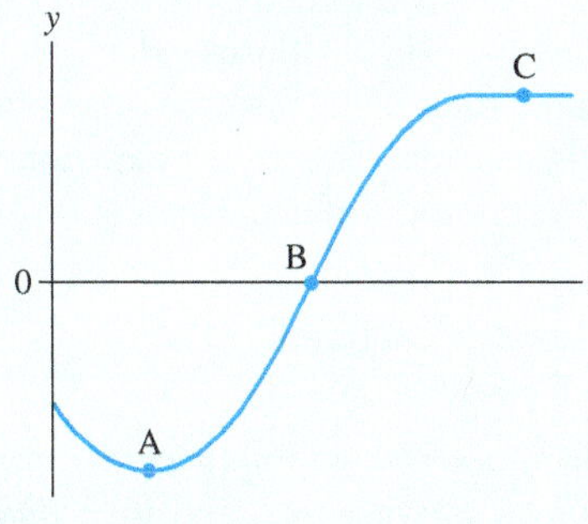

MODEL Represent the elevator as a particle.

VISUALIZE Figure 2.10 is the graphical representation.

SOLVE a. At any instant, the velocity is the slope of the position graph. FIGURE 2.11a shows that the elevator has the least speed—no speed at all!—at points A and C. At point A, the velocity is only instantaneously zero. At point C, the elevator has actually stopped and remains at rest.

FIGURE 2.11 The velocity-versus-time graph is found from the slope of the position graph.

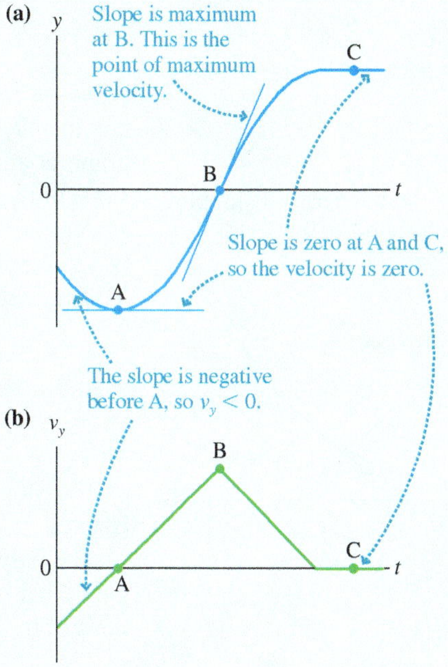

b. The elevator has maximum velocity at point B.

Continued

c. Although we cannot find an exact velocity-versus-time graph, we can see that the slope, and hence v_y, is initially negative, becomes zero at point A, rises to a maximum value at point B, decreases back to zero a little before point C, then remains at zero thereafter. Thus FIGURE 2.11b shows, at least approximately, the elevator's velocity-versus-time graph.

ASSESS Once again, the shape of the velocity graph bears no resemblance to the shape of the position graph. You must transfer *slope* information from the position graph to *value* information on the velocity graph.

Scientists and engineers must use calculus to calculate the trajectories of rockets.

A Little Calculus: Derivatives

Calculus–invented simultaneously in England by Newton and in Germany by Leibniz–is designed to deal with instantaneous quantities. In other words, it provides us with the tools for evaluating limits such as the one in Equation 2.4.

The notation ds/dt is called *the derivative of s with respect to t*, and Equation 2.4 defines it as the limiting value of a ratio. As Figure 2.9 showed, ds/dt can be interpreted graphically as the slope of the line that is tangent to the position-versus-time graph at time t.

The only functions we will use in Parts I and II of this book are powers and polynomials. Consider the function $u = ct^n$, where c and n are constants. The following result is proven in calculus:

$$\text{The derivative of } u = ct^n \text{ is } \frac{du}{dt} = nct^{n-1} \qquad (2.6)$$

NOTE ▶ The symbol u is a "dummy name." Equation 2.6 can be used to take the derivative of *any* function of the form ct^n. ◀

For example, suppose the position of a particle as a function of time is $s = 2t^2$ m where t is in s. We can find the particle's velocity by using Equation 2.6 with $c = 2$ and $n = 2$ to calculate that the derivative of $s = 2t^2$ with respect to t is

$$v_s = \frac{ds}{dt} = 2 \cdot 2t^{2-1} = 4t$$

FIGURE 2.12 shows the particle's position and velocity graphs. It is critically important to understand the relationship between these two graphs. The *value* of the velocity graph at any instant of time, which we can read directly off the vertical axis, is the *slope* of the position graph at that same time. This is illustrated at $t = 1$ s and $t = 3$ s.

FIGURE 2.12 Position-versus-time graph and the corresponding velocity-versus-time graph.

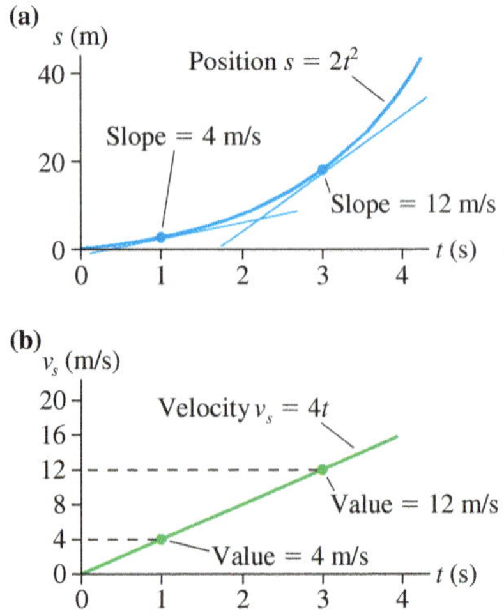

A value that doesn't change with time, such as the position of an object at rest, can be represented by the function $u = c =$ constant That is, the exponent of t^n is $n = 0$. You can see from Equation 2.6 that the derivative of a constant is zero. That is,

$$\frac{du}{dt} = 0 \text{ if } u = c = \text{constant} \qquad (2.7)$$

This makes sense. The graph of the function $u = c$ is simply a horizontal line at height c. The slope of a horizontal line—which is what the derivative du/dt measures—is zero.

The only other information we need about derivatives for now is how to evaluate the derivative of the sum of two or more functions. Let u and w be two separate functions of time. You will learn in calculus that

$$\frac{d}{dt}(u + w) = \frac{du}{dt} + \frac{dw}{dt} \qquad (2.8)$$

That is, the derivative of a sum is the sum of the derivatives.

NOTE ▶ You may have learned in calculus to take the derivative dy/dx, where y is a function of x. The derivatives we use in physics are the same; only the notation is different. We're interested in how quantities change with time, so our derivatives are with respect to t instead of x. ◀

EXAMPLE 2.5 Using calculus to find the velocity

A particle's position is given by the function $x = (-t^3 + 3t)$ m, where t is in s.

a. What are the particle's position and velocity at $t = 2$ s?
b. Draw graphs of x and v_x during the interval $-3 \text{ s} \leq t \leq 3$ s.
c. Draw a motion diagram to illustrate this motion.

SOLVE

a. We can compute the position directly from the function x:

$$x(\text{at } t = 2 \text{ s}) = -(2)^3 + (3)(2) = -8 + 6 = -2 \text{ m}$$

The velocity is $v_x = dx/dt$. The function for x is the sum of two polynomials, so

$$v_x = \frac{dx}{dt} = \frac{d}{dt}(-t^3 + 3t) = \frac{d}{dt}(-t^3) + \frac{d}{dt}(3t)$$

The first derivative is a power with $c = -1$ and $n = 3$; the second has $c = 3$ and $n = 1$. Using Equation 2.6, we have

$$v_x = (-3t^2 + 3) \text{ m/s}$$

where t is in s. Evaluating the velocity at $t = 2$ s gives

$$v_x(\text{at } t = 2 \text{ s}) = -3(2)^2 + 3 = -9 \text{ m/s}$$

The negative sign indicates that the particle, at this instant of time, is moving to the *left* at a speed of 9 m/s.

b. FIGURE 2.13 shows the position graph and the velocity graph. These were created by computing, and then graphing, the values of x and v_x at several points between -3 s and 3 s. The slope of the position-versus-time graph at $t = 2$ s is -9 m/s; this becomes the *value* that is graphed for the velocity at $t = 2$ s. Similar measurements are shown at $t = -1$ s, where the velocity is instantaneously zero.

c. Finally, we can interpret the graphs in Figure 2.13 to draw the motion diagram shown in FIGURE 2.14.

- The particle is initially to the right of the origin ($x > 0$ at $t = -3$ s) but moving to the left ($v_x < 0$). Its *speed* is slowing ($v = |v_x|$ is decreasing), so the velocity vector arrows are getting shorter.
- The particle passes the origin at $t \approx -1.5$ s, but it is still moving to the left.
- The position reaches a minimum at $t = -1$ s; the particle is as far left as it is going. The velocity is *instantaneously* $v_x = 0$ m/s as the particle reverses direction.
- The particle moves back to the right between $t = -1$ s and $t = 1$ s ($v_x > 0$).
- The particle turns around again at $t = 1$ s and begins moving back to the left ($v_x < 0$). It keeps speeding up, then disappears off to the left.

A point in the motion where a particle reverses direction is called a **turning point.** It is a point where the velocity is instantaneously zero while the position is a maximum or minimum. This particle has two turning points, at $t = -1$ s and again at $t = +1$ s. We will see many other examples of turning points.

ASSESS This example has used three different *representations* of motion: the mathematical equations, the graphs, and the motion diagram. All three describe the motion, but in different ways. Learning to move back and forth among the representations is important for solidifying your understanding of kinematics.

FIGURE 2.13 Position and velocity graphs.

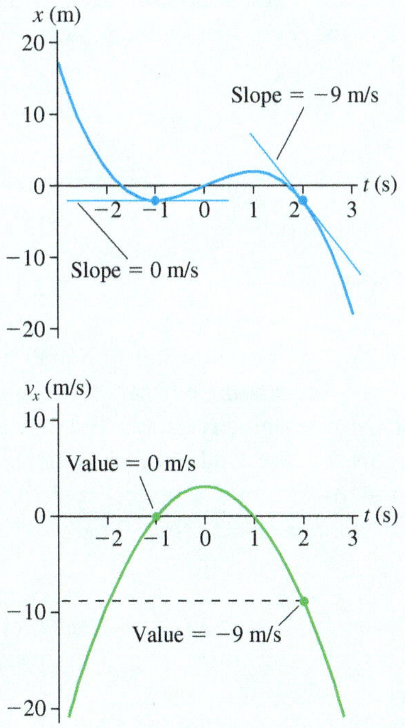

FIGURE 2.14 Motion diagram for Example 2.5.

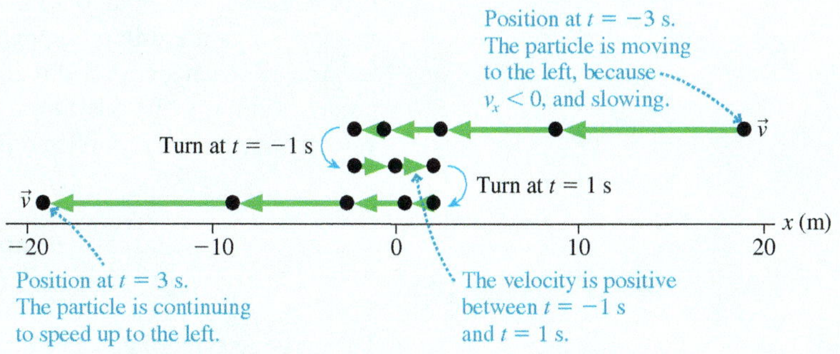

STOP TO THINK 2.2 Which velocity-versus-time graph goes with the position-versus-time graph on the left?

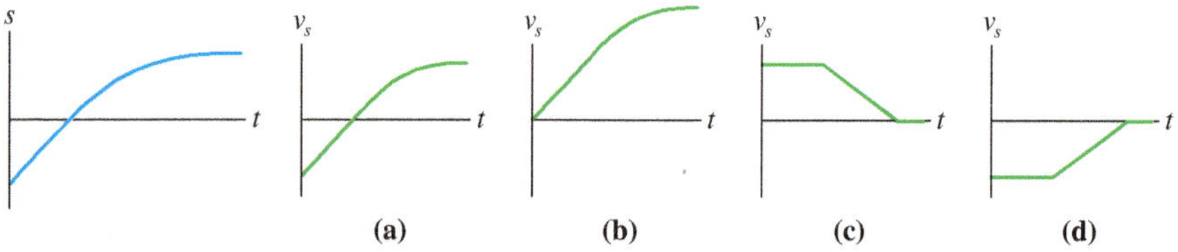

2.3 Finding Position from Velocity

Equation 2.4 provides a means of finding the instantaneous velocity v_s if we know the position s as a function of time. But what about the reverse problem? Can we use the object's velocity to calculate its position at some future time t? Equation 2.3, $s_f = s_i + v_s \Delta t$, does this for the case of uniform motion with a constant velocity. We need to find a more general expression that is valid when v_s is not constant.

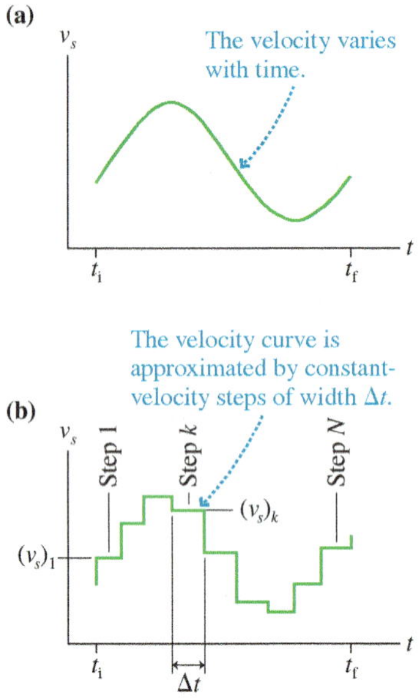

FIGURE 2.15 Approximating a velocity-versus-time graph with a series of constant-velocity steps.

FIGURE 2.15a is a velocity-versus-time graph for an object whose velocity varies with time. Suppose we know the object's position to be s_i at an initial time t_i. Our goal is to find its position s_f at a later time t_f.

Because we know how to handle constant velocities, using Equation 2.3, let's *approximate* the velocity function of Figure 2.15a as a series of constant-velocity steps of width Δt. This is illustrated in FIGURE 2.15b. During the first step, from time t_i to time $t_i + \Delta t$, the velocity has the constant value $(v_s)_1$. The velocity during step k has the constant value $(v_s)_k$. Although the approximation shown in the figure is rather rough, with only nine steps, we can easily imagine that it could be made as accurate as desired by having more and more ever-narrower steps.

The velocity during each step is constant (uniform motion), so we can apply Equation 2.3 to each step. The object's displacement Δs_1 during the first step is simply $\Delta s_1 = (v_s)_1 \Delta t$. The displacement during the second step $\Delta s_2 = (v_s)_2 \Delta t$, and during step k the displacement is $\Delta s_k = (v_s)_k \Delta t$.

The total displacement of the object between t_i and t_f can be approximated as the sum of all the individual displacements during each of the N constant-velocity steps. That is,

$$\Delta s = s_f - s_i \approx \Delta s_1 + \Delta s_2 + \cdots + \Delta s_N = \sum_{k=1}^{N} (v_s)_k \Delta t \tag{2.9}$$

where Σ (Greek sigma) is the symbol for summation. With a simple rearrangement, the particle's final position is

$$s_f \approx s_i + \sum_{k=1}^{N} (v_s)_k \Delta t \tag{2.10}$$

Our goal was to use the object's velocity to find its final position s_f. Equation 2.10 nearly reaches that goal, but Equation 2.10 is only approximate because the constant-velocity steps are only an approximation of the true velocity graph. But if we now let $\Delta t \rightarrow 0$, each step's width approaches zero while the total number of steps N approaches infinity. In this limit, the series of steps becomes a perfect replica of the velocity-versus-time graph and Equation 2.10 becomes exact. Thus

$$s_f = s_i + \lim_{\Delta t \to 0} \sum_{k=1}^{N} (v_s)_k \Delta t = s_i + \int_{t_i}^{t_f} v_s \, dt \tag{2.11}$$

The curlicue symbol is called an *integral*. The expression on the right is read, "the integral of $v_s \, dt$ from t_i to t_f." Equation 2.11 is the result that we were seeking. It allows us to predict an object's position s_f at a future time t_f.

We can give Equation 2.11 an important geometric interpretation. FIGURE 2.16 shows step k in the approximation of the velocity graph as a tall, thin rectangle of height $(v_s)_k$ and width Δt. The product $\Delta s_k = (v_s)_k \Delta t$ is the area (base $\times$ height) of this small rectangle. The sum in Equation 2.11 adds up all of these rectangular areas to give the total area enclosed between the t-axis and the tops of the steps. The limit of this sum as $\Delta t \rightarrow 0$ is the total area enclosed between the t-axis and the velocity curve. This is called the "area under the curve." Thus a graphical interpretation of Equation 2.11 is:

$$s_f = s_i + \text{area under the velocity curve } v_s \text{ between } t_i \text{ and } t_f \qquad (2.12)$$

NOTE ▶ Wait a minute! The displacement $\Delta s = s_f - s_i$ is a length. How can a length equal an area? Recall earlier, when we found that the velocity is the slope of the position graph, we made a distinction between the *actual* slope and the *physically meaningful* slope? The same distinction applies here. The velocity graph does indeed bound a certain area on the page. That is the actual area, but it is *not* the area to which we are referring. Once again, we need to measure the quantities we are using, v_s and Δt, by referring to the scales on the axes. Δt is some number of seconds while v_s is some number of meters per second. When these are multiplied together, the *physically meaningful* area has units of meters. ◀

FIGURE 2.16 The total displacement Δs is the "area under the curve."

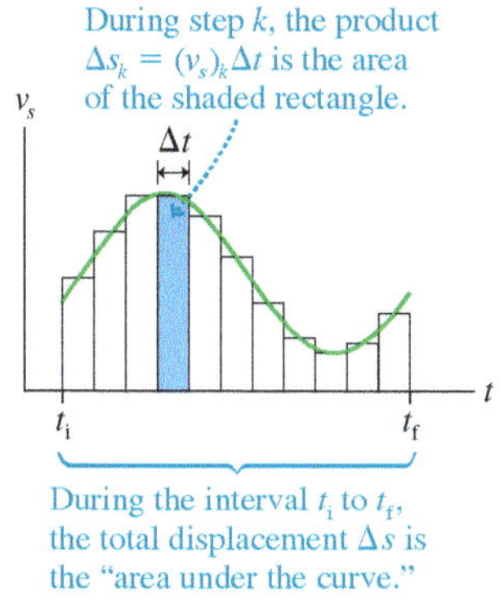

EXAMPLE 2.6 The displacement during a drag race

FIGURE 2.17 shows the velocity-versus-time graph of a drag racer. How far does the racer move during the first 3.0 s?

FIGURE 2.17 Velocity-versus-time graph for Example 2.6.

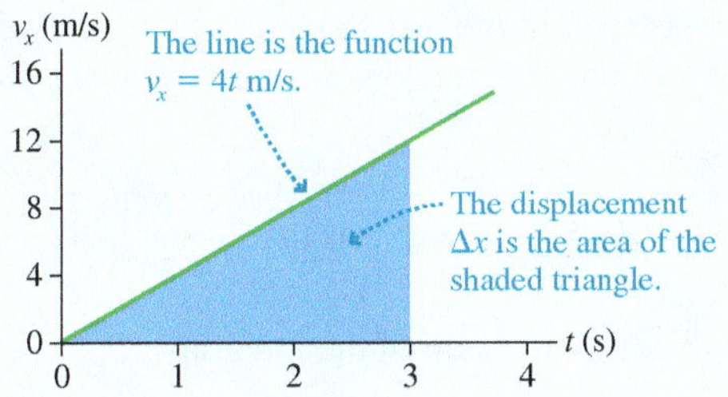

MODEL Represent the drag racer as a particle with a well-defined position at all times.

VISUALIZE Figure 2.17 is the graphical representation.

SOLVE The question "How far?" indicates that we need to find a displacement Δx rather than a position x. According to Equation 2.12, the car's displacement $\Delta x = x_f - x_i$ between $t = 0$ s and $t = 3$ s is the area under the curve from $t = 0$ s to $t = 3$ s. The curve in this case is an angled line, so the area is that of a triangle:

$$\Delta x = \text{area of triangle between } t = 0 \text{ s and } t = 3 \text{ s}$$
$$= \tfrac{1}{2} \times \text{base} \times \text{height}$$
$$= \tfrac{1}{2} \times 3 \text{ s} \times 12 \text{ m/s} = 18 \text{ m}$$

The drag racer moves 18 m during the first 3 seconds.

ASSESS The "area" is a product of s with m/s, so Δx has the proper units of m.

EXAMPLE 2.7 Finding the turning point

FIGURE 2.18 is the velocity graph for a particle that starts at $x_i = 30$ m at time $t_i = 0$ s.

a. Draw a motion diagram for the particle.
b. Where is the particle's turning point?
c. At what time does the particle reach the origin?

FIGURE 2.18 Velocity-versus-time graph for the particle of Example 2.7.

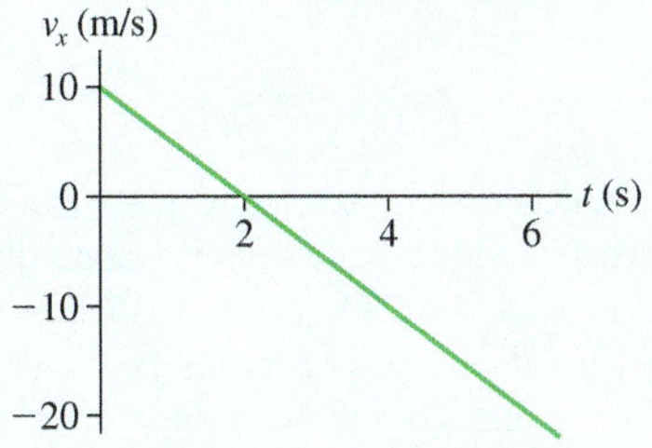

VISUALIZE The particle is initially 30 m to the right of the origin and moving *to the right* ($v_x > 0$) with a speed of 10 m/s. But v_x is decreasing, so the particle is slowing down. At $t = 2$ s the velocity, just for an instant, is zero before becoming negative. This is the turning point. The velocity is negative for $t > 2$ s, so the particle has reversed direction and moves back toward the origin. At some later time, which we want to find, the particle will pass $x = 0$ m.

SOLVE a. FIGURE 2.19 shows the motion diagram. The distance scale will be established in parts b and c but is shown here for convenience.

b. The particle reaches the turning point at $t = 2$ s. To learn *where* it is at that time we need to find the displacement during

FIGURE 2.19 Motion diagram for the particle whose velocity graph was shown in Figure 2.18.

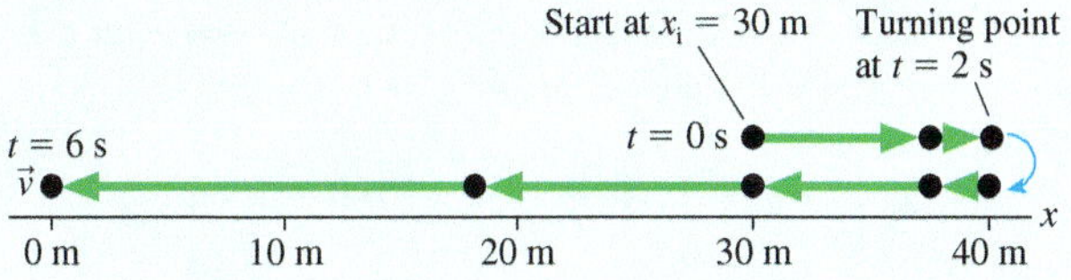

Continued

the first two seconds. We can do this by finding the area under the curve between $t = 0$ s and $t = 2$ s:

$$x(\text{at } t = 2\text{ s}) = x_i + \text{area under the curve between 0 s and 2 s}$$
$$= 30\text{ m} + \tfrac{1}{2}(2\text{ s} - 0\text{ s})(10\text{ m/s} - 0\text{ m/s})$$
$$= 40\text{ m}$$

The turning point is at $x = 40$ m.

c. The particle needs to move $\Delta x = -40$ m to get from the turning point to the origin. That is, the area under the curve from $t = 2$ s to the desired time t needs to be -40 m. Because the curve is below the axis, with negative values of v_x, the area to the right of $t = 2$ s is a *negative* area. With a bit of geometry, you will find that the triangle with a base extending from $t = 2$ s to $t = 6$ s has an area of -40 m. Thus the particle reaches the origin at $t = 6$ s.

A Little More Calculus: Integrals

Taking the derivative of a function is equivalent to finding the slope of a graph of the function. Similarly, evaluating an integral is equivalent to finding the area under a graph of the function. The graphical method is very important for building intuition about motion but is limited in its practical application. Just as derivatives of standard functions can be evaluated and tabulated, so can integrals.

The integral in Equation 2.11 is called a *definite integral* because there are two definite boundaries to the area we want to find. These boundaries are called the lower (t_i) and upper (t_f) *limits of integration*. For the important function $u = ct^n$, the essential result from calculus is that

$$\int_{t_i}^{t_f} u\,dt = \int_{t_i}^{t_f} ct^n\,dt = \left.\frac{ct^{n+1}}{n+1}\right|_{t_i}^{t_f} = \frac{ct_f^{\,n+1}}{n+1} - \frac{ct_i^{\,n+1}}{n+1} \qquad (n \neq -1) \qquad (2.13)$$

The vertical bar in the third step with subscript t_i and superscript t_f is a shorthand notation from calculus that means—as seen in the last step—the integral evaluated at the upper limit t_f *minus* the integral evaluated at the lower limit t_i. You also need to know that for two functions u and w,

$$\int_{t_i}^{t_f} (u + w)\,dt = \int_{t_i}^{t_f} u\,dt + \int_{t_i}^{t_f} w\,dt \qquad (2.14)$$

That is, the integral of a sum is equal to the sum of the integrals.

EXAMPLE 2.8 Using calculus to find the position

Use calculus to solve Example 2.7.

SOLVE Figure 2.18 is a linear graph. Its "y-intercept" is seen to be 10 m/s and its slope is -5 (m/s)/s. Thus the velocity can be described by the equation

$$v_x = (10 - 5t)\text{ m/s}$$

where t is in s. We can find the position x at time t by using Equation 2.11:

$$x = x_i + \int_0^t v_x\,dt = 30\text{ m} + \int_0^t (10 - 5t)\,dt$$
$$= 30\text{ m} + \int_0^t 10\,dt - \int_0^t 5t\,dt$$

We used Equation 2.14 for the integral of a sum to get the final expression. The first integral is a function of the form $u = ct^n$ with $c = 10$ and $n = 0$; the second is of the form $u = ct^n$ with $c = 5$ and $n = 1$. Using Equation 2.13, we have

$$\int_0^t 10\,dt = 10t\Big|_0^t = 10 \cdot t - 10 \cdot 0 = 10t\text{ m}$$

and

$$\int_0^t 5t\,dt = \tfrac{5}{2}t^2\Big|_0^t = \tfrac{5}{2}\cdot t^2 - \tfrac{5}{2}\cdot 0^2 = \tfrac{5}{2}t^2\text{ m}$$

Combining the pieces gives

$$x = (30 + 10t - \tfrac{5}{2}t^2)\text{ m}$$

This is a general result for the position at *any* time t.

The particle's turning point occurs at $t = 2$ s, and its position at that time is

$$x(\text{at } t = 2\text{ s}) = 30 + (10)(2) - \tfrac{5}{2}(2)^2 = 40\text{ m}$$

The time at which the particle reaches the origin is found by setting $x = 0$ m:

$$30 + 10t - \tfrac{5}{2}t^2 = 0$$

This quadratic equation has two solutions: $t = -2$ s or $t = 6$ s.

When we solve a quadratic equation, we cannot just arbitrarily select the root we want. Instead, we must decide which is the *meaningful* root. Here the negative root refers to a time before the problem began, so the meaningful one is the positive root, $t = 6$ s.

ASSESS The results agree with the answers we found previously from a graphical solution.

Summing Up

As you work on building intuition about motion, you need to be able to move back and forth among four different representations of the motion:

- The motion diagram;
- The position-versus-time graph;
- The velocity-versus-time graph;
- The description in words.

Given a description of a certain motion, you should be able to sketch the motion diagram and the position and velocity graphs. Given one graph, you should be able to generate the other. And given position and velocity graphs, you should be able to "interpret" them by describing the motion in words or in a motion diagram.

STOP TO THINK 2.3 Which position-versus-time graph goes with the velocity-versus-time graph on the left? The particle's position at $t_i = 0$ s is $x_i = -10$ m.

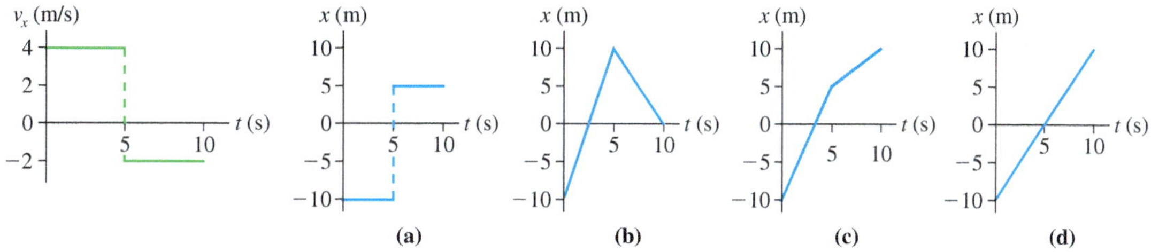

2.4 Motion with Constant Acceleration

We need one more major concept to describe one-dimensional motion: acceleration. Acceleration, as we noted in Chapter 1, is a rather abstract concept. Nonetheless, acceleration is the linchpin of mechanics. We will see very shortly that Newton's laws relate the acceleration of an object to the forces that are exerted on it.

Let's conduct a race between a Volkswagen Beetle and a Porsche to see which can achieve a velocity of 30 m/s ($\approx$60 mph) in the shortest time. Both cars are equipped with computers that will record the speedometer reading 10 times each second. This gives a nearly continuous record of the *instantaneous* velocity of each car. Table 2.1 shows some of the data. The velocity-versus-time graphs, based on these data, are shown in FIGURE 2.20.

TABLE 2.1 Velocities of a Porsche and a Volkswagen Beetle

t(s)	$v_{Porsche}$(m/s)	v_{VW} (m/s)
0.0	0.0	0.0
0.1	0.5	0.2
0.2	1.0	0.4
0.3	1.5	0.6
0.4	2.0	0.8
⋮	⋮	⋮

How can we describe the difference in performance of the two cars? It is not that one has a different velocity from the other; both achieve every velocity between 0 and 30 m/s. The distinction is how long it took each to *change* its velocity from 0 to 30 m/s. The Porsche changed velocity quickly, in 6.0 s, while the VW needed 15 s to make the same velocity change. Because the Porsche had a velocity change $\Delta v_s = 30$ m/s during a time interval $\Delta t = 6.0$ s the *rate* at which its velocity changed was

$$\text{rate of velocity change} = \frac{\Delta v_s}{\Delta t} = \frac{30 \text{ m/s}}{6.0 \text{ s}} = 5.0 \text{ (m/s)/s} \qquad (2.15)$$

FIGURE 2.20 Velocity-versus-time graphs for the Porsche and the VW Beetle.

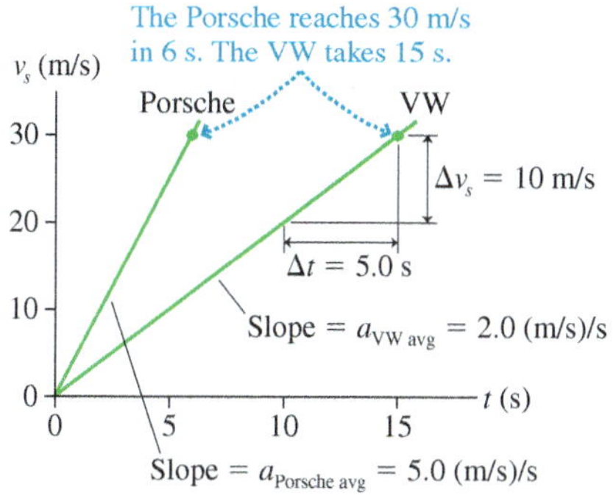

Notice the units. They are units of "velocity per second." A rate of velocity change of 5.0 "meters per second per second" means that the velocity increases by 5.0 m/s during the first second, by another 5.0 m/s during the next second, and so on. In fact, the velocity will increase by 5.0 m/s during any second in which it is changing at the rate of 5.0 (m/s)/s.

Chapter 1 introduced *acceleration* as "the rate of change of velocity." That is, acceleration measures how quickly or slowly an object's velocity changes. In parallel with our treatment of velocity, let's define the **average acceleration** a_{avg} during the time interval Δt to be

$$a_{avg} \equiv \frac{\Delta v_s}{\Delta t} \quad \text{(average acceleration)} \tag{2.16}$$

Equations 2.15 and 2.16 show that the Porsche had the rather large acceleration of 5.0 (m/s)/s.

Because Δv_s and Δt are the "rise" and "run" of a velocity-versus-time graph, we see that a_{avg} can be interpreted graphically as the *slope* of a straight-line velocity-versus-time graph. In other words,

$$a_{avg} = \text{slope of the velocity-versus-time graph} \tag{2.17}$$

Figure 2.20 uses this idea to show that the VW's average acceleration is

$$a_{\text{VW avg}} = \frac{\Delta v_s}{\Delta t} = \frac{10 \text{ m/s}}{5.0 \text{ s}} = 2.0 \text{ (m/s)/s}$$

This is less than the acceleration of the Porsche, as expected.

An object whose velocity-versus-time graph is a straight-line graph has a steady and unchanging acceleration. There's no need to specify "average" if the acceleration is constant, so we'll use the symbol a_s as we discuss motion along the s-axis with constant acceleration.

NOTE ▶ An important aspect of acceleration is its *sign*. Acceleration $\vec{a}$, like position $\vec{r}$ and velocity $\vec{v}$, is a vector. For motion in one dimension, the sign of a_x (or a_y) is positive if the vector $\vec{a}$ points to the right (or up), negative if it points to the left (or down). This was illustrated in Figure 1.19 and the very important Tactics Box 1.4, which you may wish to review. It's particularly important to emphasize that positive and negative values of a_s do *not* correspond to "speeding up" and "slowing down." ◀

EXAMPLE 2.9 Relating acceleration to velocity

a. A bicyclist has a velocity of 10 m/s and a constant acceleration of 2 (m/s)/s. What is her velocity 1 s later? 2 s later?

b. A bicyclist has a velocity of −10 m/s and a constant acceleration of 2 (m/s)/s. What is his velocity 1 s later? 2 s later?

SOLVE

a. An acceleration of 2 (m/s)/s *means* that the velocity increases by 2 m/s every 1 s. If the bicyclist's initial velocity is 10 m/s, then 1 s later her velocity will be 12 m/s. After 2 s, which is 1 additional second later, it will increase by another 2 m/s to 14 m/s. After 3 s it will be 16 m/s. Here a positive a_s is causing the bicyclist to speed up.

b. If the bicyclist's initial velocity is a *negative* −10 m/s but the acceleration is a positive +2 (m/s)/s, then 1 s later his velocity will be −8 m/s. After 2 s it will be −6 m/s, and so on. In this case, a positive a_s is causing the object to *slow down* (decreasing speed v). This agrees with the rule from Tactics Box 1.4: An object is slowing down if and only if v_s and a_s have opposite signs.

NOTE ▶ It is customary to abbreviate the acceleration units (m/s)/s as m/s^2. For example, the bicyclists in Example 2.9 had an acceleration of 2 m/s^2. We will use this notation, but keep in mind the *meaning* of the notation as "(meters per second) per second." ◀

EXAMPLE 2.10 Running the court

A basketball player starts at the left end of the court and moves with the velocity shown in FIGURE 2.21. Draw a motion diagram and an acceleration-versus-time graph for the basketball player.

FIGURE 2.21 Velocity-versus-time graph for the basketball player of Example 2.10.

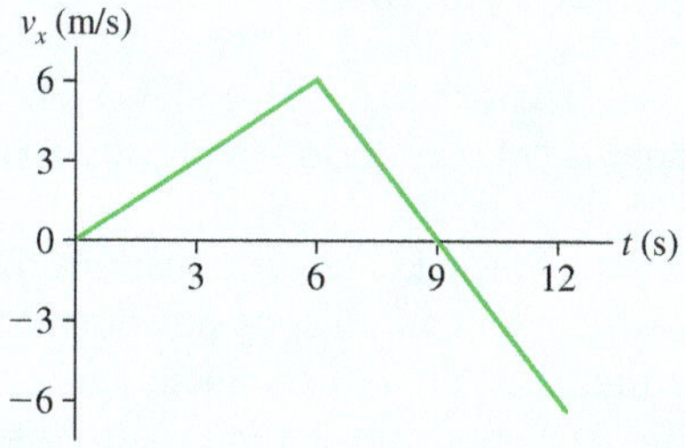

VISUALIZE The velocity is positive (motion to the right) and increasing for the first 6 s, so the velocity arrows in the motion diagram are to the right and getting longer. From $t = 6$ s to 9 s the motion is still to the right (v_x is still positive), but the arrows are getting shorter because v_x is decreasing. There's a turning point at $t = 9$ s, when $v_x = 0$, and after that the motion is to the left (v_x is negative) and getting faster. The motion diagram of FIGURE 2.22a shows the velocity and the acceleration vectors.

SOLVE Acceleration is the slope of the velocity graph. For the first 6 s, the slope has the constant value

$$a_x = \frac{\Delta v_x}{\Delta t} = \frac{6.0\text{ m/s}}{6.0\text{ s}} = 1.0\text{ m/s}^2$$

The velocity then decreases by 12 m/s during the 6 s interval from $t = 6$ s to $t = 12$ s, so

$$a_x = \frac{\Delta v_x}{\Delta t} = \frac{-12\text{ m/s}}{6.0\text{ s}} = -2.0\text{ m/s}^2$$

The acceleration graph for these 12 s is shown in FIGURE 2.22b. Notice that there is no change in the acceleration at $t = 9$ s, the turning point.

FIGURE 2.22 Motion diagram and acceleration graph for Example 2.10.

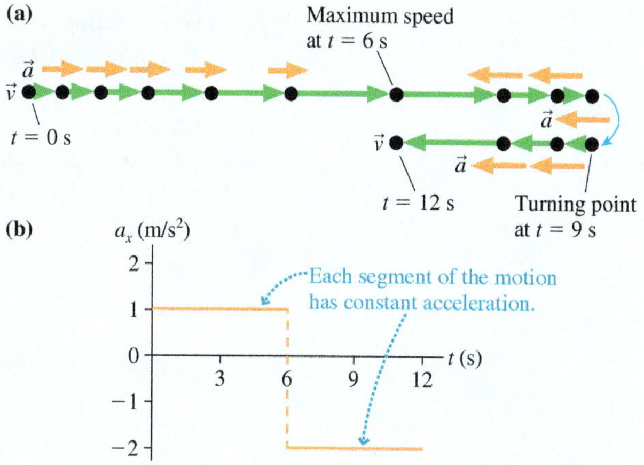

ASSESS The *sign* of a_x does *not* tell us whether the object is speeding up or slowing down. The basketball player is slowing down from $t = 6$ s to $t = 9$ s, then speeding up from $t = 9$ s to $t = 12$ s. Nonetheless, his acceleration is negative during this entire interval because his acceleration vector, as seen in the motion diagram, always points to the left.

The Kinematic Equations of Constant Acceleration

Consider an object whose acceleration a_s remains constant during the time interval $\Delta t = t_f - t_i$. At the beginning of this interval, at time t_i, the object has initial velocity v_{is} and initial position s_i. Note that t_i is often zero, but it does not have to be. We would like to predict the object's final position s_f and final velocity v_{fs} at time t_f.

The object's velocity is changing because the object is accelerating. FIGURE 2.23a shows the acceleration-versus-time graph, a horizontal line between t_i and t_f. It is not hard to find the object's velocity v_{fs} at a later time t_f. By definition,

$$a_s = \frac{\Delta v_s}{\Delta t} = \frac{v_{fs} - v_{is}}{\Delta t} \qquad (2.18)$$

which is easily rearranged to give

$$v_{fs} = v_{is} + a_s \Delta t \qquad (2.19)$$

The velocity-versus-time graph, shown in FIGURE 2.23b, is a straight line that starts at v_{is} and has slope a_s.

FIGURE 2.23 Acceleration and velocity graphs for constant acceleration.

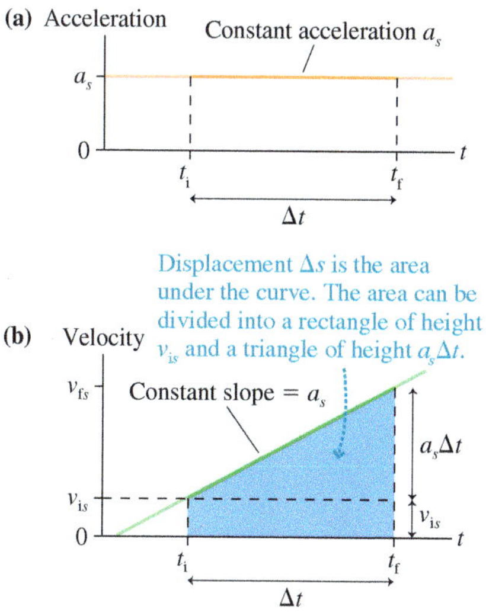

As you learned in the last section, the object's final position is

$$s_f = s_i + \text{area under the velocity curve } v_s \text{ between } t_i \text{ and } t_f \tag{2.20}$$

The shaded area in Figure 2.23b can be subdivided into a rectangle of area $v_{is}\,\Delta t$ and a triangle of area $\frac{1}{2}(a_s\Delta t)(\Delta t) = \frac{1}{2}a_s(\Delta t)^2$. Adding these gives

$$s_f = s_i + v_{is}\,\Delta t + \tfrac{1}{2}a_s(\Delta t)^2 \tag{2.21}$$

where $\Delta t = t_f - t_i$ is the elapsed time. The quadratic dependence on Δt causes the position-versus-time graph for constant-acceleration motion to have a parabolic shape, as shown below in FIGURE 2.24.

Equations 2.19 and 2.21 are two of the basic kinematic equations for motion with *constant* acceleration. They allow us to predict an object's position and velocity at a future instant of time. We need one more equation to complete our set, a direct relation between position and velocity. First use Equation 2.19 to write $\Delta t = (v_{fs} - v_{is})/a_s$. Substitute this into Equation 2.21, giving

$$s_f = s_i + v_{is}\left(\frac{v_{fs} - v_{is}}{a_s}\right) + \frac{1}{2}a_s\left(\frac{v_{fs} - v_{is}}{a_s}\right)^2 \tag{2.22}$$

With a bit of algebra, this is rearranged to read

$$v_{fs}^2 = v_{is}^2 + 2a_s\,\Delta s \tag{2.23}$$

where $\Delta s = s_f - s_i$ is the *displacement* (not the distance!).

Equations 2.19, 2.21, and 2.23, which are summarized in Table 2.2, are the key results for motion with constant acceleration.

TABLE 2.2 The kinematic equations for motion with constant acceleration

$v_{fs} = v_{is} + a_s\,\Delta t$
$s_f = s_i + v_{is}\,\Delta t + \frac{1}{2}a_s(\Delta t)^2$
$v_{fs}^2 = v_{is}^2 + 2a_s\,\Delta s$

Figure 2.24 is a comparison of motion with constant velocity (uniform motion) and motion with constant acceleration (uniformly accelerated motion). Notice that uniform motion is really a special case of uniformly accelerated motion in which the constant acceleration happens to be zero. The graphs for a negative acceleration are left as an exercise.

FIGURE 2.24 Motion with constant velocity and constant acceleration. These graphs assume $s_i = 0$, $v_{is} > 0$, and (for constant acceleration) $a_s > 0$.

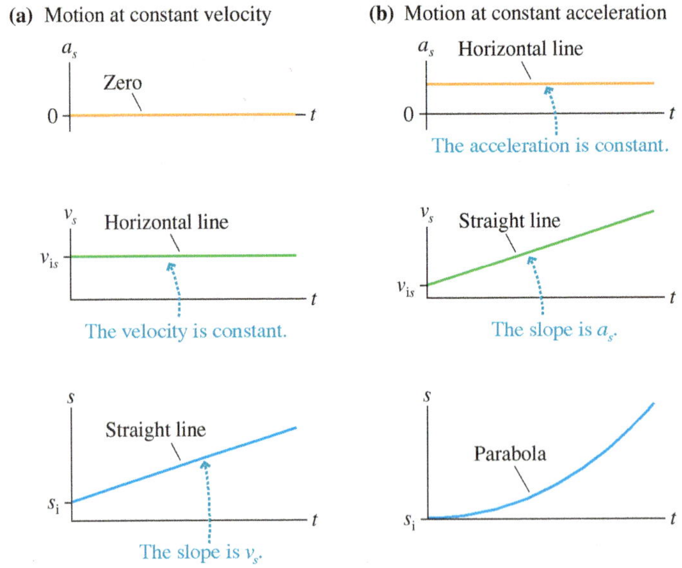

A Problem-Solving Strategy

This information can be assembled into a problem-solving strategy for kinematics with constant acceleration.

PROBLEM-SOLVING STRATEGY 2.1 Kinematics with constant acceleration MP

MODEL Use the particle model. Make simplifying assumptions.

VISUALIZE Use different representations of the information in the problem.

- Draw a *pictorial representation.* This helps you assess the information you are given and starts the process of translating the problem into symbols.
- Use a *graphical representation* if it is appropriate for the problem.
- Go back and forth between these two representations as needed.

SOLVE The mathematical representation is based on the three kinematic equations

$$v_{fs} = v_{is} + a_s \Delta t$$

$$s_f = s_i + v_{is} \Delta t + \tfrac{1}{2} a_s (\Delta t)^2$$

$$v_{fs}^2 = v_{is}^2 + 2a_s \Delta s$$

- Use x or y, as appropriate to the problem, rather than the generic s.
- Replace i and f with numerical subscripts defined in the pictorial representation.
- Uniform motion with constant velocity has $a_s = 0$.

ASSESS Is your result believable? Does it have proper units? Does it make sense?

NOTE ▶ You are strongly encouraged to solve problems on the Dynamics Worksheets found at the back of the Student Workbook. These worksheets will help you use the Problem-Solving Strategy and develop good problem-solving skills. ◀

EXAMPLE 2.11 The motion of a rocket sled

A rocket sled accelerates at 50 m/s^2 for 5.0 s, coasts for 3.0 s, then deploys a braking parachute and decelerates at 3.0 m/s^2 until coming to a halt.

a. What is the maximum velocity of the rocket sled?
b. What is the total distance traveled?

MODEL Represent the rocket sled as a particle.

VISUALIZE **FIGURE 2.25** shows the pictorial representation. Recall that we discussed the first two-thirds of this problem as Example 1.8 in Chapter 1.

FIGURE 2.25 Pictorial representation of the rocket sled.

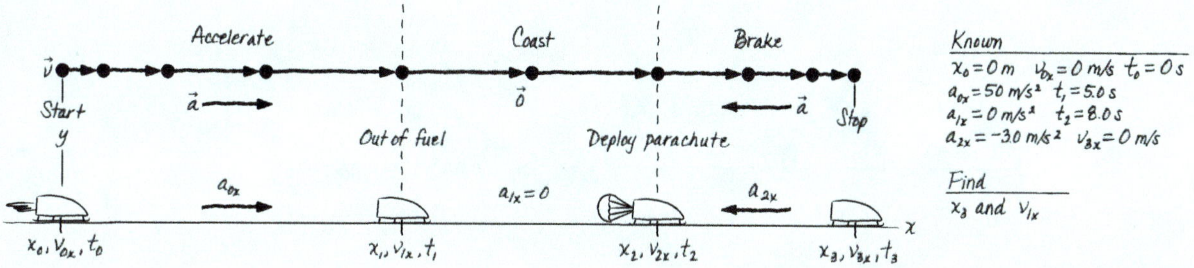

SOLVE a. The maximum velocity is identified in the pictorial representation as v_{1x}, the velocity at time t_1 when the acceleration phase ends. The first kinematic equation in Table 2.2 gives

$$v_{1x} = v_{0x} + a_{0x}(t_1 - t_0) = a_{0x} t_1$$

$$= (50 \text{ m/s}^2)(5.0 \text{ s}) = 250 \text{ m/s}$$

We started with the complete equation, then simplified by noting which terms were zero. Also notice that we found an algebraic expression for v_{1x}, *then* substituted numbers. Working algebraically

Continued

is a hallmark of good problem-solving technique, and many homework problems will ask you to do so.

b. Finding the total distance requires several steps. First, the sled's position when the acceleration ends at t_1 is found from the second equation in Table 2.2:

$$x_1 = x_0 + v_{0x}(t_1 - t_0) + \tfrac{1}{2}a_{0x}(t_1 - t_0)^2 = \tfrac{1}{2}a_{0x}t_1^2$$

$$= \tfrac{1}{2}(50\text{ m/s}^2)(5.0\text{ s})^2 = 625\text{ m}$$

During the coasting phase, which is uniform motion with no acceleration ($a_{1x} = 0$),

$$x_2 = x_1 + v_{1x}\Delta t = x_1 + v_{1x}(t_2 - t_1)$$

$$= 625\text{ m} + (250\text{ m/s})(3.0\text{ s}) = 1375\text{ m}$$

Notice that, in this case, Δt is not simply t. The braking phase is a little different because we don't know how long it lasts. But we do know that the sled ends with $v_{3x} = 0$ m/s, so we can use the third equation in Table 2.2:

$$v_{3x}^2 = v_{2x}^2 + 2a_{2x}\Delta x = v_{2x}^2 + 2a_{2x}(x_3 - x_2)$$

This can be solved for x_3:

$$x_3 = x_2 + \frac{v_{3x}^2 - v_{2x}^2}{2a_{2x}}$$

$$= 1375\text{ m} + \frac{0 - (250\text{ m/s})^2}{2(-3.0\text{ m/s}^2)} = 12{,}000\text{ m}$$

ASSESS Using the approximate conversion factor 1 m/s ≈ 2 mph from Table 1.5, we see that the top speed is ≈ 500 mph. The total distance traveled is ≈12 km ≈ 7 mi. This is reasonable because it takes a very long distance to stop from a top speed of 500 mph!

NOTE ▶ We used explicit numerical subscripts throughout the mathematical representation, each referring to a symbol that was defined in the pictorial representation. The subscripts i and f in the Table 2.2 equations are just generic "place holders" that don't have unique values. During the acceleration phase we had i = 0 and f = 1. Later, during the coasting phase, these became i = 1 and f = 2. The numerical subscripts have a clear meaning and are less likely to lead to confusion. ◀

EXAMPLE 2.12 Friday night football

Fred catches the football while standing directly on the goal line. He immediately starts running forward with an acceleration of 6 ft/s². At the moment the catch is made, Tommy is 20 yards away and heading directly toward Fred with a steady speed of 15 ft/s. If neither deviates from a straight-ahead path, where will Tommy tackle Fred?

MODEL Represent Fred and Tommy as particles.

VISUALIZE The pictorial representation is shown in FIGURE 2.26. With two moving objects we need the additional subscripts F and T to distinguish Fred's symbols and Tommy's symbols. The axes have been chosen so that Fred starts at $(x_0)_F = 0$ ft and moves to the right while Tommy starts at $(x_0)_T = 60$ ft and runs to the left with a *negative* velocity.

SOLVE We want to find *where* Fred and Tommy have the same position. The pictorial representation designates time t_1 as *when* they meet. The second equation of Table 2.2 allows us to find their positions at time t_1. These are:

$$(x_1)_F = (x_0)_F + (v_{0x})_F(t_1 - t_0) + \tfrac{1}{2}(a_x)_F(t_1 - t_0)^2$$

$$= \tfrac{1}{2}(a_x)_F t_1^2$$

$$(x_1)_T = (x_0)_T + (v_{0x})_T(t_1 - t_0) + \tfrac{1}{2}(a_x)_T(t_1 - t_0)^2$$

$$= (x_0)_T + (v_{0x})_T t_1$$

FIGURE 2.26 Pictorial representation for Example 2.12.

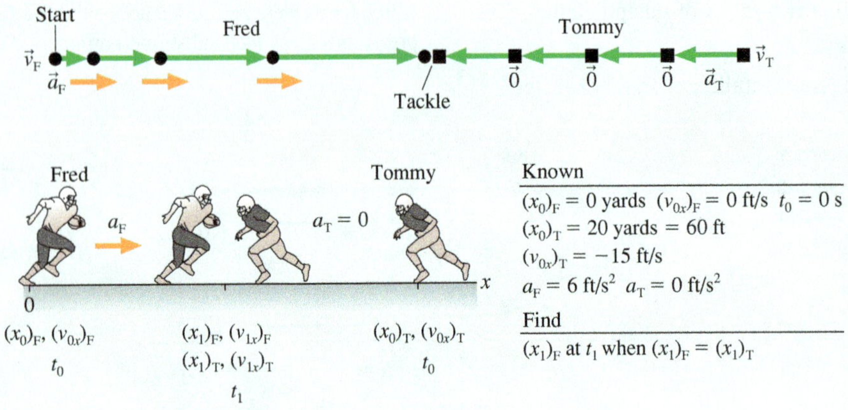

Notice that Tommy's position equation contains the term $(v_{0x})_T t_1$, not $-(v_{0x})_T t_1$. The fact that he is moving to the left has already been considered in assigning a *negative value* to $(v_{0x})_T$, hence we don't want to add any additional negative signs in the equation. If we now set $(x_1)_F$ and $(x_1)_T$ equal to each other, indicating the point of the tackle, we can solve for t_1:

$$\tfrac{1}{2}(a_x)_F t_1^2 = (x_0)_T + (v_{0x})_T t_1$$

$$\tfrac{1}{2}(a_x)_F t_1^2 - (v_{0x})_T t_1 - (x_0)_T = 0$$

$$3t_1^2 + 15t_1 - 60 = 0$$

The solutions of this quadratic equation for t_1 are $t_1 = (-7.62 \text{ s}, +2.62 \text{ s})$. The negative time is not meaningful in this problem, so the time of the tackle is $t_1 = 2.62$ s. We've kept an extra significant figure in the solution to minimize round-off error in the next step. Using this value to compute $(x_1)_F$ gives

$$(x_1)_F = \tfrac{1}{2}(a_x)_F t_1^2 = 20.6 \text{ feet} = 6.9 \text{ yards}$$

Tommy makes the tackle at just about the 7-yard line!

ASSESS The answer had to be between 0 yards and 20 yards. Because Tommy was already running, whereas Fred started from rest, it is reasonable that Fred will cover less than half the 20-yard separation before meeting Tommy. Thus 6.9 yards is a reasonable answer.

NOTE ▶ The purpose of the Assess step is not to prove that an answer must be right but to rule out answers that, with a little thought, are clearly wrong. ◀

It is worth exploring Example 2.12 graphically. FIGURE 2.27 shows position-versus-time graphs for Fred and Tommy. The curves intersect at $t = 2.62$ s, and that is where the tackle occurs. You should compare this problem to Example 2.3 and Figure 2.7 for Bob and Susan to notice the similarities and the differences.

FIGURE 2.27 Position-versus-time graphs for Fred and Tommy.

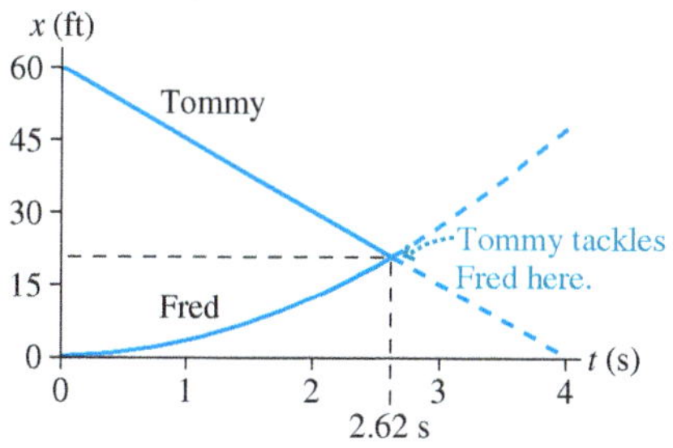

STOP TO THINK 2.4 Which velocity-versus-time graph or graphs go with the acceleration-versus-time graph? The particle is initially moving to the right.

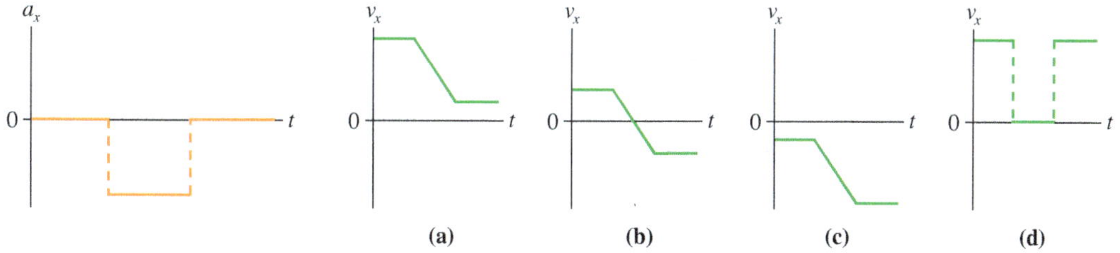

2.5 Free Fall

The motion of an object moving under the influence of gravity only, and no other forces, is called **free fall.** Strictly speaking, free fall occurs only in a vacuum, where there is no air resistance. Fortunately, the effect of air resistance is small for "heavy objects," so we'll make only a very slight error in treating these objects *as if* they were in free fall. For very light objects, such as a feather, or for objects that fall through very large distances and gain very high speeds, the effect of air resistance is *not* negligible. Motion with air resistance is a problem we will study in Chapter 6. Until then, we will restrict our attention to "heavy objects" and will make the reasonable assumption that falling objects are in free fall.

Galileo, in the 17th century, was the first to make detailed measurements of falling objects. The story of Galileo dropping different weights from the leaning bell

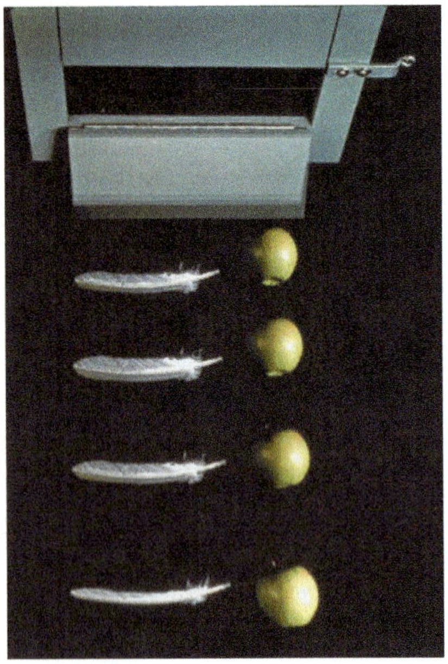

In the absence of air resistance, any two objects fall at the same rate and hit the ground at the same time. The apple and feather seen here are falling in a vacuum.

tower at the cathedral in Pisa is well known, although historians cannot confirm its truth. But bell towers were common in the Italy of Galileo's day, so he had ample opportunity to make the measurements and observations that he describes in his writings.

Galileo developed a *model* of motion—motion in the absence of air resistance—that could only be approximated by any real object. His discovery can be summarized as follows:

- Two objects dropped from the same height will, if air resistance can be neglected, hit the ground at the same time and with the same speed.
- Consequently, **any two objects in free fall, regardless of their mass, have the same acceleration $\vec{a}_{\text{free fall}}$.** This is an especially important conclusion.

FIGURE 2.28a shows the motion diagram of an object that was released from rest and falls freely. FIGURE 2.28b shows the object's velocity graph. The motion diagram and graph are identical for a falling pebble and a falling boulder. The fact that the velocity graph is a straight line tells us the motion is one of constant acceleration, and $a_{\text{free fall}}$ is easily found from the slope of the graph. Careful measurements show that the value of $\vec{a}_{\text{free fall}}$ varies ever so slightly at different places on the earth, due to the slightly nonspherical shape of the earth and to the fact that the earth is rotating. A global average, at sea level, is

$$\vec{a}_{\text{free fall}} = (9.80 \text{ m/s}^2, \text{vertically downward}) \tag{2.24}$$

FIGURE 2.28 Motion of an object in free fall.

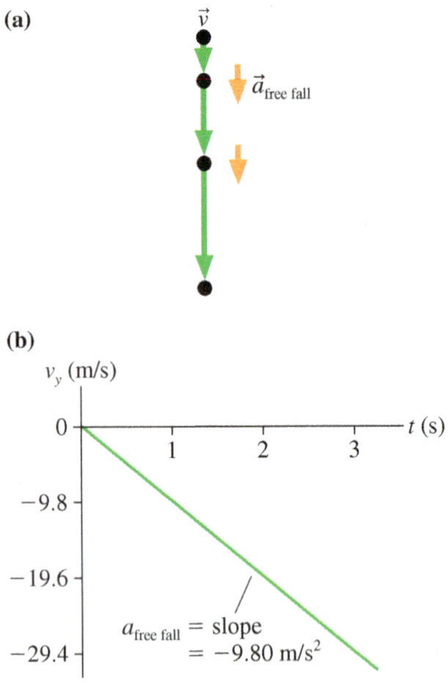

For practical purposes, *vertically downward* means along a line toward the center of the earth. However, we'll learn in Chapter 13 that the rotation of the earth has a small effect on both the size and direction of $\vec{a}_{\text{free fall}}$.

The length, or magnitude, of $\vec{a}_{\text{free fall}}$ is known as the **free-fall acceleration,** and it has the special symbol g:

$$g = 9.80 \text{ m/s}^2 \text{ (free-fall acceleration)}$$

Several points about free fall are worthy of note:

- g, by definition, is *always* positive. **There will never be a problem that will use a negative value for g.** But, you say, objects fall when you release them rather than rise, so how can g be positive?
- g is *not* the acceleration $a_{\text{free fall}}$, but simply its magnitude. Because we've chosen the y-axis to point vertically upward, the downward acceleration vector $\vec{a}_{\text{free fall}}$ has the one-dimensional acceleration

$$a_y = a_{\text{free fall}} = -g \tag{2.25}$$

  It is a_y that is negative, not g.
- Because free fall is motion with constant acceleration, we can use the kinematic equations of Table 2.2 with the acceleration being that of free fall, $a_y = -g$.
- g is not called "gravity." Gravity is a force, not an acceleration. The symbol g recognizes the influence of gravity, but g is *the free-fall acceleration.*
- $g = 9.80 \text{ m/s}^2$ only on earth. Other planets have different values of g. You will learn in Chapter 13 how to determine g for other planets.

NOTE ▶ Despite the name, free fall is not restricted to objects that are literally falling. Any object moving under the influence of gravity only, and no other forces, is in free fall. This includes objects falling straight down, objects that have been tossed or shot straight up, and projectile motion. ◀

EXAMPLE 2.13 A falling rock

A rock is released from rest at the top of a 100-m-tall building. How long does the rock take to fall to the ground, and what is its impact velocity?

MODEL Represent the rock as a particle. Assume air resistance is negligible.

VISUALIZE FIGURE 2.29 shows the pictorial representation. We have placed the origin at the ground, which makes $y_0 = 100$ m. Although the rock falls 100 m, it is important to notice that the *displacement* is $\Delta y = y_1 - y_0 = -100$ m.

FIGURE 2.29 Pictorial representation of a falling rock.

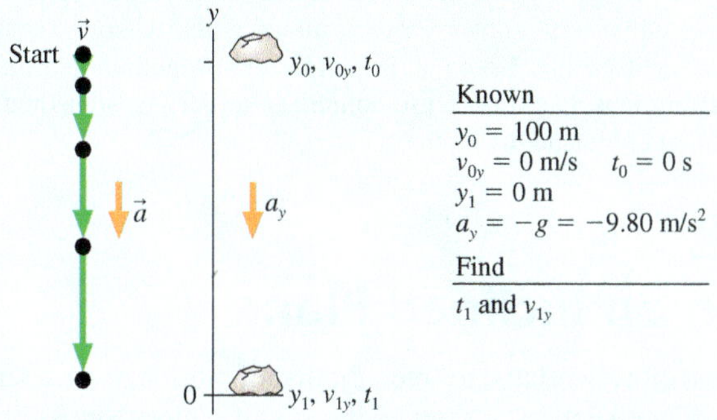

SOLVE Free fall is motion with the specific constant acceleration $a_y = -g$. The first question involves a relation between time and distance, so only the second equation in Table 2.2 is relevant. Using $v_{0y} = 0$ m/s and $t_0 = 0$ s, we find

$$y_1 = y_0 + v_{0y}\Delta t + \tfrac{1}{2}a_y \Delta t^2 = y_0 + v_{0y}\Delta t - \tfrac{1}{2}g\Delta t^2 = y_0 - \tfrac{1}{2}gt_1^2$$

We can now solve for t_1, finding

$$t_1 = \sqrt{\frac{2(y_0 - y_1)}{g}} = \sqrt{\frac{2(100\text{ m} - 0\text{ m})}{9.80\text{ m/s}^2}} = \pm 4.52\text{ s}$$

The $\pm$ sign indicates that there are two mathematical solutions; therefore we have to use physical reasoning to choose between them. A negative t_1 would refer to a time before we dropped the rock, so we select the positive root: $t_1 = 4.52$ s.

Now that we know the fall time, we can use the first kinematic equation to find v_{1y}:

$$v_{1y} = v_{0y} - g\Delta t = -gt_1 = -(9.80\text{ m/s}^2)(4.52\text{ s})$$
$$= -44.3\text{ m/s}$$

Alternatively, we could work directly from the third kinematic equation:

$$v_{1y} = \sqrt{v_{0y}^2 - 2g\Delta y} = \sqrt{-2(9.80\text{ m/s}^2)(-100\text{ m})} = \pm 44.3\text{ m/s}$$

This method is useful if you don't know Δt. However, we must again choose the correct sign of the square root. Because the velocity vector points downward, the sign of v_y has to be negative. Thus $v_{1y} = -44.3$ m/s. The importance of careful attention to the signs cannot be overemphasized!

A common error would be to say "The rock fell 100 m, so $\Delta y = 100$ m." This would have you trying to take the square root of a negative number. As noted above, Δy is not a distance. It is a *displacement,* with a carefully defined meaning of $y_f - y_i$. In this case, $\Delta y = y_1 - y_0 = -100$ m.

ASSESS Are the answers reasonable? Well, 100 m is about 300 feet, which is about the height of a 30-floor building. How long does it take something to fall 30 floors? Four or five seconds seems pretty reasonable. How fast would it be going at the bottom? Using 1 m/s $\approx$ 2 mph, we find that 44.3 m/s $\approx$ 90 mph. That also seems pretty reasonable after falling 30 floors. Had we misplaced a decimal point, though, and found 443 m/s, we would be suspicious when we converted this to $\approx$ 900 mph! The answers all seem reasonable.

EXAMPLE 2.14 Finding the height of a leap

The springbok, an antelope found in Africa, gets its name from its remarkable jumping ability. When startled, a springbok will leap straight up into the air—a maneuver called a "pronk." A springbok goes into a crouch to perform a pronk. It then extends its legs forcefully, accelerating at 35 m/s^2 for 0.70 m as its legs straighten. Legs fully extended, it leaves the ground and rises into the air. How high does it go?

MODEL Represent the springbok as a particle.

VISUALIZE FIGURE 2.30 shows the pictorial representation. This is a problem with a beginning point, an end point, and a point in between where the nature of the motion changes. We've identified these points with subscripts 0, 1, and 2. The motion from 0 to 1 is a rapid upward acceleration until the springbok's feet leave the ground at 1. Even though the springbok is moving upward from 1

FIGURE 2.30 Pictorial representation of a startled springbok.

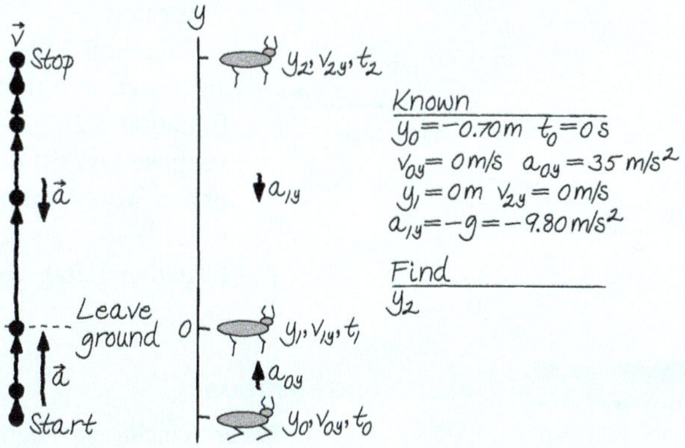

Continued

to 2, this is free-fall motion because the springbok is now moving under the influence of gravity only.

How do we put "How high?" into symbols? The clue is that the very top point of the trajectory is a *turning point*, and we've seen that the instantaneous velocity at a turning point is $v_{2y} = 0$. This was not explicitly stated but is part of our interpretation of the problem.

SOLVE For the first part of the motion, pushing off, we know a displacement but not a time interval. The third equation in Table 2.2 is perfect for this situation:

$$v_{1y}^2 = v_{0y}^2 + 2a_{0y}\Delta y = 2(35\text{ m/s}^2)(0.70\text{ m}) = 49\text{ m}^2/\text{s}^2$$

$$v_{1y} = \sqrt{49\text{ m}^2/\text{s}^2} = 7.0\text{ m/s}$$

The springbok leaves the ground with a velocity of 7.0 m/s. This is the starting point for the problem of a projectile launched straight up from the ground. One possible solution is to use the velocity equation to find how long it takes to reach maximum height, then the position equation to calculate the maximum height. But that takes two separate calculations. It is easier to make another use of the velocity-displacement equation:

$$v_{2y}^2 = 0 = v_{1y}^2 + 2a_{1y}\Delta y = v_{1y}^2 - 2g(y_2 - y_1)$$

where now the acceleration is $a_{1y} = -g$. Using $y_1 = 0$, we can solve for y_2, the height of the leap:

$$y_2 = \frac{v_{1y}^2}{2g} = \frac{(7.0\text{ m/s})^2}{2(9.80\text{ m/s}^2)} = 2.5\text{ m}$$

ASSESS 2.5 m is a bit over 8 feet, a remarkable vertical jump. But these animals are known for their jumping ability, so the answer seems reasonable. Note that it is especially important in a multi-part problem like this to use numerical subscripts to distinguish different points in the motion.

2.6 Motion on an Inclined Plane

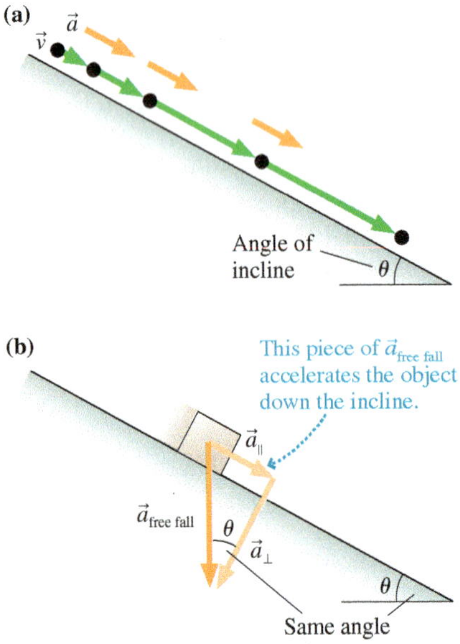

FIGURE 2.31 Acceleration on an inclined plane.

FIGURE 2.31a shows a problem closely related to free fall: that of motion down a straight, but frictionless, inclined plane, such as a skier going down a slope on frictionless snow. What is the object's acceleration? Although we're not yet prepared to give a rigorous derivation, we can deduce the acceleration with a plausibility argument.

FIGURE 2.31b shows the free-fall acceleration $\vec{a}_{\text{free fall}}$ the object would have if the incline suddenly vanished. The free-fall acceleration points straight down. This vector can be broken into two pieces: a vector $\vec{a}_{\parallel}$ that is parallel to the incline and a vector $\vec{a}_{\perp}$ that is perpendicular to the incline. The surface of the incline somehow "blocks" $\vec{a}_{\perp}$, through a process we will examine in Chapter 6, but $\vec{a}_{\parallel}$ is unhindered. It is this piece of $\vec{a}_{\text{free fall}}$, parallel to the incline, that accelerates the object.

By definition, the length, or magnitude, of $\vec{a}_{\text{free fall}}$ is g. Vector $\vec{a}_{\parallel}$ is opposite angle θ (Greek *theta*), so the length, or magnitude, of $\vec{a}_{\parallel}$ must be $g\sin\theta$. Consequently, the one-dimensional acceleration along the incline is

$$a_s = \pm g\sin\theta \qquad (2.26)$$

The correct sign depends on the direction in which the ramp is tilted. Examples will illustrate.

Equation 2.26 makes sense. Suppose the plane is perfectly horizontal. If you place an object on a horizontal surface, you expect it to stay at rest with no acceleration. Equation 2.26 gives $a_s = 0$ when $\theta = 0°$, in agreement with our expectations. Now suppose you tilt the plane until it becomes vertical, at $\theta = 90°$. Without friction, an object would simply fall, in free fall, parallel to the vertical surface. Equation 2.26 gives $a_s = -g = a_{\text{free fall}}$ when $\theta = 90°$, again in agreement with our expectations. Equation 2.26 gives the correct result in these *limiting cases*.

EXAMPLE 2.15 Measuring acceleration

In the laboratory, a 2.00-m-long track has been inclined as shown in FIGURE 2.32. Your task is to measure the acceleration of a cart on the ramp and to compare your result with what you might have expected. You have available five "photogates" that measure the cart's speed as it passes through. You place a gate every 30 cm from a line you mark near the top of the track as the starting line. One run generates the data shown in the table.

FIGURE 2.32 The experimental setup.

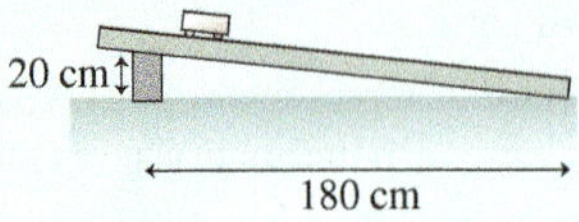

Distance (cm)	Speed (m/s)
0	0.00
30	0.75
60	1.15
90	1.38
120	1.56
150	1.76

The first entry isn't a photogate measurement, but it is a valid data point because you know the cart's speed is zero at the point where you release it.

NOTE ▶ Physics is an experimental science. Our knowledge of the universe is grounded in observations and measurements. Consequently, some examples and homework problems throughout this book will be based on data. These won't replace an actual laboratory experience, but they will provide you with an opportunity for thinking about how we make sense of the underlying theory. Data-based homework problems require the use of a spreadsheet, graphing software, or a graphing calculator in which you can "fit" data with a straight line. ◀

MODEL Represent the cart as a particle.

VISUALIZE FIGURE 2.33 shows the pictorial representation. We've chosen the x-axis to be parallel to the track, which is tilted at angle $\theta = \tan^{-1}(20\text{ cm}/180\text{ cm}) = 6.34°$. This is motion on an inclined plane, so you might expect the cart's acceleration to be $a_x = g\sin\theta = 1.08\text{ m/s}^2$. In any laboratory situation, it's good to have an idea what to expect.

FIGURE 2.33 The pictorial representation of the cart on the track.

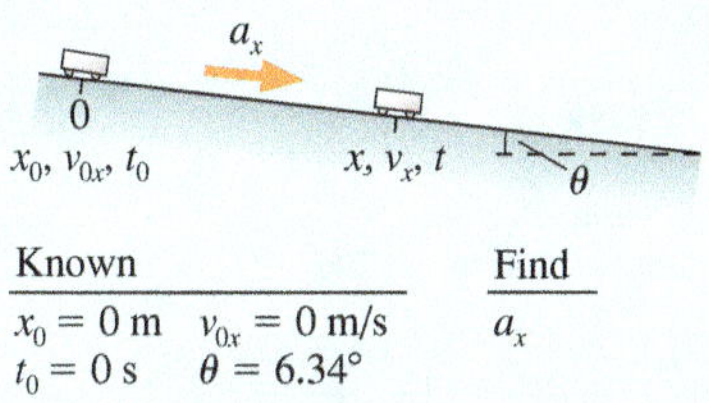

SOLVE In analyzing data, we want to use *all* the data, not just pick out one or two measurements. Further, we almost always want to use graphs when we have a series of measurements. We might start by graphing speed versus distance traveled. This is shown in FIGURE 2.34a, where—recognizing that our data table has inconsistent units—we converted distances to meters. As expected, speed increases with distance, but the graph isn't linear and that makes it hard to analyze.

FIGURE 2.34 Graphs of velocity and of velocity squared.

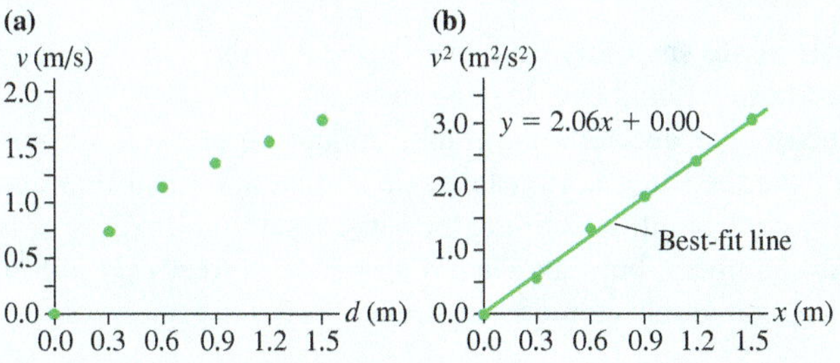

Rather than proceeding by trial and error, let's be guided by theory. We have information about speed and distance, but not about how long it took the cart to reach each photogate. *If* the cart has constant acceleration—which we don't yet know and need to confirm—the third equation of Table 2.2 tells us that velocity and displacement should be related by

$$v_x^2 = v_{0x}^2 + 2a_x\Delta x = 2a_x x$$

The last step was based on starting from rest ($v_{0x} = 0$) at the origin ($\Delta x = x - x_0 = x$). Although we measured speed, the cart is moving in the $+x$-direction, so we can interpret the speeds as velocities. And we've measured distance from the origin, so the distance values are x.

Rather than graphing v_x versus x, suppose we graphed v_x^2 versus x. If we let $y = v_x^2$, the kinematic equation reads

$$y = 2a_x x$$

This is in the form of a linear equation: $y = mx + b$, where m is the slope and b is the y-intercept. In this case, $m = 2a_x$ and $b = 0$. So if the cart really does have constant acceleration, a graph of v_x^2 versus x should be linear with a y-intercept of zero. This is a prediction that we can test.

Thus our analysis has three steps:

1. Graph v_x^2 versus x. If the graph is a straight line with a y-intercept of zero (or very close to zero), then we can conclude that the cart has constant acceleration on the ramp. If not, the acceleration is *not* constant and we cannot use the kinematic equations for constant acceleration.
2. If the graph has the correct shape, we can determine its slope m.
3. Because kinematics predicts $m = 2a_x$, the acceleration is $a_x = m/2$.

FIGURE 2.34b is the graph of v_x^2 versus x. It does turn out to be a straight line with a y-intercept of zero, and this is the evidence we need that the cart has a constant acceleration on the ramp. To proceed, we want to determine the slope by finding the straight line that is the "best fit" to the data. This is a statistical technique, justified in a statistics class, but one that is implemented in spreadsheets and graphing calculators. The solid line in Figure 2.34b is the best-fit line for this data, and its equation is shown. We see that the slope is $m = 2.06\text{ m/s}^2$. **Slopes have units**, and the units come not from the fitting procedure but by looking at the axes of the graph. Here the vertical axis is velocity squared, with units of $(\text{m/s})^2$, while the horizontal axis is position, measured in m. Thus the slope, rise over run, has units of m/s^2.

Finally, we can determine that the cart's acceleration was

$$a_x = \frac{m}{2} = 1.03\text{ m/s}^2$$

This is about 5% less than the 1.08 m/s^2 we expected. Two possibilities come to mind. Perhaps the distances used to find the tilt angle weren't measured accurately. Or, more likely, the cart rolls with a small bit of friction. The predicted acceleration $a_x = g\sin\theta$ is for a *frictionless* inclined plane; any friction would decrease the acceleration.

ASSESS How did we know to graph v_x^2 versus x rather than v_x versus x? We were guided by theory! The analysis of data requires linking measurements with theory, and we had a theoretical prediction, from kinematics, that v_x^2 is proportional to x for constant-acceleration motion that starts from rest at the origin. Thus the shape of a v_x^2-versus-x graph both tests the assertion that the acceleration is constant and, if the assertion is true, allows us to find the acceleration from the slope of the graph. We'll see this procedure over and over: Use theory to suggest a graph that should be linear if the assumptions of the theory are true, graph it, then—if the graph really is linear—match the experimentally determined slope and/or intercept with their theoretical predictions to extract useful results.

In this case, the graph was linear and we could use the slope to determine the cart's acceleration. The value was just slightly less than would be predicted for a frictionless incline, so the result is reasonable.

Thinking Graphically

Kinematics is the language of motion. The concepts we have developed in this chapter will be used extensively throughout the rest of this textbook. One of the most important ideas, summarized in Tactics Box 2.2, has been that the relationships among position, velocity, and acceleration can be expressed graphically.

TACTICS BOX 2.2 Interpreting graphical representations of motion

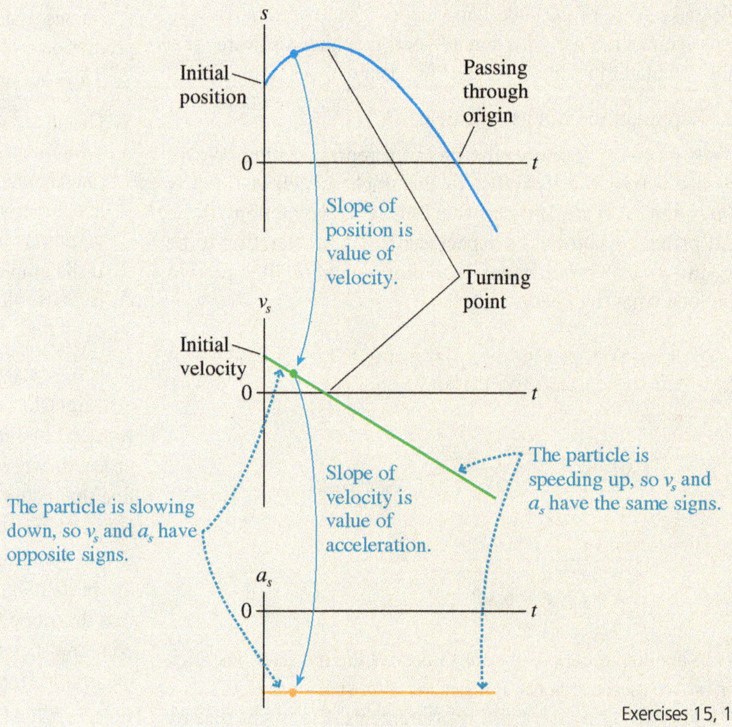

Exercises 15, 16, 22

A good way to solidify your understanding of motion graphs is to consider the problem of a hard, smooth ball rolling on a smooth track. The track is made up of several straight segments connected together. Each segment may be either horizontal or inclined. Your task is to analyze the ball's motion graphically.

There are a small number of rules to follow:

1. Assume that the ball passes smoothly from one segment of the track to the next, with no loss of speed and without ever leaving the track.
2. The position, velocity, and acceleration graphs should be stacked vertically. They should each have the same horizontal scale so that a vertical line drawn through all three connects points describing the same instant of time.
3. The graphs have no numbers, but they should show the correct *relationships*. For example, the position graph should have steeper slopes in regions of higher speed.
4. The position s is the position measured *along* the track. Similarly, v_s and a_s are the velocity and acceleration parallel to the track.

EXAMPLE 2.16 From track to graphs

Draw position, velocity, and acceleration graphs for the ball on the frictionless track of FIGURE 2.35.

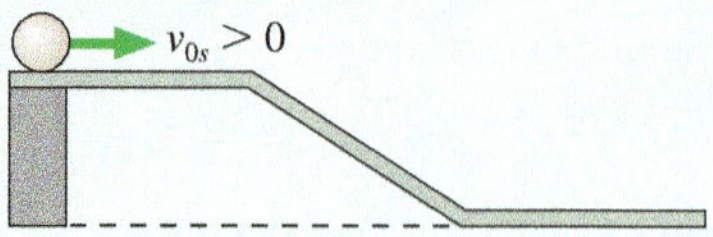

FIGURE 2.35 A ball rolling along a track.

VISUALIZE It is often easiest to begin with the velocity. Here the ball starts with an initial velocity v_{0s}. There is no acceleration on the horizontal surface ($a_s = 0$ if $\theta = 0°$), so the velocity remains constant until the ball reaches the slope. The slope is an inclined plane that, as we have learned, has constant acceleration. The velocity increases linearly with time during constant-acceleration motion. The ball returns to constant-velocity motion after reaching the bottom horizontal segment. The middle graph of FIGURE 2.36 shows the velocity.

We have enough information to draw the acceleration graph. We noted that the acceleration is zero while the ball is on the horizontal segments, and a_s has a constant positive value on the slope. These accelerations are consistent with the slope of the velocity graph: zero slope, then positive slope, then a return to zero slope. The acceleration cannot *really* change instantly from zero to a nonzero value, but the change can be so quick that we do not see it on the time scale of the graph. That is what the vertical dotted lines imply.

FIGURE 2.36 Motion graphs for the ball in Example 2.16.

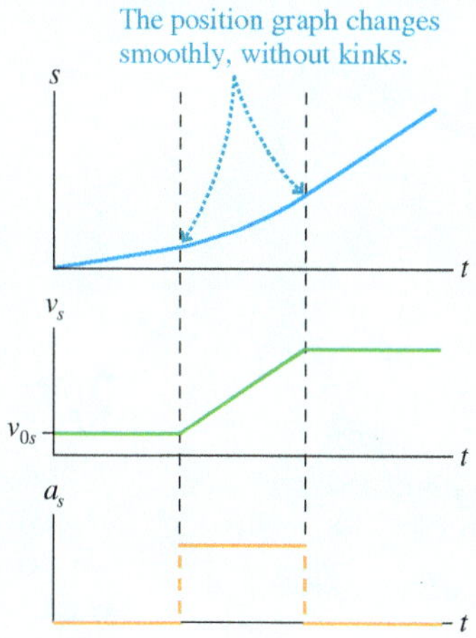

Finally, we need to find the position-versus-time graph. The position increases linearly with time during the first segment at constant velocity. It also does so during the third segment of motion, but with a steeper slope to indicate a faster velocity. In between, while the acceleration is nonzero but constant, the position graph has a *parabolic* shape.

Two points are worth noting:

1. The dotted vertical lines through the graphs show the instants when the ball moves from one segment of the track to the next. Because of rule 1, the speed does not change abruptly at these points; it changes gradually.
2. The parabolic section of the position-versus-time graph blends *smoothly* into the straight lines on either side. This is a consequence of rule 1. An abrupt change of slope (a "kink") would indicate an abrupt change in velocity and would violate rule 1.

EXAMPLE 2.17 From graphs to track

FIGURE 2.37 shows a set of motion graphs for a ball moving on a track. Draw a picture of the track and describe the ball's initial condition. Each segment of the track is *straight*, but the segments may be tilted.

FIGURE 2.37 Motion graphs of a ball rolling on a track of unknown shape.

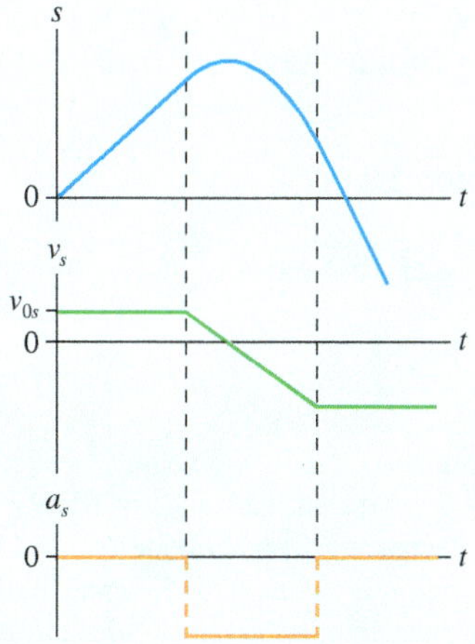

VISUALIZE Let's begin by examining the velocity graph. The ball starts with initial velocity $v_{0s} > 0$ and maintains this velocity for awhile; there's no acceleration. Thus the ball must start out rolling to the right on a horizontal track. At the end of the motion, the ball is again rolling on a horizontal track (no acceleration, constant velocity), but it's rolling to the *left* because v_s is negative. Further, the final speed ($|v_s|$) is greater than the initial speed. The middle section of the graph shows us what happens. The ball starts slowing with constant acceleration (rolling uphill), reaches a turning point (s is maximum, $v_s = 0$), then speeds up in the opposite direction (rolling downhill). This is still a negative acceleration because the ball is speeding up in the negative s-direction. It must roll farther downhill than it had rolled uphill before reaching a horizontal section of track. FIGURE 2.38 shows the track and the initial conditions that are responsible for the graphs of Figure 2.37.

FIGURE 2.38 Track responsible for the motion graphs of Figure 2.37.

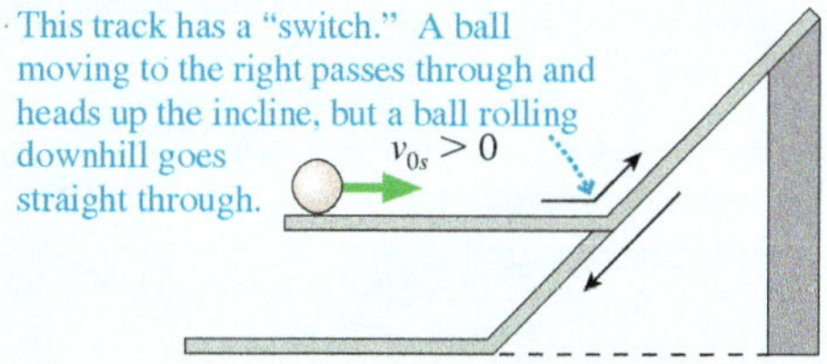

STOP TO THINK 2.5 The ball rolls up the ramp, then back down. Which is the correct acceleration graph?

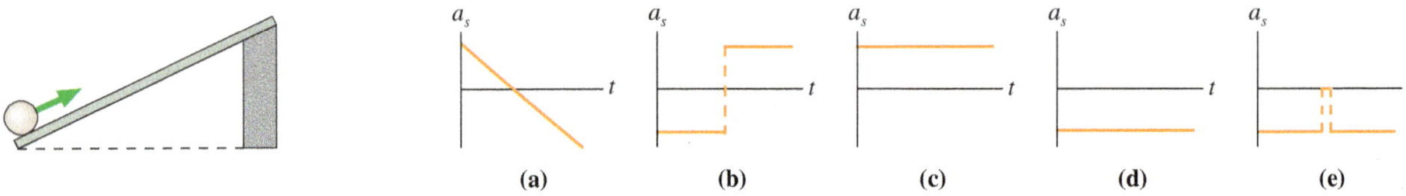

2.7 Instantaneous Acceleration

FIGURE 2.39 Velocity and acceleration graphs of a car leaving a stop sign.

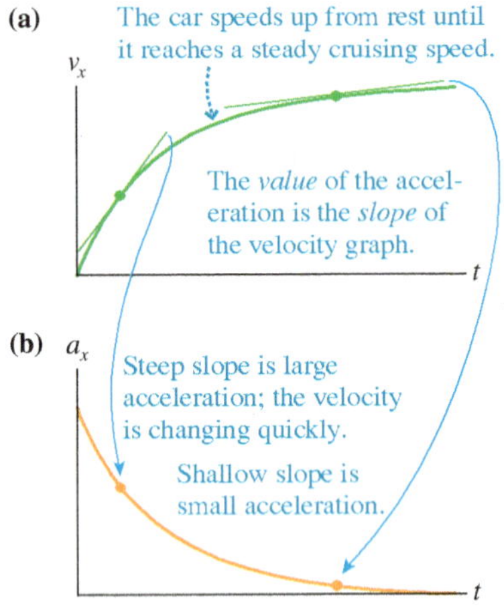

Although constant acceleration makes for straightforward problems and will often be assumed as part of a simplified model of motion, real moving objects only rarely have constant acceleration. For example, FIGURE 2.39a is a realistic velocity-versus-time graph for a car leaving a stop sign. The graph is not a straight line, so this is *not* motion with constant acceleration.

We can define an instantaneous acceleration in much the same way that we defined the instantaneous velocity. The instantaneous velocity at time t is the slope of the position-versus-time graph at that time or, mathematically, the derivative of the position with respect to time. By analogy: **The instantaneous acceleration a_s is the slope of the line that is tangent to the velocity-versus-time curve at time t.** Mathematically, this is

$$a_s = \frac{dv_s}{dt} = \text{slope of the velocity-versus-time graph at time } t \qquad (2.27)$$

The instantaneous acceleration is the rate of change of the velocity. FIGURE 2.39b applies this idea by showing the car's acceleration graph. At each instant of time, the *value* of the car's acceleration is the *slope* of its velocity graph. The initially steep slope indicates a large initial acceleration. The acceleration decreases to zero as the car reaches cruising speed.

The reverse problem—to find the velocity v_s if we know the acceleration a_s at all instants of time—is also important. Again with analogy to velocity and position, an acceleration curve can be divided into N very narrow steps so that during each step the acceleration is essentially constant. During step k, the velocity changes by $\Delta(v_s)_k = (a_s)_k \Delta t$. This is the area of the small rectangle under the step. The total velocity change between t_i and t_f is found by adding all the small $\Delta(v_s)_k$. In the limit $\Delta t \rightarrow 0$, we have

$$v_{fs} = v_{is} + \lim_{\Delta t \to 0} \sum_{k=1}^{N} (a_s)_k \Delta t = v_{is} + \int_{t_i}^{t_f} a_s \, dt \qquad (2.28)$$

The graphical interpretation of Equation 2.28 is

$$v_{fs} = v_{is} + \text{area under the acceleration curve } a_s \text{ between } t_i \text{ and } t_f \qquad (2.29)$$

EXAMPLE 2.18 A nonuniform acceleration

A particle's velocity is given by $v_s = [10 - (t - 5)^2]$ m/s, where t is in s.

a. Find an expression for the particle's acceleration a_s, then draw velocity and acceleration graphs.
b. Describe the motion.

MODEL We're told that this is a particle.

VISUALIZE FIGURE 2.40a shows the velocity graph. It is a parabola centered at $t = 5$ s with an apex $v_{max} = 10$ m/s. The slope of v_s is positive but decreasing in magnitude for $t < 5$ s. The slope is zero at $t = 5$ s, and it is negative and increasing in magnitude for $t > 5$ s. Thus the acceleration graph should start positive, decrease steadily, pass through zero at $t = 5$ s, then become increasingly negative.

SOLVE a. We can find an expression for a_s by taking the derivative of v_s. First, expand the square to give

$$v_s = (-t^2 + 10t - 15)\text{ m/s}$$

Then use the derivative rule (Equation 2.6) to find

$$a_s = \frac{dv_s}{dt} = (-2t + 10)\text{ m/s}^2$$

where t is in s. This is a linear equation that is graphed in FIGURE 2.40b. The graph meets our expectations.

b. This is a complex motion. The particle starts out moving to the left ($v_s < 0$) at 15 m/s. The positive acceleration causes the speed to decrease (slowing down because v_s and a_s have opposite signs) until the particle reaches a turning point ($v_s = 0$) just before $t = 2$ s. The particle then moves to the right ($v_s > 0$) and speeds up until reaching maximum speed at $t = 5$ s. From $t = 5$ s to just after $t = 8$ s, the particle is still moving to the right ($v_s > 0$) but slowing down. Another turning point occurs just after $t = 8$ s. Then the particle moves back to the left and gains speed as the negative a_s makes the velocity ever more negative.

FIGURE 2.40 Velocity and acceleration graphs for Example 2.18.

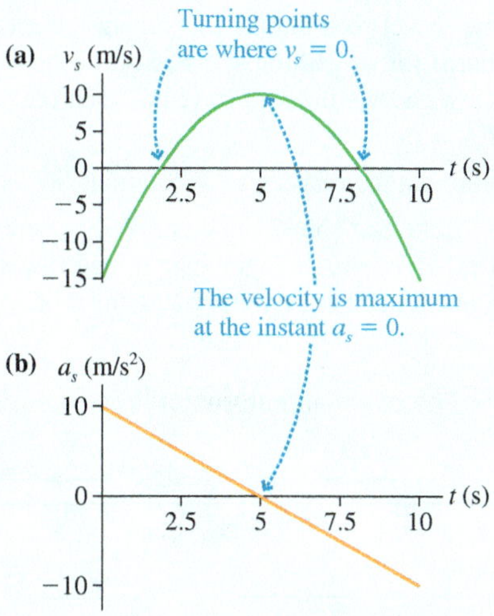

EXAMPLE 2.19 Finding velocity from acceleration

FIGURE 2.41 shows the acceleration graph for a particle with an initial velocity of 10 m/s. What is the particle's velocity at $t = 8$ s?

FIGURE 2.41 Acceleration graph for Example 2.19.

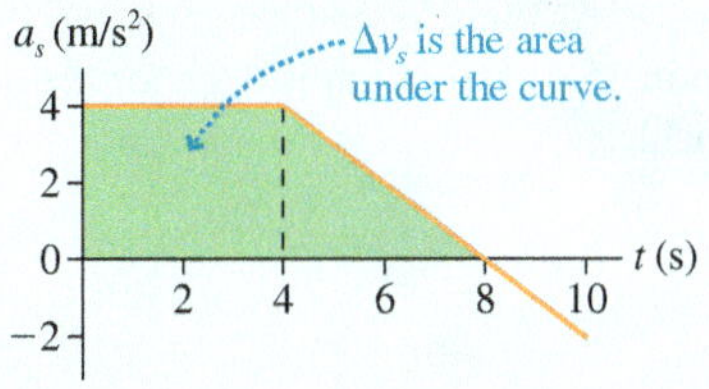

MODEL We're told this is the motion of a particle.

VISUALIZE Figure 2.41 is a graphical representation of the motion.

SOLVE The change in velocity is found as the area under the acceleration curve:

$$v_{fs} = v_{is} + \text{area under the acceleration curve } a_s \text{ between } t_i \text{ and } t_f$$

The area under the curve between $t_i = 0$ s and $t_f = 8$ s can be subdivided into a rectangle ($0\text{ s} \le t \le 4\text{ s}$) and a triangle ($4\text{ s} \le t \le 8\text{ s}$). These areas are easily computed. Thus

$$v_s(\text{at } t = 8\text{ s}) = 10\text{ m/s} + (4\text{ (m/s)/s})(4\text{ s}) + \tfrac{1}{2}(4\text{ (m/s)/s})(4\text{ s}) = 34\text{ m/s}$$

STOP TO THINK 2.6 Rank in order, from most positive to least positive, the accelerations at points A to C.

a. $a_A > a_B > a_C$
b. $a_C > a_A > a_B$
c. $a_C > a_B > a_A$
d. $a_B > a_A > a_C$

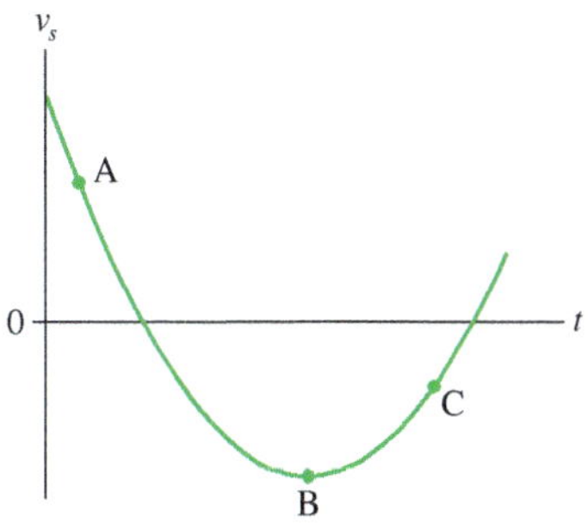

CHALLENGE EXAMPLE 2.20 Rocketing along

A rocket sled accelerates along a long, horizontal rail. Starting from rest, two rockets burn for 10 s, providing a constant acceleration. One rocket then burns out, halving the acceleration, but the other burns for an additional 5 s to boost the sled's speed to 625 m/s. How far has the sled traveled when the second rocket burns out?

MODEL Represent the rocket sled as a particle.

VISUALIZE FIGURE 2.42 shows the pictorial representation. This is a two-part problem with a beginning, an end (the second rocket burns out), and a point in between where the motion changes (the first rocket burns out).

FIGURE 2.42 The pictorial representation of the rocket sled.

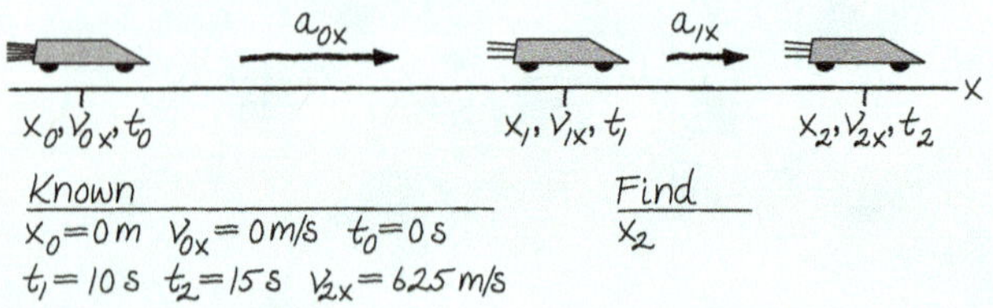

SOLVE The difficulty with this problem is that there's not enough information to completely analyze either the first or the second part of the motion. A successful solution will require combining information about both parts of the motion, and that can be done only by working algebraically, not worrying about numbers until the end of the problem. A well-drawn pictorial representation and clearly defined symbols are essential.

The first part of the motion, with both rockets firing, has acceleration a_{0x}. The sled's position and velocity when the first rocket burns out are

$$x_1 = x_0 + v_{0x}\Delta t + \tfrac{1}{2}a_{0x}(\Delta t)^2 = \tfrac{1}{2}a_{0x}t_1^2$$

$$v_{1x} = v_{0x} + a_{0x}\Delta t = a_{0x}t_1$$

where we simplified as much as possible by knowing that the sled started from rest at the origin at $t_0 = 0$ s. We can't compute numerical values, but these are valid algebraic expressions that we can carry over to the second part of the motion.

From t_1 to t_2, the acceleration is a smaller a_{1x}. The velocity when the second rocket burns out is

$$v_{2x} = v_{1x} + a_{1x}\Delta t = a_{0x}t_1 + a_{1x}(t_2 - t_1)$$

where for v_{1x} we used the algebraic result from the first part of the motion. Now we have enough information to complete the solution. We know that the acceleration is halved when the first rocket burns out, so $a_{1x} = \frac{1}{2}a_{0x}$. Thus

$$v_{2x} = 625 \text{ m/s} = a_{0x} \cdot 10 \text{ s} + \tfrac{1}{2}a_{0x} \cdot 5 \text{ s} = (12.5 \text{ s}) \cdot a_{0x}$$

Solving, we find $a_{0x} = 50 \text{ m/s}^2$.

With the acceleration now known, we can calculate the position and velocity when the first rocket burns out:

$$x_1 = \tfrac{1}{2}a_{0x}t_1^2 = \tfrac{1}{2}(50 \text{ m/s}^2)(10 \text{ s})^2 = 2500 \text{ m}$$

$$v_{1x} = a_{0x}t_1 = (50 \text{ m/s}^2)(10 \text{ s}) = 500 \text{ m/s}$$

Finally, the position when the second rocket burns out is

$$x_2 = x_1 + v_{1x}\Delta t + \tfrac{1}{2}a_{1x}(\Delta t)^2$$
$$= 2500 \text{ m} + (500 \text{ m/s})(5 \text{ s}) + \tfrac{1}{2}(25 \text{ m/s}^2)(5 \text{ s})^2 = 5300 \text{ m}$$

The sled has traveled 5300 m when it reaches 625 m/s at the burnout of the second rocket.

ASSESS 5300 m is 5.3 km, or roughly 3 miles. That's a long way to travel in 15 s! But the sled reaches incredibly high speeds. At the final speed of 625 m/s, over 1200 mph, the sled would travel nearly 10 km in 15 s. So 5.3 km in 15 s for the accelerating sled seems reasonable.

SUMMARY

The goal of Chapter 2 has been to learn how to solve problems about motion in a straight line.

General Principles

Kinematics describes motion in terms of position, velocity, and acceleration.

General kinematic relationships are given **mathematically** by:

Instantaneous velocity $v_s = ds/dt =$ slope of position graph

Instantaneous acceleration $a_s = dv_s/dt =$ slope of velocity graph

Final position $s_f = s_i + \int_{t_i}^{t_f} v_s\,dt = s_i + \begin{cases}\text{area under the velocity}\\ \text{curve from } t_i \text{ to } t_f\end{cases}$

Final velocity $v_{fs} = v_{is} + \int_{t_i}^{t_f} a_s\,dt = v_{is} + \begin{cases}\text{area under the acceleration}\\ \text{curve from } t_i \text{ to } t_f\end{cases}$

The kinematic equations for motion with constant acceleration are:

$$v_{fs} = v_{is} + a_s\,\Delta t$$

$$s_f = s_i + v_{is}\,\Delta t + \tfrac{1}{2}a_s(\Delta t)^2$$

$$v_{fs}^2 = v_{is}^2 + 2a_s\,\Delta s$$

Uniform motion is motion with constant velocity and zero acceleration:

$$s_f = s_i + v_s\,\Delta t$$

Important Concepts

Position, velocity, and acceleration are related **graphically.**

- The slope of the position-versus-time graph is the value on the velocity graph.
- The slope of the velocity graph is the value on the acceleration graph.
- s is a maximum or minimum at a turning point, and $v_s = 0$.

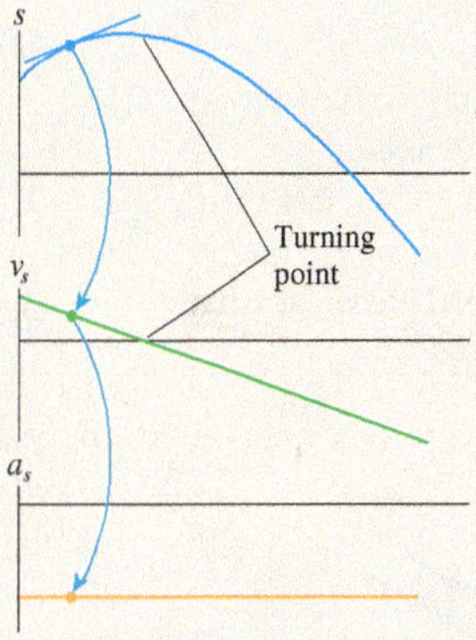

- Displacement is the area under the velocity curve.

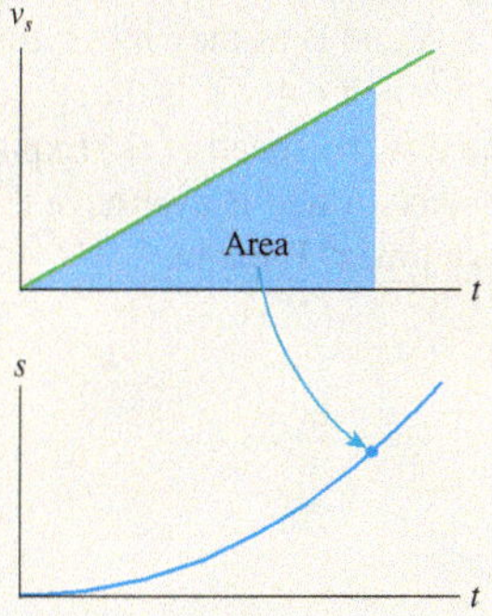

Applications

The **sign of v_s** indicates the direction of motion.

- $v_s > 0$ is motion to the right or up.
- $v_s < 0$ is motion to the left or down.

The **sign of a_s** indicates which way $\vec{a}$ points, *not* whether the object is speeding up or slowing down.

- $a_s > 0$ if $\vec{a}$ points to the right or up.
- $a_s < 0$ if $\vec{a}$ points to the left or down.
- The direction of $\vec{a}$ is found with a motion diagram.

An object is **speeding up** if and only if v_s and a_s have the same sign. An object is **slowing down** if and only if v_s and a_s have opposite signs.

Free fall is constant-acceleration motion with

$$a_y = -g = -9.80 \text{ m/s}^2$$

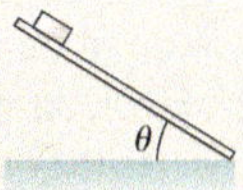

Motion on an inclined plane has $a_s = \pm g\sin\theta$. The sign depends on the direction of the tilt.

Terms and Notation

kinematics
average velocity, v_{avg}
uniform motion
speed, v
initial position, s_i
final position, s_f
instantaneous velocity, v_s
turning point
average acceleration, a_{avg}
free fall
free-fall acceleration, g
instantaneous acceleration, a_s

CONCEPTUAL QUESTIONS

For Questions 1 through 3, interpret the position graph given in each figure by writing a very short "story" of what is happening. Be creative! Have characters and situations! Simply saying that "a car moves 100 meters to the right" doesn't qualify as a story. Your stories should make *specific reference* to information you obtain from the graph, such as distance moved or time elapsed.

1.

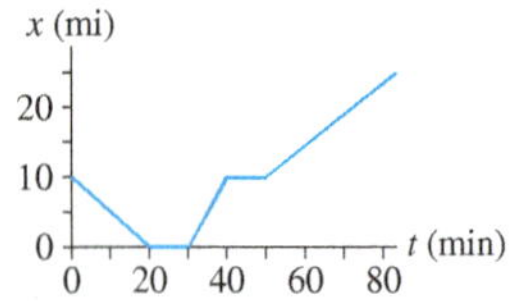

FIGURE Q2.1

2.

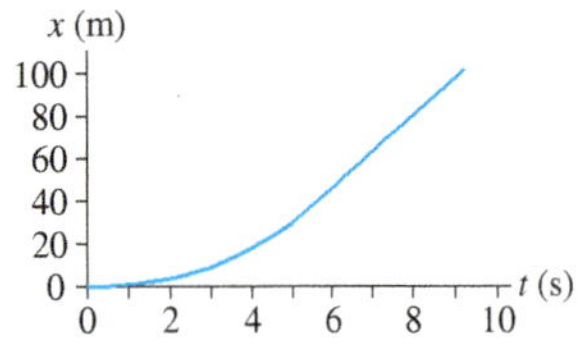

FIGURE Q2.2

3.

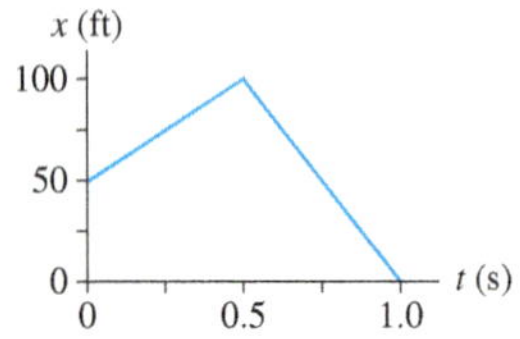

FIGURE Q2.3

4. FIGURE Q2.4 shows a position-versus-time graph for the motion of objects A and B as they move along the same axis.
   a. At the instant $t = 1$ s, is the speed of A greater than, less than, or equal to the speed of B? Explain.
   b. Do objects A and B ever have the *same* speed? If so, at what time or times? Explain.

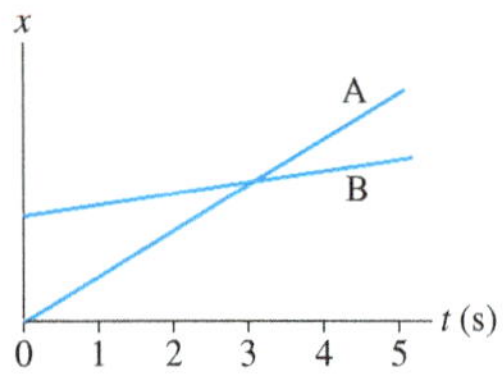

FIGURE Q2.4

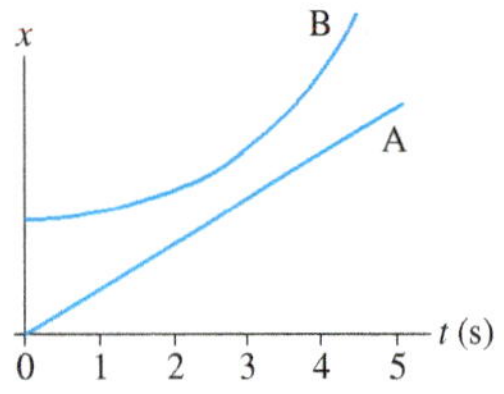

FIGURE Q2.5

5. FIGURE Q2.5 shows a position-versus-time graph for the motion of objects A and B as they move along the same axis.
   a. At the instant $t = 1$ s, is the speed of A greater than, less than, or equal to the speed of B? Explain.
   b. Do objects A and B ever have the *same* speed? If so, at what time or times? Explain.
6. FIGURE Q2.6 shows the position-versus-time graph for a moving object. At which lettered point or points:
   a. Is the object *moving* the slowest?
   b. Is the object moving the fastest?
   c. Is the object at rest?
   d. Is the object moving to the left?

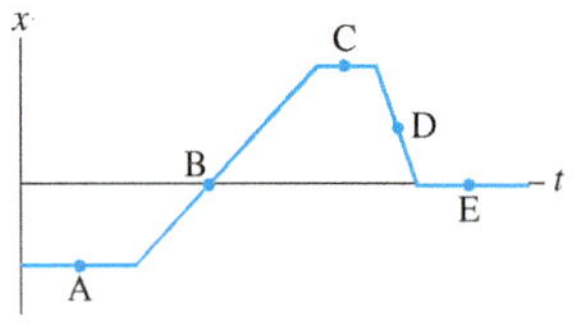

FIGURE Q2.6

7. FIGURE Q2.7 shows the position-versus-time graph for a moving object. At which lettered point or points:
   a. Is the object moving the fastest?
   b. Is the object moving to the left?
   c. Is the object speeding up?
   d. Is the object turning around?

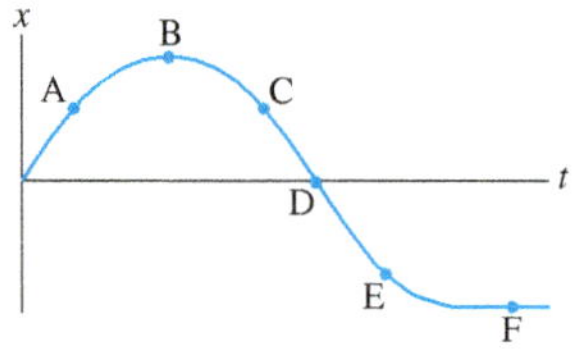

FIGURE Q2.7

8. FIGURE Q2.8 shows six frames from the motion diagrams of two moving cars, A and B.
   a. Do the two cars ever have the same position at one instant of time? If so, in which frame number (or numbers)?
   b. Do the two cars ever have the same velocity at one instant of time? If so, between which two frames?

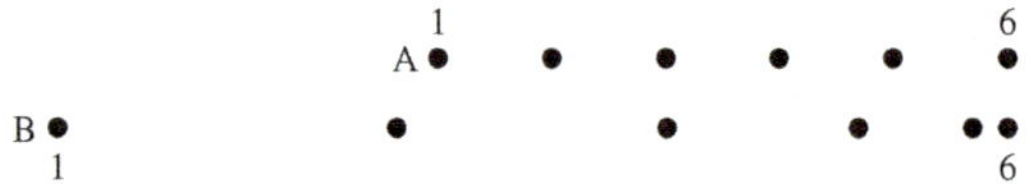

FIGURE Q2.8

9. You're driving along the highway at a steady speed of 60 mph when another driver decides to pass you. At the moment when the front of his car is exactly even with the front of your car, and you turn your head to smile at him, do the two cars have equal velocities? Explain.
10. A bicycle is traveling east. Can its acceleration vector ever point west? Explain.
11. (a) Give an example of a vertical motion with a positive velocity and a negative acceleration. (b) Give an example of a vertical motion with a negative velocity and a negative acceleration.
12. A ball is thrown straight up into the air. At each of the following instants, is the magnitude of the ball's acceleration greater than g, equal to g, less than g, or 0? Explain.
    a. Just after leaving your hand.
    b. At the very top (maximum height).
    c. Just before hitting the ground.
13. A rock is *thrown* (not dropped) straight down from a bridge into the river below. At each of the following instants, is the magnitude of the rock's acceleration greater than g, equal to g, less than g, or 0? Explain.
    a. Immediately after being released.
    b. Just before hitting the water.
14. A rubber ball dropped from a height of 2 m bounces back to a height of 1 m. Draw the ball's position, velocity, and acceleration graphs, stacked vertically, from the instant you release it until it returns to its maximum bounce height. Pay close attention to the time the ball is in contact with the ground; this is a short interval of time, but it's not zero.

EXERCISES AND PROBLEMS

Exercises

Section 2.1 Uniform Motion

1\. | Alan leaves Los Angeles at 8:00 a.m. to drive to San Francisco, 400 mi away. He travels at a steady 50 mph. Beth leaves Los Angeles at 9:00 a.m. and drives a steady 60 mph.
   a. Who gets to San Francisco first?
   b. How long does the first to arrive have to wait for the second?

2\. || Larry leaves home at 9:05 and runs at constant speed to the lamppost seen in FIGURE EX2.2. He reaches the lamppost at 9:07, immediately turns, and runs to the tree. Larry arrives at the tree at 9:10.
   a. What is Larry's average velocity, in m/min, during each of these two intervals.
   b. What is Larry's average velocity for the entire run?

FIGURE EX2.2

3\. || Julie drives 100 mi to Grandmother's house. On the way to Grandmother's, Julie drives half the distance at 40 mph and half the distance at 60 mph. On her return trip, she drives half the time at 40 mph and half the time at 60 mph.
   a. What is Julie's average speed on the way to Grandmother's house?
   b. What is her average speed on the return trip?

4\. | FIGURE EX2.4 is the position-versus-time graph of a jogger. What is the jogger's velocity at $t = 10$ s, at $t = 25$ s, and at $t = 35$ s?

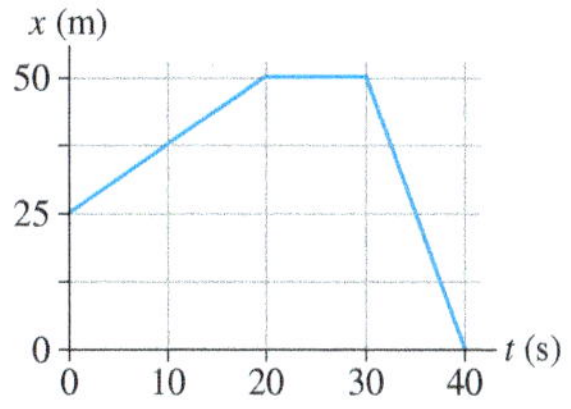

FIGURE EX2.4

Section 2.2 Instantaneous Velocity

Section 2.3 Finding Position from Velocity

5\. | FIGURE EX2.5 shows the position graph of a particle.
   a. Draw the particle's velocity graph for the interval $0 \text{ s} \leq t \leq 4 \text{ s}$.
   b. Does this particle have a turning point or points? If so, at what time or times?

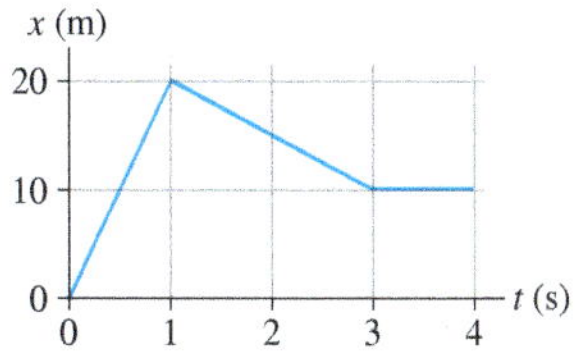

FIGURE EX2.5

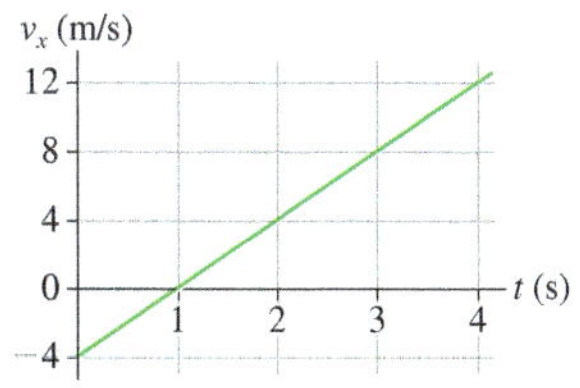

FIGURE EX2.6

6\. | A particle starts from $x_0 = 10$ m at $t_0 = 0$ s and moves with the velocity graph shown in FIGURE EX2.6.
   a. Does this particle have a turning point? If so, at what time?
   b. What is the object's position at $t = 2$ s, 3 s, and 4 s?

7\. || FIGURE EX 2.7 is a somewhat idealized graph of the velocity of blood in the ascending aorta during one beat of the heart. Approximately how far, in cm, does the blood move during one beat?
BIO

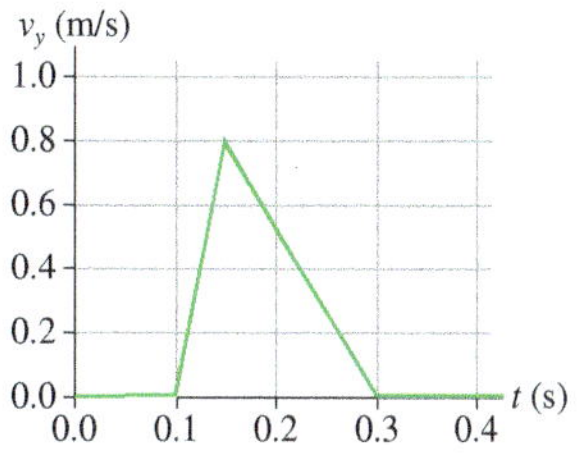

FIGURE EX2.7

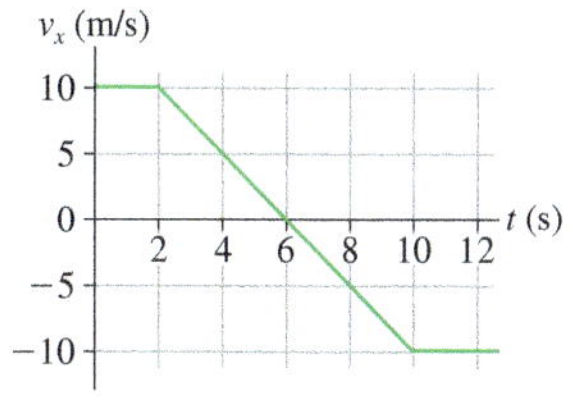

FIGURE EX2.8

8\. | FIGURE EX2.8 shows the velocity graph for a particle having initial position $x_0 = 0$ m at $t_0 = 0$ s.
   a. At what time or times is the particle found at $x = 35$ m? Work with the geometry of the graph, not with kinematic equations.
   b. Draw a motion diagram for the particle.

Section 2.4 Motion with Constant Acceleration

9\. | FIGURE EX2.9 shows the velocity graph of a particle. Draw the particle's acceleration graph for the interval $0 \text{ s} \leq t \leq 4 \text{ s}$. Give both axes an appropriate numerical scale.

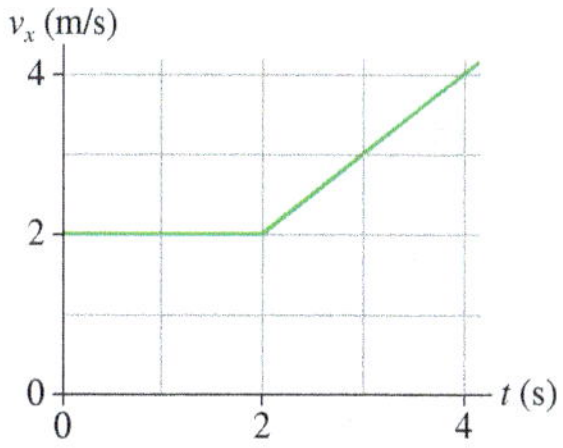

FIGURE EX2.9

10\. | FIGURE EX2.7 showed the velocity graph of blood in the aorta. Estimate the blood's acceleration during each phase of the motion, speeding up and slowing down.
BIO

11\. | FIGURE EX2.11 shows the velocity graph of a particle moving along the x-axis. Its initial position is $x_0 = 2.0$ m at $t_0 = 0$ s. At $t = 2.0$ s, what are the particle's (a) position, (b) velocity, and (c) acceleration?

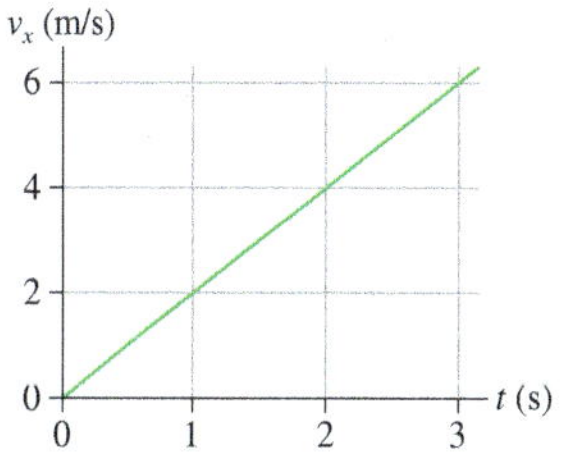

FIGURE EX2.11

12. | FIGURE EX2.12 shows the velocity-versus-time graph for a particle moving along the x-axis. Its initial position is $x_0 = 2.0$ m at $t_0 = 0$ s.
   a. What are the particle's position, velocity, and acceleration at $t = 1.0$ s?
   b. What are the particle's position, velocity, and acceleration at $t = 3.0$ s?

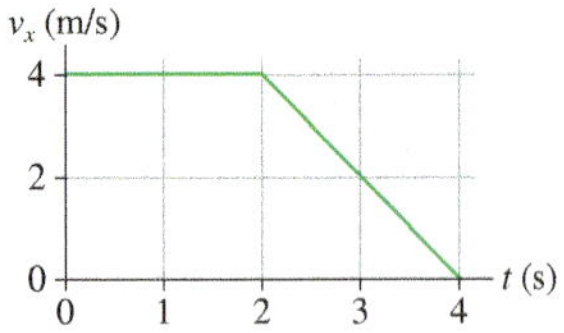

FIGURE EX2.12

13. || A jet plane is cruising at 300 m/s when suddenly the pilot turns the engines up to full throttle. After traveling 4.0 km, the jet is moving with a speed of 400 m/s. What is the jet's acceleration, assuming it to be a constant acceleration?

14. | When you sneeze, the air in your lungs accelerates from rest to 150 km/h in approximately 0.50 s. What is the acceleration of the air in m/s^2?
BIO

15. || A speed skater moving across frictionless ice at 8.0 m/s hits a 5.0-m-wide patch of rough ice. She slows steadily, then continues on at 6.0 m/s. What is her acceleration on the rough ice?

16. | A Porsche challenges a Honda to a 400 m race. Because the Porsche's acceleration of 3.5 m/s^2 is larger than the Honda's 3.0 m/s^2, the Honda gets a 1.0 s head start. Who wins?

Section 2.5 Free Fall

17. | Ball bearings are made by letting spherical drops of molten metal fall inside a tall tower—called a *shot tower*—and solidify as they fall.
   a. If a bearing needs 4.0 s to solidify enough for impact, how high must the tower be?
   b. What is the bearing's impact velocity?

18. | A ball is thrown vertically upward with a speed of 19.6 m/s.
   a. What is the ball's velocity and its height after 1.0, 2.0, 3.0, and 4.0 s?
   b. Draw the ball's velocity-versus-time graph. Give both axes an appropriate numerical scale.

19. || A student standing on the ground throws a ball straight up. The ball leaves the student's hand with a speed of 15 m/s when the hand is 2.0 m above the ground. How long is the ball in the air before it hits the ground? (The student moves her hand out of the way.)

20. || A rock is tossed straight up with a speed of 20 m/s. When it returns, it falls into a hole 10 m deep.
   a. What is the rock's velocity as it hits the bottom of the hole?
   b. How long is the rock in the air, from the instant it is released until it hits the bottom of the hole?

Section 2.6 Motion on an Inclined Plane

21. || A skier is gliding along at 3.0 m/s on horizontal, frictionless snow. He suddenly starts down a 10° incline. His speed at the bottom is 15 m/s.
   a. What is the length of the incline?
   b. How long does it take him to reach the bottom?

22. || A car traveling at 30 m/s runs out of gas while traveling up a 10° slope. How far up the hill will it coast before starting to roll back down?

Section 2.7 Instantaneous Acceleration

23. | A particle moving along the x-axis has its position described by the function $x = (2t^2 - t + 1)$ m, where t is in s. At $t = 2$ s what are the particle's (a) position, (b) velocity, and (c) acceleration?

24. || A particle moving along the x-axis has its velocity described by the function $v_x = 2t^2$ m/s, where t is in s. Its initial position is $x_0 = 1$ m at $t_0 = 0$ s. At $t = 1$ s what are the particle's (a) position, (b) velocity, and (c) acceleration?

25. || FIGURE EX2.25 shows the acceleration-versus-time graph of a particle moving along the x-axis. Its initial velocity is $v_{0x} = 8.0$ m/s at $t_0 = 0$ s. What is the particle's velocity at $t = 4.0$ s?

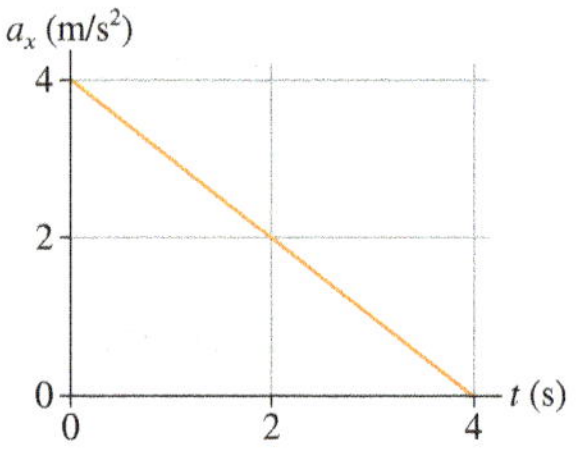

FIGURE EX2.25

Problems

26. || A particle's position on the x-axis is given by the function $x = (t^2 - 4t + 2)$ m, where t is in s.
   a. Make a position-versus-time graph for the interval $0 \text{ s} \le t \le 5$ s. Do this by calculating and plotting x every 0.5 s from 0 s to 5 s, then drawing a smooth curve through the points.
   b. Determine the particle's velocity at $t = 1.0$ s by drawing the tangent line on your graph and measuring its slope.
   c. Determine the particle's velocity at $t = 1.0$ s by evaluating the derivative at that instant. Compare this to your result from part b.
   d. Are there any turning points in the particle's motion? If so, at what position or positions?
   e. Where is the particle when $v_x = 4.0$ m/s?
   f. Draw a motion diagram for the particle.

27. || Three particles move along the x-axis, each starting with $v_{0x} = 10$ m/s at $t_0 = 0$ s. In FIGURE P2.27, the graph for A is a position-versus-time graph; the graph for B is a velocity-versus-time graph; the graph for C is an acceleration-versus-time graph. Find each particle's velocity at $t = 7.0$ s. Work with the geometry of the graphs, not with kinematic equations.

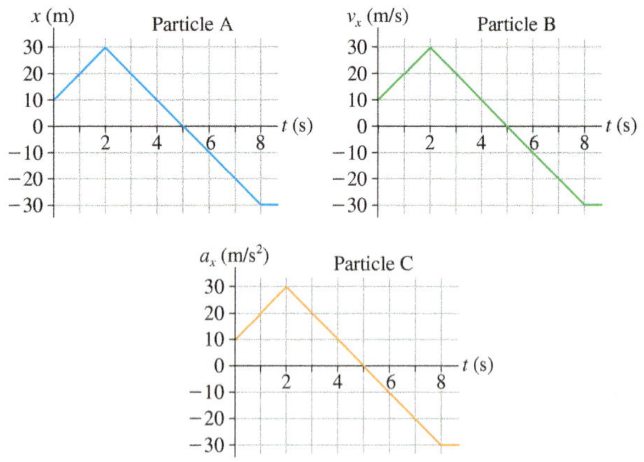

FIGURE P2.27

28. || FIGURE P2.28 shows the acceleration graph for a particle that starts from rest at $t = 0$ s. Determine the object's velocity at times $t = 0$ s, 2 s, 4 s, 6 s, and 8 s.

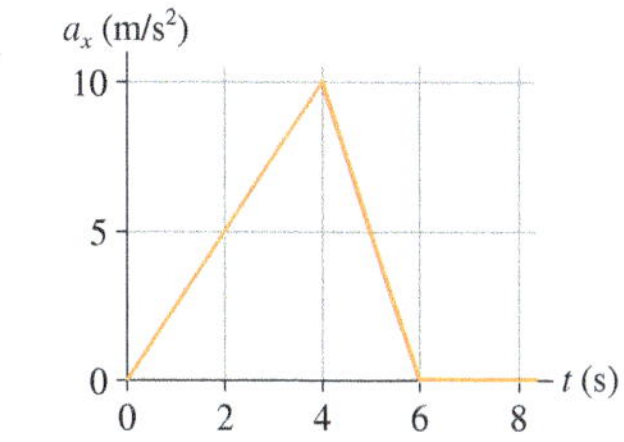

FIGURE P2.28

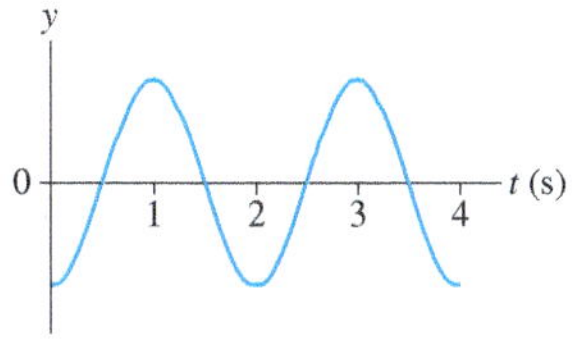

FIGURE P2.29

29. | A block is suspended from a spring, pulled down, and released. The block's position-versus-time graph is shown in FIGURE P2.29.
   a. At what times is the velocity zero? At what times is the velocity most positive? Most negative?
   b. Draw a reasonable velocity-versus-time graph.

30. || A particle's velocity is described by the function $v_x = t^2 - 7t + 10$ m/s, where t is in s.
   a. At what times does the particle reach its turning points?
   b. What is the particle's acceleration at each of the turning points?

31. | The position of a particle is given by the function $x = (2t^3 - 9t^2 + 12)$ m, where t is in s.
   a. At what time or times is $v_x = 0$ m/s?
   b. What are the particle's position and its acceleration at this time(s)?

32. | An object starts from rest at $x = 0$ m at time $t = 0$ s. Five seconds later, at $t = 5.0$ s, the object is observed to be at $x = 40$ m and to have velocity $v_x = 11$ m/s.
   a. Was the object's acceleration uniform or nonuniform? Explain your reasoning.
   b. Sketch the velocity-versus-time graph implied by these data. Is the graph a straight line or curved? If curved, is it concave upward or downward?

33. ||| A particle's velocity is described by the function $v_x = kt^2$ m/s, where k is a constant and t is in s. The particle's position at $t_0 = 0$ s is $x_0 = -9.0$ m. At $t_1 = 3.0$ s, the particle is at $x_1 = 9.0$ m. Determine the value of the constant k. Be sure to include the proper units.

34. || A particle's acceleration is described by the function $a_x = (10 - t)$ m/s^2, where t is in s. Its initial conditions are $x_0 = 0$ m and $v_{0x} = 0$ m/s at $t = 0$ s.
   a. At what time is the velocity again zero?
   b. What is the particle's position at that time?

35. || A ball rolls along the frictionless track shown in FIGURE P2.35. Each segment of the track is straight, and the ball passes smoothly from one segment to the next without changing speed or leaving the track. Draw three vertically stacked graphs showing position, velocity, and acceleration versus time. Each graph should have the same time axis, and the proportions of the graph should be qualitatively correct. Assume that the ball has enough speed to reach the top.

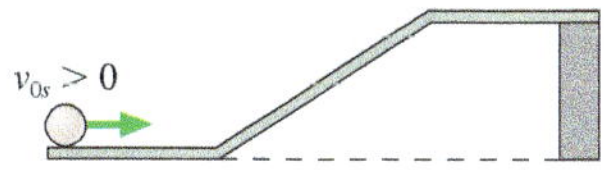

FIGURE P2.35

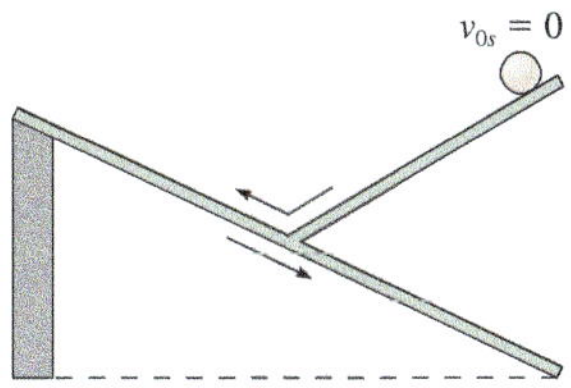

FIGURE P2.36

36. || Draw position, velocity, and acceleration graphs for the ball shown in FIGURE P2.36. See Problem 35 for more information.

37. || Draw position, velocity, and acceleration graphs for the ball shown in FIGURE P2.37. See Problem 35 for more information. The ball changes direction but not speed as it bounces from the reflecting wall.

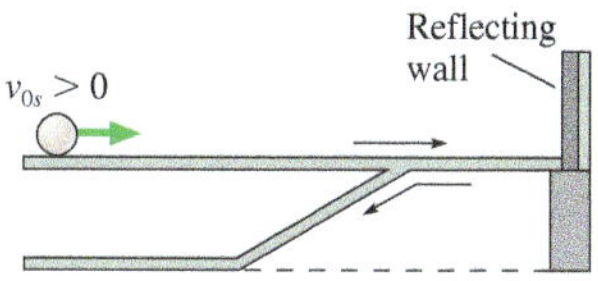

FIGURE P2.37

38. || FIGURE P2.38 shows a set of kinematic graphs for a ball rolling on a track. All segments of the track are straight lines, but some may be tilted. Draw a picture of the track and also indicate the ball's initial condition.

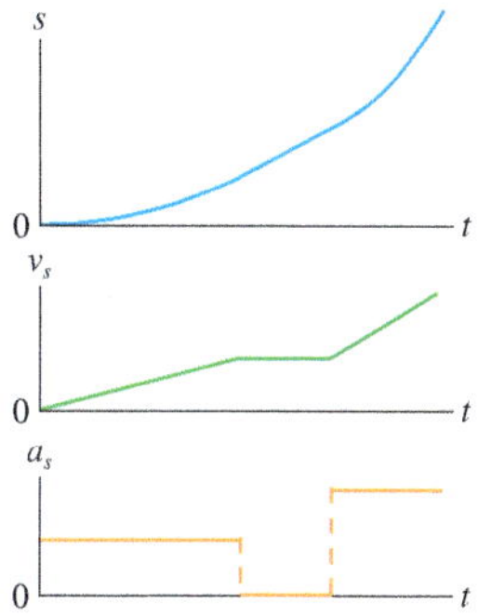

FIGURE P2.38

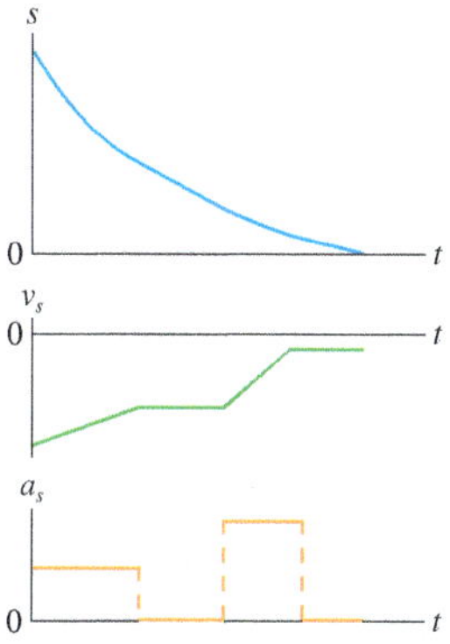

FIGURE P2.39

39. || FIGURE P2.39 shows a set of kinematic graphs for a ball rolling on a track. All segments of the track are straight lines, but some may be tilted. Draw a picture of the track and also indicate the ball's initial condition.

40. | The takeoff speed for an Airbus A320 jetliner is 80 m/s. Velocity data measured during takeoff are as shown.

t (s)	v_x (m/s)
0	0
10	23
20	46
30	69

   a. What is the takeoff speed in miles per hour?
   b. Is the jetliner's acceleration constant during takeoff? Explain.
   c. At what time do the wheels leave the ground?
   d. For safety reasons, in case of an aborted takeoff, the runway must be three times the takeoff distance. Can an A320 take off safely on a 2.5-mi-long runway?

41. | a. What constant acceleration, in SI units, must a car have to go from zero to 60 mph in 10 s?
   b. What fraction of g is this?
   c. How far has the car traveled when it reaches 60 mph? Give your answer both in SI units and in feet.

42. || a. How many days will it take a spaceship to accelerate to the speed of light (3.0×10^8 m/s) with the acceleration g?
   b. How far will it travel during this interval?
   c. What fraction of a light year is your answer to part b? A light year is the distance light travels in one year.

NOTE ▶ We know, from Einstein's theory of relativity, that no object can travel at the speed of light. So this problem, while interesting and instructive, is not realistic. ◀

43. | You are driving to the grocery store at 20 m/s. You are 110 m from an intersection when the traffic light turns red. Assume that your reaction time is 0.50 s and that your car brakes with constant acceleration.
   a. How far are you from the intersection when you begin to apply the brakes?
   b. What acceleration will bring you to rest right at the intersection?
   c. How long does it take you to stop after the light turns red?

44. || a. Suppose you are driving at speed v_0 when a sudden obstacle in the road forces you to make a quick stop. If your reaction time before applying the brakes is t_R, what constant deceleration (absolute value of a_x) do you need to stop in distance d? Assume that d is larger than the car travels during your reaction time.
   b. Suppose you are driving at 21 m/s when you suddenly see an obstacle 50 m ahead. If your reaction time is 0.50 s and if your car's maximum deceleration is 6.0 m/s^2, can you stop in time to avoid a collision?

45. || You're driving down the highway late one night at 20 m/s when a deer steps onto the road 35 m in front of you. Your reaction time before stepping on the brakes is 0.50 s, and the maximum deceleration of your car is 10 m/s^2.
   a. How much distance is between you and the deer when you come to a stop?
   b. What is the maximum speed you could have and still not hit the deer?

46. ||| The minimum stopping distance for a car traveling at a speed of 30 m/s is 60 m, including the distance traveled during the driver's reaction time of 0.50 s.
   a. What is the minimum stopping distance for the same car traveling at a speed of 40 m/s?
   b. Draw a position-versus-time graph for the motion of the car in part a. Assume the car is at $x_0 = 0$ m when the driver first sees the emergency situation ahead that calls for a rapid halt.

47. || When jumping, a flea accelerates at an astounding 1000 m/s^2,
BIO but over only the very short distance of 0.50 mm. If a flea jumps straight up, and if air resistance is neglected (a rather poor approximation in this situation), how high does the flea go?

48. || A cheetah spots a Thomson's gazelle, its preferred prey, and
BIO leaps into action, quickly accelerating to its top speed of 30 m/s, the highest of any land animal. However, a cheetah can maintain this extreme speed for only 15 s before having to let up. The cheetah is 170 m from the gazelle as it reaches top speed, and the gazelle sees the cheetah at just this instant. With negligible reaction time, the gazelle heads directly away from the cheetah, accelerating at 4.6 m/s^2 for 5.0 s, then running at constant speed. Does the gazelle escape?

49. ||| A 200 kg weather rocket is loaded with 100 kg of fuel and fired straight up. It accelerates upward at 30 m/s^2 for 30 s, then runs out of fuel. Ignore any air resistance effects.
   a. What is the rocket's maximum altitude?
   b. How long is the rocket in the air before hitting the ground?
   c. Draw a velocity-versus-time graph for the rocket from liftoff until it hits the ground.

50. || A 1000 kg weather rocket is launched straight up. The rocket motor provides a constant acceleration for 16 s, then the motor stops. The rocket altitude 20 s after launch is 5100 m. You can ignore any effects of air resistance.
   a. What was the rocket's acceleration during the first 16 s?
   b. What is the rocket's speed as it passes through a cloud 5100 m above the ground?

51. ||| A lead ball is dropped into a lake from a diving board 5.0 m above the water. After entering the water, it sinks to the bottom with a constant velocity equal to the velocity with which it hit the water. The ball reaches the bottom 3.0 s after it is released. How deep is the lake?

52. || A hotel elevator ascends 200 m with a maximum speed of 5.0 m/s. Its acceleration and deceleration both have a magnitude of 1.0 m/s^2.
   a. How far does the elevator move while accelerating to full speed from rest?
   b. How long does it take to make the complete trip from bottom to top?

53. || A car starts from rest at a stop sign. It accelerates at 4.0 m/s^2 for 6.0 s, coasts for 2.0 s, and then slows down at a rate of 3.0 m/s^2 for the next stop sign. How far apart are the stop signs?

54. || A car accelerates at 2.0 m/s^2 along a straight road. It passes two marks that are 30 m apart at times $t = 4.0$ s and $t = 5.0$ s. What was the car's velocity at $t = 0$ s?

55. || Santa loses his footing and slides down a frictionless, snowy roof that is tilted at an angle of 30°. If Santa slides 10 m before reaching the edge, what is his speed as he leaves the roof?

56. || Ann and Carol are driving their cars along the same straight road. Carol is located at $x = 2.4$ mi at $t = 0$ h and drives at a steady 36 mph. Ann, who is traveling in the same direction, is located at $x = 0.0$ mi at $t = 0.50$ h and drives at a steady 50 mph.
   a. At what time does Ann overtake Carol?
   b. What is their position at this instant?
   c. Draw a position-versus-time graph showing the motion of both Ann and Carol.

57. || a. A very slippery block of ice slides down a smooth ramp tilted at angle θ. The ice is released from rest at vertical height h above the bottom of the ramp. Find an expression for the speed of the ice at the bottom.
   b. Evaluate your answer to part a for ice released at a height of 30 cm on ramps tilted at 20° and 40°.

58. || A toy train is pushed forward and released at $x_0 = 2.0$ m with a speed of 2.0 m/s. It rolls at a steady speed for 2.0 s, then one wheel begins to stick. The train comes to a stop 6.0 m from the point at which it was released. What is the magnitude of the train's acceleration after its wheel begins to stick?

59. || Bob is driving the getaway car after the big bank robbery. He's going 50 m/s when his headlights suddenly reveal a nail strip that the cops have placed across the road 150 m in front of him. If Bob can stop in time, he can throw the car into reverse and escape. But if he crosses the nail strip, all his tires will go flat and he will be caught. Bob's reaction time before he can hit the brakes is 0.60 s, and his car's maximum deceleration is 10 m/s^2. Is Bob in jail?

60. || One game at the amusement park has you push a puck up a long, frictionless ramp. You win a stuffed animal if the puck, at its highest point, comes to within 10 cm of the end of the ramp without going off. You give the puck a push, releasing it with a speed of 5.0 m/s when it is 8.5 m from the end of the ramp. The puck's speed after traveling 3.0 m is 4.0 m/s. Are you a winner?

61. || a. Your goal in laboratory is to launch a ball of mass m straight up so that it reaches exactly height h above the top of the launching tube. You and your lab partners will earn fewer points if the ball goes too high or too low. The launch tube uses compressed air to accelerate the ball over a distance d, and you have a table of data telling you how to set the

air compressor to achieve a desired acceleration. Find an expression for the acceleration that will earn you maximum points.
b. Evaluate your answer to part a to achieve a height of 3.2 m using a 45-cm-long launch tube.

62. || Nicole throws a ball straight up. Chad watches the ball from a window 5.0 m above the point where Nicole released it. The ball passes Chad on the way up, and it has a speed of 10 m/s as it passes him on the way back down. How fast did Nicole throw the ball?

63. || A motorist is driving at 20 m/s when she sees that a traffic light 200 m ahead has just turned red. She knows that this light stays red for 15 s, and she wants to reach the light just as it turns green again. It takes her 1.0 s to step on the brakes and begin slowing. What is her speed as she reaches the light at the instant it turns green?

64. || When a 1984 Alfa Romeo Spider sports car accelerates at the maximum possible rate, its motion during the first 20 s is extremely well modeled by the simple equation

$$v_x^2 = \frac{2P}{m}t$$

where $P = 3.6 \times 10^4$ watts is the car's power output, $m = 1200$ kg is its mass, and v_x is in m/s. That is, the square of the car's velocity increases linearly with time.
a. What is the car's speed at $t = 10$ s and at $t = 20$ s?
b. Find an algebraic expression in terms of P, m, and t, for the car's acceleration at time t.
c. Evaluate the acceleration at $t = 1$ s and $t = 10$ s.
d. This simple model fails for t less than about 0.5 s. Explain how you can recognize the failure.

65. || David is driving a steady 30 m/s when he passes Tina, who is sitting in her car at rest. Tina begins to accelerate at a steady 2.0 m/s^2 at the instant when David passes.
a. How far does Tina drive before passing David?
b. What is her speed as she passes him?

66. || A cat is sleeping on the floor in the middle of a 3.0-m-wide room when a barking dog enters with a speed of 1.50 m/s. As the dog enters, the cat (as only cats can do) immediately accelerates at 0.85 m/s^2 toward an open window on the opposite side of the room. The dog (all bark and no bite) is a bit startled by the cat and begins to slow down at 0.10 m/s^2 as soon as it enters the room. Does the dog catch the cat before the cat is able to leap through the window?

67. || Jill has just gotten out of her car in the grocery store parking lot. The parking lot is on a hill and is tilted 3°. Twenty meters downhill from Jill, a little old lady lets go of a fully loaded shopping cart. The cart, with frictionless wheels, starts to roll straight downhill. Jill immediately starts to sprint after the cart with her top acceleration of 2.0 m/s^2. How far has the cart rolled before Jill catches it?

68. || As a science project, you drop a watermelon off the top of the Empire State Building, 320 m above the sidewalk. It so happens that Superman flies by at the instant you release the watermelon. Superman is headed straight down with a speed of 35 m/s. How fast is the watermelon going when it passes Superman?

69. ||| I was driving along at 20 m/s, trying to change a CD and not watching where I was going. When I looked up, I found myself 45 m from a railroad crossing. And wouldn't you know it, a train moving at 30 m/s was only 60 m from the crossing. In a split second, I realized that the train was going to beat me to the crossing and that I didn't have enough distance to stop. My only hope was to accelerate enough to cross the tracks before the train arrived. If my reaction time before starting to accelerate was 0.50 s, what minimum acceleration did my car need for me to be here today writing these words?

70. || As an astronaut visiting Planet X, you're assigned to measure the free-fall acceleration. Getting out your meter stick and stop watch, you time the fall of a heavy ball from several heights. Your data are as follows:

Height (m)	Fall time (s)
0.0	0.00
1.0	0.54
2.0	0.72
3.0	0.91
4.0	1.01
5.0	1.17

Analyze these data to determine the free-fall acceleration on Planet X. Your analysis method should involve fitting a straight line to an appropriate graph, similar to the analysis in Example 2.15.

71. || Your engineering firm has been asked to determine the deceleration of a car during hard braking. To do so, you decide to measure the lengths of the skid marks when stopping from various initial speeds. Your data are as follows:

Speed (m/s)	Skid length (m)
10	7
15	14
20	27
25	37
30	58

a. Do the data support an assertion that the deceleration is constant, independent of speed? Explain.
b. Determine an experimental value for the car's deceleration—that is, the absolute value of the acceleration. Your analysis method should involve fitting a straight line to an appropriate graph, similar to the analysis in Example 2.15.

In Problems 72 through 75, you are given the kinematic equation or equations that are used to solve a problem. For each of these, you are to:
a. Write a *realistic* problem for which this is the correct equation(s). Be sure that the answer your problem requests is consistent with the equation(s) given.
b. Draw the pictorial representation for your problem.
c. Finish the solution of the problem.

72. $64\text{ m} = 0\text{ m} + (32\text{ m/s})(4\text{ s} - 0\text{ s}) + \frac{1}{2}a_x(4\text{ s} - 0\text{ s})^2$

73. $(10\text{ m/s})^2 = v_{0y}^2 - 2(9.8\text{ m/s}^2)(10\text{ m} - 0\text{ m})$

74. $(0\text{ m/s})^2 = (5\text{ m/s})^2 - 2(9.8\text{ m/s}^2)(\sin 10°)(x_1 - 0\text{ m})$

75. $v_{1x} = 0\text{ m/s} + (20\text{ m/s}^2)(5\text{ s} - 0\text{ s})$
$x_1 = 0\text{ m} + (0\text{ m/s})(5\text{ s} - 0\text{ s}) + \frac{1}{2}(20\text{ m/s}^2)(5\text{ s} - 0\text{ s})^2$
$x_2 = x_1 + v_{1x}(10\text{ s} - 5\text{ s})$

Challenge Problems

76. The two masses in FIGURE CP2.76 slide on frictionless wires. They are connected by a pivoting rigid rod of length L. Prove that $v_{2x} = -v_{1y} \tan\theta$.

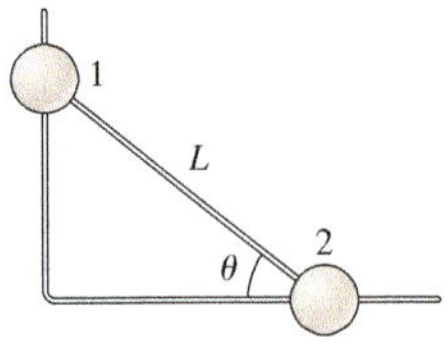

FIGURE CP2.76

77. A rocket is launched straight up with constant acceleration. Four seconds after liftoff, a bolt falls off the side of the rocket. The bolt hits the ground 6.0 s later. What was the rocket's acceleration?
78. Your school science club has devised a special event for homecoming. You've attached a rocket to the rear of a small car that has been decorated in the blue-and-gold school colors. The rocket provides a constant acceleration for 9.0 s. As the rocket shuts off, a parachute opens and slows the car at a rate of 5.0 m/s^2. The car passes the judges' box in the center of the grandstand, 990 m from the starting line, exactly 12 s after you fire the rocket. What is the car's speed as it passes the judges?
79. Careful measurements have been made of Olympic sprinters in the 100-meter dash. A simple but reasonably accurate model is that a sprinter accelerates at 3.6 m/s^2 for $3\frac{1}{3}$ s, then runs at constant velocity to the finish line.
    a. What is the race time for a sprinter who follows this model?
    b. A sprinter could run a faster race by accelerating faster at the beginning, thus reaching top speed sooner. If a sprinter's top speed is the same as in part a, what acceleration would he need to run the 100-meter dash in 9.9 s?
    c. By what percent did the sprinter need to increase his acceleration in order to decrease his time by 1%?
80. Careful measurements have been made of Olympic sprinters in the 100-meter dash. A quite realistic model is that the sprinter's velocity is given by

$$v_x = a(1 - e^{-bt})$$

where t is in s, v_x is in m/s, and the constants a and b are characteristic of the sprinter. Sprinter Carl Lewis's run at the 1987 World Championships is modeled with $a = 11.81$ m/s and $b = 0.6887$ s^{-1}.
    a. What was Lewis's acceleration at $t = 0$ s, 2.00 s, and 4.00 s?
    b. Find an expression for the distance traveled at time t.
    c. Your expression from part b is a transcendental equation, meaning that you can't solve it for t. However, it's not hard to use trial and error to find the time needed to travel a specific distance. To the nearest 0.01 s, find the time Lewis needed to sprint 100.0 m. His official time was 0.01 s more than your answer, showing that this model is very good, but not perfect.
81. A sprinter can accelerate with constant acceleration for 4.0 s before reaching top speed. He can run the 100-meter dash in 10.0 s. What is his speed as he crosses the finish line?
82. A rubber ball is shot straight up from the ground with speed v_0. Simultaneously, a second rubber ball at height h directly above the first ball is dropped from rest.
    a. At what height above the ground do the balls collide? Your answer will be an *algebraic expression* in terms of h, v_0, and g.
    b. What is the maximum value of h for which a collision occurs before the first ball falls back to the ground?
    c. For what value of h does the collision occur at the instant when the first ball is at its highest point?
83. The Starship Enterprise returns from warp drive to ordinary space with a forward speed of 50 km/s. To the crew's great surprise, a Klingon ship is 100 km directly ahead, traveling in the same direction at a mere 20 km/s. Without evasive action, the Enterprise will overtake and collide with the Klingons in just slightly over 3.0 s. The Enterprise's computers react instantly to brake the ship. What magnitude acceleration does the Enterprise need to just barely avoid a collision with the Klingon ship? Assume the acceleration is constant.
Hint: Draw a position-versus-time graph showing the motions of both the Enterprise and the Klingon ship. Let $x_0 = 0$ km be the location of the Enterprise as it returns from warp drive. How do you show graphically the situation in which the collision is "barely avoided"? Once you decide what it looks like graphically, express that situation mathematically.

STOP TO THINK ANSWERS

Stop to Think 2.1: d. The particle starts with positive x and moves to negative x.

Stop to Think 2.2: c. The velocity is the slope of the position graph. The slope is positive and constant until the position graph crosses the axis, then positive but decreasing, and finally zero when the position graph is horizontal.

Stop to Think 2.3: b. A constant positive v_x corresponds to a linearly increasing x, starting from $x_i = -10$ m. The constant negative v_x then corresponds to a linearly decreasing x.

Stop to Think 2.4: a and b. The velocity is constant while $a = 0$, it decreases linearly while a is negative. Graphs a, b, and c all have the same acceleration, but only graphs a and b have a positive initial velocity that represents a particle moving to the right.

Stop to Think 2.5: d. The acceleration vector points downhill (negative s-direction) and has the constant value $-g \sin\theta$ throughout the motion.

Stop to Think 2.6: c. Acceleration is the slope of the graph. The slope is zero at B. Although the graph is steepest at A, the slope at that point is negative, and so $a_A < a_B$. Only C has a positive slope, so $a_C > a_B$.

3 Vectors and Coordinate Systems

Wind has both a speed and a direction, hence the motion of the wind is described by a vector.

▶ **Looking Ahead** The goal of Chapter 3 is to learn how vectors are represented and used.

Vectors

A **vector** is a quantity with both a size—the *magnitude*—and a direction.

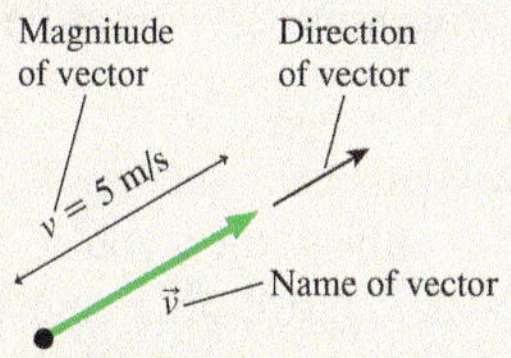

Examples of vectors that you will meet in coming chapters are:

Position	Velocity
Displacement	Acceleration
Force	Momentum

The two most basic vector operations—addition and subtraction—were introduced in Chapter 1.

◀ **Looking Back**
Tactics Box 1.1 Vector addition
Tactics Box 1.2 Vector subtraction

You may have learned in a math class to think of vectors as pairs or triplets of numbers, such as (4, −2, 5). If so, you already know a lot about vectors even though we will use a different notation in physics.

Graphical Addition and Subtraction of Vectors

You will learn to add vectors $\vec{A}$ and $\vec{B}$:

Tip-to-tail addition

Vector subtraction is addition

$$\vec{A}-\vec{B}=\vec{A}+(-\vec{B})$$

with $-\vec{B}$ defined to point opposite $\vec{B}$:

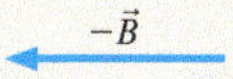

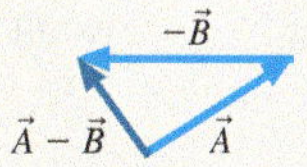

Tip-to-tail subtraction

The net displacement is the vector sum of two individual displacements.

Unit Vectors

Unit vectors define what we mean by the $+x$- and $+y$-directions in space.

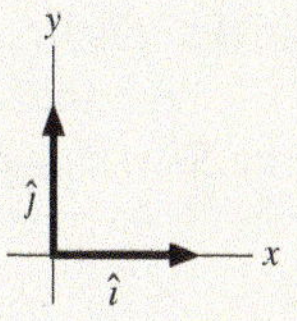

Unit vectors will be very valuable when we later use a tilted coordinate system to analyze motion on an inclined plane.

Components

You will learn how to find the *components* of vectors that are parallel to the coordinate axes. We write this as

$$\vec{E}=E_x\hat{\imath}+E_y\hat{\jmath}$$

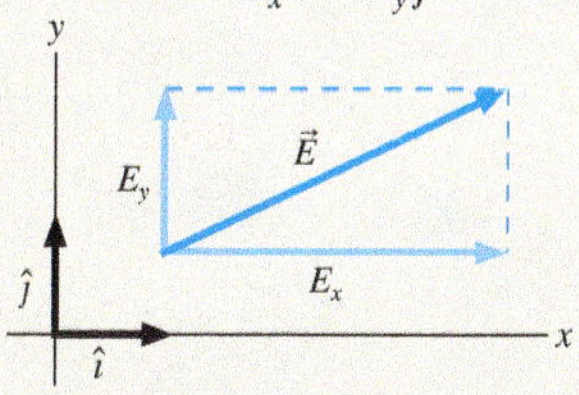

Components will simplify vector math.

3.1 Vectors

A quantity that is fully described by a single number (with units) is called a **scalar quantity.** Mass, temperature, and volume are all scalars. Other scalar quantities include pressure, density, energy, charge, and voltage. We will often use an algebraic symbol to represent a scalar quantity. Thus m will represent mass, T temperature, V volume, E energy, and so on. Notice that scalars, in printed text, are shown in italics.

Our universe has three dimensions, so some quantities also need a direction for a full description. If you ask someone for directions to the post office, the reply "Go three blocks" will not be very helpful. A full description might be, "Go three blocks south." A quantity having both a size and a direction is called a **vector quantity.**

The mathematical term for the length, or size, of a vector is **magnitude,** so we can also say that **a vector is a quantity having a magnitude and a direction.**

FIGURE 3.1 The velocity vector $\vec{v}$ has both a magnitude and a direction.

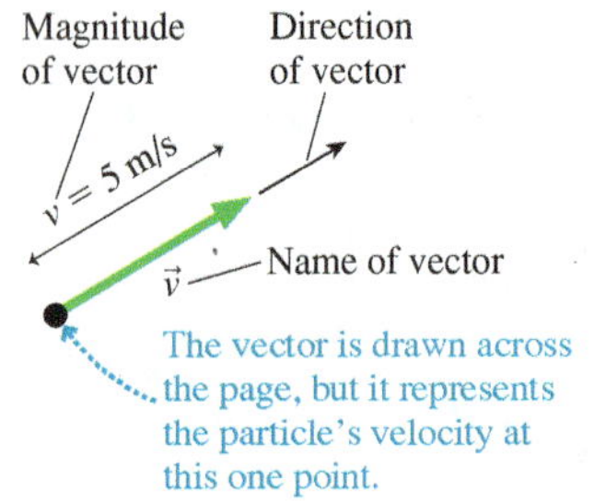

FIGURE 3.1 shows that the *geometric representation* of a vector is an arrow, with the tail of the arrow (not its tip!) placed at the point where the measurement is made. The vector then seems to radiate outward from the point to which it is attached. An arrow makes a natural representation of a vector because it inherently has both a length and a direction. As you've already seen, we label vectors by drawing a small arrow over the letter that represents the vector: $\vec{r}$ for position, $\vec{v}$ for velocity, $\vec{a}$ for acceleration, and so on.

NOTE ▶ Although the vector arrow is drawn across the page, from its tail to its tip, this does *not* indicate that the vector "stretches" across this distance. Instead, the vector arrow tells us the value of the vector quantity only at the one point where the tail of the vector is placed. ◀

The *magnitude* of a vector is sometimes shown using absolute value signs, but more frequently indicated by the letter without the arrow. For example, the magnitude of the velocity vector in Figure 3.1 is $v = |\vec{v}| = 5$ m/s. This is the object's *speed.* The magnitude of the acceleration vector $\vec{a}$ is written a. **The magnitude of a vector is a scalar quantity.**

NOTE ▶ The magnitude of a vector cannot be a negative number; it must be positive or zero, with appropriate units. ◀

It is important to get in the habit of using the arrow symbol for vectors. If you omit the vector arrow from the velocity vector $\vec{v}$ and write only v, then you're referring only to the object's speed, not its velocity. The symbols $\vec{r}$ and r, or $\vec{v}$ and v, do *not* represent the same thing, so if you omit the vector arrow from vector symbols you will soon have confusion and mistakes.

FIGURE 3.2 Displacement vectors.

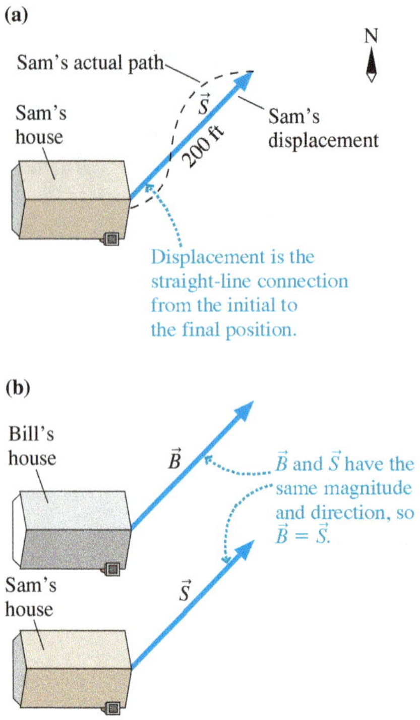

3.2 Properties of Vectors

Suppose Sam starts from his front door, walks across the street, and ends up 200 ft to the northeast of where he started. Sam's displacement, which we will label $\vec{S}$, is shown in FIGURE 3.2a. The displacement vector is a *straight-line connection* from his initial to his final position, not necessarily his actual path.

To describe a vector we must specify both its magnitude and its direction. We can write Sam's displacement as

$$\vec{S} = (200 \text{ ft, northeast})$$

where the first piece of information specifies the magnitude and the second is the direction. The magnitude of Sam's displacement is $S = |\vec{S}| = 200$ ft, the distance between his initial and final points.

Sam's next-door neighbor Bill also walks 200 ft to the northeast, starting from his own front door. Bill's displacement $\vec{B}$ = (200 ft, northeast) has the same magnitude and direction as Sam's displacement $\vec{S}$. Because vectors are defined by their magnitude and direction, **two vectors are equal if they have the same magnitude and direction.** This is true regardless of the starting points of the vectors. Thus the two displacements in FIGURE 3.2b are equal to each other, and we can write $\vec{B} = \vec{S}$.

NOTE ▶ A vector is unchanged if you move it to a different point on the page as long as you don't change its length or the direction it points. We used this idea in Chapter 1 when we moved velocity vectors around in order to find the average acceleration vector $\vec{a}$. ◀

Vector Addition

If you earn \$50 on Saturday and \$60 on Sunday, your *net* income for the weekend is the sum of \$50 and \$60. With numbers, the word *net* implies addition. The same is true with vectors. For example, FIGURE 3.3 shows the displacement of a hiker who first hikes 4 miles to the east, then 3 miles to the north. The first leg of the hike is described by the displacement $\vec{A}$ = (4 mi, east). The second leg of the hike has displacement $\vec{B}$ = (3 mi, north). Vector $\vec{C}$ is the *net displacement* because it describes the net result of the hiker's first having displacement $\vec{A}$, then displacement $\vec{B}$.

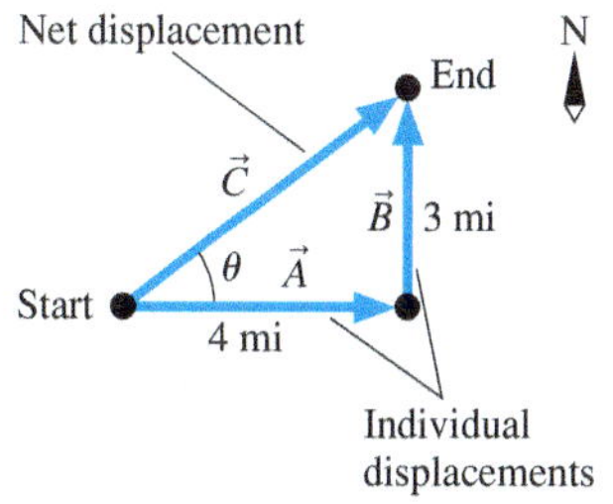

FIGURE 3.3 The net displacement $\vec{C}$ resulting from two displacements $\vec{A}$ and $\vec{B}$.

The net displacement $\vec{C}$ is an initial displacement $\vec{A}$ *plus* a second displacement $\vec{B}$, or

$$\vec{C} = \vec{A} + \vec{B} \tag{3.1}$$

The sum of two vectors is called the **resultant vector.** It's not hard to show that vector addition is commutative: $\vec{A} + \vec{B} = \vec{B} + \vec{A}$. That is, you can add vectors in any order you wish.

Look back at Tactics Box 1.1 on page 7 to see the three-step procedure for adding two vectors. This tip-to-tail method for adding vectors, which is used to find $\vec{C} = \vec{A} + \vec{B}$ in Figure 3.3, is called **graphical addition.** Any two vectors of the same type—two velocity vectors or two force vectors—can be added in exactly the same way.

The graphical method for adding vectors is straightforward, but we need to do a little geometry to come up with a complete description of the resultant vector $\vec{C}$. Vector $\vec{C}$ of Figure 3.3 is defined by its magnitude C and by its direction. Because the three vectors $\vec{A}$, $\vec{B}$, and $\vec{C}$ form a right triangle, the magnitude, or length, of $\vec{C}$ is given by the Pythagorean theorem:

$$C = \sqrt{A^2 + B^2} = \sqrt{(4\text{ mi})^2 + (3\text{ mi})^2} = 5\text{ mi} \tag{3.2}$$

Notice that Equation 3.2 uses the magnitudes A and B of the vectors $\vec{A}$ and $\vec{B}$. The angle θ, which is used in Figure 3.3 to describe the direction of $\vec{C}$, is easily found for a right triangle:

$$\theta = \tan^{-1}\left(\frac{B}{A}\right) = \tan^{-1}\left(\frac{3\text{ mi}}{4\text{ mi}}\right) = 37° \tag{3.3}$$

Altogether, the hiker's net displacement is

$$\vec{C} = \vec{A} + \vec{B} = (5\text{ mi, }37°\text{ north of east}) \tag{3.4}$$

NOTE ▶ Vector mathematics makes extensive use of geometry and trigonometry. Appendix A, at the end of this book, contains a brief review of these topics. ◀

EXAMPLE 3.1 Using graphical addition to find a displacement

A bird flies 100 m due east from a tree, then 200 m northwest (that is, 45° north of west). What is the bird's net displacement?

VISUALIZE FIGURE 3.4 shows the two individual displacements, which we've called $\vec{A}$ and $\vec{B}$. The net displacement is the vector sum $\vec{C} = \vec{A} + \vec{B}$, which is found graphically.

FIGURE 3.4 The bird's net displacement is $\vec{C} = \vec{A} + \vec{B}$.

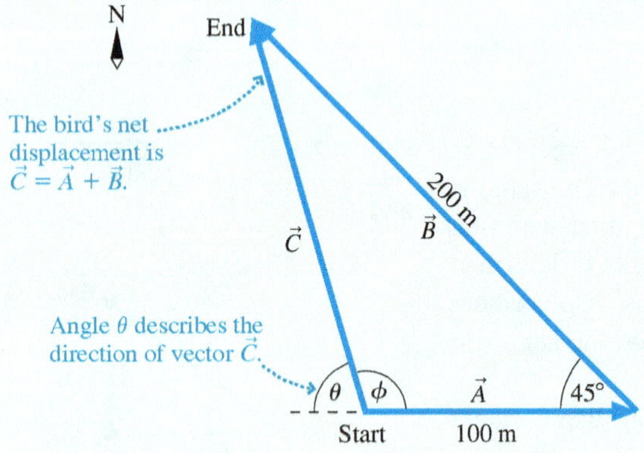

SOLVE The two displacements are $\vec{A}$ = (100 m, east) and $\vec{B}$ = (200 m, northwest). The net displacement $\vec{C} = \vec{A} + \vec{B}$ is found by drawing a vector from the initial to the final position. But describing $\vec{C}$ is a bit trickier than the example of the hiker because $\vec{A}$ and $\vec{B}$ are not at right angles. First, we can find the magnitude of $\vec{C}$ by using the law of cosines from trigonometry:

$$C^2 = A^2 + B^2 - 2AB\cos 45°$$
$$= (100\text{ m})^2 + (200\text{ m})^2 - 2(100\text{ m})(200\text{ m})\cos 45°$$
$$= 21{,}720\text{ m}^2$$

Thus $C = \sqrt{21{,}720\text{ m}^2} = 147$ m. Then a second use of the law of cosines can determine angle ϕ (the Greek letter phi):

$$B^2 = A^2 + C^2 - 2AC\cos\phi$$
$$\phi = \cos^{-1}\left[\frac{A^2 + C^2 - B^2}{2AC}\right] = 106°$$

It is easier to describe $\vec{C}$ with the angle $\theta = 180° - \phi = 74°$. The bird's net displacement is

$$\vec{C} = (147\text{ m, 74° north of west})$$

It is often convenient to draw two vectors with their tails together, as shown in FIGURE 3.5a. To evaluate $\vec{D} + \vec{E}$, you could move vector $\vec{E}$ over to where its tail is on the tip of $\vec{D}$, then use the tip-to-tail rule of graphical addition. That gives vector $\vec{F} = \vec{D} + \vec{E}$ in FIGURE 3.5b. Alternatively, FIGURE 3.5c shows that the vector sum $\vec{D} + \vec{E}$ can be found as the diagonal of the parallelogram defined by $\vec{D}$ and $\vec{E}$. This method for vector addition is called the *parallelogram rule* of vector addition.

FIGURE 3.5 Two vectors can be added using the tip-to-tail rule or the parallelogram rule.

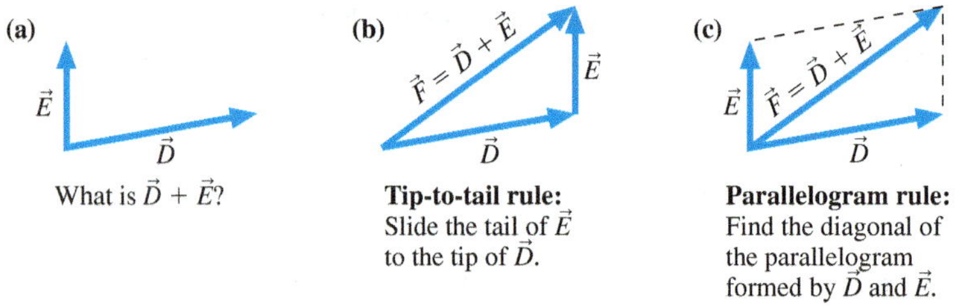

FIGURE 3.6 The net displacement after four individual displacements.

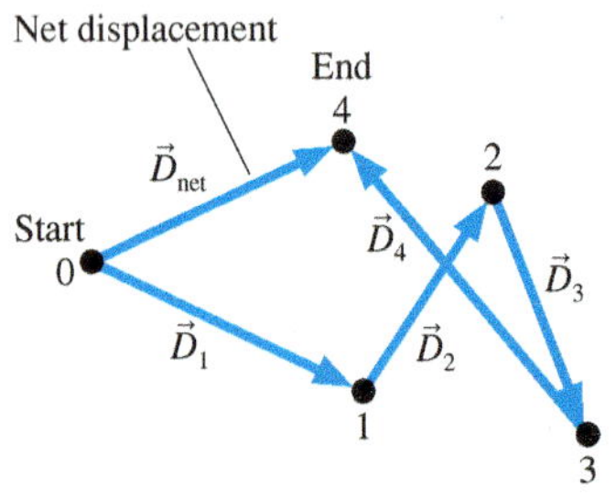

Vector addition is easily extended to more than two vectors. FIGURE 3.6 shows the path of a hiker moving from initial position 0 to position 1, then position 2, then position 3, and finally arriving at position 4. These four segments are described by displacement vectors $\vec{D}_1$, $\vec{D}_2$, $\vec{D}_3$, and $\vec{D}_4$. The hiker's *net* displacement, an arrow from position 0 to position 4, is the vector $\vec{D}_{net}$. In this case,

$$\vec{D}_{net} = \vec{D}_1 + \vec{D}_2 + \vec{D}_3 + \vec{D}_4 \tag{3.5}$$

The vector sum is found by using the tip-to-tail method three times in succession.

STOP TO THINK 3.1 Which figure shows $\vec{A}_1 + \vec{A}_2 + \vec{A}_3$?

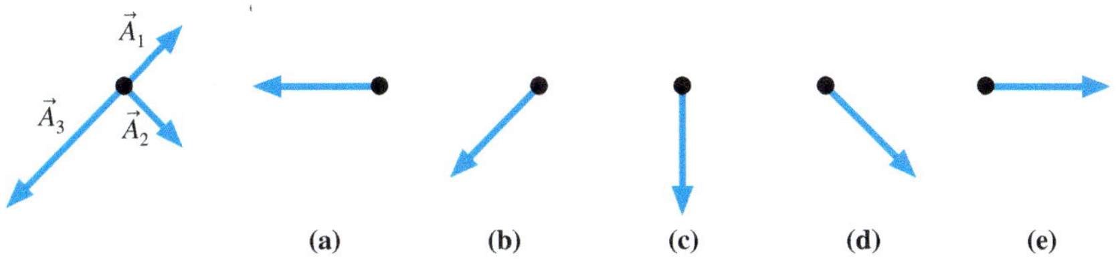

More Vector Mathematics

In addition to adding vectors, we will need to subtract vectors, multiply vectors by scalars, and understand how to interpret the negative of a vector. These operations are illustrated in FIGURE 3.7.

FIGURE 3.7 Working with vectors.

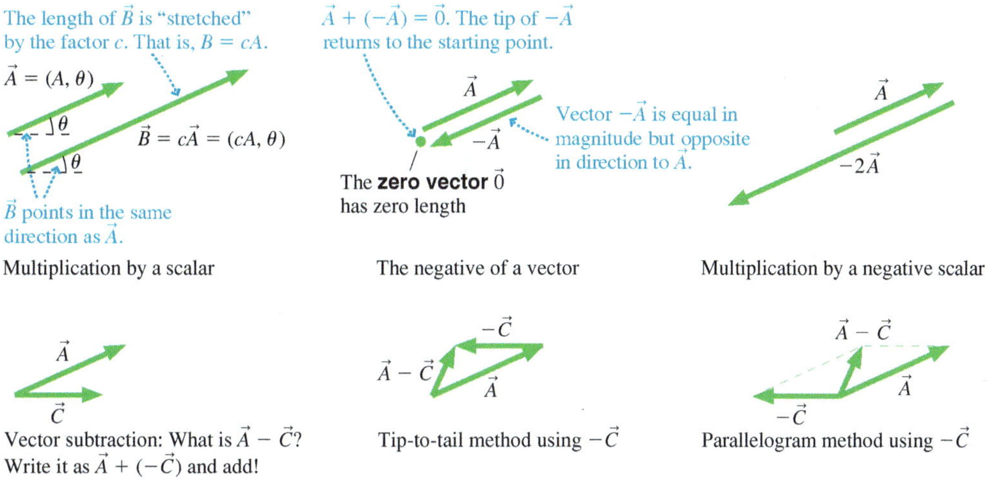

EXAMPLE 3.2 **Velocity and displacement**

Carolyn drives her car north at 30 km/h for 1 hour, east at 60 km/h for 2 hours, then north at 50 km/h for 1 hour. What is Carolyn's net displacement?

SOLVE Chapter 1 defined velocity as

$$\vec{v} = \frac{\Delta \vec{r}}{\Delta t}$$

so the displacement $\Delta \vec{r}$ during the time interval Δt is $\Delta \vec{r} = (\Delta t)\vec{v}$. This is multiplication of the vector $\vec{v}$ by the scalar Δt. Carolyn's velocity during the first hour is $\vec{v}_1 = (30 \text{ km/h, north})$, so her displacement during this interval is

$$\Delta \vec{r}_1 = (1 \text{ hour})(30 \text{ km/h, north}) = (30 \text{ km, north})$$

Similarly,

$$\Delta \vec{r}_2 = (2 \text{ hours})(60 \text{ km/h, east}) = (120 \text{ km, east})$$

$$\Delta \vec{r}_3 = (1 \text{ hour})(50 \text{ km/h, north}) = (50 \text{ km, north})$$

In this case, multiplication by a scalar changes not only the length of the vector but also its units, from km/h to km. The direction, however, is unchanged. Carolyn's net displacement is

$$\Delta \vec{r}_{net} = \Delta \vec{r}_1 + \Delta \vec{r}_2 + \Delta \vec{r}_3$$

This addition of the three vectors is shown in FIGURE 3.8, using the tip-to-tail method. $\Delta \vec{r}_{net}$ stretches from Carolyn's initial position

FIGURE 3.8 The net displacement is the vector sum $\Delta \vec{r}_{net} = \Delta \vec{r}_1 + \Delta \vec{r}_2 + \Delta \vec{r}_3$.

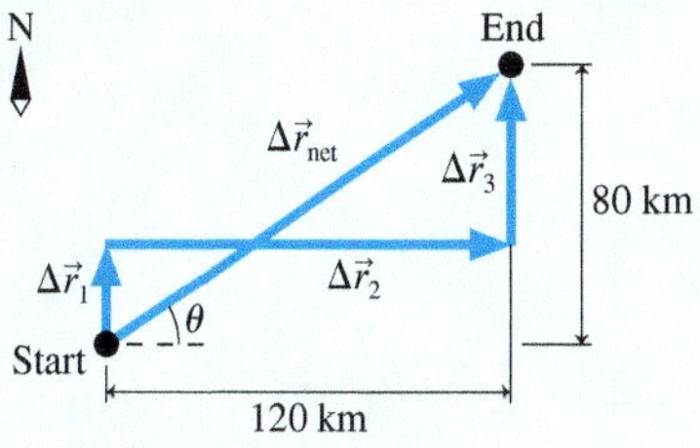

Continued

to her final position. The magnitude of her net displacement is found using the Pythagorean theorem:

$$r_{net} = \sqrt{(120\text{ km})^2 + (80\text{ km})^2} = 144\text{ km}$$

The direction of $\Delta\vec{r}_{net}$ is described by angle θ, which is

$$\theta = \tan^{-1}\left(\frac{80\text{ km}}{120\text{ km}}\right) = 34°$$

Thus Carolyn's net displacement is $\Delta\vec{r}_{net} = (144\text{ km}, 34°\text{ north of east})$.

STOP TO THINK 3.2 Which figure shows $2\vec{A} - \vec{B}$?

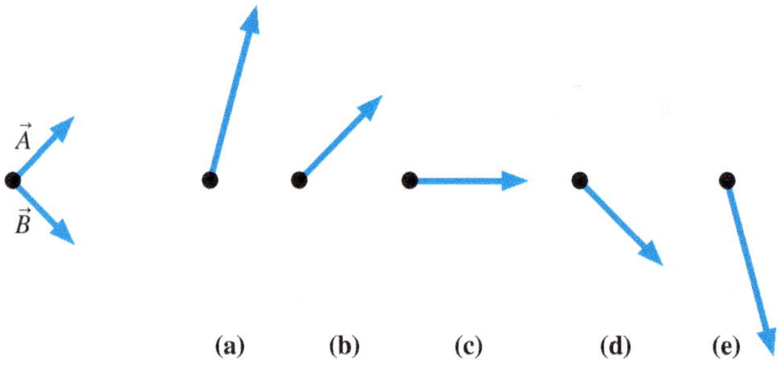

The navigator had better know which way to go, and how far, if she and the crew are to make landfall at the expected location.

3.3 Coordinate Systems and Vector Components

Vectors do not require a coordinate system. We can add and subtract vectors graphically, and we will do so frequently to clarify our understanding of a situation. But the graphical addition of vectors is not an especially good way to find quantitative results. In this section we will introduce a *coordinate representation* of vectors that will be the basis of an easier method for doing vector calculations.

Coordinate Systems

The world does not come with a coordinate system attached to it. A coordinate system is an artificially imposed grid that you place on a problem in order to make quantitative measurements. You are free to choose:

- Where to place the origin, and
- How to orient the axes.

Different problem solvers may choose to use different coordinate systems; that is perfectly acceptable. However, some coordinate systems will make a problem easier to solve. Part of our goal is to learn how to choose an appropriate coordinate system for each problem.

FIGURE 3.9 shows the xy-coordinate system we will use in this book. The placement of the axes is not entirely arbitrary. By convention, the positive y-axis is located 90° *counterclockwise* (ccw) from the positive x-axis. Figure 3.9 also identifies the four **quadrants** of the coordinate system, I through IV.

FIGURE 3.9 A conventional xy-coordinate system and the quadrants of the xy-plane.

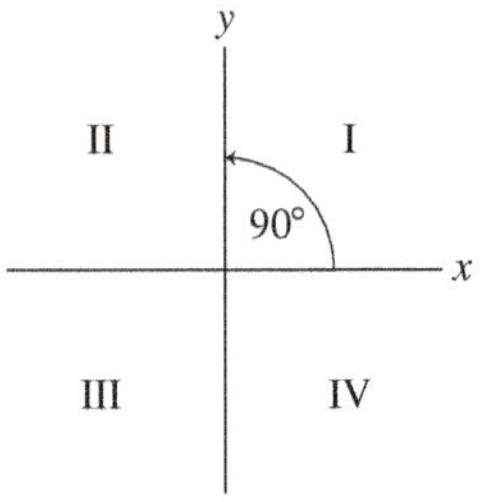

Coordinate axes have a positive end and a negative end, separated by zero at the origin where the two axes cross. When you draw a coordinate system, it is important to label the axes. This is done by placing x and y labels at the *positive* ends of the axes, as in Figure 3.9. The purpose of the labels is twofold:

- To identify which axis is which, and
- To identify the positive ends of the axes.

This will be important when you need to determine whether the quantities in a problem should be assigned positive or negative values.

Component Vectors

FIGURE 3.10 shows a vector $\vec{A}$ and an xy-coordinate system that we've chosen. Once the directions of the axes are known, we can define two new vectors *parallel to the axes* that we call the **component vectors** of $\vec{A}$. You can see, using the parallelogram rule, that $\vec{A}$ is the vector sum of the two component vectors:

$$\vec{A} = \vec{A}_x + \vec{A}_y \tag{3.6}$$

In essence, we have broken vector $\vec{A}$ into two perpendicular vectors that are parallel to the coordinate axes. This process is called the **decomposition** of vector $\vec{A}$ into its component vectors.

NOTE ▶ It is not necessary for the tail of $\vec{A}$ to be at the origin. All we need to know is the *orientation* of the coordinate system so that we can draw $\vec{A}_x$ and $\vec{A}_y$ parallel to the axes. ◀

FIGURE 3.10 Component vectors $\vec{A}_x$ and $\vec{A}_y$ are drawn parallel to the coordinate axes such that $\vec{A} = \vec{A}_x + \vec{A}_y$.

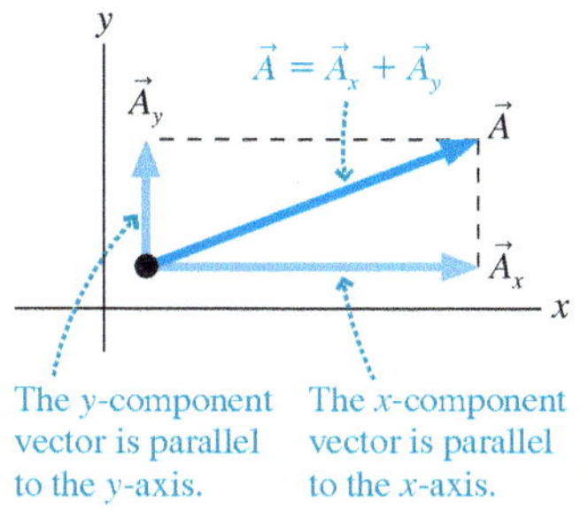

Components

You learned in Chapters 1 and 2 to give the kinematic variable v_x a positive sign if the velocity vector $\vec{v}$ points toward the positive end of the x-axis, a negative sign if $\vec{v}$ points in the negative x-direction. The basis of that rule is that v_x is what we call the *x-component* of the velocity vector. We need to extend this idea to vectors in general.

Suppose vector $\vec{A}$ has been decomposed into component vectors $\vec{A}_x$ and $\vec{A}_y$ parallel to the coordinate axes. We can describe each component vector with a single number called the **component.** The *x-component* and *y-component* of vector $\vec{A}$, denoted A_x and A_y, are determined as follows:

TACTICS BOX 3.1 Determining the components of a vector (MP)

1. The absolute value $|A_x|$ of the x-component A_x is the magnitude of the component vector $\vec{A}_x$.
2. The *sign* of A_x is positive if $\vec{A}_x$ points in the positive x-direction, negative if $\vec{A}_x$ points in the negative x-direction.
3. The y-component A_y is determined similarly.

Exercises 10–18

In other words, the component A_x tells us two things: how big $\vec{A}_x$ is and, with its sign, which end of the axis $\vec{A}_x$ points toward. FIGURE 3.11 shows three examples of determining the components of a vector.

FIGURE 3.11 Determining the components of a vector.

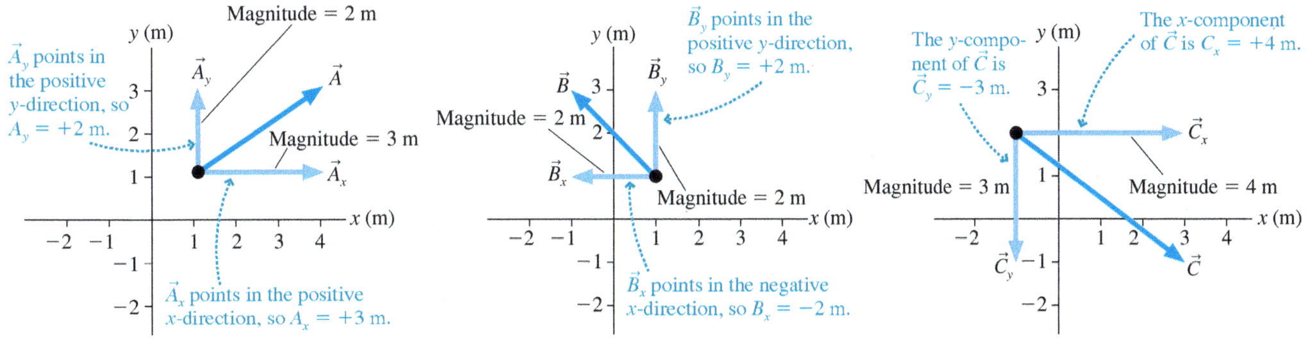

NOTE ▶ Beware of the somewhat confusing terminology. $\vec{A}_x$ and $\vec{A}_y$ are called *component vectors,* whereas A_x and A_y are simply called *components.* The components A_x and A_y are just numbers (with units), so make sure you do *not* put arrow symbols over the components. ◀

We will frequently need to decompose a vector into its components. We will also need to "reassemble" a vector from its components. In other words, we need to move back and forth between the geometric and the component representations of a vector. FIGURE 3.12 shows how this is done.

FIGURE 3.12 Moving between the geometric representation and the component representation.

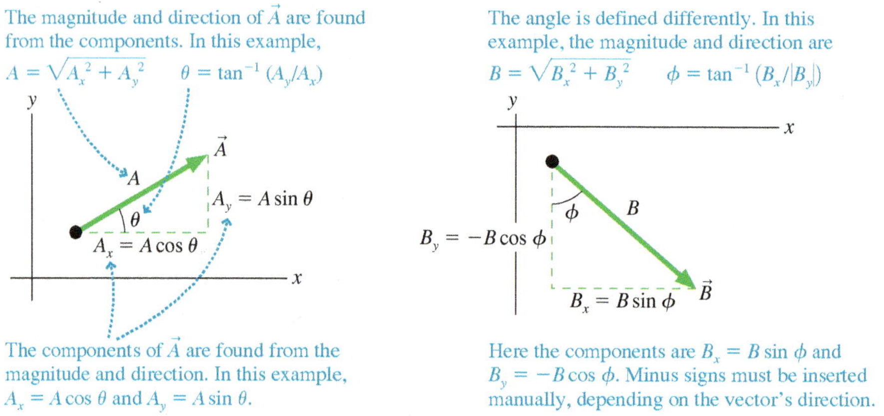

Each decomposition requires that you pay close attention to the direction in which the vector points and the angles that are defined.

- If a component vector points left (or down), you must *manually* insert a minus sign in front of the component, as was done for B_y in Figure 3.12.
- The role of sines and cosines can be reversed, depending upon which angle is used to define the direction. Compare A_x and B_x.
- The angle used to define direction is almost always between 0° and 90°, so you must take the inverse tangent of a positive number. Use absolute values of the components, as was done to find angle ϕ (Greek phi) in Figure 3.12.

EXAMPLE 3.3 Finding the components of an acceleration vector

Find the x- and y-components of the acceleration vector $\vec{a}$ shown in FIGURE 3.13.

FIGURE 3.13 The acceleration vector $\vec{a}$ of Example 3.3.

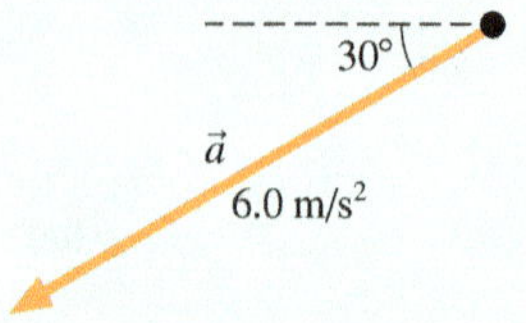

VISUALIZE It's important to *draw* vectors. FIGURE 3.14 shows the original vector $\vec{a}$ decomposed into components parallel to the axes. Notice that the axes are "acceleration axes," not xy-axes, because we're measuring an acceleration vector.

SOLVE The acceleration vector $\vec{a} = (6.0\text{ m/s}^2$, 30° below the negative x-axis) points to the left (negative x-direction) and down (negative y-direction), so the components a_x and a_y are both negative:

$$a_x = -a\cos 30° = -(6.0\text{ m/s}^2)\cos 30° = -5.2\text{ m/s}^2$$

$$a_y = -a\sin 30° = -(6.0\text{ m/s}^2)\sin 30° = -3.0\text{ m/s}^2$$

FIGURE 3.14 Decomposition of $\vec{a}$.

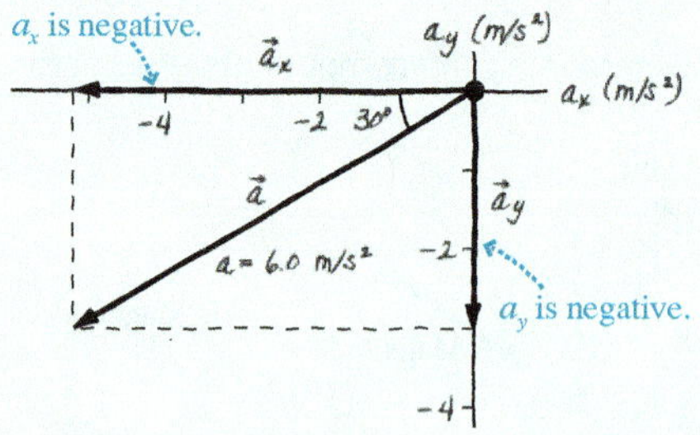

ASSESS The units of a_x and a_y are the same as the units of vector $\vec{a}$. Notice that we had to insert the minus signs manually by observing that the vector points left and down.

EXAMPLE 3.4 Finding the direction of motion

FIGURE 3.15 shows a car's velocity vector $\vec{v}$. Determine the car's speed and direction of motion.

FIGURE 3.15 The velocity vector $\vec{v}$ of Example 3.4.

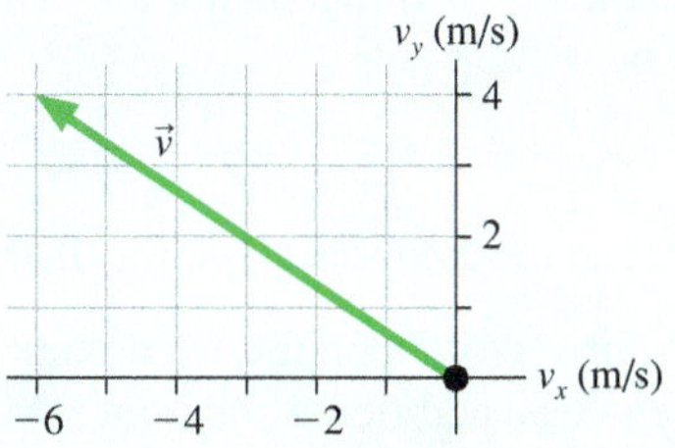

VISUALIZE FIGURE 3.16 shows the components v_x and v_y and defines an angle θ with which we can specify the direction of motion.

SOLVE We can read the components of $\vec{v}$ directly from the axes: $v_x = -6.0$ m/s and $v_y = 4.0$ m/s. Notice that v_x is negative. This is enough information to find the car's speed v, which is the magnitude of $\vec{v}$:

$$v = \sqrt{v_x^2 + v_y^2} = \sqrt{(-6.0 \text{ m/s})^2 + (4.0 \text{ m/s})^2} = 7.2 \text{ m/s}$$

FIGURE 3.16 Decomposition of $\vec{v}$.

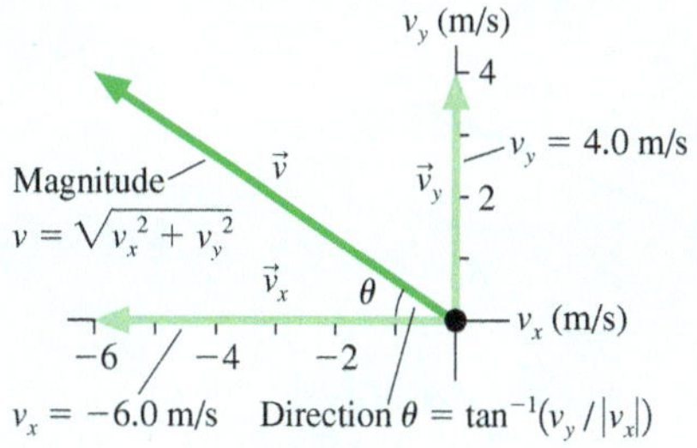

From trigonometry, angle θ is

$$\theta = \tan^{-1}\left(\frac{v_y}{|v_x|}\right) = \tan^{-1}\left(\frac{4.0 \text{ m/s}}{6.0 \text{ m/s}}\right) = 34°$$

The absolute value signs are necessary because v_x is a negative number. The velocity vector $\vec{v}$ can be written in terms of the speed and the direction of motion as

$$\vec{v} = (7.2 \text{ m/s}, 34° \text{ above the negative } x\text{-axis})$$

or, if the axes are aligned to north,

$$\vec{v} = (7.2 \text{ m/s}, 34° \text{ north of west})$$

STOP TO THINK 3.3 What are the x- and y-components C_x and C_y of vector $\vec{C}$?

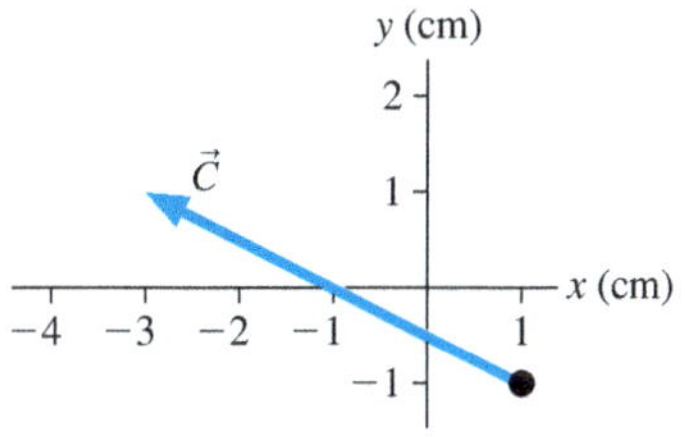

3.4 Vector Algebra

The vectors (1, $+x$-direction and (1, $+y$-direction), shown in FIGURE 3.17, have some interesting and useful properties. Each has a magnitude of 1, no units, and is parallel to a coordinate axis. A vector with these properties is called a **unit vector.** These unit vectors have the special symbols

$$\hat{\imath} \equiv (1, \text{positive } x\text{-direction})$$

$$\hat{\jmath} \equiv (1, \text{positive } y\text{-direction})$$

The notation $\hat{\imath}$ (read "i hat") and $\hat{\jmath}$ (read "j hat") indicates a unit vector with a magnitude of 1. Recall that the symbol $\equiv$ means "is defined as."

Unit vectors establish the directions of the positive axes of the coordinate system. Our choice of a coordinate system may be arbitrary, but once we decide to place a coordinate system on a problem we need something to tell us "That direction is the positive x-direction." This is what the unit vectors do.

The unit vectors provide a useful way to write component vectors. The component vector $\vec{A}_x$ is the piece of vector $\vec{A}$ that is parallel to the x-axis. Similarly, $\vec{A}_y$ is parallel

FIGURE 3.17 The unit vectors $\hat{\imath}$ and $\hat{\jmath}$.

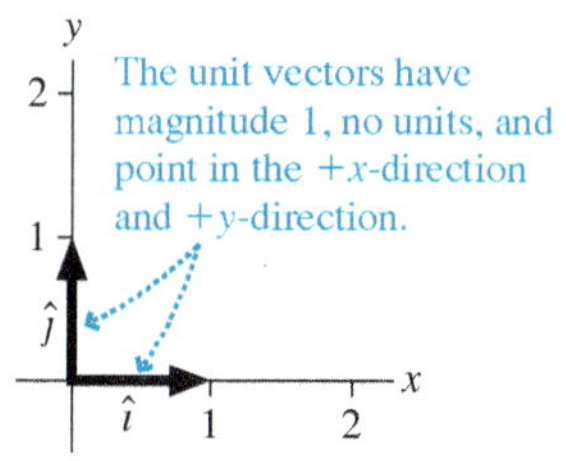

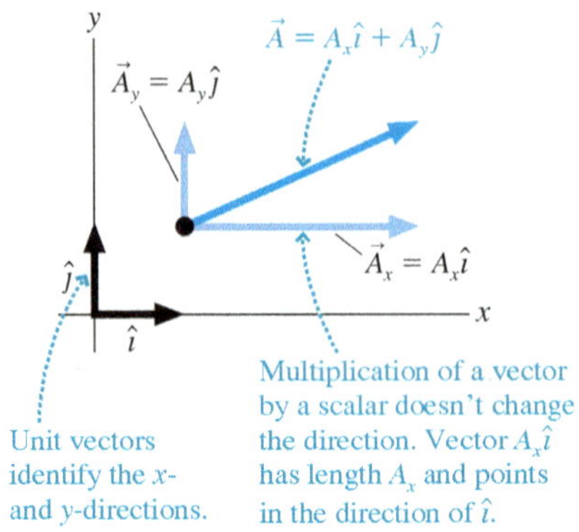

FIGURE 3.18 The decomposition of vector $\vec{A}$ is $A_x\hat{\imath} + A_y\hat{\jmath}$.

to the y-axis. Because, by definition, the vector $\hat{\imath}$ points along the x-axis and $\hat{\jmath}$ points along the y-axis, we can write

$$\vec{A}_x = A_x\hat{\imath}$$
$$\vec{A}_y = A_y\hat{\jmath} \tag{3.7}$$

Equations 3.7 separate each component vector into a length and a direction. The full decomposition of vector $\vec{A}$ can then be written

$$\vec{A} = \vec{A}_x + \vec{A}_y = A_x\hat{\imath} + A_y\hat{\jmath} \tag{3.8}$$

FIGURE 3.18 shows how the unit vectors and the components fit together to form vector $\vec{A}$.

NOTE ▶ In three dimensions, the unit vector along the $+z$-direction is called $\hat{k}$, and to describe vector $\vec{A}$ we would include an additional component vector $\vec{A}_z = A_z\hat{k}$. ◀

EXAMPLE 3.5 Run rabbit run!

A rabbit, escaping a fox, runs 40.0° north of west at 10.0 m/s. A coordinate system is established with the positive x-axis to the east and the positive y-axis to the north. Write the rabbit's velocity in terms of components and unit vectors.

VISUALIZE FIGURE 3.19 shows the rabbit's velocity vector and the coordinate axes. We're showing a velocity vector, so the axes are labeled v_x and v_y rather than x and y.

FIGURE 3.19 The velocity vector $\vec{v}$ is decomposed into components v_x and v_y.

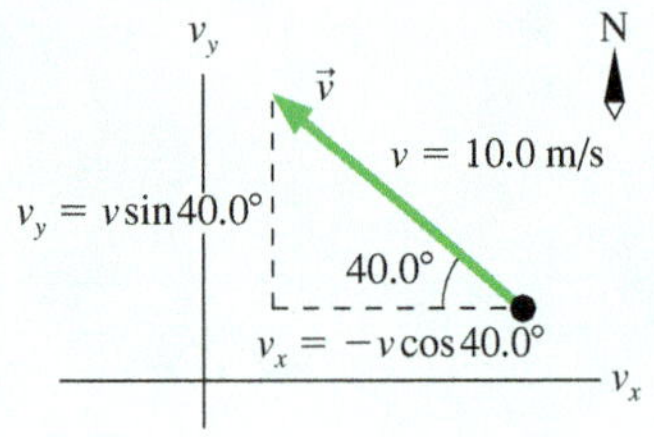

SOLVE 10.0 m/s is the rabbit's *speed,* not its velocity. The velocity, which includes directional information, is

$$\vec{v} = (10.0 \text{ m/s}, 40.0° \text{ north of west})$$

Vector $\vec{v}$ points to the left and up, so the components v_x and v_y are negative and positive, respectively. The components are

$$v_x = -(10.0 \text{ m/s})\cos 40.0° = -7.66 \text{ m/s}$$
$$v_y = +(10.0 \text{ m/s})\sin 40.0° = 6.43 \text{ m/s}$$

With v_x and v_y now known, the rabbit's velocity vector is

$$\vec{v} = v_x\hat{\imath} + v_y\hat{\jmath} = (-7.66\hat{\imath} + 6.43\hat{\jmath}) \text{ m/s}$$

Notice that we've pulled the units to the end, rather than writing them with each component.

ASSESS Notice that the minus sign for v_x was inserted manually. **Signs don't occur automatically; you have to set them after checking the vector's direction.**

Working with Vectors

You learned in Section 3.2 how to add vectors graphically, but it is a tedious problem in geometry and trigonometry to find precise values for the magnitude and direction of the resultant. The addition and subtraction of vectors become much easier if we use components and unit vectors.

To see this, let's evaluate the vector sum $\vec{D} = \vec{A} + \vec{B} + \vec{C}$. To begin, write this sum in terms of the components of each vector:

$$\begin{aligned}\vec{D} = D_x\hat{\imath} + D_y\hat{\jmath} &= \vec{A} + \vec{B} + \vec{C}\\ &= (A_x\hat{\imath} + A_y\hat{\jmath}) + (B_x\hat{\imath} + B_y\hat{\jmath}) + (C_x\hat{\imath} + C_y\hat{\jmath})\end{aligned} \tag{3.9}$$

We can group together all the x-components and all the y-components on the right side, in which case Equation 3.9 is

$$(D_x)\hat{\imath} + (D_y)\hat{\jmath} = (A_x + B_x + C_x)\hat{\imath} + (A_y + B_y + C_y)\hat{\jmath} \tag{3.10}$$

Comparing the x- and y-components on the left and right sides of Equation 3.10, we find:

$$\begin{aligned}D_x &= A_x + B_x + C_x\\ D_y &= A_y + B_y + C_y\end{aligned} \tag{3.11}$$

Stated in words, Equation 3.11 says that we can perform vector addition by adding the x-components of the individual vectors to give the x-component of the resultant and by adding the y-components of the individual vectors to give the y-component of the resultant. This method of vector addition is called **algebraic addition.**

EXAMPLE 3.6 Using algebraic addition to find a displacement

Example 3.1 was about a bird that flew 100 m to the east, then 200 m to the northwest. Use the algebraic addition of vectors to find the bird's net displacement.

VISUALIZE FIGURE 3.20 shows displacement vectors $\vec{A} = (100\text{ m, east})$ and $\vec{B} = (200\text{ m, northwest})$. We draw vectors tip-to-tail to add them graphically, but it's usually easier to draw them all from the origin if we are going to use algebraic addition.

FIGURE 3.20 The net displacement is $\vec{C} = \vec{A} + \vec{B}$.

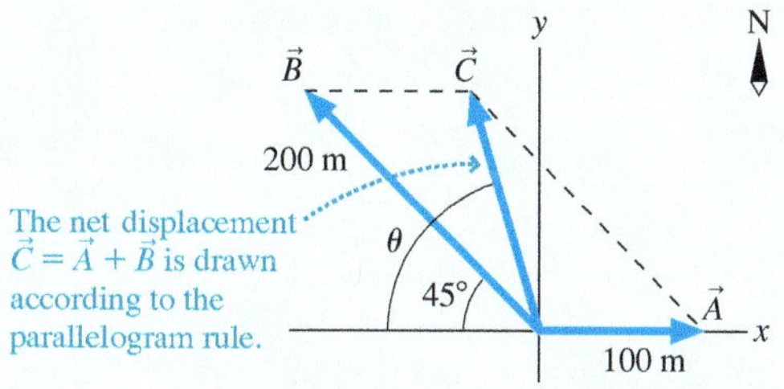

SOLVE To add the vectors algebraically we must know their components. From the figure these are seen to be

$$\vec{A} = 100\,\hat{\imath}\text{ m}$$
$$\vec{B} = (-200\cos 45°\,\hat{\imath} + 200\sin 45°\,\hat{\jmath})\text{ m}$$
$$= (-141\hat{\imath} + 141\hat{\jmath})\text{ m}$$

Notice that vector quantities must include units. Also notice, as you would expect from the figure, that $\vec{B}$ has a negative x-component. Adding $\vec{A}$ and $\vec{B}$ by components gives

$$\vec{C} = \vec{A} + \vec{B} = 100\hat{\imath}\text{ m} + (-141\hat{\imath} + 141\hat{\jmath})\text{ m}$$
$$= (100\text{ m} - 141\text{ m})\hat{\imath} + (141\text{ m})\hat{\jmath} = (-41\hat{\imath} + 141\hat{\jmath})\text{ m}$$

This would be a perfectly acceptable answer for many purposes. However, we need to calculate the magnitude and direction of $\vec{C}$ if we want to compare this result to our earlier answer. The magnitude of $\vec{C}$ is

$$C = \sqrt{C_x^2 + C_y^2} = \sqrt{(-41\text{ m})^2 + (141\text{ m})^2} = 147\text{ m}$$

The angle θ, as defined in Figure 3.20, is

$$\theta = \tan^{-1}\left(\frac{C_y}{|C_x|}\right) = \tan^{-1}\left(\frac{141\text{ m}}{41\text{ m}}\right) = 74°$$

Thus $\vec{C} = (147\text{ m, }74°\text{ north of west})$, in perfect agreement with Example 3.1.

Vector subtraction and the multiplication of a vector by a scalar, using components, are very much like vector addition. To find $\vec{R} = \vec{P} - \vec{Q}$ we would compute

$$R_x = P_x - Q_x$$
$$R_y = P_y - Q_y \qquad (3.12)$$

Similarly, $\vec{T} = c\vec{S}$ would be

$$T_x = cS_x$$
$$T_y = cS_y \qquad (3.13)$$

In other words, a vector equation is interpreted as meaning: Equate the x-components on both sides of the equals sign, then equate the y-components, and then the z-components. Vector notation allows us to write these three equations in a much more compact form.

Tilted Axes and Arbitrary Directions

As we've noted, the coordinate system is entirely your choice. It is a grid that you impose on the problem in a manner that will make the problem easiest to solve. As you've already seen in Chapter 2, it is often convenient to tilt the axes of the coordinate system, such as those shown in FIGURE 3.21. The axes are perpendicular, and the y-axis is oriented correctly with respect to the x-axis, so this is a legitimate coordinate system. There is no requirement that the x-axis has to be horizontal.

Finding components with tilted axes is no harder than what we have done so far. Vector $\vec{C}$ in Figure 3.21 can be decomposed into $\vec{C} = C_x\hat{\imath} + C_y\hat{\jmath}$, where $C_x = C\cos\theta$ and $C_y = C\sin\theta$. Note that the unit vectors $\hat{\imath}$ and $\hat{\jmath}$ correspond to the *axes,* not to "horizontal" and "vertical," so they are also tilted.

Tilted axes are useful if you need to determine component vectors "parallel to" and "perpendicular to" an arbitrary line or surface. This is illustrated in the following example.

FIGURE 3.21 A coordinate system with tilted axes.

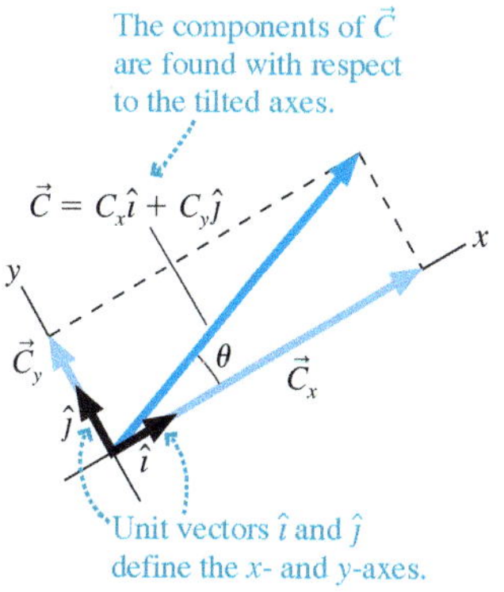

EXAMPLE 3.7 Muscle and bone

The deltoid—the rounded muscle across the top of your upper arm—allows you to lift your arm away from your side. It does so by pulling on an attachment point on the humerus, the upper arm bone, at an angle of 15° with respect to the humerus. If you hold your arm at an angle 30° below horizontal, the deltoid must pull with a force of 720 N to support the weight of your arm, as shown in FIGURE 3.22a. (You'll learn in Chapter 5 that force is a vector quantity measured in units of *newtons*, abbreviated N.) What are the components of the muscle force parallel to and perpendicular to the bone?

FIGURE 3.22 Finding the components of force parallel and perpendicular to the humerus.

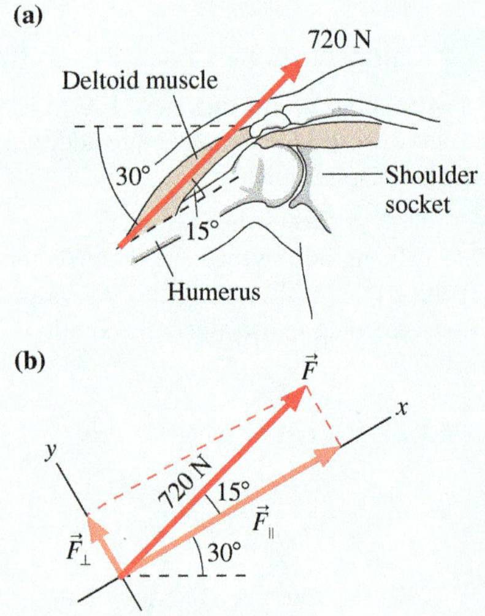

VISUALIZE FIGURE 3.22b shows a tilted coordinate system with the x-axis parallel to the humerus. The force $\vec{F}$ is shown 15° from the x-axis. The component of force parallel to the bone, which we can denote $F_{\parallel}$, is equivalent to the x-component: $F_{\parallel} = F_x$. Similarly, the component of force perpendicular to the bone is $F_{\perp} = F_y$.

SOLVE From the geometry of Figure 3.22b, we see that

$$F_{\parallel} = F\cos 15° = (720\text{ N})\cos 15° = 695\text{ N}$$

$$F_{\perp} = F\sin 15° = (720\text{ N})\sin 15° = 186\text{ N}$$

ASSESS The muscle pulls nearly parallel to the bone, so we expected $F_{\parallel} \approx 720$ N and $F_{\perp} \ll F_{\parallel}$. Thus our results seem reasonable.

STOP TO THINK 3.4 Angle ϕ that specifies the direction of $\vec{C}$ is given by

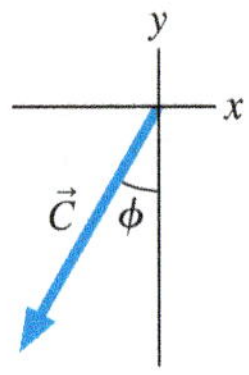

a. $\tan^{-1}(|C_x|/C_y)$
b. $\tan^{-1}(C_x/|C_y|)$
c. $\tan^{-1}(|C_x|/|C_y|)$
d. $\tan^{-1}(|C_y|/C_x)$
e. $\tan^{-1}(C_y/|C_x|)$
f. $\tan^{-1}(|C_y|/|C_x|)$

CHALLENGE EXAMPLE 3.8 Finding the net force

FIGURE 3.23 shows three forces acting at one point. What is the net force $\vec{F}_{net} = \vec{F}_1 + \vec{F}_2 + \vec{F}_3$?

FIGURE 3.23 Three forces.

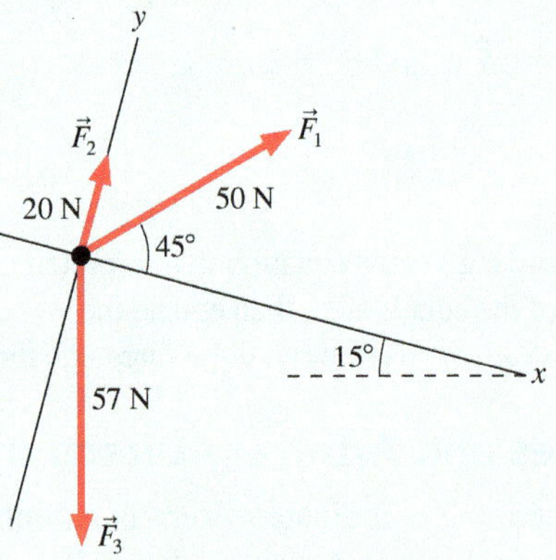

VISUALIZE Figure 3.23 show the forces and establishes a tilted coordinate system.

SOLVE The vector equation $\vec{F}_{net} = \vec{F}_1 + \vec{F}_2 + \vec{F}_3$ is really two simultaneous equations:

$$(F_{net})_x = F_{1x} + F_{2x} + F_{3x}$$

$$(F_{net})_y = F_{1y} + F_{2y} + F_{3y}$$

The components of the forces are determined with respect to the axes. Thus

$$F_{1x} = F_1\cos 45° = (50\text{ N})\cos 45° = 35\text{ N}$$

$$F_{1y} = F_1\sin 45° = (50\text{ N})\sin 45° = 35\text{ N}$$

$\vec{F}_2$ is easier. It is pointing along the y-axis, so $F_{2x} = 0$ N and $F_{2y} = 20$ N. To find the components of $\vec{F}_3$, we need to recognize—because $\vec{F}_3$ points straight down—that the angle between $\vec{F}_3$ and the x-axis is 75°. Thus

$$F_{3x} = F_3\cos 75° = (57\text{ N})\cos 75° = 15\text{ N}$$

$$F_{3y} = -F_3\sin 75° = -(57\text{ N})\sin 75° = -55\text{ N}$$

The minus sign in F_{3y} is critical, and it appears not from some formula but because we recognized—from the figure—that the y-component of $\vec{F}_3$ points in the $-y$-direction. Combining the pieces, we have

$$(F_{net})_x = 35\text{ N} + 0\text{ N} + 15\text{ N} = 50\text{ N}$$

$$(F_{net})_y = 35\text{ N} + 20\text{ N} + (-55\text{ N}) = 0\text{ N}$$

Thus the net force is $\vec{F}_{net} = 50\hat{\imath}$ N. It points along the x-axis of the tilted coordinate system.

ASSESS Notice that all work was done with reference to the axes of the coordinate system, not with respect to vertical or horizontal.

SUMMARY

The goals of Chapter 3 have been to learn how vectors are represented and used.

Important Concepts

A vector is a quantity described by both a magnitude and a direction.

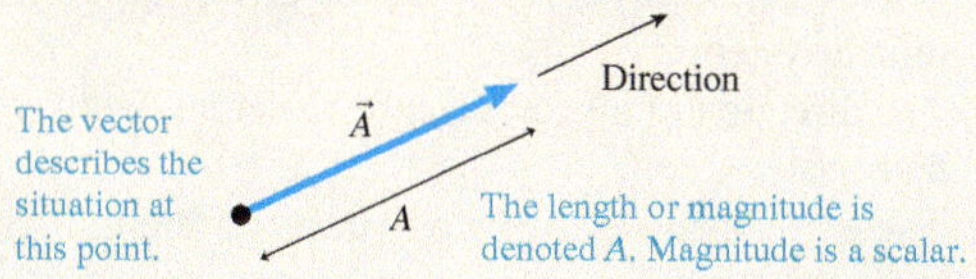

Unit Vectors

Unit vectors have magnitude 1 and no units. Unit vectors $\hat{\imath}$ and $\hat{\jmath}$ define the directions of the x- and y-axes.

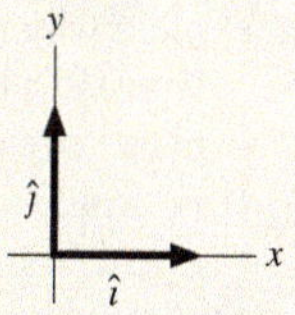

Using Vectors

Components

The component vectors are parallel to the x- and y-axes:

$$\vec{A} = \vec{A}_x + \vec{A}_y = A_x\hat{\imath} + A_y\hat{\jmath}$$

In the figure at the right, for example:

$$A_x = A\cos\theta \qquad A = \sqrt{A_x^2 + A_y^2}$$
$$A_y = A\sin\theta \qquad \theta = \tan^{-1}(A_y/A_x)$$

▶ Minus signs need to be included if the vector points down or left.

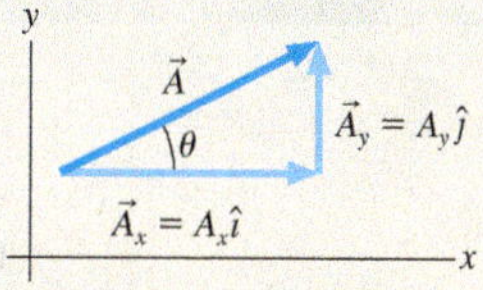

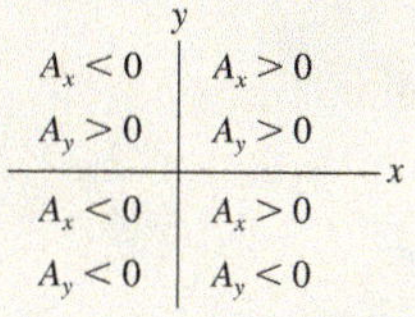

The components A_x and A_y are the magnitudes of the component vectors $\vec{A}_x$ and $\vec{A}_y$ *and* a plus or minus sign to show whether the component vector points toward the positive end or the negative end of the axis.

Working Graphically

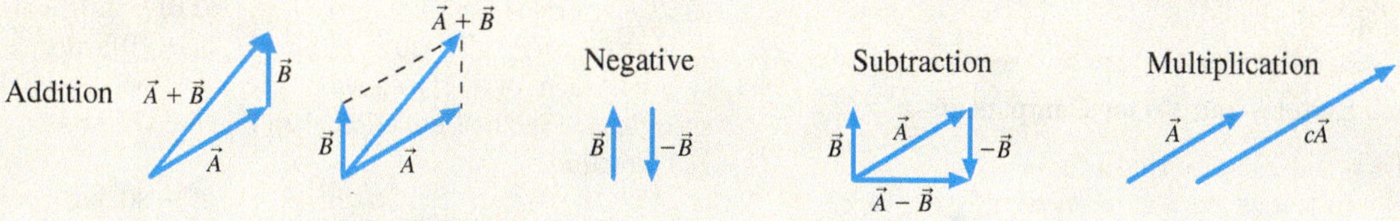

Working Algebraically

Vector calculations are done component by component:

$$\vec{C} = 2\vec{A} + \vec{B} \quad \text{means} \quad \begin{cases} C_x = 2A_x + B_x \\ C_y = 2A_y + B_y \end{cases}$$

The magnitude of $\vec{C}$ is then $C = \sqrt{C_x^2 + C_y^2}$ and its direction is found using $\tan^{-1}$.

Terms and Notation

scalar quantity
vector quantity
magnitude
resultant vector
graphical addition
zero vector, $\vec{0}$
quadrants
component vector
decomposition
component
unit vector, $\hat{\imath}$ or $\hat{\jmath}$
algebraic addition

CONCEPTUAL QUESTIONS

1. Can the magnitude of the displacement vector be more than the distance traveled? Less than the distance traveled? Explain.
2. If $\vec{C} = \vec{A} + \vec{B}$, can $C = A + B$? Can $C > A + B$? For each, show how or explain why not.
3. If $\vec{C} = \vec{A} + \vec{B}$, can $C = 0$? Can $C < 0$? For each, show how or explain why not.
4. Is it possible to add a scalar to a vector? If so, demonstrate. If not, explain why not.
5. How would you define the *zero vector* $\vec{0}$?
6. Can a vector have a component equal to zero and still have nonzero magnitude? Explain.
7. Can a vector have zero magnitude if one of its components is nonzero? Explain.
8. Suppose two vectors have unequal magnitudes. Can their sum be zero? Explain.
9. Are the following statements true or false? Explain your answer.
   a. The magnitude of a vector can be different in different coordinate systems.
   b. The direction of a vector can be different in different coordinate systems.
   c. The components of a vector can be different in different coordinate systems.

EXERCISES AND PROBLEMS

Exercises

Section 3.1 Vectors

Section 3.2 Properties of Vectors

1. | Trace the vectors in FIGURE EX3.1 onto your paper. Then find (a) $\vec{A} + \vec{B}$ and (b) $\vec{A} - \vec{B}$.

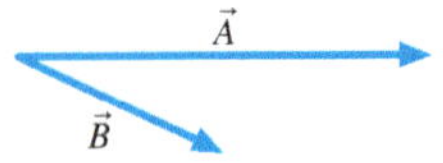

FIGURE EX3.1

FIGURE EX3.2

2. | Trace the vectors in FIGURE EX3.2 onto your paper. Then find (a) $\vec{A} + \vec{B}$ and (b) $\vec{A} - \vec{B}$.

Section 3.3 Coordinate Systems and Vector Components

3. | a. What are the x- and y-components of vector $\vec{E}$ shown in FIGURE EX3.3 in terms of the angle θ and the magnitude E?
   b. For the same vector, what are the x- and y-components in terms of the angle ϕ and the magnitude E?

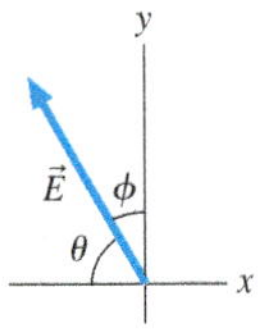

FIGURE EX3.3

4. || A velocity vector 40° below the positive x-axis has a y-component of -10 m/s. What is the value of its x-component?
5. | A position vector in the first quadrant has an x-component of 8 m and a magnitude of 10 m. What is the value of its y-component?
6. || Draw each of the following vectors, then find its x- and y-components.
   a. $\vec{r} = (100 \text{ m}, 45°$ below positive x-axis$)$
   b. $\vec{v} = (300 \text{ m/s}, 20°$ above positive x-axis$)$
   c. $\vec{a} = (5.0 \text{ m/s}^2$, negative y-direction$)$
7. || Draw each of the following vectors, then find its x- and y-components.
   a. $\vec{v} = (10 \text{ m/s}$, negative y-direction$)$
   b. $\vec{a} = (20 \text{ m/s}^2, 30°$ below positive x-axis$)$
   c. $\vec{F} = (100 \text{ N}, 36.9°$ counterclockwise from positive y-axis$)$
8. | Let $\vec{C} = (3.15 \text{ m}, 15°$ above the negative x-axis$)$ and $\vec{D} = (25.6 \text{ m}, 30°$ to the right of the negative y-axis$)$. Find the magnitude, the x-component, and the y-component of each vector.
9. | The *magnetic field* inside an instrument is $\vec{B} = (2.0\hat{\imath} - 1.0\hat{\jmath})$ T where $\vec{B}$ represents the magnetic field vector and T stands for *tesla*, the unit of the magnetic field. What are the magnitude and direction of the magnetic field?

Section 3.4 Vector Algebra

10. | Draw each of the following vectors, label an angle that specifies the vector's direction, then find its magnitude and direction.
    a. $\vec{B} = -4\hat{\imath} + 4\hat{\jmath}$
    b. $\vec{r} = (-2.0\hat{\imath} - 1.0\hat{\jmath})$ cm
    c. $\vec{v} = (-10\hat{\imath} - 100\hat{\jmath})$ m/s
    d. $\vec{a} = (20\hat{\imath} + 10\hat{\jmath})$ m/s^2
11. | Draw each of the following vectors, label an angle that specifies the vector's direction, then find the vector's magnitude and direction.
    a. $\vec{A} = 4\hat{\imath} - 6\hat{\jmath}$
    b. $\vec{r} = (50\hat{\imath} + 80\hat{\jmath})$ m
    c. $\vec{v} = (-20\hat{\imath} + 40\hat{\jmath})$ m/s
    d. $\vec{a} = (2.0\hat{\imath} - 6.0\hat{\jmath})$ m/s^2
12. | Let $\vec{A} = 2\hat{\imath} + 3\hat{\jmath}$ and $\vec{B} = 4\hat{\imath} - 2\hat{\jmath}$.
    a. Draw a coordinate system and on it show vectors $\vec{A}$ and $\vec{B}$.
    b. Use graphical vector subtraction to find $\vec{C} = \vec{A} - \vec{B}$.
13. | Let $\vec{A} = 4\hat{\imath} - 2\hat{\jmath}$, $\vec{B} = -3\hat{\imath} + 5\hat{\jmath}$, and $\vec{C} = \vec{A} + \vec{B}$.
    a. Write vector $\vec{C}$ in component form.
    b. Draw a coordinate system and on it show vectors $\vec{A}$, $\vec{B}$, and $\vec{C}$.
    c. What are the magnitude and direction of vector $\vec{C}$?
14. | Let $\vec{A} = 4\hat{\imath} - 2\hat{\jmath}$, $\vec{B} = -3\hat{\imath} + 5\hat{\jmath}$, and $\vec{D} = \vec{A} - \vec{B}$.
    a. Write vector $\vec{D}$ in component form.
    b. Draw a coordinate system and on it show vectors $\vec{A}$, $\vec{B}$, and $\vec{D}$.
    c. What are the magnitude and direction of vector $\vec{D}$?
15. | Let $\vec{A} = 4\hat{\imath} - 2\hat{\jmath}$, $\vec{B} = -3\hat{\imath} + 5\hat{\jmath}$, and $\vec{E} = 4\vec{A} + 2\vec{B}$.
    a. Write vector $\vec{E}$ in component form.
    b. Draw a coordinate system and on it show vectors $\vec{A}$, $\vec{B}$, and $\vec{E}$.
    c. What are the magnitude and direction of vector $\vec{E}$?
16. | Let $\vec{A} = 4\hat{\imath} - 2\hat{\jmath}$, $\vec{B} = -3\hat{\imath} + 5\hat{\jmath}$, and $\vec{F} = \vec{A} - 4\vec{B}$.
    a. Write vector $\vec{F}$ in component form.
    b. Draw a coordinate system and on it show vectors $\vec{A}$, $\vec{B}$, and $\vec{F}$.
    c. What are the magnitude and direction of vector $\vec{F}$?

17. | Let $\vec{B} = (5.0 \text{ m}, 60° \text{ counterclockwise from vertical})$. Find the x- and y-components of $\vec{B}$ in each of the two coordinate systems shown in FIGURE EX3.17.

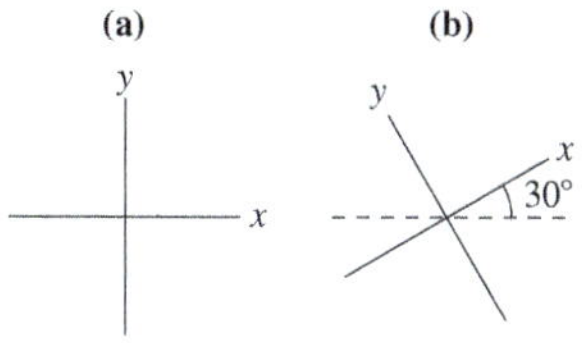

FIGURE EX3.17

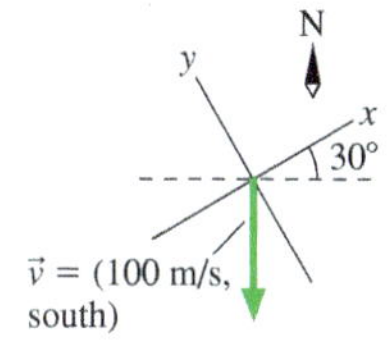

FIGURE EX3.18

18. | What are the x- and y-components of the velocity vector shown in FIGURE EX3.18?

Problems

19. || Let $\vec{A} = (3.0 \text{ m}, 20° \text{ south of east})$, $\vec{B} = (2.0 \text{ m}, \text{north})$, and $\vec{C} = (5.0 \text{ m}, 70° \text{ south of west})$.
   a. Draw and label $\vec{A}$, $\vec{B}$, and $\vec{C}$ with their tails at the origin. Use a coordinate system with the x-axis to the east.
   b. Write $\vec{A}$, $\vec{B}$, and $\vec{C}$ in component form, using unit vectors.
   c. Find the magnitude and the direction of $\vec{D} = \vec{A} + \vec{B} + \vec{C}$.

20. | Let $\vec{E} = 2\hat{\imath} + 3\hat{\jmath}$ and $\vec{F} = 2\hat{\imath} - 2\hat{\jmath}$. Find the magnitude of
   a. $\vec{E}$ and $\vec{F}$ b. $\vec{E} + \vec{F}$ c. $-\vec{E} - 2\vec{F}$

21. | The position of a particle as a function of time is given by $\vec{r} = (5.0\hat{\imath} + 4.0\hat{\jmath})t^2$ m, where t is in seconds.
   a. What is the particle's distance from the origin at $t = 0$, 2, and 5 s?
   b. Find an expression for the particle's velocity $\vec{v}$ as a function of time.
   c. What is the particle's speed at $t = 0$, 2, and 5 s?

22. || FIGURE P3.22 shows vectors $\vec{A}$ and $\vec{B}$. Let $\vec{C} = \vec{A} + \vec{B}$.

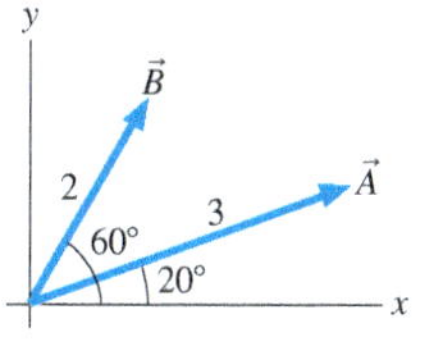

FIGURE P3.22

   a. Reproduce the figure on your page as accurately as possible, using a ruler and protractor. Draw vector $\vec{C}$ on your figure, using the graphical addition of $\vec{A}$ and $\vec{B}$. Then determine the magnitude and direction of $\vec{C}$ by *measuring* it with a ruler and protractor.
   b. Based on your figure of part a, use geometry and trigonometry to *calculate* the magnitude and direction of $\vec{C}$.
   c. Decompose vectors $\vec{A}$ and $\vec{B}$ into components, then use these to calculate algebraically the magnitude and direction of $\vec{C}$.

23. || For the three vectors shown in FIGURE P3.23, $\vec{A} + \vec{B} + \vec{C} = 1\hat{\jmath}$. What is vector $\vec{B}$?
   a. Write $\vec{B}$ in component form.
   b. Write $\vec{B}$ as a magnitude and a direction.

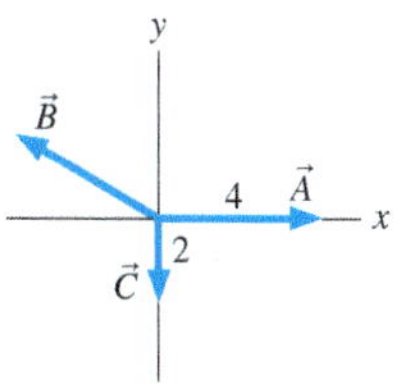

FIGURE P3.23

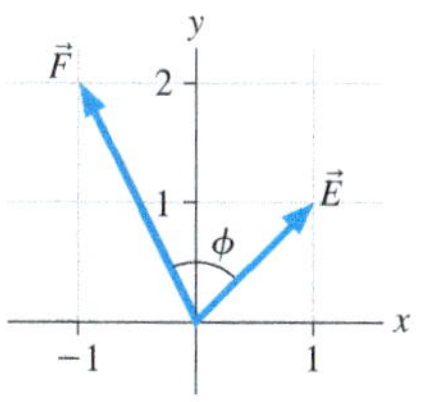

FIGURE P3.24

24. | a. What is the angle ϕ between vectors $\vec{E}$ and $\vec{F}$ in FIGURE P3.24?
   b. Use geometry and trigonometry to determine the magnitude and direction of $\vec{G} = \vec{E} + \vec{F}$.
   c. Use components to determine the magnitude and direction of $\vec{G} = \vec{E} + \vec{F}$.

25. || FIGURE P3.25 shows vectors $\vec{A}$ and $\vec{B}$. Find vector $\vec{C}$ such that $\vec{A} + \vec{B} + \vec{C} = \vec{0}$. Write your answer in component form.

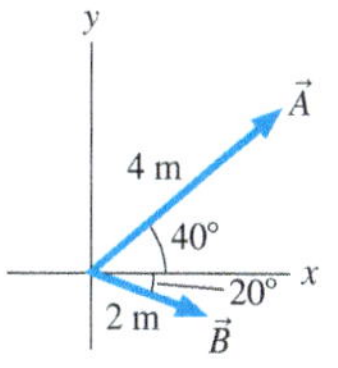

FIGURE P3.25

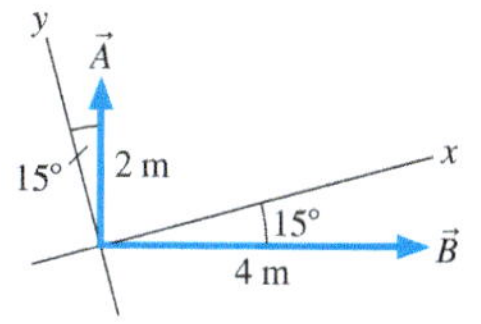

FIGURE P3.26

26. ||| FIGURE P3.26 shows vectors $\vec{A}$ and $\vec{B}$. Find $\vec{D} = 2\vec{A} + \vec{B}$. Write your answer in component form.

27. || Find a vector that points in the same direction as the vector $(\hat{\imath} + \hat{\jmath})$ and whose magnitude is 1.

28. || Carlos runs with velocity $\vec{v} = (5.0 \text{ m/s}, 25° \text{ north of east})$ for 10 minutes. How far to the north of his starting position does Carlos end up?

29. || While vacationing in the mountains you do some hiking. In the morning, your displacement is $\vec{S}_{\text{morning}} = (2000 \text{ m, east}) + (3000 \text{ m, north}) + (200 \text{ m, vertical})$. After lunch, your displacement is $\vec{S}_{\text{afternoon}} = (1500 \text{ m, west}) + (2000 \text{ m, north}) - (300 \text{ m, vertical})$.
   a. At the end of the hike, how much higher or lower are you compared to your starting point?
   b. What is the magnitude of your net displacement for the day?

30. || The minute hand on a watch is 2.0 cm in length. What is the displacement vector of the tip of the minute hand
   a. From 8:00 to 8:20 A.M.?
   b. From 8:00 to 9:00 A.M.?

31. || Bob walks 200 m south, then jogs 400 m southwest, then walks 200 m in a direction 30° east of north.
   a. Draw an accurate graphical representation of Bob's motion. Use a ruler and a protractor!
   b. Use either trigonometry or components to find the displacement that will return Bob to his starting point by the most direct route. Give your answer as a distance and a direction.
   c. Does your answer to part b agree with what you can measure on your diagram of part a?

32. || Jim's dog Sparky runs 50 m northeast to a tree, then 70 m west to a second tree, and finally 20 m south to a third tree.
   a. Draw a picture and establish a coordinate system.
   b. Calculate Sparky's net displacement in component form.
   c. Calculate Sparky's net displacement as a magnitude and an angle.

33. || A field mouse trying to escape a hawk runs east for 5.0 m, darts southeast for 3.0 m, then drops 1.0 m straight down a hole into its burrow. What is the magnitude of its net displacement?

34. | A cannon tilted upward at 30° fires a cannonball with a speed of 100 m/s. What is the component of the cannonball's velocity parallel to the ground?

35. | Jack and Jill ran up the hill at 3.0 m/s. The horizontal component of Jill's velocity vector was 2.5 m/s.
   a. What was the angle of the hill?
   b. What was the vertical component of Jill's velocity?

36. | A pine cone falls straight down from a pine tree growing on a 20° slope. The pine cone hits the ground with a speed of 10 m/s. What is the component of the pine cone's impact velocity (a) parallel to the ground and (b) perpendicular to the ground?

37. ‖ Mary needs to row her boat across a 100-m-wide river that is flowing to the east at a speed of 1.0 m/s. Mary can row the boat with a speed of 2.0 m/s relative to the water.
   a. If Mary rows straight north, how far downstream will she land?
   b. Draw a picture showing Mary's displacement due to rowing, her displacement due to the river's motion, and her net displacement.

38. ‖ The treasure map in FIGURE P3.38 gives the following directions to the buried treasure: "Start at the old oak tree, walk due north for 500 paces, then due east for 100 paces. Dig." But when you arrive, you find an angry dragon just north of the tree. To avoid the dragon, you set off along the yellow brick road at an angle 60° east of north. After walking 300 paces you see an opening through the woods. Which direction should you go, and how far, to reach the treasure?

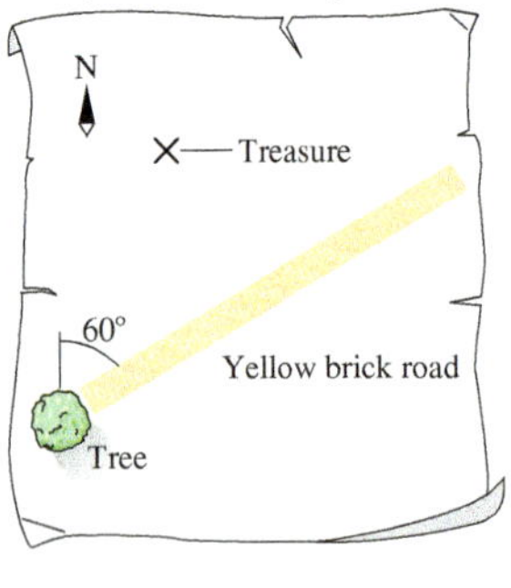

FIGURE P3.38

39. ‖ A jet plane is flying horizontally with a speed of 500 m/s over a hill that slopes upward with a 3% grade (i.e., the "rise" is 3% of the "run"). What is the component of the plane's velocity perpendicular to the ground?

40. ‖ BIO The bacterium *E. coli* is a single-cell organism that lives in the gut of healthy animals, including humans. When grown in a uniform medium in the laboratory, these bacteria swim along zigzag paths at a constant speed of 20 μm/s. FIGURE P3.40 shows the trajectory of an *E. coli* as it moves from point A to point E. What are the magnitude and direction of the bacterium's average velocity for the entire trip?

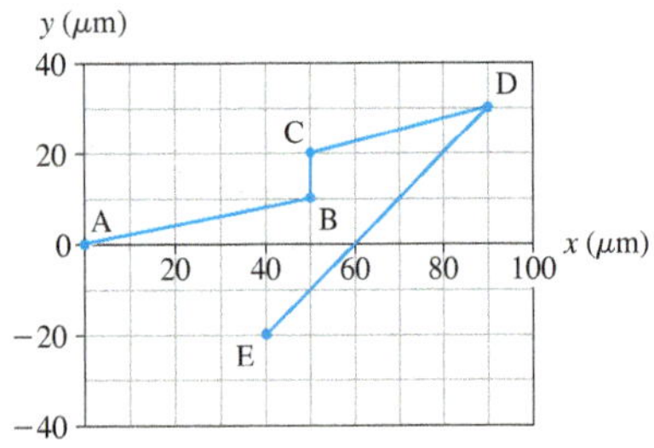

FIGURE P3.40

41. ‖ A flock of ducks is trying to migrate south for the winter, but they keep being blown off course by a wind blowing from the west at 6.0 m/s. A wise elder duck finally realizes that the solution is to fly at an angle to the wind. If the ducks can fly at 8.0 m/s relative to the air, what direction should they head in order to move directly south?

42. ‖ FIGURE P3.42 shows three ropes tied together in a knot. One of your friends pulls on a rope with 3.0 units of force and another pulls on a second rope with 5.0 units of force. How hard and in what direction must you pull on the third rope to keep the knot from moving?

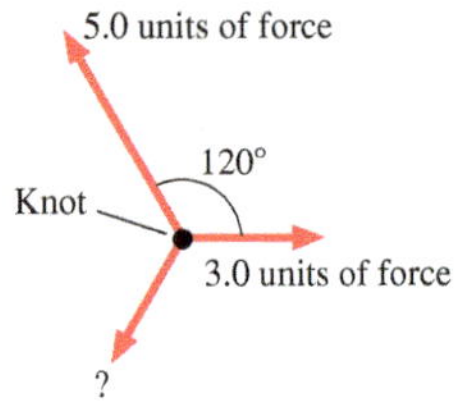

FIGURE P3.42

43. ‖ Three forces are exerted on an object placed on a tilted floor in FIGURE P3.43. The forces are measured in newtons (N). Assuming that forces are vectors,
   a. What is the component of the *net force* $\vec{F}_{net} = \vec{F}_1 + \vec{F}_2 + \vec{F}_3$ parallel to the floor?
   b. What is the component of $\vec{F}_{net}$ perpendicular to the floor?
   c. What are the magnitude and direction of $\vec{F}_{net}$?

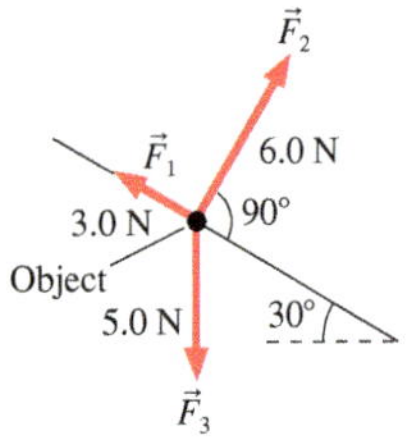

FIGURE P3.43

44. ‖ FIGURE P3.44 shows four electric charges located at the corners of a rectangle. Like charges, you will recall, repel each other while opposite charges attract. Charge B exerts a repulsive force (directly *away from* B) on charge A of 3.0 N. Charge C exerts an attractive force (directly *toward* C) on charge A of 6.0 N. Finally, charge D exerts an attractive force of 2.0 N on charge A. Assuming that forces are vectors, what are the magnitude and direction of the net force $\vec{F}_{net}$ exerted on charge A?

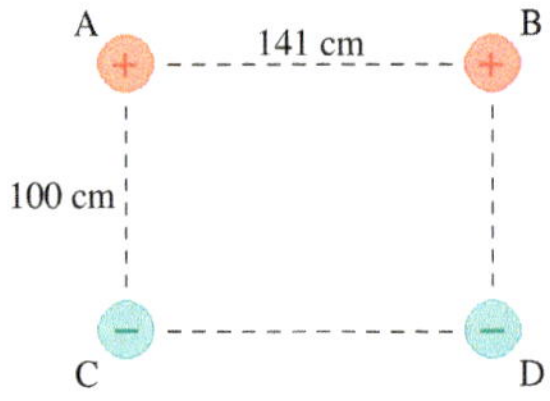

FIGURE P3.44

STOP TO THINK ANSWERS

Stop to Think 3.1: C. The graphical construction of $\vec{A}_1 + \vec{A}_2 + \vec{A}_3$ is shown at right.

Stop to Think 3.2: a. The graphical construction of $2\vec{A} - \vec{B}$ is shown at right.

Stop to Think 3.3: $C_x = -4$ **cm,** $C_y = 2$ **cm.**

Stop to Think 3.4: c. Vector $\vec{C}$ points to the left and down, so both C_x and C_y are negative. C_x is in the numerator because it is the side opposite ϕ.

STOP TO THINK 3.1

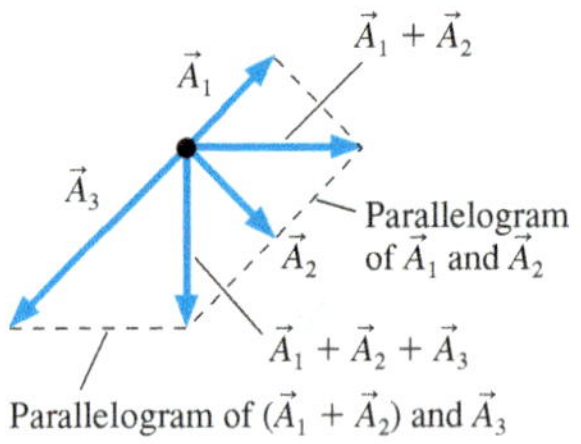

STOP TO THINK 3.2

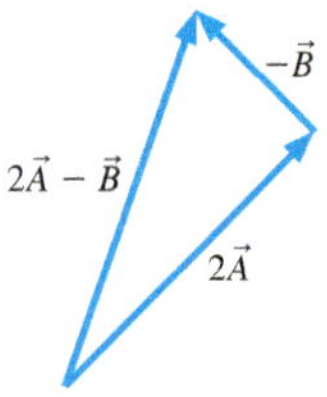

4 Kinematics in Two Dimensions

The water droplets are following the parabolic trajectories of projectile motion.

▶ **Looking Ahead** The goal of Chapter 4 is to learn how to solve problems about motion in a plane.

Two-Dimensional Motion

An object moving in two dimensions follows a **trajectory**. The object's acceleration has a component associated with changing speed *and* a component associated with changing direction. The latter is perpendicular to the direction of motion.

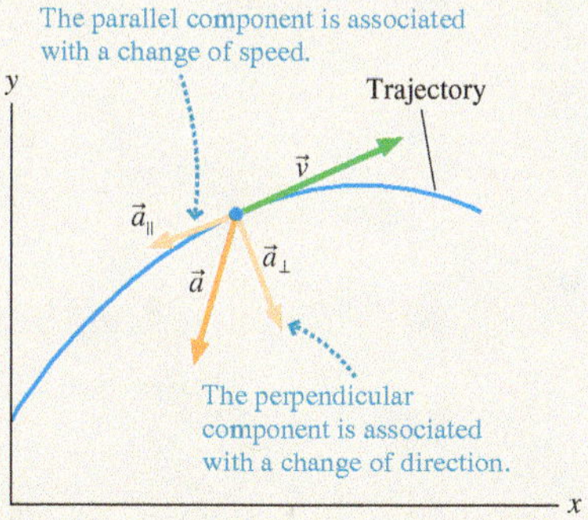

You will learn to extend the motion diagrams of Chapter 1 and the kinematics of Chapter 2 to motion in two dimensions.

◀ **Looking Back**
Section 1.5 Finding acceleration vectors on a motion diagram

◀ **Looking Back**
Sections 2.5–2.6 Constant acceleration kinematics and free fall

Projectile Motion

Projectile motion is two-dimensional motion under the influence of only gravity.

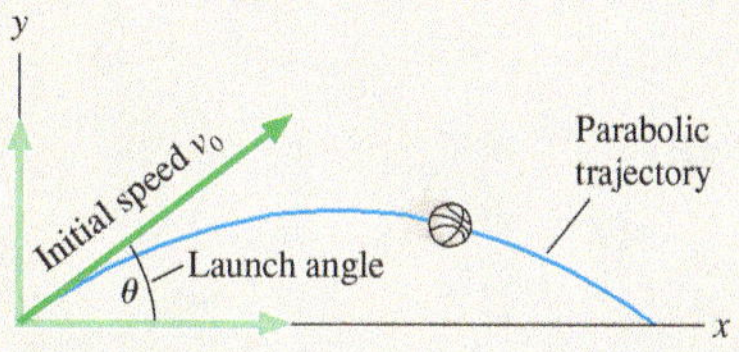

Projectile motion follows a *parabolic trajectory* characterized by the initial speed and the launch angle.

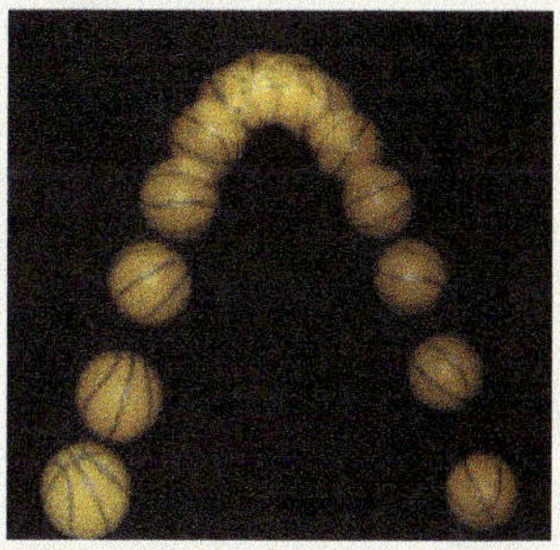

You'll learn to calculate how high and how far a projectile travels.

Circular Motion

Uniform circular motion is circular motion with constant speed. Because the direction is changing, there is a *centripetal acceleration* pointing toward the center of the circle.

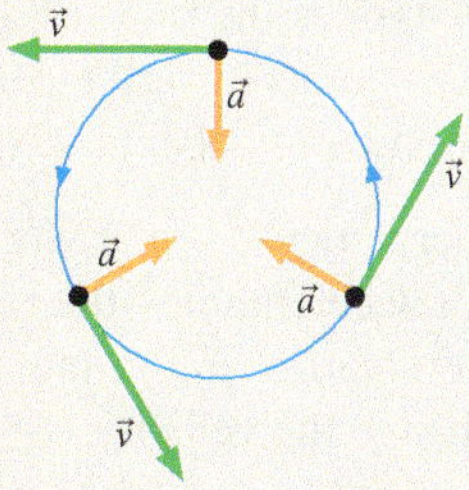

You'll learn that circular motion can be described by angular position θ, angular velocity ω, and angular acceleration α. These are analogous to the familiar linear position x, velocity v, and acceleration a.

The London Eye is a stately example of circular motion.

4.1 Acceleration

In Chapter 1 we defined the *average acceleration* $\vec{a}_{avg}$ of a moving object to be the vector

$$\vec{a}_{avg} = \frac{\Delta\vec{v}}{\Delta t} \tag{4.1}$$

From its definition, we see that $\vec{a}$ **points in the same direction as** $\Delta\vec{v}$, the change of velocity. As an object moves, its velocity vector can change in two possible ways:

1. The magnitude of $\vec{v}$ can change, indicating a change in speed, or
2. The direction of $\vec{v}$ can change, indicating that the object has changed direction.

The kinematics of Chapter 2 considered only the acceleration of changing speed. Now it's time to look at the acceleration associated with changing direction. Tactics Box 4.1 shows how we can use the velocity vectors on a motion diagram to determine the direction of the average acceleration vector. This is an extension of Tactics Box 1.3, which showed how to find $\vec{a}$ for one-dimensional motion.

FIGURE 4.1 Using Tactics Box 4.1 to find Maria's acceleration on the Ferris wheel.

(a)

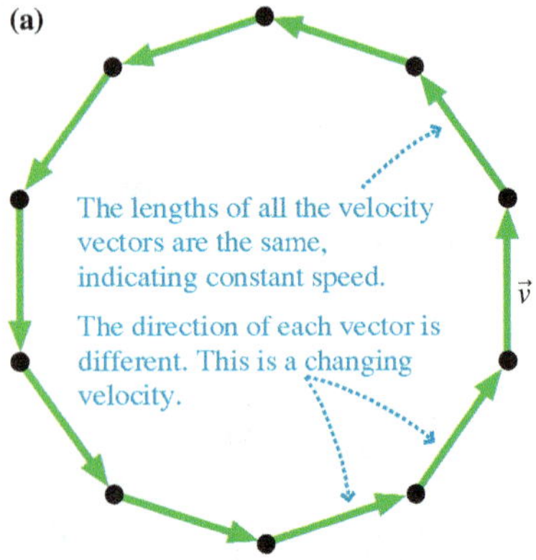

Maria moves at constant speed but *not* at constant velocity. Thus she is accelerating.

(b)

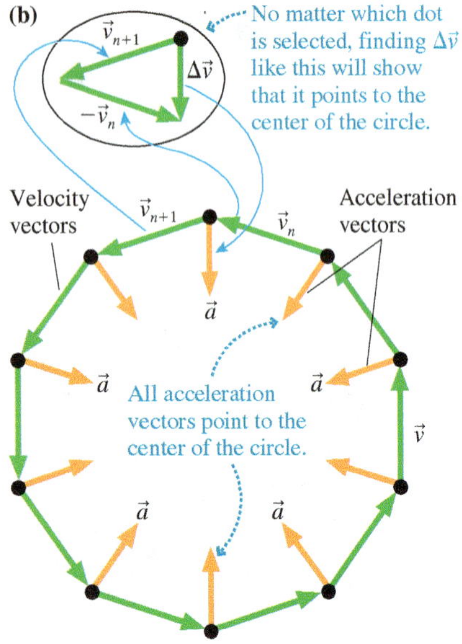

Maria's acceleration is an acceleration of changing direction, not of changing speed.

TACTICS BOX 4.1 Finding the acceleration vector

To find the acceleration between velocity $\vec{v}_n$ and velocity $\vec{v}_{n+1}$:

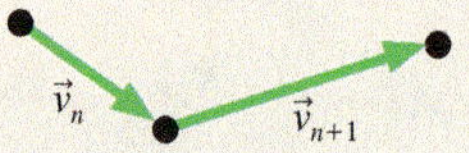

❶ Draw the velocity vector $\vec{v}_{n+1}$.

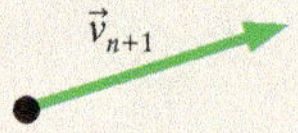

❷ Draw $-\vec{v}_n$ at the tip of $\vec{v}_{n+1}$.

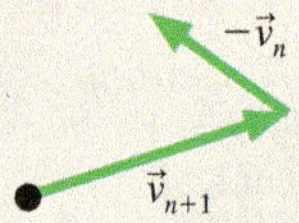

❸ Draw $\Delta\vec{v} = \vec{v}_{n+1} - \vec{v}_n$
$= \vec{v}_{n+1} + (-\vec{v}_n)$
This is the direction of $\vec{a}$.

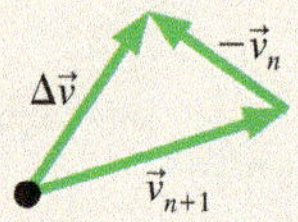

❹ Return to the original motion diagram. Draw a vector at the middle point in the direction of $\Delta\vec{v}$; label it $\vec{a}$. This is the average acceleration between $\vec{v}_n$ and $\vec{v}_{n+1}$.

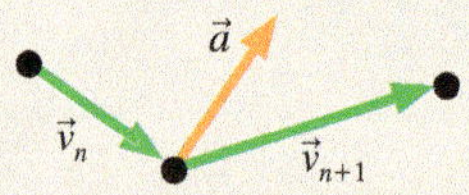

Exercises 1–4

To illustrate, FIGURE 4.1a shows a motion diagram of Maria riding a Ferris wheel at the amusement park. Maria has constant speed but *not* constant velocity, so she is accelerating. FIGURE 4.1b applies the rules of Tactics Box 4.1 to find that—at every point—Maria's acceleration points toward the center of the circle. This is an acceleration due to changing direction, not to changing speed.

NOTE ▶ Our everyday use of the word "accelerate" means "speed up." The technical definition of acceleration—the rate of change of velocity—also includes slowing down, as you learned in Chapter 2, as well as changing direction. All these are motions that change the velocity. ◀

EXAMPLE 4.1 Through the valley

A ball rolls down a long hill, through the valley, and back up the other side. Draw a complete motion diagram of the ball, showing velocity and acceleration vectors.

MODEL Model the ball as a particle.

VISUALIZE FIGURE 4.2 is the motion diagram. Where the particle moves along a *straight line,* it speeds up if $\vec{a}$ and $\vec{v}$ point in the same direction and slows down if $\vec{a}$ and $\vec{v}$ point in opposite directions. This idea was the basis for the one-dimensional kinematics we developed in Chapter 2. For linear motion, acceleration is a change of speed. When the direction of $\vec{v}$ changes, as it does when the ball goes through the valley, we need to use vector subtraction to find the direction of $\Delta\vec{v}$ and thus of $\vec{a}$. The procedure is shown at two points in the motion diagram. Notice that the point at the bottom of the valley is much like the bottom point of Maria's motion diagram in Figure 4.1b.

FIGURE 4.2 The motion diagram of the ball of Example 4.1.

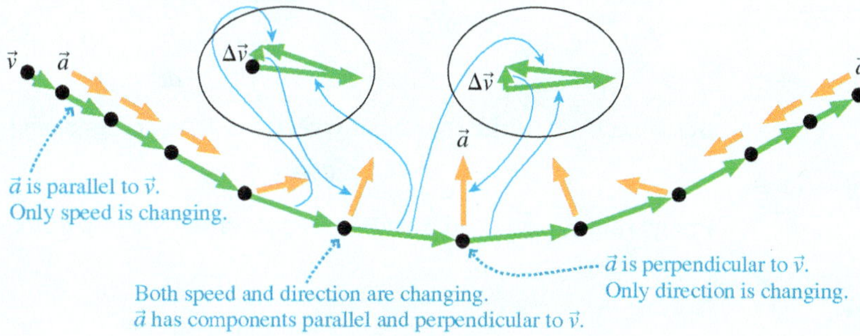

FIGURE 4.3 shows that an object's acceleration vector can be decomposed into a component parallel to the velocity—that is, parallel to the direction of motion—and a component perpendicular to the velocity. **$\vec{a}_\parallel$ is the piece of the acceleration that causes the object to change speed**, speeding up if $\vec{a}_\parallel$ points in the same direction as $\vec{v}$, slowing down if they point in opposite directions. **$\vec{a}_\perp$ is the piece of the acceleration that causes the object to change direction.** An object changing direction *always* has a component of acceleration perpendicular to the direction of motion.

Looking back at Example 4.1, we see that $\vec{a}$ is parallel to $\vec{v}$ on the straight portions of the hill where only speed is changing. At the very bottom, where the ball's direction is changing but not its speed, $\vec{a}$ is perpendicular to $\vec{v}$. The acceleration is angled with respect to velocity—having both parallel and perpendicular components—at those points where both speed and direction are changing.

FIGURE 4.3 Analyzing the acceleration vector.

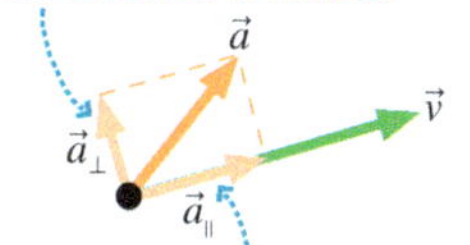

STOP TO THINK 4.1 This acceleration will cause the particle to

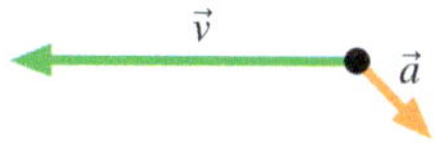

a. Speed up and curve upward.
b. Speed up and curve downward.
c. Slow down and curve upward.
d. Slow down and curve downward.
e. Move to the right and down.
f. Reverse direction.

4.2 Two-Dimensional Kinematics

Motion diagrams are an important tool for visualizing motion, but we also need to develop a mathematical description of motion in two dimensions. We're going to begin with motion in which the horizontal and vertical components of acceleration are independent of each other. For convenience, we'll say that the motion is in the *xy*-plane regardless of whether the plane of motion is horizontal or vertical.

FIGURE 4.4 shows a particle moving along a curved path—its *trajectory*—in the *xy*-plane. We can locate the particle in terms of its position vector $\vec{r} = x\hat{\imath} + y\hat{\jmath}$.

NOTE ▶ In Chapter 2 we made extensive use of position-versus-time graphs, either x versus t or y versus t. Figure 4.4, like many of the graphs we'll use in this chapter, is a graph of y versus x. In other words, it's an actual *picture* of the trajectory, not an abstract representation of the motion. ◀

FIGURE 4.4 A particle moving along a trajectory in the *xy*-plane.

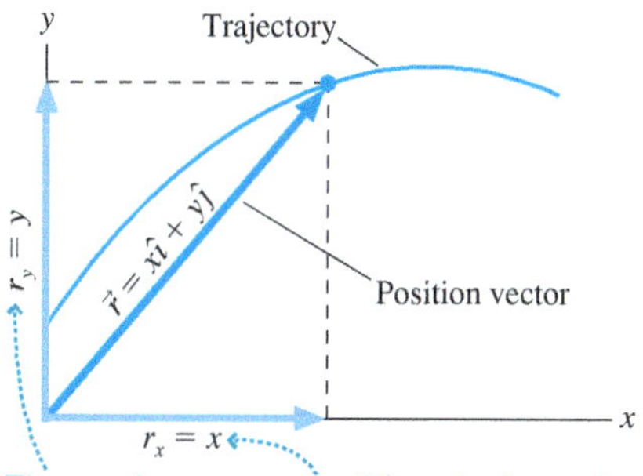

FIGURE 4.5 The instantaneous velocity vector is tangent to the trajectory.

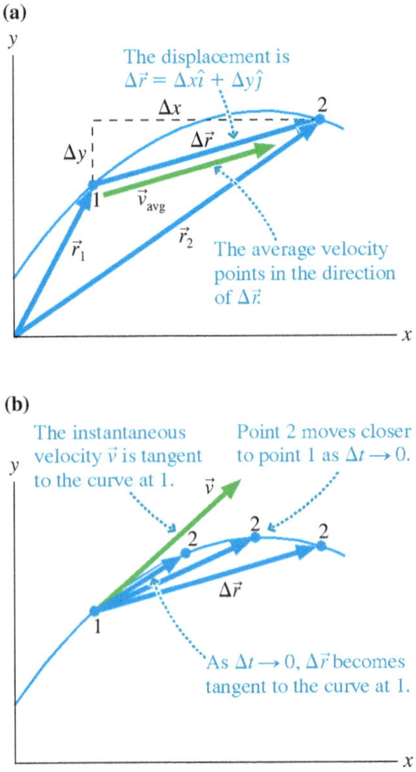

FIGURE 4.5a shows the particle moving from position $\vec{r}_1$ at time t_1 to position $\vec{r}_2$ at a later time t_2. The average velocity—pointing in the direction of the displacement $\Delta\vec{r}$—is

$$\vec{v}_{avg} = \frac{\Delta\vec{r}}{\Delta t} = \frac{\Delta x}{\Delta t}\hat{\imath} + \frac{\Delta y}{\Delta t}\hat{\jmath} \tag{4.2}$$

You learned in Chapter 2 that the instantaneous velocity is the limit of $\vec{v}_{avg}$ as $\Delta t \to 0$. As Δt decreases, point 2 moves closer to point 1 until, as FIGURE 4.5b shows, the displacement vector becomes tangent to the curve. Consequently, **the instantaneous velocity vector $\vec{v}$ is tangent to the trajectory.**

Mathematically, the limit of Equation 4.2 gives

$$\vec{v} = \lim_{\Delta t \to 0} \frac{\Delta\vec{r}}{\Delta t} = \frac{d\vec{r}}{dt} = \frac{dx}{dt}\hat{\imath} + \frac{dy}{dt}\hat{\jmath} \tag{4.3}$$

But we can also write the velocity vector in terms of its x- and y-components as

$$\vec{v} = v_x\hat{\imath} + v_y\hat{\jmath} \tag{4.4}$$

Comparing Equations 4.3 and 4.4, you can see that the velocity vector $\vec{v}$ has x- and y-components

$$v_x = \frac{dx}{dt} \quad \text{and} \quad v_y = \frac{dy}{dt} \tag{4.5}$$

That is, the x-component v_x of the velocity vector is the rate dx/dt at which the particle's x-coordinate is changing. The y-component is similar.

FIGURE 4.6 illustrates another important feature of the velocity vector. If the vector's angle θ is measured from the positive x-direction, the velocity vector components are

$$\begin{aligned} v_x &= v\cos\theta \\ v_y &= v\sin\theta \end{aligned} \tag{4.6}$$

where

$$v = \sqrt{v_x^2 + v_y^2} \tag{4.7}$$

is the particle's *speed* at that point. Speed is always a positive number (or zero), whereas the components are *signed* quantities (i.e., they can be positive or negative) to convey information about the direction of the velocity vector. Conversely, we can use the two velocity components to determine the direction of motion:

$$\tan\theta = \frac{v_y}{v_x} \tag{4.8}$$

FIGURE 4.6 Relating the components of $\vec{v}$ to the speed and direction.

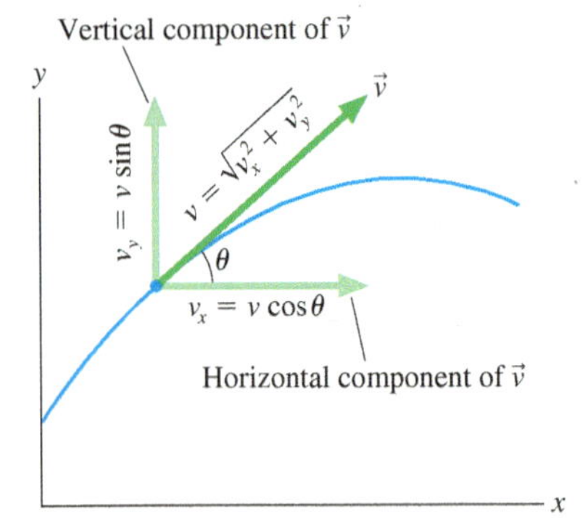

NOTE ▶ In Chapter 2, you learned that the *value* of the velocity component v_s at time t is given by the *slope* of the position-versus-time graph at time t. Now we see that the *direction* of the velocity vector $\vec{v}$ is given by the *tangent* to the y-versus-x graph of the trajectory. FIGURE 4.7 reminds you that these two graphs use different interpretations of the tangent lines. The tangent to the trajectory does not tell us anything about how fast the particle is moving, only its direction. ◀

FIGURE 4.7 Two different uses of tangent lines.

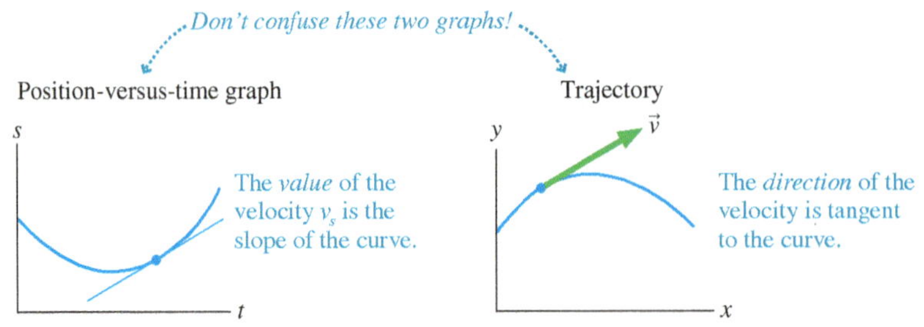

EXAMPLE 4.2 Describing the motion with graphs

A particle's motion is described by the two equations

$$x = 2t^2 \text{ m}$$
$$y = (5t + 5) \text{ m}$$

where the time t is in s.

a. Draw a graph of the particle's trajectory.
b. Draw a graph of the particle's speed as a function of time.

MODEL These are *parametric equations* that give the particle's coordinates x and y separately in terms of the parameter t.

SOLVE a. The trajectory is a curve in the xy-plane. The easiest way to proceed is to calculate x and y at several instants of time.

t (s)	x (m)	y (m)	v (m/s)
0	0	5	5.0
1	2	10	6.4
2	8	15	9.4
3	18	20	13.0
4	32	25	16.8

These points are plotted in FIGURE 4.8a; then a smooth curve is drawn through them to show the trajectory.

b. The particle's speed is given by Equation 4.7. We first need to use Equation 4.5 to find the components of the velocity vector:

$$v_x = \frac{dx}{dt} = 4t \text{ m/s} \quad \text{and} \quad v_y = \frac{dy}{dt} = 5 \text{ m/s}$$

Using these gives the particle's speed at time t:

$$v = \sqrt{v_x^2 + v_y^2} = \sqrt{16t^2 + 25} \text{ m/s}$$

FIGURE 4.8 Two motion graphs for the particle of Example 4.2.

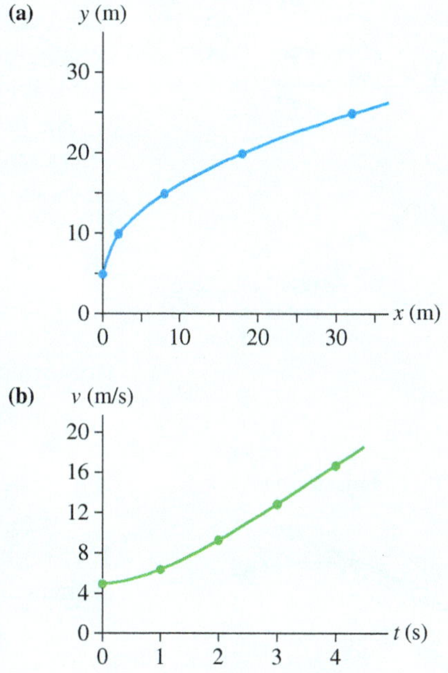

The speed was computed in the table and is graphed in FIGURE 4.8b.

ASSESS The y-versus-x graph of Figure 4.8a is a trajectory, not a position-versus-time graph. Thus the slope is *not* the particle's speed. The particle is speeding up, as you can see in the second graph, even though the slope of the trajectory is decreasing.

Acceleration

Let's return to the particle moving along a trajectory in the xy-plane. FIGURE 4.9a shows the instantaneous velocity $\vec{v}_1$ at point 1 and, a short time later, velocity $\vec{v}_2$ at point 2. These two vectors are tangent to the trajectory. We can use the vector-subtraction technique, shown in the inset, to find $\vec{a}_{avg}$ on this segment of the trajectory.

If we now take the limit $\Delta t \rightarrow 0$, the *instantaneous acceleration* is

$$\vec{a} = \lim_{\Delta t \to 0} \frac{\Delta \vec{v}}{\Delta t} = \frac{d\vec{v}}{dt} \tag{4.9}$$

As $\Delta t \rightarrow 0$, points 1 and 2 in Figure 4.9a merge, and the instantaneous acceleration $\vec{a}$ is found at the same point on the trajectory (and the same instant of time) as the instantaneous velocity $\vec{v}$. This is shown in FIGURE 4.9b.

FIGURE 4.9 The average and instantaneous acceleration vectors on a curved trajectory.

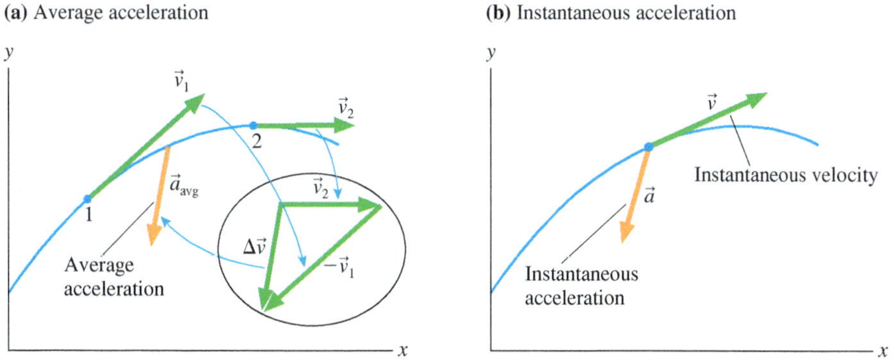

FIGURE 4.10 Decomposition of the instantaneous acceleration $\vec{a}$.

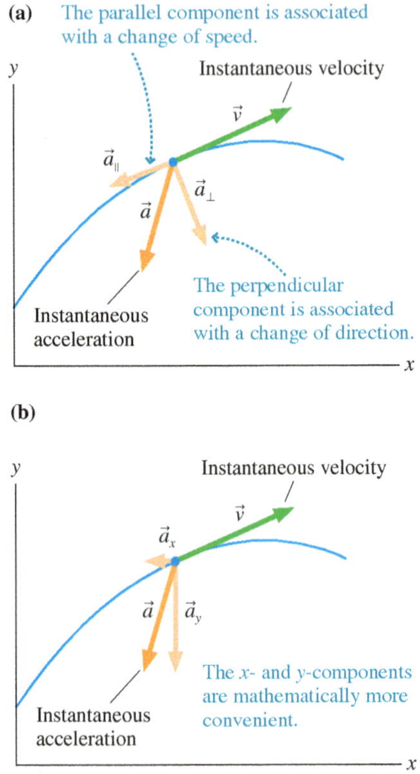

By definition, the acceleration vector $\vec{a}$ is the rate at which the velocity $\vec{v}$ is changing at that instant. To show this, FIGURE 4.10a decomposes $\vec{a}$ into components $\vec{a}_{\parallel}$ and $\vec{a}_{\perp}$ that are parallel and perpendicular to the trajectory. $\vec{a}_{\parallel}$ is associated with a change of speed, and $\vec{a}_{\perp}$ is associated with a change of direction. Both kinds of changes are accelerations. Notice that $\vec{a}_{\perp}$ always points toward the "inside" of the curve because that is the direction in which $\vec{v}$ is changing.

The parallel and perpendicular components of $\vec{a}$ convey important ideas about acceleration, but it's usually more practical to write $\vec{a}$ in terms of the x- and y-components shown in FIGURE 4.10b. Because $\vec{v} = v_x \hat{\imath} + v_y \hat{\jmath}$, we find

$$\vec{a} = a_x \hat{\imath} + a_y \hat{\jmath} = \frac{d\vec{v}}{dt} = \frac{dv_x}{dt}\hat{\imath} + \frac{dv_y}{dt}\hat{\jmath} \tag{4.10}$$

from which we see that

$$a_x = \frac{dv_x}{dt} \quad \text{and} \quad a_y = \frac{dv_y}{dt} \tag{4.11}$$

That is, the x-component of $\vec{a}$ is the rate dv_x/dt at which the x-component of velocity is changing.

Constant Acceleration

If the acceleration $\vec{a} = a_x \hat{\imath} + a_y \hat{\jmath}$ is constant, then the two components a_x and a_y are both constant. In this case, everything you learned about constant-acceleration kinematics in Chapter 2 carries over to two-dimensional motion.

Consider a particle that moves with constant acceleration from an initial position $\vec{r}_i = x_i \hat{\imath} + y_i \hat{\jmath}$, starting with initial velocity $\vec{v}_i = v_{ix} \hat{\imath} + v_{iy} \hat{\jmath}$. Its position and velocity at a final point f are

$$\begin{aligned} x_f &= x_i + v_{ix}\,\Delta t + \tfrac{1}{2}a_x(\Delta t)^2 \qquad & y_f &= y_i + v_{iy}\,\Delta t + \tfrac{1}{2}a_y(\Delta t)^2 \\ v_{fx} &= v_{ix} + a_x\,\Delta t & v_{fy} &= v_{iy} + a_y\,\Delta t \end{aligned} \tag{4.12}$$

There are *many* quantities to keep track of in two-dimensional kinematics, making the pictorial representation all the more important as a problem-solving tool.

NOTE ▶ For constant acceleration, the x-component of the motion and the y-component of the motion are independent of each other. However, they remain connected through the fact that Δt must be the same for both. ◀

EXAMPLE 4.3 Plotting the trajectory of the shuttlecraft

The up thrusters on the shuttlecraft of the starship *Enterprise* give it an upward acceleration of 5.0 m/s^2. Its forward thrusters provide a forward acceleration of 20 m/s^2. As it leaves the *Enterprise*, the shuttlecraft turns on only the up thrusters. After clearing the flight deck, 3.0 s later, it adds the forward thrusters. Plot a trajectory of the shuttlecraft for its first 6 s.

MODEL Represent the shuttlecraft as a particle. There are two segments of constant-acceleration motion.

VISUALIZE FIGURE 4.11 shows a pictorial representation. The coordinate system has been chosen so that the shuttlecraft starts at the origin and initially moves along the y-axis. The craft moves vertically for 3.0 s, then begins to acquire a forward motion. There are three points in the motion: the beginning, the end, and the point at the which forward thrusters are turned on. These points are labeled (x_0, y_0), (x_1, y_1), and (x_2, y_2). The velocities are (v_{0x}, v_{0y}), (v_{1x}, v_{1y}), and (v_{2x}, v_{2y}). This will be our standard labeling scheme for trajectories, where it is essential to keep the x-components and y-components separate.

FIGURE 4.11 Pictorial representation of the motion of the shuttlecraft.

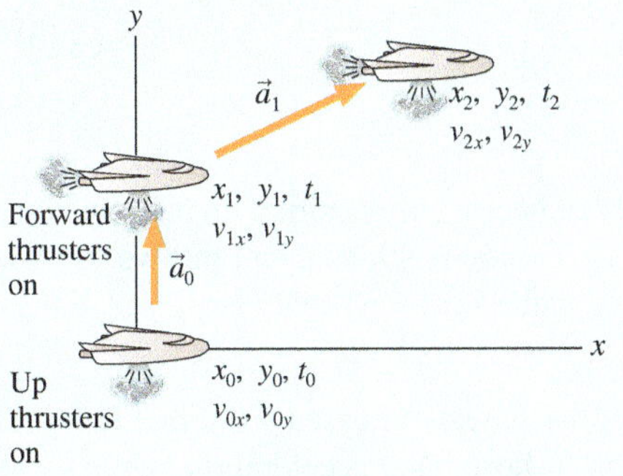

Known

$x_0 = y_0 = 0$ m $\quad v_{0x} = v_{0y} = 0$ m/s $\quad t_0 = 0$

$a_{0x} = 0$ m/s^2 $\quad a_{0y} = 5.0$ m/s^2 $\quad t_1 = 3.0$ s

$a_{1x} = 20$ m/s^2 $\quad a_{1y} = 5.0$ m/s^2 $\quad t_2 = 6.0$ s

Find

x and y at time t

FIGURE 4.12 The shuttlecraft trajectory.

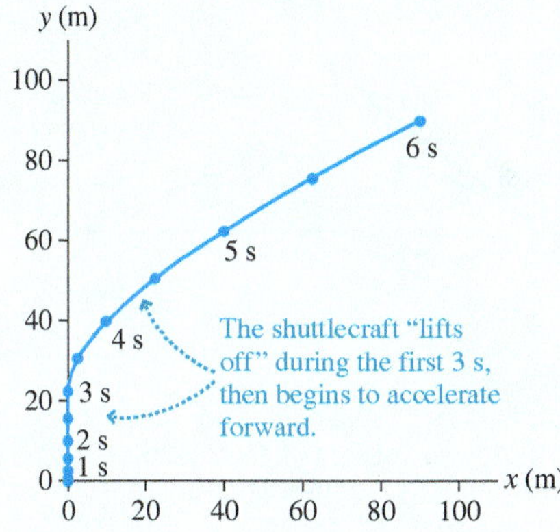

SOLVE During the first phase of the acceleration, when $a_{0x} = 0$ m/s^2 and $a_{0y} = 5.0$ m/s^2, the motion is described by

$$y = y_0 + v_{0y}(t - t_0) + \tfrac{1}{2}a_{0y}(t - t_0)^2 = 2.5t^2 \text{ m}$$

$$v_y = v_{0y} + a_{0y}(t - t_0) = 5.0t \text{ m/s}$$

where the time t is in s. These equations allow us to calculate the position and velocity at any time t. At $t_1 = 3.0$ s, when the first phase of the motion ends, we find that

$$x_1 = 0 \text{ m} \qquad v_{1x} = 0 \text{ m/s}$$

$$y_1 = 22.5 \text{ m} \qquad v_{1y} = 15 \text{ m/s}$$

During the next 3 s, when $a_{1x} = 20$ m/s^2 and $a_{1y} = 5.0$ m/s^2, the x- and y-coordinates are

$$x = x_1 + v_{1x}(t - t_1) + \tfrac{1}{2}a_{1x}(t - t_1)^2$$
$$= 10(t - 3.0)^2 \text{ m}$$
$$y = y_1 + v_{1y}(t - t_1) + \tfrac{1}{2}a_{1y}(t - t_1)^2$$
$$= \left(22.5 + 15(t - 3.0) + 2.5(t - 3.0)^2\right) \text{ m}$$

where, again, t is in s. To show the trajectory, we've calculated x and y every 0.5 s, plotted the points in FIGURE 4.12, and drawn a smooth curve through the points.

STOP TO THINK 4.2 During which time interval or intervals is the particle described by these position graphs at rest? More than one may be correct.

a. 0–1 s
b. 1–2 s
c. 2–3 s
d. 3–4 s

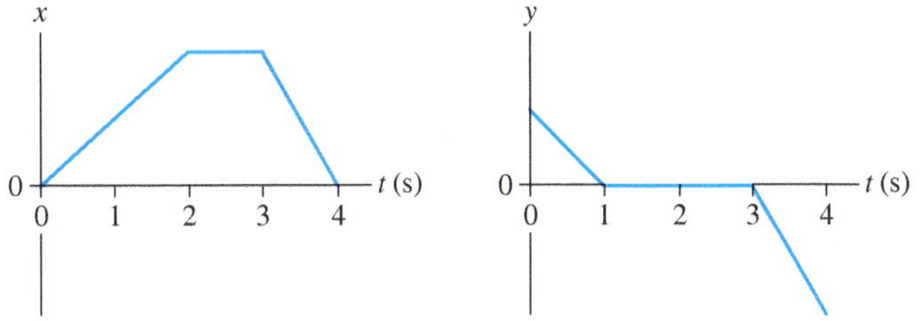

4.3 Projectile Motion

Baseballs and tennis balls flying through the air, Olympic divers, and daredevils shot from cannons all exhibit what we call *projectile motion.* A **projectile** is an object that moves in two dimensions under the influence of only gravity. Projectile motion is an extension of the free-fall motion we studied in Chapter 2. We will continue to neglect the influence of air resistance, leading to results that are a good approximation of reality for relatively heavy objects moving relatively slowly over relatively short distances. As we'll see, projectiles in two dimensions follow a *parabolic trajectory* like the one seen in FIGURE 4.13.

The start of a projectile's motion, be it thrown by hand or shot from a gun, is called the *launch,* and the angle θ of the initial velocity $\vec{v}_0$ above the horizontal (i.e., above

FIGURE 4.13 The parabolic trajectory of a bouncing ball.

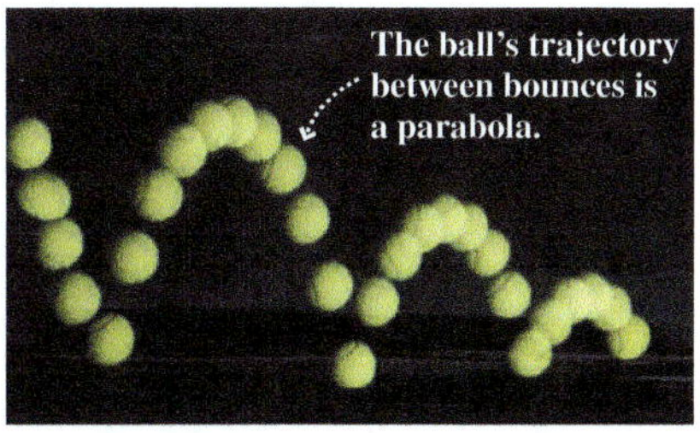

FIGURE 4.14 A projectile launched with initial velocity $\vec{v}_0$.

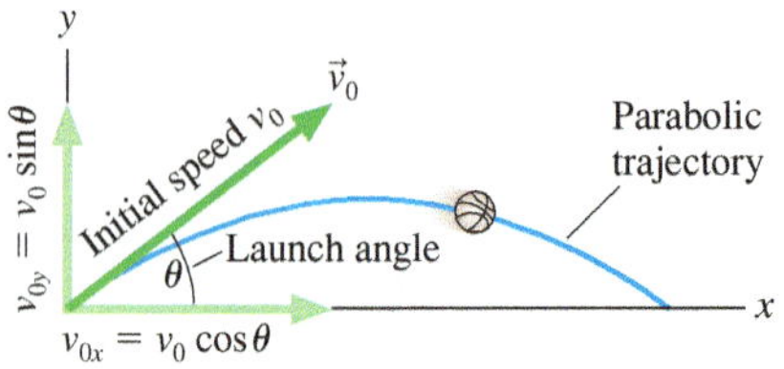

the x-axis) is called the **launch angle**. FIGURE 4.14 illustrates the relationship between the initial velocity vector $\vec{v}_0$ and the initial values of the components v_{0x} and v_{0y}. You can see that

$$v_{0x} = v_0 \cos\theta \qquad v_{0y} = v_0 \sin\theta \tag{4.13}$$

where v_0 is the initial speed.

NOTE ▶ The components v_{0x} and v_{0y} are not necessarily positive. In particular, a projectile launched at an angle *below* the horizontal (such as a ball thrown downward from the roof of a building) has *negative* values for θ and v_{0y}. However, the *speed* v_0 is always positive. ◀

FIGURE 4.15 The velocity and acceleration vectors of a projectile moving along a parabolic trajectory.

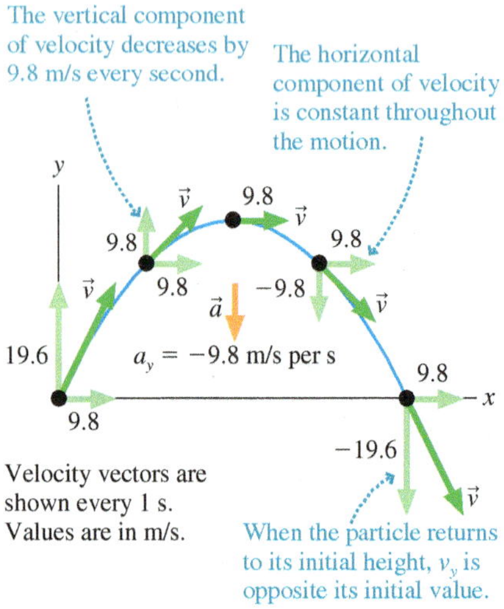

Gravity acts downward, and we know that objects released from rest fall straight down, not sideways. Hence a projectile has no horizontal acceleration, while its vertical acceleration is simply that of free fall. Thus

$$a_x = 0 \qquad a_y = -g \qquad \text{(projectile motion)} \tag{4.14}$$

In other words, **the vertical component of acceleration a_y is just the familiar $-g$ of free fall, while the horizontal component a_x is zero. Projectiles are in free fall.**

To see how these conditions influence the motion, FIGURE 4.15 shows a projectile launched from $(x_0, y_0) = (0\text{ m}, 0\text{ m})$ with an initial velocity $\vec{v}_0 = (9.8\hat{\imath} + 19.6\hat{\jmath})$ m/s. The value of v_x never changes because there's no horizontal acceleration, but v_y decreases by 9.8 m/s every second. This is what it *means* to accelerate at $a_y = -9.8\text{ m/s}^2 = (-9.8\text{ m/s})$ per second.

You can see from Figure 4.15 that **projectile motion is made up of two independent motions:** uniform motion at constant velocity in the horizontal direction and free-fall motion in the vertical direction. The kinematic equations that describe these two motions are simply Equations 4.12 with $a_x = 0$ and $a_y = -g$.

EXAMPLE 4.4 Don't try this at home!

A stunt man drives a car off a 10.0-m-high cliff at a speed of 20.0 m/s. How far does the car land from the base of the cliff?

MODEL Represent the car as a particle in free fall. Assume that the car is moving horizontally as it leaves the cliff.

VISUALIZE The pictorial representation, shown in FIGURE 4.16, is *very* important because the number of quantities to keep track of is quite large. We have chosen to put the origin at the base of the cliff. The assumption that the car is moving horizontally as it leaves the cliff leads to $v_{0x} = v_0$ and $v_{0y} = 0$ m/s.

FIGURE 4.16 Pictorial representation for the car of Example 4.4.

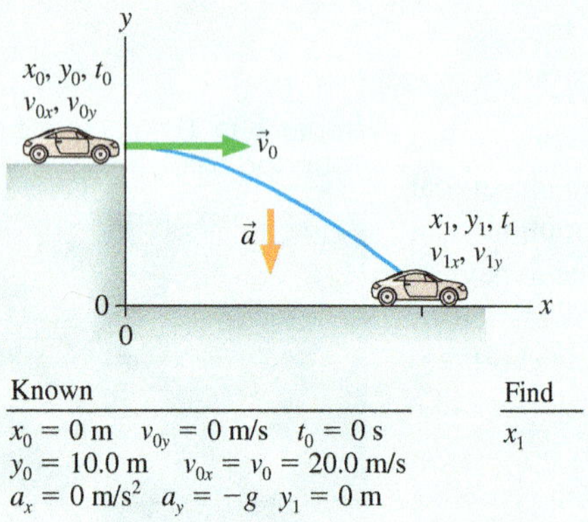

SOLVE Each point on the trajectory has x- and y-components of position, velocity, and acceleration but only *one* value of time. The time needed to move horizontally to x_1 is the *same* time needed to fall vertically through distance y_0. **Although the horizontal and vertical motions are independent, they are connected through the time t.** This is a critical observation for solving projectile motion problems. The kinematics equations with $a_x = 0$ and $a_y = -g$ are

$$x_1 = x_0 + v_{0x}(t_1 - t_0) = v_0 t_1$$
$$y_1 = 0 = y_0 + v_{0y}(t_1 - t_0) - \tfrac{1}{2}g(t_1 - t_0)^2 = y_0 - \tfrac{1}{2}g t_1^2$$

We can use the vertical equation to determine the time t_1 needed to fall distance y_0:

$$t_1 = \sqrt{\frac{2y_0}{g}} = \sqrt{\frac{2(10.0\text{ m})}{9.80\text{ m/s}^2}} = 1.43\text{ s}$$

We then insert this expression for t into the horizontal equation to find the distance traveled:

$$x_1 = v_0 t_1 = (20.0\text{ m/s})(1.43\text{ s}) = 28.6\text{ m}$$

ASSESS The cliff height is ≈ 33 ft and the initial speed is $v_0 \approx 40$ mph. Traveling $x_1 = 29\text{ m} \approx 95$ ft before hitting the ground seems reasonable.

The x- and y-equations of Example 4.4 are parametric equations. It's not hard to eliminate t and write an expression for y as a function of x. From the x_1 equation, $t_1 = x_1/v_0$. Substituting this into the y_1 equation, we find

$$y = y_0 - \frac{g}{2{v_0}^2}x^2 \tag{4.15}$$

The graph of $y = ax^2$ is a parabola, so Equation 4.15 represents an inverted parabola that starts from height y_0. This proves, as we asserted above, that a projectile follows a parabolic trajectory.

Reasoning About Projectile Motion

Think about the following question:

> A heavy ball is launched exactly horizontally at height h above a horizontal field. At the exact instant that the ball is launched, a second ball is simply dropped from height h. Which ball hits the ground first?

It may seem hard to believe, but—if air resistance is neglected—the balls hit the ground *simultaneously*. They do so because the horizontal and vertical components of projectile motion are independent of each other. The initial horizontal velocity of the first ball has *no* influence over its vertical motion. Neither ball has any initial motion in the vertical direction, so both fall distance h in the same amount of time. You can see this in FIGURE 4.17.

FIGURE 4.17 A projectile launched horizontally falls in the same time as a projectile that is released from rest.

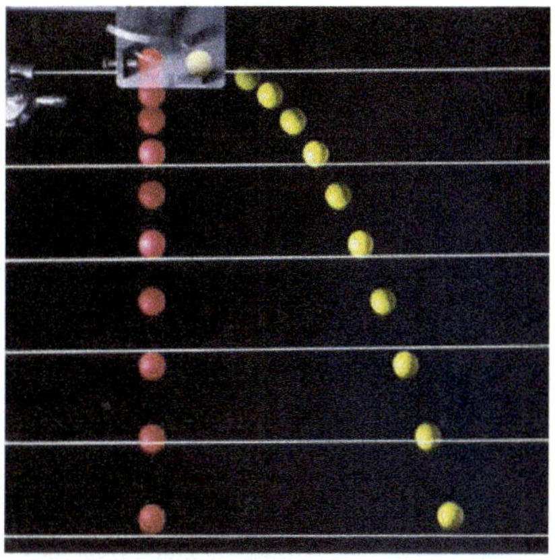

FIGURE 4.18a shows a useful way to think about the trajectory of a projectile. Without gravity, a projectile would follow a straight line. Because of gravity, the particle at time t has "fallen" a distance $\frac{1}{2}gt^2$ below this line. The separation grows as $\frac{1}{2}gt^2$, giving the trajectory its parabolic shape.

Use this idea to think about the following "classic" problem in physics:

> A hungry bow-and-arrow hunter in the jungle wants to shoot down a coconut that is hanging from the branch of a tree. He points his arrow directly at the coconut, but as luck would have it, the coconut falls from the branch at the *exact* instant the hunter releases the string. Does the arrow hit the coconut?

You might think that the arrow will miss the falling coconut, but it doesn't. Although the arrow travels very fast, it follows a slightly curved parabolic trajectory, not a straight line. Had the coconut stayed on the tree, the arrow would have curved under its target as gravity causes it to fall a distance $\frac{1}{2}gt^2$ below the straight line. But $\frac{1}{2}gt^2$ is also the distance the coconut falls while the arrow is in flight. Thus, as FIGURE 4.18b shows, the arrow and the coconut fall the same distance and meet at the same point!

FIGURE 4.18 A projectile follows a parabolic trajectory because it "falls" a distance $\frac{1}{2}gt^2$ below a straight-line trajectory.

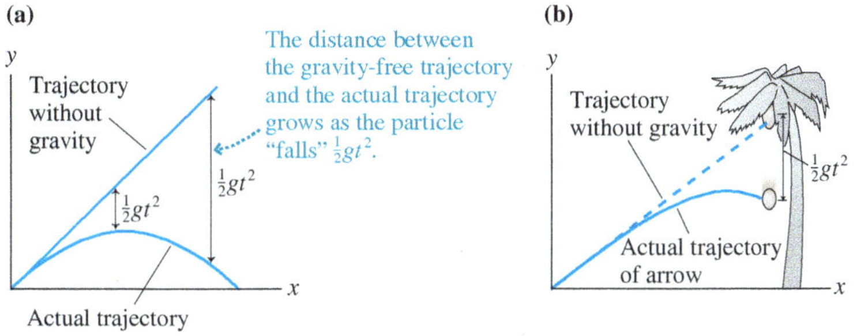

Solving Projectile Motion Problems

PROBLEM-SOLVING STRATEGY 4.1 Projectile motion problems

MODEL Make simplifying assumptions, such as treating the object as a particle. Is it reasonable to ignore air resistance?

VISUALIZE Use a pictorial representation. Establish a coordinate system with the x-axis horizontal and the y-axis vertical. Show important points in the motion on a sketch. Define symbols and identify what the problem is trying to find.

SOLVE The acceleration is known: $a_x = 0$ and $a_y = -g$. Thus the problem is one of two-dimensional kinematics. The kinematic equations are

$$x_f = x_i + v_{ix}\,\Delta t \qquad y_f = y_i + v_{iy}\,\Delta t - \tfrac{1}{2}g(\Delta t)^2$$

$$v_{fx} = v_{ix} = \text{constant} \qquad v_{fy} = v_{iy} - g\,\Delta t$$

Δt is the same for the horizontal and vertical components of the motion. Find Δt from one component, then use that value for the other component.

ASSESS Check that your result has the correct units, is reasonable, and answers the question.

EXAMPLE 4.5 Jumping frog contest

Frogs, with their long, strong legs, are excellent jumpers. And thanks to the good folks of Calaveras County, California, who have a jumping frog contest every year in honor of a Mark Twain story, we have very good data on how far a determined frog can jump.

High-speed cameras show that a good jumper goes into a crouch, then rapidly extends his legs by typically 15 cm during a 42 ms push off, leaving the ground at a 30° angle. How far does this frog leap?

MODEL Represent the frog as a particle. Model the push off as linear motion with constant acceleration. A bullfrog is fairly heavy and dense, so ignore air resistance and consider the leap to be projectile motion.

VISUALIZE This is a two-part problem: linear acceleration followed by projectile motion. A key observation is that **the final velocity for pushing off the ground becomes the initial velocity of the projectile motion.** FIGURE 4.19 shows a separate pictorial representation for each part. Notice that we've used different coordinate systems for the two parts; coordinate systems are our choice, and for each part of the motion we've chosen the coordinate system that makes the problem easiest to solve.

SOLVE While pushing off, the frog travels 15 cm = 0.15 m in 42 ms = 0.042 s. We could find his speed at the end of pushing off if we knew the acceleration. Because the initial velocity is zero, we can find the acceleration from the position-acceleration-time kinematic equation:

$$x_1 = x_0 + v_{0x}\,\Delta t + \tfrac{1}{2}a_x(\Delta t)^2 = \tfrac{1}{2}a_x(\Delta t)^2$$

$$a_x = \frac{2x_1}{(\Delta t)^2} = \frac{2(0.15\text{ m})}{(0.042\text{ s})^2} = 170\text{ m/s}^2$$

This is a substantial acceleration, but it doesn't last long. At the end of the 42 ms push off, the frog's velocity is

$$v_{1x} = v_{0x} + a_x\Delta t = (170\text{ m/s}^2)(0.042\text{ s}) = 7.14\text{ m/s}$$

We'll keep an extra significant figure here to avoid round-off error in the second half of the problem.

FIGURE 4.19 Pictorial representations of the jumping frog.

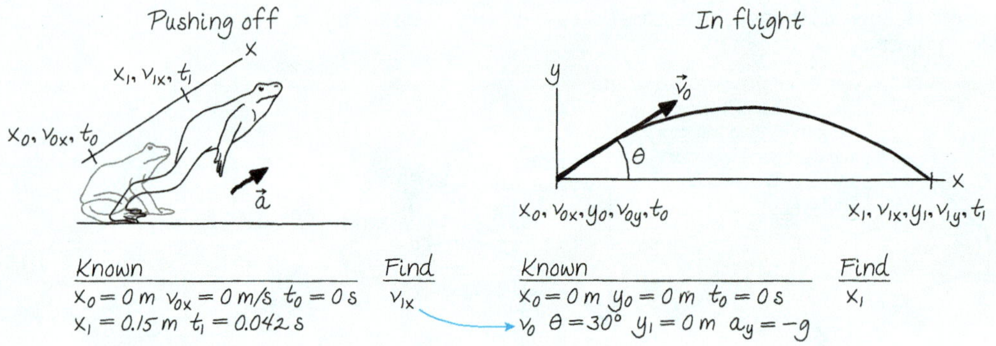

The end of the push off is the beginning of the projectile motion, so the second part of the problem is to find the distance of a projectile launched with velocity $\vec{v}_0 = (7.14 \text{ m/s}, 30°)$. The initial x- and y-components of the launch velocity are

$$v_{0x} = v_0 \cos\theta \qquad v_{0y} = v_0 \sin\theta$$

The kinematic equations of projectile motion, with $a_x = 0$ and $a_y = -g$, are

$$\begin{aligned} x_1 &= x_0 + v_{0x}\Delta t \\ &= (v_0\cos\theta)\Delta t \\ y_1 &= y_0 + v_{0y}\Delta t - \tfrac{1}{2}g(\Delta t)^2 \\ &= (v_0\sin\theta)\Delta t - \tfrac{1}{2}g(\Delta t)^2 \end{aligned}$$

We can find the time of flight from the vertical equation by setting $y_1 = 0$:

$$0 = (v_0\sin\theta)\Delta t - \tfrac{1}{2}g(\Delta t)^2 = (v_0\sin\theta - \tfrac{1}{2}g\,\Delta t)\Delta t$$

and thus

$$\Delta t = 0 \qquad \text{or} \qquad \Delta t = \frac{2v_0\sin\theta}{g}$$

Both are legitimate solutions. The first corresponds to the instant when $y = 0$ at the launch, the second to when $y = 0$ as the frog hits the ground. Clearly, we want the second solution. Substituting this expression for Δt into the equation for x_1 gives

$$x_1 = (v_0\cos\theta)\frac{2v_0\sin\theta}{g} = \frac{2v_0^2\sin\theta\cos\theta}{g}$$

We can simplify this result with the trigonometric identity $2\sin\theta\cos\theta = \sin(2\theta)$. Thus the distance traveled by the frog is

$$x_1 = \frac{v_0^2\sin(2\theta)}{g}$$

Using $v_0 = 7.14$ m/s and $\theta = 30°$, we find that the frog leaps a distance of 4.5 m.

ASSESS 4.5 m is about 15 feet. This is much farther than a human can jump from a standing start, but it seems believable. In fact, the current record holder, Rosie the Ribeter, made a leap of 6.5 m!

As Example 4.5 found, a projectile that lands at the same elevation from which it was launched travels distance

$$\text{distance} = \frac{v_0^2\sin(2\theta)}{g} \tag{4.16}$$

The maximum distance occurs for $\theta = 45°$, where $\sin(2\theta) = 1$. But there's more that we can learn from this equation. Because $\sin(180° - x) = \sin x$, it follows that $\sin(2(90° - \theta)) = \sin(2\theta)$. Consequently, a projectile launched either at angle θ *or* at angle $(90° - \theta)$ will travel the same distance *over level ground*. FIGURE 4.20 shows the trajectories of projectiles launched with the same initial speed in 15° increments of angle.

NOTE ▶ Equation 4.16 is *not* a general result. It applies *only* in situations where the projectile lands at the same elevation from which it was fired. ◀

FIGURE 4.20 Trajectories of a projectile launched at different angles with a speed of 99 m/s.

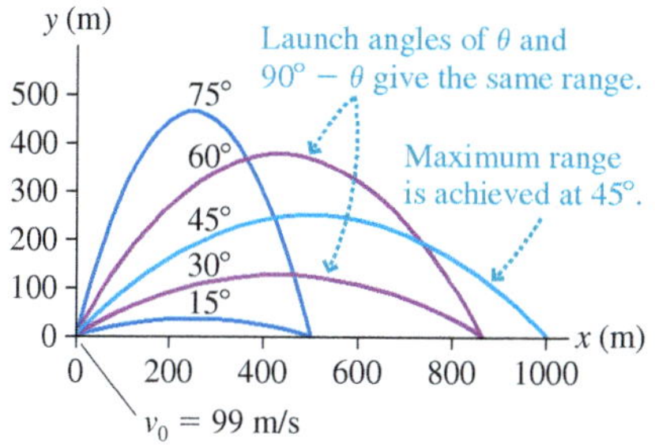

STOP TO THINK 4.3 A 50 g marble rolls off a table and lands 2 m from the base of the table. A 100 g marble rolls off the same table with the same speed. It lands at distance

a. Less than 1 m.
b. 1 m.
c. Between 1 m and 2 m.
d. 2 m.
e. Between 2 m and 4 m.
f. 4 m.

4.4 Relative Motion

FIGURE 4.21 shows Amy and Bill watching Carlos on his bicycle. According to Amy, Carlos's velocity is $v_x = 5$ m/s. Bill sees the bicycle receding in his rearview mirror, in the *negative* x-direction, getting 10 m farther away from him every second. According to Bill, Carlos's velocity is $v_x = -10$ m/s. Which is Carlos's *true* velocity?

Velocity is not a concept that can be true or false. Carlos's velocity *relative to Amy* is $(v_x)_{CA} = 5$ m/s, where the subscript notation means "C relative to A." Similarly, Carlos's velocity *relative to Bill* is $(v_x)_{CB} = -10$ m/s. These are both valid descriptions of Carlos's motion.

FIGURE 4.21 Amy and Bill each measure the velocity of Carlos on his bicycle. The velocities shown are in Amy's reference frame.

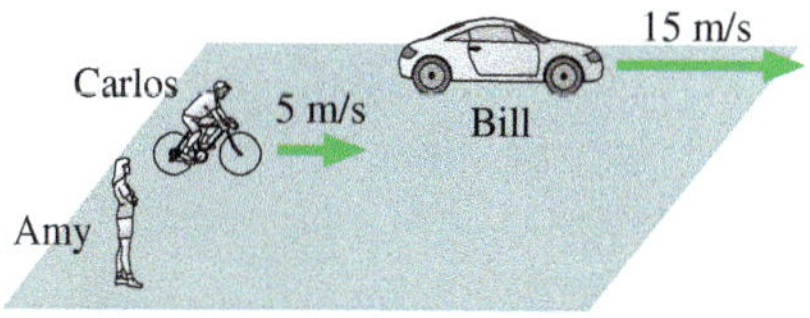

It's not hard to see how to combine the velocities for one-dimensional motion:

The first subscript is the same on both sides. The last subscript is the same on both sides.

$$(v_x)_{CB} = (v_x)_{CA} + (v_x)_{AB} \tag{4.17}$$

The inner subscripts "cancel."

We'll justify this relationship later in this section and then extend it to two-dimensional motion.

Equation 4.17 tells us that the velocity of C relative to B is the velocity of C relative to A *plus* the velocity of A relative to B. Note that

$$(v_x)_{AB} = -(v_x)_{BA} \tag{4.18}$$

because if B is moving to the right relative to A, then A is moving to the left relative to B. In Figure 4.21, Bill is moving to the right relative to Amy with $(v_x)_{BA} = 15$ m/s, so $(v_x)_{AB} = -15$ m/s. Knowing that Carlos's velocity relative to Amy is 5 m/s, we find that Carlos's velocity relative to Bill is, as expected, $(v_x)_{CB} = (v_x)_{CA} + (v_x)_{AB} = 5 \text{ m/s} + (-15) \text{ m/s} = -10$ m/s.

EXAMPLE 4.6 A speeding bullet

The police are chasing a bank robber. While driving at 50 m/s, they fire a bullet to shoot out a tire of his car. The police gun shoots bullets at 300 m/s. What is the bullet's speed as measured by a TV camera crew parked beside the road?

MODEL Assume that all motion is in the positive x-direction. The bullet is the object that is observed from both the police car and the ground.

SOLVE The bullet B's velocity relative to the gun G is $(v_x)_{BG} = 300$ m/s. The gun, inside the car, is traveling relative to the TV crew C at $(v_x)_{GC} = 50$ m/s. We can combine these values to find that the bullet's velocity relative to the TV crew on the ground is

$$(v_x)_{BC} = (v_x)_{BG} + (v_x)_{GC} = 300 \text{ m/s} + 50 \text{ m/s} = 350 \text{ m/s}$$

ASSESS It should be no surprise in this simple situation that we simply add the velocities.

Reference Frames

A coordinate system in which an experimenter (possibly with the assistance of helpers) makes position and time measurements of physical events is called a **reference frame.** In Figure 4.21, Amy and Bill each had their own reference frame (where they were at rest) in which they measured Carlos's velocity.

More generally, FIGURE 4.22 shows two reference frames, A and B, and an object C. It is assumed that the reference frames are moving with respect to each other. At this instant of time, the position vector of C in reference frame A is $\vec{r}_{CA}$, meaning "the position of C relative to the origin of frame A." Similarly, $\vec{r}_{CB}$ is the position vector of C in reference frame B. Using vector addition, you can see that

$$\vec{r}_{CB} = \vec{r}_{CA} + \vec{r}_{AB} \tag{4.19}$$

where $\vec{r}_{AB}$ locates the origin of A relative to the origin of B.

FIGURE 4.22 Object C is measured from two different reference frames.

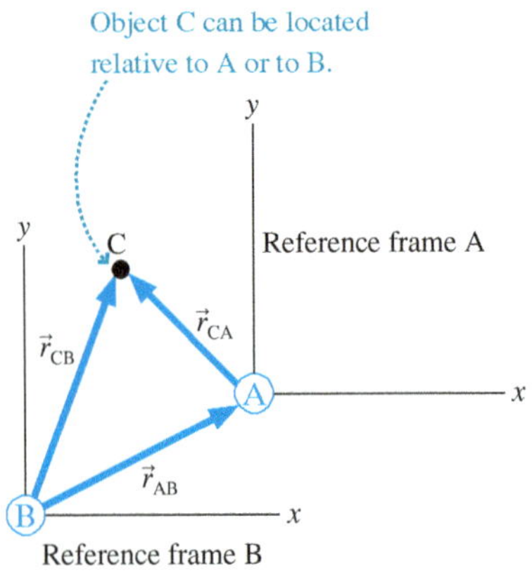

In general, object C is moving relative to both reference frames. To find its velocity in each reference frame, take the time derivative of Equation 4.19:

$$\frac{d\vec{r}_{CB}}{dt} = \frac{d\vec{r}_{CA}}{dt} + \frac{d\vec{r}_{AB}}{dt} \tag{4.20}$$

By definition, $d\vec{r}/dt$ is a velocity. The first derivative is $\vec{v}_{CB}$, the velocity of C relative to B. Similarly, the second derivative is the velocity of C relative to A, $\vec{v}_{CA}$. The last derivative is slightly different because it doesn't refer to object C. Instead, this is the velocity $\vec{v}_{AB}$ of reference frame A relative to reference frame B. As we noted in one dimension, $\vec{v}_{AB} = -\vec{v}_{BA}$.

Writing Equation 4.20 in terms of velocities, we have

$$\vec{v}_{CB} = \vec{v}_{CA} + \vec{v}_{AB} \tag{4.21}$$

This relationship between velocities in different reference frames was recognized by Galileo in his pioneering studies of motion, hence it is known as the **Galilean transformation of velocity.** If you know an object's velocity in one reference frame, you can *transform* it into the velocity that would be measured in a different reference frame. Just as in one dimension, the velocity of C relative to B is the velocity of C relative to A plus the velocity of A relative to B, *but* you must add the velocities as vectors for two-dimensional motion.

As we've seen, the Galilean velocity transformation is pretty much common sense for one-dimensional motion. The real usefulness appears when an object travels in a *medium* moving with respect to the earth. For example, a boat moves relative to the water. What is the boat's net motion if the water is a flowing river? Airplanes fly relative to the air, but the air at high altitudes often flows at high speed. Navigation of boats and planes requires knowing both the motion of the vessel in the medium and the motion of the medium relative to the earth.

EXAMPLE 4.7 Flying to Cleveland I

Cleveland is 300 miles east of Chicago. A plane leaves Chicago flying due east at 500 mph. The pilot forgot to check the weather and doesn't know that the wind is blowing to the south at 50 mph. What is the plane's ground speed? Where is the plane 0.60 h later, when the pilot expects to land in Cleveland?

MODEL Establish a coordinate system with the x-axis pointing east and the y-axis north. The plane P flies in the air, so its velocity relative to the air A is $\vec{v}_{PA} = 500\hat{\imath}$ mph. Meanwhile, the air is moving relative to the ground G at $\vec{v}_{AG} = -50\hat{\jmath}$ mph.

FIGURE 4.23 The wind causes a plane flying due east in the air to move to the southeast relative to the ground.

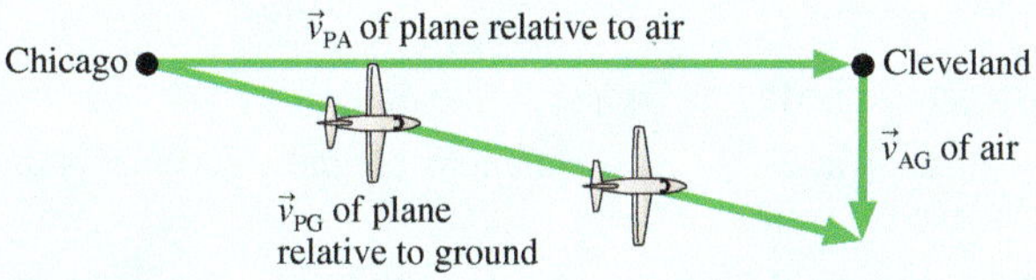

SOLVE The velocity equation $\vec{v}_{PG} = \vec{v}_{PA} + \vec{v}_{AG}$ is a vector-addition equation. FIGURE 4.23 shows graphically what happens. Although the nose of the plane points east, the wind carries the plane in a direction somewhat south of east. The plane's velocity relative to the ground is

$$\vec{v}_{PG} = \vec{v}_{PA} + \vec{v}_{AG} = (500\hat{\imath} - 50\hat{\jmath})\text{ mph}$$

The plane's ground speed is

$$v = \sqrt{(v_x)_{PG}^2 + (v_y)_{PG}^2} = 502\text{ mph}$$

After flying for 0.60 h at this velocity, the plane's location (relative to Chicago) is

$$x = (v_x)_{PG}t = (500\text{ mph})(0.60\text{ h}) = 300\text{ mi}$$

$$y = (v_y)_{PG}t = (-50\text{ mph})(0.60\text{ h}) = -30\text{ mi}$$

The plane is 30 mi due south of Cleveland! Although the pilot thought he was flying to the east, his actual heading has been $\tan^{-1}(50\text{ mph}/500\text{ mph}) = \tan^{-1}(0.10) = 5.71°$ south of east.

EXAMPLE 4.8 Flying to Cleveland II

A wiser pilot flying from Chicago to Cleveland on the same day plots a course that will take her directly to Cleveland. In which direction does she fly the plane? How long does it take to reach Cleveland?

MODEL Establish a coordinate system with the x-axis pointing east and the y-axis north. The air is moving relative to the ground at $\vec{v}_{AG} = -50\hat{\jmath}$ mph.

SOLVE The objective of navigation is to move between two points on the earth's surface. The wiser pilot, who knows that the wind will affect her plane, draws the vector picture of FIGURE 4.24. She sees that she'll need $(v_y)_{PG} = 0$, in order to fly due east to Cleveland. This will require turning the nose of the plane at an angle θ north of east, making $\vec{v}_{PA} = (500\cos\theta\,\hat{\imath} + 500\sin\theta\,\hat{\jmath})$ mph.

The velocity equation is $\vec{v}_{PG} = \vec{v}_{PA} + \vec{v}_{AG}$. The desired heading is found from setting the y-component of this equation to zero:

$$(v_y)_{PG} = (v_y)_{PA} + (v_y)_{AG} = (500\sin\theta - 50)\text{ mph} = 0\text{ mph}$$

$$\theta = \sin^{-1}\left(\frac{50\text{ mph}}{500\text{ mph}}\right) = 5.74°$$

FIGURE 4.24 To travel due east in a south wind, a pilot has to point the plane somewhat to the northeast.

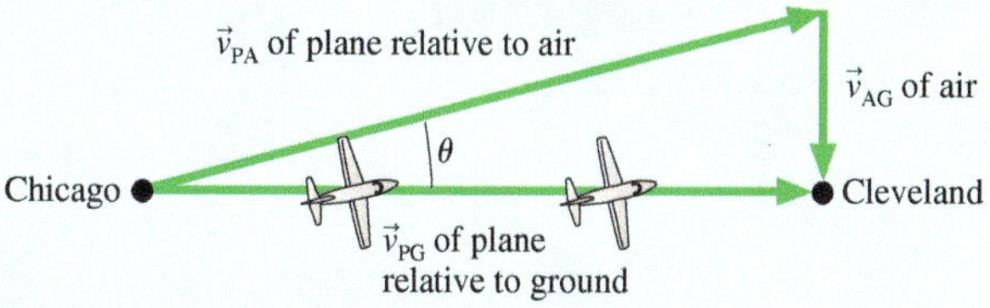

The plane's velocity relative to the ground is then $\vec{v}_{PG} = (500\text{ mph}) \times \cos 5.74°\,\hat{\imath} = 497\hat{\imath}$ mph. This is slightly slower than the speed relative to the air. The time needed to fly to Cleveland at this speed is

$$t = \frac{300\text{ mi}}{497\text{ mph}} = 0.604\text{ h}$$

It takes 0.004 h = 14 s longer to reach Cleveland than it would on a day without wind.

ASSESS A boat crossing a river or an ocean current faces the same difficulties. These are exactly the kinds of calculations performed by pilots of boats and planes as part of navigation.

STOP TO THINK 4.4 A plane traveling horizontally to the right at 100 m/s flies past a helicopter that is going straight up at 20 m/s. From the helicopter's perspective, the plane's direction and speed are

a. Right and up, less than 100 m/s.
b. Right and up, 100 m/s.
c. Right and up, more than 100 m/s.
d. Right and down, less than 100 m/s.
e. Right and down, 100 m/s.
f. Right and down, more than 100 m/s.

4.5 Uniform Circular Motion

FIGURE 4.25 A particle in uniform circular motion.

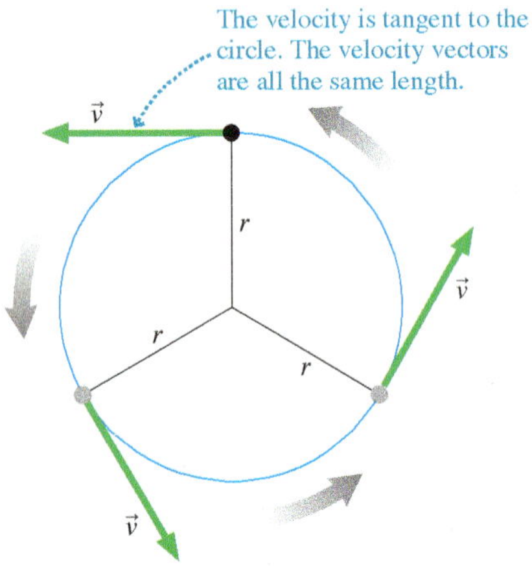

FIGURE 4.25 shows a particle moving around a circle of radius r. The particle might be a satellite in an orbit, a ball on the end of a string, or even just a dot painted on the side of a rotating wheel. Circular motion is another example of motion in a plane, but it is quite different from projectile motion.

To begin the study of circular motion, consider a particle that moves at *constant speed* around a circle of radius r. This is called **uniform circular motion**. Regardless of what the particle represents, its velocity vector $\vec{v}$ is always tangent to the circle. The particle's speed v is constant, so vector $\vec{v}$ is always the same length.

The time interval it takes the particle to go around the circle once, completing one revolution (abbreviated rev), is called the **period** of the motion. Period is represented by the symbol T. It's easy to relate the particle's period T to its speed v. For a particle moving with constant speed, speed is simply distance/time. In one period, the particle moves once around a circle of radius r and travels the circumference $2\pi r$. Thus

$$v = \frac{1 \text{ circumference}}{1 \text{ period}} = \frac{2\pi r}{T} \tag{4.22}$$

Circular motion is one of the most common types of motion.

EXAMPLE 4.9 A rotating crankshaft

A 4.0-cm-diameter crankshaft turns at 2400 rpm (revolutions per minute). What is the speed of a point on the surface of the crankshaft?

SOLVE We need to determine the time it takes the crankshaft to make 1 rev. First, we convert 2400 rpm to revolutions per second:

$$\frac{2400 \text{ rev}}{1 \text{ min}} \times \frac{1 \text{ min}}{60 \text{ s}} = 40 \text{ rev/s}$$

If the crankshaft turns 40 times in 1 s, the time for 1 rev is

$$T = \frac{1}{40} \text{ s} = 0.025 \text{ s}$$

Thus the speed of a point on the surface, where $r = 2.0 \text{ cm} = 0.020 \text{ m}$, is

$$v = \frac{2\pi r}{T} = \frac{2\pi(0.020 \text{ m})}{0.025 \text{ s}} = 5.0 \text{ m/s}$$

Angular Position

FIGURE 4.26 A particle's position is described by distance r and angle θ.

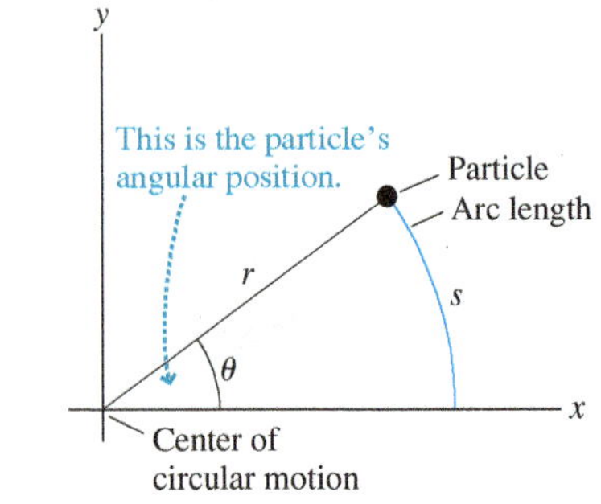

Rather than using xy-coordinates, it will be more convenient to describe the position of a particle in circular motion by its distance r from the center of the circle and its angle θ from the positive x-axis. This is shown in FIGURE 4.26. The angle θ is the **angular position** of the particle.

We can distinguish a position above the x-axis from a position that is an equal angle below the x-axis by *defining* θ to be positive when measured *counterclockwise* (ccw) from the positive x-axis. An angle measured clockwise (cw) from the positive x-axis has a negative value. "Clockwise" and "counterclockwise" in circular motion are analogous, respectively, to "left of the origin" and "right of the origin" in linear motion, which we associated with negative and positive values of x. A particle 30° below the positive x-axis is equally well described by either $\theta = -30°$ or $\theta = +330°$. We could also describe this particle by $\theta = \frac{11}{12}$ rev, where *revolutions* are another way to measure the angle.

Although degrees and revolutions are widely used measures of angle, mathematicians and scientists usually find it more useful to measure the angle θ in Figure 4.26 by using the **arc length** s that the particle travels along the edge of a circle of radius r. We define the angular unit of **radians** such that

$$\theta(\text{radians}) \equiv \frac{s}{r} \tag{4.23}$$

The radian, which is abbreviated rad, is the SI unit of an angle. An angle of 1 rad has an arc length s exactly equal to the radius r.

The arc length completely around a circle is the circle's circumference $2\pi r$. Thus the angle of a full circle is

$$\theta_{\text{full circle}} = \frac{2\pi r}{r} = 2\pi \text{ rad}$$

This relationship is the basis for the well-known conversion factors

$$1 \text{ rev} = 360° = 2\pi \text{ rad}$$

As a simple example of converting between radians and degrees, let's convert an angle of 1 rad to degrees:

$$1 \text{ rad} = 1 \text{ rad} \times \frac{360°}{2\pi \text{ rad}} = 57.3°$$

Thus a rough approximation is 1 rad ≈ 60°. We will often specify angles in degrees, but keep in mind that the SI unit is the radian.

An important consequence of Equation 4.23 is that the arc length spanning angle θ is

$$s = r\theta \qquad (\text{with } \theta \text{ in rad}) \tag{4.24}$$

This is a result that we will use often, but it is valid *only* if θ is measured in radians and not in degrees. This very simple relationship between angle and arc length is one of the primary motivations for using radians.

NOTE ▶ Units of angle are often troublesome. Unlike the kilogram or the second, for which we have standards, the radian is a *defined* unit. Further, its definition as a ratio of two lengths makes it a *pure number* without dimensions. Thus the unit of angle, be it radians or degrees or revolutions, is really just a *name* to remind us that we're dealing with an angle. Consequently, the radian unit sometimes appears or disappears without warning. This seems rather mysterious until you get used to it. This textbook will call your attention to such behavior the first few times it occurs. With a little practice, you'll soon learn when the rad unit is needed and when it's not. ◀

Angular Velocity

FIGURE 4.27 shows a particle moving in a circle from an initial angular position θ_i at time t_i to a final angular position θ_f at a later time t_f. The change $\Delta\theta = \theta_f - \theta_i$ is called the **angular displacement.** We can measure the particle's circular motion in terms of the rate of change of θ, just as we measured the particle's linear motion in terms of the rate of change of its position s.

In analogy with linear motion, let's define the *average angular velocity* to be

$$\text{average angular velocity} \equiv \frac{\Delta\theta}{\Delta t} \tag{4.25}$$

As the time interval Δt becomes very small, $\Delta t \to 0$, we arrive at the definition of the instantaneous **angular velocity**

$$\omega \equiv \lim_{\Delta t \to 0} \frac{\Delta\theta}{\Delta t} = \frac{d\theta}{dt} \qquad (\text{angular velocity}) \tag{4.26}$$

FIGURE 4.27 A particle moves with angular velocity ω.

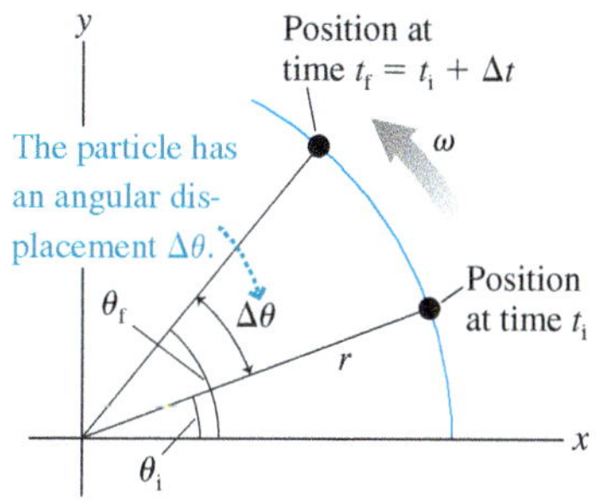

The symbol ω is a lowercase Greek omega, *not* an ordinary w. The SI unit of angular velocity is rad/s, but °/s, rev/s, and rev/min are also common units. Revolutions per minute is abbreviated rpm.

Angular velocity is the *rate* at which a particle's angular position is changing as it moves around a circle. A particle that starts from $\theta = 0$ rad with an angular velocity of 0.5 rad/s will be at angle $\theta = 0.5$ rad after 1 s, at $\theta = 1.0$ rad after 2 s, at $\theta = 1.5$ rad after 3 s, and so on. Its angular position is increasing at the *rate* of 0.5 radian per second. **A particle moves with uniform circular motion if and only if its angular velocity ω is constant and unchanging.**

FIGURE 4.28 Positive and negative angular velocities.

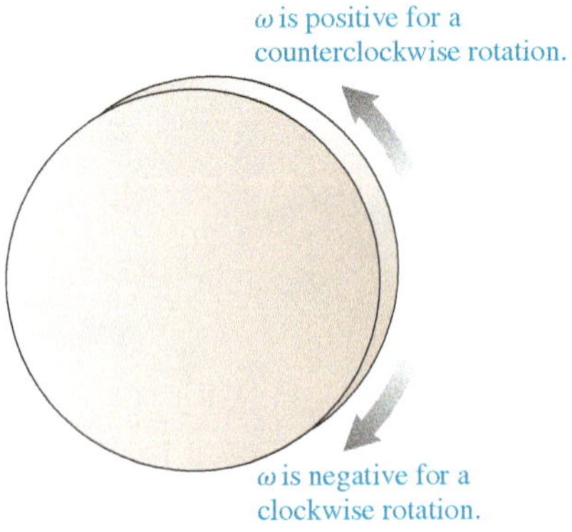

Angular velocity, like the velocity v_s of one-dimensional motion, can be positive or negative. The signs shown in FIGURE 4.28 are based on the fact that θ was defined to be positive for a counterclockwise rotation. Because the definition $\omega = d\theta/dt$ for circular motion parallels the definition $v_s = ds/dt$ for linear motion, the graphical relationships we found between v_s and s in Chapter 2 apply equally well to ω and θ:

$$\begin{aligned} \omega &= \text{slope of the } \theta\text{-versus-}t \text{ graph at time } t \\ \theta_f &= \theta_i + \text{area under the } \omega\text{-versus-}t \text{ curve between } t_i \text{ and } t_f \qquad (4.27) \\ &= \theta_i + \omega \Delta t \end{aligned}$$

You will see many more instances where circular motion is analogous to linear motion with angular variables replacing linear variables. Thus much of what you learned about linear kinematics carries over to circular motion.

EXAMPLE 4.10 A graphical representation of circular motion

FIGURE 4.29 shows the angular position of a painted dot on the edge of a rotating wheel. Describe the wheel's motion and draw an ω-versus-t graph.

FIGURE 4.29 Angular position graph for the wheel of Example 4.10.

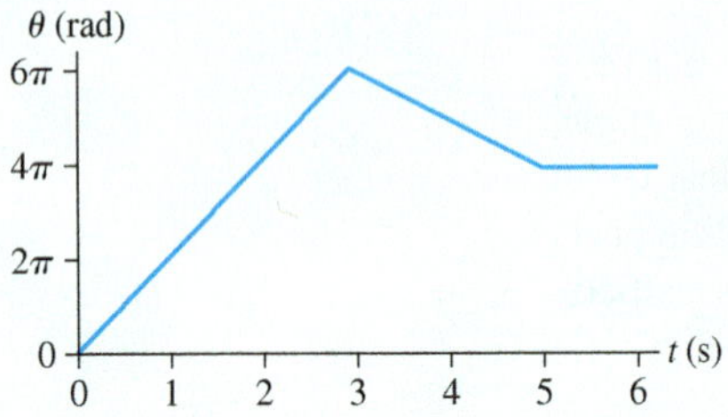

SOLVE Although circular motion seems to "start over" every revolution (every 2π rad), the angular position θ continues to increase. $\theta = 6\pi$ rad corresponds to three revolutions. This wheel makes 3 ccw rev (because θ is getting more positive) in 3 s, immediately reverses direction and makes 1 cw rev in 2 s, then stops at $t = 5$ s and holds the position $\theta = 4\pi$ rad. The angular velocity is found by measuring the slope of the graph:

$t = 0\text{–}3$ s slope $= \Delta\theta/\Delta t = 6\pi \text{ rad}/3 \text{ s} = 2\pi$ rad/s

$t = 3\text{–}5$ s slope $= \Delta\theta/\Delta t = -2\pi \text{ rad}/2 \text{ s} = -\pi$ rad/s

$t > 5$ s slope $= \Delta\theta/\Delta t = 0$ rad/s

These results are shown as an ω-versus-t graph in FIGURE 4.30. For the first 3 s, the motion is uniform circular motion with $\omega = 2\pi$ rad/s. The wheel then changes to a different uniform circular motion with $\omega = -\pi$ rad/s for 2 s, then stops.

FIGURE 4.30 ω-versus-t graph for the wheel of Example 4.10.

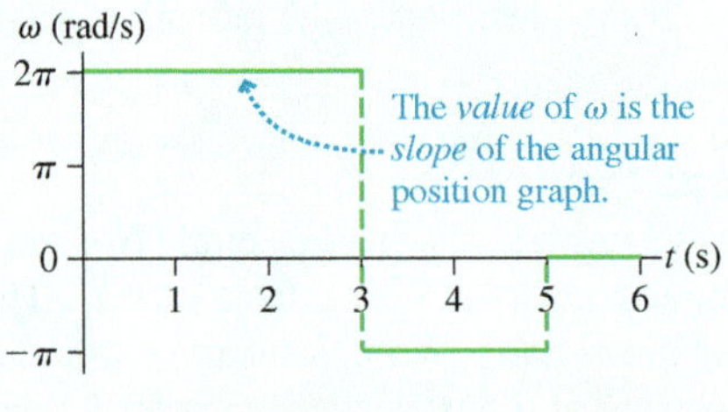

NOTE ▶ In physics, we nearly always want to give results as numerical values. Example 4.9 had a π in the equation, but we used its numerical value to compute $v = 5.0$ m/s. However, angles in radians are an exception to this rule. It's okay to leave a π in the value of θ or ω, and we have done so in Example 4.10. ◀

Not surprisingly, the angular velocity ω is closely related to the *period* T of the motion. As a particle goes around a circle one time, its angular displacement is

$\Delta\theta = 2\pi$ rad during the interval $\Delta t = T$. Thus, using the definition of angular velocity, we find

$$|\omega| = \frac{2\pi \text{ rad}}{T} \quad \text{or} \quad T = \frac{2\pi \text{ rad}}{|\omega|} \tag{4.28}$$

The period alone gives only the absolute value of $|\omega|$. You need to know the direction of motion to determine the sign of ω.

EXAMPLE 4.11 At the roulette wheel

A small steel roulette ball rolls ccw around the inside of a 30-cm-diameter roulette wheel. The ball completes 2.0 rev in 1.20 s.

a. What is the ball's angular velocity?
b. What is the ball's position at $t = 2.0$ s? Assume $\theta_i = 0$.

MODEL Model the ball as a particle in uniform circular motion.

SOLVE a. The period of the ball's motion, the time for 1 rev, is $T = 0.60$ s. Angular velocity is positive for ccw motion, so

$$\omega = \frac{2\pi \text{ rad}}{T} = \frac{2\pi \text{ rad}}{0.60 \text{ s}} = 10.47 \text{ rad/s}$$

b. The ball starts at $\theta_i = 0$ rad. After $\Delta t = 2.0$ s, its position is

$$\theta_f = 0 \text{ rad} + (10.47 \text{ rad/s})(2.0 \text{ s}) = 20.94 \text{ rad}$$

where we've kept an extra significant figure to avoid round-off error. Although this is a mathematically acceptable answer, an observer would say that the ball is always located somewhere between 0° and 360°. Thus it is common practice to subtract an integer number of 2π rad, representing the completed revolutions. Because $20.94/2\pi = 3.333$, we can write

$$\begin{aligned}\theta_f &= 20.94 \text{ rad} = 3.333 \times 2\pi \text{ rad}\\ &= 3 \times 2\pi \text{ rad} + 0.333 \times 2\pi \text{ rad}\\ &= 3 \times 2\pi \text{ rad} + 2.09 \text{ rad}\end{aligned}$$

In other words, at $t = 2.0$ s the ball has completed 3 rev and is 2.09 rad = 120° into its fourth revolution. An observer would say that the ball's position is $\theta_f = 120°$.

STOP TO THINK 4.5 A particle moves cw around a circle at constant speed for 2.0 s. It then reverses direction and moves ccw at half the original speed until it has traveled through the same angle. Which is the particle's angle-versus-time graph?

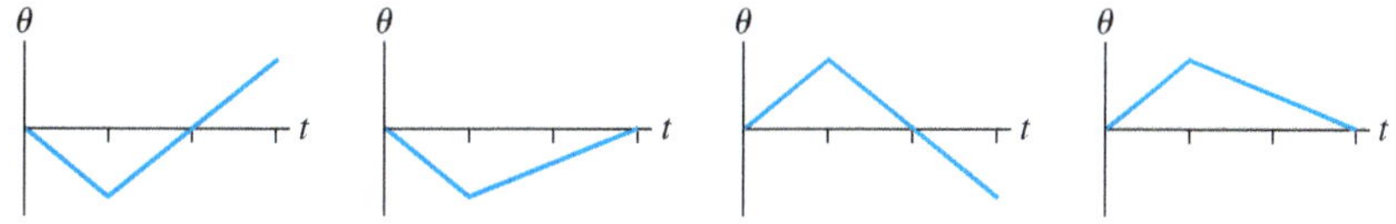

(a) (b) (c) (d)

4.6 Velocity and Acceleration in Uniform Circular Motion

For a particle in circular motion, such as the one in FIGURE 4.31, the velocity vector $\vec{v}$ is always tangent to the circle. In other words, the velocity vector has only a *tangential component*, which we will designate v_t.

The tangential velocity component v_t is the rate ds/dt at which the particle moves *around* the circle, where s is the arc length measured from the positive x-axis. From Equation 4.24, the arc length is $s = r\theta$. Taking the derivative, we find

$$v_t = \frac{ds}{dt} = r\frac{d\theta}{dt}$$

But $d\theta/dt$ is the angular velocity ω. Thus the tangential velocity and the angular velocity are related by

$$v_t = \omega r \quad \text{(with } \omega \text{ in rad/s)} \tag{4.29}$$

NOTE ▶ ω is restricted to rad/s because the relationship $s = r\theta$ is the definition of radians. While it may be convenient in some problems to measure ω in rev/s or rpm, you must convert to SI units of rad/s before using Equation 4.29. ◀

FIGURE 4.31 Velocity and acceleration of uniform circular motion.

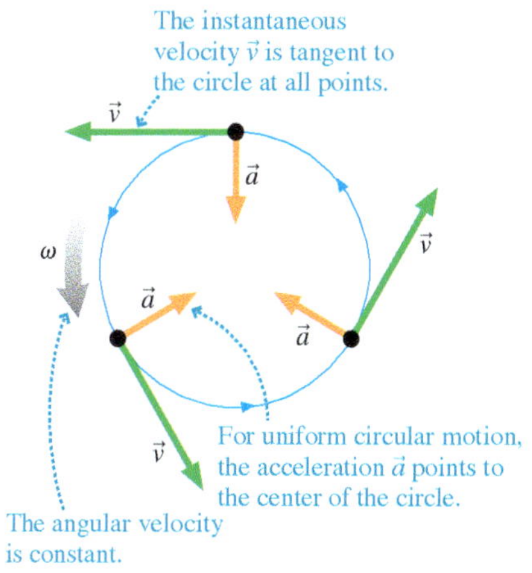

The tangential velocity v_t is positive for ccw motion, negative for cw motion. Because v_t is the only nonzero component of $\vec{v}$, the particle's speed is $v = |v_t| = |\omega|r$. We'll sometimes write this as $v = \omega r$ if there's no ambiguity about the sign of ω.

As a simple example, a particle moving cw at 2.0 m/s in a circle of radius 40 cm has angular velocity

$$\omega = \frac{v_t}{r} = \frac{-2.0\ \text{m/s}}{0.40\ \text{m}} = -5.0\ \text{rad/s}$$

where v_t and ω are negative because the motion is clockwise. Notice the units. Velocity divided by distance has units of s^{-1}. But because the division, in this case, gives us an angular quantity, we've inserted the *dimensionless* unit rad to give ω the appropriate units of rad/s.

Acceleration

The centripetal acceleration is enormous in a high-speed centrifuge.

Figure 4.1 at the beginning of this chapter looked at the uniform circular motion of a Ferris wheel. You are strongly encouraged to review that figure. There we found that a particle in uniform circular motion, although moving with constant speed, has an acceleration because the *direction* of the velocity vector $\vec{v}$ is always changing. The motion-diagram analysis showed that the **acceleration $\vec{a}$ points toward the center of the circle.** The instantaneous velocity is tangent to the circle, so $\vec{v}$ and $\vec{a}$ are perpendicular to each other at all points on the circle, as Figure 4.31 shows.

The acceleration of uniform circular motion is called **centripetal acceleration,** a term from a Greek root meaning "center seeking." Centripetal acceleration is not a new type of acceleration; all we are doing is *naming* an acceleration that corresponds to a particular type of motion. The magnitude of the centripetal acceleration is constant because each successive $\Delta\vec{v}$ in the motion diagram has the same length.

The motion diagram tells us the direction of $\vec{a}$, but it doesn't give us a value for a. To complete our description of uniform circular motion, we need to find a quantitative relationship between a and the particle's speed v. FIGURE 4.32 shows the velocity $\vec{v}_i$ at one instant of motion and the velocity $\vec{v}_f$ an infinitesimal amount of time dt later. During this small interval of time, the particle has moved through the infinitesimal angle $d\theta$ and traveled distance $ds = r\,d\theta$.

FIGURE 4.32 Finding the acceleration of circular motion.

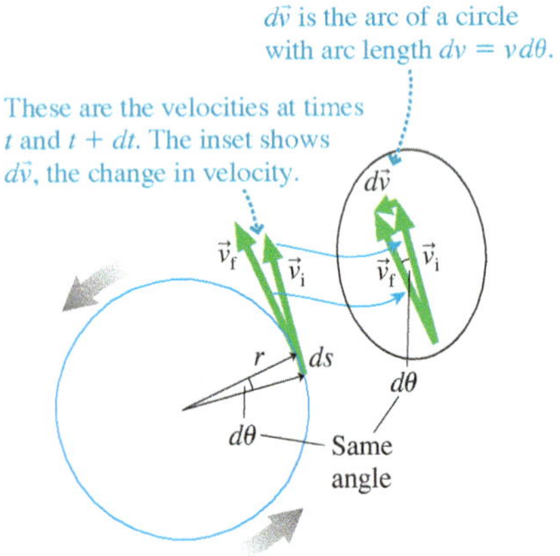

By definition, the acceleration is $\vec{a} = d\vec{v}/dt$. We can see from the inset to Figure 4.32 that $d\vec{v}$ points toward the center of the circle—that is, $\vec{a}$ is a centripetal acceleration. To find the magnitude of $\vec{a}$, we can see from the isosceles triangle of velocity vectors that, if $d\theta$ is in radians,

$$dv = |d\vec{v}| = v\,d\theta \tag{4.30}$$

For uniform circular motion at constant speed, $v = ds/dt = r\,d\theta/dt$ and thus the time to rotate through angle $d\theta$ is

$$dt = \frac{r\,d\theta}{v} \tag{4.31}$$

Combining Equations 4.30 and 4.31, we see that the acceleration has magnitude

$$a = |\vec{a}| = \frac{|d\vec{v}|}{dt} = \frac{v\,d\theta}{r\,d\theta/v} = \frac{v^2}{r}$$

In vector notation, we can write

$$\vec{a} = \left(\frac{v^2}{r},\ \text{toward center of circle}\right) \qquad \text{(centripetal acceleration)} \tag{4.32}$$

Using Equation 4.29, $v = \omega r$, we can also express the magnitude of the centripetal acceleration in terms of the angular velocity ω as

$$a = \omega^2 r \tag{4.33}$$

NOTE ▶ Centripetal acceleration is not a constant acceleration. The magnitude of the centripetal acceleration is constant during uniform circular motion, but the direction of $\vec{a}$ is constantly changing. **Thus the constant-acceleration kinematics equations of Chapter 2 do *not* apply to circular motion.** ◀

EXAMPLE 4.12 The acceleration of a Ferris wheel

A typical carnival Ferris wheel has a radius of 9.0 m and rotates 4.0 times per minute. What magnitude acceleration do the riders experience?

MODEL Model the rider as a particle in uniform circular motion.

SOLVE The period is $T = \frac{1}{4}$ min = 15 s. From Equation 4.22, a rider's speed is

$$v = \frac{2\pi r}{T} = \frac{2\pi(9.0\text{ m})}{15\text{ s}} = 3.77\text{ m/s}$$

Consequently, the centripetal acceleration is

$$a = \frac{v^2}{r} = \frac{(3.77\text{ m/s})^2}{9.0\text{ m}} = 1.6\text{ m/s}^2$$

ASSESS This was not intended to be a profound problem, merely to illustrate how centripetal acceleration is computed. The acceleration is enough to be noticed and make the ride interesting, but not enough to be scary.

STOP TO THINK 4.6 Rank in order, from largest to smallest, the centripetal accelerations a_a to a_e of particles a to e.

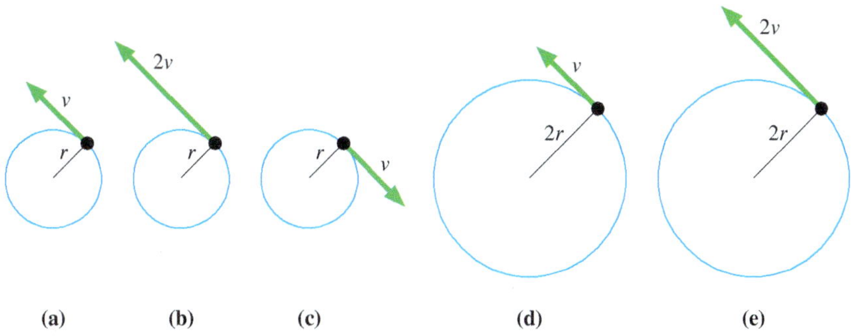

4.7 Nonuniform Circular Motion and Angular Acceleration

A roller coaster car doing a loop-the-loop slows down as it goes up one side, speeds up as it comes back down the other. The ball in a roulette wheel gradually slows until it stops. Circular motion with a changing speed is called **nonuniform circular motion.**

To begin our analysis of nonuniform circular motion, FIGURE 4.33 shows a wheel rotating on an axle. Notice that two points on the wheel, marked with dots, turn through the *same angle* as the wheel rotates, even though their radii may differ. That is, $\Delta\theta_1 = \Delta\theta_2$ during some time interval Δt. As a consequence, any two points on a rotating object have equal angular velocities, $\omega_1 = \omega_2$, and we can refer to ω as the angular velocity *of the wheel.*

Suppose the wheel's rotation is speeding up or slowing down—that is, points on the wheel have nonuniform circular motion. For linear motion, we defined acceleration as $a_x = dv_x/dt$. By analogy, let's define the **angular acceleration** α (Greek alpha) of a rotating object, or a point on the object, to be

$$\alpha \equiv \frac{d\omega}{dt} \quad \text{(angular acceleration)} \tag{4.34}$$

The units of angular acceleration are rad/s^2. Angular acceleration is the *rate* at which the angular velocity ω changes, just as linear acceleration is the rate at which the linear velocity v_x changes. FIGURE 4.34 on the next page illustrates this idea.

FIGURE 4.33 All points on the wheel rotate with the same angular velocity.

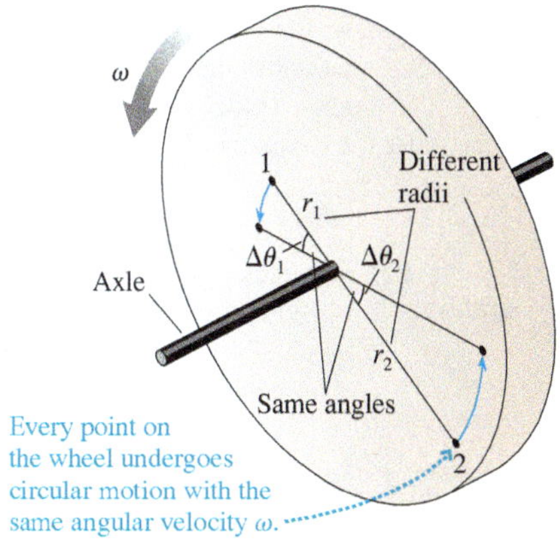

FIGURE 4.34 A wheel with angular acceleration $\alpha = 2\,\text{rad/s}^2$.

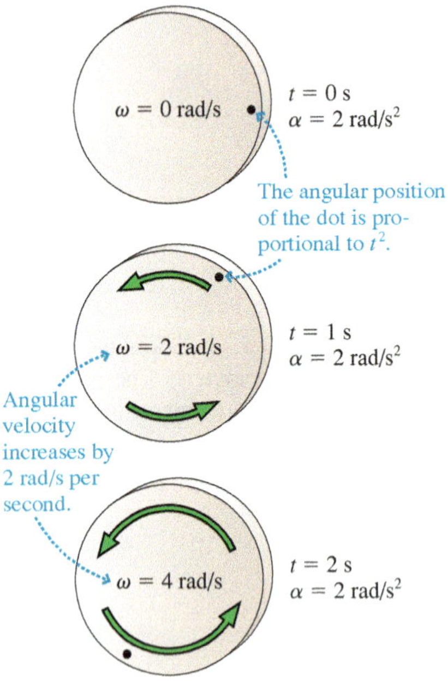

For linear acceleration, positive a_x means that v_x is increasing to the right or decreasing to the left; negative a_x means that v_x is increasing to the left or decreasing to the right. For rotational motion, α is positive if ω is increasing ccw (the direction of positive angle) or decreasing cw, negative if ω is increasing cw or decreasing ccw. These ideas are illustrated in FIGURE 4.35.

FIGURE 4.35 The signs of angular velocity and acceleration. The rotation is speeding up if ω and α have the same sign, slowing down if they have opposite signs.

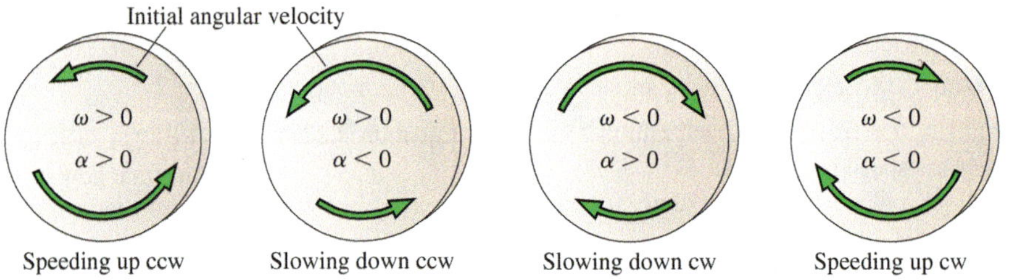

NOTE ▶ Be careful with the sign of α. You learned in Chapter 2 that positive and negative values of the acceleration can't be interpreted as simply "speeding up" and "slowing down." Similarly, positive and negative values of angular acceleration can't be interpreted as a rotation that is speeding up or slowing down. ◀

Because α is the time derivative of ω, we can use exactly the same graphical relationships that we found for linear motion:

$$\alpha = \text{slope of the } \omega\text{-versus-}t \text{ graph at time } t$$
$$\omega_f = \omega_i + \text{area under the } \alpha\text{-versus-}t \text{ curve between } t_i \text{ and } t_f \qquad (4.35)$$

These relationships involving slopes and areas are illustrated in the following example.

EXAMPLE 4.13 A rotating wheel

FIGURE 4.36a is a graph of angular velocity versus time for a rotating wheel. Describe the motion and draw a graph of angular acceleration versus time.

SOLVE This is a wheel that starts from rest, gradually speeds up *counterclockwise* until reaching top speed at t_1, maintains a constant angular velocity until t_2, then gradually slows down until stopping at t_3. The motion is always ccw because ω is always positive. The angular acceleration graph of FIGURE 4.36b is based on the fact that α is the slope of the ω-versus-t graph.

Conversely, the initial linear increase of ω can be seen as the increasing area under the α-versus-t graph as t increases from 0 to t_1. The angular velocity doesn't change from t_1 to t_2 when the area under the α-versus-t is zero.

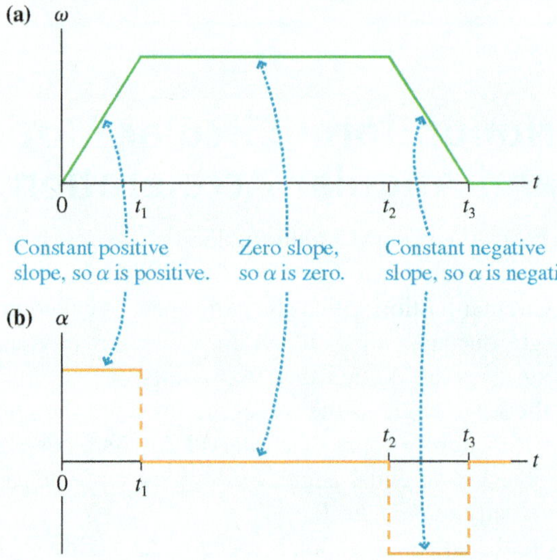

FIGURE 4.36 ω-versus-t graph and the corresponding α-versus-t graph for a rotating wheel.

Table 4.1 shows the kinematic equations for rotational motion with constant angular acceleration. These equations apply to a particle in circular motion or to the rotation of a rigid object. **The rotational kinematic equations are exactly analogous to the linear kinematic equations,** as they must be since the mathematical relationships among θ, ω, and α are identical to the relationships among x, v_x, and a_x. Thus all the problem-solving techniques you learned in Chapter 2 for linear motion carry over to circular and rotational motion.

TABLE 4.1 Rotational and linear kinematics for constant acceleration

Rotational kinematics	Linear kinematics
$\omega_f = \omega_i + \alpha\,\Delta t$	$v_{fs} = v_{is} + a_s\,\Delta t$
$\theta_f = \theta_i + \omega_i\,\Delta t + \frac{1}{2}\alpha(\Delta t)^2$	$s_f = s_i + v_{is}\,\Delta t + \frac{1}{2}a_s(\Delta t)^2$
$\omega_f^2 = \omega_i^2 + 2\alpha\,\Delta\theta$	$v_{fs}^2 = v_{is}^2 + 2a_s\,\Delta s$

EXAMPLE 4.14 **Back to the roulette wheel**

A small steel roulette ball rolls around the inside of a 30-cm-diameter roulette wheel. It is spun at 150 rpm, but it slows to 60 rpm over an interval of 5.0 s. How many revolutions does the ball make during these 5.0 s?

MODEL The ball is a particle in nonuniform circular motion. Assume constant angular acceleration as it slows.

SOLVE During these 5.0 s the ball rotates through angle

$$\Delta\theta = \theta_f - \theta_i = \omega_i\,\Delta t + \tfrac{1}{2}\alpha(\Delta t)^2$$

where $\Delta t = 5.0$ s. We can find the angular acceleration from the initial and final angular velocities, but first they must be converted to SI units:

$$\omega_i = 150\,\frac{\text{rev}}{\text{min}} \times \frac{1\text{ min}}{60\text{ s}} \times \frac{2\pi\text{ rad}}{1\text{ rev}} = 15.71\text{ rad/s}$$

$$\omega_f = 60\,\frac{\text{rev}}{\text{min}} = 0.40\omega_i = 6.28\text{ rad/s}$$

The angular acceleration α is

$$\alpha = \frac{\Delta\omega}{\Delta t} = \frac{6.28\text{ rad/s} - 15.71\text{ rad/s}}{5.0\text{ s}} = -1.89\text{ rad/s}^2$$

Thus the ball rotates through angle

$$\Delta\theta = (15.71\text{ rad/s})(5.0\text{ s}) + \tfrac{1}{2}(-1.89\text{ rad/s}^2)(5.0\text{ s})^2 = 54.9\text{ rad}$$

Because $54.9/2\pi = 8.75$, the ball completes $8\frac{3}{4}$ revolutions as it slows to 60 rpm.

ASSESS This problem is solved just like the linear kinematics problems you learned to solve in Chapter 2.

Tangential Acceleration

FIGURE 4.37 shows a particle in nonuniform circular motion. *Any* circular motion, whether uniform or nonuniform, has a centripetal acceleration because the particle is changing direction; this was the acceleration component $\vec{a}_\perp$ of Figure 4.10. The centripetal acceleration, which points radially toward the center of the circle, will now be called the **radial acceleration** a_r. The centripetal expression $a_r = v_t^2/r = \omega^2 r$ is still valid in nonuniform circular motion.

For a particle to speed up or slow down as it moves around a circle, it needs—in addition to the centripetal acceleration—an acceleration parallel to the trajectory or, equivalently, parallel to $\vec{v}$. This is the acceleration component $\vec{a}_\parallel$ associated with changing speed. We'll call this the **tangential acceleration** a_t because, like the velocity v_t, it is always tangent to the circle. Because of the tangential acceleration, **the acceleration vector $\vec{a}$ of a particle in nonuniform circular motion does not point toward the center of the circle.** It points "ahead" of center for a particle that is speeding up, as in Figure 4.37, but it would point "behind" center for a particle slowing down. You can see from Figure 4.37 that the magnitude of the acceleration is

$$a = \sqrt{a_r^2 + a_t^2} \tag{4.36}$$

If a_t is constant, then the arc length s traveled by the particle around the circle and the tangential velocity v_t are found from constant-acceleration kinematics:

$$\begin{aligned} s_f &= s_i + v_{it}\,\Delta t + \tfrac{1}{2}a_t(\Delta t)^2 \\ v_{ft} &= v_{it} + a_t\,\Delta t \end{aligned} \tag{4.37}$$

Because tangential acceleration is the rate at which the tangential velocity changes, $a_t = dv_t/dt$, and we already know that the tangential velocity is related to the angular velocity by $v_t = \omega r$, it follows that

$$a_t = \frac{dv_t}{dt} = \frac{d(\omega r)}{dt} = \frac{d\omega}{dt}r = \alpha r \tag{4.38}$$

FIGURE 4.37 Acceleration in nonuniform circular motion.

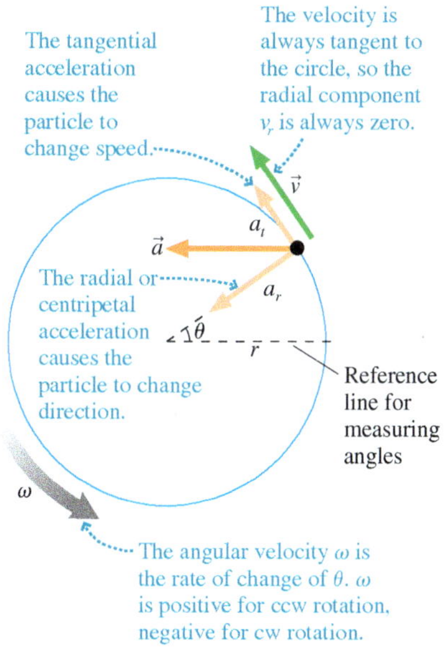

Thus $v_t = \omega r$ and $a_t = \alpha r$ are analogous equations for the tangential velocity and acceleration. In Example 4.14, where we found the roulette ball to have angular acceleration $\alpha = -1.89\ \text{rad/s}^2$, its tangential acceleration was

$$a_t = \alpha r = (-1.89\ \text{rad/s}^2)(0.15\ \text{m}) = -0.28\ \text{m/s}^2$$

EXAMPLE 4.15 Analyzing rotational data

You've been assigned the task of measuring the start-up characteristics of a large industrial motor. After several seconds, when the motor has reached full speed, you know that the angular acceleration will be zero, but you hypothesize that the angular acceleration may be constant during the first couple of seconds as the motor speed increases. To find out, you attach a shaft encoder to the 3.0-cm-diameter axle. A shaft encoder is a device that converts the angular position of a shaft or axle to a signal that can be read by a computer. After setting the computer program to read four values a second, you start the motor and acquire the following data:

Time (s)	Angle(°)
0.00	0
0.25	16
0.50	69
0.75	161
1.00	267
1.25	428
1.50	620

a. Do the data support your hypothesis of a constant angular acceleration? If so, what is the angular acceleration? If not, is the angular acceleration increasing or decreasing with time?
b. A 76-cm-diameter blade is attached to the motor shaft. At what time does the acceleration of the tip of the blade reach $10\ \text{m/s}^2$?

MODEL The axle is rotating with nonuniform circular motion. Model the tip of the blade as a particle.

VISUALIZE FIGURE 4.38 shows that the blade tip has both a tangential and a radial acceleration.

FIGURE 4.38 Pictorial representation of the axle and blade.

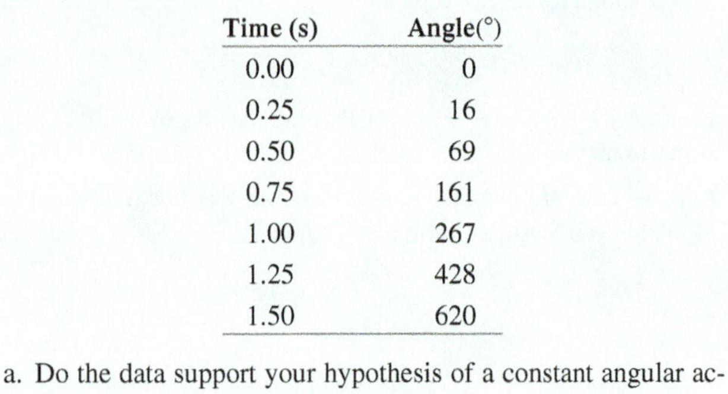

SOLVE

a. *If* the motor starts up with constant angular acceleration, with $\theta_i = 0$ and $\omega_i = 0$ rad/s, the angle-time equation of rotational kinematics is $\theta = \frac{1}{2}\alpha t^2$. This can be written as a linear equation $y = mx + b$ if we let $\theta = y$ and $t^2 = x$. That is, constant angular acceleration predicts that a graph of θ versus t^2 should be a straight line with slope $m = \frac{1}{2}\alpha$ and y-intercept $b = 0$. We can test this. If the graph turns out to be a straight line with zero y-intercept, it will confirm the hypothesis of constant angular acceleration and we can then use its slope to determine the angular acceleration: $\alpha = 2m$. If the graph is not a straight line, our observation of whether it curves upward or downward will tell us whether the angular acceleration us increasing or decreasing.

FIGURE 4.39 is the graph of θ versus t^2, and it confirms our hypothesis that the motor starts up with constant angular acceleration. The best-fit line, found using a spreadsheet, gives a slope of $274.6°/\text{s}^2$. The units come not from the spreadsheet but by looking at the units of rise (°) over run (s^2 because we're graphing t^2 on the x-axis). Thus the angular acceleration is

$$\alpha = 2m = 549.2°/\text{s}^2 \times \frac{\pi\ \text{rad}}{180°} = 9.6\ \text{rad/s}^2$$

where we used $180° = \pi$ rad to convert to SI units of rad/s^2.

FIGURE 4.39 Graph of θ versus t^2 for the motor shaft.

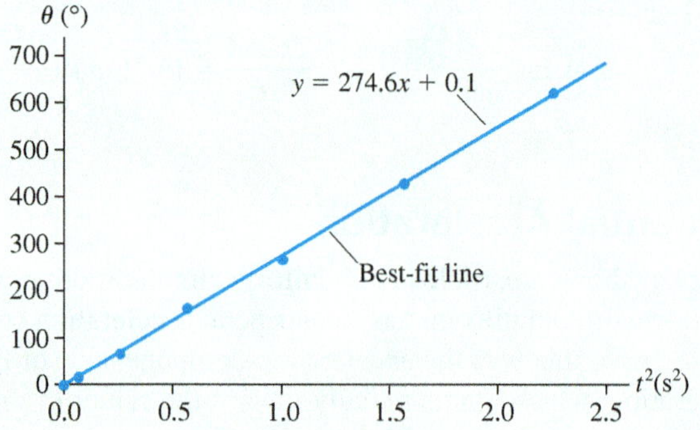

b. The magnitude of the linear acceleration is

$$a = \sqrt{a_r^2 + a_t^2}$$

Constant angular acceleration implies constant tangential acceleration, and the tangential acceleration of the blade tip is

$$a_t = \alpha r = (9.6\ \text{rad/s}^2)(0.38\ \text{m}) = 3.65\ \text{m/s}^2$$

We were careful to use the blade's radius, not its diameter, and we kept an extra significant figure to avoid round-off error. The radial (centripetal) acceleration increases as the rotation speed increases, and the total acceleration reaches $10\ \text{m/s}^2$ when

$$a_r = \sqrt{a^2 - a_t^2} = \sqrt{(10\ \text{m/s}^2)^2 - (3.65\ \text{m/s}^2)^2} = 9.31\ \text{m/s}^2$$

Radial acceleration is $a_r = \omega^2 r$, so the corresponding angular velocity is

$$\omega = \sqrt{\frac{a_r}{r}} = \sqrt{\frac{9.31\ \text{m/s}^2}{0.38\ \text{m}}} = 4.95\ \text{rad/s}$$

For constant angular acceleration, $\omega = \alpha t$, so this angular velocity is achieved at

$$t = \frac{\omega}{\alpha} = \frac{4.95\ \text{rad/s}}{9.6\ \text{rad/s}^2} = 0.52\ \text{s}$$

Thus it takes 0.52 s for the acceleration of the blade tip to reach $10\ \text{m/s}^2$.

ASSESS The motor has not completed 2 full revolutions in 1.5 s, so it has a slow start and modest accelerations. A tangential acceleration of $3.65\ \text{m/s}^2$ seems reasonable, so we have confidence in our final answer of 0.52 s.

STOP TO THINK 4.7 The fan blade is slowing down. What are the signs of ω and α?

a. ω is positive and α is positive.
b. ω is positive and α is negative.
c. ω is negative and α is positive.
d. ω is negative and α is negative.

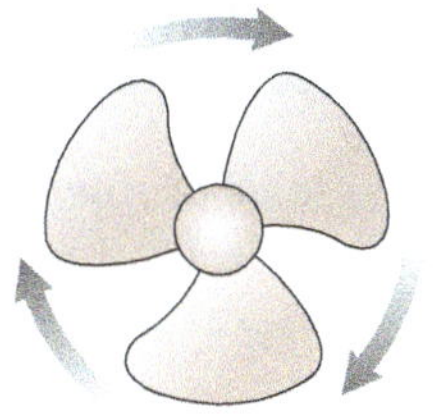

CHALLENGE EXAMPLE 4.16 Hit the target!

One day when you come into lab, you see a spring-loaded wheel that can launch a ball straight up. To do so, you place the ball in a cup on the rim of the wheel, turn the wheel to stretch the spring, then release. The wheel rotates through an angle $\Delta\theta$, then hits a stop when the cup is level with the axle and pointing straight up. The cup stops, but the ball flies out and keeps going. You're told that the wheel has been designed to have constant angular acceleration as it rotates through $\Delta\theta$. The lab assignment is first to measure the wheel's angular acceleration. Then the lab instructor is going to place a target at height h above the point where the ball is launched. Your task will be to launch the ball so that it just barely hits the target. You'll lose points if the ball doesn't reach the target or if it slams into the target.

a. Find an expression in terms of quantities that you can measure for the angle $\Delta\theta$ that launches the ball at the correct speed.
b. Evaluate $\Delta\theta$ if you've determined that the wheel's diameter is 62 cm, its angular acceleration is 200 rad/s^2, the mass of the ball is 25 g, and the instructor places the target 190 cm above the launch point.

MODEL Model the ball as a particle. It first undergoes nonuniform circular motion. We'll then ignore air resistance and treat the vertical motion as free fall.

VISUALIZE FIGURE 4.40 is a pictorial representation. This is a two-part problem, with the speed at the end of the angular acceleration being the launch speed for the vertical motion. We've chosen to call the wheel radius R and the target height h. These and the angular acceleration α are considered "known" because we will measure them, but we don't have numerical values at this time.

FIGURE 4.40 Pictorial representation of the ball launcher.

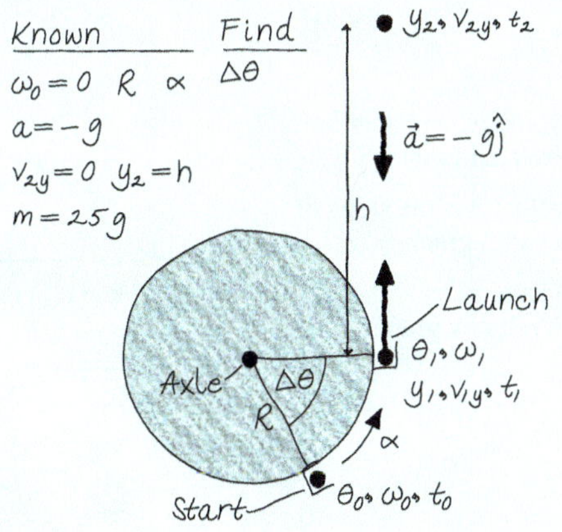

SOLVE

a. The circular motion problem and the vertical motion problem are connected through the ball's speed: The final speed of the angular acceleration is the launch speed of the vertical motion. We don't know anything about time intervals, which suggests using the kinematic equations that relate distance and acceleration (for the vertical motion) and angle and angular acceleration (for the circular motion). For the angular acceleration, with $\omega_0 = 0$ rad/s,

$$\omega_1^2 = \omega_0^2 + 2\alpha\,\Delta\theta = 2\alpha\,\Delta\theta$$

The final speed of the ball and cup, when the wheel hits the stop, is

$$v_1 = \omega_1 R = R\sqrt{2\alpha\,\Delta\theta}$$

Thus the vertical-motion problem begins with the ball being shot upward with velocity $v_{1y} = R\sqrt{2\alpha\,\Delta\theta}$. How high does it go? The highest point is the point where $v_{2y} = 0$, so the free-fall equation is

$$v_{2y}^2 = 0 = v_{1y}^2 - 2g\Delta y = R^2 \cdot 2\alpha\,\Delta\theta - 2gh$$

Rather than solve for height h, we need to solve for the angle that produces a given height. This is

$$\Delta\theta = \frac{gh}{\alpha R^2}$$

Once we've determined the properties of the wheel and then measured the height at which our instructor places the target, we'll quickly be able to calculate the angle through which we should pull back the wheel to launch the ball.

b. For the values given in the problem statement, $\Delta\theta = 0.969$ rad $= 56°$. Don't forget that equations involving angles need values in radians and return values in radians.

ASSESS The angle needed to be less than 90° or else the ball would fall out of the cup before launch. And an angle of only a few degrees would seem suspiciously small. Thus 56° seems to be reasonable. Notice that the mass was not needed in this problem. Part of becoming a better problem solver is evaluating the information you have to see what is relevant. Some homework problems will help you develop this skill by providing information that isn't necessary.

SUMMARY

The goal of Chapter 4 has been to learn how to solve problems about motion in a plane.

General Principles

The instantaneous velocity

$$\vec{v} = d\vec{r}/dt$$

is a vector tangent to the trajectory.

The instantaneous acceleration is

$$\vec{a} = d\vec{v}/dt$$

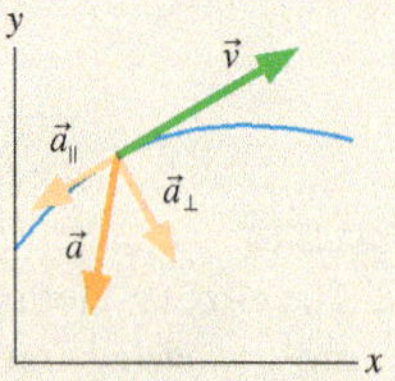

$\vec{a}_\parallel$, the component of $\vec{a}$ parallel to $\vec{v}$, is responsible for change of *speed*. $\vec{a}_\perp$, the component of $\vec{a}$ perpendicular to $\vec{v}$, is responsible for change of *direction*.

Relative motion

If object C moves relative to reference frame A with velocity $\vec{v}_{CA}$, then it moves relative to a different reference frame B with velocity

$$\vec{v}_{CB} = \vec{v}_{CA} + \vec{v}_{AB}$$

where $\vec{v}_{AB}$ is the velocity of A relative to B. This is the Galilean transformation of velocity.

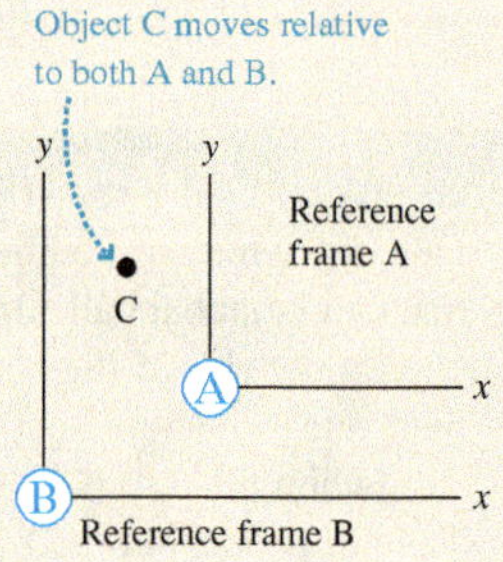

Important Concepts

Uniform Circular Motion

Angular velocity $\omega = d\theta/dt$.

v_t and ω are constant:

$$v_t = \omega r$$

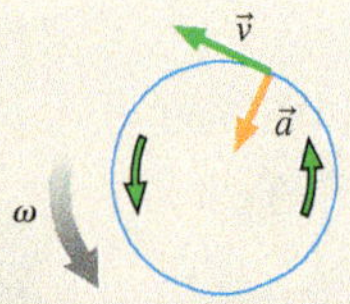

The centripetal acceleration points toward the center of the circle:

$$a = \frac{v^2}{r} = \omega^2 r$$

It changes the particle's direction but not its speed.

Nonuniform Circular Motion

Angular acceleration $\alpha = d\omega/dt$.

The radial acceleration

$$a_r = \frac{v^2}{r} = \omega^2 r$$

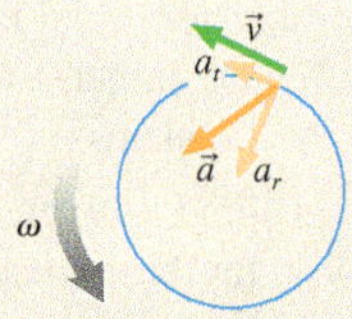

changes the particle's direction. The tangential component

$$a_t = \alpha r$$

changes the particle's speed.

Applications

Kinematics in two dimensions

If $\vec{a}$ is constant, then the x- and y-components of motion are independent of each other.

$$x_f = x_i + v_{ix}\,\Delta t + \tfrac{1}{2}a_x(\Delta t)^2$$

$$y_f = y_i + v_{iy}\,\Delta t + \tfrac{1}{2}a_y(\Delta t)^2$$

$$v_{fx} = v_{ix} + a_x\,\Delta t$$

$$v_{fy} = v_{iy} + a_y\,\Delta t$$

Projectile motion occurs if the object moves under the influence of only gravity. The motion is a parabola.

- Uniform motion in the horizontal direction with $v_{0x} = v_0\cos\theta$.
- Free-fall motion in the vertical direction with $a_y = -g$ and $v_{0y} = v_0\sin\theta$.
- The x and y kinematic equations have the *same* value for Δt.

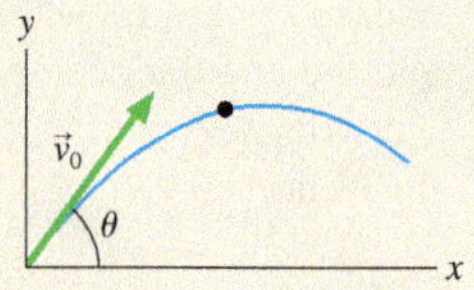

Circular motion kinematics

Period $T = \dfrac{2\pi r}{v} = \dfrac{2\pi}{\omega}$

Angular position $\theta = \dfrac{s}{r}$

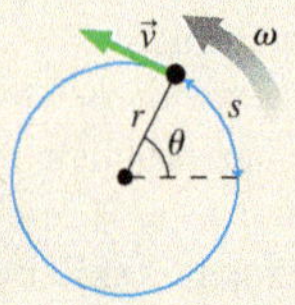

$$\omega_f = \omega_i + \alpha\,\Delta t$$

$$\theta_f = \theta_i + \omega_i\,\Delta t + \tfrac{1}{2}\alpha(\Delta t)^2$$

$$\omega_f^2 = \omega_i^2 + 2\alpha\,\Delta\theta$$

Angle, angular velocity, and angular acceleration are related graphically.

- The angular velocity is the slope of the angular position graph.
- The angular acceleration is the slope of the angular velocity graph.

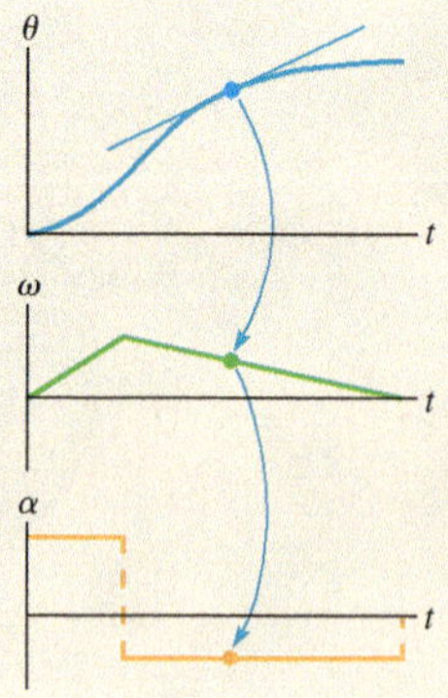

Terms and Notation

projectile	uniform circular motion	angular displacement, $\Delta\theta$	radial acceleration, a_r
launch angle, θ	period, T	angular velocity, ω	tangential acceleration, a_t
reference frame	angular position, θ	centripetal acceleration, a	
Galilean transformation of velocity	arc length, s	nonuniform circular motion	
	radians	angular acceleration, α	

CONCEPTUAL QUESTIONS

1. a. At this instant, is the particle in FIGURE Q4.1 speeding up, slowing down, or traveling at constant speed?
   b. Is this particle curving to the right, curving to the left, or traveling straight?

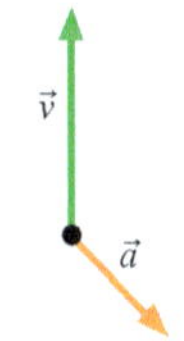

FIGURE Q4.1

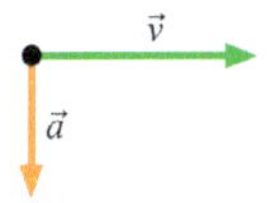

FIGURE Q4.2

2. a. At this instant, is the particle in FIGURE Q4.2 speeding up, slowing down, or traveling at constant speed?
   b. Is this particle curving upward, curving downward, or traveling straight?
3. Tarzan swings through the jungle by hanging from a vine.
   a. Immediately after stepping off a branch to swing over to another tree, is Tarzan's acceleration $\vec{a}$ zero or not zero? If not zero, which way does it point? Explain.
   b. Answer the same question at the lowest point in Tarzan's swing.
4. A projectile is launched at an angle of 30°.
   a. Is there any point on the trajectory where $\vec{v}$ and $\vec{a}$ are parallel to each other? If so, where?
   b. Is there any point where $\vec{v}$ and $\vec{a}$ are perpendicular to each other? If so, where?
5. For a projectile, which of the following quantities are constant during the flight: x, y, r, v_x, v_y, v, a_x, a_y? Which of these quantities are zero throughout the flight?
6. A cart that is rolling at constant velocity on a level table fires a ball straight up.
   a. When the ball comes back down, will it land in front of the launching tube, behind the launching tube, or directly in the tube? Explain.
   b. Will your answer change if the cart is accelerating in the forward direction? If so, how?
7. A rock is thrown from a bridge at an angle 30° below horizontal. Immediately after the rock is released, is the magnitude of its acceleration greater than, less than, or equal to g? Explain.
8. Anita is running to the right at 5 m/s in FIGURE Q4.8. Balls 1 and 2 are thrown toward her by friends standing on the ground. According to Anita, both balls are approaching her at 10 m/s. Which ball was thrown at a faster speed? Or were they thrown with the same speed? Explain.

FIGURE Q4.8

9. An electromagnet on the ceiling of an airplane holds a steel ball. When a button is pushed, the magnet releases the ball. The experiment is first done while the plane is parked on the ground, and the point where the ball hits the floor is marked with an X. Then the experiment is repeated while the plane is flying level at a steady 500 mph. Does the ball land slightly in front of the X (toward the nose of the plane), on the X, or slightly behind the X (toward the tail of the plane)? Explain.
10. Zack is driving past his house in FIGURE Q4.10. He wants to toss his physics book out the window and have it land in his driveway. If he lets go of the book exactly as he passes the end of the driveway, should he direct his throw outward and toward the front of the car (throw 1), straight outward (throw 2), or outward and toward the back of the car (throw 3)? Explain.

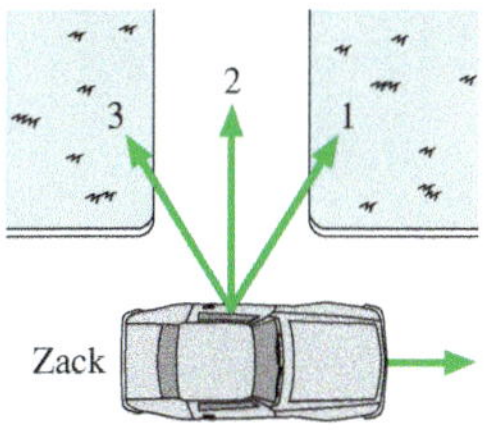

FIGURE Q4.10

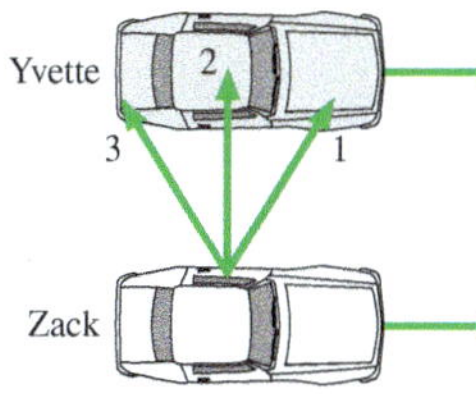

FIGURE Q4.11

11. In FIGURE Q4.11, Yvette and Zack are driving down the freeway side by side with their windows down. Zack wants to toss his physics book out the window and have it land in Yvette's front seat. Ignoring air resistance, should he direct his throw outward and toward the front of the car (throw 1), straight outward (throw 2), or outward and toward the back of the car (throw 3)? Explain.
12. In uniform circular motion, which of the following quantities are constant: speed, instantaneous velocity, tangential velocity, radial acceleration, tangential acceleration? Which of these quantities are zero throughout the motion?

13. FIGURE Q4.13 shows three points on a steadily rotating wheel.
    a. Rank in order, from largest to smallest, the angular velocities ω_1, ω_2, and ω_3 of these points. Explain.
    b. Rank in order, from largest to smallest, the speeds v_1, v_2, and v_3 of these points. Explain.

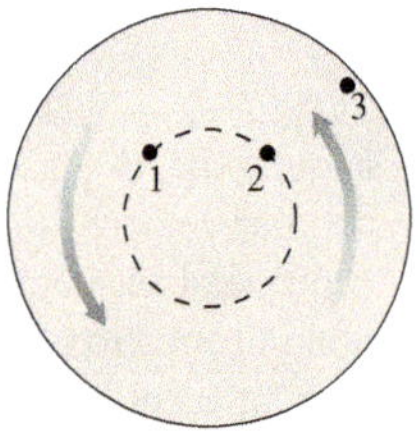

FIGURE Q4.13

14. FIGURE Q4.14 shows four rotating wheels. For each, determine the signs (+ or −) of ω and α.

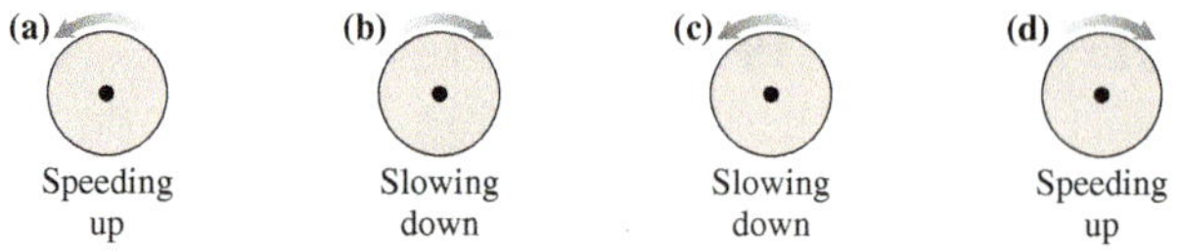

FIGURE Q4.14

15. FIGURE Q4.15 shows a pendulum at one end point of its arc.
    a. At this point, is ω positive, negative, or zero? Explain.
    b. At this point, is α positive, negative, or zero? Explain.

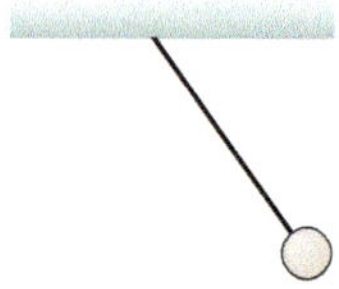

FIGURE Q4.15

EXERCISES AND PROBLEMS

Exercises

Section 4.1 Acceleration

Problems 1 and 2 show a partial motion diagram. For each:
a. Complete the motion diagram by adding acceleration vectors.
b. Write a physics *problem* for which this is the correct motion diagram. Be imaginative! Don't forget to include enough information to make the problem complete and to state clearly what is to be found.

1. |

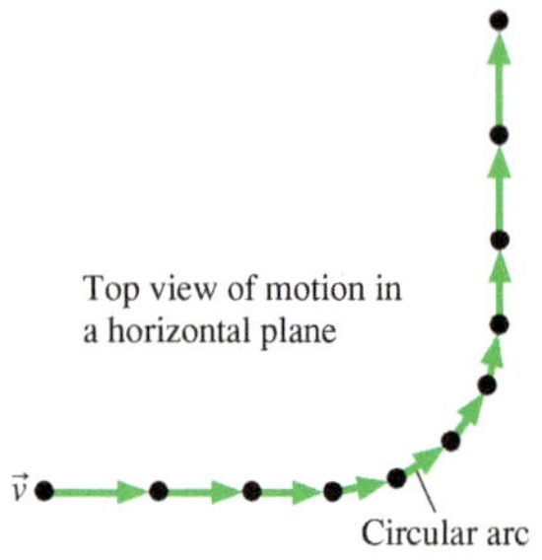

FIGURE EX4.1

2. |

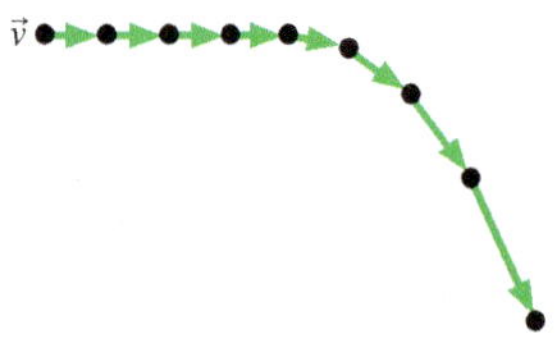

FIGURE EX4.2

Answer Problems 3 through 5 by choosing one of the eight labeled acceleration vectors or selecting option I: $\vec{a} = \vec{0}$.

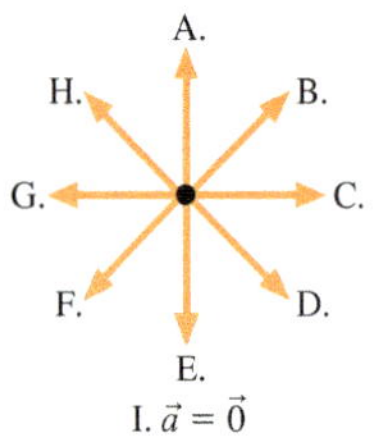

3. | At this instant, the particle is slowing and curving upward. What is the direction of its acceleration?

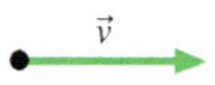

FIGURE EX4.3

4. | At this instant, the particle has steady speed and is curving to the right. What is the direction of its acceleration?

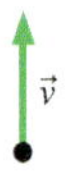

FIGURE EX4.4

5. | At this instant, the particle is speeding up and curving downward. What is the direction of its acceleration?

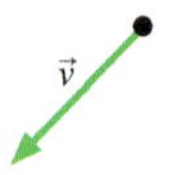

FIGURE EX4.5

Section 4.2 Two-Dimensional Kinematics

6. || A sailboat is traveling east at 5.0 m/s. A sudden gust of wind gives the boat an acceleration $\vec{a} = (0.80\ \text{m/s}^2,\ 40°\ \text{north of east})$. What are the boat's speed and direction 6.0 s later when the gust subsides?
7. || A model rocket is launched from rest with an upward acceleration of 6.00 m/s^2 and, due to a strong wind, a horizontal acceleration of 1.50 m/s^2 How far is the rocket from the launch pad 6.00 s later when the rocket engine runs out of fuel?
8. || A particle's trajectory is described by $x = \left(\frac{1}{2}t^3 - 2t^2\right)$ m and $y = \left(\frac{1}{2}t^2 - 2t\right)$ m, where t is in s.
    a. What are the particle's position and speed at $t = 0$ s and $t = 4$ s?
    b. What is the particle's direction of motion, measured as an angle from the x-axis, at $t = 0$ s and $t = 4$ s?
9. || A rocket-powered hockey puck moves on a horizontal frictionless table. FIGURE EX4.9 shows graphs of v_x and v_y, the x- and y-components of the puck's velocity. The puck starts at the origin.
    a. In which direction is the puck moving at $t = 2$ s? Give your answer as an angle from the x-axis.
    b. How far from the origin is the puck at $t = 5$ s?

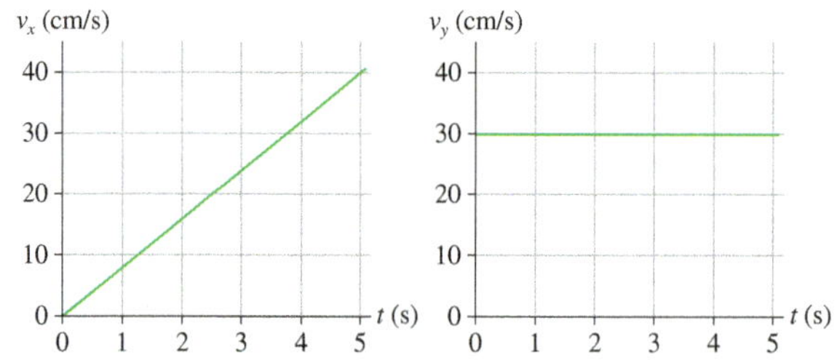

FIGURE EX4.9

10. || A rocket-powered hockey puck moves on a horizontal frictionless table. FIGURE EX4.10 shows graphs of v_x and v_y, the x- and y-components of the puck's velocity. The puck starts at the origin. What is the magnitude of the puck's acceleration at $t = 5$ s?

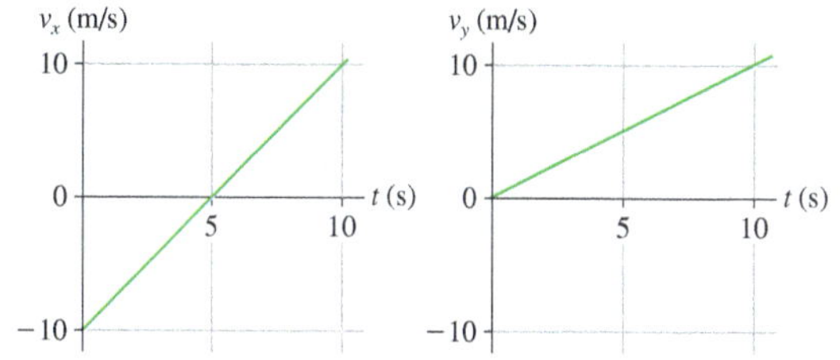

FIGURE EX4.10

Section 4.3 Projectile Motion

11. | A physics student on Planet Exidor throws a ball, and it follows the parabolic trajectory shown in FIGURE EX4.11. The ball's position is shown at 1 s intervals until $t = 3$ s. At $t = 1$ s, the ball's velocity is $\vec{v} = (2.0\hat{\imath} + 2.0\hat{\jmath})$ m/s.
    a. Determine the ball's velocity at $t = 0$ s, 2 s, and 3 s.
    b. What is the value of g on Planet Exidor?
    c. What was the ball's launch angle?

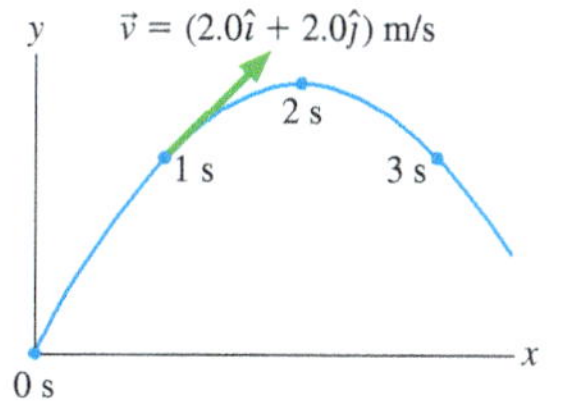

FIGURE EX4.11

12. | A ball thrown horizontally at 25 m/s travels a horizontal distance of 50 m before hitting the ground. From what height was the ball thrown?
13. || A rifle is aimed horizontally at a target 50 m away. The bullet hits the target 2.0 cm below the aim point.
    a. What was the bullet's flight time?
    b. What was the bullet's speed as it left the barrel?
14. || A supply plane needs to drop a package of food to scientists working on a glacier in Greenland. The plane flies 100 m above the glacier at a speed of 150 m/s. How far short of the target should it drop the package?

Section 4.4 Relative Motion

15. || A boat takes 3.0 hours to travel 30 km down a river, then 5.0 hours to return. How fast is the river flowing?
16. || When the moving sidewalk at the airport is broken, as it often seems to be, it takes you 50 s to walk from your gate to baggage claim. When it is working and you stand on the moving sidewalk the entire way, without walking, it takes 75 s to travel the same distance. How long will it take you to travel from the gate to baggage claim if you walk while riding on the moving sidewalk?
17. || Mary needs to row her boat across a 100-m-wide river that is flowing to the east at a speed of 1.0 m/s. Mary can row with a speed of 2.0 m/s.
    a. If Mary points her boat due north, how far from her intended landing spot will she be when she reaches the opposite shore?
    b. What is her speed with respect to the shore?
18. || Susan, driving north at 60 mph, and Trent, driving east at 45 mph, are approaching an intersection. What is Trent's speed relative to Susan's reference frame?

Section 4.5 Uniform Circular Motion

19. | FIGURE EX4.19 shows the angular-position-versus-time graph for a particle moving in a circle. What is the particle's angular velocity at (a) $t = 1$ s, (b) $t = 4$ s, and (c) $t = 7$ s?

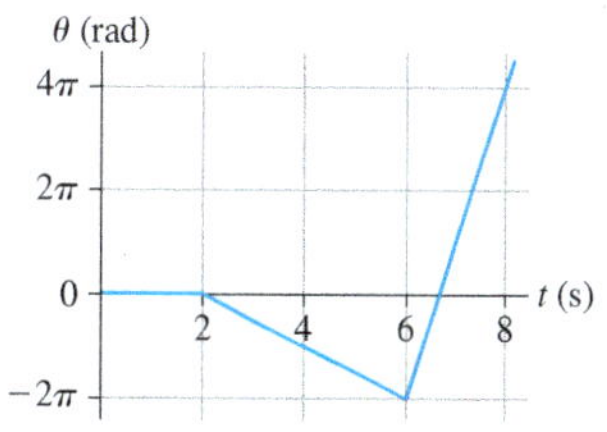

FIGURE EX4.19

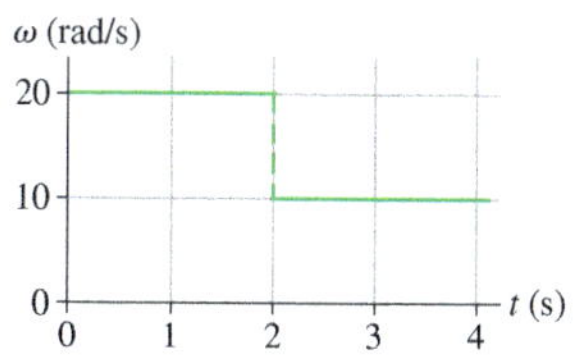

FIGURE EX4.20

20. || FIGURE EX4.20 shows the angular-velocity-versus-time graph for a particle moving in a circle. How many revolutions does the object make during the first 4 s?
21. | FIGURE EX4.21 shows the angular-velocity-versus-time graph for a particle moving in a circle, starting from $\theta_0 = 0$ rad at $t = 0$ s. Draw the angular-position-versus-time graph. Include an appropriate scale on both axes.

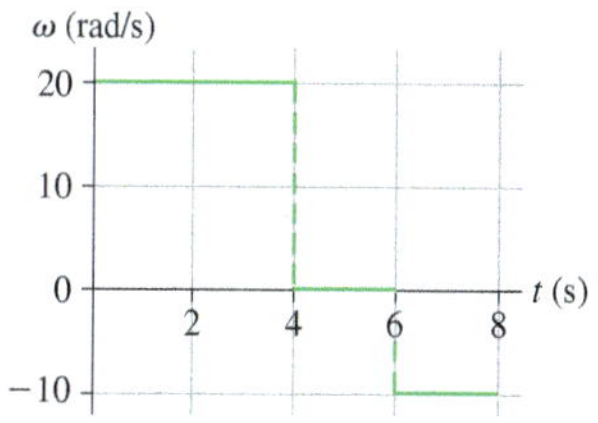

FIGURE EX4.21

22. | An old-fashioned single-play vinyl record rotates on a turntable at 45 rpm. What are (a) the angular velocity in rad/s and (b) the period of the motion?
23. || The earth's radius is about 4000 miles. Kampala, the capital of Uganda, and Singapore are both nearly on the equator. The distance between them is 5000 miles. The flight from Kampala to Singapore takes 9.0 hours. What is the plane's angular velocity with respect to the earth's surface? Give your answer in °/h.

Section 4.6 Velocity and Acceleration in Uniform Circular Motion

24. || A 3000-m-high mountain is located on the equator. How much faster does a climber on top of the mountain move than a surfer at a nearby beach? The earth's radius is 6400 km.
25. | How fast must a plane fly along the earth's equator so that the sun stands still relative to the passengers? In which direction must the plane fly, east to west or west to east? Give your answer in both km/h and mph. The earth's radius is 6400 km.
26. | BIO To withstand "g-forces" of up to 10 g's, caused by suddenly pulling out of a steep dive, fighter jet pilots train on a "human centrifuge." 10 g's is an acceleration of 98 m/s^2. If the length of the centrifuge arm is 12 m, at what speed is the rider moving when she experiences 10 g's?

27. | The radius of the earth's very nearly circular orbit around the sun is 1.5×10^{11} m. Find the magnitude of the earth's (a) velocity, (b) angular velocity, and (c) centripetal acceleration as it travels around the sun. Assume a year of 365 days.

28. || Your roommate is working on his bicycle and has the bike upside down. He spins the 60-cm-diameter wheel, and you notice that a pebble stuck in the tread goes by three times every second. What are the pebble's speed and acceleration?

Section 4.7 Nonuniform Circular Motion and Angular Acceleration

29. | FIGURE EX4.29 shows the angular velocity graph of the crankshaft in a car. What is the crankshaft's angular acceleration at (a) $t = 1$ s, (b) $t = 3$ s, and (c) $t = 5$ s?

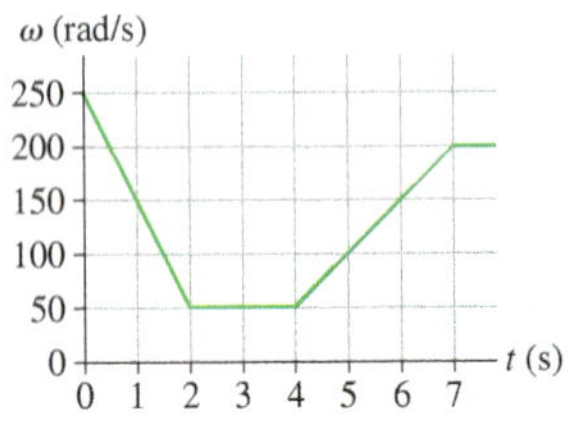

FIGURE EX4.29

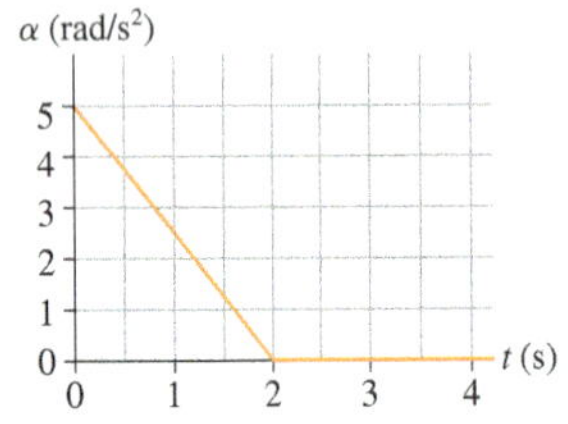

FIGURE EX4.30

30. || FIGURE EX4.30 shows the angular acceleration graph of a turntable that starts from rest. What is the turntable's angular velocity at (a) $t = 1$ s, (b) $t = 2$ s, and (c) $t = 3$ s?

31. || FIGURE EX4.31 shows the angular-velocity-versus-time graph for a particle moving in a circle. How many revolutions does the object make during the first 4 s?

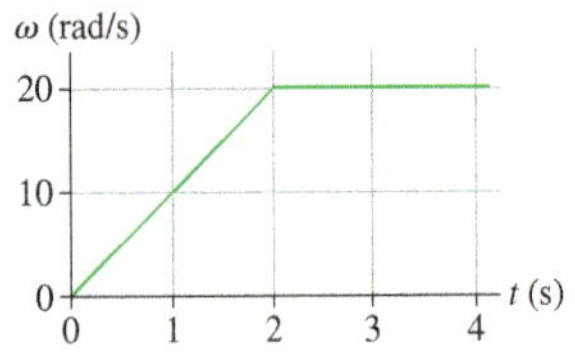

FIGURE EX4.31

32. || A 5.0-m-diameter merry-go-round is initially turning with a 4.0 s period. It slows down and stops in 20 s.
a. Before slowing, what is the speed of a child on the rim?
b. How many revolutions does the merry-go-round make as it stops?

33. || An electric fan goes from rest to 1800 rpm in 4.0 s. What is its angular acceleration?

34. ||| A bicycle wheel is rotating at 50 rpm when the cyclist begins to pedal harder, giving the wheel a constant angular acceleration of 0.50 rad/s^2.
a. What is the wheel's angular velocity, in rpm, 10 s later?
b. How many revolutions does the wheel make during this time?

35. ||| A 3.0-cm-diameter crankshaft that is rotating at 2500 rpm comes to a halt in 1.5 s.
a. What is the tangential acceleration of a point on the surface?
b. How many revolutions does the crankshaft make as it stops?

Problems

36. || A particle starts from rest at $\vec{r}_0 = 9.0\hat{j}$ m and moves in the xy-plane with the velocity shown in FIGURE P4.36. The particle passes through a wire hoop located at $\vec{r}_1 = 20\hat{i}$ m, then continues onward.
a. At what time does the particle pass through the hoop?
b. What is the value of v_{4y}, the y-component of the particle's velocity at $t = 4$ s?

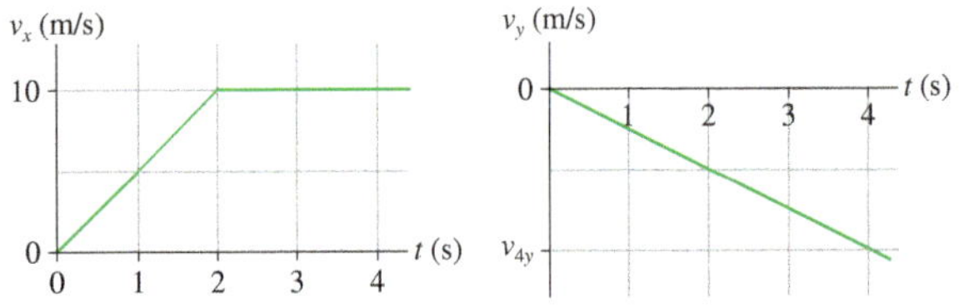

FIGURE EX4.36

37. || A spaceship maneuvering near Planet Zeta is located at $\vec{r} = (600\hat{i} - 400\hat{j} + 200\hat{k}) \times 10^3$ km, relative to the planet, and traveling at $\vec{v} = 9500\hat{i}$ m/s. It turns on its thruster engine and accelerates with $\vec{a} = (40\hat{i} - 20\hat{k})$ m/s^2 for 35 min. Where is the spaceship located when the engine shuts off? Give your answer as a vector measured in km.

38. || A projectile's horizontal range on level ground is $R = v_0^2 \sin 2\theta/g$. At what launch angle or angles will the projectile land at half of its maximum possible range?

39. || a. A projectile is launched with speed v_0 and angle θ. Derive an expression for the projectile's maximum height h.
b. A baseball is hit with a speed of 33.6 m/s. Calculate its height and the distance traveled if it is hit at angles of 30.0°, 45.0°, and 60.0°.

40. || A gray kangaroo can bound across level ground with each jump carrying it 10 m from the takeoff point. Typically the kangaroo leaves the ground at a 20° angle. If this is so:
BIO
a. What is its takeoff speed?
b. What is its maximum height above the ground?

41. || A projectile is fired with an initial speed of 30 m/s at an angle of 60° above the horizontal. The object hits the ground 7.5 s later.
a. How much higher or lower is the launch point relative to the point where the projectile hits the ground?
b. To what maximum height above the launch point does the projectile rise?

42. || In the Olympic shotput event, an athlete throws the shot with an initial speed of 12.0 m/s at a 40.0° angle from the horizontal. The shot leaves her hand at a height of 1.80 m above the ground.
a. How far does the shot travel?
b. Repeat the calculation of part (a) for angles 42.5°, 45.0°, and 47.5°. Put all your results, including 40.0°, in a table. At what angle of release does she throw the farthest?

43. || On the Apollo 14 mission to the moon, astronaut Alan Shepard hit a golf ball with a 6 iron. The free-fall acceleration on the moon is 1/6 of its value on earth. Suppose he hit the ball with a speed of 25 m/s at an angle 30° above the horizontal.
a. How much farther did the ball travel on the moon than it would have on earth?
b. For how much more time was the ball in flight?

44. || A ball is thrown toward a cliff of height h with a speed of 30 m/s and an angle of 60° above horizontal. It lands on the edge of the cliff 4.0 s later.
a. How high is the cliff?
b. What was the maximum height of the ball?
c. What is the ball's impact speed?

45. || A tennis player hits a ball 2.0 m above the ground. The ball leaves his racquet with a speed of 20.0 m/s at an angle 5.0° above the horizontal. The horizontal distance to the net is 7.0 m, and the net is 1.0 m high. Does the ball clear the net? If so, by how much? If not, by how much does it miss?

46. ||| A baseball player friend of yours wants to determine his pitching speed. You have him stand on a ledge and throw the ball horizontally from an elevation 4.0 m above the ground. The ball lands 25 m away.
   a. What is his pitching speed?
   b. As you think about it, you're not sure he threw the ball exactly horizontally. As you watch him throw, the pitches seem to vary from 5° below horizontal to 5° above horizontal. What are the lowest and highest speeds with which the ball might have left his hand?

47. || You are playing right field for the baseball team. Your team is up by one run in the bottom of the last inning of the game when a ground ball slips through the infield and comes straight toward you. As you pick up the ball 65 m from home plate, you see a runner rounding third base and heading for home with the tying run. You throw the ball at an angle of 30° above the horizontal with just the right speed so that the ball is caught by the catcher, standing on home plate, at the same height as you threw it. As you release the ball, the runner is 20 m from home plate and running full speed at 8.0 m/s. Will the ball arrive in time for your team's catcher to make the tag and win the game?

48. || You're 6.0 m from one wall of the house seen in FIGURE P4.48. You want to toss a ball to your friend who is 6.0 m from the opposite wall. The throw and catch each occur 1.0 m above the ground.
   a. What minimum speed will allow the ball to clear the roof?
   b. At what angle should you toss the ball?

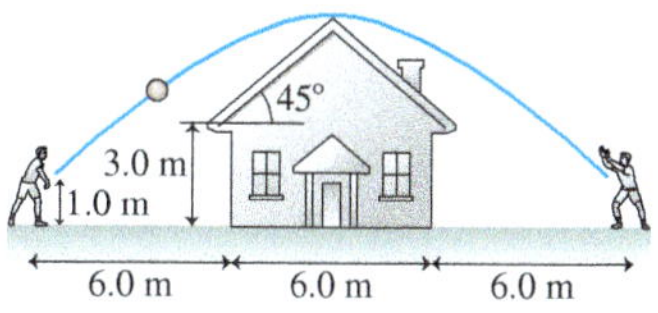

FIGURE P4.48

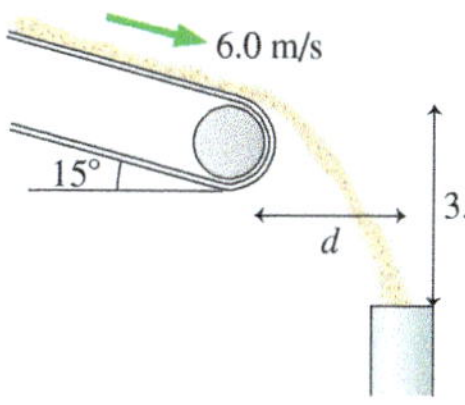

FIGURE P4.49

49. || Sand moves without slipping at 6.0 m/s down a conveyer that is tilted at 15°. The sand enters a pipe 3.0 m below the end of the conveyer belt, as shown in FIGURE P4.49. What is the horizontal distance *d* between the conveyer belt and the pipe?

50. || A stunt man drives a car at a speed of 20 m/s off a 30-m-high cliff. The road leading to the cliff is inclined upward at an angle of 20°.
   a. How far from the base of the cliff does the car land?
   b. What is the car's impact speed?

51. || BIO A javelin thrower standing at rest holds the center of the javelin behind her head, then accelerates it through a distance of 70 cm as she throws. She releases the javelin 2.0 m above the ground traveling at an angle of 30° above the horizontal. Top-rated javelin throwers do throw at about a 30° angle, not the 45° you might have expected, because the biomechanics of the arm allow them to throw the javelin much faster at 30° than they would be able to at 45°. In this throw, the javelin hits the ground 62 m away. What was the acceleration of the javelin during the throw? Assume that it has a constant acceleration.

52. || Ships A and B leave port together. For the next two hours, ship A travels at 20 mph in a direction 30° west of north while the ship B travels 20° east of north at 25 mph.
   a. What is the distance between the two ships two hours after they depart?
   b. What is the speed of ship A as seen by ship B?

53. || A kayaker needs to paddle north across a 100-m-wide harbor. The tide is going out, creating a tidal current that flows to the east at 2.0 m/s. The kayaker can paddle with a speed of 3.0 m/s.
   a. In which direction should he paddle in order to travel straight across the harbor?
   b. How long will it take him to cross?

54. || Mike throws a ball upward and toward the east at a 63° angle with a speed of 22 m/s. Nancy drives east past Mike at 30 m/s at the instant he releases the ball.
   a. What is the ball's initial angle in Nancy's reference frame?
   b. Find and graph the ball's trajectory as seen by Nancy.

55. || While driving north at 25 m/s during a rainstorm you notice that the rain makes an angle of 38° with the vertical. While driving back home moments later at the same speed but in the opposite direction, you see that the rain is falling straight down. From these observations, determine the speed and angle of the raindrops relative to the ground.

56. || You've been assigned the task of using a shaft encoder—a device that measures the angle of a shaft or axle and provides a signal to a computer—to analyze the rotation of an engine crankshaft under certain conditions. The table lists the crankshaft's angles over a 0.6 s interval.

Time (s)	Angle (rad)
0.0	0.0
0.1	2.0
0.2	3.2
0.3	4.3
0.4	5.3
0.5	6.1
0.6	7.0

Is the crankshaft rotating with uniform circular motion? If so, what is its angular velocity in rpm? If not, is the angular acceleration positive or negative?

57. || A speck of dust on a spinning DVD has a centripetal acceleration of 20 m/s^2.
   a. What is the acceleration of a different speck of dust that is twice as far from the center of the disk?
   b. What would be the acceleration of the first speck of dust if the disk's angular velocity was doubled?

58. || A typical laboratory centrifuge rotates at 4000 rpm. Test tubes have to be placed into a centrifuge very carefully because of the very large accelerations.
   a. What is the acceleration at the end of a test tube that is 10 cm from the axis of rotation?
   b. For comparison, what is the magnitude of the acceleration a test tube would experience if dropped from a height of 1.0 m and stopped in a 1.0-ms-long encounter with a hard floor?

59. || BIO Astronauts use a centrifuge to simulate the acceleration of a rocket launch. The centrifuge takes 30 s to speed up from rest to its top speed of 1 rotation every 1.3 s. The astronaut is strapped into a seat 6.0 m from the axis.
   a. What is the astronaut's tangential acceleration during the first 30 s?
   b. How many g's of acceleration does the astronaut experience when the device is rotating at top speed? Each 9.8 m/s^2 of acceleration is 1 g.

60. || Peregrine falcons are known for their maneuvering ability. In
BIO a tight circular turn, a falcon can attain a centripetal acceleration 1.5 times the free-fall acceleration. What is the radius of the turn if the falcon is flying at 25 m/s?

61. || As the earth rotates, what is the speed of (a) a physics student in Miami, Florida, at latitude 26°, and (b) a physics student in Fairbanks, Alaska, at latitude 65°? Ignore the revolution of the earth around the sun. The radius of the earth is 6400 km.

62. || Communications satellites are placed in a circular orbit where they stay directly over a fixed point on the equator as the earth rotates. These are called *geosynchronous orbits.* The radius of the earth is 6.37×10^6 m, and the altitude of a geosynchronous orbit is 3.58×10^7 m ($\approx$ 22,000 miles). What are (a) the speed and (b) the magnitude of the acceleration of a satellite in a geosynchronous orbit?

63. || A computer hard disk 8.0 cm in diameter is initially at rest. A small dot is painted on the edge of the disk. The disk accelerates at 600 rad/s^2 for $\frac{1}{2}$ s, then coasts at a steady angular velocity for another $\frac{1}{2}$ s.
   a. What is the speed of the dot at $t = 1.0$ s?
   b. Through how many revolutions has the disk turned?

64. || a. A turbine spinning with angular velocity ω_0 rad/s comes to a halt in T seconds. Find an expression for the angle $\Delta\theta$ through which the turbine turns while stopping.
   b. A turbine is spinning at 3800 rpm. Friction in the bearings is so low that it takes 10 min to coast to a stop. How many revolutions does the turbine make while stopping?

65. || A high-speed drill rotating ccw at 2400 rpm comes to a halt in 2.5 s.
   a. What is the drill's angular acceleration?
   b. How many revolutions does it make as it stops?

66. || A wheel initially rotating at 60 rpm experiences the angular acceleration shown in FIGURE P4.66. What is the wheel's angular velocity, in rpm, at $t = 3.0$ s?

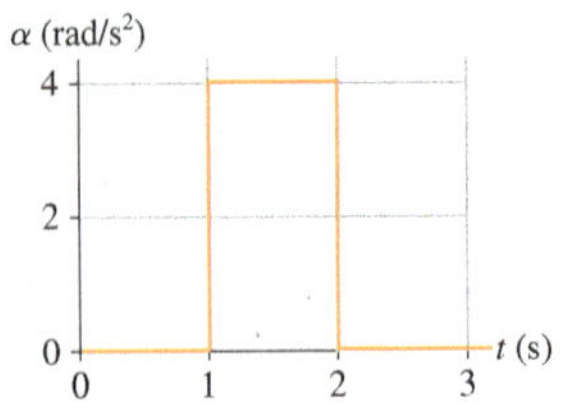

FIGURE P4.66

67. || Your car tire is rotating at 3.5 rev/s when suddenly you press down hard on the accelerator. After traveling 200 m, the tire's rotation has increased to 6.0 rev/s. What was the tire's angular acceleration? Give your answer in rad/s^2.

68. || The angular velocity of a process control motor is $\omega = (20 - \frac{1}{2}t^2)$ rad/s, where t is in seconds.
   a. At what time does the motor reverse direction?
   b. Through what angle does the motor turn between $t = 0$ s and the instant at which it reverses direction?

69. || A Ferris wheel of radius R speeds up with angular acceleration α starting from rest. Find an expression for the (a) velocity and (b) centripetal acceleration of a rider after the Ferris wheel has rotated through angle $\Delta\theta$.

70. || A 6.0-cm-diameter gear rotates with angular velocity $\omega = (2.0 + \frac{1}{2}t^2)$ rad/s, where t is in seconds. At $t = 4.0$ s, what are:
   a. The gear's angular acceleration?
   b. The tangential acceleration of a tooth on the gear?

71. || On a lonely highway, with no other cars in sight, you decide to measure the angular acceleration of your engine's crankshaft while braking gently. Having excellent memory, you are able to read the tachometer every 1.0 s and remember seven values long enough to later write them down. The table shows your data:

Time (s)	rpm
0.0	3010
1.0	2810
2.0	2450
3.0	2250
4.0	1940
5.0	1810
6.0	1510

What is the *magnitude* of the crankshaft's angular acceleration? Give your result in rad/s^2.

72. ||| A car starts from rest on a curve with a radius of 120 m and accelerates at 1.0 m/s^2. Through what angle will the car have traveled when the magnitude of its total acceleration is 2.0 m/s^2?

73. ||| A long string is wrapped around a 6.0-cm-diameter cylinder, initially at rest, that is free to rotate on an axle. The string is then pulled with a constant acceleration of 1.5 m/s^2 until 1.0 m of string has been unwound. If the string unwinds without slipping, what is the cylinder's angular speed, in rpm, at this time?

In Problems 74 through 76 you are given the equations that are used to solve a problem. For each of these, you are to
   a. Write a realistic problem for which these are the correct equations. Be sure that the answer your problem requests is consistent with the equations given.
   b. Finish the solution of the problem, including a pictorial representation.

74. $100 \text{ m} = 0 \text{ m} + (50\cos\theta \text{ m/s})t_1$
   $0 \text{ m} = 0 \text{ m} + (50\sin\theta \text{ m/s})t_1 - \frac{1}{2}(9.80 \text{ m/s}^2){t_1}^2$

75. $v_x = -(6.0\cos 45°) \text{ m/s} + 3.0 \text{ m/s}$
   $v_y = (6.0\sin 45°) \text{ m/s} + 0 \text{ m/s}$
   $100 \text{ m} = v_y t_1,\ x_1 = v_x t_1$

76. $2.5 \text{ rad} = 0 \text{ rad} + \omega_i(10 \text{ s}) + \big((1.5 \text{ m/s}^2)/2(50 \text{ m})\big)(10 \text{ s})^2$
   $\omega_f = \omega_i + \big((1.5 \text{ m/s}^2)/(50 \text{ m})\big)(10 \text{ s})$

Challenge Problems

77. You are asked to consult for the city's research hospital,
BIO where a group of doctors is investigating the bombardment of cancer tumors with high-energy ions. The ions are fired directly toward the center of the tumor at speeds of 5.0×10^6 m/s. To cover the entire tumor area, the ions are deflected sideways by passing them between two charged metal plates that accelerate the ions perpendicular to the direction of their initial motion. The acceleration region is 5.0 cm long, and the ends of the acceleration plates are 1.5 m from the patient. What sideways acceleration is required to deflect an ion 2.0 cm to one side?

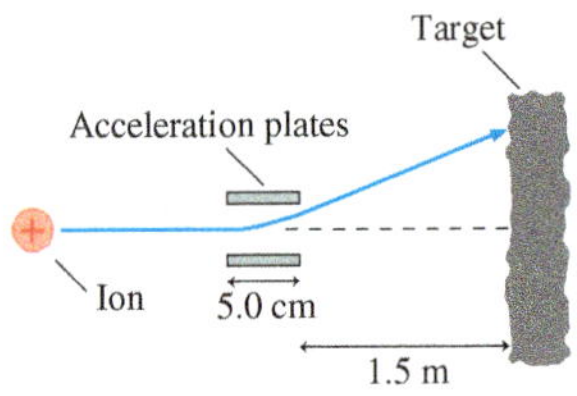

FIGURE CP4.77

78. In one contest at the county fair, seen in FIGURE CP4.78, a spring-loaded plunger launches a ball at a speed of 3.0 m/s from one corner of a smooth, flat board that is tilted up at a 20° angle. To win, you must make the ball hit a small target at the adjacent corner, 2.50 m away. At what angle θ should you tilt the ball launcher?

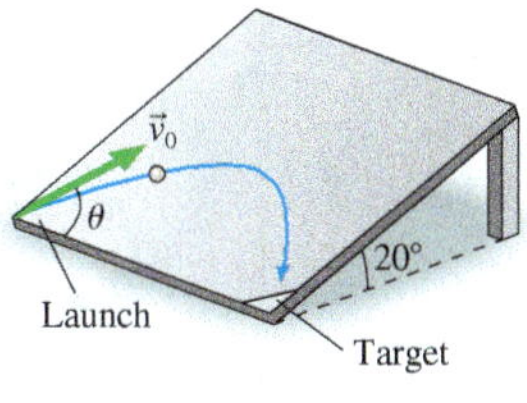

FIGURE CP4.78

79. You are watching an archery tournament when you start wondering how fast an arrow is shot from the bow. Remembering your physics, you ask one of the archers to shoot an arrow parallel to the ground. You find the arrow stuck in the ground 60 m away, making a 3.0° angle with the ground. How fast was the arrow shot?
80. An archer standing on a 15° slope shoots an arrow 20° above the horizontal, as shown in FIGURE CP4.80. How far down the slope does the arrow hit if it is shot with a speed of 50 m/s from 1.75 m above the ground?

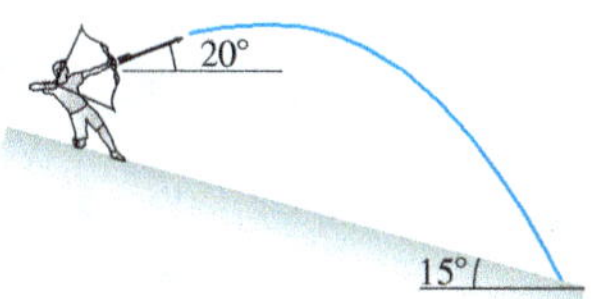

FIGURE CP4.80

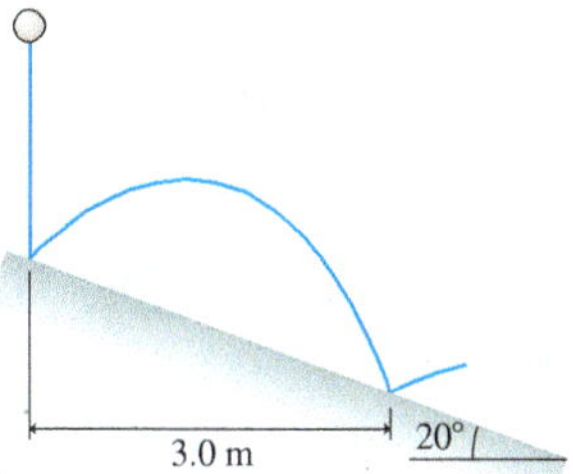

FIGURE CP4.81

81. A rubber ball is dropped onto a ramp that is tilted at 20°, as shown in FIGURE CP4.81. A bouncing ball obeys the "law of reflection," which says that the ball leaves the surface at the same angle it approached the surface. The ball's next bounce is 3.0 m to the right of its first bounce. What is the ball's rebound speed on its first bounce?
82. A skateboarder starts up a 1.0-m-high, 30° ramp at a speed of 7.0 m/s. The skateboard wheels roll without friction. How far from the end of the ramp does the skateboarder touch down?
83. A motorcycle daredevil wants to set a record for jumping over burning school buses. He has hired you to help with the design. He intends to ride off a horizontal platform at 40 m/s, cross the burning buses in a pit below him, then land on a ramp sloping down at 20°. It's very important that he not bounce when he hits the landing ramp because that could cause him to lose control and crash. You immediately recognize that he won't bounce if his velocity is parallel to the ramp as he touches down. This can be accomplished if the ramp is tangent to his trajectory *and* if he lands right on the front edge of the ramp. There's no room for error! Your task is to determine where to place the landing ramp. That is, how far from the edge of the launching platform should the front edge of the landing ramp be horizontally and how far below it? There's a clause in your contract that requires you to test your design before the hero goes on national television to set the record.
84. A cannon on a train car fires a projectile to the right with speed v_0, relative to the train, from a barrel elevated at angle θ. The cannon fires just as the train, which had been cruising to the right along a level track with speed v_{train}, begins to accelerate with acceleration a. Find an expression for the angle at which the projectile should be fired so that it lands as far as possible from the cannon. You can ignore the small height of the cannon above the track.
85. A child in danger of drowning in a river is being carried downstream by a current that flows uniformly with a speed of 2.0 m/s. The child is 200 m from the shore and 1500 m upstream of the boat dock from which the rescue team sets out. If their boat speed is 8.0 m/s with respect to the water, at what angle from the shore should the pilot leave the shore to go directly to the child?
86. An amusement park game, shown in FIGURE CP4.86, launches a marble toward a small cup. The marble is placed directly on top of a spring-loaded wheel and held with a clamp. When released, the wheel spins around clockwise at constant angular acceleration, opening the clamp and releasing the marble after making $\frac{11}{12}$ revolution. What angular acceleration is needed for the ball to land in the cup? The top of the cup is level with the center of the wheel.

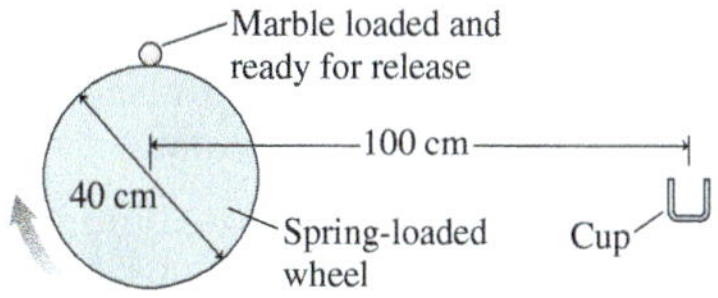

FIGURE CP4.86

STOP TO THINK ANSWERS

Stop to Think 4.1: d. The parallel component of $\vec{a}$ is opposite $\vec{v}$ and will cause the particle to slow down. The perpendicular component of $\vec{a}$ will cause the particle to change direction downward.

Stop to Think 4.2: c. $v = 0$ requires both $v_x = 0$ and $v_y = 0$. Neither x nor y can be changing.

Stop to Think 4.3: d. A projectile's acceleration $\vec{a} = -g\hat{j}$ does not depend on its mass. The second marble has the same initial velocity and the same acceleration, so it follows the same trajectory and lands at the same position.

Stop to Think 4.4: f. The plane's velocity relative to the helicopter is $\vec{v}_{PH} = \vec{v}_{PG} + \vec{v}_{GH} = \vec{v}_{PG} - \vec{v}_{HG}$, where G is the ground. The vector addition shows that $\vec{v}_{PH}$ is to the right and down with a magnitude greater than the 100 m/s of $\vec{v}_{PG}$.

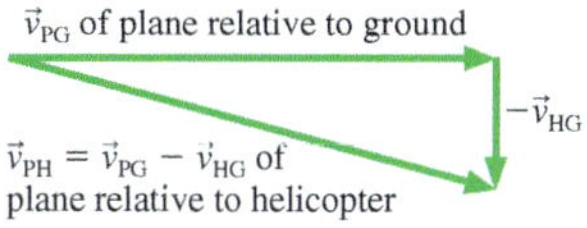

Stop to Think 4.5: b. An initial cw rotation causes the particle's angular position to become increasingly negative. The speed drops to half after reversing direction, so the slope becomes positive and is half as steep as the initial slope. Turning through the same angle returns the particle to $\theta = 0°$.

Stop to Think 4.6: $a_b > a_e > a_a = a_c > a_d$. Centripetal acceleration is v^2/r. Doubling r decreases a_r by a factor of 2. Doubling v increases a_r by a factor of 4. Reversing direction doesn't change a_r.

Stop to Think 4.7: c. ω is negative because the rotation is cw. Because ω is negative but becoming *less* negative, the change $\Delta\omega$ is *positive*. So α is positive.

5 Force and Motion

These ice boats are a memorable example of the connection between force and motion.

▶ **Looking Ahead** The goal of Chapter 5 is to establish a connection between force and motion.

What Causes Motion?

Kinematics describes *how* an object moves. For the more fundamental issue of understanding *why* an object moves, we now turn our attention to **dynamics**.

Dynamics joins with kinematics to form **mechanics**, the science of motion.

◀ **Looking Back**
Section 1.5 Acceleration
Section 3.2 Vector addition

Force

The fundamental concept of dynamics is that of *force*.

- A force is a push or a pull.
- A force acts on an object.
- A force is a vector.
- A force can be a contact force or a long-range force.

Some important forces that we'll study in this chapter are

Gravity

Tension

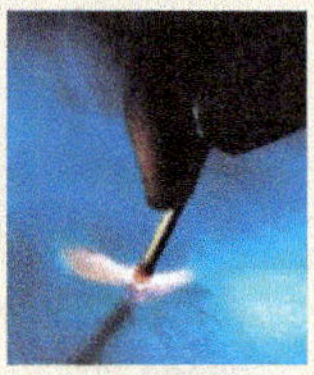
Friction

Drag

Force and Motion

Force causes an object to *accelerate!*

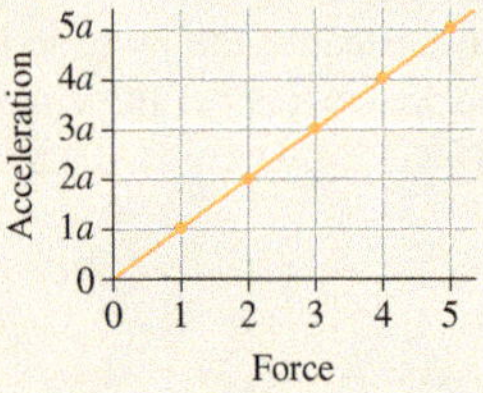

You'll learn that the acceleration of an object is directly proportional to the force exerted on it.

Newton's Laws

You've likely seen Newton's second law, the famous equation $F = ma$. This is the first of several chapters in which you'll learn to use Newton's *three* laws of motion to solve dynamics problems.

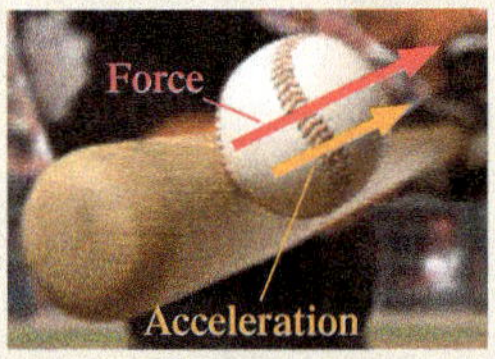

An object accelerates in the same direction as the net force on the object.

Identifying Forces

In this chapter you will learn to *identify* forces and then to represent them on a **free-body diagram.**

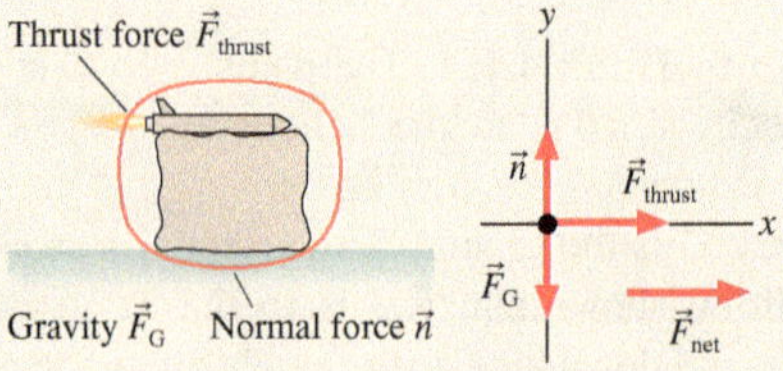

Except for the long-range force of gravity, forces act at points of contact.

5.1 Force

The two major issues that this chapter will examine are:

- What is a force?
- What is the connection between force and motion?

We begin with the first of these questions in the table below.

What is a force?

A force is a push or a pull.

Our commonsense idea of a **force** is that it is a *push* or a *pull.* We will refine this idea as we go along, but it is an adequate starting point. Notice our careful choice of words: We refer to "*a* force," rather than simply "force." We want to think of a force as a very specific *action,* so that we can talk about a single force or perhaps about two or three individual forces that we can clearly distinguish. Hence the concrete idea of "a force" acting on an object.

A force acts on an object.

Implicit in our concept of force is that a **force acts on an object.** In other words, pushes and pulls are applied *to* something—an object. From the object's perspective, it has a force *exerted* on it. Forces do not exist in isolation from the object that experiences them.

A force requires an agent.

Every force has an **agent,** something that acts or exerts power. That is, a force has a specific, identifiable *cause.* As you throw a ball, it is your hand, while in contact with the ball, that is the agent or the cause of the force exerted on the ball. *If* a force is being exerted on an object, you must be able to identify a specific cause (i.e., the agent) of that force. Conversely, a force is not exerted on an object *unless* you can identify a specific cause or agent. Although this idea may seem to be stating the obvious, you will find it to be a powerful tool for avoiding some common misconceptions about what is and is not a force.

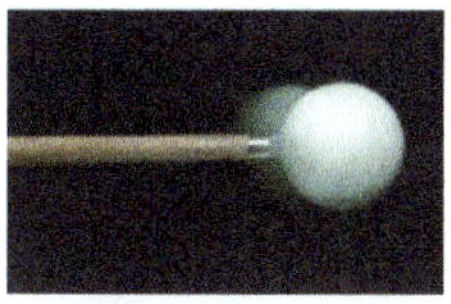

A force is a vector.

If you push an object, you can push either gently or very hard. Similarly, you can push either left or right, up or down. To quantify a push, we need to specify both a magnitude *and* a direction. It should thus come as no surprise that a force is a vector quantity. The general symbol for a force is the vector symbol $\vec{F}$. The size or strength of a force is its magnitude F.

A force can be either a contact force . . .

There are two basic classes of forces, depending on whether the agent touches the object or not. **Contact forces** are forces that act on an object by touching it at a point of contact. The bat must touch the ball to hit it. A string must be tied to an object to pull it. The majority of forces that we will examine are contact forces.

. . . or a long-range force.

Long-range forces are forces that act on an object without physical contact. Magnetism is an example of a long-range force. You have undoubtedly held a magnet over a paper clip and seen the paper clip leap up to the magnet. A coffee cup released from your hand is pulled to the earth by the long-range force of gravity.

NOTE ▶ In the particle model, objects cannot exert forces on themselves. A force on an object will always have an agent or cause external to the object. Now, there are certainly objects that have internal forces (think of all the forces inside the engine of your car!), but the particle model is not valid if you need to consider those internal forces. If you are going to treat your car as a particle and look only at the overall motion of the car as a whole, that motion will be a consequence of external forces acting on the car. ◀

Force Vectors

We can use a simple diagram to visualize how forces are exerted on objects.

TACTICS BOX 5.1 Drawing force vectors

1. Represent the object as a particle.
2. Place the *tail* of the force vector on the particle.
3. Draw the force vector as an arrow pointing in the proper direction and with a length proportional to the size of the force.
4. Give the vector an appropriate label.

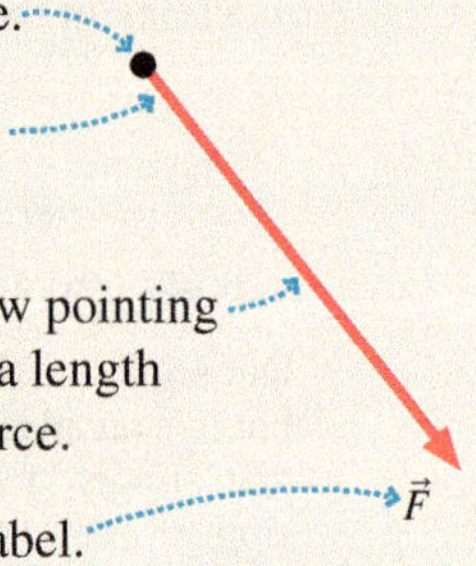

Step 2 may seem contrary to what a "push" should do, but recall that moving a vector does not change it as long as the length and angle do not change. The vector $\vec{F}$ is the same regardless of whether the tail or the tip is placed on the particle. FIGURE 5.1 shows three examples of force vectors.

FIGURE 5.1 Three examples of forces and their vector representations.

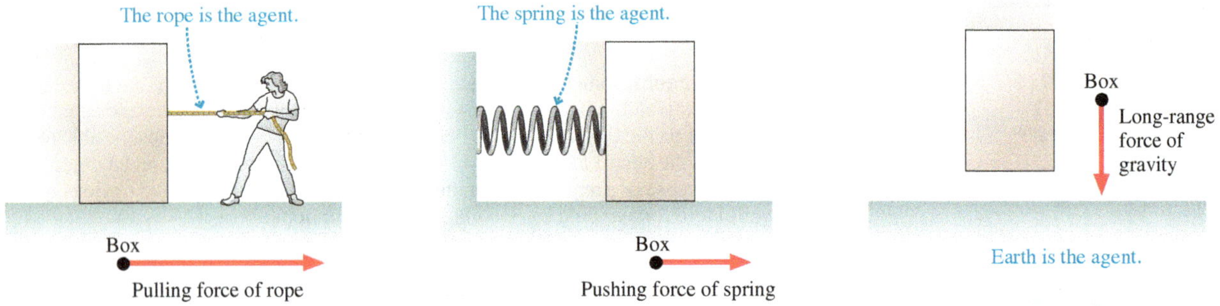

Combining Forces

FIGURE 5.2a shows a box being pulled by two ropes, each exerting a force on the box. How will the box respond? Experimentally, we find that when several forces $\vec{F}_1$, $\vec{F}_2$, $\vec{F}_3$, . . . are exerted on an object, they combine to form a **net force** given by the *vector* sum of *all* the forces:

$$\vec{F}_{net} \equiv \sum_{i=1}^{N} \vec{F}_i = \vec{F}_1 + \vec{F}_2 + \cdots + \vec{F}_N \qquad (5.1)$$

Recall that ≡ is the symbol meaning "is defined as." Mathematically, this summation is called a **superposition of forces.** FIGURE 5.2b shows the net force on the box.

FIGURE 5.2 Two forces applied to a box.

(a)

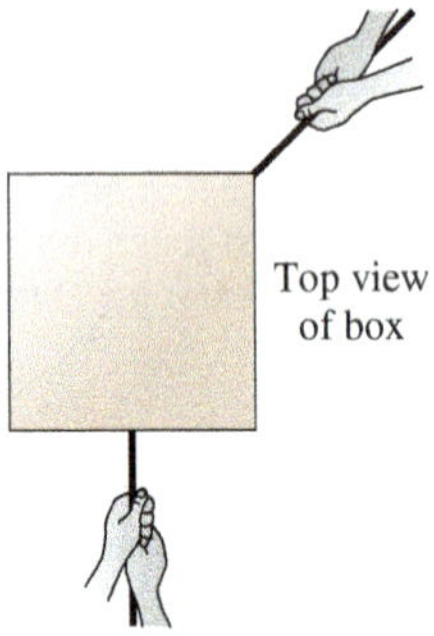

(b)

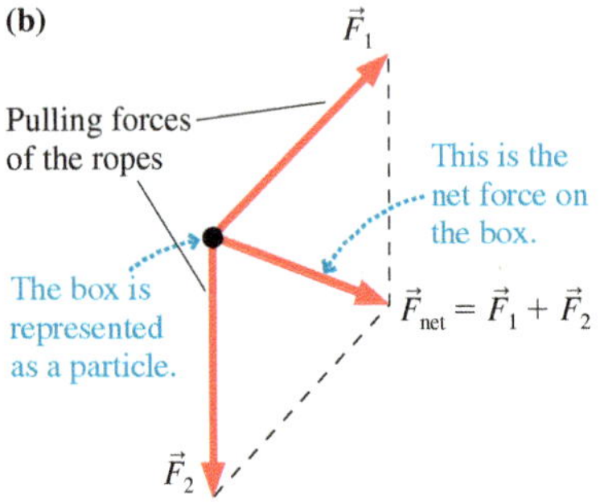

STOP TO THINK 5.1 Two of the three forces exerted on an object are shown. The net force points to the left. Which is the missing third force?

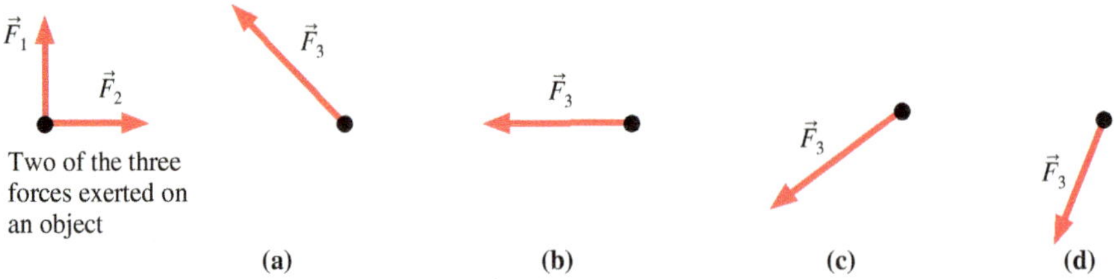

5.2 A Short Catalog of Forces

There are many forces we will deal with over and over. This section will introduce you to some of them. Many of these forces have special symbols. As you learn the major forces, be sure to learn the symbol for each.

Gravity

Gravity—the only long-range force we will encounter in the next few chapters—keeps you in your chair, and the planets in their orbits around the sun. We'll have a thorough look at gravity in Chapter 13. For now we'll concentrate on objects on or near the surface of the earth (or other planet).

The pull of a planet on an object on or near the surface is called the **gravitational force.** The agent for the gravitational force is the *entire planet*. Gravity acts on *all* objects, whether moving or at rest. The symbol for gravitational force is $\vec{F}_G$. The gravitational force vector always points vertically downward, as shown in FIGURE 5.3.

FIGURE 5.3 Gravity.

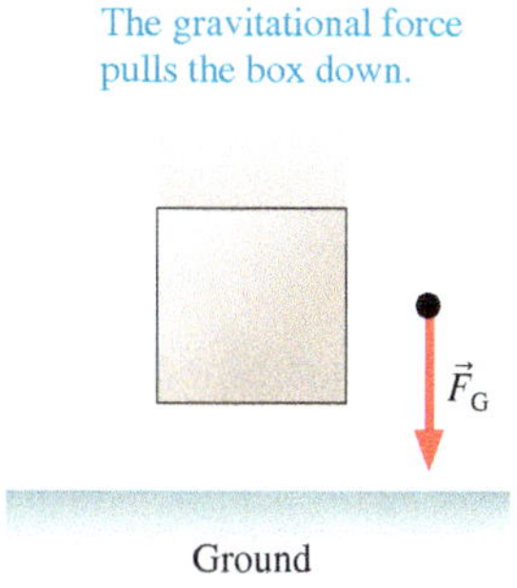

NOTE ▶ We often refer to "the weight" of an object. For an object at rest on the surface of a planet, its weight is simply the magnitude F_G of the gravitational force. However, weight and gravitational force are not the same thing, nor is weight the same as mass. We will briefly examine mass later in the chapter, and we'll explore the rather subtle connections among gravity, weight, and mass in Chapter 6. ◀

Spring Force

Springs exert one of the most common contact forces. A spring can either push (when compressed) or pull (when stretched). FIGURE 5.4 shows the **spring force,** for which we use the symbol $\vec{F}_{sp}$. In both cases, pushing and pulling, the tail of the force vector is placed on the particle in the force diagram.

FIGURE 5.4 The spring force.

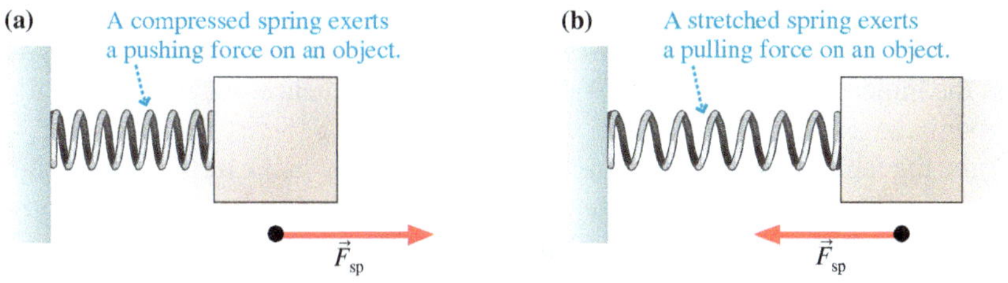

Although you may think of a spring as a metal coil that can be stretched or compressed, this is only one type of spring. Hold a ruler, or any other thin piece of wood or metal, by the ends and bend it slightly. It flexes. When you let go, it "springs" back to its original shape. This is just as much a spring as is a metal coil.

Tension Force

When a string or rope or wire pulls on an object, it exerts a contact force that we call the **tension force,** represented by a capital $\vec{T}$. The direction of the tension force is always in the direction of the string or rope, as you can see in FIGURE 5.5. The commonplace reference to "the tension" in a string is an informal expression for T, the size or magnitude of the tension force.

FIGURE 5.5 Tension.

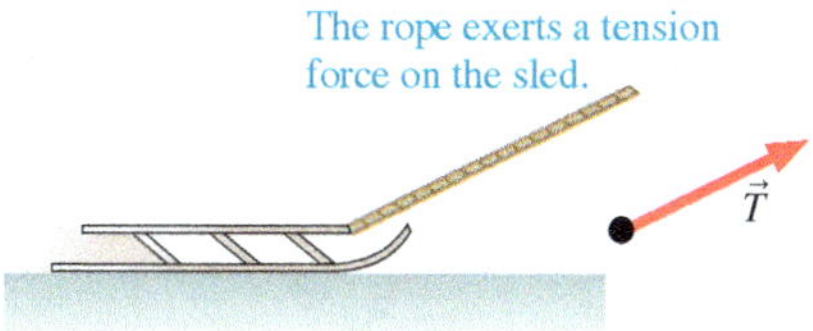

NOTE ▶ Tension is represented by the symbol T. This is logical, but there's a risk of confusing the tension T with the identical symbol T for the period of a particle in circular motion. The number of symbols used in science and engineering is so large that some letters are used several times to represent different quantities. The use of T is the first time we've run into this problem, but it won't be the last. You must be alert to the *context* of a symbol's use to deduce its meaning. ◀

FIGURE 5.6 An atomic model of tension.

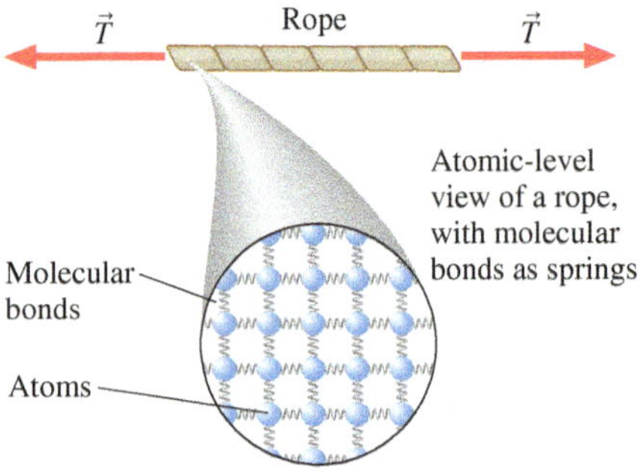

If you were to use a very powerful microscope to look inside a rope, you would "see" that it is made of *atoms* joined together by *molecular bonds*. Molecular bonds are not rigid connections between the atoms. They are more accurately thought of as tiny *springs* holding the atoms together, as in FIGURE 5.6. Pulling on the ends of a string or rope stretches the molecular springs ever so slightly. The tension within a rope and the tension force experienced by an object at the end of the rope are really the net spring force being exerted by billions and billions of microscopic springs.

This atomic-level view of tension introduces a new idea: a microscopic **atomic model** for understanding the behavior and properties of macroscopic objects. It is a *model* because atoms and molecular bonds aren't really little balls and springs. We're using macroscopic concepts—balls and springs—to understand atomic-scale phenomena that we cannot directly see or sense. This is a good model for explaining the elastic properties of materials, but it would not necessarily be a good model for explaining other phenomena. We will frequently use atomic models to obtain a deeper understanding of our observations.

FIGURE 5.7 An atomic model of the force exerted by a table.

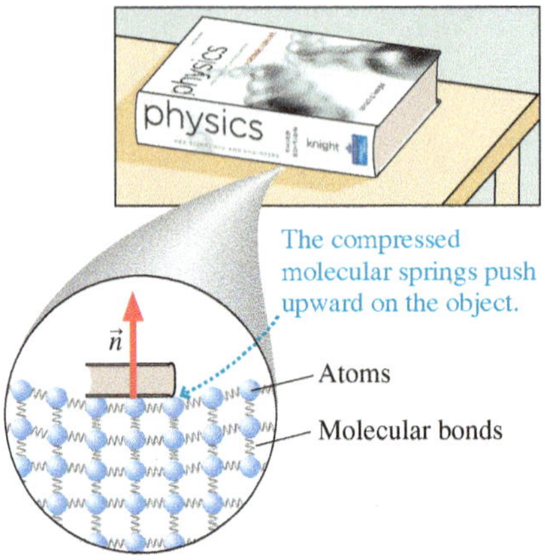

Normal Force

If you sit on a bed, the springs in the mattress compress and, as a consequence of the compression, exert an upward force on you. Stiffer springs would show less compression but still exert an upward force. The compression of extremely stiff springs might be measurable only by sensitive instruments. Nonetheless, the springs would compress ever so slightly and exert an upward spring force on you.

FIGURE 5.7 shows an object resting on top of a sturdy table. The table may not visibly flex or sag, but—just as you do to the bed—the object compresses the molecular springs in the table. The size of the compression is very small, but it is not zero. As a consequence, the compressed molecular springs *push upward* on the object. We say that "the table" exerts the upward force, but it is important to understand that the pushing is *really* done by molecular springs. Similarly, an object resting on the ground compresses the molecular springs holding the ground together and, as a consequence, the ground pushes up on the object.

We can extend this idea. Suppose you place your hand on a wall and lean against it, as shown in FIGURE 5.8. Does the wall exert a force on your hand? As you lean, you compress the molecular springs in the wall and, as a consequence, they push outward against your hand. So the answer is yes, the wall does exert a force on you.

FIGURE 5.8 The wall pushes outward.

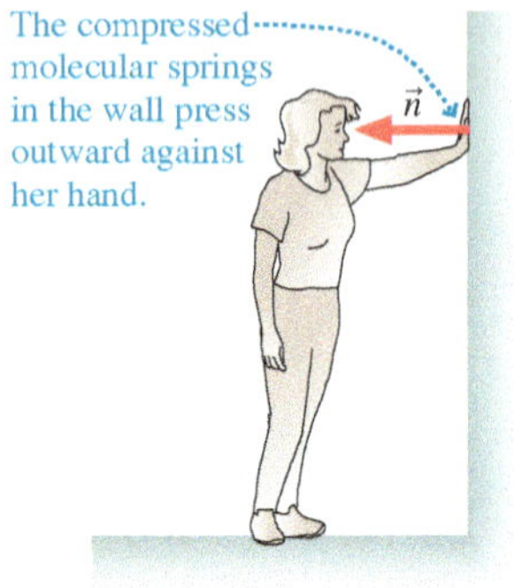

The force the table surface exerts is vertical; the force the wall exerts is horizontal. In all cases, the force exerted on an object that is pressing against a surface is in a direction *perpendicular* to the surface. Mathematicians refer to a line that is perpendicular to a surface as being *normal* to the surface. In keeping with this terminology, we define the **normal force** as the force exerted by a surface (the agent) against an object that is pressing against the surface. The symbol for the normal force is $\vec{n}$.

We're not using the word *normal* to imply that the force is an "ordinary" force or to distinguish it from an "abnormal force." A surface exerts a force *perpendicular* (i.e., normal) to itself as the molecular springs press *outward*. FIGURE 5.9 shows an object on an inclined surface, a common situation.

In essence, the normal force is just a spring force, but one exerted by a vast number of microscopic springs acting at once. The normal force is responsible for the "solidness" of solids. It is what prevents you from passing right through the chair you are sitting in and what causes the pain and the lump if you bang your head into a door.

FIGURE 5.9 The normal force.

Friction

Friction, like the normal force, is exerted by a surface. But whereas the normal force is perpendicular to the surface, the friction force is always *tangent* to the surface. It is useful to distinguish between two kinds of friction:

- *Kinetic friction*, denoted $\vec{f}_k$, appears as an object slides across a surface. This is a force that "opposes the motion," meaning that the friction force vector $\vec{f}_k$ points in a direction opposite the velocity vector $\vec{v}$ (i.e., "the motion").

- *Static friction,* denoted $\vec{f}_s$, is the force that keeps an object "stuck" on a surface and prevents its motion. Finding the direction of $\vec{f}_s$ is a little trickier than finding it for $\vec{f}_k$. Static friction points opposite the direction in which the object *would* move if there were no friction. That is, it points in the direction necessary to *prevent* motion.

FIGURE 5.10 shows examples of kinetic and static friction.

FIGURE 5.10 Kinetic and static friction.

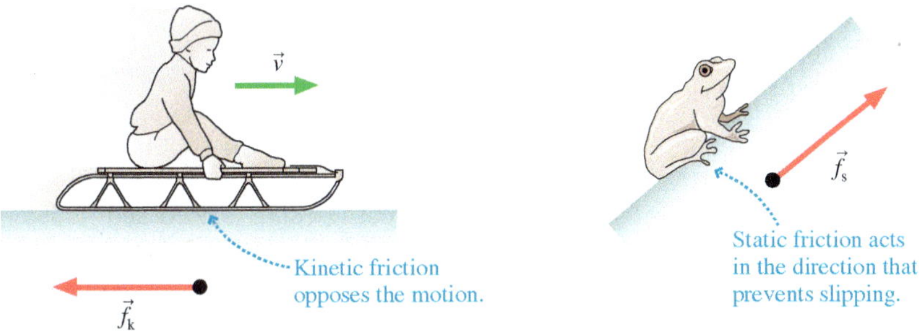

NOTE ▶ A surface exerts a kinetic friction force when an object moves *relative to* the surface. A package on a conveyor belt is in motion, but it does not experience a kinetic friction force because it is not moving relative to the belt. So to be precise, we should say that the kinetic friction force points opposite to an object's motion *relative to* a surface. ◀

Drag

Friction at a surface is one example of a *resistive force,* a force that opposes or resists motion. Resistive forces are also experienced by objects moving through fluids—gases and liquids. The resistive force of a fluid is called **drag,** with symbol $\vec{D}$. Drag, like kinetic friction, points opposite the direction of motion. FIGURE 5.11 shows an example.

Drag can be a significant force for objects moving at high speeds or in dense fluids. Hold your arm out the window as you ride in a car and feel how the air resistance against it increases rapidly as the car's speed increases. Drop a lightweight object into a beaker of water and watch how slowly it settles to the bottom.

For objects that are heavy and compact, that move in air, and whose speed is not too great, the drag force of air resistance is fairly small. To keep things as simple as possible, **you can neglect air resistance in all problems unless a problem explicitly asks you to include it.**

FIGURE 5.11 Air resistance is an example of drag.

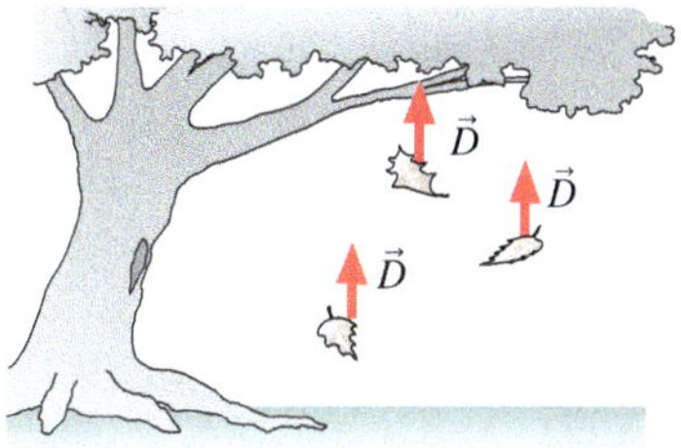

Thrust

A jet airplane obviously has a force that propels it forward during takeoff. Likewise for the rocket being launched in FIGURE 5.12. This force, called **thrust,** occurs when a jet or rocket engine expels gas molecules at high speed. Thrust is a contact force, with the exhaust gas being the agent that pushes on the engine. The process by which thrust is generated is rather subtle, and we will postpone a full discussion until we study Newton's third law in Chapter 7. For now, we will treat thrust as a force opposite the direction in which the exhaust gas is expelled. There's no special symbol for thrust, so we will call it $\vec{F}_{\text{thrust}}$.

FIGURE 5.12 Thrust force on a rocket.

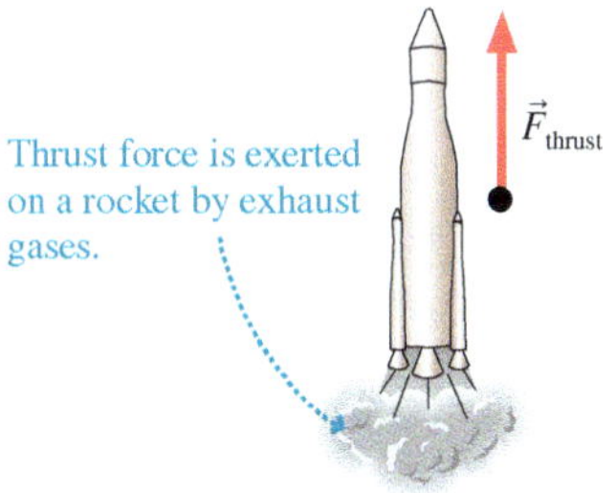

Electric and Magnetic Forces

Electricity and magnetism, like gravity, exert long-range forces. We will study electric and magnetic forces in detail in Part VI. For now, it is worth noting that the forces holding molecules together—the molecular bonds—are not actually tiny springs. Atoms and molecules are made of charged particles—electrons and protons—and what we call a molecular bond is really an electric force between these particles. So when we say that the normal force and the tension force are due to "molecular springs," or that friction is due to atoms running into each other, what we're really saying is that these forces, at the most fundamental level, are actually electric forces between the charged particles in the atoms.

5.3 Identifying Forces

Force	Notation
General force	$\vec{F}$
Gravitational force	$\vec{F}_G$
Spring force	$\vec{F}_{sp}$
Tension	$\vec{T}$
Normal force	$\vec{n}$
Static friction	$\vec{f}_s$
Kinetic friction	$\vec{f}_k$
Drag	$\vec{D}$
Thrust	$\vec{F}_{thrust}$

Force and motion problems generally have two basic steps:

1. Identify all of the forces acting on an object.
2. Use Newton's laws and kinematics to determine the motion.

Understanding the first step is the primary goal of this chapter. We'll turn our attention to step 2 in the next chapter.

A typical physics problem describes an object that is being pushed and pulled in various directions. Some forces are given explicitly; others are only implied. In order to proceed, it is necessary to determine all the forces that act on the object. The procedure for identifying forces will become part of the *pictorial representation* of the problem.

TACTICS BOX 5.2 Identifying forces

MP

1. **Identify the object of interest.** This is the object whose motion you wish to study.
2. **Draw a picture of the situation.** Show the object of interest and all other objects—such as ropes, springs, or surfaces—that touch it.
3. **Draw a closed curve around the object.** Only the object of interest is inside the curve; everything else is outside.
4. **Locate every point on the boundary of this curve where other objects touch the object of interest.** These are the points where *contact forces* are exerted on the object.
5. **Name and label each contact force acting on the object.** There is at least one force at each point of contact; there may be more than one. When necessary, use subscripts to distinguish forces of the same type.
6. **Name and label each long-range force acting on the object.** For now, the only long-range force is the gravitational force.

Exercises 3–8

EXAMPLE 5.1 Forces on a bungee jumper

A bungee jumper has leapt off a bridge and is nearing the bottom of her fall. What forces are being exerted on the jumper?

VISUALIZE

FIGURE 5.13 Forces on a bungee jumper.

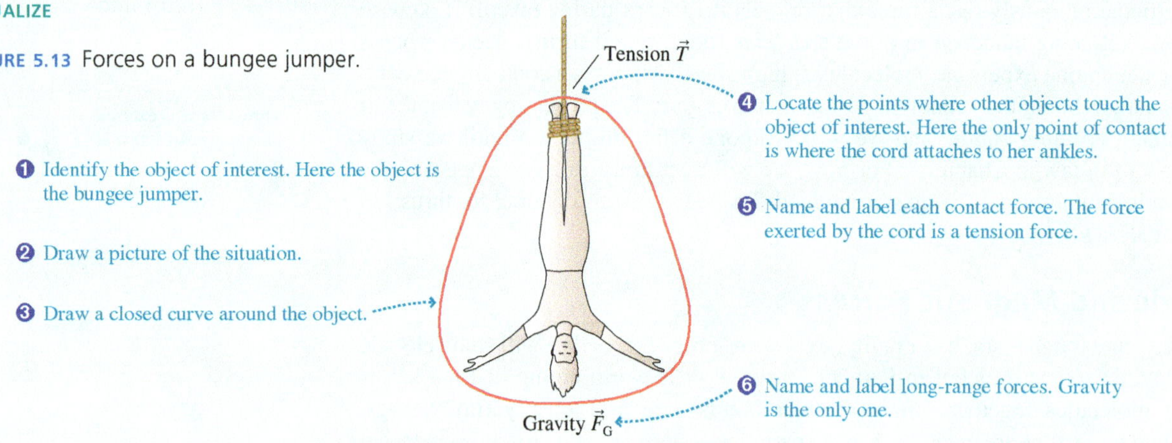

EXAMPLE 5.2 Forces on a skier

A skier is being towed up a snow-covered hill by a tow rope. What forces are being exerted on the skier?

VISUALIZE

FIGURE 5.14 Forces on a skier.

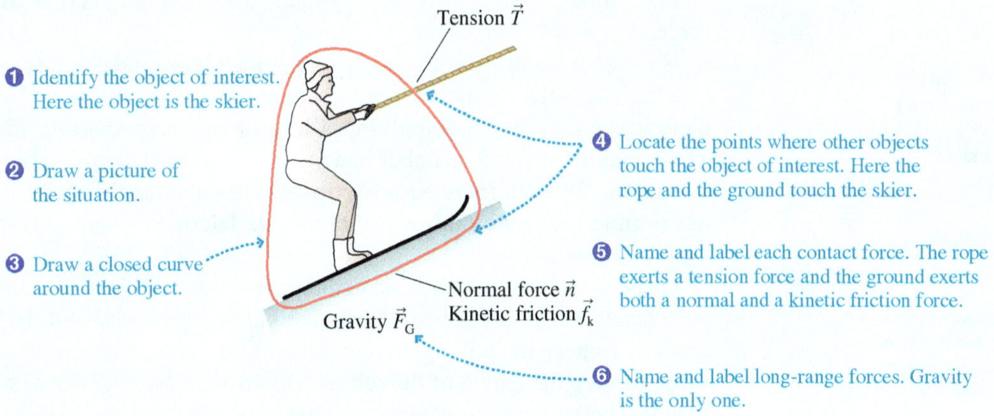

NOTE ▶ You might have expected two friction forces and two normal forces in Example 5.2, one on each ski. Keep in mind, however, that we're working within the particle model, which represents the skier by a single point. A particle has only one contact with the ground, so there is one normal force and one friction force. ◀

EXAMPLE 5.3 Forces on a rocket

A rocket is being launched to place a new satellite in orbit. Air resistance is not negligible. What forces are being exerted on the rocket?

VISUALIZE This drawing is much more like the sketch you would make when identifying forces as part of solving a problem.

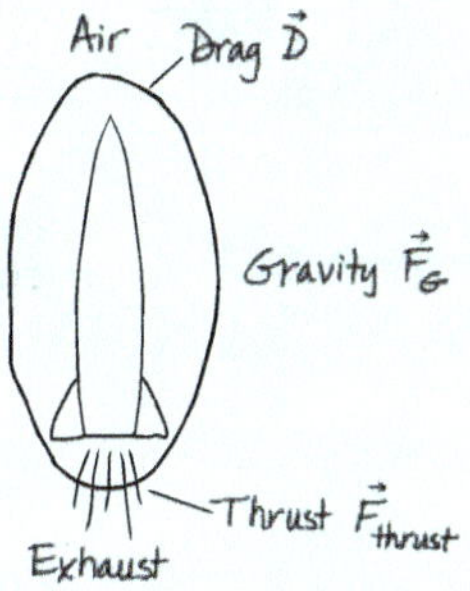

FIGURE 5.15 Forces on a rocket.

STOP TO THINK 5.2 You've just kicked a rock, and it is now sliding across the ground about 2 meters in front of you. Which of these forces act on the rock? List all that apply.

a. Gravity, acting downward.
b. The normal force, acting upward.
c. The force of the kick, acting in the direction of motion.
d. Friction, acting opposite the direction of motion.

5.4 What Do Forces Do? A Virtual Experiment

Having learned to identify forces, we ask the next question: How does an object move when a force is exerted on it? The only way to answer this question is to do experiments. Let's conduct a "virtual experiment," one you can easily visualize. Imagine using your fingers to stretch a rubber band to a certain length—say

FIGURE 5.16 A reproducible force.

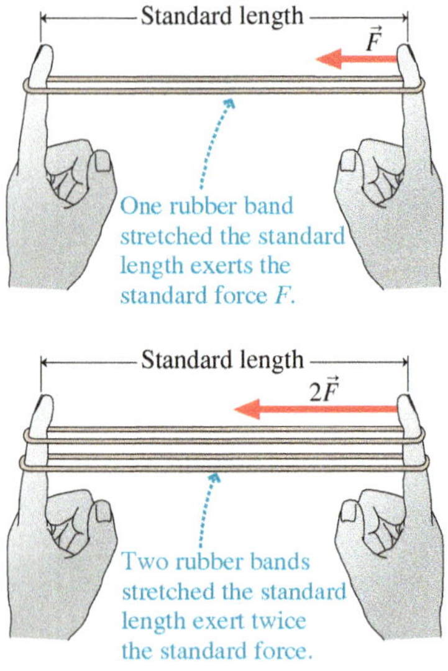

10 centimeters—that you can measure with a ruler, as shown in FIGURE 5.16. You know that a stretched rubber band exerts a force—a spring force—because your fingers *feel* the pull. Furthermore, this is a reproducible force; the rubber band exerts the same force every time you stretch it to this length. We'll call this the *standard force* F. Not surprisingly, two identical rubber bands exert twice the pull of one rubber band, and N side-by-side rubber bands exert N times the standard force: $F_{net} = NF$.

Now attach one rubber band to a 1 kg block and stretch it to the standard length. The object experiences the same force F as did your finger. The rubber band gives us a way of applying a known and reproducible force to an object. Then imagine using the rubber band to pull the block across a horizontal, frictionless table. (We can imagine a frictionless table since this is a virtual experiment, but in practice you could nearly eliminate friction by supporting the object on a cushion of air.)

If you stretch the rubber band and then release the object, the object moves toward your hand. But as it does so, the rubber band gets shorter and the pulling force decreases. To keep the pulling force constant, you must *move your hand* at just the right speed to keep the length of the rubber band from changing! FIGURE 5.17a shows the experiment being carried out. Once the motion is complete, you can use motion diagrams and kinematics to analyze the object's motion.

FIGURE 5.17 Measuring the motion of an object that is pulled with a constant force.

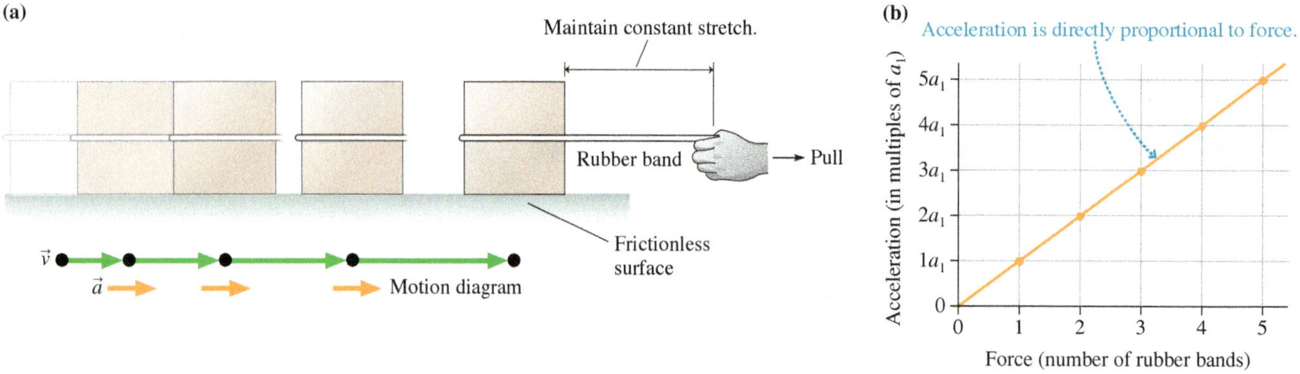

The first important finding of this experiment is that **an object pulled with a constant force moves with a constant acceleration.** That is, the answer to the question What does a force do? is: A force causes an object to accelerate, and a constant force produces a constant acceleration. This finding could not have been anticipated in advance. It's conceivable that the object would speed up for a while, then move with a steady speed. Or that it would speed up, but that the *rate* of increase, the acceleration, would steadily decline. These are conceivable motions, but they're not what happens. Instead, the object accelerates *with a constant acceleration* a_1 for as long as you pull it with a constant force F.

What happens if you increase the force by using several rubber bands? To find out, use two rubber bands, then three rubber bands, then four, and so on. With N rubber bands, the force on the block is NF. FIGURE 5.17b shows the results of this experiment. You can see that doubling the force causes twice the acceleration, tripling the force causes three times the acceleration, and so on. The graph reveals our second important finding: **The acceleration is directly proportional to the force.** This result can be written as

$$a = cF \tag{5.2}$$

where c, called the *proportionality constant,* is the slope of the graph.

MATHEMATICAL ASIDE Proportionality and proportional reasoning

The concept of **proportionality** arises frequently in physics. A quantity symbolized by u is *proportional* to another quantity symbolized by v if

$$u = cv$$

where c (which might have units) is called the **proportionality constant.** This relationship between u and v is often written

$$u \propto v$$

where the symbol $\propto$ means "is proportional to."

If v is doubled to $2v$, then u is doubled to $c(2v) = 2(cv) = 2u$. In general, if v is changed by any factor f, then u changes by the same factor. This is the essence of what we *mean* by proportionality.

A graph of u versus v is a straight line *passing through the origin* (i.e., the y-intercept is zero) with slope $= c$. Notice that proportionality is a much more specific relationship between u and v than mere linearity. The linear equation $u = cv + b$ has a straight-line graph, but it doesn't pass through the origin (unless b happens to be zero) and doubling v does not double u.

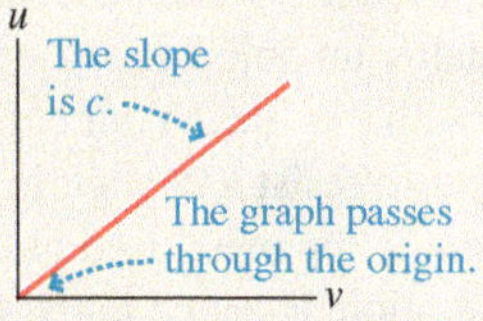

u is proportional to v.

If $u \propto v$, then $u_1 = cv_1$ and $u_2 = cv_2$. Dividing the second equation by the first, we find

$$\frac{u_2}{u_1} = \frac{v_2}{v_1}$$

By working with *ratios,* we can deduce information about u without needing to know the value of c. (This would not be true if the relationship were merely linear.) This is called **proportional reasoning.**

Proportionality is not limited to being linearly proportional. The graph on the left below shows that u is clearly not proportional to w. But a graph of u versus $1/w^2$ *is* a straight line passing through the origin, thus, in this case, u is proportional to $1/w^2$, or $u \propto 1/w^2$. We would say that "u is proportional to the inverse square of w."

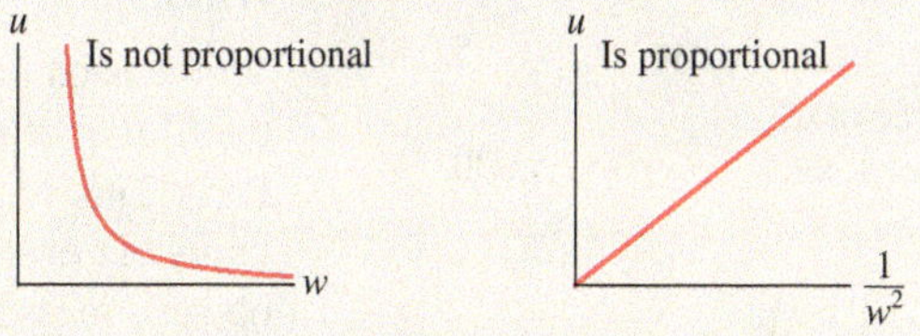

u is proportional to the inverse square of w.

EXAMPLE u is proportional to the inverse square of w. By what factor does u change if w is tripled?

SOLUTION This is an opportunity for proportional reasoning; we don't need to know the proportionality constant. If u is proportional to $1/w^2$, then

$$\frac{u_2}{u_1} = \frac{1/w_2^2}{1/w_1^2} = \frac{w_1^2}{w_2^2} = \left(\frac{w_1}{w_2}\right)^2$$

Tripling w, with $w_2/w_1 = 3$, and thus $w_1/w_2 = \frac{1}{3}$, changes u to

$$u_2 = \left(\frac{w_1}{w_2}\right)^2 u_1 = \left(\frac{1}{3}\right)^2 u_1 = \frac{1}{9}u_1$$

Tripling w causes u to become $\frac{1}{9}$ of its original value.

Many *Student Workbook* and end-of-chapter homework questions will require proportional reasoning. It's an important skill to learn.

The final question for our virtual experiment is: How does the acceleration depend on the mass of the object being pulled? To find out, apply the *same force*—for example, the standard force of one rubber band—to a 2 kg block, then a 3 kg block, and so on, and for each measure the acceleration. Doing so gives you the results shown in FIGURE 5.18. An object with twice the mass of the original block has only half the acceleration when both are subjected to the same force.

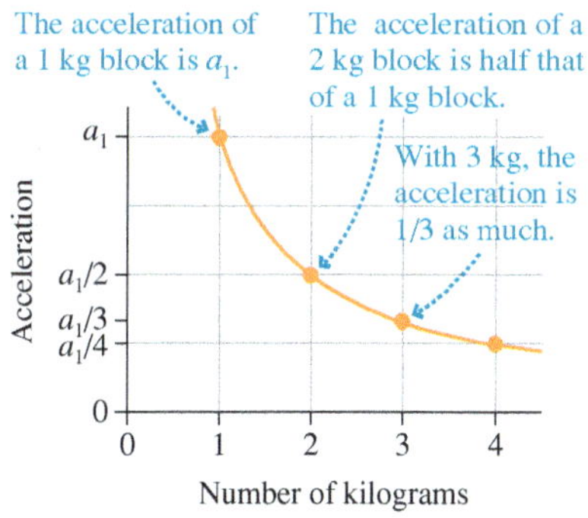

FIGURE 5.18 Acceleration is inversely proportional to mass.

Mathematically, the graph of Figure 5.18 is one of *inverse proportionality*. That is, **the acceleration is inversely proportional to the object's mass,** which we can write as

$$a = \frac{c'}{m} \tag{5.3}$$

where c' is another proportionality constant.

Force causes an object to *accelerate!* The results of our experiment are that the acceleration is directly proportional to the force applied and inversely proportional to the object's mass. We can combine these into the single statement

$$a = \frac{F}{m} \tag{5.4}$$

if we define the basic unit of force as the force that causes a 1 kg mass to accelerate at 1 m/s². That is,

$$1 \text{ basic unit of force} \equiv 1 \text{ kg} \times 1 \frac{\text{m}}{\text{s}^2} = 1 \frac{\text{kg m}}{\text{s}^2}$$

TABLE 5.1 Approximate magnitude of some typical forces

Force	Approximate magnitude (newtons)
Weight of a U.S. quarter	0.05
Weight of a 1 pound object	5
Weight of a 110 pound person	500
Propulsion force of a car	5,000
Thrust force of a rocket motor	5,000,000

This basic unit of force is called a newton:

> One **newton** is the force that causes a 1 kg mass to accelerate at 1 m/s^2. The abbreviation for newton is N. Mathematically, $1\ N = 1\ kg\ m/s^2$.

Table 5.1 lists some typical forces. As you can see, "typical" forces on "typical" objects are likely to be in the range 0.01–10,000 N.

Mass

We've been using the term *mass* without a clear definition. As we learned in Chapter 1, the SI unit of mass, the kilogram, is based on a particular metal block kept in a vault in Paris. This suggests that *mass* is the amount of matter an object contains, and that is certainly our everyday concept of mass. Now we see that a more precise way of defining an object's mass is in terms of its acceleration in response to a force. Figure 5.18 shows that an object with twice the amount of matter accelerates only half as much in response to the same force. The more matter an object has, the more it *resists* accelerating in response to a force. You're familiar with this idea: Your car is much harder to push than your bicycle. The tendency of an object to resist a *change* in its velocity (i.e., to resist acceleration) is called **inertia.** Consequently, the mass used in Equation 5.4, a measure of an object's resistance to changing its motion, is called **inertial mass.** We'll meet a different concept of mass, *gravitational mass,* when we study Newton's law of gravity in Chapter 13.

STOP TO THINK 5.3 Two rubber bands stretched to the standard length cause an object to accelerate at 2 m/s^2. Suppose another object with twice the mass is pulled by four rubber bands stretched to the standard length. The acceleration of this second object is

a. 1 m/s^2 b. 2 m/s^2 c. 4 m/s^2 d. 8 m/s^2 e. 16 m/s^2

Hint: Use proportional reasoning.

5.5 Newton's Second Law

Equation 5.4 is an important finding, but our experiment was limited to looking at an object's response to a single applied force. Realistically, an object is likely to be subjected to several distinct forces $\vec{F}_1$, $\vec{F}_2$, $\vec{F}_3$, . . . that may point in different directions. What happens then? In that case, it is found experimentally that the acceleration is determined by the *net* force.

Newton was the first to recognize the connection between force and motion. This relationship is known today as Newton's second law.

> **Newton's second law** An object of mass m subjected to forces $\vec{F}_1$, $\vec{F}_2$, $\vec{F}_3$, . . . will undergo an acceleration $\vec{a}$ given by
>
> $$\vec{a} = \frac{\vec{F}_{net}}{m} \tag{5.5}$$
>
> where the net force $\vec{F}_{net} = \vec{F}_1 + \vec{F}_2 + \vec{F}_3 + \cdots$ is the vector sum of all forces acting on the object. The acceleration vector $\vec{a}$ points in the same direction as the net force vector $\vec{F}_{net}$.

The significance of Newton's second law cannot be overstated. There was no reason to suspect that there should be any simple relationship between force and acceleration. Yet there it is, a simple but exceedingly powerful equation relating the two. The critical idea is that **an object accelerates in the direction of the net force vector $\vec{F}_{net}$.**

We can rewrite Newton's second law in the form

$$\vec{F}_{net} = m\vec{a} \tag{5.6}$$

which is how you'll see it presented in many textbooks. Equations 5.5 and 5.6 are mathematically equivalent, but Equation 5.5 better describes the central idea of Newtonian mechanics: A force applied to an object causes the object to accelerate.

It's also worth noting that **the object responds only to the forces acting on it *at this instant.*** The object has no memory of forces that may have been exerted at earlier times. This idea is sometimes called **Newton's zeroth law.**

NOTE ▶ Be careful not to think that one force "overcomes" the others to determine the motion. Forces are not in competition with each other! It is $\vec{F}_{net}$, the sum of *all* the forces, that determines the acceleration $\vec{a}$. ◀

As an example, FIGURE 5.19a shows a box being pulled by two ropes. The ropes exert tension forces $\vec{T}_1$ and $\vec{T}_2$ on the box. FIGURE 5.19b represents the box as a particle, shows the forces acting on the box, and adds them graphically to find the net force $\vec{F}_{net}$. The box will accelerate in the direction of $\vec{F}_{net}$ with acceleration

$$\vec{a} = \frac{\vec{F}_{net}}{m} = \frac{\vec{T}_1 + \vec{T}_2}{m}$$

NOTE ▶ The acceleration is *not* $(T_1 + T_2)/m$. You must add the forces as *vectors,* not merely add their magnitudes as scalars. ◀

FIGURE 5.19 Acceleration of a pulled box.

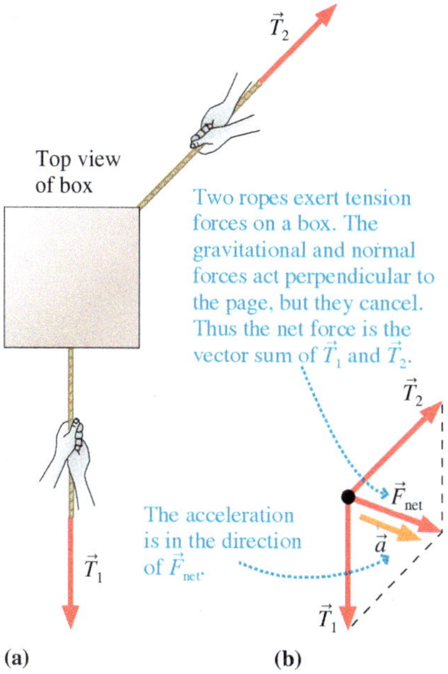

Forces Are Interactions

There's one more important aspect of forces. If you push against a door (the object) to close it, the door pushes back against your hand (the agent). If a tow rope pulls on a car (the object), the car pulls back on the rope (the agent). In general, if an agent exerts a force on an object, the object exerts a force on the agent. We really need to think of a force as an *interaction* between two objects. This idea is captured in Newton's third law—that for every action there is an equal but opposite reaction.

Although the interaction perspective is a more exact way to view forces, it adds complications that we would like to avoid for now. Our approach will be to start by focusing on how a single object responds to forces exerted on it. Then, in Chapter 7, we'll return to Newton's third law and the larger issue of how two or more objects interact with each other.

STOP TO THINK 5.4 Three forces act on an object. In which direction does the object accelerate?

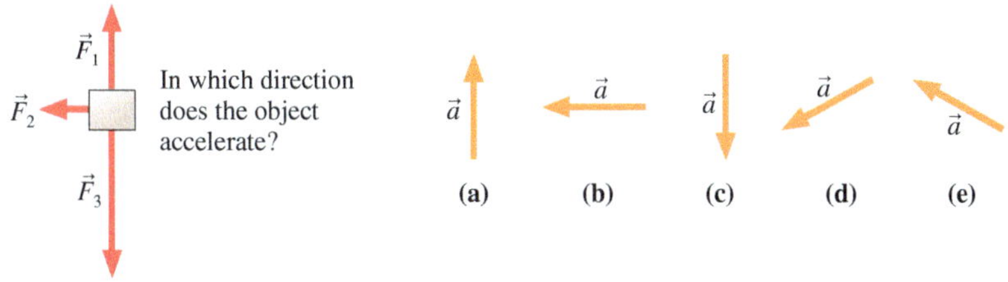

5.6 Newton's First Law

Aristotle and his contemporaries in the world of ancient Greece were very interested in motion. One question they asked was: What is the "natural state" of an object if left to itself? It is easy to see that every moving object on earth, if left to itself, eventually comes to rest. Aristotle concluded that the natural state of an earthly object is to be at rest. An object at rest requires no explanation. A moving object, though, is not in its natural state and thus requires an explanation: Why is this object moving? What keeps it going and prevents it from being in its natural state?

Galileo reopened the question of the "natural state" of objects. He suggested focusing on the *limiting case* in which resistance to the motion (e.g., friction or air resistance) is zero. Many careful experiments in which he minimized the influence of friction led Galileo to a conclusion that was in sharp contrast to Aristotle's belief that rest is an object's natural state.

Galileo found that an external influence (i.e., a force) is needed to make an object accelerate—to *change* its velocity. In particular, a force is needed to put an object in motion. In the absence of friction or air resistance, a moving object would continue to move along a straight line forever with no loss of speed. In other words, the natural state of an object—its behavior if free of external influences—is *uniform motion* with constant velocity! This does not happen in practice because friction or air resistance prevents the object from being left alone. "At rest" has no special significance in Galileo's view of motion; it is simply uniform motion that happens to have $\vec{v} = \vec{0}$.

It was left to Newton to generalize this result, and today we call it Newton's first law of motion.

Newton's first law An object that is at rest will remain at rest, or an object that is moving will continue to move in a straight line with constant velocity, if and only if the net force acting on the object is zero.

FIGURE 5.20 Two examples of mechanical equilibrium.

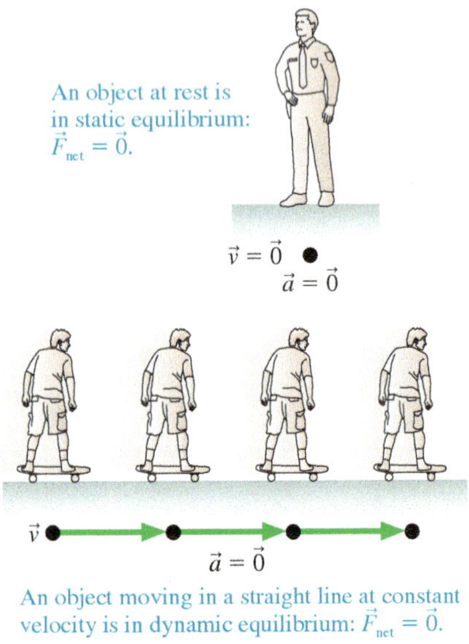

Newton's first law is also known as the *law of inertia.* If an object is at rest, it has a tendency to stay at rest. If it is moving, it has a tendency to continue moving with the *same velocity.*

NOTE ▶ The first law refers to *net* force. An object can remain at rest, or can move in a straight line with constant velocity, even though forces are exerted on it as long as the *net* force is zero. ◀

Notice the "if and only if" aspect of Newton's first law. If an object is at rest or moves with constant velocity, then we can conclude that there is no net force acting on it. Conversely, if no net force is acting on it, we can conclude that the object will have constant velocity, not just constant speed. The direction remains constant, too!

An object on which the net force is zero, $\vec{F}_{net} = \vec{0}$, is said to be in **mechanical equilibrium.** There are two distinct forms of mechanical equilibrium:

1. The object is at rest. This is **static equilibrium.**
2. The object is moving in a straight line with constant velocity. This is **dynamic equilibrium.**

Two examples of mechanical equilibrium are shown in FIGURE 5.20. Both share the common feature that the acceleration is zero: $\vec{a} = \vec{0}$.

What Good Is Newton's First Law?

The first law completes our definition of force. It answers the question: What is a force? If an "influence" on an object disturbs a state of equilibrium by causing the object's velocity to change, the influence is a force.

Newton's first law changes the question the ancient Greeks were trying to answer: What causes an object to move? Newton's first law says **no cause is needed for an object to move!** Uniform motion is the object's natural state. Nothing at all is required for it to remain in that state. The proper question, according to Newton, is: What causes an object to *change* its velocity? Newton, with Galileo's help, also gave us the answer. **A *force* is what causes an object to change its velocity.**

The preceding paragraph contains the essence of Newtonian mechanics. This new perspective on motion, however, is often contrary to our common experience. We all know perfectly well that you must keep pushing an object—exerting a force on it—to keep it moving. Newton is asking us to change our point of view and to consider motion *from the object's perspective* rather than from our personal perspective. As far as the object is concerned, our push is just one of several forces acting on it. Others might include friction, air resistance, or gravity. Only by knowing the *net* force can we determine the object's motion.

Newton's first law may seem to be merely a special case of Newton's second law. After all, the equation $\vec{F}_{net} = m\vec{a}$ tells us that an object moving with constant velocity ($\vec{a} = \vec{0}$) has $\vec{F}_{net} = \vec{0}$. The difficulty is that the second law assumes that we already know what force is. The purpose of the first law is to *identify* a force as something that disturbs a state of equilibrium. The second law then describes how the object responds to this force. Thus from a *logical* perspective, the first law really is a separate statement that must precede the second law. But this is a rather formal distinction. From a pedagogical perspective it is better—as we have done—to use a commonsense understanding of force and start with Newton's second law.

Inertial Reference Frames

If a car stops suddenly, you may be "thrown" into the windshield if you're not wearing your seat belt. You have a very real forward acceleration *relative to the car,* but is there a force pushing you forward? A force is a push or a pull caused by an identifiable agent in contact with the object. Although you *seem* to be pushed forward, there's no agent to do the pushing.

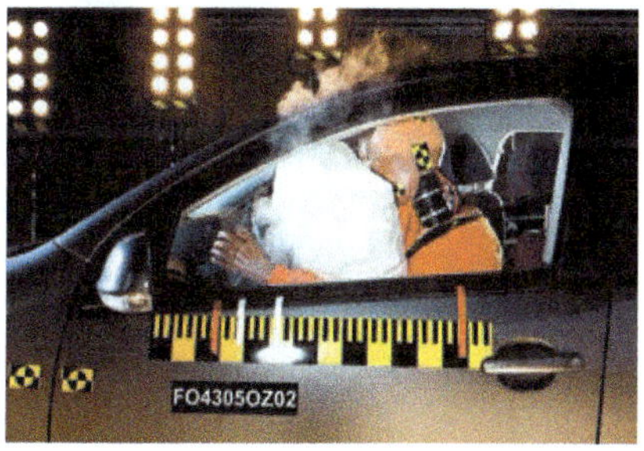

This guy thinks there's a force hurling him into the windshield. What a dummy!

The difficulty—an acceleration without an apparent force—comes from using an inappropriate reference frame. Your acceleration measured in a reference frame attached to the car is not the same as your acceleration measured in a reference frame attached to the ground. Newton's second law says $\vec{F}_{net} = m\vec{a}$. But which $\vec{a}$? Measured in which reference frame?

We define an **inertial reference frame** as a reference frame in which Newton's laws are valid. The first law provides a convenient way to test whether a reference frame is inertial. If $\vec{a} = \vec{0}$ (an object is at rest or moving with constant velocity) only when $\vec{F}_{net} = \vec{0}$, then the reference frame in which $\vec{a}$ is measured is an inertial reference frame.

Not all reference frames are inertial reference frames. FIGURE 5.21a shows a physics student cruising at constant velocity in an airplane. If the student places a ball on the floor, it stays there. There are no horizontal forces, and the ball remains at rest relative to the airplane. That is, $\vec{a} = \vec{0}$ in the airplane's reference frame when $\vec{F}_{net} = \vec{0}$. Newton's first law is satisfied, so this airplane is an inertial reference frame.

FIGURE 5.21 Reference frames.

(a)

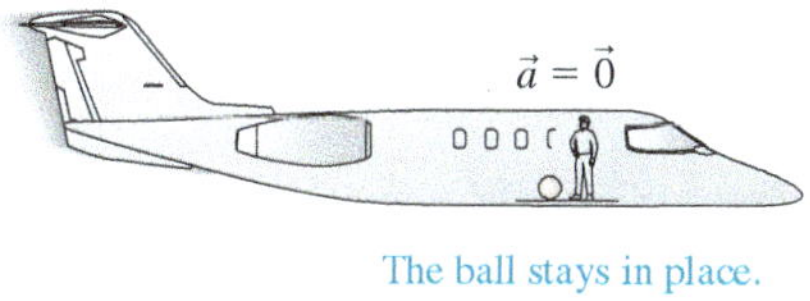

A ball with no horizontal forces stays at rest in an airplane cruising at constant velocity. The airplane is an inertial reference frame.

(b)

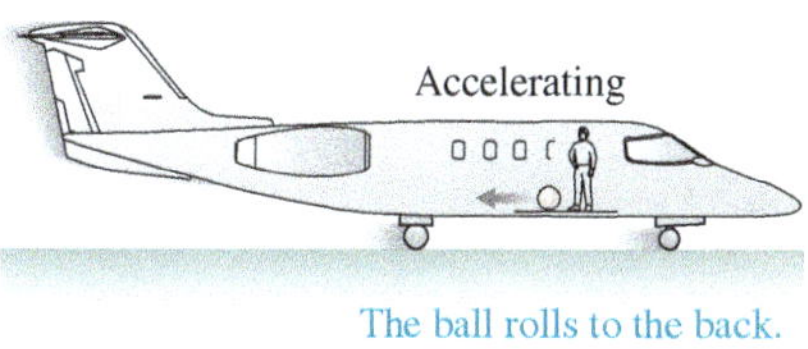

The ball rolls to the back of the plane during takeoff. An accelerating plane is not an inertial reference frame.

The physics student in FIGURE 5.21b conducts the same experiment during takeoff. He carefully places the ball on the floor just as the airplane starts to accelerate down the runway. You can imagine what happens. The ball rolls to the back of the plane as the passengers are being pressed back into their seats. Nothing exerts a horizontal contact force on the ball, yet the ball accelerates *in the plane's reference frame.* This violates Newton's first law, so the plane is *not* an inertial reference frame during takeoff.

In the first example, the plane is traveling with constant velocity. In the second, the plane is accelerating. **Accelerating reference frames are not inertial reference frames.** Consequently, Newton's laws are not valid in an accelerating reference frame.

The earth is not exactly an inertial reference frame because the earth rotates on its axis and orbits the sun. However, the earth's acceleration is so small that violations of Newton's laws can be measured only in high-precision experiments. We will treat the earth and laboratories attached to the earth as inertial reference frames, an approximation that is exceedingly well justified.

To understand the motion of the passengers in a braking car, you need to measure velocities and accelerations *relative to the ground.* From the perspective of an observer on the ground, the body of a passenger in a braking car tries to continue moving forward with constant velocity, exactly as we would expect on the basis of Newton's first law, while his immediate surroundings are decelerating. The passenger is not "thrown" into the windshield. Instead, the windshield runs into the passenger!

Thinking About Force

It is important to identify correctly all the forces acting on an object. It is equally important not to include forces that do not really exist. We have established a number of criteria for identifying forces; the three critical ones are:

- A force has an agent. Something tangible and identifiable causes the force.
- Forces exist at the point of contact between the agent and the object experiencing the force (except for the few special cases of long-range forces).
- Forces exist due to interactions happening *now*, not due to what happened in the past.

There's no "force of motion" or any other forward force on this arrow. It continues to move because of inertia.

We all have had many experiences suggesting that a force is necessary to keep something moving. Consider a bowling ball rolling along on a smooth floor. It is very tempting to think that a horizontal "force of motion" keeps it moving in the forward direction. But *nothing contacts the ball* except the floor. No agent is giving the ball a forward push. According to our definition, then, there is *no* forward "force of motion" acting on the ball. So what keeps it going? Recall our discussion of the first law: *No* cause is needed to keep an object moving at constant velocity. It continues to move forward simply because of its inertia.

One reason for wanting to include a "force of motion" is that we tend to view the problem from our perspective as one of the agents of force. You certainly have to keep pushing to move a box across the floor at constant velocity. If you stop, it stops. Newton's laws, though, require that we adopt the object's perspective. The box experiences your pushing force in one direction *and* a friction force in the opposite direction. The box moves at constant velocity if the *net* force is zero. This will be true as long as your pushing force exactly balances the friction force. When you stop pushing, the friction force causes an acceleration that slows and stops the box.

A related problem occurs if you throw a ball. A pushing force was indeed required to accelerate the ball *as it was thrown.* But that force disappears the instant the ball loses contact with your hand. The force does not stick with the ball as the ball travels through the air. Once the ball has acquired a velocity, *nothing* is needed to keep it moving with that velocity.

5.7 Free-Body Diagrams

Having discussed at length what is and is not a force, we are ready to assemble our knowledge about force and motion into a single diagram called a *free-body diagram.* You will learn in the next chapter how to write the equations of motion directly from the free-body diagram. Solution of the equations is a mathematical exercise—possibly a difficult one, but nonetheless an exercise that could be done by a computer. The *physics* of the problem, as distinct from the purely calculational aspects, are the steps that lead to the free-body diagram.

A **free-body diagram,** part of the *pictorial representation* of a problem, represents the object as a particle and shows *all* of the forces acting on the object.

TACTICS BOX 5.3 Drawing a free-body diagram

1. **Identify all forces acting on the object.** This step was described in Tactics Box 5.2.
2. **Draw a coordinate system.** Use the axes defined in your pictorial representation.
3. **Represent the object as a dot at the origin of the coordinate axes.** This is the particle model.
4. **Draw vectors representing each of the identified forces.** This was described in Tactics Box 5.1. Be sure to label each force vector.
5. **Draw and label the *net force* vector $\vec{F}_{net}$.** Draw this vector beside the diagram, not on the particle. Or, if appropriate, write $\vec{F}_{net} = \vec{0}$. Then check that $\vec{F}_{net}$ points in the same direction as the acceleration vector $\vec{a}$ on your motion diagram.

Exercises 24–29

EXAMPLE 5.4 An elevator accelerates upward

An elevator, suspended by a cable, speeds up as it moves upward from the ground floor. Identify the forces and draw a free-body diagram of the elevator.

MODEL Treat the elevator as a particle.

VISUALIZE

FIGURE 5.22 Free-body diagram of an elevator accelerating upward.

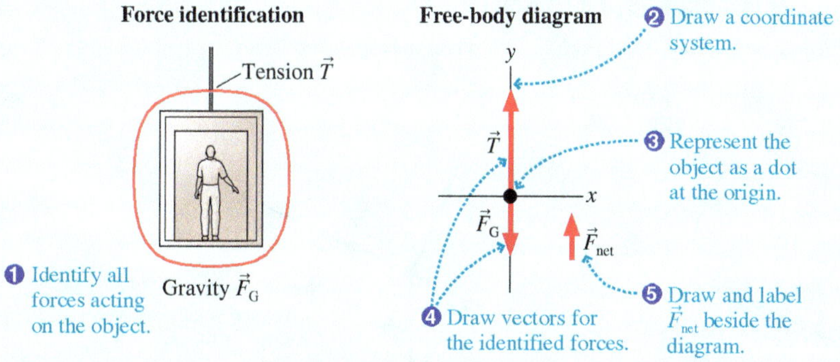

ASSESS The coordinate axes, with a vertical y-axis, are the ones we would use in a pictorial representation of the motion. The elevator is accelerating upward, so $\vec{F}_{net}$ must point upward. For this to be true, the magnitude of $\vec{T}$ must be larger than the magnitude of $\vec{F}_G$. The diagram has been drawn accordingly.

EXAMPLE 5.5 An ice block shoots across a frozen lake

Bobby straps a small model rocket to a block of ice and shoots it across the smooth surface of a frozen lake. Friction is negligible. Draw a pictorial representation of the block of ice.

MODEL Treat the block of ice as a particle. The pictorial representation consists of a motion diagram to determine $\vec{a}$, a force-identification picture, and a free-body diagram. The statement of the situation implies that friction is negligible.

VISUALIZE

FIGURE 5.23 Pictorial representation for a block of ice shooting across a frictionless frozen lake.

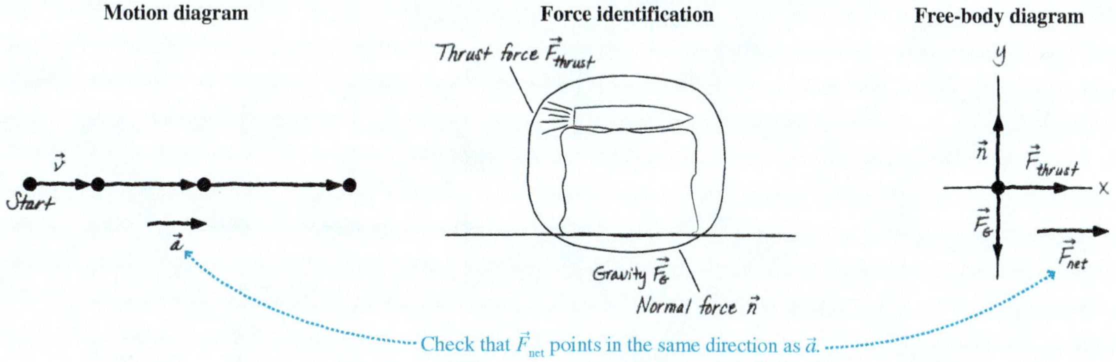

ASSESS The motion diagram tells us that the acceleration is in the positive x-direction. According to the rules of vector addition, this can be true only if the upward-pointing $\vec{n}$ and the downward-pointing $\vec{F}_G$ are equal in magnitude and thus cancel each other $\left((F_G)_y = -n_y\right)$. The vectors have been drawn accordingly, and this leaves the net force vector pointing toward the right, in agreement with $\vec{a}$ from the motion diagram.

EXAMPLE 5.6 **A skier is pulled up a hill**

A tow rope pulls a skier up a snow-covered hill at a constant speed. Draw a pictorial representation of the skier.

MODEL This is Example 5.2 again with the additional information that the skier is moving at constant speed. The skier will be treated as a particle in *dynamic equilibrium.* If we were doing a kinematics problem, the pictorial representation would use a tilted coordinate system with the x-axis parallel to the slope, so we use these same tilted coordinate axes for the free-body diagram.

VISUALIZE

FIGURE 5.24 Pictorial representation for a skier being towed at a constant speed.

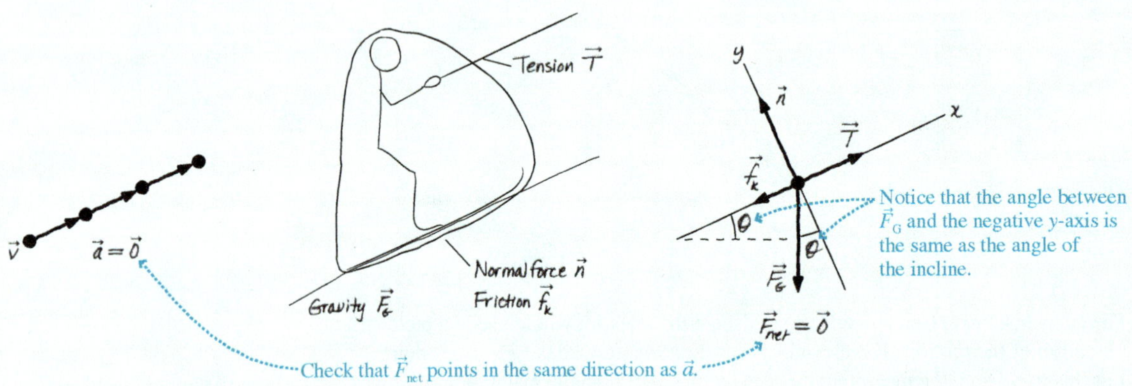

ASSESS We have shown $\vec{T}$ pulling parallel to the slope and $\vec{f}_k$, which opposes the direction of motion, pointing down the slope. $\vec{n}$ is perpendicular to the surface and thus along the y-axis. Finally, and this is important, the gravitational force $\vec{F}_G$ is *vertically* downward, *not* along the negative y-axis. In fact, you should convince yourself from the geometry that the angle θ between the $\vec{F}_G$ vector and the negative y-axis is the same as the angle θ of the incline above the horizontal. The skier moves in a straight line with constant speed, so $\vec{a} = \vec{0}$ and, from Newton's first law, $\vec{F}_{net} = \vec{0}$. Thus we have drawn the vectors such that the y-component of $\vec{F}_G$ is equal in magnitude to $\vec{n}$. Similarly, $\vec{T}$ must be large enough to match the negative x-components of both $\vec{f}_k$ and $\vec{F}_G$.

Free-body diagrams will be our major tool for the next several chapters. Careful practice with the workbook exercises and homework in this chapter will pay immediate benefits in the next chapter. Indeed, it is not too much to assert that a problem is half solved, or even more, when you complete the free-body diagram.

STOP TO THINK 5.5 An elevator suspended by a cable is moving upward and slowing to a stop. Which free-body diagram is correct?

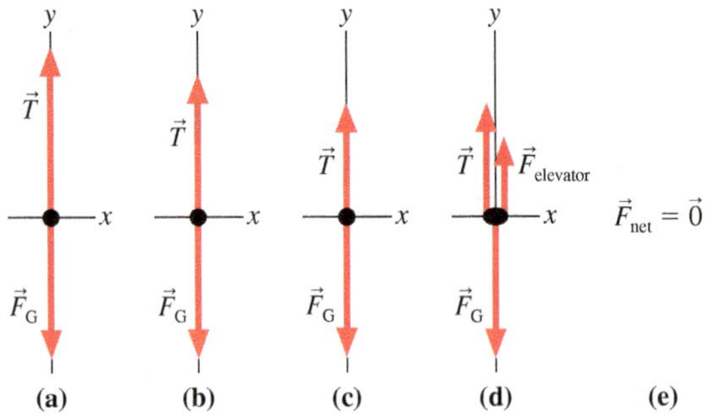

SUMMARY

The goal of Chapter 5 has been to establish a connection between force and motion.

General Principles

Newton's First Law

An object at rest will remain at rest, or an object that is moving will continue to move in a straight line with constant velocity, if and only if the net force on the object is zero.

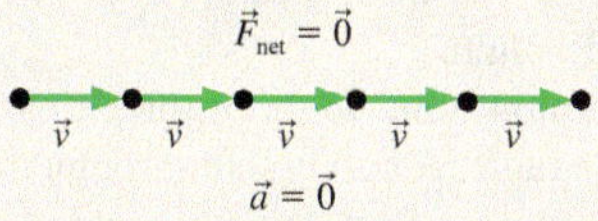

The first law tells us that no "cause" is needed for motion. Uniform motion is the "natural state" of an object.

Newton's laws are valid only in inertial reference frames.

Newton's Second Law

An object with mass m will undergo acceleration

$$\vec{a} = \frac{1}{m}\vec{F}_{net}$$

where $\vec{F}_{net} = \vec{F}_1 + \vec{F}_2 + \vec{F}_3 + \cdots$ is the vector sum of all the individual forces acting on the object.

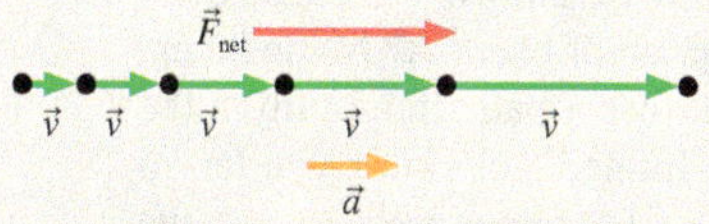

The second law tells us that a net force causes an object to accelerate. This is the connection between force and motion that we are seeking.

Important Concepts

Acceleration is the link to kinematics.

From $\vec{F}_{net}$, find $\vec{a}$.
From a, find v and x.

$\vec{a} = \vec{0}$ is the condition for equilibrium.

Static equilibrium if $\vec{v} = \vec{0}$.
Dynamic equilibrium if $\vec{v} =$ constant.

Equilibrium occurs if and only if $\vec{F}_{net} = \vec{0}$.

Mass is the resistance of an object to acceleration. It is an intrinsic property of an object.

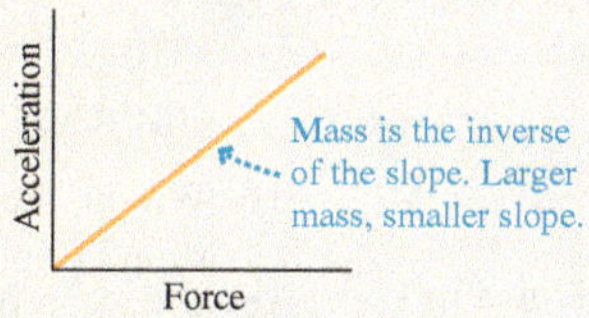

Force is a push or a pull on an object.

- Force is a vector, with a magnitude and a direction.
- Force requires an agent.
- Force is either a contact force or a long-range force.

Key Skills

Identifying Forces

Forces are identified by locating the points where other objects touch the object of interest. These are points where contact forces are exerted. In addition, objects with mass feel a long-range gravitational force.

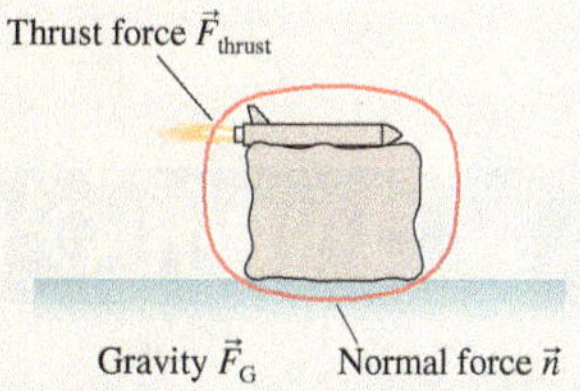

Free-Body Diagrams

A free-body diagram represents the object as a particle at the origin of a coordinate system. Force vectors are drawn with their tails on the particle. The net force vector is drawn beside the diagram.

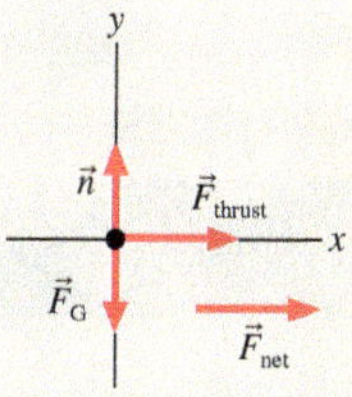

Terms and Notation

dynamics
mechanics
force, $\vec{F}$
agent
contact force
long-range force
net force, $\vec{F}_{net}$
superposition of forces
gravitational force, $\vec{F}_G$
spring force, $\vec{F}_{sp}$
tension force, $\vec{T}$
atomic model
normal force, $\vec{n}$
friction, $\vec{f}_k$ or $\vec{f}_s$
drag, $\vec{D}$
thrust, $\vec{F}_{thrust}$
proportionality
proportionality constant
proportional reasoning
newton, N
inertia
inertial mass, m
Newton's second law
Newton's zeroth law
Newton's first law
mechanical equilibrium
static equilibrium
dynamic equilibrium
inertial reference frame
free-body diagram

CONCEPTUAL QUESTIONS

1. An elevator suspended by a cable is descending at constant velocity. How many force vectors would be shown on a free-body diagram? Name them.
2. A compressed spring is pushing a block across a rough horizontal table. How many force vectors would be shown on a free-body diagram? Name them.
3. A brick is falling from the roof of a three-story building. How many force vectors would be shown on a free-body diagram? Name them.
4. In FIGURE Q5.4, block B is falling and dragging block A across a table. How many force vectors would be shown on a free-body diagram of block A? Name them.

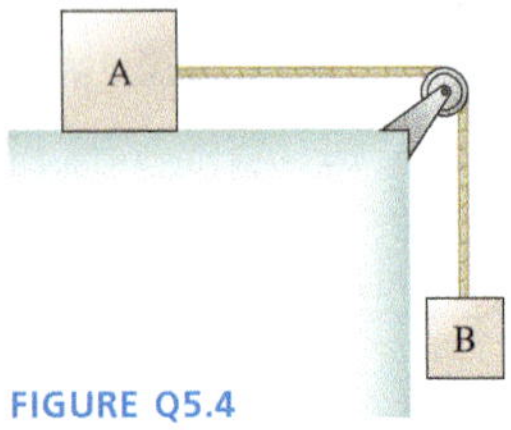

FIGURE Q5.4

5. You toss a ball straight up in the air. Immediately after you let go of it, what forces are acting on the ball? For each force you name, (a) state whether it is a contact force or a long-range force and (b) identify the agent of the force.
6. A constant force applied to A causes A to accelerate at 5 m/s^2. The same force applied to B causes an acceleration of 3 m/s^2. Applied to C, it causes an acceleration of 8 m/s^2.
   a. Which object has the largest mass? Explain.
   b. Which object has the smallest mass?
   c. What is the ratio m_A/m_B of the mass of A to the mass of B?
7. An object experiencing a constant force accelerates at 10 m/s^2. What will the acceleration of this object be if
   a. The force is doubled? Explain.
   b. The mass is doubled?
   c. The force is doubled *and* the mass is doubled?
8. An object experiencing a constant force accelerates at 8 m/s^2. What will the acceleration of this object be if
   a. The force is halved? Explain.
   b. The mass is halved?
   c. The force is halved *and* the mass is halved?
9. If an object is at rest, can you conclude that there are no forces acting on it? Explain.
10. If a force is exerted on an object, is it possible for that object to be moving with constant velocity? Explain.
11. Is the statement "An object always moves in the direction of the net force acting on it" true or false? Explain.
12. Newton's second law says $\vec{F}_{net} = m\vec{a}$. So is $m\vec{a}$ a force? Explain.
13. Is it possible for the friction force on an object to be in the direction of motion? If so, give an example. If not, why not?
14. Suppose you press your physics book against a wall hard enough to keep it from moving. Does the friction force on the book point (a) into the wall, (b) out of the wall, (c) up, (d) down, or (e) is there no friction force? Explain.
15. FIGURE Q5.15 shows a hollow tube forming three-quarters of a circle. It is lying flat on a table. A ball is shot through the tube at high speed. As the ball emerges from the other end, does it follow path A, path B, or path C? Explain.

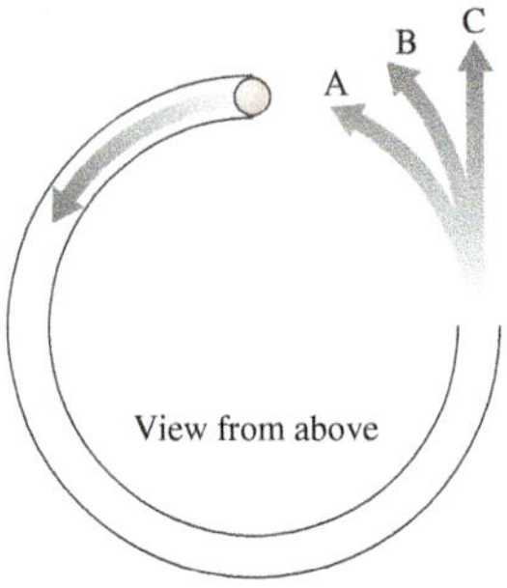

FIGURE Q5.15

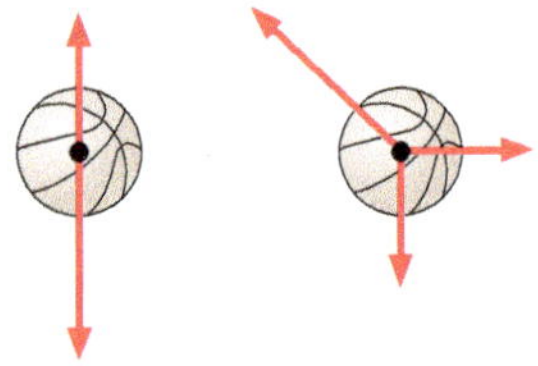
FIGURE Q5.16

16. Which, if either, of the basketballs in FIGURE Q5.16 are in equilibrium? Explain.
17. Which of the following are inertial reference frames? Explain.
   a. A car driving at steady speed on a straight and level road.
   b. A car driving at steady speed up a 10° incline.
   c. A car speeding up after leaving a stop sign.
   d. A car driving at steady speed around a curve.

EXERCISES AND PROBLEMS

Exercises

Section 5.3 Identifying Forces

1. | A chandelier hangs from a chain in the middle of a dining room. Identify the forces on the chandelier.
2. | A car is parked on a steep hill. Identify the forces on the car.
3. || A jet plane is speeding down the runway during takeoff. Air resistance is not negligible. Identify the forces on the jet.
4. | A baseball player is sliding into second base. Identify the forces on the baseball player.
5. || A bullet has just been shot from a gun and is now traveling horizontally. Air resistance is not negligible. Identify the forces on the bullet.

Section 5.4 What Do Forces Do? A Virtual Experiment

6. | Two rubber bands cause an object to accelerate with acceleration a. How many rubber bands are needed to cause an object with half the mass to accelerate three times as quickly?
7. | Two rubber bands pulling on an object cause it to accelerate at 1.2 m/s^2.
   a. What will be the object's acceleration if it is pulled by four rubber bands?
   b. What will be the acceleration of two of these objects glued together if they are pulled by two rubber bands?

8. | FIGURE EX5.8 shows an acceleration-versus-force graph for three objects pulled by rubber bands. The mass of object 2 is 0.20 kg. What are the masses of objects 1 and 3? Explain your reasoning.

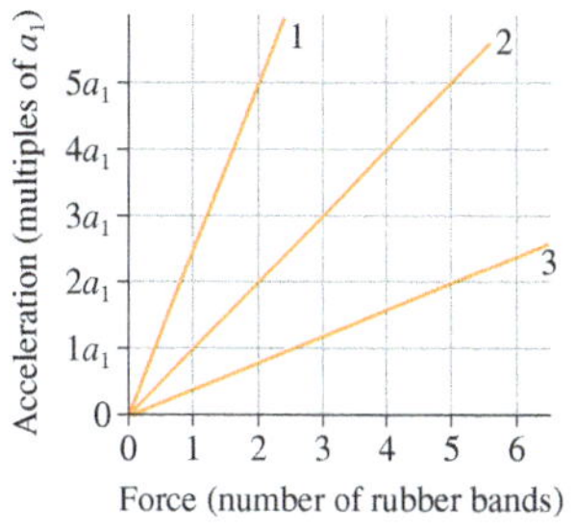

FIGURE EX5.8

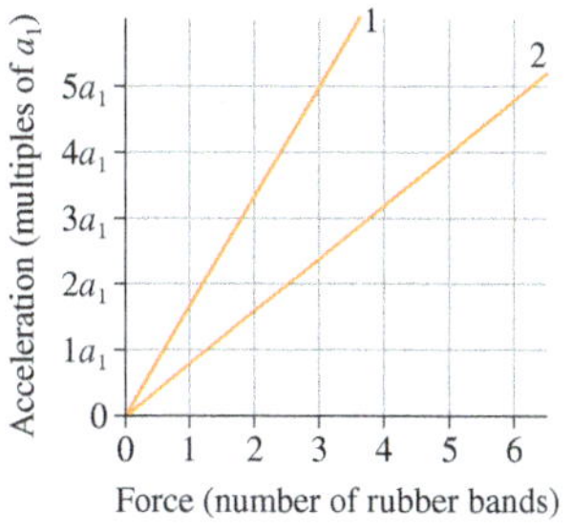

FIGURE EX5.9

9. | FIGURE EX5.9 shows acceleration-versus-force graphs for two objects pulled by rubber bands. What is the mass ratio m_1/m_2?

10. | For an object starting from rest and accelerating with constant acceleration, distance traveled is proportional to the square of the time. If an object travels 2.0 furlongs in the first 2.0 s, how far will it travel in the first 4.0 s?

11. | The period of a pendulum is proportional to the square root of its length. A 2.0-m-long pendulum has a period of 3.0 s. What is the period of a 3.0-m-long pendulum?

Section 5.5 Newton's Second Law

12. | FIGURE EX5.12 shows an acceleration-versus-force graph for a 500 g object. What acceleration values go in the blanks on the vertical scale?

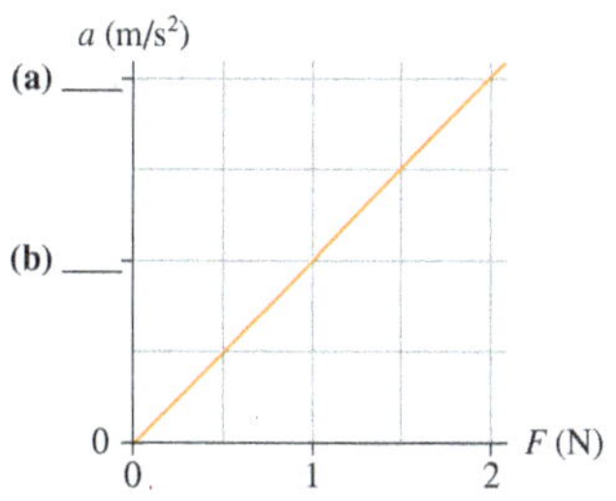

FIGURE EX5.12

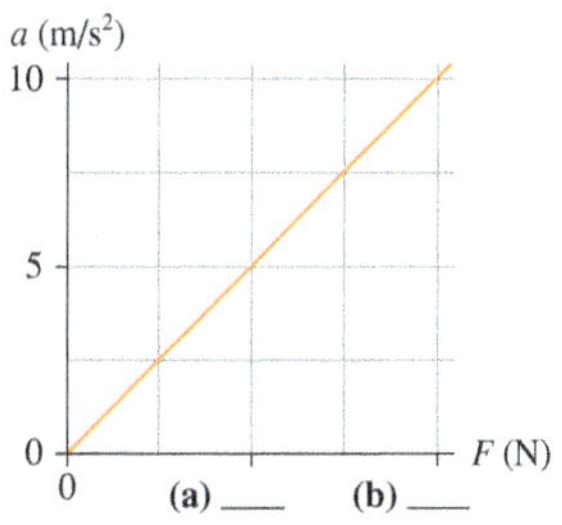

FIGURE EX5.13

13. | FIGURE EX5.13 shows an acceleration-versus-force graph for a 200 g object. What force values go in the blanks on the horizontal scale?

14. | FIGURE EX5.14 shows an object's acceleration-versus-force graph. What is the object's mass?

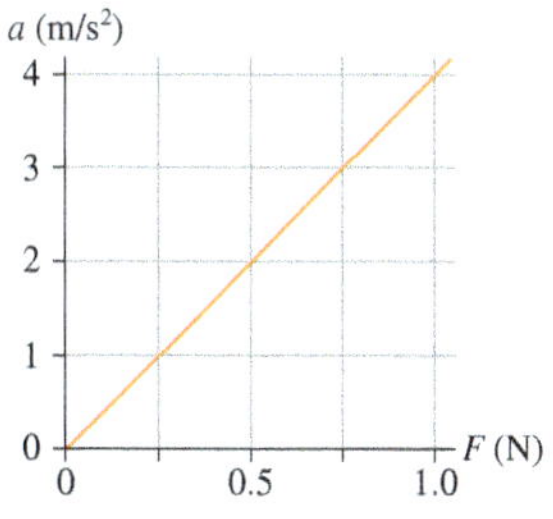

FIGURE EX5.14

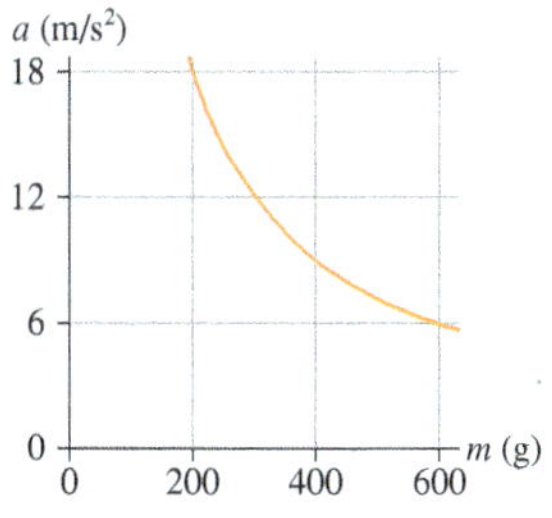

FIGURE EX5.15

15. | FIGURE EX5.15 shows the acceleration of objects of different mass that experience the same force. What is the magnitude of the force?

16. | Based on the information in Table 5.1, *estimate*
   a. The weight of a laptop computer.
   b. The propulsion force of a bicycle.

17. | Based on the information in Table 5.1, *estimate*
   a. The weight of a pencil.
   b. The propulsion force of a sprinter.

Section 5.6 Newton's First Law

Exercises 18 through 20 show two of the three forces acting on an object in equilibrium. Redraw the diagram, showing all three forces. Label the third force $\vec{F}_3$.

18. ||

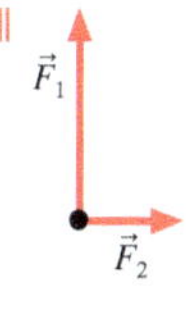

FIGURE EX5.18

19. ||

FIGURE EX5.19

20. ||

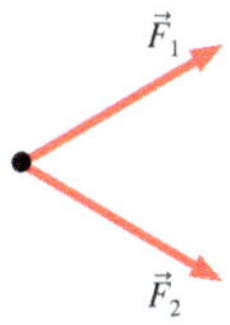

FIGURE EX5.20

Section 5.7 Free-Body Diagrams

Exercises 21 through 23 show a free-body diagram. For each:
   a. Redraw the free-body diagram.
   b. Write a short description of a real object for which this is the correct free-body diagram. Use Examples 5.4, 5.5, and 5.6 as models of what a description should be like.

21. |

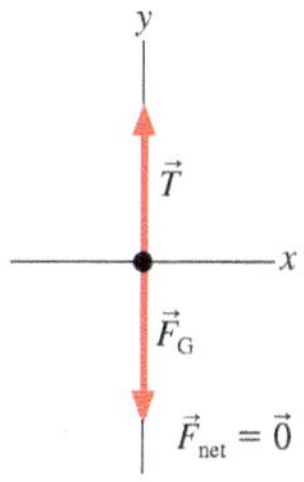

FIGURE EX5.21

22. |

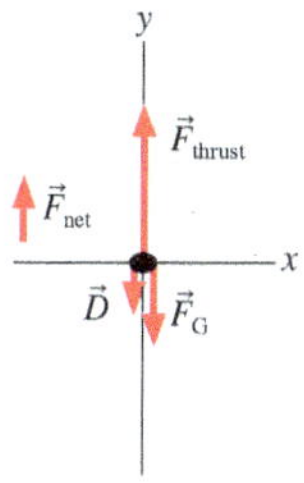

FIGURE EX5.22

23. |

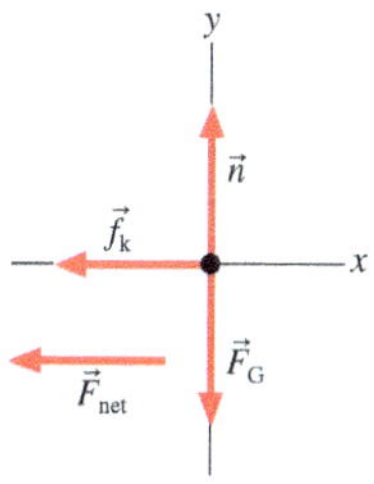

FIGURE EX5.23

Exercises 24 through 27 describe a situation. For each, identify all forces acting on the object and draw a free-body diagram of the object.

24. | A cat is sitting on a window sill.

25. | An ice hockey puck glides across frictionless ice.

26. | Your physics textbook is sliding across the table.

27. | A steel beam is being lowered at steady speed by a crane.

Problems

28. | Redraw the two motion diagrams shown in FIGURE P5.28, then draw a vector beside each one to show the direction of the net force acting on the object. Explain your reasoning.

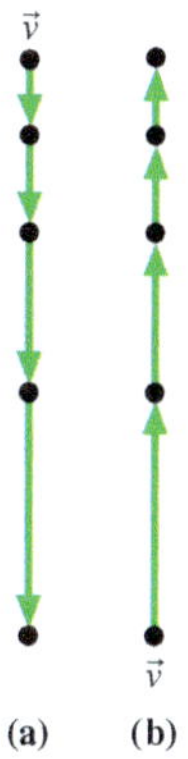

FIGURE P5.28

29. | Redraw the two motion diagrams shown in FIGURE P5.29, then draw a vector beside each one to show the direction of the net force acting on the object. Explain your reasoning.

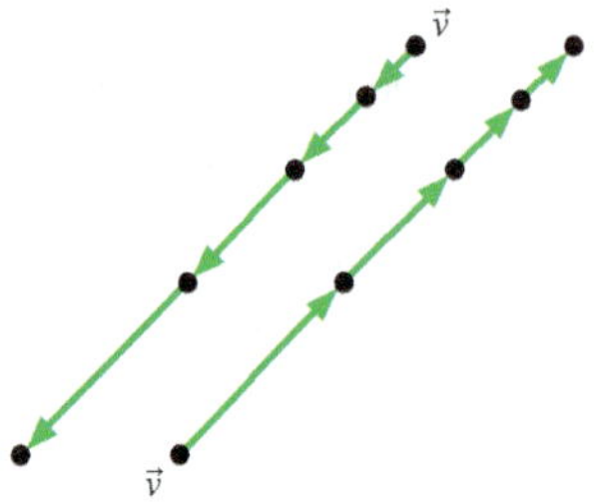
(a) (b)

FIGURE P5.29

30. | A single force with x-component F_x acts on a 2.0 kg object as it moves along the x-axis. The object's acceleration graph (a_x versus t) is shown in FIGURE P5.30. Draw a graph of F_x versus t.

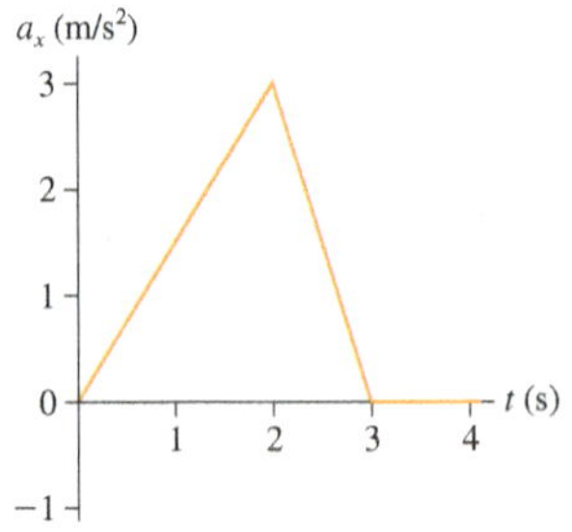

FIGURE P5.30

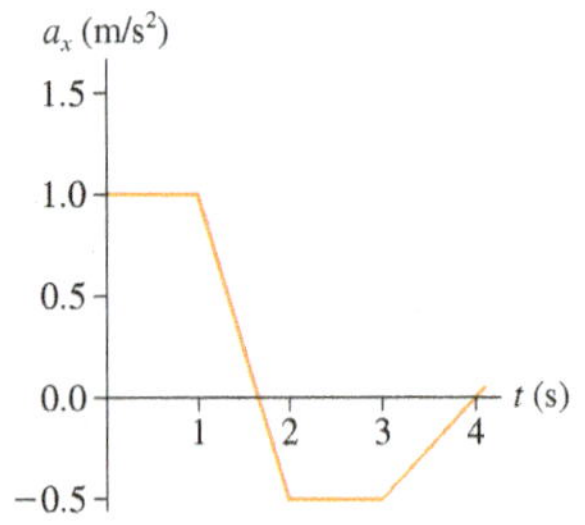

FIGURE P5.31

31. | A single force with x-component F_x acts on a 500 g object as it moves along the x-axis. The object's acceleration graph (a_x versus t) is shown in FIGURE P5.31. Draw a graph of F_x versus t.

32. | A single force with x-component F_x acts on a 2.0 kg object as it moves along the x-axis. A graph of F_x versus t is shown in FIGURE P5.32. Draw an acceleration graph (a_x versus t) for this object.

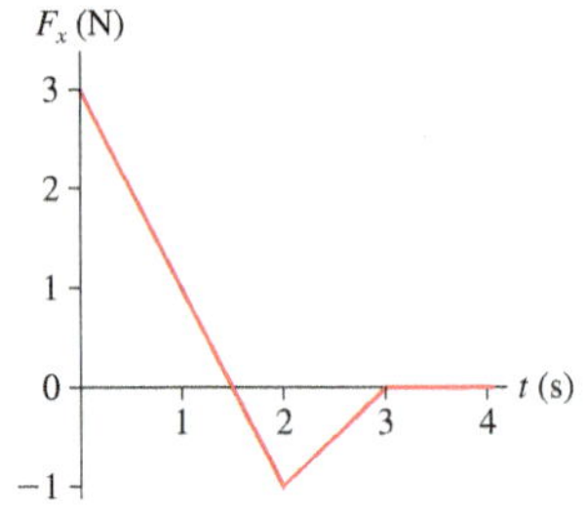

FIGURE P5.32

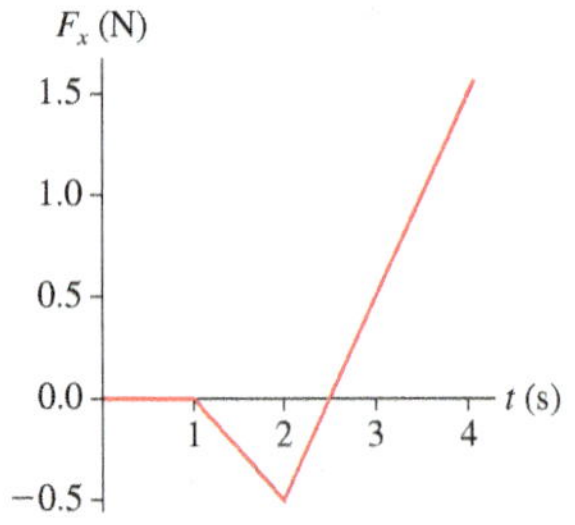

FIGURE P5.33

33. | A single force with x-component F_x acts on a 500 g object as it moves along the x-axis. A graph of F_x versus t is shown in FIGURE P5.33. Draw an acceleration graph (a_x versus t) for this object.

34. | A constant force is applied to an object, causing the object to accelerate at 8.0 m/s^2. What will the acceleration be if
   a. The force is doubled?
   b. The object's mass is doubled?
   c. The force and the object's mass are both doubled?
   d. The force is doubled and the object's mass is halved?

35. | A constant force is applied to an object, causing the object to accelerate at 10 m/s^2. What will the acceleration be if
   a. The force is halved?
   b. The object's mass is halved?
   c. The force and the object's mass are both halved?
   d. The force is halved and the object's mass is doubled?

Problems 36 through 41 show a free-body diagram. For each:
   a. Redraw the diagram.
   b. Identify the direction of the acceleration vector $\vec{a}$ and show it as a vector next to your diagram. Or, if appropriate, write $\vec{a} = \vec{0}$.
   c. If possible, identify the direction of the velocity vector $\vec{v}$ and show it as a labeled vector.
   d. Write a short description of a real object for which this is the correct free-body diagram. Use Examples 5.4, 5.5, and 5.6 as models of what a description should be like.

36. |

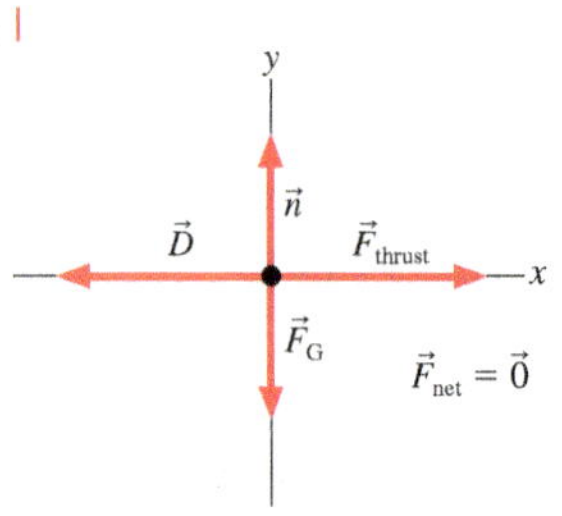

FIGURE P5.36

37. |

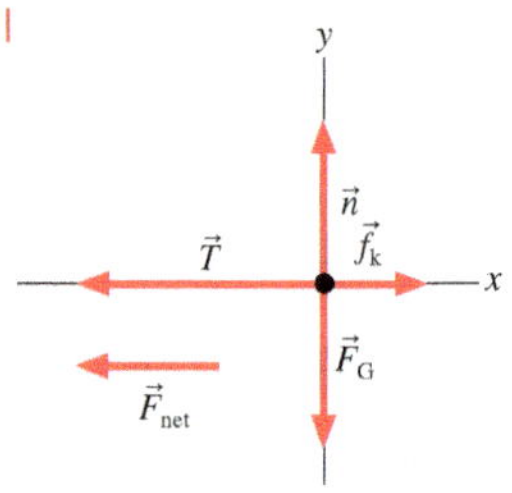

FIGURE P5.37

38. |

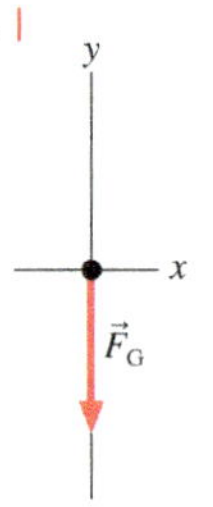

FIGURE P5.38

39. |

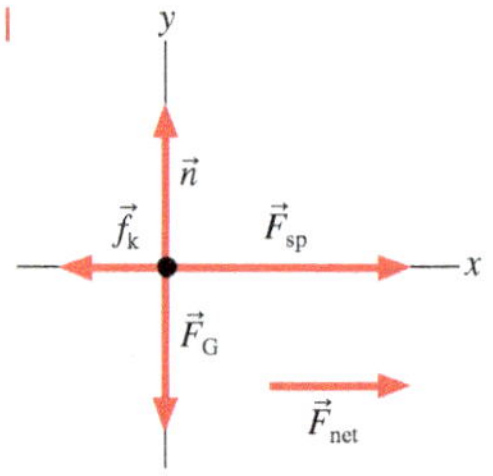

FIGURE P5.39

40. ||

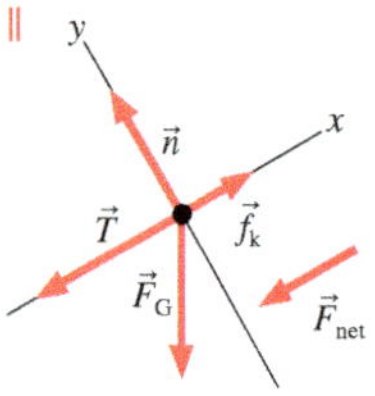

FIGURE P5.40

41. ||

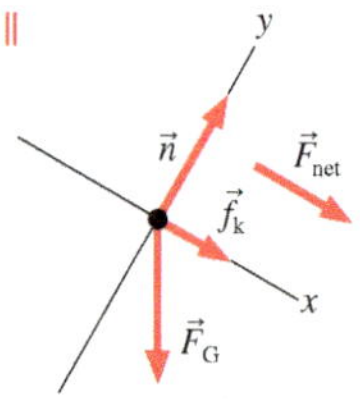

FIGURE P5.41

42. || In lab, you propel a cart with four known forces while using an ultrasonic motion detector to measure the cart's acceleration. Your data are as follows:

Force (N)	Acceleration (m/s^2)
0.25	0.5
0.50	0.8
0.75	1.3
1.00	1.8

   a. How should you graph these data so as to determine the mass of the cart from the slope of the line? That is, what values should you graph on the horizontal axis and what on the vertical axis?
   b. Is there another data point that would be reasonable to add, even though you made no measurements? If so, what is it?
   c. What is your best determination of the cart's mass?

Problems 43 through 52 describe a situation. For each, draw a motion diagram, a force-identification diagram, and a free-body diagram.

43. | An elevator, suspended by a single cable, has just left the tenth floor and is speeding up as it descends toward the ground floor.
44. || A rocket is being launched straight up. Air resistance is not negligible.
45. | A jet plane is speeding down the runway during takeoff. Air resistance is not negligible.
46. | You've slammed on the brakes and your car is skidding to a stop while going down a 20° hill.
47. || A skier is going down a 20° slope. A *horizontal* headwind is blowing in the skier's face. Friction is small, but not zero.
48. || You've just kicked a rock on the sidewalk and it is now sliding along the concrete.
49. | A Styrofoam ball has just been shot straight up. Air resistance is not negligible.
50. | A spring-loaded gun shoots a plastic ball. The trigger has just been pulled and the ball is starting to move down the barrel. The barrel is horizontal.
51. || A person on a bridge throws a rock straight down toward the water. The rock has just been released.
52. | A gymnast has just landed on a trampoline. She's still moving downward as the trampoline stretches.
53. || The leaf hopper, champion jumper of the insect world, can
BIO jump straight up at 4 m/s^2. The jump itself lasts a mere 1 ms before the insect is clear of the ground.
    a. Draw a free-body diagram of this mighty leaper while the jump is taking place.
    b. While the jump is taking place, is the force of the ground on the leaf hopper greater than, less than, or equal to the force of gravity on the leaf hopper? Explain.

Challenge Problems

54. A heavy box is in the back of a truck. The truck is accelerating to the right. Draw a motion diagram, a force-identification diagram, and a free-body diagram for the box.
55. A bag of groceries is on the seat of your car as you stop for a stop light. The bag does not slide. Draw a motion diagram, a force-identification diagram, and a free-body diagram for the bag.
56. A rubber ball bounces. We'd like to understand *how* the ball bounces.
    a. A rubber ball has been dropped and is bouncing off the floor. Draw a motion diagram of the ball during the brief time interval that it is in contact with the floor. Show 4 or 5 frames as the ball compresses, then another 4 or 5 frames as it expands. What is the direction of $\vec{a}$ during each of these parts of the motion?
    b. Draw a picture of the ball in contact with the floor and identify all forces acting on the ball.
    c. Draw a free-body diagram of the ball during its contact with the ground. Is there a net force acting on the ball? If so, in which direction?
    d. Write a paragraph in which you describe what you learned from parts a to c and in which you answer the question: How does a ball bounce?
57. If a car stops suddenly, you feel "thrown forward." We'd like to understand what happens to the passengers as a car stops. Imagine yourself sitting on a *very* slippery bench inside a car. This bench has no friction, no seat back, and there's nothing for you to hold onto.
    a. Draw a picture and identify all of the forces acting on you as the car travels at a perfectly steady speed on level ground.
    b. Draw your free-body diagram. Is there a net force on you? If so, in which direction?
    c. Repeat parts a and b with the car slowing down.
    d. Describe what happens to you as the car slows down.
    e. Use Newton's laws to explain why you seem to be "thrown forward" as the car stops. Is there really a force pushing you forward?
    f. Suppose now that the bench is not slippery. As the car slows down, you stay on the bench and don't slide off. What force is responsible for your deceleration? In which direction does this force point? Include a free-body diagram as part of your answer.

STOP TO THINK ANSWERS

Stop to Think 5.1: c.

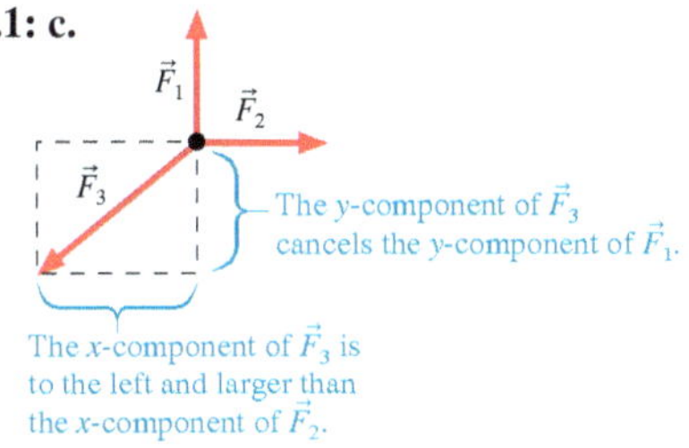

Stop to Think 5.2: a, b, and d. Friction and the normal force are the only contact forces. Nothing is touching the rock to provide a "force of the kick."

Stop to Think 5.3: b. Acceleration is proportional to force, so doubling the number of rubber bands doubles the acceleration of the original object from 2 m/s^2 to 4 m/s^2. But acceleration is also inversely proportional to mass. Doubling the mass cuts the acceleration in half, back to 2 m/s^2.

Stop to Think 5.4: d.

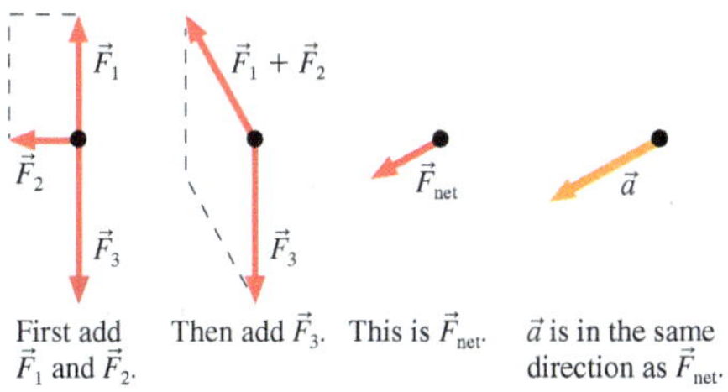

Stop to Think 5.5: c. The acceleration vector points downward as the elevator slows. $\vec{F}_{net}$ points in the same direction as $\vec{a}$, so $\vec{F}_{net}$ also points down. This will be true if the tension is less than the gravitational force: $T < F_G$.

6 Dynamics I: Motion Along a Line

The powerful thrust of the jet engines accelerates this enormous plane to a speed of over 150 mph in less than a mile.

▶ **Looking Ahead** The goal of Chapter 6 is to learn how to solve linear force-and-motion problems.

Forces

In Chapter 5 you learned what a force is and how force and motion are related through Newton's second law.

In this chapter, you'll study in more detail some of the forces introduced in Chapter 5. You'll also learn to solve **equilibrium** problems (with zero net force) and then **dynamics** problems.

The problem-solving procedures developed in this chapter will be used throughout the remainder of the book.

◀ **Looking Back**
Sections 5.3 and 5.7 Identifying forces, drawing free-body diagrams

Equilibrium

An object at rest or moving in a straight line with constant velocity is in equilibrium. The net force is zero.

What tension is needed to tow the car at constant velocity?

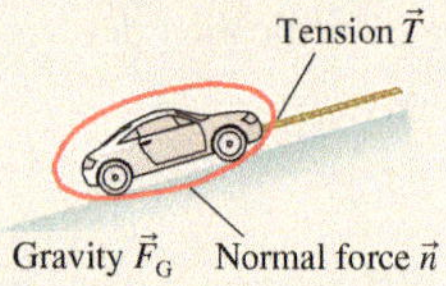

To solve equilibrium problems, you must be able to identify and work with forces.

◀ **Looking Back**
Section 5.2 A catalog of forces

Dynamics

A net force on an object causes the object to accelerate. This is Newton's second law.

What is the acceleration of the file cabinet?

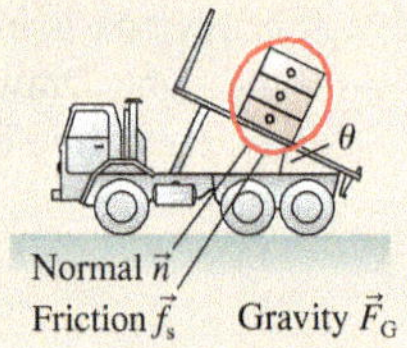

To solve dynamics problems, you will need to use constant-acceleration kinematics.

◀ **Looking Back**
Sections 2.4–2.6 Constant-acceleration kinematics

Mass and Weight

You'll learn how mass and weight are different.

- **Mass** is the amount of matter in an object. It is the same everywhere.
- **Weight** is the result of weighing an object on a scale. It depends on gravity and acceleration.

This astronaut on the moon weighs only 1/6 of what he does on earth, but his mass is the same.

Friction and Drag

We'll expand our understanding of friction and drag by developing a model of each.

- Static and kinetic friction depend on the **coefficient of friction**, but not on the object's speed.
- Drag depends on the *square* of the speed and also on the object's cross-section area.

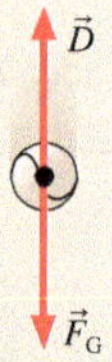

A falling object reaches terminal speed when the drag force balances the gravitational force.

Problem Solving

We'll develop a *strategy* for solving force and motion problems, one based on a set of *procedures* rather than a memorized set of equations.

This chapter focuses on motion in a straight line, the motion of bicycles, cars, planes, and rockets.

6.1 Equilibrium

An object on which the net force is zero is in *equilibrium.* The object might be at rest in *static equilibrium,* or it might be moving along a straight line with constant velocity in *dynamic equilibrium.* Both are identical from a Newtonian perspective because $\vec{F}_{\text{net}} = \vec{0}$ and $\vec{a} = \vec{0}$.

Newton's first law is the basis for a four-step *strategy* for solving equilibrium problems.

PROBLEM-SOLVING STRATEGY 6.1 **Equilibrium problems**

MODEL Make simplifying assumptions. When appropriate, represent the object as a particle.

VISUALIZE

- Establish a coordinate system, define symbols, and identify what the problem is asking you to find. This is the process of translating words into symbols.
- Identify *all* forces acting on the object and show them on a free-body diagram.
- These elements form the **pictorial representation** of the problem.

SOLVE The mathematical representation is based on Newton's first law:

$$\vec{F}_{\text{net}} = \sum_i \vec{F}_i = \vec{0}$$

The vector sum of the forces is found directly from the free-body diagram.

ASSESS Check that your result has the correct units, is reasonable, and answers the question.

Newton's laws are *vector equations.* The requirement for equilibrium, $\vec{F}_{\text{net}} = \vec{0}$, is a shorthand way of writing two simultaneous equations:

$$(F_{\text{net}})_x = \sum_i (F_i)_x = 0$$
$$(F_{\text{net}})_y = \sum_i (F_i)_y = 0 \qquad (6.1)$$

The concept of equilibrium is essential for the engineering analysis of stationary objects such as bridges.

In other words, each component of $\vec{F}_{\text{net}}$ must simultaneously be zero. Although real-world situations often have forces pointing in three dimensions, thus requiring a third equation for the z-component of $\vec{F}_{\text{net}}$, we will restrict ourselves for now to problems that can be analyzed in two dimensions.

NOTE ▶ The equilibrium condition of Equations 6.1 applies only to particles, which cannot rotate. Equilibrium of an extended object, which can rotate, requires an additional condition that we will study in Chapter 12. ◀

Equilibrium problems occur frequently, especially in engineering applications. Let's look at a couple of examples.

Static Equilibrium

EXAMPLE 6.1 Finding the force on the kneecap

Your kneecap (patella) is attached by a tendon to your quadriceps muscle. This tendon pulls at a 10° angle relative to the femur, the bone of your upper leg. The patella is also attached to your lower leg (tibia) by a tendon that pulls parallel to the leg. To balance these forces, the lower end of your femur pushes outward on the patella. Bending your knee increases the tension in the tendons, and both have a tension of 60 N when the knee is bent to make a 70° angle between the upper and lower leg. What force does the femur exert on the kneecap in this position?

MODEL Model the kneecap as a particle in static equilibrium.

FIGURE 6.1 Pictorial representation of the kneecap in static equilibrium.

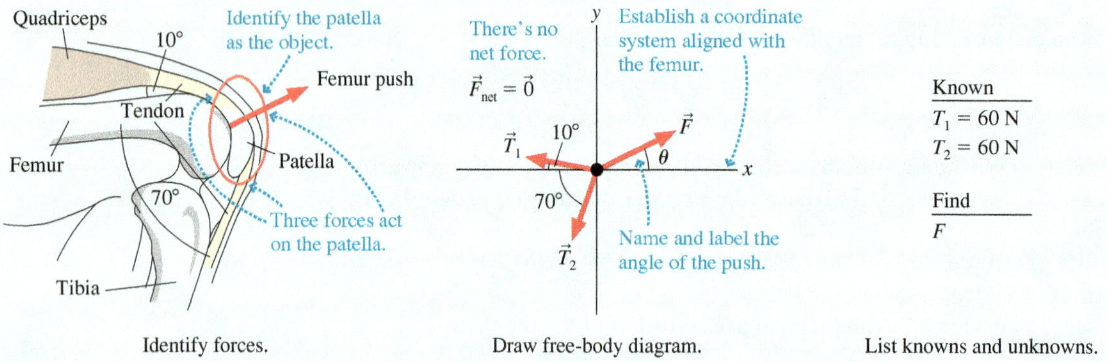

VISUALIZE **FIGURE 6.1** shows how to draw a pictorial representation. We've chosen to align the *x*-axis with the femur. The three forces—shown on the free-body diagram—are labeled $\vec{T}_1$ and $\vec{T}_2$ for the tensions and $\vec{F}$ for the femur's push. Notice that we've *defined* angle θ to indicate the direction of the femur's force on the kneecap.

SOLVE This is a static-equilibrium problem, with three forces on the kneecap that must sum to zero. Newton's first law, written in component form, is

$$(F_{net})_x = \sum_i (F_i)_x = T_{1x} + T_{2x} + F_x = 0$$

$$(F_{net})_y = \sum_i (F_i)_y = T_{1y} + T_{2y} + F_y = 0$$

NOTE ▶ You might have been tempted to write $-T_{1x}$ in the equation since $\vec{T}_1$ points to the left. But the net force, by definition, is the *sum* of all the individual forces. That fact that $\vec{T}_1$ points to the left will be taken into account when we *evaluate* the components. ◀

The components of the force vectors can be evaluated directly from the free-body diagram:

$$T_{1x} = -T_1 \cos 10° \qquad T_{1y} = T_1 \sin 10°$$
$$T_{2x} = -T_2 \cos 70° \qquad T_{2y} = -T_2 \sin 70°$$
$$F_x = F\cos\theta \qquad F_y = F\sin\theta$$

This is where signs enter, with T_{1x} being assigned a negative value because $\vec{T}_1$ points to the left. Similarly, $\vec{T}_2$ points both to the left and down, so both T_{2x} and T_{2y} are negative. With these components, Newton's first law becomes

$$-T_1 \cos 10° - T_2 \cos 70° + F\cos\theta = 0$$

$$T_1 \sin 10° - T_2 \sin 70° + F\sin\theta = 0$$

These are two simultaneous equations for the two unknowns F and θ. We will encounter equations of this form on many occasions, so make a note of the method of solution. First, rewrite the two equations as

$$F\cos\theta = T_1 \cos 10° + T_2 \cos 70°$$

$$F\sin\theta = -T_1 \sin 10° + T_2 \sin 70°$$

Next, divide the second equation by the first to eliminate F:

$$\frac{F\sin\theta}{F\cos\theta} = \tan\theta = \frac{-T_1 \sin 10° + T_2 \sin 70°}{T_1 \cos 10° + T_2 \cos 70°}$$

Then solve for θ:

$$\theta = \tan^{-1}\left(\frac{-T_1 \sin 10° + T_2 \sin 70°}{T_1 \cos 10° + T_2 \cos 70°}\right)$$

$$= \tan^{-1}\left(\frac{-(60\text{ N})\sin 10° + (60\text{ N})\sin 70°}{(60\text{ N})\cos 10° + (60\text{ N})\cos 70°}\right) = 30°$$

Finally, use θ to find F:

$$F = \frac{T_1 \cos 10° + T_2 \cos 70°}{\cos\theta}$$

$$= \frac{(60\text{ N})\cos 10° + (60\text{ N})\cos 70°)}{\cos 30°} = 92\text{ N}$$

The question asked What force? and force is a vector, so we must specify both the magnitude and the direction. With the knee in this position, the femur exerts a force $\vec{F} = (92\text{ N}, 30° \text{ above horizontal})$ on the kneecap.

ASSESS The magnitude of the force would be 0 N if the leg were straight, 120 N if the knee could be bent 180° so that the two tendons pull in parallel. The knee is closer to fully bent than to straight, so we would expect a femur force between 60 N and 120 N. Thus the calculated magnitude of 92 N seems reasonable.

Dynamic Equilibrium

EXAMPLE 6.2 Towing a car up a hill

A car with a weight of 15,000 N is being towed up a 20° slope at constant velocity. Friction is negligible. The tow rope is rated at 6000 N maximum tension. Will it break?

MODEL We'll treat the car as a particle in dynamic equilibrium.

VISUALIZE This problem asks for a yes or no answer, not a number, but we still need a quantitative analysis. Part of our analysis of the problem statement is to determine which quantity or quantities allow us to answer the question. In this case the answer is clear: We need to calculate the tension in the rope. FIGURE 6.2 shows the pictorial representation. Note the similarities to Examples 5.2 and 5.6 in Chapter 5, which you may want to review.

We noted in Chapter 5 that the weight of an object at rest is the magnitude F_G of the gravitational force acting on it, and that information has been listed as known.

FIGURE 6.2 Pictorial representation of a car being towed up a hill.

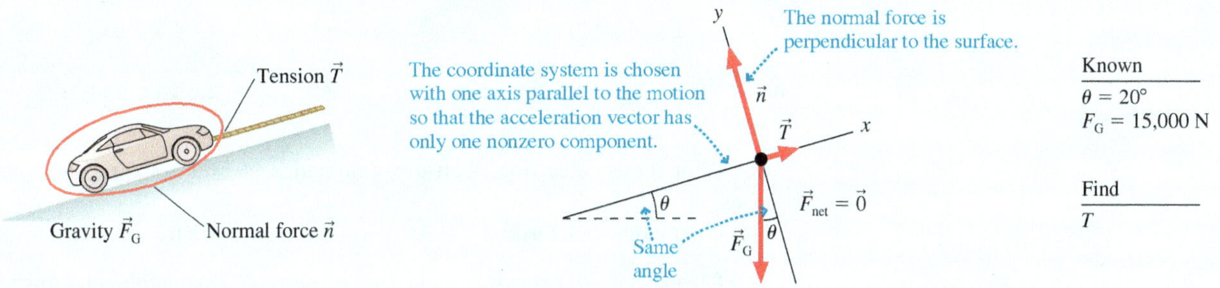

SOLVE The free-body diagram shows forces $\vec{T}$, $\vec{n}$, and $\vec{F}_G$ acting on the car. Newton's first law is

$$(F_{net})_x = \sum F_x = T_x + n_x + (F_G)_x = 0$$

$$(F_{net})_y = \sum F_y = T_y + n_y + (F_G)_y = 0$$

From here on, we'll use $\sum F_x$ and $\sum F_y$, without the label i, as a simple shorthand notation to indicate that we're adding all the x-components and all the y-components of the forces.

We can deduce the components directly from the free-body diagram:

$$T_x = T \qquad T_y = 0$$
$$n_x = 0 \qquad n_y = n$$
$$(F_G)_x = -F_G \sin\theta \qquad (F_G)_y = -F_G \cos\theta$$

NOTE ▶ The gravitational force has both x- and y-components in this coordinate system, both of which are negative due to the direction of the vector $\vec{F}_G$. You'll see this situation often, so be sure you understand where $(F_G)_x$ and $(F_G)_y$ come from. ◀

With these components, the first law becomes

$$T - F_G \sin\theta = 0$$

$$n - F_G \cos\theta = 0$$

The first of these can be rewritten as

$$T = F_G \sin\theta = (15{,}000\ \text{N})\sin 20° = 5100\ \text{N}$$

Because $T < 6000$ N, we conclude that the rope will *not* break. It turned out that we did not need the y-component equation in this problem.

ASSESS Because there's no friction, it would not take *any* tension force to keep the car rolling along a horizontal surface ($\theta = 0°$). At the other extreme, $\theta = 90°$, the tension force would need to equal the car's weight ($T = 15{,}000$ N) to lift the car straight up at constant velocity. The tension force for a 20° slope should be somewhere in between, and 5100 N is a little less than half the weight of the car. That our result is reasonable doesn't prove it's right, but we have at least ruled out careless errors that give unreasonable results.

6.2 Using Newton's Second Law

The essence of Newtonian mechanics can be expressed in two steps:

- The forces acting on an object determine its acceleration $\vec{a} = \vec{F}_{net}/m$.
- The object's trajectory can be determined by using $\vec{a}$ in the equations of kinematics.

These two ideas are the basis of a strategy for solving dynamics problems.

PROBLEM-SOLVING STRATEGY 6.2 Dynamics problems

MP

MODEL Make simplifying assumptions.

VISUALIZE Draw a **pictorial representation.**

- Show important points in the motion with a sketch, establish a coordinate system, define symbols, and identify what the problem is trying to find.
- Use a motion diagram to determine the object's acceleration vector $\vec{a}$.
- Identify all forces acting on the object *at this instant* and show them on a free-body diagram.
- It's OK to go back and forth between these steps as you visualize the situation.

SOLVE The mathematical representation is based on Newton's second law:

$$\vec{F}_{\text{net}} = \sum_i \vec{F}_i = m\vec{a}$$

The vector sum of the forces is found directly from the free-body diagram. Depending on the problem, either

- Solve for the acceleration, then use kinematics to find velocities and positions; or
- Use kinematics to determine the acceleration, then solve for unknown forces.

ASSESS Check that your result has the correct units, is reasonable, and answers the question.

Exercise 22

Newton's second law is a vector equation. To apply the step labeled Solve, you must write the second law as two simultaneous equations:

$$\begin{aligned}(F_{\text{net}})_x &= \sum F_x = ma_x \\ (F_{\text{net}})_y &= \sum F_y = ma_y\end{aligned} \tag{6.2}$$

The primary goal of this chapter is to illustrate the use of this strategy.

EXAMPLE 6.3 Speed of a towed car

A 1500 kg car is pulled by a tow truck. The tension in the tow rope is 2500 N, and a 200 N friction force opposes the motion. If the car starts from rest, what is its speed after 5.0 seconds?

MODEL We'll treat the car as an accelerating particle. We'll assume, as part of our *interpretation* of the problem, that the road is horizontal and that the direction of motion is to the right.

VISUALIZE FIGURE 6.3 on the next page shows the pictorial representation. We've established a coordinate system and defined symbols to represent kinematic quantities. We've identified the speed v_1, rather than the velocity v_{1x}, as what we're trying to find.

SOLVE We begin with Newton's second law:

$$(F_{\text{net}})_x = \sum F_x = T_x + f_x + n_x + (F_G)_x = ma_x$$
$$(F_{\text{net}})_y = \sum F_y = T_y + f_y + n_y + (F_G)_y = ma_y$$

All four forces acting on the car have been included in the vector sum. The equations are perfectly general, with + signs everywhere, because the four vectors are *added* to give $\vec{F}_{\text{net}}$. We can now "read" the vector components from the free-body diagram:

$$T_x = +T \qquad T_y = 0 \qquad n_x = 0 \qquad n_y = +n$$
$$f_x = -f \qquad f_y = 0 \qquad (F_G)_x = 0 \qquad (F_G)_y = -F_G$$

The signs depend on which way the vectors point. Substituting these into the second-law equations and dividing by m give

$$a_x = \frac{1}{m}(T - f)$$
$$= \frac{1}{1500\text{ kg}}(2500\text{ N} - 200\text{ N}) = 1.53\text{ m/s}^2$$
$$a_y = \frac{1}{m}(n - F_G)$$

NOTE ▶ Newton's second law has allowed us to determine a_x exactly but has given only an algebraic expression for a_y. However, we know *from the motion diagram* that $a_y = 0$! That is, the motion is purely along the x-axis, so there is *no* acceleration along the y-axis. The requirement $a_y = 0$ allows us to conclude that $n = F_G$. Although we do not need n for this problem, it will be important in many future problems. ◀

FIGURE 6.3 Pictorial representation of a car being towed.

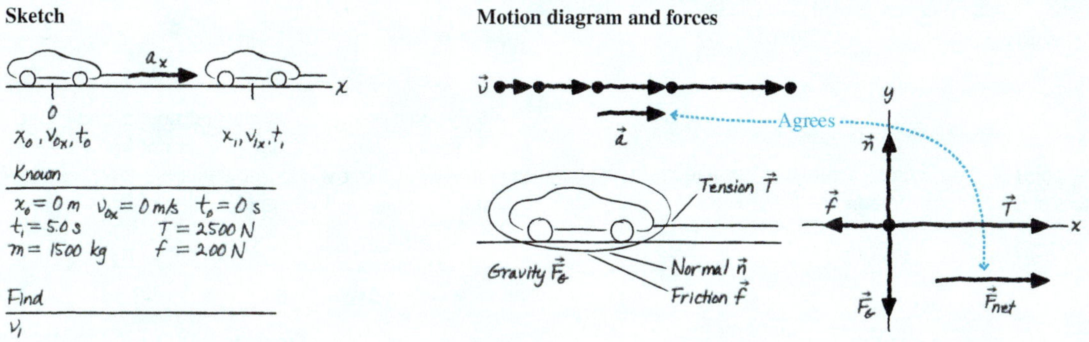

Because a_x is a constant 1.53 m/s^2, we can finish by using constant-acceleration kinematics to find the velocity:

$$v_{1x} = v_{0x} + a_x \Delta t$$
$$= 0 + (1.53 \text{ m/s}^2)(5.0 \text{ s}) = 7.7 \text{ m/s}$$

The problem asked for the *speed* after 5.0 s, which is $v_1 = 7.7$ m/s.

ASSESS 7.7 m/s $\approx$ 15 mph, a quite reasonable speed after 5 s of acceleration.

EXAMPLE 6.4 Altitude of a rocket

A 500 g model rocket with a weight of 4.90 N is launched straight up. The small rocket motor burns for 5.00 s and has a steady thrust of 20.0 N. What maximum altitude does the rocket reach? Ignore the mass loss of the burned fuel.

MODEL We'll treat the rocket as an accelerating particle. Air resistance will be neglected.

VISUALIZE The pictorial representation of **FIGURE 6.4** finds that this is a two-part problem. First, the rocket accelerates straight up. Second, the rocket continues going up as it slows down, a free-fall situation. The maximum altitude is at the end of the second part of the motion.

SOLVE We now know what the problem is asking, have established relevant symbols and coordinates, and know what the forces are. We begin the mathematical representation by writing Newton's second law, in component form, as the rocket accelerates upward. The free-body diagram shows two forces, so

$$(F_{\text{net}})_x = \sum F_x = (F_{\text{thrust}})_x + (F_G)_x = ma_{0x}$$
$$(F_{\text{net}})_y = \sum F_y = (F_{\text{thrust}})_y + (F_G)_y = ma_{0y}$$

The fact that vector $\vec{F}_G$ points downward—and which might have tempted you to use a minus sign in the y-equation—will be taken into account when we *evaluate* the components. None of

FIGURE 6.4 Pictorial representation of a rocket launch.

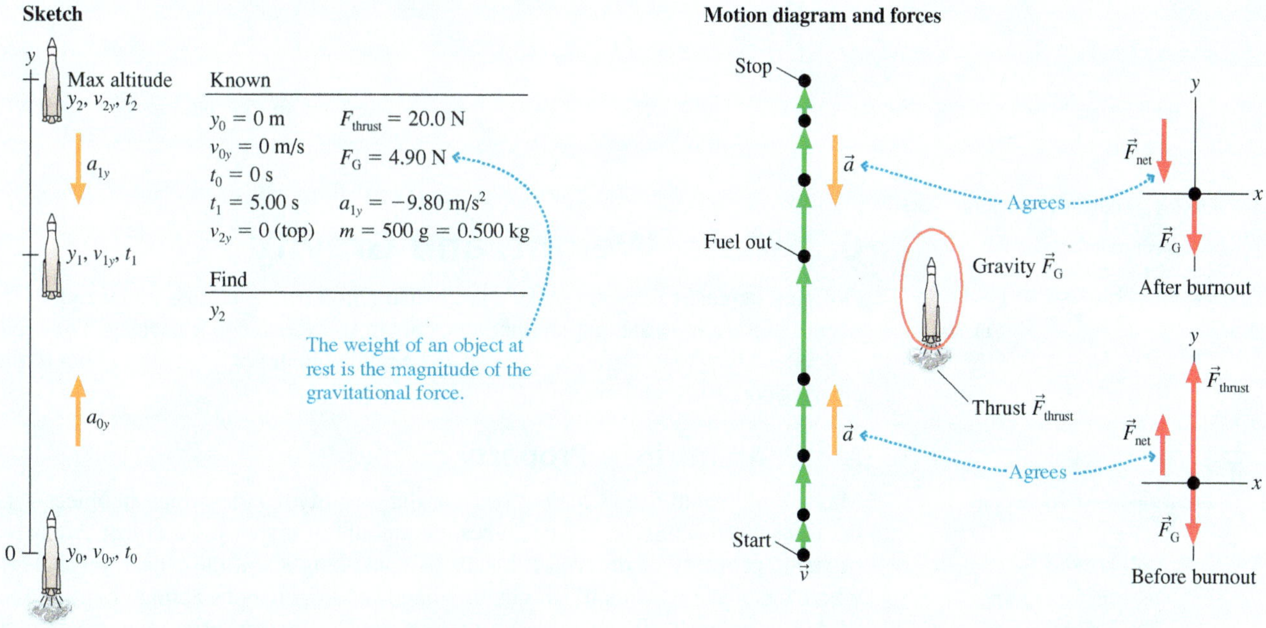

Continued

the vectors in this problem has an x-component, so only the y-component of the second law is needed. We can use the free-body diagram to see that

$$(F_{\text{thrust}})_y = +F_{\text{thrust}}$$
$$(F_G)_y = -F_G$$

This is the point at which the directional information about the force vectors enters. The y-component of the second law is then

$$a_{0y} = \frac{1}{m}(F_{\text{thrust}} - F_G)$$
$$= \frac{20.0\text{ N} - 4.90\text{ N}}{0.500\text{ kg}} = 30.2\text{ m/s}^2$$

Notice that we converted the mass to SI units of kilograms before doing any calculations and that, because of the definition of the newton, the division of newtons by kilograms automatically gives the correct SI units of acceleration.

The acceleration of the rocket is constant until it runs out of fuel, so we can use constant-acceleration kinematics to find the altitude and velocity at burnout ($\Delta t = t_1 = 5.00$ s):

$$y_1 = y_0 + v_{0y}\,\Delta t + \tfrac{1}{2}a_{0y}(\Delta t)^2$$
$$= \tfrac{1}{2}a_{0y}(\Delta t)^2 = 377\text{ m}$$
$$v_{1y} = v_{0y} + a_{0y}\,\Delta t = a_{0y}\,\Delta t = 151\text{ m/s}$$

The only force on the rocket after burnout is gravity, so the second part of the motion is free fall. We do not know how long it takes to reach the top, but we do know that the final velocity is $v_{2y} = 0$. Constant-acceleration kinematics with $a_{1y} = -g$ gives

$$v_{2y}^2 = 0 = v_{1y}^2 - 2g\,\Delta y = v_{1y}^2 - 2g(y_2 - y_1)$$

which we can solve to find

$$y_2 = y_1 + \frac{v_{1y}^2}{2g} = 377\text{ m} + \frac{(151\text{ m/s})^2}{2(9.80\text{ m/s}^2)}$$
$$= 1540\text{ m} = 1.54\text{ km}$$

ASSESS The maximum altitude reached by this rocket is 1.54 km, or just slightly under one mile. While this does not seem unreasonable for a high-acceleration rocket, the neglect of air resistance was probably not a terribly realistic assumption.

The solutions to these first few examples have been quite detailed. Our purpose has been to show how the problem-solving strategy is put into practice. Future examples will be briefer, but the basic *procedure* will remain the same.

STOP TO THINK 6.1 A Martian lander is approaching the surface. It is slowing its descent by firing its rocket motor. Which is the correct free-body diagram?

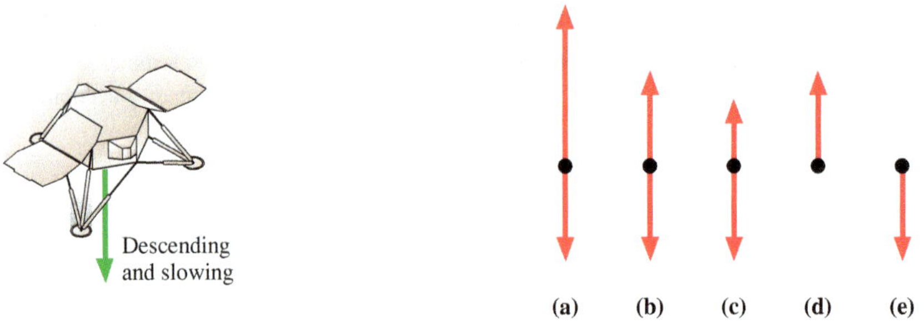

6.3 Mass, Weight, and Gravity

Ordinary language does not make a large distinction between mass and weight. However, these are separate and distinct concepts in science and engineering. We need to understand how they differ, and how they're related to gravity, if we're going to think clearly about force and motion.

Mass: An Intrinsic Property

Mass, you'll recall from Chapter 5, is a scalar quantity that describes an object's inertia. Loosely speaking, it also describes the amount of matter in an object. **Mass is an intrinsic property of an object.** It tells us something about the object, regardless of where the object is, what it's doing, or whatever forces may be acting on it.

A *pan balance,* shown in FIGURE 6.5, is a device for measuring mass. Although a pan balance requires gravity to function, it does not depend on the strength of gravity. Consequently, the pan balance would give the same result on another planet.

FIGURE 6.5 A pan balance measures mass.

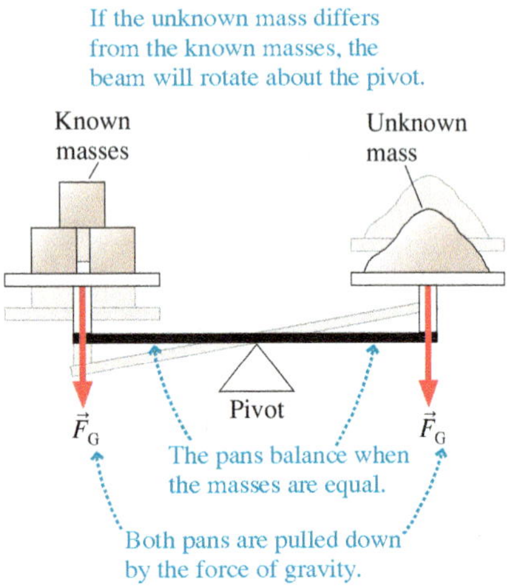

Gravity: A Force

The idea of gravity has a long and interesting history intertwined with our evolving ideas about the solar system. It was Newton who—along with discovering his three laws of motion—first recognized that **gravity is an attractive, long-range force between *any* two objects.**

FIGURE 6.6 shows two objects with masses m_1 and m_2 separated by distance r. Each object pulls on the other with a force given by *Newton's law of gravity:*

$$F_{1\text{ on }2} = F_{2\text{ on }1} = \frac{Gm_1 m_2}{r^2} \quad \text{(Newton's law of gravity)} \qquad (6.3)$$

where $G = 6.67 \times 10^{-11}\ \text{N m}^2/\text{kg}^2$, called the *gravitational constant,* is one of the basic constants of nature. Notice that the force gets weaker as the distance between the objects increases.

The gravitational force between two human-sized objects is minuscule, completely insignificant in comparison with other forces. That's why you're not aware of being tugged toward everything around you. Only when one or both objects is planet-sized or larger does gravity become an important force. Indeed, Chapter 13 will explore in detail the application of Newton's law of gravity to the orbits of satellites and planets.

For objects moving near the surface of the earth (or other planet), things like balls and cars and planes that we'll be studying in the next few chapters, we can make the **flat-earth approximation** shown in FIGURE 6.7. That is, if the height above the surface is very small in comparison with the size of the planet, then the curvature of the surface is not noticeable and there's virtually no difference between r and the planet's radius R. Consequently, a very good approximation for the gravitational force of the planet on mass m is simply

$$\vec{F}_G = \vec{F}_{\text{planet on }m} = \left(\frac{GMm}{R^2}, \text{straight down}\right) = (mg, \text{straight down}) \qquad (6.4)$$

The magnitude or size of the gravitational force is $F_G = mg$, where the quantity g—a property of the planet—is defined to be

$$g = \frac{GM}{R^2} \qquad (6.5)$$

In addition, the direction of the gravitational force defines what we *mean* by "straight down."

But why did we choose to call it g, a symbol we've already used for free-fall acceleration? To see the connection, recall that free fall is motion under the influence of gravity only. FIGURE 6.8 shows the free-body diagram of an object in free fall near the surface of a planet. With $\vec{F}_{\text{net}} = \vec{F}_G$, Newton's second law predicts the acceleration to be

$$\vec{a}_{\text{free fall}} = \frac{\vec{F}_{\text{net}}}{m} = \frac{\vec{F}_G}{m} = (g, \text{straight down}) \qquad (6.6)$$

Because g is a property of the planet, independent of the object, **all objects on the same planet, regardless of mass, have the same free-fall acceleration.** We introduced this idea in Chapter 2 as an experimental discovery of Galileo, but now we see that the mass independence of $\vec{a}_{\text{free fall}}$ is a prediction of Newton's law of gravity.

But does Newton's law predict the correct value, which we know from experiment to be $g = |a_{\text{free fall}}| = 9.80\ \text{m/s}^2$? We can use the average radius ($R_{\text{earth}} = 6.37 \times 10^6$ m) and mass ($M_{\text{earth}} = 5.98 \times 10^{24}$ kg) of the earth to calculate

$$g_{\text{earth}} = \frac{GM_{\text{earth}}}{(R_{\text{earth}})^2} = \frac{(6.67 \times 10^{-11}\ \text{N m}^2/\text{kg}^2)(5.98 \times 10^{24}\ \text{kg})}{(6.37 \times 10^6\ \text{m})^2} = 9.83\ \text{N/kg}$$

You should convince yourself that N/kg is equivalent to m/s², so $g_{\text{earth}} = 9.83\ \text{m/s}^2$. (Data for other astronomical objects, which you may need for homework, are provided inside the back cover of the book.)

FIGURE 6.6 Newton's law of gravity.

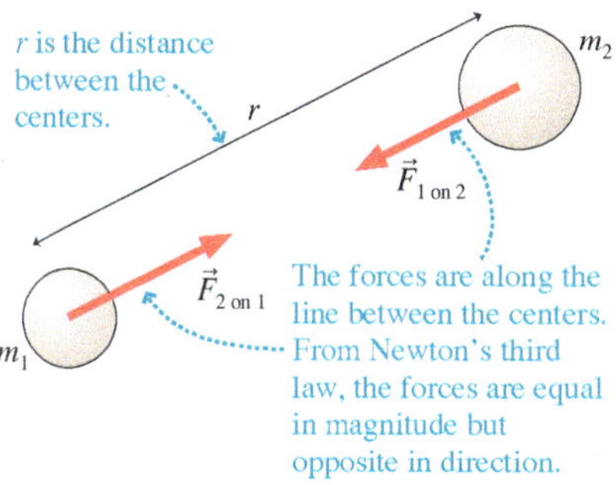

FIGURE 6.7 Gravity near the surface of a planet.

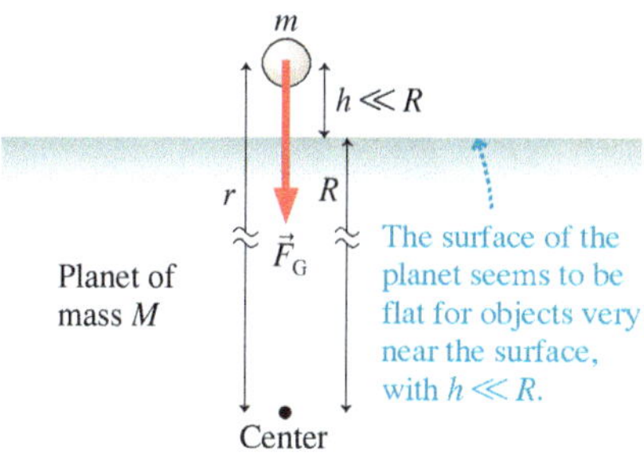

FIGURE 6.8 The free-body diagram of an object in free fall.

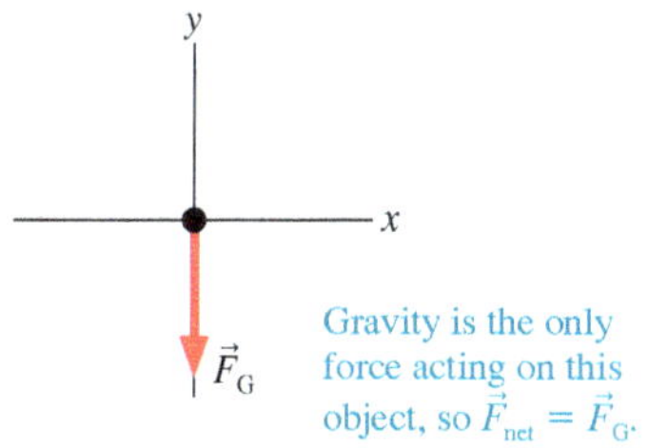

Newton's prediction is very close, but it's not quite right. The free-fall acceleration *would* be 9.83 m/s^2 on a stationary earth, but, in reality, the earth is rotating on its axis. The "missing" 0.03 m/s^2 is due to the earth's rotation, a claim we'll justify when we study circular motion in Chapter 8. Because we're on the outside of a rotating sphere, rather like being on the outside edge of a merry-go-round, the effect of rotation is to "weaken" gravity.

Our goal is to analyze motion from within our own reference frame, a reference frame attached to the earth. Strictly speaking, Newton's laws of motion are not valid in our reference frame because it is rotating and thus is not an inertial reference frame. Fortunately, we can use Newton's laws to analyze motion near the earth's surface, and we can use $F_G = mg$ for the gravitational force *if* we use $g = |a_{\text{free fall}}| = 9.80 \text{ m/s}^2$ rather than $g = g_{\text{earth}}$. (This assertion is proved in more advanced classes.) In our rotating reference frame, $\vec{F}_G$ is the *effective gravitational force*, the true gravitational force given by Newton's law of gravity plus a small correction due to our rotation. This is the force to show on free-body diagrams and use in calculations.

Weight: A Measurement

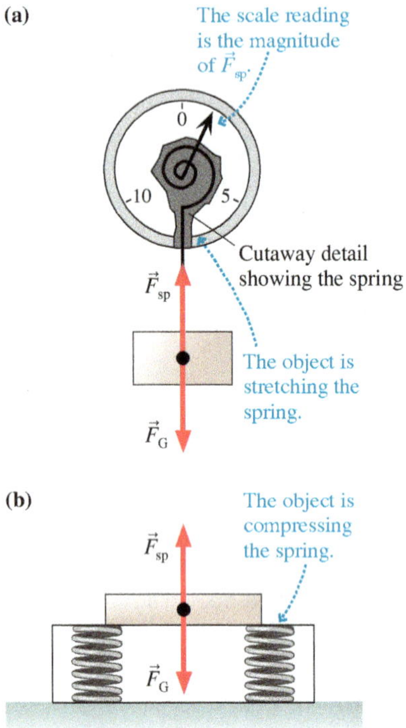

FIGURE 6.9 A spring scale measures weight.

When you weigh yourself, you stand on a *spring scale* and compress a spring. You weigh apples in the grocery store by placing them in a spring scale and stretching a spring. The reading of a spring scale, such as the two shown in FIGURE 6.9, is F_{sp}, the magnitude of the force the spring is exerting.

With that in mind, let's define the **weight** of an object as the reading F_{sp} of a calibrated spring scale on which the object is stationary. That is, **weight is a measurement, the result of "weighing" an object.** Because F_{sp} is a force, weight is measured in newtons.

Suppose the scales in Figure 6.9 are at rest relative to the earth. Then the object being weighed is in static equilibrium, with $\vec{F}_{\text{net}} = \vec{0}$. The stretched spring *pulls* up, the compressed spring *pushes* up, but in both cases $\vec{F}_{\text{net}} = \vec{0}$ only if the upward spring force exactly balances the downward gravitational force of magnitude mg:

$$F_{sp} = F_G = mg \tag{6.7}$$

Because we defined weight as the reading F_{sp} of a spring scale, the weight of a stationary object is

$$w = mg \qquad \text{(weight of a stationary object)} \tag{6.8}$$

The scale does not "know" the weight of the object. All it can do is to measure how much its spring is stretched or compressed. On earth, a student with a mass of 70 kg has weight $w = (70 \text{ kg})(9.80 \text{ m/s}^2) = 686 \text{ N}$ *because* he compresses a spring until the spring pushes upward with 686 N. On a different planet, with a different value for g, the expansion or compression of the spring would be different and the student's weight would be different.

NOTE ▶ **Mass and weight are not the same thing.** Mass, in kg, is an intrinsic property of an object; its value is unique and always the same. Weight, in N, depends on the object's mass, but it also depends on the situation—the strength of gravity and, as we will see, whether or not the object is accelerating. Weight is *not* a property of the object, and thus weight does not have a unique value. ◀

Surprisingly, you cannot directly feel or sense gravity. Your *sensation*—how heavy you feel—is due to contact forces pressing against you, forces that touch you and activate nerve endings in your skin. As you read this, your sensation of weight is due to the normal force exerted on you by the chair in which you are sitting. When you stand, you feel the contact force of the floor pushing against your feet.

But recall the sensations you feel while accelerating. You feel "heavy" when an elevator suddenly accelerates upward, but this sensation vanishes as soon as the elevator reaches a steady speed. Your stomach seems to rise a little and you feel lighter than normal as the upward-moving elevator brakes to a halt or a roller coaster goes over the top. Has your weight actually changed?

To answer this question, FIGURE 6.10 shows a man weighing himself on a spring scale in an accelerating elevator. The only forces acting on the man are the upward spring force of the scale and the downward gravitational force. This seems to be the same situation as Figure 6.9b, but there's one big difference: The man is accelerating, hence there must be a net force on the man in the direction of $\vec{a}$.

For the net force $\vec{F}_{net}$ to point upward, the magnitude of the spring force must be *greater* than the magnitude of the gravitational force. That is, $F_{sp} > mg$. Looking at the free-body diagram in Figure 6.10, we see that the y-component of Newton's second law is

$$(F_{net})_y = (F_{sp})_y + (F_G)_y = F_{sp} - mg = ma_y \tag{6.9}$$

where m is the man's mass.

We defined weight as the reading F_{sp} of a calibrated spring scale *on which the object is stationary*. That is the case here as the scale and man accelerate upward together. Thus the man's weight as he accelerates vertically is

$$w = \text{scale reading } F_{sp} = mg + ma_y = mg\left(1 + \frac{a_y}{g}\right) \tag{6.10}$$

If an object is either at rest or moving with constant velocity, then $a_y = 0$ and $w = mg$. That is, the weight of an object at rest is the magnitude of the (effective) gravitational force acting on it. But its weight differs if it has a vertical acceleration.

You *do* weigh more as an elevator accelerates upward ($a_y > 0$) because the reading of a scale—a weighing—increases. Similarly, your weight is less when the acceleration vector $\vec{a}$ points downward ($a_y < 0$) because the scale reading goes down. Weight, as we've defined it, corresponds to your sensation of heaviness or lightness.*

We found Equation 6.10 by considering a person in an accelerating elevator, but it applies to any object with a vertical acceleration. Further, an object doesn't really have to be on a scale to have a weight; an object's weight is the magnitude of the contact force supporting it. It makes no difference whether this is the spring force of the scale or simply the normal force of the floor.

NOTE ▶ Informally, we sometimes say "This object weighs such and such" or "The weight of this object is. . . ." We'll interpret these expressions as meaning mg, the weight of an object of mass m at rest ($a_y = 0$) on the surface of the earth or some other astronomical body. ◀

FIGURE 6.10 A man weighing himself in an accelerating elevator.

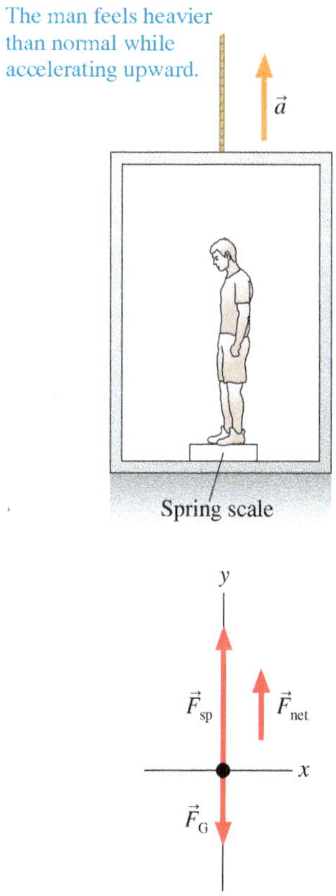

Weightlessness

Suppose the elevator cable breaks and the elevator, along with the man and his scale, plunges straight down in free fall! What will the scale read? When the free-fall acceleration $a_y = -g$ is used in Equation 6.10, we find $w = 0$. In other words, *the man has no weight!*

Suppose, as the elevator falls, the man inside releases a ball from his hand. In the absence of air resistance, as Galileo discovered, both the man and the ball would fall at the same rate. From the man's perspective, the ball would appear to "float" beside him. Similarly, the scale would float beneath him and not press against his feet. He is what we call *weightless*. Gravity is still pulling down on him—that's why he's falling—but he has no *sensation* of weight as everything floats around him in free fall.

But isn't this exactly what happens to astronauts orbiting the earth? If an astronaut tries to stand on a scale, it does not exert any force against her feet and reads zero. She is said to be weightless. But if the criterion to be weightless is to be in free fall, and if astronauts orbiting the earth are weightless, does this mean that they are in free fall? This is a very interesting question to which we shall return in Chapter 8.

Astronauts are weightless as they orbit the earth.

*Surprisingly, there is no universally agreed-upon definition of *weight*. Some textbooks define weight as the gravitational force on an object, $\vec{w} = (mg, \text{down})$. In that case, the scale reading of an accelerating object, and your sensation of weight, is often called *apparent weight*. This textbook prefers the definition of *weight* as being what a scale reads, the result of a weighing measurement.

FIGURE 6.11 Static friction keeps an object from slipping.

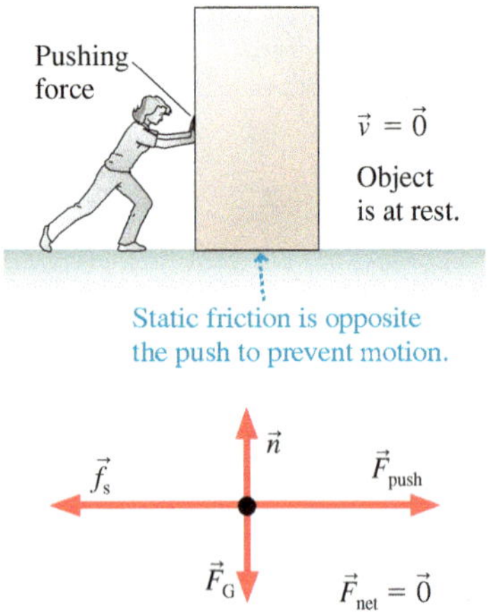

FIGURE 6.12 Static friction acts in *response* to an applied force.

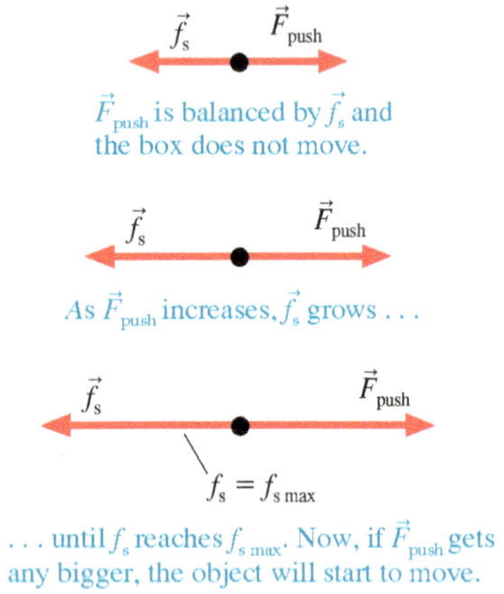

FIGURE 6.13 The kinetic friction force is opposite the direction of motion.

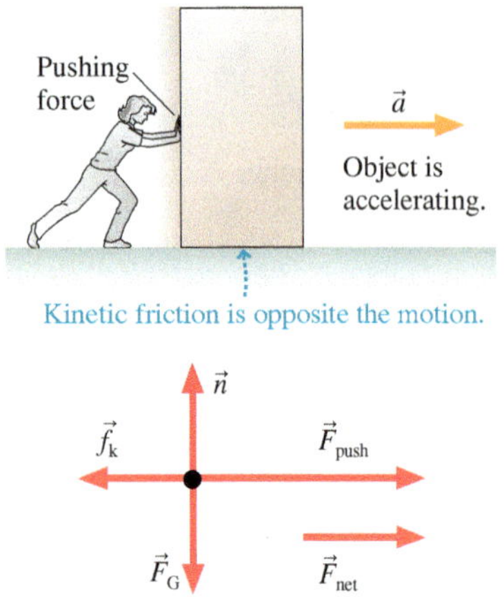

STOP TO THINK 6.2 An elevator that has descended from the 50th floor is coming to a halt at the 1st floor. As it does, your weight is

a. More than mg. b. Less than mg. c. Equal to mg. d. Zero.

6.4 Friction

Friction is absolutely essential for many things we do. Without friction you could not walk, drive, or even sit down (you would slide right off the chair!). Although friction is a complicated force, many aspects of friction can be described with a simple model.

Static Friction

Chapter 5 defined *static friction* $\vec{f}_s$ as the force on an object that keeps it from slipping. FIGURE 6.11 shows a person pushing on a box that, due to static friction, isn't moving. The box is in static equilibrium, so the static friction force must exactly balance the pushing force:

$$f_s = F_{push} \tag{6.11}$$

To determine the direction of $\vec{f}_s$, decide which way the object would move if there were no friction. The static friction force $\vec{f}_s$ points in the *opposite* direction to prevent the motion.

Unlike the gravitational force, which has the precise and unambiguous magnitude $F_G = mg$, the size of the static friction force depends on how hard you push. The harder the person in Figure 6.11 pushes, the harder the floor pushes back. Reduce the pushing force, and the static friction force will automatically be reduced to match. Static friction acts in *response* to an applied force. FIGURE 6.12 illustrates this idea.

But there's clearly a limit to how big f_s can get. If you push hard enough, the object slips and starts to move. In other words, the static friction force has a *maximum* possible size $f_{s\,max}$.

- An object remains at rest as long as $f_s < f_{s\,max}$.
- The object slips when $f_s = f_{s\,max}$.
- A static friction force $f_s > f_{s\,max}$ is not physically possible.

Experiments with friction show that $f_{s\,max}$ is proportional to the magnitude of the normal force. That is,

$$f_{s\,max} = \mu_s n \tag{6.12}$$

where the proportionality constant μ_s is called the **coefficient of static friction**. The coefficient is a dimensionless number that depends on the materials of which the object and the surface are made. Table 6.1 on the next page shows some typical coefficients of friction. It is to be emphasized that these are only approximate. The exact value of the coefficient depends on the roughness, cleanliness, and dryness of the surfaces.

Kinetic Friction

Once the box starts to slide, as in FIGURE 6.13, the static friction force is replaced by a kinetic friction force $\vec{f}_k$. Experiments show that kinetic friction, unlike static friction, has a nearly *constant* magnitude. Furthermore, the size of the kinetic friction force is *less* than the maximum static friction, $f_k < f_{s\,max}$, which explains why it is easier to keep the box moving than it was to start it moving. The direction of $\vec{f}_k$ is always opposite to the direction in which an object slides across the surface.

The kinetic friction force is also proportional to the magnitude of the normal force:

$$f_k = \mu_k n \tag{6.13}$$

where μ_k is called the **coefficient of kinetic friction.** Table 6.1 includes typical values of μ_k. You can see that $\mu_k < \mu_s$, causing the kinetic friction to be less than the maximum static friction.

Rolling Friction

If you slam on the brakes hard enough, your car tires slide against the road surface and leave skid marks. This is kinetic friction. A wheel *rolling* on a surface also experiences friction, but not kinetic friction. The portion of the wheel that contacts the surface is stationary with respect to the surface, not sliding. To see this, roll a wheel slowly and watch how it touches the ground.

No wheel is perfectly round and thus, as FIGURE 6.14 shows, a wheel has an area of contact with the ground. Molecular bonds are quickly established where the wheel presses against the surface. These bonds have to be broken as the wheel rolls forward, and the effort needed to break them causes **rolling friction**. (Think how it is to walk with a wad of chewing gum stuck to the sole of your shoe!) The force of rolling friction can be calculated in terms of a **coefficient of rolling friction** μ_r:

$$f_r = \mu_r n \qquad (6.14)$$

Rolling friction acts very much like kinetic friction, but values of μ_r (see Table 6.1) are much lower than values of μ_k. This is why it is easier to roll an object on wheels than to slide it.

FIGURE 6.14 Rolling friction is due to the contact area between a wheel and the surface.

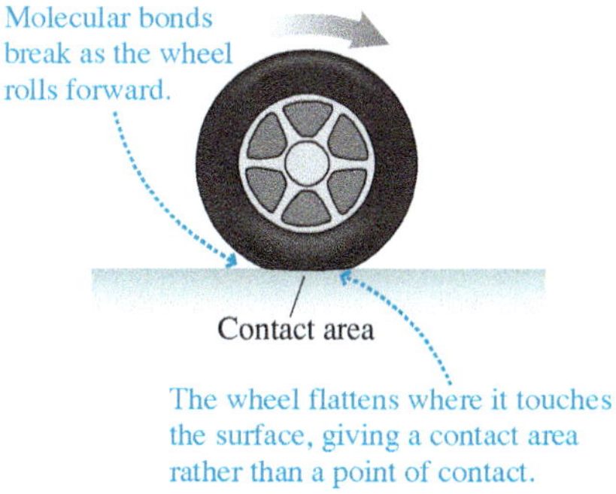

A Model of Friction

These ideas can be summarized in a *model* of friction:

$$\begin{aligned}\text{Static: } \vec{f}_s &\le (\mu_s n, \text{ direction as necessary to prevent motion})\\ \text{Kinetic: } \vec{f}_k &= (\mu_k n, \text{ direction opposite the motion})\\ \text{Rolling: } \vec{f}_r &= (\mu_r n, \text{ direction opposite the motion})\end{aligned} \qquad (6.15)$$

Here "motion" means "motion relative to the surface." The maximum value of static friction $f_{s\,max} = \mu_s n$ occurs at the point where the object slips and begins to move.

NOTE ▶ Equations 6.15 are a "model" of friction, not a "law" of friction. These equations—a simplification of reality—provide a reasonably accurate, but not perfect, description of how friction forces act. They are not a "law of nature" on a level with Newton's laws. ◀

TABLE 6.1 Coefficients of friction

Materials	Static μ_s	Kinetic μ_k	Rolling μ_r
Rubber on concrete	1.00	0.80	0.02
Steel on steel (dry)	0.80	0.60	0.002
Steel on steel (lubricated)	0.10	0.05	
Wood on wood	0.50	0.20	
Wood on snow	0.12	0.06	
Ice on ice	0.10	0.03	

FIGURE 6.15 summarizes these ideas graphically by showing how the friction force changes as the magnitude of an applied force $\vec{F}_{push}$ increases.

FIGURE 6.15 The friction force response to an increasing applied force.

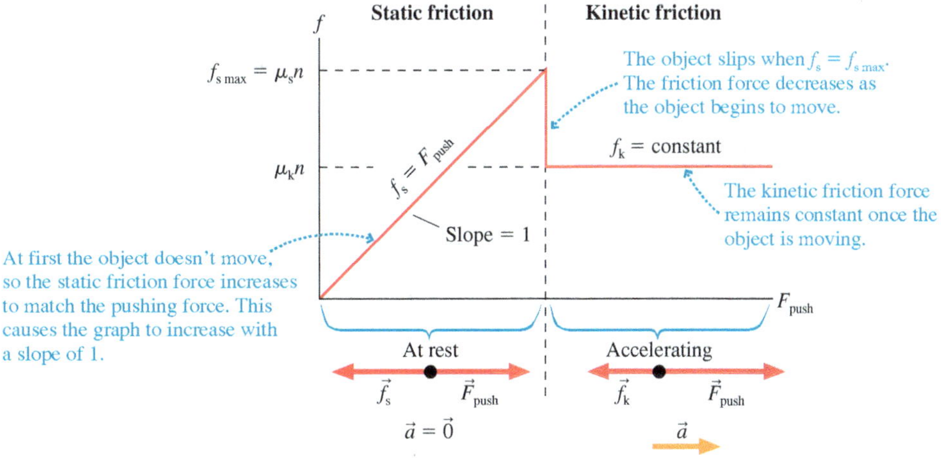

STOP TO THINK 6.3 Rank in order, from largest to smallest, the sizes of the friction forces $\vec{f}_a$ to $\vec{f}_e$ in these 5 different situations. The box and the floor are made of the same materials in all situations.

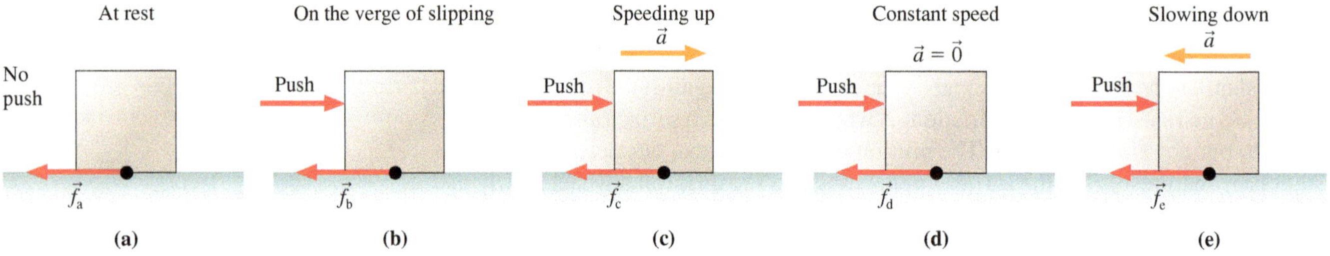

EXAMPLE 6.5 How far does a box slide?

Carol pushes a 50 kg wood box across a wood floor at a steady speed of 2.0 m/s. How much force does Carol exert on the box? If she stops pushing, how far will the box slide before coming to rest?

MODEL We model the box as a particle and we describe the friction forces with the model of static and kinetic friction. This is a two-part problem: first while Carol is pushing the box, then as it slides after she releases it.

VISUALIZE This is a fairly complex situation, one that calls for careful visualization. FIGURE 6.16 shows the pictorial representation both while Carol pushes, when $\vec{a} = \vec{0}$, and after she stops. We've placed $x = 0$ at the point where she stops pushing because this is the point where the kinematics calculation for "How far?" will begin. Notice that each part of the motion needs its own free-body diagram. The box is moving until the very instant that the problem ends, so only kinetic friction is relevant.

SOLVE We'll start by finding how hard Carol has to push to keep the box moving at a steady speed. The box is in dynamic equilibrium ($\vec{a} = \vec{0}$), and Newton's first law is

$$\sum F_x = F_{push} - f_k = 0$$

$$\sum F_y = n - F_G = n - mg = 0$$

where we've used $F_G = mg$ for the gravitational force. The negative sign occurs in the first equation because $\vec{f}_k$ points to the left and thus the *component* is negative: $(f_k)_x = -f_k$. Similarly, $(F_G)_y = -F_G$ because the gravitational force vector—with magnitude mg—points down. In addition to Newton's laws, we also have our model of kinetic friction:

$$f_k = \mu_k n$$

Altogether we have three simultaneous equations in the three unknowns F_{push}, f_k, and n. Fortunately, these equations are easy to solve. The y-component of Newton's law tells us that $n = mg$. We can then find the friction force to be

$$f_k = \mu_k mg$$

We substitute this into the x-component of the first law, giving

$$F_{push} = f_k = \mu_k mg$$
$$= (0.20)(50\text{ kg})(9.80\text{ m/s}^2) = 98\text{ N}$$

This is how hard Carol pushes to keep the box moving at a steady speed.

The box is not in equilibrium after Carol stops pushing it. Our strategy for the second half of the problem is to use Newton's second law to find the acceleration, then use constant-acceleration kinematics to find how far the box moves before stopping. We

FIGURE 6.16 Pictorial representation of a box sliding across a floor.

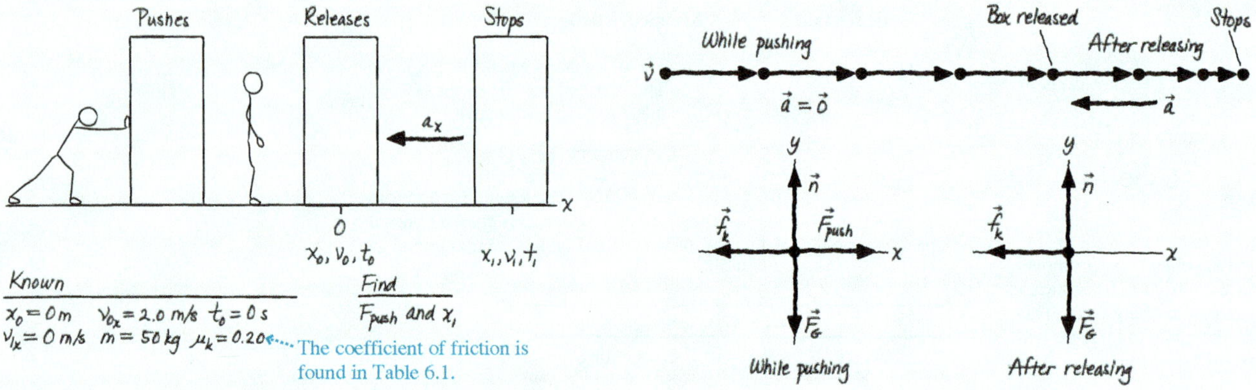

know from the motion diagram that $a_y = 0$. Newton's second law, applied to the second free-body diagram of Figure 6.16, is

$$\sum F_x = -f_k = ma_x$$

$$\sum F_y = n - mg = ma_y = 0$$

We also have our model of friction,

$$f_k = \mu_k n$$

We see from the y-component equation that $n = mg$, and thus $f_k = \mu_k mg$. Using this in the x-component equation gives

$$ma_x = -f_k = -\mu_k mg$$

This is easily solved to find the box's acceleration:

$$a_x = -\mu_k g = -(0.20)(9.80 \text{ m/s}^2) = -1.96 \text{ m/s}^2$$

The acceleration component a_x is negative because the acceleration vector $\vec{a}$ points to the left, as we see from the motion diagram.

Now we are left with a problem of constant-acceleration kinematics. We are interested in a distance, rather than a time interval, so the easiest way to proceed is

$$v_{1x}{}^2 = 0 = v_{0x}{}^2 + 2a_x\,\Delta x = v_{0x}{}^2 + 2a_x x_1$$

from which the distance that the box slides is

$$x_1 = \frac{-v_{0x}{}^2}{2a_x} = \frac{-(2.0 \text{ m/s})^2}{2(-1.96 \text{ m/s}^2)} = 1.0 \text{ m}$$

ASSESS Carol was pushing at 2 m/s ≈ 4 mph, which is fairly fast. The box slides 1.0 m, which is slightly over 3 feet. That sounds reasonable.

NOTE ▶ We needed both the horizontal and the vertical components of the second law even though the motion was entirely horizontal. This need is typical when friction is involved because we must find the normal force before we can evaluate the friction force. ◀

EXAMPLE 6.6 Dumping a file cabinet

A 50 kg steel file cabinet is in the back of a dump truck. The truck's bed, also made of steel, is slowly tilted. What is the size of the static friction force on the cabinet when the bed is tilted 20°? At what angle will the file cabinet begin to slide?

MODEL We'll model the file cabinet as a particle. We'll also use the model of static friction. The file cabinet will slip when the static friction force reaches its maximum value $f_{s\,max}$.

VISUALIZE FIGURE 6.17 shows the pictorial representation when the truck bed is tilted at angle θ. We can make the analysis easier if we tilt the coordinate system to match the bed of the truck. To prevent the file cabinet from slipping, the static friction force must point *up* the slope.

SOLVE The file cabinet is in static equilibrium. Newton's first law is

$$(F_{net})_x = \sum F_x = n_x + (F_G)_x + (f_s)_x = 0$$

$$(F_{net})_y = \sum F_y = n_y + (F_G)_y + (f_s)_y = 0$$

From the free-body diagram we see that f_s has only a *negative* x-component and that n has only a positive y-component. The gravitational force vector can be written $\vec{F}_G = +F_G \sin\theta\,\hat{\imath} - F_G\cos\theta\,\hat{\jmath}$, so $\vec{F}_G$ has both x- and y-components in this coordinate system. Thus the first law becomes

$$\sum F_x = F_G \sin\theta - f_s = mg\sin\theta - f_s = 0$$

$$\sum F_y = n - F_G\cos\theta = n - mg\cos\theta = 0$$

where we've used $F_G = mg$.

You might be tempted to solve the y-component equation for n, then to use Equation 6.12 to calculate the static friction force as $\mu_s n$. **But Equation 6.12 does *not* say $f_s = \mu_s n$.** Equation 6.12 gives only the maximum possible static friction force $f_{s\,max}$, the point at which the object slips. In nearly all situations, the actual static friction force is less than $f_{s\,max}$. In this problem, we can use the x-component equation—which tells us that static friction has to exactly balance the component of the gravitational force along the incline—to find the size of the static friction force when $\theta = 20°$:

$$f_s = mg\sin\theta = (50 \text{ kg})(9.80 \text{ m/s}^2)\sin 20°$$

$$= 170 \text{ N}$$

FIGURE 6.17 The pictorial representation of a file cabinet in a tilted dump truck.

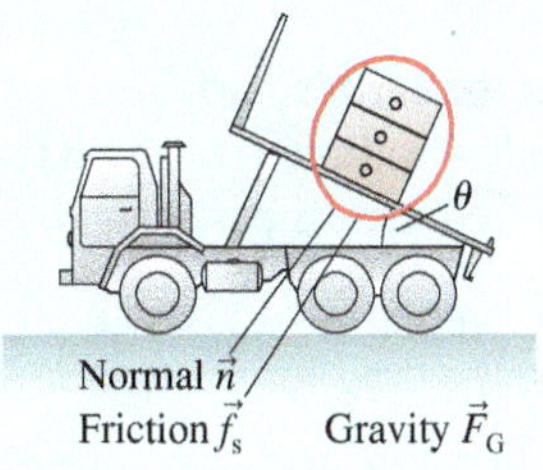

Known

$\mu_s = 0.80$ $m = 50$ kg
$\mu_k = 0.60$

Find

f_s where $\theta = 20°$
θ where cabinet slips

The coefficients of friction are found in Table 6.1.

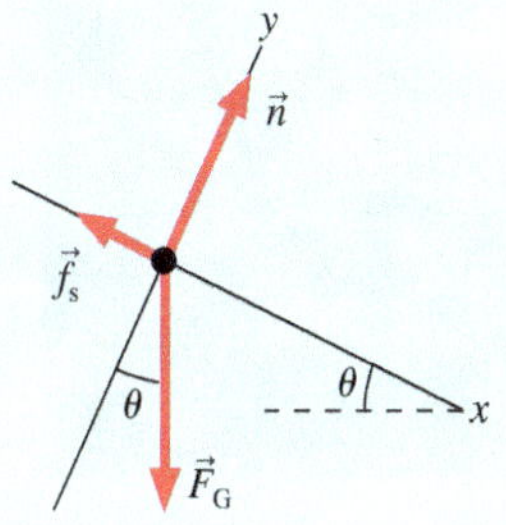

Continued

Slipping occurs when the static friction reaches its maximum value

$$f_s = f_{s\,max} = \mu_s n$$

From the y-component of Newton's law we see that $n = mg\cos\theta$. Consequently,

$$f_{s\,max} = \mu_s mg\cos\theta$$

NOTE ▶ A common error is to use simply $n = mg$. Be sure to evaluate the normal force within the context of each specific problem. In this example, $n = mg\cos\theta$. ◀

Substituting this into the x-component of the first law gives

$$mg\sin\theta - \mu_s mg\cos\theta = 0$$

The mg in both terms cancels, and we find

$$\frac{\sin\theta}{\cos\theta} = \tan\theta = \mu_s$$

$$\theta = \tan^{-1}\mu_s = \tan^{-1}(0.80) = 39°$$

ASSESS Steel doesn't slide all that well on unlubricated steel, so a fairly large angle is not surprising. The answer seems reasonable.

FIGURE 6.18 An atomic-level view of friction.

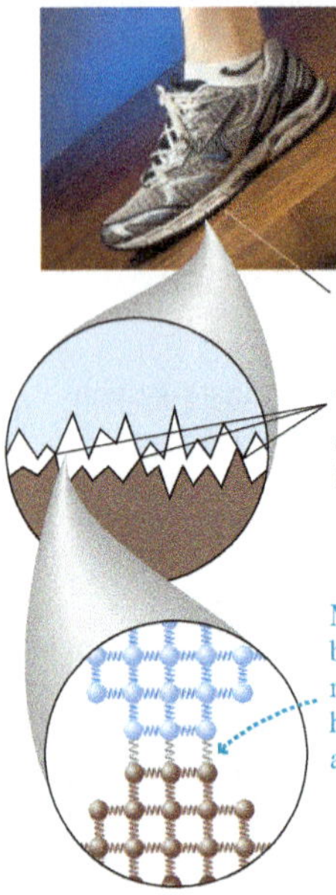

Causes of Friction

It is worth a brief pause to look at the *causes* of friction. All surfaces, even those quite smooth to the touch, are very rough on a microscopic scale. When two objects are placed in contact, they do not make a smooth fit. Instead, as FIGURE 6.18 shows, the high points on one surface become jammed against the high points on the other surface, while the low points are not in contact at all. The amount of contact depends on how hard the surfaces are pushed together, which is why friction forces are proportional to n.

At the points of actual contact, the atoms in the two materials are pressed closely together and molecular bonds are established between them. These bonds are the "cause" of the static friction force. For an object to slip, you must push it hard enough to break these molecular bonds between the surfaces. Once they are broken, and the two surfaces are sliding against each other, there are still attractive forces between the atoms on the opposing surfaces as the high points of the materials push past each other. However, the atoms move past each other so quickly that they do not have time to establish the tight bonds of static friction. That is why the kinetic friction force is smaller. Friction can be minimized with lubrication, a very thin film of liquid between the surfaces that allows them to "float" past each other with many fewer points in actual contact.

6.5 Drag

The air exerts a drag force on objects as they move through the air. You experience drag forces every day as you jog, bicycle, ski, or drive your car. The drag force $\vec{D}$

- Is opposite in direction to $\vec{v}$.
- Increases in magnitude as the object's speed increases.

FIGURE 6.19 The drag force on a high-speed motorcyclist is significant.

FIGURE 6.19 illustrates the drag force.

Drag is a more complex force than ordinary friction because drag depends on the object's speed. Drag also depends on the object's shape and on the density of the medium through which it moves. Fortunately, we can use a fairly simple *model* of drag if the following three conditions are met:

- The object is moving through the air near the earth's surface.
- The object's size (diameter) is between a few millimeters and a few meters.
- The object's speed is less than a few hundred meters per second.

These conditions are usually satisfied for balls, people, cars, and many other objects in our everyday world. Under these conditions, the drag force on an object moving with speed v can be written

$$\vec{D} = (\tfrac{1}{2}C\rho Av^2, \text{direction opposite the motion}) \tag{6.16}$$

Notice that the drag force is proportional to the *square* of the object's speed. The symbols in Equation 6.16 are:

- A is the *cross-section area* of the object as it "faces into the wind," as illustrated in FIGURE 6.20.
- ρ is the density of the air, which is 1.2 kg/m^3 at atmospheric pressure and room temperature.
- C is the **drag coefficient.** It is smaller for aerodynamically shaped objects, larger for objects presenting a flat face to the wind. Figure 6.20 gives approximate values for a sphere and two cylinders.

This model of drag fails for objects that are very small (such as dust particles), very fast (such as bullets), or that move in liquids (such as water). We'll leave those situations to more advanced textbooks.

FIGURE 6.20 Cross-section areas for objects of different shape.

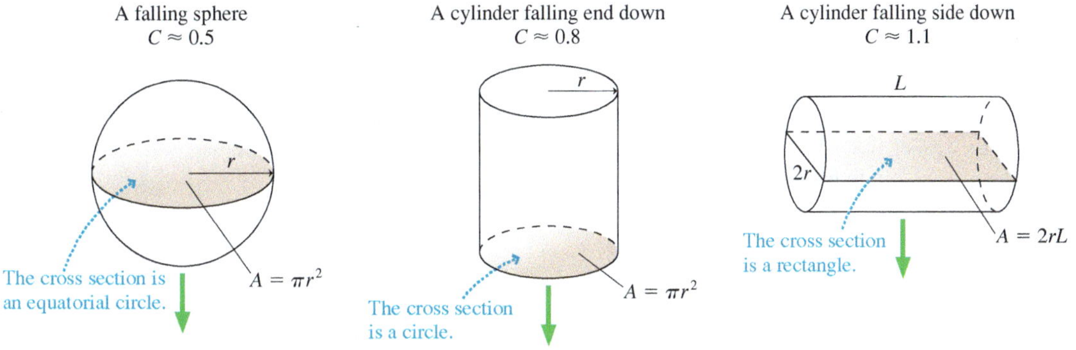

EXAMPLE 6.7 Air resistance compared to rolling friction

The profile of a typical 1500 kg passenger car, as seen from the front, is 1.6 m wide and 1.4 m high. Aerodynamic body shaping gives a drag coefficient of 0.35. At what speed does the magnitude of the drag equal the magnitude of the rolling friction?

MODEL Treat the car as a particle. Use the models of rolling friction and drag.

VISUALIZE FIGURE 6.21 shows the car and a free-body diagram. A full pictorial representation is not needed because we won't be doing any kinematics calculations.

FIGURE 6.21 A car experiences both rolling friction and drag.

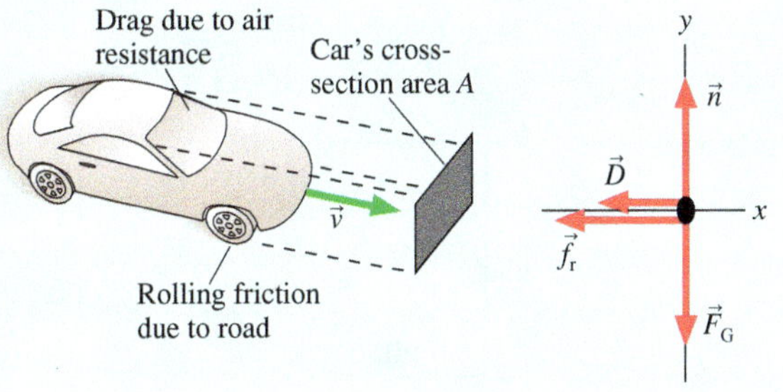

SOLVE Drag is less than friction at low speeds, where air resistance is negligible. But drag increases as v increases, so there will be a speed at which the two forces are equal in size. Above this speed, drag is more important than rolling friction.

There's no motion and no acceleration in the vertical direction, so we can see from the free-body diagram that $n = F_G = mg$. Thus $f_r = \mu_r mg$. Equating friction and drag, we have

$$\tfrac{1}{2}C\rho Av^2 = \mu_r mg$$

Solving for v, we find

$$v = \sqrt{\frac{2\mu_r mg}{C\rho A}} = \sqrt{\frac{2(0.02)(1500\ \text{kg})(9.80\ \text{m/s}^2)}{(0.35)(1.2\ \text{kg/m}^3)\,(1.4\ \text{m} \times 1.6\ \text{m})}} = 25\ \text{m/s}$$

where the value of μ_r for rubber on concrete was taken from Table 6.1.

ASSESS 25 m/s is approximately 50 mph, a reasonable result. This calculation shows that we can reasonably ignore air resistance for car speeds less than 30 or 40 mph. Calculations that neglect drag will be increasingly inaccurate as speeds go above 50 mph.

Terminal Speed

FIGURE 6.22 An object falling at terminal speed.

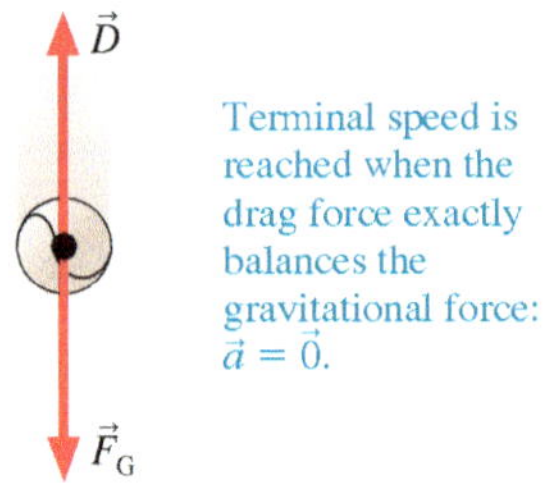

The drag force increases as an object falls and gains speed. If the object falls far enough, it will eventually reach a speed, shown in FIGURE 6.22, at which $D = F_G$. That is, the drag force will be equal and opposite to the gravitational force. The net force at this speed is $\vec{F}_{net} = \vec{0}$, so there is no further acceleration and the object falls with a *constant* speed. The speed at which the exact balance between the upward drag force and the downward gravitational force causes an object to fall without acceleration is called the **terminal speed** v_{term}. Once an object has reached terminal speed, it will continue falling at that speed until it hits the ground.

It's not hard to compute the terminal speed. It is the speed, by definition, at which $D = F_G$ or, equivalently, $\frac{1}{2}C\rho Av^2 = mg$. This speed is

$$v_{term} = \sqrt{\frac{2mg}{C\rho A}} \tag{6.17}$$

A more massive object has a larger terminal speed than a less massive object of equal size and shape. A 10-cm-diameter lead ball, with a mass of 6 kg, has a terminal speed of 160 m/s, while a 10-cm-diameter Styrofoam ball, with a mass of 50 g, has a terminal speed of only 15 m/s.

A popular use of Equation 6.17 is to find the terminal speed of a skydiver. A skydiver is rather like the cylinder of Figure 6.20 falling "side down," for which we see that $C \approx 1.1$. A typical skydiver is 1.8 m long and 0.40 m wide ($A = 0.72\ \text{m}^2$) and has a mass of 75 kg. His terminal speed is

$$v_{term} = \sqrt{\frac{2mg}{C\rho A}} = \sqrt{\frac{2(75\ \text{kg})(9.8\ \text{m/s}^2)}{(1.1)(1.2\ \text{kg/m}^3)(0.72\ \text{m}^2)}} = 39\ \text{m/s}$$

FIGURE 6.23 The velocity-versus-time graph of a falling object with and without drag.

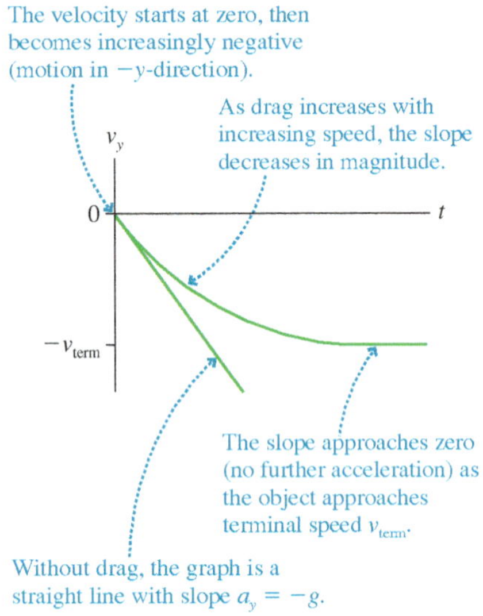

This is roughly 90 mph. A higher speed can be reached by falling feet first or head first, which reduces the area A and the drag coefficient.

FIGURE 6.23 shows the results of a more detailed calculation for a falling object. Without drag, the velocity graph is a straight line with slope $= a_y = -g$. When drag is included, the slope steadily decreases in magnitude and approaches zero (no further acceleration) as the object reaches terminal speed.

Although we've focused our analysis on objects moving vertically, the same ideas apply to objects moving horizontally. If an object is thrown or shot horizontally, $\vec{D}$ causes the object to slow down. An airplane reaches its maximum speed, which is analogous to the terminal speed, when the drag is equal and opposite to the thrust: $D = F_{thrust}$. The net force is then zero and the plane cannot go any faster. The maximum speed of a passenger jet is about 550 mph.

STOP TO THINK 6.4 The terminal speed of a Styrofoam ball is 15 m/s. Suppose a Styrofoam ball is shot straight down with an initial speed of 30 m/s. Which velocity graph is correct?

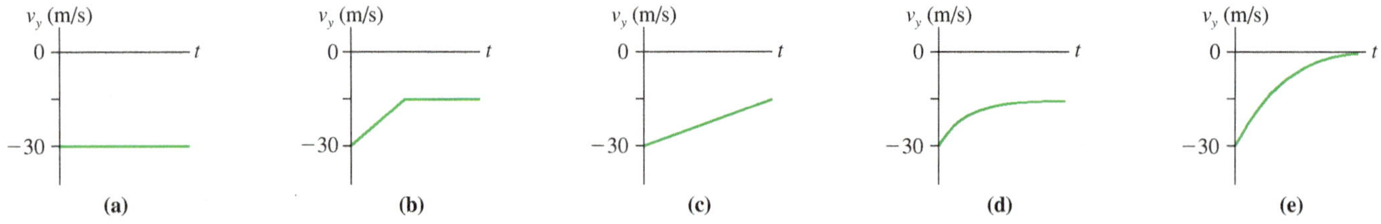

6.6 More Examples of Newton's Second Law

We will finish this chapter with four additional examples in which we use the problem-solving strategy in more complex scenarios.

EXAMPLE 6.8 Stopping distances

A 1500 kg car is traveling at a speed of 30 m/s when the driver slams on the brakes and skids to a halt. Determine the stopping distance if the car is traveling up a 10° slope, down a 10° slope, or on a level road.

MODEL We'll represent the car as a particle and we'll use the model of kinetic friction. We want to solve the problem only once, not three separate times, so we'll leave the slope angle θ unspecified until the end.

VISUALIZE **FIGURE 6.24** shows the pictorial representation. We've shown the car sliding uphill, but these representations work equally well for a level or downhill slide if we let θ be zero or negative, respectively. We've used a tilted coordinate system so that the motion is along one of the axes. We've *assumed* that the car is traveling to the right, although the problem didn't state this. You could equally well make the opposite assumption, but you would have to be careful with negative values of x and v_x. The car *skids* to a halt, so we've taken the coefficient of *kinetic* friction for rubber on concrete from Table 6.1.

SOLVE Newton's second law and the model of kinetic friction are

$$\sum F_x = n_x + (F_G)_x + (f_k)_x = -mg\sin\theta - f_k = ma_x$$

$$\sum F_y = n_y + (F_G)_y + (f_k)_y = n - mg\cos\theta = ma_y = 0$$

$$f_k = \mu_k n$$

We've written these equations by "reading" the motion diagram and the free-body diagram. Notice that both components of the gravitational force vector $\vec{F}_G$ are negative. $a_y = 0$ because the motion is entirely along the x-axis.

The second equation gives $n = mg\cos\theta$. Using this in the friction model, we find $f_k = \mu_k mg\cos\theta$. Inserting this result back into the first equation then gives

$$\begin{aligned} ma_x &= -mg\sin\theta - \mu_k mg\cos\theta \\ &= -mg(\sin\theta + \mu_k\cos\theta) \\ a_x &= -g(\sin\theta + \mu_k\cos\theta) \end{aligned}$$

This is a constant acceleration. Constant-acceleration kinematics gives

$$v_{1x}^2 = 0 = v_{0x}^2 + 2a_x(x_1 - x_0) = v_{0x}^2 + 2a_x x_1$$

which we can solve for the stopping distance x_1:

$$x_1 = -\frac{v_{0x}^2}{2a_x} = \frac{v_{0x}^2}{2g(\sin\theta + \mu_k\cos\theta)}$$

Notice how the minus sign in the expression for a_x canceled the minus sign in the expression for x_1. Evaluating our result at the three different angles gives the stopping distances:

$$x_1 = \begin{cases} 48\text{ m} & \theta = 10^\circ & \text{uphill} \\ 57\text{ m} & \theta = 0^\circ & \text{level} \\ 75\text{ m} & \theta = -10^\circ & \text{downhill} \end{cases}$$

The implications are clear about the danger of driving downhill too fast!

ASSESS 30 m/s ≈ 60 mph and 57 m ≈ 180 feet on a level surface. This is similar to the stopping distances you learned when you got your driver's license, so the results seem reasonable. Additional confirmation comes from noting that the expression for a_x becomes $-g\sin\theta$ if $\mu_k = 0$. This is what you learned in Chapter 2 for the acceleration on a frictionless inclined plane.

FIGURE 6.24 Pictorial representation of a skidding car.

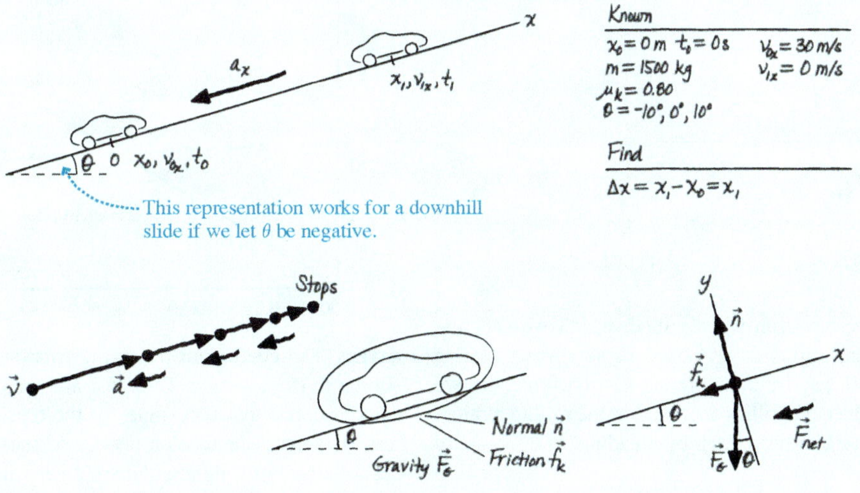

EXAMPLE 6.9 Measuring the tension pulling a cart

Your instructor has set up a lecture demonstration in which a 250 g cart can roll along a level, 2.00-m-long track while its velocity is measured with a motion detector. First, the instructor simply gives the cart a push and measures its velocity as it rolls down the track. The data below show that the cart slows slightly before reaching the end of the track. Then, as FIGURE 6.25 shows, the instructor attaches a string to the cart and uses a falling weight to pull the cart. She then asks you to determine the tension in the string. For extra credit, find the coefficient of rolling friction.

Time (s)	Rolled velocity (m/s)	Pulled velocity (m/s)
0.00	1.20	0.00
0.25	1.17	0.36
0.50	1.15	0.80
0.75	1.12	1.21
1.00	1.08	1.52
1.25	1.04	1.93
1.50	1.02	2.33

FIGURE 6.25 The experimental arrangement.

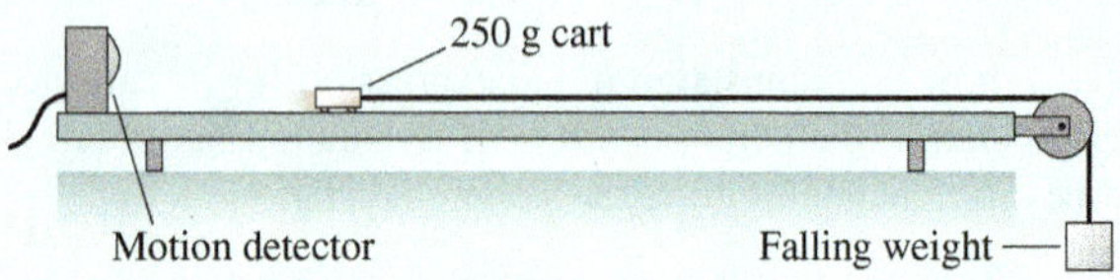

MODEL Model the cart as a particle.

VISUALIZE The cart changes velocity—it accelerates—when both pulled and rolled. Consequently, there must be a net force for both motions. For rolling, force identification finds that the only horizontal force is rolling friction, a force that opposes the motion and slows the cart. There is no "force of motion" or "force of the hand" because the hand is no longer in contact with the cart. (Recall Newton's "zeroth law": The cart responds only to forces applied *at this instant.*) Pulling adds a tension force in the direction of motion. The two free-body diagrams are shown in FIGURE 6.26.

FIGURE 6.26 Pictorial representations of the cart.

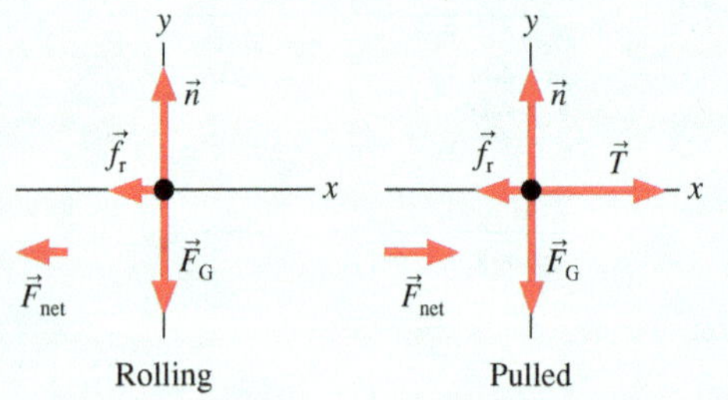

SOLVE The cart's acceleration when pulled, which we can find from the velocity data, will allow us to find the net force. Isolating the tension force will require knowing the friction force, but we can determine that from the rolling motion. For the rolling motion, Newton's second law can be written by "reading" the free-body diagram on the left:

$$\sum F_x = (f_r)_x = -f_r = ma_x = ma_{roll}$$

$$\sum F_y = n_y + (F_G)_y = n - mg = 0$$

Make sure you understand where the signs come from and how we used our knowledge that $\vec{a}$ has only an x-component, which we called a_{roll}. The magnitude of the friction force, which is all we'll need to determine the tension, is found from the x-component equation:

$$f_r = -ma_{roll} = -m \times \text{slope of the rolling-velocity graph}$$

But we'll need to do a bit more analysis to get the coefficient of rolling friction. The y-component equation tells us that $n = mg$. Using this in the model of rolling friction, $f_r = \mu_r n = \mu_r mg$, we see that the coefficient of rolling friction is

$$\mu_r = \frac{f_r}{mg}$$

The x-component equation of Newton's second law when the cart is pulled is

$$\sum F_x = T + (f_r)_x = T - f_r = ma_x = ma_{pulled}$$

Thus the tension that we seek is

$$T = f_r + ma_{pulled} = f_r + m \times \text{slope of the pulled-velocity graph}$$

FIGURE 6.27 shows the graphs of the velocity data. The accelerations are the slopes of these lines, and from the equations of the best-fit lines we find $a_{roll} = -0.124\text{ m/s}^2$ and $a_{pulled} = 1.55\text{ m/s}^2$. Thus the friction force is

$$f_r = -ma_{roll} = -(0.25\text{ kg})(-0.124\text{ m/s}^2) = 0.031\text{ N}$$

Knowing this, we find that the string tension pulling the cart is

$$T = f_r + ma_{pulled} = 0.031\text{ N} + (0.25\text{ kg})(1.55\text{ m/s}^2) = 0.42\text{ N}$$

and the coefficient of rolling friction is

$$\mu_r = \frac{f_r}{mg} = \frac{0.031\text{ N}}{(0.25\text{ kg})(9.80\text{ m/s}^2)} = 0.013$$

FIGURE 6.27 The velocity graphs of the rolling and pulled motion. The slopes of these graphs are the cart's acceleration.

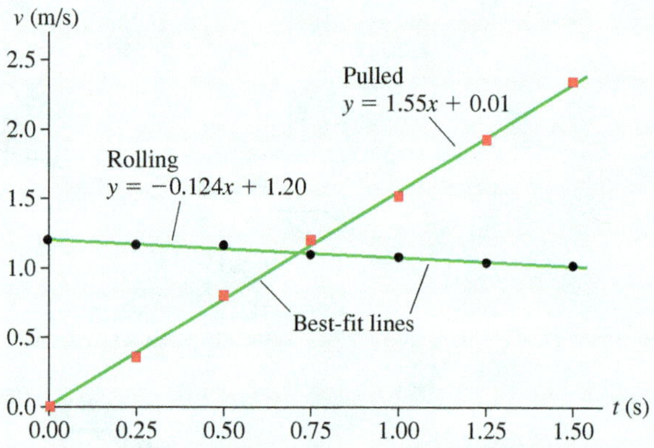

ASSESS The coefficient of rolling friction is very small, but it's similar to the values in Table 6.1 and thus believable. That gives us confidence that our value for the tension is also correct. It's reasonable that the tension needed to accelerate the cart is small because the cart is light and there's very little friction.

EXAMPLE 6.10 **Make sure the cargo doesn't slide**

A 100 kg box of dimensions 50 cm × 50 cm × 50 cm is in the back of a flatbed truck. The coefficients of friction between the box and the bed of the truck are $\mu_s = 0.40$ and $\mu_k = 0.20$. What is the maximum acceleration the truck can have without the box slipping?

MODEL This is a somewhat different problem from any we have looked at thus far. Let the box, which we'll model as a particle, be the object of interest. It contacts other objects only where it touches the truck bed, so only the truck can exert contact forces on the box. If the box does *not* slip, then there is no motion of the box *relative to the truck* and the box must accelerate *with the truck:* $a_{box} = a_{truck}$. As the box accelerates, it must, according to Newton's second law, have a net force acting on it. But from what?

Imagine, for a moment, that the truck bed is frictionless. The box would slide backward (as seen in the truck's reference frame) as the truck accelerates. The force that prevents sliding is *static friction,* so the truck must exert a static friction force on the box to "pull" the box along with it and prevent the box from sliding *relative to the truck.*

VISUALIZE This situation is shown in FIGURE 6.28. There is only one horizontal force on the box, $\vec{f}_s$, and it points in the *forward* direction to accelerate the box. Notice that we're solving the problem with the ground as our reference frame. Newton's laws are not valid in the accelerating truck because it is not an inertial reference frame.

SOLVE Newton's second law, which we can "read" from the free-body diagram, is

$$\sum F_x = f_s = ma_x$$
$$\sum F_y = n - F_G = n - mg = ma_y = 0$$

Now, static friction, you will recall, can be *any* value between 0 and $f_{s\,max}$. If the truck accelerates slowly, so that the box doesn't slip, then $f_s < f_{s\,max}$. However, we're interested in the acceleration a_{max} at which the box begins to slip. This is the acceleration at which f_s reaches its maximum possible value

$$f_s = f_{s\,max} = \mu_s n$$

The y-equation of the second law and the friction model combine to give $f_{s\,max} = \mu_s mg$. Substituting this into the x-equation, and noting that a_x is now a_{max}, we find

$$a_{max} = \frac{f_{s\,max}}{m} = \mu_s g = 3.9\ \text{m/s}^2$$

The truck must keep its acceleration less than 3.9 m/s^2 if slipping is to be avoided.

ASSESS 3.9 m/s^2 is about one-third of g. You may have noticed that items in a car or truck are likely to *tip over* when you start or stop, but they slide only if you really floor it and accelerate very quickly. So this answer seems reasonable. Notice that the dimensions of the crate were not needed. Real-world situations rarely have exactly the information you need, no more and no less. Many problems in this textbook will require you to assess the information in the problem statement in order to learn which is relevant to the solution.

FIGURE 6.28 Pictorial representation for the box in a flatbed truck.

Known

$m = 100$ kg
Box dimensions 50 cm × 50 cm × 50 cm
$\mu_s = 0.40$ $\mu_k = 0.20$

Find

Acceleration at which box slips

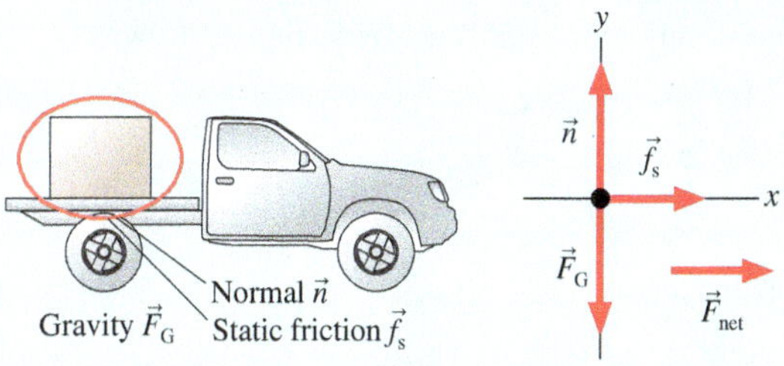

The mathematical representation of this last example was quite straightforward. The challenge was in the analysis that preceded the mathematics—that is, in the *physics* of the problem rather than the mathematics. It is here that our analysis tools—motion diagrams, force identification, and free-body diagrams—prove their value.

CHALLENGE EXAMPLE 6.11 **Acceleration from a variable force**

Force $F_x = c\sin(\pi t/T)$, where c and T are constants, is applied to an object of mass m that moves on a horizontal, frictionless surface. The object is at rest at the origin at $t = 0$.

a. Find an expression for the object's velocity. Graph your result for $0 \le t \le T$.

b. What is the maximum velocity of a 500 g object if $c = 2.5$ N and $T = 1.0$ s?

MODEL Model the object as a particle.

VISUALIZE The sine function is 0 at $t = 0$ and again at $t = T$, when the value of the argument is π rad. Over the interval $0 \le t \le T$, the force grows from 0 to c and back to 0, always pointing in the positive x-direction. FIGURE 6.29 on the next page shows a graph of the force and a pictorial representation.

Continued

FIGURE 6.29 Pictorial representation for a variable force.

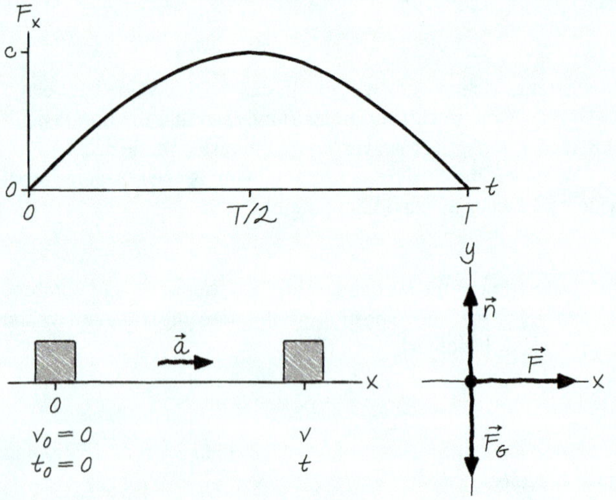

SOLVE The object's acceleration increases between 0 and $T/2$ as the force increases. You might expect the object to slow down between $T/2$ and T as the force decreases. However, *there's still a net force in the positive x-direction, so there must be an acceleration in the positive x-direction.* The object continues to speed up, only more slowly as the acceleration decreases. Maximum velocity is reached at $t = T$.

a. This is not constant-acceleration motion, so we cannot use the familiar equations of constant-acceleration kinematics. Instead, we must use the definition of acceleration as the rate of change—the time derivative—of velocity. With no friction, we need only the x-component equation of Newton's second law:

$$a_x = \frac{dv_x}{dt} = \frac{F_{\text{net}}}{m} = \frac{c}{m}\sin\left(\frac{\pi t}{T}\right)$$

First we rewrite this as

$$dv_x = \frac{c}{m}\sin\left(\frac{\pi t}{T}\right)dt$$

Then we integrate both sides from the initial conditions ($v_x = v_{0x} = 0$ at $t = t_0 = 0$) to the final conditions (v_x at the later time t):

$$\int_0^{v_x} dv_x = \frac{c}{m}\int_0^t \sin\left(\frac{\pi t}{T}\right)dt$$

The fraction c/m is a constant that we could take outside the integral. The integral on the right side is of the form

$$\int \sin(ax)\,dx = -\frac{1}{a}\cos(ax)$$

Using this, and integrating both sides of the equation, we find

$$v_x\Big|_0^{v_x} = v_x - 0 = -\frac{cT}{\pi m}\cos\left(\frac{\pi t}{T}\right)\Bigg|_0^t = -\frac{cT}{\pi m}\left(\cos\left(\frac{\pi t}{T}\right) - 1\right)$$

Simplifying, we find the object's velocity at time t is

$$v_x = \frac{cT}{\pi m}\left(1 - \cos\left(\frac{\pi t}{T}\right)\right)$$

This expression is graphed in FIGURE 6.30, where we see that, as predicted, maximum velocity is reached at $t = T$.

FIGURE 6.30 The object's velocity as a function of time.

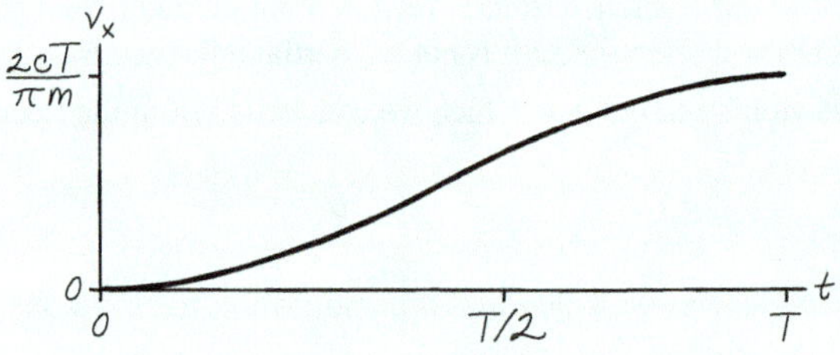

b. Maximum velocity, at $t = T$, is

$$v_{\max} = \frac{cT}{\pi m}(1 - \cos\pi) = \frac{2cT}{\pi m} = \frac{2(2.5\text{ N})(1.0\text{ s})}{\pi(0.50\text{ kg})} = 3.2\text{ m/s}$$

ASSESS A steady 2.5 N force would cause a 0.5 kg object to accelerate at 5 m/s^2 and reach a speed of 5 m/s in 1 s. A variable force with a maximum of 2.5 N will produce less acceleration, so a top speed of 3.2 m/s seems reasonable.

SUMMARY

The goal of Chapter 6 has been to learn how to solve linear force-and-motion problems.

General Strategy

All examples in this chapter follow a four-part strategy. You'll become a better problem solver if you adhere to it as you do the homework problems. The *Dynamics Worksheets* in the *Student Workbook* will help you structure your work in this way.

Equilibrium Problems

Object at rest or moving with constant velocity.

MODEL Make simplifying assumptions.

VISUALIZE

- Translate words into symbols.
- Identify forces.
- Draw a free-body diagram.

SOLVE Use Newton's first law:

$$\vec{F}_{\text{net}} = \sum_i \vec{F}_i = \vec{0}$$

"Read" the vectors from the free-body diagram.

ASSESS Is the result reasonable?

Go back and forth between these steps as needed.

Dynamics Problems

Object accelerating.

MODEL Make simplifying assumptions.

VISUALIZE

- Translate words into symbols.
- Draw a sketch to define the situation.
- Draw a motion diagram.
- Identify forces.
- Draw a free-body diagram.

SOLVE Use Newton's second law:

$$\vec{F}_{\text{net}} = \sum_i \vec{F}_i = m\vec{a}$$

"Read" the vectors from the free-body diagram.
Use kinematics to find velocities and positions.

ASSESS Is the result reasonable?

Important Concepts

Specific information about three important forces:

Gravity $\vec{F}_G = (mg, \text{downward})$

Friction $\vec{f}_s = (0 \text{ to } \mu_s n, \text{direction as necessary to prevent motion})$

$\vec{f}_k = (\mu_k n, \text{direction opposite the motion})$

$\vec{f}_r = (\mu_r n, \text{direction opposite the motion})$

Drag $\vec{D} = (\tfrac{1}{2}C\rho A v^2, \text{direction opposite the motion})$

Newton's laws are vector expressions. You must write them out by **components**:

$$(F_{\text{net}})_x = \sum F_x = ma_x$$

$$(F_{\text{net}})_y = \sum F_y = ma_y$$

The acceleration is zero in equilibrium and also along an axis perpendicular to the motion.

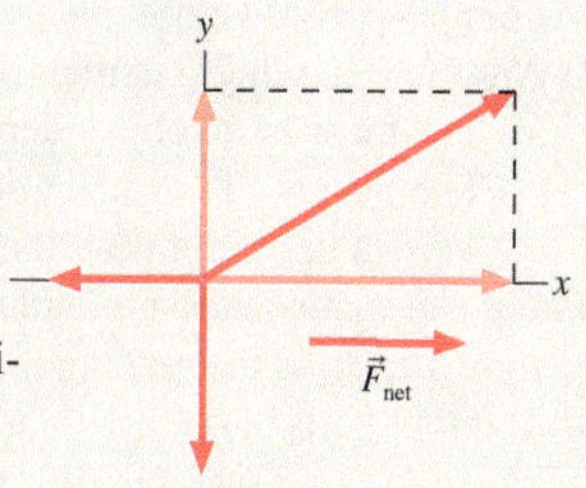

Applications

Mass is an intrinsic property of an object that describes the object's inertia and, loosely speaking, its quantity of matter.

The **weight** of an object is the reading of a calibrated spring scale on which the object is stationary. Weight is the result of weighing. An object's weight depends on its mass, its acceleration, and the strength of gravity. An object in free fall is weightless.

A falling object reaches **terminal speed**

$$v_{\text{term}} = \sqrt{\frac{2mg}{C\rho A}}$$

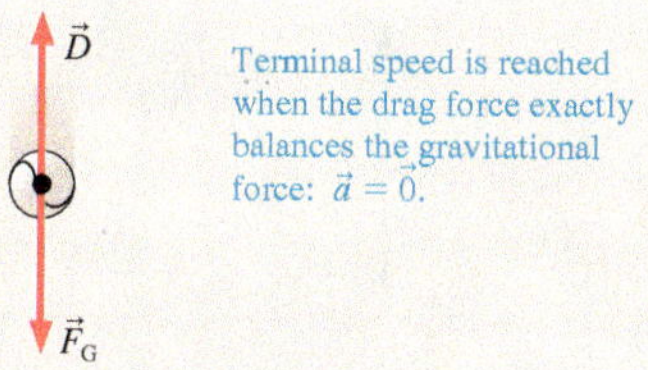

Terms and Notation

flat-earth approximation
weight
coefficient of static friction, μ_s
coefficient of kinetic friction, μ_k
rolling friction
coefficient of rolling friction, μ_r
drag coefficient, C
terminal speed, v_{term}

CONCEPTUAL QUESTIONS

1. Are the objects described here in static equilibrium, dynamic equilibrium, or not in equilibrium at all? Explain.
   a. A 200 pound barbell is held over your head.
   b. A girder is lifted at constant speed by a crane.
   c. A girder is being lowered into place. It is slowing down.
   d. A jet plane has reached its cruising speed and altitude.
   e. A box in the back of a truck doesn't slide as the truck stops.
2. A ball tossed straight up has $v = 0$ at its highest point. Is it in equilibrium? Explain.
3. Kat, Matt, and Nat are arguing about why a physics book on a table doesn't fall. According to Kat, "Gravity pulls down on it, but the table is in the way so it can't fall." "Nonsense," says Matt. "There are all kinds of forces acting on the book, but the upward forces overcome the downward forces to prevent it from falling." "But what about Newton's first law?" counters Nat. "It's not moving, so there can't be any forces acting on it." None of the statements is exactly correct. Who comes closest, and how would you change his or her statement to make it correct?
4. If you know all of the forces acting on a moving object, can you tell the direction the object is moving? If yes, explain how. If no, give an example.
5. An elevator, hanging from a single cable, moves upward at constant speed. Friction and air resistance are negligible. Is the tension in the cable greater than, less than, or equal to the gravitational force on the elevator? Explain. Include a free-body diagram as part of your explanation.
6. An elevator, hanging from a single cable, moves downward and is slowing. Friction and air resistance are negligible. Is the tension in the cable greater than, less than, or equal to the gravitational force on the elevator? Explain. Include a free-body diagram as part of your explanation.
7. Are the following statements true or false? Explain.
   a. The mass of an object depends on its location.
   b. The weight of an object depends on its location.
   c. Mass and weight describe the same thing in different units.
8. An astronaut takes his bathroom scale to the moon and then stands on it. Is the reading of the scale his weight? Explain.
9. The four balls in FIGURE Q6.9 have been thrown straight up. They have the same size, but different masses. Air resistance is negligible. Rank in order, from largest to smallest, the magnitude of the net force acting on each ball. Some may be equal. Give your answer in the form a > b = c > d and explain your ranking.

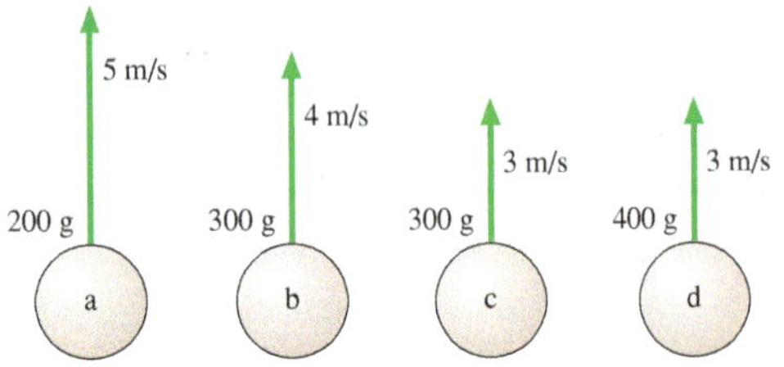

FIGURE Q6.9

10. Suppose you attempt to pour out 100 g of salt, using a pan balance for measurements, while in a rocket accelerating upward. Will the quantity of salt be too much, too little, or the correct amount? Explain.
11. A box with a passenger inside is launched straight up into the air by a giant rubber band. Before launch, the passenger stood on a scale and weighed 750 N. Once the box has left the rubber band but is still moving upward, is the passenger's weight more than 750 N, 750 N, less than 750 N but not zero, or zero? Explain.
12. An astronaut orbiting the earth is handed two balls that have identical outward appearances. However, one is hollow while the other is filled with lead. How can the astronaut determine which is which? Cutting or altering the balls is not allowed.
13. A hand presses down on the book in FIGURE Q6.13. Is the normal force of the table on the book larger than, smaller than, or equal to mg?

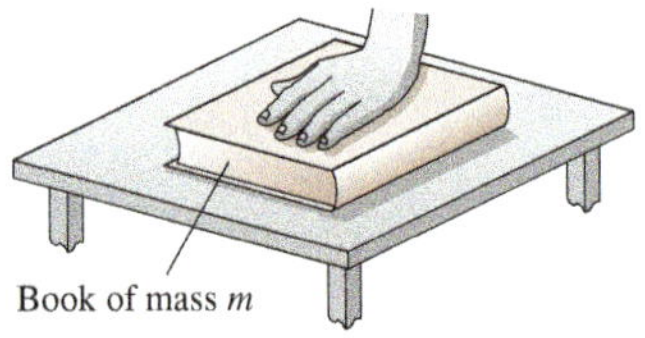

FIGURE Q6.13

14. Suppose you push a hockey puck of mass m across frictionless ice for a time Δt, starting from rest, giving the puck speed v after traveling distance d. If you repeat the experiment with a puck of mass $2m$,
    a. How long will you have to push for the puck to reach the same speed v?
    b. How long will you have to push for the puck to travel the same distance d?
15. A block pushed along the floor with velocity v_{0x} slides a distance d after the pushing force is removed.
    a. If the mass of the block is doubled but its initial velocity is not changed, what distance does the block slide before stopping?
    b. If the initial velocity is doubled to $2v_{0x}$ but the mass is not changed, what distance does the block slide before stopping?
16. Can the friction force on an object ever point in the direction of the object's motion? If yes, give an example. If no, why not?
17. A crate of fragile dishes is in the back of a pickup truck. The truck accelerates north from a stop sign, and the crate moves without slipping. Does the friction force on the crate point north or south? Or is the friction force zero? Explain.
18. Five balls move through the air as shown in FIGURE Q6.18. All five have the same size and shape. Air resistance is not negligible. Rank in order, from largest to smallest, the magnitudes of the accelerations a_a to a_e. Some may be equal. Give your answer in the form a > b = c > d > e and explain your ranking.

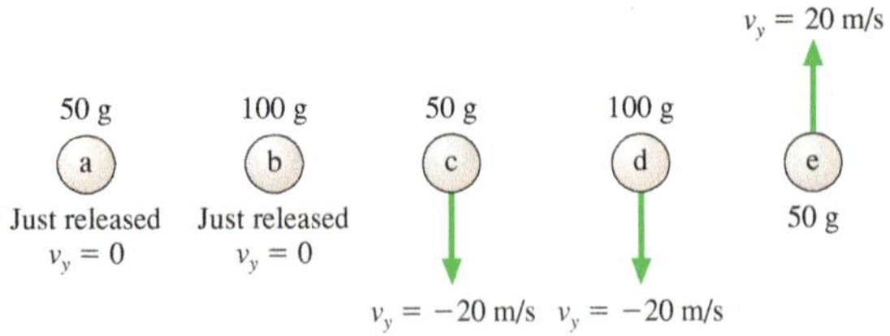

FIGURE Q6.18

EXERCISES AND PROBLEMS

Exercises

Section 6.1 Equilibrium

1. | The three ropes in FIGURE EX6.1 are tied to a small, very light ring. Two of the ropes are anchored to walls at right angles, and the third rope pulls as shown. What are T_1 and T_2, the magnitudes of the tension forces in the first two ropes?

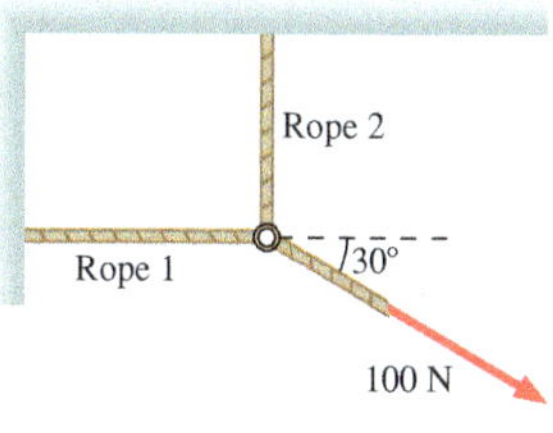

FIGURE EX6.1

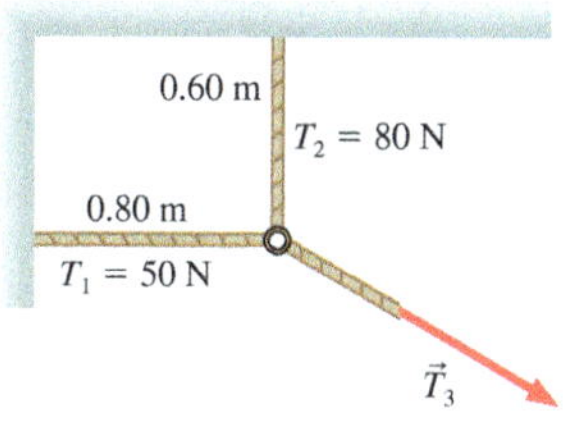

FIGURE EX6.2

2. | The three ropes in FIGURE EX6.2 are tied to a small, very light ring. Two of these ropes are anchored to walls at right angles with the tensions shown in the figure. What are the magnitude and direction of the tension $\vec{T}_3$ in the third rope?
3. || A 20 kg loudspeaker is suspended 2.0 m below the ceiling by two 3.0-m-long cables that angle outward at equal angles. What is the tension in the cables?
4. || A football coach sits on a sled while two of his players build their strength by dragging the sled across the field with ropes. The friction force on the sled is 1000 N and the angle between the two ropes is 20°. How hard must each player pull to drag the coach at a steady 2.0 m/s?
5. | A construction worker with a weight of 850 N stands on a roof that is sloped at 20°. What is the magnitude of the normal force of the roof on the worker?

Section 6.2 Using Newton's Second Law

6. | In each of the two free-body diagrams in FIGURE EX6.6, the forces are acting on a 2.0 kg object. For each diagram, find the values of a_x and a_y, the x- and y-components of the acceleration.

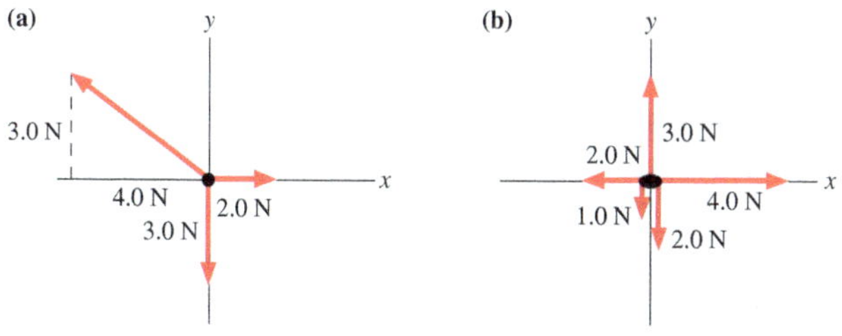

FIGURE EX6.6

7. || In each of the two free-body diagrams in FIGURE EX6.7, the forces are acting on a 2.0 kg object. For each diagram, find the values of a_x and a_y, the x- and y-components of the acceleration.

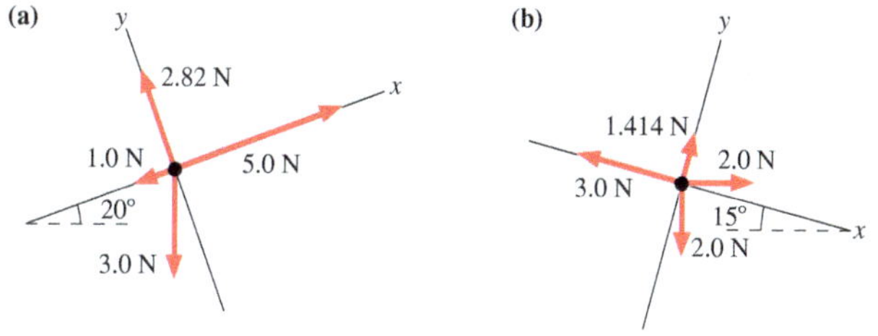

FIGURE EX6.7

8. | FIGURE EX6.8 shows the velocity graph of a 2.0 kg object as it moves along the x-axis. What is the net force acting on this object at $t = 1$ s? At 4 s? At 7 s?

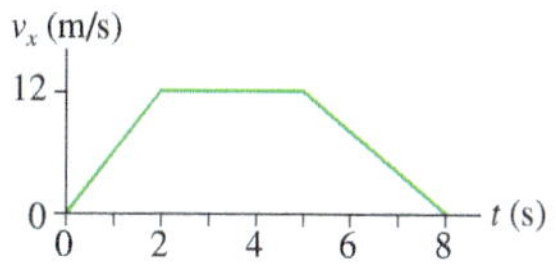

FIGURE EX6.8

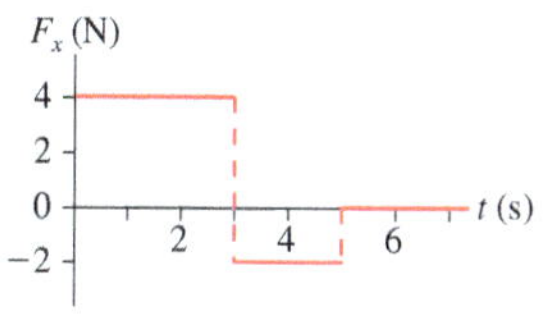

FIGURE EX6.9

9. | FIGURE EX6.9 shows the force acting on a 2.0 kg object as it moves along the x-axis. The object is at rest at the origin at $t = 0$ s. What are its acceleration and velocity at $t = 6$ s?
10. | A horizontal rope is tied to a 50 kg box on frictionless ice. What is the tension in the rope if:
    a. The box is at rest?
    b. The box moves at a steady 5.0 m/s?
    c. The box has $v_x = 5.0$ m/s and $a_x = 5.0$ m/s^2?
11. | A 50 kg box hangs from a rope. What is the tension in the rope if:
    a. The box is at rest?
    b. The box moves up at a steady 5.0 m/s?
    c. The box has $v_y = 5.0$ m/s and is speeding up at 5.0 m/s^2?
    d. The box has $v_y = 5.0$ m/s and is slowing down at 5.0 m/s^2?
12. || a. The block in FIGURE EX6.12 floats on a cushion of air. It is pushed to the right with a force that remains constant as the block moves from 0 to 1. The block
    A. Speeds up from 0 to 1.
    B. Speeds up at first, then has constant speed.
    C. Moves with constant speed from 0 to 1.

    b. From 1 to 2, the size of the force steadily decreases until it reaches half of its initial value. The block
    A. Continues to speed up from 1 to 2.
    B. Moves with constant speed from 1 to 2.
    C. Slows down.

FIGURE EX6.12

Section 6.3 Mass, Weight, and Gravity

13. | A woman has a mass of 55 kg.
    a. What is her weight while standing on earth?
    b. What are her mass and her weight on Mars, where $g = 3.76$ m/s^2?

14. | It takes the elevator in a skyscraper 4.0 s to reach its cruising speed of 10 m/s. A 60 kg passenger gets aboard on the ground floor. What is the passenger's weight
   a. Before the elevator starts moving?
   b. While the elevator is speeding up?
   c. After the elevator reaches its cruising speed?
15. || FIGURE EX6.15 shows the velocity graph of a 75 kg passenger in an elevator. What is the passenger's weight at $t = 1$ s? At 5 s? At 9 s?

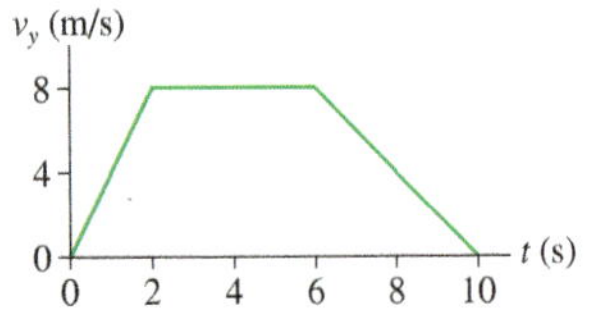

FIGURE EX6.15

16. || What thrust does a 200 g model rocket need in order to have a vertical acceleration of 10 m/s^2
   a. On earth?
   b. On the moon, where $g = 1.62$ m/s^2?

Section 6.4 Friction

17. || Bonnie and Clyde are sliding a 300 kg bank safe across the floor to their getaway car. The safe slides with a constant speed if Clyde pushes from behind with 385 N of force while Bonnie pulls forward on a rope with 350 N of force. What is the safe's coefficient of kinetic friction on the bank floor?
18. | A stubborn, 120 kg mule sits down and refuses to move. To drag the mule to the barn, the exasperated farmer ties a rope around the mule and pulls with his maximum force of 800 N. The coefficients of friction between the mule and the ground are $\mu_s = 0.8$ and $\mu_k = 0.5$. Is the farmer able to move the mule?
19. || A 10 kg crate is placed on a horizontal conveyor belt. The materials are such that $\mu_s = 0.5$ and $\mu_k = 0.3$.
   a. Draw a free-body diagram showing all the forces on the crate if the conveyer belt runs at constant speed.
   b. Draw a free-body diagram showing all the forces on the crate if the conveyer belt is speeding up.
   c. What is the maximum acceleration the belt can have without the crate slipping?
20. | Bob is pulling a 30 kg filing cabinet with a force of 200 N, but the filing cabinet refuses to move. The coefficient of static friction between the filing cabinet and the floor is 0.80. What is the magnitude of the friction force on the filing cabinet?
21. || A 4000 kg truck is parked on a 15° slope. How big is the friction force on the truck? The coefficient of static friction between the tires and the road is 0.90.
22. | A 1500 kg car skids to a halt on a wet road where $\mu_k = 0.50$. How fast was the car traveling if it leaves 65-m-long skid marks?
23. || A 50,000 kg locomotive is traveling at 10 m/s when its engine and brakes both fail. How far will the locomotive roll before it comes to a stop? Assume the track is level.

Section 6.5 Drag

24. || A 75 kg skydiver can be modeled as a rectangular "box" with dimensions 20 cm × 40 cm × 180 cm. What is his terminal speed if he falls feet first? Use 0.8 for the drag coefficient.
25. || A 6.5-cm-diameter tennis ball has a terminal speed of 26 m/s. What is the ball's mass?

Problems

26. || A 5.0 kg object initially at rest at the origin is subjected to the time-varying force shown in FIGURE P6.26. What is the object's velocity at $t = 6$ s?

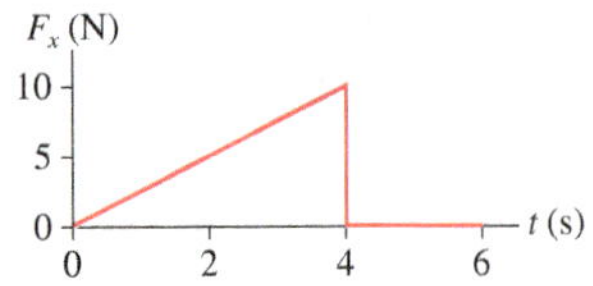

FIGURE P6.26

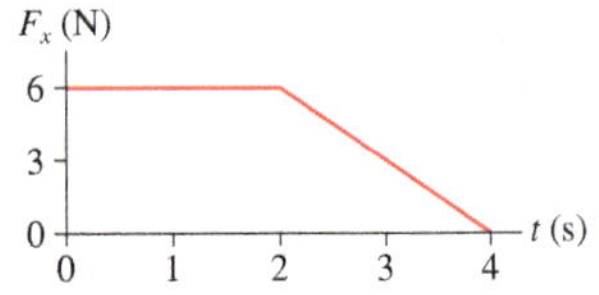

FIGURE P6.27

27. || A 2.0 kg object initially at rest at the origin is subjected to the time-varying force shown in FIGURE P6.27. What is the object's velocity at $t = 4$ s?
28. || The 1000 kg steel beam in FIGURE P6.28 is supported by two ropes. What is the tension in each?

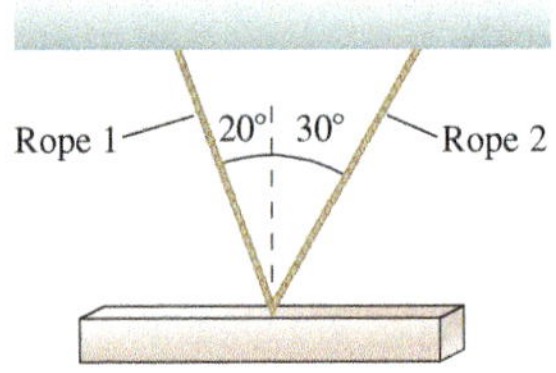

FIGURE P6.28

29. || In an electricity experiment, a 1.0 g plastic ball is suspended on a 60-cm-long string and given an electric charge. A charged rod brought near the ball exerts a horizontal electrical force $\vec{F}_{elec}$ on it, causing the ball to swing out to a 20° angle and remain there.
   a. What is the magnitude of $\vec{F}_{elec}$?
   b. What is the tension in the string?
30. | A 500 kg piano is being lowered into position by a crane while two people steady it with ropes pulling to the sides. Bob's rope pulls to the left, 15° below horizontal, with 500 N of tension. Ellen's rope pulls toward the right, 25° below horizontal.
   a. What tension must Ellen maintain in her rope to keep the piano descending at a steady speed?
   b. What is the tension in the main cable supporting the piano?
31. || Henry gets into an elevator on the 50th floor of a building and it begins moving at $t = 0$ s. FIGURE P6.31 shows his weight over the next 12 s.
   a. Is the elevator's initial direction up or down? Explain how you can tell.
   b. What is Henry's mass?
   c. How far has Henry traveled at $t = 12$ s?

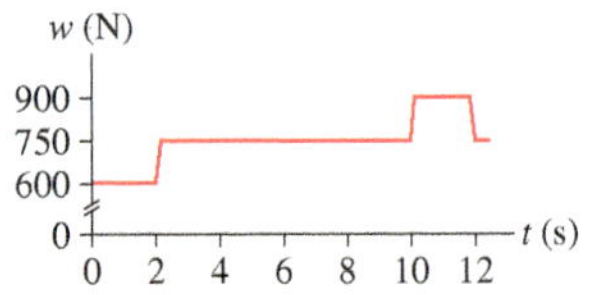

FIGURE P6.31

32. | Zach, whose mass is 80 kg, is in an elevator descending at 10 m/s. The elevator takes 3.0 s to brake to a stop at the first floor.
   a. What is Zach's weight before the elevator starts braking?
   b. What is Zach's weight while the elevator is braking?
33. || An accident victim with a broken leg is being placed in traction. The patient wears a special boot with a pulley attached to the sole. The foot and boot together have a mass of 4.0 kg, and the
BIO

doctor has decided to hang a 6.0 kg mass from the rope. The boot is held suspended by the ropes, as shown in FIGURE P6.33, and does not touch the bed.

a. Determine the amount of tension in the rope by using Newton's laws to analyze the hanging mass.
b. The net traction force needs to pull straight out on the leg. What is the proper angle θ for the upper rope?
c. What is the net traction force pulling on the leg?

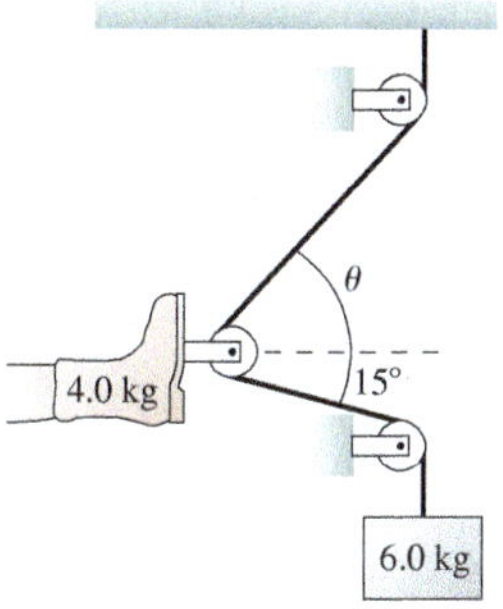

FIGURE P6.33

Hint: If the pulleys are frictionless, which we will assume, the tension in the rope is constant from one end to the other.

34. || Seat belts and air bags save lives by reducing the forces exerted on the driver and passengers in an automobile collision. Cars are designed with a "crumple zone" in the front of the car. In the event of an impact, the passenger compartment decelerates over a distance of about 1 m as the front of the car crumples. An occupant restrained by seat belts and air bags decelerates with the car. By contrast, an unrestrained occupant keeps moving forward with no loss of speed (Newton's first law!) until hitting the dashboard or windshield. These are unyielding surfaces, and the unfortunate occupant then decelerates over a distance of only about 5 mm.
BIO
a. A 60 kg person is in a head-on collision. The car's speed at impact is 15 m/s. Estimate the net force on the person if he or she is wearing a seat belt and if the air bag deploys.
b. Estimate the net force that ultimately stops the person if he or she is not restrained by a seat belt or air bag.
c. How do these two forces compare to the person's weight?

35. || The position of a 2.0 kg mass is given by $x = (2t^3 - 3t^2)$ m, where t is in seconds. What is the net horizontal force on the mass at (a) $t = 0$ s and (b) $t = 1$ s?

36. || The piston of a machine exerts a constant force on a ball as it moves horizontally through a distance of 15 cm. You use a motion detector to measure the speed of five different balls as they come off the piston; the data are shown below. Use theory to find two quantities that, when graphed, should give a straight line. Then use the graph to find the size of the piston's force.

Mass (g)	Speed (m/s)
200	9.4
400	6.3
600	5.2
800	4.9
1000	4.0

37. || Compressed air is used to fire a 50 g ball vertically upward from a 1.0-m-tall tube. The air exerts an upward force of 2.0 N on the ball as long as it is in the tube. How high does the ball go above the top of the tube?

38. || a. A rocket of mass m is launched straight up with thrust $\vec{F}_{\text{thrust}}$. Find an expression for the rocket's speed at height h if air resistance is neglected.
b. The motor of a 350 g model rocket generates 9.5 N thrust. If air resistance can be neglected, what will be the rocket's speed as it reaches a height of 85 m?

39. || A rifle with a barrel length of 60 cm fires a 10 g bullet with a horizontal speed of 400 m/s. The bullet strikes a block of wood and penetrates to a depth of 12 cm.
a. What resistive force (assumed to be constant) does the wood exert on the bullet?
b. How long does it take the bullet to come to rest?
c. Draw a velocity-versus-time graph for the bullet in the wood.

40. || A 20,000 kg rocket has a rocket motor that generates 3.0×10^5 N of thrust.
a. What is the rocket's initial upward acceleration?
b. At an altitude of 5000 m the rocket's acceleration has increased to 6.0 m/s^2. What mass of fuel has it burned?

41. || a. An object of mass m is at rest at the top of a smooth slope of height h and length L. The coefficient of kinetic friction between the object and the surface, μ_k, is small enough that the object will slide down the slope if given a very small push to get it started. Find an expression for the object's speed at the bottom of the slope.
b. Sam, whose mass is 75 kg, stands at the top of a 12-m-high, 100-m-long snow-covered slope. His skis have a coefficient of kinetic friction on snow of 0.07. If he uses his poles to get started, then glides down, what is his speed at the bottom?

42. || Sam, whose mass is 75 kg, takes off across level snow on his jet-powered skis. The skis have a thrust of 200 N and a coefficient of kinetic friction on snow of 0.10. Unfortunately, the skis run out of fuel after only 10 s.
a. What is Sam's top speed?
b. How far has Sam traveled when he finally coasts to a stop?

43. ||| Sam, whose mass is 75 kg, takes off down a 50-m-high, 10° slope on his jet-powered skis. The skis have a thrust of 200 N. Sam's speed at the bottom is 40 m/s. What is the coefficient of kinetic friction of his skis on snow?

44. || A baggage handler drops your 10 kg suitcase onto a conveyor belt running at 2.0 m/s. The materials are such that $\mu_s = 0.50$ and $\mu_k = 0.30$. How far is your suitcase dragged before it is riding smoothly on the belt?

45. ||| You and your friend Peter are putting new shingles on a roof pitched at 25°. You're sitting on the very top of the roof when Peter, who is at the edge of the roof directly below you, 5.0 m away, asks you for the box of nails. Rather than carry the 2.5 kg box of nails down to Peter, you decide to give the box a push and have it slide down to him. If the coefficient of kinetic friction between the box and the roof is 0.55, with what speed should you push the box to have it gently come to rest right at the edge of the roof?

46. || It's moving day, and you need to push a 100 kg box up a 20° ramp into the truck. The coefficients of friction for the box on the ramp are $\mu_s = 0.90$ and $\mu_k = 0.60$. Your largest pushing force is 1000 N. Can you get the box into the truck without assistance if you get a running start at the ramp? If you stop on the ramp, will you be able to get the box moving again?

47. || An Airbus A320 jetliner has a takeoff mass of 75,000 kg. It reaches its takeoff speed of 82 m/s (180 mph) in 35 s. What is the thrust of the engines? You can neglect air resistance but not rolling friction.

48. || A 2.0 kg wood block is launched up a wooden ramp that is inclined at a 30° angle. The block's initial speed is 10 m/s.
a. What vertical height does the block reach above its starting point?
b. What speed does it have when it slides back down to its starting point?

49. It's a snowy day and you're pulling a friend along a level road on a sled. You've both been taking physics, so she asks what you think the coefficient of friction between the sled and the snow is. You've been walking at a steady 1.5 m/s, and the rope pulls up on the sled at a 30° angle. You estimate that the mass of the sled, with your friend on it, is 60 kg and that you're pulling with a force of 75 N. What answer will you give?

50. a. A large box of mass M is pulled across a horizontal, frictionless surface by a horizontal rope with tension T. A small box of mass m sits on top of the large box. The coefficients of static and kinetic friction between the two boxes are μ_s and μ_k, respectively. Find an expression for the maximum tension T_{max} for which the small box rides on top of the large box without slipping.
b. A horizontal rope pulls a 10 kg wood sled across frictionless snow. A 5.0 kg wood box rides on the sled. What is the largest tension force for which the box doesn't slip?

51. a. A large box of mass M is moving on a horizontal surface at speed v_0. A small box of mass m sits on top of the large box. The coefficients of static and kinetic friction between the two boxes are μ_s and μ_k, respectively. Find an expression for the shortest distance d_{min} in which the large box can stop without the small box slipping.
b. A pickup truck with a steel bed is carrying a steel file cabinet. If the truck's speed is 15 m/s, what is the shortest distance in which it can stop without the file cabinet sliding?

52. Your assignment in lab is to measure the coefficient of kinetic friction between a 350 g block and a smooth metal table. To do so, you decide to launch the block at various speeds and measure how far it slides; your data are listed in the table. Use a graph to determine the value of μ_k.

Speed (m/s)	Distance (cm)
0.5	5
1.0	24
1.5	41
2.0	83
2.5	130

53. You're driving along at 25 m/s with your aunt's valuable antiques in the back of your pickup truck when suddenly you see a giant hole in the road 55 m ahead of you. Fortunately, your foot is right beside the brake and your reaction time is zero! Will the antiques be as fortunate?
a. Can you stop the truck before it falls into the hole?
b. If your answer to part a is yes, can you stop without the antiques sliding and being damaged? Their coefficients of friction are $\mu_s = 0.60$ and $\mu_k = 0.30$.

Hint: You're not trying to stop in the shortest possible distance. What's your best strategy for avoiding damage to the antiques?

54. The 2.0 kg wood box in FIGURE P6.54 slides down a vertical wood wall while you push on it at a 45° angle. What magnitude of force should you apply to cause the box to slide down at a constant speed?

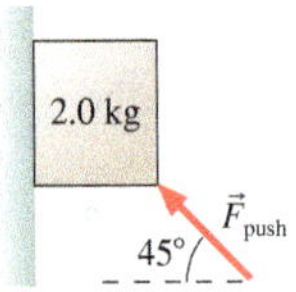

FIGURE P6.54

55. A 1.0 kg wood block is pressed against a vertical wood wall by the 12 N force shown in FIGURE P6.55. If the block is initially at rest, will it move upward, move downward, or stay at rest?

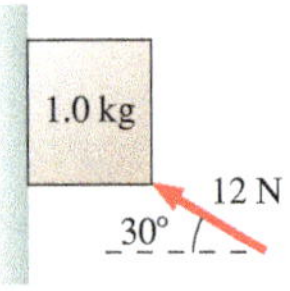

FIGURE P6.55

56. BIO A person with compromised pinch strength in his fingers can exert a force of only 6.0 N to either side of a pinch-held object, such as the book shown in FIGURE P6.56. What is the heaviest book he can hold vertically before it slips out of his fingers? The coefficient of static friction between his fingers and the book cover is 0.80.

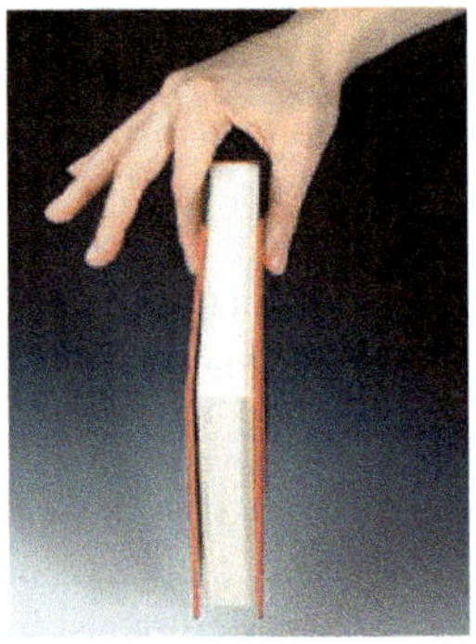
FIGURE P6.56

57. What is the terminal speed for an 80 kg skier going down a 40° snow-covered slope on wooden skis? Assume that the skier is 1.8 m tall and 0.40 m wide.

58. A ball is shot from a compressed-air gun at twice its terminal speed.
a. What is the ball's initial acceleration, as a multiple of g, if it is shot straight up?
b. What is the ball's initial acceleration, as a multiple of g, if it is shot straight down?

59. An artist friend of yours needs help hanging a 500 lb sculpture from the ceiling. For artistic reasons, she wants to use just two ropes. One will be 30° from vertical, the other 60°. She needs you to determine the smallest diameter rope that can safely support this expensive piece of art. On a visit to the hardware store you find that rope is sold in increments of $\frac{1}{8}$-inch diameter and that the safety rating is 4000 pounds per square inch of cross section. What diameter rope should you buy?

60. You've entered a "slow ski race" where the winner is the skier who takes the *longest* time to go down a 15° slope without ever stopping. You need to choose the best wax to apply to your skis. Red wax has a coefficient of kinetic friction 0.25, yellow is 0.20, green is 0.15, and blue is 0.10. Having just finished taking physics, you realize that a wax too slippery will cause you to accelerate down the slope and lose the race. But a wax that's too sticky will cause you to stop and be disqualified. You know that a strong headwind will apply a 50 N horizontal force against you as you ski, and you know that your mass is 82 kg. Which wax do you choose?

61. Astronauts in space "weigh" themselves by oscillating on a spring. Suppose the position of an oscillating 75 kg astronaut is given by $x = (0.30\text{ m})\sin((\pi\text{ rad/s})\cdot t)$, where t is in s. What force does the spring exert on the astronaut at (a) $t = 1.0$ s and (b) 1.5 s? Note that the angle of the sine function is in radians.

62. A particle of mass m moving along the x-axis experiences the net force $F_x = ct$, where c is a constant. The particle has velocity v_{0x} at $t = 0$. Find an algebraic expression for the particle's velocity v_x at a later time t.

63. || At $t = 0$, an object of mass m is at rest at $x = 0$ on a horizontal, frictionless surface. A horizontal force $F_x = F_0(1 - t/T)$, which decreases from F_0 at $t = 0$ to zero at $t = T$, is exerted on the object. Find an expression for the object's (a) velocity and (b) position at time T.

64. ||| At $t = 0$, an object of mass m is at rest at $x = 0$ on a horizontal, frictionless surface. Starting at $t = 0$, a horizontal force $F_x = F_0 e^{-t/T}$ is exerted on the object.
   a. Find and graph an expression for the object's velocity at an arbitrary lateer time t.
   b. What is the object's velocity after a very long time has elapsed?

65. ||| Large objects have inertia and tend to keep moving—Newton's first law. Life is very different for small microorganisms that swim through water. For them, drag forces are so large that they instantly stop, without coasting, if they cease their swimming motion. To swim at constant speed, they must exert a constant propulsion force by rotating corkscrew-like flagella or beating hair-like cilia. The quadratic model of drag of Equation 6.16 fails for very small particles. Instead, a small object moving in a liquid experiences a *linear* drag force, $\vec{D} = (bv$, direction opposite the motion), where b is a constant. For a sphere of radius R, the drag constant can be shown to be $b = 6\pi\eta R$, where η is the *viscosity* of the liquid. Water at 20°C has viscosity 1.0×10^{-3} N s/m^2.
BIO
   a. A *paramecium* is about 100 μm long. If it's modeled as a sphere, how much propulsion force must it exert to swim at a typical speed of 1.0 mm/s? How about the propulsion force of a 2.0-μm-diameter *E. coli* bacterium swimming at 30 μm/s?
   b. The propulsion forces are very small, but so are the organisms. To judge whether the propulsion force is large or small *relative to the organism*, compute the acceleration that the propulsion force could give each organism if there were no drag. The density of both organisms is the same as that of water, 1000 kg/m^3.

66. ||| Very small objects, such as dust particles, experience a *linear* drag force, $\vec{D} = (bv$, direction opposite the motion), where b is a constant. That is, the quadratic model of drag of Equation 6.16 fails for very small particles. For a sphere of radius R, the drag constant can be shown to be $b = 6\pi\eta R$, where η is the *viscosity* of the gas.
   a. Find an expression for the terminal speed v_{term} of a spherical particle of radius R and mass m falling through a gas of viscosity η.
   b. Suppose a gust of wind has carried a 50-μm-diameter dust particle to a height of 300 m. If the wind suddenly stops, how long will it take the dust particle to settle back to the ground? Dust has a density of 2700 kg/m^3, the viscosity of 25°C air is 2.0×10^{-5} N s/m^2, and you can assume that the falling dust particle reaches terminal speed almost instantly.

Problems 67 and 68 show a free-body diagram. For each:
   a. Write a realistic dynamics problem for which this is the correct free-body diagram. Your problem should ask a question that can be answered with a value of position or velocity (such as "How far?" or "How fast?"), and should give sufficient information to allow a solution.
   b. Solve your problem!

67.

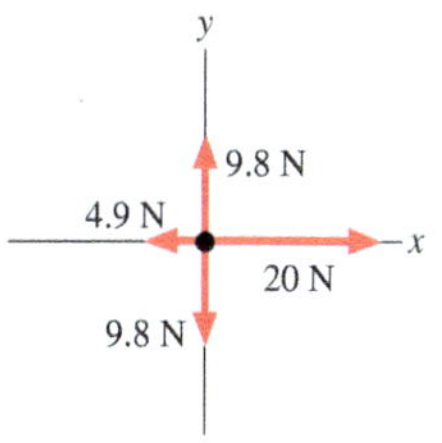

FIGURE P6.67

68.

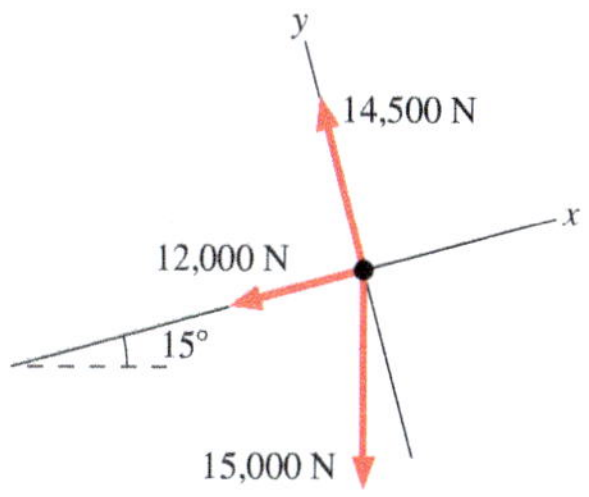

FIGURE P6.68

In Problems 69 through 71 you are given the dynamics equations that are used to solve a problem. For each of these, you are to
   a. Write a realistic problem for which these are the correct equations.
   b. Draw the free-body diagram and the pictorial representation for your problem.
   c. Finish the solution of the problem.

69. $-0.80n = (1500 \text{ kg})a_x$
$n - (1500 \text{ kg})(9.80 \text{ m/s}^2) = 0$

70. $T - 0.20n - (20 \text{ kg})(9.80 \text{ m/s}^2)\sin 20° = (20 \text{ kg})(2.0 \text{ m/s}^2)$
$n - (20 \text{ kg})(9.80 \text{ m/s}^2)\cos 20° = 0$

71. $(100 \text{ N})\cos 30° - f_k = (20 \text{ kg})a_x$
$n + (100 \text{ N})\sin 30° - (20 \text{ kg})(9.80 \text{ m/s}^2) = 0$
$f_k = 0.20n$

Challenge Problems

72. A block of mass m is at rest at the origin at $t = 0$. It is pushed with constant force F_0 from $x = 0$ to $x = L$ across a horizontal surface whose coefficient of kinetic friction is $\mu_k = \mu_0(1 - x/L)$. That is, the coefficient of friction decreases from μ_0 at $x = 0$ to zero at $x = L$.
   a. Use what you've learned in calculus to prove that
$$a_x = v_x \frac{dv_x}{dx}$$
   b. Find an expression for the block's speed as it reaches position L.

73. The machine in FIGURE CP6.73 has an 800 g steel shuttle that is pulled along a square steel rail by an elastic cord. The shuttle is released when the elastic cord has 20 N tension at a 45° angle. What is the initial acceleration of the shuttle?

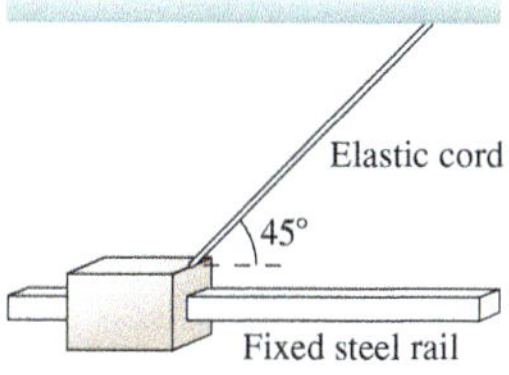

FIGURE CP6.73

74. FIGURE CP6.74 shows an *accelerometer*, a device for measuring the horizontal acceleration of cars and airplanes. A ball is free to roll on a parabolic track described by the equation $y = x^2$, where both x and y are in meters. A scale along the bottom is used to measure the ball's horizontal position x.

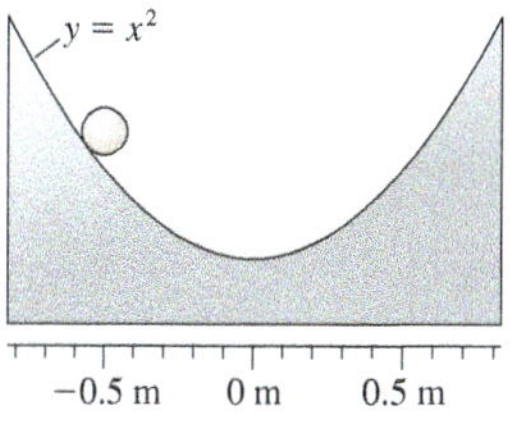

FIGURE CP6.74

   a. Find an expression that allows you to use a measured position x (in m) to compute the acceleration a_x (in m/s^2). (For example, $a_x = 3x$ is a possible expression.)
   b. What is the acceleration if $x = 20$ cm?

75. An object moving in a liquid experiences a *linear* drag force: $\vec{D} = (bv$, direction opposite the motion), where b is a constant called the *drag coefficient.* For a sphere of radius R, the drag constant can be computed as $b = 6\pi\eta R$, where η is the *viscosity* of the liquid.
   a. Find an algebraic expression for $v_x(t)$, the x-component of velocity as a function of time, for a spherical particle of radius R and mass m that is shot horizontally with initial speed v_0 through a liquid of viscosity η.
   b. Water at 20°C has viscosity $\eta = 1.0 \times 10^{-3}$ N s/m^2. Suppose a 4.0-cm-diameter, 33 g ball is shot horizontally into a tank of 20°C water. How long will it take for the horizontal speed to decrease to 50% of its initial value?

76. An object moving in a liquid experiences a *linear* drag force: $\vec{D} = (bv$, direction opposite the motion), where b is a constant called the *drag coefficient.* For a sphere of radius R, the drag constant can be computed as $b = 6\pi\eta R$, where η is the *viscosity* of the liquid.
   a. Use what you've learned in calculus to prove that

$$a_x = v_x \frac{dv_x}{dx}$$

   b. Find an algebraic expression for $v_x(x)$, the x-component of velocity as a function of distance traveled, for a spherical particle of radius R and mass m that is shot horizontally with initial speed v_0 through a liquid of viscosity η.
   c. Water at 20°C has viscosity $\eta = 1.0 \times 10^{-3}$ N s/m^2. Suppose a 1.0-cm-diameter, 1.0 g marble is shot horizontally into a tank of 20°C water at 10 cm/s. How far will it travel before stopping?

77. An object with cross section A is shot horizontally across frictionless ice. Its initial velocity is v_{0x} at $t_0 = 0$ s. Air resistance is not negligible.
   a. Show that the velocity at time t is given by the expression

$$v_x = \frac{v_{0x}}{1 + C\rho A v_{0x} t/2m}$$

   b. A 1.6-m-wide, 1.4-m-high, 1500 kg car with a drag coefficient of 0.35 hits a very slick patch of ice while going 20 m/s. If friction is neglected, how long will it take until the car's speed drops to 10 m/s? To 5 m/s?
   c. Assess whether or not it is reasonable to neglect kinetic friction.

STOP TO THINK ANSWERS

Stop to Think 6.1: a. The lander is descending and slowing. The acceleration vector points upward, and so $\vec{F}_{net}$ points upward. This can be true only if the thrust has a larger magnitude than the weight.

Stop to Think 6.2: a. You are descending and slowing, so your acceleration vector points upward and there is a net upward force on you. The floor pushes up against your feet harder than gravity pulls down.

Stop to Think 6.3: $f_b > f_c = f_d = f_e > f_a$. Situations c, d, and e are all kinetic friction, which does not depend on either velocity or acceleration. Kinetic friction is smaller than the maximum static friction that is exerted in b. $f_a = 0$ because no friction is needed to keep the object at rest.

Stop to Think 6.4: d. The ball is shot *down* at 30 m/s, so $v_{0y} = -30$ m/s. This exceeds the terminal speed, so the upward drag force is *larger* than the downward weight force. Thus the ball *slows down* even though it is "falling." It will slow until $v_y = -15$ m/s, the terminal velocity, then maintain that velocity.

7 Newton's Third Law

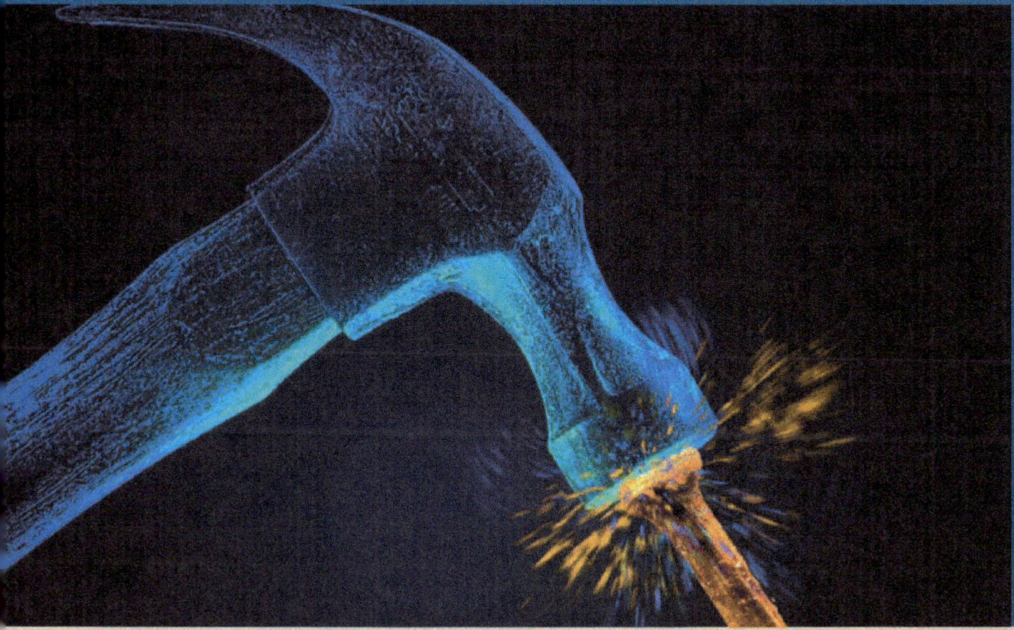

The hammer and nail are interacting. The forces of the hammer on the nail and the nail on the hammer are an action/reaction pair of forces.

▶ **Looking Ahead** The goal of Chapter 7 is to use Newton's third law to understand how objects interact.

Interactions

Newton's second law treats an object as an isolated entity acted upon by external forces.

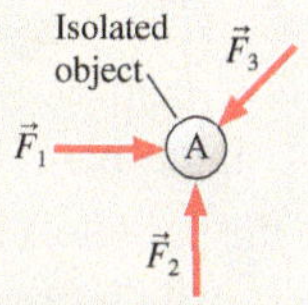

For example, we've looked at the normal force of a table on a book. But what about the book's effect on the table?

Whenever two or more objects exert forces on each other, by touching, being tied together, or via long-range forces, we say that they *interact*.

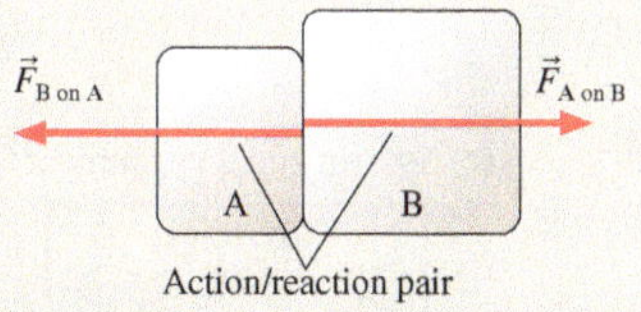

You'll learn that if object A exerts a force on object B, then object B exerts a force on object A. These two forces form what is called an **action/reaction pair.**

◀ **Looking Back**
Sections 5.1–5.3 Basic concepts of force and the atomic model of tension

Newton's Third Law

Interactions are described by **Newton's third law:**

- *Every* force occurs as one member of an action/reaction pair.
- The two members of a pair act on two *different* objects.
- The two members of a pair are equal in magnitude but opposite in direction.

Thrust and propulsion are two important applications of Newton's third law.

Interaction Diagrams

You'll learn how to draw **interaction diagrams** to show the action/reaction pairs of forces between interacting objects.

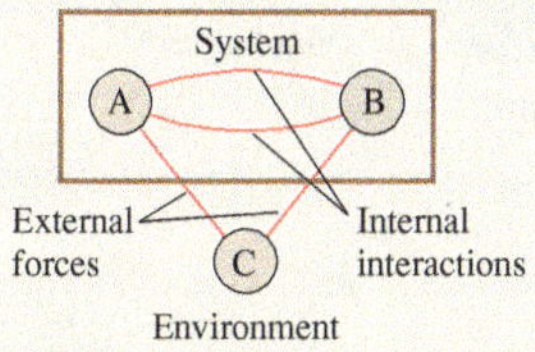

The **system** consists of those objects whose motion we wish to analyze. Objects that exert forces but whose motion is not of interest—such as the earth—form the **environment**.

Ropes and Pulleys

A common way that two objects interact is via ropes or cables or strings. Pulleys can be used to change the direction of the tension force.

You'll learn that:

- Objects that are connected together must have the same acceleration.
- *Tension is constant* throughout a rope or string if we can model it as being massless and pulleys as frictionless.

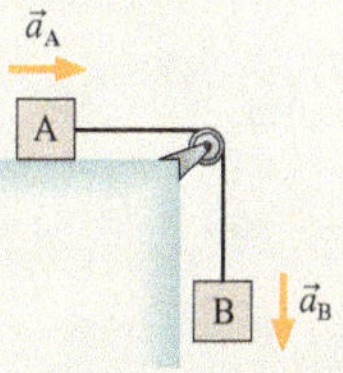

Problem-Solving Strategy

We will expand the problem-solving strategy that we began in Chapter 6.

- Draw an interaction diagram.
- Identify the system.
- Draw a separate free-body diagram for each object in the system.
- Write Newton's second law for each object.
- Use Newton's third law to relate action/reaction pairs of forces.

◀ **Looking Back**
Sections 6.1–6.2 Problem-solving strategies for force and motion

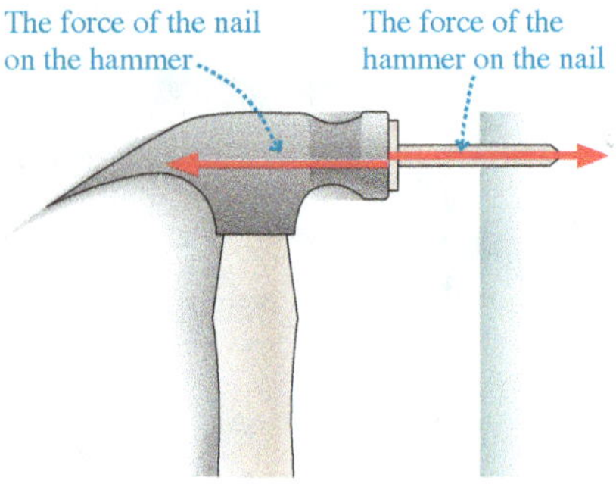

FIGURE 7.1 The hammer and nail are interacting with each other.

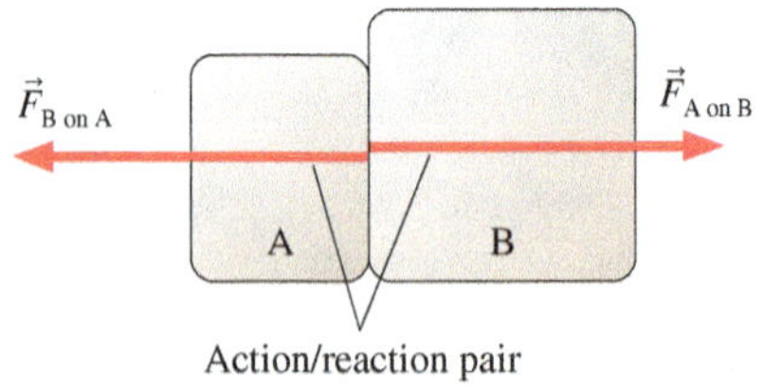

FIGURE 7.2 An action/reaction pair of forces.

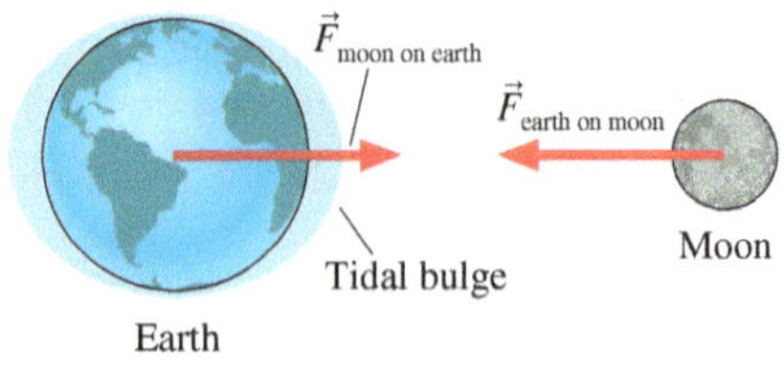

FIGURE 7.3 The ocean tides are an indication of the long-range gravitational interaction of the earth and the moon.

7.1 Interacting Objects

FIGURE 7.1 shows a hammer hitting a nail. The hammer exerts a force on the nail as it drives the nail forward. At the same time, the nail exerts a force on the hammer. If you're not sure that it does, imagine hitting the nail with a glass hammer. It's the force of the nail on the hammer that would cause the glass to shatter.

In fact, any time an object A pushes or pulls on another object B, B pushes or pulls back on A. When you pull someone with a rope in a tug-of-war, that person pulls back on you. Your chair pushes up on you (the normal force) as you push down on the chair. These are examples of an **interaction,** the mutual influence of two objects on each other.

To be more specific, if object A exerts a force $\vec{F}_{\text{A on B}}$ on object B, then object B exerts a force $\vec{F}_{\text{B on A}}$ on object A. This pair of forces, shown in FIGURE 7.2, is called an **action/reaction pair.** Two objects interact by exerting an action/reaction pair of forces on each other. Notice the very explicit subscripts on the force vectors. The first letter is the *agent,* the second letter is the object on which the force acts. $\vec{F}_{\text{A on B}}$ is a force exerted *by* A *on* B.

NOTE ▶ The name "action/reaction pair" is somewhat misleading. The forces occur simultaneously, and we cannot say which is the "action" and which the "reaction." **An action/reaction pair of forces exists as a pair, or not at all.** In identifying action/reaction pairs, the labels are the key. Force $\vec{F}_{\text{A on B}}$ is paired with force $\vec{F}_{\text{B on A}}$. ◀

The hammer and nail interact through contact forces. The same idea holds true for long-range forces such as gravity. If you release a ball, it falls because the earth's gravity exerts a downward force $\vec{F}_{\text{earth on ball}}$. But does the ball really pull upward on the earth with a force $\vec{F}_{\text{ball on earth}}$?

Newton was the first to realize that, indeed, the ball *does* pull upward on the earth. His evidence was the tides. Astronomers had known since antiquity that the tides depend on the phase of the moon, but Newton was the first to understand that tides are the ocean's response to the gravitational pull of the moon on the earth. As FIGURE 7.3 shows, the flexible water bulges toward the moon while the relatively inflexible crust remains behind.

Objects, Systems, and the Environment

Chapters 5 and 6 considered forces acting on a single object that we modeled as a particle. FIGURE 7.4a shows a diagrammatic representation of single-particle dynamics. We can use Newton's second law, $\vec{a} = \vec{F}_{\text{net}}/m$, to determine the particle's acceleration.

We now want to extend the particle model to situations in which two or more objects, each represented as a particle, interact with each other. For example, FIGURE 7.4b shows three objects interacting via action/reaction pairs of forces. The forces can be given labels such as $\vec{F}_{\text{A on B}}$ and $\vec{F}_{\text{B on A}}$. How do these particles move?

We will often be interested in the motion of some of the objects, say objects A and B, but not of others. For example, objects A and B might be the hammer and the nail, while object C is the earth. The earth interacts with both the hammer and the nail via

FIGURE 7.4 Single-particle dynamics and a model of interacting objects.

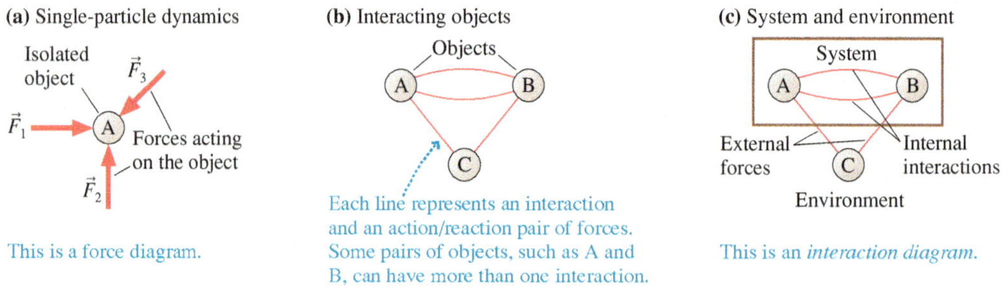

gravity, but in a practical sense the earth remains "at rest" while the hammer and nail move. Let's define the **system** as those objects whose motion we want to analyze and the **environment** as objects external to the system.

FIGURE 7.4c is a new kind of diagram, an **interaction diagram,** in which we've enclosed the objects of the system in a box and represented interactions as lines connecting objects. This is a rather abstract, schematic diagram, but it captures the essence of the interactions. Notice that interactions with objects in the environment are called **external forces.** For the hammer and nail, the gravitational force on each—an interaction with the earth—is an external force.

The bat and the ball are interacting with each other.

NOTE ▶ The system–environment distinction is a practical matter, not a fundamental distinction. If object A pushes or pulls on object B, then B pushes or pulls on A. *Every* force is one member of an action/reaction pair, and there is no such thing as a true "external force." What we call an external force is an interaction between an object of interest, one we've chosen to place inside the system, and an object whose motion is not of interest. ◀

7.2 Analyzing Interacting Objects

TACTICS BOX 7.1 Analyzing interacting objects

MP

❶ **Represent each object as a circle.** Place each in the correct position relative to other objects.

- Give each a name and a label.
- The surface of the earth (contact forces) and the entire earth (long-range forces) should be considered separate objects. Label the **e**ntire **e**arth EE.
- Ropes and pulleys often need to be considered objects.

❷ **Identify interactions.** Draw connecting lines between the circles to represent interactions.

- Draw *one* line for each interaction. Label it with the type of force.
- Every interaction line connects two and only two objects.
- There can be at most two interactions at a surface: a force parallel to the surface (e.g., friction) and a force perpendicular to the surface (e.g., a normal force).
- The entire earth interacts only by the long-range gravitational force.

❸ **Identify the system.** Identify the objects of interest; draw and label a box enclosing them. This completes the interaction diagram.

❹ **Draw a free-body diagram for each object in the system.** Include only the forces acting *on* each object, not forces exerted by the object.

- Every interaction line crossing the system boundary is one external force acting on the object. The usual force symbols, such as $\vec{n}$ and $\vec{T}$, can be used.
- Every interaction line within the system represents an action/reaction pair of forces. There is one force vector on *each* of the objects, and these forces always point in opposite directions. Use labels like $\vec{F}_{\text{A on B}}$ and $\vec{F}_{\text{B on A}}$.
- Connect the two action/reaction forces—which must be on *different* free-body diagrams—with a dashed line.

Exercises 1–7

We'll illustrate these ideas with two concrete examples. The first example will be much longer than usual because we'll go carefully through all the steps in the reasoning.

EXAMPLE 7.1 Pushing a crate

FIGURE 7.5 shows a person pushing a large crate across a rough surface. Identify all interactions, show them on an interaction diagram, then draw free-body diagrams of the person and the crate.

FIGURE 7.5 A person pushes a crate across a rough floor.

VISUALIZE The interaction diagram of FIGURE 7.6 starts by representing every object as a circle in the correct position but separated from all other objects. The person and the crate are obvious objects. The earth is also an object that both exerts and experiences forces, but it's necessary to distinguish between the surface, which exerts contact forces, and the entire earth, which exerts the long-range gravitational force.

FIGURE 7.6 The interaction diagram.

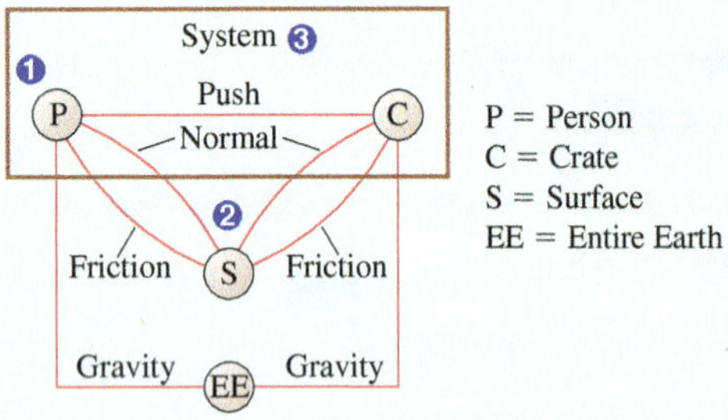

Figure 7.6 also identifies the various interactions. Some, like the pushing interaction between the person and the crate, are fairly obvious. The interactions with the earth are a little trickier. Gravity, a long-range force, is an interaction between each object and the earth as a whole. Friction forces and normal forces are contact interactions between each object and the earth's surface. These are two different interactions, so two interaction lines connect the crate to the surface and the person to the surface. Altogether, there are seven interactions. Finally, we've enclosed the person and crate in a box labeled System. These are the objects whose motion we wish to analyze.

NOTE ▶ Interactions are between two *different* objects. None of the interactions are between an object and itself. ◀

We can now draw free-body diagrams for the objects in the system, the crate and the person. FIGURE 7.7 correctly locates the crate's free-body diagram to the right of the person's free-body diagram. For each, three interaction lines cross the system boundary and thus represent external forces. These are the gravitational force from the entire earth, the upward normal force from the surface, and a friction force from the surface. We can use familiar labels such as $\vec{n}_P$ and $\vec{f}_C$, but **it's very important to distinguish different forces with subscripts.** There's now more than one normal force. If you call both simply $\vec{n}$, you're almost certain to make mistakes when you start writing out the second-law equations.

FIGURE 7.7 Free-body diagrams of the person and the crate.

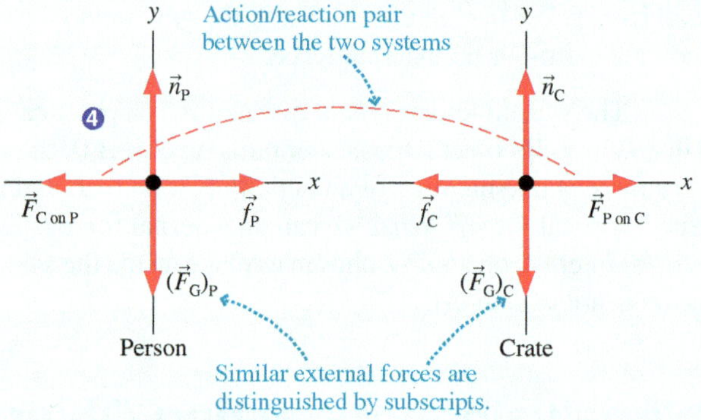

The directions of the normal forces and the gravitational forces are clear, but we have to be careful with friction. Friction force $\vec{f}_C$ is kinetic friction of the crate sliding across the surface, so it points left, opposite the motion. But what about friction between the person and the surface? It is tempting to draw force $\vec{f}_P$ pointing to the left. After all, friction forces are supposed to be in the direction opposite the motion. But if we did so, the person would have two forces to the left, $\vec{F}_{C\text{ on }P}$ and $\vec{f}_P$, and none to the right, causing the person to accelerate *backward!* That is clearly not what happens, so what is wrong?

Imagine pushing a crate to the right across loose sand. Each time you take a step, you tend to kick the sand to the *left,* behind you. Thus friction force $\vec{f}_{P\text{ on }S}$, the force of the person pushing against the earth's surface, is to the *left.* In reaction, the force of the earth's surface against the person is a friction force to the *right.* It is force $\vec{f}_{S\text{ on }P}$, which we've shortened to $\vec{f}_P$, that causes the person to accelerate in the forward direction. Further, as we'll discuss more below, this is a *static* friction force; your foot is planted on the ground, not sliding across the surface.

Finally, we have one internal interaction. The crate is pushed with force $\vec{F}_{P\text{ on }C}$. If A pushes or pulls on B, then B pushes or pulls back on A. The reaction to force $\vec{F}_{P\text{ on }C}$ is $\vec{F}_{C\text{ on }P}$, the crate pushing back against the person's hands. Force $\vec{F}_{P\text{ on }C}$ is a force exerted on the crate, so it's shown on the crate's free-body diagram. Force $\vec{F}_{C\text{ on }P}$ is exerted on the person, so it is drawn on the person's free-body diagram. **The two forces of an action/ reaction pair never occur on the same object.** Notice that forces $\vec{F}_{P\text{ on }C}$ and $\vec{F}_{C\text{ on }P}$ are pointing in opposite directions. We've connected them with a dashed line to show that they are an action/ reaction pair.

ASSESS The completed free-body diagrams of Figure 7.7 could now be the basis for a quantitative analysis.

Propulsion

The friction force $\vec{f}_P$ (force of surface on person) is an example of **propulsion.** It is the force that a system with an internal source of energy uses to drive itself forward. Propulsion is an important feature not only of walking or running but also of the forward motion of cars, jets, and rockets. Propulsion is somewhat counterintuitive, so it is worth a closer look.

If you try to walk across a frictionless floor, your foot slips and slides *backward.* In order for you to walk, the floor needs to have friction so that your foot *sticks* to the floor as you straighten your leg, moving your body forward. The friction that prevents slipping is *static* friction. Static friction, you will recall, acts in the direction that prevents slipping. The static friction force $\vec{f}_P$ has to point in the *forward* direction to prevent your foot from slipping backward. It is this forward-directed static friction force that propels you forward! The force of your foot on the floor, the other half of the action/reaction pair, is in the opposite direction.

The distinction between you and the crate is that you have an *internal source of energy* that allows you to straighten your leg by pushing backward against the surface. In essence, you walk by pushing the earth away from you. The earth's surface responds by pushing you forward. These are static friction forces. In contrast, all the crate can do is slide, so *kinetic* friction opposes the motion of the crate.

What force causes this sprinter to accelerate?

FIGURE 7.8 shows how propulsion works. A car uses its motor to spin the tires, causing the tires to push backward against the ground. This is why dirt and gravel are kicked backward, not forward. The earth's surface responds by pushing the car forward. These are also *static* friction forces. The tire is rolling, but the bottom of the tire, where it contacts the road, is instantaneously at rest. If it weren't, you would leave one giant skid mark as you drove and would burn off the tread within a few miles.

FIGURE 7.8 Examples of propulsion.

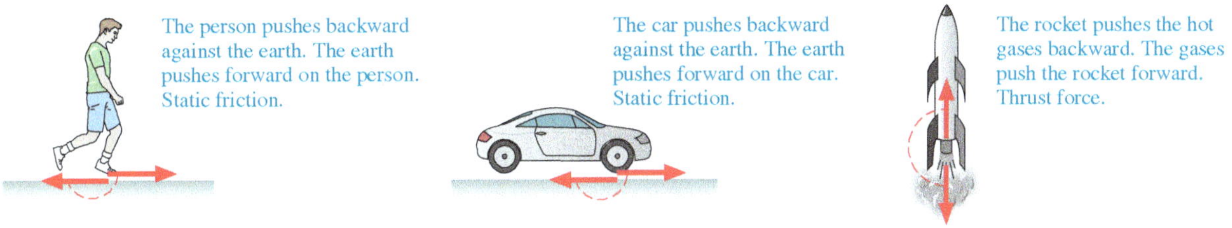

EXAMPLE 7.2 Towing a car

A tow truck uses a rope to pull a car along a horizontal road, as shown in FIGURE 7.9. Identify all interactions, show them on an interaction diagram, then draw free-body diagrams of each object in the system.

FIGURE 7.9 A truck towing a car.

VISUALIZE The interaction diagram of FIGURE 7.10 represents the objects as separate circles, but with the correct relative positions. The rope is shown as a separate object. Many of the interactions are identical to those in Example 7.1. The system—the objects in motion—consists of the truck, the rope, and the car.

FIGURE 7.10 The interaction diagram.

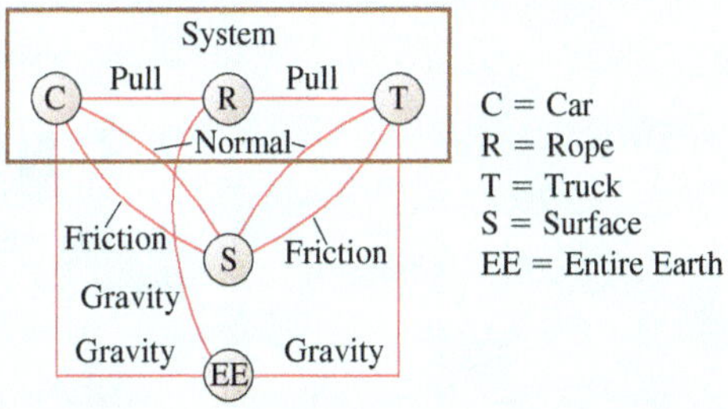

The three objects in the system require three free-body diagrams, shown in FIGURE 7.11 on the next page. Gravity, friction, and normal forces at the surface are all interactions that cross the system boundary and are shown as external forces. The car is an inert object rolling along. It would slow and stop if the rope were cut, so the surface must exert a rolling friction force $\vec{f}_C$ to the left. The truck, however, has an internal source of energy. The truck's drive wheels

Continued

FIGURE 7.11 Free-body diagrams of Example 7.2.

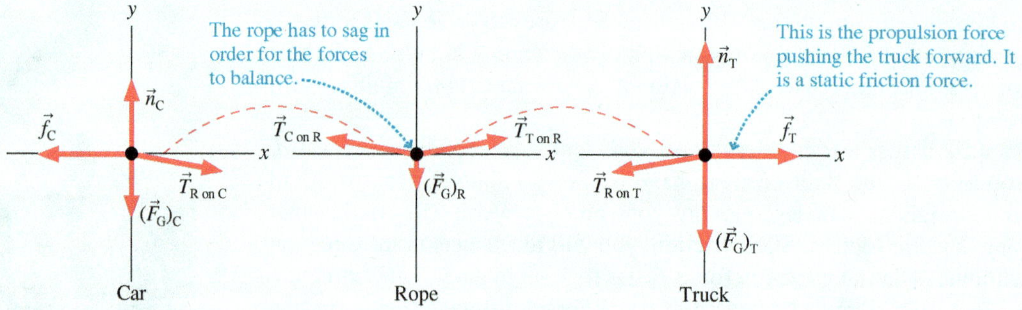

push the ground to the left with force $\vec{f}_{\text{T on S}}$. In reaction, the ground propels the truck forward, to the right, with force $\vec{f}_{\text{T}}$.

We next need to identify the horizontal forces between the car, the truck, and the rope. The rope pulls on the car with a tension force $\vec{T}_{\text{R on C}}$. You might be tempted to put the reaction force on the truck because we say that "the truck pulls the car," but the truck is not in contact with the car. The truck pulls on the rope, then the rope pulls on the car. Thus the reaction to $\vec{T}_{\text{R on C}}$ is a force on the *rope:* $\vec{T}_{\text{C on R}}$. These are an action/reaction pair. At the other end, $\vec{T}_{\text{T on R}}$ and $\vec{T}_{\text{R on T}}$ are also an action/reaction pair.

NOTE ▶ Drawing an interaction diagram helps you avoid mistakes because it shows very clearly what is interacting with what. ◀

Notice that the tension forces of the rope *cannot* be horizontal. If they were, the rope's free-body diagram would show a net downward force, because of its weight, and the rope would accelerate downward. The tension forces $\vec{T}_{\text{T on R}}$ and $\vec{T}_{\text{C on R}}$ have to angle slightly upward to balance the gravitational force, so any real rope has to sag at least a little in the center.

ASSESS Make sure you avoid the common error of considering $\vec{n}$ and $\vec{F}_{\text{G}}$ to be an action/reaction pair. These are both forces on the *same* object, whereas the two forces of an action/reaction pair are always on two *different* objects that are interacting with each other. The normal and gravitational forces are often equal in magnitude, as they are in this example, but that doesn't make them an action/reaction pair of forces.

STOP TO THINK 7.1 A rope of negligible mass pulls a crate across the floor. The rope and crate are the system; the hand is part of the environment. What, if anything, is wrong with the free-body diagrams?

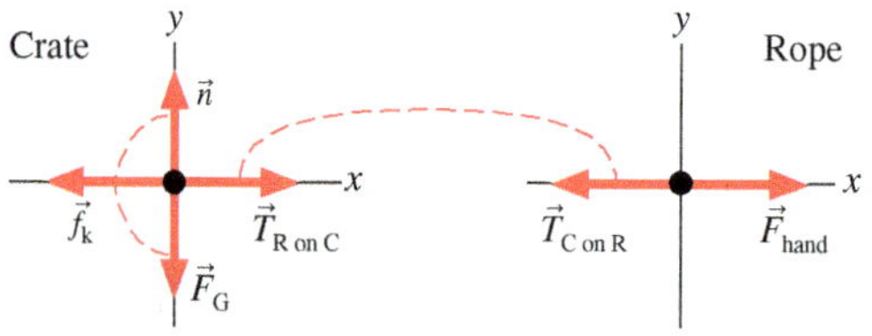

7.3 Newton's Third Law

Newton was the first to recognize how the two members of an action/reaction pair of forces are related to each other. Today we know this as Newton's third law:

Newton's third law Every force occurs as one member of an action/reaction pair of forces.

- The two members of an action/reaction pair act on two *different* objects.
- The two members of an action/reaction pair are equal in magnitude but opposite in direction: $\vec{F}_{\text{A on B}} = -\vec{F}_{\text{B on A}}$.

We deduced most of the third law in Section 7.2. There we found that the two members of an action/reaction pair are always opposite in direction (see Figures 7.7 and 7.11). According to the third law, this will always be true. But the most significant portion of the third law, which is by no means obvious, is that the two members of an action/reaction pair have *equal* magnitudes. That is, $F_{\text{A on B}} = F_{\text{B on A}}$. This is the quantitative relationship that will allow you to solve problems of interacting objects.

Newton's third law is frequently stated as "For every action there is an equal but opposite reaction." While this is indeed a catchy phrase, it lacks the preciseness of our preferred version. In particular, it fails to capture an essential feature of action/reaction pairs—that they each act on a *different* object.

NOTE ▶ Newton's third law extends and completes our concept of *force.* We can now recognize force as an *interaction* between objects rather than as some "thing" with an independent existence of its own. The concept of an interaction will become increasingly important as we begin to study the laws of momentum and energy. ◀

Reasoning with Newton's Third Law

Newton's third law is easy to state but harder to grasp. For example, consider what happens when you release a ball. Not surprisingly, it falls down. But if the ball and the earth exert equal and opposite forces on each other, as Newton's third law alleges, why doesn't the earth "fall up" to meet the ball?

The key to understanding this and many similar puzzles is that **the forces are equal but the accelerations are not.** Equal causes can produce very unequal effects. FIGURE 7.12 shows equal-magnitude forces on the ball and the earth. The force on ball B is simply the gravitational force of Chapter 6:

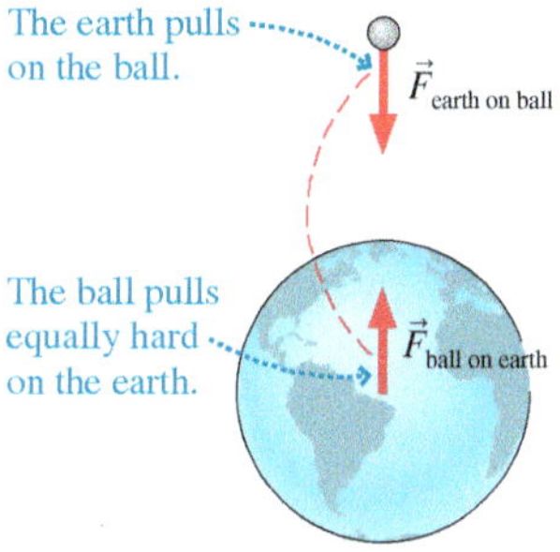

FIGURE 7.12 The action/reaction forces of a ball and the earth are equal in magnitude.

$$\vec{F}_{\text{earth on ball}} = (\vec{F}_G)_B = -m_B g\hat{j} \tag{7.1}$$

where m_B is the mass of the ball. According to Newton's second law, this force gives the ball an acceleration

$$\vec{a}_B = \frac{(\vec{F}_G)_B}{m_B} = -g\hat{j} \tag{7.2}$$

This is just the familiar free-fall acceleration.

According to Newton's third law, the ball pulls up on the earth with force $\vec{F}_{\text{ball on earth}}$. Because $\vec{F}_{\text{earth on ball}}$ and $\vec{F}_{\text{ball on earth}}$ are an action/reaction pair, $\vec{F}_{\text{ball on earth}}$ must be equal in magnitude and opposite in direction to $\vec{F}_{\text{earth on ball}}$. That is,

$$\vec{F}_{\text{ball on earth}} = -\vec{F}_{\text{earth on ball}} = -(\vec{F}_G)_B = +m_B g\hat{j} \tag{7.3}$$

Using this result in Newton's second law, we find the upward acceleration of the earth as a whole is

$$\vec{a}_E = \frac{\vec{F}_{\text{ball on earth}}}{m_E} = \frac{m_B g\hat{j}}{m_E} = \left(\frac{m_B}{m_E}\right) g\hat{j} \tag{7.4}$$

The upward acceleration of the earth is less than the downward acceleration of the ball by the factor m_B/m_E. If we assume a 1 kg ball, we can estimate the magnitude of $\vec{a}_E$:

$$a_E = \frac{1\ \text{kg}}{6 \times 10^{24}\ \text{kg}} g \approx 2 \times 10^{-24}\ \text{m/s}^2$$

With this incredibly small acceleration, it would take the earth 8×10^{15} years, approximately 500,000 times the age of the universe, to reach a speed of 1 mph! So we certainly would not expect to see or feel the earth "fall up" after we drop a ball.

NOTE ▶ Newton's third law equates the size of two forces, not two accelerations. The acceleration continues to depend on the mass, as Newton's second law states. **In an interaction between two objects of different mass, the lighter mass will do essentially all of the accelerating even though the forces exerted on the two objects are equal.** ◀

EXAMPLE 7.3 The forces on accelerating boxes

The hand shown in FIGURE 7.13 pushes boxes A and B to the right across a frictionless table. The mass of B is larger than the mass of A.

a. Draw free-body diagrams of A, B, and the hand H, showing only the *horizontal* forces. Connect action/reaction pairs with dashed lines.
b. Rank in order, from largest to smallest, the horizontal forces shown on your free-body diagrams.

FIGURE 7.13 Hand H pushes boxes A and B.

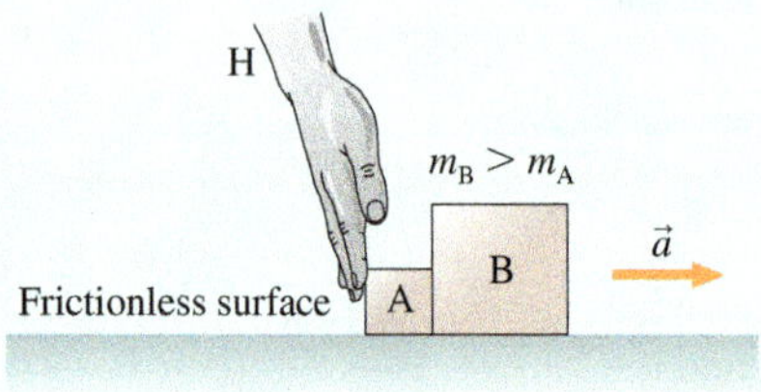

VISUALIZE a. The hand H pushes on box A, and A pushes back on H. Thus $\vec{F}_{H\text{ on }A}$ and $\vec{F}_{A\text{ on }H}$ are an action/reaction pair. Similarly, A pushes on B and B pushes back on A. **The hand H does not touch box B, so there is no interaction between them.** There is no friction. FIGURE 7.14 shows the four horizontal forces and identifies two action/reaction pairs. Notice that each force is shown on the free-body diagram of the object that it acts *on*.

b. According to Newton's third law, $F_{A\text{ on }H} = F_{H\text{ on }A}$ and $F_{A\text{ on }B} = F_{B\text{ on }A}$. But the third law is not our only tool. The boxes are *accelerating* to the right, because there's no friction, so Newton's *second* law tells us that box A must have a net force to the right. Consequently, $F_{H\text{ on }A} > F_{B\text{ on }A}$. Thus

$$F_{A\text{ on }H} = F_{H\text{ on }A} > F_{B\text{ on }A} = F_{A\text{ on }B}$$

ASSESS You might have expected $F_{A\text{ on }B}$ to be larger than $F_{H\text{ on }A}$ because $m_B > m_A$. It's true that the *net* force on B is larger than the *net* force on A, but we have to reason more closely to judge the individual forces. Notice how we used both the second and the third laws to answer this question.

FIGURE 7.14 The free-body diagrams, showing only the horizontal forces.

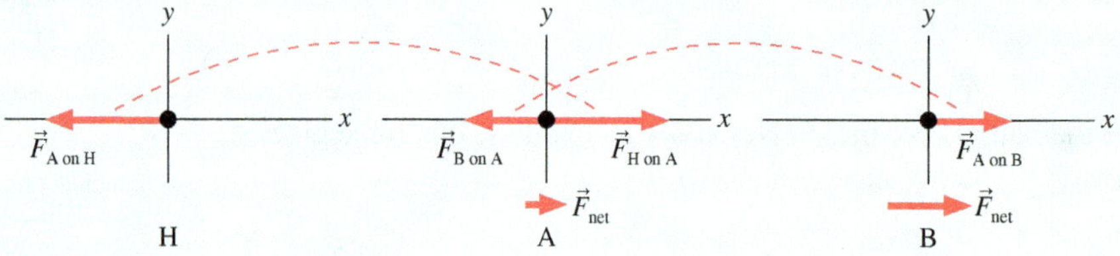

STOP TO THINK 7.2 Car B is stopped for a red light. Car A, which has the same mass as car B, doesn't see the red light and runs into the back of B. Which of the following statements is true?

a. B exerts a force on A, but A doesn't exert a force on B.
b. B exerts a larger force on A than A exerts on B.
c. B exerts the same amount of force on A as A exerts on B.
d. A exerts a larger force on B than B exerts on A.
e. A exerts a force on B, but B doesn't exert a force on A.

Acceleration Constraints

Newton's third law is one quantitative relationship you can use to solve problems of interacting objects. In addition, we frequently have other information about the motion in a problem. For example, if two objects A and B move together, their accelerations are *constrained* to be equal: $\vec{a}_A = \vec{a}_B$. A well-defined relationship between the accelerations of two or more objects is called an **acceleration constraint.** It is an independent piece of information that can help solve a problem.

FIGURE 7.15 The car and the truck have the same acceleration.

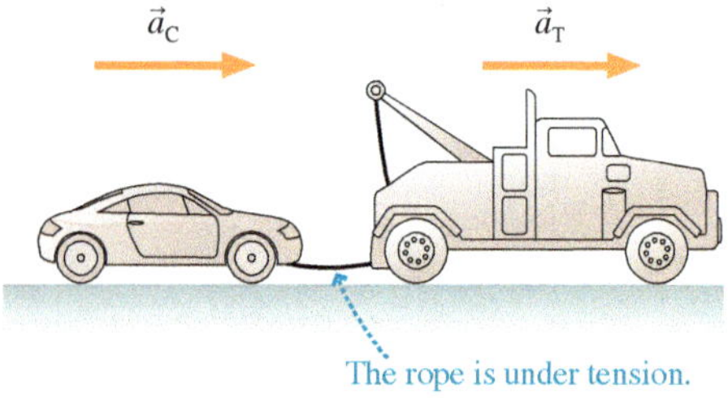

In practice, we'll express acceleration constraints in terms of the x- and y-components of $\vec{a}$. Consider the car being towed in FIGURE 7.15. This is one-dimensional motion, so we can write the acceleration constraint as

$$a_{Cx} = a_{Tx} = a_x$$

Because the accelerations of both objects are equal, we can drop the subscripts C and T and call both of them a_x.

Don't assume the accelerations of A and B will always have the same sign. Consider blocks A and B in FIGURE 7.16. The blocks are connected by a string, so they are constrained to move together and their accelerations have equal magnitudes. But A has a positive acceleration (to the right) in the x-direction while B has a negative acceleration (downward) in the y-direction. Thus the acceleration constraint is

$$a_{Ax} = -a_{By}$$

This relationship does *not* say that a_{Ax} is a negative number. It is simply a relational statement, saying that a_{Ax} is (-1) times whatever a_{By} happens to be. The acceleration a_{By} in Figure 7.16 is a negative number, so a_{Ax} is positive. In some problems, the signs of a_{Ax} and a_{By} may not be known until the problem is solved, but the *relationship* is known from the beginning.

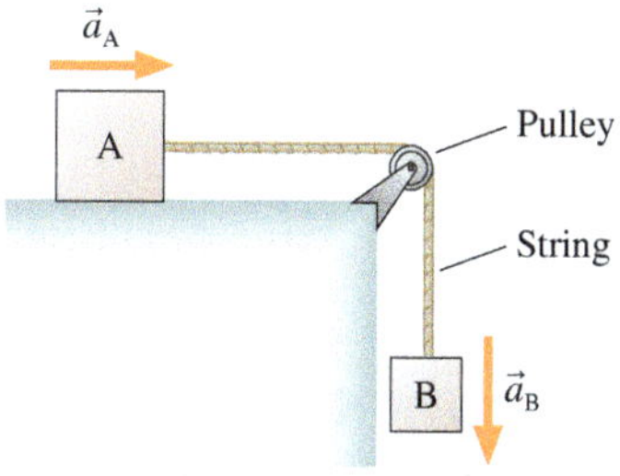

FIGURE 7.16 The string constrains the two objects to accelerate together.

A Revised Strategy for Interacting-Objects Problems

Problems of interacting objects can be solved with a few modifications to the basic problem-solving strategy we developed in Chapter 6. A revised problem-solving strategy follows.

PROBLEM-SOLVING STRATEGY 7.1 Interacting-objects problems

MP

MODEL Identify which objects are part of the system and which are part of the environment. Make simplifying assumptions.

VISUALIZE Draw a pictorial representation.

- Show important points in the motion with a sketch. You may want to give each object a separate coordinate system. Define symbols and identify what the problem is trying to find.
- Identify acceleration constraints.
- Draw an interaction diagram to identify the forces on each object and all action/reaction pairs.
- Draw a *separate* free-body diagram for each object. Each shows only the forces acting *on* that object, not forces exerted by the object.
- Connect the force vectors of action/reaction pairs with dashed lines. Use subscript labels to distinguish forces that act independently on more than one object.

SOLVE Use Newton's second and third laws.

- Write the equations of Newton's second law for *each* object, using the force information from the free-body diagrams.
- Equate the magnitudes of action/reaction pairs.
- Include the acceleration constraints, the friction model, and other quantitative information relevant to the problem.
- Solve for the acceleration, then use kinematics to find velocities and positions.

ASSESS Check that your result has the correct units, is reasonable, and answers the question.

You might be puzzled that the Solve step calls for the use of the third law to equate just the *magnitudes* of action/reaction forces. What about the "opposite in direction" part of the third law? You have already used it! Your free-body diagrams should show the two members of an action/reaction pair to be opposite in direction, and that information will have been utilized in writing the second-law equations. Because the directional information has already been used, all that is left is the magnitude information.

NOTE ▶ Two steps are especially important when drawing the free-body diagrams. First, draw a *separate* diagram for each object. The diagrams need not have the same coordinate system. Second, show only the forces acting *on* that object. The force $\vec{F}_{\text{A on B}}$ goes on the free-body diagram of object B, but $\vec{F}_{\text{B on A}}$ goes on the diagram of object A. The two members of an action/reaction pair *always* appear on two different free-body diagrams—*never* on the same diagram. ◀

EXAMPLE 7.4 Keep the crate from sliding

You and a friend have just loaded a 200 kg crate filled with priceless art objects into the back of a 2000 kg truck. As you press down on the accelerator, force $\vec{F}_{\text{surface on truck}}$ propels the truck forward. To keep things simple, call this just $\vec{F}_T$. What is the maximum magnitude $\vec{F}_T$ can have without the crate sliding? The static and kinetic coefficients of friction between the crate and the bed of the truck are 0.80 and 0.30. Rolling friction of the truck is negligible.

MODEL The crate and the truck are separate objects that form the system. We'll model them as particles. The earth and the road surface are part of the environment.

VISUALIZE The sketch in FIGURE 7.17 establishes a coordinate system, lists the known information, and—new to problems of interacting objects—identifies the acceleration constraint. As long as the crate doesn't slip, it must accelerate *with* the truck. Both accelerations are in the positive x-direction, so the acceleration constraint in this problem is

$$a_{Cx} = a_{Tx} = a_x$$

The interaction diagram of Figure 7.17 shows the crate interacting twice with the truck—a friction force parallel to the surface of the truck bed and a normal force perpendicular to this surface. The truck interacts similarly with the road surface, but notice that the crate does not interact with the ground; there's no contact between them. The two interactions within the system are each an action/reaction pair, so this is a total of four forces. You can also see four external forces crossing the system boundary, so the free-body diagrams should show a total of eight forces.

Finally, the interaction information is transferred to the free-body diagrams, where we see friction between the crate and truck as an action/reaction pair and the normal forces (the truck pushes up on the crate, the crate pushes down on the truck) as another action/reaction pair. It's easy to overlook forces such as $\vec{f}_{\text{C on T}}$, but you won't make this mistake if you first identify action/reaction pairs on an interaction diagram. Note that $\vec{f}_{\text{C on T}}$ and $\vec{f}_{\text{T on C}}$ are *static* friction forces because they are forces that prevent slipping; force $\vec{f}_{\text{T on C}}$ must point forward to prevent the crate from sliding out the back of the truck.

SOLVE Now we're ready to write Newton's second law. For the crate:

$$\sum (F_{\text{on crate}})_x = f_{\text{T on C}} = m_C a_{Cx} = m_C a_x$$

$$\sum (F_{\text{on crate}})_y = n_{\text{T on C}} - (F_G)_C = n_{\text{T on C}} - m_C g = 0$$

For the truck:

$$\sum (F_{\text{on truck}})_x = F_T - f_{\text{C on T}} = m_T a_{Tx} = m_T a_x$$

$$\sum (F_{\text{on truck}})_y = n_T - (F_G)_T - n_{\text{C on T}}$$

$$= n_T - m_T g - n_{\text{C on T}} = 0$$

Be sure you agree with all the signs, which are based on the free-body diagrams. The net force in the y-direction is zero because there's no motion in the y-direction. It may seem like a lot of effort to write all the subscripts, but it is very important in problems with more than one object.

Notice that we've already used the acceleration constraint $a_{Cx} = a_{Tx} = a_x$. Another important piece of information is Newton's third law, which tells us that $f_{\text{C on T}} = f_{\text{T on C}}$ and $n_{\text{C on T}} = n_{\text{T on C}}$. Finally, we know that the maximum value of F_T will occur when the static friction on the crate reaches its maximum value:

$$f_{\text{T on C}} = f_{\text{s max}} = \mu_s n_{\text{T on C}}$$

FIGURE 7.17 Pictorial representation of the crate and truck in Example 7.4.

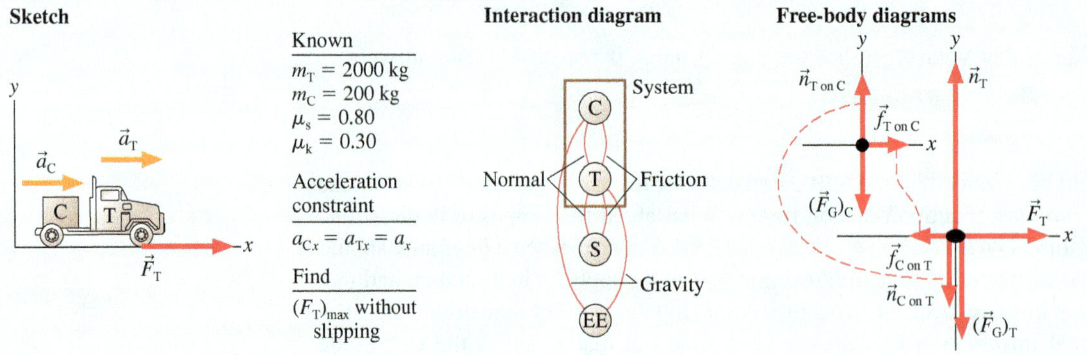

The friction depends on the normal force on the crate, not the normal force on the truck.

Now we can assemble all the pieces. From the y-equation of the crate, $n_{\text{T on C}} = m_C g$. Thus

$$f_{\text{T on C}} = \mu_s n_{\text{T on C}} = \mu_s m_C g$$

Using this in the x-equation of the crate, we find that the acceleration is

$$a_x = \frac{f_{\text{T on C}}}{m_C} = \mu_s g$$

This is the crate's maximum acceleration without slipping. Now use this acceleration *and* the fact that $f_{\text{C on T}} = f_{\text{T on C}} = \mu_s m_C g$ in the x-equation of the truck to find

$$F_T - f_{\text{C on T}} = F_T - \mu_s m_C g = m_T a_x = m_T \mu_s g$$

Solving for F_T, we find the maximum propulsion without the crate sliding is

$$(F_T)_{\max} = \mu_s (m_T + m_C) g$$
$$= (0.80)(2200\ \text{kg})(9.80\ \text{m/s}^2) = 17{,}000\ \text{N}$$

ASSESS This is a hard result to assess. Few of us have any intuition about the size of forces that propel cars and trucks. Even so, the fact that the forward force on the truck is a significant fraction (80%) of the combined weight of the truck and the crate seems plausible. We might have been suspicious if F_T had been only a tiny fraction of the weight or much greater than the weight.

As you can see, there are many equations and many pieces of information to keep track of when solving a problem of interacting objects. These problems are not inherently harder than the problems you learned to solve in Chapter 6, but they do require a high level of organization. Using the systematic approach of the problem-solving strategy will help you solve similar problems successfully.

STOP TO THINK 7.3 Boxes A and B are sliding to the right across a frictionless table. The hand H is slowing them down. The mass of A is larger than the mass of B. Rank in order, from largest to smallest, the *horizontal* forces on A, B, and H.

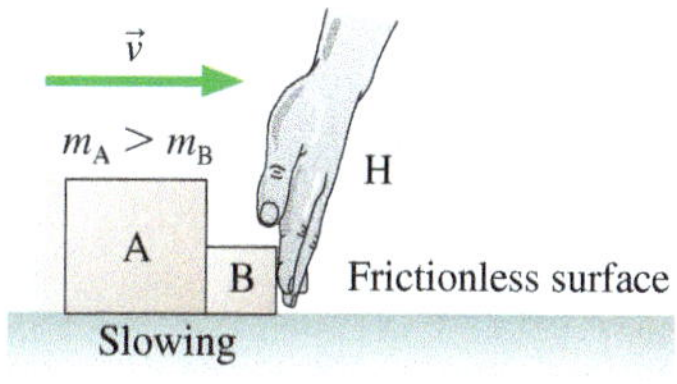

a. $F_{\text{B on H}} = F_{\text{H on B}} = F_{\text{A on B}} = F_{\text{B on A}}$
b. $F_{\text{B on H}} = F_{\text{H on B}} > F_{\text{A on B}} = F_{\text{B on A}}$
c. $F_{\text{B on H}} = F_{\text{H on B}} < F_{\text{A on B}} = F_{\text{B on A}}$
d. $F_{\text{H on B}} = F_{\text{H on A}} > F_{\text{A on B}}$

7.4 Ropes and Pulleys

Many objects are connected by strings, ropes, cables, and so on. In single-particle dynamics, we defined *tension* as the force exerted on an object by a rope or string. Now we need to think more carefully about the string itself. Just what do we mean when we talk about the tension "in" a string?

Tension Revisited

FIGURE 7.18a shows a heavy safe hanging from a rope, placing the rope under tension. If you cut the rope, the safe and the lower portion of the rope will fall. Thus there must be a force *within* the rope by which the upper portion of the rope pulls upward on the lower portion to prevent it from falling.

Chapter 5 introduced an atomic-level model in which tension is due to the stretching of spring-like molecular bonds within the rope. Stretched springs exert pulling forces, and the combined pulling force of billions of stretched molecular springs in a string or rope is what we call *tension.*

An important aspect of tension is that it pulls equally *in both directions.* FIGURE 7.18b is a very thin cross section through the rope. This small piece of rope is in equilibrium, so it must be pulled equally from both sides. To gain a mental picture, imagine holding your arms outstretched and having two friends pull on them. You'll remain at rest—but "in tension"—as long as they pull with equal strength in opposite directions. But if one lets go, analogous to the breaking of molecular bonds if a rope breaks or is cut, you'll fly off in the other direction!

FIGURE 7.18 Tension forces within the rope are due to stretching the spring-like molecular bonds.

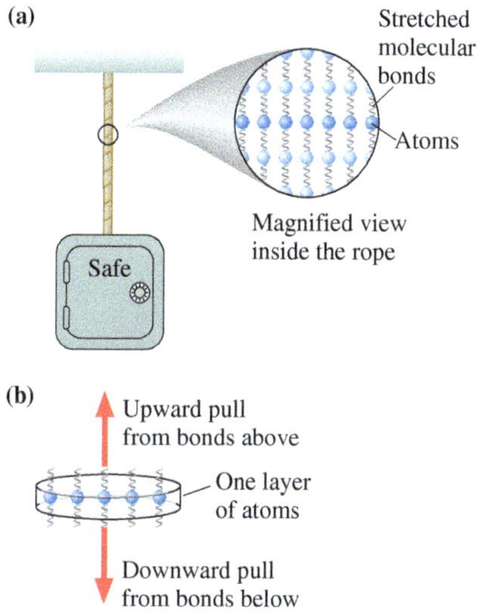

EXAMPLE 7.5 Pulling a rope

FIGURE 7.19a shows a student pulling horizontally with a 100 N force on a rope that is attached to a wall. In FIGURE 7.19b, two students in a tug-of-war pull on opposite ends of a rope with 100 N each. Is the tension in the second rope larger than, smaller than, or the same as that in the first rope?

FIGURE 7.19 Pulling on a rope. Which produces a larger tension?

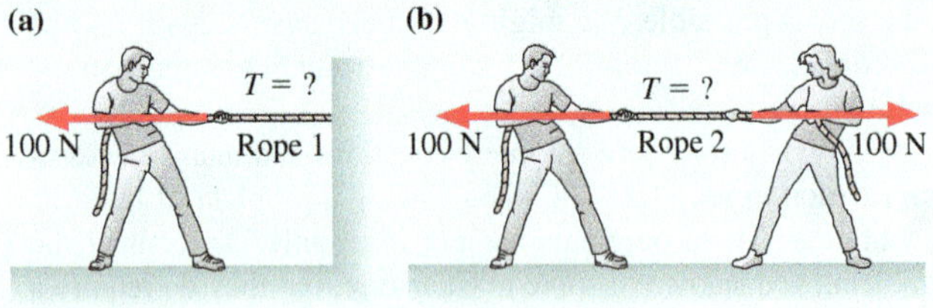

SOLVE Surely pulling on a rope from both ends causes more tension than pulling on one end. Right? Before jumping to conclusions, let's analyze the situation carefully.

FIGURE 7.20a shows the first student, the rope, and the wall as separate, interacting objects. Force $\vec{F}_{\text{S on R}}$ is the student pulling on the rope, so it has magnitude 100 N. Forces $\vec{F}_{\text{S on R}}$ and $\vec{F}_{\text{R on S}}$ are an action/reaction pair and must have equal magnitudes. Similarly for forces $\vec{F}_{\text{W on R}}$ and $\vec{F}_{\text{R on W}}$. Finally, because the rope is in static equilibrium, force $\vec{F}_{\text{W on R}}$ has to balance force $\vec{F}_{\text{S on R}}$. Thus

$$F_{\text{R on W}} = F_{\text{W on R}} = F_{\text{S on R}} = F_{\text{R on S}} = 100\text{ N}$$

The first and third equalities are Newton's third law; the second equality is Newton's first law for the rope.

Forces $\vec{F}_{\text{R on S}}$ and $\vec{F}_{\text{R on W}}$ are the pulling forces exerted by the rope and are what we *mean* by "the tension in the rope." Thus the tension in the first rope is 100 N.

FIGURE 7.20b repeats the analysis for the rope pulled by two students. Each student pulls with 100 N, so $F_{\text{S1 on R}} = 100$ N and $F_{\text{S2 on R}} = 100$ N. Just as before, there are two action/reaction pairs and the rope is in static equilibrium. Thus

$$F_{\text{R on S2}} = F_{\text{S2 on R}} = F_{\text{S1 on R}} = F_{\text{R on S1}} = 100\text{ N}$$

The tension in the rope—the pulling forces $\vec{F}_{\text{R on S1}}$ and $\vec{F}_{\text{R on S2}}$—is still 100 N!

You may have *assumed* that the student on the right in Figure 7.19b is doing something to the rope that the wall in Figure 7.19a does not do. But our analysis finds that the wall, just like the student, pulls to the right with 100 N. The rope doesn't care whether it's pulled by a wall or a hand. It experiences the same forces in both cases, so the rope's tension is the same in both.

ASSESS Ropes and strings exert forces at *both* ends. The force with which they pull—and thus the force pulling on them at each end—*is* the tension in the rope. Tension is not the sum of the pulling forces.

FIGURE 7.20 Analysis of tension forces.

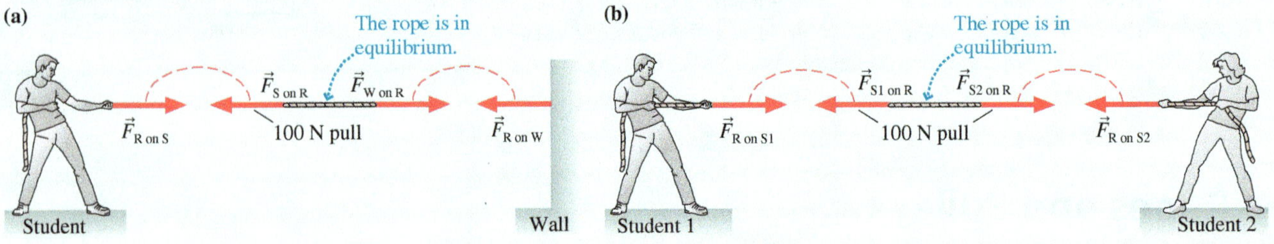

STOP TO THINK 7.4 All three 50 kg blocks are at rest. Is the tension in rope 2 greater than, less than, or equal to the tension in rope 1?

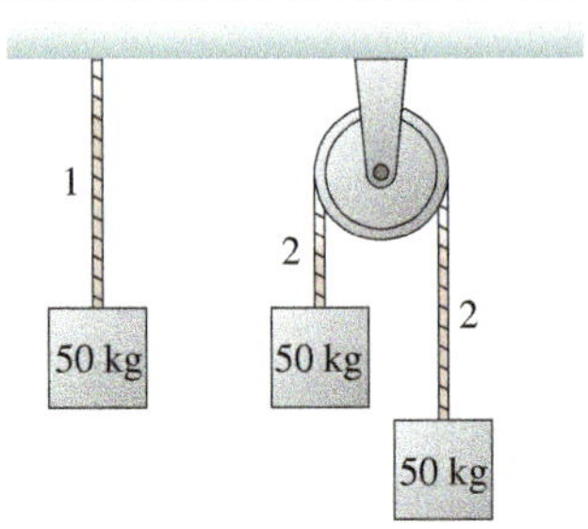

The Massless String Approximation

The tension is constant throughout a rope that is in equilibrium, but what happens if the rope is accelerating? For example, FIGURE 7.21a shows two connected blocks being pulled by force $\vec{F}$. Is the string's tension at the right end, where it pulls back on B, the same as the tension at the left end, where it pulls on A?

FIGURE 7.21b shows the horizontal forces acting on the blocks and the string. The only horizontal forces acting on the string are $\vec{T}_{\text{A on S}}$ and $\vec{T}_{\text{B on S}}$, so Newton's second law *for the string* is

$$(F_{\text{net}})_x = T_{\text{B on S}} - T_{\text{A on S}} = m_s a_x \tag{7.5}$$

where m_s is the mass of the string. If the string is accelerating, then the tensions at the two ends can *not* be the same. The tension at the "front" of the string must be higher than the tension at the "back" in order to accelerate the string!

Often in physics and engineering problems the mass of the string or rope is much less than the masses of the objects that it connects. In such cases, we can adopt the **massless string approximation.** In the limit $m_s \to 0$, Equation 7.5 becomes

$$T_{\text{B on S}} = T_{\text{A on S}} \quad \text{(massless string approximation)} \tag{7.6}$$

In other words, **the tension in a massless string is constant.** This is nice, but it isn't the primary justification for the massless string approximation.

Look again at Figure 7.21b. If $T_{\text{B on S}} = T_{\text{A on S}}$, then

$$\vec{T}_{\text{S on A}} = -\vec{T}_{\text{S on B}} \tag{7.7}$$

That is, the force on block A is equal and opposite to the force on block B. Forces $\vec{T}_{\text{S on A}}$ and $\vec{T}_{\text{S on B}}$ act *as if* they are an action/reaction pair of forces. Thus we can draw the simplified diagram of FIGURE 7.22 in which the string is missing and blocks A and B interact directly with each other through forces that we can call $\vec{T}_{\text{A on B}}$ and $\vec{T}_{\text{B on A}}$.

In other words, **if objects A and B interact with each other through a massless string, we can omit the string and treat forces $\vec{F}_{\text{A on B}}$ and $\vec{F}_{\text{B on A}}$ as if they are an action/reaction pair.** This is not literally true because A and B are not in contact. Nonetheless, all a massless string does is transmit a force from A to B without changing the magnitude of that force. This is the real significance of the massless string approximation.

NOTE ▶ For problems in this book, you can assume that any strings or ropes are massless unless the problem explicitly states otherwise. The simplified view of Figure 7.22 is appropriate under these conditions. But if the string has a mass, it must be treated as a separate object. ◀

FIGURE 7.21 The string's tension pulls forward on block A, backward on block B.

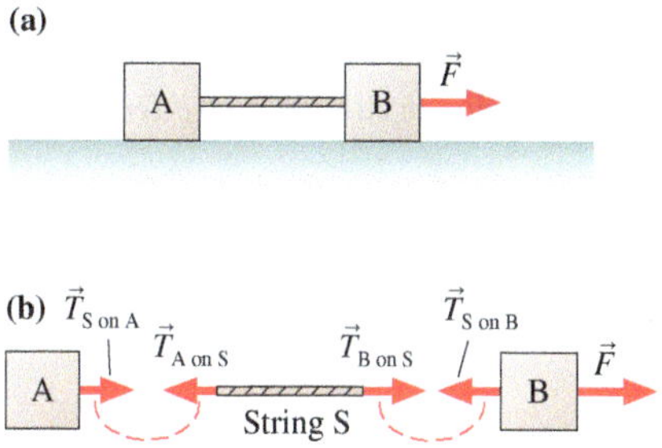

FIGURE 7.22 The massless string approximation allows objects A and B to act *as if* they are directly interacting.

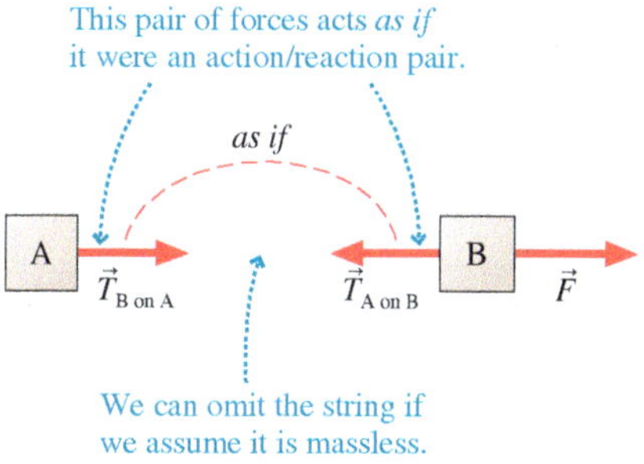

EXAMPLE 7.6 Comparing two tensions

Blocks A and B in FIGURE 7.23 are connected by massless string 2 and pulled across a frictionless table by massless string 1. B has a larger mass than A. Is the tension in string 2 larger than, smaller than, or equal to the tension in string 1?

FIGURE 7.23 Blocks A and B are pulled across a frictionless table by massless strings.

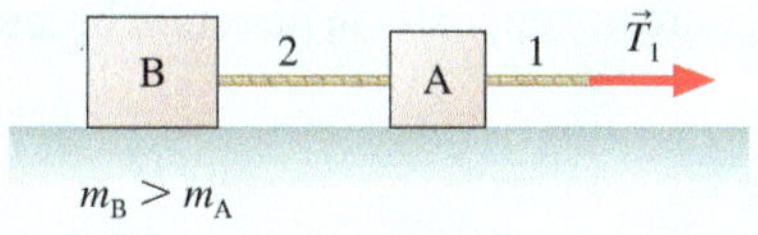

MODEL The massless string approximation allows us to treat A and B *as if* they interact directly with each other. The blocks are accelerating because there's a force to the right and no friction.

SOLVE B has a larger mass, so it may be tempting to conclude that string 2, which pulls B, has a greater tension than string 1, which pulls A. The flaw in this reasoning is that Newton's second law tells us only about the *net* force. The net force on B *is* larger than the net force on A, but the net force on A is *not* just the tension $\vec{T}_1$ in the forward direction. The tension in string 2 also pulls *backward* on A!

FIGURE 7.24 shows the horizontal forces in this frictionless situation. Forces $\vec{T}_{\text{A on B}}$ and $\vec{T}_{\text{B on A}}$ act *as if* they are an action/reaction pair.

From Newton's third law,

$$T_{\text{A on B}} = T_{\text{B on A}} = T_2$$

where T_2 is the tension in string 2. From Newton's second law, the net force on A is

$$(F_{\text{A net}})_x = T_1 - T_{\text{B on A}} = T_1 - T_2 = m_A a_{Ax}$$

FIGURE 7.24 The horizontal forces on blocks A and B.

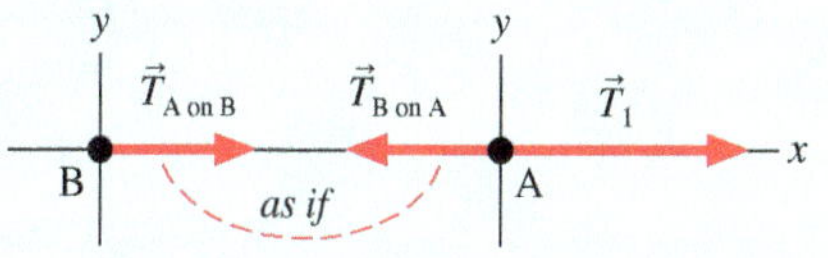

Continued

The net force on A is the *difference* in tensions. The blocks are accelerating to the right, making $a_{Ax} > 0$, so

$$T_1 > T_2$$

The tension in string 2 is *smaller* than the tension in string 1.

ASSESS This is not an intuitively obvious result. A careful study of the reasoning in this example is worthwhile. An alternative analysis would note that $\vec{T}_1$ accelerates *both* blocks, of combined mass $(m_A + m_B)$, whereas $\vec{T}_2$ accelerates only block B. Thus string 1 must have the larger tension.

Pulleys

Strings and ropes often pass over pulleys. The application might be as simple as lifting a heavy weight or as complex as the internal cable-and-pulley arrangement that precisely moves a robot arm.

FIGURE 7.25a shows a simple situation in which block B drags block A across a frictionless table as it falls. FIGURE 7.25b shows the objects separately as well as the forces. As the string moves, static friction between the string and pulley causes the pulley to turn. If we assume that

- The string *and* the pulley are both massless, and
- There is no friction where the pulley turns on its axle,

then no net force is needed to accelerate the string or turn the pulley. In this case,

$$T_{A\text{ on }S} = T_{B\text{ on }S}$$

In other words, **the tension in a massless string remains constant as it passes over a massless, frictionless pulley.**

FIGURE 7.25 Blocks A and B are connected by a string that passes over a pulley.

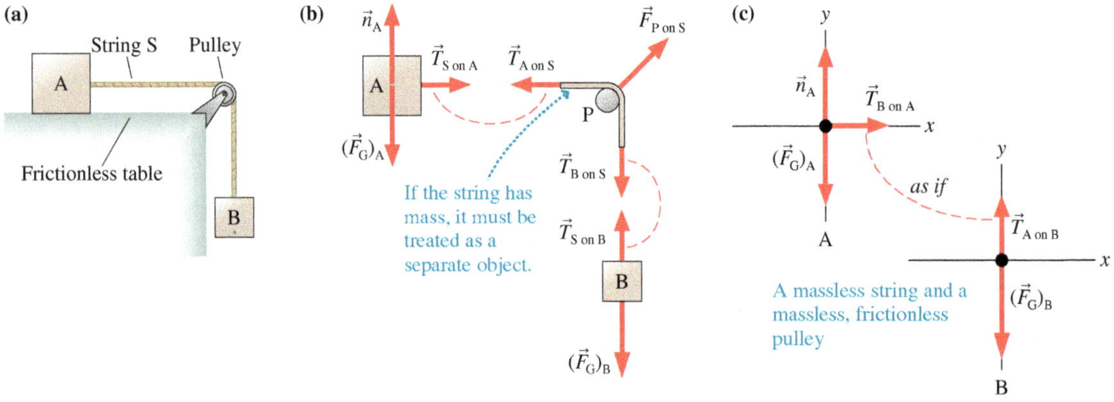

Because of this, we can draw the simplified free-body diagram of FIGURE 7.25c, in which the string and pulley are omitted. Forces $\vec{T}_{A\text{ on }B}$ and $\vec{T}_{B\text{ on }A}$ act *as if* they are an action/reaction pair, even though they are not opposite in direction because the tension force gets "turned" by the pulley.

TACTICS BOX 7.2 Working with ropes and pulleys

For massless ropes or strings and massless, frictionless pulleys:

- If a force pulls on one end of a rope, the tension in the rope equals the magnitude of the pulling force.
- If two objects are connected by a rope, the tension is the same at both ends.
- If the rope passes over a pulley, the tension in the rope is unaffected.

Exercises 17–22

STOP TO THINK 7.5 In Figure 7.25, is the tension in the string greater than, less than, or equal to the gravitational force acting on block B?

7.5 Examples of Interacting-Objects Problems

We will conclude this chapter with three extended examples. Although the mathematics will be more involved than in any of our work up to this point, we will continue to emphasize the *reasoning* one uses in approaching problems such as these. The solutions will be based on Problem-Solving Strategy 7.1. In fact, these problems are now reaching such a level of complexity that, for all practical purposes, it becomes impossible to work them unless you are following a well-planned strategy. Our earlier emphasis on identifying forces and using free-body diagrams will now really begin to pay off!

EXAMPLE 7.7 Placing a leg in traction

Serious fractures of the leg often need a stretching force to keep contracting leg muscles from forcing the broken bones together too hard. This is done using *traction*, an arrangement of a rope, a weight, and pulleys as shown in FIGURE 7.26. The rope must make the same angle on both sides of the pulley so that the net force on the leg is horizontal, but the angle can be adjusted to control the amount of traction. The doctor has specified 50 N of traction for this patient with a 4.2 kg hanging mass. What is the proper angle?

FIGURE 7.26 A leg in traction.

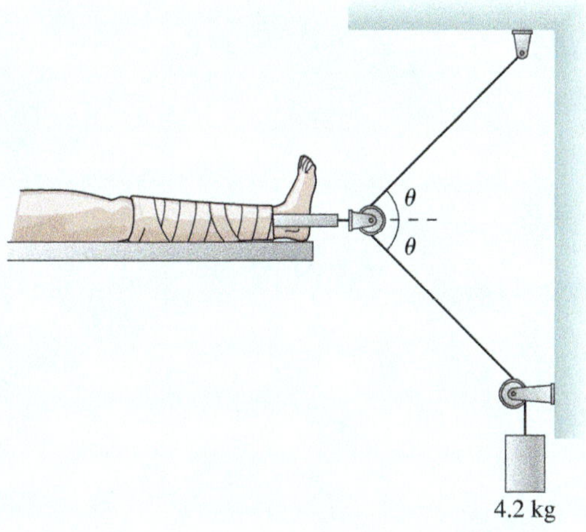

MODEL Model the leg and the weight as particles. The other point where forces are applied is the pulley attached to the patient's foot, which we'll treat as a separate object. We'll assume massless ropes and a massless, frictionless pulley.

VISUALIZE FIGURE 7.27 shows three free-body diagrams. Forces $\vec{T}_{P1}$ and $\vec{T}_{P2}$ are the tension forces of the rope as it pulls on the pulley. The pulley is in static equilibrium, so these forces are balanced by $\vec{F}_{L\text{ on }P}$, which forms an action/reaction pair with the 50 N traction force $\vec{F}_{P\text{ on }L}$. Our model of the rope and pulley makes the tension force constant, $T_{P1} = T_{P2} = T_W$, so we'll call it simply T.

FIGURE 7.27 The free-body diagrams.

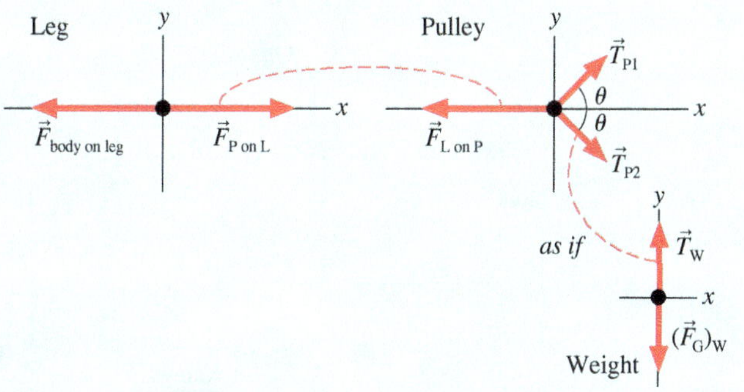

SOLVE The x-component equation of Newton's first law for the pulley is

$$\sum (F_{\text{on }P})_x = T_{P1}\cos\theta + T_{P2}\cos\theta - F_{L\text{ on }P}$$
$$= 2T\cos\theta - F_{L\text{ on }P} = 0$$

Thus the correct angle for the ropes is

$$\theta = \cos^{-1}\left(\frac{F_{L\text{ on }P}}{2T}\right)$$

We know, from Newton's third law, that $F_{L\text{ on }P} = F_{P\text{ on }L} = 50$ N. We can determine the tension force by analyzing the weight. It also is in static equilibrium, so the upward tension force exactly balances the downward gravitational force:

$$T = (F_G)_W = m_W g = (4.2\text{ kg})(9.80\text{ m/s}^2) = 41\text{ N}$$

Thus the proper angle is

$$\theta = \cos^{-1}\left(\frac{50\text{ N}}{2(41\text{ N})}\right) = 52°$$

ASSESS The traction force would approach 82 N if angle θ approached zero because the two ropes would pull in parallel. Conversely, the traction would approach 0 N if θ approached 90°. The desired traction is roughly midway between these two extremes, so an angle near 45° seems reasonable.

EXAMPLE 7.8 The show must go on!

A 200 kg set used in a play is stored in the loft above the stage. The rope holding the set passes up and over a pulley, then is tied backstage. The director tells a 100 kg stagehand to lower the set. When he unties the rope, the set falls and the unfortunate man is hoisted into the loft. What is the stagehand's acceleration?

MODEL The system is the stagehand M and the set S, which we will model as particles. Assume a massless rope and a massless, frictionless pulley.

VISUALIZE FIGURE 7.28 shows the pictorial representation. The man's acceleration a_{My} is positive, while the set's acceleration a_{Sy} is negative. These two accelerations have the same magnitude because the two objects are connected by a rope, but they have opposite signs. Thus the acceleration constraint is $a_{Sy} = -a_{My}$. Forces $\vec{T}_{\text{M on S}}$ and $\vec{T}_{\text{S on M}}$ are not literally an action/reaction pair, but they act *as if* they are because the rope is massless and the pulley is massless and frictionless. Notice that the pulley has "turned" the tension force so that $\vec{T}_{\text{M on S}}$ and $\vec{T}_{\text{S on M}}$ are *parallel* to each other rather than opposite, as members of a true action/reaction pair would have to be.

SOLVE Newton's second law for the man and the set are

$$\sum (F_{\text{on M}})_y = T_{\text{S on M}} - m_M g = m_M a_{My}$$

$$\sum (F_{\text{on S}})_y = T_{\text{M on S}} - m_S g = m_S a_{Sy} = -m_S a_{My}$$

Only the y-equations are needed. Notice that we used the acceleration constraint in the last step. Newton's third law is

$$T_{\text{M on S}} = T_{\text{S on M}} = T$$

where we can drop the subscripts and call the tension simply T. With this substitution, the two second-law equations can be written

$$T - m_M g = m_M a_{My}$$

$$T - m_S g = -m_S a_{My}$$

These are simultaneous equations in the two unknowns T and a_{My}. We can eliminate T by subtracting the second equation from the first to give

$$(m_S - m_M)g = (m_S + m_M)a_{My}$$

Finally, we can solve for the hapless stagehand's acceleration:

$$a_{My} = \frac{m_S - m_M}{m_S + m_M} g = \frac{100 \text{ kg}}{300 \text{ kg}} 9.80 \text{ m/s}^2 = 3.27 \text{ m/s}^2$$

This is also the acceleration with which the set falls. If the rope's tension was needed, we could now find it from $T = m_M a_{My} + m_M g$.

ASSESS If the stagehand weren't holding on, the set would fall with free-fall acceleration g. The stagehand acts as a *counterweight* to reduce the acceleration.

FIGURE 7.28 Pictorial representation for Example 7.8.

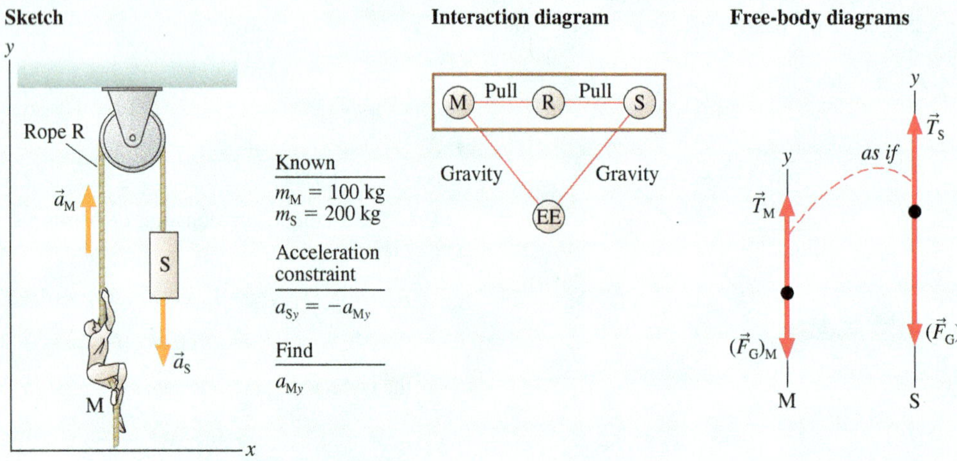

STOP TO THINK 7.6 A small car is pushing a larger truck that has a dead battery. The mass of the truck is larger than the mass of the car. Which of the following statements is true?

a. The car exerts a force on the truck, but the truck doesn't exert a force on the car.
b. The car exerts a larger force on the truck than the truck exerts on the car.
c. The car exerts the same amount of force on the truck as the truck exerts on the car.
d. The truck exerts a larger force on the car than the car exerts on the truck.
e. The truck exerts a force on the car, but the car doesn't exert a force on the truck.

CHALLENGE EXAMPLE 7.9 A not-so-clever bank robbery

Bank robbers have pushed a 1000 kg safe to a second-story floor-to-ceiling window. They plan to break the window, then lower the safe 3.0 m to their truck. Not being too clever, they stack up 500 kg of furniture, tie a rope between the safe and the furniture, and place the rope over a pulley. Then they push the safe out the window. What is the safe's speed when it hits the truck? The coefficient of kinetic friction between the furniture and the floor is 0.50.

MODEL This is a continuation of the situation that we analyzed in Figures 7.16 and 7.25, which are worth reviewing. The system is the safe S and the furniture F, which we will model as particles. We will assume a massless rope and a massless, frictionless pulley.

VISUALIZE The safe and the furniture are tied together, so their accelerations have the same magnitude. The safe has a y-component of acceleration a_{Sy} that is negative because the safe accelerates in the negative y-direction. The furniture has an x-component a_F that is positive. Thus the acceleration constraint is

$$a_{Fx} = -a_{Sy}$$

The free-body diagrams shown in FIGURE 7.29 are modeled after Figure 7.25 but now include a kinetic friction force on the furniture. Forces $\vec{T}_{F\text{ on }S}$ and $\vec{T}_{S\text{ on }F}$ act *as if* they are an action/reaction pair, so they have been connected with a dashed line.

SOLVE We can write Newton's second law directly from the free-body diagrams. For the furniture,

$$\sum(F_{\text{on F}})_x = T_{\text{S on F}} - f_k = T - f_k = m_F a_{Fx} = -m_F a_{Sy}$$

$$\sum(F_{\text{on F}})_y = n - m_F g = 0$$

And for the safe,

$$\sum(F_{\text{on S}})_y = T - m_S g = m_S a_{Sy}$$

Notice how we used the acceleration constraint in the first equation. We also went ahead and made use of Newton's third law: $T_{\text{F on S}} = T_{\text{S on F}} = T$. We have one additional piece of information, the model of kinetic friction:

$$f_k = \mu_k n = \mu_k m_F g$$

where we used the y-equation of the furniture to deduce that $n = m_F g$. Substitute this result for f_k into the x-equation of the furniture, then rewrite the furniture's x-equation and the safe's y-equation:

$$T - \mu_k m_F g = -m_F a_{Sy}$$

$$T - m_S g = m_S a_{Sy}$$

We have succeeded in reducing our knowledge to two simultaneous equations in the two unknowns a_{Sy} and T. Subtract the second equation from the first to eliminate T:

$$(m_S - \mu_k m_F)g = -(m_S + m_F)a_{Sy}$$

Finally, solve for the safe's acceleration:

$$a_{Sy} = -\left(\frac{m_S - \mu_k m_F}{m_S + m_F}\right)g$$

$$= -\frac{1000\text{ kg} - (0.50)(500\text{ kg})}{1000\text{ kg} + 500\text{ kg}}9.80\text{ m/s}^2 = -4.9\text{ m/s}^2$$

Now we need to calculate the kinematics of the falling safe. Because the time of the fall is not known or needed, we can use

$$v_{1y}{}^2 = v_{0y}{}^2 + 2a_{Sy}\,\Delta y = 0 + 2a_{Sy}(y_1 - y_0) = -2a_{Sy}y_0$$

$$v_1 = \sqrt{-2a_{Sy}y_0} = \sqrt{-2(-4.9\text{ m/s}^2)(3.0\text{ m})} = 5.4\text{ m/s}$$

The value of v_{1y} is negative, but we only needed to find the speed so we took the absolute value. This is about 12 mph, so it seems unlikely that the truck will survive the impact of the 1000 kg safe!

FIGURE 7.29 Pictorial representation for Challenge Example 7.9.

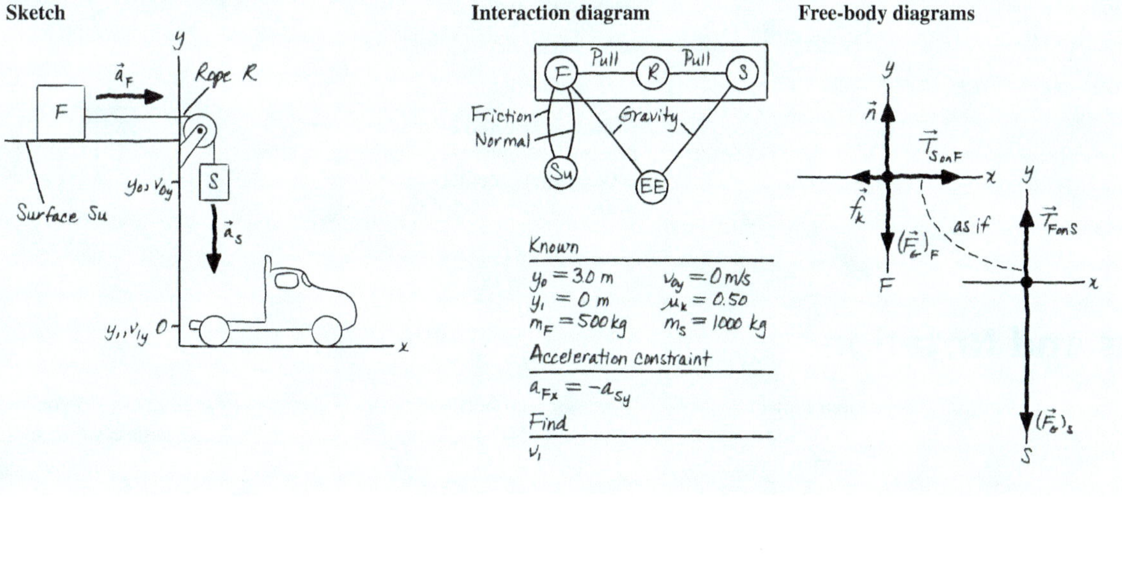

SUMMARY

The goal of Chapter 7 has been to use Newton's third law to understand how objects interact.

General Principles

Newton's Third Law

Every force occurs as one member of an **action/reaction pair** of forces. The two members of an action/reaction pair:

- Act on two *different* objects.
- Are equal in magnitude but opposite in direction:

$$\vec{F}_{\text{A on B}} = -\vec{F}_{\text{B on A}}$$

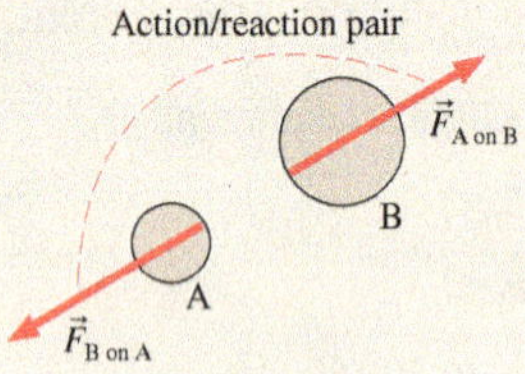

Solving Interacting-Objects Problems

MODEL Choose the objects of interest.

VISUALIZE
Draw a pictorial representation.
Sketch and define coordinates.
Identify acceleration constraints.
Draw an interaction diagram.
Draw a separate free-body diagram for each object.
Connect action/reaction pairs with dashed lines.

SOLVE Write Newton's second law for each object.
Include *all* forces acting *on* each object.
Use Newton's third law to equate the magnitudes of action/reaction pairs.
Include acceleration constraints and friction.

ASSESS Is the result reasonable?

Important Concepts

Objects, systems, and the environment

Objects whose motion is of interest are the system.
Objects whose motion is not of interest form the environment.
The objects of interest interact with the environment, but those interactions can be considered external forces.

Interaction diagram

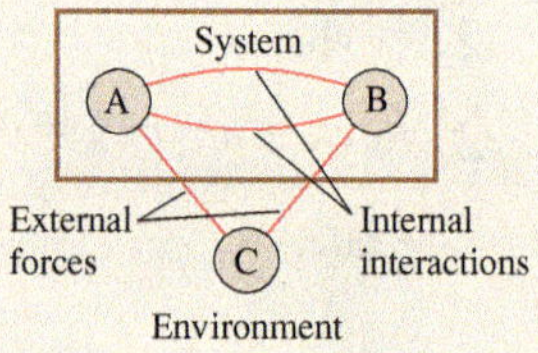

Applications

Acceleration constraints

Objects that are constrained to move together must have accelerations of equal magnitude: $a_A = a_B$.
This must be expressed in terms of components, such as $a_{Ax} = -a_{By}$.

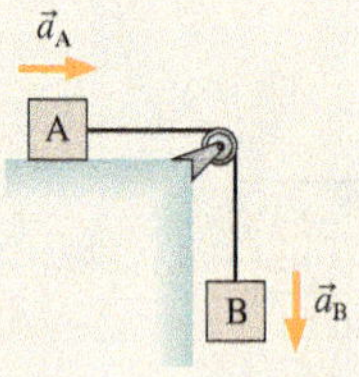

Strings and pulleys

The tension in a string or rope pulls in both directions. The tension is constant in a string if the string is:

- Massless, or
- In equilibrium

Objects connected by massless strings passing over massless, frictionless pulleys act *as if* they interact via an action/reaction pair of forces.

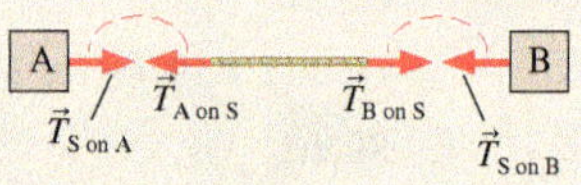

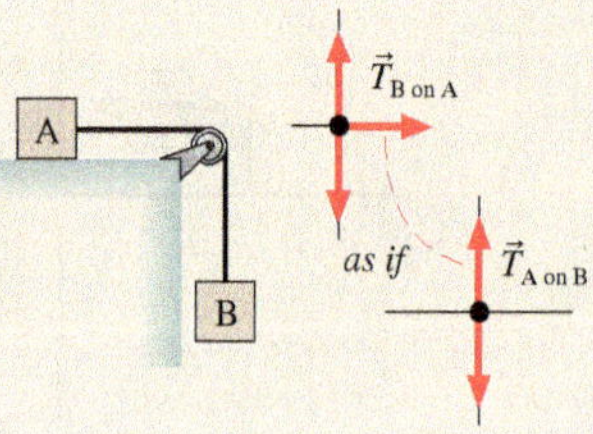

Terms and Notation

interaction	environment	propulsion	acceleration constraint
action/reaction pair	interaction diagram	Newton's third law	massless string approximation
system	external force		

CONCEPTUAL QUESTIONS

1. You find yourself in the middle of a frozen lake with a surface so slippery ($\mu_s = \mu_k = 0$) you cannot walk. However, you happen to have several rocks in your pocket. The ice is extremely hard. It cannot be chipped, and the rocks slip on it just as much as your feet do. Can you think of a way to get to shore? Use pictures, forces, and Newton's laws to explain your reasoning.
2. How do you paddle a canoe in the forward direction? Explain. Your explanation should include diagrams showing forces on the water and forces on the paddle.
3. How does a rocket take off? What is the upward force on it? Your explanation should include diagrams showing forces on the rocket and forces on the parcel of hot gas that was just expelled from the rocket's exhaust.
4. How do basketball players jump straight up into the air? Your explanation should include pictures showing forces on the player and forces on the ground.
5. A mosquito collides head-on with a car traveling 60 mph. Is the force of the mosquito on the car larger than, smaller than, or equal to the force of the car on the mosquito? Explain.
6. A mosquito collides head-on with a car traveling 60 mph. Is the magnitude of the mosquito's acceleration larger than, smaller than, or equal to the magnitude of the car's acceleration? Explain.
7. A small car is pushing a large truck. They are speeding up. Is the force of the truck on the car larger than, smaller than, or equal to the force of the car on the truck?
8. A very smart 3-year-old child is given a wagon for her birthday. She refuses to use it. "After all," she says, "Newton's third law says that no matter how hard I pull, the wagon will exert an equal but opposite force on me. So I will never be able to get it to move forward." What would you say to her in reply?
9. Teams red and blue are having a tug-of-war. According to Newton's third law, the force with which the red team pulls on the blue team exactly equals the force with which the blue team pulls on the red team. How can one team ever win? Explain.
10. Will hanging a magnet in front of the iron cart in FIGURE Q7.10 make it go? Explain.

FIGURE Q7.10

11. FIGURE Q7.11 shows two masses at rest. The string is massless and the pulley is frictionless. The spring scale reads in kg. What is the reading of the scale?

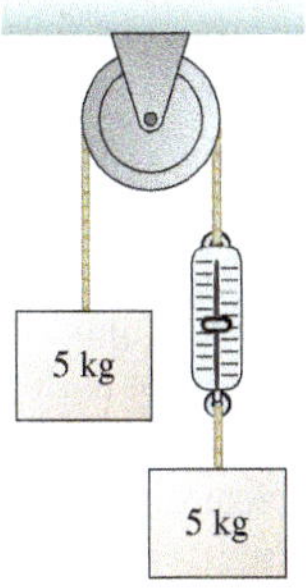

FIGURE Q7.11

12. FIGURE Q7.12 shows two masses at rest. The string is massless and the pulley is frictionless. The spring scale reads in kg. What is the reading of the scale?

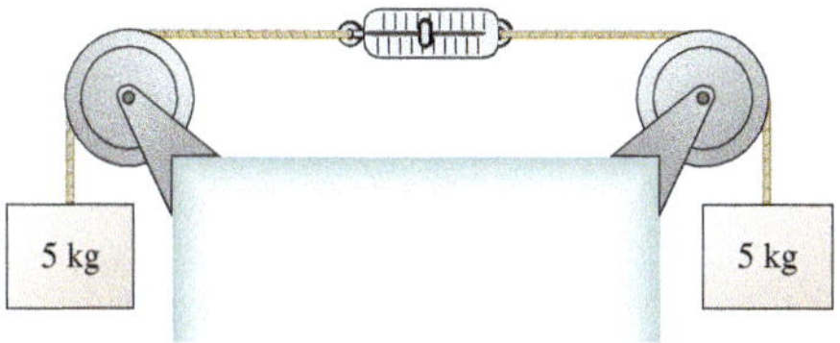

FIGURE Q7.12

13. The hand in FIGURE Q7.13 is pushing on the back of block A. Blocks A and B, with $m_B > m_A$, are connected by a massless string and slide on a frictionless surface. Is the force of the string on B larger than, smaller than, or equal to the force of the hand on A? Explain.

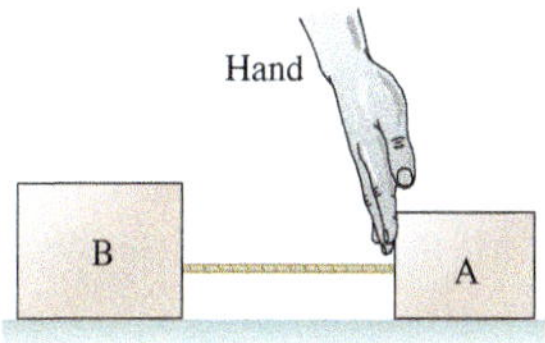

FIGURE Q7.13

14. Blocks A and B in FIGURE Q7.14 are connected by a massless string over a massless, frictionless pulley. The blocks have just been released from rest. Will the pulley rotate clockwise, counterclockwise, or not at all? Explain.

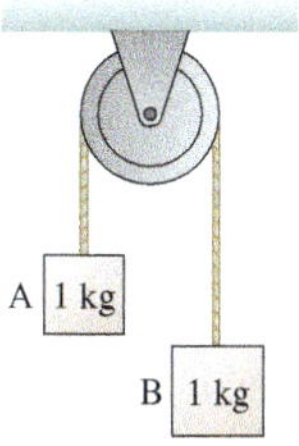

FIGURE Q7.14

15. In case a in FIGURE Q7.15, block A is accelerated across a frictionless table by a hanging 10 N weight (1.02 kg). In case b, block A is accelerated across a frictionless table by a steady 10 N tension in the string. The string is massless, and the pulley is massless and frictionless. Is A's acceleration in case b greater than, less than, or equal to its acceleration in case a? Explain.

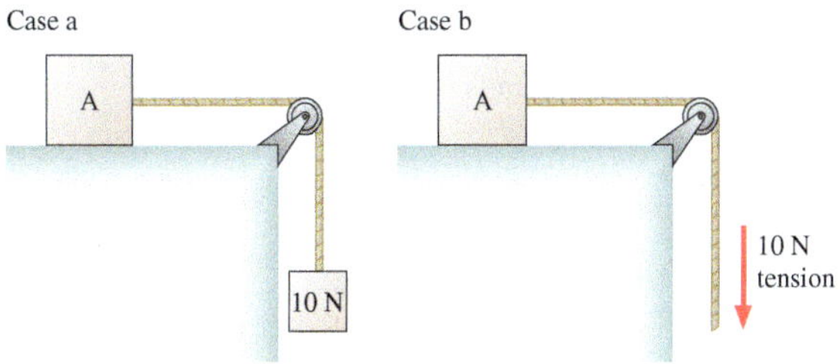

FIGURE Q7.15

EXERCISES AND PROBLEMS

Exercises

Section 7.2 Analyzing Interacting Objects

Exercises 1 through 6 describe a situation. For each:

a. Draw an interaction diagram, following the steps of Tactics Box 7.1.
b. Identify the "system" on your interaction diagram.
c. Draw a free-body diagram for each object in the system. Use dashed lines to connect the members of an action/reaction pair.

1. | A weightlifter stands up at constant speed from a squatting position while holding a heavy barbell across his shoulders.
2. | A soccer ball and a bowling ball have a head-on collision at this instant. Rolling friction is negligible.
3. | A mountain climber is using a rope to pull a bag of supplies up a 45° slope. The rope is not massless.
4. || A battery-powered toy car pushes a stuffed rabbit across the floor.
5. || Block A in FIGURE EX7.5 is heavier than block B and is sliding down the incline. All surfaces have friction. The rope is massless, and the massless pulley turns on frictionless bearings. The rope and the pulley are among the interacting objects, but you'll have to decide if they're part of the system.

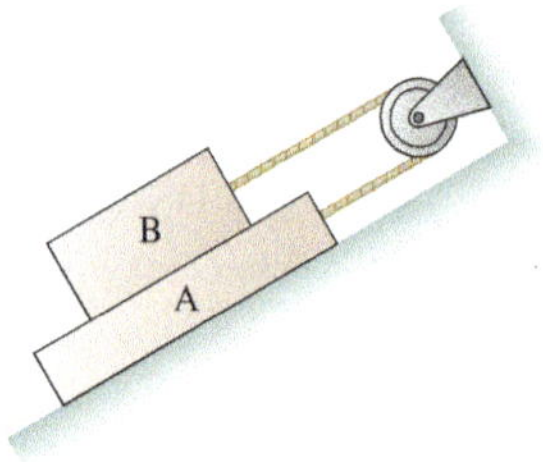

FIGURE EX7.5

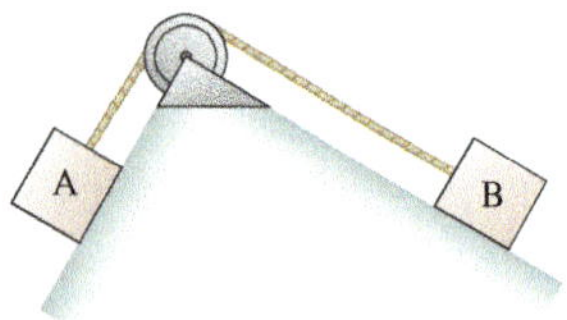

FIGURE EX7.6

6. || Block A in FIGURE EX7.6 is sliding down the incline. The rope is massless, and the massless pulley turns on frictionless bearings, but the surface is not frictionless. The rope and the pulley are among the interacting objects, but you'll have to decide if they're part of the system.

Section 7.3 Newton's Third Law

7. | a. How much force does an 80 kg astronaut exert on his chair while sitting at rest on the launch pad?
   b. How much force does the astronaut exert on his chair while accelerating straight up at 10 m/s^2?
8. || Block B in FIGURE EX7.8 rests on a surface for which the static and kinetic coefficients of friction are 0.60 and 0.40, respectively. The ropes are massless. What is the maximum mass of block A for which the system is in equilibrium?

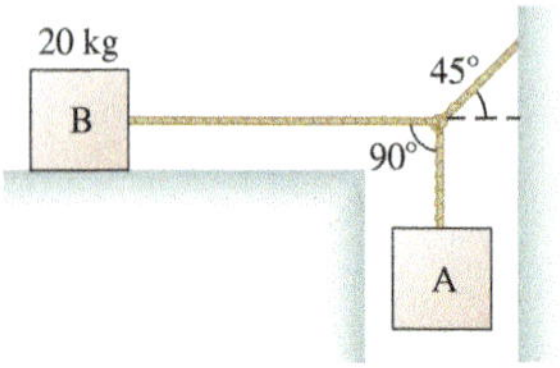

FIGURE EX7.8

9. || A 1000 kg car pushes a 2000 kg truck that has a dead battery. When the driver steps on the accelerator, the drive wheels of the car push against the ground with a force of 4500 N. Rolling friction can be neglected.
   a. What is the magnitude of the force of the car on the truck?
   b. What is the magnitude of the force of the truck on the car?
10. || Blocks with masses of 1 kg, 2 kg, and 3 kg are lined up in a row on a frictionless table. All three are pushed forward by a 12 N force applied to the 1 kg block.
   a. How much force does the 2 kg block exert on the 3 kg block?
   b. How much force does the 2 kg block exert on the 1 kg block?
11. || A massive steel cable drags a 20 kg block across a horizontal, frictionless surface. A 100 N force applied to the cable causes the block to reach a speed of 4.0 m/s in a distance of 2.0 m. What is the mass of the cable?

Section 7.4 Ropes and Pulleys

12. || What is the tension in the rope of FIGURE EX7.12?

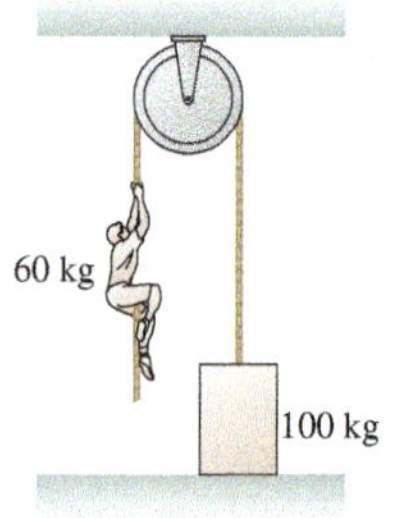

FIGURE EX7.12

13. || FIGURE EX7.13 shows two 1.0 kg blocks connected by a rope. A second rope hangs beneath the lower block. Both ropes have a mass of 250 g. The entire assembly is accelerated upward at 3.0 m/s^2 by force $\vec{F}$.
   a. What is F?
   b. What is the tension at the top end of rope 1?
   c. What is the tension at the bottom end of rope 1?
   d. What is the tension at the top end of rope 2?

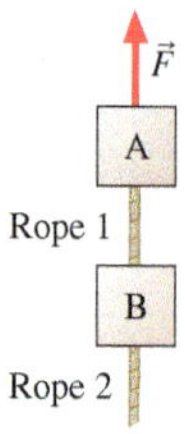

FIGURE EX7.13

14. || Jimmy has caught two fish in Yellow Creek. He has tied the line holding the 3.0 kg steelhead trout to the tail of the 1.5 kg carp. To show the fish to a friend, he lifts upward on the carp with a force of 60 N.
   a. Draw separate free-body diagrams for the trout and the carp. Label all forces, then use dashed lines to connect action/reaction pairs or forces that act as if they are a pair.
   b. Rank in order, from largest to smallest, the magnitudes of all the forces shown on your free-body diagrams. Explain your reasoning.
15. || A 2.0-m-long, 500 g rope pulls a 10 kg block of ice across a horizontal, frictionless surface. The block accelerates at 2.0 m/s^2. How much force pulls forward on (a) the ice, (b) the rope?
16. || The cable cars in San Francisco are pulled along their tracks by an underground steel cable that moves along at 9.5 mph. The cable is driven by large motors at a central power station and extends, via an intricate pulley arrangement, for several miles beneath the city streets. The length of a cable stretches by up

to 100 ft during its lifetime. To keep the tension constant, the cable passes around a 1.5-m-diameter "tensioning pulley" that rolls back and forth on rails, as shown in FIGURE EX7.16. A 2000 kg block is attached to the tensioning pulley's cart, via a rope and pulley, and is suspended in a deep hole. What is the tension in the cable car's cable?

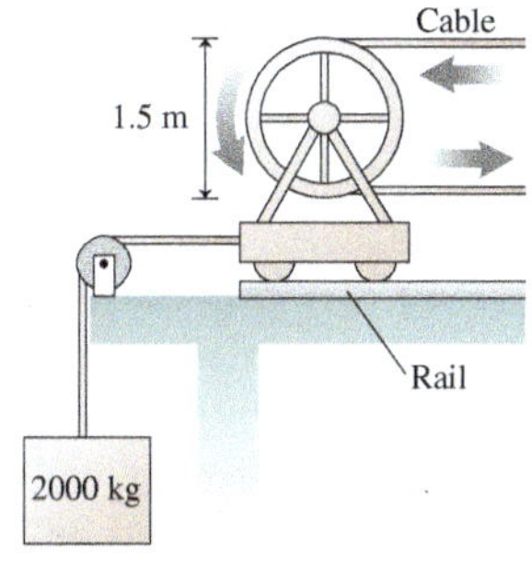

FIGURE EX7.16

17. || A 2.0 kg rope hangs from the ceiling. What is the tension at the midpoint of the rope?

18. || A mobile at the art museum has a 2.0 kg steel cat and a 4.0 kg steel dog suspended from a lightweight cable, as shown in FIGURE EX7.18. It is found that $\theta_1 = 20°$ when the center rope is adjusted to be perfectly horizontal. What are the tension and the angle of rope 3?

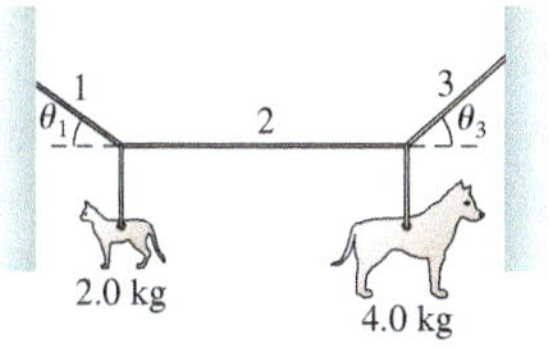

FIGURE EX7.18

Problems

19. ||| FIGURE P7.19 shows two strong magnets on opposite sides of a small table. The long-range attractive force between the magnets keeps the lower magnet in place.
    a. Draw an interaction diagram and draw free-body diagrams for both magnets and the table. Use dashed lines to connect the members of an action/reaction pair.
    b. Suppose the weight of the table is 20 N, the weight of each magnet is 2.0 N, and the magnetic force on the lower magnet is three times its weight. Find the magnitude of each of the forces shown on your free-body diagrams.

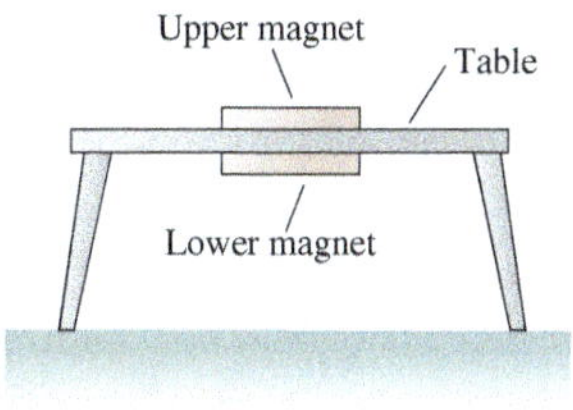

FIGURE P7.19

20. || An 80 kg spacewalking astronaut pushes off a 640 kg satellite, exerting a 100 N force for the 0.50 s it takes him to straighten his arms. How far apart are the astronaut and the satellite after 1.0 min?

21. || A massive steel cable drags a 20 kg block across a horizontal, frictionless surface. A 100 N force applied to the cable causes the block to reach a speed of 4.0 m/s in 2.0 s. What is the difference in tension between the two ends of the cable?

22. || FIGURE P7.22 shows a 6.0 N force pushing two gliders along an air track. The 200 g spring between the gliders is compressed. How much force does the spring exert on (a) glider A and (b) glider B?

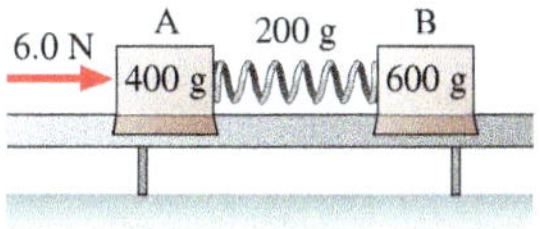

FIGURE P7.22

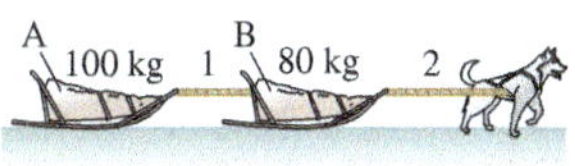

FIGURE P7.23

23. || The sled dog in FIGURE P7.23 drags sleds A and B across the snow. The coefficient of friction between the sleds and the snow is 0.10. If the tension in rope 1 is 150 N, what is the tension in rope 2?

24. || A rope of length L and mass m is suspended from the ceiling. Find an expression for the tension in the rope at position y, measured upward from the free end of the rope.

25. || While driving to work last year, I was holding my coffee mug in my left hand while changing the CD with my right hand. Then the cell phone rang, so I placed the mug on the flat part of my dashboard. Then, believe it or not, a deer ran out of the woods and on to the road right in front of me. Fortunately, my reaction time was zero, and I was able to stop from a speed of 20 m/s in a mere 50 m, just barely avoiding the deer. Later tests revealed that the static and kinetic coefficients of friction of the coffee mug on the dash are 0.50 and 0.30, respectively; the coffee and mug had a mass of 0.50 kg; and the mass of the deer was 120 kg. Did my coffee mug slide?

26. || Two-thirds of the weight of a 1500 kg car rests on the drive wheels. What is the maximum acceleration of this car on a concrete surface?

27. ||| A Federation starship (2.0×10^6 kg) uses its tractor beam to pull a shuttlecraft (2.0×10^4 kg) aboard from a distance of 10 km away. The tractor beam exerts a constant force of 4.0×10^4 N on the shuttlecraft. Both spacecraft are initially at rest. How far does the starship move as it pulls the shuttlecraft aboard?

28. || Your forehead can withstand a force of about 6.0 kN before fracturing, while your cheekbone can withstand only about 1.3 kN. Suppose a 140 g baseball traveling at 30 m/s strikes your head and stops in 1.5 ms.
BIO
    a. What is the magnitude of the force that stops the baseball?
    b. What force does the baseball exert on your head? Explain.
    c. Are you in danger of a fracture if the ball hits you in the forehead? On the cheek?

29. || Bob, who has a mass of 75 kg, can throw a 500 g rock with a speed of 30 m/s. The distance through which his hand moves as he accelerates the rock from rest until he releases it is 1.0 m.
    a. What constant force must Bob exert on the rock to throw it with this speed?
    b. If Bob is standing on frictionless ice, what is his recoil speed after releasing the rock?

30. || You see the boy next door trying to push a crate down the sidewalk. He can barely keep it moving, and his feet occasionally slip. You start to wonder how heavy the crate is. You call to ask the boy his mass, and he replies "50 kg." From your recent physics class you estimate that the static and kinetic coefficients of friction are 0.8 and 0.4 for the boy's shoes, and 0.5 and 0.2 for the crate. Estimate the mass of the crate.

31. ||| Two packages at UPS start sliding down the 20° ramp shown in FIGURE P7.31. Package A has a mass of 5.0 kg and a coefficient of friction of 0.20. Package B has a mass of 10 kg and a coefficient of friction of 0.15. How long does it take package A to reach the bottom?

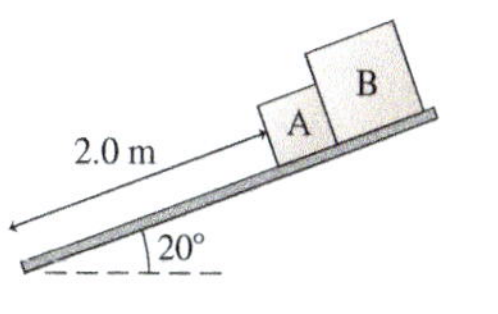

FIGURE P7.31

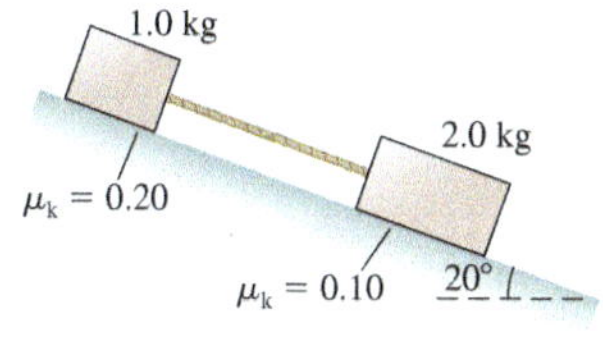

FIGURE P7.32

32. ||| The two blocks in FIGURE P7.32 are sliding down the incline. What is the tension in the massless string?

33. || The 1.0 kg block in FIGURE P7.33 is tied to the wall with a rope. It sits on top of the 2.0 kg block. The lower block is pulled to the right with a tension force of 20 N. The coefficient of kinetic friction at both the lower and upper surfaces of the 2.0 kg block is $\mu_k = 0.40$.
   a. What is the tension in the rope holding the 1.0 kg block to the wall?
   b. What is the acceleration of the 2.0 kg block?

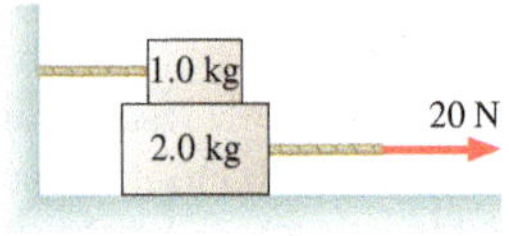

FIGURE P7.33

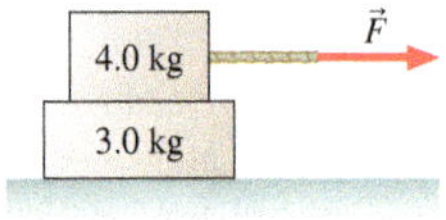

FIGURE P7.34

34. || The coefficient of static friction is 0.60 between the two blocks in FIGURE P7.34. The coefficient of kinetic friction between the lower block and the floor is 0.20. Force $\vec{F}$ causes both blocks to cross a distance of 5.0 m, starting from rest. What is the least amount of time in which this motion can be completed without the top block sliding on the lower block?

35. ||| The lower block in FIGURE P7.35 is pulled on by a rope with a tension force of 20 N. The coefficient of kinetic friction between the lower block and the surface is 0.30. The coefficient of kinetic friction between the lower block and the upper block is also 0.30. What is the acceleration of the 2.0 kg block?

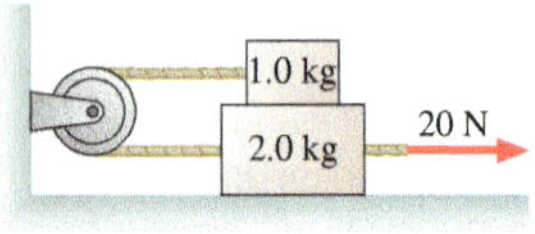

FIGURE P7.35

36. || The block of mass M in FIGURE P7.36 slides on a frictionless surface. Find an expression for the tension in the string.

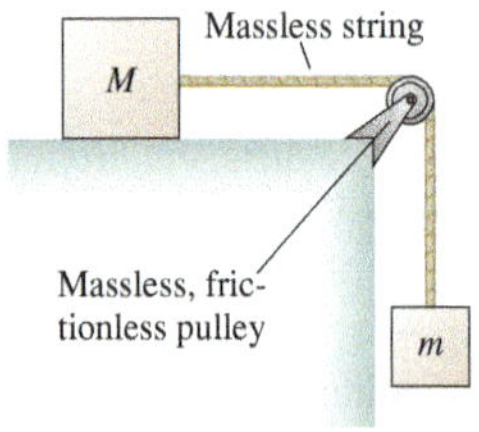

FIGURE P7.36

37. ||| A rope attached to a 20 kg wood sled pulls the sled up a 20° snow-covered hill. A 10 kg wood box rides on top of the sled. If the tension in the rope steadily increases, at what value of the tension does the box slip?

38. || The 100 kg block in FIGURE P7.38 takes 6.0 s to reach the floor after being released from rest. What is the mass of the block on the left? The pulley is massless and frictionless.

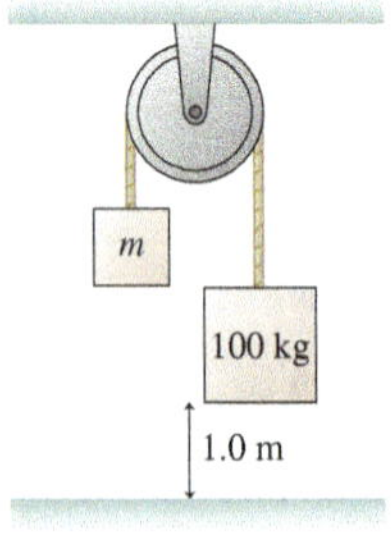

FIGURE P7.38

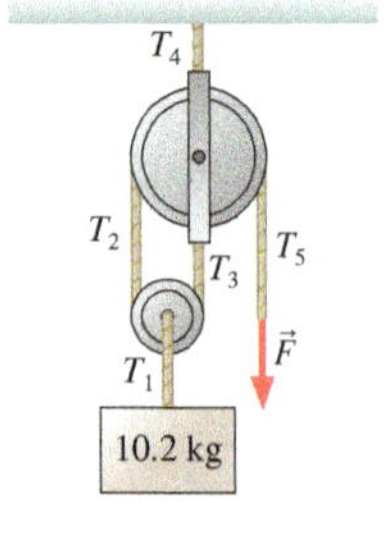

FIGURE P7.39

39. ||| The 10.2 kg block in FIGURE P7.39 is held in place by a force applied to a rope passing over two massless, frictionless pulleys. Find the tensions T_1 to T_5 and the magnitude of force $\vec{F}$.

40. ||| The coefficient of kinetic friction between the 2.0 kg block in FIGURE P7.40 and the table is 0.30. What is the acceleration of the 2.0 kg block?

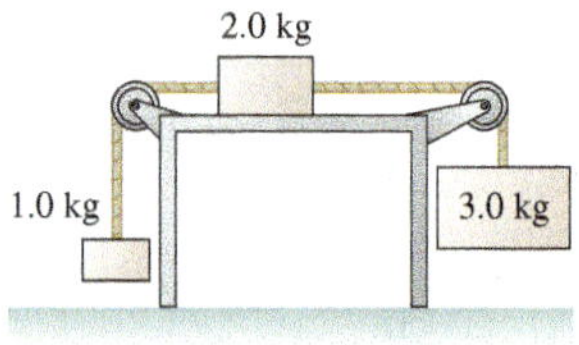

FIGURE P7.40

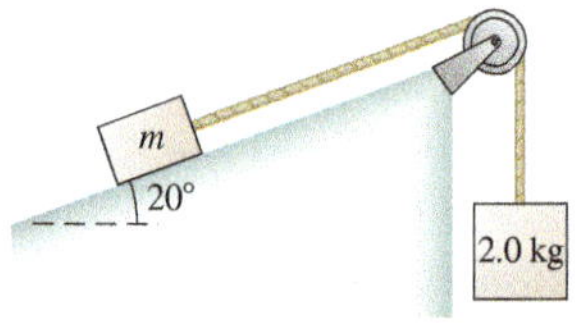

FIGURE P7.41

41. || FIGURE P7.41 shows a block of mass m resting on a 20° slope. The block has coefficients of friction $\mu_s = 0.80$ and $\mu_k = 0.50$ with the surface. It is connected via a massless string over a massless, frictionless pulley to a hanging block of mass 2.0 kg.
   a. What is the minimum mass m that will stick and not slip?
   b. If this minimum mass is nudged ever so slightly, it will start being pulled up the incline. What acceleration will it have?

42. || A 4.0 kg box is on a frictionless 35° slope and is connected via a massless string over a massless, frictionless pulley to a hanging 2.0 kg weight. The picture for this situation is similar to FIGURE P7.41.
   a. What is the tension in the string if the 4.0 kg box is *held* in place, so that it cannot move?
   b. If the box is then released, which way will it move on the slope?
   c. What is the tension in the string once the box begins to move?

43. || The 1.0 kg physics book in FIGURE P7.43 is connected by a string to a 500 g coffee cup. The book is given a push up the slope and released with a speed of 3.0 m/s. The coefficients of friction are $\mu_s = 0.50$ and $\mu_k = 0.20$.
   a. How far does the book slide?
   b. At the highest point, does the book stick to the slope, or does it slide back down?

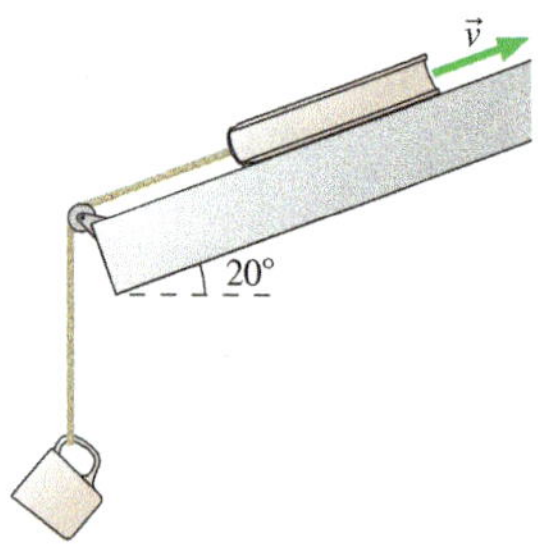

FIGURE P7.43

44. || The 2000 kg cable car shown in FIGURE P7.44 descends a 200-m-high hill. In addition to its brakes, the cable car controls its speed by pulling an 1800 kg counterweight up the other side of the hill. The rolling friction of both the cable car and the counterweight are negligible.
   a. How much braking force does the cable car need to descend at constant speed?
   b. One day the brakes fail just as the cable car leaves the top on its downward journey. What is the runaway car's speed at the bottom of the hill?

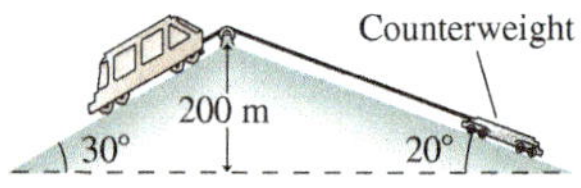

FIGURE P7.44

45. ||| The century-old *ascensores* in Valparaiso, Chile, are small cable cars that go up and down the steep hillsides. As FIGURE P7.45 shows, one car ascends as the other descends. The cars use a two-cable arrangement to compensate for friction; one cable passing around a large pulley connects the cars, the second is pulled by a small motor. Suppose the mass of both cars (with passengers) is 1500 kg, the coefficient of rolling friction is 0.020, and the cars move at constant speed. What is the tension in (a) the connecting cable and (b) the cable to the motor?

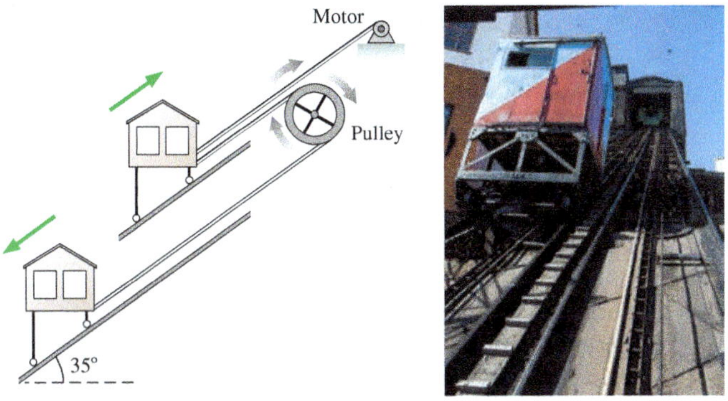

FIGURE P7.45

46. ‖ A house painter uses the chair-and-pulley arrangement of FIGURE P7.46 to lift himself up the side of a house. The painter's mass is 70 kg and the chair's mass is 10 kg. With what force must he pull down on the rope in order to accelerate upward at 0.20 m/s^2?

FIGURE P7.46

FIGURE P7.47

47. ||| Jorge, with mass m, is wearing roller skates whose coefficient of friction with the floor is μ_r. He ties a massless rope around his waist, passes it around a frictionless pulley, and grabs hold of the other end, as shown in FIGURE P7.47. Jorge then pulls hand over hand on the rope with a constant force F. Find an expression for Jorge's acceleration toward the wall.

48. ||| A 70 kg tightrope walker stands at the center of a rope. The rope supports are 10 m apart and the rope sags 10° at each end. The tightrope walker crouches down, then leaps straight up with an acceleration of 8.0 m/s^2 to catch a passing trapeze. What is the tension in the rope as he jumps?

49. ‖ Find an expression for the magnitude of the horizontal force F in FIGURE P7.49 for which m_1 does not slip either up or down along the wedge. All surfaces are frictionless.

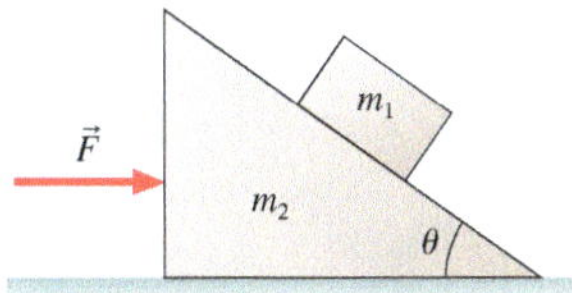

FIGURE P7.49

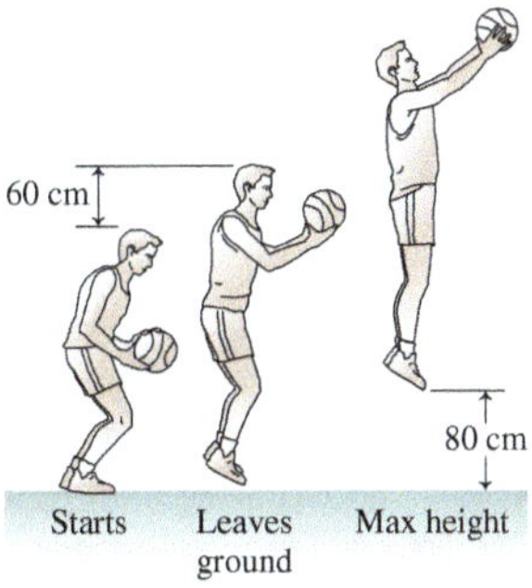

FIGURE P7.50

50. ‖ A 100 kg basketball player can leap straight up in the air to a height of 80 cm, as shown in FIGURE P7.50. You can understand how by analyzing the situation as follows:

a. The player bends his legs until the upper part of his body has dropped by 60 cm, then he begins his jump. Draw separate free-body diagrams for the player and for the floor *as* he is jumping, but before his feet leave the ground.
b. Is there a net force on the player as he jumps (before his feet leave the ground)? How can that be? Explain.
c. With what speed must the player leave the ground to reach a height of 80 cm?
d. What was his acceleration, assumed to be constant, as he jumped?
e. Suppose the player jumps while standing on a bathroom scale that reads in newtons. What does the scale read before he jumps, as he is jumping, and after his feet leave the ground?

Problems 51 and 52 show the free-body diagrams of two interacting systems. For each of these, you are to

a. Write a realistic problem for which these are the correct free-body diagrams. Be sure that the answer your problem requests is consistent with the diagrams shown.
b. Finish the solution of the problem.

51.

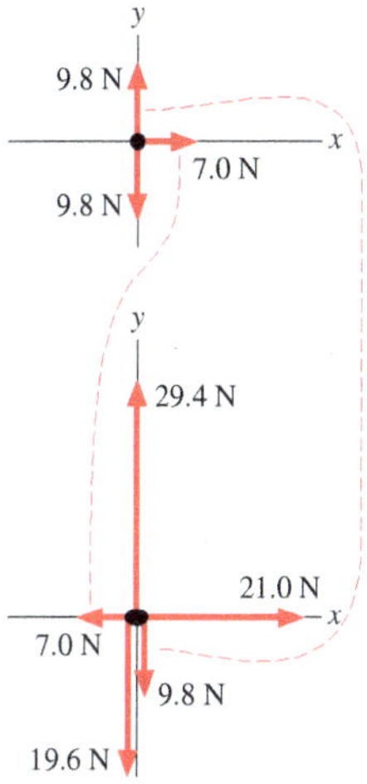

FIGURE P7.51

52.

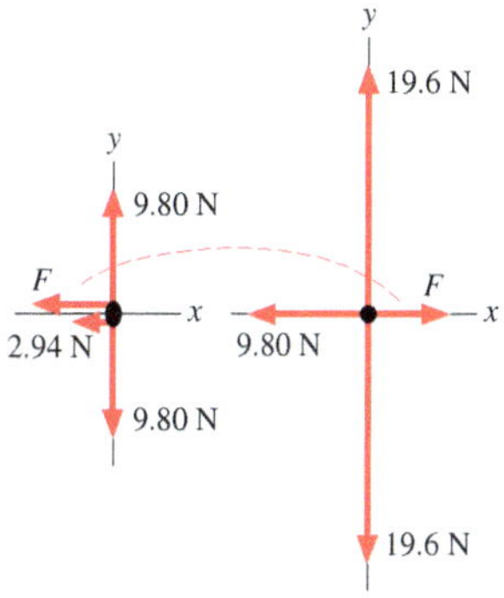

FIGURE P7.52

Challenge Problems

53. A 100 g ball of clay is thrown horizontally with a speed of 10 m/s toward a 900 g block resting on a frictionless surface. It hits the block and sticks. The clay exerts a constant force on the block during the 10 ms it takes the clay to come to rest relative to the block. After 10 ms, the block and the clay are sliding along the surface as a single system.

a. What is their speed after the collision?
b. What is the force of the clay on the block during the collision?
c. What is the force of the block on the clay?

NOTE ▶ This problem can be worked using the conservation laws you will be learning in the next few chapters. However, here you're asked to solve the problem using Newton's laws. ◀

54. In FIGURE CP7.54, find an expression for the acceleration of m_1. The pulleys are massless and frictionless. **Hint:** Think carefully about the acceleration constraint.

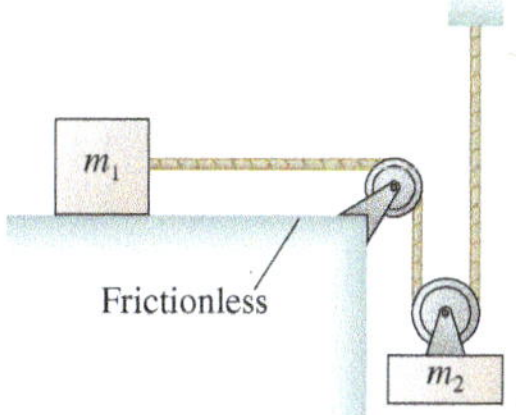

FIGURE CP7.54

55. What is the acceleration of the 2.0 kg block in FIGURE CP7.55 across the frictionless table?
Hint: Think carefully about the acceleration constraint.

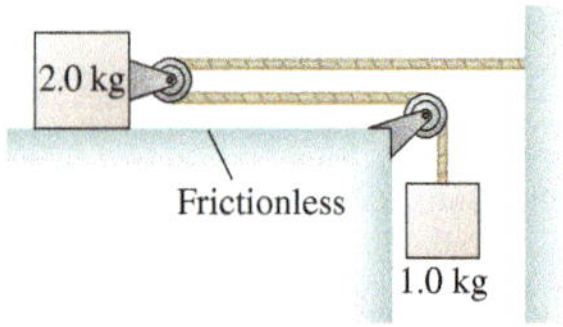

FIGURE CP7.55

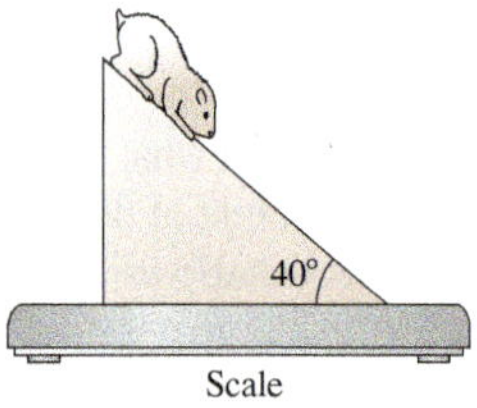

FIGURE CP7.56

56. FIGURE CP7.56 shows a 200 g hamster sitting on an 800 g wedge-shaped block. The block, in turn, rests on a spring scale. An extra-fine lubricating oil having $\mu_s = \mu_k = 0$ is sprayed on the top surface of the block, causing the hamster to slide down. Friction between the block and the scale is large enough that the block does *not* slip on the scale. What does the scale read, in grams, as the hamster slides down?

57. FIGURE CP7.57 shows three hanging masses connected by massless strings over two massless, frictionless pulleys.
   a. Find the acceleration constraint for this system. It is a single equation relating a_{1y}, a_{2y}, and a_{3y}.
   Hint: y_A isn't constant.
   b. Find an expression for the tension in string A.
   Hint: You should be able to write four second-law equations. These, plus the acceleration constraint, are five equations in five unknowns.
   c. Suppose: $m_1 = 2.5$ kg, $m_2 = 1.5$ kg, and $m_3 = 4.0$ kg. Find the acceleration of each.
   d. The 4.0 kg mass would appear to be in equilibrium. Explain why it accelerates.

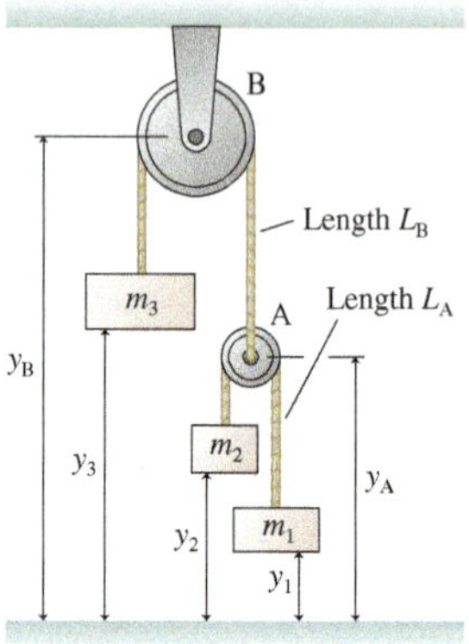

FIGURE CP7.57

STOP TO THINK ANSWERS

Stop to Think 7.1: The crate's gravitational force and the normal force are incorrectly identified as an action/reaction pair. The normal force should be paired with a downward force of the crate on the ground. Gravity is the pull of the entire earth, so $\vec{F}_G$ should be paired with a force pulling up on the entire earth.

Stop to Think 7.2: c. Newton's third law says that the force of A on B is *equal* and opposite to the force of B on A. This is always true. The speed of the objects isn't relevant.

Stop to Think 7.3: b. $F_{\text{B on H}} = F_{\text{H on B}}$ and $F_{\text{A on B}} = F_{\text{B on A}}$ because these are action/reaction pairs. Box B is slowing down and therefore must have a net force to the left. So from Newton's second law we also know that $F_{\text{H on B}} > F_{\text{A on B}}$.

Stop to Think 7.4: Equal to. Each block is hanging in equilibrium, with no net force, so the upward tension force is mg.

Stop to Think 7.5: Less than. Block B is *accelerating* downward, so the net force on B must point down. The only forces acting on B are the tension and gravity, so $T_{\text{S on B}} < (F_G)_B$.

Stop to Think 7.6: c. Newton's third law says that the force of A on B is *equal* and opposite to the force of B on A. This is always true. The mass of the objects isn't relevant.

8 Dynamics II: Motion in a Plane

Why doesn't the roller coaster fall off the track at the top of the loop?

▶ **Looking Ahead** The goal of Chapter 8 is to learn how to solve problems about motion in a plane.

Newton's Laws in 2D

This chapter extends Newton's laws to two-dimensional motion in a plane.

One important application is circular motion. You studied the kinematics in Chapter 4; now we want to look at the forces of circular motion.

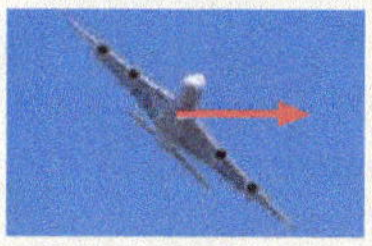

You'll learn that the net force on this turning plane is directed toward the center of the circle.

You'll learn to analyze circular motion using a coordinate system with *radial* and *tangential* components—what we'll call the *rtz*-coordinate system.

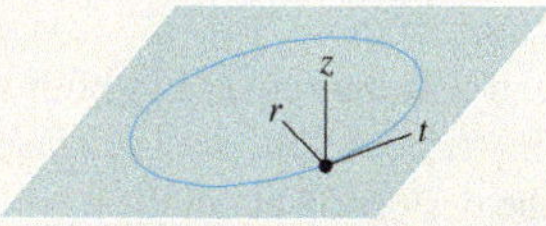

The kinematics of projectile motion was another important topic of Chapter 4. We'll justify those equations and learn how to handle situations where there are forces in addition to gravity.

◀ **Looking Back**
Chapter 4 Kinematics of planar and circular motion

◀ **Looking Back**
Section 6.2 Solving dynamics problems with Newton's second law

Dynamics of Circular Motion

For uniform circular motion, there must be a net force toward the center of the circle to create the centripetal acceleration of changing direction.

Acceleration points toward the center for uniform circular motion at constant speed. You'll learn that non-uniform circular motion has a tangential component of acceleration.

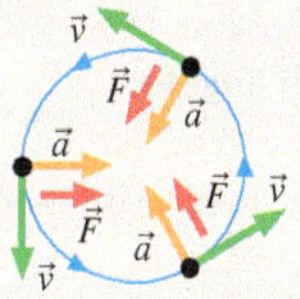

Dynamics in a Plane

Two-dimensional motion with acceleration along both axes often can be analyzed writing Newton's second law in terms of its *x*- and *y*-components.

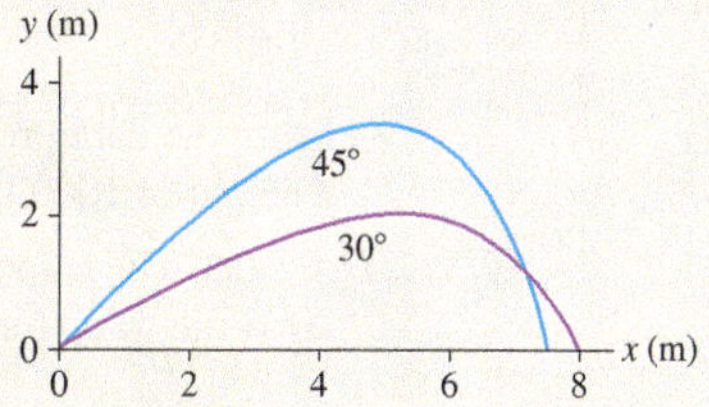

An example is projectile motion with air resistance, for which maximum range no longer occurs for a 45° launch.

Gravity and Orbits

You'll see that an **orbit** can be thought of as projectile motion that never gets any closer to the ground because the ground curves away as fast as the object falls.

An orbiting projectile is in free fall!

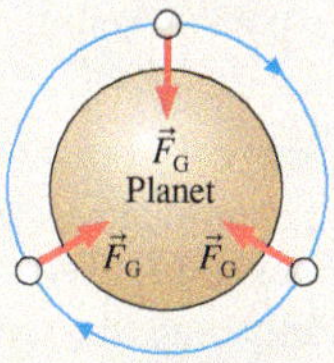

◀ **Looking Back**
Section 6.3 Gravity and weight

Reasoning About Circular Motion

Water in a bucket swung over your head has a downward gravitation pull, but it doesn't fall out.

These riders at the carnival feel pressed against the wall, yet the force on them points inward, toward the center.

You will learn to understand and explain the physics of these seemingly odd effects.

8.1 Dynamics in Two Dimensions

Newton's second law, $\vec{a} = \vec{F}_{net}/m$, determines an object's acceleration. It makes no distinction between linear motion and two-dimensional motion in a plane. In general, the x- and y-components of the acceleration vector are given by

$$a_x = \frac{(F_{net})_x}{m} \quad \text{and} \quad a_y = \frac{(F_{net})_y}{m} \tag{8.1}$$

Suppose the x- and y-components of acceleration are *independent* of each other. That is, a_x does not depend on either y or v_y, and similarly a_y does not depend on x or v_x. Then Problem-Solving Strategy 6.2 for dynamics problems, on page 142, is still valid. As a quick review, you should

1. Draw a pictorial representation—a sketch and a free-body diagram.
2. Use Newton's second law in component form:

$$(F_{net})_x = \sum F_x = ma_x \quad \text{and} \quad (F_{net})_y = \sum F_y = ma_y$$

The force components (including proper signs) are found from the free-body diagram.

3. Solve for the acceleration. If the acceleration is constant, use the two-dimensional kinematic equations of Chapter 4 to find velocities and positions.

EXAMPLE 8.1 Rocketing in the wind

A small rocket for gathering weather data has a mass of 30 kg and generates 1500 N of thrust. On a windy day, the wind exerts a 20 N horizontal force on the rocket. If the rocket is launched straight up, what is the shape of its trajectory, and by how much has it been deflected sideways when it reaches a height of 1.0 km? Because the rocket goes much higher than this, assume there's no significant mass loss during the first 1.0 km of flight.

MODEL Model the rocket as a particle. We need to find the *function* $y(x)$ describing the curve the rocket follows. Because rockets have pointy, aerodynamic shapes, we'll assume no vertical air resistance.

VISUALIZE FIGURE 8.1 shows a pictorial representation. We've chosen a coordinate system with a vertical y-axis. Three forces act on the rocket: two vertical and one horizontal. The wind force is essentially drag (the rocket is moving sideways relative to the wind), so we've labeled it $\vec{D}$.

SOLVE The vertical and horizontal forces are independent of each other, so we can follow the problem-solving strategy summarized above. Newton's second law is

$$a_x = \frac{(F_{net})_x}{m} = \frac{D}{m}$$

$$a_y = \frac{(F_{net})_y}{m} = \frac{F_{thrust} - mg}{m}$$

Both accelerations are constant, so we can use kinematics to find

$$x = \tfrac{1}{2}a_x(\Delta t)^2 = \frac{D}{2m}(\Delta t)^2$$

$$y = \tfrac{1}{2}a_y(\Delta t)^2 = \frac{F_{thrust} - mg}{2m}(\Delta t)^2$$

where we used the fact that all initial positions and velocities are zero. From the x-equation, $(\Delta t)^2 = 2mx/D$. Substituting this into the y-equation, we find

$$y(x) = \left[\frac{F_{thrust} - mg}{D}\right]x$$

This is the equation of the rocket's trajectory. It is a linear equation. Somewhat surprisingly, given that the rocket has both vertical and horizontal accelerations, its trajectory is a *straight line.* We can rearrange this result to find the deflection at height y:

$$x = \left[\frac{D}{F_{thrust} - mg}\right]y$$

From the data provided, we can calculate a deflection of 17 m at a height of 1000 m.

ASSESS The solution depended on the fact that the time parameter Δt is the *same* for both components of the motion.

FIGURE 8.1 Pictorial representation of the rocket launch.

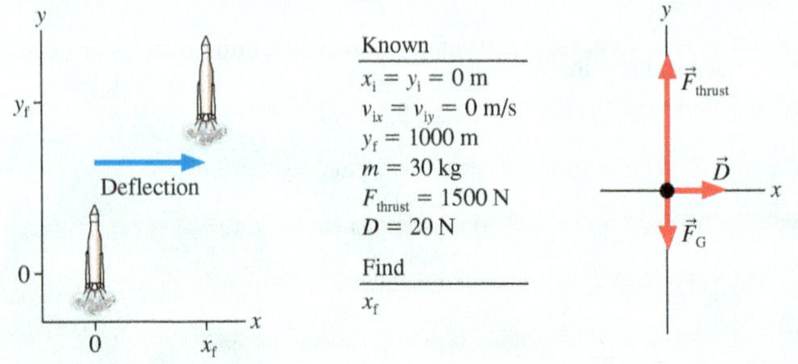

Projectile Motion

We found in Chapter 6 that the gravitational force on an object near the surface of a planet is $\vec{F}_G = (mg, \text{down})$. If we choose a coordinate system with a vertical y-axis, then

$$\vec{F}_G = -mg\hat{j} \tag{8.2}$$

Consequently, from Newton's second law, the acceleration is

$$a_x = \frac{(F_G)_x}{m} = 0$$
$$a_y = \frac{(F_G)_y}{m} = -g \tag{8.3}$$

Equations 8.3 justify the accelerations of Chapter 4—a downward acceleration $a_y = -g$ with no horizontal acceleration—that led to the parabolic motion of a drag-free projectile. The vertical motion is free fall, while the horizontal motion is one of constant velocity.

However, the situation is quite different for a low-mass projectile, where the effects of drag are too large to ignore. We'll leave it as a homework problem for you to show that the acceleration of a projectile subject to drag is

$$a_x = -\frac{\rho CA}{2m} v_x \sqrt{v_x^2 + v_y^2}$$
$$a_y = -g - \frac{\rho CA}{2m} v_y \sqrt{v_x^2 + v_y^2} \tag{8.4}$$

Here the components of acceleration are *not* independent of each other because a_x depends on v_y and vice versa. It turns out that these two equations cannot be solved exactly for the trajectory, but they can be solved numerically. FIGURE 8.2 shows the numerical solution for the motion of a 5 g plastic ball that's been hit with an initial speed of 25 m/s. It doesn't travel very far (the maximum distance would be more than 60 m in a vacuum), and the maximum range is no longer reached for a launch angle of 45°. In this case, maximum distance is achieved by hitting the ball at a 30° angle. A 60° launch angle, which gives the same distance as 30° in vacuum, travels only ≈ 75% as far. Notice that the trajectories are not at all parabolic.

When drag is included, the angle for maximum range of a projectile depends both on its size and mass. The optimum angle is roughly 35° for baseballs. The flight of a golf ball is even more complex because the dimples and the high rate of spin greatly affect its aerodynamics. Professional golfers achieve their maximum distance at launch angles of barely 15°.

FIGURE 8.2 A projectile is affected by drag. This example shows trajectories of a plastic ball launched at different angles.

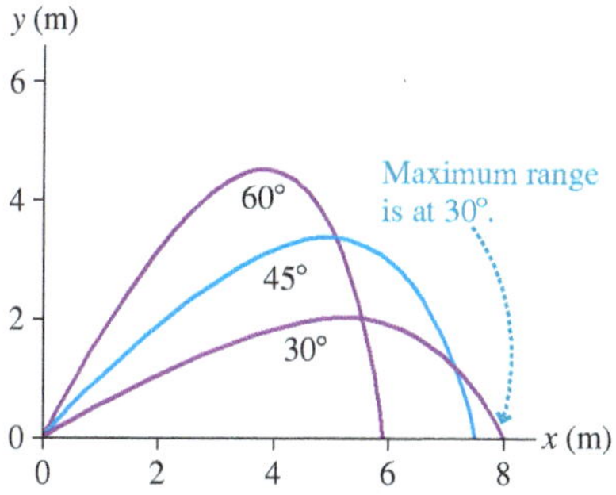

STOP TO THINK 8.1 This acceleration will cause the particle to

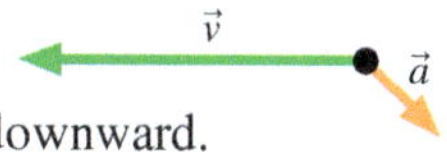

a. Speed up and curve upward.
b. Speed up and curve downward.
c. Slow down and curve upward.
d. Slow down and curve downward.
e. Move to the right and down.
f. Reverse direction.

8.2 Uniform Circular Motion

We studied the mathematics of circular motion in Chapter 4, and a review is *highly* recommended. Recall that a particle in uniform circular motion with angular velocity ω has speed $v = \omega r$ and centripetal acceleration

$$\vec{a} = \left(\frac{v^2}{r}, \text{toward center of circle}\right) = (\omega^2 r, \text{toward center of circle}) \tag{8.5}$$

Now we're ready to study *dynamics*—how forces *cause* circular motion.

The xy-coordinate system we've been using for linear motion and projectile motion is not the best coordinate system for circular dynamics. FIGURE 8.3 shows a circular

FIGURE 8.3 The rtz-coordinate system.

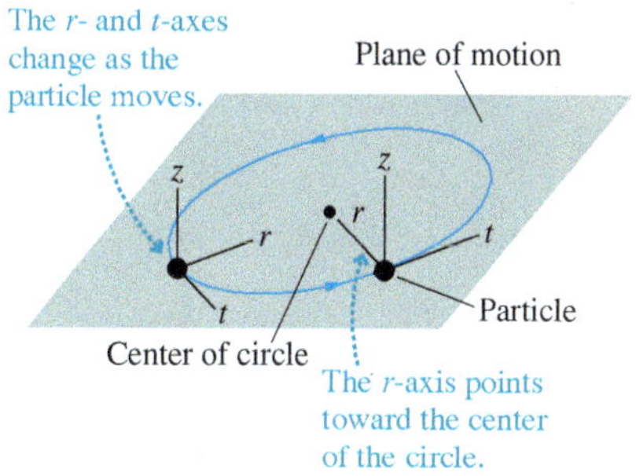

trajectory and the plane in which the circle lies. Let's establish a coordinate system with its origin at the point where the particle is located. The axes are defined as follows:

- The r-axis (radial axis) points *from* the particle *toward* the center of the circle.
- The t-axis (tangential axis) is tangent to the circle, pointing in the ccw direction.
- The z-axis is perpendicular to the plane of motion.

The three axes of this rtz-coordinate system are mutually perpendicular, just like the axes of the familiar xyz-coordinate system. Notice how the axes move with the particle so that the r-axis always points to the center of the circle. It will take a little getting used to, but you will soon see that circular-motion problems are most easily described in these coordinates.

FIGURE 8.4 The velocity and acceleration vectors in the rtz-coordinate system.

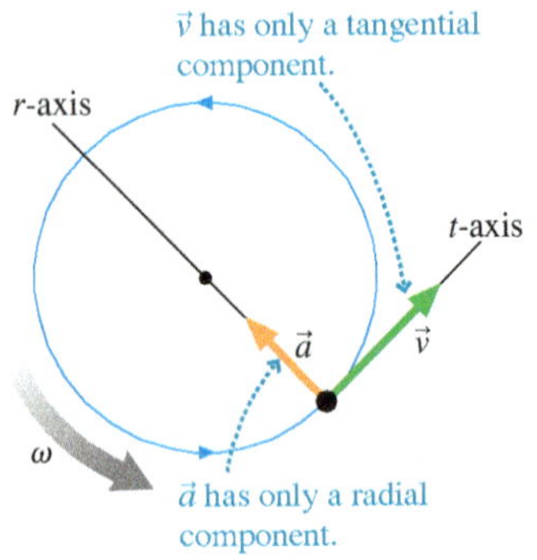

FIGURE 8.4, from Chapter 4, reminds you that a particle in uniform circular motion has a velocity tangential to the circle and an acceleration—the centripetal acceleration—pointing toward the center of the circle. Thus the rtz-components of $\vec{v}$ and $\vec{a}$ are

$$\begin{aligned} v_r &= 0 & a_r &= \frac{v^2}{r} = \omega^2 r \\ v_t &= \omega r & a_t &= 0 \\ v_z &= 0 & a_z &= 0 \end{aligned} \tag{8.6}$$

where $\omega = d\theta/dt$, the angular velocity, must be in rad/s. In other words, **the velocity vector has only a tangential component, the acceleration vector has only a radial component.** Now you can begin to see the advantages of the rtz-coordinate system. For convenience, we'll generally refer to a_r as "the centripetal acceleration" rather than "the radial acceleration."

NOTE ▶ Recall that ω is positive for a counterclockwise (ccw) rotation, negative for a clockwise (cw) rotation. Hence the tangential velocity v_t is positive/negative for ccw/cw rotations. Because v_t is the only nonzero component of velocity, the particle's speed is $v = |v_t| = |\omega| r$. We'll sometimes write this as $v = \omega r$ if there's no ambiguity about the signs. ◀

EXAMPLE 8.2 The ultracentrifuge

A 17-cm-diameter ultracentrifuge produces an extraordinarily large acceleration of 600,000g, where g is the free-fall acceleration. What is the rotational frequency in rpm? What is the speed of a bacterium at the bottom of the centrifuge tube?

SOLVE The radius of the circular motion is 8.5 cm, or 0.085 m. From Equations 8.6, we see that the angular velocity is

$$\omega = \sqrt{\frac{a_r}{r}} = \sqrt{\frac{(600{,}000)(9.8 \text{ m/s}^2)}{0.085 \text{ m}}} = 8320 \text{ rad/s}$$

Converting to rpm, we find

$$\omega = 8320 \frac{\text{rad}}{\text{s}} \times \frac{1 \text{ rev}}{2\pi \text{ rad}} \times \frac{60 \text{ s}}{1 \text{ min}} = 80{,}000 \text{ rpm}$$

This incredibly fast rotation rate is why it's called an *ultra*centrifuge. The tubes pivot outward as the centrifuge spins, so a bacterium at the "bottom" of the tube is rotating at the end of an arm of radius 8.5 cm. Its speed is

$$v = \omega r = (8320 \text{ rad/s})(0.085 \text{ m}) = 710 \text{ m/s}$$

This is more than twice the speed of sound!

STOP TO THINK 8.2 Rank in order, from largest to smallest, the centripetal accelerations $(a_r)_a$ to $(a_r)_e$ of particles a to e.

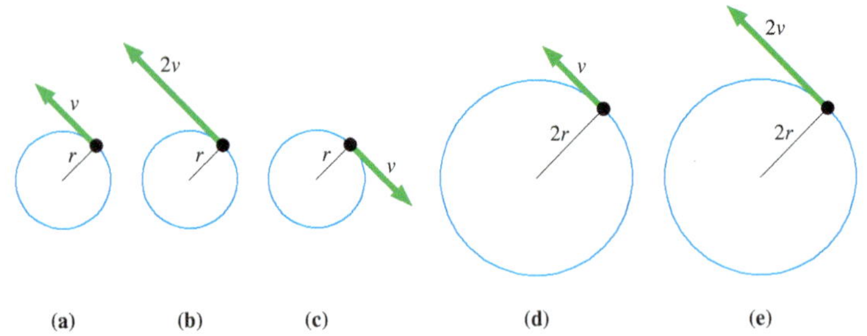

Dynamics of Uniform Circular Motion

A particle in uniform circular motion is clearly not traveling at constant velocity in a straight line. Consequently, according to Newton's first law, the particle *must* have a net force acting on it. We've already determined the acceleration of a particle in uniform circular motion—the centripetal acceleration of Equation 8.5. Newton's second law tells us exactly how much net force is needed to cause this acceleration:

$$\vec{F}_{\text{net}} = m\vec{a} = \left(\frac{mv^2}{r}, \text{ toward center of circle}\right) \tag{8.7}$$

In other words, a particle of mass m moving at constant speed v around a circle of radius r *must* have a net force of magnitude mv^2/r pointing toward the center of the circle. Without such a force, the particle would move off in a straight line tangent to the circle.

Highway and racetrack curves are banked to allow the normal force of the road to provide the centripetal acceleration of the turn.

FIGURE 8.5 shows the net force $\vec{F}_{\text{net}}$ acting on a particle as it undergoes uniform circular motion. You can see that **$\vec{F}_{\text{net}}$ points along the radial axis of the *rtz*-coordinate system, toward the center of the circle.** The tangential and perpendicular components of $\vec{F}_{\text{net}}$ are zero.

FIGURE 8.5 The net force points in the radial direction, toward the center of the circle.

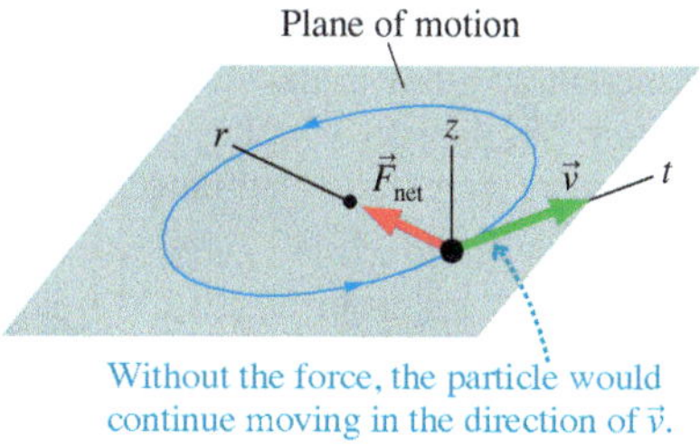

NOTE ▶ The force described by Equation 8.7 is not a *new* force. Our rules for identifying forces have not changed. What we are saying is that a particle moves with uniform circular motion *if and only if* a net force always points toward the center of the circle. The force itself must have an identifiable agent and will be one of our familiar forces, such as tension, friction, or the normal force. Equation 8.7 simply tells us how the force needs to act—how strongly and in which direction—to cause the particle to move with speed v in a circle of radius r. ◀

The usefulness of the *rtz*-coordinate system becomes apparent when we write Newton's second law, Equation 8.7, in terms of the r-, t-, and z-components:

$$\begin{aligned}(F_{\text{net}})_r &= \sum F_r = ma_r = \frac{mv^2}{r} = m\omega^2 r\\ (F_{\text{net}})_t &= \sum F_t = ma_t = 0\\ (F_{\text{net}})_z &= \sum F_z = ma_z = 0\end{aligned} \tag{8.8}$$

Notice that we've used our explicit knowledge of the acceleration, as given in Equations 8.6, to write the right-hand sides of these equations. **For uniform circular motion, the sum of the forces along the *t*-axis and along the *z*-axis *must* equal zero, and the sum of the forces along the *r*-axis must equal ma_r, where a_r is the centripetal acceleration.**

A few examples will clarify these ideas and show how some of the forces you've come to know can be involved in circular motion.

EXAMPLE 8.3 Spinning in a circle

An energetic father places his 20 kg child on a 5.0 kg cart to which a 2.0-m-long rope is attached. He then holds the end of the rope and spins the cart and child around in a circle, keeping the rope parallel to the ground. If the tension in the rope is 100 N, how many revolutions per minute (rpm) does the cart make? Rolling friction between the cart's wheels and the ground is negligible.

MODEL Model the child in the cart as a particle in uniform circular motion.

VISUALIZE FIGURE 8.6 on the next page shows the pictorial representation. A circular-motion problem usually does not have starting and ending points like a projectile problem, so numerical subscripts such as x_1 or y_2 are usually not needed. Here we need to define the cart's speed v and the radius r of the circle. Further, a motion diagram is not needed for uniform circular motion because we already know the acceleration $\vec{a}$ points to the center of the circle.

Continued

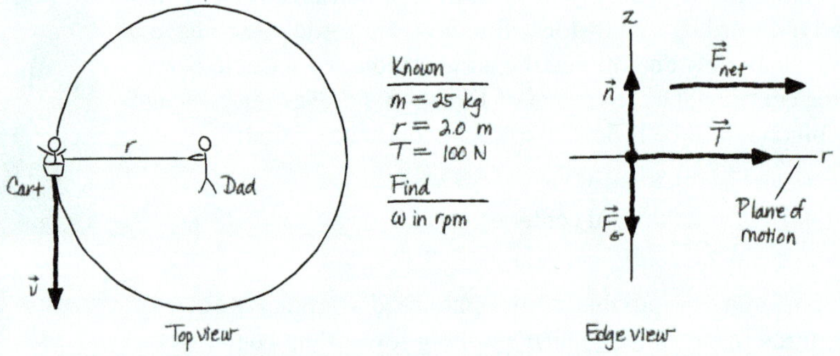

FIGURE 8.6 Pictorial representation of a cart spinning in a circle.

The essential part of the pictorial representation is the free-body diagram. **For uniform circular motion we'll draw the free-body diagram in the rz-plane, looking at the edge of the circle, because this is the plane of the forces.** The contact forces acting on the cart are the normal force of the ground and the tension force of the rope. The normal force is perpendicular to the plane of the motion and thus in the z-direction. The direction of $\vec{T}$ is determined by the statement that the rope is parallel to the ground. In addition, there is the long-range gravitational force $\vec{F}_G$.

SOLVE We defined the r-axis to point toward the center of the circle, so $\vec{T}$ points in the positive r-direction and has r-component $T_r = T$. Newton's second law, using the rtz-components of Equations 8.8, is

$$\sum F_r = T = \frac{mv^2}{r}$$

$$\sum F_z = n - mg = 0$$

We've taken the r- and z-components of the forces directly from the free-body diagram, as you learned to do in Chapter 6. Then we've *explicitly* equated the sums to $a_r = v^2/r$ and $a_z = 0$. This is the basic strategy for all uniform circular-motion problems. From the z-equation we can find that $n = mg$. This would be useful if we needed to determine a friction force, but it's not needed in this problem. From the r-equation, the speed of the cart is

$$v = \sqrt{\frac{rT}{m}} = \sqrt{\frac{(2.0\text{ m})(100\text{ N})}{25\text{ kg}}} = 2.83\text{ m/s}$$

The cart's angular velocity is then found from Equations 8.6:

$$\omega = \frac{v_t}{r} = \frac{v}{r} = \frac{2.83\text{ m/s}}{2.0\text{ m}} = 1.41\text{ rad/s}$$

This is another case where we inserted the radian unit because ω is specifically an *angular* velocity. Finally, we need to convert ω to rpm:

$$\omega = \frac{1.41\text{ rad}}{1\text{ s}} \times \frac{1\text{ rev}}{2\pi\text{ rad}} \times \frac{60\text{ s}}{1\text{ min}} = 14\text{ rpm}$$

ASSESS 14 rpm corresponds to a period $T = 4.3$ s. This result is reasonable.

EXAMPLE 8.4 Turning the corner I

What is the maximum speed with which a 1500 kg car can make a left turn around a curve of radius 50 m on a level (unbanked) road without sliding?

MODEL Although the car turns only a quarter of a circle, we can model the car as a particle in uniform circular motion as it goes around the turn. Assume that rolling friction is negligible.

VISUALIZE FIGURE 8.7 shows the pictorial representation. The issue we must address is *how* a car turns a corner. What force or forces cause the direction of the velocity vector to change? Imagine driving on a completely frictionless road, such as a very icy road. You would not be able to turn a corner. Turning the steering wheel would be of no use; the car would slide straight ahead, in accordance with both Newton's first law and the experience of anyone who has ever driven on ice! So it must be *friction* that somehow allows the car to turn.

Figure 8.7 shows the top view of a tire as it turns a corner. If the road surface were frictionless, the tire would slide straight ahead. The force that prevents an object from sliding across a surface is *static friction.* Static friction $\vec{f}_s$ pushes *sideways* on the tire, toward the center of the circle. How do we know the direction is sideways? If $\vec{f}_s$ had a component either parallel to $\vec{v}$ or opposite to $\vec{v}$, it would cause the car to speed up or slow down. Because the car changes direction but not speed, static friction must be perpendicular to $\vec{v}$. $\vec{f}_s$ causes the centripetal acceleration of circular motion around the curve, and thus the free-body diagram, drawn from behind the car, shows the static friction force pointing toward the center of the circle.

FIGURE 8.7 Pictorial representation of a car turning a corner.

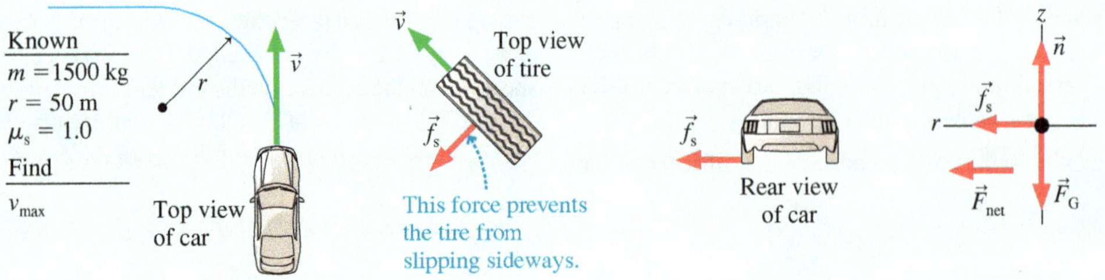

SOLVE The maximum turning speed is reached when the static friction force reaches its maximum $f_{s\,max} = \mu_s n$. If the car enters the curve at a speed higher than the maximum, static friction will not be large enough to provide the necessary centripetal acceleration and the car will slide.

The static friction force points in the positive r-direction, so its radial component is simply the magnitude of the vector: $(f_s)_r = f_s$. Newton's second law in the rtz-coordinate system is

$$\sum F_r = f_s = \frac{mv^2}{r}$$

$$\sum F_z = n - mg = 0$$

The only difference from Example 8.3 is that the tension force toward the center has been replaced by a static friction force toward the center. From the radial equation, the speed is

$$v = \sqrt{\frac{rf_s}{m}}$$

The speed will be a maximum when f_s reaches its maximum value:

$$f_s = f_{s\,max} = \mu_s n = \mu_s mg$$

where we used $n = mg$ from the z-equation. At that point,

$$v_{max} = \sqrt{\frac{rf_{s\,max}}{m}} = \sqrt{\mu_s rg}$$

$$= \sqrt{(1.0)(50\text{ m})(9.80\text{ m/s}^2)} = 22\text{ m/s}$$

where the coefficient of static friction was taken from Table 6.1.

ASSESS 22 m/s ≈ 45 mph, a reasonable answer for how fast a car can take an unbanked curve. Notice that the car's mass canceled out and that the final equation for v_{max} is quite simple. This is another example of why it pays to work algebraically until the very end.

Because μ_s depends on road conditions, the maximum safe speed through turns can vary dramatically. Wet roads, in particular, lower the value of μ_s and thus lower the speed of turns. Icy conditions are even worse. The corner you turn every day at 45 mph will require a speed of no more than 15 mph if the coefficient of static friction drops to 0.1.

EXAMPLE 8.5 Turning the corner II

A highway curve of radius 70 m is banked at a 15° angle. At what speed v_0 can a car take this curve without assistance from friction?

MODEL The car is a particle in uniform circular motion.

VISUALIZE Having just discussed the role of friction in turning corners, it is perhaps surprising to suggest that the same turn can also be accomplished without friction. Example 8.4 considered a level roadway, but real highway curves are *banked* by being tilted up at the outside edge of the curve. The angle is modest on ordinary highways, but it can be quite large on high-speed racetracks. The purpose of banking becomes clear if you look at the free-body diagram in FIGURE 8.8. The normal force $\vec{n}$ is perpendicular to the road, so tilting the road causes $\vec{n}$ to have a component toward the center of the circle. **The radial component n_r is the inward force that causes the centripetal acceleration needed to turn the car.** Notice that we are *not* using a tilted coordinate system, although this looks rather like an inclined-plane problem. The center of the circle is in the same horizontal plane as the car, and for circular-motion problems we need the r-axis to pass through the center. Tilted axes are for *linear* motion along an incline.

FIGURE 8.8 Pictorial representation of a car on a banked curve.

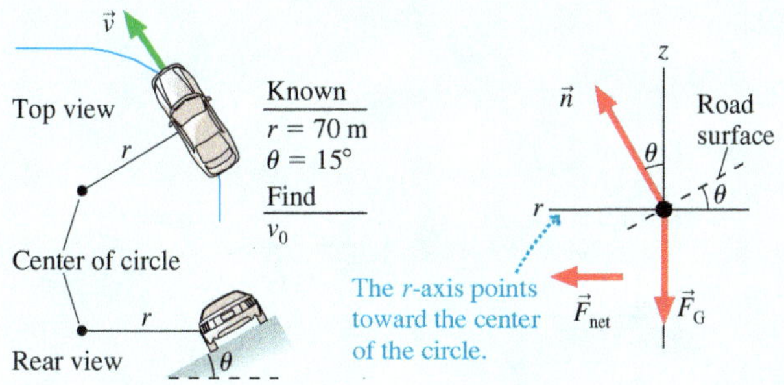

SOLVE Without friction, $n_r = n\sin\theta$ is the only component of force in the radial direction. It is this inward component of the normal force on the car that causes it to turn the corner. Newton's second law is

$$\sum F_r = n\sin\theta = \frac{mv_0^2}{r}$$

$$\sum F_z = n\cos\theta - mg = 0$$

where θ is the angle at which the road is banked and we've assumed that the car is traveling at the correct speed v_0. From the z-equation,

$$n = \frac{mg}{\cos\theta}$$

Substituting this into the r-equation and solving for v_0 give

$$\frac{mg}{\cos\theta}\sin\theta = mg\tan\theta = \frac{mv_0^2}{r}$$

$$v_0 = \sqrt{rg\tan\theta} = 14\text{ m/s}$$

ASSESS This is ≈ 28 mph, a reasonable speed. Only at this very specific speed can the turn be negotiated without reliance on friction forces.

It's interesting to explore what happens at other speeds on a banked curve. FIGURE 8.9 shows that the car will need to rely on both the banking *and* friction if it takes the curve at a speed higher or lower than v_0.

FIGURE 8.9 Free-body diagrams for a car going around a banked curve at speeds higher and lower than the friction-free speed v_0.

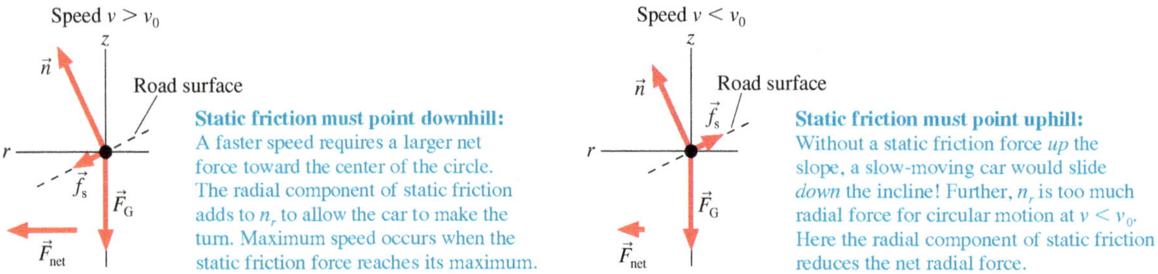

EXAMPLE 8.6 A rock in a sling

A Stone Age hunter places a 1.0 kg rock in a sling and swings it in a horizontal circle around his head on a 1.0-m-long vine. If the vine breaks at a tension of 200 N, what is the maximum angular speed, in rpm, with which he can swing the rock?

MODEL Model the rock as a particle in uniform circular motion.

VISUALIZE This problem appears, at first, to be essentially the same as Example 8.3, where the father spun his child around on a rope. However, the lack of a normal force from a supporting surface makes a *big* difference. In this case, the *only* contact force on the rock is the tension in the vine. Because the rock moves in a horizontal circle, you may be tempted to draw a free-body diagram like FIGURE 8.10a, where $\vec{T}$ is directed along the r-axis. You will quickly run into trouble, however, because this diagram has a net force in the z-direction and it is impossible to satisfy $\sum F_z = 0$. The gravitational force $\vec{F}_G$ certainly points vertically downward, so the difficulty must be with $\vec{T}$.

As an experiment, tie a small weight to a string, swing it over your head, and check the *angle* of the string. You will quickly discover that the string is *not* horizontal but, instead, is angled downward. The sketch of FIGURE 8.10b labels the angle θ. Notice that the rock moves in a *horizontal* circle, so the center of the circle is *not* at his hand. The r-axis points to the center of the circle, but the tension force is directed along the vine. Thus the correct free-body diagram is the one in Figure 8.10b.

SOLVE The free-body diagram shows that the downward gravitational force is balanced by an upward component of the tension, leaving the radial component of the tension to cause the centripetal acceleration. Newton's second law is

$$\sum F_r = T\cos\theta = \frac{mv^2}{r}$$

$$\sum F_z = T\sin\theta - mg = 0$$

where θ is the angle of the vine below horizontal. From the z-equation we find

$$\sin\theta = \frac{mg}{T}$$

$$\theta = \sin^{-1}\left(\frac{(1.0\text{ kg})(9.8\text{ m/s}^2)}{200\text{ N}}\right) = 2.81°$$

where we've evaluated the angle at the maximum tension of 200 N. The vine's angle of inclination is small but not zero.

Turning now to the r-equation, we find the rock's speed is

$$v = \sqrt{\frac{rT\cos\theta}{m}}$$

Careful! The radius r of the circle is *not* the length L of the vine. You can see in Figure 8.10b that $r = L\cos\theta$. Thus

$$v = \sqrt{\frac{LT\cos^2\theta}{m}} = \sqrt{\frac{(1.0\text{ m})(200\text{ N})(\cos 2.81°)^2}{1.0\text{ kg}}} = 14.1\text{ m/s}$$

We can now find the maximum angular speed, the value of ω that brings the tension to the breaking point:

$$\omega_{max} = \frac{v}{r} = \frac{v}{L\cos\theta} = \frac{14.1\text{ rad}}{1\text{ s}} \times \frac{1\text{ rev}}{2\pi\text{ rad}} \times \frac{60\text{ s}}{1\text{ min}} = 135\text{ rpm}$$

FIGURE 8.10 Pictorial representation of a rock in a sling.

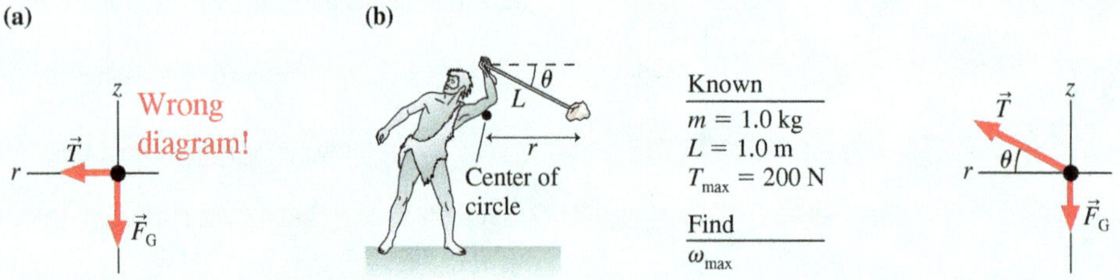

STOP TO THINK 8.3 A block on a string spins in a horizontal circle on a frictionless table. Rank order, from largest to smallest, the tensions T_a to T_e acting on blocks a to e.

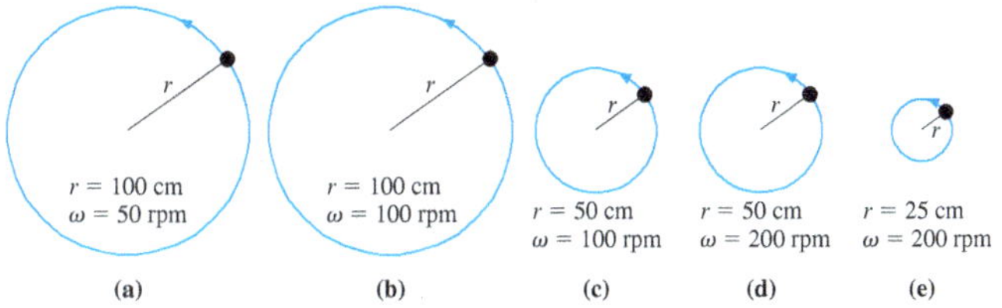

8.3 Circular Orbits

Satellites orbit the earth, the earth orbits the sun, and our entire solar system orbits the center of the Milky Way galaxy. Not all orbits are circular, but in this section we'll limit our analysis to circular orbits. We'll look at the elliptical orbits of satellites and planets in Chapter 13.

How does a satellite orbit the earth? What forces act on it? Why does it move in a circle? To answer these important questions, let's return, for a moment, to projectile motion. Projectile motion occurs when the only force on an object is gravity. Our analysis of projectiles assumed that the earth is flat and that the acceleration due to gravity is everywhere straight down. This is an acceptable approximation for projectiles of limited range, such as baseballs or cannon balls, but there comes a point where we can no longer ignore the curvature of the earth.

FIGURE 8.11 shows a perfectly smooth, spherical, airless planet with one tower of height h. A projectile is launched from this tower parallel to the ground ($\theta = 0°$) with speed v_0. If v_0 is very small, as in trajectory A, the "flat-earth approximation" is valid and the problem is identical to Example 4.4 in which a car drove off a cliff. The projectile simply falls to the ground along a parabolic trajectory.

As the initial speed v_0 is increased, the projectile begins to notice that the ground is curving out from beneath it. It is falling the entire time, always getting closer to the ground, but the distance that the projectile travels before finally reaching the ground—that is, its range—increases because the projectile must "catch up" with the ground that is curving away from it. Trajectories B and C are of this type. The actual calculation of these trajectories is beyond the scope of this textbook, but you should be able to understand the factors that influence the trajectory.

If the launch speed v_0 is sufficiently large, there comes a point where the curve of the trajectory and the curve of the earth are parallel. In this case, the projectile "falls" but it never gets any closer to the ground! This is the situation for trajectory D. A closed trajectory around a planet or star, such as trajectory D, is called an **orbit**.

The most important point of this qualitative analysis is that **an orbiting projectile is in free fall.** This is, admittedly, a strange idea, but one worth careful thought. An orbiting projectile is really no different from a thrown baseball or a car driving off a cliff. The only force acting on it is gravity, but its tangential velocity is so large that the curvature of its trajectory matches the curvature of the earth. When this happens, the projectile "falls" under the influence of gravity but never gets any closer to the surface, which curves away beneath it.

In the flat-earth approximation, shown in FIGURE 8.12a, the gravitational force acting on an object of mass m is

$$\vec{F}_G = (mg, \text{ vertically downward}) \quad \text{(flat-earth approximation)} \quad (8.9)$$

But since stars and planets are actually spherical (or very close to it), the "real" force of gravity acting on an object is directed toward the *center* of the planet, as shown in FIGURE 8.12b. In this case the gravitational force is

$$\vec{F}_G = (mg, \text{ toward center}) \quad \text{(spherical planet)} \quad (8.10)$$

FIGURE 8.11 Projectiles being launched at increasing speeds from height h on a smooth, airless planet.

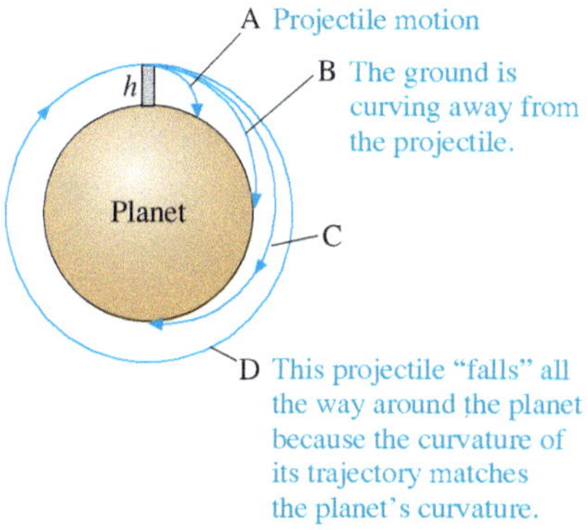

FIGURE 8.12 The "real" gravitational force is always directed toward the center of the planet.

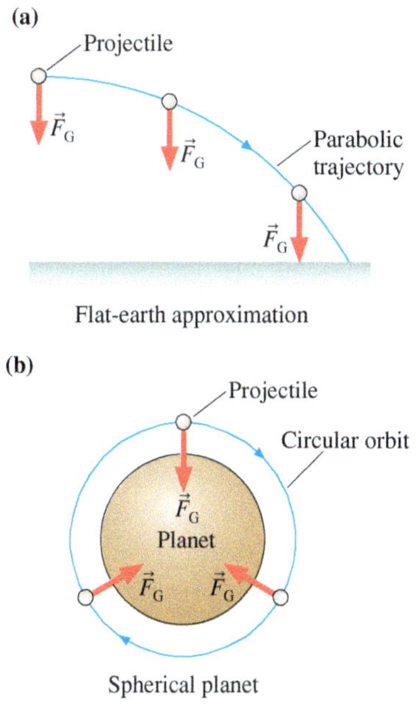

The orbiting space shuttle is in free fall.

As you have learned, a force of constant magnitude that always points toward the center of a circle causes the centripetal acceleration of uniform circular motion. Thus the gravitational force of Equation 8.10 on the object in Figure 8.12b causes it to have acceleration

$$\vec{a} = \frac{\vec{F}_{net}}{m} = (g, \text{toward center}) \tag{8.11}$$

An object moving in a circle of radius r at speed v_{orbit} will have this centripetal acceleration if

$$a_r = \frac{(v_{orbit})^2}{r} = g \tag{8.12}$$

That is, if an object moves parallel to the surface with the speed

$$v_{orbit} = \sqrt{rg} \tag{8.13}$$

then the free-fall acceleration provides exactly the centripetal acceleration needed for a circular orbit of radius r. An object with any other speed will not follow a circular orbit.

The earth's radius is $r = R_e = 6.37 \times 10^6$ m. (A table of useful astronomical data is inside the back cover of this book.) The orbital speed of a projectile just skimming the surface of an airless, bald earth is

$$v_{orbit} = \sqrt{rg} = \sqrt{(6.37 \times 10^6 \text{ m})(9.80 \text{ m/s}^2)} = 7900 \text{ m/s} \approx 16{,}000 \text{ mph}$$

Even if there were no trees and mountains, a real projectile moving at this speed would burn up from the friction of air resistance.

Suppose, however, that we launched the projectile from a tower of height $h = 200 \text{ mi} \approx 3.2 \times 10^5$ m, just above the earth's atmosphere. This is approximately the height of low-earth-orbit satellites, such as the space shuttle. Note that $h \ll R_e$, so the radius of the orbit $r = R_e + h = 6.69 \times 10^6$ m is only 5% greater than the earth's radius. Many people have a mental image that satellites orbit far above the earth, but in fact many satellites come pretty close to skimming the surface. Our calculation of v_{orbit} thus turns out to be quite a good estimate of the speed of a satellite in low earth orbit.

We can use v_{orbit} to calculate the period of a satellite orbit:

$$T = \frac{2\pi r}{v_{orbit}} = 2\pi\sqrt{\frac{r}{g}} \tag{8.14}$$

For a low earth orbit, with $r = R_e + 200$ miles, we find $T = 5190 \text{ s} = 87$ min. The period of the space shuttle at an altitude of 200 mi is, indeed, close to 87 minutes. (The actual period of the shuttle at this elevation is 91 min. The difference, you'll learn in Chapter 13, arises because g is slightly less at a satellite's altitude.)

When we discussed *weightlessness* in Chapter 6, we discovered that it occurs during free fall. We asked the question, at the end of Section 6.3, whether astronauts and their spacecraft were in free fall. We can now give an affirmative answer: They are, indeed, in free fall. They are falling continuously around the earth, under the influence of only the gravitational force, but never getting any closer to the ground because the earth's surface curves beneath them. Weightlessness in space is no different from the weightlessness in a free-falling elevator. It does *not* occur from an absence of gravity. Instead, the astronaut, the spacecraft, and everything in it are weightless because they are all falling together.

Gravity

We can leave this section with a glance ahead, where we will look at the gravitational force more closely. If a satellite is simply "falling" around the earth, with the gravitational force causing a centripetal acceleration, then what about the moon? Is it obeying the same laws of physics? Or do celestial objects obey laws that we cannot discover by experiments here on earth?

The radius of the moon's orbit around the earth is $r = R_m = 3.84 \times 10^8$ m. If we use Equation 8.14 to calculate the period of the moon's orbit, the time it takes the moon to circle the earth once, we get

$$T = 2\pi\sqrt{\frac{r}{g}} = 2\pi\sqrt{\frac{3.84 \times 10^8 \text{ m}}{9.80 \text{ m/s}^2}} = 655 \text{ min} \approx 11 \text{ h}$$

This is clearly wrong. As you probably know, the full moon occurs roughly once a month. More exactly, we know from astronomical measurements that the period of the moon's orbit is $T = 27.3$ days $= 2.36 \times 10^6$ s, a factor of 60 longer than we calculated it to be.

Newton believed that the laws of motion he had discovered were *universal.* That is, they should apply to the motion of the moon as well as to the motion of objects in the laboratory. But why should we assume that the free-fall acceleration g is the same at the distance of the moon as it is on or near the earth's surface? If gravity is the force of the earth pulling on an object, it seems plausible that the size of that force, and thus the size of g, should diminish with increasing distance from the earth.

If the moon orbits the earth because of the earth's gravitational pull, what value of g would be needed to explain the moon's period? We can calculate $g_{\text{at moon}}$ from Equation 8.14 and the observed value of the moon's period:

$$g_{\text{at moon}} = \frac{4\pi^2 R_m}{T_{\text{moon}}^{\;2}} = 0.00272 \text{ m/s}^2$$

This is much less than the earth-bound value of 9.80 m/s^2.

As you learned in Chapter 6, Newton proposed the idea that the earth's force of gravity decreases inversely with the square of the distance from the earth. In Chapter 13, we'll use Newton's law of gravity, the mass of the earth, and the distance to the moon to *predict* that $g_{\text{at moon}} = 0.00272$ m/s^2, exactly as expected. The moon, just like the space shuttle, is simply "falling" around the earth!

Saturn's beautiful rings consist of dust particles and small rocks orbiting the planet.

8.4 Fictitious Forces

If you are riding in a car that makes a sudden stop, you may feel as if a force "throws" you forward toward the windshield. But there really is no such force. You cannot identify any agent that does the throwing. An observer watching from beside the road would simply see you continuing forward as the car stops.

The decelerating car is not an inertial reference frame. You learned in Chapter 5 that Newton's laws are valid only in inertial reference frames. The roadside observer is in the earth's inertial reference frame. His observations of the car decelerating relative to the earth while you continue forward with constant velocity are in accord with Newton's laws.

Nonetheless, the fact that you *seem* to be hurled forward relative to the car is a very real experience. You can describe your experience in terms of what are called **fictitious forces**. These are not real forces because no agent is exerting them, but they describe your motion *relative to a noninertial reference frame.* FIGURE 8.13 shows the situation from both reference frames.

FIGURE 8.13 The forces are properly identified only in an inertial reference frame.

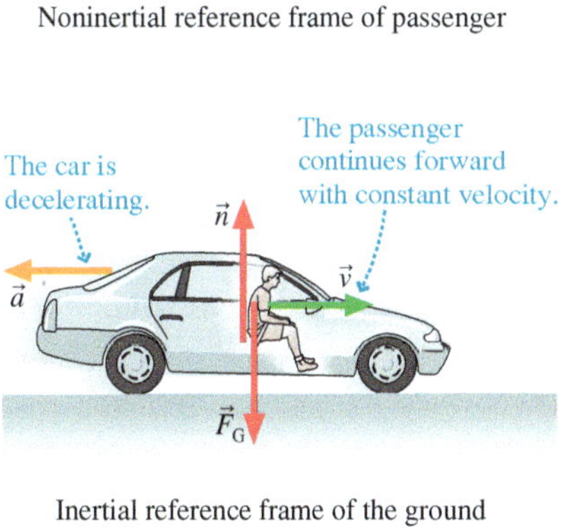

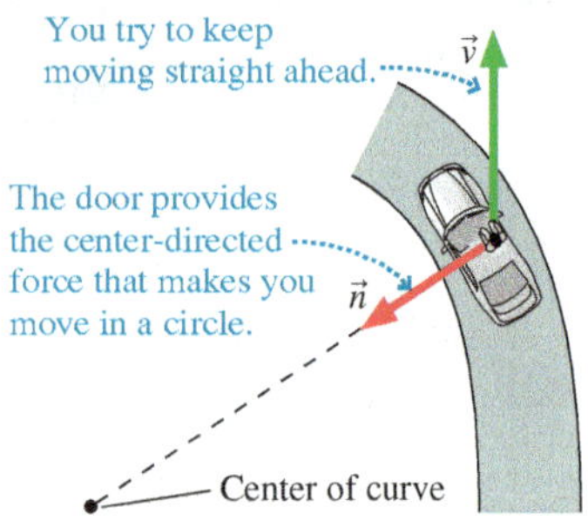

FIGURE 8.14 Bird's-eye view of a passenger as a car turns a corner.

Centrifugal Force?

If the car turns a corner quickly, you feel "thrown" against the door. But is there really such a force? **FIGURE 8.14** shows a bird's-eye view of you riding in a car as it makes a left turn. You try to continue moving in a straight line, obeying Newton's first law, when—without having been provoked—the door suddenly turns in front of you and runs into you! You do, indeed, then feel the force of the door because it is now the normal force of the door, pointing *inward* toward the center of the curve, causing you to turn the corner. But you were not "thrown" into the door; the door ran into you. The bird's-eye view, from an inertial reference frame, gives the proper perspective of what happens.

The "force" that seems to push an object to the outside of a circle is called the *centrifugal force.* Despite having a name, the centrifugal force is a fictitious force. It describes your experience *relative to a noninertial reference frame,* but there really is no such force. **You must always use Newton's laws in an inertial reference frame.** There are no centrifugal forces in an inertial reference frame.

NOTE ▶ You might wonder if the *rtz*-coordinate system is an inertial reference frame. It is, and Newton's laws apply, although the reason is rather subtle. We're using the *rtz*-coordinates to establish directions for decomposing vectors, but we're not making measurements in the *rtz*-system. That is, velocities and accelerations are measured in the laboratory reference frame. The particle would always be at rest ($\vec{v} = \vec{0}$) if we measured velocities in a reference frame attached to the particle. Thus the analysis of this chapter really is in an inertial reference frame. ◀

Gravity on a Rotating Earth

There is one small problem with the admonition that you must use Newton's laws in an inertial reference frame: A reference frame attached to the ground isn't truly inertial because of the earth's rotation. Fortunately, we can make a simple correction that allows us to continue using Newton's laws on the earth's surface.

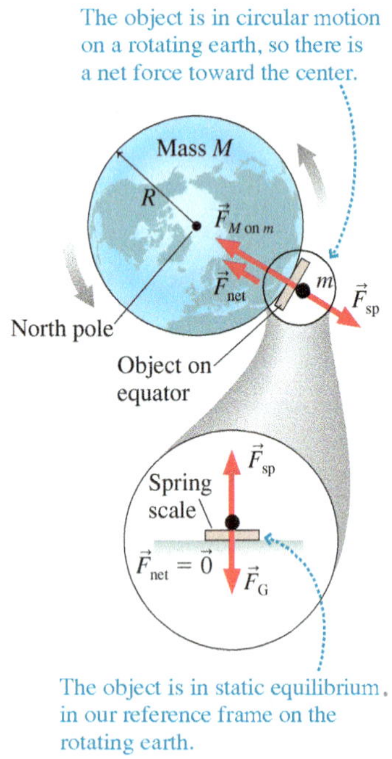

FIGURE 8.15 The earth's rotation affects the measured value of g.

FIGURE 8.15 shows an object being weighed by a spring scale on the earth's equator. An observer hovering in an inertial reference frame above the north pole sees two forces on the object: the gravitational force $\vec{F}_{M \text{ on } m}$, given by Newton's law of gravity, and the outward spring force $\vec{F}_{\text{sp}}$. The object moves in a circle as the earth rotates—it's accelerating—and circular motion *requires* a net force directed toward the center of the circle. The gravitational force points toward the center of the circle, the spring force points away, so Newton's second law is

$$\sum F_r = F_{M \text{ on } m} - F_{\text{sp}} = m\omega^2 R$$

where ω is the angular speed of the rotating earth. The spring-scale reading $F_{\text{sp}} = F_{M \text{ on } m} - m\omega^2 R$ is *less* than it would be on a nonrotating earth.

The blow-up in Figure 8.15 shows how we see things in a noninertial, flat-earth reference frame. Here the object is at rest, in static equilibrium. If we insist on using Newton's laws, we have to conclude that $\vec{F}_{\text{net}} = \vec{0}$ and hence the upward spring force must be exactly balanced by a downward gravitational force $\vec{F}_{\text{G}}$. Thus the spring-scale reading is $F_{\text{sp}} = F_{\text{G}}$.

Now both we and the hovering, inertial observer measure the same spring compression and read the same number on the scale. If F_{sp} is the same for both of us, then

$$F_{\text{G}} = F_{M \text{ on } m} - m\omega^2 R \tag{8.15}$$

In other words, force $\vec{F}_{\text{G}}$—what we called the *effective* gravitational force in Chapter 6—is slightly less than the true gravitational force $\vec{F}_{M \text{ on } m}$ because of the earth's rotation. In essence, $m\omega^2 R$ is the centrifugal force, a fictitious force trying—from our perspective in a noninertial reference frame—to "throw" us off the rotating platform.

This has the effect of "weakening" gravity. There really is no such force, but—this is the important point—**we can continue to use Newton's laws in our rotating reference frame if we pretend there is.**

Because $F_G = mg$ for an object at rest, the effect of the centrifugal term in Equation 8.15 is to make g a little smaller than it would be on a nonrotating earth:

$$g = \frac{F_G}{m} = \frac{F_{M\text{ on }m} - m\omega^2 R}{m} = \frac{GM}{R^2} - \omega^2 R = g_{\text{earth}} - \omega^2 R \qquad (8.16)$$

We calculated $g_{\text{earth}} = 9.83\ \text{m/s}^2$ in Chapter 6. Using $\omega = 1$ rev/day (which must be converted to SI units) and $R = 6370$ km, we find $\omega^2 R = 0.033\ \text{m/s}^2$ at the equator. Thus the free-fall acceleration—what we actually measure in our rotating reference frame—is about $9.80\ \text{m/s}^2$. The purely gravitational acceleration g_{earth} has been reduced by the centripetal acceleration of our rotation.

Things are a little more complicated at other latitudes, but the bottom line is that we can safely use Newton's laws in our rotating, noninertial reference frame on the earth's surface if we calculate the gravitational force—as we've been doing—as $F_G = mg$ with g the measured free-fall value, a value that compensates for our rotation, rather than the purely gravitational g_{earth}.

Why Does the Water Stay in the Bucket?

If you swing a bucket of water over your head quickly, the water stays in, but you'll get a shower if you swing too slowly. Why? We'll answer this question by starting with an equivalent situation, a roller coaster doing a loop-the-loop.

FIGURE 8.16a shows a roller-coaster car going around a vertical loop-the-loop of radius r. Why doesn't the car fall off at the top of the circle? FIGURE 8.16b shows the car's free-body diagrams at the top and bottom of the loop. Now, motion in a vertical circle is *not* uniform circular motion; the car slows down as it goes up one side and speeds up as it comes back down the other. But at the very top and very bottom points, only the car's direction is changing, not its speed, so at those points the acceleration is purely centripetal.

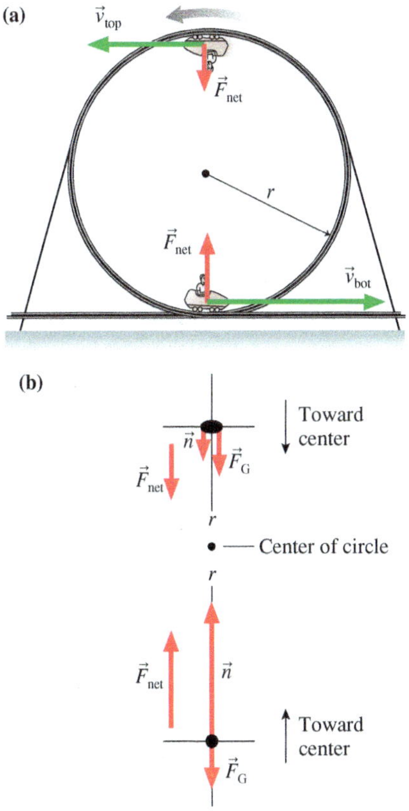

FIGURE 8.16 A roller-coaster car going around a loop-the-loop.

Because the car is moving in a circle, there must be a net force toward the center of the circle. First consider the very bottom of the loop. The only forces acting on the car are the gravitational force $\vec{F}_G$ and the normal force $\vec{n}$ of the track pushing up on it, so a net force toward the center—upward at this point—requires $n > F_G$. The normal force has to *exceed* the gravitational force to provide the net force needed to "turn the corner" at the bottom of the circle. This is why you "feel heavy" at the bottom of the circle or at the bottom of a valley on a roller coaster.

We can analyze the situation quantitatively by writing the r-component of Newton's second law. At the bottom of the circle, with the r-axis pointing upward, we have

$$\sum F_r = n_r + (F_G)_r = n - mg = ma_r = \frac{m(v_{\text{bot}})^2}{r} \qquad (8.17)$$

From Equation 8.17 we find

$$n = mg + \frac{m(v_{\text{bot}})^2}{r} \qquad (8.18)$$

The normal force at the bottom is *larger* than mg.

Things are a little trickier as the roller-coaster car crosses the top of the loop. Whereas the normal force of the track pushes up when the car is at the bottom of the circle, it *presses down* when the car is at the top and the track is above the car. Think about the free-body diagram to make sure you agree.

The car is still moving in a circle, so there *must* be a net force toward the center of the circle to provide the centripetal acceleration. The r-axis, which points toward the center of the circle, now points *downward*. Consequently, both forces have *positive* components. Newton's second law at the top of the circle is

$$\sum F_r = n_r + (F_G)_r = n + mg = \frac{m(v_{top})^2}{r} \qquad (8.19)$$

Thus at the top the normal force of the track on the car is

$$n = \frac{m(v_{top})^2}{r} - mg \qquad (8.20)$$

The normal force at the top can exceed mg if v_{top} is large enough. Our interest, however, is in what happens as the car goes slower and slower. As v_{top} decreases, there comes a point when n reaches zero. "No normal force" means "no contact," so at that speed, the track is *not* pushing against the car. Instead, the car is able to complete the circle because gravity alone provides sufficient centripetal acceleration.

The speed at which $n = 0$ is called the *critical speed* v_c:

$$v_c = \sqrt{\frac{rmg}{m}} = \sqrt{rg} \qquad (8.21)$$

FIGURE 8.17 A roller-coaster car at the top of the loop.

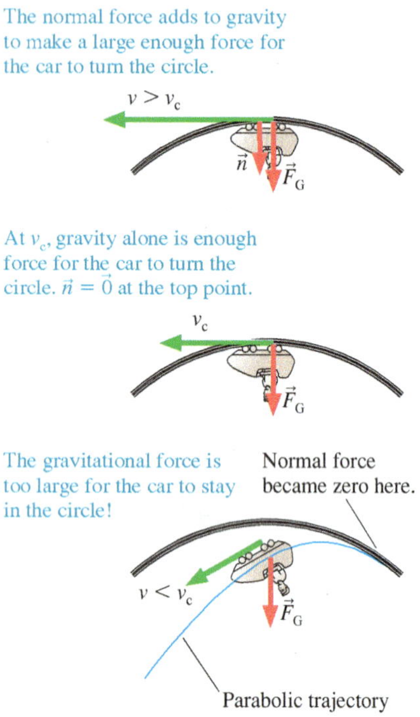

The critical speed is the slowest speed at which the car can complete the circle. Equation 8.20 would give a negative value for n if $v < v_c$, but that is physically impossible. The track can push against the wheels of the car ($n > 0$), but it can't pull on them. If $v < v_c$, the car cannot turn the full loop but, instead, comes off the track and becomes a projectile! FIGURE 8.17 summarizes this reasoning for the car on the loop-the-loop.

Water stays in a bucket swung over your head for the same reason: Circular motion requires a net force toward the center of the circle. At the top of the circle—if you swing the bucket fast enough—the bucket adds to the force of gravity by pushing *down* on the water, just like the downward normal force of the track on the roller-coaster car. As long as the bucket is pushing against the water, the bucket and the water are in contact and thus the water is "in" the bucket. As you swing slower and slower, requiring the water to have less and less centripetal acceleration, the bucket-on-water normal force decreases until it becomes zero at the critical speed. At the critical speed, gravity alone provides sufficient centripetal acceleration. Below the critical speed, gravity provides *too much* downward force for circular motion, so the water leaves the bucket and becomes a projectile following a parabolic trajectory toward your head!

Notice the similarity to the car making the left turn in Figure 8.14. The passenger feels like he's being "hurled" into the door by a centrifugal force, but it's actually the pushing force from the door, toward the center of the circle, that causes the passenger to turn the corner instead of moving straight ahead. Here, while it seems like the water is being "pinned" against the bottom of the bucket by a centrifugal force, it's really the pushing force from the bottom of the bucket causing the water to move in a circle instead of following a free-fall parabola.

STOP TO THINK 8.4 An out-of-gas car is rolling over the top of a hill at speed v. At this instant,

a. $n > F_G$
b. $n < F_G$
c. $n = F_G$
d. We can't tell about n without knowing v.

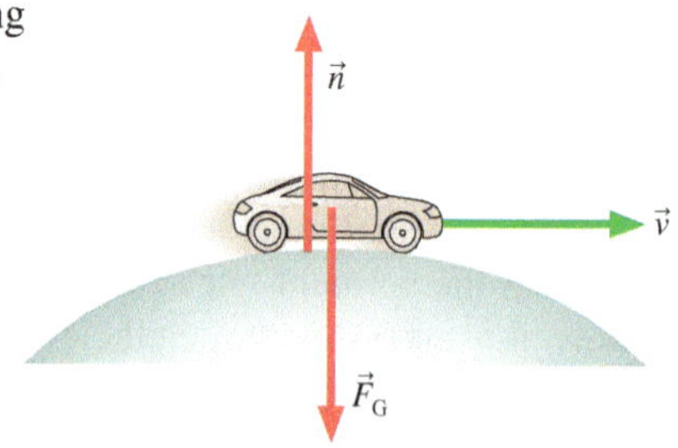

8.5 Nonuniform Circular Motion

Many interesting examples of circular motion involve objects whose speed changes. As we've already noted, a roller-coaster car doing a loop-the-loop slows down as it goes up one side, speeds up as it comes back down the other side. Circular motion with a changing speed is called *nonuniform circular motion.*

FIGURE 8.18, which is borrowed from Chapter 4, reminds you of the key ideas. Here the particle is speeding up or slowing down as it moves around the circle. In addition to centripetal acceleration, a particle in nonuniform circular motion has a tangential acceleration a_t. Tangential acceleration, parallel to $\vec{v}$, is the acceleration of changing speed. Mathematically, the tangential acceleration is simply the rate at which the tangential velocity changes:

$$a_t = \frac{dv_t}{dt} \tag{8.22}$$

It is usually most convenient to write the kinematic equations for circular motion in terms of the angular velocity ω and the angular acceleration $\alpha = d\omega/dt$. In Chapter 4, we found the connection between the tangential and angular accelerations to be

$$a_t = r\alpha \tag{8.23}$$

This is analogous to the similar equation $v_t = r\omega$ for tangential and angular velocity. In terms of angular quantities, the equations of constant-acceleration kinematics are

$$\begin{aligned} \theta_f &= \theta_i + \omega_i\,\Delta t + \tfrac{1}{2}\alpha(\Delta t)^2 \\ \omega_f &= \omega_i + \alpha\,\Delta t \\ \omega_f^2 &= \omega_i^2 + 2\alpha\,\Delta\theta \end{aligned} \tag{8.24}$$

In addition, the centripetal acceleration equation $a_r = v^2/r = \omega^2 r$ is still valid.

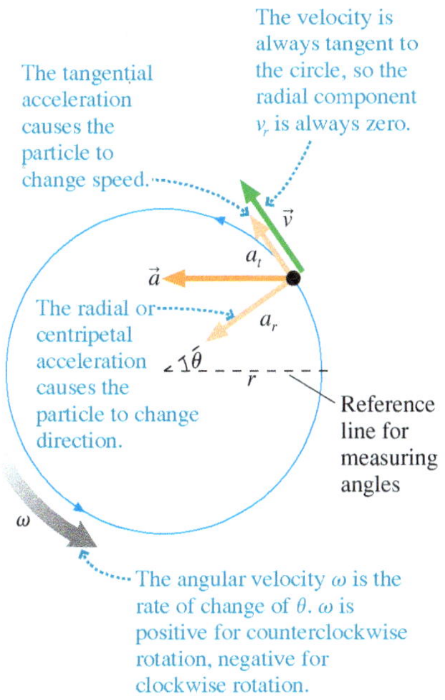

FIGURE 8.18 Nonuniform circular motion.

Dynamics of Nonuniform Circular Motion

FIGURE 8.19 shows a net force $\vec{F}_{net}$ acting on a particle as it moves around a circle of radius r. $\vec{F}_{net}$ is likely to be a superposition of several forces, such as a tension force in a string, a thrust force, a friction force, and so on.

We can decompose the force vector $\vec{F}_{net}$ into a *tangential* component $(F_{net})_t$ and a radial component $(F_{net})_r$. The component $(F_{net})_t$ is positive for a tangential force in the ccw direction, negative for a tangential force in the cw direction. Because of our definition of the r-axis, the component $(F_{net})_r$ is positive for a radial force *toward* the center, negative for a radial force away from the center. For example, the particular force illustrated in Figure 8.19 has positive values for both $(F_{net})_t$ and $(F_{net})_r$.

The force component $(F_{net})_r$ perpendicular to the trajectory creates a centripetal acceleration and causes the particle to change directions. It is the component $(F_{net})_t$ parallel to the trajectory that creates a tangential acceleration and causes the particle to change speed. Force and acceleration are related to each other through Newton's second law:

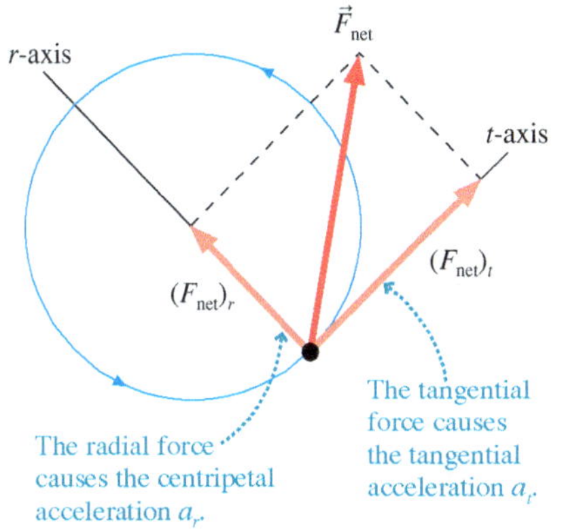

FIGURE 8.19 Net force $\vec{F}_{net}$ is applied to a particle moving in a circle.

$$\begin{aligned} (F_{net})_r &= \sum F_r = ma_r = \frac{mv^2}{r} = m\omega^2 r \\ (F_{net})_t &= \sum F_t = ma_t \\ (F_{net})_z &= \sum F_z = 0 \end{aligned} \tag{8.25}$$

NOTE ▶ Equations 8.25 differ from Equations 8.8 for uniform circular motion only in the fact that a_t is no longer constrained to be zero. ◀

EXAMPLE 8.7 Circular motion of a grinding machine

A machine for grinding small samples down to thin slabs of uniform thickness consists of a 2.0 kg steel block that spins around on a table at the end of an 80-cm-long arm. Samples are glued to the bottom of the heavy block, then dragged across the sandpaper-like surface of the table as the block spins. A tachometer attached to the motor shaft gives the following readings after the motor is disengaged at $t = 0$ s:

Time (s)	rpm
0.0	156
0.5	114
1.0	88
1.5	52
2.0	17

What is the coefficient of kinetic friction between the sample and the surface? How many revolutions does the block make as it comes to a halt?

MODEL Model the block and sample as a particle in nonuniform circular motion. Assume that the mass of the sample is negligible compared to the mass of the block and that the axle is frictionless.

VISUALIZE FIGURE 8.20 shows a pictorial representation. For the first time, we need a free-body diagram showing forces in three dimensions.

FIGURE 8.20 Pictorial representation of the circular motion.

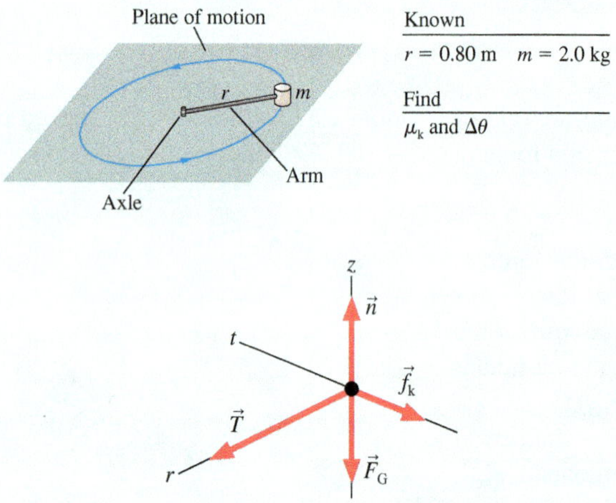

SOLVE The block slows down because kinetic friction between the sample and the table's surface exerts a retarding force $\vec{f}_k$. Kinetic friction is always opposite the direction of motion, so $\vec{f}_k$ is *tangent* to the circle and has magnitude $f_k = \mu_k n$.

There's no net force in the vertical direction, so the z-component of the second law is

$$\sum F_z = n - F_G = 0$$

from which we can conclude that $n = F_G = mg$ and thus $f_k = \mu_k mg$. The friction force is the only tangential component of force, so we can use the t-component of Newton's second law to find the tangential acceleration:

$$\sum F_t = (f_k)_t = -f_k = ma_t$$

$$a_t = \frac{-f_k}{m} = \frac{-\mu_k mg}{m} = -\mu_k g$$

Thus the angular acceleration is $\alpha = a_t/r = -\mu_k g/r$.

We can find α experimentally as the slope of the ω-versus-t graph. FIGURE 8.21 shows a graph of the data (after conversion of rpm to rad/s) and a best-fit line. We see that the angular acceleration is $\alpha = -7.12$ rad/s^2, and with this value we can calculate the coefficient of kinetic friction:

$$\mu_k = -\frac{\alpha r}{g} = -\frac{(-7.12\text{ rad/s}^2)(0.80\text{ m})}{9.80\text{ m/s}^2} = 0.58$$

We can now use the kinematic equation $\omega_f^2 = 0 = \omega_i^2 + 2\alpha\,\Delta\theta$ to find how many revolutions the block makes as it comes to rest. But what is ω_i? The first data entry is 156 rpm = 16.3 rad/s, but clearly—from the graph—the data have uncertainties. The first entry has no more claim to being "perfect" than any other entry. It's better to use the y-intercept of our best-fit line, 16.1 rad/s. That is, a statistical analysis of *all* the data tells us that 16.1 rad/s (154 rpm) is our best estimate of the angular velocity ω_i when the motor was disengaged at $t = 0$ s. With this:

$$\Delta\theta = -\frac{\omega_i^2}{2\alpha} = -\frac{(16.1\text{ rad/s})^2}{2(-7.12\text{ rad/s}^2)} = 18.2\text{ rad} = 2.9\text{ rev}$$

The block makes 2.9 revolutions while stopping.

FIGURE 8.21 Graph of angular velocity versus time. Angular acceleration is the slope of the best-fit line.

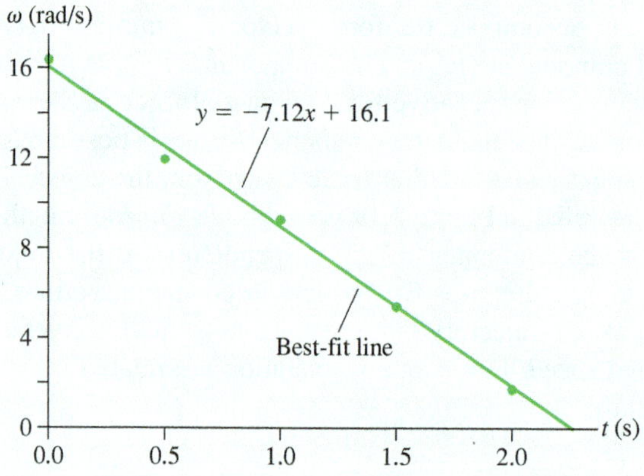

ASSESS A coefficient of kinetic friction of 0.58 is reasonable for a sandpaper-like surface. And even though the friction is fairly large, it's reasonable that the block would make several revolutions before stopping. The purpose of the Assess step, as always, is not to prove that the answer is right but to rule out obviously unreasonable answers that have been reached by mistake.

We've come a long way since our first dynamics problems in Chapter 6, but our basic strategy has not changed.

PROBLEM-SOLVING STRATEGY 8.1 Circular-motion problems

MODEL Make simplifying assumptions.

VISUALIZE Draw a pictorial representation. Use *rtz*-coordinates.

- Establish a coordinate system with the *r*-axis pointing toward the center of the circle.
- Show important points in the motion on a sketch. Define symbols and identify what the problem is trying to find.
- Identify the forces and show them on a free-body diagram.

SOLVE Newton's second law is

$$(F_{net})_r = \sum F_r = ma_r = \frac{mv^2}{r} = m\omega^2 r$$
$$(F_{net})_t = \sum F_t = ma_t$$
$$(F_{net})_z = \sum F_z = 0$$

- Determine the force components from the free-body diagram. Be careful with signs.
- The tangential acceleration for uniform circular motion is $a_t = 0$.
- Solve for the acceleration, then use kinematics to find velocities and positions.

ASSESS Check that your result has the correct units, is reasonable, and answers the question.

Exercise 17

STOP TO THINK 8.5 A ball on a string is swung in a vertical circle. The string happens to break when it is parallel to the ground and the ball is moving up. Which trajectory does the ball follow?

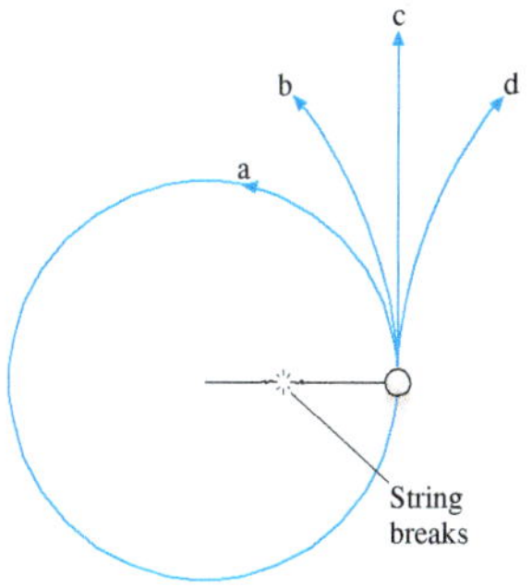

CHALLENGE EXAMPLE 8.8 Circular motion inside a cone

A small ball of mass m rolls in a horizontal circle around the inside of the inverted cone shown in FIGURE 8.22. The walls of the cone make an angle θ with a horizontal plane. The coefficient of static friction between the ball and the cone is μ_s; rolling friction is negligible. What minimum speed v_{min} must the ball maintain to remain at a constant height h?

MODEL Model the ball as a particle in uniform circular motion.

FIGURE 8.22 Pictorial representation of the circular motion.

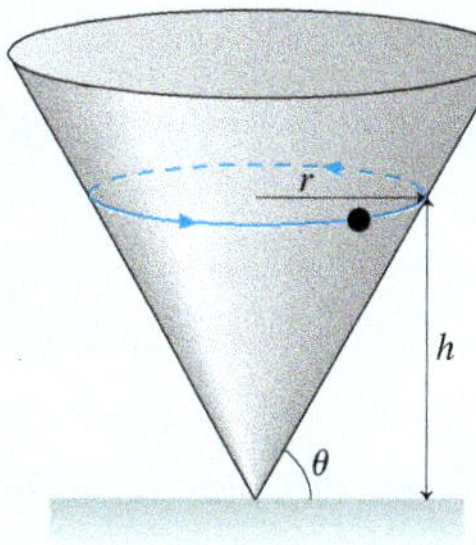

Continued

VISUALIZE The forces on the ball are a normal force, perpendicular to the surface; a static friction force parallel to the surface that keeps the ball from sliding up or down; and gravity. FIGURE 8.23 shows a free-body diagram with the r-axis pointing toward the center of the circle. Notice that the situation is very similar to that of a car on a banked curve. Figure 8.9 showed that the static friction force must point up the slope to keep a slow-moving car from sliding down the slope, and that information was used in drawing Figure 8.23.

FIGURE 8.23 Free-body diagram of the ball.

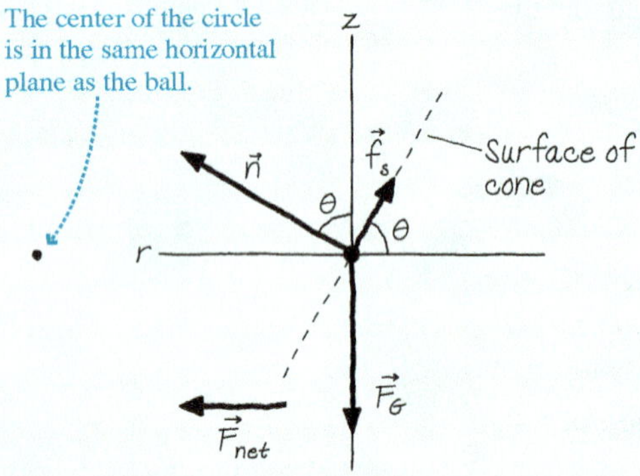

SOLVE This is uniform circular motion, so we need to consider only the r- and z-component equations of Newton's second law. All the information is on the free-body diagram, but considerable care is needed to write down all the components correctly. The two equations are

$$\sum F_r = n_r + (f_s)_r + (F_G)_r = n\sin\theta - f_s\cos\theta = \frac{mv^2}{r}$$

$$\sum F_z = n_z + (f_s)_z + (F_G)_z = n\cos\theta + f_s\sin\theta - mg = 0$$

The r-axis points toward the center of the circle, here on the left, so $\vec{n}$ has a positive r-component while $\vec{f}_s$ has a negative r-component.

There is one specific speed at which the ball would roll around the inside without friction, just like the car in Example 8.5. For slower speeds, some amount of static friction is needed to keep the ball from sliding down. Our task is to find the *minimum* speed for maintaining motion in a horizontal plane, and that occurs when the static friction force reaches its *maximum* value: $f_{s\,max} = \mu_s n$. Then the r-component equation becomes

$$\frac{m{v_{min}}^2}{r} = n\sin\theta - \mu_s n\cos\theta = n(\sin\theta - \mu_s\cos\theta)$$

To find n, we use the z-component equation with $f_{s\,max} = \mu_s n$:

$$n\cos\theta + \mu_s n\sin\theta = n(\cos\theta + \mu_s\sin\theta) = mg$$

$$n = \frac{mg}{\cos\theta + \mu_s\sin\theta}$$

Substituting this into the equation for v_{min} and taking the square root, we find

$$v_{min} = \sqrt{rg\left(\frac{\sin\theta - \mu_s\cos\theta}{\cos\theta + \mu_s\sin\theta}\right)}$$

This is a complicated answer, but we can check it because without friction the ball should roll around at the same speed as the car turning a banked curve without friction. If we set $\mu_s = 0$, we find

$$v_{frictionless} = \sqrt{rg\tan\theta}$$

which, indeed, was the answer to Example 8.5.

The information we have is the ball's height h, not the radius of the circle, so the final step, which we can see from Figure 8.22, is to substitute $r = h/\tan\theta$. Thus the minimum speed for the ball to circle at height h is

$$v_{min} = \sqrt{\frac{hg}{\tan\theta}\left(\frac{\sin\theta - \mu_s\cos\theta}{\cos\theta + \mu_s\sin\theta}\right)}$$

ASSESS An important problem-solving skill to learn is checking new results by comparing them to previously known results. In this case, we recognized that the ball rolling around the inside of the cone without the aid of friction is equivalent to a car turning a banked curve without friction. The fact that we could reproduce that earlier result gives us confidence in our answer.

SUMMARY

The goal of Chapter 8 has been to learn how to solve problems about motion in a plane.

General Principles

Newton's Second Law

Expressed in x- and y-component form:

$$(F_{net})_x = \sum F_x = ma_x$$

$$(F_{net})_y = \sum F_y = ma_y$$

Expressed in rtz-component form:

$$(F_{net})_r = \sum F_r = ma_r = \frac{mv^2}{r} = m\omega^2 r$$

$$(F_{net})_t = \sum F_t = \begin{cases} 0 & \text{uniform circular motion} \\ ma_t & \text{nonuniform circular motion} \end{cases}$$

$$(F_{net})_z = \sum F_z = 0$$

Uniform Circular Motion

- v is constant.
- $\vec{F}_{net}$ points toward the center of the circle.
- The **centripetal acceleration** $\vec{a}$ points toward the center of the circle. It changes the particle's direction but not its speed.

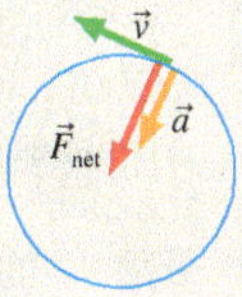

Nonuniform Circular Motion

- v changes.
- $\vec{a}$ is parallel to $\vec{F}_{net}$.
- The radial component a_r changes the particle's direction.
- The tangential component a_t changes the particle's speed.

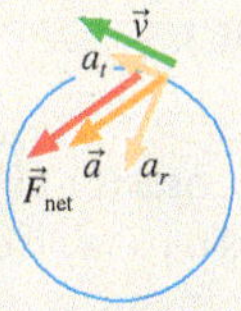

Important Concepts

rtz-coordinates

- The r-axis points toward the center of the circle.
- The t-axis is tangent, pointing counterclockwise.

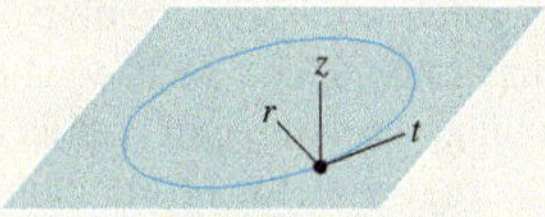

Angular velocity

$\omega = d\theta/dt$

$v_t = \omega r$

Angular acceleration

$\alpha = d\omega/dt$

$a_t = \alpha r$

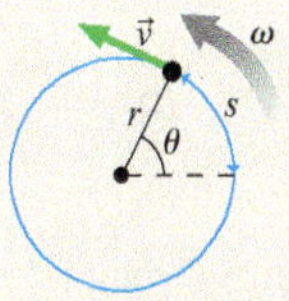

Applications

Orbits

A circular orbit has radius r if

$$v = \sqrt{rg}$$

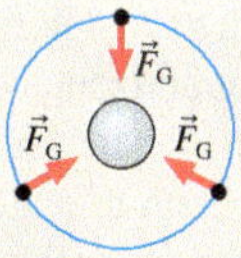

Hills and valleys

Circular motion requires a net force pointing to the center. n must be > 0 for the object to be in contact with a surface.

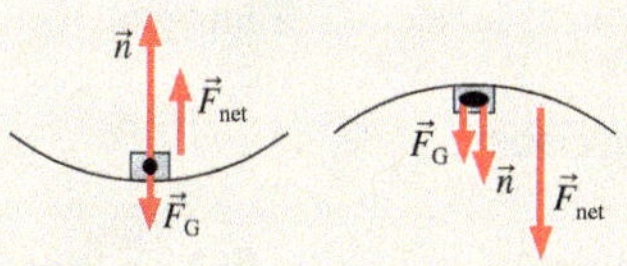

Terms and Notation

orbit
fictitious force

CONCEPTUAL QUESTIONS

1. In uniform circular motion, which of the following are constant: speed, velocity, angular velocity, centripetal acceleration, magnitude of the net force?
2. A car runs out of gas while driving down a hill. It rolls through the valley and starts up the other side. At the very bottom of the valley, which of the free-body diagrams in FIGURE Q8.2 is correct? The car is moving to the right, and drag and rolling friction are negligible.

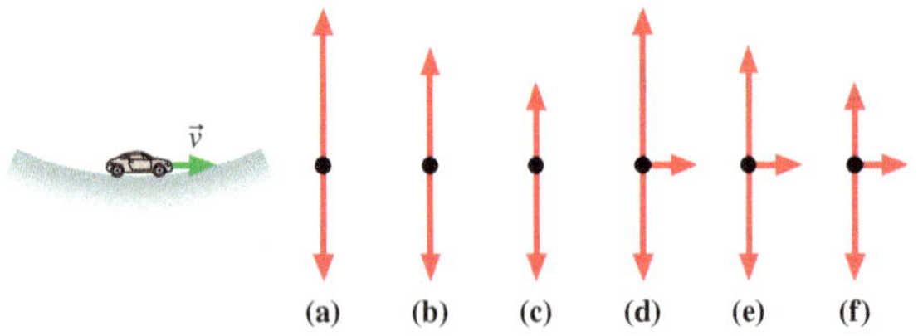

FIGURE Q8.2

3. FIGURE Q8.3 is a bird's-eye view of particles moving in horizontal circles on a tabletop. All are moving at the same speed. Rank in order, from largest to smallest, the tensions T_a to T_d. Give your answer in the form a > b = c > d and explain your ranking.

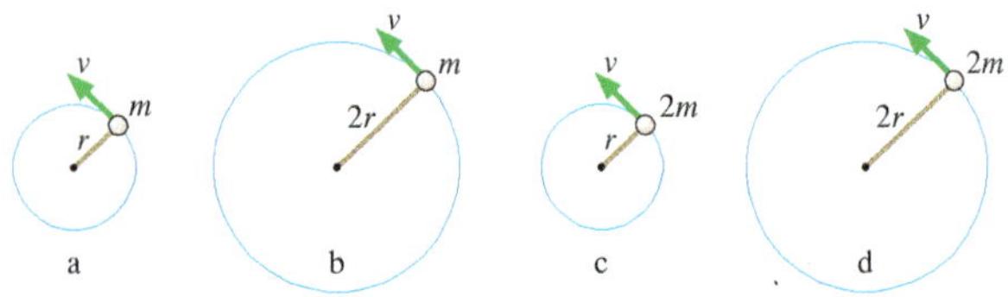

FIGURE Q8.3

4. Tarzan swings through the jungle on a vine. At the lowest point of his swing, is the tension in the vine greater than, less than, or equal to the gravitational force on Tarzan? Explain.
5. FIGURE Q8.5 shows two balls of equal mass moving in vertical circles. Is the tension in string A greater than, less than, or equal to the tension in string B if the balls travel over the top of the circle (a) with equal speed and (b) with equal angular velocity?

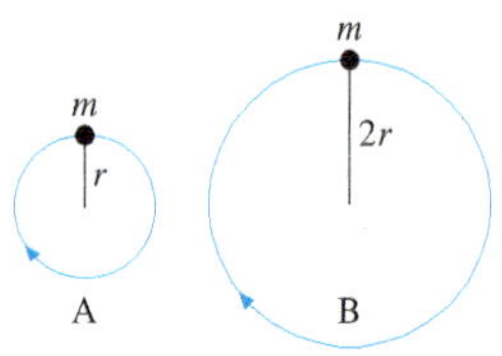

FIGURE Q8.5

6. Ramon and Sally are observing a toy car speed up as it goes around a circular track. Ramon says, "The car's speeding up, so there must be a net force parallel to the track." "I don't think so," replies Sally. "It's moving in a circle, and that requires centripetal acceleration. The net force has to point to the center of the circle." Do you agree with Ramon, Sally, or neither? Explain.
7. A jet plane is flying on a level course at constant speed. The engines are at full throttle.
   a. What is the net force on the plane? Explain.
   b. Draw a free-body diagram of the plane as seen from the side with the plane flying to the right. Name (don't just label) any and all forces shown on your diagram.
   c. Airplanes bank when they turn. Draw a free-body diagram of the plane as seen from behind as it makes a right turn.
   d. *Why* do planes bank as they turn? Explain.
8. A small projectile is launched parallel to the ground at height $h = 1$ m with sufficient speed to orbit a completely smooth, airless planet. A bug rides inside a small hole inside the projectile. Is the bug weightless? Explain.
9. You can swing a ball on a string in a vertical circle if you swing it fast enough. But if you swing too slowly, the string goes slack as the ball nears the top. Explain *why* there's a minimum speed to keep the ball moving in a circle.
10. A golfer starts with the club over her head and swings it to reach maximum speed as it contacts the ball. Halfway through her swing, when the golf club is parallel to the ground, does the acceleration vector of the club head point (a) straight down, (b) parallel to the ground, approximately toward the golfer's shoulders, (c) approximately toward the golfer's feet, or (d) toward a point above the golfer's head? Explain.

EXERCISES AND PROBLEMS

Problems labeled integrate material from earlier chapters.

Exercises

Section 8.1 Dynamics in Two Dimensions

1. || As a science fair project, you want to launch an 800 g model rocket straight up and hit a horizontally moving target as it passes 30 m above the launch point. The rocket engine provides a constant thrust of 15.0 N. The target is approaching at a speed of 15 m/s. At what horizontal distance between the target and the rocket should you launch?
2. || A 500 g model rocket is on a cart that is rolling to the right at a speed of 3.0 m/s. The rocket engine, when it is fired, exerts an 8.0 N thrust on the rocket. Your goal is to have the rocket pass through a small horizontal hoop that is 20 m above the launch point. At what horizontal distance left of the hoop should you launch?
3. || A 4.0×10^{10} kg asteroid is heading directly toward the center of the earth at a steady 20 km/s. To save the planet, astronauts strap a giant rocket to the asteroid perpendicular to its direction of travel. The rocket generates 5.0×10^9 N of thrust. The rocket is fired when the asteroid is 4.0×10^6 km away from earth. You can ignore the earth's gravitational force on the asteroid and their rotation about the sun.
   a. If the mission fails, how many hours is it until the asteroid impacts the earth?
   b. The radius of the earth is 6400 km. By what minimum angle must the asteroid be deflected to just miss the earth?
   c. The rocket fires at full thrust for 300 s before running out of fuel. Is the earth saved?

Section 8.2 Uniform Circular Motion

4. | A 1500 kg car drives around a flat 200-m-diameter circular track at 25 m/s. What are the magnitude and direction of the net force on the car? What causes this force?
5. | A 1500 kg car takes a 50-m-radius unbanked curve at 15 m/s. What is the size of the friction force on the car?
6. || A 200 g block on a 50-cm-long string swings in a circle on a horizontal, frictionless table at 75 rpm.
   a. What is the speed of the block?
   b. What is the tension in the string?
7. || In the Bohr model of the hydrogen atom, an electron (mass $m = 9.1 \times 10^{-31}$ kg) orbits a proton at a distance of 5.3×10^{-11} m. The proton pulls on the electron with an electric force of 8.2×10^{-8} N. How many revolutions per second does the electron make?
8. || A highway curve of radius 500 m is designed for traffic moving at a speed of 90 km/h. What is the correct banking angle of the road?
9. || Suppose the moon were held in its orbit not by gravity but by a massless cable attached to the center of the earth. What would be the tension in the cable? Use the table of astronomical data inside the back cover of the book.
10. | It is proposed that future space stations create an artificial gravity by rotating. Suppose a space station is constructed as a 1000-m-diameter cylinder that rotates about its axis. The inside surface is the deck of the space station. What rotation period will provide "normal" gravity?

Section 8.3 Circular Orbits

11. | A satellite orbiting the moon very near the surface has a period of 110 min. What is free-fall acceleration on the surface of the moon?
12. || What is free-fall acceleration toward the sun at the distance of the earth's orbit? Astronomical data are inside the back cover of the book.

Section 8.4 Fictitious Forces

13. | A car drives over the top of a hill that has a radius of 50 m. What maximum speed can the car have at the top without flying off the road?
14. || The weight of passengers on a roller coaster increases by 50% as the car goes through a dip with a 30 m radius of curvature. What is the car's speed at the bottom of the dip?
15. || A roller coaster car crosses the top of a circular loop-the-loop at twice the critical speed. What is the ratio of the normal force to the gravitational force?
16. || The normal force equals the magnitude of the gravitational force as a roller coaster car crosses the top of a 40-m-diameter loop-the-loop. What is the car's speed at the top?
17. || A student has 65-cm-long arms. What is the minimum angular velocity (in rpm) for swinging a bucket of water in a vertical circle without spilling any? The distance from the handle to the bottom of the bucket is 35 cm.
18. | While at the county fair, you decide to ride the Ferris wheel. Having eaten too many candy apples and elephant ears, you find the motion somewhat unpleasant. To take your mind off your stomach, you wonder about the motion of the ride. You estimate the radius of the big wheel to be 15 m, and you use your watch to find that each loop around takes 25 s.
   a. What are your speed and the magnitude of your acceleration?
   b. What is the ratio of your weight at the top of the ride to your weight while standing on the ground?
   c. What is the ratio of your weight at the bottom of the ride to your weight while standing on the ground?

Section 8.5 Nonuniform Circular Motion

19. || A new car is tested on a 200-m-diameter track. If the car speeds up at a steady 1.5 m/s^2, how long after starting is the magnitude of its centripetal acceleration equal to the tangential acceleration?
20. || A toy train rolls around a horizontal 1.0-m-diameter track. The coefficient of rolling friction is 0.10.
   a. What is the magnitude of the train's angular acceleration after it is released?
   b. How long does it take the train to stop if it's released with an angular speed of 30 rpm?

Problems

21. || A popular pastime is to see who can push an object closest to the edge of a table without its going off. You push the 100 g object and release it 2.0 m from the table edge. Unfortunately, you push a little too hard. The object slides across, sails off the edge, falls 1.0 m to the floor, and lands 30 cm from the edge of the table. If the coefficient of kinetic friction is 0.50, what was the object's speed as you released it?
22. || A motorcycle daredevil plans to ride up a 2.0-m-high, 20° ramp, sail across a 10-m-wide pool filled with hungry crocodiles, and land at ground level on the other side. He has done this stunt many times and approaches it with confidence. Unfortunately, the motorcycle engine dies just as he starts up the ramp. He is going 11 m/s at that instant, and the rolling friction of his rubber tires (coefficient 0.02) is not negligible. Does he survive, or does he become crocodile food?
23. ||| Sam (75 kg) takes off up a 50-m-high, 10° frictionless slope on his jet-powered skis. The skis have a thrust of 200 N. He keeps his skis tilted at 10° after becoming airborne, as shown in FIGURE P8.23. How far does Sam land from the base of the cliff?

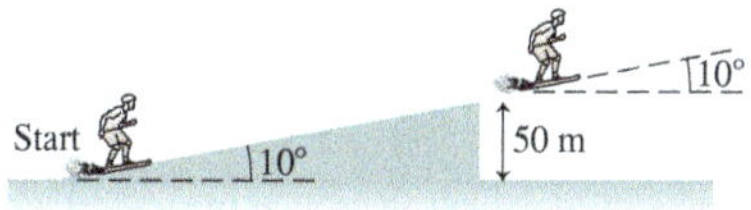

FIGURE P8.23

24. || Derive Equations 8.4 for the acceleration of a projectile subject to drag.
25. || A 5000 kg interceptor rocket is launched at an angle of 44.7°. The thrust of the rocket motor is 140,700 N.
   a. Find an equation $y(x)$ that describes the rocket's trajectory.
   b. What is the shape of the trajectory?
   c. At what elevation does the rocket reach the speed of sound, 330 m/s?
26. ||| A rocket-powered hockey puck has a thrust of 2.0 N and a total mass of 1.0 kg. It is released from rest on a frictionless table, 4.0 m from the edge of a 2.0 m drop. The front of the rocket is pointed directly toward the edge. How far does the puck land from the base of the table?

27. || A 500 g model rocket is resting horizontally at the top edge of a 40-m-high wall when it is accidentally bumped. The bump pushes it off the edge with a horizontal speed of 0.5 m/s and at the same time causes the engine to ignite. When the engine fires, it exerts a constant 20 N horizontal thrust away from the wall.
   a. How far from the base of the wall does the rocket land?
   b. Describe the trajectory of the rocket while it travels to the ground.
28. || An experimental aircraft begins its takeoff at $t = 0$ s. Every second, an onboard GPS measures and records the plane's distances east and north of a reference marker. The following data are downloaded to your computer:

Time (s)	East (m)	North (m)
0.0	91	0
1.0	86	4
2.0	77	18
3.0	65	39
4.0	39	63
5.0	19	101

   Analyze these data to determine the magnitude of the aircraft's takeoff acceleration.
29. || Communications satellites are placed in circular orbits where they stay directly over a fixed point on the equator as the earth rotates. These are called *geosynchronous orbits.* The altitude of a geosynchronous orbit is 3.58×10^7 m ($\approx$22,000 miles).
   a. What is the period of a satellite in a geosynchronous orbit?
   b. Find the value of g at this altitude.
   c. What is the weight of a 2000 kg satellite in a geosynchronous orbit?
30. || A 75 kg man weighs himself at the north pole and at the equator. Which scale reading is higher? By how much?
31. || A 500 g ball swings in a vertical circle at the end of a 1.5-m-long string. When the ball is at the bottom of the circle, the tension in the string is 15 N. What is the speed of the ball at that point?
32. ||| A concrete highway curve of radius 70 m is banked at a 15° angle. What is the maximum speed with which a 1500 kg rubber-tired car can take this curve without sliding?
33. || a. An object of mass m swings in a horizontal circle on a string of length L that tilts downward at angle θ. Find an expression for the angular velocity ω.
   b. A student ties a 500 g rock to a 1.0-m-long string and swings it around her head in a horizontal circle. At what angular speed, in rpm, does the string tilt down at a 10° angle?
34. || A 5.0 g coin is placed 15 cm from the center of a turntable. The coin has static and kinetic coefficients of friction with the turntable surface of $\mu_s = 0.80$ and $\mu_k = 0.50$. The turntable very slowly speeds up to 60 rpm. Does the coin slide off?
35. || You've taken your neighbor's young child to the carnival to ride the rides. She wants to ride The Rocket. Eight rocket-shaped cars hang by chains from the outside edge of a large steel disk. A vertical axle through the center of the ride turns the disk, causing the cars to revolve in a circle. You've just finished taking physics, so you decide to figure out the speed of the cars while you wait. You estimate that the disk is 5 m in diameter and the chains are 6 m long. The ride takes 10 s to reach full speed, then the cars swing out until the chains are 20° from vertical. What is the cars' speed?
36. || A conical pendulum is formed by attaching a ball of mass m to a string of length L, then allowing the ball to move in a horizontal circle of radius r. FIGURE P8.36 shows that the string traces out the surface of a cone, hence the name.
   a. Find an expression for the tension T in the string.
   b. Find an expression for the ball's angular speed ω.
   c. What are the tension and angular speed (in rpm) for a 500 g ball swinging in a 20-cm-radius circle at the end of a 1.0-m-long string?

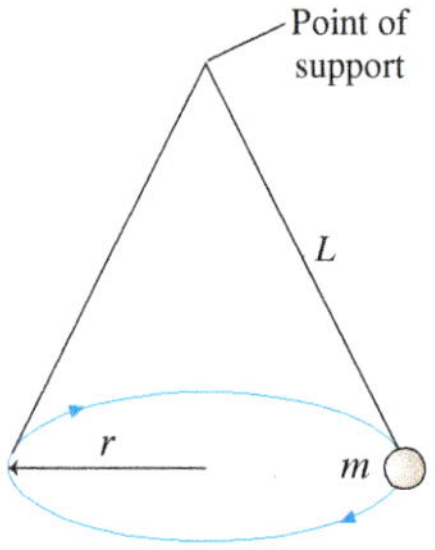

FIGURE P8.36

37. ||| Two wires are tied to the 2.0 kg sphere shown in FIGURE P8.37. The sphere revolves in a horizontal circle at constant speed.
   a. For what speed is the tension the same in both wires?
   b. What is the tension?

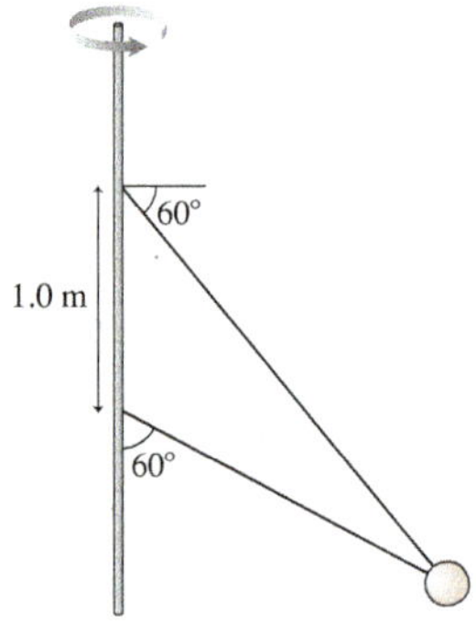

FIGURE P8.37

38. ||| In an old-fashioned amusement park ride, passengers stand inside a 5.0-m-diameter hollow steel cylinder with their backs against the wall. The cylinder begins to rotate about a vertical axis. Then the floor on which the passengers are standing suddenly drops away! If all goes well, the passengers will "stick" to the wall and not slide. Clothing has a static coefficient of friction against steel in the range 0.60 to 1.0 and a kinetic coefficient in the range 0.40 to 0.70. A sign next to the entrance says "No children under 30 kg allowed." What is the minimum angular speed, in rpm, for which the ride is safe?
39. || A 10 g steel marble is spun so that it rolls at 150 rpm around the *inside* of a vertically oriented steel tube. The tube, shown in FIGURE P8.39, is 12 cm in diameter. Assume that the rolling resistance is small enough for the marble to maintain 150 rpm for several seconds. During this time, will the marble spin in a horizontal circle, at constant height, or will it spiral down the inside of the tube?

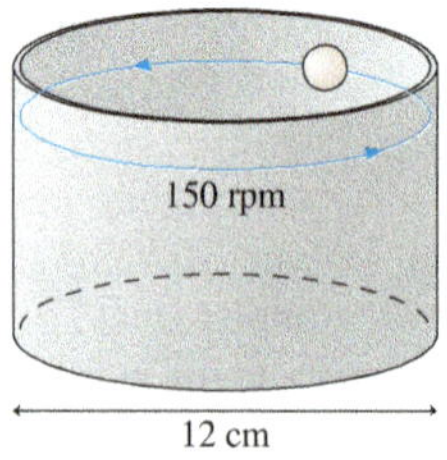

FIGURE P8.39

40. || The ultracentrifuge is an important tool for separating and analyzing proteins. Because of the enormous centripetal accelerations, the centrifuge must be carefully balanced, with each sample matched by a sample of identical mass on the opposite side. Any difference in the masses of opposing samples creates a net force on the shaft of the rotor, potentially leading to a catastrophic failure of the apparatus. Suppose a scientist makes a slight error in sample preparation and one sample has a mass 10 mg larger than the opposing sample. If the samples are 12 cm from the axis of the rotor and the ultracentrifuge spins at 70,000 rpm, what is the magnitude of the net force on the rotor due to the unbalanced samples?
BIO

41. || Three cars are driving at 25 m/s along the road shown in FIGURE P8.41. Car B is at the bottom of a hill and car C is at the top. Both hills have a 200 m radius of curvature. Suppose each car suddenly brakes hard and starts to skid. What is the tangential acceleration (i.e., the acceleration parallel to the road) of each car? Assume $\mu_k = 1.0$.

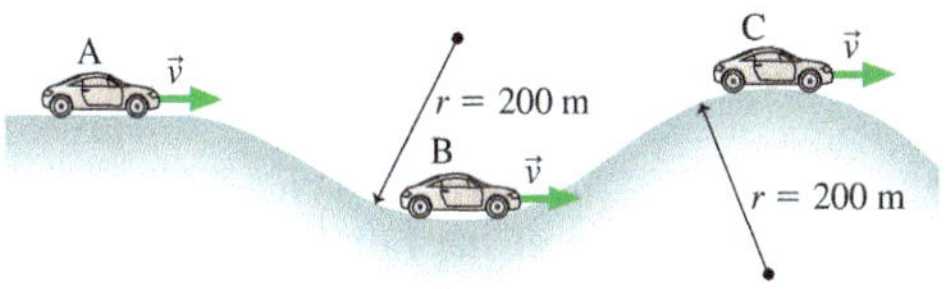

FIGURE P8.41

42. || A 500 g ball moves in a vertical circle on a 102-cm-long string. If the speed at the top is 4.0 m/s, then the speed at the bottom will be 7.5 m/s. (You'll learn how to show this in Chapter 10.)
   a. What is the gravitational force acting on the ball?
   b. What is the tension in the string when the ball is at the top?
   c. What is the tension in the string when the ball is at the bottom?

43. || In an amusement park ride called The Roundup, passengers stand inside a 16-m-diameter rotating ring. After the ring has acquired sufficient speed, it tilts into a vertical plane, as shown in FIGURE P8.43.

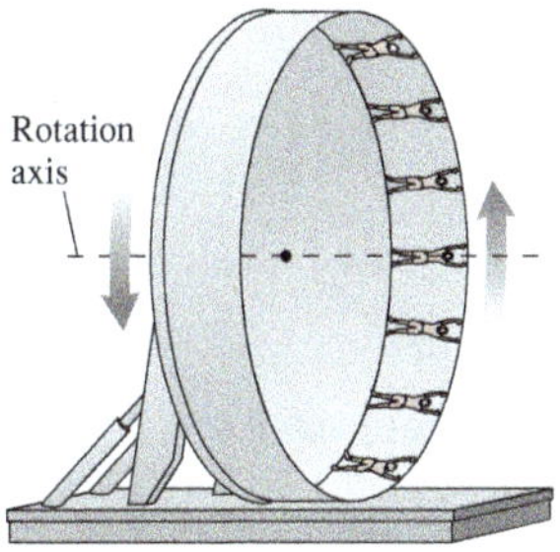

FIGURE P8.43

   a. Suppose the ring rotates once every 4.5 s. If a rider's mass is 55 kg, with how much force does the ring push on her at the top of the ride? At the bottom?
   b. What is the longest rotation period of the wheel that will prevent the riders from falling off at the top?

44. || You have a new job designing rides for an amusement park. In one ride, the rider's chair is attached by a 9.0-m-long chain to the top of a tall rotating tower. The tower spins the chair and rider around at the rate of 1.0 rev every 4.0 s. In your design, you've assumed that the maximum possible combined weight of the chair and rider is 150 kg. You've found a great price for chain at the local discount store, but your supervisor wonders if the chain is strong enough. You contact the manufacturer and learn that the chain is rated to withstand a tension of 3000 N. Will this chain be strong enough for the ride?

45. || Suppose you swing a ball of mass m in a vertical circle on a string of length L. As you probably know from experience, there is a minimum angular velocity ω_{min} you must maintain if you want the ball to complete the full circle without the string going slack at the top.
   a. Find an expression for ω_{min}.
   b. Evaluate ω_{min} in rpm for a 65 g ball tied to a 1.0-m-long string.

46. || A heavy ball with a weight of 100 N ($m = 10.2$ kg) is hung from the ceiling of a lecture hall on a 4.5-m-long rope. The ball is pulled to one side and released to swing as a pendulum, reaching a speed of 5.5 m/s as it passes through the lowest point. What is the tension in the rope at that point?

47. || A 30 g ball rolls around a 40-cm-diameter L-shaped track, shown in FIGURE P8.47, at 60 rpm. What is the magnitude of the *net* force that the track exerts on the ball? Rolling friction can be neglected.

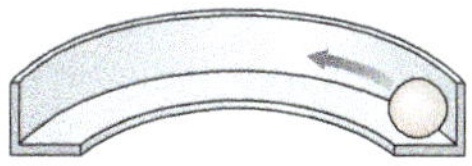

FIGURE P8.47

Hint: The track exerts more than one force on the ball.

48. || Mass m_1 on the frictionless table of FIGURE P8.48 is connected by a string through a hole in the table to a hanging mass m_2. With what speed must m_1 rotate in a circle of radius r if m_2 is to remain hanging at rest?

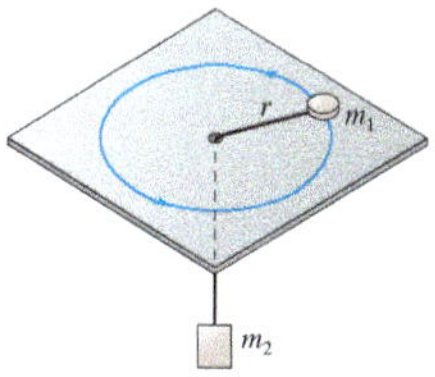

FIGURE P8.48

49. || The physics of circular motion sets an upper limit to the speed of human walking. (If you need to go faster, your gait changes from a walk to a run.) If you take a few steps and watch what's happening, you'll see that your body pivots in circular motion over your forward foot as you bring your rear foot forward for the next step. As you do so, the normal force of the ground on your foot decreases and your body tries to "lift off" from the ground.
BIO
   a. A person's center of mass is very near the hips, at the top of the legs. Model a person as a particle of mass m at the top of a leg of length L. Find an expression for the person's maximum walking speed v_{max}.
   b. Evaluate your expression for the maximum walking speed of a 70 kg person with a typical leg length of 70 cm. Give your answer in both m/s and mph, then comment, based on your experience, as to whether this is a reasonable result. A "normal" walking speed is about 3 mph.

50. || A 100 g ball on a 60-cm-long string is swung in a vertical circle about a point 200 cm above the floor. The tension in the string when the ball is at the very bottom of the circle is 5.0 N. A very sharp knife is suddenly inserted, as shown in FIGURE P8.50, to cut the string directly below the point of support. How far to the right of where the string was cut does the ball hit the floor?

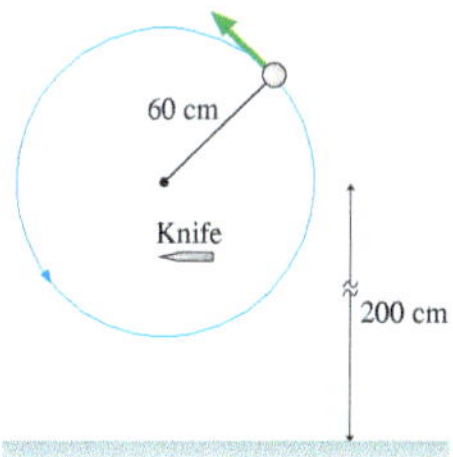

FIGURE P8.50

51. || A 60 g ball is tied to the end of a 50-cm-long string and swung in a vertical circle. The center of the circle, as shown in FIGURE P8.51, is 150 cm above the floor. The ball is swung at the minimum speed necessary to make it over the top without the string going slack. If the string is released at the instant the ball is at the top of the loop, how far to the right does the ball hit the ground?

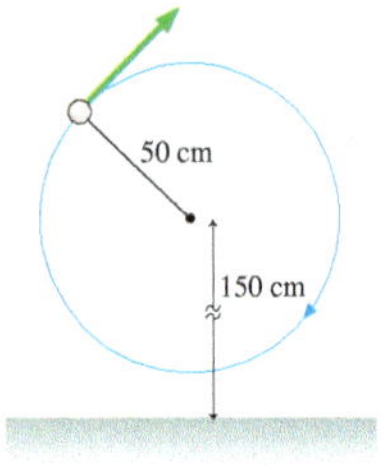

FIGURE P8.51

52. || Elm Street has a pronounced dip at the bottom of a steep hill before going back uphill on the other side. Your science teacher has asked everyone in the class to measure the radius of curvature of the dip. Some of your classmates are using surveying equipment, but you decide to base your measurement on what you've learned in physics. To do so, you sit on a spring scale, drive through the dip at different speeds, and for each speed record the scale's reading as you pass through the bottom of the dip. Your data are as follows:

Speed (m/s)	Scale reading (N)
5	599
10	625
15	674
20	756
25	834

Sitting on the scale while the car is parked gives a reading of 588 N. Analyze your data, using a graph, to determine the dip's radius of curvature.

53. || A 100 g ball on a 60-cm-long string is swung in a vertical circle about a point 200 cm above the floor. The string suddenly breaks when it is parallel to the ground and the ball is moving upward. The ball reaches a height 600 cm above the floor. What was the tension in the string an instant before it broke?

54. ||| A 1500 kg car starts from rest and drives around a flat 50-m-diameter circular track. The forward force provided by the car's drive wheels is a constant 1000 N.
   a. What are the magnitude and direction of the car's acceleration at $t = 10$ s? Give the direction as an angle from the r-axis.
   b. If the car has rubber tires and the track is concrete, at what time does the car begin to slide out of the circle?

55. || A 500 g steel block rotates on a steel table while attached to a 2.0-m-long massless rod. Compressed air fed through the rod is ejected from a nozzle on the back of the block, exerting a thrust force of 3.5 N. The nozzle is 70° from the radial line, as shown in FIGURE P8.55. The block starts from rest.
   a. What is the block's angular velocity after 10 rev?
   b. What is the tension in the rod after 10 rev?

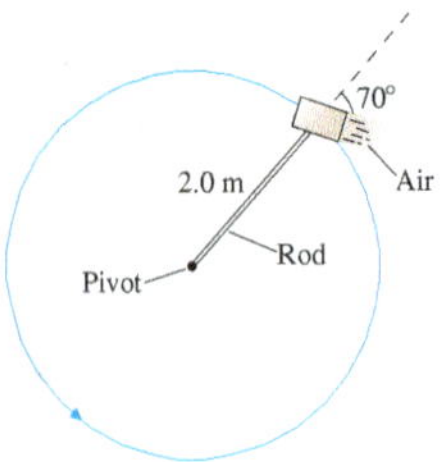

FIGURE P8.55

56. || A 2.0 kg ball swings in a vertical circle on the end of an 80-cm-long string. The tension in the string is 20 N when its angle from the highest point on the circle is $\theta = 30°$.
   a. What is the ball's speed when $\theta = 30°$?
   b. What are the magnitude and direction of the ball's acceleration when $\theta = 30°$?

In Problems 57 and 58 you are given the equation used to solve a problem. For each of these, you are to
   a. Write a realistic problem for which this is the correct equation. Be sure that the answer your problem requests is consistent with the equation given.
   b. Finish the solution of the problem.

57. $60\text{ N} = (0.30\text{ kg})\omega^2(0.50\text{ m})$
58. $(1500\text{ kg})(9.8\text{ m/s}^2) - 11{,}760\text{ N} = (1500\text{ kg})\,v^2/(200\text{ m})$

Challenge Problems

59. In the absence of air resistance, a projectile that lands at the elevation from which it was launched achieves maximum range when launched at a 45° angle. Suppose a projectile of mass m is launched with speed v_0 into a headwind that exerts a constant, horizontal retarding force $\vec{F}_{wind} = -F_{wind}\hat{\imath}$.
   a. Find an expression for the angle at which the range is maximum.
   b. By what percentage is the maximum range of a 0.50 kg ball reduced if $F_{wind} = 0.60$ N?

60. The father of Example 8.3 stands at the summit of a conical hill as he spins his 20 kg child around on a 5.0 kg cart with a 2.0-m-long rope. The sides of the hill are inclined at 20°. He again keeps the rope parallel to the ground, and friction is negligible. What rope tension will allow the cart to spin with the same 14 rpm it had in the example?

61. A small ball rolls around a horizontal circle at height y inside the cone shown in FIGURE CP8.61. Find an expression for the ball's speed in terms of a, h, y, and g.

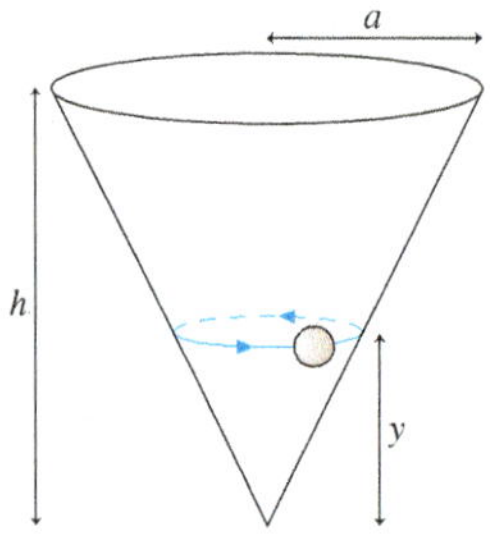

FIGURE CP8.61

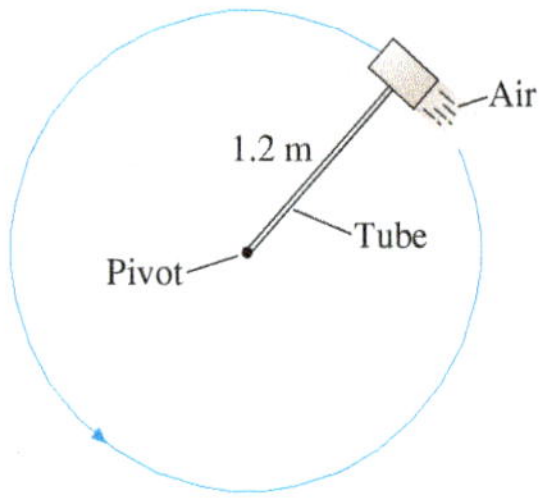

FIGURE CP8.62

62. A 500 g steel block rotates on a steel table while attached to a 1.2-m-long hollow tube as shown in FIGURE CP8.62. Compressed air fed through the tube and ejected from a nozzle on the back of the block exerts a thrust force of 4.0 N perpendicular to the tube. The maximum tension the tube can withstand without breaking is 50 N. If the block starts from rest, how many revolutions does it make before the tube breaks?

63. Two wires are tied to the 300 g sphere shown in FIGURE CP8.63. The sphere revolves in a horizontal circle at a constant speed of 7.5 m/s. What is the tension in each of the wires?

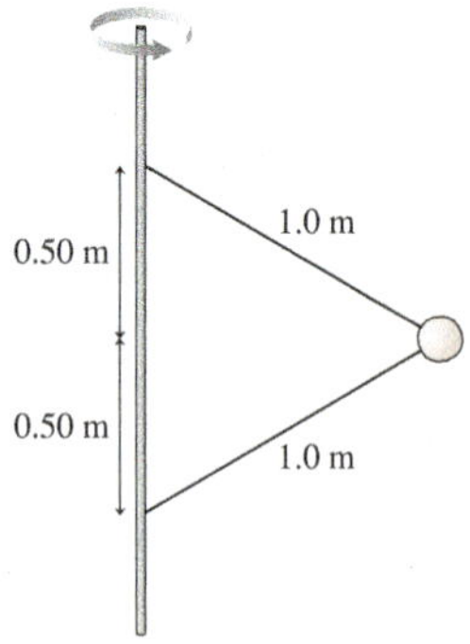

FIGURE CP8.63

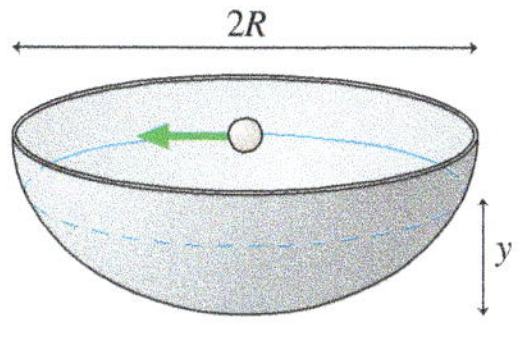

FIGURE CP8.64

64. A small ball rolls around a horizontal circle at height y inside a frictionless hemispherical bowl of radius R, as shown in FIGURE CP8.64.
   a. Find an expression for the ball's angular velocity in terms of R, y, and g.
   b. What is the minimum value of ω for which the ball can move in a circle?
   c. What is ω in rpm if $R = 20$ cm and the ball is halfway up?

65. You are flying to New York. You've been reading the in-flight magazine, which has an article about the physics of flying. You learned that the airflow over the wings creates a *lift force* that is always perpendicular to the wings. In level flight, the upward lift force exactly balances the downward gravitational force. The pilot comes on to say that, because of heavy traffic, the plane is going to circle the airport for a while. She says that you'll maintain a speed of 400 mph at an altitude of 20,000 ft. You start to wonder what the diameter of the plane's circle around the airport is. You notice that the pilot has banked the plane so that the wings are 10° from horizontal. The safety card in the seatback pocket informs you that the plane's wing span is 250 ft. What can you learn about the diameter?

66. If a vertical cylinder of water (or any other liquid) rotates about its axis, as shown in FIGURE CP8.66, the surface forms a smooth curve. Assuming that the water rotates as a unit (i.e., all the water rotates with the same angular velocity), show that the shape of the surface is a parabola described by the equation $z = (\omega^2/2g)r^2$. **Hint:** Each particle of water on the surface is subject to only two forces: gravity and the normal force due to the water underneath it. The normal force, as always, acts perpendicular to the surface.

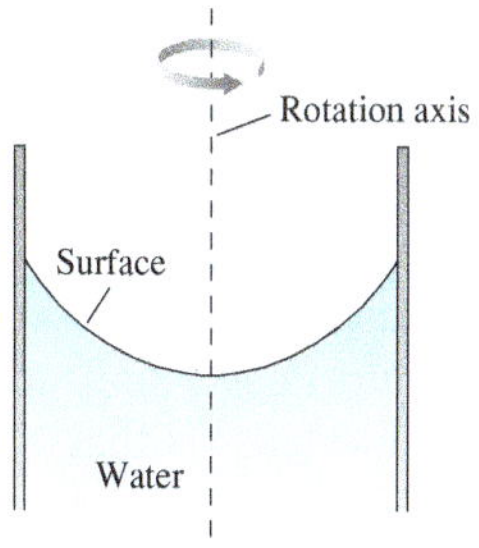

FIGURE CP8.66

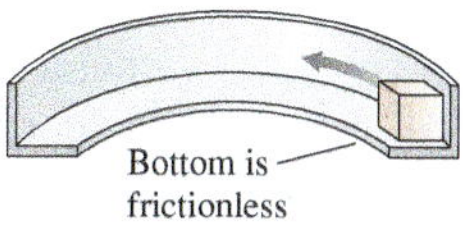

FIGURE CP8.67

67. FIGURE CP8.67 shows a small block of mass m sliding around the inside of an L-shaped track of radius r. The bottom of the track is frictionless; the coefficient of kinetic friction between the block and the wall of the track is μ_k. The block's speed is v_0 at $t_0 = 0$. Find an expression for the block's speed at a later time t.

STOP TO THINK ANSWERS

Stop to Think 8.1: d. The parallel component of $\vec{a}$ is opposite $\vec{v}$ and will cause the particle to slow down. The perpendicular component of $\vec{a}$ will cause the particle to change directions in a downward direction.

Stop to Think 8.2: $(a_r)_b > (a_r)_e > (a_r)_a = (a_r)_c > (a_r)_d$. Centripetal acceleration is v^2/r. Doubling r decreases a_r by a factor of 2. Doubling v increases a_r by a factor of 4. Reversing direction doesn't change a_r.

Stop to Think 8.3: $T_d > T_b = T_e > T_c > T_a$. The center-directed force is $m\omega^2 r$. Changing r by a factor of 2 changes the tension by a factor of 2, but changing ω by a factor of 2 changes the tension by a factor of 4.

Stop to Think 8.4: b. The car is moving in a circle, so there must be a net force toward the center of the circle. The circle is below the car, so the net force must point downward. This can be true only if $F_G > n$.

Stop to Think 8.5: c. The ball does not have a "memory" of its previous motion. The velocity $\vec{v}$ is straight up at the instant the string breaks. The only force on the ball after the string breaks is the gravitational force, straight down. This is just like tossing a ball straight up.

PART I SUMMARY

Newton's Laws

The goal of Part I has been to discover the connection between force and motion. We started with *kinematics,* which is the mathematical description of motion; then we proceeded to *dynamics,* which is the explanation of motion in terms of forces. Newton's three laws of motion form the basis of our explanation. All of the examples we have studied so far are applications of Newton's laws.

The table below is called a *knowledge structure* for Newton's laws. A knowledge structure summarizes the essential concepts, the general principles, and the primary applications of a theory. The first section of the table tells us that Newtonian mechanics is concerned with how *particles* respond to *forces.* The second section indicates that we have introduced only three general principles, Newton's three laws of motion.

You use this knowledge structure by working your way through it, from top to bottom. Once you recognize a problem as a dynamics problem, you immediately know to start with Newton's laws. You can then determine the category of motion and apply Newton's second law in the appropriate form. Newton's third law will help you identify the forces acting on particles as they interact. Finally, the kinematic equations for that category of motion allow you to reach the solution you seek.

The knowledge structure provides the *procedural knowledge* for solving dynamics problems, but it does not represent the total knowledge required. You must add to it knowledge about what position and velocity are, about how forces are identified, about action/reaction pairs, about drawing and using free-body diagrams, and so on. These are specific *tools* for problem solving. The problem-solving strategies of Chapters 5 through 8 combine the procedures and the tools into a powerful method for thinking about and solving problems.

KNOWLEDGE STRUCTURE I Newton's Laws

ESSENTIAL CONCEPTS	Particle, acceleration, force, interaction
BASIC GOALS	How does a particle respond to a force? How do objects interact?

GENERAL PRINCIPLES	**Newton's first law**	An object will remain at rest or will continue to move with constant velocity (equilibrium) if and only if $\vec{F}_{net} = \vec{0}$.
	Newton's second law	$\vec{F}_{net} = m\vec{a}$
	Newton's third law	$\vec{F}_{A\ on\ B} = -\vec{F}_{B\ on\ A}$

BASIC PROBLEM-SOLVING STRATEGY Use Newton's second law for each particle or object. Use Newton's third law to equate the magnitudes of the two members of an action/reaction pair.

Linear motion

$\sum F_x = ma_x$, $\sum F_y = 0$ or $\sum F_x = 0$, $\sum F_y = ma_y$

Trajectory motion

$\sum F_x = ma_x$

$\sum F_y = ma_y$

Circular motion

$\sum F_r = mv^2/r = m\omega^2 r$

$\sum F_t = 0$ or ma_t

$\sum F_z = 0$

Linear and trajectory kinematics

Uniform acceleration: (a_s = constant)

$$v_{fs} = v_{is} + a_s \Delta t$$

$$s_f = s_i + v_{is} \Delta t + \tfrac{1}{2} a_s (\Delta t)^2$$

$$v_{fs}^2 = v_{is}^2 + 2a_s \Delta s$$

Trajectories: The same equations are used for both x and y.

Uniform motion: ($a = 0$, v_s = constant)

$$s_f = s_i + v_s \Delta t$$

General case

$v_s = ds/dt =$ slope of the position graph

$a_s = dv_s/dt =$ slope of the velocity graph

$$v_{fs} = v_{is} + \int_{t_i}^{t_f} a_s\, dt = v_{is} + \text{area under the acceleration curve}$$

$$s_f = s_i + \int_{t_i}^{t_f} v_s\, dt = s_i + \text{area under the velocity curve}$$

Circular kinematics

Uniform circular motion:

$$T = 2\pi r/v = 2\pi/\omega$$

$$\theta_f = \theta_i + \omega \Delta t$$

$$a_r = v^2/r = \omega^2 r$$

$$v_t = \omega r$$

Nonuniform circular motion:

$$\omega_f = \omega_i + \alpha \Delta t$$

$$\theta_f = \theta_i + \omega_i \Delta t + \tfrac{1}{2}\alpha(\Delta t)^2$$

$$\omega_f^2 = \omega_i^2 + 2\alpha \Delta\theta$$

ONE STEP BEYOND

The Forces of Nature

What are the fundamental forces of nature? That is, what set of distinct, irreducible forces can explain everything we know about nature? This is a question that has long intrigued physicists. For example, friction is not a fundamental force because it can be reduced to electric forces between atoms. What about other forces?

Physicists have long recognized three basic forces: the gravitational force, the electric force, and the magnetic force. The gravitational force is an inherent attraction between two masses. The electric force is a force between charges. The magnetic force, which is a bit more mysterious, causes compass needles to point north and holds your shopping list on the refrigerator door.

In the 1860s, the Scottish physicist James Clerk Maxwell developed a theory that *unified* the electric and magnetic forces into a single *electromagnetic force.* Where there had appeared to be two separate forces, Maxwell found there to be a single force that, under appropriate conditions, exhibits "electric behavior" or "magnetic behavior." Maxwell used his theory to predict the existence of *electromagnetic waves,* including light. Our entire telecommunications industry is testimony to Maxwell's genius.

Maxwell's electromagnetic force was soon found to be the "glue" holding atoms, molecules, and solids together. With the exception of gravity, *every* force we have considered so far can be traced to electromagnetic forces between atoms.

The discovery of the atomic nucleus, about 1910, presented difficulties that could not be explained by either gravitational or electromagnetic forces. The atomic nucleus is an unimaginably dense ball of protons and neutrons. But what holds it together against the repulsive electric forces between the protons? There must be an attractive force inside the nucleus that is stronger than the repulsive electric force. This force, called the *strong force,* is the force that holds atomic nuclei together. The strong force is a *short-range* force, extending only about 10^{-14} m. It is completely negligible outside the nucleus. The subatomic particles called *quarks,* of which you have likely heard, are part of our understanding of how the strong force works.

In the 1930s, physicists found that the nuclear radioactivity called *beta decay* could not be explained by either the electromagnetic or the strong force. Careful experiments established that the decay is due to a previously undiscovered force within the nucleus. The strength of this force is less than either the strong force or the electromagnetic force, so this new force was named the *weak force.* Although discovered in conjunction with radioactivity, it is now known to play an important role in the fusion reactions that power the stars.

By 1940, the recognized forces of nature were four: the gravitational force, the electromagnetic force, the strong force, and the weak force. Physicists were understandably curious whether all four of these were truly fundamental or if some of them could be further unified. Indeed, innovative work in the 1960s and 1970s produced a theory that unified the electromagnetic force and the weak force.

Predictions of this new theory were confirmed during the 1980s at some of the world's largest particle accelerators, and we now speak of the *electroweak force.* Under appropriate conditions, the electroweak force exhibits either "electromagnetic behavior" or "weak behavior." But under other conditions, new phenomena appear that are consequences of the full electroweak force. These conditions appear on earth only in the largest and most energetic particle accelerators, which is why we were not previously aware of the unified nature of these two forces. However, the earliest moments of the Big Bang provided the right conditions for the electroweak force to play a significant role. Thus a theory developed to help us understand the workings of nature on the smallest subatomic scale has unexpectedly given us powerful new insights into the origin of the universe.

The success of the electroweak theory has prompted efforts to unify the electroweak force and the strong force into a *grand unified theory.* Only time will tell if the strong force and the electroweak force are really just two different aspects of a single force, or if they are truly distinct. Some physicists even envision a day when all the forces of nature will be unified in a single theory, the so-called *Theory of Everything!* For today, however, our understanding of the forces of nature is in terms of three fundamental forces: the gravitational force, the electroweak force, and the strong force.

FIGURE I.1 A historical progression of our understanding of the fundamental forces of nature.

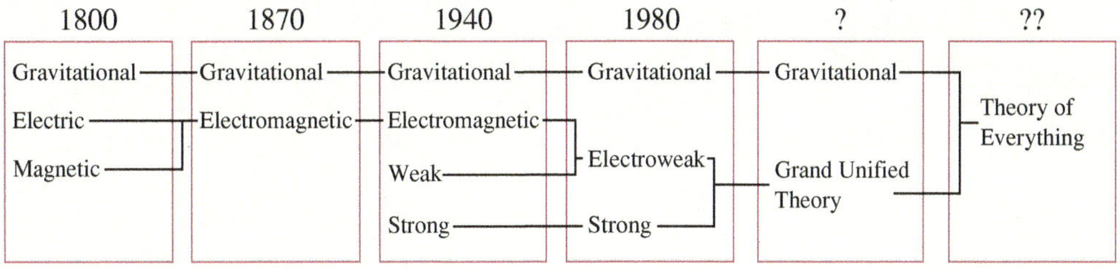

PART

II Conservation Laws

Energy is the lifeblood of modern society. This power plant in the Mojave Desert transforms solar energy into electrical energy and, unavoidably, increased thermal energy.

OVERVIEW

Why Some Things Don't Change

Part I of this textbook was about *change*. One particular type of change—motion—is governed by Newton's second law. Although Newton's second law is a very powerful statement, it isn't the whole story. Part II will now focus on things that *stay the same* as other things around them change.

Consider, for example, an explosive chemical reaction taking place inside a closed, sealed box. No matter how violent the explosion, the total mass of the products—the final mass M_f —is the same as the initial mass M_i of the reactants. In other words, matter cannot be created or destroyed, only rearranged. This is an important and powerful statement about nature.

A quantity that *stays the same* throughout an interaction is said to be *conserved*. Our knowledge about mass can be stated as a *conservation law*:

Law of conservation of mass The total mass in a closed system is constant. Mathematically, $M_f = M_i$.*

The qualification "in a closed system" is important. Mass certainly won't be conserved if you open the box halfway through and remove some of the matter. Other conservation laws we'll discover also have qualifications stating the circumstances under which they apply.

A system of interacting objects has another curious property. Each system is characterized by a certain number, and no matter how complex the interactions, the value of this number never changes. This number is called the *energy* of the system, and the fact that it never changes is called the *law of conservation of energy*. It is, perhaps, the single most important physical law ever discovered.

But what is energy? How do you determine the energy number for a system? These are not easy questions. Energy is an abstract idea, not as tangible or easy to picture as mass or force. Our modern concept of energy wasn't fully formulated until the middle of the 19th century, two hundred years after Newton, when the relationship between *energy* and *heat* was finally understood. That is a topic we will take up in Part IV, where the concept of energy will be found to be the basis of thermodynamics. But all that in due time. In Part II we will be content to introduce the concept of energy and show how energy can be a useful problem-solving tool. We'll also meet another quantity—*momentum*—that is conserved under the proper circumstances.

Conservation laws give us a new and different perspective on motion. This is not insignificant. You've seen optical illusions where a figure appears first one way, then another, even though the information has not changed. Likewise with motion. Some situations are most easily analyzed from the perspective of Newton's laws; others make more sense from a conservation-law perspective. An important goal of Part II is to learn which is better for a given problem.

*Surprisingly, Einstein's 1905 theory of relativity showed that there are circumstances in which mass is *not* conserved but can be converted to energy in accordance with his famous formula $E = mc^2$. Nonetheless, conservation of mass is an exceedingly good approximation in nearly all applications of science and engineering.

9 Impulse and Momentum

An exploding firework is a dramatic event. Nonetheless, the explosion obeys some simple laws of physics.

▶ **Looking Ahead** The goals of Chapter 9 are to understand and apply the new concepts of impulse and momentum.

Momentum

An object's **momentum** is the product of its mass and velocity: $\vec{p} = m\vec{v}$.

An object can have a large momentum by having a large mass or a large velocity.

Momentum is a vector. Paying attention to the *signs* of the components of momentum will be especially important.

You'll learn to write Newton's second law in terms of momentum.

Impulse

A force of short duration is an *impulsive force.* The **impulse** J_x is the area under the force-versus-time curve.

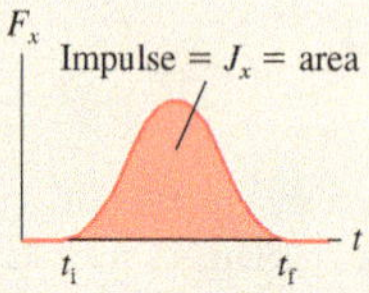

We say that the bat delivers an impulse to the ball.

The **impulse-momentum theorem** says that an impulse changes a particle's momentum: $\Delta p_x = J_x$.

Conservation Laws

Part I of this textbook was about how interactions cause things to change. Part II will explore how some things are *not* changed by the interactions. We say they are *conserved.*

The particles of an **isolated system** interact with each other—perhaps very intensely—but not with the external environment.

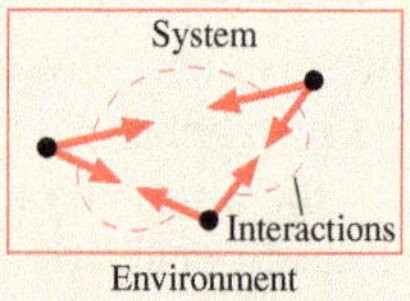

The mass, the momentum, and the energy of an isolated system are conserved. Conservation laws will be the basis of a new and powerful problem solving strategy:

final value = initial value

Conservation of momentum for an isolated system is a consequence of Newton's third law.

◀ **Looking Back**
Sections 7.1–7.3 Action/reaction force pairs and Newton's third law

Representations

Conservation laws require new visual tools. You will learn to draw and use:

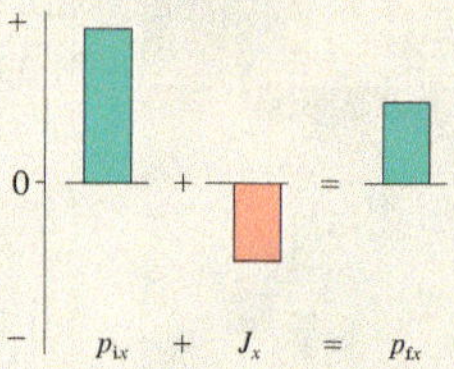

Momentum bar charts to show how impulse changes momentum.

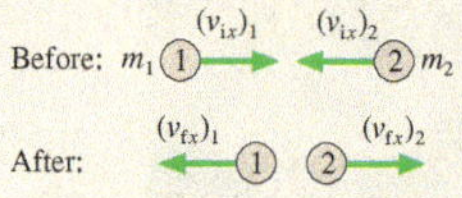

Before-and-after pictorial representations to compare quantities before and after an interaction.

Collisions and Explosions

You will learn to apply conservation of momentum to the analysis of *collisions* and *explosions*.

A **collision** is when two or more particles come together for a short but intense interaction.

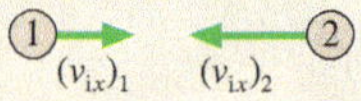

An **explosion** is when a short but intense interaction causes two or more particles to move apart.

9.1 Momentum and Impulse

A **collision** is a short-duration interaction between two objects. The collision between a tennis ball and a racket, or a baseball and a bat, may seem instantaneous to your eye, but that is a limitation of your perception. A careful look at the photograph reveals that the right side of the ball is flattened and pressed up against the strings of the racket. It takes time to compress the ball, and more time for the ball to re-expand as it leaves the racket.

A tennis ball collides with a racket. Notice that the right side of the ball is flattened.

The duration of a collision depends on the materials from which the objects are made, but 1 to 10 ms (0.001 to 0.010 s) is fairly typical. This is the time during which the two objects are in contact with each other. The harder the objects, the shorter the contact time. A collision between two steel balls lasts less than 1 ms.

FIGURE 9.1 shows a microscopic view of a collision in which object A bounces off object B. The spring-like molecular bonds—the same bonds that cause normal forces and tension forces—compress during the collision, then re-expand as A bounces back. The forces $\vec{F}_{\text{A on B}}$ and $\vec{F}_{\text{B on A}}$ are an action/reaction pair and, according to Newton's third law, have equal magnitudes: $F_{\text{A on B}} = F_{\text{B on A}}$. The force increases rapidly as the bonds compress, reaches a maximum at the instant A is at rest (point of maximum compression), then decreases as the bonds re-expand.

FIGURE 9.1 Atomic model of a collision.

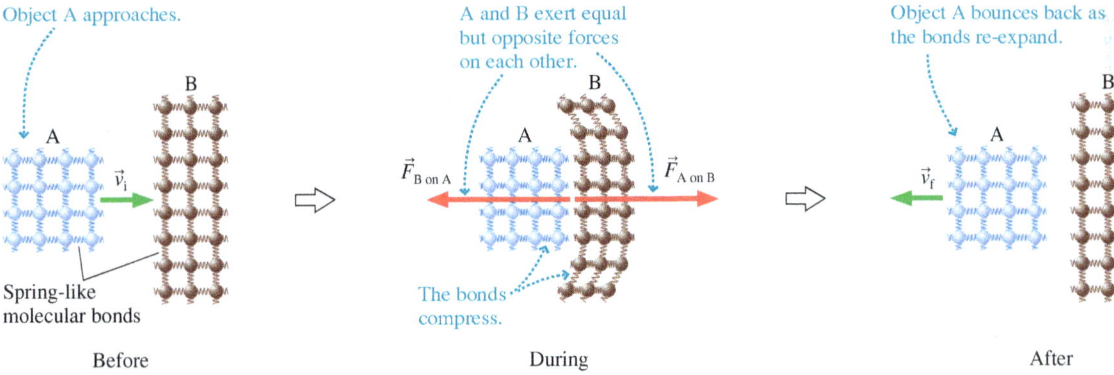

A large force exerted for a small interval of time is called an **impulsive force.** FIGURE 9.2 shows that a particle undergoing a collision enters with initial velocity $\vec{v}_i$, experiences an impulsive force of short duration Δt, then leaves with final velocity $\vec{v}_f$. The graph shows how a typical impulsive force behaves, growing to a maximum and then decreasing back to zero. Because an impulsive force is a function of time, we will write it as $F_x(t)$.

FIGURE 9.2 A particle undergoes a collision.

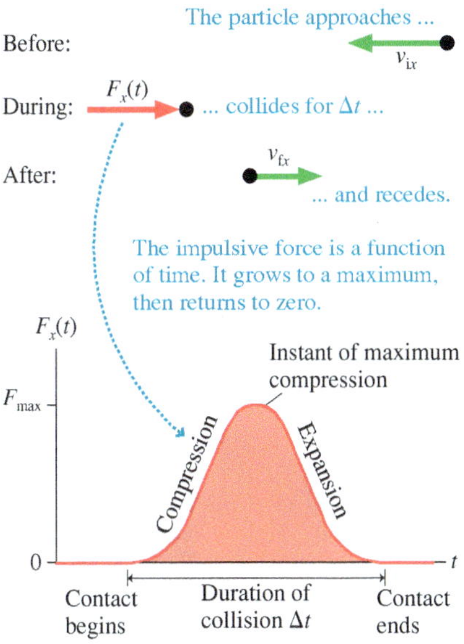

NOTE ▶ Both v_x and F_x are components of vectors and thus have *signs* indicating which way the vectors point. ◀

We can use Newton's second law to find the final velocity. Acceleration in one dimension is $a_x = dv_x/dt$, so the second law is

$$ma_x = m\frac{dv_x}{dt} = F_x(t)$$

After multiplying both sides by dt, we can write the second law as

$$m\,dv_x = F_x(t)\,dt \tag{9.1}$$

The force is nonzero only during the interval of time from t_i to t_f, so let's integrate Equation 9.1 over this interval. The velocity changes from v_{ix} to v_{fx} during the collision; thus

$$m\int_{v_i}^{v_f} dv_x = mv_{fx} - mv_{ix} = \int_{t_i}^{t_f} F_x(t)\,dt \tag{9.2}$$

We need some new tools to help us make sense of Equation 9.2.

Momentum

The product of a particle's mass and velocity is called the *momentum* of the particle:

$$\textbf{momentum} = \vec{p} \equiv m\vec{v} \tag{9.3}$$

Momentum, like velocity, is a vector. The units of momentum are kg m/s. The plural of "momentum" is "momenta," from its Latin origin.

The momentum vector $\vec{p}$ is parallel to the velocity vector $\vec{v}$. FIGURE 9.3 shows that $\vec{p}$, like any vector, can be decomposed into x- and y-components. Equation 9.3, which is a vector equation, is a shorthand way to write the simultaneous equations

$$p_x = mv_x$$
$$p_y = mv_y$$

FIGURE 9.3 A particle's momentum vector $\vec{p}$ can be decomposed into x- and y-components.

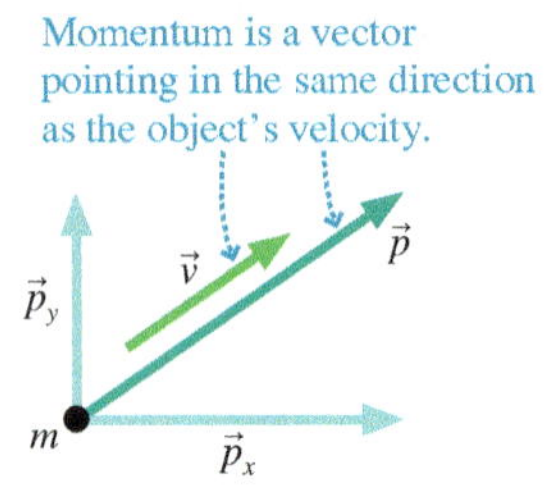

An object can have a large momentum by having either a small mass but a large velocity (a bullet fired from a rifle) or a small velocity but a large mass (a large truck rolling at a slow 1 mph).

NOTE ▶ One of the most common errors in momentum problems is a failure to use the appropriate signs. The momentum component p_x has the same sign as v_x. Momentum is *negative* for a particle moving to the left (on the x-axis) or down (on the y-axis). ◀

Newton actually formulated his second law in terms of momentum rather than acceleration:

$$\vec{F} = m\vec{a} = m\frac{d\vec{v}}{dt} = \frac{d(m\vec{v})}{dt} = \frac{d\vec{p}}{dt} \tag{9.4}$$

This statement of the second law, saying that **force is the rate of change of momentum,** is more general than our earlier version $\vec{F} = m\vec{a}$. It allows for the possibility that the mass of the object might change, such as a rocket that is losing mass as it burns fuel.

Returning to Equation 9.2, you can see that mv_{ix} and mv_{fx} are p_{ix} and p_{fx}, the x-component of the particle's momentum before and after the collision. Further, $p_{fx} - p_{ix}$ is Δp_x, the *change* in the particle's momentum. In terms of momentum, Equation 9.2 is

$$\Delta p_x = p_{fx} - p_{ix} = \int_{t_i}^{t_f} F_x(t)\,dt \tag{9.5}$$

Now we need to examine the right-hand side of Equation 9.5.

Impulse

Equation 9.5 tells us that the particle's change in momentum is related to the time integral of the force. Let's define a quantity J_x called the *impulse* to be

$$\textbf{impulse} = J_x \equiv \int_{t_i}^{t_f} F_x(t)\,dt$$
$$= \text{area under the } F_x(t) \text{ curve between } t_i \text{ and } t_f \tag{9.6}$$

FIGURE 9.4 Looking at the impulse graphically.

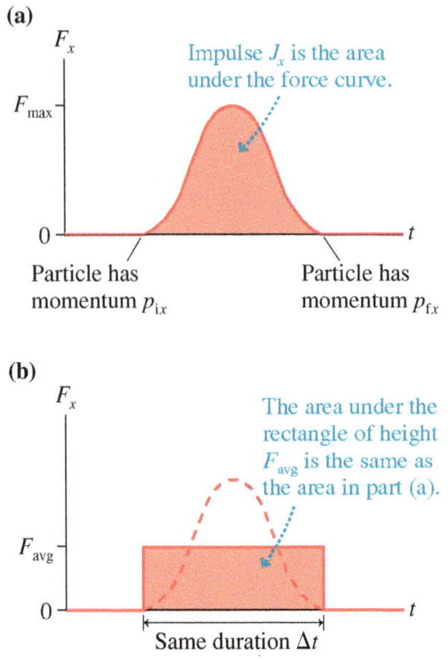

Strictly speaking, impulse has units of N s, but you should be able to show that N s are equivalent to kg m/s, the units of momentum.

The interpretation of the integral in Equation 9.6 as an area under a curve is especially important. FIGURE 9.4a portrays the impulse graphically. Because the force changes in a complicated way during a collision, it is often useful to describe the collision in terms of an *average* force F_{avg}. As FIGURE 9.4b shows, F_{avg} is the height of a rectangle that has the same area, and thus the same impulse, as the real force curve. The impulse exerted during the collision is

$$J_x = F_{avg}\,\Delta t \tag{9.7}$$

Equation 9.2, which we found by integrating Newton's second law, can now be rewritten in terms of impulse and momentum as

$$\Delta p_x = J_x \qquad \text{(impulse-momentum theorem)} \tag{9.8}$$

This result is called the **impulse-momentum theorem.** The name is rather unusual, but it's not the name that is important. The important new *idea* is that **an impulse delivered to a particle changes the particle's momentum.** The momentum p_{fx} "after" an interaction, such as a collision or an explosion, is equal to the momentum p_{ix} "before" the interaction *plus* the impulse that arises from the interaction:

$$p_{fx} = p_{ix} + J_x \quad (9.9)$$

FIGURE 9.5 illustrates the impulse-momentum theorem for a rubber ball bouncing off a wall. Notice the signs; they are very important. The ball is initially traveling toward the right, so v_{ix} and p_{ix} are positive. After the bounce, v_{fx} and p_{fx} are negative. The force *on the ball* is toward the left, so F_x is also negative. The graphs show how the force and the momentum change with time.

Although the interaction is very complex, the impulse—the area under the force graph—is all we need to know to find the ball's velocity as it rebounds from the wall. The final momentum is

$$p_{fx} = p_{ix} + J_x = p_{ix} + \text{area under the force curve}$$

and the final velocity is $v_{fx} = p_{fx}/m$. In this example, the area has a negative value.

FIGURE 9.5 The impulse-momentum theorem helps us understand a rubber ball bouncing off a wall.

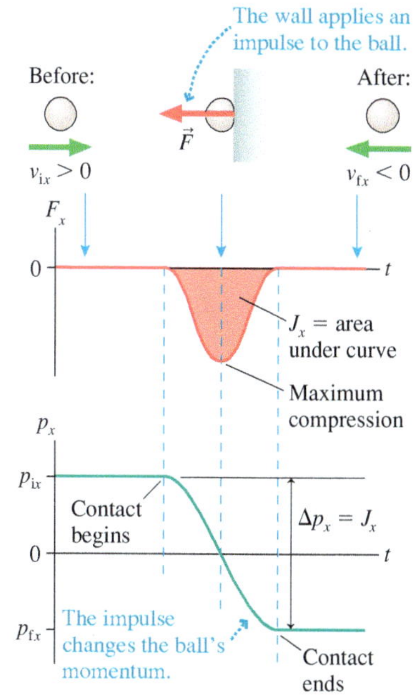

Momentum Bar Charts

The impulse-momentum theorem tells us that **impulse transfers momentum to an object.** If an object has 2 kg m/s of momentum, a 1 kg m/s impulse exerted on the object increases its momentum to 3 kg m/s. That is, $p_{fx} = p_{ix} + J_x$.

We can represent this "momentum accounting" with a **momentum bar chart.** FIGURE 9.6a shows a bar chart in which one unit of impulse adds to an initial two units of momentum to give three units of momentum. The bar chart of FIGURE 9.6b represents the ball colliding with a wall in Figure 9.5. Momentum bar charts are a tool for visualizing an interaction.

FIGURE 9.6 Two examples of momentum bar charts.

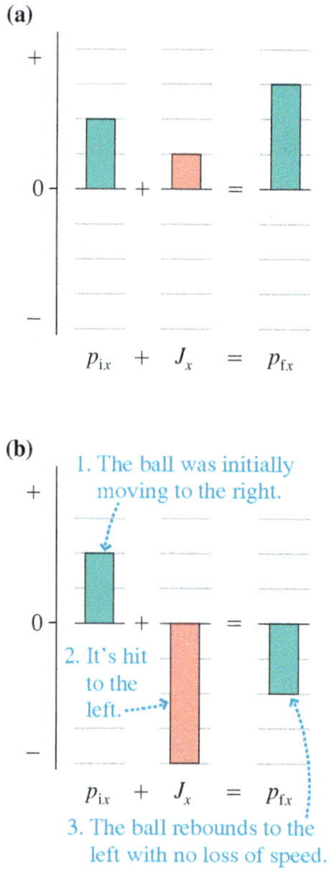

NOTE ▶ The vertical scale of a momentum bar chart has no numbers; it can be adjusted to match any problem. However, be sure that all bars in a given problem use a consistent scale. ◀

STOP TO THINK 9.1 The cart's change of momentum is

a. −30 kg m/s
b. −20 kg m/s
c. 0 kg m/s
d. 10 kg m/s
e. 20 kg m/s
f. 30 kg m/s

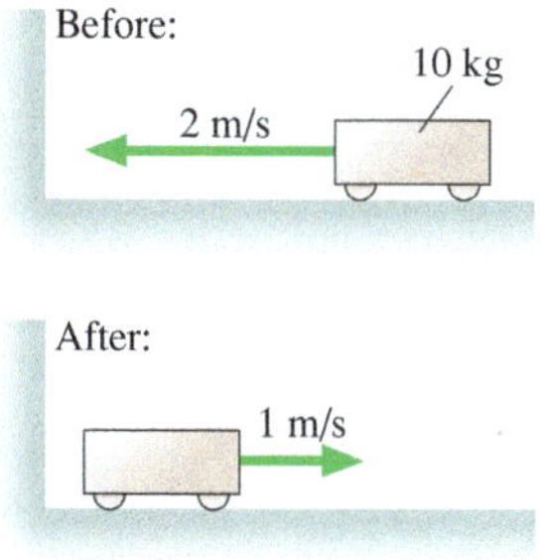

9.2 Solving Impulse and Momentum Problems

Pictorial representations have become an important problem-solving tool. The pictorial representations you learned to draw in Part I were oriented toward the use of Newton's laws and a subsequent kinematic analysis. For conservation-law problems we need a new representation, the **before-and-after pictorial representation.**

TACTICS BOX 9.1 Drawing a before-and-after pictorial representation

1. **Sketch the situation.** Use two drawings, labeled "Before" and "After," to show the objects *before* they interact and again *after* they interact.
2. **Establish a coordinate system.** Select your axes to match the motion.
3. **Define symbols.** Define symbols for the masses and for the velocities before and after the interaction. Position and time are not needed.
4. **List known information.** Give the values of quantities that are known from the problem statement or that can be found quickly with simple geometry or unit conversions. Before-and-after pictures are simpler than the pictures for dynamics problems, so listing known information on the sketch is adequate.
5. **Identify the desired unknowns.** What quantity or quantities will allow you to answer the question? These should have been defined in step 3.
6. If appropriate, **draw a momentum bar chart** to clarify the situation and establish appropriate signs.

Exercises 17–19

NOTE ▶ The generic subscripts i and f, for *initial* and *final*, are adequate in equations for a simple problem, but using numerical subscripts, such as v_{1x} and v_{2x}, will help keep all the symbols straight in more complex problems. ◀

EXAMPLE 9.1 Hitting a baseball

A 150 g baseball is thrown with a speed of 20 m/s. It is hit straight back toward the pitcher at a speed of 40 m/s. The interaction force between the ball and the bat is shown in FIGURE 9.7. What *maximum* force F_{max} does the bat exert on the ball? What is the *average* force of the bat on the ball?

FIGURE 9.7 The interaction force between the baseball and the bat.

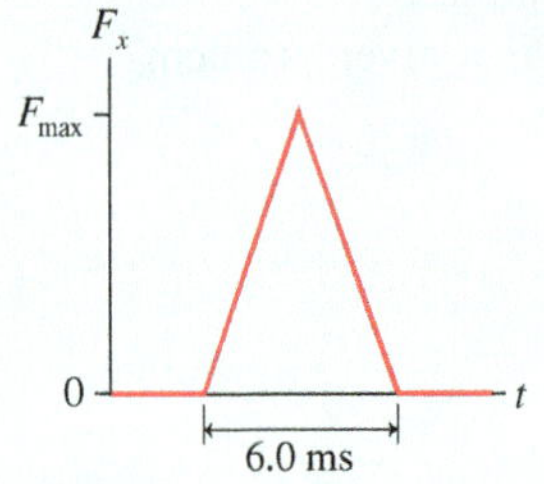

MODEL Model the baseball as a particle and the interaction as a collision.

VISUALIZE FIGURE 9.8 is a before-and-after pictorial representation. The steps from Tactics Box 9.1 are explicitly noted. Because F_x is positive (a force to the right), we know the ball was initially moving toward the left and is hit back toward the right. Thus we converted the statements about *speeds* into information about *velocities*, with v_{ix} negative.

SOLVE Until now we've consistently started the mathematical representation with Newton's second law. Now we want to use the impulse-momentum theorem:

$$\Delta p_x = J_x = \text{area under the force curve}$$

We know the velocities before and after the collision, so we can calculate the ball's momenta:

$$p_{ix} = mv_{ix} = (0.15\text{ kg})(-20\text{ m/s}) = -3.0\text{ kg m/s}$$

$$p_{fx} = mv_{fx} = (0.15\text{ kg})(40\text{ m/s}) = 6.0\text{ kg m/s}$$

FIGURE 9.8 A before-and-after pictorial representation.

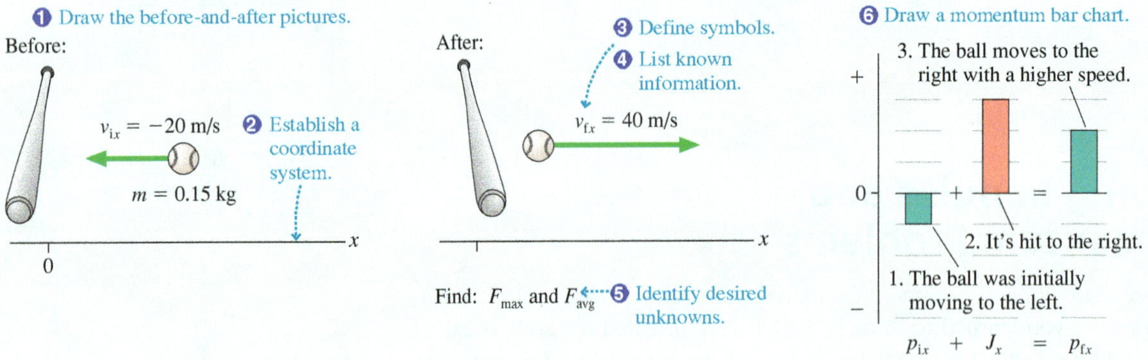

Thus the *change* in momentum is

$$\Delta p_x = p_{fx} - p_{ix} = 9.0\text{ kg m/s}$$

The force curve is a triangle with height F_{max} and width 6.0 ms. The area under the curve is

$$J_x = \text{area} = \frac{1}{2} \times F_{max} \times (0.0060\text{ s}) = (F_{max})(0.0030\text{ s})$$

According to the impulse-momentum theorem,

$$9.0\text{ kg m/s} = (F_{max})(0.0030\text{ s})$$

Thus the *maximum* force is

$$F_{max} = \frac{9.0\text{ kg m/s}}{0.0030\text{ s}} = 3000\text{ N}$$

The *average* force, which depends on the collision duration $\Delta t = 0.0060$ s, has the smaller value:

$$F_{avg} = \frac{J_x}{\Delta t} = \frac{\Delta p_x}{\Delta t} = \frac{9.0\text{ kg m/s}}{0.0060\text{ s}} = 1500\text{ N}$$

ASSESS F_{max} is a large force, but quite typical of the impulsive forces during collisions. The main thing to focus on is our new perspective: An impulse changes the momentum of an object.

Other forces often act on an object during a collision or other brief interaction. In Example 9.1, for instance, the baseball is also acted on by gravity. Usually these other forces are *much* smaller than the interaction forces. The 1.5 N weight of the ball is vastly less than the 3000 N force of the bat on the ball. We can reasonably neglect these small forces *during* the brief time of the impulsive force by using what is called the **impulse approximation.**

When we use the impulse approximation, p_{ix} and p_{fx} (and v_{ix} and v_{fx}) are the momenta (and velocities) *immediately* before and *immediately* after the collision. For example, the velocities in Example 9.1 are those of the ball just before and after it collides with the bat. We could then do a follow-up problem, including gravity and drag, to find the ball's speed a second later as the second baseman catches it. We'll look at some two-part examples later in the chapter.

EXAMPLE 9.2 A bouncing ball

A 100 g rubber ball is dropped from a height of 2.00 m onto a hard floor. FIGURE 9.9 shows the force that the floor exerts on the ball. How high does the ball bounce?

FIGURE 9.9 The force of the floor on a bouncing rubber ball.

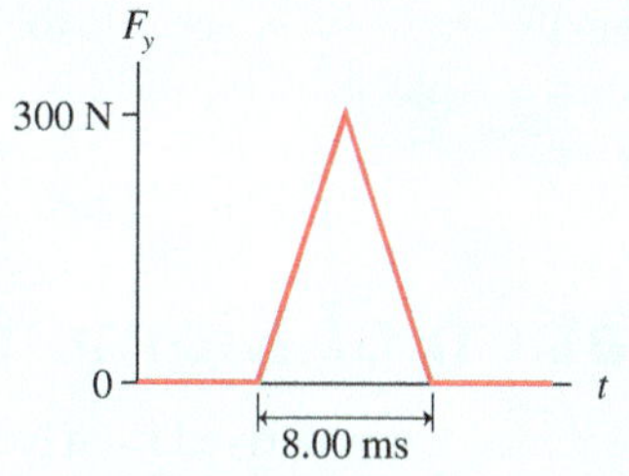

MODEL Model the ball as a particle subjected to an impulsive force while in contact with the floor. Using the impulse approximation, we'll neglect gravity during these 8.00 ms. The fall and subsequent rise are free-fall motion.

VISUALIZE FIGURE 9.10 is a pictorial representation. Here we have a three-part problem (downward free fall, impulsive collision, upward free fall), so the pictorial motion includes both the before and after of the collision (v_{1y} changing to v_{2y}) and the beginning and end of the free-fall motion. The bar chart shows the momentum change during the brief collision. Note that p is negative for downward motion.

FIGURE 9.10 Pictorial representation of the ball and a momentum bar chart of the collision with the floor.

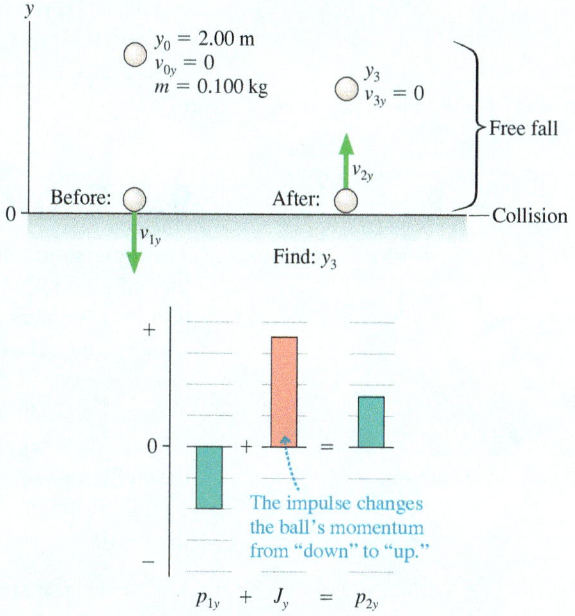

Continued

SOLVE Velocity v_{1y}, the ball's velocity *immediately* before the collision, is found using free-fall kinematics with $\Delta y = -2.0$ m:

$$v_{1y}^2 = v_{0y}^2 - 2g\Delta y = 0 - 2g\Delta y$$

$$v_{1y} = \sqrt{-2g\Delta y} = \sqrt{-2(9.80\text{ m/s}^2)(-2.00\text{ m})} = -6.26\text{ m/s}$$

We've chosen the negative root because the ball is moving in the negative y-direction.

The impulse-momentum theorem is $p_{2y} = p_{1y} + J_y$. The initial momentum, just before the collision, is $p_{1y} = mv_{1y} = -0.626$ kg m/s. The force of the floor is upward, so J_y is positive. From Figure 9.9, the impulse J_y is

$$J_y = \text{area under the force curve} = \frac{1}{2} \times (300\text{ N}) \times (0.00800\text{ s})$$
$$= 1.200\text{ N s}$$

Thus

$$p_{2y} = p_{1y} + J_y = (-0.626\text{ kg m/s}) + 1.200\text{ N s} = 0.574\text{ kg m/s}$$

and the post-collision velocity is

$$v_{2y} = \frac{p_{2y}}{m} = \frac{0.574\text{ kg m/s}}{0.100\text{ kg}} = 5.74\text{ m/s}$$

The rebound speed is less than the impact speed, as expected. Finally a second use of free-fall kinematics yields

$$v_{3y}^2 = 0 = v_{2y}^2 - 2g\Delta y = v_{2y}^2 - 2gy_3$$

$$y_3 = \frac{v_{2y}^2}{2g} = \frac{(5.74\text{ m/s})^2}{2(9.80\text{ m/s}^2)} = 1.68\text{ m}$$

The ball bounces back to a height of 1.68 m.

ASSESS The ball bounces back to less than its initial height, which is realistic.

NOTE ▶ Example 9.2 illustrates an important point: The impulse-momentum theorem applies *only* during the brief interval in which an impulsive force is applied. Many problems will have segments of the motion that must be analyzed with kinematics or Newton's laws. The impulse-momentum theorem is a new and useful tool, but it doesn't replace all that you've learned up until now. ◀

STOP TO THINK 9.2 A 10 g rubber ball and a 10 g clay ball are thrown at a wall with equal speeds. The rubber ball bounces, the clay ball sticks. Which ball exerts a larger impulse on the wall?

a. The clay ball exerts a larger impulse because it sticks.
b. The rubber ball exerts a larger impulse because it bounces.
c. They exert equal impulses because they have equal momenta.
d. Neither exerts an impulse on the wall because the wall doesn't move.

FIGURE 9.11 A collision between two objects.

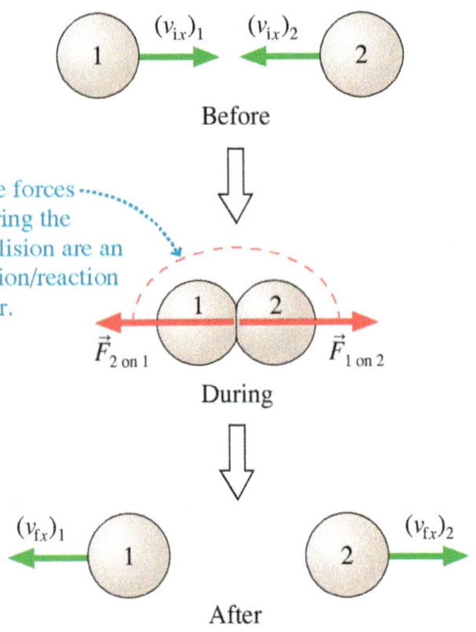

9.3 Conservation of Momentum

The impulse-momentum theorem was derived from Newton's second law and is really just an alternative way of looking at single-particle dynamics. To discover the real power of momentum for problem solving, we need also to invoke Newton's third law, which will lead us to one of the most important principles in physics: conservation of momentum.

FIGURE 9.11 shows two objects with initial velocities $(v_{ix})_1$ and $(v_{ix})_2$. The objects collide, then bounce apart with final velocities $(v_{fx})_1$ and $(v_{fx})_2$. The forces during the collision, as the objects are interacting, are the action/reaction pair $\vec{F}_{1\text{ on }2}$ and $\vec{F}_{2\text{ on }1}$. For now, we'll continue to assume that the motion is one dimensional along the x-axis.

NOTE ▶ The notation, with all the subscripts, may seem excessive. But there are two objects, and each has an initial and a final velocity, so we need to distinguish among four different velocities. ◀

Newton's second law for each object *during* the collision is

$$\frac{d(p_x)_1}{dt} = (F_x)_{2\text{ on }1}$$
$$\frac{d(p_x)_2}{dt} = (F_x)_{1\text{ on }2} = -(F_x)_{2\text{ on }1} \qquad (9.10)$$

We made explicit use of Newton's third law in the second equation.

Although Equations 9.10 are for two different objects, suppose—just to see what happens—we were to *add* these two equations. If we do, we find that

$$\frac{d(p_x)_1}{dt} + \frac{d(p_x)_2}{dt} = \frac{d}{dt}\Big[(p_x)_1 + (p_x)_2\Big] = (F_x)_{2\text{ on }1} + (-(F_x)_{2\text{ on }1}) = 0 \qquad (9.11)$$

If the time derivative of the quantity $(p_x)_1 + (p_x)_2$ is zero, it must be the case that

$$(p_x)_1 + (p_x)_2 = \text{constant} \qquad (9.12)$$

Equation 9.12 is a conservation law! If $(p_x)_1 + (p_x)_2$ is a constant, then the sum of the momenta *after* the collision equals the sum of the momenta *before* the collision. That is,

$$(p_{fx})_1 + (p_{fx})_2 = (p_{ix})_1 + (p_{ix})_2 \qquad (9.13)$$

Furthermore, this equality is independent of the interaction force. We don't need to know anything about $\vec{F}_{1\text{ on }2}$ and $\vec{F}_{2\text{ on }1}$ to make use of Equation 9.13.

As an example, FIGURE 9.12 is a before-and-after pictorial representation of two equal-mass train cars colliding and coupling. Equation 9.13 relates the momenta of the cars after the collision to their momenta before the collision:

$$m_1(v_{fx})_1 + m_2(v_{fx})_2 = m_1(v_{ix})_1 + m_2(v_{ix})_2$$

Initially, car 1 is moving with velocity $(v_{ix})_1 = v_i$ while car 2 is at rest. Afterward, they roll together with the common final velocity v_f. Furthermore, $m_1 = m_2 = m$. With this information, the sum of the momenta is

$$mv_f + mv_f = 2mv_f = mv_i + 0$$

The mass cancels, and we find that the train cars' final velocity is $v_f = \frac{1}{2}v_i$. That is, we can make the very simple prediction that the final speed is exactly half the initial speed of car 1 without knowing anything at all about the very complex interaction between the two cars as they collide.

FIGURE 9.12 Two colliding train cars.

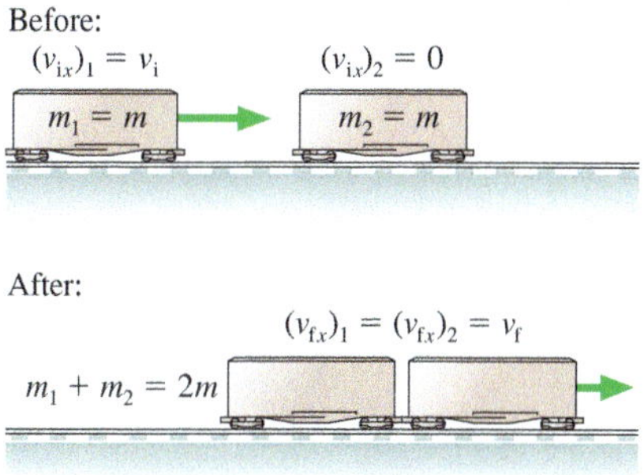

Law of Conservation of Momentum

Equation 9.13 illustrates the idea of a conservation law for momentum, but it was derived for the specific case of two particles colliding in one dimension. Our goal is to develop a more general law of conservation of momentum, a law that will be valid in three dimensions and that will work for any type of interaction. The next few paragraphs are fairly mathematical, so you might want to begin by looking ahead to Equations 9.21 and the statement of the law of conservation of momentum to see where we're heading.

Consider a system consisting of N particles. FIGURE 9.13 shows a simple case where $N = 3$. The particles might be large entities (cars, baseballs, etc.), or they might be the microscopic atoms in a gas. We can identify each particle by an identification number k. Every particle in the system *interacts* with every other particle via action/reaction pairs of forces $\vec{F}_{j\text{ on }k}$ and $\vec{F}_{k\text{ on }j}$. In addition, every particle is subjected to possible *external forces* $\vec{F}_{\text{ext on }k}$ from agents outside the system.

If particle k has velocity $\vec{v}_k$, its momentum is $\vec{p}_k = m_k\vec{v}_k$. We define the **total momentum** $\vec{P}$ of the system as the vector sum

$$\vec{P} = \text{total momentum} = \vec{p}_1 + \vec{p}_2 + \vec{p}_3 + \cdots + \vec{p}_N = \sum_{k=1}^{N}\vec{p}_k \qquad (9.14)$$

FIGURE 9.13 A system of particles.

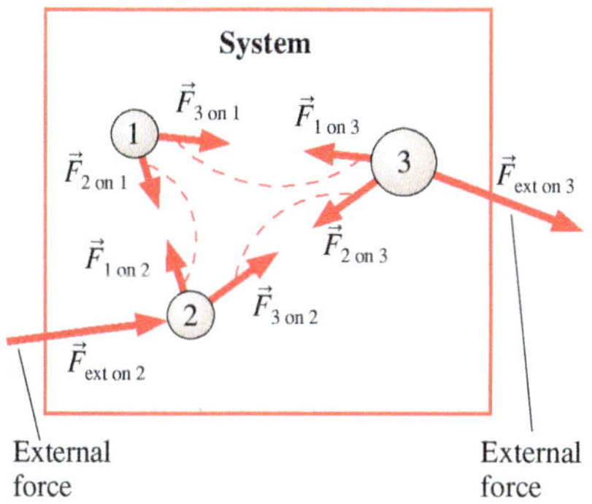

The total momentum of the rocket + gases system is conserved, so the rocket accelerates forward as the gases are expelled backward.

In other words, the total momentum *of the system* is the vector sum of all the individual momenta.

The time derivative of $\vec{P}$ tells us how the total momentum of the system changes with time:

$$\frac{d\vec{P}}{dt} = \sum_k \frac{d\vec{p}_k}{dt} = \sum_k \vec{F}_k \tag{9.15}$$

where we used Newton's second law for each particle in the form $\vec{F}_k = d\vec{p}_k/dt$, which was Equation 9.4.

The net force acting on particle k can be divided into *external forces,* from outside the system, and *interaction forces* due to all the other particles in the system:

$$\vec{F}_k = \sum_{j \neq k} \vec{F}_{j \text{ on } k} + \vec{F}_{\text{ext on } k} \tag{9.16}$$

The restriction $j \neq k$ expresses the fact that particle k does not exert a force on itself. Using this in Equation 9.15 gives the rate of change of the total momentum $\vec{P}$ of the system:

$$\frac{d\vec{P}}{dt} = \sum_k \sum_{j \neq k} \vec{F}_{j \text{ on } k} + \sum_k \vec{F}_{\text{ext on } k} \tag{9.17}$$

The double sum on $\vec{F}_{j \text{ on } k}$ adds *every* interaction force within the system. But the interaction forces come in action/reaction pairs, with $\vec{F}_{k \text{ on } j} = -\vec{F}_{j \text{ on } k}$, so $\vec{F}_{k \text{ on } j} + \vec{F}_{j \text{ on } k} = \vec{0}$. Consequently, **the sum of all the interaction forces is zero.** As a result, Equation 9.17 becomes

$$\frac{d\vec{P}}{dt} = \sum_k \vec{F}_{\text{ext on } k} = \vec{F}_{\text{net}} \tag{9.18}$$

where $\vec{F}_{\text{net}}$ is the net force exerted on the system by agents outside the system. But this is just Newton's second law written for the system as a whole! That is, **the rate of change of the total momentum of the system is equal to the net force applied to the system.**

Equation 9.18 has two very important implications. First, we can analyze the motion of the system as a whole without needing to consider interaction forces between the particles that make up the system. In fact, we have been using this idea all along as an *assumption* of the particle model. When we treat cars and rocks and baseballs as particles, we assume that the internal forces between the atoms—the forces that hold the object together—do not affect the motion of the object as a whole. Now we have *justified* that assumption.

The second implication of Equation 9.18, and the more important one from the perspective of this chapter, applies to what we call an *isolated system.* An **isolated system** is a system for which the *net* external force is zero: $\vec{F}_{\text{net}} = \vec{0}$. That is, an isolated system is one on which there are *no* external forces or for which the external forces are balanced and add to zero.

For an isolated system, Equation 9.18 is simply

$$\frac{d\vec{P}}{dt} = \vec{0} \quad \text{(isolated system)} \tag{9.19}$$

In other words, **the *total* momentum of an isolated system does not change.** The total momentum $\vec{P}$ remains constant, *regardless* of whatever interactions are going on *inside* the system. The importance of this result is sufficient to elevate it to a law of nature, alongside Newton's laws.

Law of conservation of momentum The total momentum $\vec{P}$ of an isolated system is a constant. Interactions within the system do not change the system's total momentum.

Mathematically, the law of conservation of momentum for an isolated system is

$$\vec{P}_f = \vec{P}_i \tag{9.20}$$

The total momentum after an interaction is equal to the total momentum before the interaction. Because Equation 9.20 is a vector equation, the equality is true for each of the components of the momentum vector. That is,

$$\begin{aligned}(p_{fx})_1 + (p_{fx})_2 + (p_{fx})_3 + \cdots &= (p_{ix})_1 + (p_{ix})_2 + (p_{ix})_3 + \cdots \\ (p_{fy})_1 + (p_{fy})_2 + (p_{fy})_3 + \cdots &= (p_{iy})_1 + (p_{iy})_2 + (p_{iy})_3 + \cdots\end{aligned} \tag{9.21}$$

The x-equation is an extension of Equation 9.13 to N interacting particles.

NOTE ▶ It is worth emphasizing the critical role of Newton's third law. The law of conservation of momentum is a direct consequence of the fact that interactions within an isolated system are action/reaction pairs. ◀

EXAMPLE 9.3 A glider collision

A 250 g air-track glider is pushed across a level track toward a 500 g glider that is at rest. FIGURE 9.14 shows a position-versus-time graph of the 250 g glider as recorded by a motion detector. Best-fit lines have been found. What is the speed of the 500 g glider after the collision?

FIGURE 9.14 Position graph of the 250 g glider.

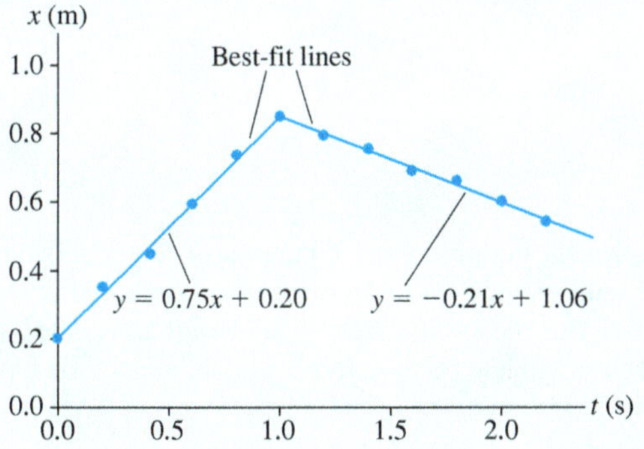

MODEL The two gliders, modeled as particles, are the system. The gliders interact with each other, but the external forces (normal force and gravity) balance to make $\vec{F}_{net} = \vec{0}$. Thus the gliders form an isolated system and their total momentum is conserved.

VISUALIZE FIGURE 9.15 is a before-and-after pictorial representation. The graph of Figure 9.14 tells us that the 250 g glider initially moves to the right, collides at $t = 1.0$ s, then rebounds to the left (decreasing x).

FIGURE 9.15 Before-and-after pictorial representation of a glider collision.

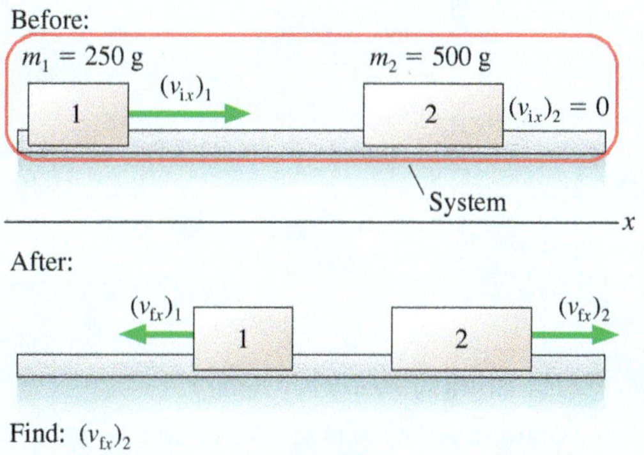

SOLVE Conservation of momentum for this one-dimensional problem requires that the final momentum equal the initial momentum: $P_{fx} = P_{ix}$. In terms of the individual components, conservation of momentum is

$$(p_{fx})_1 + (p_{fx})_2 = (p_{ix})_1 + (p_{ix})_2$$

Each momentum is mv_x, so conservation of momentum in terms of velocities is

$$m_1(v_{fx})_1 + m_2(v_{fx})_2 = m_1(v_{ix})_1 + m_2(v_{ix})_2 = m_1(v_{ix})_1$$

where, in the last step, we used $(v_{ix})_2 = 0$ for the 500 g glider. Solving for the heavier glider's final velocity gives

$$(v_{fx})_2 = \frac{m_1}{m_2}\left[(v_{ix})_1 - (v_{fx})_1\right]$$

From Chapter 2 kinematics, the velocities of the 250 g glider before and after the collision are the slopes of the position-versus-time graph. Referring to Figure 9.14, we see that $(v_{ix})_1 = 0.75$ m/s and $(v_{fx})_1 = -0.21$ m/s. The latter is negative because the rebound motion is to the left. Thus

$$(v_{fx})_2 = \frac{250\text{ g}}{500\text{ g}}\left[0.75\text{ m/s} - (-0.21\text{ m/s})\right] = 0.48\text{ m/s}$$

The 500 g glider moves away from the collision at 0.48 m/s.

ASSESS The 500 g glider has twice the mass of the glider that was pushed, so a somewhat smaller speed seems reasonable. Paying attention to the *signs*—which are positive and which negative—was very important for reaching a correct answer. We didn't convert the masses to kilograms because only the mass *ratio* of 0.50 was needed.

A Strategy for Conservation of Momentum Problems

PROBLEM-SOLVING STRATEGY 9.1 Conservation of momentum

MP

MODEL Clearly define *the system.*

- If possible, choose a system that is isolated ($\vec{F}_{net} = \vec{0}$) or within which the interactions are sufficiently short and intense that you can ignore external forces for the duration of the interaction (the impulse approximation). Momentum is conserved.
- If it's not possible to choose an isolated system, try to divide the problem into parts such that momentum is conserved during one segment of the motion. Other segments of the motion can be analyzed using Newton's laws or, as you'll learn in Chapters 10 and 11, conservation of energy.

VISUALIZE Draw a before-and-after pictorial representation. Define symbols that will be used in the problem, list known values, and identify what you're trying to find.

SOLVE The mathematical representation is based on the law of conservation of momentum: $\vec{P}_f = \vec{P}_i$. In component form, this is

$$(p_{fx})_1 + (p_{fx})_2 + (p_{fx})_3 + \cdots = (p_{ix})_1 + (p_{ix})_2 + (p_{ix})_3 + \cdots$$

$$(p_{fy})_1 + (p_{fy})_2 + (p_{fy})_3 + \cdots = (p_{iy})_1 + (p_{iy})_2 + (p_{iy})_3 + \cdots$$

ASSESS Check that your result has the correct units, is reasonable, and answers the question.

Exercise 16

EXAMPLE 9.4 Rolling away

Bob sees a stationary cart 8.0 m in front of him. He decides to run to the cart as fast as he can, jump on, and roll down the street. Bob has a mass of 75 kg and the cart's mass is 25 kg. If Bob accelerates at a steady 1.0 m/s^2, what is the cart's speed just after Bob jumps on?

MODEL This is a two-part problem. First Bob accelerates across the ground. Then Bob lands on and sticks to the cart, a "collision" between Bob and the cart. The interaction forces between Bob and the cart (i.e., friction) act only over the fraction of a second it takes Bob's feet to become stuck to the cart. Using the impulse approximation allows the system Bob + cart to be treated as an isolated system during the brief interval of the "collision," and thus the total momentum of Bob + cart is conserved during this interaction. But the system Bob + cart is *not* an isolated system for the entire problem because Bob's initial acceleration has nothing to do with the cart.

VISUALIZE Our strategy is to divide the problem into an *acceleration* part, which we can analyze using kinematics, and a *collision* part, which we can analyze with momentum conservation. The pictorial representation of FIGURE 9.16 includes information about both parts. Notice that Bob's velocity $(v_{1x})_B$ at the end of his run is his "before" velocity for the collision.

FIGURE 9.16 Pictorial representation of Bob and the cart.

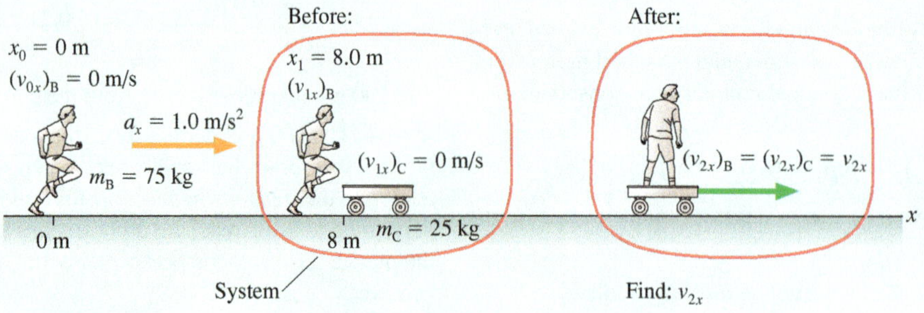

SOLVE The first part of the mathematical representation is kinematics. We don't know how long Bob accelerates, but we do know his acceleration and the distance. Thus

$$(v_{1x})_B^2 = (v_{0x})_B^2 + 2a_x\,\Delta x = 2a_x x_1$$

His velocity after accelerating for 8.0 m is

$$(v_{1x})_B = \sqrt{2a_x x_1} = 4.0 \text{ m/s}$$

The second part of the problem, the collision, uses conservation of momentum: $P_{2x} = P_{1x}$. Equation 9.21 is

$$m_B(v_{2x})_B + m_C(v_{2x})_C = m_B(v_{1x})_B + m_C(v_{1x})_C = m_B(v_{1x})_B$$

where we've used $(v_{1x})_C = 0$ m/s because the cart starts at rest. In this problem, Bob and the cart move together at the end with a common velocity, so we can replace both $(v_{2x})_B$ and $(v_{2x})_C$ with simply v_{2x}. Solving for v_{2x}, we find

$$v_{2x} = \frac{m_B}{m_B + m_C}(v_{1x})_B = \frac{75 \text{ kg}}{100 \text{ kg}} \times 4.0 \text{ m/s} = 3.0 \text{ m/s}$$

The cart's speed is 3.0 m/s immediately after Bob jumps on.

Notice how easy this was! No forces, no acceleration constraints, no simultaneous equations. Why didn't we think of this before? Conservation laws are indeed powerful, but they can answer only certain questions. Had we wanted to know how far Bob slid across the cart before sticking to it, how long the slide took, or what the cart's acceleration was during the collision, we would not have been able to answer such questions on the basis of the conservation law. There is a price to pay for finding a simple connection between before and after, and that price is the loss of information about the details of the interaction. If we are satisfied with knowing only about before and after, then conservation laws are a simple and straightforward way to proceed. But many problems *do* require us to understand the interaction, and for these there is no avoiding Newton's laws.

It Depends on the System

The first step in the problem-solving strategy asks you to clearly define *the system.* This is worth emphasizing because many problem-solving errors arise from trying to apply momentum conservation to an inappropriate system. **The goal is to choose a system whose momentum will be conserved.** Even then, it is the *total* momentum of the system that is conserved, not the momenta of the individual particles within the system.

As an example, consider what happens if you drop a rubber ball and let it bounce off a hard floor. Is momentum conserved during the collision of the ball with the floor? You might be tempted to answer yes because the ball's rebound speed is very nearly equal to its impact speed. But there are two errors in this reasoning.

First, momentum depends on *velocity,* not speed. The ball's velocity and momentum just before the collision are negative. They are positive after the collision. Even if their magnitudes are equal, the ball's momentum after the collision is *not* equal to its momentum before the collision.

But more important, we haven't defined the system. The momentum of what? Whether or not momentum is conserved depends on the system. FIGURE 9.17 shows two different choices of systems. In FIGURE 9.17a, where the ball itself is chosen as the system, the gravitational force of the earth on the ball is an external force. This force causes the ball to accelerate toward the earth, changing the ball's momentum. The force of the floor on the ball is also an external force. The impulse of $\vec{F}_{\text{floor on ball}}$ changes the ball's momentum from "down" to "up" as the ball bounces. The momentum of this system is most definitely *not* conserved.

FIGURE 9.17b shows a different choice. Here the system is ball + earth. Now the gravitational forces and the impulsive forces of the collision are interactions *within* the system. This is an isolated system, so the *total* momentum $\vec{P} = \vec{p}_{\text{ball}} + \vec{p}_{\text{earth}}$ is conserved.

In fact, the total momentum is $\vec{P} = \vec{0}$. Before you release the ball, both the ball and the earth are at rest (in the earth's reference frame). The total momentum is zero before

FIGURE 9.17 Whether or not momentum is conserved as a ball falls to earth depends on your choice of the system.

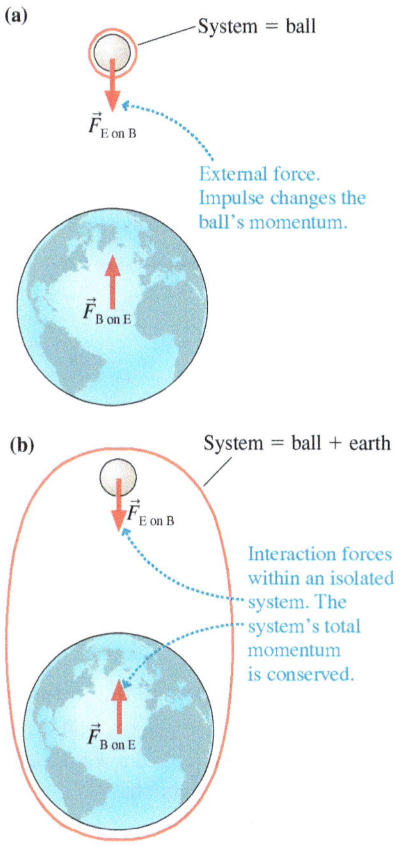

you release the ball, so it will *always* be zero. Just before the ball hits the floor with velocity v_{By}, it must be the case that $m_B v_{By} + m_E v_{Ey} = 0$ and thus

$$v_{Ey} = -\frac{m_B}{m_E} v_{By}$$

In other words, as the ball is pulled down toward the earth, the ball pulls up on the earth (action/reaction pair of forces) until the entire earth reaches velocity v_{Ey}. The earth's momentum is equal and opposite to the ball's momentum.

Why don't we notice the earth "leaping up" toward us each time we drop something? Because of the earth's enormous mass relative to everyday objects. A typical rubber ball has a mass of 60 g and hits the ground with a velocity of about −5 m/s. The earth's upward velocity is thus

$$v_{Ey} \approx -\frac{6 \times 10^{-2}\ \text{kg}}{6 \times 10^{24}\ \text{kg}}(-5\ \text{m/s}) = 5 \times 10^{-26}\ \text{m/s}$$

The earth does, indeed, have a momentum equal and opposite to that of the ball, but the earth is so massive that it needs only an infinitesimal velocity to match the ball's momentum. At this speed, it would take the earth 300 million years to move the diameter of an atom!

STOP TO THINK 9.3 Objects A and C are made of different materials, with different "springiness," but they have the same mass and are initially at rest. When ball B collides with object A, the ball ends up at rest. When ball B is thrown with the same speed and collides with object C, the ball rebounds to the left. Compare the velocities of A and C after the collisions. Is v_A greater than, equal to, or less than v_C?

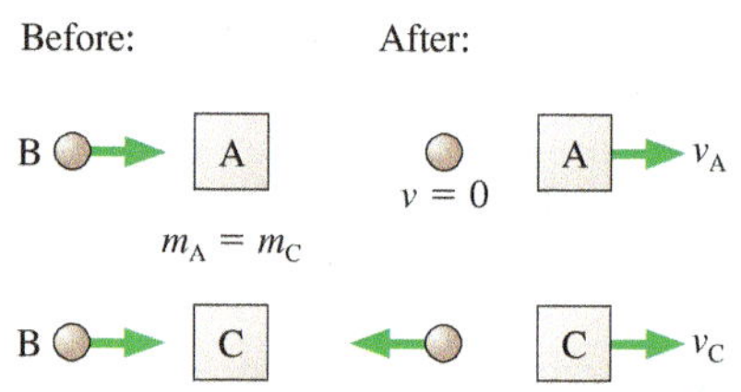

9.4 Inelastic Collisions

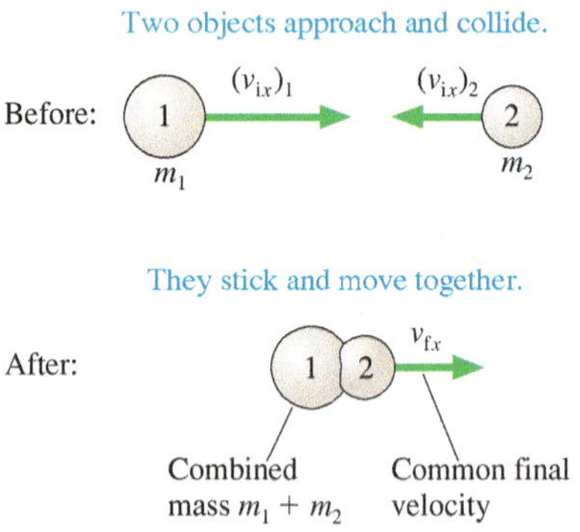

FIGURE 9.18 An inelastic collision.

Collisions can have different possible outcomes. A rubber ball dropped on the floor bounces, but a ball of clay sticks to the floor without bouncing. A golf club hitting a golf ball causes the ball to rebound away from the club, but a bullet striking a block of wood embeds itself in the block.

A collision in which the two objects stick together and move with a common final velocity is called a **perfectly inelastic collision.** The clay hitting the floor and the bullet embedding itself in the wood are examples of perfectly inelastic collisions. Other examples include railroad cars coupling together upon impact and darts hitting a dart board. FIGURE 9.18 emphasizes the fact that the two objects have a common final velocity after they collide.

In an *elastic collision,* by contrast, the two objects bounce apart. We've looked at some examples of elastic collisions, but a full analysis requires ideas about energy. We will return to elastic collisions in Chapter 10.

EXAMPLE 9.5 An inelastic glider collision

In a laboratory experiment, a 200 g air-track glider and a 400 g air-track glider are pushed toward each other from opposite ends of the track. The gliders have Velcro tabs on the front and will stick together when they collide. The 200 g glider is pushed with an initial speed of 3.0 m/s. The collision causes it to reverse direction at 0.40 m/s. What was the initial speed of the 400 g glider?

MODEL Model the gliders as particles. Define the two gliders together as the system. This is an isolated system, so its total momentum is conserved in the collision. The gliders stick together, so this is a perfectly inelastic collision.

VISUALIZE **FIGURE 9.19** shows a pictorial representation. We've chosen to let the 200 g glider (glider 1) start out moving to the right, so $(v_{ix})_1$ is a positive 3.0 m/s. The gliders move to the left after the collision, so their common final velocity is $v_{fx} = -0.40$ m/s.

SOLVE The law of conservation of momentum, $P_{fx} = P_{ix}$, is

$$(m_1 + m_2)v_{fx} = m_1(v_{ix})_1 + m_2(v_{ix})_2$$

where we made use of the fact that the combined mass $m_1 + m_2$ moves together after the collision. We can easily solve for the initial velocity of the 400 g glider:

$$(v_{ix})_2 = \frac{(m_1 + m_2)v_{fx} - m_1(v_{ix})_1}{m_2}$$
$$= \frac{(0.60\text{ kg})(-0.40\text{ m/s}) - (0.20\text{ kg})(3.0\text{ m/s})}{0.40\text{ kg}}$$
$$= -2.1\text{ m/s}$$

FIGURE 9.19 The before-and-after pictorial representation of an inelastic collision.

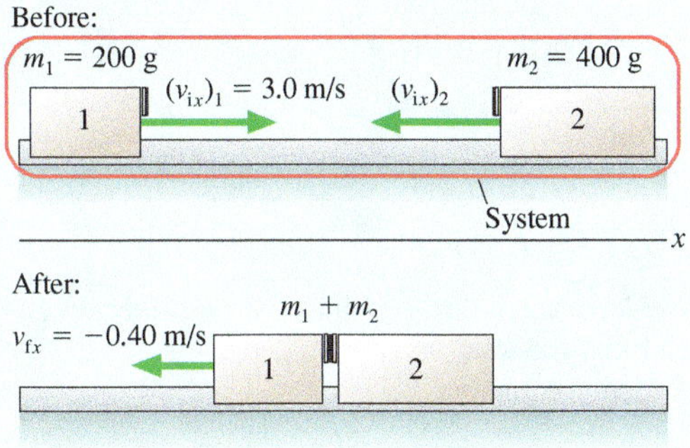

The negative sign indicates that the 400 g glider started out moving to the left. The initial *speed* of the glider, which we were asked to find, is 2.1 m/s.

EXAMPLE 9.6 Momentum in a car crash

A 2000 kg Cadillac had just started forward from a stop sign when it was struck from behind by a 1000 kg Volkswagen. The bumpers became entangled, and the two cars skidded forward together until they came to rest. Officer Tom, responding to the accident, measured the skid marks to be 3.0 m long. He also took testimony from the driver that the Cadillac's speed just before the impact was 5.0 m/s. Officer Tom charged the Volkswagen driver with reckless driving. Should the Volkswagen driver also be charged with exceeding the 50 km/h speed limit? The judge calls you as an "expert witness" to analyze the evidence. What is your conclusion?

MODEL This is really *two* problems. First, there is an inelastic collision. The two cars are not a perfectly isolated system because of external friction forces, but during the brief collision the external impulse delivered by friction will be negligible. Within the impulse approximation, the momentum of the Volkswagen + Cadillac system will be conserved in the collision. Then we have a second problem, a dynamics problem of the two cars sliding.

VISUALIZE **FIGURE 9.20a** is a pictorial representation showing both the before and after of the collision and the more familiar picture for the dynamics of the skidding. We do not need to consider forces during the collision because we will use the law of conservation of momentum, but we do need a free-body diagram of the cars during the subsequent skid. This is shown in **FIGURE 9.20b**.

The cars have a common velocity v_{1x} just after the collision. This is the *initial* velocity for the dynamics problem. Our goal is to find $(v_{0x})_{VW}$, the Volkswagen's velocity at the moment of impact. The 50 km/h speed limit has been converted to 14 m/s.

SOLVE First, the inelastic collision. The law of conservation of momentum is

$$(m_{VW} + m_C)v_{1x} = m_{VW}(v_{0x})_{VW} + m_C(v_{0x})_C$$

Solving for the initial velocity of the Volkswagen, we find

$$(v_{0x})_{VW} = \frac{(m_{VW} + m_C)v_{1x} - m_C(v_{0x})_C}{m_{VW}}$$

FIGURE 9.20 Pictorial representation and a free-body diagram of the cars as they skid.

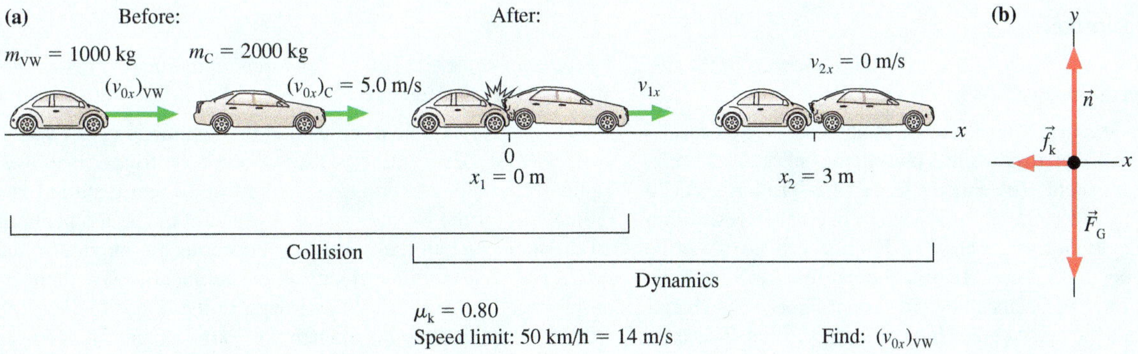

Continued

To evaluate $(v_{0x})_{VW}$, we need to know v_{1x}, the velocity *immediately* after the collision as the cars begin to skid. This information will come out of the dynamics of the skid. Newton's second law and the model of kinetic friction are

$$\sum F_x = -f_k = (m_{VW} + m_C)a_x$$
$$\sum F_y = n - (m_{VW} + m_C)g = 0$$
$$f_k = \mu_k n$$

where we have noted that $\vec{f}_k$ points to the left (negative x-component) and that the total mass is $m_{VW} + m_C$. From the y-equation and the friction equation,

$$f_k = \mu_k(m_{VW} + m_C)g$$

Using this in the x-equation gives

$$a_x = \frac{-f_k}{m_{VW} + m_C} = -\mu_k g = -7.84 \text{ m/s}^2$$

where the coefficient of kinetic friction for rubber on concrete is taken from Table 6.1. With the acceleration determined, we can move on to the kinematics. This is constant acceleration, so

$$v_{2x}^2 = 0 = v_{1x}^2 + 2a_x(\Delta x) = v_{1x}^2 + 2a_x x_2$$

Hence the skid starts with velocity

$$v_{1x} = \sqrt{-2a_x x_2} = \sqrt{-2(-7.84 \text{ m/s}^2)(3.0 \text{ m})} = 6.9 \text{ m/s}$$

As we have noted, this is the final velocity of the collision. Inserting v_{1x} back into the momentum conservation equation, we finally determine that

$$(v_{0x})_{VW} = \frac{(3000 \text{ kg})(6.9 \text{ m/s}) - (2000 \text{ kg})(5.0 \text{ m/s})}{1000 \text{ kg}}$$
$$= 11 \text{ m/s}$$

On the basis of your testimony, the Volkswagen driver is *not* charged with speeding!

NOTE ▶ Momentum is conserved only for an isolated system. In this example, momentum was conserved during the collision (isolated system) but *not* during the skid (not an isolated system). In practice, it is not unusual for momentum to be conserved in one part or one aspect of a problem but not in others. ◀

STOP TO THINK 9.4 The two particles are both moving to the right. Particle 1 catches up with particle 2 and collides with it. The particles stick together and continue on with velocity v_f. Which of these statements is true?

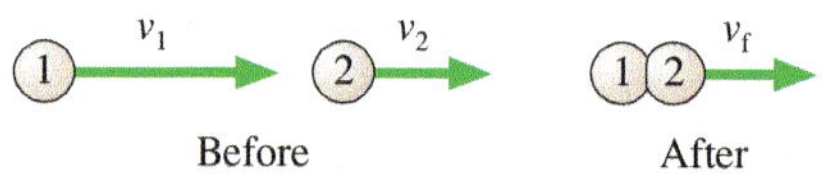

a. v_f is greater than v_1.
b. $v_f = v_1$
c. v_f is greater than v_2 but less than v_1.
d. $v_f = v_2$
e. v_f is less than v_2.
f. Can't tell without knowing the masses.

9.5 Explosions

An **explosion,** where the particles of the system move apart from each other after a brief, intense interaction, is the opposite of a collision. The explosive forces, which could be from an expanding spring or from expanding hot gases, are *internal* forces. If the system is isolated, its total momentum during the explosion will be conserved.

EXAMPLE 9.7 Recoil

A 10 g bullet is fired from a 3.0 kg rifle with a speed of 500 m/s. What is the recoil speed of the rifle?

MODEL The rifle causes a small mass of gunpowder to explode, and the expanding gas then exerts forces on *both* the bullet and the rifle. Let's define the system to be bullet + gas + rifle. The forces due to the expanding gas during the explosion are internal forces, within the system. Any friction forces between the bullet and the rifle as the bullet travels down the barrel are also internal forces. Gravity, the only external force, is balanced by the normal forces of the barrel on the bullet and the person holding the rifle, so $\vec{F}_{net} = \vec{0}$. This is an isolated system and the law of conservation of momentum applies.

VISUALIZE FIGURE 9.21 shows a pictorial representation before and after the bullet is fired.

SOLVE The x-component of the total momentum is $P_x = (p_x)_B + (p_x)_R + (p_x)_{gas}$. Everything is at rest before the trigger is pulled, so the initial momentum is zero. After the trigger is pulled, the momentum of the expanding gas is the sum of the momenta of all the molecules in the gas. For every molecule moving in the forward direction with velocity v and momentum mv there is, on average, another molecule moving in the opposite direction with velocity $-v$ and thus momentum $-mv$. When summed over the enormous number of molecules in the gas, we will be left

FIGURE 9.21 Before-and-after pictorial representation of a rifle firing a bullet.

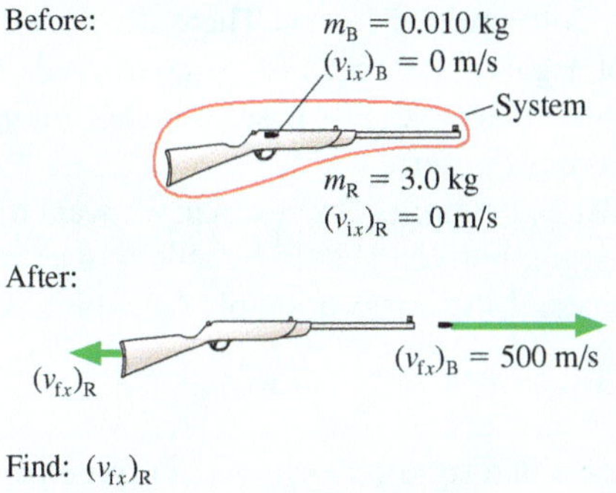

with $p_{gas} \approx 0$. In addition, the mass of the gas is much less than that of the rifle or bullet. For both reasons, we can reasonably neglect the momentum of the gas. The law of conservation of momentum is thus

$$P_{fx} = m_B(v_{fx})_B + m_R(v_{fx})_R = P_{ix} = 0$$

Solving for the rifle's velocity, we find

$$(v_{fx})_R = -\frac{m_B}{m_R}(v_{fx})_B = -\frac{0.010 \text{ kg}}{3.0 \text{ kg}} \times 500 \text{ m/s} = -1.7 \text{ m/s}$$

The minus sign indicates that the rifle's recoil is to the left. The recoil *speed* is 1.7 m/s.

We would not know where to begin to solve a problem such as this using Newton's laws. But Example 9.7 is a simple problem when approached from the before-and-after perspective of a conservation law. The selection of bullet + gas + rifle as "the system" was the critical step. For momentum conservation to be a useful principle, we had to select a system in which the complicated forces due to expanding gas and friction were all internal forces. The rifle by itself is *not* an isolated system, so its momentum is *not* conserved.

EXAMPLE 9.8 Radioactivity

A ^{238}U uranium nucleus is radioactive. It spontaneously disintegrates into a small fragment that is ejected with a measured speed of 1.50×10^7 m/s and a "daughter nucleus" that recoils with a measured speed of 2.56×10^5 m/s. What are the atomic masses of the ejected fragment and the daughter nucleus?

MODEL The notation ^{238}U indicates the isotope of uranium with an atomic mass of 238 u, where u is the abbreviation for the *atomic mass unit.* The nucleus contains 92 protons (uranium is atomic number 92) and 146 neutrons. The disintegration of a nucleus is, in essence, an explosion. Only *internal* nuclear forces are involved, so the total momentum is conserved in the decay.

VISUALIZE FIGURE 9.22 shows the pictorial representation. The mass of the daughter nucleus is m_1 and that of the ejected fragment is m_2. Notice that we converted the speed information to velocity information, giving $(v_{fx})_1$ and $(v_{fx})_2$ opposite signs.

FIGURE 9.22 Before-and-after pictorial representation of the decay of a ^{238}U nucleus.

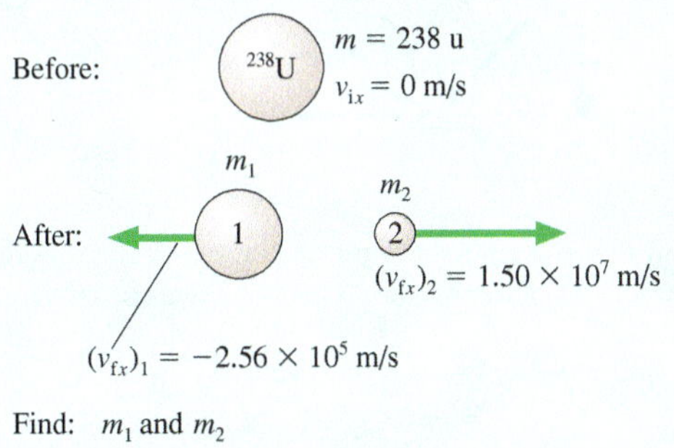

SOLVE The nucleus was initially at rest, hence the total momentum is zero. The momentum after the decay is still zero if the two pieces fly apart in opposite directions with momenta equal in magnitude but opposite in sign. That is,

$$P_{fx} = m_1(v_{fx})_1 + m_2(v_{fx})_2 = P_{ix} = 0$$

Although we know both final velocities, this is not enough information to find the two unknown masses. However, we also have another conservation law, conservation of mass, that requires

$$m_1 + m_2 = 238 \text{ u}$$

Combining these two conservation laws gives

$$m_1(v_{fx})_1 + (238 \text{ u} - m_1)(v_{fx})_2 = 0$$

The mass of the daughter nucleus is

$$m_1 = \frac{(v_{fx})_2}{(v_{fx})_2 - (v_{fx})_1} \times 238 \text{ u}$$

$$= \frac{1.50 \times 10^7 \text{ m/s}}{(1.50 \times 10^7 - (-2.56 \times 10^5)) \text{ m/s}} \times 238 \text{ u} = 234 \text{ u}$$

With m_1 known, the mass of the ejected fragment is $m_2 = 238 - m_1 = 4$ u.

ASSESS All we learn from a momentum analysis is the masses. Chemical analysis shows that the daughter nucleus is the element thorium, atomic number 90, with two fewer protons than uranium. The ejected fragment carried away two protons as part of its mass of 4 u, so it must be a particle with two protons and two neutrons. This is the nucleus of a helium atom, ^{4}He, which in nuclear physics is called an *alpha particle* α. Thus the radioactive decay of ^{238}U can be written as $^{238}\text{U} \rightarrow {}^{234}\text{Th} + \alpha$.

FIGURE 9.23 Rocket propulsion is an example of conservation of momentum.

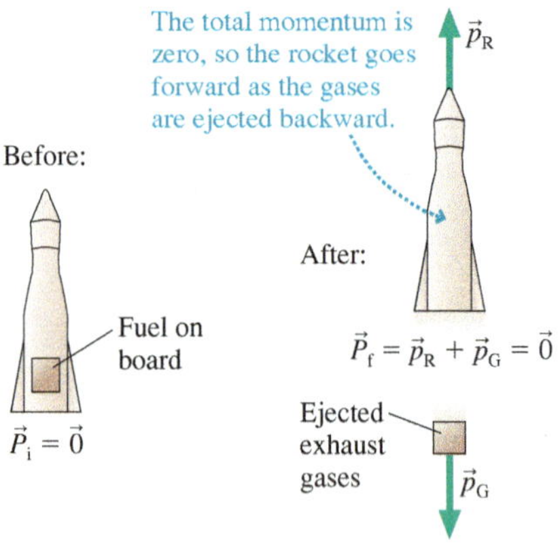

Much the same reasoning explains how a rocket or jet aircraft accelerates. FIGURE 9.23 shows a rocket with a parcel of fuel on board. Burning converts the fuel to hot gases that are expelled from the rocket motor. If we choose rocket + gases to be the system, the burning and expulsion are both internal forces. There are no other forces, so the total momentum of the rocket + gases system must be conserved. The rocket gains forward velocity and momentum as the exhaust gases are shot out the back, but the *total* momentum of the system remains zero.

The details of rocket propulsion are more complex than we want to handle, because the mass of the rocket is changing, but you should be able to use the law of conservation of momentum to understand the basic principle by which rocket propulsion occurs.

STOP TO THINK 9.5 An explosion in a rigid pipe shoots out three pieces. A 6 g piece comes out the right end. A 4 g piece comes out the left end with twice the speed of the 6 g piece. From which end, left or right, does the third piece emerge?

Collisions and explosions often involve motion in two dimensions.

9.6 Momentum in Two Dimensions

Our examples thus far have been confined to motion along a one-dimensional axis. Many practical examples of momentum conservation involve motion in a plane. The total momentum $\vec{P}$ is a *vector* sum of the momenta $\vec{p} = m\vec{v}$ of the individual particles. Consequently, as we found in Section 9.3, momentum is conserved only if each component of $\vec{P}$ is conserved:

$$
\begin{aligned}
(p_{fx})_1 + (p_{fx})_2 + (p_{fx})_3 + \cdots &= (p_{ix})_1 + (p_{ix})_2 + (p_{ix})_3 + \cdots \\
(p_{fy})_1 + (p_{fy})_2 + (p_{fy})_3 + \cdots &= (p_{iy})_1 + (p_{iy})_2 + (p_{iy})_3 + \cdots
\end{aligned}
\tag{9.22}
$$

In this section we'll apply momentum conservation to motion in two dimensions.

EXAMPLE 9.9 A peregrine falcon strike

Peregrine falcons often grab their prey from above while both falcon and prey are in flight. A 0.80 kg falcon, flying at 18 m/s, swoops down at a 45° angle from behind a 0.36 kg pigeon flying horizontally at 9.0 m/s. What are the speed and direction of the falcon (now holding the pigeon) immediately after impact?

MODEL The two birds, modeled as particles, are the system. This is a perfectly inelastic collision because after the collision the falcon and pigeon move at a common final velocity. The birds are not a perfectly isolated system because of external forces of the air, but during the brief collision the external impulse delivered by the air resistance will be negligible. Within this approximation, the total momentum of the falcon + pigeon system is conserved during the collision.

VISUALIZE FIGURE 9.24 is a before-and-after pictorial representation. We've used angle ϕ to label the post-collision direction.

FIGURE 9.24 Pictorial representation of a falcon catching a pigeon.

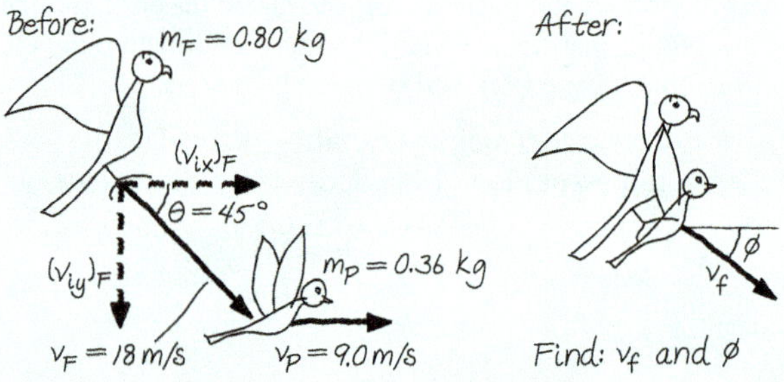

SOLVE The initial velocity components of the falcon are $(v_{ix})_F = v_F \cos\theta$ and $(v_{iy})_F = -v_F \sin\theta$. The pigeon's initial velocity is entirely along the x-axis. After the collision, when the falcon and pigeon have the common velocity $\vec{v}_f$, the components are $v_{fx} = v_f \cos\phi$ and $v_{fy} = -v_f \sin\phi$. Conservation of momentum in two dimensions requires conservation of both the x- and y-components of momentum. This gives two conservation equations:

$$
\begin{aligned}
(m_F + m_P)v_{fx} &= (m_F + m_P)v_f \cos\phi \\
&= m_F(v_{ix})_F + m_P(v_{ix})_P = m_F v_F \cos\theta + m_P v_P \\
(m_F + m_P)v_{fy} &= -(m_F + m_P)v_f \sin\phi \\
&= m_F(v_{iy})_F + m_P(v_{iy})_P = -m_F v_F \sin\theta
\end{aligned}
$$

The unknowns are v_f and ϕ. Dividing both equations by the total mass gives

$$v_f \cos\phi = \frac{m_F v_F \cos\theta + m_P v_P}{m_F + m_P} = 11.6 \text{ m/s}$$

$$v_f \sin\phi = \frac{m_F v_F \sin\theta}{m_F + m_P} = 8.78 \text{ m/s}$$

We can eliminate v_f by dividing the second equation by the first to give

$$\frac{v_f \sin\phi}{v_f \cos\phi} = \tan\phi = \frac{8.78 \text{ m/s}}{11.6 \text{ m/s}} = 0.757$$

$$\phi = \tan^{-1}(0.757) = 37°$$

Then $v_f = (11.6 \text{ m/s})/\cos(37°) = 15$ m/s. Immediately after impact, the falcon, with its meal, is traveling at 15 m/s at an angle 37° below the horizontal.

ASSESS It makes sense that the falcon would slow down after grabbing the slower-moving pigeon. And Figure 9.24 tells us that the total momentum is at an angle between 0° (the pigeon's momentum) and 45° (the falcon's momentum). Thus our answer seems reasonable.

CHALLENGE EXAMPLE 9.10 A three-piece explosion

A 10.0 g projectile is traveling east at 2.0 m/s when it suddenly explodes into three pieces. A 3.0 g fragment is shot due west at 10 m/s while another 3.0 g fragment travels 40° north of east at 12 m/s. What are the speed and direction of the third fragment?

MODEL Although many complex forces are involved in the explosion, they are all internal to the system. There are no external forces, so this is an isolated system and its total momentum is conserved.

VISUALIZE **FIGURE 9.25** shows a before-and-after pictorial representation. We'll use uppercase M and V to distinguish the initial object from the three pieces into which it explodes.

FIGURE 9.25 Before-and-after pictorial representation of the three-piece explosion.

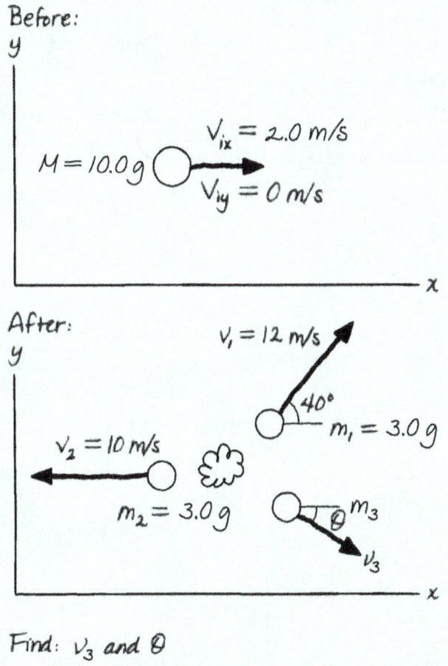

SOLVE The system is the initial object and the subsequent three pieces. Conservation of momentum requires

$$m_1(v_{fx})_1 + m_2(v_{fx})_2 + m_3(v_{fx})_3 = MV_{ix}$$

$$m_1(v_{fy})_1 + m_2(v_{fy})_2 + m_3(v_{fy})_3 = MV_{iy}$$

Conservation of mass implies that

$$m_3 = M - m_1 - m_2 = 4.0 \text{ g}$$

Neither the original object nor m_2 has any momentum along the y-axis. We can use Figure 9.25 to write out the x- and y-components of $\vec{v}_1$ and $\vec{v}_3$, leading to

$$m_1 v_1 \cos 40° - m_2 v_2 + m_3 v_3 \cos\theta = MV$$

$$m_1 v_1 \sin 40° - m_3 v_3 \sin\theta = 0$$

where we used $(v_{fx})_2 = -v_2$ because m_2 is moving in the negative x-direction. Inserting known values in these equations gives us

$$-2.42 + 4v_3 \cos\theta = 20$$

$$23.14 - 4v_3 \sin\theta = 0$$

We can leave the masses in grams in this situation because the conversion factor to kilograms appears on both sides of the equation and thus cancels out. To solve, first use the second equation to write $v_3 = 5.79/\sin\theta$. Substitute this result into the first equation, noting that $\cos\theta/\sin\theta = 1/\tan\theta$, to get

$$-2.42 + 4\left(\frac{5.79}{\sin\theta}\right)\cos\theta = -2.42 + \frac{23.14}{\tan\theta} = 20$$

Now solve for θ:

$$\tan\theta = \frac{23.14}{20 + 2.42} = 1.03$$

$$\theta = \tan^{-1}(1.03) = 45.8°$$

Finally, use this result in the earlier expression for v_3 to find

$$v_3 = \frac{5.79}{\sin 45.8°} = 8.1 \text{ m/s}$$

The third fragment, with a mass of 4.0 g, is shot 46° south of east at a speed of 8.1 m/s.

SUMMARY

The goals of Chapter 9 have been to understand and apply the new concepts of impulse and momentum.

General Principles

Law of Conservation of Momentum

The total momentum $\vec{P} = \vec{p}_1 + \vec{p}_2 + \cdots$ of an isolated system is a constant. Thus

$$\vec{P}_f = \vec{P}_i$$

Newton's Second Law

In terms of momentum, Newton's second law is

$$\vec{F} = \frac{d\vec{p}}{dt}$$

Solving Momentum Conservation Problems

MODEL Choose an isolated system or a system that is isolated during at least part of the problem.

VISUALIZE Draw a pictorial representation of the system before and after the interaction.

SOLVE Write the law of conservation of momentum in terms of vector components:

$$(p_{fx})_1 + (p_{fx})_2 + \cdots = (p_{ix})_1 + (p_{ix})_2 + \cdots$$
$$(p_{fy})_1 + (p_{fy})_2 + \cdots = (p_{iy})_1 + (p_{iy})_2 + \cdots$$

ASSESS Is the result reasonable?

Important Concepts

Momentum $\vec{p} = m\vec{v}$

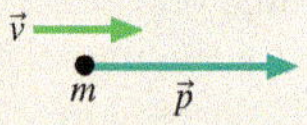

Impulse $J_x = \int_{t_i}^{t_f} F_x(t)\,dt$ = area under force curve

Impulse and momentum are related by the impulse-momentum theorem

$$\Delta p_x = J_x$$

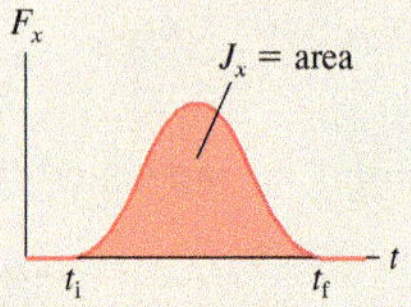

The impulse delivered to a particle causes the particle's momentum to change. This is an alternative statement of Newton's second law.

System A group of interacting particles.

Isolated system A system on which there are no external forces or the net external force is zero.

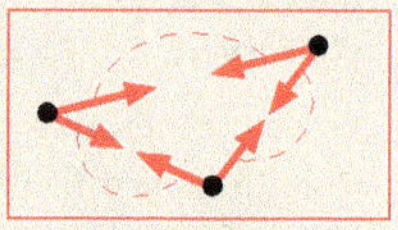

Before-and-after pictorial representation

- Define the system.
- Use two drawings to show the system *before* and *after* the interaction.
- List known information and identify what you are trying to find.

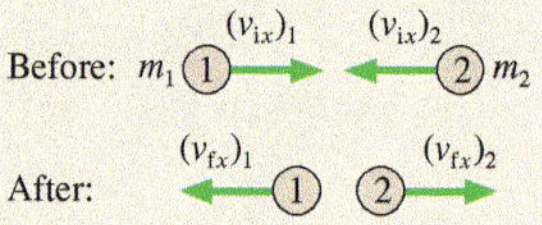

Applications

Collisions Two or more particles come together. In a perfectly inelastic collision, they stick together and move with a common final velocity.

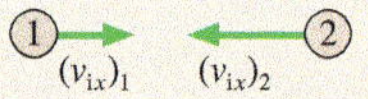

Explosions Two or more particles move away from each other.

Two dimensions No new ideas, but both the x- and y-components of $\vec{P}$ must be conserved, giving two simultaneous equations.

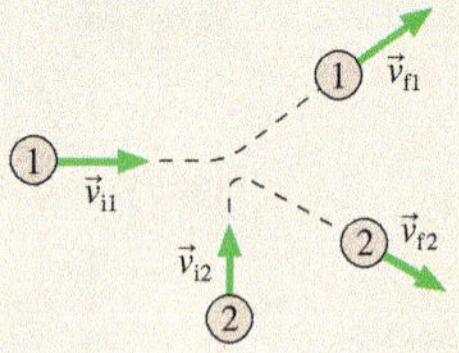

Momentum bar charts display the impulse-momentum theorem $p_{fx} = p_{ix} + J_x$ in graphical form.

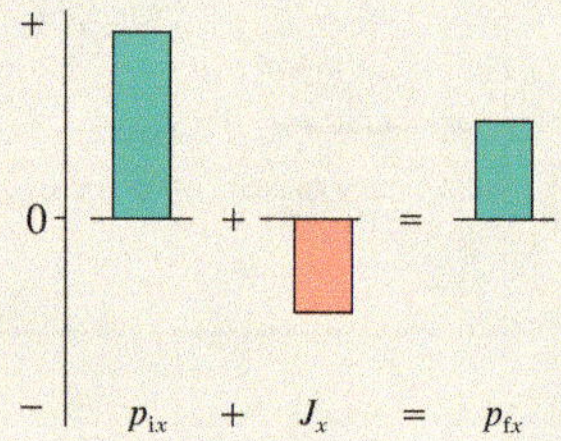

Terms and Notation

collision
impulsive force
momentum, $\vec{p}$
impulse, J_x
impulse-momentum theorem
momentum bar chart
before-and-after pictorial representation
impulse approximation
total momentum, $\vec{P}$
isolated system
law of conservation of momentum
perfectly inelastic collision
explosion

CONCEPTUAL QUESTIONS

1. Rank in order, from largest to smallest, the momenta $(p_x)_a$ to $(p_x)_e$ of the objects in FIGURE Q9.1.

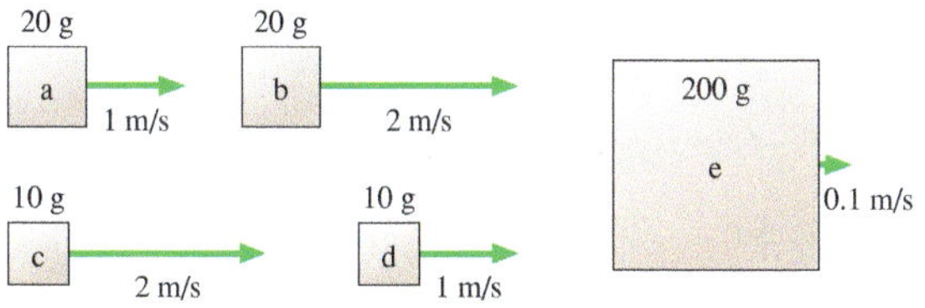

FIGURE Q9.1

2. Explain the concept of *impulse* in nonmathematical language. That is, don't simply put the equation in words to say that "impulse is the time integral of force." Explain it in terms that would make sense to an educated person who had never heard of it.
3. Explain the concept of *isolated system* in nonmathematical language that would make sense to an educated person who had never heard of it.
4. A 0.2 kg plastic cart and a 20 kg lead cart can both roll without friction on a horizontal surface. Equal forces are used to push both carts forward for a time of 1 s, starting from rest. After the force is removed at $t = 1$ s, is the momentum of the plastic cart greater than, less than, or equal to the momentum of the lead cart? Explain.
5. A 0.2 kg plastic cart and a 20 kg lead cart can both roll without friction on a horizontal surface. Equal forces are used to push both carts forward for a distance of 1 m, starting from rest. After traveling 1 m, is the momentum of the plastic cart greater than, less than, or equal to the momentum of the lead cart? Explain.
6. Angie, Brad, and Carlos are discussing a physics problem in which two identical bullets are fired with equal speeds at equal-mass wood and steel blocks resting on a frictionless table. One bullet bounces off the steel block while the second becomes embedded in the wood block. "All the masses and speeds are the same," says Angie, "so I think the blocks will have equal speeds after the collisions." "But what about momentum?" asks Brad. "The bullet hitting the wood block transfers all its momentum and energy to the block, so the wood block should end up going faster than the steel block." "I think the bounce is an important factor," replies Carlos. "The steel block will be faster because the bullet bounces off it and goes back the other direction." Which of these three do you agree with, and why?
7. It feels better to catch a hard ball while wearing a padded glove than to catch it bare handed. Use the ideas of this chapter to explain why.
8. Automobiles are designed with "crumple zones" intended to collapse in a collision. Use the ideas of this chapter to explain why.
9. A 2 kg object is moving to the right with a speed of 1 m/s when it experiences an impulse of 4 N s. What are the object's speed and direction after the impulse?
10. A 2 kg object is moving to the right with a speed of 1 m/s when it experiences an impulse of -4 N s. What are the object's speed and direction after the impulse?
11. A golf club continues forward after hitting the golf ball. Is momentum conserved in the collision? Explain, making sure you are careful to identify "the system."
12. Suppose a rubber ball collides head-on with a steel ball of equal mass traveling in the opposite direction with equal speed. Which ball, if either, receives the larger impulse? Explain.
13. Two particles collide, one of which was initially moving and the other initially at rest.
    a. Is it possible for *both* particles to be at rest after the collision? Give an example in which this happens, or explain why it can't happen.
    b. Is it possible for *one* particle to be at rest after the collision? Give an example in which this happens, or explain why it can't happen.
14. Two ice skaters, Paula and Ricardo, push off from each other. Ricardo weighs more than Paula.
    a. Which skater, if either, has the greater momentum after the push-off? Explain.
    b. Which skater, if either, has the greater speed after the push-off? Explain.

EXERCISES AND PROBLEMS

Problems labeled [ ] integrate material from earlier chapters.

Exercises

Section 9.1 Momentum and Impulse

1. | What is the magnitude of the momentum of
   a. A 3000 kg truck traveling at 15 m/s?
   b. A 200 g baseball thrown at 40 m/s?
2. | At what speed do a bicycle and its rider, with a combined mass of 100 kg, have the same momentum as a 1500 kg car traveling at 5.0 m/s?
3. || What impulse does the force shown in FIGURE EX9.3 exert on a 250 g particle?

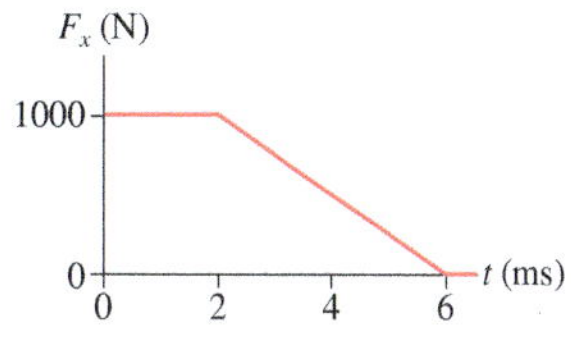

FIGURE EX9.3

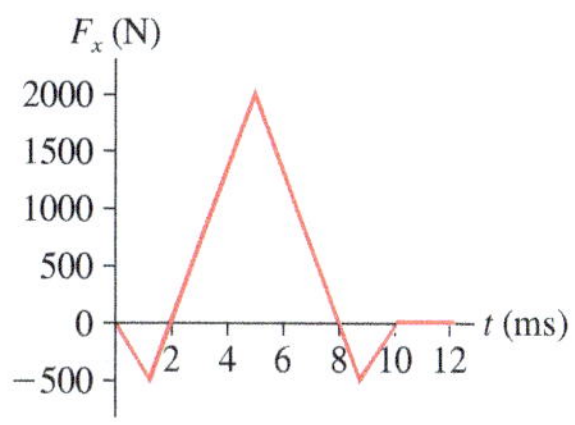

FIGURE EX9.4

4. || What is the impulse on a 3.0 kg particle that experiences the force shown in FIGURE EX9.4?

5. || In FIGURE EX9.5, what value of F_{max} gives an impulse of 6.0 N s?

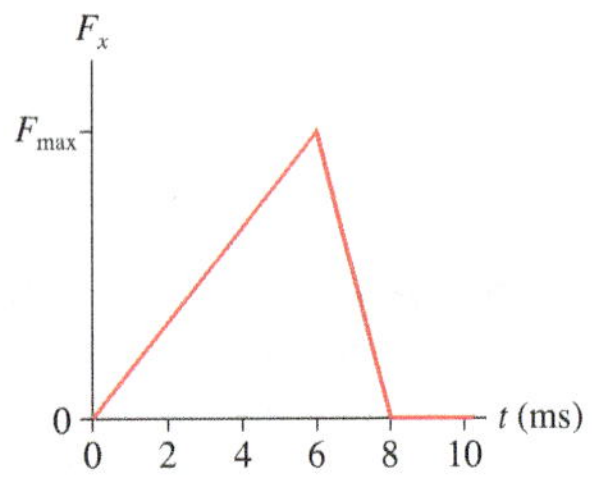

FIGURE EX9.5

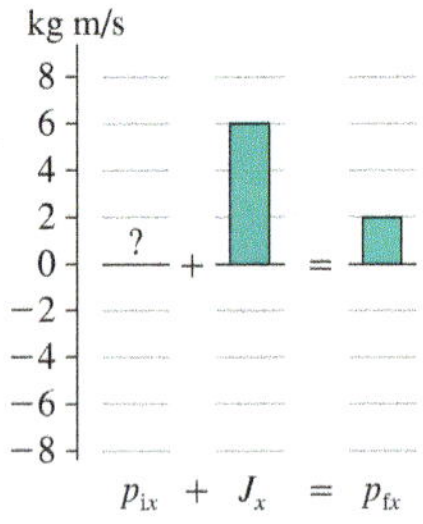

FIGURE EX9.6

6. || FIGURE EX9.6 is an incomplete momentum bar chart for a 50 g particle that experiences an impulse lasting 10 ms. What were the speed and direction of the particle before the impulse?

7. || FIGURE EX9.7 is an incomplete momentum bar chart for a collision that lasts 10 ms. What are the magnitude and direction of the average collision force exerted on the object?

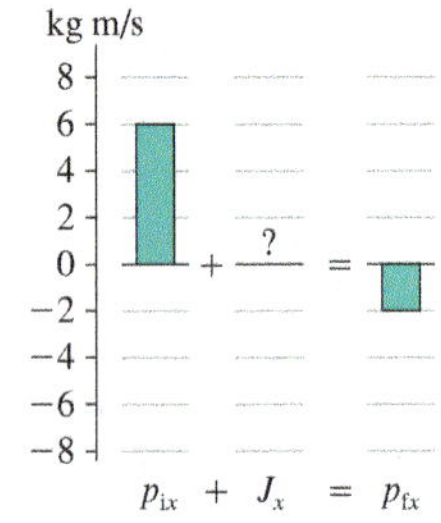

FIGURE EX9.7

Section 9.2 Solving Impulse and Momentum Problems

8. | A 2.0 kg object is moving to the right with a speed of 1.0 m/s when it experiences the force shown in FIGURE EX9.8. What are the object's speed and direction after the force ends?

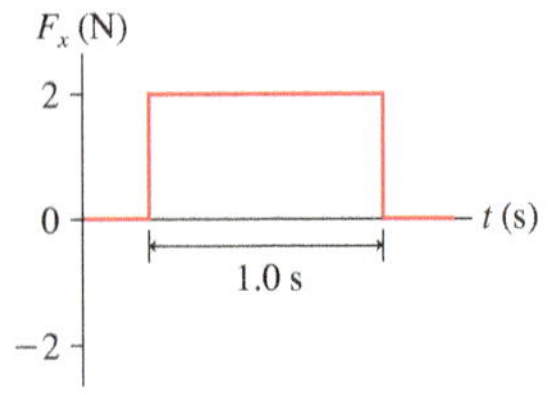

FIGURE EX9.8

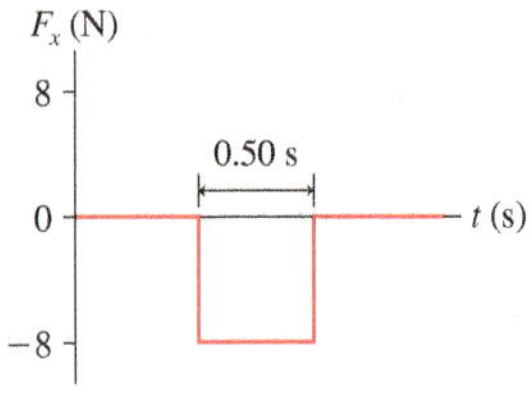

FIGURE EX9.9

9. | A 2.0 kg object is moving to the right with a speed of 1.0 m/s when it experiences the force shown in FIGURE EX9.9. What are the object's speed and direction after the force ends?

10. | A sled slides along a horizontal surface on which the coefficient of kinetic friction is 0.25. Its velocity at point A is 8.0 m/s and at point B is 5.0 m/s. Use the impulse-momentum theorem to find how long the sled takes to travel from A to B.

11. | Far in space, where gravity is negligible, a 425 kg rocket traveling at 75 m/s fires its engines. FIGURE EX9.11 shows the thrust force as a function of time. The mass lost by the rocket during these 30 s is negligible.
   a. What impulse does the engine impart to the rocket?
   b. At what time does the rocket reach its maximum speed? What is the maximum speed?

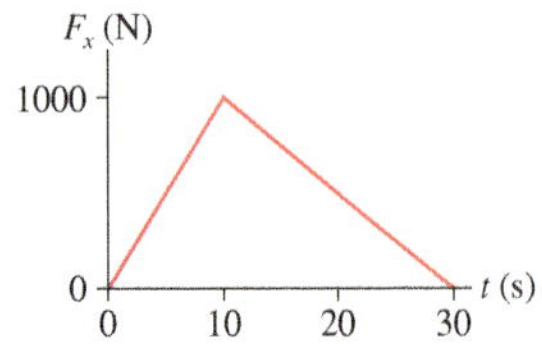

FIGURE EX9.11

12. || A 250 g ball collides with a wall. FIGURE EX9.12 shows the ball's velocity and the force exerted on the ball by the wall. What is v_{fx}, the ball's rebound velocity?

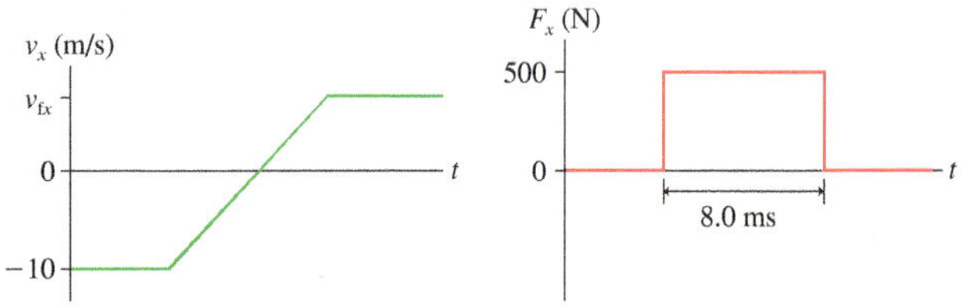

FIGURE EX9.12

13. || A 600 g air-track glider collides with a spring at one end of the track. FIGURE EX9.13 shows the glider's velocity and the force exerted on the glider by the spring. How long is the glider in contact with the spring?

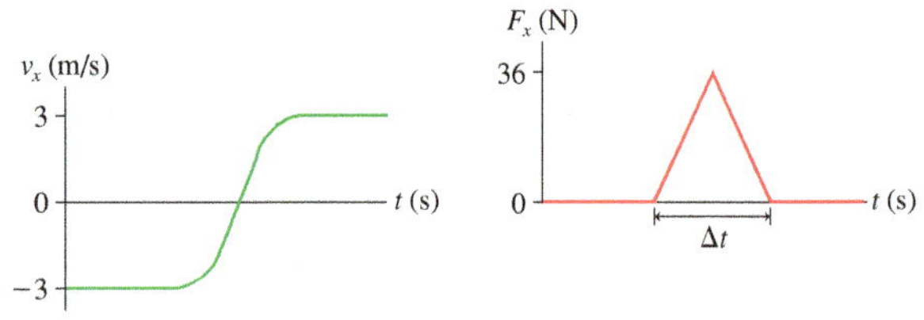

FIGURE EX9.13

Section 9.3 Conservation of Momentum

14. | A 10,000 kg railroad car is rolling at 2.0 m/s when a 4000 kg load of gravel is suddenly dropped in. What is the car's speed just after the gravel is loaded?

15. | A 5000 kg open train car is rolling on frictionless rails at 22 m/s when it starts pouring rain. A few minutes later, the car's speed is 20 m/s. What mass of water has collected in the car?

16. || A 10-m-long glider with a mass of 680 kg (including the passengers) is gliding horizontally through the air at 30 m/s when a 60 kg skydiver drops out by releasing his grip on the glider. What is the glider's velocity just after the skydiver lets go?

Section 9.4 Inelastic Collisions

17. | A 300 g bird flying along at 6.0 m/s sees a 10 g insect heading straight toward it with a speed of 30 m/s. The bird opens its mouth wide and enjoys a nice lunch. What is the bird's speed immediately after swallowing?

18. | The parking brake on a 2000 kg Cadillac has failed, and it is rolling slowly, at 1.0 mph, toward a group of small children. Seeing the situation, you realize you have just enough time to drive your 1000 kg Volkswagen head-on into the Cadillac and save the children. With what speed should you impact the Cadillac to bring it to a halt?

19. | A 1500 kg car is rolling at 2.0 m/s. You would like to stop the car by firing a 10 kg blob of sticky clay at it. How fast should you fire the clay?

Section 9.5 Explosions

20. | A 50 kg archer, standing on frictionless ice, shoots a 100 g arrow at a speed of 100 m/s. What is the recoil speed of the archer?

21. | Dan is gliding on his skateboard at 4.0 m/s. He suddenly jumps backward off the skateboard, kicking the skateboard forward at 8.0 m/s. How fast is Dan going as his feet hit the ground? Dan's mass is 50 kg and the skateboard's mass is 5.0 kg?

22. | A 70.0 kg football player is gliding across very smooth ice at 2.00 m/s. He throws a 0.450 kg football straight forward. What is the player's speed afterward if the ball is thrown at
   a. 15.0 m/s relative to the ground?
   b. 15.0 m/s relative to the player?

Section 9.6 Momentum in Two Dimensions

23. | Two particles collide and bounce apart. FIGURE EX9.23 shows the initial momenta of both and the final momentum of particle 2. What is the final momentum of particle 1? Write your answer in component form.

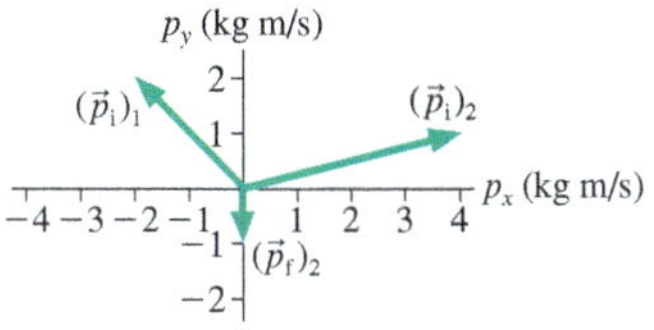

FIGURE EX9.23

24. | An object at rest explodes into three fragments. FIGURE EX9.24 shows the momentum vectors of two of the fragments. What are p_x and p_y of the third fragment?

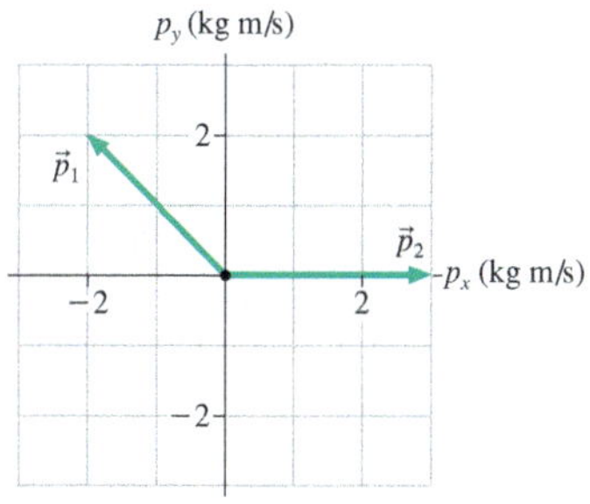

FIGURE EX9.24

25. | A 20 g ball of clay traveling east at 3.0 m/s collides with a 30 g ball of clay traveling north at 2.0 m/s. What are the speed and the direction of the resulting 50 g ball of clay?

Problems

26. | A 60 g tennis ball with an initial speed of 32 m/s hits a wall and rebounds with the same speed. FIGURE P9.26 shows the force of the wall on the ball during the collision. What is the value of F_{max}, the maximum value of the contact force during the collision?

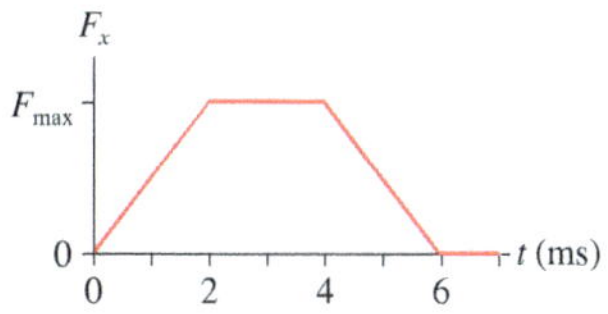

FIGURE P9.26

27. | A tennis player swings her 1000 g racket with a speed of 10 m/s. She hits a 60 g tennis ball that was approaching her at a speed of 20 m/s. The ball rebounds at 40 m/s.
   a. How fast is her racket moving immediately after the impact? You can ignore the interaction of the racket with her hand for the brief duration of the collision.
   b. If the tennis ball and racket are in contact for 10 ms, what is the average force that the racket exerts on the ball? How does this compare to the gravitational force on the ball?

28. ||| A 200 g ball is dropped from a height of 2.0 m, bounces on a hard floor, and rebounds to a height of 1.5 m. FIGURE P9.28 shows the impulse received from the floor. What maximum force does the floor exert on the ball?

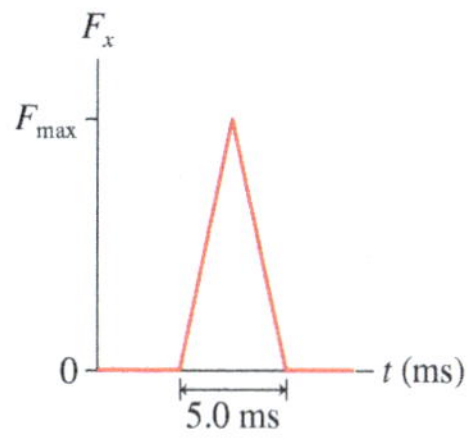

FIGURE P9.28

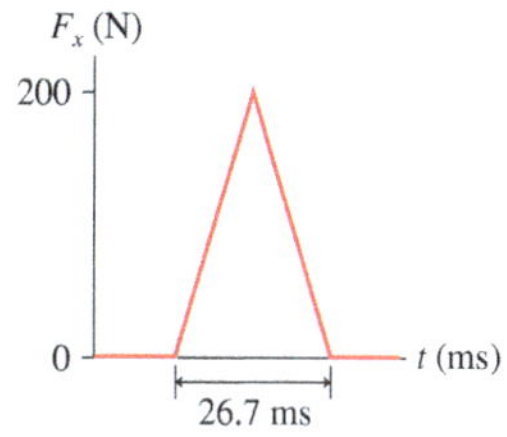

FIGURE P9.29

29. | A 500 g cart is released from rest 1.00 m from the bottom of a frictionless, 30.0° ramp. The cart rolls down the ramp and bounces off a rubber block at the bottom. FIGURE P9.29 shows the force during the collision. After the cart bounces, how far does it roll back up the ramp?

30. | One week in lab, you're given a spring-loaded bar that can be used to strike a metal ball. Your assignment is to measure what size impulse the bar delivers to the ball. You and your lab partner decide to place several balls of different mass on the edge of the lab table, use the striker to launch them horizontally, and measure the horizontal distance to where each ball hits the floor.
   a. Let the table height be h and the horizontal distance traveled by the ball be its range R. Find an expression for the range. The range depends on h, the ball's mass m, and the impulse J.
   b. What should you graph the measured range against to get a linear graph whose slope is related to J?
   c. After measuring the table height to be 1.5 m, you and your partner acquire the following data:

Mass (g)	Range (cm)
100	247
150	175
200	129
250	98

   Draw an appropriate graph of the data and, from the slope of the best-fit line, determine the impulse.

31. | The flowers of the bunchberry plant open with astonishing
BIO force and speed, causing the pollen grains to be ejected out of the flower in a mere 0.30 ms at an acceleration of 2.5×10^4 m/s^2. If the acceleration is constant, what impulse is delivered to a pollen grain with a mass of 1.0×10^{-7} g?

32. | A particle of mass m is at rest at $t = 0$. Its momentum for $t > 0$ is given by $p_x = 6t^2$ kg m/s, where t is in s. Find an expression for $F_x(t)$, the force exerted on the particle as a function of time.

33. | A small rocket to gather weather data is launched straight up. Several seconds into the flight, its velocity is 120 m/s and it is accelerating at 18 m/s^2. At this instant, the rocket's mass is 48 kg and it is losing mass at the rate of 0.50 kg/s as it burns fuel. What is the net force on the rocket? **Hint:** Newton's second law was presented in a new form in this chapter.

34. | Three identical train cars, coupled together, are rolling east at speed v_0. A fourth car traveling east at $2v_0$ catches up with the three and couples to make a four-car train. A moment later, the train cars hit a fifth car that was at rest on the tracks, and it couples to make a five-car train. What is the speed of the five-car train?

35. ‖ A clay blob of mass m_1, initially at rest, is pushed across a frictionless surface with constant force F for a distance d. It then hits and sticks to a second clay blob of mass m_2 that is at rest. Find an expression for their speed after the collision.

36. ‖ Air-track gliders with masses 300 g, 400 g, and 200 g are lined up and held in place with lightweight springs compressed between them. All three are released at once. The 200 g glider flies off to the right while the 300 g glider goes left. Their position-versus-time graphs, as measured by motion detectors, are shown in FIGURE P9.36. What are the direction (right or left) and speed of the 400 g glider that was in the middle?

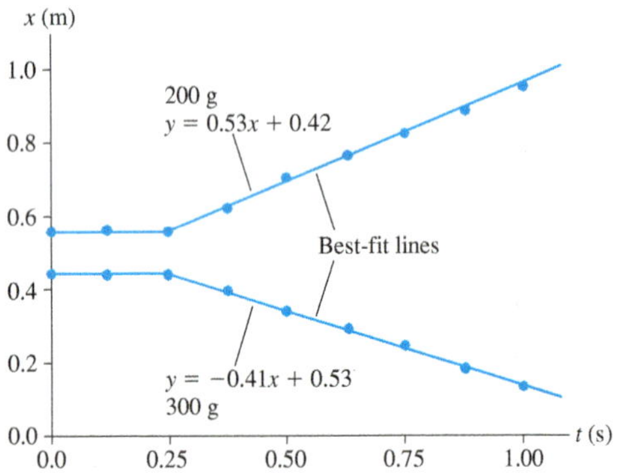

FIGURE P9.36

37. ‖ Most geologists believe that the dinosaurs became extinct 65 million years ago when a large comet or asteroid struck the earth, throwing up so much dust that the sun was blocked out for a period of many months. Suppose an asteroid with a diameter of 2.0 km and a mass of 1.0×10^{13} kg hits the earth with an impact speed of 4.0×10^4 m/s.
   a. What is the earth's recoil speed after such a collision? (Use a reference frame in which the earth was initially at rest.)
   b. What percentage is this of the earth's speed around the sun? (Use the astronomical data inside the back cover.)

38. ‖ At the center of a 50-m-diameter circular ice rink, a 75 kg skater traveling north at 2.5 m/s collides with and holds onto a 60 kg skater who had been heading west at 3.5 m/s.
   a. How long will it take them to glide to the edge of the rink?
   b. Where will they reach it? Give your answer as an angle north of west.

39. ‖ Squids rely on jet propulsion to move around. A 1.5 kg squid drifting at 0.40 m/s suddenly expels 0.10 kg of water backward to quickly get itself moving forward at 2.5 m/s. If drag is ignored over the small interval of time needed to expel the water (the impulse approximation), with what speed relative to itself does the squid eject the water?
BIO

40. | Two ice skaters, with masses of 50 kg and 75 kg, are at the center of a 60-m-diameter circular rink. The skaters push off against each other and glide to opposite edges of the rink. If the heavier skater reaches the edge in 20 s, how long does the lighter skater take to reach the edge?

41. ‖ A firecracker in a coconut blows the coconut into three pieces. Two pieces of equal mass fly off south and west, perpendicular to each other, at speed v_0. The third piece has twice the mass as the other two. What are the speed and direction of the third piece? Give the direction as an angle east of north.

42. ‖ One billiard ball is shot east at 2.0 m/s. A second, identical billiard ball is shot west at 1.0 m/s. The balls have a glancing collision, not a head-on collision, deflecting the second ball by 90° and sending it north at 1.41 m/s. What are the speed and direction of the first ball after the collision? Give the direction as an angle south of east.

43. ‖ a. A bullet of mass m is fired into a block of mass M that is at rest. The block, with the bullet embedded, slides distance d across a horizontal surface. The coefficient of kinetic friction is μ_k. Find an expression for the bullet's speed v_{bullet}.
   b. What is the speed of a 10 g bullet that, when fired into a 10 kg stationary wood block, causes the block to slide 5.0 cm across a wood table?

44. ‖ Fred (mass 60 kg) is running with the football at a speed of 6.0 m/s when he is met head-on by Brutus (mass 120 kg), who is moving at 4.0 m/s. Brutus grabs Fred in a tight grip, and they fall to the ground. Which way do they slide, and how far? The coefficient of kinetic friction between football uniforms and Astroturf is 0.30.

45. | You are part of a search-and-rescue mission that has been called out to look for a lost explorer. You've found the missing explorer, but, as FIGURE P9.45 shows, you're separated from him by a 200-m-high cliff and a 30-m-wide raging river. To save his life, you need to get a 5.0 kg package of emergency supplies across the river. Unfortunately, you can't throw the package hard enough to make it across. Fortunately, you happen to have a 1.0 kg rocket intended for launching flares. Improvising quickly, you attach a sharpened stick to the front of the rocket, so that it will impale itself into the package of supplies, then fire the rocket at ground level toward the supplies. What minimum speed must the rocket have just before impact in order to save the explorer's life?

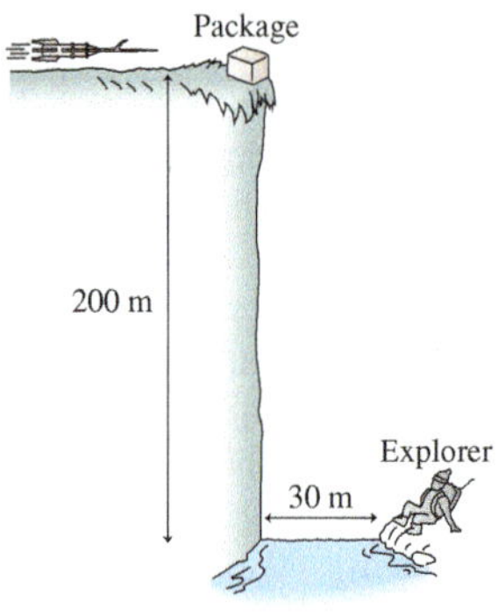

FIGURE P9.45

46. ‖ An object at rest on a flat, horizontal surface explodes into two fragments, one seven times as massive as the other. The heavier fragment slides 8.2 m before stopping. How far does the lighter fragment slide? Assume that both fragments have the same coefficient of kinetic friction.

47. ‖ A 1500 kg weather rocket accelerates upward at 10 m/s^2. It explodes 2.0 s after liftoff and breaks into two fragments, one twice as massive as the other. Photos reveal that the lighter fragment traveled straight up and reached a maximum height of 530 m. What were the speed and direction of the heavier fragment just after the explosion?

48. ‖ In a ballistics test, a 25 g bullet traveling horizontally at 1200 m/s goes through a 30-cm-thick 350 kg stationary target and emerges with a speed of 900 m/s. The target is free to slide on a smooth horizontal surface. What is the target's speed just after the bullet emerges?

49. | Two 500 g blocks of wood are 2.0 m apart on a frictionless table. A 10 g bullet is fired at 400 m/s toward the blocks. It passes all the way through the first block, then embeds itself in the second block. The speed of the first block immediately afterward is 6.0 m/s. What is the speed of the second block after the bullet stops in it?

50. ‖ The skiing duo of Brian (80 kg) and Ashley (50 kg) is always a crowd pleaser. In one routine, Brian, wearing wood skis, starts at the top of a 200-m-long, 20° slope. Ashley waits for him halfway down. As he skis past, she leaps into his arms and he carries her

the rest of the way down. What is their speed at the bottom of the slope?

51. || The stoplight had just changed and a 2000 kg Cadillac had entered the intersection, heading north at 3.0 m/s, when it was struck by a 1000 kg eastbound Volkswagen. The cars stuck together and slid to a halt, leaving skid marks angled 35° north of east. How fast was the Volkswagen going just before the impact?

52. || Ann (mass 50 kg) is standing at the left end of a 15-m-long, 500 kg cart that has frictionless wheels and rolls on a frictionless track. Initially both Ann and the cart are at rest. Suddenly, Ann starts running along the cart at a speed of 5.0 m/s relative to the cart. How far will Ann have run *relative to the ground* when she reaches the right end of the cart?

53. || A ball of mass m and another ball of mass $3m$ are placed inside a smooth metal tube with a massless spring compressed between them. When the spring is released, the heavier ball flies out of one end of the tube with speed v_0. With what speed does the lighter ball emerge from the other end?

54. ||| Force $F_x = (10\text{ N})\sin(2\pi t/4.0\text{ s})$ is exerted on a 250 g particle during the interval $0\text{ s} \leq t \leq 2.0\text{ s}$. If the particle starts from rest, what is its speed at $t = 2.0$ s?

55. ||| A 500 g particle has velocity $v_x = -5.0$ m/s at $t = -2$ s. Force $F_x = (4 - t^2)$ N is exerted on the particle between $t = -2$ s and $t = 2$ s. This force increases from 0 N at $t = -2$ s to 4 N at $t = 0$ s and then back to 0 N at $t = 2$ s. What is the particle's velocity at $t = 2$ s?

56. || A 30 ton rail car and a 90 ton rail car, initially at rest, are connected together with a giant but massless compressed spring between them. When released, the 30 ton car is pushed away at a speed of 4.0 m/s relative to the 90 ton car. What is the speed of the 30 ton car relative to the ground?

57. || A 75 kg shell is fired with an initial speed of 125 m/s at an angle 55° above horizontal. Air resistance is negligible. At its highest point, the shell explodes into two fragments, one four times more massive than the other. The heavier fragment lands directly below the point of the explosion. If the explosion exerts forces only in the horizontal direction, how far from the launch point does the lighter fragment land?

58. || A proton (mass 1 u) is shot at a speed of 5.0×10^7 m/s toward a gold target. The nucleus of a gold atom (mass 197 u) repels the proton and deflects it straight back toward the source with 90% of its initial speed. What is the recoil speed of the gold nucleus?

59. || A proton (mass 1 u) is shot toward an unknown target nucleus at a speed of 2.50×10^6 m/s. The proton rebounds with its speed reduced by 25% while the target nucleus acquires a speed of 3.12×10^5 m/s. What is the mass, in atomic mass units, of the target nucleus?

60. || The nucleus of the polonium isotope ^{214}Po (mass 214 u) is radioactive and decays by emitting an alpha particle (a helium nucleus with mass 4 u). Laboratory experiments measure the speed of the alpha particle to be 1.92×10^7 m/s. Assuming the polonium nucleus was initially at rest, what is the recoil speed of the nucleus that remains after the decay?

61. || A neutron is an electrically neutral subatomic particle with a mass just slightly greater than that of a proton. A free neutron is radioactive and decays after a few minutes into other subatomic particles. In one experiment, a neutron at rest was observed to decay into a proton (mass 1.67×10^{-27} kg) and an electron (mass 9.11×10^{-31} kg). The proton and electron were shot out back-to-back. The proton speed was measured to be 1.0×10^5 m/s and the electron speed was 3.0×10^7 m/s. No other decay products were detected.
   a. Was momentum conserved in the decay of this neutron?

   NOTE ▶ Experiments such as this were first performed in the 1930s and seemed to indicate a failure of the law of conservation of momentum. In 1933, Wolfgang Pauli postulated that the neutron might have a *third* decay product that is virtually impossible to detect. Even so, it can carry away just enough momentum to keep the total momentum conserved. This proposed particle was named the *neutrino,* meaning "little neutral one." Neutrinos were, indeed, discovered nearly 20 years later. ◀

   b. If a neutrino was emitted in the above neutron decay, in which direction did it travel? Explain your reasoning.
   c. How much momentum did this neutrino "carry away" with it?

62. || A 20 g ball of clay traveling east at 2.0 m/s collides with a 30 g ball of clay traveling 30° south of west at 1.0 m/s. What are the speed and direction of the resulting 50 g blob of clay?

63. || FIGURE P9.63 shows a collision between three balls of clay. The three hit simultaneously and stick together. What are the speed and direction of the resulting blob of clay?

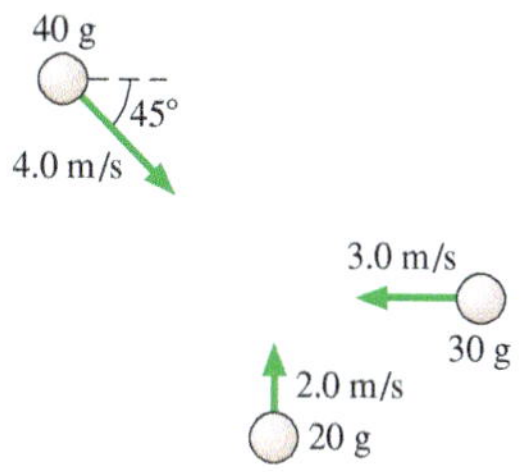

FIGURE P9.63

64. || A 2100 kg truck is traveling east through an intersection at 2.0 m/s when it is hit simultaneously from the side and the rear. (Some people have all the luck!) One car is a 1200 kg compact traveling north at 5.0 m/s. The other is a 1500 kg midsize traveling east at 10 m/s. The three vehicles become entangled and slide as one body. What are their speed and direction just after the collision?

65. || The carbon isotope ^{14}C is used for carbon dating of archeological artifacts. ^{14}C (mass 2.34×10^{-26} kg) decays by the process known as *beta decay* in which the nucleus emits an electron (the beta particle) and a subatomic particle called a neutrino. In one such decay, the electron and the neutrino are emitted at right angles to each other. The electron (mass 9.11×10^{-31} kg) has a speed of 5.0×10^7 m/s and the neutrino has a momentum of 8.0×10^{-24} kg m/s. What is the recoil speed of the nucleus?

In Problems 66 through 69 you are given the equation used to solve a problem. For each of these, you are to
   a. Write a realistic problem for which this is the correct equation.
   b. Finish the solution of the problem, including a pictorial representation.

66. $(0.10\text{ kg})(40\text{ m/s}) - (0.10\text{ kg})(-30\text{ m/s}) = \frac{1}{2}(1400\text{ N})\,\Delta t$

67. $(600\text{ g})(4.0\text{ m/s}) = (400\text{ g})(3.0\text{ m/s}) + (200\text{ g})(v_{ix})_2$

68. $(3000\text{ kg})v_{fx} = (2000\text{ kg})(5.0\text{ m/s}) + (1000\text{ kg})(-4.0\text{ m/s})$

69. $(50\text{ g})(v_{fx})_1 + (100\text{ g})(7.5\text{ m/s}) = (150\text{ g})(1.0\text{ m/s})$

Challenge Problems

70. A 1000 kg cart is rolling to the right at 5.0 m/s. A 70 kg man is standing on the right end of the cart. What is the speed of the cart if the man suddenly starts running to the left with a speed of 10 m/s relative to the cart?

71. A spaceship of mass 2.0×10^6 kg is cruising at a speed of 5.0×10^6 m/s when the antimatter reactor fails, blowing the ship into three pieces. One section, having a mass of 5.0×10^5 kg, is blown straight backward with a speed of 2.0×10^6 m/s. A second piece, with mass 8.0×10^5 kg, continues forward at 1.0×10^6 m/s. What are the direction and speed of the third piece?
72. A 20 kg wood ball hangs from a 2.0-m-long wire. The maximum tension the wire can withstand without breaking is 400 N. A 1.0 kg projectile traveling horizontally hits and embeds itself in the wood ball. What is the greatest speed this projectile can have without causing the cable to break?
73. A two-stage rocket is traveling at 1200 m/s with respect to the earth when the first stage runs out of fuel. Explosive bolts release the first stage and push it backward with a speed of 35 m/s relative to the second stage. The first stage is three times as massive as the second stage. What is the speed of the second stage after the separation?
74. You are the ground-control commander of a 2000 kg scientific rocket that is approaching Mars at a speed of 25,000 km/h. It needs to quickly slow to 15,000 km/h to begin a controlled descent to the surface. If the rocket enters the Martian atmosphere too fast it will burn up, and if it enters too slowly, it will use up its maneuvering fuel before reaching the surface and will crash. The rocket has a new braking system: Several 5.0 kg "bullets" on the front of the rocket can be fired straight ahead. Each has a high-explosive charge that fires it at a speed of 139,000 m/s relative to the rocket. You need to send the rocket an instruction to tell it how many bullets to fire. Success will bring you fame and glory, but failure of this $500,000,000 mission will ruin your career. How many bullets will you tell the rocket to fire?
75. You are a world-famous physicist-lawyer defending a client who has been charged with murder. It is alleged that your client, Mr. Smith, shot the victim, Mr. Wesson. The detective who investigated the scene of the crime found a second bullet, from a shot that missed Mr. Wesson, that had embedded itself into a chair. You arise to cross-examine the detective.

You: In what type of chair did you find the bullet?
Det: A wooden chair.
You: How massive was this chair?
Det: It had a mass of 20 kg.
You: How did the chair respond to being struck with a bullet?
Det: It slid across the floor.
You: How far?
Det: A good three centimeters. The slide marks on the dusty floor are quite distinct.
You: What kind of floor was it?
Det: A wood floor, very nice oak planks.
You: What was the mass of the bullet you retrieved from the chair?
Det: Its mass was 10 g.
You: And how far had it penetrated into the chair?
Det: A distance of 1.5 cm.
You: Have you tested the gun you found in Mr. Smith's possession?
Det: I have.
You: What is the muzzle velocity of bullets fired from that gun?
Det: The muzzle velocity is 450 m/s.
You: And the barrel length?
Det: The gun has a barrel length of 16 cm.

With only a slight hesitation, you turn confidently to the jury and proclaim, "My client's gun did not fire these shots!" How are you going to convince the jury and the judge?

STOP TO THINK ANSWERS

Stop to Think 9.1: f. The cart is initially moving in the negative x-direction, so $p_{ix} = -20$ kg m/s. After it bounces, $p_{fx} = 10$ kg m/s. Thus $\Delta p = (10 \text{ kg m/s}) - (-20 \text{ kg m/s}) = 30$ kg m/s.

Stop to Think 9.2: b. The clay ball goes from $v_{ix} = v$ to $v_{fx} = 0$, so $J_{clay} = \Delta p_x = -mv$. The rubber ball rebounds, going from $v_{ix} = v$ to $v_{fx} = -v$ (same speed, opposite direction). Thus $J_{rubber} = \Delta p_x = -2mv$. The rubber ball has a larger momentum change, and this requires a larger impulse.

Stop to Think 9.3: Less than. The ball's momentum $m_B v_B$ is the same in both cases. Momentum is conserved, so the *total* momentum is the same after both collisions. The ball that rebounds from C has *negative* momentum, so C must have a larger momentum than A.

Stop to Think 9.4: c. Momentum conservation requires $(m_1 + m_2) \times v_f = m_1 v_1 + m_2 v_2$. Because $v_1 > v_2$, it must be that $(m_1 + m_2) \times v_f = m_1 v_1 + m_2 v_2 > m_1 v_2 + m_2 v_2 = (m_1 + m_2) v_2$. Thus $v_f > v_2$. Similarly, $v_2 < v_1$ so $(m_1 + m_2) v_f = m_1 v_1 + m_2 v_2 < m_1 v_1 + m_2 v_1 = (m_1 + m_2) v_1$. Thus $v_f < v_1$. The collision causes m_1 to slow down and m_2 to speed up.

Stop to Think 9.5: Right end. The pieces started at rest, so the total momentum of the system is zero. It's an isolated system, so the total momentum after the explosion is still zero. The 6 g piece has momentum $6v$. The 4 g piece, with velocity $-2v$, has momentum $-8v$. The combined momentum of these two pieces is $-2v$. In order for P to be zero, the third piece must have a *positive* momentum $(+2v)$ and thus a positive velocity.

10 Energy

These photovoltaic panels are transforming solar energy into electrical energy.

▶ **Looking Ahead** The goal of Chapter 10 is to introduce the concept of energy and the basic energy model.

Basic Energy Model

Energy is one of the most important concepts in physics. Chapters 10 and 11 will develop the **basic energy model**, a powerful set of ideas for solving problems in mechanics.

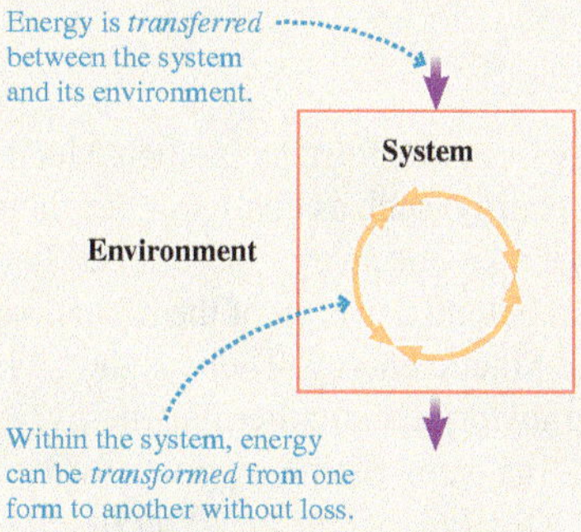

This chapter focuses on energy transformations within the system as one kind of energy is converted to another. Chapter 11 will explore energy transfers to and from the system. For mechanical systems, that transfer is called *work*. Part IV will expand our understanding of energy even further by incorporating the concepts of *heat* and thermodynamics.

Forms of Energy

- **Kinetic energy** is energy associated with an object's motion.
- **Potential energy** is stored energy. Potential energy is associated with an object's position.
- **Thermal energy** is the energy of the random motions of atoms within an object. Thermal energy is associated with temperature.

You will learn about *gravitational potential energy,* the *elastic potential energy* of a stretched or compressed spring, and how these potential energies can be transformed into kinetic energy.

Conservation of Mechanical Energy

Mechanical energy, the sum of kinetic and potential energies, is *conserved* in a system that is both isolated and frictionless. As you learned with momentum, conservation means that

$$\text{final value} = \text{initial value}$$

This will be the basis for a new problem-solving strategy.

◀ **Looking Back**
Sections 9.2–9.3 Before-and-after pictorial representations and conservation of momentum

Energy Diagrams

You'll learn how to interpret an **energy diagram**, a graphical representation for understanding how the speed of a particle changes as it moves through space.

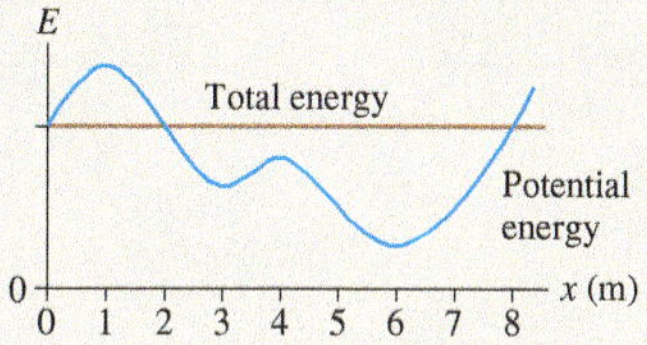

As you'll see, maxima and minima are points of unstable and stable equilibrium, respectively.

Elastic Collisions

A collision that conserves both momentum and mechanical energy is a **perfectly elastic collision.**

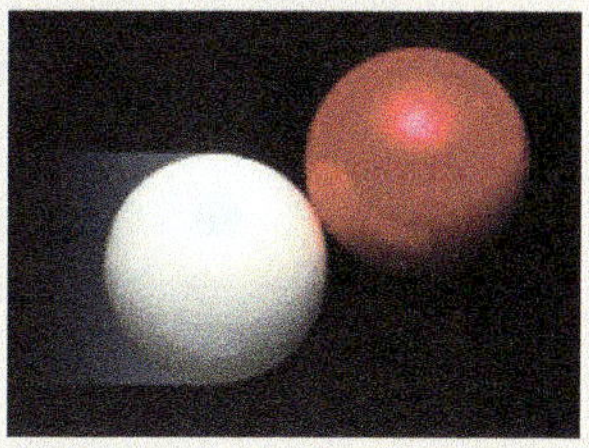

Collisions between two billiard balls or two steel balls come very close to being perfectly elastic.

10.1 The Basic Energy Model

Energy. It's a word you hear all the time. We use chemical energy to heat our homes, electrical energy to power our lights and computers, and solar energy to grow our crops and forests. We're told to use energy wisely and not to waste it.

But just what is energy? The concept of energy has grown and changed with time, and it is not easy to define in a general way just what energy is. Rather than starting with a formal definition, we're going to let the concept of energy expand slowly over the course of several chapters. This chapter introduces three of the most fundamental forms of energy: kinetic energy, potential energy, and thermal energy. Our goal is to understand the characteristics of energy, how energy is used, and, especially important, how energy is transformed from one form to another.

Ultimately we will discover a very powerful conservation law for energy. Some scientists consider the law of conservation of energy to be the most important of all the laws of nature. But all that in due time; first we have to start with the basic ideas.

Some important forms of energy

Kinetic energy K

Kinetic energy is the energy of motion. All moving objects have kinetic energy. The more massive an object or the faster it moves, the larger its kinetic energy.

Potential energy U

Potential energy is stored energy associated with an object's position. The roller coaster's gravitational potential energy depends on its height above the ground.

Thermal energy E_{th}

Thermal energy is the sum of the microscopic kinetic and potential energies of all the atoms and bonds that make up the object. An object has more thermal energy when hot than when cold.

Energy Transfer and Transformation

This chapter focuses on the *transformation* of energy from one form to another. Much of modern technology is concerned with transforming energy, such as changing the chemical energy of oil molecules to electrical energy or to the kinetic energy of your car. In the pictures above, you can imagine that the gravitational potential energy of the roller coaster at the top of the hill will soon be transformed into kinetic energy. Then, as the brakes are applied and get hot, that kinetic energy will be transformed into thermal energy.

Remarkably, the total energy of the system—the sum of the various forms of energy—is not changed by these transformations. This *law of conservation of energy* was not recognized until the mid-19th century, long after Newton. The belated discovery of such an important idea was because it took scientists a long time to realize how many types of energy there are and the various ways that energy can be converted from one form into another. As you'll learn, energy ideas go well beyond Newtonian mechanics to include concepts about heat, about chemical energy, and about the energy of the individual atoms that make up a system. All of these forms of energy ultimately have to be included in the law of energy conservation.

Energy not only can be transformed from one kind to another but also can be transferred from one system to another. For example, the roller coaster at the top of the hill acquired its potential energy not through an energy transformation but because an outside force—a chain powered by a motor—dragged it up the hill and in the process *transferred* energy to the roller coaster. This mechanical transfer of energy to a system via forces is called *work*, a topic we'll explore in detail in the next chapter.

As we use energy concepts, we will be "accounting" for energy that is transferred into or out of a system or that is transformed from one form to another within a system. FIGURE 10.1 shows a **basic energy model** that illustrates these ideas. There are many details that must be added to this model, but it's a good starting point. The fact that nature "balances the books" for energy, so that energy is never created or destroyed, is one of the most profound discoveries of science.

There's a lot to say about energy, and energy is an abstract idea, so we'll take it one step at a time. This chapter focuses on the transformations that take place inside the system, especially idealized transformations that don't change the thermal energy. Then, after you've had some practice using the basic concepts of energy, Chapter 11 will introduce energy transfers between the system and the environment and will establish a more rigorous way of defining potential energy. We will extend these ideas even further in Part IV when we reach the study of thermodynamics.

FIGURE 10.1 The basic energy model.

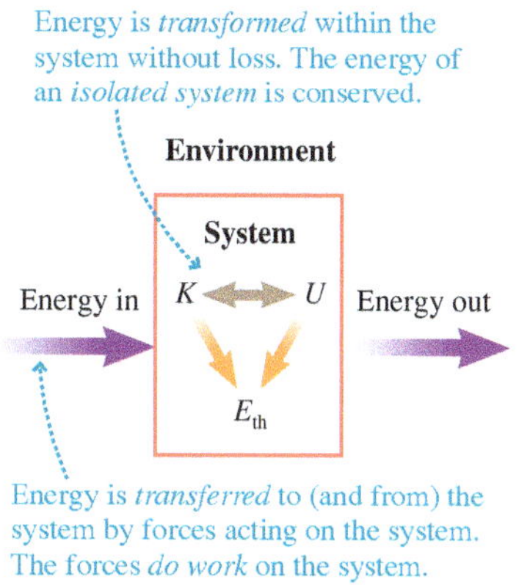

10.2 Kinetic Energy and Gravitational Potential Energy

FIGURE 10.2 is a before-and-after pictorial representation of an object in free fall, as you learned to draw in Chapter 9. We didn't call attention to it in Chapter 2, but one of the free-fall equations also relates "before" and "after." In particular, the free-fall kinematic equation with $a_y = -g$

$$v_{fy}^2 = v_{iy}^2 + 2a_y\,\Delta y = v_{iy}^2 - 2g(y_f - y_i) \qquad (10.1)$$

can easily be rewritten as

$$v_{fy}^2 + 2gy_f = v_{iy}^2 + 2gy_i \qquad (10.2)$$

Equation 10.2 is a conservation law for free-fall motion. It tells us that the quantity $v_y^2 + 2gy$ has the same value *after* free fall (regardless of whether the motion is upward or downward) that it had *before* free fall.

Let's introduce a more general technique to arrive at the same result, but a technique that can be extended to other types of motion. Newton's second law for one-dimensional motion along the y-axis is

$$(F_{net})_y = ma_y = m\frac{dv_y}{dt} \qquad (10.3)$$

The net force on an object in free fall is $(F_{net})_y = -mg$, so Equation 10.3 becomes

$$m\frac{dv_y}{dt} = -mg \qquad (10.4)$$

Recall, from calculus, that we can use the chain rule to write

$$\frac{dv_y}{dt} = \frac{dv_y}{dy}\frac{dy}{dt} = v_y\frac{dv_y}{dy} \qquad (10.5)$$

where we used $v_y = dy/dt$. Substituting this into Equation 10.4 gives

$$mv_y\frac{dv_y}{dy} = -mg \qquad (10.6)$$

The chain rule has allowed us to change from a derivative of v_y with respect to time to a derivative of v_y with respect to position.

We can rewrite Equation 10.6 as

$$mv_y\,dv_y = -mg\,dy \qquad (10.7)$$

Now we can integrate both sides of the equation. However, we have to be careful to make sure the limits of integration match. We want to integrate from "before," when the object

FIGURE 10.2 The before-and-after representation of an object in free fall.

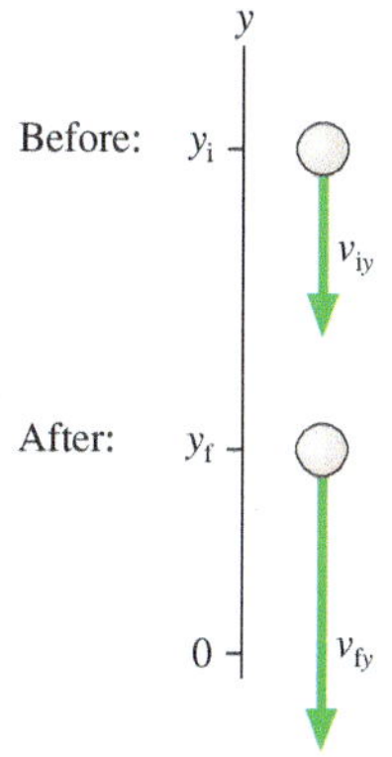

is at position y_i and has velocity v_{iy}, to "after," when the object is at position y_f and has velocity v_{fy}. Figure 10.2 shows these points in the motion. With these limits, the integrals are

$$\int_{v_{iy}}^{v_{fy}} mv_y\, dv_y = -\int_{y_i}^{y_f} mg\, dy \qquad (10.8)$$

Carrying out the integrations, with m and g as constants, we find

$$\frac{1}{2}mv_y^2\Big|_{v_{iy}}^{v_{fy}} = \frac{1}{2}mv_{fy}^2 - \frac{1}{2}mv_{iy}^2 = -mgy\Big|_{y_i}^{y_f} = -mgy_f + mgy_i \qquad (10.9)$$

Because v_y is squared wherever it appears in Equation 10.9, the sign of v_y is not relevant. All we need to know are the initial and final *speeds* v_i and v_f. With this, Equation 10.9 can be written

$$\frac{1}{2}mv_f^2 + mgy_f = \frac{1}{2}mv_i^2 + mgy_i \qquad (10.10)$$

You should recognize that Equation 10.10, other than a constant factor of $\frac{1}{2}m$, is the same as Equation 10.2. This seems like a lot of effort to get to a result we already knew. However, our purpose was not to get the answer but to introduce a *procedure* that will turn out to have other valuable applications.

Kinetic and Potential Energy

The quantity

$$K = \frac{1}{2}mv^2 \quad \text{(kinetic energy)} \qquad (10.11)$$

is called the **kinetic energy** of the object. **Kinetic energy is an energy of motion.** It depends on the object's speed but not its location. The quantity

$$U_g = mgy \quad \text{(gravitational potential energy)} \qquad (10.12)$$

is the object's **gravitational potential energy. Potential energy is an energy of position.** It depends on the object's position but not its speed.

The unit of kinetic energy is mass multiplied by velocity squared. In the SI system of units, this is $kg\,m^2/s^2$. The unit of energy is so important that it has been given its own name, the **joule.** We define:

$$1 \text{ joule} = 1\text{ J} \equiv 1 \text{ kg m}^2/\text{s}^2$$

The unit of potential energy, $kg \times m/s^2 \times m = kg\,m^2/s^2$, is also the joule.

To give you an idea about the size of a joule, consider a 0.5 kg mass (weight on earth $\approx$ 1 lb) moving at 4 m/s ($\approx$ 10 mph). Its kinetic energy is

$$K = \frac{1}{2}mv^2 = \frac{1}{2}(0.5 \text{ kg})(4 \text{ m/s})^2 = 4 \text{ J}$$

Its gravitational potential energy at a height of 1 m ($\approx$ 3 ft) is

$$U_g = mgy = (0.5 \text{ kg})(9.8 \text{ m/s}^2)(1 \text{ m}) \approx 5 \text{ J}$$

This suggests that ordinary-sized objects moving at ordinary speeds will have energies of a fraction of a joule up to, perhaps, a few thousand joules (a running person has $K \approx 1000$ J). A high-speed truck might have $K \approx 10^6$ J.

NOTE ▶ You *must* have masses in kg and velocities in m/s before doing energy calculations. ◀

In terms of energy, Equation 10.10 says that for an object in free fall,

$$K_f + U_{gf} = K_i + U_{gi} \qquad (10.13)$$

In other words, the sum $K + U_g$ of kinetic energy and gravitational potential energy is not changed by free fall. Its value *after* free fall (regardless of whether the motion is upward or downward) is the same as *before* free fall. FIGURE 10.3 illustrates this important idea.

FIGURE 10.3 Kinetic energy and gravitational potential energy.

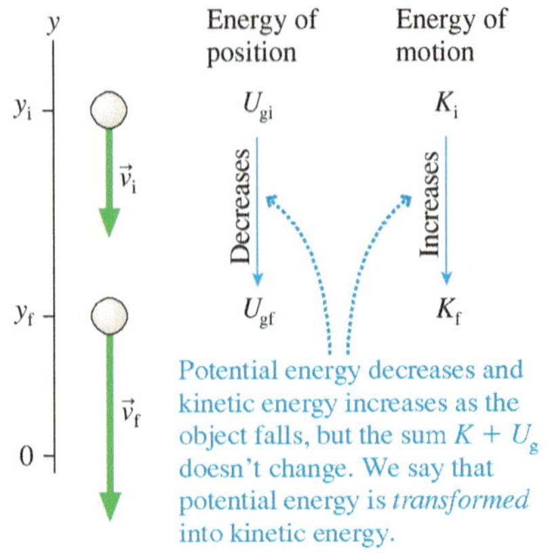

EXAMPLE 10.1 Launching a pebble

Bob uses a slingshot to shoot a 20 g pebble straight up with a speed of 25 m/s. How high does the pebble go?

MODEL This is free-fall motion, so the sum of the kinetic and gravitational potential energy does not change as the pebble rises.

VISUALIZE FIGURE 10.4 shows a before-and-after pictorial representation. The pictorial representation for energy problems is essentially the same as the pictorial representation you learned in Chapter 9 for momentum problems. We'll use numerical subscripts 0 and 1 for the initial and final points.

FIGURE 10.4 Pictorial representation of a pebble shot upward from a slingshot.

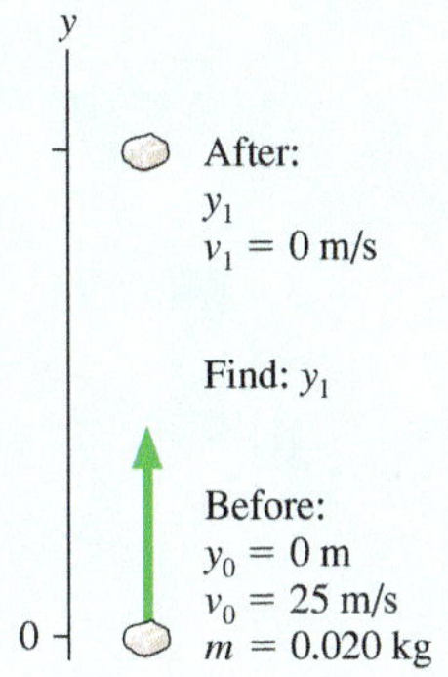

SOLVE Equation 10.13,

$$K_1 + U_{g1} = K_0 + U_{g0}$$

tells us that the sum $K + U_g$ is not changed by the motion. Using the definitions of K and U_g, we have

$$\frac{1}{2}mv_1^2 + mgy_1 = \frac{1}{2}mv_0^2 + mgy_0$$

Here $y_0 = 0$ m and $v_1 = 0$ m/s, so the equation simplifies to

$$mgy_1 = \frac{1}{2}mv_0^2$$

This is easily solved for the height y_1:

$$y_1 = \frac{v_0^2}{2g} = \frac{(25\text{ m/s})^2}{2(9.80\text{ m/s}^2)} = 32\text{ m}$$

ASSESS Notice that the mass canceled and wasn't needed, a fact about free fall that you should remember from Chapter 2.

One of the most important characteristics of energy is that it is a scalar, not a vector. Kinetic energy depends on an object's speed v but *not* on the direction of motion. The kinetic energy of a particle is the same whether it moves up or down or left or right. Consequently, the mathematics of using energy is often much easier than the vector mathematics required by force and acceleration.

NOTE ▶ By its definition, kinetic energy can never be a negative number. If you find, in the course of solving a problem, that K is negative—stop! You have made an error somewhere. Don't just "lose" the minus sign and hope that everything turns out OK. ◀

Energy Bar Charts

The pebble of Example 10.1 started with all kinetic energy, an energy of motion. As the pebble ascends, kinetic energy is converted into gravitational potential energy, *but the sum of the two doesn't change.* At the top, the pebble's energy is entirely potential energy. The simple bar chart in FIGURE 10.5 shows graphically how kinetic energy is transformed into gravitational potential energy as a pebble rises. The potential energy

FIGURE 10.5 Simple energy bar chart for a pebble tossed into the air.

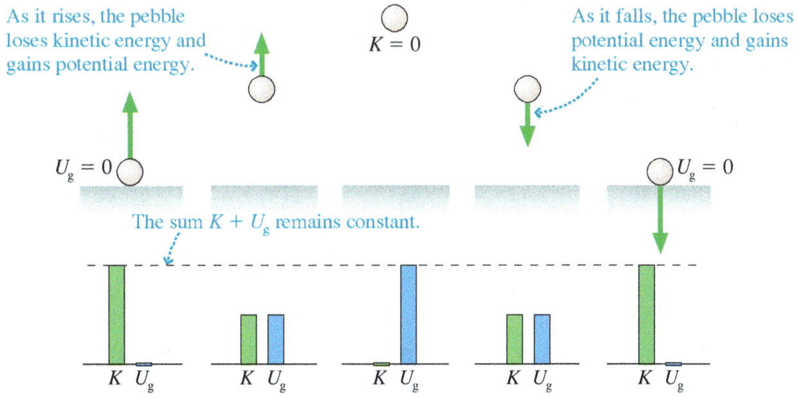

is then transformed back into kinetic energy as the pebble falls. The sum $K + U_g$ remains constant throughout the motion.

FIGURE 10.6a is an energy bar chart more suitable to problem solving. The chart is a graphical representation of the energy equation $K_f + U_{gf} = K_i + U_{gi}$. FIGURE 10.6b applies this to the pebble of Example 10.1. The initial kinetic energy is transformed entirely into potential energy as the pebble reaches its highest point. There are no numerical scales on a bar chart, but you should draw the bar heights proportional to the amount of each type of energy.

FIGURE 10.6 An energy bar chart suitable for problem solving.

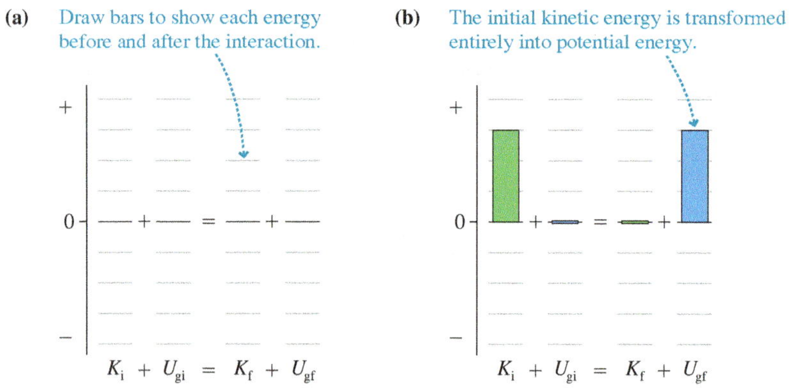

STOP TO THINK 10.1 Rank in order, from largest to smallest, the gravitational potential energies of balls a to d.

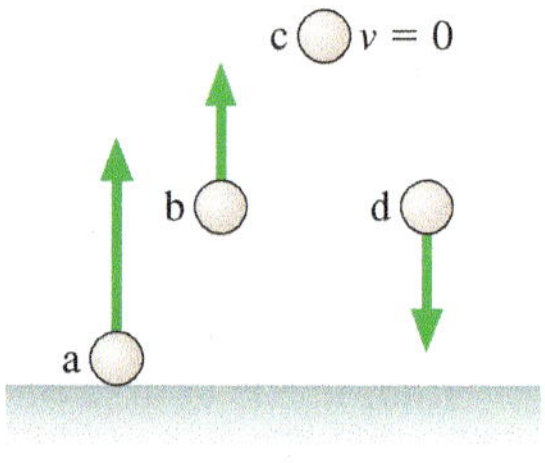

The Zero of Potential Energy

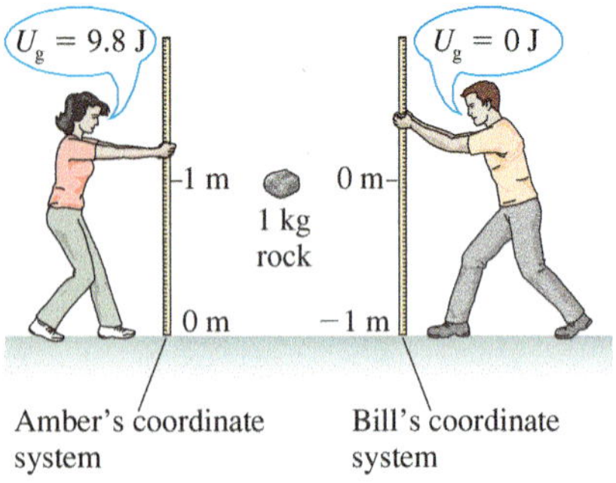

FIGURE 10.7 Amber and Bill use coordinate systems with different origins to determine the potential energy of a rock.

Our expression for the gravitational potential energy $U_g = mgy$ seems straightforward. But you might notice, on further reflection, that the value of U_g depends on where you choose to put the origin of your coordinate system. Consider FIGURE 10.7, where Amber and Bill are attempting to determine the potential energy of a 1 kg rock that is 1 m above the ground. Amber chooses to put the origin of her coordinate system on the ground, measures $y_{rock} = 1$ m, and quickly computes $U_g = mgy = 9.8$ J. Bill, on the other hand, read Chapter 1 very carefully and recalls that it is entirely up to him where to locate the origin of his coordinate system. So he places his origin next to the rock, measures $y_{rock} = 0$ m, and declares that $U_g = mgy = 0$ J!

How can the potential energy of one rock at one position in space have two different values? The source of this apparent difficulty comes from our interpretation of Equation 10.9. The integral of $-mg\,dy$ resulted in the expression $-mg(y_f - y_i)$, and this led us to propose that $U_g = mgy$. But all we are *really* justified in concluding is that the potential energy *changes* by $\Delta U = -mg(y_f - y_i)$. To go beyond this and claim $U_g = mgy$ is consistent with $\Delta U = -mg(y_f - y_i)$, but so also would be a claim that $U_g = mgy + C$, where C is any constant.

No matter where the rock is located, Amber's value of y will always equal Bill's value plus 1 m. Consequently, her value of the potential energy will always equal Bill's value plus 9.8 J. That is, their values of U_g differ by a constant. Nonetheless, both will calculate exactly the *same* value for ΔU if the rock changes position.

EXAMPLE 10.2 The speed of a falling rock

The 1.0 kg rock shown in Figure 10.7 is released from rest. Use both Amber's and Bill's perspectives to calculate its speed just before it hits the ground.

MODEL This is free-fall motion, so the sum of the kinetic and gravitational potential energy does not change as the rock falls.

VISUALIZE FIGURE 10.8 shows a before-and-after pictorial representation using both Amber's and Bill's coordinate systems.

FIGURE 10.8 The before-and-after pictorial representation of a falling rock.

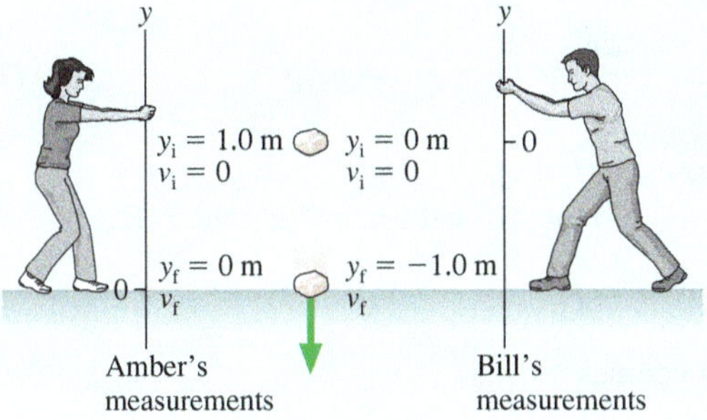

SOLVE The energy equation is $K_f + U_{gf} = K_i + U_{gi}$. Bill and Amber both agree that $K_i = 0$ because the rock was released from rest, so we have

$$K_f = \frac{1}{2}mv_f^2 = -(U_{gf} - U_{gi}) = -\Delta U$$

According to Amber, $U_{gi} = mgy_i = 9.8$ J and $U_{gf} = mgy_f = 0$ J. Thus

$$\Delta U_{Amber} = U_{gf} - U_{gi} = -9.8 \text{ J}$$

The rock *loses* potential energy as it falls. According to Bill, $U_{gi} = mgy_i = 0$ J and $U_{gf} = mgy_f = -9.8$ J. Thus

$$\Delta U_{Bill} = U_{gf} - U_{gi} = -9.8 \text{ J}$$

Bill has different values for U_{gi} and U_{gf} but the *same* value for ΔU. Thus they both agree that the rock hits the ground with speed

$$v_f = \sqrt{\frac{-2\Delta U}{m}} = \sqrt{\frac{-2(-9.8 \text{ J})}{1.0 \text{ kg}}} = 4.4 \text{ m/s}$$

FIGURE 10.9 shows energy bar charts for Amber and Bill. Despite their disagreement over the value of U_g, Amber and Bill arrive at the same value for v_f and their K_f bars are the same height. The reason is that only ΔU has physical significance, not U_g itself, and Amber and Bill found the same value for ΔU. **You can place the origin of your coordinate system, and thus the "zero of potential energy," wherever you choose and be assured of getting the correct answer to a problem.**

FIGURE 10.9 Amber's and Bill's energy bar charts for the falling rock.

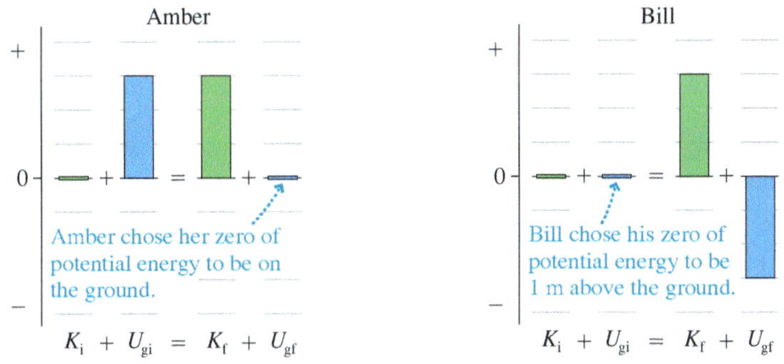

NOTE ▶ Gravitational potential energy can be negative, as U_{gf} is for Bill. **A negative value for U_g means that the particle has less potential for motion at that point than it does at $y = 0$.** But there's nothing wrong with that. Contrast this with kinetic energy, which *cannot* be negative. ◀

10.3 A Closer Look at Gravitational Potential Energy

The concept of energy would be of little interest or use if it applied only to free fall. Let's begin to expand the idea. FIGURE 10.10a on the next page shows an object of mass m sliding along a frictionless surface. The only forces acting on the object are gravity

FIGURE 10.10 A particle moving along a frictionless surface of arbitrary shape.

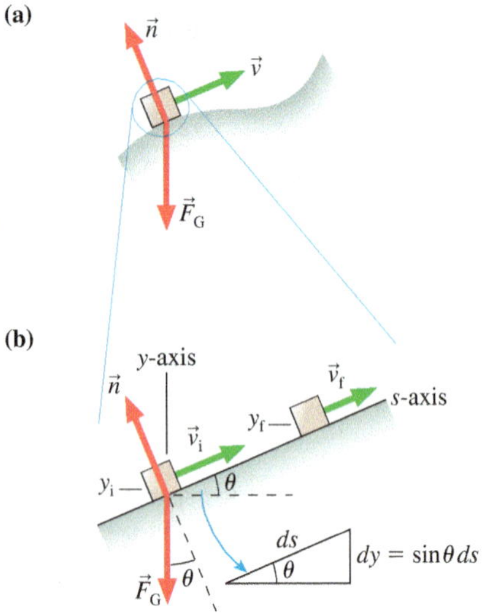

and the normal force from the surface. If the surface is curved, you know from calculus that we can subdivide the surface into many small (perhaps infinitesimal) straight-line segments. **FIGURE 10.10b** shows a magnified segment of the surface that, over some small distance, is a straight line at angle θ.

We can analyze the motion along this small segment using the procedure of Equations 10.3 through 10.10. We define an s-axis parallel to the direction of motion. Newton's second law along this axis is

$$(F_{net})_s = ma_s = m\frac{dv_s}{dt} \tag{10.14}$$

Using the chain rule, we can write Equation 10.14 as

$$(F_{net})_s = m\frac{dv_s}{dt} = m\frac{dv_s}{ds}\frac{ds}{dt} = mv_s\frac{dv_s}{ds} \tag{10.15}$$

where, in the last step, we used $ds/dt = v_s$.

You can see from Figure 10.10b that the net force along the s-axis is

$$(F_{net})_s = -F_G \sin\theta = -mg\sin\theta \tag{10.16}$$

Thus Newton's second law becomes

$$-mg\sin\theta = mv_s\frac{dv_s}{ds} \tag{10.17}$$

Multiplying both sides by ds gives

$$mv_s dv_s = -mg\sin\theta\, ds \tag{10.18}$$

You can see from the figure that $\sin\theta\, ds$ is dy, so Equation 10.18 becomes

$$mv_s\, dv_s = -mg\, dy \tag{10.19}$$

This is *identical* to Equation 10.7, which we found for free fall. Consequently, integrating this equation from "before" to "after" leads again to Equation 10.10:

$$\frac{1}{2}mv_f^2 + mgy_f = \frac{1}{2}mv_i^2 + mgy_i \tag{10.20}$$

where v_i^2 and v_f^2 are the squares of the *speeds* at the beginning and end of this segment of the motion.

We previously defined the kinetic energy $K = \frac{1}{2}mv^2$ and the gravitational potential energy $U_g = mgy$. Equation 10.20 shows that

$$K_f + U_{gf} = K_i + U_{gi} \tag{10.21}$$

for a particle moving along *any* frictionless surface, regardless of the shape.

NOTE ▶ For energy calculations, the y-axis is specifically a *vertical* axis. Gravitational potential energy depends on the *height* above the earth's surface. A tilted coordinate system, such as we often used in dynamics problems, doesn't work for problems with gravitational potential energy. ◀

STOP TO THINK 10.2 A small child slides down the four frictionless slides a–d. Each has the same height. Rank in order, from largest to smallest, her speeds v_a to v_d at the bottom.

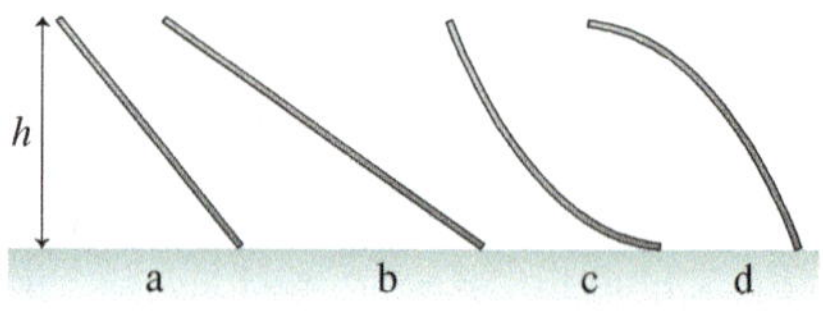

EXAMPLE 10.3 The speed of a sled

Christine runs forward with her sled at 2.0 m/s. She hops onto the sled at the top of a 5.0-m-high, very slippery slope. What is her speed at the bottom?

MODEL Model Christine and the sled as a particle. Assume the slope is frictionless. In that case, the sum of her kinetic and gravitational potential energy does not change as she slides down.

VISUALIZE FIGURE 10.11a shows a before-and-after pictorial representation. We are not told the angle of the slope, or even if it is a straight slope, but the *change* in potential energy depends only on the height Christine descends and *not* on the shape of the hill. **FIGURE 10.11b** is an energy bar chart in which we see an initial kinetic *and* potential energy being transformed into entirely kinetic energy as she goes down the slope.

SOLVE The quantity $K + U_g$ is the same at the bottom of the hill as it was at the top. Thus

$$\frac{1}{2}mv_1^2 + mgy_1 = \frac{1}{2}mv_0^2 + mgy_0$$

This is easily solved for Christine's speed at the bottom:

$$v_1 = \sqrt{v_0^2 + 2g(y_0 - y_1)} = \sqrt{v_0^2 + 2gh} = 10 \text{ m/s}$$

ASSESS We did not need the mass of either Christine or the sled.

FIGURE 10.11 Pictorial representation and energy bar chart of Christine sliding down the hill.

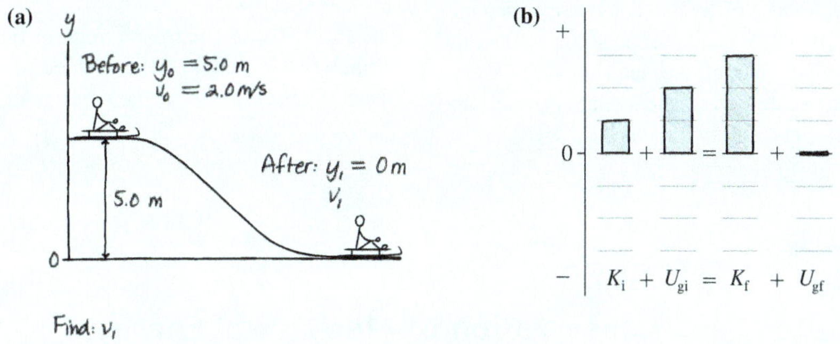

Notice that the normal force $\vec{n}$ doesn't enter an energy analysis. The equation $K_f + U_{gf} = K_i + U_{gi}$ is a statement about how the particle's speed changes as it changes position. $\vec{n}$ does not have a component in the direction of motion, so it cannot change the particle's speed.

The same is true for an object tied to a string and moving in a circle. The tension in the string causes the direction to change, but $\vec{T}$ does not have a component in the direction of motion and does not change the speed of the object. Hence Equation 10.21 also applies to a *pendulum.*

EXAMPLE 10.4 A ballistic pendulum

A 10 g bullet is fired into a 1200 g wood block hanging from a 150-cm-long string. The bullet embeds itself into the block, and the block then swings out to an angle of 40°. What was the speed of the bullet? (This is called a *ballistic pendulum.*)

MODEL This is a two-part problem. The impact of the bullet with the block is an inelastic collision. We haven't done any analysis to let us know what happens to energy during a collision, but you learned in Chapter 9 that *momentum* is conserved in an inelastic collision. After the collision is over, the block swings out as a pendulum. The sum of the kinetic and gravitational potential energy does not change as the block swings to its largest angle.

VISUALIZE FIGURE 10.12 is a pictorial representation in which we've identified before-and-after quantities for both the collision and the swing.

SOLVE The momentum conservation equation $P_f = P_i$ applied to the inelastic collision gives

$$(m_W + m_B)v_{1x} = m_W(v_{0x})_W + m_B(v_{0x})_B$$

FIGURE 10.12 A ballistic pendulum is used to measure the speed of a bullet.

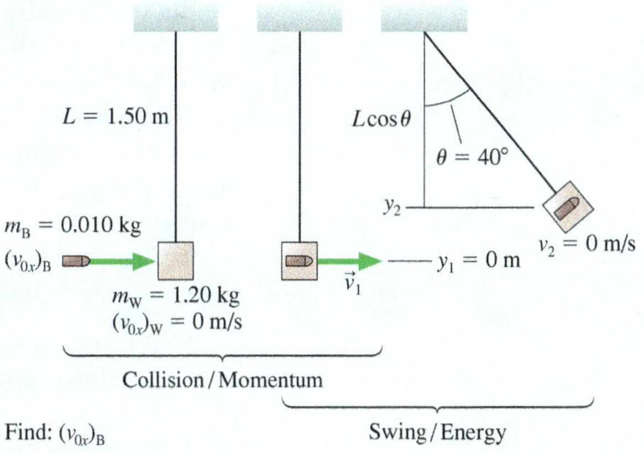

Continued

The wood block is initially at rest, with $(v_{0x})_W = 0$, so the bullet's velocity is

$$(v_{0x})_B = \frac{m_W + m_B}{m_B} v_{1x}$$

where v_{1x} is the velocity of the block + bullet *immediately* after the collision, as the pendulum begins to swing. If we can determine v_{1x} from an analysis of the swing, then we will be able to calculate the speed of the bullet. Turning our attention to the swing, the energy equation $K_f + U_{gf} = K_i + U_{gi}$ is

$$\frac{1}{2}(m_W + m_B)v_2^2 + (m_W + m_B)gy_2 = \frac{1}{2}(m_W + m_B)v_1^2 + (m_W + m_B)gy_1$$

We used the *total* mass $(m_W + m_B)$ of the block and embedded bullet, but notice that it cancels out. We also dropped the x-subscript on v_1 because for energy calculations we need only speed, not velocity. The speed is zero at the top of the swing $(v_2 = 0)$, and we've defined the y-axis such that $y_1 = 0$ m. Thus

$$v_1 = \sqrt{2gy_2}$$

The initial speed is found simply from the maximum height of the swing. You can see from the geometry of Figure 10.12 that

$$y_2 = L - L\cos\theta = L(1 - \cos\theta) = 0.351 \text{ m}$$

With this, the initial velocity of the pendulum, immediately after the collision, is

$$v_{1x} = v_1 = \sqrt{2gy_2} = \sqrt{2(9.80 \text{ m/s}^2)(0.351 \text{ m})} = 2.62 \text{ m/s}$$

Having found v_{1x} from an energy analysis of the swing, we can now calculate that the speed of the bullet was

$$(v_{0x})_B = \frac{m_W + m_B}{m_B} v_{1x} = \frac{1.210 \text{ kg}}{0.010 \text{ kg}} \times 2.62 \text{ m/s} = 320 \text{ m/s}$$

ASSESS It would have been very difficult to solve this problem using Newton's laws, but it yielded to a straightforward analysis based on the concepts of momentum and energy.

Conservation of Mechanical Energy

The sum of the kinetic energy and the potential energy of a system is called the **mechanical energy:**

$$E_{mech} = K + U \tag{10.22}$$

Here K is the *total* kinetic energy of all the particles in the system and U is the potential energy stored in the system. Our examples thus far suggest that a particle's mechanical energy does not change as it moves under the influence of only gravity. The kinetic energy and the potential energy change, as they are transformed back and forth into each other, but their sum remains constant. We can express the unchanging value of E_{mech} as

$$K_f + U_f = K_i + U_i \tag{10.23}$$

This statement is called the **law of conservation of mechanical energy.**

But is this really a law of nature? Consider shoving a box that then slides along the floor until it stops. The box gains kinetic energy, but it comes from the shove, an outside force, rather than from a transformation of potential energy. The box then loses kinetic energy as it slows down, but in this case kinetic energy is transformed into thermal energy (the box and the floor get hotter) rather than into potential energy. Mechanical energy is conserved neither as the box speeds up nor as it slows down.

In Chapter 9 you learned that momentum is conserved only for an isolated system. Similarly, mechanical energy is conserved only if two requirements are satisfied:

1. The system is isolated, meaning that no external forces transfer energy into or out of the system.
2. There is no friction or drag that would transform kinetic or potential energy into thermal energy.

Fortunately, enough realistic situations satisfy the restrictions, or come very close, that the law of conservation of mechanical energy is an important problem-solving strategy.

PROBLEM-SOLVING STRATEGY 10.1 **Conservation of mechanical energy**

MODEL Choose a system that is isolated and has no friction or other losses of mechanical energy.

VISUALIZE Draw a before-and-after pictorial representation. Define symbols, list known values, and identify what you're trying to find.

SOLVE The mathematical representation is based on the law of conservation of mechanical energy:

$$K_f + U_f = K_i + U_i$$

ASSESS Check that your result has the correct units, is reasonable, and answers the question.

Exercise 8

STOP TO THINK 10.3 A box slides along the frictionless surface shown in the figure. It is released from rest at the position shown. Is the highest point the box reaches on the other side at level a, level b, or level c?

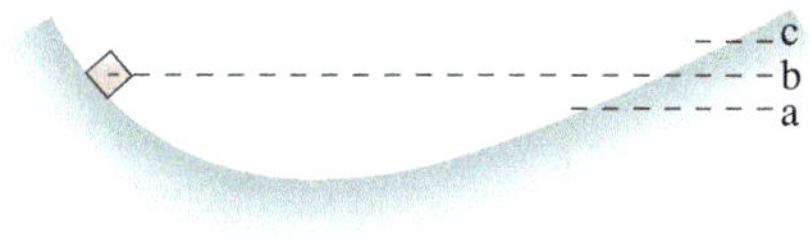

10.4 Restoring Forces and Hooke's Law

If you stretch a rubber band, a force tries to pull the rubber band back to its equilibrium, or unstretched, length. A force that restores a system to an equilibrium position is called a **restoring force.** Systems that exhibit restoring forces are called **elastic.** The most basic examples of elasticity are things like springs and rubber bands. If you stretch a spring, a tension-like force pulls back. Similarly, a compressed spring tries to re-expand to its equilibrium length. Other examples of elasticity and restoring forces abound. The steel beams bend slightly as you drive your car over a bridge, but they are restored to equilibrium after your car passes by. Nearly everything that stretches, compresses, flexes, bends, or twists exhibits a restoring force and can be called elastic.

We're going to use a simple spring as a prototype of elasticity. Suppose you have a spring whose **equilibrium length** is L_0. This is the length of the spring when it is neither pushing nor pulling. If you now stretch the spring to length L, how hard does it pull back? One way to find out is to attach the spring to a bar, as shown in FIGURE 10.13, then to hang a mass m from the spring. The mass stretches the spring to length L. Lengths L_0 and L are easily measured with a meter stick.

The mass hangs in static equilibrium, so the upward spring force $\vec{F}_{sp}$ exactly balances the downward gravitational force $\vec{F}_G$ to give $\vec{F}_{net} = \vec{0}$. That is,

$$F_{sp} = F_G = mg \qquad (10.24)$$

By using different masses to stretch the spring to different lengths, we can determine how F_{sp}, the magnitude of the spring's restoring force, depends on the length L.

FIGURE 10.14 shows measured data for the restoring force of a real spring. Notice that the quantity graphed along the horizontal axis is $\Delta s = L - L_0$. This is the distance that the end of the spring has moved, which we call the **displacement from equilibrium.** The graph shows that the restoring force is proportional to the displacement. That is, the data fall along the straight line

$$F_{sp} = k\,\Delta s \qquad (10.25)$$

The proportionality constant k, the slope of the force-versus-displacement graph, is called the **spring constant.** The units of the spring constant are N/m.

FIGURE 10.13 A hanging mass stretches a spring of equilibrium length L_0 to length L.

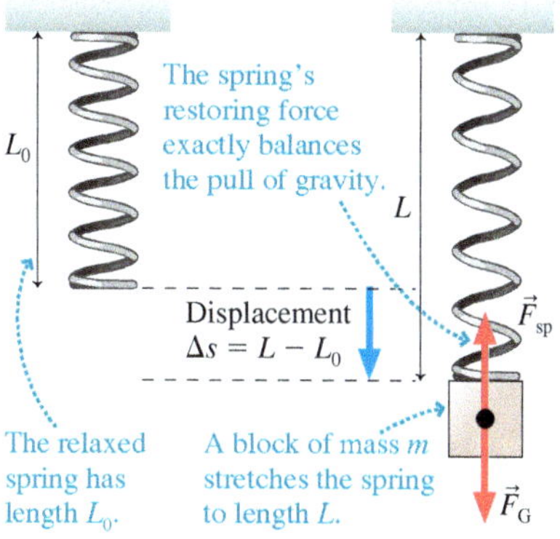

FIGURE 10.14 Measured data for the restoring force of a real spring.

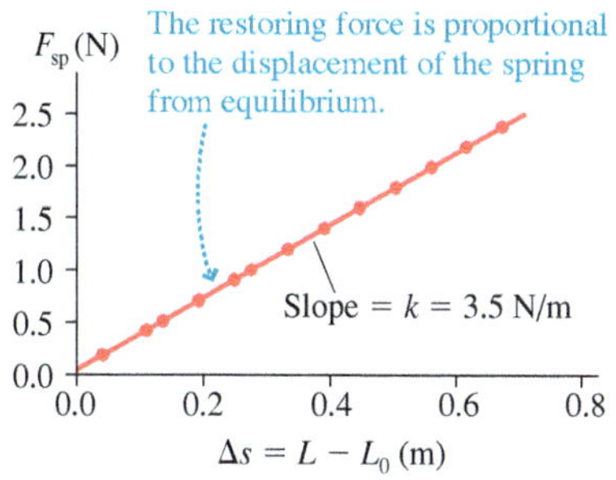

NOTE ▶ The force does not depend on the spring's physical length L but, instead, on the *displacement* Δs of the end of the spring. ◀

The spring constant k is a property that characterizes a spring, just as mass m characterizes a particle. If k is large, it takes a large pull to cause a significant stretch, and we call the spring a "stiff" spring. A spring with small k can be stretched with very little force, and we call it a "soft" spring. The spring constant for the spring in Figure 10.14 can be determined from the slope of the straight line to be $k = 3.5$ N/m.

NOTE ▶ Just as we used massless strings, we will adopt the idealization of a *massless spring*. While not a perfect description, it is a good approximation if the mass attached to a spring is much larger than the mass of the spring itself. ◀

Hooke's Law

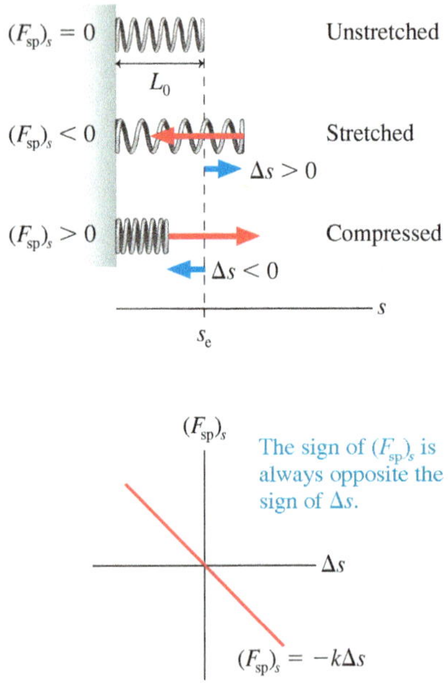

FIGURE 10.15 The direction of $\vec{F}_{sp}$ is always opposite the displacement $\Delta\vec{s}$.

FIGURE 10.15 shows a spring along a generic s-axis. The equilibrium position of the end of the spring is denoted s_e. This is the *position,* or coordinate, of the free end of the spring, *not* the spring's equilibrium length L_0.

When the spring is stretched, the displacement from equilibrium $\Delta s = s - s_e$ is *positive* while $(F_{sp})_s$, the s-component of the restoring force pointing to the left, is *negative.* If the spring is compressed, the displacement from equilibrium Δs is negative while the s-component of $\vec{F}_{sp}$, which now points to the right, is positive. Either way, the sign of the force component $(F_{sp})_s$ is always opposite to the sign of the displacement Δs. We can write this mathematically as

$$(F_{sp})_s = -k\,\Delta s \qquad \text{(Hooke's law)} \qquad (10.26)$$

where $\Delta s = s - s_e$ is the displacement of the end of the spring from equilibrium. The minus sign is the mathematical indication of a *restoring* force.

Equation 10.26 for the restoring force of a spring is called **Hooke's law.** This "law" was first suggested by Robert Hooke, a contemporary (and sometimes bitter rival) of Newton. Hooke's law is not a true "law of nature," in the sense that Newton's laws are, but is actually just a *model* of a restoring force. It works extremely well for some springs, as in Figure 10.14, but less well for others. Hooke's law will fail for any spring that is compressed or stretched too far.

NOTE ▶ Some of you, in an earlier physics course, may have learned Hooke's law as $F_{sp} = -kx$ (for a spring along the x-axis), rather than as $-k\,\Delta x$. This can be misleading, and it is a common source of errors. The restoring force will be $-kx$ *only* if the coordinate system in the problem is chosen such that the origin is at the equilibrium position of the free end of the spring. That is, $\Delta x = x$ only if $x_e = 0$. This is often done, but in some problems it will be more convenient to locate the origin of the coordinate system elsewhere. So make sure you learn Hooke's law as $(F_{sp})_s = -k\,\Delta s$. ◀

EXAMPLE 10.5 Pull until it slips

FIGURE 10.16 shows a spring attached to a 2.0 kg block. The other end of the spring is pulled by a motorized toy train that moves forward at 5.0 cm/s. The spring constant is 50 N/m, and the coefficient of static friction between the block and the surface is 0.60. The spring is at its equilibrium length at $t = 0$ s when the train starts to move. When does the block slip?

FIGURE 10.16 A toy train stretches the spring until the block slips.

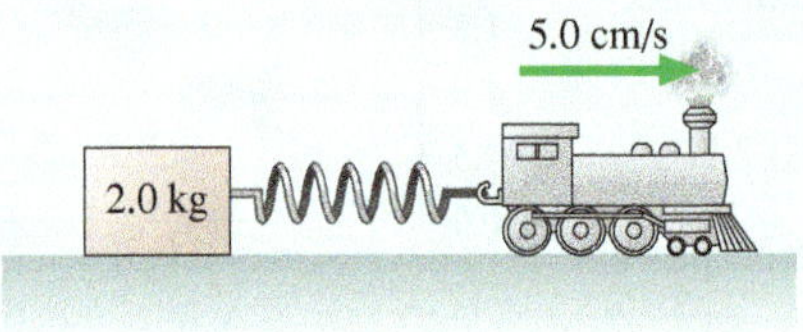

MODEL Model the block as a particle and the spring as an ideal spring obeying Hooke's law.

VISUALIZE **FIGURE 10.17** is a free-body diagram for the block.

FIGURE 10.17 The free-body diagram.

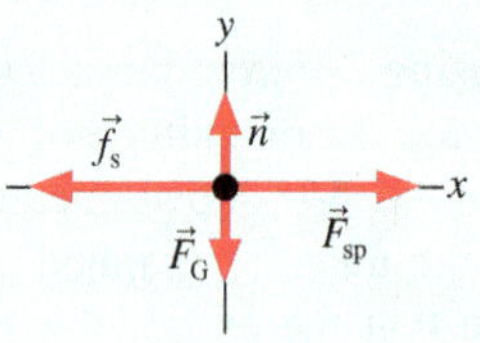

SOLVE Recall that the tension in a massless string pulls equally at *both* ends of the string. The same is true for the spring force: It pulls (or pushes) equally at *both* ends. This is the key to solving the problem. As the right end of the spring moves, stretching the spring, the spring pulls backward on the train *and* forward on the block with equal strength. As the spring stretches, the static friction force on the block increases in magnitude to keep the block at rest. The block is in static equilibrium, so

$$\sum (F_{net})_x = (F_{sp})_x + (f_s)_x = F_{sp} - f_s = 0$$

where F_{sp} is the *magnitude* of the spring force. The magnitude is $F_{sp} = k\,\Delta x$, where $\Delta x = v_x t$ is the distance the train has moved. Thus

$$f_s = F_{sp} = k\,\Delta x$$

The block slips when the static friction force reaches its maximum value $f_{s\,max} = \mu_s n = \mu_s mg$. This occurs when the train has moved

$$\Delta x = \frac{f_{s\,max}}{k} = \frac{\mu_s mg}{k} = \frac{(0.60)(2.0\text{ kg})(9.80\text{ m/s}^2)}{50\text{ N/m}}$$

$$= 0.235\text{ m} = 23.5\text{ cm}$$

The time at which the block slips is

$$t = \frac{\Delta x}{v_x} = \frac{23.5\text{ cm}}{5.0\text{ cm/s}} = 4.7\text{ s}$$

This example illustrates a class of motion called *stick-slip motion*. Once the block slips, it will shoot forward some distance, then stop and stick again. As the train continues, there will be a recurring sequence of stick, slip, stick, slip, stick. . . .

Earthquakes are an important example of stick-slip motion. The large tectonic plates making up the earth's crust are attempting to slide past each other, but friction causes the edges of the plates to stick together. You may think of rocks as rigid and brittle, but large masses of rock are somewhat elastic and can be "stretched." Eventually the elastic force of the deformed rocks exceeds the friction force between the plates. An earthquake occurs as the plates slip and lurch forward. Once the tension is released, the plates stick together again and the process starts all over.

The slip can range from a few centimeters in a relatively small earthquake to several meters in a very large earthquake.

STOP TO THINK 10.4 The graph shows force versus displacement for three springs. Rank in order, from largest to smallest, the spring constants k_a, k_b, and k_c.

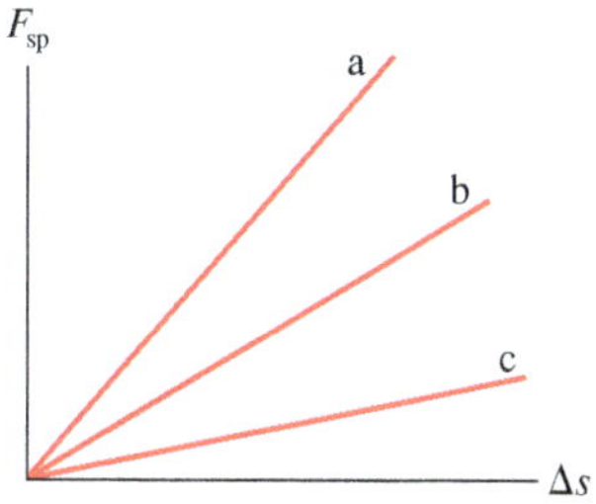

10.5 Elastic Potential Energy

The forces we have worked with thus far—gravity, friction, tension—have been constant forces. That is, their magnitudes do not change as an object moves. That feature has been important because the kinematic equations we developed in Chapter 2 are for motion with constant acceleration. But a spring exerts a *variable* force. The force is zero if $\Delta s = 0$ (no displacement), and it steadily increases as the stretching increases. The "natural motion" of a mass on a spring—think of pulling down on a spring and then releasing it—is an *oscillation*. This is *not* constant-acceleration motion, and we haven't yet developed the kinematics to handle oscillatory motion.

But suppose we're interested not in the time dependence of motion, only in before-and-after situations. For example, FIGURE 10.18 shows a before-and-after situation in which a spring launches a ball. Asking how the compression of the spring (the "before") affects the speed of the ball (the "after") is very different from wanting to know the ball's position as a function of time as the spring expands.

You certainly have a sense that a compressed spring has "stored energy," and Figure 10.18 shows clearly that the stored energy is transformed into the kinetic energy of the ball. Let's analyze this process with the same method we developed for motion under the influence of gravity. Newton's second law for the ball is

$$(F_{net})_s = ma_s = m\frac{dv_s}{dt} \tag{10.27}$$

FIGURE 10.18 Before and after a spring launches a ball.

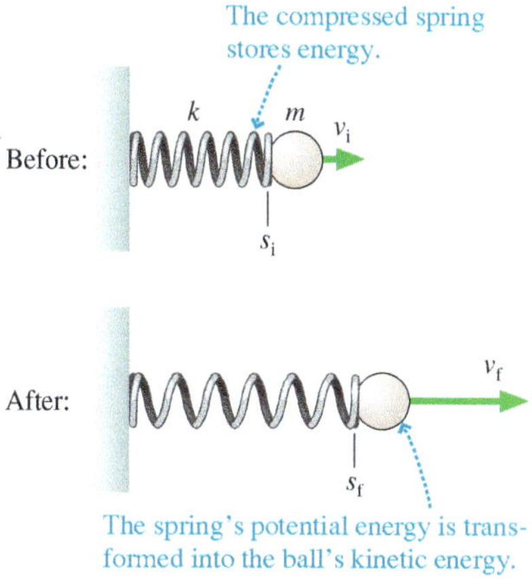

Springs and rubber bands store energy—potential energy—that can be transformed into kinetic energy.

The net force on the ball is given by Hooke's law, $(F_{net})_s = -k(s - s_e)$. Thus

$$m\frac{dv_s}{dt} = -k(s - s_e) \tag{10.28}$$

We'll use a generic s-axis, although it is better in actual problem solving to use x or y, depending on whether the motion is horizontal or vertical.

As before, we use the chain rule to write

$$\frac{dv_s}{dt} = \frac{dv_s}{ds}\frac{ds}{dt} = v_s\frac{dv_s}{ds} \tag{10.29}$$

We substitute this into Equation 10.28 and then multiply both sides by ds to get

$$mv_s\,dv_s = -k(s - s_e)\,ds \tag{10.30}$$

We can integrate both sides of the equation from the initial conditions i to the final conditions f—that is, integrate "from before to after"—to give

$$\int_{v_i}^{v_f} mv_s\,dv_s = \frac{1}{2}mv_f^2 - \frac{1}{2}mv_i^2 = -k\int_{s_i}^{s_f}(s - s_e)\,ds \tag{10.31}$$

The integral on the right is not difficult, but many of you are new to calculus so we'll proceed step by step. The easiest way to get the answer in the most useful form is to make a change of variables. Define $u = (s - s_e)$, in which case $ds = du$. This changes the integrand from $(s - s_e)\,ds$ to $u\,du$.

When we change variables, we also must change the limits of integration. In particular, $s = s_i$ at the lower integration limit makes $u = s_i - s_e = \Delta s_i$, where Δs_i is the initial displacement of the spring from equilibrium. Likewise, $s = s_f$ makes $u = s_f - s_e = \Delta s_f$ at the upper limit. FIGURE 10.19 clarifies the meanings of Δs_i and Δs_f.

FIGURE 10.19 The initial and final displacements of the spring.

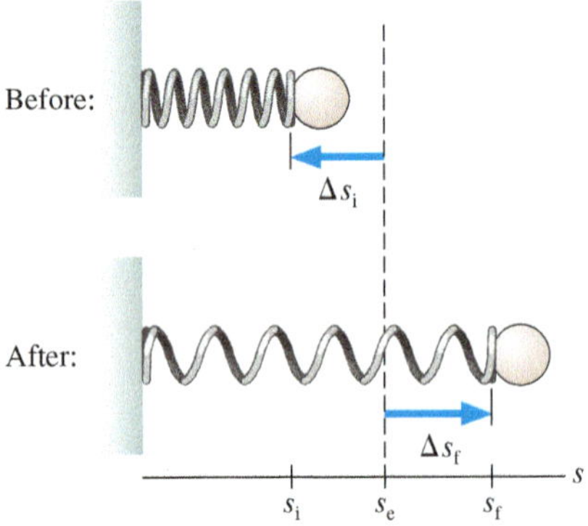

With this change of variables, the integral is

$$-k\int_{s_i}^{s_f}(s - s_e)\,ds = -k\int_{\Delta s_i}^{\Delta s_f} u\,du = -\frac{1}{2}ku^2\Big|_{\Delta s_i}^{\Delta s_f}$$
$$= -\frac{1}{2}k(\Delta s_f)^2 + \frac{1}{2}k(\Delta s_i)^2 \tag{10.32}$$

Using this result makes Equation 10.31 become

$$\frac{1}{2}mv_f^2 - \frac{1}{2}mv_i^2 = -\frac{1}{2}k(\Delta s_f)^2 + \frac{1}{2}k(\Delta s_i)^2 \tag{10.33}$$

which can be rewritten as

$$\frac{1}{2}mv_f^2 + \frac{1}{2}k(\Delta s_f)^2 = \frac{1}{2}mv_i^2 + \frac{1}{2}k(\Delta s_i)^2 \tag{10.34}$$

We've succeeded in our goal of relating before and after. In particular, the quantity

$$\frac{1}{2}mv^2 + \frac{1}{2}k(\Delta s)^2 \tag{10.35}$$

does not change as the spring compresses or expands. You recognize $\frac{1}{2}mv^2$ as the kinetic energy K. Let's define the **elastic potential energy** U_s of a spring to be

$$U_s = \frac{1}{2}k(\Delta s)^2 \quad \text{(elastic potential energy)} \tag{10.36}$$

Then Equation 10.34 tells us that an object moving on a spring obeys

$$K_f + U_{sf} = K_i + U_{si} \tag{10.37}$$

In other words, the mechanical energy $E_{mech} = K + U_s$ is conserved for an object moving *without friction* on an ideal spring.

NOTE ▶ Because Δs is squared, the elastic potential energy is positive for a spring that is either stretched or compressed. U_s is zero when the spring is at its equilibrium length L_0 and $\Delta s = 0$. ◀

EXAMPLE 10.6 **A spring-launched plastic ball**

A spring-loaded toy gun launches a 10 g plastic ball. The spring, with spring constant 10 N/m, is compressed by 10 cm as the ball is pushed into the barrel. When the trigger is pulled, the spring is released and shoots the ball back out. What is the ball's speed as it leaves the barrel? Assume friction is negligible.

MODEL Assume an ideal spring that obeys Hooke's law. Also assume that the gun is held firmly enough to prevent recoil. There's no friction; hence the mechanical energy $K + U_s$ is conserved.

VISUALIZE FIGURE 10.20a shows a before-and-after pictorial representation. We have chosen to put the origin of the coordinate system at the equilibrium position of the free end of the spring. The bar chart of FIGURE 10.20b shows the potential energy stored in the compressed spring being entirely transformed into the kinetic energy of the ball.

SOLVE The energy conservation equation is $K_2 + U_{s2} = K_1 + U_{s1}$. We can use the elastic potential energy of the spring, Equation 10.36, to write this as

$$\frac{1}{2}mv_2^2 + \frac{1}{2}k(x_2 - x_e)^2 = \frac{1}{2}mv_1^2 + \frac{1}{2}k(x_1 - x_e)^2$$

Notice that we used x, rather than the generic s, and that we explicitly wrote out the meaning of Δx_1 and Δx_2. Using $x_2 = x_e = 0$ m and $v_1 = 0$ m/s simplifies this to

$$\frac{1}{2}mv_2^2 = \frac{1}{2}kx_1^2$$

FIGURE 10.20 Pictorial representation and energy bar chart of a ball being shot from a spring-loaded toy gun.

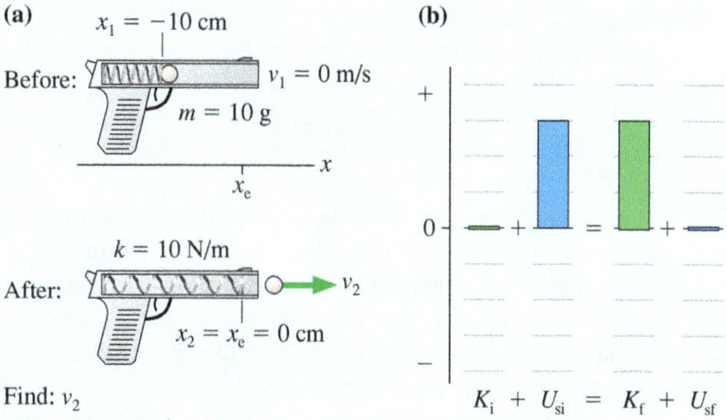

It is now straightforward to solve for the ball's speed:

$$v_2 = \sqrt{\frac{kx_1^2}{m}} = \sqrt{\frac{(10\ \text{N/m})(-0.10\ \text{m})^2}{0.010\ \text{kg}}} = 3.2\ \text{m/s}$$

ASSESS This is a problem that we could *not* have solved with Newton's laws. The acceleration is not constant, and we have not learned how to handle the kinematics of nonconstant acceleration. But with conservation of energy—it's easy! The result, 3.2 m/s, seems reasonable for a toy gun.

EXAMPLE 10.7 **A spring-launched projectile**

Your lab assignment for the week is to devise a method to determine the spring constant of a spring. You notice several small blocks of different mass lying around, so you decide to measure how high the compressed spring will launch each of the blocks. You and your lab partners quickly realize that you need to compress the spring the same amount each time, so that only the mass is varying, and you choose to use a compression of 4.0 cm. Measuring the height from where you place the mass on the compressed spring generates the following data:

Mass (g)	Height (m)
50	2.07
100	1.11
150	0.65
200	0.51

What value will you report for the spring constant?

MODEL Assume an ideal spring that obeys Hooke's law. There's no friction, and we'll assume no drag; hence the mechanical energy $K + U$ is conserved. However, this system has both elastic *and* gravitational potential energy—two distinct ways of storing energy—and we need to include them both. Thus $U = U_g + U_s$.

VISUALIZE FIGURE 10.21 is a before-and-after pictorial representation. We've chosen to place the origin of the coordinate system at the point of launch, so in this problem the equilibrium position of the spring is *not* $y_e = 0$. The projectile reaches height $y_2 = h$, at which point $v_2 = 0$ m/s.

FIGURE 10.21 Pictorial representation of a spring-launched projectile.

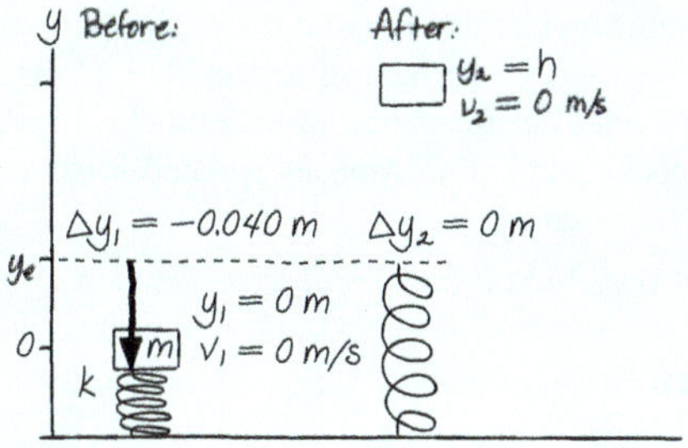

SOLVE Mechanical energy is now $K + U_g + U_s$, so the conservation equation is

$$\frac{1}{2}mv_2^2 + mgy_2 + \frac{1}{2}k(\Delta y_2)^2 = \frac{1}{2}mv_1^2 + mgy_1 + \frac{1}{2}k(\Delta y_1)^2$$

It is important to distinguish between the *position* of the projectile and the *compression* of the spring. While the projectile moves to position y_2, the end of the spring stops at y_e. Thus $\Delta y_2 = 0$, not $\Delta y_2 = y_2$. The initial and final speeds are zero, as is the initial position, so the equation simplifies to

$$mgh = \frac{1}{2}k(\Delta y_1)^2$$

Continued

This equation tells us that the net effect of the launch is to transform the potential energy initially stored in the spring entirely into gravitational potential energy. Kinetic energy is zero at the beginning and again zero at the highest point. The projectile does have kinetic energy as it comes off the spring, but we don't need to know that. Solving for the height, we find

$$h = \frac{k(\Delta y_1)^2}{2mg} = \frac{k(\Delta y_1)^2}{2g} \cdot \frac{1}{m}$$

The first expression for h is correct as an algebraic expression, but here we want to use the result to analyze an experiment in which we measure h as m is varied. By isolating the mass term, we see that plotting h versus $1/m$ (that is, using $1/m$ as the x-variable) should yield a straight line with slope $k(\Delta y_1)^2/2g$. Thus we can use the experimentally determined slope to find k.

FIGURE 10.22 is a graph of h versus $1/m$, with masses first converted to kg. The graph is linear and the best-fit line has a y-intercept very near zero, confirming our analysis of the situation. The experimentally determined slope is 0.105 m kg, with the units determined by rise over run. Thus the experimental value of the spring constant is

$$k = \frac{2g}{(\Delta y_1)^2} \times \text{slope} = 1290 \text{ N/m}$$

FIGURE 10.22 Graph of the projectile height versus the inverse of its mass.

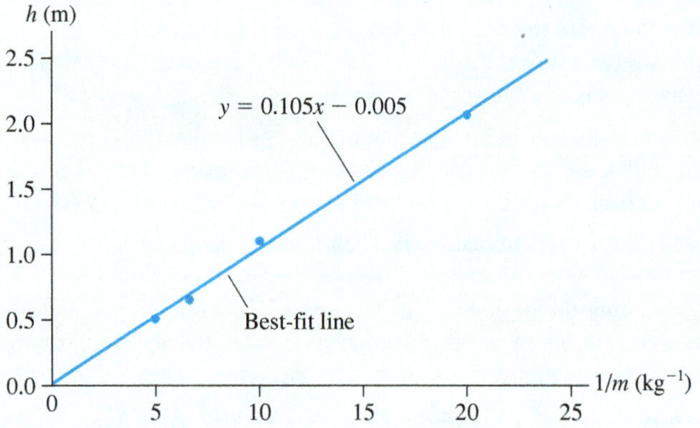

ASSESS A spring with spring constant 1290 N/m has potential energy $U_s = \frac{1}{2}k(\Delta y)^2 = 1.0$ J when compressed 4 cm. A 100 g mass has potential energy $U_g = mgy = 1.1$ J at a height of 1.1 m. That these energies are very nearly equal (perfect equality isn't expected with experimental data) gives us confidence in our value for the spring constant.

EXAMPLE 10.8 Pushing apart

A spring with spring constant 2000 N/m is sandwiched between a 1.0 kg block and a 2.0 kg block on a frictionless table. The blocks are pushed together to compress the spring by 10 cm, then released. What are the velocities of the blocks as they fly apart?

MODEL Assume an ideal spring that obeys Hooke's law. There's no friction; hence the mechanical energy $K + U_s$ is conserved. Here K is the *total* kinetic energy of both blocks. In addition, because the blocks and spring form an isolated system, their total momentum is conserved.

VISUALIZE FIGURE 10.23 is a pictorial representation.

FIGURE 10.23 Pictorial representation of the blocks and spring.

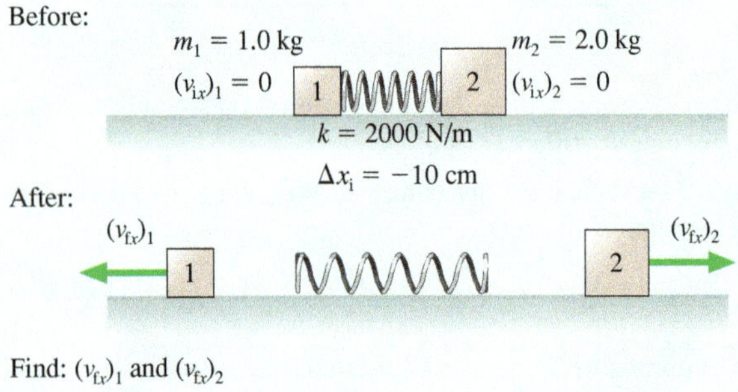

SOLVE The initial energy, with the spring compressed, is entirely potential. The final energy is entirely kinetic. The energy conservation equation $K_f + U_{sf} = K_i + U_{si}$ is

$$\frac{1}{2}m_1(v_f)_1^2 + \frac{1}{2}m_2(v_f)_2^2 + 0 = 0 + 0 + \frac{1}{2}k(\Delta x_i)^2$$

Notice that *both* blocks contribute to the kinetic energy. The energy equation has two unknowns, $(v_f)_1$ and $(v_f)_2$, and one equation is not enough to solve the problem. Fortunately, momentum is also conserved. The initial momentum is zero because both blocks are at rest, so the momentum equation is

$$m_1(v_{fx})_1 + m_2(v_{fx})_2 = 0$$

which can be solved to give

$$(v_{fx})_1 = -\frac{m_2}{m_1}(v_{fx})_2$$

The minus sign indicates that the blocks move in opposite directions. The speed $(v_f)_1 = (m_2/m_1)(v_f)_2$ is all we need to calculate the kinetic energy. Substituting $(v_f)_1$ into the energy equation gives

$$\frac{1}{2}m_1\left(\frac{m_2}{m_1}(v_f)_2\right)^2 + \frac{1}{2}m_2(v_f)_2^2 = \frac{1}{2}k(\Delta x_i)^2$$

which simplifies to

$$m_2\left(1 + \frac{m_2}{m_1}\right)(v_f)_2^2 = k(\Delta x_i)^2$$

Solving for $(v_f)_2$, we find

$$(v_f)_2 = \sqrt{\frac{k(\Delta x_i)^2}{m_2(1 + m_2/m_1)}} = 1.8 \text{ m/s}$$

Finally, we can go back to find

$$(v_{fx})_1 = -\frac{m_2}{m_1}(v_{fx})_2 = -3.6 \text{ m/s}$$

The 2.0 kg block moves to the right at 1.8 m/s while the 1.0 kg block goes left at 3.6 m/s.

ASSESS Speeds of a few m/s seem reasonable.

STOP TO THINK 10.5 A spring-loaded gun shoots a plastic ball with a speed of 4 m/s. If the spring is compressed twice as far, the ball's speed will be

a. 2 m/s.
b. 4 m/s.
c. 8 m/s.
d. 16 m/s.

10.6 Energy Diagrams

Potential energy is an energy of position. The gravitational potential energy depends on the height of an object, and the elastic potential energy depends on a spring's displacement. Other potential energies you will meet in the future will depend in some way on position. Functions of position are easy to represent as graphs. A graph showing a system's potential energy and total energy as a function of position is called an **energy diagram.** Energy diagrams allow you to visualize motion based on energy considerations.

FIGURE 10.24 is the energy diagram of a particle in free fall. The gravitational potential energy $U_g = mgy$ is graphed as a line through the origin with slope mg. The *potential-energy curve* is labeled PE. The line labeled TE is the *total energy line,* $E = K + U_g$. It is horizontal because mechanical energy is conserved, meaning that the object's mechanical energy E has the same value at every position.

Suppose the particle is at position y_1. By definition, the distance from the axis to the potential-energy curve is the particle's potential energy U_{g1} at that position. Because $K_1 = E - U_{g1}$, the distance between the potential-energy curve and the total energy line is the particle's kinetic energy.

The four-frame "movie" of FIGURE 10.25 illustrates how an energy diagram is used to visualize motion. The first frame shows a particle projected upward from $y_a = 0$ with kinetic energy K_a. Initially the energy is entirely kinetic, with $U_{ga} = 0$. A pictorial representation and an energy bar chart help to illustrate what the energy diagram is showing.

FIGURE 10.24 The energy diagram of a particle in free fall.

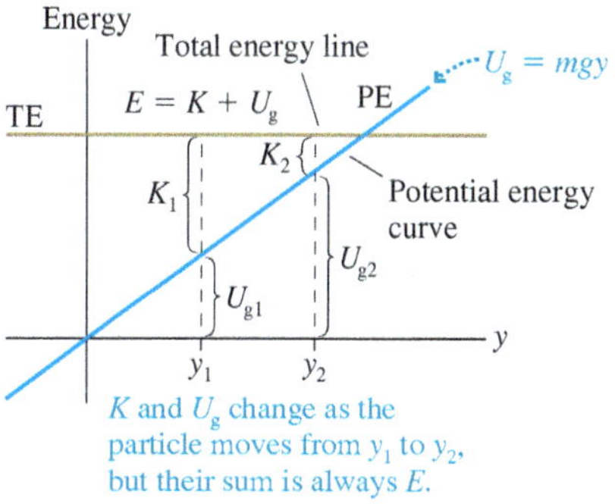

FIGURE 10.25 A four-frame movie of a particle in free fall.

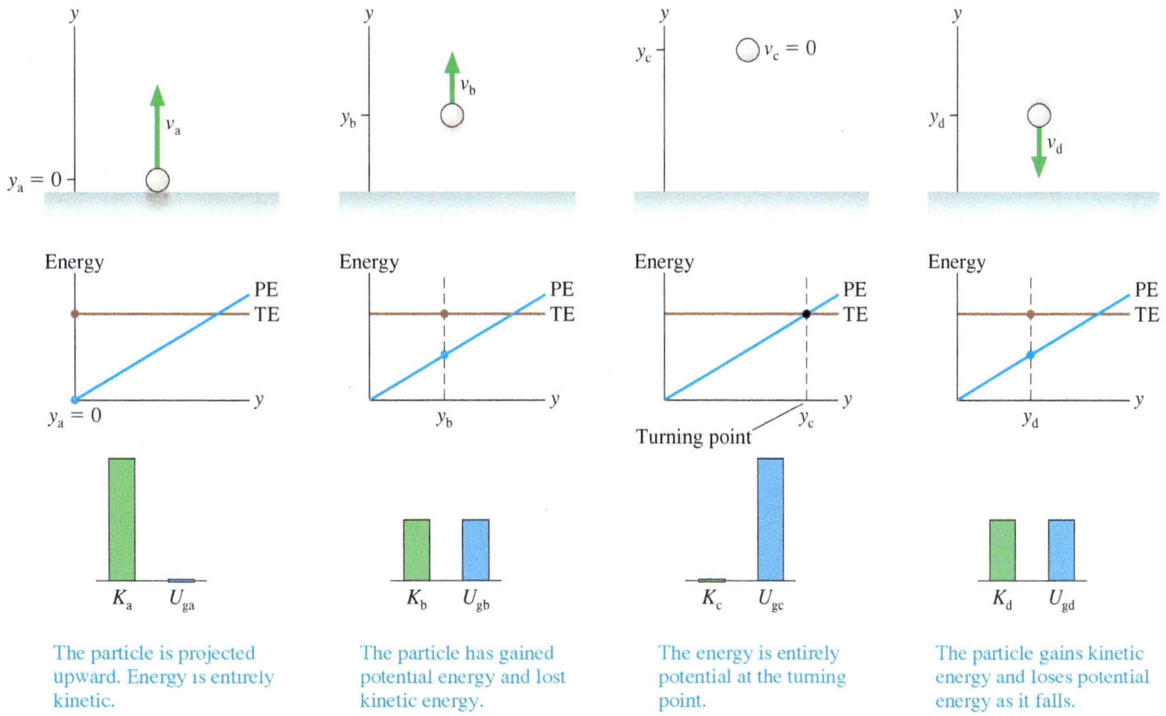

In the second frame, the particle has gained height but lost speed. The potential energy U_{gb} is larger, and the distance K_b between the potential-energy curve and the total energy line is less. The particle continues rising and slowing until, in the third frame, it reaches the y-value where the total energy line crosses the potential-energy curve. This point, where $K = 0$ and the energy is entirely potential, is a *turning point* where the particle reverses direction. Finally, we see the particle speeding up as it falls.

A particle with this amount of total energy would need negative kinetic energy to be to the right of the point, at y_c, where the total energy line crosses the potential-energy curve. Negative K is not physically possible, so **the particle cannot be at positions with $U > E$.** Now, it's certainly true that you could make the particle reach a larger value of y simply by throwing it harder. But that would increase E and move the total energy line higher.

NOTE ▶ The TE line is under your control. You can move the TE line as far up or down as you wish by changing the initial conditions, such as projecting the particle upward with a different speed or dropping it from a different height. Once you've determined the initial conditions, you can use the energy diagram to analyze the motion for that amount of total energy. ◀

FIGURE 10.26 The energy diagram of a mass on a horizontal spring.

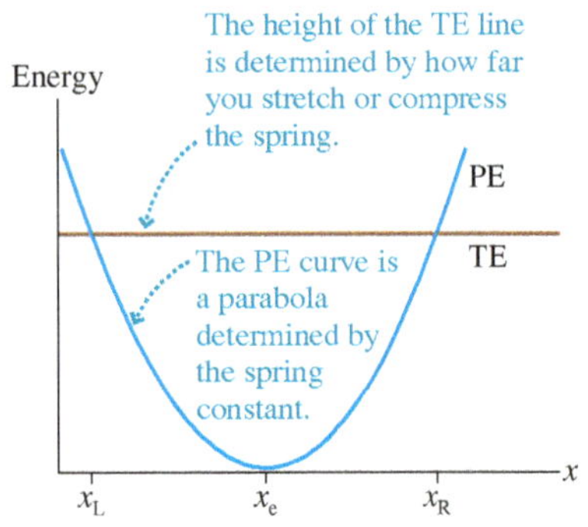

FIGURE 10.26 shows the energy diagram of a mass on a horizontal spring. The potential-energy curve $U_s = \frac{1}{2}k(x - x_e)^2$ is a parabola centered at the equilibrium position x_e. The PE curve is determined by the spring constant; you can't change it. But you can set the TE to any height you wish simply by stretching the spring to the proper length. The figure shows one possible TE line.

Suppose you pull the mass out to position x_R and release it. FIGURE 10.27 is a four-frame movie of the subsequent motion. Initially, the energy is entirely potential. The restoring force of the spring pulls the mass toward x_e, increasing the kinetic energy as the potential energy decreases. The mass has maximum speed at position x_e, where $U_s = 0$, and then it slows down as the spring starts to compress.

If the movie were to continue, you should be able to visualize that position x_L is a turning point. The mass will instantaneously have $v_L = 0$ and $K_L = 0$, then reverse direction as the spring starts to expand. The mass will speed up until x_e, then slow

FIGURE 10.27 A four-frame movie of a mass oscillating on a spring.

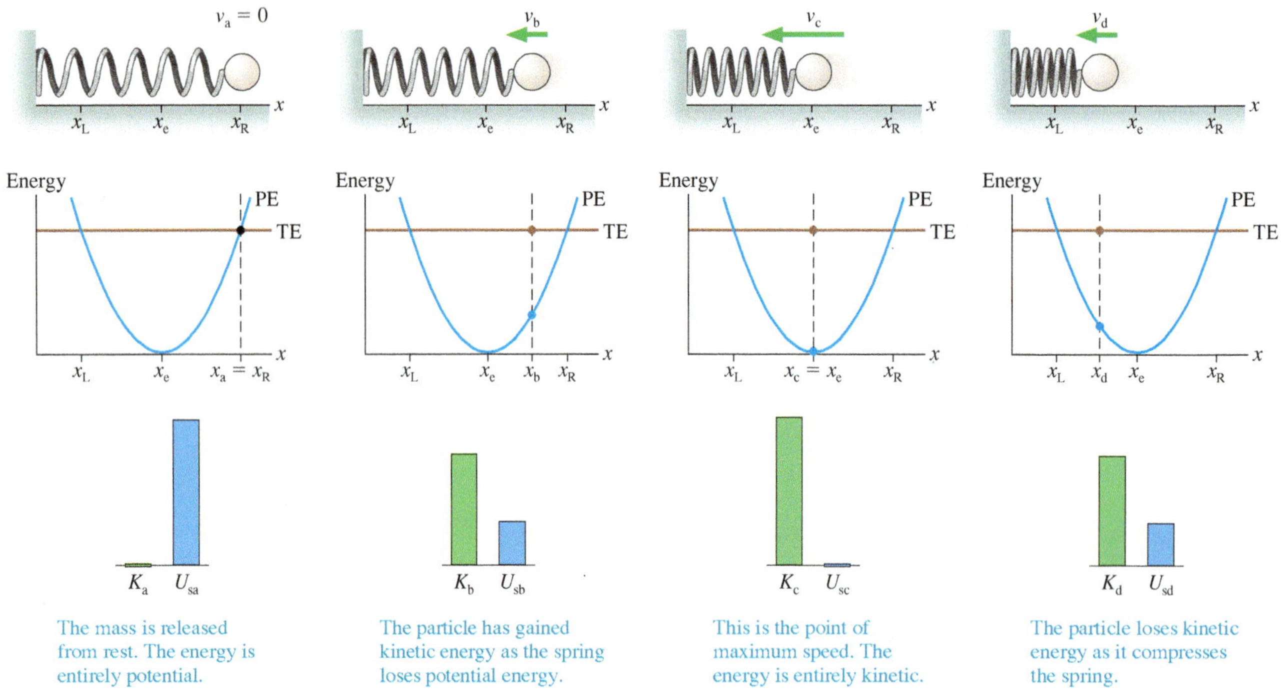

down until reaching x_R, where it started. This is another turning point. It will reverse direction again and start the process over. In other words, the mass will *oscillate* back and forth between the left and right turning points at x_L and x_R where the TE line crosses the PE curve.

FIGURE 10.28 applies these ideas to a more general energy diagram. We don't know how this potential energy was created, but we can visualize the motion of a particle that has this potential energy. Suppose the particle is released from rest at position x_1. How will it then move?

The particle's kinetic energy at x_1 is zero; hence the TE line must cross the PE curve at this point. The particle cannot move to the left because $U > E$, so it begins to move toward the right. The particle speeds up from x_1 to x_2 as U decreases and K increases, then slows down from x_2 to x_3 as it goes up the "potential-energy hill." The particle doesn't stop at x_3 because it still has kinetic energy. It speeds up from x_3 to x_4, reaching its maximum speed at x_4, then slows down between x_4 and x_5. Position x_5 is a turning point, a point where the TE line crosses the PE curve. The particle is instantaneously at rest, then reverses direction. The particle will oscillate back and forth between x_1 and x_5, following the pattern of slowing down and speeding up that we've outlined.

FIGURE 10.28 A more general energy diagram.

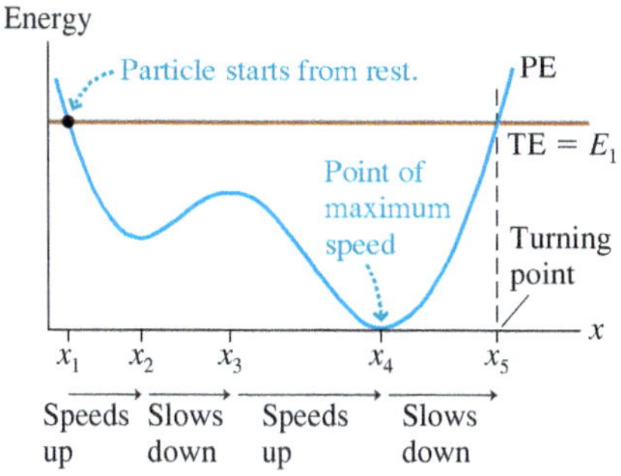

Equilibrium Positions

Positions x_2, x_3, and x_4 in Figure 10.28, where the potential energy has a local minimum or maximum, are special positions. Consider a particle with the total energy E_2 shown in FIGURE 10.29. The particle can be at rest at x_2, with $K = 0$, but it cannot move away from x_2. In other words, a particle with energy E_2 is in *static equilibrium* at x_2. If you disturb the particle, giving it a small kinetic energy and a total energy just *slightly* larger than E_2, the particle will undergo a very small oscillation centered on x_2, like a marble in the bottom of a bowl. An equilibrium for which small disturbances cause small oscillations is called a point of **stable equilibrium.** You should recognize that *any* minimum in the PE curve is a point of stable equilibrium. Position x_4 is also a point of stable equilibrium, in this case for a particle with $E = 0$.

Figure 10.29 also shows a particle with energy E_3 that is tangent to the curve at x_3. If a particle is placed *exactly* at x_3, it will stay there at rest ($K = 0$). But if you disturb the particle at x_3, giving it an energy only slightly more than E_3, it will speed up as it moves away from x_3. This is like trying to balance a marble on top of a hill. The slightest displacement will cause the marble to roll down the hill. A point of equilibrium for which a small disturbance causes the particle to move away is called a point of **unstable equilibrium.** Any maximum in the PE curve, such as x_3, is a point of unstable equilibrium.

We can summarize these lessons as follows:

FIGURE 10.29 Points of stable and unstable equilibrium.

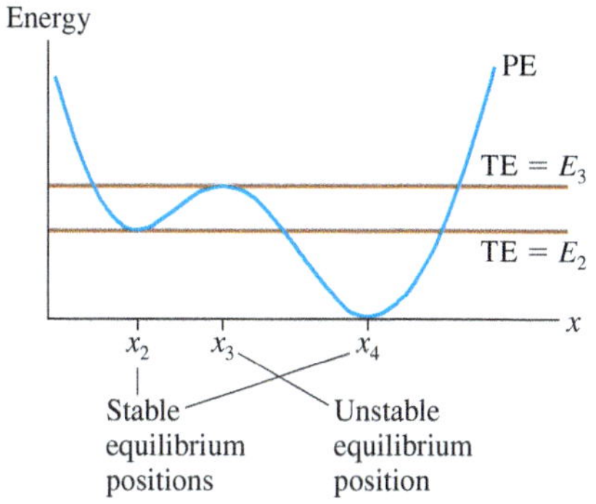

TACTICS BOX 10.1 Interpreting an energy diagram

MP

1. The distance from the axis to the PE curve is the particle's potential energy. The distance from the PE curve to the TE line is its kinetic energy. These are transformed as the position changes, causing the particle to speed up or slow down, but the sum $K + U$ doesn't change.
2. A point where the TE line crosses the PE curve is a turning point. The particle reverses direction.
3. The particle cannot be at a point where the PE curve is above the TE line.
4. The PE curve is determined by the properties of the system—mass, spring constant, and the like. You cannot change the PE curve. However, you can raise or lower the TE line simply by changing the initial conditions to give the particle more or less total energy.
5. A minimum in the PE curve is a point of stable equilibrium. A maximum in the PE curve is a point of unstable equilibrium.

Exercises 18–20

EXAMPLE 10.9 Balancing a mass on a spring

A spring of length L_0 and spring constant k is standing on one end. A block of mass m is placed on the spring, compressing it. What is the length of the compressed spring?

MODEL Assume an ideal spring obeying Hooke's law. The block + spring system has both gravitational potential energy U_g *and* elastic potential energy U_s. The block sitting on top of the spring is at a point of stable equilibrium (small disturbances cause the block to oscillate slightly around the equilibrium position), so we can solve this problem by looking at the energy diagram.

VISUALIZE FIGURE 10.30a is a pictorial representation. We've used a coordinate system with the origin at ground level, so the equilibrium position of the uncompressed spring is $y_e = L_0$.

SOLVE FIGURE 10.30b shows the two potential energies separately and also shows the total potential energy:

$$U_{tot} = U_g + U_s = mgy + \frac{1}{2}k(y - L_0)^2$$

The equilibrium position (the minimum of U_{tot}) has shifted from L_0 to a smaller value of y, closer to the ground. We can find the equilibrium by locating the position of the minimum in the PE curve. You know from calculus that the minimum of a function is at the point where the derivative (or slope) is zero. The derivative of U_{tot} is

$$\frac{dU_{tot}}{dy} = mg + k(y - L_0)$$

The derivative is zero at the point y_{eq}, so we can easily find

$$mg + k(y_{eq} - L_0) = 0$$

$$y_{eq} = L_0 - \frac{mg}{k}$$

The block compresses the spring by the length mg/k from its original length L_0, giving it a new equilibrium length $L_0 - mg/k$.

FIGURE 10.30 The block + spring system has both gravitational and elastic potential energy.

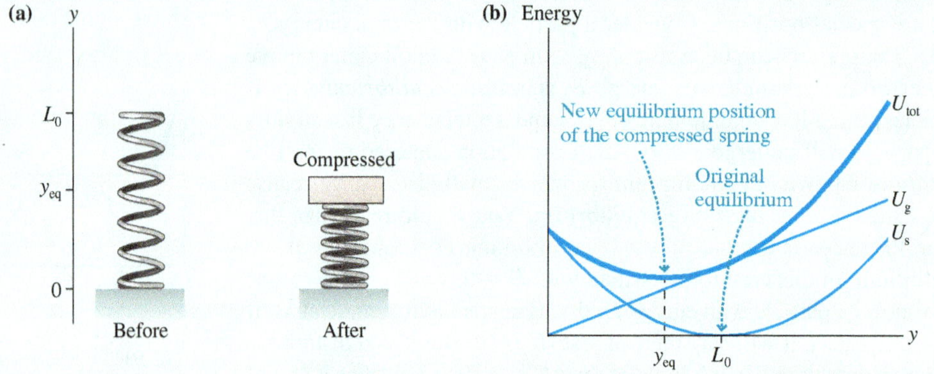

STOP TO THINK 10.6 A particle with the potential energy shown in the graph is moving to the right. It has 1 J of kinetic energy at $x = 1$ m. Where is the particle's turning point?

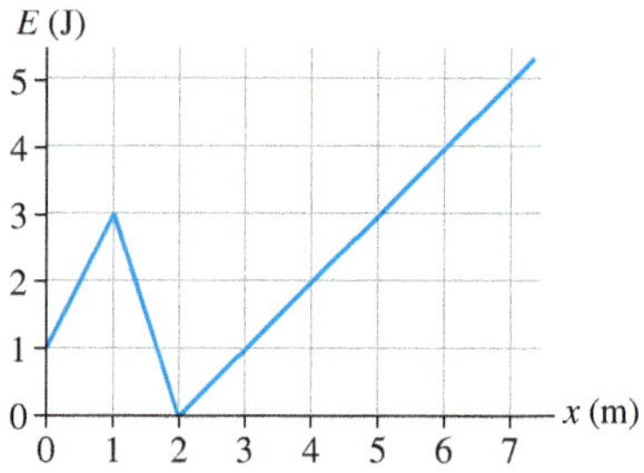

FIGURE 10.31 The energy diagram of the diatomic molecule HCl.

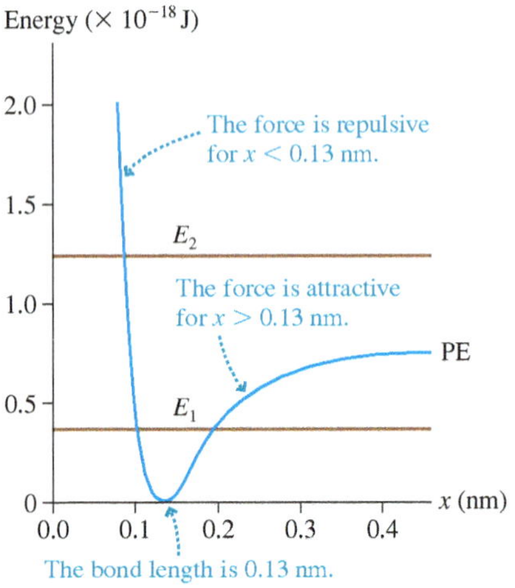

Molecular Bonds

Molecular bonds cause the "springiness" in our atomic models of tension, the normal force, and collisions. A *molecular bond* that holds two atoms together is an electric interaction between the charged electrons and nuclei. FIGURE 10.31 shows the potential-energy diagram for the diatomic molecule HCl (hydrogen chloride) as it has been experimentally determined. Distance x is the *atomic separation,* the distance between the hydrogen and the chlorine atoms. Note the very tiny distances: 1 nm = 10^{-9} m.

Although the potential energy is an electric energy, we can *interpret* the diagram using the steps in Tactics Box 10.1. The molecule has a stable equilibrium at an atomic separation of $x_{eq} = 0.13$ nm. This is the *bond length* of HCl, and you can find this value listed in chemistry books. If we try to push the atoms closer together (smaller x), the potential energy rises very rapidly. Physically, this is the repulsive electric force between the electrons orbiting each atom, preventing the atoms from getting too close.

There is also an attractive force between the atoms, called the *polarization force.* It is similar to the static electricity force by which a comb that has been brushed through your hair attracts small pieces of paper. If you try to pull the atoms apart (larger x), the attractive polarization force resists and is responsible for the increasing potential energy for $x > x_{eq}$. The equilibrium position is where the repulsive force between the electrons and the attractive polarization force are exactly balanced.

The repulsive force keeps getting stronger as you push the atoms together, and thus the potential-energy curve keeps getting steeper on the left. But the attractive polarization force gets *weaker* as the atoms get farther apart. This is why the potential-energy curve becomes *less* steep as the atomic separation increases. This difference between the repulsive and attractive forces leads to an asymmetric curve.

It turns out that, for quantum physics reasons, a molecule cannot have $E = 0$ and thus cannot simply rest at the equilibrium position. By requiring the molecule to have some energy, such as E_1, we see that the atoms oscillate back and forth between two turning points. This is a *molecular vibration,* and atoms held together by molecular bonds are constantly vibrating. For a molecule having an energy $E_1 = 0.35 \times 10^{-18}$ J, as illustrated in Figure 10.31, the bond oscillates in length between roughly 0.10 nm and 0.18 nm.

Suppose we increase the molecule's energy to $E_2 = 1.25 \times 10^{-18}$ J. This could happen if the molecule absorbs some light. You can see from the energy diagram that atoms with this energy are not bound together at large values of x. There is no turning point on the right, so the atoms will keep moving apart. By raising the molecule's energy to E_2 we have *broken the molecular bond.* The breaking of molecular bonds through the absorption of light is called *photodissociation.* It is an important process in making integrated circuits.

10.7 Elastic Collisions

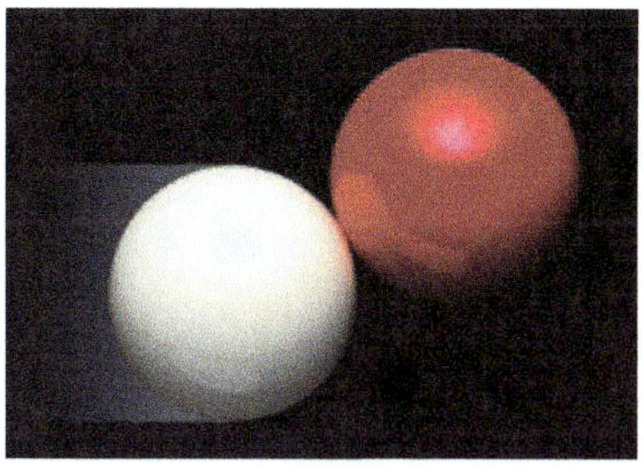

A perfectly elastic collision conserves both momentum and mechanical energy.

Figure 9.1 showed a molecular-level view of a collision. Billions of spring-like molecular bonds are compressed as two objects collide, then the bonds expand and push the objects apart. In the language of energy, the kinetic energy of the objects is transformed into the elastic potential energy of molecular bonds, then back into kinetic energy as the two objects spring apart.

In some cases, such as the inelastic collisions of Chapter 9, some of the mechanical energy is dissipated inside the objects as thermal energy and not all of the kinetic energy is recovered. We're now interested in collisions in which *all* of the kinetic energy is stored as elastic potential energy in the bonds, and then *all* of the stored energy is transformed back into the post-collision kinetic energy of the objects. A collision in which mechanical energy is conserved is called a **perfectly elastic collision.** Collisions between two very hard objects, such as two billiard balls or two steel balls, come close to being perfectly elastic.

FIGURE 10.32 shows a head-on, perfectly elastic collision of a ball of mass m_1, having initial velocity $(v_{ix})_1$, with a ball of mass m_2 that is initially at rest. The balls' velocities after the collision are $(v_{fx})_1$ and $(v_{fx})_2$. These are velocities, not speeds, and have signs. Ball 1, in particular, might bounce backward and have a negative value for $(v_{fx})_1$.

FIGURE 10.32 A perfectly elastic collision.

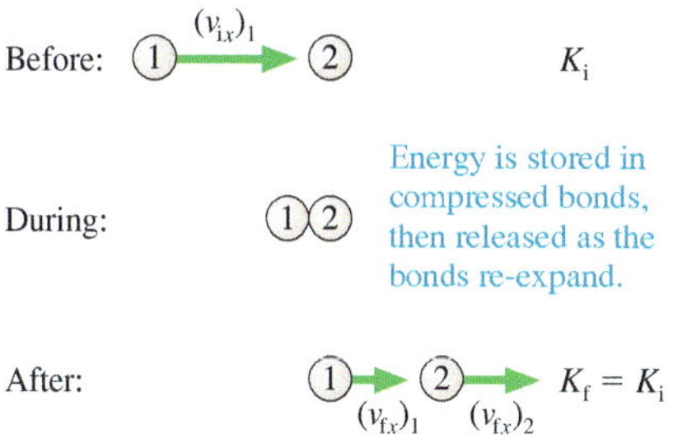

The collision must obey two conservation laws: conservation of momentum (obeyed in any collision) and conservation of mechanical energy (because the collision is perfectly elastic). Although the energy is transformed into potential energy during the collision, the mechanical energy before and after the collision is purely kinetic energy. Thus

$$\text{momentum conservation:} \quad m_1(v_{fx})_1 + m_2(v_{fx})_2 = m_1(v_{ix})_1 \tag{10.38}$$

$$\text{energy conservation:} \quad \frac{1}{2}m_1(v_{fx})_1^2 + \frac{1}{2}m_2(v_{fx})_2^2 = \frac{1}{2}m_1(v_{ix})_1^2 \tag{10.39}$$

Momentum conservation alone is not sufficient to analyze the collision because there are two unknowns: the two final velocities. That is why we did not consider perfectly elastic collisions in Chapter 9. Energy conservation gives us another condition. Isolating $(v_{fx})_1$ in Equation 10.38 gives

$$(v_{fx})_1 = (v_{ix})_1 - \frac{m_2}{m_1}(v_{fx})_2 \tag{10.40}$$

We substitute this into Equation 10.39:

$$\frac{1}{2}m_1\left((v_{ix})_1 - \frac{m_2}{m_1}(v_{fx})_2\right)^2 + \frac{1}{2}m_2(v_{fx})_2^2 = \frac{1}{2}m_1(v_{ix})_1^2$$

With a bit of algebra, this can be rearranged to give

$$(v_{fx})_2\left[\left(1+\frac{m_2}{m_1}\right)(v_{fx})_2 - 2(v_{ix})_1\right] = 0 \tag{10.41}$$

One possible solution to this equation is seen to be $(v_{fx})_2 = 0$. However, this solution is of no interest; it is the case where ball 1 misses ball 2. The other solution is

$$(v_{fx})_2 = \frac{2m_1}{m_1+m_2}(v_{ix})_1$$

which, finally, can be substituted back into Equation 10.40 to yield $(v_{fx})_1$. The complete solution is

$$\begin{aligned}(v_{fx})_1 &= \frac{m_1-m_2}{m_1+m_2}(v_{ix})_1 \\ (v_{fx})_2 &= \frac{2m_1}{m_1+m_2}(v_{ix})_1\end{aligned} \quad \text{(perfectly elastic collision with ball 2 initially at rest)} \tag{10.42}$$

FIGURE 10.33 Three special elastic collisions.

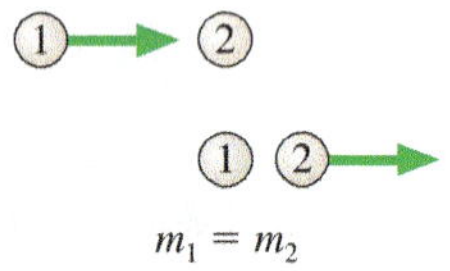

Ball 1 stops. Ball 2 goes forward with $v_{f2} = v_{i1}$.

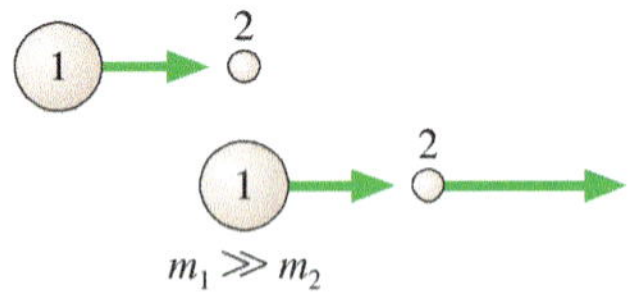

Ball 1 hardly slows down. Ball 2 is knocked forward at $v_{f2} \approx 2v_{i1}$.

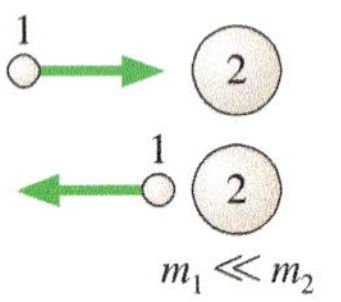

Ball 1 bounces off ball 2 with almost no loss of speed. Ball 2 hardly moves.

Equations 10.42 allow us to compute the final velocity of each ball. These equations are a little difficult to interpret, so let us look at the three special cases shown in FIGURE 10.33.

Case 1: $m_1 = m_2$. This is the case of one billiard ball striking another of equal mass. For this case, Equations 10.42 give

$$\begin{aligned}v_{f1} &= 0 \\ v_{f2} &= v_{i1}\end{aligned}$$

Case 2: $m_1 \gg m_2$. This is the case of a bowling ball running into a Ping-Pong ball. We do not want an exact solution here, but an approximate solution for the limiting case that $m_1 \to \infty$. Equations 10.42 in this limit give

$$\begin{aligned}v_{f1} &\approx v_{i1} \\ v_{f2} &\approx 2v_{i1}\end{aligned}$$

Case 3: $m_1 \ll m_2$. Now we have the reverse case of a Ping-Pong ball colliding with a bowling ball. Here we are interested in the limit $m_1 \to 0$, in which case Equations 10.42 become

$$\begin{aligned}v_{f1} &\approx -v_{i1} \\ v_{f2} &\approx 0\end{aligned}$$

These cases agree well with our expectations and give us confidence that Equations 10.42 accurately describe a perfectly elastic collision.

Using Reference Frames

Equations 10.42 assumed that ball 2 was at rest prior to the collision. Suppose, however, you need to analyze the perfectly elastic collision that is just about to take place in FIGURE 10.34. What are the direction and speed of each ball after the collision? You could solve the simultaneous momentum and energy equations, but the mathematics becomes quite messy when both balls have an initial velocity. Fortunately, there's an easier way.

You already know the answer—Equations 10.42—when ball 2 is initially at rest. And in Chapter 4 you learned the Galilean transformation of velocity. This transformation relates an object's velocity as measured in one reference frame to its velocity in a different reference frame that moves with respect to the first. The Galilean transformation provides an elegant and straightforward way to analyze the collision of Figure 10.34.

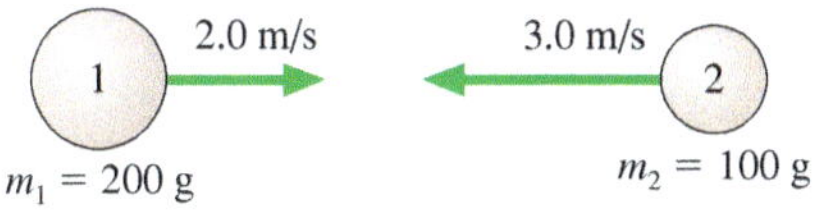

FIGURE 10.34 A perfectly elastic collision in which both balls have an initial velocity.

TACTICS BOX 10.2 Analyzing elastic collisions

❶ Use the Galilean transformation to transform the initial velocities of balls 1 and 2 from the "lab frame" to a reference frame in which ball 2 is at rest.
❷ Use Equations 10.42 to determine the outcome of the collision in the frame where ball 2 is initially at rest.
❸ Transform the final velocities back to the "lab frame."

FIGURE 10.35a shows the situation, just before the collision, in the lab frame L. Ball 1 has initial velocity $(v_{ix})_{1L} = 2.0$ m/s. Recall from Chapter 4 that the subscript notation means "velocity of ball 1 relative to the lab frame L." Because ball 2 is moving to the left, it has $(v_{ix})_{2L} = -3.0$ m/s. We would like to observe the collision from a reference frame in which ball 2 is at rest. That will be true if we choose a moving reference frame M that travels alongside ball 2 with the same velocity: $(v_x)_{ML} = -3.0$ m/s.

FIGURE 10.35 The collision seen in two reference frames: the lab frame L and a moving frame M in which ball 2 is initially at rest.

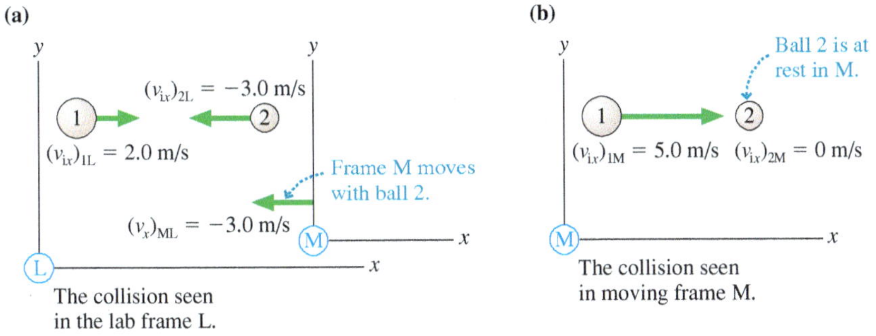

We first need to transform the balls' velocities from the lab frame to the moving reference frame. From Chapter 4, the Galilean transformation of velocity for an object O is

$$(v_x)_{OM} = (v_x)_{OL} + (v_x)_{LM} \tag{10.43}$$

That is, O's velocity in reference frame M is its velocity in reference frame L plus the velocity of frame L relative to frame M. Because reference frame M is moving to the left relative to L with $(v_x)_{ML} = -3.0$ m/s, reference frame L is moving to the right relative to M with $(v_x)_{LM} = +3.0$ m/s. Applying the transformation to the two initial velocities gives

$$\begin{aligned}(v_{ix})_{1M} &= (v_{ix})_{1L} + (v_x)_{LM} = 2.0\text{ m/s} + 3.0\text{ m/s} = 5.0\text{ m/s}\\ (v_{ix})_{2M} &= (v_{ix})_{2L} + (v_x)_{LM} = -3.0\text{ m/s} + 3.0\text{ m/s} = 0\text{ m/s}\end{aligned} \tag{10.44}$$

$(v_{ix})_{2M} = 0$ m/s, as expected, because we chose a moving reference frame in which ball 2 would be at rest.

FIGURE 10.35b now shows a situation—with ball 2 initially at rest—in which we can use Equations 10.42 to find the post-collision velocities in frame M:

$$(v_{fx})_{1M} = \frac{m_1 - m_2}{m_1 + m_2}(v_{ix})_{1M} = 1.7 \text{ m/s}$$
$$(v_{fx})_{2M} = \frac{2m_1}{m_1 + m_2}(v_{ix})_{1M} = 6.7 \text{ m/s} \tag{10.45}$$

Reference frame M hasn't changed—it's still moving to the left in the lab frame at 3.0 m/s—but the collision has changed both balls' velocities in frame M.

To finish, we need to transform the post-collision velocities in frame M back to the lab frame L. We can do so with another application of the Galilean transformation:

$$(v_{fx})_{1L} = (v_{fx})_{1M} + (v_x)_{ML} = 1.7 \text{ m/s} + (-3.0 \text{ m/s}) = -1.3 \text{ m/s}$$
$$(v_{fx})_{2L} = (v_{fx})_{2M} + (v_x)_{ML} = 6.7 \text{ m/s} + (-3.0 \text{ m/s}) = 3.7 \text{ m/s} \tag{10.46}$$

FIGURE 10.36 shows the outcome of the collision in the lab frame. It's not hard to confirm that these final velocities do, indeed, conserve both momentum and energy.

FIGURE 10.36 The post-collision velocities in the lab frame.

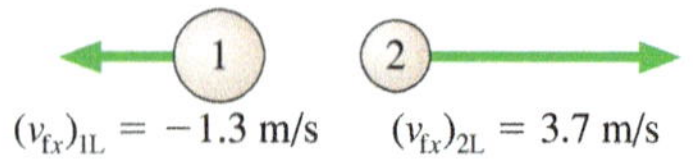

CHALLENGE EXAMPLE 10.10 A rebounding pendulum

A 200 g steel ball hangs on a 1.0-m-long string. The ball is pulled sideways so that the string is at a 45° angle, then released. At the very bottom of its swing the ball strikes a 500 g steel paperweight that is resting on a frictionless table. To what angle does the ball rebound?

MODEL We can divide this problem into three parts. First the ball swings down as a pendulum. Second, the ball and paperweight have a collision. Steel balls bounce off each other very well, so we will assume that the collision is perfectly elastic. Third, the ball, after it bounces off the paperweight, swings back up as a pendulum.

VISUALIZE FIGURE 10.37 shows four distinct moments of time: as the ball is released, an instant before the collision, an instant after the collision but before the ball and paperweight have had time to move, and as the ball reaches its highest point on the rebound. Call the ball A and the paperweight B, so $m_A = 0.20$ kg and $m_B = 0.50$ kg.

FIGURE 10.37 Four moments in the collision of a pendulum with a paperweight.

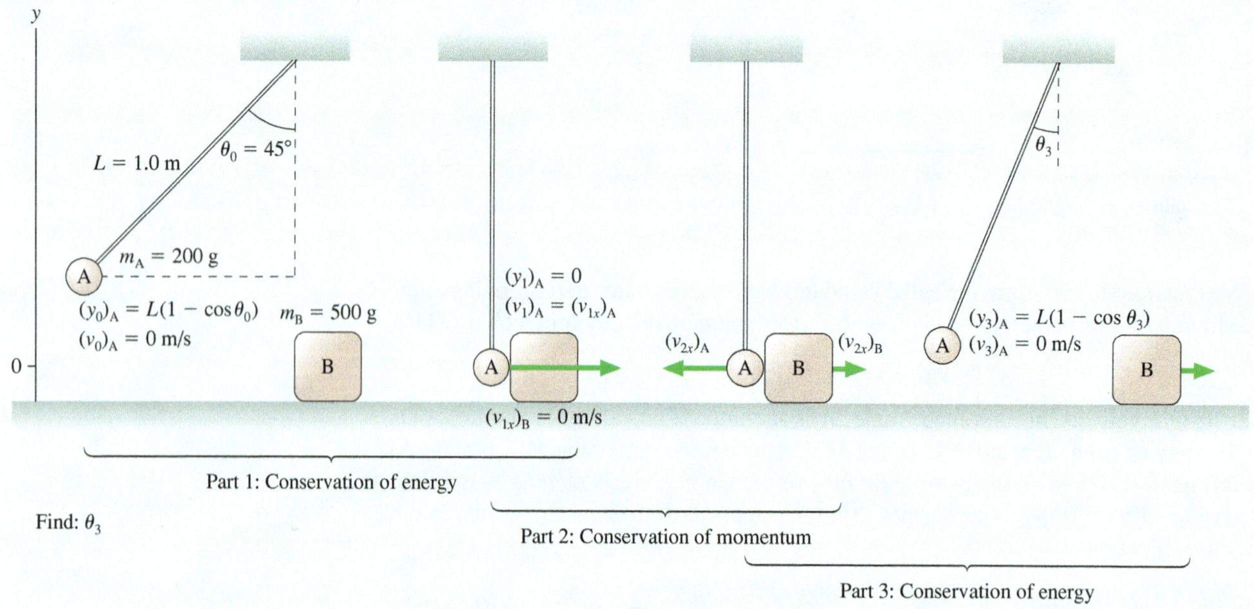

SOLVE Part 1: The first part involves the ball only. Its initial height is

$$(y_0)_A = L - L\cos\theta_0 = L(1 - \cos\theta_0) = 0.293 \text{ m}$$

We can use conservation of mechanical energy to find the ball's velocity at the bottom, just before impact on the paperweight:

$$\frac{1}{2}m_A(v_1)_A^2 + m_A g(y_1)_A = \frac{1}{2}m_A(v_0)_A^2 + m_A g(y_0)_A$$

We know $(v_0)_A = 0$. Solving for the velocity at the bottom, where $(y_1)_A = 0$, gives

$$(v_1)_A = \sqrt{2g(y_0)_A} = 2.40 \text{ m/s}$$

Part 2: The ball and paperweight undergo a perfectly elastic collision in which the paperweight is initially at rest. These are the conditions for which Equations 10.42 were derived. The velocities *immediately* after the collision, prior to any further motion, are

$$(v_{2x})_A = \frac{m_A - m_B}{m_A + m_B}(v_{1x})_A = -1.03 \text{ m/s}$$

$$(v_{2x})_B = \frac{2m_A}{m_A + m_B}(v_{1x})_A = +1.37 \text{ m/s}$$

The ball rebounds toward the left with a speed of 1.03 m/s while the paperweight moves to the right at 1.37 m/s. Kinetic energy has been conserved (you might want to check this), but it is now shared between the ball and the paperweight.

Part 3: Now the ball is a pendulum with an initial speed of 1.03 m/s. Mechanical energy is again conserved, so we can find its maximum height at the point where $(v_3)_A = 0$:

$$\frac{1}{2}m_A(v_3)_A^2 + m_A g(y_3)_A = \frac{1}{2}m_A(v_2)_A^2 + m_A g(y_2)_A$$

Solving for the maximum height gives

$$(y_3)_A = \frac{(v_2)_A^2}{2g} = 0.0541 \text{ m}$$

The height $(y_3)_A$ is related to angle θ_3 by $(y_3)_A = L(1 - \cos\theta_3)$. This can be solved to find the angle of rebound:

$$\theta_3 = \cos^{-1}\left(1 - \frac{(y_3)_A}{L}\right) = 19°$$

The paperweight speeds away at 1.37 m/s and the ball rebounds to an angle of 19°.

ASSESS The ball and the paperweight aren't hugely different in mass, so we expect the ball to transfer a significant fraction of its energy to the paperweight when they collide. Thus a rebound to roughly half the initial angle seems reasonable.

SUMMARY

The goals of Chapter 10 have been to introduce the concept of energy and the basic energy model.

General Principles

Law of Conservation of Mechanical Energy

If a system is isolated and frictionless, then the mechanical energy $E_{mech} = K + U$ of the system is conserved. Thus

$$K_f + U_f = K_i + U_i$$

- K is the sum of the kinetic energies of all particles.
- U is the sum of all potential energies.

Solving Energy Conservation Problems

MODEL Choose an isolated system without friction or other losses of mechanical energy.

VISUALIZE Draw a before-and-after pictorial representation.

SOLVE Use the law of conservation of energy:

$$K_f + U_f = K_i + U_i$$

ASSESS Is the result reasonable?

Important Concepts

Kinetic energy is an energy of motion: $K = \frac{1}{2}mv^2$.

Potential energy is an energy of position.

- **Gravitational:** $U_g = mgy$
- **Elastic:** $U_s = \frac{1}{2}k(\Delta s)^2$

Thermal energy is due to atomic motions. Hotter objects have more thermal energy.

Basic Energy Model

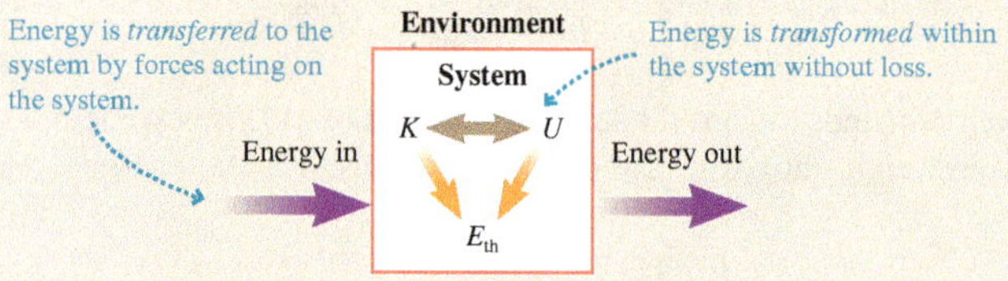

Energy diagrams

These diagrams show the potential-energy curve PE and the total mechanical energy line TE.

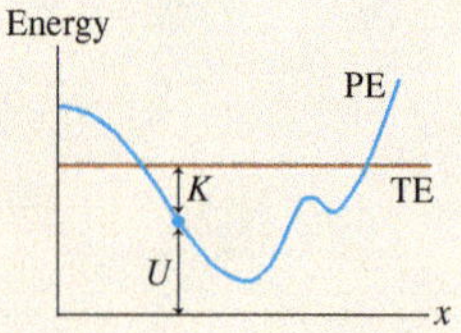

- The distance from the axis to the curve is PE.
- The distance from the curve to the TE line is KE.
- A point where the TE line crosses the PE curve is a **turning point.**
- Minima in the PE curve are points of **stable equilibrium.** Maxima are points of **unstable equilibrium.**
- Regions where PE is greater than TE are forbidden.

Applications

Hooke's law

The restoring force of an ideal spring is

$$(F_{sp})_s = -k\,\Delta s$$

where k is the spring constant and $\Delta s = s - s_e$ is the displacement from equilibrium.

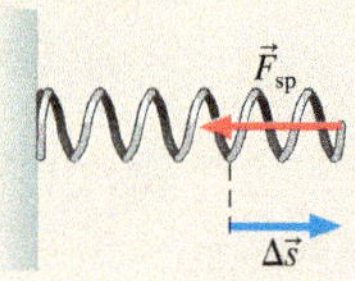

Perfectly elastic collisions

Both mechanical energy and momentum are conserved.

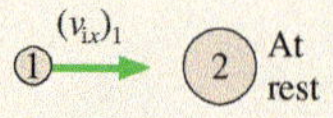

$$(v_{fx})_1 = \frac{m_1 - m_2}{m_1 + m_2}(v_{ix})_1 \qquad (v_{fx})_2 = \frac{2m_1}{m_1 + m_2}(v_{ix})_1$$

If ball 2 is moving, transform to a reference frame in which ball 2 is at rest.

Terms and Notation

energy
basic energy model
kinetic energy, K
gravitational potential energy, U_g
joule, J
mechanical energy
law of conservation of mechanical energy
restoring force
elastic
equilibrium length, L_0
displacement from equilibrium, Δs
spring constant, k
Hooke's law
elastic potential energy, U_s
energy diagram
stable equilibrium
unstable equilibrium
perfectly elastic collision

CONCEPTUAL QUESTIONS

1. Upon what basic quantity does kinetic energy depend? Upon what basic quantity does potential energy depend?
2. Can kinetic energy ever be negative? Can gravitational potential energy ever be negative? For each, give a plausible *reason* for your answer without making use of any equations.
3. If a particle's speed increases by a factor of 3, by what factor does its kinetic energy change?
4. Particle A has half the mass and eight times the kinetic energy of particle B. What is the speed ratio v_A/v_B?
5. A roller-coaster car rolls down a frictionless track, reaching speed v_0 at the bottom. If you want the car to go twice as fast at the bottom, by what factor must you increase the height of the track? Explain.
6. The three balls in FIGURE Q10.6, which have equal masses, are fired with equal speeds from the same height above the ground. Rank in order, from largest to smallest, their speeds v_a, v_b, and v_c as they hit the ground. Explain.

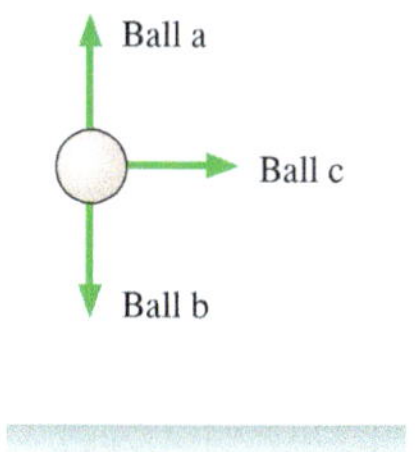

FIGURE Q10.6

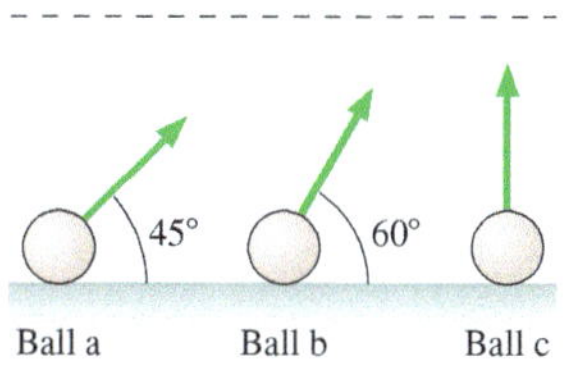

FIGURE Q10.7

7. The three balls in FIGURE Q10.7, which have equal masses, are fired with equal speeds at the angles shown. Rank in order, from largest to smallest, their speeds v_a, v_b, and v_c as they cross the dashed horizontal line. Explain. (All three are fired with sufficient speed to reach the line.)
8. A spring has an unstretched length of 10 cm. It exerts a restoring force F when stretched to a length of 11 cm.
   a. For what length of the spring is its restoring force $3F$?
   b. At what compressed length is the restoring force $2F$?
9. The left end of a spring is attached to a wall. When Bob pulls on the right end with a 200 N force, he stretches the spring by 20 cm. The same spring is then used for a tug-of-war between Bob and Carlos. Each pulls on his end of the spring with a 200 N force. How far does the spring stretch? Explain.
10. Rank in order, from most to least, the elastic potential energy $(U_s)_a$ to $(U_s)_d$ stored in the springs of FIGURE Q10.10. Explain.

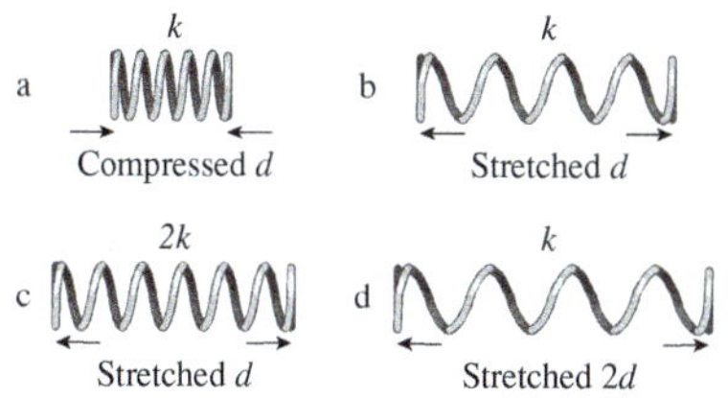

FIGURE Q10.10

11. A spring is compressed 1.0 cm. How far must you compress a spring with twice the spring constant to store the same amount of energy?
12. A spring gun shoots out a plastic ball at speed v_0. The spring is then compressed twice the distance it was on the first shot. By what factor is the ball's speed increased? Explain.
13. A particle with the potential energy shown in FIGURE Q10.13 is moving to the right at $x = 5$ m with total energy E.
   a. At what value or values of x is this particle's speed a maximum?
   b. Does this particle have a turning point or points in the range of x covered by the graph? If so, where?
   c. If E is changed appropriately, could the particle remain at rest at any point or points in the range of x covered by the graph? If so, where?

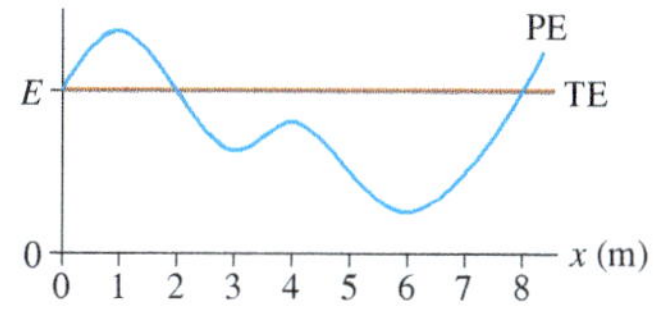

FIGURE Q10.13

14. Two balls of clay of known masses hang from the ceiling on massless strings of equal length. They barely touch when both hang at rest. One ball is pulled back until its string is at 45°, then released. It swings down, collides with the second ball, and they stick together. To determine the angle to which the balls swing on the opposite side, would you invoke (a) conservation of momentum, (b) conservation of mechanical energy, (c) both, (d) either but not both, or (e) these laws alone are not sufficient to find the angle? Explain.

EXERCISES AND PROBLEMS

Problems labeled integrate material from earlier chapters.

Exercises

Section 10.2 Kinetic Energy and Gravitational Potential Energy

1. | Which has the larger kinetic energy, a 10 g bullet fired at 500 m/s or a 75 kg student running at 5.5 m/s?
2. | The lowest point in Death Valley is 85 m below sea level. The summit of nearby Mt. Whitney has an elevation of 4420 m. What is the change in potential energy of an energetic 65 kg hiker who makes it from the floor of Death Valley to the top of Mt. Whitney?
3. | At what speed does a 1000 kg compact car have the same kinetic energy as a 20,000 kg truck going 25 km/h?

4. | a. What is the kinetic energy of a 1500 kg car traveling at a speed of 30 m/s ($\approx$ 65 mph)?
   b. From what height would the car have to be dropped to have this same amount of kinetic energy just before impact?
   c. Does your answer to part b depend on the car's mass?
5. | A boy reaches out of a window and tosses a ball straight up with a speed of 10 m/s. The ball is 20 m above the ground as he releases it. Use energy to find
   a. The ball's maximum height above the ground.
   b. The ball's speed as it passes the window on its way down.
   c. The speed of impact on the ground.
6. | a. With what minimum speed must you toss a 100 g ball straight up to just touch the 10-m-high roof of the gymnasium if you release the ball 1.5 m above the ground? Solve this problem using energy.
   b. With what speed does the ball hit the ground?
7. || A mother has four times the mass of her young son. Both are running with the same kinetic energy. What is the ratio v_{son}/v_{mother} of their speeds?

Section 10.3 A Closer Look at Gravitational Potential Energy

8. | A 55 kg skateboarder wants to just make it to the upper edge of a "quarter pipe," a track that is one-quarter of a circle with a radius of 3.0 m. What speed does he need at the bottom?
9. || What minimum speed does a 100 g puck need to make it to the top of a 3.0-m-long, 20° frictionless ramp?
10. || A pendulum is made by tying a 500 g ball to a 75-cm-long string. The pendulum is pulled 30° to one side, then released.
    a. What is the ball's speed at the lowest point of its trajectory?
    b. To what angle does the pendulum swing on the other side?
11. || A 20 kg child is on a swing that hangs from 3.0-m-long chains. What is her maximum speed if she swings out to a 45° angle?
12. || A 1500 kg car traveling at 10 m/s suddenly runs out of gas while approaching the valley shown in FIGURE EX10.12. The alert driver immediately puts the car in neutral so that it will roll. What will be the car's speed as it coasts into the gas station on the other side of the valley?

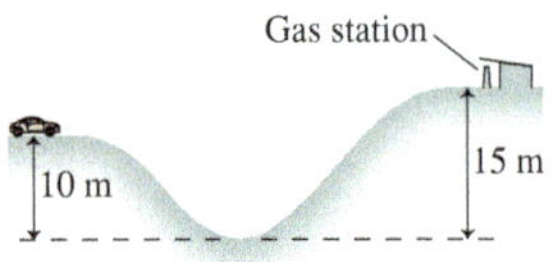

FIGURE EX10.12

Section 10.4 Restoring Forces and Hooke's Law

13. | You need to make a spring scale for measuring mass. You want each 1.0 cm length along the scale to correspond to a mass difference of 100 g. What should be the value of the spring constant?
14. || A 10-cm-long spring is attached to the ceiling. When a 2.0 kg mass is hung from it, the spring stretches to a length of 15 cm.
    a. What is the spring constant k?
    b. How long is the spring when a 3.0 kg mass is suspended from it?
15. || A 60 kg student is standing atop a spring in an elevator as it accelerates upward at 3.0 m/s^2. The spring constant is 2500 N/m. By how much is the spring compressed?
16. || A spring hanging from the ceiling has equilibrium length L_0. Hanging mass m from the spring stretches its length to L_1. Find an expression for the spring's length L_3 when mass $3m$ hangs from it.
17. || A 5.0 kg mass hanging from a spring scale is slowly lowered onto a vertical spring, as shown in FIGURE EX10.17. The scale reads in newtons.
    a. What does the spring scale read just before the mass touches the lower spring?
    b. The scale reads 20 N when the lower spring has been compressed by 2.0 cm. What is the value of the spring constant for the lower spring?
    c. At what compression length will the scale read zero?

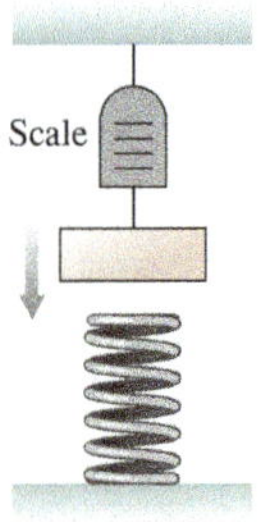

FIGURE EX10.17

Section 10.5 Elastic Potential Energy

18. | How far must you stretch a spring with $k = 1000$ N/m to store 200 J of energy?
19. | A stretched spring stores 2.0 J of energy. How much energy will be stored if the spring is stretched three times as far?
20. || A student places her 500 g physics book on a frictionless table. She pushes the book against a spring, compressing the spring by 4.0 cm, then releases the book. What is the book's speed as it slides away? The spring constant is 1250 N/m.
21. | A block sliding along a horizontal frictionless surface with speed v collides with a spring and compresses it by 2.0 cm. What will be the compression if the same block collides with the spring at a speed of $2v$?
22. || A 10 kg runaway grocery cart runs into a spring with spring constant 250 N/m and compresses it by 60 cm. What was the speed of the cart just before it hit the spring?
23. | The desperate contestants on a TV survival show are very hungry. The only food they can see is some fruit hanging on a branch high in a tree. Fortunately, they have a spring they can use to launch a rock. The spring constant is 1000 N/m, and they can compress the spring a maximum of 30 cm. All the rocks on the island seem to have a mass of 400 g.
    a. With what speed does the rock leave the spring?
    b. If the fruit hangs 15 m above the ground, will they feast or go hungry?
24. || As a 15,000 kg jet plane lands on an aircraft carrier, its tail hook snags a cable to slow it down. The cable is attached to a spring with spring constant 60,000 N/m. If the spring stretches 30 m to stop the plane, what was the plane's landing speed?

Section 10.6 Energy Diagrams

25. | FIGURE EX10.25 is the potential-energy diagram for a 20 g particle that is released from rest at $x = 1.0$ m.
    a. Will the particle move to the right or to the left? How can you tell?
    b. What is the particle's maximum speed? At what position does it have this speed?
    c. Where are the turning points of the motion?

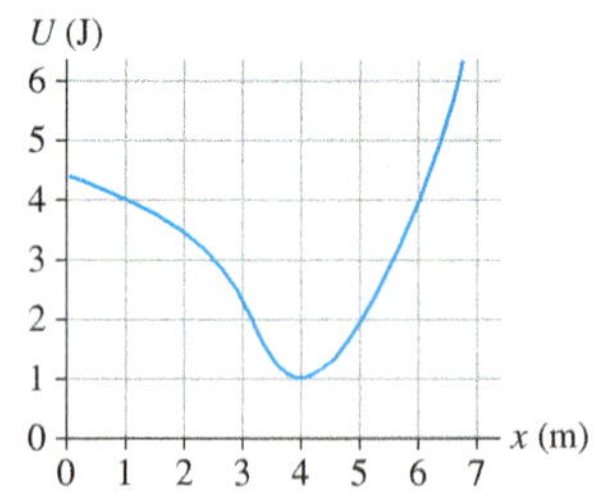

FIGURE EX10.25

26. || FIGURE EX10.26 is the potential-energy diagram for a 500 g particle that is released from rest at A. What are the particle's speeds at B, C, and D?

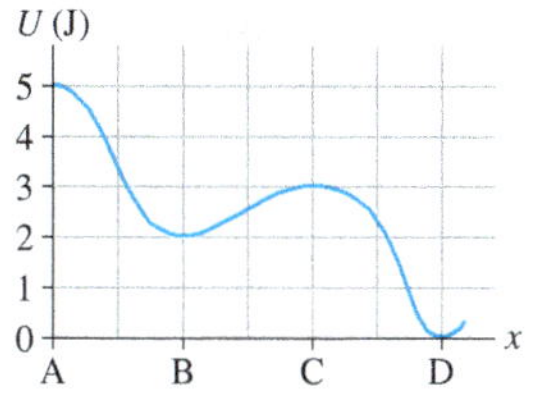

FIGURE EX10.26

27. | a. In FIGURE EX10.27, what minimum speed does a 100 g particle need at point A to reach point B?
b. What minimum speed does a 100 g particle need at point B to reach point A?

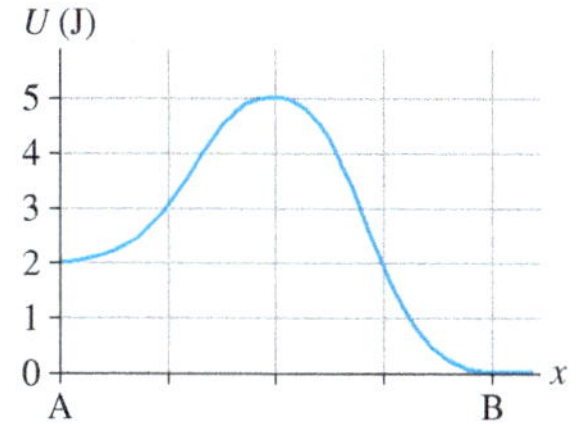

FIGURE EX10.27

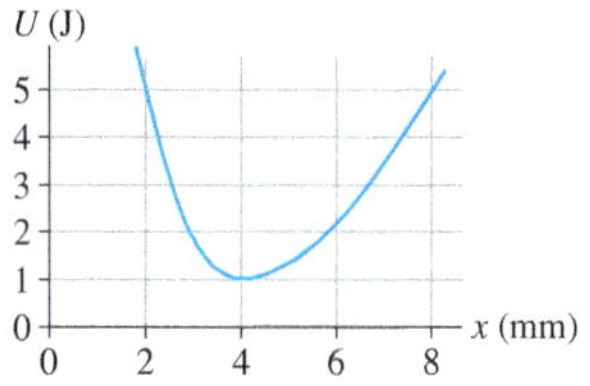

FIGURE EX10.28

28. || In FIGURE EX10.28, what is the maximum speed of a 2.0 g particle that oscillates between $x = 2.0$ mm and $x = 8.0$ mm?

Section 10.7 Elastic Collisions

29. | A 50 g marble moving at 2.0 m/s strikes a 20 g marble at rest. What is the speed of each marble immediately after the collision?

30. | A proton is traveling to the right at 2.0×10^7 m/s. It has a head-on perfectly elastic collision with a carbon atom. The mass of the carbon atom is 12 times the mass of the proton. What are the speed and direction of each after the collision?

31. | Ball 1, with a mass of 100 g and traveling at 10 m/s, collides head-on with ball 2, which has a mass of 300 g and is initially at rest. What is the final velocity of each ball if the collision is (a) perfectly elastic? (b) perfectly inelastic?

32. || A 50 g ball of clay traveling at speed v_0 hits and sticks to a 1.0 kg brick sitting at rest on a frictionless surface.
a. What is the speed of the brick after the collision?
b. What percentage of the mechanical energy is lost in this collision?

Problems

33. | BIO The maximum energy a bone can absorb without breaking is surprisingly small. Experimental data show that the leg bones of a healthy, 60 kg human can absorb about 200 J.
a. From what maximum height could a 60 kg person jump and land rigidly upright on both feet without breaking his legs? Assume that all energy is absorbed by the leg bones in a rigid landing.
b. People jump safely from much greater heights than this. Explain how this is possible.

34. | You're driving at 35 km/h when the road suddenly descends 15 m into a valley. You take your foot off the accelerator and coast down the hill. Just as you reach the bottom you see the policeman hiding behind the speed limit sign that reads "70 km/h." Are you going to get a speeding ticket?

35. || A cannon tilted up at a 30° angle fires a cannon ball at 80 m/s from atop a 10-m-high fortress wall. What is the ball's impact speed on the ground below?

36. || You have a ball of unknown mass, a spring with spring constant 950 N/m, and a meter stick. You use various compressions of the spring to launch the ball vertically, then use the meter stick to measure the ball's maximum height above the launch point. Your data are as follows:

Compression (cm)	Height (cm)
2.0	32
3.0	65
4.0	115
5.0	189

Use an appropriate graph of the data to determine the ball's mass.

37. || A very slippery ice cube slides in a *vertical* plane around the inside of a smooth, 20-cm-diameter horizontal pipe. The ice cube's speed at the bottom of the circle is 3.0 m/s.
a. What is the ice cube's speed at the top?
b. Find an algebraic expression for the ice cube's speed when it is at angle θ, where the angle is measured counterclockwise from the bottom of the circle. Your expression should give 3.0 m/s for $\theta = 0°$ and your answer to part a for $\theta = 180°$.

38. | A 50 g rock is placed in a slingshot and the rubber band is stretched. The force of the rubber band on the rock is shown by the graph in FIGURE P10.38.
a. Is the rubber band stretched to the right or to the left? How can you tell?
b. Does this rubber band obey Hooke's law? Explain.
c. What is the rubber band's spring constant k?
d. The rubber band is stretched 30 cm and then released. What is the speed of the rock?

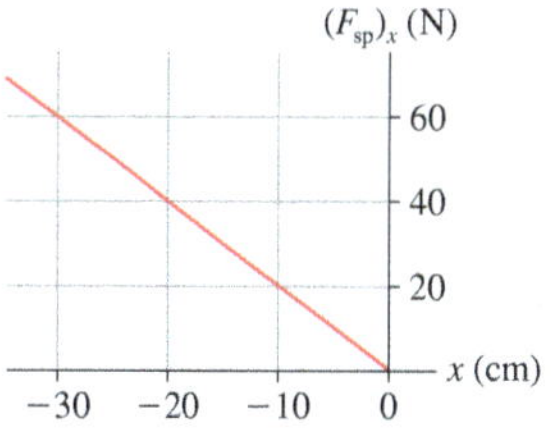

FIGURE P10.38

39. | BIO The elastic energy stored in your tendons can contribute up to 35% of your energy needs when running. Sports scientists find that (on average) the knee extensor tendons in sprinters stretch 41 mm while those of nonathletes stretch only 33 mm. The spring constant of the tendon is the same for both groups, 33 N/mm. What is the difference in maximum stored energy between the sprinters and the nonathletes?

40. || The spring in FIGURE P10.40a is compressed by Δx. It launches the block across a frictionless surface with speed v_0. The two springs in FIGURE P10.40b are identical to the spring of Figure P10.40a. They are compressed by the same Δx and used to launch the same block. What is the block's speed now?

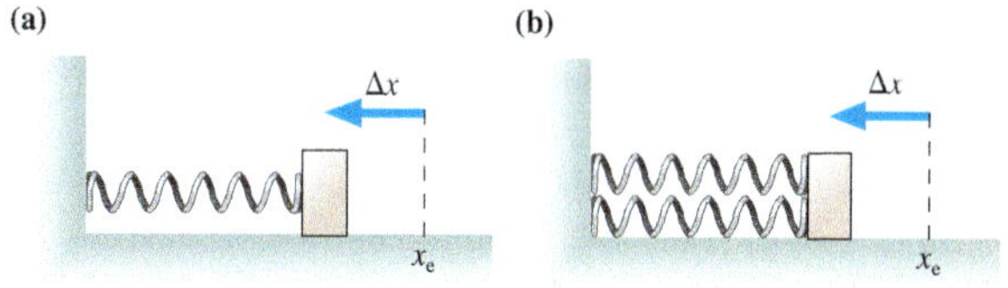

FIGURE P10.40

41. ||| The spring in FIGURE P10.41a is compressed by Δx. It launches the block across a frictionless surface with speed v_0. The two springs in FIGURE P10.41b are identical to the spring of Figure P10.41a. They are compressed the same *total* Δx and used to launch the same block. What is the block's speed now?

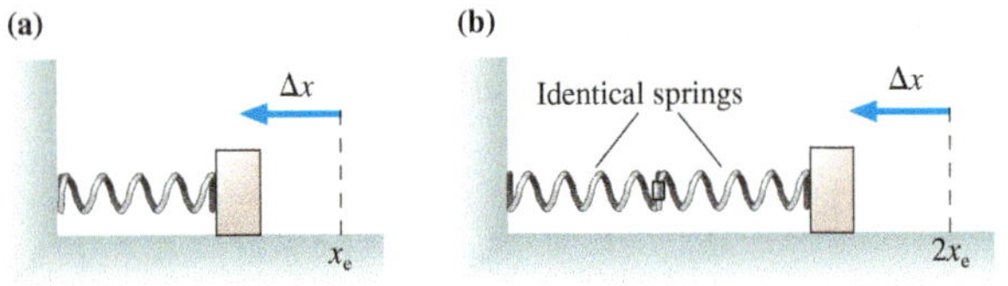

FIGURE P10.41

42. || a. A block of mass m can slide up and down a frictionless slope tilted at angle θ. The block is pressed against a spring at the bottom of the slope, compressing the spring (with spring constant k) by Δx, then released. Find an expression for the block's maximum height h above its starting point.
    b. A 50 g ice cube can slide up and down a frictionless 30° slope. At the bottom, a spring with spring constant 25 N/m is compressed 10 cm and used to launch the ice cube up the slope. How high does it go?

43. || A package of mass m is released from rest at a warehouse loading dock and slides down the 3.0-m-high, frictionless chute of FIGURE P10.43 to a waiting truck. Unfortunately, the truck driver went on a break without having removed the previous package, of mass $2m$, from the bottom of the chute.
    a. Suppose the packages stick together. What is their common speed after the collision?
    b. Suppose the collision between the packages is perfectly elastic. To what height does the package of mass m rebound?

FIGURE P10.43

44. ||| A 100 g granite cube slides down a 40° frictionless ramp. At the bottom, just as it exits onto a horizontal table, it collides with a 200 g steel cube at rest. How high above the table should the granite cube be released to give the steel cube a speed of 150 cm/s?

45. || A 1000 kg safe is 2.0 m above a heavy-duty spring when the rope holding the safe breaks. The safe hits the spring and compresses it 50 cm. What is the spring constant of the spring?

46. || A vertical spring with $k = 490$ N/m is standing on the ground. You are holding a 5.0 kg block just above the spring, not quite touching it.
    a. How far does the spring compress if you let go of the block suddenly?
    b. How far does the spring compress if you slowly lower the block to the point where you can remove your hand without disturbing it?
    c. Why are your two answers different?

47. || You have been hired to design a spring-launched roller coaster that will carry two passengers per car. The car goes up a 10-m-high hill, then descends 15 m to the track's lowest point. You've determined that the spring can be compressed a maximum of 2.0 m and that a loaded car will have a maximum mass of 400 kg. For safety reasons, the spring constant should be 10% larger than the minimum needed for the car to just make it over the top.
    a. What spring constant should you specify?
    b. What is the maximum speed of a 350 kg car if the spring is compressed the full amount?

48. || It's been a great day of new, frictionless snow. Julie starts at the top of the 60° slope shown in FIGURE P10.48. At the bottom, a circular arc carries her through a 90° turn, and she then launches off a 3.0-m-high ramp. How far horizontally is her touchdown point from the end of the ramp?

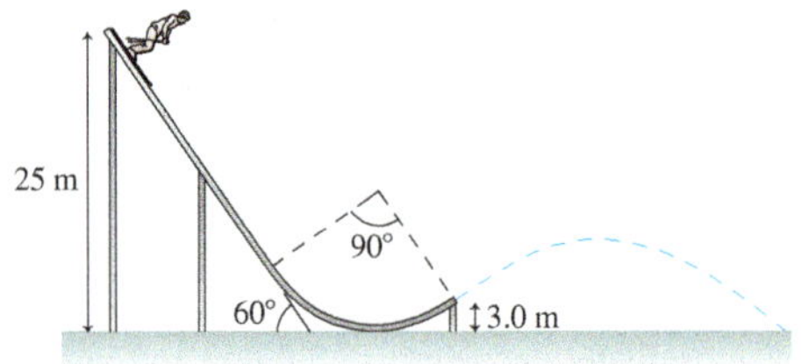

FIGURE P10.48

49. || A 100 g block on a frictionless table is firmly attached to one end of a spring with $k = 20$ N/m. The other end of the spring is anchored to the wall. A 20 g ball is thrown horizontally toward the block with a speed of 5.0 m/s.
    a. If the collision is perfectly elastic, what is the ball's speed immediately after the collision?
    b. What is the maximum compression of the spring?
    c. Repeat parts a and b for the case of a perfectly inelastic collision.

50. || You have been asked to design a "ballistic spring system" to measure the speed of bullets. A bullet of mass m is fired into a block of mass M. The block, with the embedded bullet, then slides across a frictionless table and collides with a horizontal spring whose spring constant is k. The opposite end of the spring is anchored to a wall. The spring's maximum compression d is measured.
    a. Find an expression for the bullet's speed v_B in terms of m, M, k, and d.
    b. What was the speed of a 5.0 g bullet if the block's mass is 2.0 kg and if the spring, with $k = 50$ N/m, was compressed by 10 cm?
    c. What fraction of the bullet's energy is "lost"? Where did it go?

51. ||| You have been asked to design a "ballistic spring system" to measure the speed of bullets. A spring whose spring constant is k is suspended from the ceiling. A block of mass M hangs from the spring. A bullet of mass m is fired vertically upward into the bottom of the block and stops in the block. The spring's maximum compression d is measured.
    a. Find an expression for the bullet's speed v_B in terms of m, M, k, and d.
    b. What was the speed of a 10 g bullet if the block's mass is 2.0 kg and if the spring, with $k = 50$ N/m, was compressed by 45 cm?

52. || In FIGURE P10.52, a block of mass m slides along a frictionless track with speed v_m. It collides with a stationary block of mass M. Find an expression for the minimum value of v_m that will allow the second block to circle the loop-the-loop without falling off if the collision is (a) perfectly inelastic or (b) perfectly elastic.

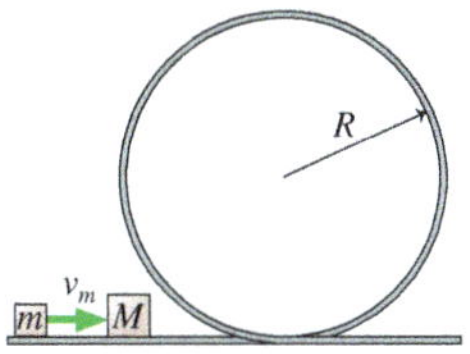

FIGURE P10.52

53. || A block of mass m slides down a frictionless track, then around the inside of a circular loop-the-loop of radius R. From what minimum height h must the block start to make it around without falling off? Give your answer as a multiple of R.

54. || A new event has been proposed for the Winter Olympics. As seen in FIGURE P10.54, an athlete will sprint 100 m, starting from rest, then leap onto a 20 kg bobsled. The person and bobsled will then slide down a 50-m-long ice-covered ramp, sloped at 20°, and into a spring with a carefully calibrated spring constant of 2000 N/m. The athlete who compresses the spring the farthest wins the gold medal. Lisa, whose mass is 40 kg, has been training for this event. She can reach a maximum speed of 12 m/s in the 100 m dash.

a. How far will Lisa compress the spring?
b. The Olympic committee has very exact specifications about the shape and angle of the ramp. Is this necessary? What factors about the ramp are important?

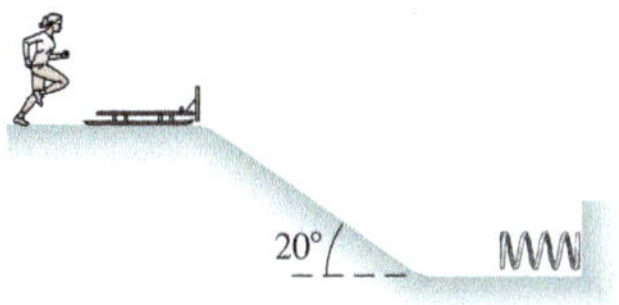

FIGURE P10.54

55. ||| A 20 g ball is fired horizontally with speed v_0 toward a 100 g ball hanging motionless from a 1.0-m-long string. The balls undergo a head-on, perfectly elastic collision, after which the 100 g ball swings out to a maximum angle $\theta_{max} = 50°$. What was v_0?

56. || A 100 g ball moving to the right at 4.0 m/s collides head-on with a 200 g ball that is moving to the left at 3.0 m/s.

a. If the collision is perfectly elastic, what are the speed and direction of each ball after the collision?
b. If the collision is perfectly inelastic, what are the speed and direction of the combined balls after the collision?

57. || A 100 g ball moving to the right at 4.0 m/s catches up and collides with a 400 g ball that is moving to the right at 1.0 m/s. If the collision is perfectly elastic, what are the speed and direction of each ball after the collision?

58. || FIGURE P10.58 shows the potential energy of a 500 g particle as it moves along the x-axis. Suppose the particle's mechanical energy is 12 J.

a. Where are the particle's turning points?
b. What is the particle's speed when it is at $x = 6.0$ m?
c. What is the particle's maximum speed? At what position or positions does this occur?
d. Write a description of the motion of the particle as it moves from the left turning point to the right turning point.
e. Suppose the particle's energy is lowered to 4.0 J. Describe the possible motions.

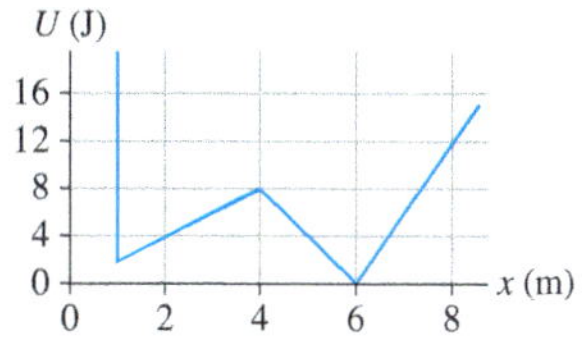

FIGURE P10.58

59. || A particle has potential energy

$$U(x) = x + \sin((2\ \text{rad/m})x)$$

over the range $0\ \text{m} \le x \le \pi$ m.

a. Where are the equilibrium positions in this range?
b. For each, is it a point of stable or unstable equilibrium?

60. ||| Protons and neutrons (together called *nucleons*) are held together in the nucleus of an atom by a force called the *strong force.* At very small separations, the strong force between two nucleons is larger than the repulsive electrical force between two protons—hence its name. But the strong force quickly weakens as the distance between the protons increases. A well-established model for the potential energy of two nucleons interacting via the strong force is

$$U = U_0[1 - e^{-x/x_0}]$$

where x is the distance between the centers of the two nucleons, x_0 is a constant having the value $x_0 = 2.0 \times 10^{-15}$ m, and $U_0 = 6.0 \times 10^{-11}$ J.

a. Calculate and draw an accurate potential-energy curve from $x = 0$ m to $x = 10 \times 10^{-15}$ m. Either use your calculator to compute the value at several points or use computer software.
b. Quantum effects are essential for a proper understanding of how nucleons behave. Nonetheless, let us innocently consider two neutrons *as if* they were small, hard, electrically neutral spheres of mass 1.67×10^{-27} kg and diameter 1.0×10^{-15} m. (We will consider neutrons rather than protons so as to avoid complications from the electric forces between protons.) You are going to hold two neutrons 5.0×10^{-15} m apart, measured between their centers, then release them. Draw the total energy line for this situation on your diagram of part a.
c. What is the speed of each neutron as they crash together? Keep in mind that *both* neutrons are moving.

61. ||| A 50 g air-track glider is repelled by a post fixed at one end of the track. It is hypothesized that the glider's potential energy is $U = c/x$, where x is the distance from the post and c is an unknown constant. To test this hypothesis, you launch the glider with the same speed at various distances from the post and then use a motion detector to measure its speed when it is 1.0 m from the post. Your data are as follows:

Initial distance (cm)	Speed at 1.0 m (m/s)
2.0	1.40
4.0	0.98
6.0	0.79
8.0	0.68

a. Do the data support the hypothesis? To find out, you'll need to compare the shape of an appropriate graph to a theoretical prediction.
b. Find an experimental value for c. Don't forget to determine the appropriate units.
Hint: Both the slope *and* the y-intercept of the graph are important.

62. Write a realistic problem for which the energy bar chart shown in FIGURE P10.62 correctly shows the energy at the beginning and end of the problem.

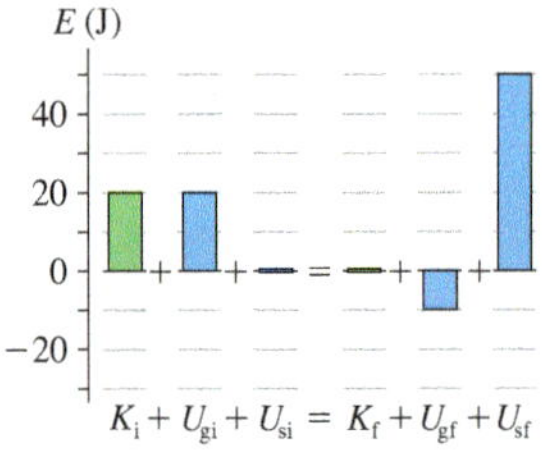

FIGURE P10.62

In Problems 63 through 66 you are given the equation used to solve a problem. For each of these, you are to

a. Write a realistic problem for which this is the correct equation.
b. Draw the before-and-after pictorial representation.
c. Finish the solution of the problem.

63. $\frac{1}{2}(1500\text{ kg})(5.0\text{ m/s})^2 + (1500\text{ kg})(9.80\text{ m/s}^2)(10\text{ m})$

$= \frac{1}{2}(1500\text{ kg})(v_i)^2 + (1500\text{ kg})(9.80\text{ m/s}^2)(0\text{ m})$

64. $\frac{1}{2}(0.20\text{ kg})(2.0\text{ m/s})^2 + \frac{1}{2}k(0\text{ m})^2$

$= \frac{1}{2}(0.20\text{ kg})(0\text{ m/s})^2 + \frac{1}{2}k(-0.15\text{ m})^2$

65. $(0.10\text{ kg} + 0.20\text{ kg})v_{1x} = (0.10\text{ kg})(3.0\text{ m/s})$

$\frac{1}{2}(0.30\text{ kg})(0\text{ m/s})^2 + \frac{1}{2}(3.0\text{ N/m})(\Delta x_2)^2$

$= \frac{1}{2}(0.30\text{ kg})(v_{1x})^2 + \frac{1}{2}(3.0\text{ N/m})(0\text{ m})^2$

66. $\frac{1}{2}(0.50\text{ kg})(v_f)^2 + (0.50\text{ kg})(9.80\text{ m/s}^2)(0\text{ m})$

$+ \frac{1}{2}(400\text{ N/m})(0\text{ m})^2 = \frac{1}{2}(0.50\text{ kg})(0\text{ m/s})^2$

$+ (0.50\text{ kg})(9.80\text{ m/s}^2)((-0.10\text{ m})\sin 30°)$

$+ \frac{1}{2}(400\text{ N/m})(-0.10\text{ m})^2$

Challenge Problems

67. A massless pan hangs from a spring that is suspended from the ceiling. When empty, the pan is 50 cm below the ceiling. If a 100 g clay ball is placed gently on the pan, the pan hangs 60 cm below the ceiling. Suppose the clay ball is dropped from the ceiling onto an empty pan. What is the pan's distance from the ceiling when the spring reaches its maximum length?

68. A pendulum is formed from a small ball of mass m on a string of length L. As FIGURE CP10.68 shows, a peg is height $h = L/3$ above the pendulum's lowest point. From what minimum angle θ must the pendulum be released in order for the ball to go over the top of the peg without the string going slack?

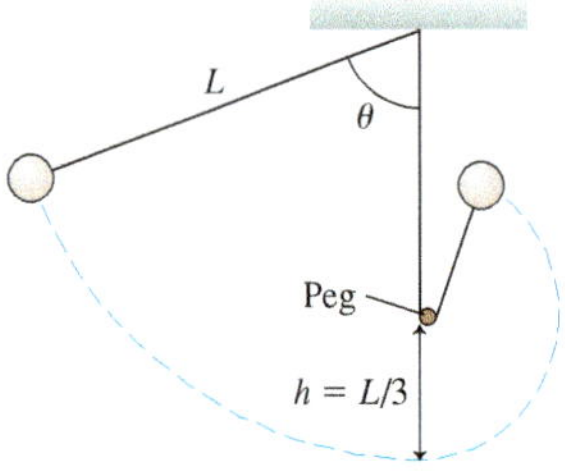

FIGURE CP10.68

69. In a physics lab experiment, a compressed spring launches a 20 g metal ball at a 30° angle. Compressing the spring 20 cm causes the ball to hit the floor 1.5 m below the point at which it leaves the spring after traveling 5.0 m horizontally. What is the spring constant?

70. It's your birthday, and to celebrate you're going to make your first bungee jump. You stand on a bridge 100 m above a raging river and attach a 30-m-long bungee cord to your harness. A bungee cord, for practical purposes, is just a long spring, and this cord has a spring constant of 40 N/m. Assume that your mass is 80 kg. After a long hesitation, you dive off the bridge. How far are you above the water when the cord reaches its maximum elongation?

71. A 10 kg box slides 4.0 m down the frictionless ramp shown in FIGURE CP10.71, then collides with a spring whose spring constant is 250 N/m.
a. What is the maximum compression of the spring?
b. At what compression of the spring does the box have its maximum speed?

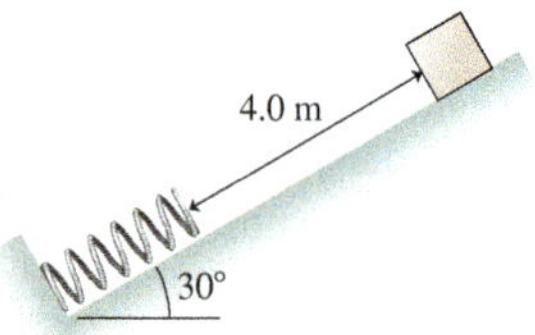

FIGURE CP10.71

72. Old naval ships fired 10 kg cannon balls from a 200 kg cannon. It was very important to stop the recoil of the cannon, since otherwise the heavy cannon would go careening across the deck of the ship. In one design, a large spring with spring constant 20,000 N/m was placed behind the cannon. The other end of the spring braced against a post that was firmly anchored to the ship's frame. What was the speed of the cannon ball if the spring compressed 50 cm when the cannon was fired?

73. A 2.0 kg cart has a spring with $k = 5000$ N/m attached to its front, parallel to the ground. This cart rolls at 4.0 m/s toward a stationary 1.0 kg cart.
a. What is the maximum compression of the spring during the collision?
b. What is the speed of each cart after the collision?

74. The air-track carts in FIGURE CP10.74 are sliding to the right at 1.0 m/s. The spring between them has a spring constant of 120 N/m and is compressed 4.0 cm. The carts slide past a flame that burns through the string holding them together. Afterward, what are the speed and direction of each cart?

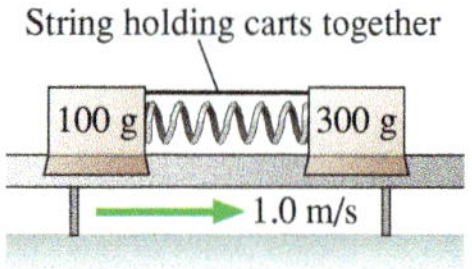

FIGURE CP10.74

75. A 100 g steel ball and a 200 g steel ball each hang from 1.0-m-long strings. At rest, the balls hang side by side, barely touching. The 100 g ball is pulled to the left until the angle between its string and vertical is 45°. The 200 g ball is pulled to a 45° angle on the right. The balls are released so as to collide at the very bottom of their swings. To what angle does each ball rebound?

76. A sled starts from rest at the top of the frictionless, hemispherical, snow-covered hill shown in FIGURE CP10.76.
   a. Find an expression for the sled's speed when it is at angle ϕ.
   b. Use Newton's laws to find the maximum speed the sled can have at angle ϕ without leaving the surface.
   c. At what angle ϕ_{max} does the sled "fly off" the hill?

FIGURE CP10.76

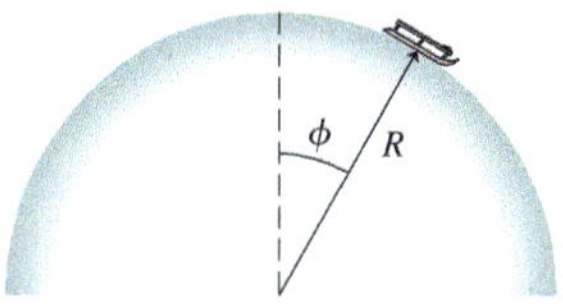

STOP TO THINK ANSWERS

Stop to Think 10.1: $(U_g)_c > (U_g)_b = (U_g)_d > (U_g)_a$. Gravitational potential energy depends only on height, not on speed.

Stop to Think 10.2: $v_a = v_b = v_c = v_d$. Her increase in kinetic energy depends only on the vertical height through which she falls, not the shape of the slide.

Stop to Think 10.3: b. Mechanical energy is conserved on a frictionless surface. Because $K_i = 0$ and $K_f = 0$, it must be true that $U_f = U_i$ and thus $y_f = y_i$. The final height matches the initial height.

Stop to Think 10.4: $k_a > k_b > k_c$. The spring constant is the slope of the force-versus-displacement graph.

Stop to Think 10.5: c. U_s depends on $(\Delta s)^2$, so doubling the compression increases U_s by a factor of 4. All the potential energy is converted to kinetic energy, so K increases by a factor of 4. But K depends on v^2, so v increases by only a factor of $(4)^{1/2} = 2$.

Stop to Think 10.6: $x = 6$ m. From the graph, the particle's potential energy at $x = 1$ m is $U = 3$ J. Its total energy is thus $E = K + U = 4$ J. A TE line at 4 J crosses the PE curve at $x = 6$ m.

11 Work

The bow may be very contemporary, but it's still the bow string doing work on the arrow that makes the arrow fly.

▶ **Looking Ahead** The goal of Chapter 11 is to develop a more complete understanding of energy and its conservation.

Unanswered Questions

Chapter 10 introduced energy but left many questions unanswered:

- How does a system gain or lose energy?
- When is a system's energy conserved?
- What role does friction play?

In Chapter 11 we'll answer these questions and introduce powerful new problem-solving tools.

◀ **Looking Back**
Chapter 10 Kinetic energy, potential energy, and energy diagrams

Work

When a force pushes or pulls a particle through a distance, we say that the force does **work** on the particle. The work W changes the particle's kinetic energy by $\Delta K = W$. One of the most important ways to change a system's energy is to do work on the system.

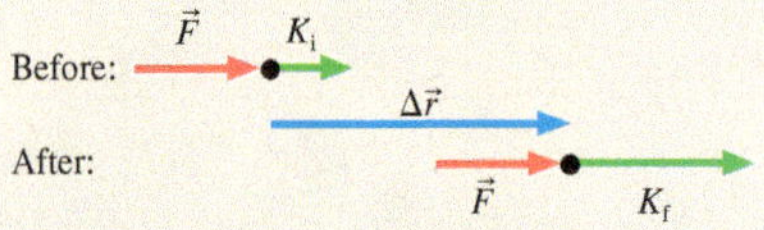

You'll learn a simple relationship among work, the force, and the displacement.

Work and Potential Energy

You'll learn that only certain kinds of forces, called **conservative forces**, can be associated with a potential energy. For these forces, the work W done by the force changes the potential energy by $\Delta U = W$.

Gravity is a conservative force. Lifting an object increases its gravitational potential energy.

The Basic Energy Model

We'll expand our basic energy model to include

- Work as an energy *transfer* between the system and the environment, and
- Energy *transformations* within the system.

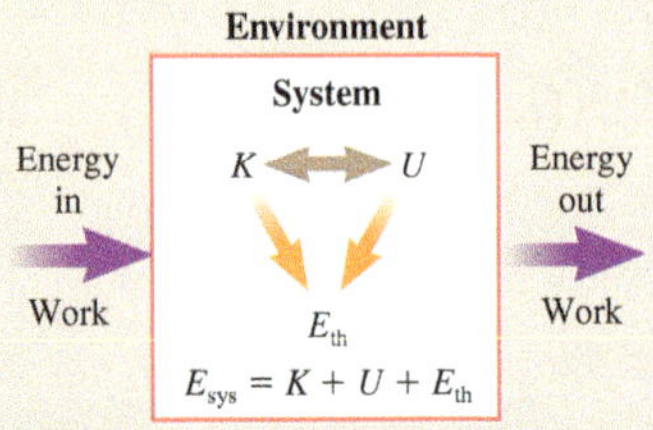

The Energy Equation

The ideas of the basic energy model—transfer and transformation—are captured in the **energy equation**:

$$\Delta E_{sys} = \Delta K + \Delta U + \Delta E_{th} = W_{ext}$$

- An isolated system has $W_{ext} = 0$. The total energy E_{sys} is conserved.
- An isolated system with no friction also has $\Delta E_{th} = 0$, so the mechanical energy $E_{mech} = K + U$ is conserved.

Power

Power is the *rate* at which energy is transferred or transformed. Power is measured in **watts**, where 1 watt is a rate of 1 joule per second.

A 100 W lightbulb transforms electrical energy into light and thermal energy at a rate of 100 J/s.

11.1 The Basic Energy Model Revisited

Chapter 10 introduced the *basic energy model* of FIGURE 11.1 but then focused on isolated systems. We found that kinetic and potential energy could be transformed back and forth without loss in an isolated, frictionless system, which led to the *law of conservation of mechanical energy:* The mechanical energy $E_{mech} = K + U$ is conserved in an isolated, frictionless system.

That was a good start, with many applications, but we need to expand our understanding of energy beyond ideal, isolated systems. Consider the following:

- A speeding car skids to a halt.
- A hand places a book on a high shelf.

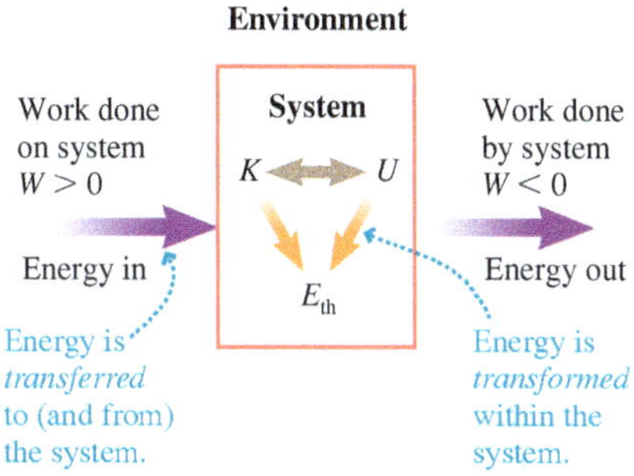

FIGURE 11.1 The basic energy model of a system interacting with its environment.

Neither of these situations conserves mechanical energy. In the first, kinetic energy is transformed not into potential energy but into **thermal energy** E_{th}, the energy of the random, microscopic motions of the atoms inside the tires, the brakes, and the road. We'll use an arrow $\rightarrow$ as a shorthand way to indicate an energy transformation, writing the energy transformation of the skidding car as $K \rightarrow E_{th}$.

Thermal energy is associated with the system's temperature. Friction raises the temperature—think of rubbing your hands together briskly—so a system with friction transforms kinetic or potential energy into thermal energy. This is usually a one-way process, as we'll discuss when we get to thermodynamics. That is, you can't get your car moving again by transforming the thermal energy of the hot brakes back into the car's kinetic energy. Thus the transformation arrows in Figure 11.1 point only to, not from, E_{th}.

Nonetheless, the energy remains inside the system if we define the system to be all objects that interact—the car and the road in this case. **System energy** E_{sys} is defined to be the sum of the mechanical energy of the objects plus the thermal energy of the atoms *inside* the objects. That is,

$$E_{sys} = E_{mech} + E_{th} = K + U + E_{th} \tag{11.1}$$

Energy transformations within the system do not change the value of E_{sys}, so we can say that **the total energy of an isolated system is conserved.** This generalizes the law of conservation of mechanical energy to include thermal energy.

In the second situation above, an external force—the hand—acts on the book to increase the book's gravitational potential energy. The book is not an isolated system but, instead, is surrounded by a larger environment with which it can exchange energy. Such an exchange is called an **energy transfer.** There are two energy-transfer processes. The first is due to forces—pushes and pulls—exerted on the system by the environment. In this case, the hand gives the book potential energy by lifting it upward. This *mechanical* transfer of energy to or from the system is called **work.** The symbol for work is W.

The second means of transferring energy between a system and its environment is a *nonmechanical* process called *heat*. Heat is a crucial idea that we will add to the energy model when we study thermodynamics, but for now we want to concentrate on the mechanical transfer of energy via work.

As the arrows in Figure 11.1 show, energy can both enter and leave the system. We'll distinguish between the two directions of energy flow by allowing the work W to be either positive or negative. The sign of W is interpreted as follows:

$W > 0$ **The environment does work on the system and the system's energy increases.**

$W < 0$ **The system does work on the environment and the system's energy decreases.**

What is the relationship among the quantities of the basic energy model? Our hypothesis, which is confirmed by experiment, is that

$$\Delta E_{sys} = \Delta K + \Delta U + \Delta E_{th} = W \tag{11.2}$$

The two essential ideas of the basic energy model and Equation 11.2 are:

1. Energy can be *transferred* to or from a system by doing work on the system. This process changes the energy of the system: $\Delta E_{sys} = W$.
2. Energy can be *transformed* within the system among K, U, and E_{th}. These processes don't change the energy of the system: $\Delta E_{sys} = 0$.

This is the essence of the basic energy model. The rest of Chapter 11 will substantiate Equation 11.2 and look at its many implications.

STOP TO THINK 11.1 A child slides down a playground slide at constant speed. The energy transformation is

a. $U \rightarrow K$ b. $K \rightarrow U$
c. There is no transformation because energy is conserved.
d. $U \rightarrow E_{th}$ e. $K \rightarrow E_{th}$

11.2 Work and Kinetic Energy

One dictionary defines "work" as:

1. Physical or mental effort; labor.
2. The activity by which one makes a living.
3. A task or duty.
4. Something produced as a result of effort, such as a *work of art.*
5. Plural *works:* A factory or plant where industry is carried on, such as *steel works.*
6. Plural *works:* The essential or operating parts of a mechanism.
7. The transfer of energy to a body by application of a force.

"Work" is a common word in the English language, with many meanings. When you first think of work, you probably think of the first two definitions in this list. After all, we talk about "working out," or we say, "I just got home from work." But that is *not* what work means in physics.

The basic energy model uses "work" in the sense of definition 7: energy transferred to or from a body or system by the application of force. The critical question we must answer is: *How much energy* does a force transfer?

We can answer this question by following the procedure we used in Chapter 10 to find the potential energy of gravity and of a spring. We'll begin, in FIGURE 11.2, with a force $\vec{F}$ acting on a particle of mass m as the particle moves along an s-axis from an initial position s_i, with kinetic energy K_i, to a final position s_f where the kinetic energy is K_f.

FIGURE 11.2 Force $\vec{F}$ does work as the particle moves from s_i to s_f.

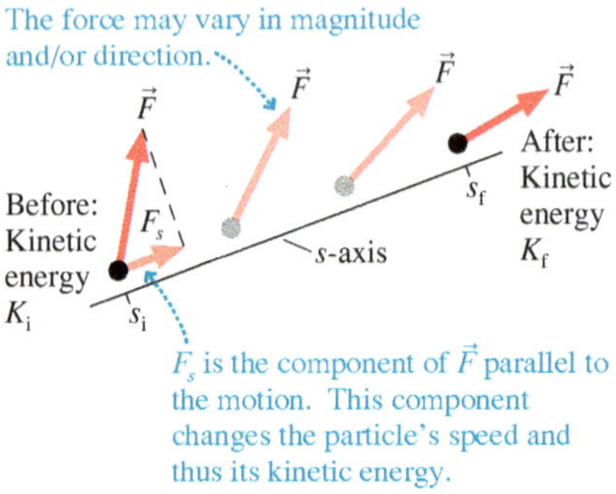

The force component F_s parallel to the s-axis causes the particle to speed up or slow down, thus transferring energy to or from the particle. We say that force $\vec{F}$ *does work* on the particle. Our goal is to find a relationship between F_s and ΔK. The s-component of Newton's second law is

$$F_s = ma_s = m\frac{dv_s}{dt} \tag{11.3}$$

where the v_s is the s-component of $\vec{v}$. As we did in Chapter 10, we can use the chain rule to write

$$m\frac{dv_s}{dt} = m\frac{dv_s}{ds}\frac{ds}{dt} = mv_s\frac{dv_s}{ds} \tag{11.4}$$

where $ds/dt = v_s$. Substituting Equation 11.4 into Equation 11.3 gives

$$F_s = mv_s\frac{dv_s}{ds} \tag{11.5}$$

The crucial step here, as it was in Chapter 10, was changing from a derivative with respect to time to a derivative with respect to position. We're going to want to integrate, so we first multiply through by ds to get

$$mv_s\,dv_s = F_s\,ds \tag{11.6}$$

Now we can integrate both sides from "before," where the position is s_i and the speed is v_i, to "after," giving

$$\int_{v_i}^{v_f} mv_s\,dv_s = \frac{1}{2}mv_s^2\Big|_{v_i}^{v_f} = \frac{1}{2}mv_f^2 - \frac{1}{2}mv_i^2 = \int_{s_i}^{s_f} F_s\,ds \tag{11.7}$$

The left side of Equation 11.7 is ΔK, the change in the particle's kinetic energy as it moves from s_i to s_f. The integral on the right apparently specifies the extent to which the applied force changes the particle's kinetic energy. We define the *work* done by force $\vec{F}$ as the particle moves from s_i to s_f as

$$W = \int_{s_i}^{s_f} F_s \, ds \tag{11.8}$$

The unit of work, that of force multiplied by distance, is the N m. Using the definition of the newton gives

$$1 \text{ N m} = 1 \text{ (kg m/s}^2\text{) m} = 1 \text{ kg m}^2\text{/s}^2 = 1 \text{ J}$$

Thus the unit of work is really the unit of energy. This is consistent with the idea that work is a transfer of energy. Rather than use N m, we will measure work in joules.

Using Equation 11.8 as the definition of work, we can write Equation 11.7 as

$$\Delta K = W \tag{11.9}$$

Equation 11.9 tells us that a force transfers kinetic energy to a particle by pushing or pulling on it. Furthermore, **Equation 11.8 gives us a specific method to calculate *how much* energy is transferred by the push or pull.** This energy transfer, by mechanical means, is what we mean by the term "work."

Notice that *no* work is done if there is no displacement ($s_f = s_i$) because an integral that spans no interval is zero. **A force does work on a particle only if the particle is displaced.** If you were to hold a 200 lb weight over your head, you might break out in a sweat and your arms would tire. You might "feel" that you had done a lot of work, but you would have done *zero* work in the physics sense because the weight was not displaced while you were holding it and thus you transferred no energy to it.

This pitcher is increasing the ball's kinetic energy by doing work on it.

The Work-Kinetic Energy Theorem

Equation 11.8 is the work done by one force. Because $\vec{F}_{\text{net}} = \sum \vec{F}_i$, it's easy to see that the net work done on a particle by several forces is $W_{\text{net}} = \sum W_i$, where W_i is the work done by force $\vec{F}_i$. In that case, Equation 11.9 becomes

$$\Delta K = W_{\text{net}} \tag{11.10}$$

This basic idea—that the net work done on a particle causes the particle's kinetic energy to change—is a general principle, one worth giving a name:

The work-kinetic energy theorem When one or more forces act on a particle as it is displaced from an initial position to a final position, the net work done on the particle by these forces causes the particle's kinetic energy to *change* by $\Delta K = W_{\text{net}}$.

An Analogy with the Impulse-Momentum Theorem

You might have noticed that there is a similarity between the work-kinetic energy theorem and the impulse-momentum theorem of Chapter 9:

$$\begin{aligned} &\text{Work-kinetic energy theorem:} \quad && \Delta K = W = \int_{s_i}^{s_f} F_s \, ds \\ &\text{Impulse-momentum theorem:} \quad && \Delta p_s = J_s = \int_{t_i}^{t_f} F_s \, dt \end{aligned} \tag{11.11}$$

In both cases, a force acting on a particle changes the state of the system. If the force acts over a time interval from t_i to t_f, it creates an *impulse* that changes the particle's momentum. If the force acts over the spatial interval from s_i to s_f, it does *work* that

FIGURE 11.3 Impulse and work are both the area under a force graph, but it's very important to know what the horizontal axis is.

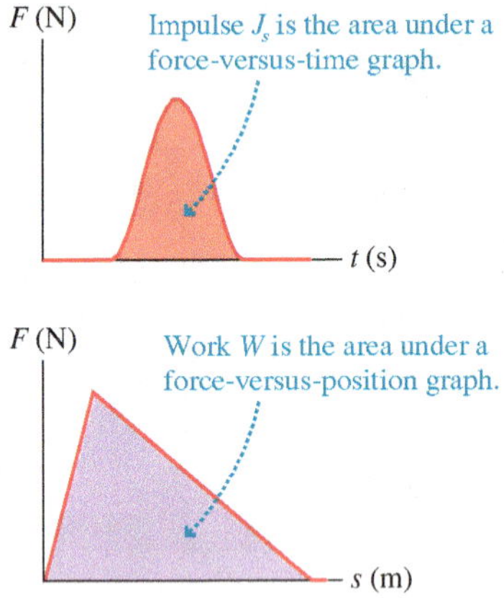

changes the particle's kinetic energy. FIGURE 11.3 shows that the geometric interpretation of impulse as the area under the F-versus-t graph applies equally well to an interpretation of work as the area under the F-versus-s graph.

This does not mean that a force *either* creates an impulse *or* does work but does not do both. Quite the contrary. A force acting on a particle *both* creates an impulse *and* does work, changing both the momentum and the kinetic energy of the particle. Whether you use the work-kinetic energy theorem or the impulse-momentum theorem depends on the question you are trying to answer.

We can, in fact, express the kinetic energy in terms of the momentum as

$$K = \frac{1}{2}mv^2 = \frac{(mv)^2}{2m} = \frac{p^2}{2m} \tag{11.12}$$

You cannot change a particle's kinetic energy without also changing its momentum.

STOP TO THINK 11.2 A particle moving along the x-axis experiences the force shown in the graph. If the particle has 2.0 J of kinetic energy as it passes $x = 0$ m, what is its kinetic energy when it reaches $x = 4$ m?

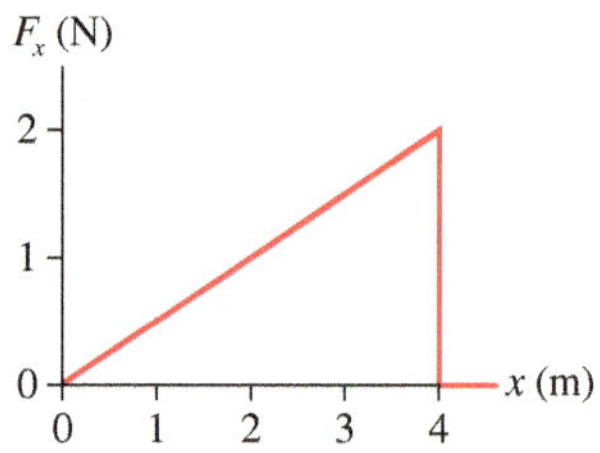

11.3 Calculating and Using Work

In this section we'll practice calculating work and using the work-kinetic energy theorem. We'll also introduce a new mathematical idea, the *dot product* of two vectors, that will allow us to write the work in a compact notation.

Constant Force

FIGURE 11.4 Work being done by a constant force as a particle moves through displacement $\Delta\vec{r}$.

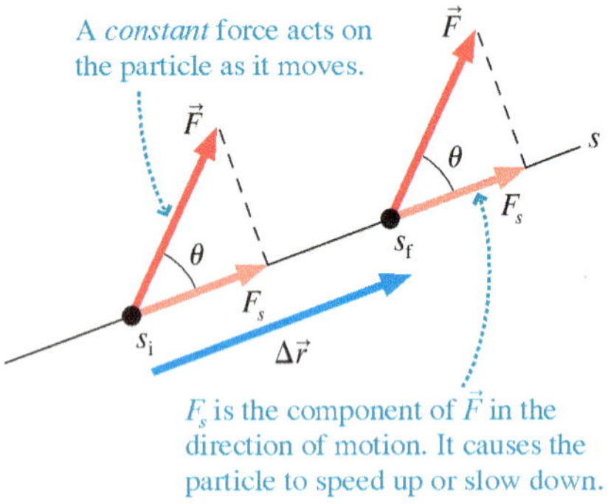

We'll begin by calculating the work done by a force $\vec{F}$ that acts with a *constant* strength and in a *constant* direction as a particle moves along a straight line through a displacement $\Delta\vec{r}$. FIGURE 11.4 shows the force acting on the particle as it moves along the s-axis. The force vector $\vec{F}$ makes an angle θ with respect to the displacement $\Delta\vec{r}$, so the component of the force vector along the direction of motion is $F_s = F\cos\theta$. According to Equation 11.8, the work done on the particle by this force is

$$W = \int_{s_i}^{s_f} F_s\, ds = \int_{s_i}^{s_f} F\cos\theta\, ds$$

Both F and θ are constant, so they can be taken outside the integral. Thus

$$W = F\cos\theta \int_{s_i}^{s_f} ds = F\cos\theta(s_f - s_i) = F(\Delta r)\cos\theta \tag{11.13}$$

where we used $s_f - s_i = \Delta r$, the magnitude of the particle's displacement. We can use Equation 11.13 to calculate the work done by a constant force if we know the magnitude F of the force, the angle θ of the force from the line of motion, and the distance Δr through which the particle is displaced.

NOTE ▶ You may have learned in an earlier physics course that work is "force times distance." This is *not* the definition of work, merely a special case. Work is "force times distance" only if the force is constant *and* parallel to the displacement (i.e., $\theta = 0°$). ◀

EXAMPLE 11.1 Pulling a suitcase

A rope inclined upward at a 45° angle pulls a suitcase through the airport. The tension in the rope is 20 N. How much work does the tension do if the suitcase is pulled 100 m?

MODEL Model the suitcase as a particle.

VISUALIZE FIGURE 11.5 shows a pictorial representation.

SOLVE The motion is along the x-axis, so in this case $\Delta r = \Delta x$. We can use Equation 11.13 to find that the tension does work:

$$W = T(\Delta x)\cos\theta = (20\text{ N})(100\text{ m})\cos 45° = 1400\text{ J}$$

FIGURE 11.5 Pictorial representation of a suitcase pulled by a rope.

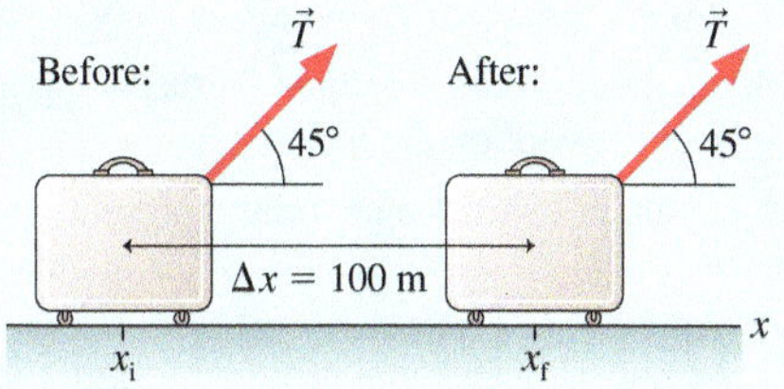

ASSESS Because a person pulls the rope, we would say informally that the person does 1400 J of work on the suitcase.

According to the basic energy model, work can be either positive or negative to indicate energy transfer into or out of the system. The quantities F and Δr are always positive, so **the sign of W is determined entirely by the angle θ between the force $\vec{F}$ and the displacement $\Delta\vec{r}$.**

TACTICS BOX 11.1 Calculating the work done by a constant force

Force and displacement	θ	Work W	Sign	Energy transfer
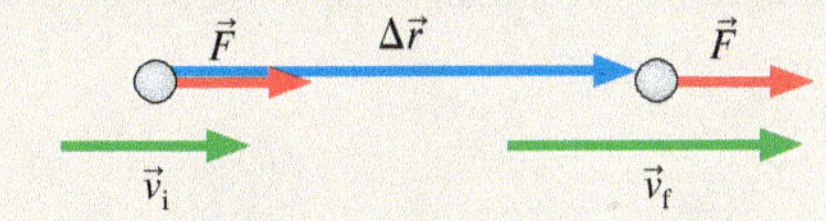	0°	$F(\Delta r)$	+	Energy is transferred into the system. The particle speeds up. K increases.
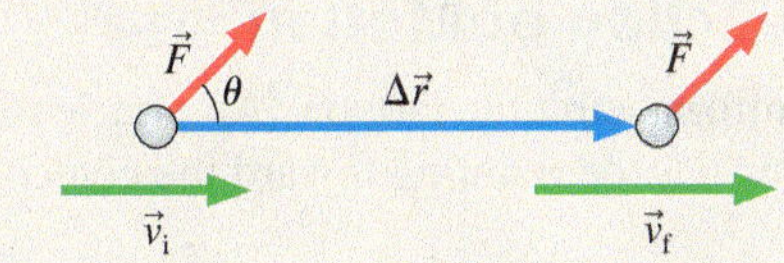	<90°	$F(\Delta r)\cos\theta$	+	
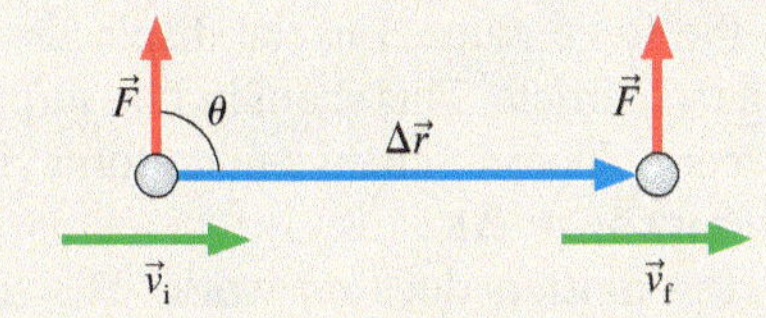	90°	0	0	No energy is transferred. Speed and K are constant.
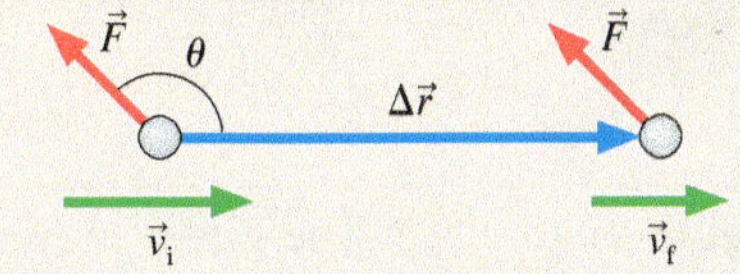	>90°	$F(\Delta r)\cos\theta$	−	Energy is transferred out of the system. The particle slows down. K decreases.
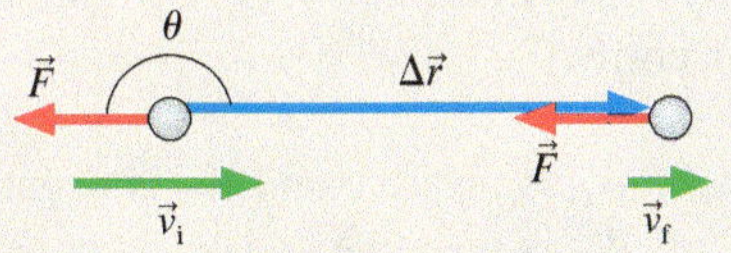	180°	$-F(\Delta r)$	−	

Exercises 3–10

NOTE ▶ The sign of W depends on the angle between the force vector and the displacement vector, *not* on the coordinate axes. A force to the left does *positive* work if it pushes a particle to the left (the force and the displacement are in the same direction, so this is a $\theta = 0°$ situation) even though the force component F_x is negative. Think about whether the force is trying to increase the particle's speed ($W > 0$) or decrease the particle's speed ($W < 0$). ◀

EXAMPLE 11.2 **Work during a rocket launch**

A 150,000 kg rocket is launched straight up. The rocket motor generates a thrust of 4.0×10^6 N. What is the rocket's speed at a height of 500 m? Ignore air resistance and any slight mass loss.

MODEL Model the rocket as a particle. Thrust and gravity are constant forces that do work on the rocket.

VISUALIZE **FIGURE 11.6** shows a pictorial representation and a free-body diagram.

FIGURE 11.6 Pictorial representation and free-body diagram of a rocket launch.

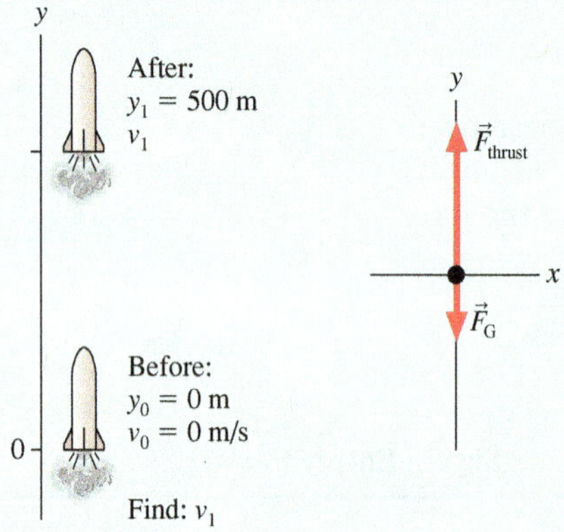

SOLVE We can solve this problem with the work-kinetic energy theorem $\Delta K = W_{net}$. Both forces do work on the rocket. The thrust is in the direction of motion, with $\theta = 0°$, and thus

$$W_{thrust} = F_{thrust}(\Delta r) = (4.0 \times 10^6 \text{ N})(500 \text{ m}) = 2.00 \times 10^9 \text{ J}$$

The gravitational force points downward, opposite the displacement $\Delta\vec{r}$, so $\theta = 180°$. Thus the work done by gravity is

$$W_{grav} = -F_G(\Delta r) = -mg(\Delta r)$$
$$= -(1.5 \times 10^5 \text{ kg})(9.8 \text{ m/s}^2)(500 \text{ m}) = -0.74 \times 10^9 \text{ J}$$

The work done by the thrust is positive. By itself, the thrust would cause the rocket to speed up. The work done by gravity is negative, not because $\vec{F}_G$ points down but because $\vec{F}_G$ is opposite the displacement. By itself, gravity would cause the rocket to slow down. The work-kinetic energy theorem, using $v_0 = 0$ m/s, is

$$\Delta K = \frac{1}{2}mv_1^2 - 0 = W_{net} = W_{thrust} + W_{grav} = 1.26 \times 10^9 \text{ J}$$

This is easily solved for the speed:

$$v_1 = \sqrt{\frac{2W_{net}}{m}} = 130 \text{ m/s}$$

ASSESS The net work is positive, meaning that energy is transferred *to* the rocket. In response, the rocket speeds up.

Force Perpendicular to the Direction of Motion

FIGURE 11.7 A perpendicular force does no work.

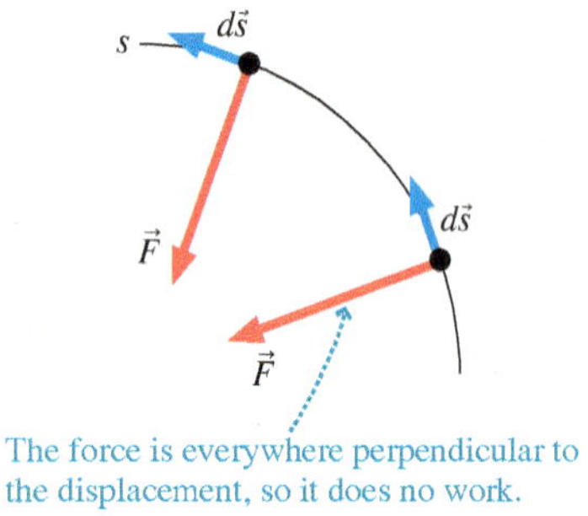

FIGURE 11.7 shows a particle moving in uniform circular motion. As you learned in Chapter 8, uniform circular motion requires a force pointing toward the center of the circle. How much work does this force do?

Zero! You can see that the force is everywhere perpendicular to the small displacement $d\vec{s}$. Thus F_s, the component of the force parallel to the displacement, is everywhere zero. The force does *no* work on the particle. This shouldn't be surprising. The particle's speed, and hence its kinetic energy, doesn't change in uniform circular motion, so the work-kinetic energy theorem says $W = \Delta K = 0$.

A force everywhere perpendicular to the motion does no work. The friction force on a car turning a corner does no work. Neither does the tension force when a mass on a string is in circular motion.

STOP TO THINK 11.3 A crane lowers a steel girder into place. The girder moves with constant speed. Consider the work W_G done by gravity and the work W_T done by the tension in the cable. Which of the following is correct?

a. W_G is positive and W_T is positive.
b. W_G is positive and W_T is negative.
c. W_G is negative and W_T is positive.
d. W_G is negative and W_T is negative.
e. W_G and W_T are both zero.

The Dot Product of Two Vectors

There's something different about the quantity $F(\Delta r)\cos\theta$ in Equation 11.13. We've spent many chapters adding vectors, but this is the first time we've *multiplied* two vectors. Multiplying vectors is not like multiplying scalars. In fact, there is more than one way to multiply vectors. We will introduce one way now, the *dot product*.

FIGURE 11.8 shows two vectors, $\vec{A}$ and $\vec{B}$, with angle α between them. We define the **dot product** of $\vec{A}$ and $\vec{B}$ as

$$\vec{A} \cdot \vec{B} = AB\cos\alpha \tag{11.14}$$

A dot product *must have* the dot symbol • between the vectors. The notation $\vec{A}\vec{B}$, without the dot, is *not* the same thing as $\vec{A} \cdot \vec{B}$. The dot product is also called the **scalar product** because the value is a scalar. Later, when we need it, we'll introduce a different way to multiply vectors called the *cross product*.

The dot product of two vectors depends on the orientation of the vectors. FIGURE 11.9 shows five different situations, including the three "special cases" where $\alpha = 0°$, 90°, and 180°.

NOTE ▶ The dot product of a vector with itself is well defined. If $\vec{B} = \vec{A}$ (i.e., $\vec{B}$ is a copy of $\vec{A}$), then $\alpha = 0°$. Thus $\vec{A} \cdot \vec{A} = A^2$. ◀

FIGURE 11.8 Vectors $\vec{A}$ and $\vec{B}$, with angle α between them.

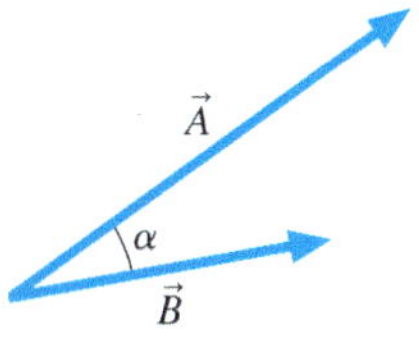

FIGURE 11.9 The dot product $\vec{A} \cdot \vec{B}$ as α ranges from 0° to 180°.

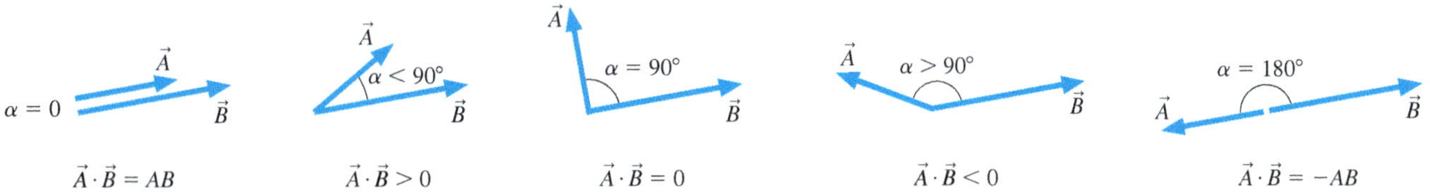

EXAMPLE 11.3 Calculating a dot product

Compute the dot product of the two vectors in FIGURE 11.10

FIGURE 11.10 Vectors $\vec{A}$ and $\vec{B}$ of Example 11.3.

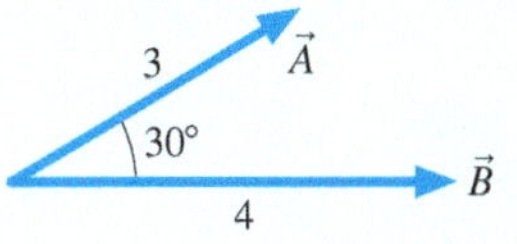

SOLVE The angle between the vectors is $\alpha = 30°$, so

$$\vec{A} \cdot \vec{B} = AB\cos\alpha = (3)(4)\cos 30° = 10.4$$

Like vector addition and subtraction, calculating the dot product of two vectors is often performed most easily using vector components. FIGURE 11.11 reminds you of the unit vectors $\hat{\imath}$ and $\hat{\jmath}$ that point in the positive x-direction and positive y-direction. The two unit vectors are perpendicular to each other, so their dot product is $\hat{\imath} \cdot \hat{\jmath} = 0$. Furthermore, because the magnitudes of $\hat{\imath}$ and $\hat{\jmath}$ are 1, $\hat{\imath} \cdot \hat{\imath} = 1$ and $\hat{\jmath} \cdot \hat{\jmath} = 1$.

FIGURE 11.11 The unit vectors $\hat{\imath}$ and $\hat{\jmath}$.

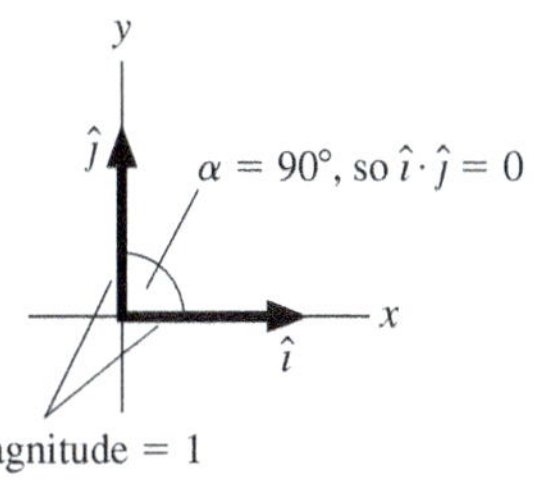

In terms of components, we can write the dot product of vectors $\vec{A}$ and $\vec{B}$ as

$$\vec{A} \cdot \vec{B} = (A_x\hat{\imath} + A_y\hat{\jmath}) \cdot (B_x\hat{\imath} + B_y\hat{\jmath})$$

Multiplying this out, and using the results for the dot products of the unit vectors:

$$\begin{aligned}\vec{A} \cdot \vec{B} &= A_x B_x \hat{\imath} \cdot \hat{\imath} + (A_x B_y + A_y B_x)\hat{\imath} \cdot \hat{\jmath} + A_y B_y \hat{\jmath} \cdot \hat{\jmath} \\ &= A_x B_x + A_y B_y\end{aligned} \tag{11.15}$$

That is, the **dot product is the sum of the products of the components.**

EXAMPLE 11.4 Calculating a dot product using components

Compute the dot product of $\vec{A} = 3\hat{\imath} + 3\hat{\jmath}$ and $\vec{B} = 4\hat{\imath} - \hat{\jmath}$.

FIGURE 11.12 Vectors $\vec{A}$ and $\vec{B}$.

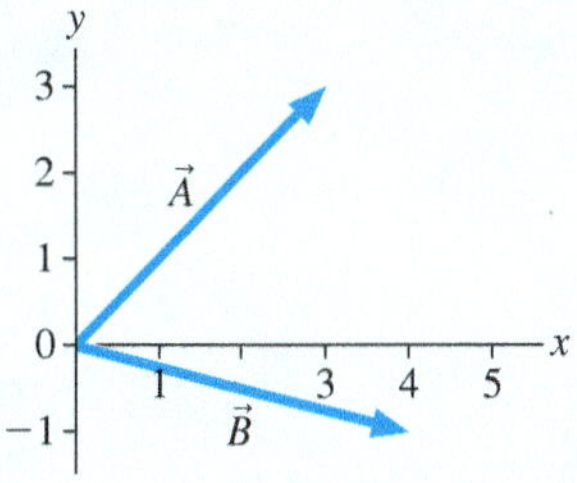

SOLVE FIGURE 11.12 shows vectors $\vec{A}$ and $\vec{B}$. We could calculate the dot product by first doing the geometry needed to find the angle between the vectors and then using Equation 11.14. But calculating the dot product from the vector components is much easier. It is

$$\vec{A} \cdot \vec{B} = A_x B_x + A_y B_y = (3)(4) + (3)(-1) = 9$$

Looking at Equation 11.13, the work done by a constant force, you should recognize that it is the dot product of the force vector and the displacement vector:

$$W = \vec{F} \cdot \Delta\vec{r} \quad \text{(work done by a constant force)} \qquad (11.16)$$

This definition of work is valid for a constant force.

EXAMPLE 11.5 Calculating work using the dot product

A 70 kg skier is gliding at 2.0 m/s when he starts down a very slippery 50-m-long, 10° slope. What is his speed at the bottom?

MODEL Model the skier as a particle and interpret "very slippery" to mean frictionless. Use the work-kinetic energy theorem to find his final speed.

VISUALIZE FIGURE 11.13 shows a pictorial representation.

FIGURE 11.13 Pictorial representation of the skier.

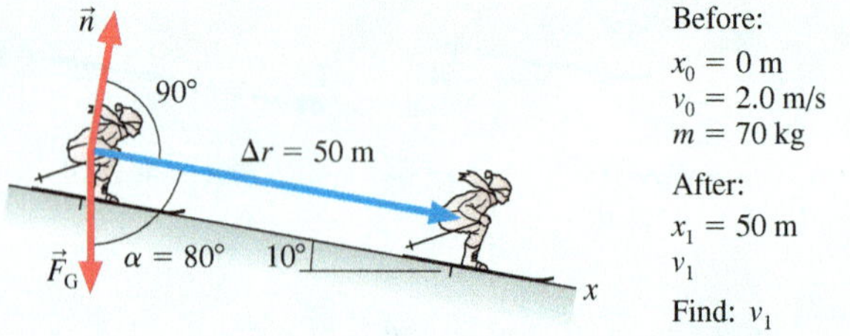

SOLVE The only forces on the skier are $\vec{F}_G$ and $\vec{n}$. The normal force is perpendicular to the motion and thus does no work. The work done by gravity is easily calculated as a dot product:

$$W = \vec{F}_G \cdot \Delta\vec{r} = mg(\Delta r)\cos\alpha$$

$$= (70 \text{ kg})(9.8 \text{ m/s}^2)(50 \text{ m})\cos 80° = 5960 \text{ J}$$

Notice that the angle *between* the vectors is 80°, not 10°. Then, from the work-kinetic energy theorem, we find

$$\Delta K = \frac{1}{2}mv_1^2 - \frac{1}{2}mv_0^2 = W$$

$$v_1 = \sqrt{v_0^2 + \frac{2W}{m}} = \sqrt{(2.0 \text{ m/s})^2 + \frac{2(5960 \text{ J})}{70 \text{ kg}}} = 13 \text{ m/s}$$

NOTE ▶ While in the midst of the mathematics of calculating work, do not lose sight of what the work-kinetic energy theorem is all about. It is a statement about *energy transfer:* Work causes a particle's kinetic energy to either increase or decrease. ◀

STOP TO THINK 11.4 Which force does the most work as a particle undergoes displacement $\Delta\vec{r}$?

a. The 10 N force.
b. The 8 N force.
c. The 6 N force.
d. They all do the same amount of work.

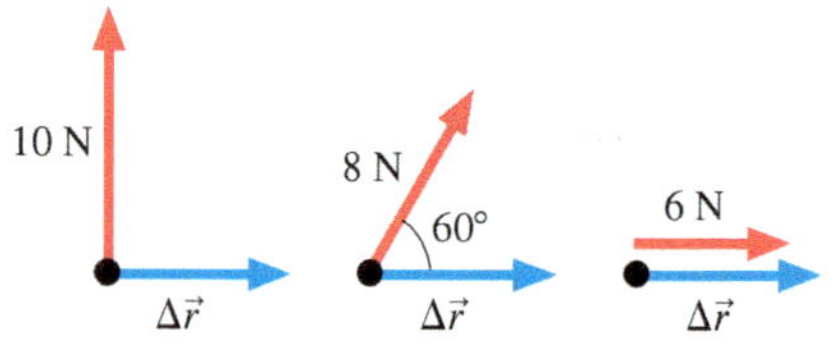

11.4 The Work Done by a Variable Force

We've learned how to calculate the work done on an object by a constant force, but what about a force that changes in either magnitude or direction as the object moves? Equation 11.8, the definition of work, is all we need:

$$W = \int_{s_i}^{s_f} F_s\, ds = \text{area under the force-versus-position graph} \qquad (11.17)$$

The integral sums up the small amounts of work $F_s\,ds$ done in each step along the trajectory. The only new feature, because F_s now varies with position, is that we cannot take F_s outside the integral. We must evaluate the integral either geometrically, by finding the area under the curve, or by actually doing the integration.

EXAMPLE 11.6 **Using work to find the speed of a car**

A 1500 kg car accelerates from rest. FIGURE 11.14 shows the net force on the car (propulsion force minus any drag forces) as it travels from $x = 0$ m to $x = 200$ m. What is the car's speed after traveling 200 m?

FIGURE 11.14 Force-versus-position graph for a car.

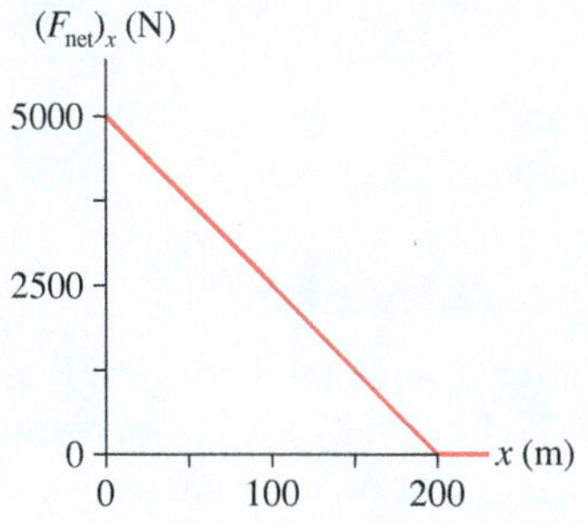

SOLVE The acceleration $a_x = (F_{net})_x/m$ is high as the car starts but decreases as the car picks up speed because of increasing drag. Figure 11.14 is a realistic portrayal of the net force on a car. But a variable force means that we cannot use the familiar constant-acceleration kinematics. Instead, we can use the work-kinetic energy theorem. Because $v_i = 0$ m/s, we have

$$\Delta K = \frac{1}{2}mv_f^2 - 0 = W_{net}$$

Starting from $x_i = 0$ m, the work is

$$W_{net} = \int_{0\,m}^{x_f} (F_{net})_x\,dx$$

$$= \text{area under the } (F_{net})_x\text{-versus-}x \text{ graph from 0 m to } x_f$$

The area under the curve of Figure 11.14 is that of a triangle of width 200 m. Thus

$$W_{net} = \text{area} = \frac{1}{2}(5000\text{ N})(200\text{ m}) = 500{,}000\text{ J}$$

The work-kinetic energy theorem then gives

$$v_f = \sqrt{\frac{2W_{net}}{m}} = \sqrt{\frac{2(500{,}000\text{ J})}{1500\text{ kg}}} = 26\text{ m/s}$$

ASSESS 26 m/s ≈ 55 mph is a reasonable speed after accelerating for 200 m, so we can have confidence in our answer.

EXAMPLE 11.7 **Using the work-kinetic energy theorem for a spring**

The "pincube machine" was an ill-fated predecessor of the pinball machine. A 100 g cube is launched by pulling a spring back 20 cm and releasing it. What is the cube's launch speed, as it leaves the spring, if the spring constant is 20 N/m and the surface is frictionless?

MODEL Model the spring as an ideal spring obeying Hooke's law. Use the work-kinetic energy theorem to find the launch speed.

VISUALIZE FIGURE 11.15 shows a before-and-after pictorial representation and a free-body diagram. We've placed the origin of the x-axis at the equilibrium position of the spring.

FIGURE 11.15 Pictorial representation and free-body diagram for Example 11.7.

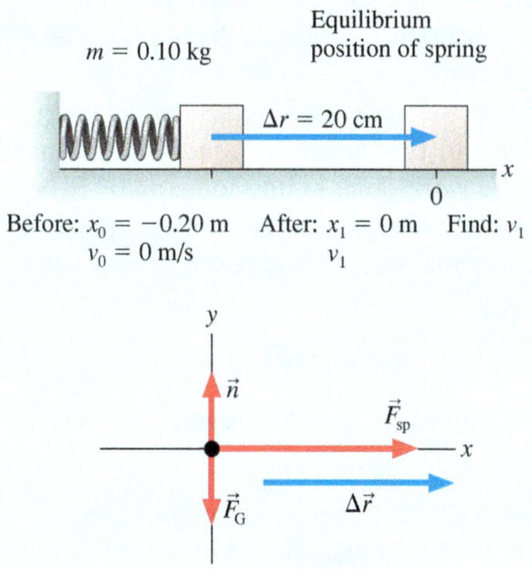

SOLVE The normal force and gravity are perpendicular to the motion and do no work. We can use the work-kinetic energy theorem, with $v_0 = 0$ m/s, to find the launch speed:

$$\Delta K = \frac{1}{2}mv_1^2 - 0 = W_{sp}$$

The spring force is a variable force: $(F_{sp})_x = -k\Delta x = -kx$, where $\Delta x = x - x_e = x$ because we chose a coordinate system with $x_e = 0$ m. Despite the minus sign, $(F_{sp})_x$ is a positive quantity (force pointing to the right) because x is negative throughout the motion. The spring force points in the direction of motion, so W_{sp} is positive. We can use Equation 11.17 to evaluate W_{sp}:

$$W_{sp} = \int_{x_0}^{x_1} (F_{sp})_x\,dx = -k\int_{x_0}^{x_1} x\,dx = -\frac{1}{2}kx^2\Big|_{x_0}^{x_1}$$

$$= -\left(\frac{1}{2}kx_1^2 - \frac{1}{2}kx_0^2\right)$$

Evaluating W_{sp} for $x_0 = -0.20$ m and $x_1 = 0$ m gives

$$W_{sp} = \frac{1}{2}(20\text{ N/m})(-0.20\text{ m})^2 = 0.400\text{ J}$$

We can now solve for the launch speed, finding

$$v_1 = \sqrt{\frac{2W_{sp}}{m}} = \sqrt{\frac{2(0.400\text{ J})}{0.100\text{ kg}}} = 2.8\text{ m/s}$$

ASSESS You might have noticed that the work done by the spring looks a lot like the spring's potential energy $U_{sp} = \frac{1}{2}k(\Delta x)^2$. The next section will find a connection between work and potential energy.

11.5 Work and Potential Energy

It's time to look more closely at the concept of potential energy. In Chapter 10, the new concept of gravitational potential energy was associated with the gravitational force. Then, after introducing Hooke's law, we used the spring force to "discover" elastic potential energy. In both cases, a force was associated with a potential energy, and we found that it is often easier to solve problems with energy laws rather than force laws.

But that raises the question: Is there a potential energy associated with every force? Is there a "tension potential energy" and a "friction potential energy"? If not, what's special about the gravitational force and the spring force? What conditions must a force meet in order to have an associated potential energy?

Conservative and Nonconservative Forces

FIGURE 11.16 A particle can move from A to B along either path 1 or path 2.

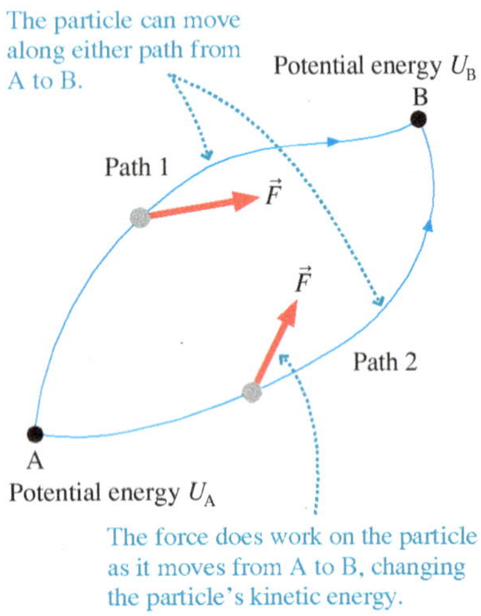

FIGURE 11.16 shows a particle that can move from point A to point B along two possible paths while a force $\vec{F}$ is exerted on it. The force may vary from point to point in space, so the force experienced along path 1 may not be the same as the force experienced along path 2. The force changes the particle's speed, as well as its direction, so the particle's speed and kinetic energy when it arrives at B will differ from the speed and kinetic energy it had when it left A.

Let's assume that there is a potential energy associated with force $\vec{F}$, just as the gravitational potential energy $U_g = mgy$ is associated with the gravitational force $\vec{F}_G = -mg\hat{j}$. What restrictions does this assumption place on $\vec{F}$? There are three steps in the logic:

1. Potential energy is an energy of position. The system has one value of potential energy when the particle is at A, a different value when the particle is at B. Thus the overall change in potential energy $\Delta U = U_B - U_A$ is the same whether the particle moves along path 1 or path 2.
2. Potential energy is transformed into kinetic energy, with $\Delta K = -\Delta U$. If ΔU is independent of the path, then ΔK is also independent of the path. The transformation of energy causes the particle to have the same kinetic energy at B no matter which path it follows.
3. The change in a particle's kinetic energy is related to the work done on the particle by force $\vec{F}$. According to the work-kinetic energy theorem, $\Delta K = W$. Because ΔK is independent of the path, it *must* be the case that **the work done by force $\vec{F}$ as the particle moves from A to B is independent of the path followed.**

A force for which the work done on a particle as it moves from an initial to a final position is independent of the path followed is called a **conservative force.** (The name, as you'll soon see, is related to the conditions under which mechanical energy is conserved.) The importance of conservative forces is that **a potential energy can be associated with any conservative force.**

To establish a general connection between work and potential energy, suppose an object moves from initial position i to final position f under the influence of a conservative force $\vec{F}$. We'll denote the work done by the force as $W_c(\text{i} \rightarrow \text{f})$, where the notation $\text{i} \rightarrow \text{f}$ means "as the object moves from position i to position f." Because $\Delta K = W$ and $\Delta K = -\Delta U$, the potential energy difference between these two points must be

$$\Delta U = U_f - U_i = -W_c(\text{i} \rightarrow \text{f}) \tag{11.18}$$

Equation 11.18 is a general definition of the potential energy associated with a conservative force.

NOTE ▶ Equation 11.18 defines only the *change* in potential energy ΔU. We can add a constant to both U_f and U_i without changing ΔU. This was the basis for our discussion in Chapter 10 about the zero of potential energy. ◀

For example, Chapter 10 showed that the kinetic energy gained by an object sliding down a frictionless slope depends only on the vertical distance change Δy and is independent of the shape of the slope. It follows, because $W = \Delta K$, that the work done by gravity does not depend on the path followed from initial height y_i to final height y_f. That is, the gravitational force is a conservative force, and *that* is why we were able to establish a gravitational potential energy. (If you look back, you'll see that the analysis of Chapter 10 that led to $U_g = mgy$ was really a calculation of the work done by the gravitational force, although we didn't call it that at the time.)

What about springs? A homework problem will let you show that Hooke's law is also a conservative force. In Example 11.7 we showed that the work done by a spring is

$$W_{sp}(i \rightarrow f) = \int_{x_i}^{x_f} F_{sp}\, dx = -\left(\frac{1}{2}kx_f^2 - \frac{1}{2}kx_i^2\right)$$

from which it follows that $U_s = \frac{1}{2}kx^2$. Example 11.7 was a "special case" in that we defined the coordinate system to make $x_e = 0$. A more general analysis would give $U_s = \frac{1}{2}k(\Delta s)^2$, as you learned in Chapter 10.

Not all forces are conservative forces. For example, FIGURE 11.17 is a bird's-eye view of two particles sliding across a surface. The friction force always points opposite the direction of motion, 180° from $d\vec{s}$, hence the small amount of work done during displacement $d\vec{s}$ is $dW_{fric} = \vec{f}_k \cdot d\vec{s} = -\mu_k mg\, ds$. Summed over the entire path, the work done by friction as a particle travels total distance Δs is $W_{fric} = -\mu_k mg \Delta s$. We see that the work done by friction depends on Δs, the distance traveled. More work is done on the particle traveling the longer path, so the work done by friction is *not* independent of the path followed.

FIGURE 11.17 Top view of two particles sliding across a surface.

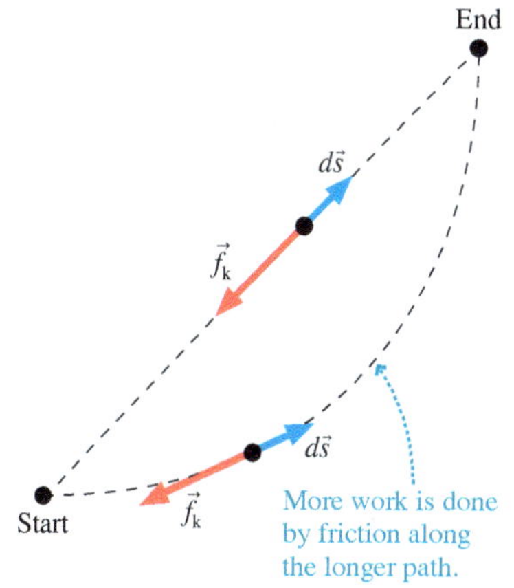

NOTE ▶ This analysis applies only to the motion of a particle, which has no internal structure and thus no thermal energy. These ideas will be applied to extended objects—such as a car skidding to a halt—in Section 11.7 on thermal energy. The particle equation $W_{fric} = -\mu_k mg \Delta s$ should *not* be used in problem solving. ◀

A force for which the work is *not* independent of the path is called a **nonconservative force.** It is not possible to define a potential energy for a nonconservative force. Friction is a nonconservative force, so we cannot define a potential energy of friction.

This makes sense. If you toss a ball straight up, kinetic energy is transformed into gravitational potential energy. The ball has the potential to transform this energy back into kinetic energy, and it does so as the ball falls. But you cannot recover the kinetic energy lost to friction as a box slides to a halt. There's no "potential" that can be transformed back into kinetic energy.

Mechanical Energy

Consider a system of objects interacting via both conservative forces and nonconservative forces. The conservative forces do work W_c as the particles move from initial positions i to final positions f. The nonconservative forces do work W_{nc}. The total work done by *all* forces is $W_{net} = W_c + W_{nc}$. The change in the system's kinetic energy ΔK, as determined by the work-kinetic energy theorem, is

$$\Delta K = W_{net} = W_c(i \rightarrow f) + W_{nc}(i \rightarrow f) \quad (11.19)$$

The work done by the conservative forces can now be associated with a potential energy U. According to Equation 11.18, $W_c(i \rightarrow f) = -\Delta U$. With this definition, Equation 11.19 becomes

$$\Delta K + \Delta U = \Delta E_{mech} = W_{nc} \quad (11.20)$$

where, as in Chapter 10, the *mechanical energy* is $E_{mech} = K + U$.

Now we can see that **mechanical energy is conserved if there are no nonconservative forces.** That is,

$$\Delta E_{mech} = 0 \text{ if } W_{nc} = 0 \quad (11.21)$$

Mechanical energy isn't always conserved. Here, mechanical energy is being transformed into thermal energy.

This important conclusion is what we called the law of conservation of mechanical energy in Chapter 10. There we saw that friction prevents E_{mech} from being conserved, but we really didn't know why. Equation 11.20 tells us that the work done by any nonconservative force causes the mechanical energy to change. Friction and other "dissipative forces" lead to a loss of mechanical energy. Other outside forces, such as the pull of a rope, might increase the mechanical energy.

Equally important, Equation 11.20 tells us what to do if the mechanical energy isn't conserved. You can still use energy concepts to analyze the motion if you compute the work done by the nonconservative forces.

EXAMPLE 11.8 Using work and potential energy

The skier from Example 11.5 repeats his run after the wind comes up. Recall that the 70 kg skier was gliding at 2.0 m/s when he started down a 50-m-long, 10°, frictionless slope. What is his speed at the bottom if the wind exerts a steady 50 N retarding force opposite his motion?

MODEL This time let the system be the skier and the earth.

VISUALIZE Figure 11.13 showed the pictorial representation and free-body diagram.

SOLVE In solving this problem with the work-kinetic energy theorem, we had to explicitly calculate the work done by gravity. Now let's use Equation 11.20. Gravity is a conservative force that we can associate with the gravitational potential energy U_g. The retarding force of the wind is nonconservative. Thus

$$\Delta K + \Delta U_g = W_{nc} = W_{wind}$$

The gravitational potential energy is $U_g = mgy$. Because the wind force is opposite the skier's motion, with $\theta = 180°$, it does work $W_{wind} = \vec{F}_{wind} \cdot \Delta\vec{r} = -F_{wind}\,\Delta r$. Thus the energy equation becomes

$$\frac{1}{2}mv_1^2 - \frac{1}{2}mv_0^2 + mgy_1 - mgy_0 = -F_{wind}\,\Delta r$$

Using the values given, we find

$$v_1 = \sqrt{v_0^2 + 2gy_0 - 2F_{wind}\,\Delta r/m} = 10 \text{ m/s}$$

ASSESS What appeared to be a difficult problem, with both gravity and a retarding force, turned out to be straightforward when analyzed with energy and work. The skier's final speed is about 25% slower when the wind is blowing.

Example 11.8 illustrates an important idea. When we associate a potential energy with a conservative force, we

- Enlarge the system to include all objects that interact via conservative forces.
- "Precompute" the work. We can do this because we don't need to know what paths the objects are going to follow. This precomputed work becomes a potential energy and moves from the right side of $\Delta K = W$ to the left side of Equation 11.20.

NOTE ▶ When you use a potential energy, you've already taken the work of that force into account. Don't compute the work explicitly, or you'll double count it! ◀

In Example 11.5, the system consisted of just the skier. We treated the gravitational force as a force from the environment doing work on the system. In Example 11.8, where we revisited the same problem, we brought the earth into the system and represented the conservative earth-skier interaction with a potential energy.

In summary, to analyze a problem using work and energy, you can either

1. Use the work-kinetic energy theorem $\Delta K = W$ and explicitly compute the work done by *every* force. This was the method of Example 11.5. Or
2. Represent the work done by conservative forces as potential energies, then use $\Delta K + \Delta U = W_{nc}$. The only work that must be computed is the work of any nonconservative forces. This was the method of Example 11.8.

In practice, **method 2 is always easier and is the preferred method.**

11.6 Finding Force from Potential Energy

We now know how to find the potential energy due to a conservative force. Can we reverse the procedure and find the force from the potential energy?

FIGURE 11.18a shows an object moving through a *small* displacement Δs while being acted on by a conservative force $\vec{F}$. If Δs is sufficiently small, the force component F_s in the direction of motion is essentially constant during the displacement. The work done on the object as it moves from s to $s + \Delta s$ is

$$W(s \rightarrow s + \Delta s) = F_s \, \Delta s \tag{11.22}$$

Because $\vec{F}$ is a conservative force, the object's potential energy as it moves through Δs changes by

$$\Delta U = -W(s \rightarrow s + \Delta s) = -F_s \, \Delta s$$

which we can rewrite as

$$F_s = -\frac{\Delta U}{\Delta s} \tag{11.23}$$

In the limit $\Delta s \rightarrow 0$, we find that the force at position s is

$$F_s = \lim_{\Delta s \to 0}\left(-\frac{\Delta U}{\Delta s}\right) = -\frac{dU}{ds} \tag{11.24}$$

We see that the force on the object is the *negative* of the derivative of the potential energy with respect to position. FIGURE 11.18b shows that we can interpret this result graphically by saying

$$F_s = \text{the negative of the slope of the } U\text{-versus-}s \text{ graph at } s \tag{11.25}$$

In practice, of course, we will usually use either $F_x = -dU/dx$ or $F_y = -dU/dy$.

As an example, consider the gravitational potential energy $U_g = mgy$. FIGURE 11.19a shows the potential-energy diagram U_g versus y. It is simply a straight-line graph passing through the origin. The force on the object at position y, according to Equations 11.24 and 11.25, is simply

$$(F_G)_y = -\frac{dU_g}{dy} = -(\text{slope of } U_g) = -mg$$

The negative sign, as always, indicates that the force points in the negative y-direction. FIGURE 11.19b shows the corresponding F-versus-y graph. At each point, the *value* of F is equal to the negative of the *slope* of the U-versus-y graph. This is similar to position and velocity graphs, where the value of v_x at any time t is equal to the slope of the x-versus-t graph.

We already knew that $(F_G)_y = -mg$, of course, so the point of this particular example was to illustrate the meaning of Equation 11.25 rather than to find out anything new. Had we *not* known the gravitational force, we see that it is possible to find it from the potential energy.

FIGURE 11.20a is a more interesting example. The slope of the potential-energy graph is negative between x_1 and x_2. This means that the force on the object, which is the negative of the slope of U, is *positive*. An object between x_1 and x_2 experiences a force toward the right. The force decreases as the magnitude of the slope decreases until, at x_2, $F_x = 0$. This is consistent with our prior identification of x_2 as a point of stable *equilibrium*. The slope is positive (force negative and thus to the left) between x_2 and x_3, zero (zero force) at the unstable equilibrium point x_3, and so on. Point x_4, where the slope is most negative, is the point of maximum force.

FIGURE 11.20b is a plausible graph of F versus x. We don't know the exact shape because we don't have an exact expression for U, but the force graph must look very much like this.

FIGURE 11.18 Relating force and potential energy.

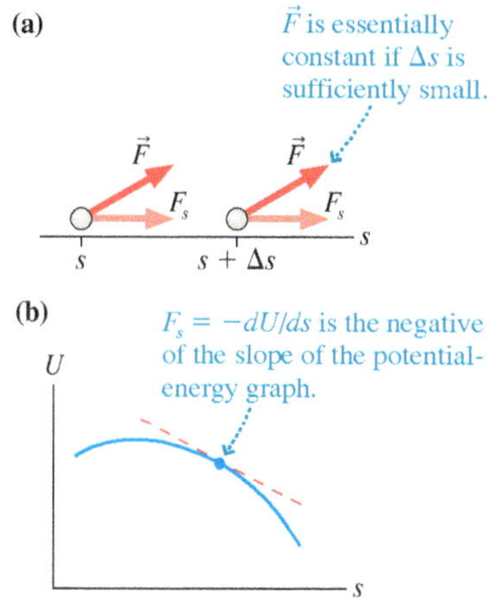

FIGURE 11.19 Gravitational potential energy and force diagrams.

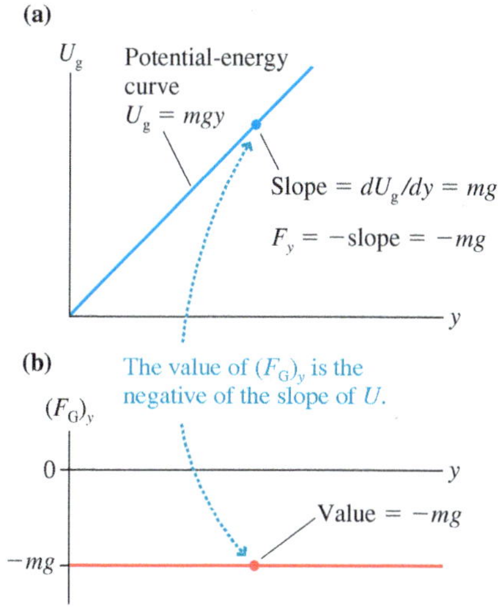

FIGURE 11.20 A potential-energy diagram and the corresponding force diagram.

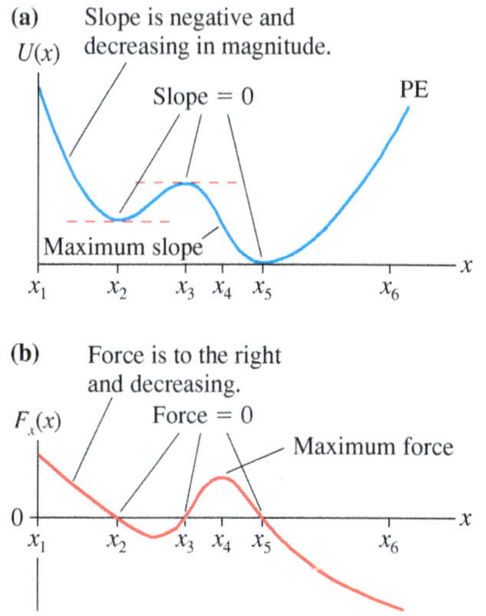

STOP TO THINK 11.5 A particle moves along the x-axis with the potential energy shown. The x-component of the force on the particle when it is at $x = 4$ m is

a. 4 N.
b. 2 N.
c. 1 N.
d. −4 N.
e. −2 N.
f. −1 N.

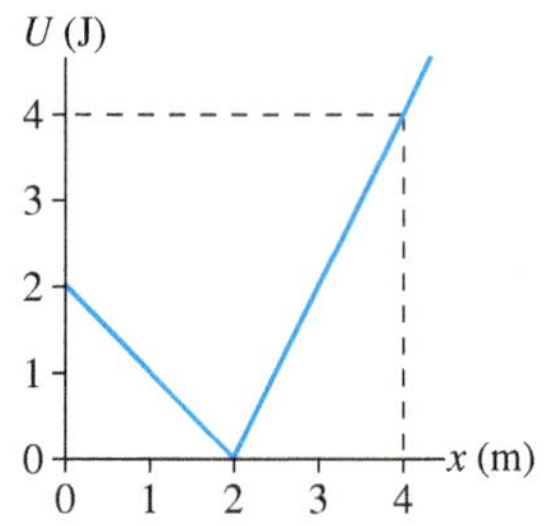

11.7 Thermal Energy

All of the objects we handle and use every day consist of vast numbers of particle-like atoms. We will use the terms **macrophysics** to refer to the motion and dynamics of the object as a whole and **microphysics** to refer to the motion of atoms. You recognize the prefix *micro,* meaning "small." You may not be familiar with *macro,* which means "large" or "large-scale."

Kinetic and Potential Energy at the Microscopic Level

FIGURE 11.21 Two perspectives of motion and energy.

(a) The macroscopic motion of the system as a whole

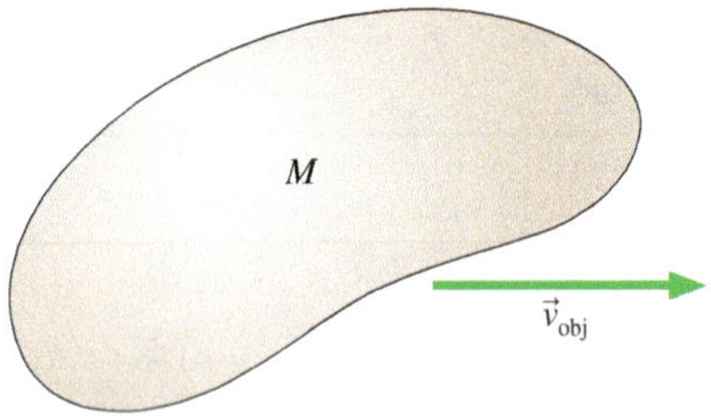

(b) The microscopic motion of the atoms inside

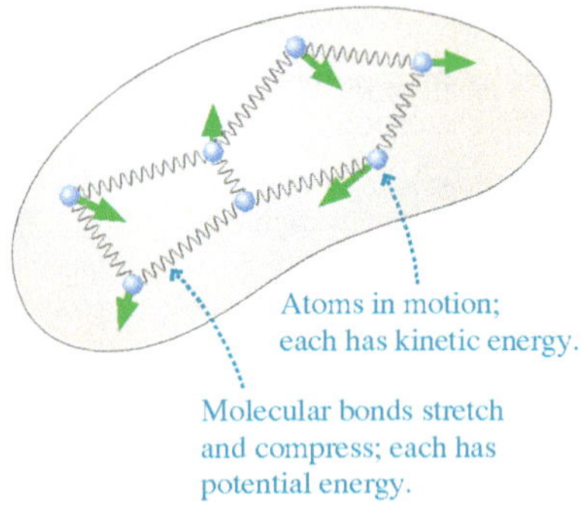

Figure 11.21 shows two different perspectives of an object. In the macrophysics perspective of FIGURE 11.21a you see an object of mass M moving as a whole with velocity v_{obj}. As a consequence of its motion, the object has macroscopic kinetic energy $K_{macro} = \frac{1}{2}Mv_{obj}^2$.

FIGURE 11.21b is a microphysics view of the same object, where now we see a *system of particles.* Each of these atoms is moving about, and in doing so they stretch and compress the spring-like bonds between them. Consequently, there is a *microscopic* kinetic and potential energy associated with the motion of atoms and bonds.

The kinetic energy of one atom is exceedingly small, but there are enormous numbers of atoms in a macroscopic object. The total kinetic energy of all the atoms is what we call the *microscopic kinetic energy* K_{micro}. The total potential energy of all the bonds is the *microscopic potential energy* U_{micro}. These microscopic energies are quite distinct from the energies K_{macro} and U_{macro} of the object as a whole.

Is the microscopic energy worth worrying about? To see, consider a 500 g ($\approx$1 lb) iron ball moving at the respectable speed of 20 m/s ($\approx$45 mph). Its macroscopic kinetic energy is

$$K_{macro} = \frac{1}{2}Mv_{obj}^2 = 100 \text{ J}$$

A periodic table of the elements shows that iron has atomic mass 56. Recall from chemistry that 56 g of iron is 1 gram-molecular weight and has Avogadro's number ($N_A = 6.02 \times 10^{23}$) of atoms. Thus 500 g of iron is $\approx$9 gram-molecular weights and contains $N \approx 9N_A \approx 5.4 \times 10^{24}$ iron atoms. The mass of each atom is

$$m = \frac{M}{N} \approx \frac{0.50 \text{ kg}}{5.4 \times 10^{24}} \approx 9 \times 10^{-26} \text{ kg}$$

How fast do atoms move? In Part IV you'll learn that $v \approx 500$ m/s is a reasonable estimate. The kinetic energy of one iron atom at this speed is

$$K_{atom} = \frac{1}{2}mv^2 \approx 1.1 \times 10^{-20} \text{ J}$$

This is very tiny, but there are a great many atoms. If all atoms move at roughly this speed, the microscopic kinetic energy is

$$K_{micro} \approx NK_{atom} \approx 60{,}000 \text{ J}$$

We'll later see that, on average, U_{micro} for a solid is equal to K_{micro}, so the total microscopic energy is $\approx 120{,}000$ J. The microscopic energy is much larger than the macroscopic kinetic energy of the object as a whole!

The combined microscopic kinetic and potential energy of the atoms is called the *thermal energy* of the system:

$$E_{\text{th}} = K_{\text{micro}} + U_{\text{micro}} \tag{11.26}$$

This energy is usually hidden from view in our macrophysics perspective, but it is quite real. We will discover later, when we reach thermodynamics, that the thermal energy is related to the *temperature* of the system. Raising the temperature causes the atoms to move faster and the bonds to stretch more, giving the system more thermal energy.

NOTE ▶ The microscopic energy of atoms is *not* called "heat." The word "heat," like the word "work," has a narrow and precise meaning in physics that is much more restricted than its use in everyday language.We will introduce the concept of heat later, when we need it. For the time being we want to use the correct term "thermal energy" to describe the random, thermal motion of the particles in a system. If the temperature of a system goes up (i.e., it gets hotter), it is because the system's thermal energy has increased. ◀

FIGURE 11.22 The atomic-level view of friction.

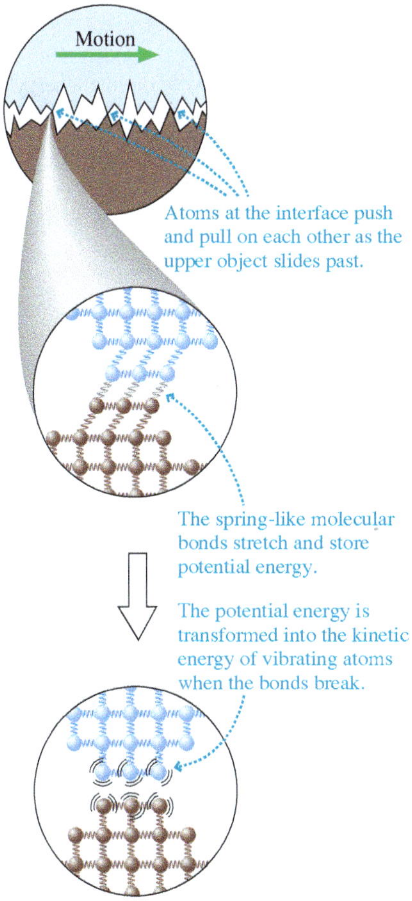

Dissipative Forces

If you shove a book across the table, it gradually slows down and stops. Where did the energy go? The common answer "It went into heat" isn't quite right.

FIGURE 11.22, the atomic model of friction from Chapter 6, shows why. As two objects slide against each other, atomic interactions at the boundary transform the kinetic energy K_{macro} of the moving object—it's slowing down—into microscopic kinetic and potential energy of vibrating atoms and stretched bonds. The energy transformation is $K \rightarrow E_{\text{th}}$, and we perceive this as an increased temperature of *both* objects. Thus the correct answer to What happens to the energy? is "It is transformed into thermal energy."

Forces such as friction and drag cause the macroscopic kinetic energy of a system to be "dissipated" as thermal energy. Hence these are called **dissipative forces.** Dissipative forces are always nonconservative forces. The energy analysis of dissipative forces is a bit subtle. Because friction causes *both* objects to get warmer, we must define the system to include both objects—both the book *and* the table, or both the car *and* the road.

FIGURE 11.23 shows a box being pulled at constant speed across a horizontal surface with friction. As you can imagine, both the surface and the box are getting warmer—increasing thermal energy. But neither the kinetic nor the potential energy of the box is changing, so where is the thermal energy coming from? Recall, from the basic energy model, that work is energy transferred to a system by forces from the environment. If we define the system to be box + surface, then the increasing thermal energy of the system is entirely due to the work being done on the system by tension in the rope: $\Delta E_{\text{th}} = W_{\text{tension}}$.

The work done by tension in pulling the box a distance Δs is simply $W_{\text{tension}} = T\Delta s$; thus $\Delta E_{\text{th}} = T\Delta s$. Because the box is moving with constant velocity, Newton's first law $\vec{F}_{\text{net}} = \vec{0}$ requires the tension to exactly balance the friction force: $T = f_k$. Consequently, the increase in thermal energy due to the dissipative force of friction is

$$\Delta E_{\text{th}} = f_k \Delta s \text{ (increased thermal energy due to friction)} \tag{11.27}$$

FIGURE 11.23 Work done by tension is dissipated as increased thermal energy.

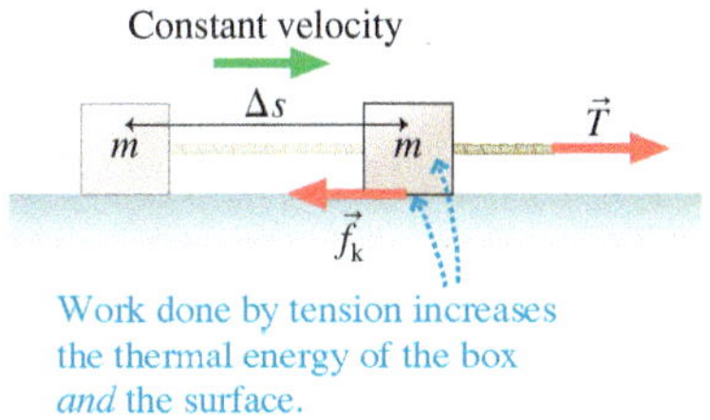

Notice that the increase in thermal energy is directly proportional to the total distance of sliding. **Dissipative forces always increase the thermal energy; they never decrease it.**

You might wonder why we didn't simply calculate the work done by friction. The rather subtle reason is that work is defined only for forces acting on a *particle*.

A particle has no internal structure and thus cannot have thermal energy. Thermal energy appears when we have to deal with extended objects, nonrigid systems of many particle-like atoms.

There is work being done on individual atoms at the boundary as they are pulled this way and that, but we would need a detailed knowledge of atomic-level friction forces to calculate this work. The friction force $\vec{f}_k$ is an average force on the object as a whole; it is not a force on any particular particle, so we cannot use it to calculate work. Further, increasing thermal energy is not an energy transfer from the book to the surface or from the surface to the book. Both book *and* surface are gaining thermal energy at the expense of the macroscopic kinetic energy.

NOTE ▶ The analysis of thermal energy is rather subtle, as we noted above. The considerations that led to Equation 11.27 do allow us to calculate the total increase in thermal energy of the entire system, but we cannot determine what fraction of ΔE_{th} goes to the book and what fraction goes to the surface. ◀

EXAMPLE 11.9 Calculating the increase in thermal energy

A rope pulls a 10 kg wooden crate 3.0 m across a wood floor. What is the change in thermal energy? The coefficient of kinetic friction is 0.20.

MODEL Let the system be crate + floor. Assume the floor is horizontal.

SOLVE The friction force on an object moving on a horizontal surface is $F_k = \mu_k n = \mu_k mg$. Thus the change in thermal energy, given by Equation 11.27, is

$$\Delta E_{th} = f_k \Delta s = \mu_k mg \Delta s$$
$$= (0.20)(10 \text{ kg})(9.80 \text{ m/s}^2)(3.0 \text{ m}) = 59 \text{ J}$$

ASSESS The thermal energy of the crate *and* floor increases by 59 J. We cannot determine ΔE_{th} for the crate (or floor) alone.

11.8 Conservation of Energy

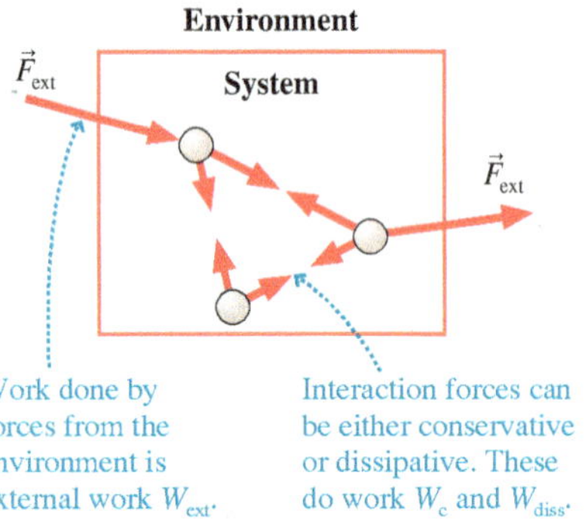

FIGURE 11.24 A system with both internal interaction forces and external forces.

Let's return to the basic energy model and start pulling together the many ideas introduced in this chapter. FIGURE 11.24 shows a general system consisting of several macroscopic objects. These objects interact with each other, and they may be acted on by external forces from the environment. Both the interaction forces and the external forces do work on the objects. The change in the system's kinetic energy is given by the work-kinetic energy theorem, $\Delta K = W_{net}$.

We previously divided W_{net} into the work W_c done by conservative forces and the work W_{nc} done by nonconservative forces. The work done by the conservative forces can be represented by a potential energy U. Let's now make a further distinction by dividing the nonconservative forces into *dissipative forces* and *external forces*. That is,

$$W_{nc} = W_{diss} + W_{ext} \tag{11.28}$$

To illustrate what we mean by an external force, suppose you pick up a box at rest on the floor and place it at rest on a table. The box gains gravitational potential energy, but $\Delta K = 0$. Or consider pulling the box across the table with a string. The box gains kinetic energy, but not by transforming potential energy. The force of your hand and the tension of the string are forces that "reach in" from the environment to change the system. Thus they are *external forces.*

We have to be careful choosing the system if we want this distinction to be valid. As you can imagine, we're going to associate W_{diss} with ΔE_{th}. We want the thermal energy E_{th} to be an energy *of the system.* Otherwise, it wouldn't make sense to talk about transforming kinetic energy into thermal energy. But for E_{th} to be an energy of the system, *both* objects involved in a dissipative interaction must be part of the system. The book sliding across the table raises the temperature of both the book *and the table.* Consequently, we must include both the book *and the table* in the system. The dissipative forces, like the conservative forces, are atomic-level interaction forces *inside* the system.

With this distinction, the work-kinetic energy theorem is

$$\Delta K = W_c + W_{diss} + W_{ext} \tag{11.29}$$

As before, we define the potential energy U such that $\Delta U = -W_c$. Remember that potential energy is really just the precomputed work of a conservative force. We've also seen that the work done by dissipative forces—the forces stretching the bonds at the boundary—increases the system's thermal energy: $\Delta E_{th} = -W_{diss}$. With these substitutions, the work-kinetic energy theorem becomes

$$\Delta K = -\Delta U - \Delta E_{th} + W_{ext}$$

We can write this more profitably as

$$\Delta K + \Delta U + \Delta E_{th} = \Delta E_{mech} + \Delta E_{th} = \Delta E_{sys} = W_{ext} \tag{11.30}$$

where $E_{sys} = E_{mech} + E_{th}$ is the total energy of the system. Equation 11.30 is the **energy equation** of the system.

Equation 11.30 is our most general statement about how the energy of a system changes, but we still need to give a clear interpretation as to what it says. In Chapter 9 we defined an *isolated system* as a system for which the *net* external force is zero. It follows that no external work is done on an isolated system: $W_{ext} = 0$. Thus one conclusion from Equation 11.30 is that **the total energy E_{sys} of an isolated system is conserved**. That is, $\Delta E_{sys} = 0$ for an isolated system. If, in addition, the system is nondissipative (i.e., no friction forces), then $\Delta E_{th} = 0$. In that case, the mechanical energy E_{mech} is conserved.

These conclusions about energy can be summarized as the *law of conservation of energy:*

Law of conservation of energy The total energy $E_{sys} = E_{mech} + E_{th}$ of an isolated system is a constant. The kinetic, potential, and thermal energy within the system can be transformed into each other, but their sum cannot change. Further, the mechanical energy $E_{mech} = K + U$ is conserved if the system is both isolated and nondissipative.

The law of conservation of energy is one of the most powerful statements in physics. FIGURE 11.25a redraws the basic energy model of Figure 11.1 Now you can see that this is a pictorial representation of Equation 11.30. E_{sys}, the total energy of the system, changes only if external forces transfer energy into or out of the system by doing work on the system. The kinetic, potential, and thermal energy within the system can be transformed into each other by interaction forces within the system. As FIGURE 11.25b shows, $E_{sys} = K + U + E_{th}$ remains constant if the system is isolated. The *transfer* and *transformation* of energy are what the basic energy model is all about.

FIGURE 11.25 The basic energy model is a pictorial representation of the energy equation.

(a) A system interacting with its environment

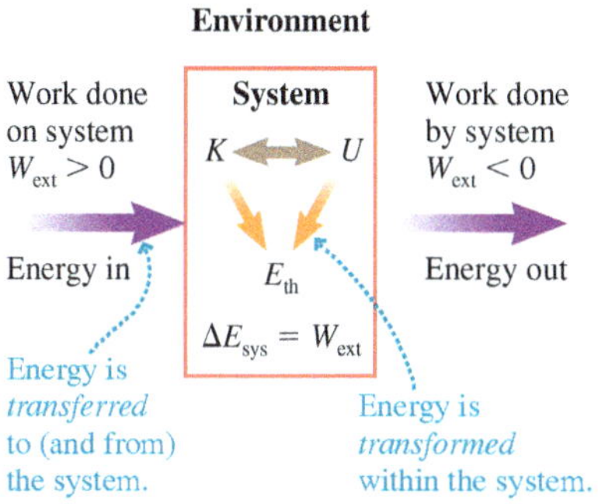

(b) An isolated system

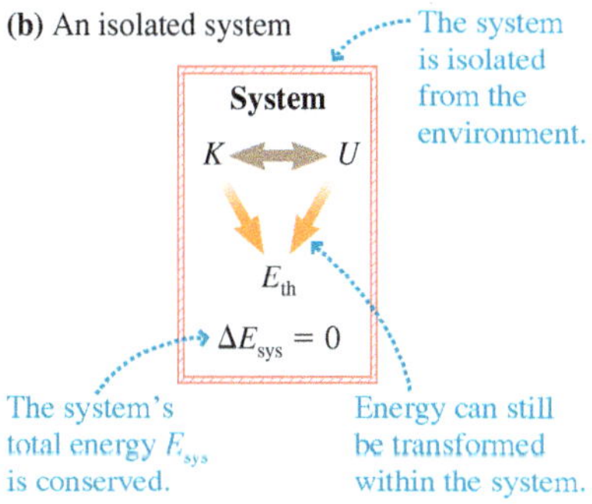

Energy Bar Charts

The energy bar charts of Chapter 10 can now be expanded to include the thermal energy and the work done by external forces. The energy equation, Equation 11.30, can be written

$$K_i + U_i + W_{ext} = K_f + U_f + \Delta E_{th} \tag{11.31}$$

The left side is the "before" condition ($K_i + U_i$) plus any energy that is added to or removed from the system. The right side is the "after" situation. The "energy accounting" of Equation 11.31 can be represented by the bar chart of FIGURE 11.26 on the next page.

NOTE ▶ We don't have any way to determine $(E_{th})_i$ or $(E_{th})_f$, but ΔE_{th} is always positive whenever the system contains dissipative forces. ◀

FIGURE 11.26 An energy bar chart shows how all the energy is accounted for.

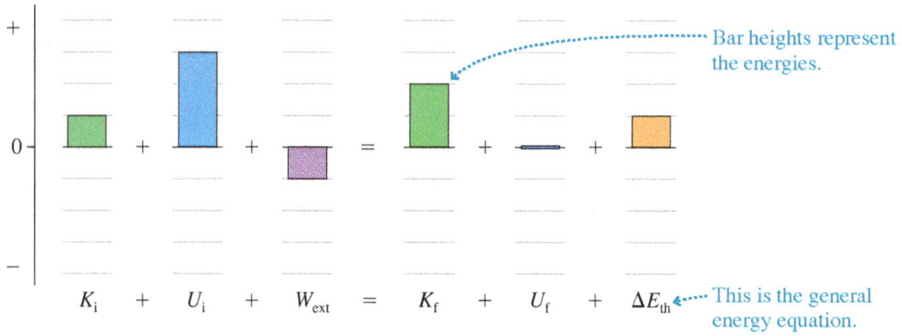

Let's look at a few examples.

EXAMPLE 11.10 **Energy bar chart I**

A speeding car skids to a halt. Show the energy transfers and transformations on an energy bar chart.

SOLVE The car has an initial kinetic energy K_i. That energy is transformed into the thermal energy of the car and the road. The potential energy doesn't change and no work is done by external forces, so the process is an energy transformation $K_i \rightarrow E_{th}$. This is shown in FIGURE 11.27. E_{sys} is conserved but E_{mech} is not.

FIGURE 11.27 Energy bar chart for Example 11.10.

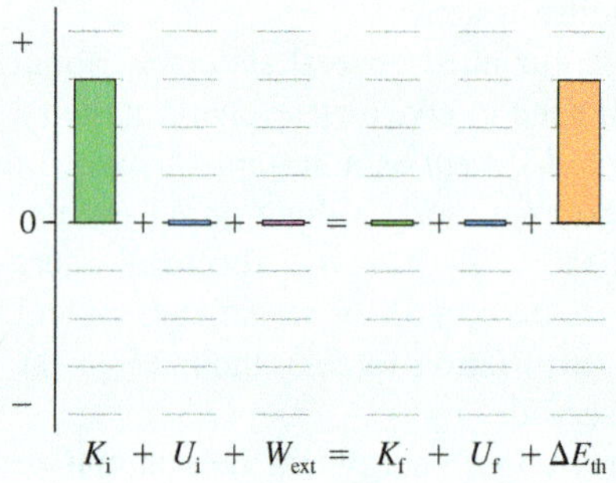

EXAMPLE 11.11 **Energy bar chart II**

A rope lifts a box at constant speed. Show the energy transfers and transformations on an energy bar chart.

SOLVE The tension in the rope is an external force that does work on the box, increasing the potential energy of the box. The kinetic energy is unchanged because the speed is constant. The process is an energy transfer $W_{ext} \rightarrow U_f$, as FIGURE 11.28 shows. This is not an isolated system, so E_{sys} is not conserved.

FIGURE 11.28 Energy bar chart for Example 11.11.

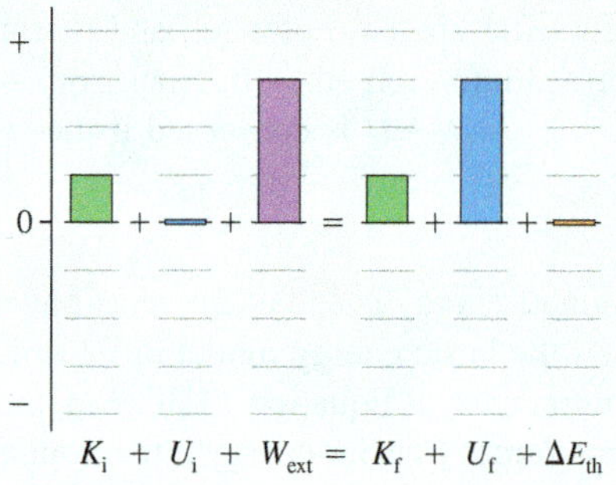

EXAMPLE 11.12 **Energy bar chart III**

The box that was lifted in Example 11.11 falls at a steady speed as the rope spins a generator and causes a lightbulb to glow. Air resistance is negligible. Show the energy transfers and transformations on an energy bar chart.

SOLVE The initial potential energy decreases, but K does not change and $\Delta E_{th} = 0$. The tension in the rope is an external force that does work, but W_{ext} is negative in this case because $\vec{T}$ points up while the displacement $\Delta\vec{r}$ is down. Negative work means that energy is transferred from the system to the environment or, in more informal terms, that the *system does work on the environment.* The falling box does work on the generator to spin it and light the bulb. Energy is transferred out of the system and eventually ends up in the lightbulb as electrical energy. The process is $U_i \rightarrow W_{ext}$. This is shown in FIGURE 11.29.

FIGURE 11.29 Energy bar chart for Example 11.12.

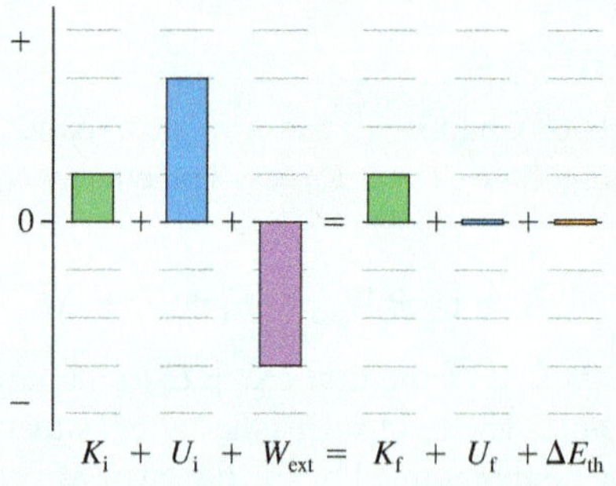

Strategy for Energy Problems

This is a good place to summarize the strategy we have been developing for using the concept of energy.

PROBLEM-SOLVING STRATEGY 11.1 Solving energy problems

MODEL Identify which objects are part of the system and which are in the environment. When possible, choose a system without friction or other dissipative forces. Some problems may need to be subdivided into two or more parts.

VISUALIZE Draw a before-and-after pictorial representation and an energy bar chart. A free-body diagram is helpful if you're going to calculate work.

SOLVE If the system is both isolated and nondissipative, then the mechanical energy is conserved:

$$K_f + U_f = K_i + U_i$$

If there are external or dissipative forces, calculate W_{ext} and ΔE_{th}. Then use the more general energy equation

$$K_f + U_f + \Delta E_{th} = K_i + U_i + W_{ext}$$

Kinematics and/or other conservation laws may be needed for some problems.

ASSESS Check that your result has the correct units, is reasonable, and answers the question.

STOP TO THINK 11.6 A child at the playground slides down a pole at constant speed. This is a situation in which

a. $U \rightarrow K$. E_{mech} is not conserved but E_{sys} is.
b. $U \rightarrow E_{th}$. E_{mech} is conserved.
c. $U \rightarrow E_{th}$. E_{mech} is not conserved but E_{sys} is.
d. $K \rightarrow E_{th}$. E_{mech} is not conserved but E_{sys} is.
e. $U \rightarrow W_{ext}$. Neither E_{mech} nor E_{sys} is conserved.

11.9 Power

Work is a transfer of energy between the environment and a system. In many situations we would like to know *how quickly* the energy is transferred. Does the force act quickly and transfer the energy very rapidly, or is it a slow and lazy transfer of energy? If you need to buy a motor to lift 2000 lb of bricks up 50 ft, it makes a *big* difference whether the motor has to do this in 30 s or 30 min!

The question How quickly? implies that we are talking about a *rate*. For example, the velocity of an object—how quickly it is moving—is the *rate of change* of position. So when we raise the issue of how quickly the energy is transferred, we are talking about the *rate of transfer* of energy. The rate at which energy is transferred or transformed is called the **power** P, and it is defined as

$$P \equiv \frac{dE_{sys}}{dt} \tag{11.32}$$

The unit of power is the **watt,** which is defined as 1 watt $= 1\text{ W} \equiv 1\text{ J/s}$.

The English unit of power is the *horsepower*. The conversion factor to watts is

$$1\text{ horsepower} = 1\text{ hp} = 746\text{ W}$$

Many common appliances, such as motors, are rated in hp.

A force that is doing work (i.e., transferring energy) at a rate of 3 J/s has an "output power" of 3 W. The system gaining energy at the rate of 3 J/s is said to "consume" 3 W of power. Common prefixes used with power are mW (milliwatts), kW (kilowatts), and MW (megawatts).

EXAMPLE 11.13 Choosing a motor

What power motor is needed to lift a 2000 kg elevator at a steady 3.0 m/s?

SOLVE The tension in the cable does work on the elevator to lift it. Because the cable is pulled by the motor, we say that the motor does the work of lifting the elevator. The net force is zero because the elevator moves at constant velocity, so the tension is simply $T = mg = 19{,}600$ N. The energy gained by the elevator is

$$\Delta E_{sys} = W_{ext} = T\,\Delta y$$

The power required to give the system this much energy in a time interval Δt is

$$P = \frac{\Delta E_{sys}}{\Delta t} = \frac{T\,\Delta y}{\Delta t}$$

But $\Delta y = v\Delta t$, so

$$P = Tv = (19{,}600\text{ N})(3.0\text{ m/s}) = 58{,}800\text{ W}$$
$$= 79\text{ hp}$$

Highly trained athletes have a tremendous power output.

The idea of power as a *rate* of energy transfer applies no matter what the form of energy. FIGURE 11.30 shows three examples of the idea of power. For now, we want to focus primarily on *work* as the source of energy transfer. Within this more limited scope, power is simply the **rate of doing work**: $P = dW/dt$. If a particle moves through a small displacement $d\vec{r}$ while acted on by force $\vec{F}$, the force does a small amount of work dW given by

$$dW = \vec{F}\cdot d\vec{r}$$

Dividing both sides by dt, to give a rate of change, yields

$$\frac{dW}{dt} = \vec{F}\cdot\frac{d\vec{r}}{dt}$$

But $d\vec{r}/dt$ is the velocity $\vec{v}$, so we can write the power as

$$P = \vec{F}\cdot\vec{v} = Fv\cos\theta \qquad (11.33)$$

In other words, the power delivered to a particle by a force acting on it is the dot product of the force and the particle's velocity. These ideas will become clearer with some examples.

FIGURE 11.30 Examples of power.

Electrical energy ⟶ light and thermal energy at 100 J/s

Chemical energy of glucose and fat ⟶ mechanical energy at ≈350 J/s ≈ $\frac{1}{2}$ hp

Chemical energy of gas ⟶ thermal energy at 20,000 J/s

EXAMPLE 11.14 **Power output of a motor**

A factory uses a motor and a cable to drag a 300 kg machine to the proper place on the factory floor. What power must the motor supply to drag the machine at a speed of 0.50 m/s? The coefficient of friction between the machine and the floor is 0.60.

SOLVE The force applied by the motor, through the cable, is the tension force $\vec{T}$. This force does work on the machine with power $P = Tv$. The machine is in dynamic equilibrium because the motion is at constant velocity, hence the tension in the rope balances the friction and is

$$T = f_k = \mu_k mg$$

The motor's power output is

$$P = Tv = \mu_k mgv = 882 \text{ W}$$

EXAMPLE 11.15 **Power output of a car engine**

A 1500 kg car has a front profile that is 1.6 m wide and 1.4 m high and a drag coefficient of 0.50. The coefficient of rolling friction is 0.02. What power must the engine provide to drive at a steady 30 m/s (≈65 mph) if 25% of the power is "lost" before reaching the drive wheels?

SOLVE The net force on a car moving at a steady speed is zero. The motion is opposed both by rolling friction and by air resistance. The forward force on the car $\vec{F}_{car}$ (recall that this is really $\vec{F}_{ground\ on\ car}$, a reaction to the drive wheels pushing backward on the ground with $\vec{F}_{car\ on\ ground}$) exactly balances the two opposing forces:

$$F_{car} = f_r + D$$

where $\vec{D}$ is the drag due to the air. Using the results of Chapter 6, where both rolling friction and drag were introduced, this becomes

$$F_{car} = \mu_r mg + \frac{1}{2}C\rho Av^2 = 294 \text{ N} + 605 \text{ N} = 899 \text{ N}$$

Here $A = (1.6 \text{ m}) \times (1.4 \text{ m})$ is the front cross-section area of the car, and we used 1.2 kg/m^3 as the density of air. The power required to push the car forward at 30 m/s is

$$P_{car} = F_{car}v = (899 \text{ N})(30 \text{ m/s}) = 27{,}000 \text{ W} = 36 \text{ hp}$$

This is the power *needed* at the drive wheels to push the car against the dissipative forces of friction and air resistance. The power output of the engine is larger because some energy is used to run the water pump, the power steering, and other accessories. In addition, energy is lost to friction in the drive train. If 25% of the power is lost (a typical value), leading to $P_{car} = 0.75P_{engine}$, the engine's power output is

$$P_{engine} = \frac{P_{car}}{0.75} = 31{,}900 \text{ W} = 48 \text{ hp}$$

ASSESS Automobile engines are typically rated at ≈200 hp. Most of that power is reserved for fast acceleration and climbing hills.

STOP TO THINK 11.7 Four students run up the stairs in the time shown. Rank in order, from largest to smallest, their power outputs P_a to P_d.

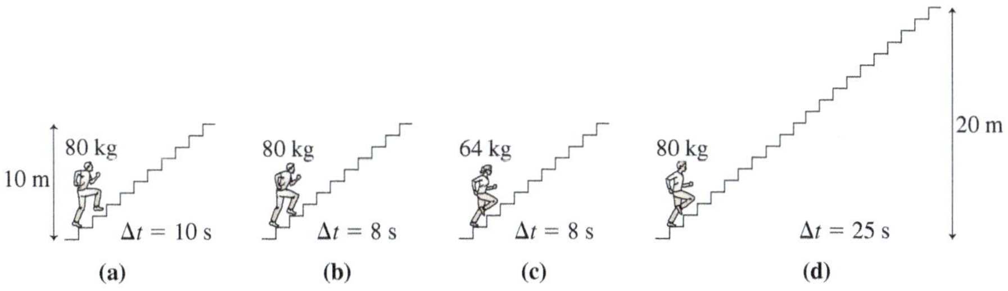

CHALLENGE EXAMPLE 11.16 Stretching a spring

A 5.0 kg box is attached to one end of a spring with spring constant 80 N/m. The other end of the spring is anchored to a wall. Initially the box is at rest at the spring's equilibrium position. A rope with a constant tension of 100 N then pulls the box away from the wall. The coefficient of friction between the box and the floor is 0.30. How much power is being supplied by the hand or motor pulling the rope when the box has moved 50 cm?

MODEL This is a complex situation, but one that we can analyze. First, identify the box, the spring, and the floor as the system. We need the floor inside the system because friction increases the temperature of the box *and* the floor. The tension in the rope is an external force. The work W_{ext} done by the rope's tension transfers energy into the system, causing K, U_s, and E_{th} all to increase.

VISUALIZE FIGURE 11.31a is a before-and-after pictorial representation. The energy transfers and transformations are shown in the energy bar chart of FIGURE 11.31b.

SOLVE The power supplied by the rope's tension—the rate at which energy is being delivered to the system—is $P = Tv$. We know the rope's tension, so we need to use energy considerations to find the speed v_1 after the box has moved to $x_1 = 50$ cm. The energy equation $K_f + U_f + \Delta E_{th} = K_i + U_i + W_{ext}$ is

$$\frac{1}{2}mv_1^2 + \frac{1}{2}kx_1^2 + \Delta E_{th} = \frac{1}{2}mv_0^2 + \frac{1}{2}kx_0^2 + W_{ext}$$

We know that $x_0 = 0$ m and $v_0 = 0$ m/s, so the energy equation simplifies to

$$\frac{1}{2}mv_1^2 = W_{ext} - \Delta E_{th} - \frac{1}{2}kx_1^2$$

The external work done by the rope's tension is

$$W_{ext} = \vec{T} \cdot \Delta\vec{r} = T(\Delta x)\cos 0° = (100\text{ N})(0.50\text{ m}) = 50.0\text{ J}$$

The increase in thermal energy is given by Equation 11.27:

$$\Delta E_{th} = f_k \Delta x = \mu_k mg\Delta x$$
$$= (0.30)(5.0\text{ kg})(9.80\text{ m/s}^2)(0.50\text{ m}) = 7.4\text{ J}$$

Solving for the speed v_1 at $x_1 = 50$ cm $= 0.50$ m gives

$$v_1 = \sqrt{\frac{2(W_{ext} - \Delta E_{th} - \frac{1}{2}kx_1^2)}{m}} = 3.6\text{ m/s}$$

The power being supplied at this instant is

$$P_1 = Tv_1 = (100\text{ N})(3.6\text{ m/s}) = 360\text{ W}$$

ASSESS The work done by the rope's tension is energy transferred to the system. Part of the energy increases the speed of the box, part increases the potential energy stored in the spring, and part is transformed into increased thermal energy, increasing the temperature. We had to bring all the energy ideas together to solve this problem.

FIGURE 11.31 Pictorial representation and energy bar chart for Challenge Example 11.16.

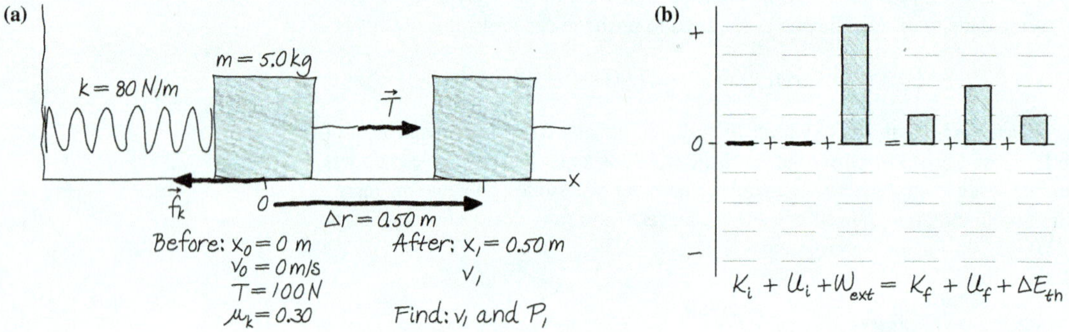

SUMMARY

The goal of Chapter 11 has been to develop a more complete understanding of energy and its conservation.

General Principles

Basic Energy Model

- Energy is *transferred* to or from the system by work.
- Energy is *transformed* within the system.

Two versions of the energy equation are

$$\Delta E_{sys} = \Delta K + \Delta U + \Delta E_{th} = W_{ext}$$

$$K_f + U_f + \Delta E_{th} = K_i + U_i + W_{ext}$$

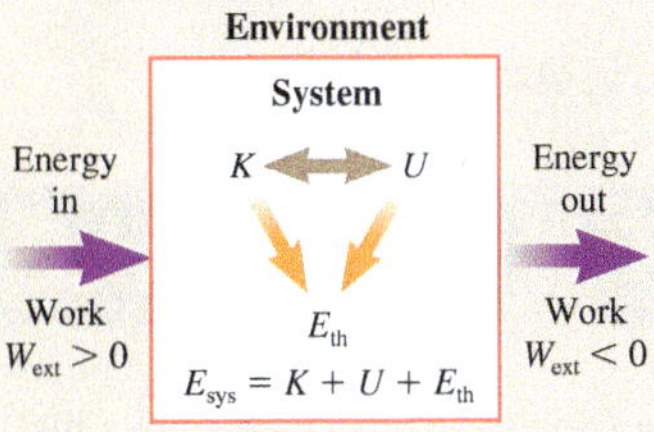

Solving Energy Problems

MODEL Identify objects in the system.

VISUALIZE Draw a before-and-after pictorial representation and an energy bar chart.

SOLVE Use the energy equation

$$K_f + U_f + \Delta E_{th} = K_i + U_i + W_{ext}$$

ASSESS Is the result reasonable?

Law of Conservation of Energy

- **Isolated system:** $W_{ext} = 0$. The total energy $E_{sys} = E_{mech} + E_{th}$ is conserved. $\Delta E_{sys} = 0$.
- **Isolated, nondissipative system:** $W_{ext} = 0$ and $W_{diss} = 0$. The mechanical energy E_{mech} is conserved.

$$\Delta E_{mech} = 0 \text{ or } K_f + U_f = K_i + U_i$$

Important Concepts

The **work-kinetic energy theorem** is

$$\Delta K = W_{net} = W_c + W_{diss} + W_{ext}$$

With $W_c = -\Delta U$ for conservative forces and $W_{diss} = -\Delta E_{th}$ for dissipative forces, this becomes the energy equation.

The **work** done by a force on a particle as it moves from s_i to s_f is

$$W = \int_{s_i}^{s_f} F_s\, ds = \text{area under the force curve}$$

$$= \vec{F} \cdot \Delta\vec{r} \text{ if } \vec{F} \text{ is a constant force}$$

Conservative forces are forces for which the work is independent of the path followed. The work done by a conservative force can be represented as a **potential energy:**

$$\Delta U = U_f - U_i = -W_c(\mathrm{i} \rightarrow \mathrm{f})$$

A conservative force is found from the potential energy by

$$F_s = -dU/ds = \text{negative of the slope of the PE curve}$$

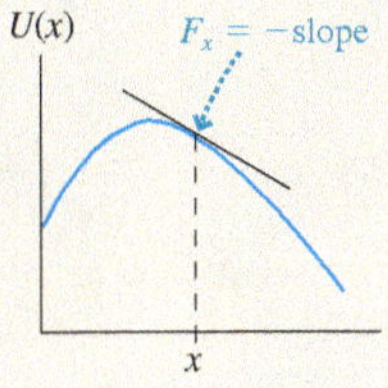

Dissipative forces transform **macroscopic energy** into thermal energy, which is the **microscopic energy** of the atoms and molecules. For friction:

$$\Delta E_{th} = f_k \Delta s$$

Applications

Power is the rate at which energy is transferred or transformed:

$$P = \frac{dE_{sys}}{dt}$$

For a particle moving with velocity $\vec{v}$, the power delivered to the particle by force $\vec{F}$ is $P = \vec{F} \cdot \vec{v} = Fv\cos\theta$.

Dot product

$$\vec{A} \cdot \vec{B} = AB\cos\alpha = A_xB_x + A_yB_y$$

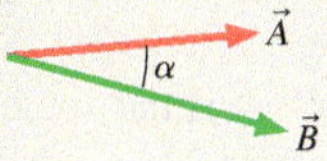

Energy bar charts display the energy equation $K_f + U_f + \Delta E_{th} = K_i + U_i + W_{ext}$ in graphical form.

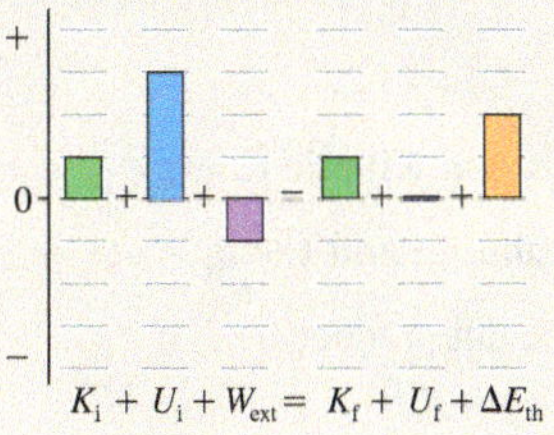

Terms and Notation

thermal energy, E_{th}
system energy, E_{sys}
energy transformation
energy transfer
work, W
work-kinetic energy theorem
dot product
scalar product
conservative force
nonconservative force
macrophysics
microphysics
dissipative force
energy equation
law of conservation of energy
power, P
watt, W

CONCEPTUAL QUESTIONS

1. A process occurs in which a system's potential energy decreases while the system does work on the environment. Does the system's kinetic energy increase, decrease, or stay the same? Or is there not enough information to tell? Explain.
2. A process occurs in which a system's potential energy increases while the environment does work on the system. Does the system's kinetic energy increase, decrease, or stay the same? Or is there not enough information to tell? Explain.
3. The kinetic energy of a system decreases while its potential energy and thermal energy are unchanged. Does the environment do work on the system, or does the system do work on the environment? Explain.
4. You drop a ball from a high balcony and it falls freely. Does the ball's kinetic energy increase by equal amounts in equal time intervals, or by equal amounts in equal distances? Explain.
5. A particle moves in a vertical plane along the *closed* path seen in FIGURE Q11.5, starting at A and eventually returning to its starting point. How much work is done on the particle by gravity? Explain.

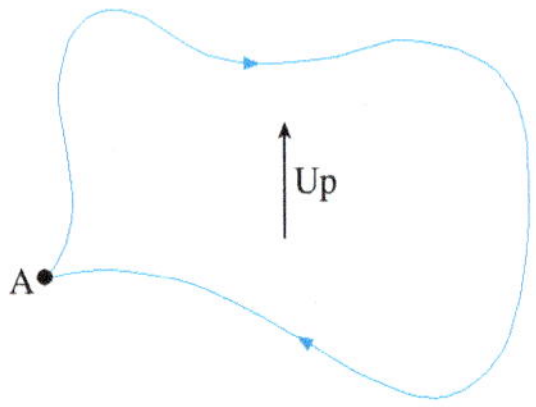

FIGURE Q11.5

6. A 0.2 kg plastic cart and a 20 kg lead cart both roll without friction on a horizontal surface. Equal forces are used to push both carts forward a distance of 1 m, starting from rest. After traveling 1 m, is the kinetic energy of the plastic cart greater than, less than, or equal to the kinetic energy of the lead cart? Explain.
7. You need to raise a heavy block by pulling it with a massless rope. You can either (a) pull the block straight up height h, or (b) pull it up a long, frictionless plane inclined at a 15° angle until its height has increased by h. Assume you will move the block at constant speed either way. Will you do more work in case a or case b? Or is the work the same in both cases? Explain.
8. a. If the force on a particle at some point in space is zero, must its potential energy also be zero at that point? Explain.
   b. If the potential energy of a particle at some point in space is zero, must the force on it also be zero at that point? Explain.
9. A car traveling at 60 mph slams on its brakes and skids to a halt. What happened to the kinetic energy the car had just before stopping?
10. What energy transformations occur as a skier glides down a gentle slope at constant speed?
11. Give a *specific* example of a situation in which
    a. $W_{ext} \rightarrow K$ with $\Delta U = 0$ and $\Delta E_{th} = 0$.
    b. $W_{ext} \rightarrow E_{th}$ with $\Delta K = 0$ and $\Delta U = 0$.
12. The motor of a crane uses power P to lift a steel beam. By what factor must the motor's power increase to lift the beam twice as high in half the time?

EXERCISES AND PROBLEMS

Problems labeled [shaded box] integrate material from earlier chapters.

Exercises

Section 11.2 Work and Kinetic Energy

Section 11.3 Calculating and Using Work

1. | Evaluate the dot product $\vec{A} \cdot \vec{B}$ if
   a. $\vec{A} = 3\hat{\imath} + 4\hat{\jmath}$ and $\vec{B} = 2\hat{\imath} - 6\hat{\jmath}$.
   b. $\vec{A} = 3\hat{\imath} - 2\hat{\jmath}$ and $\vec{B} = 6\hat{\imath} + 4\hat{\jmath}$.
2. | Evaluate the dot product $\vec{A} \cdot \vec{B}$ if
   a. $\vec{A} = 4\hat{\imath} - 2\hat{\jmath}$ and $\vec{B} = -2\hat{\imath} - 3\hat{\jmath}$.
   b. $\vec{A} = -4\hat{\imath} + 2\hat{\jmath}$ and $\vec{B} = 2\hat{\imath} + 4\hat{\jmath}$.
3. || What is the angle θ between vectors $\vec{A}$ and $\vec{B}$ in each part of Exercise 1?
4. || What is the angle θ between vectors $\vec{A}$ and $\vec{B}$ in each part of Exercise 2?

5. | Evaluate the dot product of the three pairs of vectors in FIGURE EX11.5.

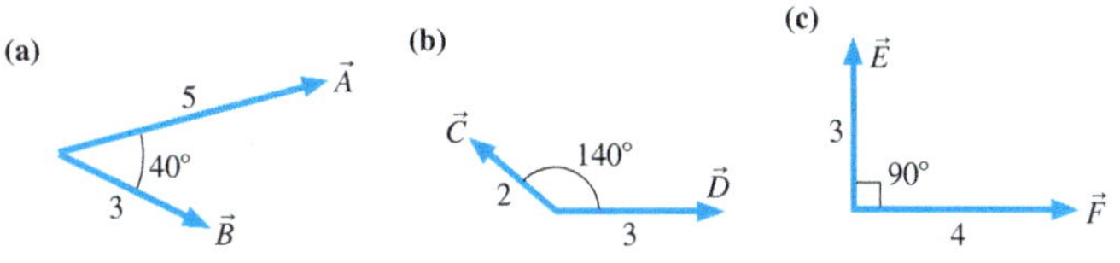

FIGURE EX11.5

6. | Evaluate the dot product of the three pairs of vectors in FIGURE EX11.6.

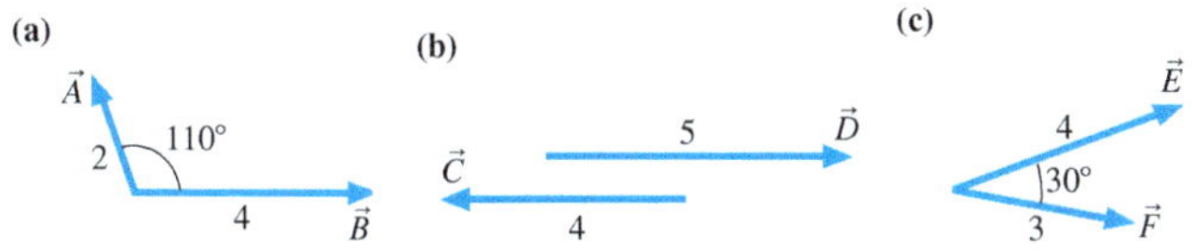

FIGURE EX11.6

7. | How much work is done by the force $\vec{F} = (-3.0\hat{\imath} + 6.0\hat{\jmath})$ N on a particle that moves through displacement (a) $\Delta\vec{r} = 2.0\hat{\imath}$ m and (b) $\Delta\vec{r} = 2.0\hat{\jmath}$ m?

8. | How much work is done by the force $\vec{F} = (-4.0\hat{\imath} - 6.0\hat{\jmath})$ N on a particle that moves through displacement (a) $\Delta\vec{r} = -3.0\hat{\imath}$ m and (b) $\Delta\vec{r} = (3.0\hat{\imath} - 2.0\hat{\jmath})$ m?

9. || A 20 g particle is moving to the left at 30 m/s. How much net work must be done on the particle to cause it to move to the right at 30 m/s?

10. | A 2.0 kg book is lying on a 0.75-m-high table. You pick it up and place it on a bookshelf 2.25 m above the floor.
   a. How much work does gravity do on the book?
   b. How much work does your hand do on the book?

11. || The two ropes seen in FIGURE EX11.11 are used to lower a 255 kg piano 5.00 m from a second-story window to the ground. How much work is done by each of the three forces?

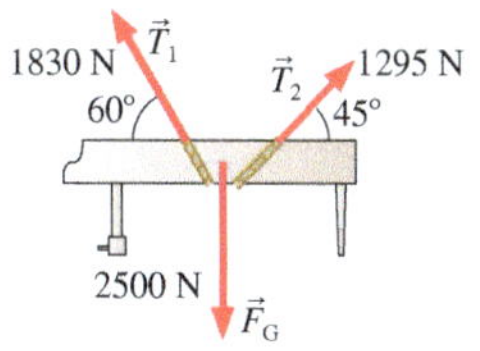

FIGURE EX11.11

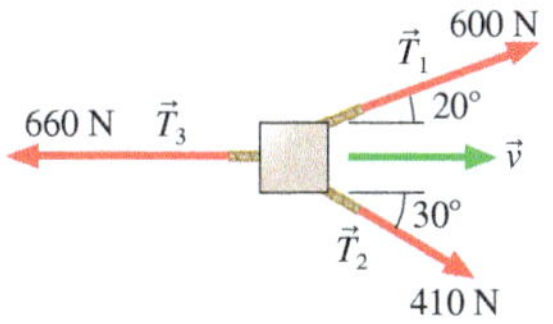

FIGURE EX11.12

12. || The three ropes shown in the bird's-eye view of FIGURE EX11.12 are used to drag a crate 3.0 m across the floor. How much work is done by each of the three forces?

13. || FIGURE EX11.13 is the velocity-versus-time graph for a 2.0 kg object moving along the x-axis. Determine the work done on the object during each of the four intervals AB, BC, CD, and DE.

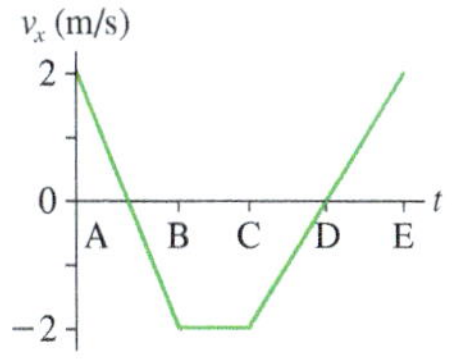

FIGURE EX11.13

Section 11.4 The Work Done by a Variable Force

14. | FIGURE EX11.14 is the force-versus-position graph for a particle moving along the x-axis. Determine the work done on the particle during each of the three intervals 0–1 m, 1–2 m, and 2–3 m.

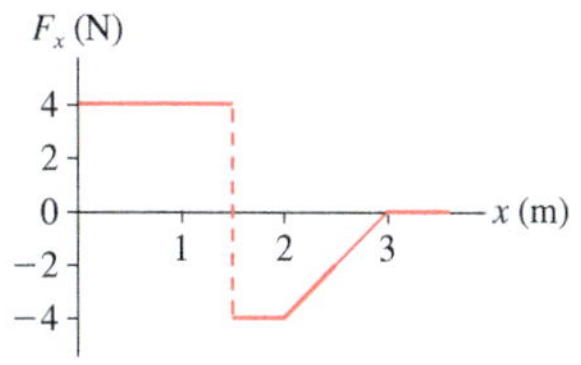

FIGURE EX11.14

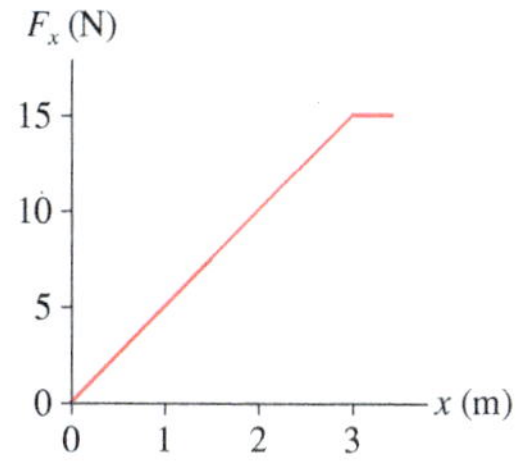

FIGURE EX11.15

15. || A 500 g particle moving along the x-axis experiences the force shown in FIGURE EX11.15. The particle's velocity is 2.0 m/s at $x = 0$ m. What is its velocity at $x = 1$ m, 2 m, and 3 m?

16. || A 2.0 kg particle moving along the x-axis experiences the force shown in FIGURE EX11.16. The particle's velocity is 4.0 m/s at $x = 0$ m. What is its velocity at $x = 2$ m and 4 m?

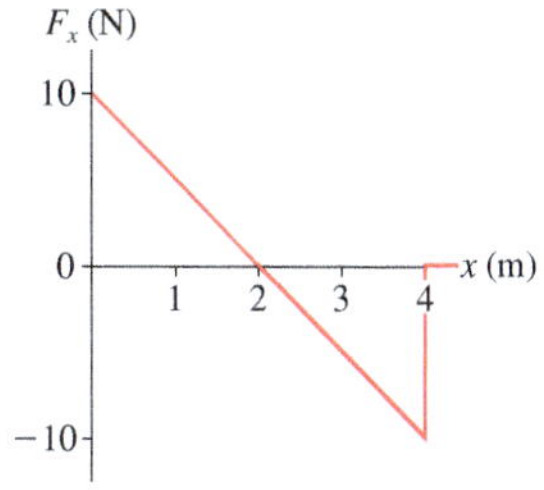

FIGURE EX11.16

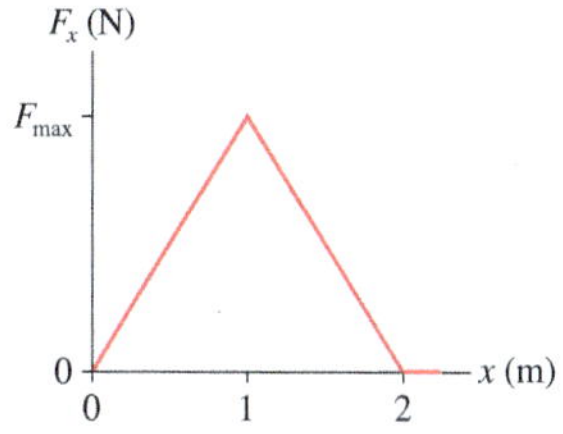

FIGURE EX11.17

17. || A 500 g particle moving along the x-axis experiences the force shown in FIGURE EX11.17. The particle goes from $v_x = 2.0$ m/s at $x = 0$ m to $v_x = 6.0$ m/s at $x = 2$ m. What is F_{max}?

Section 11.5 Work and Potential Energy

Section 11.6 Finding Force from Potential Energy

18. || A particle has the potential energy shown in FIGURE EX11.18. What is the x-component of the force on the particle at $x = 5$, 15, 25, and 35 cm?

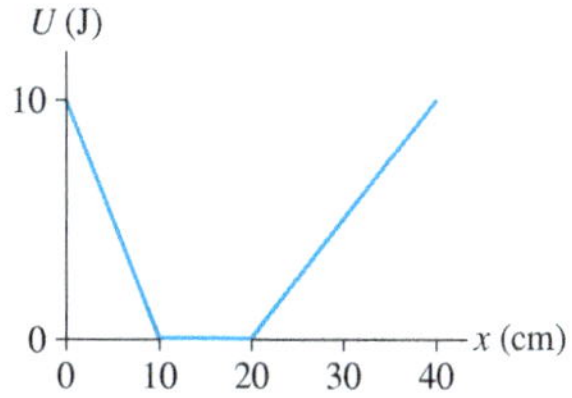

FIGURE EX11.18

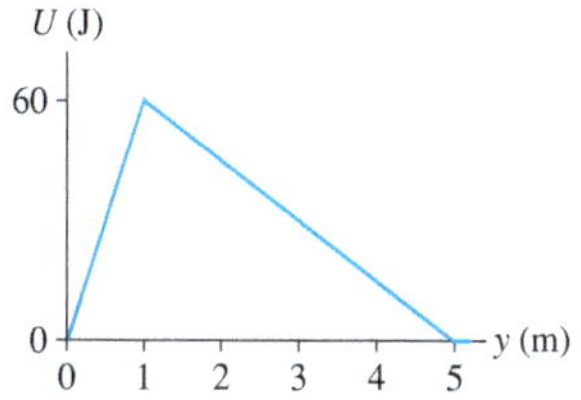

FIGURE EX11.19

19. || A particle has the potential energy shown in FIGURE EX11.19. What is the y-component of the force on the particle at $y = 0.5$ m and 4 m?

20. || A particle moving along the y-axis has the potential energy $U = 4y^3$ J, where y is in m. What is the y-component of the force on the particle at $y = 0$ m, 1 m, and 2 m?

21. || A particle moving along the x-axis has the potential energy $U = 10/x$ J, where x is in m. What is the x-component of the force on the particle at $x = 2$ m, 5 m, and 8 m?

Section 11.7 Thermal Energy

22. || The mass of a carbon atom is 2.0×10^{-26} kg.
   a. What is the kinetic energy of a carbon atom moving with a speed of 500 m/s?

b. Two carbon atoms are joined by a spring-like carbon-carbon bond. The potential energy stored in the bond has the value you calculated in part a if the bond is stretched 0.050 nm. What is the bond's spring constant?

23. || In Part IV you'll learn to calculate that 1 mole (6.02×10^{23} atoms) of helium atoms in the gas phase has 3700 J of microscopic kinetic energy at room temperature. If we assume that all atoms move with the same speed, what is that speed? The mass of a helium atom is 6.68×10^{-27} kg.

24. || A 20 kg child slides down a 3.0-m-high playground slide. She starts from rest, and her speed at the bottom is 2.0 m/s.
a. Describe the energy transfers and transformations occurring during the slide.
b. What is the change in the combined thermal energy of the slide and the seat of her pants?

Section 11.8 Conservation of Energy

25. || A system loses 400 J of potential energy. In the process, it does 400 J of work on the environment and the thermal energy increases by 100 J. Show this process on an energy bar chart.

26. || A system loses 500 J of kinetic energy while gaining 200 J of potential energy. The thermal energy increases 100 J. Show this process on an energy bar chart.

27. | How much work is done by the environment in the process shown in FIGURE EX11.27? Is energy transferred from the environment to the system or from the system to the environment?

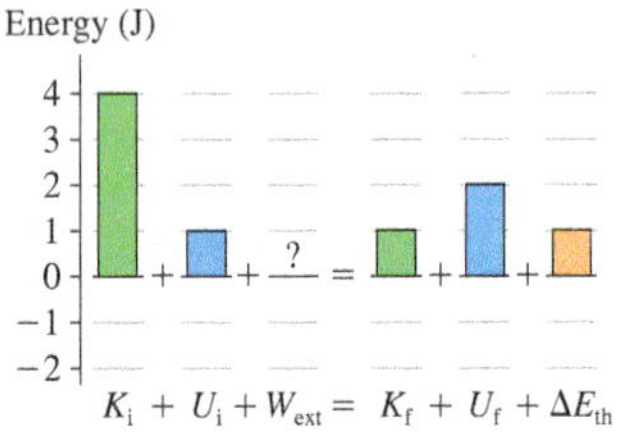

FIGURE EX11.27

28. || A cable with 20.0 N of tension pulls straight up on a 1.02 kg block that is initially at rest. What is the block's speed after being lifted 2.00 m? Solve this problem using work and energy.

Section 11.9 Power

29. | a. How much work does an elevator motor do to lift a 1000 kg elevator a height of 100 m?
b. How much power must the motor supply to do this in 50 s at constant speed?

30. | a. How much work must you do to push a 10 kg block of steel across a steel table at a steady speed of 1.0 m/s for 3.0 s?
b. What is your power output while doing so?

31. || At midday, solar energy strikes the earth with an intensity of about 1 kW/m^2. What is the area of a solar collector that could collect 150 MJ of energy in 1 h? This is roughly the energy content of 1 gallon of gasoline.

32. || Which consumes more energy, a 1.2 kW hair dryer used for 10 min or a 10 W night light left on for 24 h?

33. | A 2.0 hp electric motor on a water well pumps water from 10 m below the surface. The density of water is 1.0 kg per liter. How many liters of water does the motor pump in 1 h?

34. || A 50 kg sprinter, starting from rest, runs 50 m in 7.0 s at constant acceleration.
a. What is the magnitude of the horizontal force acting on the sprinter?
b. What is the sprinter's power output at 2.0 s, 4.0 s, and 6.0 s?

35. | BIO a. Estimate the height in meters of the two flights of stairs that go from the first to the third floor of a building.
b. Estimate how long it takes you to *run* up these two flights of stairs.
c. Estimate your power output in both watts and horsepower while running up the stairs.

36. | BIO A 70 kg human sprinter can accelerate from rest to 10 m/s in 3.0 s. During the same time interval, a 30 kg greyhound can go from rest to 20 m/s. What is the average power output of each? *Average* power over a time interval Δt is $\Delta E / \Delta t$.

Problems

37. | A particle moves from A to D in FIGURE P11.37 while experiencing force $\vec{F} = (6\hat{\imath} + 8\hat{\jmath})$ N. How much work does the force do if the particle follows path (a) ABD, (b) ACD, and (c) AD? Is this a conservative force? Explain.

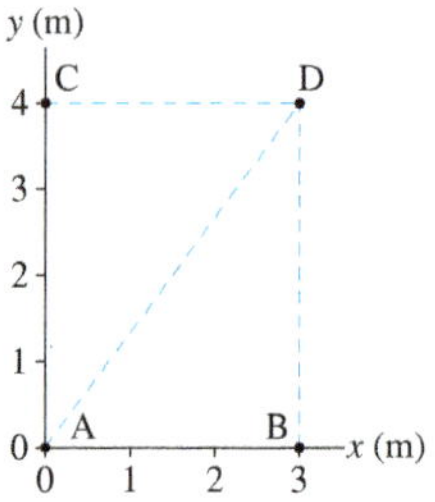

FIGURE P11.37

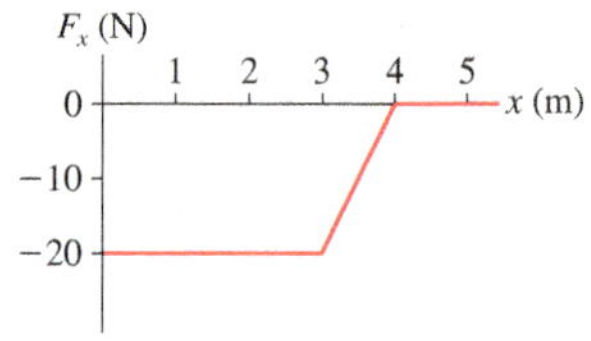

FIGURE P11.38

38. || A 100 g particle experiences the one-dimensional, conservative force F_x shown in FIGURE P11.38.
a. Draw a graph of the potential energy U from $x = 0$ m to $x = 5$ m. Let the zero of the potential energy be at $x = 0$ m. **Hint:** Think about the definition of potential energy *and* the geometric interpretation of the work done by a varying force.
b. The particle is shot toward the right from $x = 1.0$ m with a speed of 25 m/s. What is the particle's mechanical energy?
c. Draw the particle's total energy line on your graph of part a.
d. Where is the particle's turning point?

39. || A 10 g particle has the potential energy shown in FIGURE P11.39.
a. Draw a force-versus-position graph from $x = 0$ cm to $x = 8$ cm.
b. How much work does the force do as the particle moves from $x = 2$ cm to $x = 6$ cm?
c. What speed does the particle need at $x = 2$ cm to arrive at $x = 6$ cm with a speed of 10 m/s?

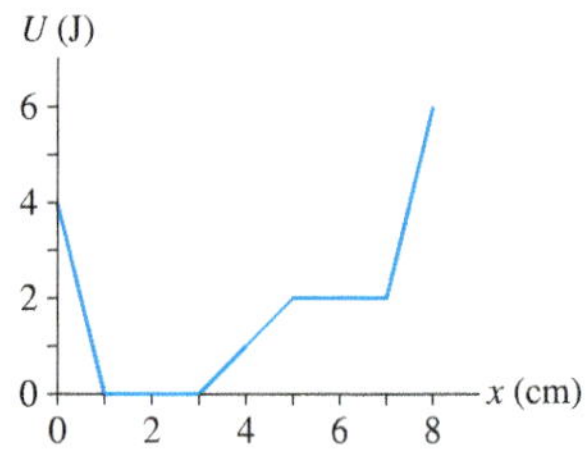

FIGURE P11.39

40. ||| a. FIGURE P11.40a shows the force F_x exerted on a particle that moves along the x-axis. Draw a graph of the particle's potential energy as a function of position x. Let U be zero at $x = 0$ m.
b. FIGURE P11.40b shows the potential energy U of a particle that moves along the x-axis. Draw a graph of the force F_x as a function of position x.

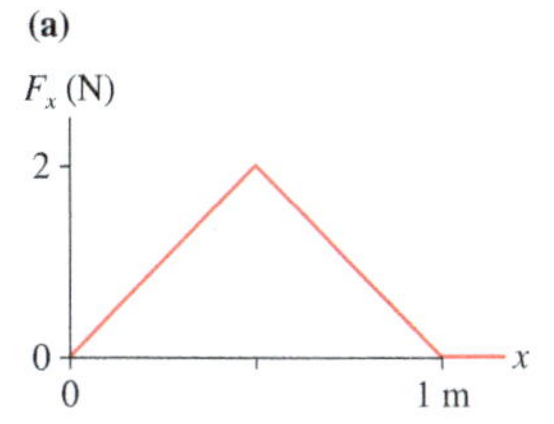

FIGURE P11.40

41. || FIGURE P11.41 is the velocity-versus-time graph of a 500 g particle that starts at $x = 0$ m and moves along the x-axis. Draw graphs of the following by calculating and plotting numerical values at $t = 0$, 1, 2, 3, and 4 s. Then sketch lines or curves of the appropriate shape between the points. Make sure you include appropriate scales on both axes of each graph.

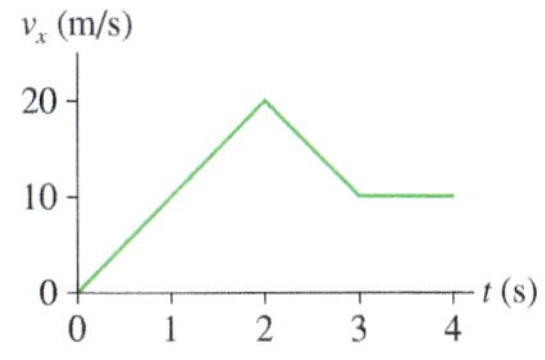

FIGURE P11.41

a. Acceleration versus time.
b. Position versus time.
c. Kinetic energy versus time.
d. Force versus time.
e. Use your F_x-versus-t graph to determine the *impulse* delivered to the particle during the time interval 0–2 s and also the interval 2–4 s.
f. Use the impulse-momentum theorem to determine the particle's velocity at $t = 2$ s and at $t = 4$ s. Do your results agree with the velocity graph?
g. Now draw a graph of force versus *position*. This requires no calculations; just think carefully about what you learned in parts a to d.
h. Use your F_x-versus-x graph to determine the *work* done on the particle during the time interval 0–2 s and also the interval 2–4 s.
i. Use the work-kinetic energy theorem to determine the particle's velocity at $t = 2$ s and at $t = 4$ s. Do your results agree with the velocity graph?

42. || A 1000 kg elevator accelerates upward at 1.0 m/s^2 for 10 m, starting from rest.
a. How much work does gravity do on the elevator?
b. How much work does the tension in the elevator cable do on the elevator?
c. Use the work-kinetic energy theorem to find the kinetic energy of the elevator as it reaches 10 m.
d. What is the speed of the elevator as it reaches 10 m?

43. | Bob can throw a 500 g rock with a speed of 30 m/s. He moves his hand forward 1.0 m while doing so.
a. How much work does Bob do on the rock?
b. How much force, assumed to be constant, does Bob apply to the rock?
c. What is Bob's maximum power output as he throws the rock?

44. || a. Starting from rest, a crate of mass m is pushed up a frictionless slope of angle θ by a *horizontal* force of magnitude F. Use work and energy to find an expression for the crate's speed v when it is at height h above the bottom of the slope.
b. Doug uses a 25 N horizontal force to push a 5.0 kg crate up a 2.0-m-high, 20° frictionless slope. What is the speed of the crate at the top of the slope?

45. || Sam, whose mass is 75 kg, straps on his skis and starts down a 50-m-high, 20° frictionless slope. A strong headwind exerts a *horizontal* force of 200 N on him as he skis. Use work and energy to find Sam's speed at the bottom.

46. ||| Susan's 10 kg baby brother Paul sits on a mat. Susan pulls the mat across the floor using a rope that is angled 30° above the floor. The tension is a constant 30 N and the coefficient of friction is 0.20. Use work and energy to find Paul's speed after being pulled 3.0 m.

47. || A horizontal spring with spring constant 100 N/m is compressed 20 cm and used to launch a 2.5 kg box across a frictionless, horizontal surface. After the box travels some distance, the surface becomes rough. The coefficient of kinetic friction of the box on the surface is 0.15. Use work and energy to find how far the box slides across the rough surface before stopping.

48. || a. A box of mass m and initial speed v_0 slides distance d across a horizontal floor before coming to rest. Use work and energy to find an expression for the coefficient of kinetic friction.
b. A baggage handler throws a 15 kg suitcase along the floor of an airplane luggage compartment with a speed of 1.2 m/s. The suitcase slides 2.0 m before stopping. What is the suitcase's coefficient of kinetic friction on the floor?

49. || Truck brakes can fail if they get too hot. In some mountainous areas, ramps of loose gravel are constructed to stop runaway trucks that have lost their brakes. The combination of a slight upward slope and a large coefficient of rolling resistance as the truck tires sink into the gravel brings the truck safely to a halt. Suppose a gravel ramp slopes upward at 6.0° and the coefficient of rolling friction is 0.40. Use work and energy to find the length of a ramp that will stop a 15,000 kg truck that enters the ramp at 35 m/s ($\approx$75 mph).

50. || A freight company uses a compressed spring to shoot 2.0 kg packages up a 1.0-m-high frictionless ramp into a truck, as FIGURE P11.50 shows. The spring constant is 500 N/m and the spring is compressed 30 cm.
a. What is the speed of the package when it reaches the truck?
b. A careless worker spills his soda on the ramp. This creates a 50-cm-long sticky spot with a coefficient of kinetic friction 0.30. Will the next package make it into the truck?

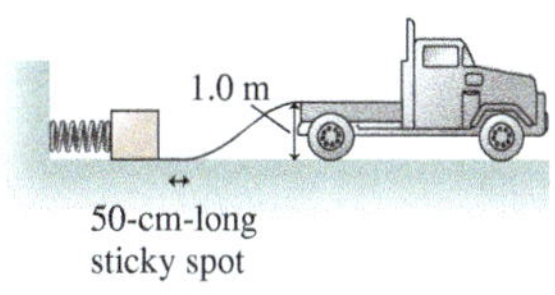

FIGURE P11.50

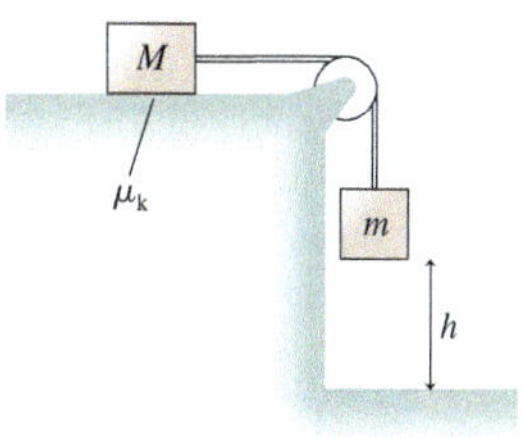

FIGURE P11.51

51. ||| Use work and energy to find an expression for the speed of the block in FIGURE P11.51 just before it hits the floor if (a) the coefficient of kinetic friction for the block on the table is μ_k and (b) the table is frictionless.

52. ||| An 8.0 kg crate is pulled 5.0 m up a 30° incline by a rope angled 18° above the incline. The tension in the rope is 120 N, and the crate's coefficient of kinetic friction on the incline is 0.25.
a. How much work is done by tension, by gravity, and by the normal force?
b. What is the increase in thermal energy of the crate and incline?

53. || You've taken a summer job at a water park. In one stunt, a water skier is going to glide up the 2.0-m-high frictionless ramp shown in FIGURE P11.53, then sail over a 5.0-m-wide tank filled with hungry sharks. You will be driving the boat that pulls her to the ramp. She'll drop the tow rope at the base of the ramp just as you veer away. What minimum speed must you have as you reach the ramp in order for her to live to do this again tomorrow?

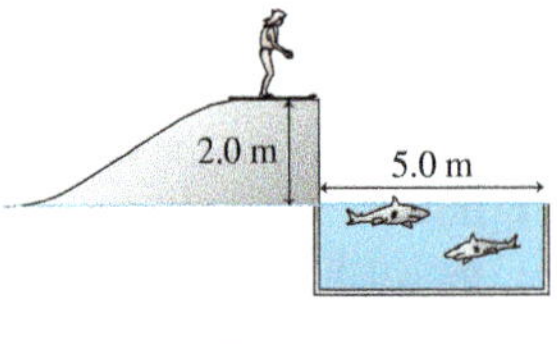

FIGURE P11.53

54. || A 50 kg ice skater is gliding along the ice, heading due north at 4.0 m/s. The ice has a small coefficient of static friction, to prevent the skater from slipping sideways, but $\mu_k = 0$. Suddenly, a wind from the northeast exerts a force of 4.0 N on the skater.
   a. Use work and energy to find the skater's speed after gliding 100 m in this wind.
   b. What is the minimum value of μ_s that allows her to continue moving straight north?

55. || a. A 50 g ice cube can slide without friction up and down a 30° slope. The ice cube is pressed against a spring at the bottom of the slope, compressing the spring 10 cm. The spring constant is 25 N/m. When the ice cube is released, what total distance will it travel up the slope before reversing direction?
   b. The ice cube is replaced by a 50 g plastic cube whose coefficient of kinetic friction is 0.20. How far will the plastic cube travel up the slope? Use work and energy.

56. || A 5.0 kg box slides down a 5.0-m-high frictionless hill, starting from rest, across a 2.0-m-wide horizontal surface, then hits a horizontal spring with spring constant 500 N/m. The other end of the spring is anchored against a wall. The ground under the spring is frictionless, but the 2.0-m-wide horizontal surface is rough. The coefficient of kinetic friction of the box on this surface is 0.25.
   a. What is the speed of the box just before reaching the rough surface?
   b. What is the speed of the box just before hitting the spring?
   c. How far is the spring compressed?
   d. Including the first crossing, how many *complete* trips will the box make across the rough surface before coming to rest?

57. || The spring shown in FIGURE P11.57 is compressed 50 cm and used to launch a 100 kg physics student. The track is frictionless until it starts up the incline. The student's coefficient of kinetic friction on the 30° incline is 0.15.
   a. What is the student's speed just after losing contact with the spring?
   b. How far up the incline does the student go?

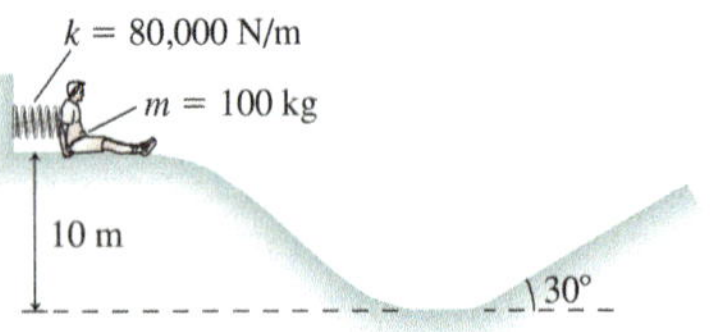

FIGURE P11.57

58. || A block of mass m starts from rest at height h. It slides down a frictionless incline, across a rough horizontal surface of length L, then up a frictionless incline. The coefficient of kinetic friction on the rough surface is μ_k.
   a. What is the block's speed at the bottom of the first incline?
   b. How high does the block go on the second incline?
   Give your answers in terms of m, h, L, μ_k, and g.

59. || Show that Hooke's law for an ideal spring is a conservative force. To do so, first calculate the work done by the spring as it expands from A to B. Then calculate the work done by the spring as it expands from A to point C, which is beyond B, then returns from C to B.

60. || A clever engineer designs a "sprong" that obeys the force law $F_x = -q(x - x_e)^3$, where x_e is the equilibrium position of the end of the sprong and q is the sprong constant. For simplicity, we'll let $x_e = 0$ m. Then $F_x = -qx^3$.
   a. What are the units of q?
   b. Find an expression for the potential energy of a stretched or compressed sprong.
   c. A sprong-loaded toy gun shoots a 20 g plastic ball. What is the launch speed if the sprong constant is 40,000, with the units you found in part a, and the sprong is compressed 10 cm? Assume the barrel is frictionless.

61. || A particle of mass m starts from $x_0 = 0$ m with $v_0 > 0$ m/s. The particle experiences the variable force $F_x = F_0 \sin(cx)$ as it moves to the right along the x-axis, where F_0 and c are constants.
   a. What are the units of F_0?
   b. What are the units of c?
   c. At what position x_{max} does the force first reach a maximum value? Your answer will be in terms of the constants F_0 and c and perhaps other numerical constants.
   d. Sketch a graph of F versus x from x_0 to x_{max}.
   e. What is the particle's velocity as it reaches x_{max}? Give your answer in terms of m, v_0, F_0, and c.

62. || A 5.0 kg cat leaps from the floor to the top of a 95-cm-high table. If the cat pushes against the floor for 0.20 s to accomplish this feat, what was her average power output during the pushoff period?

63. || The human heart pumps the average adult's 6.0 L (6000 cm^3) BIO of blood through the body every minute. The heart must do work to overcome frictional forces that resist blood flow. The average adult blood pressure is 1.3×10^4 N/m^2.
   a. How much work does the heart do to move the 6.0 L of blood completely through the body?
   b. What power output must the heart have to do this task once a minute?

Hint: When the heart contracts, it applies force to the blood. Pressure is force/area. Model the circulatory system as a single closed tube, with cross-section area A and volume $V = 6.0$ L, filled with blood to which the heart applies a force.

64. || When you ride a bicycle at constant speed, nearly all the en- BIO ergy you expend goes into the work you do against the drag force of the air. Model a cyclist as having cross-section area 0.45 m^2 and, because the human body is not aerodynamically shaped, a drag coefficient of 0.90.
   a. What is the cyclist's power output while riding at a steady 7.3 m/s (16 mph)?
   b. *Metabolic power* is the rate at which your body "burns" fuel to power your activities. For many activities, your body is roughly 25% efficient at converting the chemical energy of food into mechanical energy. What is the cyclist's metabolic power while cycling at 7.3 m/s?
   c. The food calorie is equivalent to 4190 J. How many calories does the cyclist burn if he rides over level ground at 7.3 m/s for 1 h?

65. || In a hydroelectric dam, water falls 25 m and then spins a turbine to generate electricity.

a. What is ΔU of 1.0 kg of water?
b. Suppose the dam is 80% efficient at converting the water's potential energy to electrical energy. How many kilograms of water must pass through the turbines each second to generate 50 MW of electricity? This is a typical value for a small hydroelectric dam.

66. || The force required to tow a water skier at speed v is proportional to the speed. That is, $F_{tow} = Av$, where A is a proportionality constant. If a speed of 2.5 mph requires 2 hp, how much power is required to tow a water skier at 7.5 mph?

67. || A Porsche 944 Turbo has a rated engine power of 217 hp. 30% of the power is lost in the drive train, and 70% reaches the wheels. The total mass of the car and driver is 1480 kg, and two-thirds of the weight is over the drive wheels.
a. What is the maximum acceleration of the Porsche on a concrete surface where $\mu_s = 1.00$?
Hint: What force pushes the car forward?
b. If the Porsche accelerates at a_{max}, what is its speed when it reaches maximum power output?
c. How long does it take the Porsche to reach the maximum power output?

In Problems 68 through 71 you are given the equation(s) used to solve a problem. For each of these, you are to
a. Write a realistic problem for which this is the correct equation(s).
b. Draw a pictorial representation.
c. Finish the solution of the problem.

68. $\frac{1}{2}(2.0\text{ kg})(4.0\text{ m/s})^2 + 0$

$+ (0.15)(2.0\text{ kg})(9.8\text{ m/s}^2)(2.0\text{ m}) = 0 + 0 + T(2.0\text{ m})$

69. $\frac{1}{2}(20\text{ kg})v_1^2 + 0$

$+ (0.15)(20\text{ kg})(9.8\text{ m/s}^2)\cos 40°((2.5\text{ m})/\sin 40°)$

$= 0 + (20\text{ kg})(9.8\text{ m/s}^2)(2.5\text{ m}) + 0$

70. $F_{push} - (0.20)(30\text{ kg})(9.8\text{ m/s}^2) = 0$

$75\text{ W} = F_{push}v$

71. $T - (1500\text{ kg})(9.8\text{ m/s}^2) = (1500\text{ kg})(1.0\text{ m/s}^2)$

$P = T(2.0\text{ m/s})$

Challenge Problems

72. A 10.2 kg weather rocket generates a thrust of 200 N. The rocket, pointing upward, is clamped to the top of a vertical spring. The bottom of the spring, whose spring constant is 500 N/m, is anchored to the ground.
a. Initially, before the engine is ignited, the rocket sits at rest on top of the spring. How much is the spring compressed?
b. After the engine is ignited, what is the rocket's speed when the spring has stretched 40 cm? For comparison, what would be the rocket's speed after traveling this distance if it weren't attached to the spring?

73. The spring in FIGURE CP11.73 has a spring constant of 1000 N/m. It is compressed 15 cm, then launches a 200 g block. The horizontal surface is frictionless, but the block's coefficient of kinetic friction on the incline is 0.20. What distance d does the block sail through the air?

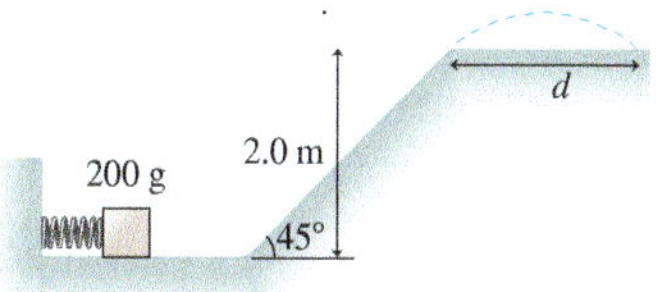

FIGURE CP11.73

74. The equation mgy for gravitational potential energy is valid only for objects near the surface of a planet. Consider two very large objects of mass m_1 and m_2, such as stars or planets, whose centers are separated by the large distance r. These two large objects exert gravitational forces on each other. You'll learn in Chapter 13 that the gravitational potential energy is

$$U = -\frac{Gm_1 m_2}{r}$$

where $G = 6.67 \times 10^{-11}\text{ N m}^2/\text{kg}^2$ is the *gravitational constant*.
a. Sketch a graph of U versus r. The mathematical difficulty at $r = 0$ is not a physically significant difficulty because the masses will collide before they get that close together.
b. What separation r has been chosen as the point of zero potential energy? Does this make sense? Explain.
c. Two stars are at rest 1.0×10^{14} m apart. This is about 10 times the diameter of the solar system. The first star is the size of our sun, with a mass of 2.0×10^{30} kg and a radius of 7.0×10^8 m. The second star has mass 8.0×10^{30} kg and radius of 11.0×10^8 m. Gravitational forces pull the two stars together. What is the speed of each star at the moment of impact?

75. A gardener pushes a 12 kg lawnmower whose handle is tilted up 37° above horizontal. The lawnmower's coefficient of rolling friction is 0.15. How much power does the gardener have to supply to push the lawnmower at a constant speed of 1.2 m/s? Assume his push is parallel to the handle.

STOP TO THINK ANSWERS

Stop to Think 11.1: d. Constant speed means $\Delta K = 0$. Gravitational potential energy is lost, and friction heats up the slide and the child's pants.

Stop to Think 11.2: 6.0 J. $K_f = K_i + W$. W is the area under the curve, which is 4.0 J.

Stop to Think 11.3: b. The gravitational force $\vec{F}_G$ is in the same direction as the displacement. It does positive work. The tension force $\vec{T}$ is opposite the displacement. It does negative work.

Stop to Think 11.4: c. $W = F(\Delta r)\cos\theta$. The 10 N force at 90° does no work at all. $\cos 60° = \frac{1}{2}$, so the 8 N force does less work than the 6 N force.

Stop to Think 11.5: e. Force is the negative of the slope of the potential-energy diagram. At $x = 4$ m the potential energy has risen by 4 J over a distance of 2 m, so the slope is 2 J/m = 2 N.

Stop to Think 11.6: c. Constant speed means $\Delta K = 0$. Gravitational potential energy is lost, and friction heats up the pole and the child's hands.

Stop to Think 11.7: $P_b > P_a = P_c > P_d$. The work done is $mg\Delta y$, so the power is $mg\Delta y/\Delta t$. Runner b does the same work as a but in less time. The ratio $m/\Delta t$ is the same for a and c. Runner d does twice the work of a but takes more than twice as long.

PART II

SUMMARY
Conservation Laws

In Part II we have discovered that we don't need to know all the details of an interaction to relate the properties of a system "before" an interaction to the system's properties "after" the interaction. Along the way, we found two important quantities, momentum and energy, that characterize a system of particles.

Momentum and energy have specific conditions under which they are conserved. In particular, the total momentum $\vec{P}$ and the total energy E_{sys} are conserved for an *isolated system,* one on which the net external force is zero. Further, the system's mechanical energy is conserved if the system is both isolated and nondissipative (i.e., no friction forces). These ideas are captured in the two most important conservation laws, the law of conservation of momentum and the law of conservation of energy.

Of course, not all systems are isolated. For both momentum and energy, it was useful to develop a *model* of a system interacting with its environment. Interactions between the system and the environment change the system's momentum and energy. In particular,

- Impulse is the transfer of momentum to or from the system: $\Delta\vec{P} = \vec{J}$.
- Work is the transfer of energy to or from the system: $\Delta E_{sys} = W_{ext}$.

Interactions within the system do not change $\vec{P}$ or E_{sys}. The kinetic, potential, and thermal energy within the system can be transformed without changing E_{sys}. The basic energy model is built around the twin ideas of the transfer and the transformation of energy.

The table below is a knowledge structure of conservation laws. You should compare this with the knowledge structure of Newtonian mechanics in the Part I Summary. Add the problem-solving strategies, and you now have a very powerful set of tools for understanding motion.

KNOWLEDGE STRUCTURE II Conservation Laws

ESSENTIAL CONCEPTS	Impulse, momentum, work, energy
BASIC GOALS	How is the system "after" an interaction related to the system "before"? What quantities are conserved, and under what conditions?
GENERAL PRINCIPLES	**Impulse-momentum theorem** $\Delta p_s = J_s$ **Work-kinetic energy theorem** $\Delta K = W_{net} = W_c + W_{diss} + W_{ext}$ **Energy equation** $\Delta E_{sys} = \Delta K + \Delta U + \Delta E_{th} = W_{ext}$
CONSERVATION LAWS	For an isolated system, with $\vec{F}_{net} = \vec{0}$ and $W_{net} = 0$ • The total momentum $\vec{P}$ is conserved. • The total energy $E_{sys} = E_{mech} + E_{th}$ is conserved. For an isolated and nondissipative system, with $W_{diss} = 0$ • The mechanical energy $E_{mech} = K + U$ is conserved.

BASIC PROBLEM-SOLVING STRATEGY Draw a before-and-after pictorial representation, then use the momentum or energy equations to relate "before" to "after." Where possible, choose a system for which momentum and/or energy are conserved. If necessary, calculate impulse and/or work.

Basic model of momentum and energy

Impulse and momentum

$\vec{p} = m\vec{v}$

$$J_s = \int_{t_i}^{t_f} F_s(t)\,dt$$

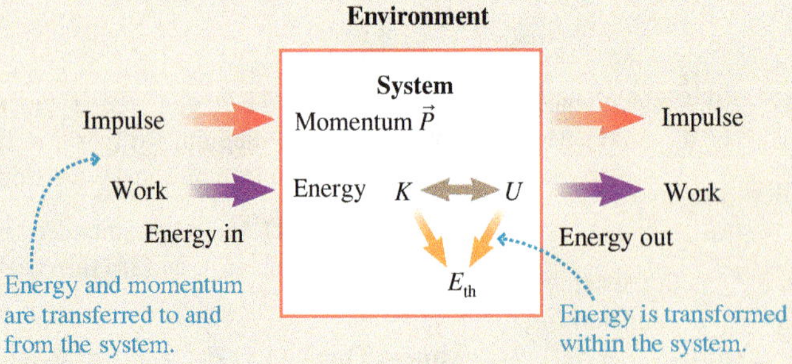

Work and energy

$K = \frac{1}{2}mv^2$

$$W = \int_{s_i}^{s_f} F_s\,ds = \vec{F}\cdot\Delta\vec{r}$$

$U_g = mgy$

$U_s = \frac{1}{2}k(\Delta s)^2$

ONE STEP BEYOND

Energy Conservation

You hear it all the time. Turn off lights. Buy a more fuel-efficient car. Conserve energy. But why conserve energy if energy is already conserved? Consider the earth as a whole. No work is done on the earth. And while heat energy flows from the sun to the earth, the earth radiates an equal amount of heat back into space. With no work and no net heat flow, the earth's total energy E_{earth} is conserved.

Pumping oil, driving your car, running a nuclear reactor, and turning on the lights are all interactions *within* the earth system. They transform energy from one type to another, but they don't affect the value of E_{earth}. Consider two examples.

- Crude oil, stored in the earth, has chemical energy E_{chem}. Chemical energy, a form of microscopic potential energy, is released when chemical reactions rearrange the bonds. As you burn gasoline in your car engine, the chemical energy is transformed into the kinetic energy of the moving pistons. This kinetic energy, in turn, is transformed into the car's kinetic energy. The car's kinetic energy is ultimately dissipated as thermal energy in the brakes, air, tires, and road because of friction and drag. Overall, the energy process of driving looks like

$$E_{chem} \rightarrow K_{piston} \rightarrow K_{car} \rightarrow E_{th}$$

- Water stored behind a dam has gravitational potential energy U_g. Potential energy is transformed into kinetic energy as the water falls, then into the spinning turbine's kinetic energy. The turbine converts mechanical energy into electric energy E_{elec}. The electric energy reaches a lightbulb where it is transformed partly into thermal energy (lightbulbs are hot!) and partly into light energy. The light is absorbed by surfaces, heating them slightly and thus transforming the light energy into thermal energy. The overall energy process is

$$U_g \rightarrow K_{water} \rightarrow K_{turbine} \rightarrow E_{elec} \rightarrow E_{light} \rightarrow E_{th}$$

Do you notice a trend? Stored energy (fossil fuel, water behind a dam) is transformed through a series of steps, some of which are considered "useful," until the energy is ultimately dissipated as thermal energy. **The total energy has not changed, but its "usefulness" has.**

Thermal energy is rarely "useful" energy. A room full of moving air molecules has a huge thermal energy, but you can't run your lights or your air conditioner with it. You can't turn the thermal energy of your hot brakes back into the kinetic energy of the car. Energy may be conserved, but there's a one-way characteristic of the transformations.

The energy stored in fuels and the energy of the sun are "high-quality energy" because of their potential to be transformed into such useful forms of energy as moving your car and heating your house. But as FIGURE PSII.1 shows, high-quality energy becomes "degraded" into thermal energy, where it is no longer useful. Thus the phrase "conserve energy" isn't used literally. Instead, it means to conserve or preserve the earth's sources of high-quality energy.

Conserving high-quality energy is important because fossil fuels are a finite resource. Experts may disagree as to how long fossil fuels will last, but all agree that it won't be forever. Oil and natural gas will likely become scarce during your lifetime. In addition, burning fossil fuel generates carbon dioxide, a major contributor to global warming. Energy conservation helps fuels last longer and minimizes their side effects.

There are two paths to conserving energy. One is to use less high-quality energy. Turning off lights and bicycling rather than driving are actions that preserve high-quality energy. A second path is to use energy more efficiently. That is, get more of the useful activity (miles driven, rooms lit) for the same amount of high-quality energy.

Lightbulbs offer a good example. A 100 W incandescent lightbulb actually produces only about 10 W of light energy. Ninety watts of the high-quality electric energy is immediately degraded as thermal energy without doing anything useful. By contrast, a 25 W compact fluorescent bulb generates the same 10 W of light but only 15 W of thermal energy. The same amount of high-quality energy can light four times as many rooms if 100 W incandescent bulbs are replaced by 25 W compact fluorescent bulbs.

So why conserve energy if energy is already conserved? Because technological society needs a dependable and sustainable supply of high-quality energy. Both technology improvement and lifestyle choices will help us achieve a sustainable energy future.

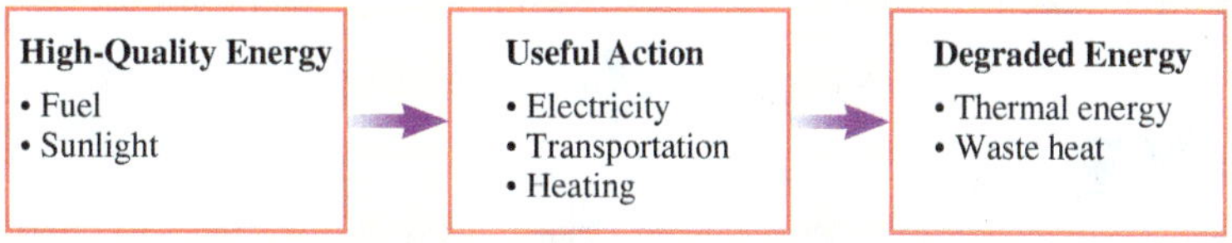

FIGURE PSII.1 "Using" energy transforms high-quality energy into thermal energy.

PART

III

Applications of Newtonian Mechanics

Hurricane Ivan approaches the United States in 2004. A hurricane is a fluid—the air—moving on a rotating sphere—the earth—under the influence of gravity. Understanding hurricanes is very much an application of Newtonian mechanics.

OVERVIEW

Power Over Our Environment

Early humans had to endure whatever nature provided. Only within the last few thousand years have agriculture and technology provided some level of control over the environment. And it has been a mere couple of centuries since machines, and later electronics, began to do much of our work and provide us with "creature comforts."

It's no coincidence that machines began to appear about a century after Galileo, Newton, and others ignited what we now call the *scientific revolution.* The machines and other devices we take for granted today are direct consequences of scientific knowledge and the scientific method.

Parts I and II have established Newton's theory of motion, the foundation of modern science. Most of the applications will be developed in other science and engineering courses, but we're now in a good position to examine a few of the more practical aspects of Newtonian mechanics.

Our goal for Part III is to apply our newfound theory to four important topics:

- **Rotation.** Rotation is a very important form of motion, but to understand rotational motion we'll need to introduce a new model—the *rigid-body model.* We'll then be able to study rolling wheels and spinning space stations. Rotation will also lead to the law of conservation of angular momentum.
- **Gravity.** By adding one more law, Newton's law of gravity, we'll be able to understand much about the physics of the space shuttle, communication satellites, the solar system, and interplanetary travel.
- **Oscillations.** Oscillations are seen in systems ranging from the pendulum in a grandfather clock to the quartz crystal oscillator providing the timing signals in sophisticated electronic circuits. The physics and mathematics of oscillations will later be the starting point for our study of waves.
- **Fluids.** Liquids and gases *flow.* Surprisingly, it takes no new physics to understand the basic mechanical properties of fluids. By applying our understanding of force, we'll be able to understand what pressure is, how a steel ship can float, and how fluids flow through pipes.

Newton's laws of motion and the conservation laws, especially conservation of energy, will be the tools that allow us to analyze and understand a variety of interesting and practical applications.

Science has given us the power to control our environment, but science and engineering are a two-edged sword. Much of the progress of the last two hundred years has come at the expense of the environment. We humans have deforested much of the world, polluted our air and water, and driven many of our fellow travelers on Spaceship Earth to extinction. Now, at the beginning of the 21st century, the evidence is increasingly clear that humans are altering the earth's climate and causing other global changes.

Fortunately, science also gives us the ability to understand the consequences of our actions and to develop better techniques and procedures. It is more important than ever that scientists and engineers in the 21st century distinguish control that is beneficial from control that is harmful. We'll return to some of these ideas in the Summary to Part III.

12 Rotation of a Rigid Body

Not all motion can be described as that of a particle. Rotation requires the idea of an extended object.

▶ **Looking Ahead** The goal of Chapter 12 is to understand the physics of rotating objects.

Rigid Bodies

A **rigid body** is an object whose size and shape don't change as it moves.

This chapter focuses primarily on the rotation of rigid bodies. We'll emphasize two types of rotation:

- Rotation about a fixed axle.
- Rolling without slipping.

The mathematics of circular motion—angular velocity and angular acceleration— will be very important. A review is highly recommended.

◀ **Looking Back**
Sections 4.5–4.7 Kinematics of circular motion

We'll also consider the conditions under which an extended object is in static equilibrium, neither translating nor rotating. Static equilibrium has many important applications.

◀ **Looking Back**
Section 6.1 Equilibrium

Properties of Rigid Bodies

An extended object's motion and stability depend on how its mass is distributed. You'll learn how to calculate an object's **center of mass** and its **moment of inertia.** Moment of inertia is the rotational equivalent of mass.

Torque

Torque is the tendency or ability of a force to rotate an object around a pivot point.

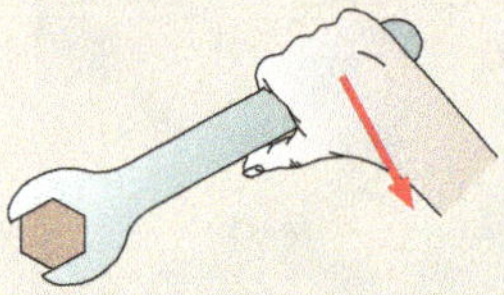

You will learn to calculate torque and will find that the torque depends not only on the magnitude of the force but also on where the force is applied relative to the pivot point. A longer wrench gives a larger torque.

Conservation Laws

Kinetic energy and linear momentum have their rotational equivalents. A rotating object's **rotational kinetic energy** and **angular momentum** depend on its moment of inertia and its angular velocity.

You'll learn to solve problems using

- Conservation of energy for frictionless, rotating systems, and
- Conservation of angular momentum for isolated systems.

◀ **Looking Back**
Section 9.3 Momentum conservation

Newton's Second Law

Torque is to rotation what force is to linear motion. Torque τ causes an object with moment of inertia I to undergo angular acceleration $\alpha = \tau/I$.

You'll learn to use Newton's second law to solve problems of rotational dynamics.

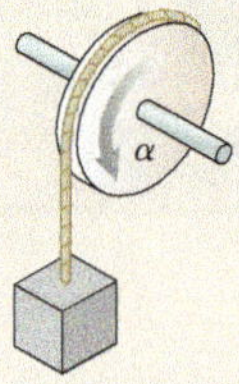

◀ **Looking Back**
Section 6.2 Newton's second law

12.1 Rotational Motion

Thus far, our study of physics has focused almost exclusively on the *particle model* in which an object is represented as a mass at a single point in space. The particle model is a perfectly good description of the physics in a vast number of situations, but there are other situations for which we need to consider an *extended object*—a system of particles for which the size and shape *do* make a difference and cannot be neglected.

A **rigid body** is an extended object whose size and shape do not change as it moves. For example, a bicycle wheel can be thought of as a rigid body. FIGURE 12.1 shows a rigid body as a collection of atoms held together by the rigid "massless rods" of molecular bonds.

Real molecular bonds are, of course, not perfectly rigid. That's why an object seemingly as rigid as a bicycle wheel can flex and bend. Thus Figure 12.1 is really a simplified *model* of an extended object, the **rigid-body model.** The rigid-body model is a very good approximation of many real objects of practical interest, such as wheels and axles. Even nonrigid objects can often be modeled as rigid bodies during parts of their motion. For example, a diver is well described as a rotating rigid body while she's in the tuck position.

FIGURE 12.2 illustrates the three basic types of motion of a rigid body: **translational motion, rotational motion,** and **combination motion.**

FIGURE 12.1 The rigid-body model.

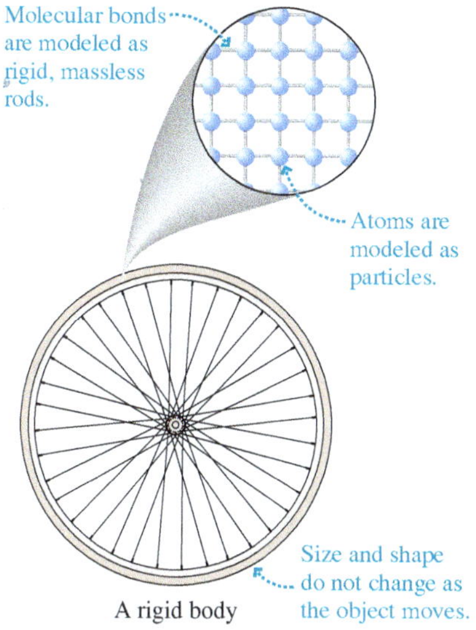

FIGURE 12.2 Three basic types of motion of a rigid body.

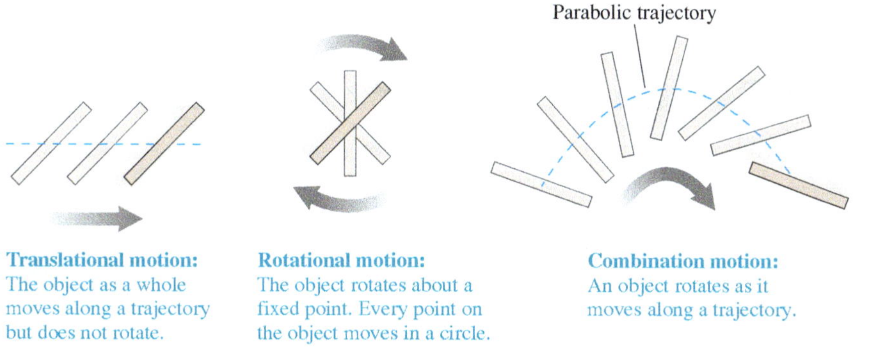

Brief Review of Rotational Kinematics

Rotation is an extension of circular motion, so we begin with a brief summary of Chapter 4. **A review of Sections 4.5–4.7 is highly recommended.** FIGURE 12.3 shows a wheel rotating on an axle. Its angular velocity

$$\omega = \frac{d\theta}{dt} \tag{12.1}$$

is the rate at which the wheel rotates. The SI units of ω are radians per second (rad/s), but revolutions per second (rev/s) and revolutions per minute (rpm) are frequently used. Notice that all points have equal angular velocities, so we can refer to the angular velocity ω *of the wheel.*

If the wheel is speeding up or slowing down, its angular acceleration is

$$\alpha = \frac{d\omega}{dt} \tag{12.2}$$

The units of angular acceleration are rad/s^2. Angular acceleration is the *rate* at which the angular velocity ω changes, just as the linear acceleration is the rate at which the linear velocity v changes. Table 12.1 on the next page summarizes the kinematic equations for rotation with constant angular acceleration.

FIGURE 12.3 Two points on a wheel rotate with the same angular velocity.

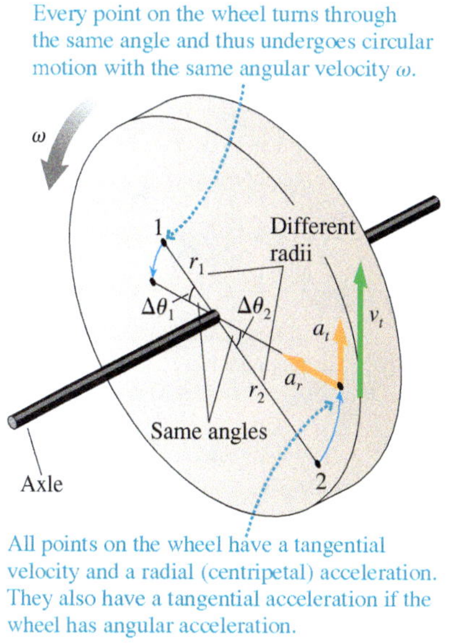

TABLE 12.1 Rotational kinematics for constant angular acceleration

$\omega_f = \omega_i + \alpha\,\Delta t$
$\theta_f = \theta_i + \omega_i\,\Delta t + \frac{1}{2}\alpha(\Delta t)^2$
$\omega_f^2 = \omega_i^2 + 2\alpha\,\Delta\theta$

FIGURE 12.4 reminds you of the sign conventions for angular velocity and acceleration. They will be especially important in the present chapter. Be careful with the sign of α. Just as with linear acceleration, positive and negative values of α can't be interpreted as simply "speeding up" and "slowing down."

FIGURE 12.4 The signs of angular velocity and angular acceleration.

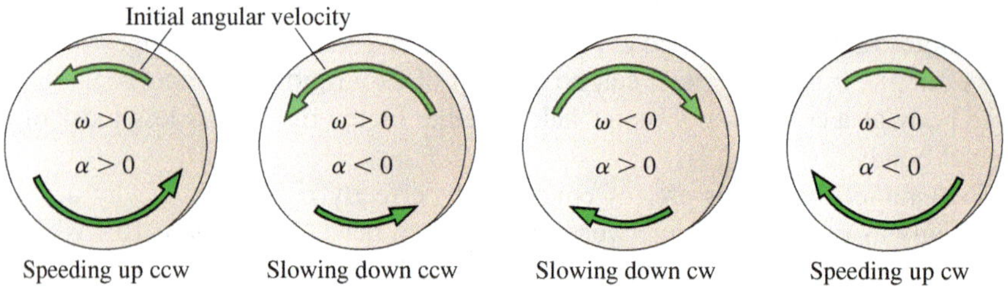

A point at distance r from the rotation axis has instantaneous velocity and acceleration, shown in Figure 12.3, given by

$$v_r = 0 \qquad a_r = \frac{v_t^2}{r} = \omega^2 r$$
$$v_t = r\omega \qquad a_t = r\alpha \tag{12.3}$$

The sign convention for ω implies that v_t and a_t are positive if they point in the counterclockwise (ccw) direction, negative if they point in the clockwise (cw) direction.

12.2 Rotation About the Center of Mass

Imagine yourself floating in a space capsule deep in space. Suppose you take an object like that shown in FIGURE 12.5a, spin it, then let go. The object will rotate, but it will have no translational motion as it floats beside you. *About what point does it rotate?* That is the question we need to answer.

An unconstrained object (i.e., one not on an axle or a pivot) on which there is no net force rotates about a point called the **center of mass.** The center of mass remains motionless while every other point in the object undergoes circular motion around it. You need not go deep into space to demonstrate rotation about the center of mass. If you have an air table, a flat object rotating on the air table rotates about its center of mass.

To locate the center of mass, FIGURE 12.5b models the object as if it were constructed from particles numbered $i = 1, 2, 3, \ldots$. Particle i has mass m_i and is located at position (x_i, y_i). We'll prove later in this section that the center of mass is located at position

$$x_{cm} = \frac{1}{M}\sum_i m_i x_i = \frac{m_1x_1 + m_2x_2 + m_3x_3 + \cdots}{m_1 + m_2 + m_3 + \cdots}$$
$$y_{cm} = \frac{1}{M}\sum_i m_i y_i = \frac{m_1y_1 + m_2y_2 + m_3y_3 + \cdots}{m_1 + m_2 + m_3 + \cdots} \tag{12.4}$$

where $M = m_1 + m_2 + m_3 + \cdots$ is the object's total mass.

FIGURE 12.5 Rotation about the center of mass.

(a)

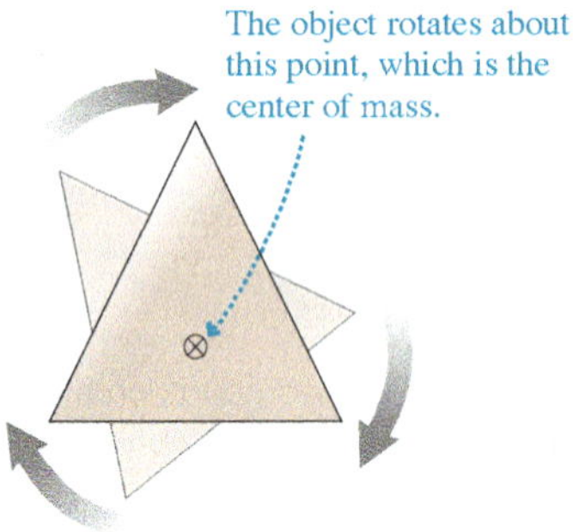

(b)

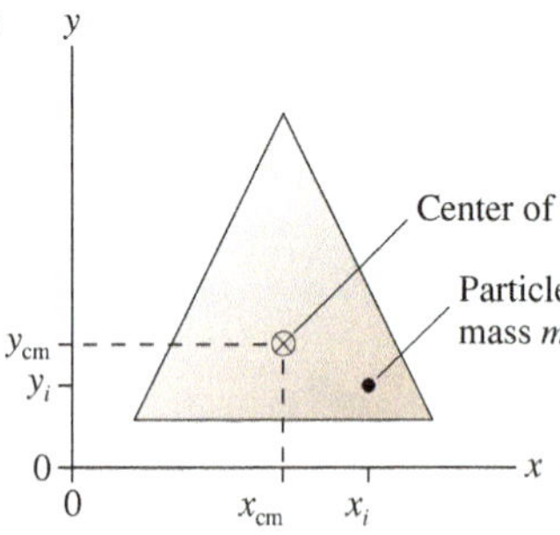

Let's see if Equations 12.4 make sense. Suppose you have an object consisting of N particles, all with the same mass m. That is, $m_1 = m_2 = \cdots = m_N = m$. We can factor the m out of the numerator, and the denominator becomes simply Nm. The m cancels, and the x-coordinate of the center of mass is

$$x_{cm} = \frac{x_1 + x_2 + \cdots + x_N}{N} = x_{average}$$

In this case, x_{cm} is simply the *average* x-coordinate of all the particles. Likewise, y_{cm} will be the average of all the y-coordinates.

This *does* make sense! If the particle masses are all the same, the center of mass should be at the center of the object. And the "center of the object" is the average position of all the particles. To allow for *unequal* masses, Equations 12.4 are called *weighted averages.* Particles of higher mass count more than particles of lower mass, but the basic idea remains the same. **The center of mass is the mass-weighted center of the object.**

EXAMPLE 12.1 The center of mass

A 500 g ball and a 2.0 kg ball are connected by a massless 50-cm-long rod.

a. Where is the center of mass?
b. What is the speed of each ball if they rotate about the center of mass at 40 rpm?

MODEL Model each ball as a particle.

VISUALIZE FIGURE 12.6 shows the two masses. We've chosen a coordinate system in which the masses are on the x-axis with the 2.0 kg mass at the origin.

SOLVE a We can use Equations 12.4 to calculate that the center of mass is

$$x_{cm} = \frac{m_1x_1 + m_2x_2}{m_1 + m_2}$$

$$= \frac{(2.0\text{ kg})(0.0\text{ m}) + (0.50\text{ kg})(0.50\text{ m})}{2.0\text{ kg} + 0.50\text{ kg}} = 0.10\text{ m}$$

$y_{cm} = 0$ because all the masses are on the x-axis. The center of mass is 20% of the way from the 2.0 kg ball to the 0.50 kg ball.

FIGURE 12.6 Finding the center of mass.

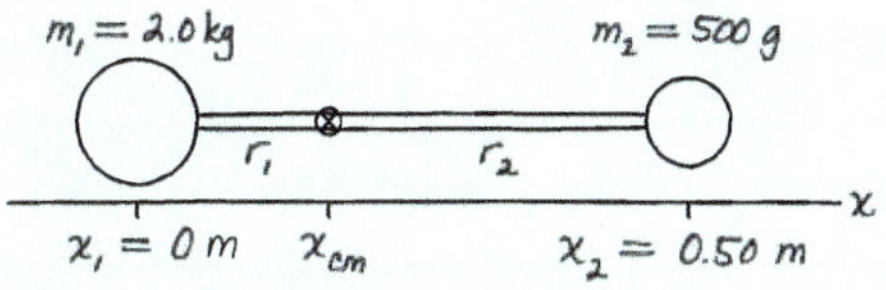

b. Each ball rotates about the center of mass. The radii of the circles are $r_1 = 0.10$ m and $r_2 = 0.40$ m. The tangential velocities are $(v_i)_t = r_i\omega$, but this equation requires ω to be in rad/s. The conversion is

$$\omega = 40\,\frac{\text{rev}}{\text{min}} \times \frac{1\text{ min}}{60\text{ s}} \times \frac{2\pi\text{ rad}}{1\text{ rev}} = 4.19\text{ rad/s}$$

Consequently,

$$(v_1)_t = r_1\omega = (0.10\text{ m})(4.19\text{ rad/s}) = 0.42\text{ m/s}$$

$$(v_2)_t = r_2\omega = (0.40\text{ m})(4.19\text{ rad/s}) = 1.68\text{ m/s}$$

ASSESS The center of mass is closer to the heavier ball than to the lighter ball. We expected this because x_{cm} is a mass-weighted average of the positions. But the lighter mass moves faster because it is farther from the rotation axis.

For any realistic object, carrying out the summations of Equations 12.4 over all the atoms in the object is not practical. Instead, as FIGURE 12.7 shows, we can divide an extended object into many small cells or boxes, each with the very small mass Δm. We will number the cells 1, 2, 3, . . . , just as we did the particles. Cell i has coordinates (x_i, y_i) and mass $m_i = \Delta m$. The center-of-mass coordinates are then

$$x_{cm} = \frac{1}{M}\sum_i x_i\,\Delta m \qquad \text{and} \qquad y_{cm} = \frac{1}{M}\sum_i y_i\,\Delta m$$

FIGURE 12.7 Calculating the center of mass of an extended object.

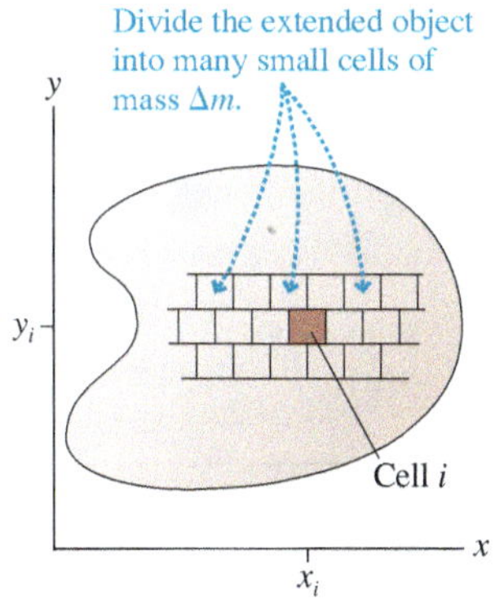

Now, as you might expect, we'll let the cells become smaller and smaller, with the total number increasing. As each cell becomes infinitesimally small, we can replace Δm with dm and the sum by an integral. Then

$$x_{cm} = \frac{1}{M}\int x\,dm \qquad \text{and} \qquad y_{cm} = \frac{1}{M}\int y\,dm \qquad (12.5)$$

Equations 12.5 are a formal definition of the center of mass, but they are *not* ready to integrate in this form. First, integrals are carried out over *coordinates,* not over masses. Before we can integrate, we must replace dm by an equivalent expression involving a coordinate differential such as dx or dy. Second, no limits of integration have been specified. The procedure for using Equations 12.5 is best shown with an example.

EXAMPLE 12.2 The center of mass of a rod

Find the center of mass of a thin, uniform rod of length L and mass M. Use this result to find the tangential acceleration of one tip of a 1.60-m-long rod that rotates about its center of mass with an angular acceleration of 6.0 rad/s^2.

VISUALIZE FIGURE 12.8 shows the rod. We've chosen a coordinate system such that the rod lies along the x-axis from 0 to L. Because the rod is "thin," we'll assume that $y_{cm} = 0$.

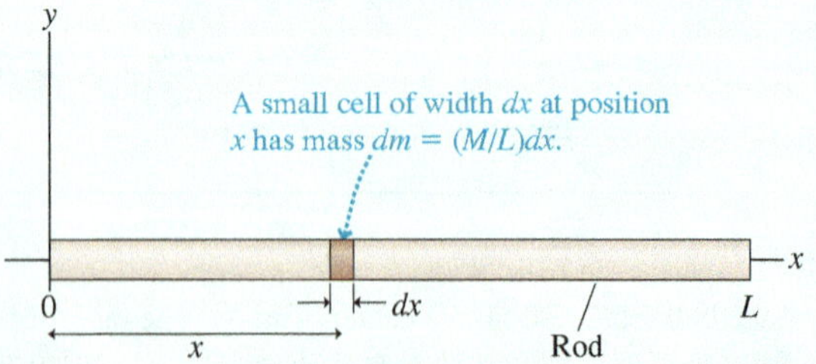

FIGURE 12.8 Finding the center of mass of a long, thin rod.

SOLVE Our first task is to find x_{cm}, which lies somewhere on the x-axis. To do this, we divide the rod into many small cells of mass dm. One such cell, at position x, is shown. The cell's width is dx. Because the rod is *uniform*, the mass of this little cell is the *same fraction* of the total mass M that dx is of the total length L. That is,

$$\frac{dm}{M} = \frac{dx}{L}$$

Consequently, we can express dm in terms of the coordinate differential dx as

$$dm = \frac{M}{L}dx$$

NOTE ▶ The change of variables from dm to the differential of a coordinate is *the* key step in calculating the center of mass. ◀

With this expression for dm, Equation 12.5 for x_{cm} becomes

$$x_{cm} = \frac{1}{M}\left(\frac{M}{L}\int x\,dx\right) = \frac{1}{L}\int_0^L x\,dx$$

where in the last step we've noted that summing "all the mass in the rod" means integrating from $x = 0$ to $x = L$. This is a straightforward integral to carry out, giving

$$x_{cm} = \frac{1}{L}\left[\frac{x^2}{2}\right]_0^L = \frac{1}{L}\left[\frac{L^2}{2} - 0\right] = \frac{1}{2}L$$

The center of mass is at the center of the rod. For a 1.60-m-long rod, each tip of the rod rotates in a circle with $r = \frac{1}{2}L = 0.80$ m. The tangential acceleration, the rate at which the tip is speeding up, is

$$a_t = r\alpha = (0.80 \text{ m})(6.0 \text{ rad/s}^2) = 4.8 \text{ m/s}^2$$

ASSESS You could have guessed that the center of mass is at the center of the rod, but now we've shown it rigorously.

NOTE ▶ For any symmetrical object of uniform density, the center of mass is at the physical center of the object. ◀

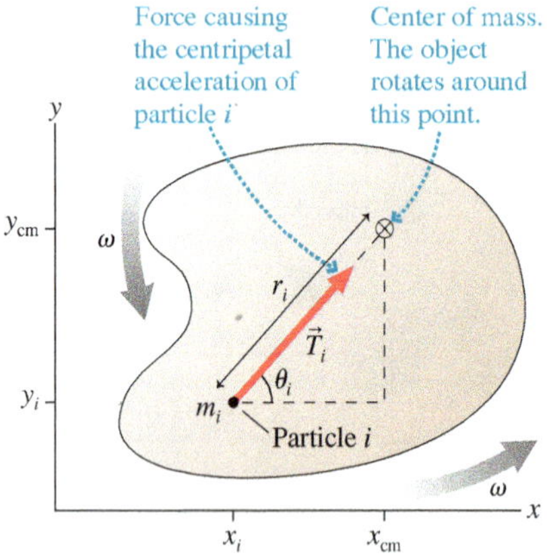

FIGURE 12.9 Finding the center of mass.

To see where the center-of-mass equations come from, FIGURE 12.9 shows an object rotating about its center of mass. Particle i is moving in a circle, so it *must* have a centripetal acceleration. Acceleration requires a force, and this force is due to tension in the molecular bonds that hold the object together. Force $\vec{T}_i$ on particle i has magnitude

$$T_i = m_i(a_i)_r = m_i r_i \omega^2 \tag{12.6}$$

where r_i is the distance of particle i from the center of mass and we used Equation 12.3 for a_r. All points in a rigid rotating object have the *same* angular velocity, so ω doesn't need a subscript.

At every instant of time, the internal tension forces are all paired as action/reaction forces, equal in magnitude but opposite in direction, so the sum of all the tension forces must be zero. That is, $\sum \vec{T}_i = \vec{0}$. The x-component of this sum is

$$\sum_i (T_i)_x = \sum_i T_i \cos\theta_i = \sum_i (m_i r_i \omega^2)\cos\theta_i = 0 \tag{12.7}$$

You can see from Figure 12.9 that $\cos\theta_i = (x_{cm} - x_i)/r_i$. Thus

$$\sum_i (T_i)_x = \sum_i (m_i r_i \omega^2)\frac{x_{cm} - x_i}{r_i} = \left(\sum_i m_i x_{cm} - \sum_i m_i x_i\right)\omega^2 = 0 \tag{12.8}$$

This equation will be true if the term in parentheses is zero. x_{cm} is a constant, so we can bring it outside the summation to write

$$\sum_i m_i x_{cm} - \sum_i m_i x_i = \left(\sum_i m_i\right)x_{cm} - \sum_i m_i x_i = Mx_{cm} - \sum_i m_i x_i = 0 \tag{12.9}$$

where we used the fact that $\sum m_i$ is simply the object's total mass M. Solving for x_{cm}, we find the x-coordinate of the object's center of mass to be

$$x_{cm} = \frac{1}{M}\sum_i m_i x_i = \frac{m_1x_1 + m_2x_2 + m_3x_3 + \cdots}{m_1 + m_2 + m_3 + \cdots} \quad (12.10)$$

This was Equation 12.4. The y-equation is found similarly.

12.3 Rotational Energy

A rotating rigid body has kinetic energy because all atoms in the object are in motion. The kinetic energy due to rotation is called **rotational kinetic energy.**

FIGURE 12.10 shows a few of the particles making up a solid object that rotates with angular velocity ω. Particle i, which rotates in a circle of radius r_i, moves with speed $v_i = r_i\omega$. The object's rotational kinetic energy is the sum of the kinetic energies of each of the particles:

$$\begin{aligned} K_{rot} &= \frac{1}{2}m_1v_1^2 + \frac{1}{2}m_2v_2^2 + \cdots \\ &= \frac{1}{2}m_1r_1^2\omega^2 + \frac{1}{2}m_2r_2^2\omega^2 + \cdots = \frac{1}{2}\left(\sum_i m_ir_i^2\right)\omega^2 \end{aligned} \quad (12.11)$$

The quantity $\sum m_ir_i^2$ is called the object's **moment of inertia** I:

$$I = m_1r_1^2 + m_2r_2^2 + m_3r_3^2 + \cdots = \sum_i m_ir_i^2 \quad (12.12)$$

The units of moment of inertia are kg m^2. **An object's moment of inertia depends on the axis of rotation.** Once the axis is specified, allowing the values of r_i to be determined, the moment of inertia *about that axis* can be calculated from Equation 12.12.

Written using the moment of inertia I, the rotational kinetic energy is

$$K_{rot} = \frac{1}{2}I\omega^2 \quad (12.13)$$

Rotational kinetic energy is *not* a new form of energy. This is the familiar kinetic energy of motion, only now expressed in a form that is especially convenient for rotational motion. Notice the analogy with the familiar $\frac{1}{2}mv^2$.

FIGURE 12.10 Rotational kinetic energy is due to the motion of the particles.

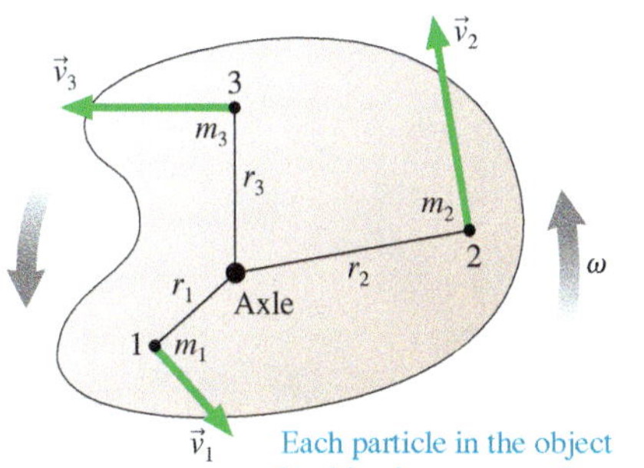

EXAMPLE 12.3 A rotating widget

Students participating in an engineering project design the triangular widget seen in FIGURE 12.11. The three masses, held together by lightweight plastic rods, rotate in the plane of the page about an axle passing through the right-angle corner. At what angular velocity does the widget have 100 mJ of rotational energy?

FIGURE 12.11 The rotating widget.

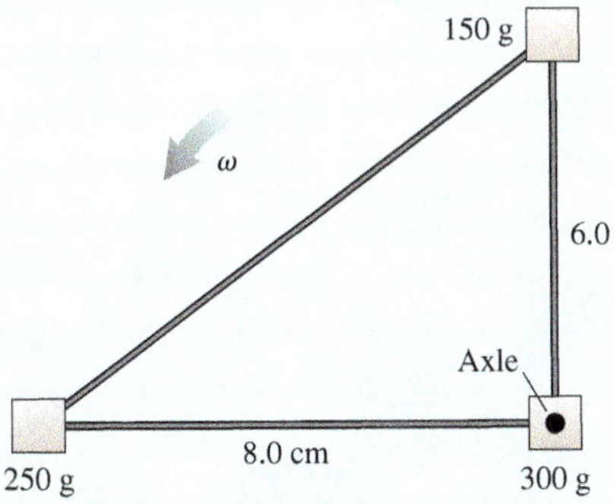

MODEL The widget can be modeled as three particles connected by massless rods.

SOLVE Rotational energy is $K = \frac{1}{2}I\omega^2$. The moment of inertia is measured about the rotation axis, thus

$$\begin{aligned} I = \sum_i m_ir_i^2 &= (0.25\text{ kg})(0.080\text{ m})^2 + (0.15\text{ kg})(0.060\text{ m})^2 \\ &\quad + (0.30\text{ kg})(0\text{ m})^2 \\ &= 2.14 \times 10^{-3}\text{ kg m}^2 \end{aligned}$$

The largest mass makes no contribution to I because it is right on the rotation axis with $r = 0$. With I known, the desired angular velocity is

$$\begin{aligned} \omega &= \sqrt{\frac{2K}{I}} = \sqrt{\frac{2(0.10\text{ J})}{2.14 \times 10^{-3}\text{ kg m}^2}} \\ &= 9.67\text{ rad/s} \times \frac{1\text{ rev}}{2\pi\text{ rad}} = 1.54\text{ rev/s} = 92\text{ rpm} \end{aligned}$$

ASSESS The moment of inertia depends on the distance of each mass from the rotation axis. The moment of inertia would be different for an axle passing through either of the other two masses, and thus the required angular velocity would be different.

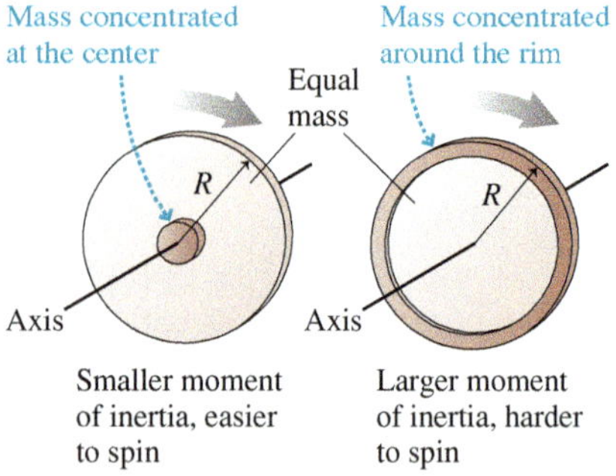

FIGURE 12.12 Moment of inertia depends on both the mass and how the mass is distributed.

Before rushing to calculate moments of inertia, let's get a better understanding of the meaning. First, notice that **moment of inertia is the rotational equivalent of mass.** It plays the same role in Equation 12.13 as mass m in the now-familiar $K = \frac{1}{2}mv^2$. Recall that the quantity we call *mass* was actually defined as the *inertial mass.* Objects with larger mass have a larger *inertia,* meaning that they're harder to accelerate. Similarly, an object with a larger moment of inertia is harder to rotate. The fact that *moment of inertia* retains the word "inertia" reminds us of this.

But why does the moment of inertia depend on the distances r_i from the rotation axis? Think about the two wheels shown in FIGURE 12.12. They have the same total mass M and the same radius R. As you probably know from experience, it's much easier to spin the wheel whose mass is concentrated at the center than to spin the one whose mass is concentrated around the rim. This is because having the mass near the center (smaller values of r_i) lowers the moment of inertia.

Moments of inertia for many solid objects are tabulated and found in various science and engineering handbooks. You would need to compute I yourself only for an object of unusual shape. Table 12.2 is a short list of common moments of inertia. We'll see in the next section where these come from, but do notice how I depends on the rotation axis.

If the rotation axis is not through the center of mass, then rotation may cause the center of mass to move up or down. In that case, the object's gravitational potential energy $U_g = Mgy_{cm}$ will change. If there are no dissipative forces (i.e., if the axle is frictionless) and if no work is done by external forces, then the mechanical energy

$$E_{mech} = K_{rot} + U_g = \frac{1}{2}I\omega^2 + Mgy_{cm} \tag{12.14}$$

is a conserved quantity.

TABLE 12.2 Moments of inertia of objects with uniform density

Object and axis	Picture	I	Object and axis	Picture	I
Thin rod, about center	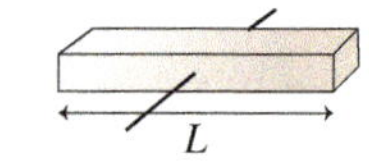	$\frac{1}{12}ML^2$	Cylinder or disk, about center	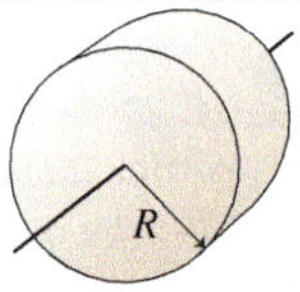	$\frac{1}{2}MR^2$
Thin rod, about end	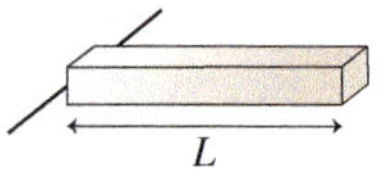	$\frac{1}{3}ML^2$	Cylindrical hoop, about center	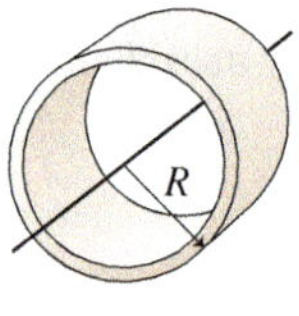	MR^2
Plane or slab, about center	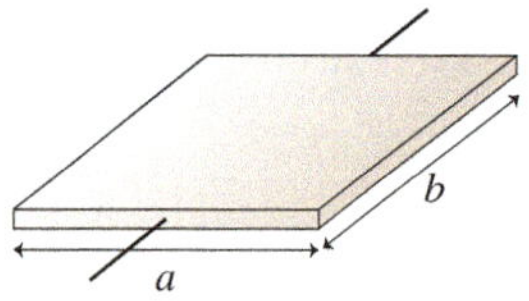	$\frac{1}{12}Ma^2$	Solid sphere, about diameter	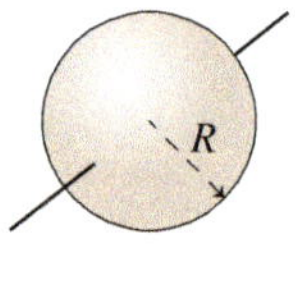	$\frac{2}{5}MR^2$
Plane or slab, about edge	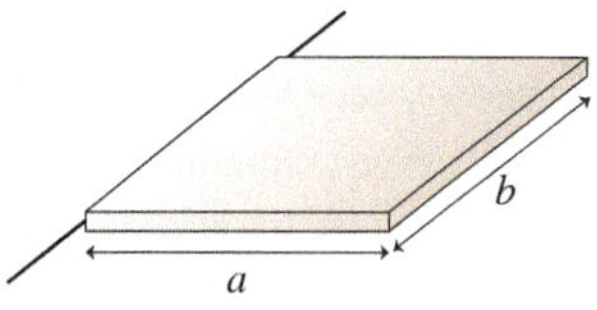	$\frac{1}{3}Ma^2$	Spherical shell, about diameter	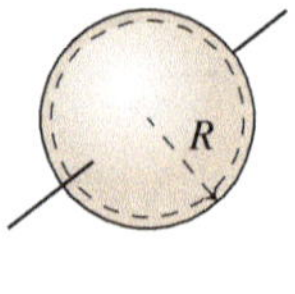	$\frac{2}{3}MR^2$

EXAMPLE 12.4 The speed of a rotating rod

A 1.0-m-long, 200 g rod is hinged at one end and connected to a wall. It is held out horizontally, then released. What is the speed of the tip of the rod as it hits the wall?

MODEL The mechanical energy is conserved if we assume the hinge is frictionless. The rod's gravitational potential energy is transformed into rotational kinetic energy as it "falls."

FIGURE 12.13 A before-and-after pictorial representation of the rod.

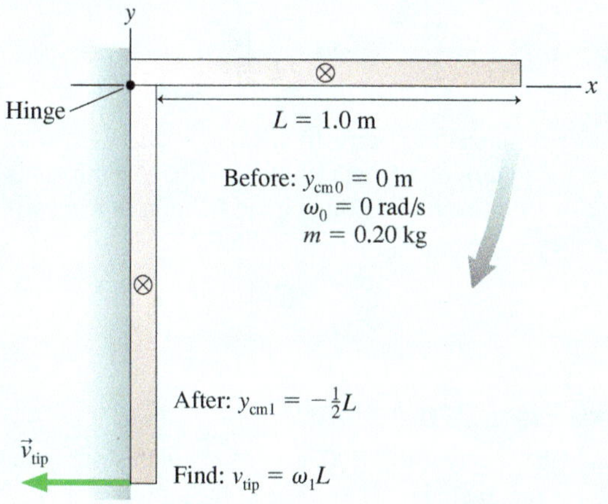

VISUALIZE FIGURE 12.13 is a familiar before-and-after pictorial representation of the rod. We've placed the origin of the coordinate system at the pivot point.

SOLVE Mechanical energy is conserved, so we can equate the rod's final mechanical energy to its initial mechanical energy:

$$\frac{1}{2}I\omega_1^2 + Mgy_{cm1} = \frac{1}{2}I\omega_0^2 + Mgy_{cm0}$$

The initial conditions are $\omega_0 = 0$ and $y_{cm0} = 0$. The center of mass moves to $y_{cm1} = -\frac{1}{2}L$ as the rod hits the wall. From Table 12.2 we find $I = \frac{1}{3}ML^2$ for a rod rotating about one end. Thus

$$\frac{1}{2}I\omega_1^2 + Mgy_{cm1} = \frac{1}{6}ML^2\omega_1^2 - \frac{1}{2}MgL = 0$$

We can solve this for the rod's angular velocity as it hits the wall:

$$\omega_1 = \sqrt{\frac{3g}{L}}$$

The tip of the rod is moving in a circle with radius $r = L$. Its final speed is

$$v_{tip} = \omega_1 L = \sqrt{3gL} = 5.4 \text{ m/s}$$

ASSESS Energy conservation is a powerful tool for rotational motion, just as it was for translational motion.

12.4 Calculating Moment of Inertia

The equation for rotational energy is easy to write, but we can't make use of it without knowing an object's moment of inertia. Unlike mass, we can't measure moment of inertia by putting an object on a scale. And while we can guess that the center of mass of a symmetrical object is at the physical center of the object, we can *not* guess the moment of inertia of even a simple object. To find I, we really must carry through the calculation.

Equation 12.12 defines the moment of inertia as a sum over all the particles in the system. As we did for the center of mass, we can replace the individual particles with cells 1, 2, 3, . . . of mass Δm. Then the moment of inertia summation can be converted to an integration:

$$I = \sum_i r_i^2 \Delta m \xrightarrow[\Delta m \to 0]{} I = \int r^2\, dm \tag{12.15}$$

where r is the distance from the rotation axis. If we let the rotation axis be the z-axis, then we can write the moment of inertia as

$$I = \int (x^2 + y^2)\, dm \tag{12.16}$$

NOTE ▶ You *must* replace dm by an equivalent expression involving a coordinate differential such as dx or dy before you can carry out the integration of Equation 12.16. ◀

You can use any coordinate system to calculate the coordinates x_{cm} and y_{cm} of the center of mass. But the moment of inertia is defined for rotation about a particular axis, and r is measured from that axis. Thus the coordinate system used for moment-of-inertia calculations *must* have its origin at the pivot point. Two examples will illustrate these ideas.

EXAMPLE 12.5 Moment of inertia of a rod about a pivot at one end

Find the moment of inertia of a thin, uniform rod of length L and mass M that rotates about a pivot at one end.

MODEL An object's moment of inertia depends on the axis of rotation. In this case, the rotation axis is at the end of the rod.

VISUALIZE FIGURE 12.14 defines an x-axis with the origin at the pivot point.

FIGURE 12.14 Finding the moment of inertia about one end of a long, thin rod.

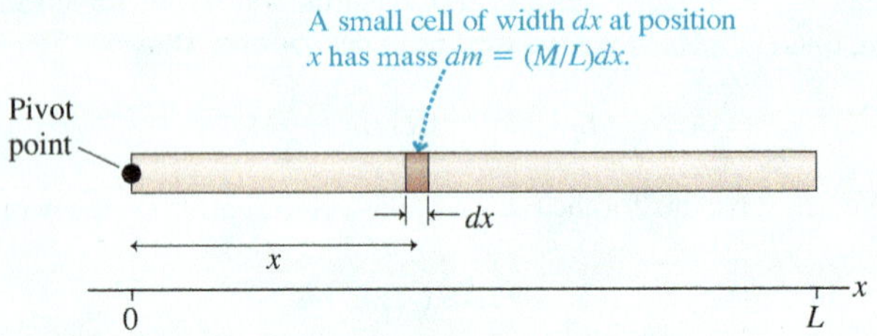

SOLVE Because the rod is thin, we can assume that $y \approx 0$ for all points on the rod. Thus

$$I = \int x^2\, dm$$

The small amount of mass dm in the small length dx is $dm = (M/L)\,dx$, as we found in Example 12.2. The rod extends from $x = 0$ to $x = L$, so the moment of inertia for a rod about one end is

$$I_{\text{end}} = \frac{M}{L}\int_0^L x^2\, dx = \frac{M}{L}\left[\frac{x^3}{3}\right]_0^L = \frac{1}{3}ML^2$$

ASSESS The moment of inertia involves a product of the total mass M with the *square* of a length, in this case L. All moments of inertia have a similar form, although the fraction in front will vary. This is the result shown earlier in Table 12.2.

EXAMPLE 12.6 Moment of inertia of a circular disk about an axis through the center

Find the moment of inertia of a circular disk of radius R and mass M that rotates on an axis passing through its center.

VISUALIZE FIGURE 12.15 shows the disk and defines distance r from the axis.

FIGURE 12.15 Finding the moment of inertia of a disk about an axis through the center.

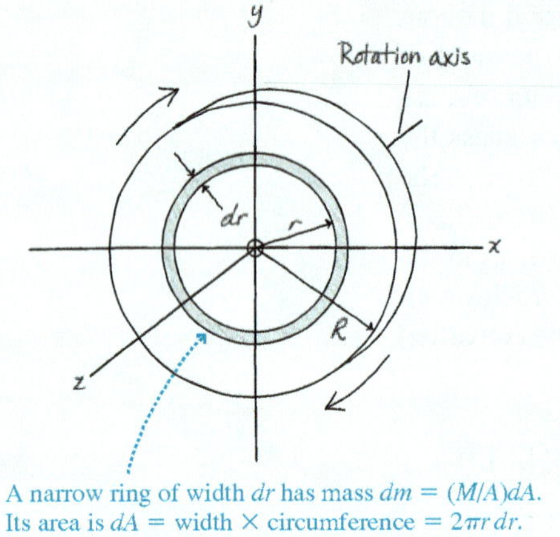

SOLVE This is a situation of great practical importance. To solve this problem, we need to use a two-dimensional integration scheme that you learned in calculus. Rather than dividing the disk into little boxes, let's divide it into narrow *rings* of mass dm. Figure 12.15 shows one such ring, of radius r and width dr. Let dA represent the area of this ring. The mass dm in this ring is the same fraction of the total mass M as dA is of the total area A. That is,

$$\frac{dm}{M} = \frac{dA}{A}$$

Thus the mass in the small area dA is

$$dm = \frac{M}{A}dA$$

This is the reasoning we used to find the center of mass of the rod in Example 12.2, only now we're using it in two dimensions.

The total area of the disk is $A = \pi R^2$, but what is dA? If we imagine unrolling the little ring, it would form a long, thin rectangle of length $2\pi r$ and height dr. Thus the *area* of this little ring is $dA = 2\pi r\, dr$. With this information we can write

$$dm = \frac{M}{\pi R^2}(2\pi r dr) = \frac{2M}{R^2} r\, dr$$

Now we have an expression for dm in terms of a coordinate differential dr, so we can proceed to carry out the integration for I. Using Equation 12.15, we find

$$I_{\text{disk}} = \int r^2\, dm = \int r^2\left(\frac{2M}{R^2} r\, dr\right) = \frac{2M}{R^2}\int_0^R r^3\, dr$$

where in the last step we have used the fact that the disk extends from $r = 0$ to $r = R$. Performing the integration gives

$$I_{\text{disk}} = \frac{2M}{R^2}\left[\frac{r^4}{4}\right]_0^R = \frac{1}{2}MR^2$$

ASSESS Once again, the moment of inertia involves a product of the total mass M with the *square* of a length, in this case R.

If a complex object can be divided into simpler pieces 1, 2, 3, ... whose moments of inertia I_1, I_2, I_3 ... are already known, the moment of inertia of the entire object is

$$I_{\text{object}} = I_1 + I_2 + I_3 + \cdots \tag{12.17}$$

This follows from the fact that the sum $I = \sum m_i r_i^2$ can be broken into smaller sums over the simpler objects. Equation 12.17 is useful for solving many problems.

The Parallel-Axis Theorem

The moment of inertia depends on the rotation axis. Suppose you need to know the moment of inertia for rotation about the off-center axis in FIGURE 12.16. You can find this quite easily if you know the moment of inertia for rotation around a *parallel axis* through the center of mass.

If the axis of interest is distance d from a parallel axis through the center of mass, the moment of inertia is

$$I = I_{cm} + Md^2 \qquad (12.18)$$

Equation 12.18 is called the **parallel-axis theorem.** We'll give a proof for the one-dimensional object shown in FIGURE 12.17.

The x-axis has its origin at the rotation axis, and the x'-axis has its origin at the center of mass. You can see that the coordinates of dm along these two axes are related by $x = x' + d$. By definition, the moment of inertia about the rotation axis is

$$I = \int x^2\,dm = \int (x' + d)^2\,dm = \int (x')^2\,dm + 2d\int x'\,dm + d^2\int dm \qquad (12.19)$$

The first of the three integrals on the right, by definition, is the moment of inertia I_{cm} about the center of mass. The third is simply Md^2 because adding up (integrating) all the dm gives the total mass M.

If you refer back to Equations 12.5, the definition of the center of mass, you'll see that the middle integral on the right is equal to Mx'_{cm}. But $x'_{cm} = 0$ because we specifically chose the x'-axis to have its origin at the center of mass. Thus the second integral is zero and we end up with Equation 12.18. The proof in two dimensions is similar.

FIGURE 12.16 Rotation about an off-center axis.

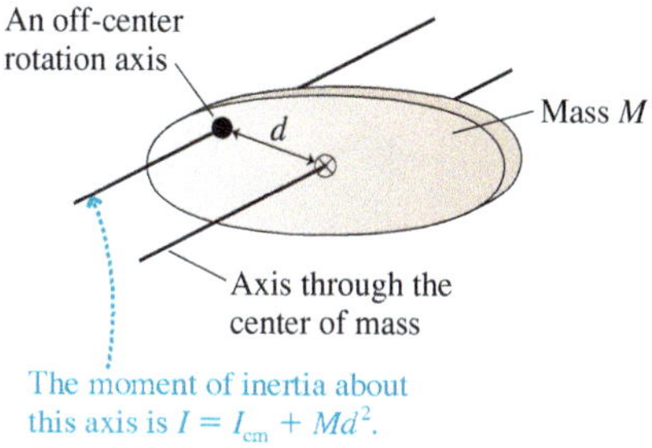

FIGURE 12.17 Proving the parallel-axis theorem.

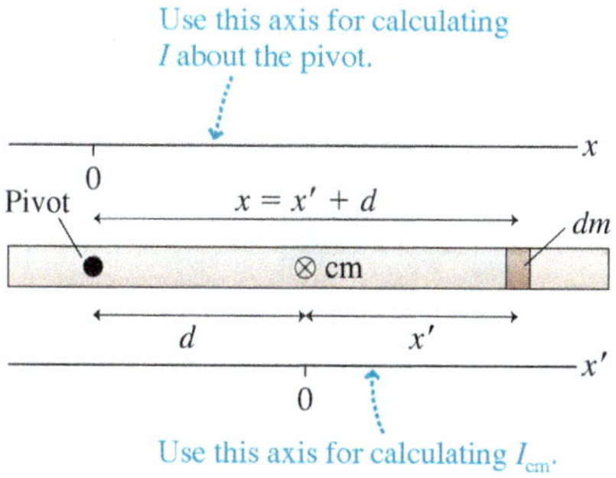

EXAMPLE 12.7 The moment of inertia of a thin rod

Find the moment of inertia of a thin rod with mass M and length L about an axis one-third of the length from one end.

SOLVE From Table 12.2 we know the moment of inertia about the center of mass is $\frac{1}{12}ML^2$. The center of mass is at the center of the rod. An axis $\frac{1}{3}L$ from one end is $d = \frac{1}{6}L$ from the center of mass. Using the parallel-axis theorem, we have

$$I = I_{cm} + Md^2 = \frac{1}{12}ML^2 + M\left(\frac{1}{6}L\right)^2 = \frac{1}{9}ML^2$$

STOP TO THINK 12.1 Four Ts are made from two identical rods of equal mass and length. Rank in order, from largest to smallest, the moments of inertia I_a to I_d for rotation about the dashed line.

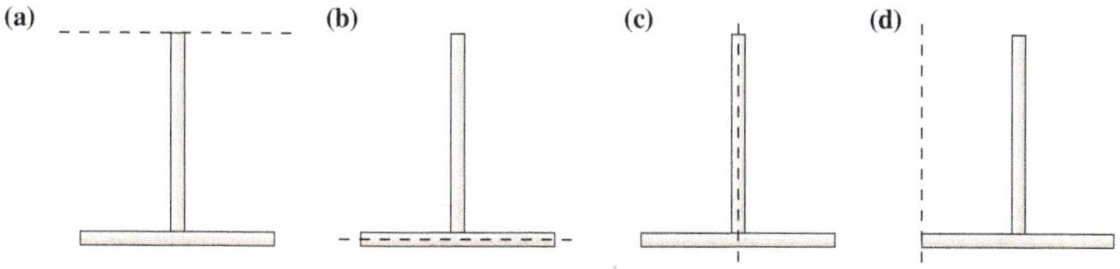

12.5 Torque

Consider the common experience of pushing open a door. FIGURE 12.18 is a top view of a door hinged on the left. Four pushing forces are shown, all of equal strength. Which of these will be most effective at opening the door?

Force $\vec{F}_1$ will open the door, but force $\vec{F}_2$, which pushes straight at the hinge, will not. Force $\vec{F}_3$ will open the door, but not as easily as $\vec{F}_1$. What about $\vec{F}_4$? It is perpendicular to the door, it has the same magnitude as $\vec{F}_1$, but you know from experience that pushing close to the hinge is not as effective as pushing at the outer edge of the door.

FIGURE 12.18 The four forces have different effects on the swinging door.

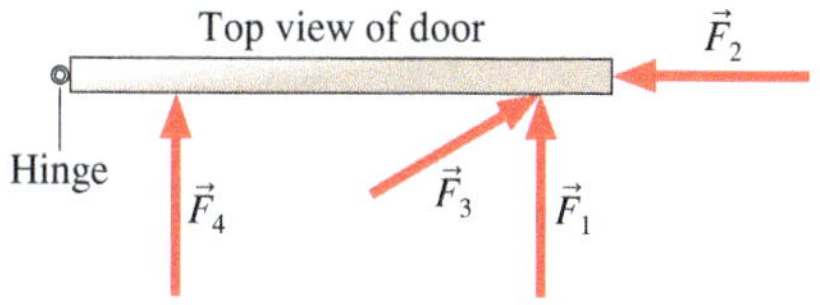

FIGURE 12.19 Force $\vec{F}$ exerts a torque about the pivot point.

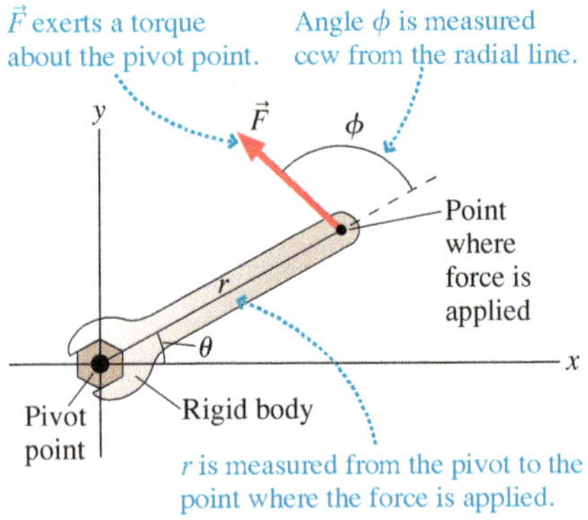

Torque is to rotational motion as force is to linear motion.

The ability of a force to cause a rotation depends on three factors:

1. The magnitude F of the force.
2. The distance r from the point of application to the pivot.
3. The angle at which the force is applied.

To make these ideas specific, FIGURE 12.19 shows a force $\vec{F}$ applied at one point on a rigid body. For example, a string might be pulling on the object at that point, in which case the force would be a tension force.

NOTE ▶ Angle ϕ is measured *counterclockwise* from the dashed line that extends outward along the radial line. This is consistent with our sign convention for the angular position θ. ◀

Let's define a new quantity called the **torque τ** (Greek tau) as

$$\tau \equiv rF\sin\phi \tag{12.20}$$

Torque depends on the three properties we just listed: the magnitude of the force, its distance from the pivot, and its angle. Loosely speaking, τ measures the "effectiveness" of the force at causing an object to rotate about a pivot. **Torque is the rotational equivalent of force.**

The SI units of torque are newton-meters, abbreviated N m. Although we defined 1 N m = 1 J during our study of energy, torque is not an energy-related quantity and so we do *not* use joules as a measure of torque.

Torque, like force, has a sign. A torque that tries to rotate the object in a ccw direction is positive while a negative torque gives a cw rotation. FIGURE 12.20 summarizes the signs. Notice that a force pushing straight toward the pivot or pulling straight out from the pivot exerts *no* torque.

FIGURE 12.20 Signs and strengths of the torque.

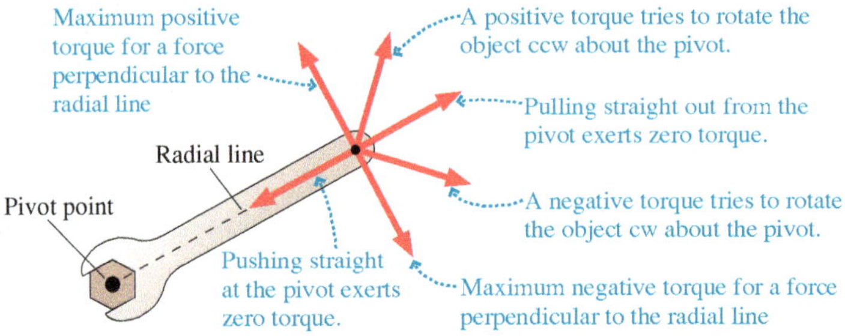

NOTE ▶ Torque differs from force in a very important way. Torque is calculated or measured *about a pivot point.* To say that a torque is 20 N m is meaningless. You need to say that the torque is 20 N m about a particular point. Torque can be calculated about any pivot point, but its value depends on the point chosen. In practice, we measure or calculate torques about the same point from which we measure an object's angular position θ (and thus its angular velocity ω and angular acceleration α). This assumption is built into the equations of rotational dynamics. ◀

Returning to the door of Figure 12.18, you can see that $\vec{F}_1$ is most effective at opening the door because $\vec{F}_1$ exerts the largest torque *about the pivot point.* $\vec{F}_3$ has equal magnitude, but it is applied at an angle less than 90° and thus exerts less torque. $\vec{F}_2$, pushing straight at the hinge with $\phi = 0°$, exerts no torque at all. And $\vec{F}_4$, with a smaller value for r, exerts less torque than $\vec{F}_1$.

Interpreting Torque

Torque can be interpreted from two perspectives. First, FIGURE 12.21a shows that the quantity $F \sin \phi$ is the tangential force component F_t. Consequently, the torque is

$$\tau = rF_t \qquad (12.21)$$

In other words, torque is the product of r with the force component F_t that is *perpendicular* to the radial line. This interpretation makes sense because the radial component of $\vec{F}$ points straight at the pivot point and cannot exert a torque.

FIGURE 12.21 Two useful interpretations of the torque.

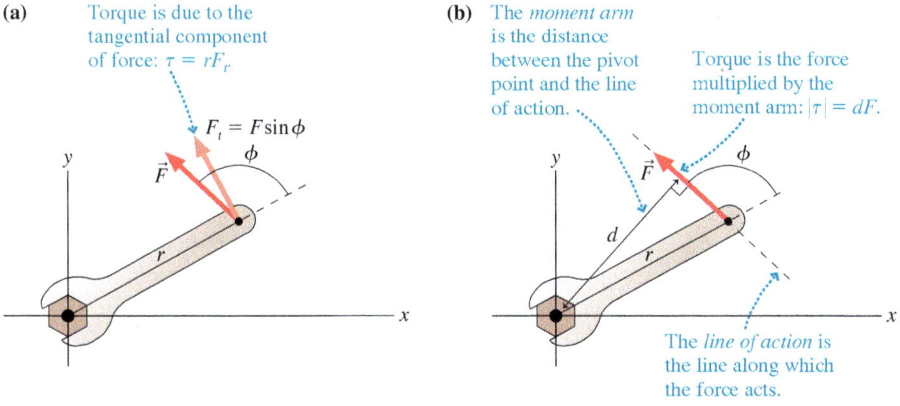

A second perspective, widely used in applications, is based on the idea of a *moment arm*. FIGURE 12.21b shows the **line of action,** the line along which the force acts. The minimum distance between the pivot point and the line of action—the length of a line drawn *perpendicular to the line of action*—is called the **moment arm** (or the *lever arm*) d. Because $\sin(180° - \theta) = \sin\phi$, it is easy to see that $d = r\sin\phi$. Thus the torque $rF\sin\theta$ can also be written

$$|\tau| = dF \qquad (12.22)$$

NOTE ▶ Equation 12.22 gives only $|\tau|$, the magnitude of the torque; the sign has to be supplied by observing the direction in which the torque acts. ◀

EXAMPLE 12.8 Applying a torque

Luis uses a 20-cm-long wrench to turn a nut. The wrench handle is tilted 30° above the horizontal, and Luis pulls straight down on the end with a force of 100 N. How much torque does Luis exert on the nut?

VISUALIZE FIGURE 12.22 shows the situation. The angle is a negative $\phi = -120°$ because it is *clockwise* from the radial line.

FIGURE 12.22 A wrench being used to turn a nut.

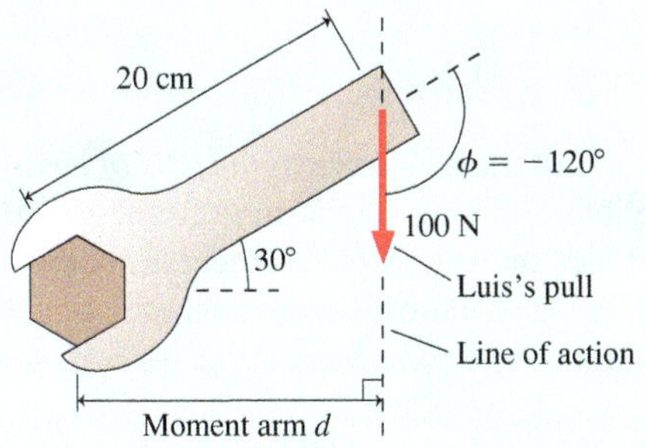

SOLVE The tangential component of the force is

$$F_t = F\sin\phi = -86.6 \text{ N}$$

According to our sign convention, F_t is negative because it points in a cw direction. The torque, from Equation 12.21, is

$$\tau = rF_t = (0.20 \text{ m})(-86.6 \text{ N}) = -17 \text{ N m}$$

Alternatively, Figure 12.22 has drawn the *line of action* by extending the force vector forward and backward. The *moment arm*, the distance between the pivot point and the line of action, is

$$d = r\sin(60°) = 0.17 \text{ m}$$

Inserting the moment arm in Equation 12.22 gives

$$|\tau| = dF = (0.17 \text{ m})(100 \text{ N}) = 17 \text{ N m}$$

The torque acts to give a cw rotation, so we insert a minus sign to end up with

$$\tau = -17 \text{ N m}$$

ASSESS Luis could increase the torque by changing the angle so that his pull is perpendicular to the wrench ($\phi = -90°$).

STOP TO THINK 12.2 Rank in order, from largest to smallest, the five torques τ_a to τ_e. The rods all have the same length and are pivoted at the dot.

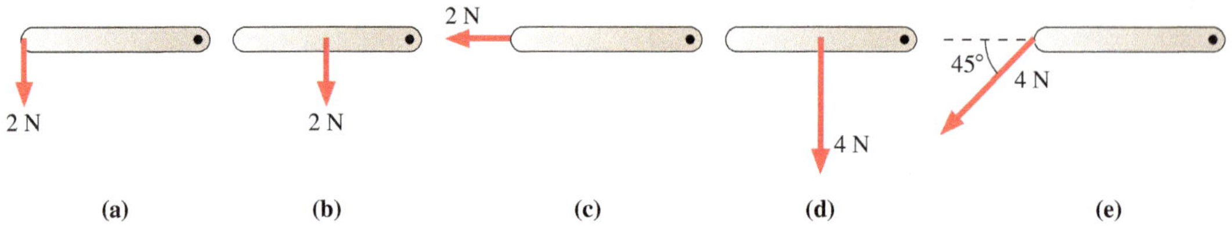

Net Torque

FIGURE 12.23 The forces exert a net torque about the pivot point.

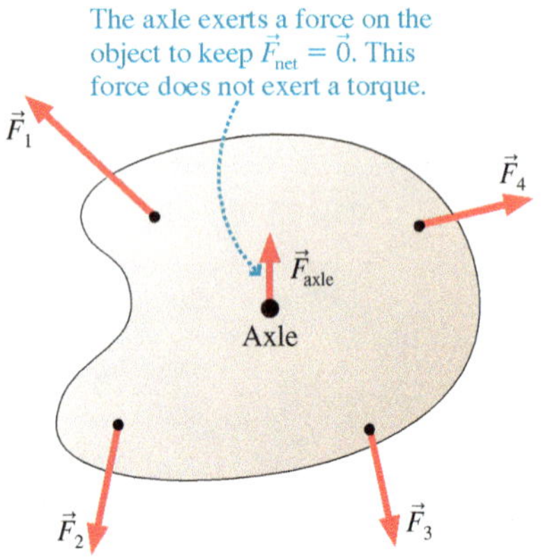

FIGURE 12.23 shows forces $\vec{F}_1, \vec{F}_2, \vec{F}_3, \ldots$ applied to an extended object. The object is free to rotate about the axle, but the axle prevents the object from having any translational motion. It does so by exerting force $\vec{F}_{axle}$ on the object to balance the other forces and keep $\vec{F}_{net} = \vec{0}$.

Forces $\vec{F}_1, \vec{F}_2, \vec{F}_3, \ldots$ exert torques $\tau_1, \tau_2, \tau_3, \ldots$ on the object, but $\vec{F}_{axle}$ does *not* exert a torque because it is applied at the pivot point and has zero moment arm. Thus the *net* torque about the axle is the sum of the torques due to the applied forces:

$$\tau_{net} = \tau_1 + \tau_2 + \tau_3 + \cdots = \sum_i \tau_i \qquad (12.23)$$

Gravitational Torque

FIGURE 12.24 Gravitational torque.

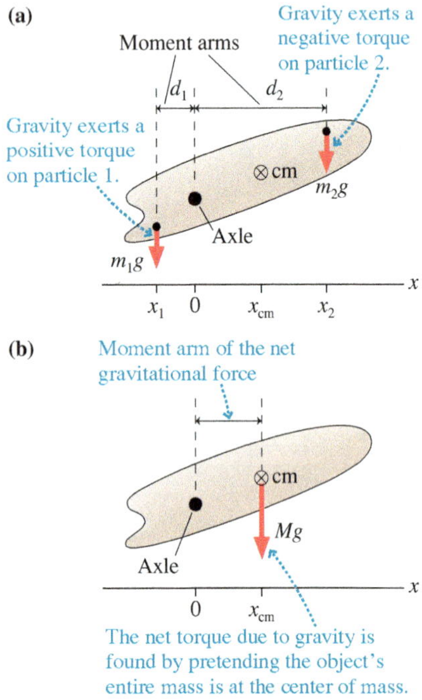

Gravity exerts a torque on many objects. If the object in FIGURE 12.24 is released, a torque due to gravity will cause it to rotate around the axle. To calculate the torque about the axle, we start with the fact that gravity acts on *every* particle in the object, exerting a downward force of magnitude $F_i = m_i g$ on particle *i*. The *magnitude* of the gravitational torque on particle *i* is $|\tau_i| = d_i m_i g$, where d_i is the moment arm. But we need to be careful with signs.

A moment arm must be a positive number because it's a distance. If we establish a coordinate system with the origin at the axle, then you can see from FIGURE 12.24a that the moment arm d_i of particle *i* is $|x_i|$. A particle to the right of the axle (positive x_i) experiences a *negative* torque because gravity tries to rotate this particle in a clockwise direction. Similarly, a particle to the left of the axle (negative x_i) has a positive torque. The torque is opposite in sign to x_i, so we can get the sign right by writing

$$\tau_i = -x_i m_i g = -(m_i x_i)g \qquad (12.24)$$

The net torque due to gravity is found by summing Equation 12.24 over all particles:

$$\tau_{grav} = \sum_i \tau_i = \sum_i (-m_i x_i g) = -\left(\sum_i m_i x_i\right) g \qquad (12.25)$$

But according to the definition of center of mass, Equations 12.4, $\sum m_i x_i = M x_{cm}$. Thus the torque due to gravity is

$$\tau_{grav} = -Mgx_{cm} \qquad (12.26)$$

where x_{cm} is the position of the center of mass *relative to the axis of rotation.*

Equation 12.26 has the simple interpretation shown in FIGURE 12.24b. Mg is the net gravitational force on the entire object, and x_{cm} is the moment arm between the rotation axis and the center of mass. The gravitational torque on an extended object of mass M is equivalent to the torque of a *single* force vector $\vec{F}_{grav} = -Mg\hat{j}$ acting at the object's center of mass.

In other words, **the gravitational torque is found by treating the object as if all its mass were concentrated at the center of mass.** This is the basis for the well-known technique of finding an object's center of mass by balancing it. An object will balance on a pivot, as shown in FIGURE 12.25, only if the center of mass is directly above the pivot point. If the pivot is *not* under the center of mass, the gravitational torque will cause the object to rotate.

NOTE ▶ The point at which gravity acts is also called the *center of gravity*. As long as gravity is uniform over the object—always true for earthbound objects—there's no difference between center of mass and center of gravity. ◀

FIGURE 12.25 An object balances on a pivot that is directly under the center of mass.

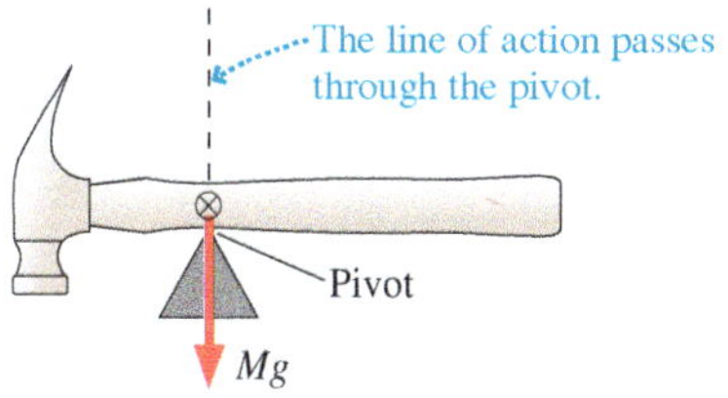

EXAMPLE 12.9 The gravitational torque on a beam

The 4.00-m-long, 500 kg steel beam shown in FIGURE 12.26 is supported 1.20 m from the right end. What is the gravitational torque about the support?

FIGURE 12.26 A steel beam supported at one point.

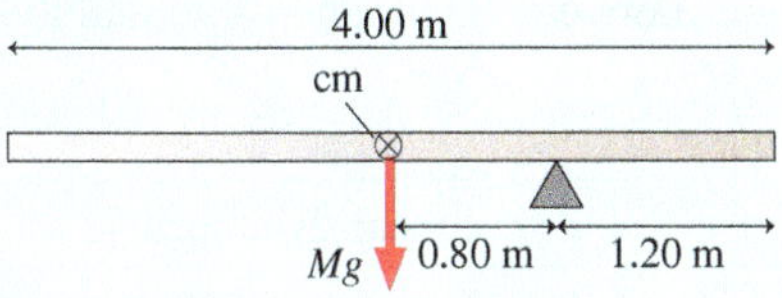

MODEL The center of mass of the beam is at the midpoint. $x_{cm} = -0.80$ m is measured from the pivot point.

SOLVE This is a straightforward application of Equation 12.26. The gravitational torque is

$$\tau_{grav} = -Mgx_{cm} = -(500 \text{ kg})(9.80 \text{ m/s}^2)(-0.80 \text{ m})$$
$$= 3920 \text{ N m}$$

ASSESS The torque is positive because gravity tries to rotate the beam ccw around the point of support. Notice that the beam in Figure 12.26 is *not* in equilibrium. It will fall over unless other forces, not shown, are supporting it.

12.6 Rotational Dynamics

What does a torque do? **A torque causes an angular acceleration.** To see why, FIGURE 12.27 shows a rigid body undergoing *pure rotational motion* about a fixed and unmoving axis. This might be an unconstrained rotation about the object's center of mass, such as we considered in Section 12.2. Or it might be an object, such as a pulley or a turbine, rotating on an axle.

FIGURE 12.27 The external forces on a rigid body exert a torque about the rotation axis and thus cause an angular acceleration.

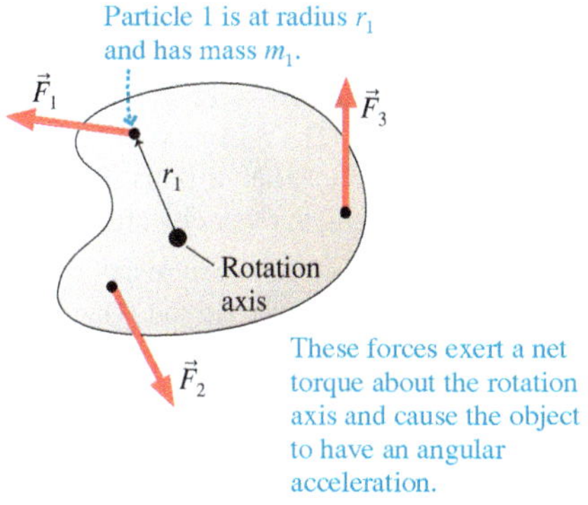

The forces $\vec{F}_1$, $\vec{F}_2$, $\vec{F}_3$, ... in Figure 12.27 are external forces acting on particles of masses m_1, m_2, m_3, ... that are part of the rigid body. These forces exert torques τ_1, τ_2, τ_3, ... about the rotation axis. The *net* torque on the object is the sum of the torques on all the individual particles in the object:

$$\tau_{net} = \sum_i \tau_i \qquad (12.27)$$

Focus on particle i, which is acted on by force $\vec{F}_i$ and undergoes circular motion with radius r_i. In Chapter 8, we found that the radial component of $\vec{F}_i$ is responsible for the centripetal acceleration of circular motion, while the tangential component $(F_i)_t$ causes the particle to speed up or slow down with a tangential acceleration $(a_i)_t$. Newton's second law is

$$(F_i)_t = m_i(a_i)_t = m_i r_i \alpha \qquad (12.28)$$

where in the last step we used the relationship between tangential and angular acceleration: $a_t = r\alpha$. The angular acceleration α does not have a subscript because *all particles in the object have the same angular acceleration.* That is, α is the angular acceleration of the entire object.

Multiplying both sides by r_i gives

$$r_i(F_i)_t = m_i r_i^2 \alpha \qquad (12.29)$$

But $r_i(F_i)_t$ is the torque τ_i on particle i; hence Newton's second law for a single particle in the object is

$$\tau_i = m_i r_i^2 \alpha \qquad (12.30)$$

Returning now to Equation 12.27, we see that the net torque on the object in Figure 12.27 is

$$\tau_{net} = \sum_i \tau_i = \sum_i m_i r_i^2 \alpha = \left(\sum_i m_i r_i^2\right)\alpha \qquad (12.31)$$

In the last step, we factored out α by using the key idea that every particle in a rotating rigid body has the *same* angular acceleration.

You'll recognize the quantity in parentheses as the moment of inertia I. Substituting the moment of inertia into Equation 12.31 puts the final piece of the puzzle into place. An object that experiences a net torque τ_{net} about the axis of rotation undergoes an angular acceleration

$$\alpha = \frac{\tau_{net}}{I} \quad \text{(Newton's second law for rotational motion)} \qquad (12.32)$$

where I is the object's moment of inertia *about the rotation axis.* This result, Newton's second law for rotation, is the fundamental equation of rigid-body dynamics.

In practice we often write $\tau_{net} = I\alpha$, but Equation 12.32 better conveys the idea that **torque is the cause of angular acceleration.** In the absence of a net torque ($\tau_{net} = 0$), the object either does not rotate ($\omega = 0$) or rotates with *constant* angular velocity ($\omega =$ constant).

Table 12.3 summarizes the analogies between linear and rotational dynamics.

TABLE 12.3 Rotational and linear dynamics

Rotational dynamics		Linear dynamics	
torque	τ_{net}	force	$\vec{F}_{net}$
moment of inertia	I	mass	m
angular acceleration	α	acceleration	$\vec{a}$
second law	$\alpha = \tau_{net}/I$	second law	$\vec{a} = \vec{F}_{net}/m$

EXAMPLE 12.10 Rotating rockets

Far out in space, a 100,000 kg rocket and a 200,000 kg rocket are docked at opposite ends of a motionless 90-m-long connecting tunnel. The tunnel is rigid and its mass is much less than that of either rocket. The rockets start their engines simultaneously, each generating 50,000 N of thrust in opposite directions. What is the structure's angular velocity after 30 s?

MODEL The entire structure can be modeled as two masses at the ends of a massless, rigid rod. There's no net force, so the structure does not undergo translational motion, but the thrusts do create torques that will give the structure angular acceleration and cause it to rotate. We'll assume the thrust forces are perpendicular to the connecting tunnel. This is an unconstrained rotation, so the structure will rotate about its center of mass.

VISUALIZE FIGURE 12.28 shows the rockets and defines distances r_1 and r_2 from the center of mass.

FIGURE 12.28 The thrusts exert a torque on the structure.

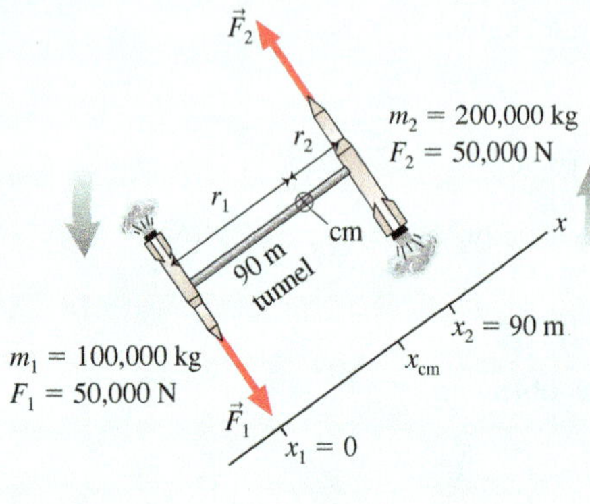

SOLVE Our strategy will be to use Newton's second law to find the angular acceleration, followed by rotational kinematics to find ω. We'll need to determine the moment of inertia, and that requires knowing the distances of the two rockets from the rotation axis. As we did in Example 12.1, we choose a coordinate system in which the masses are on the x-axis and in which m_1 is at the origin. Then

$$x_{cm} = \frac{m_1x_1 + m_2x_2}{m_1 + m_2}$$

$$= \frac{(100{,}000\text{ kg})(0\text{ m}) + (200{,}000\text{ kg})(90\text{ m})}{100{,}000\text{ kg} + 200{,}000\text{ kg}} = 60\text{ m}$$

The structure's center of mass is $r_1 = 60$ m from the 100,000 kg rocket and $r_2 = 30$ m from the 200,000 kg rocket. The moment of inertia about the center of mass is

$$I = m_1r_1^2 + m_2r_2^2 = 540{,}000{,}000\text{ kg m}^2$$

The two rocket thrusts exert net torque

$$\tau_{net} = r_1F_1 + r_2F_2 = (60\text{ m})(50{,}000\text{ N}) + (30\text{ m})(50{,}000\text{ N})$$
$$= 4{,}500{,}000\text{ N m}$$

With I and τ_{net} now known, we can use Newton's second law to find the angular acceleration:

$$\alpha = \frac{\tau}{I} = \frac{4{,}500{,}000\text{ N m}}{540{,}000{,}000\text{ kg m}^2} = 0.00833\text{ rad/s}^2$$

After 30 seconds, the structure's angular velocity is

$$\omega = \alpha \Delta t = 0.25 \text{ rad/s}$$

ASSESS Few of us have the experience to judge whether or not 0.25 rad/s is a reasonable answer to this problem. The significance of the example is to demonstrate the approach to a rotational dynamics problem.

STOP TO THINK 12.3 Rank in order, from largest to smallest, the angular accelerations α_a to α_e.

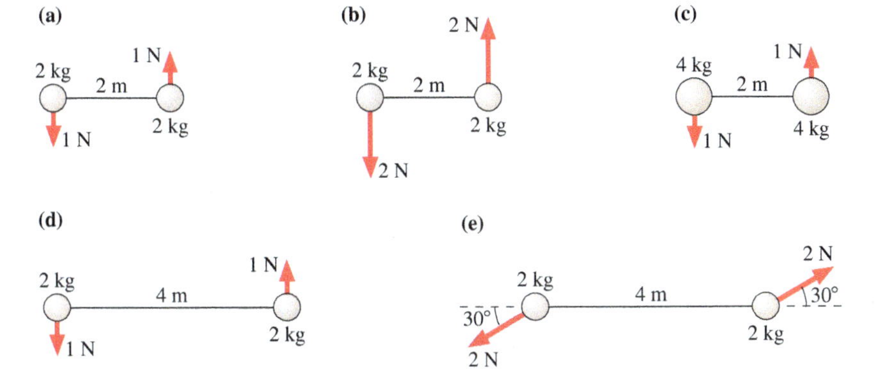

12.7 Rotation About a Fixed Axis

In this section we'll look at rigid bodies that rotate about a fixed axis. The problem-solving strategy for rotational dynamics is very similar to that for linear dynamics.

PROBLEM-SOLVING STRATEGY 12.1 Rotational dynamics problems

MODEL Model the object as a simple shape.

VISUALIZE Draw a pictorial representation to clarify the situation, define coordinates and symbols, and list known information.

- Identify the axis about which the object rotates.
- Identify forces and determine their distances from the axis. For most problems it will be useful to draw a free-body diagram.
- Identify any torques caused by the forces and the signs of the torques.

SOLVE The mathematical representation is based on Newton's second law for rotational motion:

$$\tau_{\text{net}} = I\alpha \qquad \text{or} \qquad \alpha = \frac{\tau_{\text{net}}}{I}$$

- Find the moment of inertia in Table 12.2 or, if needed, calculate it as an integral or by using the parallel-axis theorem.
- Use rotational kinematics to find angles and angular velocities.

ASSESS Check that your result has the correct units, is reasonable, and answers the question.

Exercise 26

EXAMPLE 12.11 Starting an airplane engine

The engine in a small airplane is specified to have a torque of 60 N m. This engine drives a 2.0-m-long, 40 kg propeller. On start-up, how long does it take the propeller to reach 200 rpm?

MODEL The propeller can be modeled as a rod that rotates about its center. The engine exerts a torque on the propeller.

VISUALIZE FIGURE 12.29 shows the propeller and the rotation axis.

FIGURE 12.29 A rotating airplane propeller.

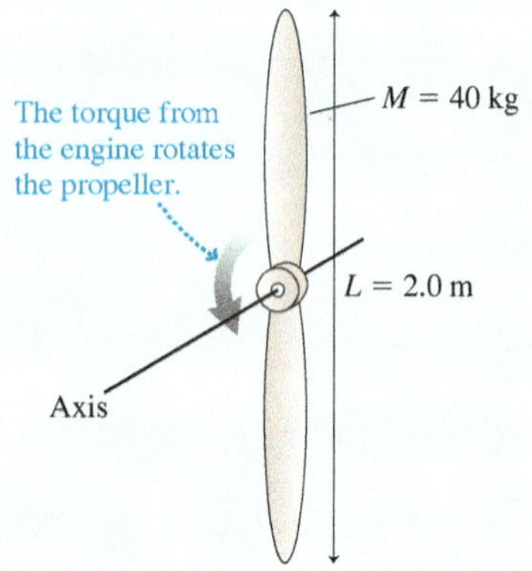

SOLVE The moment of inertia of a rod rotating about its center is found from Table 12.2:

$$I = \frac{1}{12}ML^2 = \frac{1}{12}(40\text{ kg})(2.0\text{ m})^2 = 13.33\text{ kg m}^2$$

The 60 N m torque of the engine causes an angular acceleration

$$\alpha = \frac{\tau}{I} = \frac{60\text{ N m}}{13.33\text{ kg m}^2} = 4.50\text{ rad/s}^2$$

The time needed to reach $\omega_f = 200\text{ rpm} = 3.33\text{ rev/s} = 20.9\text{ rad/s}$ is

$$\Delta t = \frac{\Delta\omega}{\alpha} = \frac{\omega_f - \omega_i}{\alpha} = \frac{20.9\text{ rad/s} - 0\text{ rad/s}}{4.5\text{ rad/s}^2} = 4.6\text{ s}$$

ASSESS We've assumed a constant angular acceleration, which is reasonable for the first few seconds while the propeller is still turning slowly. Eventually, air resistance and friction will cause opposing torques and the angular acceleration will decrease. At full speed, the negative torque due to air resistance and friction cancels the torque of the engine. Then $\tau_{net} = 0$ and the propeller turns at *constant* angular velocity with no angular acceleration.

EXAMPLE 12.12 An off-center disk

FIGURE 12.30 shows a piece of a large machine. A 10.0-cm-diameter, 5.0 kg disk turns on an axle. A vertical cable attached to the edge of the disk exerts a 100 N force but, initially, a pin keeps the disk from rotating. What is the initial angular acceleration of the disk when the pin is removed?

FIGURE 12.30 A disk rotates on an off-center axle after the pin is removed.

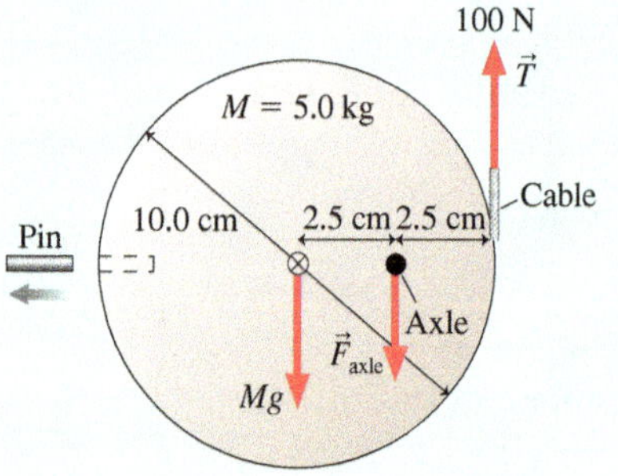

MODEL The disk has an off-center axle. Gravity and tension exert torques about the axle.

VISUALIZE Both the cable tension and gravity rotate the disk ccw, so their torques are positive.

SOLVE After the pin is removed, the forces on the disk are a downward gravitational force, an upward force from the cable, and a force exerted by the axle. The axle force, which is exerted at the pivot, does not contribute to the torque and doesn't affect the rotation. The center of mass is to the *left* of the axle, at $x_{cm} = -\frac{1}{2}R$; thus the gravitational torque is

$$\tau_{grav} = -Mgx_{cm} = \frac{1}{2}MgR$$

This is a positive torque, as expected. The net torque, including the cable tension, is

$$\tau_{net} = \tau_{grav} + \tau_{cable} = \frac{1}{2}MgR + \frac{1}{2}RT = 3.73\text{ N m}$$

To find the angular acceleration, we need to know the moment of inertia about the axle. This is where the parallel-axis theorem is useful. We know the moment of inertia about an axis through the center from Table 12.2. The axle is offset by $d = \frac{1}{2}R$. Thus

$$I = I_{cm} + Md^2 = \frac{1}{2}MR^2 + M\left(\frac{1}{2}R\right)^2 = \frac{3}{4}MR^2$$
$$= 9.38 \times 10^{-3}\text{ kg m}^2$$

The torque causes an angular acceleration

$$\alpha = \frac{\tau_{net}}{I} = \frac{3.73\text{ N m}}{9.38 \times 10^{-3}\text{ kg m}^2} = 400\text{ rad/s}^2$$

The angular acceleration is positive, indicating that the disk begins rotating in a ccw direction.

ASSESS As the disk rotates, τ_{net} will change as the moment arms change. Consequently, the disk will *not* have constant angular acceleration. This is simply the *initial* value of α.

Constraints Due to Ropes and Pulleys

Many important applications of rotational dynamics involve objects, such as pulleys, that are connected via ropes or belts to other objects. FIGURE 12.31 shows a rope passing over a pulley and connected to an object in linear motion. If the rope does not slip as the pulley rotates, then the rope's speed v_{rope} must exactly match the speed of the rim of the pulley, which is $v_{rim} = |\omega|R$. If the pulley has an angular acceleration, the rope's acceleration a_{rope} must match the *tangential* acceleration of the rim of the pulley, $a_t = |\alpha|R$.

The object attached to the other end of the rope has the same speed and acceleration as the rope. Consequently, an object connected to a pulley of radius R by a rope that does not slip must obey the constraints

$$v_{obj} = |\omega|R \qquad \text{(motion constraints for a nonslipping rope)} \qquad (12.33)$$
$$a_{obj} = |\alpha|R$$

These constraints are very similar to the acceleration constraints introduced in Chapter 7 for two objects connected by a string or rope.

NOTE ▶ The constraints are given as magnitudes. Specific problems will need to introduce signs that depend on the direction of motion and on the choice of coordinate system. ◀

FIGURE 12.31 The rope's motion must match the motion of the rim of the pulley.

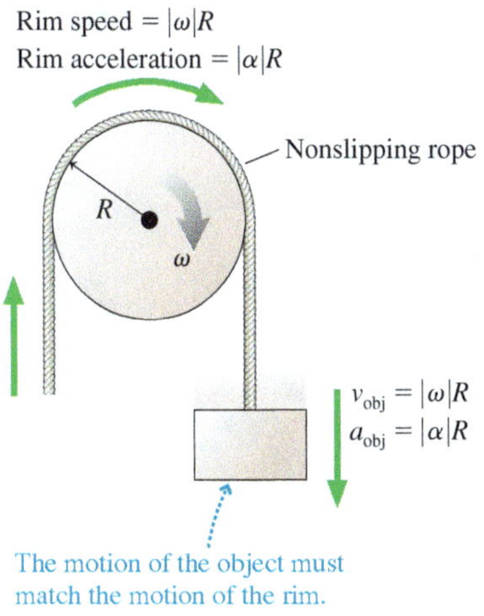

EXAMPLE 12.13 Lowering a bucket

A 2.0 kg bucket is attached to a massless string that is wrapped around a 1.0 kg, 4.0-cm-diameter cylinder, as shown in FIGURE 12.32a. The cylinder rotates on an axle through the center. The bucket is released from rest 1.0 m above the floor. How long does it take to reach the floor?

FIGURE 12.32 The falling bucket turns the cylinder.

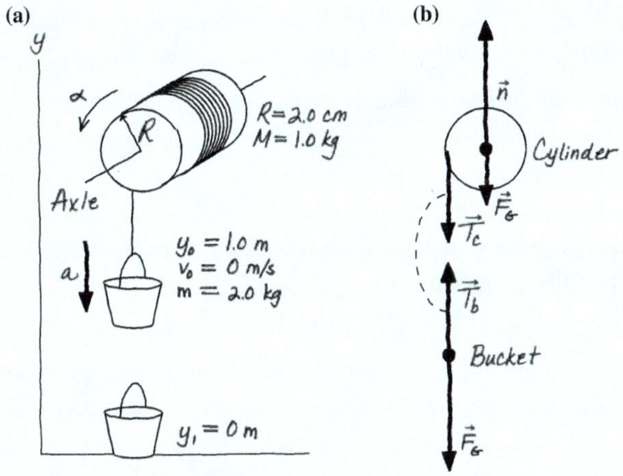

MODEL Assume the string does not slip.

VISUALIZE FIGURE 12.32b shows the free-body diagram for the cylinder and the bucket. The string tension exerts an upward force on the bucket and a downward force on the outer edge of the cylinder. The string is massless, so these two tension forces act as if they are an action/reaction pair: $T_b = T_c = T$.

SOLVE Newton's second law applied to the linear motion of the bucket is

$$ma_y = T - mg$$

where, as usual, the y-axis points upward. What about the cylinder? The only torque comes from the string tension. The moment arm for the tension is $d = R$, and the torque is positive because the string turns the cylinder ccw. Thus $\tau_{string} = TR$ and Newton's second law for the rotational motion is

$$\alpha = \frac{\tau_{net}}{I} = \frac{TR}{\frac{1}{2}MR^2} = \frac{2T}{MR}$$

The moment of inertia of a cylinder rotating about a center axis was taken from Table 12.2.

The last piece of information we need is the constraint due to the fact that the string doesn't slip. Equation 12.33 relates only the absolute values, but in this problem α is positive (ccw acceleration) while a_y is negative (downward acceleration). Hence

$$a_y = -\alpha R$$

Using α from the cylinder's equation in the constraint, we find

$$a_y = -\alpha R = -\frac{2T}{MR}R = -\frac{2T}{M}$$

Thus the tension is $T = -\frac{1}{2}Ma_y$. If we use this value of the tension in the bucket's equation, we can solve for the acceleration:

$$ma_y = -\frac{1}{2}Ma_y - mg$$
$$a_y = -\frac{g}{(1 + M/2m)} = -7.84 \text{ m/s}^2$$

The time to fall through $\Delta y = -1.0$ m is found from kinematics:

$$\Delta y = \frac{1}{2}a_y(\Delta t)^2$$
$$\Delta t = \sqrt{\frac{2\Delta y}{a_y}} = \sqrt{\frac{2(-1.0 \text{ m})}{-7.84 \text{ m/s}^2}} = 0.50 \text{ s}$$

ASSESS The expression for the acceleration gives $a_y = -g$ if $M = 0$. This makes sense because the bucket would be in free fall if there were no cylinder. When the cylinder has mass, the downward force of gravity on the bucket has to accelerate the bucket *and* spin the cylinder. Consequently, the acceleration is reduced and the bucket takes longer to fall.

12.8 Static Equilibrium

Structures such as bridges are analyzed in engineering statics.

We now have two versions of Newton's second law: $\vec{F}_{net} = M\vec{a}$ for translational motion and $\tau_{net} = I\alpha$ for rotational motion. The condition for a rigid body to be in *static equilibrium* is both $\vec{F}_{net} = \vec{0}$ *and* $\tau_{net} = 0$. That is, no net force *and* no net torque. An important branch of engineering called *statics* analyzes buildings, dams, bridges, and other structures in total static equilibrium.

No matter which pivot point you choose, an object that is not rotating is not rotating about that point. This would seem to be a trivial statement, but it has an important implication: **For a rigid body in total equilibrium, there is no net torque about any point.** This is the basis of a problem-solving strategy.

PROBLEM-SOLVING STRATEGY 12.2 **Static equilibrium problems**

MP

MODEL Model the object as a simple shape.

VISUALIZE Draw a pictorial representation showing all forces and distances. List known information.

- Pick any point you wish as a pivot point. The net torque about this point is zero.
- Determine the moment arms of all forces about this pivot point.
- Determine the sign of each torque about this pivot point.

SOLVE The mathematical representation is based on the fact that an object in total equilibrium has no net force and no net torque:

$$\vec{F}_{net} = \vec{0} \quad \text{and} \quad \tau_{net} = 0$$

- Write equations for $\sum F_x = 0$, $\sum F_y = 0$, and $\sum \tau = 0$.
- Solve the three simultaneous equations.

ASSESS Check that your result is reasonable and answers the question.

Although you can pick any point you wish as a pivot point, some choices make the problem easier than others. Often the best choice is a point at which several forces act because the torques exerted by those forces will be zero.

EXAMPLE 12.14 Lifting weights

Weightlifting can exert extremely large forces on the body's joints and tendons. In the *strict curl* event, a standing athlete uses both arms to lift a barbell by moving only his forearms, which pivot at the elbows. The record weight lifted in the strict curl is over 200 pounds (about 900 N). FIGURE 12.33 shows the arm bones and the biceps, the main lifting muscle when the forearm is horizontal. What is the tension in the tendon connecting the biceps muscle to the bone while a 900 N barbell is held stationary in this position?

FIGURE 12.33 An arm holding a barbell.

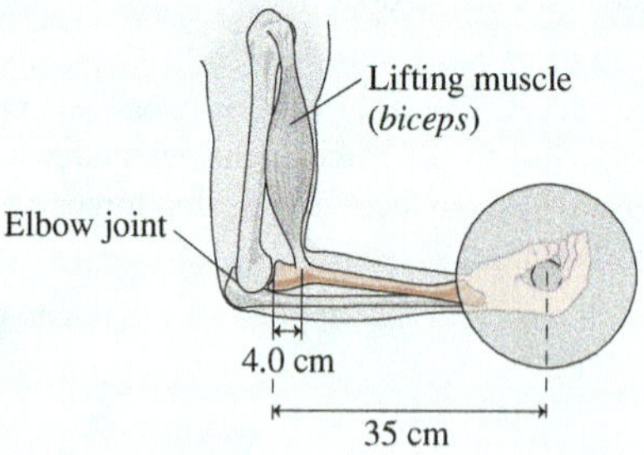

MODEL Model the arm as two rigid rods connected by a hinge. We'll ignore the arm's weight because it is so much less than that of the barbell. Although the tendon pulls at a slight angle, it is close enough to vertical that we'll treat it as such.

VISUALIZE **FIGURE 12.34** shows the forces acting on our simplified model of the forearm. The biceps pulls the forearm up against the upper arm at the elbow, so the force $\vec{F}_{elbow}$ *on* the forearm at the elbow—a force due to the upper arm—is a downward force.

FIGURE 12.34 A pictorial representation of the forces involved.

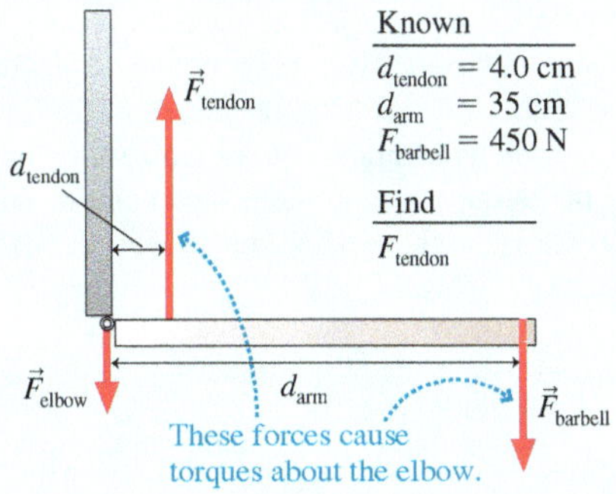

SOLVE Static equilibrium requires both the net force *and* the net torque on the forearm to be zero. Only the y-component of force is relevant, and setting it to zero gives a first equation:

$$\sum F_y = F_{tendon} - F_{elbow} - F_{barbell} = 0$$

Because each arm supports half the weight of the barbell, $F_{barbell} = 450$ N. We don't know either F_{tendon} or F_{elbow}, nor does the force equation give us enough information to find them. But the fact that the net torque also must be zero gives us that extra information. The torque is zero about *every* point, so we can choose any point we wish to calculate the torque. The elbow joint is a convenient point because force $\vec{F}_{elbow}$ exerts no torque about this point; its moment arm is zero. Thus the torque equation is

$$\tau_{net} = d_{tendon}F_{tendon} - d_{arm}F_{barbell} = 0$$

The tension in the tendon tries to rotate the arm ccw, so it produces a positive torque. Similarly, the torque due to the barbell is negative. We can solve the torque equation for F_{tendon} to find

$$F_{tendon} = F_{barbell}\frac{d_{arm}}{d_{tendon}} = (450\text{ N})\frac{35\text{ cm}}{4.0\text{ cm}} = 3900\text{ N}$$

ASSESS The short distance d_{tendon} from the tendon to the elbow joint means that the force supplied by the biceps has to be very large to counter the torque generated by a force applied at the opposite end of the forearm. Although we ended up not needing the force equation in this problem, we could now use it to calculate that the force exerted at the elbow is $F_{elbow} = 3450$ N. These large forces can easily damage the tendon or the elbow.

EXAMPLE 12.15 Walking the plank

Adrienne (50 kg) and Bo (90 kg) are playing on a 100 kg rigid plank resting on the supports seen in **FIGURE 12.35**. If Adrienne stands on the left end, can Bo walk all the way to the right end without the plank tipping over? If not, how far can he get past the support on the right?

FIGURE 12.35 Adrienne and Bo on the plank.

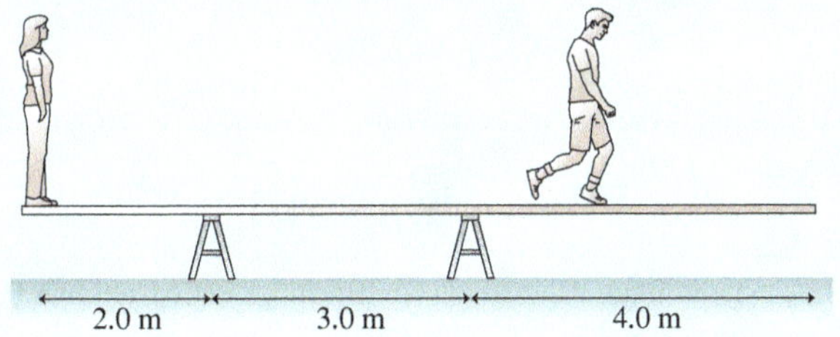

FIGURE 12.36 A pictorial representation of the forces on the plank.

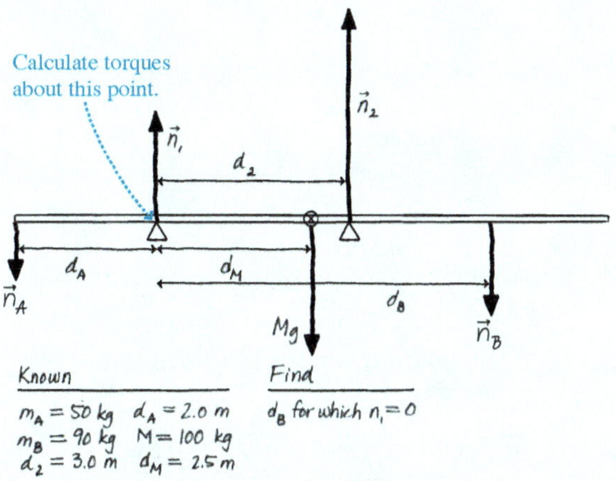

MODEL Model Adrienne and Bo as particles. Assume the plank is uniform, with its center of mass at the center.

VISUALIZE **FIGURE 12.36** shows the forces acting on the plank. Both supports exert upward forces. $\vec{n}_A$ and $\vec{n}_B$ are the normal forces of Adrienne's and Bo's feet pushing down on the board.

SOLVE Because the plank is resting on the supports, not held down, forces $\vec{n}_1$ and $\vec{n}_2$ must point upward. (The supports could pull down if the plank were nailed to them, but that's not the case here.) Force $\vec{n}_1$ will decrease as Bo moves to the right, and the tipping point occurs when $n_1 = 0$. The plank remains in static equilibrium right up to the tipping point, so both the net force and the net torque on it are zero. The force equation is

$$\sum F_y = n_1 + n_2 - n_A - n_B - Mg$$
$$= n_1 + n_2 - m_A g - m_B g - Mg = 0$$

Continued

Adrienne is at rest, with zero net force, so her downward force on the board, an action/reaction pair with the upward normal force of the board on her, equals her weight: $n_A = m_A g$. Bo's center of mass oscillates up and down as he walks, so he's *not* in equilibrium and, strictly speaking, $n_B \neq m_B g$. But we'll assume that he edges out onto the board slowly, with minimal bouncing, in which case $n_B = m_B g$ is a reasonable approximation.

We can again choose any point we wish for calculating torque. Let's use the support on the left. Adrienne and the support on the right exert positive torques about this point; the other forces exert negative torques. Force $\vec{n}_1$ exerts no torque, since it acts at the pivot point. Thus the torque equation is

$$\tau_{net} = d_A m_A g - d_B m_B g - d_M M g + d_2 n_2 = 0$$

At the tipping point, where $n_1 = 0$, the force equation gives $n_2 = (m_A + m_B + M)g$. Substituting this into the torque equation and then solving for Bo's position give

$$d_B = \frac{d_A m_A - d_M M + d_2(m_A + m_B + M)}{m_B} = 6.3 \text{ m}$$

Bo doesn't quite make it to the end. The plank tips when he's 6.3 m past the left support, our pivot point, and thus 3.3 m past the support on the right.

ASSESS We could have solved this problem somewhat more simply had we chosen the support on the right for calculating the torques. However, you might not recognize the "best" point for calculating the torques in a problem. The point of this example is that it doesn't matter which point you choose.

EXAMPLE 12.16 Will the ladder slip?

A 3.0-m-long ladder leans against a frictionless wall at an angle of 60°. What is the minimum value of μ_s, the coefficient of static friction with the ground, that prevents the ladder from slipping?

MODEL The ladder is a rigid rod of length L. To not slip, it must be in both translational equilibrium ($\vec{F}_{net} = \vec{0}$) and rotational equilibrium ($\tau_{net} = 0$).

VISUALIZE FIGURE 12.37 shows the ladder and the forces acting on it.

FIGURE 12.37 A ladder in total equilibrium.

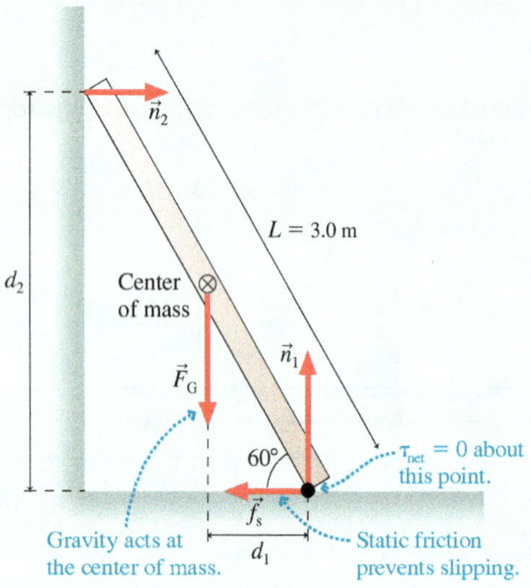

SOLVE The x- and y-components of $\vec{F}_{net} = \vec{0}$ are

$$\sum F_x = n_2 - f_s = 0$$
$$\sum F_y = n_1 - Mg = 0$$

The net torque is zero about *any* point, so which should we choose? The bottom corner of the ladder is a good choice because two forces pass through this point and have no torque about it. The torque about the bottom corner is

$$\tau_{net} = d_1 F_G - d_2 n_2 = \frac{1}{2}(L\cos 60°)Mg - (L\sin 60°)n_2 = 0$$

The signs are based on the observation that $\vec{F}_G$ would cause the ladder to rotate ccw while $\vec{n}_2$ would cause it to rotate cw. All together, we have three equations in the three unknowns n_1, n_2, and f_s. If we solve the third for n_2,

$$n_2 = \frac{\frac{1}{2}(L\cos 60°)Mg}{L\sin 60°} = \frac{Mg}{2\tan 60°}$$

we can then substitute this into the first to find

$$f_s = \frac{Mg}{2\tan 60°}$$

Our model of friction is $f_s \leq f_{s\,max} = \mu_s n_1$. We can find n_1 from the second equation: $n_1 = Mg$. Using this, the model of static friction tells us that

$$f_s \leq \mu_s Mg$$

Comparing these two expressions for f_s, we see that μ_s must obey

$$\mu_s \geq \frac{1}{2\tan 60°} = 0.29$$

Thus the minimum value of the coefficient of static friction is 0.29.

ASSESS You know from experience that you can lean a ladder or other object against a wall if the ground is "rough," but it slips if the surface is too smooth. 0.29 is a "medium" value for the coefficient of static friction, which is reasonable.

Balance and Stability

If you tilt a box up on one edge by a small amount and let go, it falls back down. If you tilt it too much, it falls over. And if you tilt "just right," you can get the box to balance on its edge. What determines these three possible outcomes?

FIGURE 12.38 illustrates the idea with a car, but the results are general and apply in many situations. As long as the object's center of mass remains over the base of

FIGURE 12.38 Stability depends on the position of the center of mass.

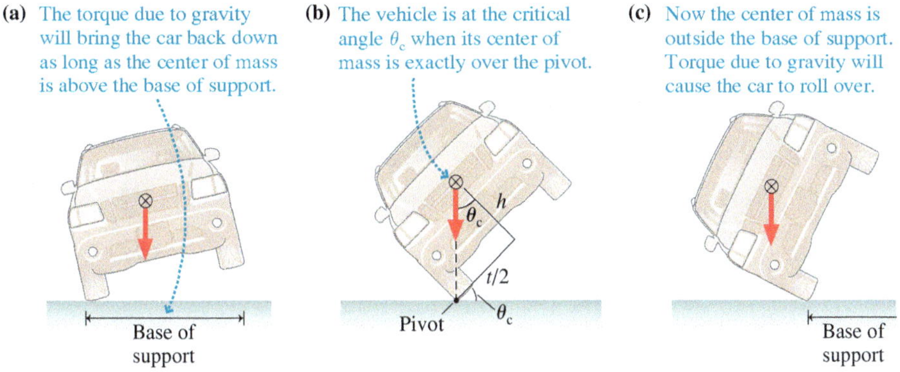

support, torque due to gravity will rotate the object back toward its stable equilibrium position. But if the center of mass gets outside the base of support, the torque due to gravity causes a rotation in the opposite direction. Now the box falls over or the car rolls over.

A *critical angle* θ_c is reached when the center of mass is directly over the pivot point. This is the point of balance, with no net torque. For vehicles, the distance between the tires is called the track width t. If the height of the center of mass is h, you can see from FIGURE 12.38b that the critical angle is

$$\theta_c = \tan^{-1}\left(\frac{t}{2h}\right)$$

For passenger cars with $h \approx 0.33t$, the critical angle is $\theta_c \approx 57°$. But for a sport utility vehicle (SUV) with $h \approx 0.47t$, a higher center of mass, the critical angle is only $\theta_c \approx 47°$. Loading an SUV with cargo further raises the center of gravity, especially if the roof rack is used, thus reducing θ_c even more. Various automobile safety groups have determined that a vehicle with $\theta_c > 50°$ is unlikely to roll over in an accident. A rollover becomes increasingly likely when θ_c is reduced below 50°. The general rule is that **a wider base of support and/or a lower center of mass improve stability.**

This dancer balances *en pointe* by having her center of mass directly over her toes, her base of support.

EXAMPLE 12.17 Tilting cans

A typical can of food is 7.5 cm in diameter. What is the tallest can of food that can rest on a 30° incline without falling over?

MODEL Assume the food inside is uniformly distributed so that the center of mass is at the center of the can.

FIGURE 12.39 A can balanced at the critical angle.

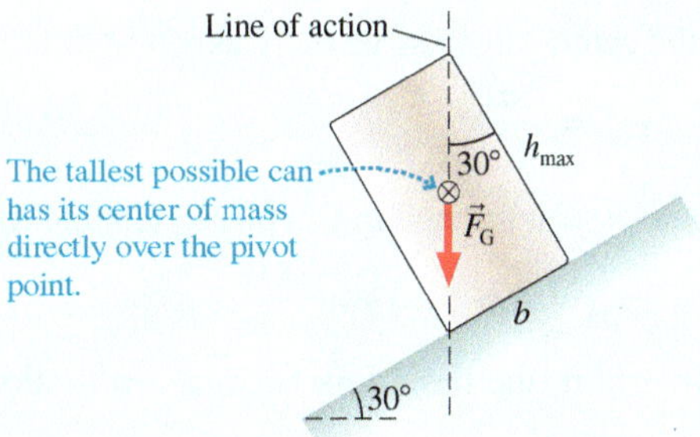

VISUALIZE FIGURE 12.39 shows a can at the critical angle. This is the tallest possible can. A shorter can would have its center of mass inside the base of support and would be stable; a taller can would have its center of mass outside the base of support and would fall over.

SOLVE For a can whose height puts it at the critical angle, the line of action is a diagonal through the can. If the height is h_{max} and the diameter of the base b, we see from the figure that $\tan 30° = b/h_{max}$ and thus

$$h_{max} = \frac{b}{\tan 30°} = \frac{7.5\text{ cm}}{\tan 30°} = 13\text{ cm}$$

ASSESS A typical can of soup is just under 13 cm tall. It will stand on a 30° incline—try it!—but anything taller will fall over.

STOP TO THINK 12.4 A student holds a meter stick straight out with one or more masses dangling from it. Rank in order, from most difficult to least difficult, how hard it will be for the student to keep the meter stick from rotating.

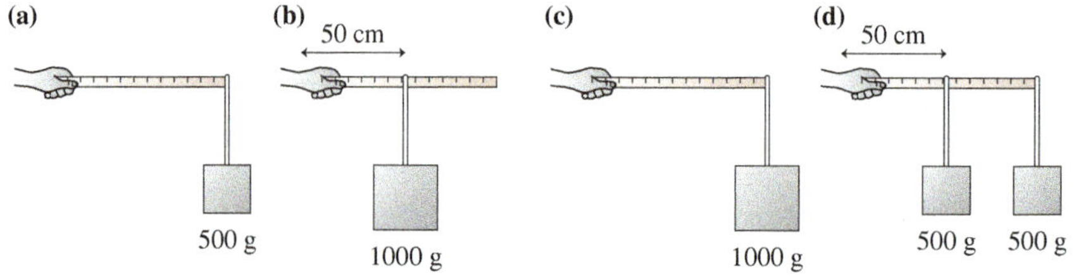

12.9 Rolling Motion

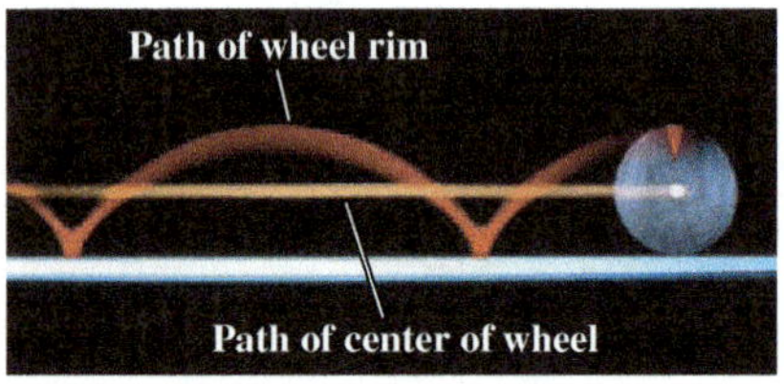

FIGURE 12.40 The trajectories of the center of a wheel and of a point on the rim are seen in a time-exposure photograph.

Rolling is a *combination motion* in which an object rotates about an axis that is moving along a straight-line trajectory. For example, FIGURE 12.40 is a time-exposure photo of a rolling wheel with one lightbulb on the axis and a second lightbulb at the edge. The axis light moves straight ahead, but the edge light follows a curve called a *cycloid.* Let's see if we can understand this interesting motion. We'll consider only objects that roll without slipping.

FIGURE 12.41 shows a round object—a wheel or a sphere—that rolls forward exactly one revolution. The point that had been on the bottom follows the cycloid, the curve you saw in Figure 12.40, to the top and back to the bottom. *Because the object doesn't slip,* the center of mass moves forward exactly one circumference: $\Delta x_{cm} = 2\pi R$.

FIGURE 12.41 An object rolling through one revolution.

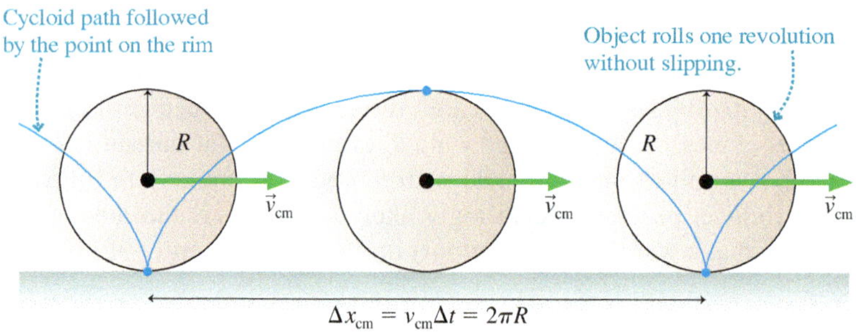

We can also write the distance traveled in terms of the velocity of the center of mass: $\Delta x_{cm} = v_{cm}\,\Delta t$. But Δt, the time it takes the object to make one complete revolution, is nothing other than the rotation period T. In other words, $\Delta x_{cm} = v_{cm}T$.

These two expressions for Δx_{cm} come from two perspectives on the motion: one looking at the rotation and the other looking at the translation of the center of mass. But it's the same distance no matter how you look at it, so these two expressions must be equal. Consequently,

$$\Delta x_{cm} = 2\pi R = v_{cm}T \tag{12.34}$$

If we divide by T, we can write the center-of-mass velocity as

$$v_{cm} = \frac{2\pi}{T}R \tag{12.35}$$

But $2\pi/T$ is the angular velocity ω, as you learned in Chapter 4, leading to

$$v_{cm} = R\omega \tag{12.36}$$

Equation 12.36 is the **rolling constraint,** the basic link between translation and rotation for objects that roll without slipping.

NOTE ▶ The rolling constraint is equivalent to Equation 12.33 for the speed of a rope that doesn't slip as it passes over a pulley. ◀

Let's look carefully at a particle in the rolling object. As FIGURE 12.42a shows, the position vector $\vec{r}_i$ for particle i is the vector sum $\vec{r}_i = \vec{r}_{cm} + \vec{r}_{i,\text{rel}}$. Taking the time derivative of this equation, we can write the velocity of particle i as

$$\vec{v}_i = \vec{v}_{cm} + \vec{v}_{i,\text{rel}} \tag{12.37}$$

In other words, the velocity of particle i can be divided into two parts: the velocity $\vec{v}_{cm}$ of the object as a whole plus the velocity $\vec{v}_{i,\text{rel}}$ of particle i relative to the center of mass (i.e., the velocity that particle i would have if the object were only rotating and had no translational motion).

FIGURE 12.42b applies this idea to point P at the very bottom of the rolling object, the point of contact between the object and the surface. This point is moving around the center of the object at angular velocity ω, so $v_{i,\text{rel}} = -R\omega$. The negative sign indicates that the motion is cw. At the same time, the center-of-mass velocity, Equation 12.36, is $v_{cm} = R\omega$. Adding these, we find that the velocity of point P, the lowest point, is $v_i = 0$. In other words, **the point on the bottom of a rolling object is instantaneously at rest.**

Although this seems surprising, it is really what we mean by "rolling without slipping." If the bottom point had a velocity, it would be moving horizontally relative to the surface. In other words, it would be slipping or sliding across the surface. To roll without slipping, the bottom point, the point touching the surface, must be at rest.

FIGURE 12.43 shows how the velocity vectors at the top, center, and bottom of a rotating wheel are found by adding the rotational velocity vectors to the center-of-mass velocity. You can see that $v_{\text{bottom}} = 0$ and that $v_{\text{top}} = 2R\omega = 2v_{cm}$.

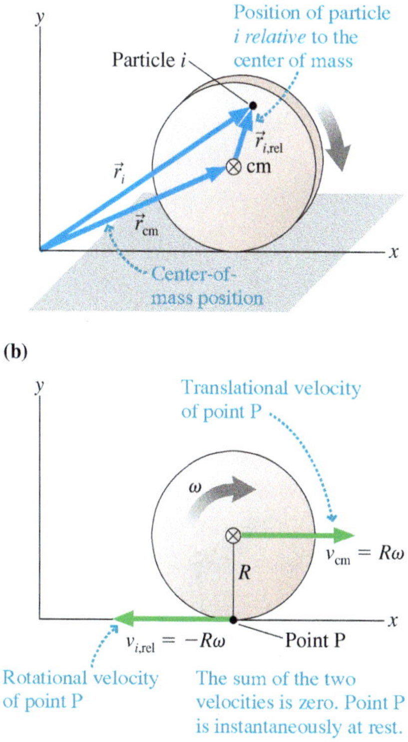

FIGURE 12.42 The motion of a particle in the rolling object.

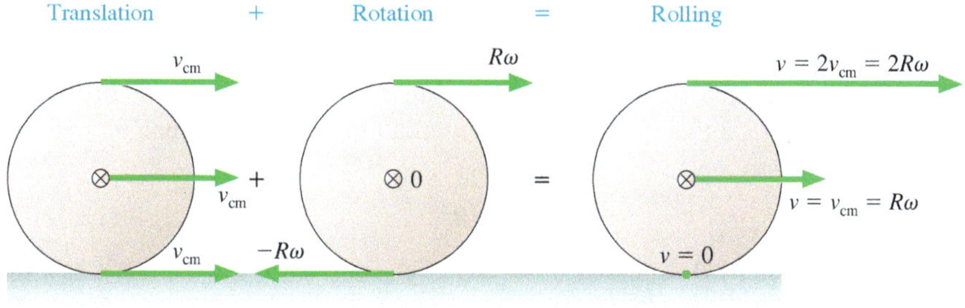

FIGURE 12.43 Rolling without slipping is a combination of translation and rotation.

Kinetic Energy of a Rolling Object

We found earlier that the rotational kinetic energy of a rigid body in pure rotational motion is $K_{\text{rot}} = \frac{1}{2}I\omega^2$. Now we would like to find the kinetic energy of an object that rolls without slipping, a combination of rotational and translation motion.

We begin with the observation that the bottom point in FIGURE 12.44 is instantaneously at rest. Consequently, we can think of an axis through P as an *instantaneous axis of rotation.* The idea of an instantaneous axis of rotation seems a little far-fetched, but it is confirmed by looking at the instantaneous velocities of the center point and the top point. We found these in Figure 12.43 and they are shown again in Figure 12.44. They are exactly what you would expect as the tangential velocity $v_t = r\omega$ for rotation about P at distances R and $2R$.

From this perspective, the object's motion is pure rotation about point P. Thus the kinetic energy is that of pure rotation:

$$K = K_{\text{rotation about P}} = \frac{1}{2}I_P\omega^2 \tag{12.38}$$

I_P is the moment of inertia for rotation about point P. We can use the parallel-axis theorem to write I_P in terms of the moment of inertia I_{cm} about the center of mass. Point P is displaced by distance $d = R$; thus

$$I_P = I_{cm} + MR^2$$

FIGURE 12.44 Rolling motion is an instantaneous rotation about point P.

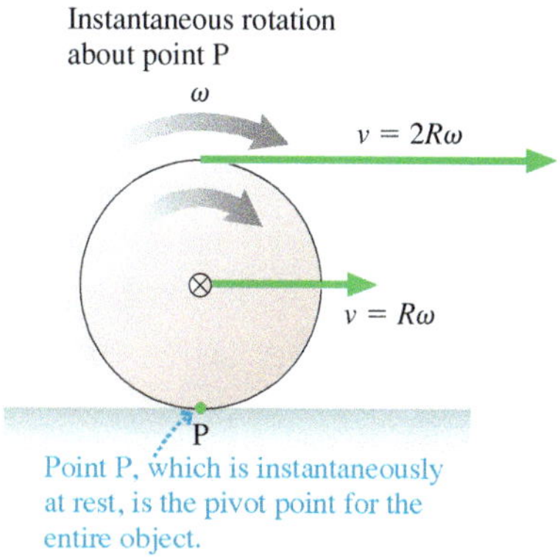

Using this expression in Equation 12.38 gives us the kinetic energy:

$$K = \frac{1}{2}I_{cm}\omega^2 + \frac{1}{2}M(R\omega)^2 \tag{12.39}$$

We know from the rolling constraint that $R\omega$ is the center-of-mass velocity v_{cm}. Thus the kinetic energy of a rolling object is

$$K_{rolling} = \frac{1}{2}I_{cm}\omega^2 + \frac{1}{2}Mv_{cm}^2 = K_{rot} + K_{cm} \tag{12.40}$$

In other words, **the rolling motion of a rigid body can be described as a translation of the center of mass (with kinetic energy K_{cm}) plus a rotation about the center of mass (with kinetic energy K_{rot}).**

The Great Downhill Race

FIGURE 12.45 Which will win the downhill race?

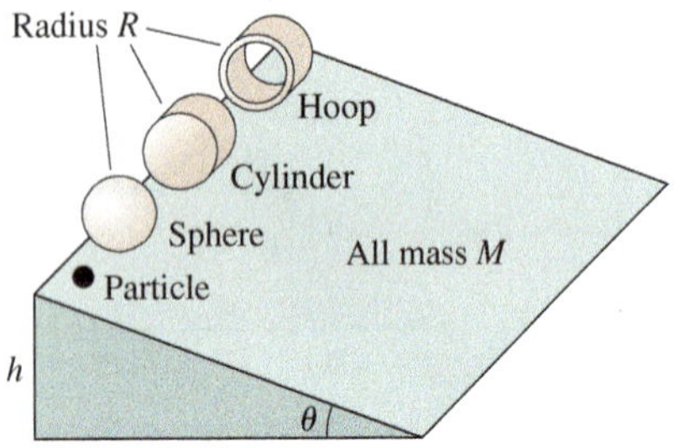

FIGURE 12.45 shows a contest in which a sphere, a cylinder, and a circular hoop, all of mass M and radius R, are placed at height h on a slope of angle θ. All three are released from rest at the same instant of time and roll down the ramp without slipping. To make things more interesting, they are joined by a particle of mass M that slides down the ramp without friction. Which one will win the race to the bottom of the hill? Does rotation affect the outcome?

An object's initial gravitational potential energy is transformed into kinetic energy as it rolls (or slides, in the case of the particle). The kinetic energy, as we just discovered, is a combination of translational and rotational kinetic energy. If we choose the bottom of the ramp as the zero point of potential energy, the statement of energy conservation $K_f = U_i$ can be written

$$\frac{1}{2}I_{cm}\omega^2 + \frac{1}{2}Mv_{cm}^2 = Mgh \tag{12.41}$$

The translational and rotational velocities are related by $\omega = v_{cm}/R$. In addition, notice from Table 12.2 that the moments of inertia of all the objects can be written in the form

$$I_{cm} = cMR^2 \tag{12.42}$$

where c is a constant that depends on the object's geometry. For example, $c = \frac{2}{5}$ for a sphere but $c = 1$ for a circular hoop. Even the particle can be represented by $c = 0$, which eliminates the rotational kinetic energy.

With this information, Equation 12.41 becomes

$$\frac{1}{2}(cMR^2)\left(\frac{v_{cm}}{R}\right)^2 + \frac{1}{2}Mv_{cm}^2 = \frac{1}{2}M(1+c)v_{cm}^2 = Mgh$$

Thus the finishing speed of an object with $I = cMR^2$ is

$$v_{cm} = \sqrt{\frac{2gh}{1+c}} \tag{12.43}$$

The final speed is independent of both M and R, but it does depend on the *shape* of the rolling object. The particle, with the smallest value of c, will finish with the highest speed, while the circular hoop, with the largest c, will be the slowest. In other words, the rolling aspect of the motion *does* matter!

We can use Equation 12.43 to find the acceleration a_{cm} of the center of mass. The objects move through distance $\Delta x = h/\sin\theta$, so we can use constant-acceleration kinematics to find

$$\begin{aligned} v_{cm}^2 &= 2a_{cm}\Delta x \\ a_{cm} &= \frac{v_{cm}^2}{2\Delta x} = \frac{2gh/(1+c)}{2h/\sin\theta} = \frac{g\sin\theta}{1+c} \end{aligned} \tag{12.44}$$

Recall, from Chapter 2, that $a_{particle} = g\sin\theta$ is the acceleration of a particle sliding down a frictionless incline. We can use this fact to write Equation 12.44 in an interesting form:

$$a_{cm} = \frac{a_{particle}}{1 + c} \tag{12.45}$$

This analysis leads us to the conclusion that **the acceleration of a rolling object is less—in some cases significantly less—than the acceleration of a particle.** The reason is that the energy has to be shared between translational kinetic energy and rotational kinetic energy. A particle, by contrast, can put all its energy into translational kinetic energy.

FIGURE 12.46 shows the results of the race. The simple particle wins by a fairly wide margin. Of the solid objects, the sphere has the largest acceleration. Even so, its acceleration is only 71% the acceleration of a particle. The acceleration of the circular hoop, which comes in last, is a mere 50% that of a particle.

NOTE ▶ The objects having the largest acceleration are those whose mass is most concentrated near the center. Placing the mass far from the center, as in the hoop, increases the moment of inertia. Thus it requires a larger effort to get a hoop rolling than to get a sphere of equal mass rolling. ◀

FIGURE 12.46 And the winner is...

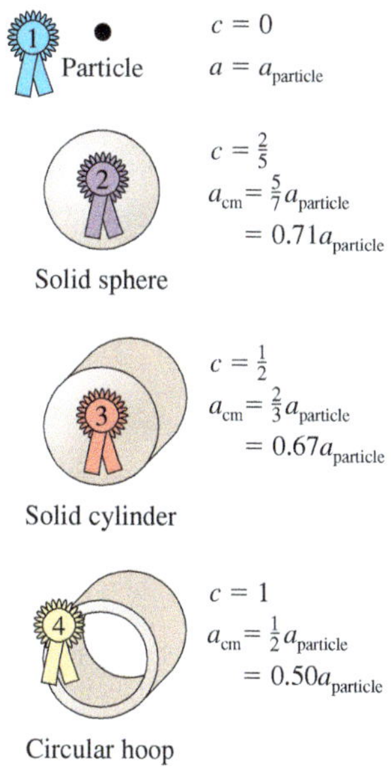

12.10 The Vector Description of Rotational Motion

Rotation about a fixed axis, such as an axle, can be described in terms of a scalar angular velocity ω and a scalar torque τ, using a plus or minus sign to indicate the direction of rotation. This is very much analogous to the one-dimensional kinematics of Chapter 2. For more general rotational motion, angular velocity, torque, and other quantities must be treated as *vectors*. We won't go into much detail because the subject rapidly gets very complicated, but we will sketch some important basic ideas.

The Angular Velocity Vector

FIGURE 12.47 shows a rotating rigid body. We can define an angular velocity vector $\vec{\omega}$ as follows:

- The magnitude of $\vec{\omega}$ is the object's angular velocity ω.
- $\vec{\omega}$ points along the axis of rotation in the direction given by the *right-hand rule* illustrated in Figure 12.47.

If the object rotates in the xy-plane, the vector $\vec{\omega}$ points along the z-axis. The scalar angular velocity $\omega = v_t/r$ that we've been using is now seen to be ω_z, the z-component of the vector $\vec{\omega}$. You should convince yourself that the sign convention for ω (positive for ccw rotation, negative for cw rotation) is equivalent to having the vector $\vec{\omega}$ pointing in the positive z-direction or the negative z-direction.

FIGURE 12.47 The angular velocity vector $\vec{\omega}$ is found using the right-hand rule.

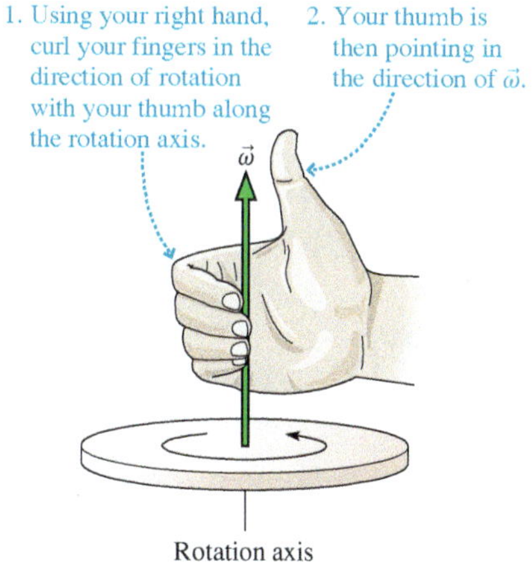

The Cross Product of Two Vectors

We defined the torque exerted by force $\vec{F}$ to be $\tau = rF\sin\phi$. The quantity F is the magnitude of the force vector $\vec{F}$, and the distance r is really the magnitude of the position vector $\vec{r}$. Hence torque looks very much like a product of the two vectors $\vec{r}$ and $\vec{F}$. Previously, in conjunction with the definition of work, we introduced the dot product of two vectors: $\vec{A}\cdot\vec{B} = AB\cos\alpha$, where α is the angle between the vectors. $\tau = rF\sin\phi$ is a different way of multiplying vectors that depends on the *sine* of the angle between them.

FIGURE 12.48 The cross product $\vec{A} \times \vec{B}$, is a vector perpendicular to the plane of vectors $\vec{A}$ and $\vec{B}$.

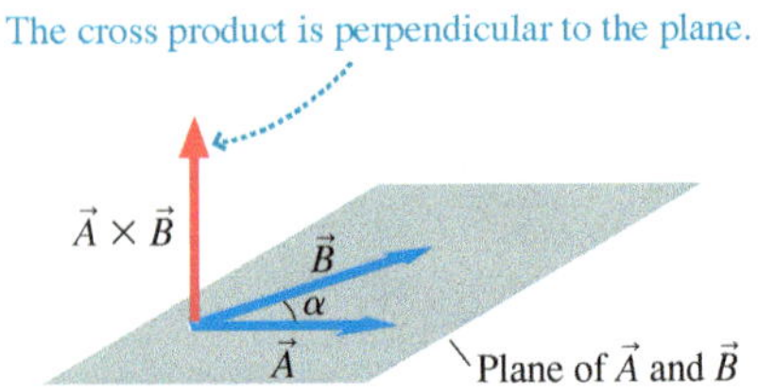

FIGURE 12.48 shows two vectors, $\vec{A}$ and $\vec{B}$, with angle α between them. We define the **cross product** of $\vec{A}$ and $\vec{B}$ as the vector

$$\vec{A} \times \vec{B} \equiv (AB\sin\alpha, \text{in the direction given by the right-hand rule}) \qquad (12.46)$$

The symbol $\times$ between the vectors is *required* to indicate a cross product. The cross product is also called the **vector product** because the result is a vector.

The **right-hand rule,** which specifies the direction of $\vec{A} \times \vec{B}$, can be stated in three different ways:

Using the right-hand rule

Spread your *right* thumb and index finger apart by angle α. Bend your middle finger so that it is *perpendicular* to your thumb and index finger. Orient your hand so that your thumb points in the direction of $\vec{A}$ and your index finger in the direction of $\vec{B}$. Your middle finger now points in the direction of $\vec{A} \times \vec{B}$.

Make a loose fist with your *right* hand with your thumb extended outward. Orient your hand so that your thumb is perpendicular to the plane of $\vec{A}$ and $\vec{B}$ and your fingers are curling *from* the line of vector $\vec{A}$ *toward* the line of vector $\vec{B}$. Your thumb now points in the direction of $\vec{A} \times \vec{B}$.

Imagine using a screwdriver to turn the slot in the head of a screw from the direction of $\vec{A}$ to the direction of $\vec{B}$. The screw will move either "in" or "out." The direction in which the screw moves is the direction of $\vec{A} \times \vec{B}$.

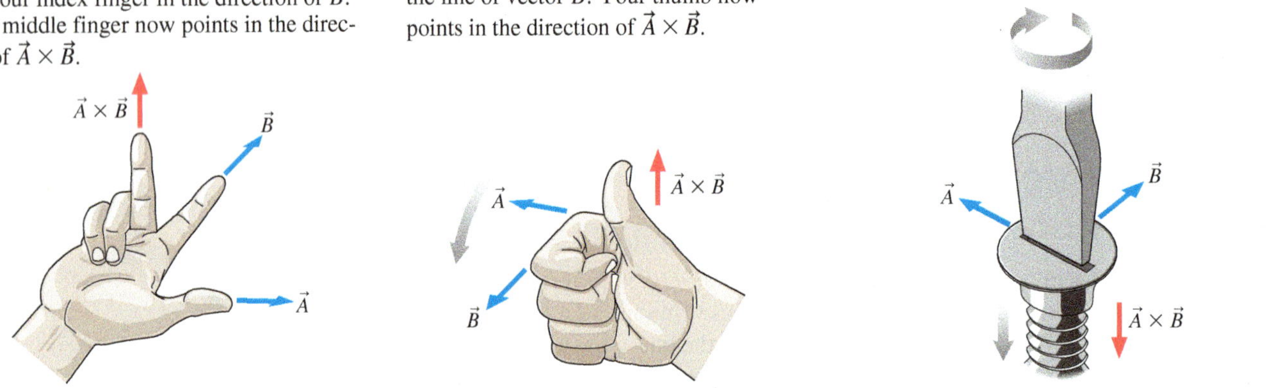

These methods are easier to demonstrate than to describe in words! Your instructor will show you how they work. Some individuals find one method of thinking about the direction of the cross product easier than the others, but they all work, and you'll soon find the method that works best for you.

Referring back to Figure 12.48, you should use the right-hand rule to convince yourself that the cross product $\vec{A} \times \vec{B}$ is a vector that points *upward,* perpendicular to the plane of $\vec{A}$ and $\vec{B}$. FIGURE 12.49 shows that the cross product, like the dot product, depends on the angle between the two vectors. Notice the two special cases: $\vec{A} \times \vec{B} = \vec{0}$ when $\alpha = 0°$ (parallel vectors) and $\vec{A} \times \vec{B}$ has its maximum magnitude AB when $\alpha = 90°$ (perpendicular vectors).

FIGURE 12.49 The magnitude of the cross-product vector increases from 0 to AB as α increases from 0° to 90°.

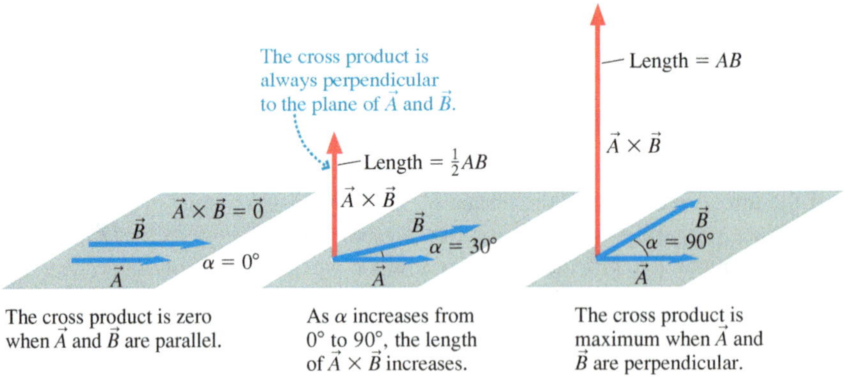

EXAMPLE 12.18 Calculating a cross product

FIGURE 12.50 shows vectors $\vec{C}$ and $\vec{D}$ in the plane of the page. What is the cross product $\vec{E} = \vec{C} \times \vec{D}$?

FIGURE 12.50 Vectors $\vec{C}$ and $\vec{D}$.

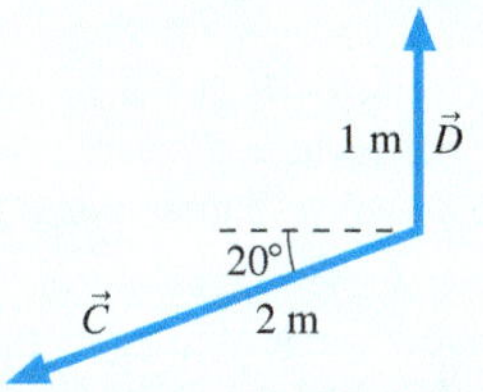

SOLVE The angle between the two vectors is $\alpha = 110°$. Consequently, the magnitude of the cross product is

$$E = CD\sin\alpha = (2\text{ m})(1\text{ m})\sin(110°) = 1.88\text{ m}^2$$

The direction of $\vec{E}$ is given by the right-hand rule. To curl your right fingers from $\vec{C}$ to $\vec{D}$, you have to point your thumb *into* the page. Alternatively, if you turned a screwdriver from $\vec{C}$ to $\vec{D}$ you would be driving a screw *into* the page. Thus

$$\vec{E} = (1.88\text{ m}^2,\text{ into page})$$

ASSESS Notice that $\vec{E}$ has units of m^2.

The cross product has three important properties:

1. The product $\vec{A} \times \vec{B}$ is *not* equal to the product $\vec{B} \times \vec{A}$. That is, the cross product does not obey the commutative rule $ab = ba$ that you know from arithmetic. In fact, you can see from the right-hand rule that the product $\vec{B} \times \vec{A}$ points in exactly the opposite direction from $\vec{A} \times \vec{B}$. Thus, as FIGURE 12.51a shows,
$$\vec{B} \times \vec{A} = -\vec{A} \times \vec{B}$$
2. In a *right-handed coordinate system,* which is the standard coordinate system of science and engineering, the z-axis is oriented relative to the xy-plane such that the unit vectors obey $\hat{\imath} \times \hat{\jmath} = \hat{k}$. This is shown in FIGURE 12.51b. You can also see from this figure that $\hat{\jmath} \times \hat{k} = \hat{\imath}$ and $\hat{k} \times \hat{\imath} = \hat{\jmath}$.
3. The derivative of a cross product is
$$\frac{d}{dt}(\vec{A} \times \vec{B}) = \frac{d\vec{A}}{dt} \times \vec{B} + \vec{A} \times \frac{d\vec{B}}{dt} \tag{12.47}$$

FIGURE 12.51 Properties of the cross product.

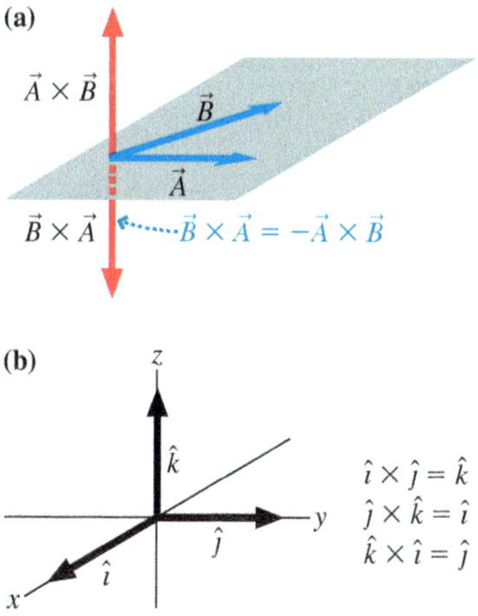

Torque

Now let's return to torque. As a concrete example, FIGURE 12.52 shows a long wrench being used to loosen the nuts holding a car wheel on. Force $\vec{F}$ exerts a torque about the origin. Let's define a *torque vector*

$$\vec{\tau} \equiv \vec{r} \times \vec{F} \tag{12.48}$$

If we place the vector tails together in order to use the right-hand rule, we see that the torque vector is perpendicular to the plane of $\vec{r}$ and $\vec{F}$. The angle between the vectors is ϕ, so the magnitude of the torque is $\tau = rF|\sin\phi|$.

FIGURE 12.52 The torque vector.

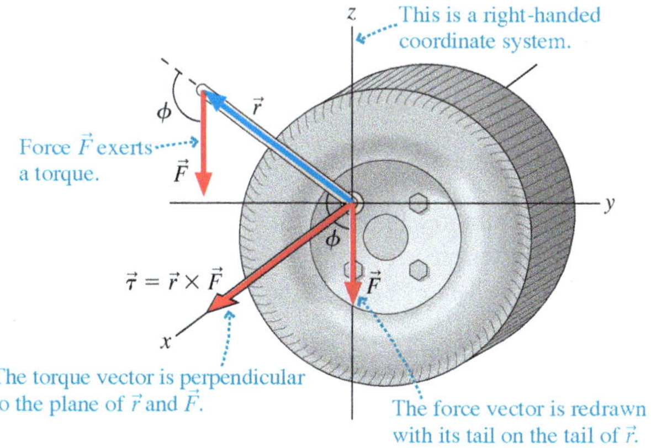

You can see that the scalar torque $\tau = rF\sin\phi$ we've been using is really the component along the rotation axis—in this case τ_x—of the vector $\vec{\tau}$. This is the basis for our earlier sign convention for τ. In Figure 12.52, where the force causes a ccw rotation, the torque vector points in the positive x-direction, and thus τ_x is positive.

EXAMPLE 12.19 Wrench torque revisited

Example 12.8 found the torque that Luis exerts on a nut by pulling on the end of a wrench. What is the torque vector?

VISUALIZE FIGURE 12.53 shows the position vector $\vec{r}$, drawn from the pivot point to the point where the force is applied. The figure also redraws the force vector $\vec{F}$ at the pivot point, not because force is applied there but because it's easiest to use the right-hand rule if the vectors are drawn with their tails together.

FIGURE 12.53 Calculating the torque vector.

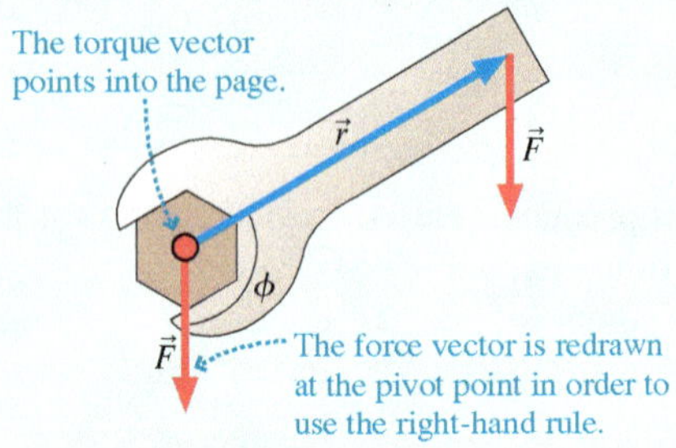

SOLVE We already know the magnitude of the torque, 17 N m, from Example 12.8. Now we need to apply the right-hand rule. If you place your right thumb along $\vec{r}$ and your index finger along $\vec{F}$, which is somewhat awkward, you'll see that your middle finger points into the page. Alternatively, make a loose fist of your right hand, then orient your fist so that your fingers curl *from $\vec{r}$ toward $\vec{F}$*. Doing so requires your thumb to point into the page. Using either method, we conclude that

$$\vec{\tau} = (17 \text{ N m, into page})$$

12.11 Angular Momentum

FIGURE 12.54 shows a particle that, at this instant, is located at position $\vec{r}$ and is moving with momentum $\vec{p} = m\vec{v}$. Together, $\vec{r}$ and $\vec{p}$ define the *plane of motion*. We define the particle's **angular momentum** $\vec{L}$ relative to the origin to be the vector

$$\vec{L} \equiv \vec{r} \times \vec{p} = (mrv\sin\beta, \text{ direction of right-hand rule}) \tag{12.49}$$

Because of the cross product, **the angular momentum vector is perpendicular to the plane of motion.** The units of angular momentum are kg m^2/s.

FIGURE 12.54 The angular momentum vector $\vec{L}$.

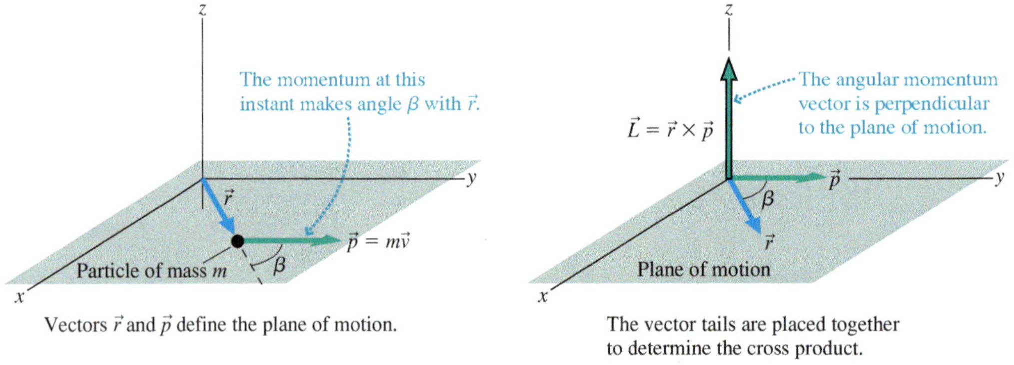

NOTE ▶ Angular momentum is the rotational equivalent of linear momentum in much the same way that torque is the rotational equivalent of force. Notice that the vector definitions are parallel: $\vec{\tau} \equiv \vec{r} \times \vec{F}$ and $\vec{L} \equiv \vec{r} \times \vec{p}$. ◀

Angular momentum, like torque, is *about* the point from which $\vec{r}$ is measured. A different origin would yield a different angular momentum. Angular momentum is especially simple for a particle in circular motion. As FIGURE 12.55 shows, the angle β between $\vec{p}$ (or $\vec{v}$) and $\vec{r}$ is always 90° if we make the obvious choice of measuring $\vec{r}$ from the center of the circle. For motion in the xy-plane, the angular momentum vector $\vec{L}$—which must be perpendicular to the plane of motion—is entirely along the z-axis:

$$L_z = mrv_t \qquad \text{(particle in circular motion)} \tag{12.50}$$

where v_t is the tangential component of velocity. Our sign convention for v_t makes L_z, like ω, positive for a ccw rotation, negative for a cw rotation.

In Chapter 9, we found that Newton's second law for a particle can be written $\vec{F}_{\text{net}} = d\vec{p}/dt$. There's a similar connection between torque and angular momentum. To show this, we take the time derivative of $\vec{L}$:

$$\begin{aligned}\frac{d\vec{L}}{dt} &= \frac{d}{dt}(\vec{r}\times\vec{p}) = \frac{d\vec{r}}{dt}\times\vec{p} + \vec{r}\times\frac{d\vec{p}}{dt} \\ &= \vec{v}\times\vec{p} + \vec{r}\times\vec{F}_{\text{net}}\end{aligned} \tag{12.51}$$

where we used Equation 12.47 for the derivative of a cross product. We also used the definitions $\vec{v} = d\vec{r}/dt$ and $\vec{F}_{\text{net}} = d\vec{p}/dt$.

Vectors $\vec{v}$ and $\vec{p}$ are parallel, and the cross product of two parallel vectors is $\vec{0}$. Thus the first term in Equation 12.51 vanishes. The second term $\vec{r}\times\vec{F}_{\text{net}}$ is the net torque, $\vec{\tau}_{\text{net}} = \vec{\tau}_1 + \vec{\tau}_2 + \cdots$, so we arrive at

$$\frac{d\vec{L}}{dt} = \vec{\tau}_{\text{net}} \tag{12.52}$$

Equation 12.52, which says **a net torque causes the particle's angular momentum to change,** is the rotational equivalent of $d\vec{p}/dt = \vec{F}_{\text{net}}$.

FIGURE 12.55 Angular momentum of circular motion.

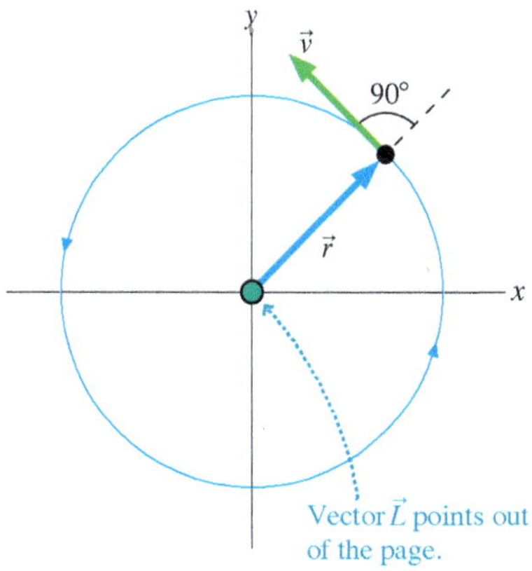

Angular Momentum of a Rigid Body

Equation 12.52 is the angular momentum of a single particle. The angular momentum of a rigid body composed of particles with individual angular momenta $\vec{L}_1, \vec{L}_2, \vec{L}_3, \ldots$ is the vector sum

$$\vec{L} = \vec{L}_1 + \vec{L}_2 + \vec{L}_3 + \cdots = \sum_i \vec{L}_i \tag{12.53}$$

We can combine Equations 12.52 and 12.53 to find the rate of change of the system's angular momentum:

$$\frac{d\vec{L}}{dt} = \sum_i \frac{d\vec{L}_i}{dt} = \sum_i \vec{\tau}_i = \vec{\tau}_{\text{net}} \tag{12.54}$$

Because any internal forces are action/reaction pairs of forces, acting with the same strength in opposite directions, the net torque due to internal forces is zero. Thus the only forces that contribute to the net torque are external forces exerted on the system by the environment.

For a system of particles, **the rate of change of the system's angular momentum is the net torque on the system.** Equation 12.54 is analogous to the Chapter 9 result $d\vec{P}/dt = \vec{F}_{\text{net}}$, which says that the rate of change of a system's total linear momentum is the net force on the system. Table 12.4 summarizes the analogies between linear and angular momentum and energy.

The spin of an ice skater is determined by her angular momentum.

TABLE 12.4 Angular and linear momentum and energy

Angular motion	Linear motion
$K_{\text{rot}} = \frac{1}{2}I\omega^2$	$K_{\text{cm}} = \frac{1}{2}Mv_{\text{cm}}{}^2$
$\vec{L} = I\vec{\omega}$ *	$\vec{P} = M\vec{v}_{\text{cm}}$
$d\vec{L}/dt = \vec{\tau}_{\text{net}}$	$d\vec{P}/dt = \vec{F}_{\text{net}}$
The angular momentum of a system is conserved if there is no net torque.	The linear momentum of a system is conserved if there is no net force.

*Rotation about an axis of symmetry.

Conservation of Angular Momentum

A net torque on a rigid body causes its angular momentum to change. Conversely, the angular momentum does *not* change—it is *conserved*—for a system with no net torque. This is the basis of the law of conservation of angular momentum.

Law of conservation of angular momentum The angular momentum $\vec{L}$ of an isolated system ($\vec{\tau}_{net} = \vec{0}$) is conserved. The final angular momentum $\vec{L}_f$ is equal to the initial angular momentum $\vec{L}_i$. Both the magnitude *and* the direction of $\vec{L}$ are unchanged.

EXAMPLE 12.20 An expanding rod

Two equal masses are at the ends of a massless 50-cm-long rod. The rod spins at 2.0 rev/s about an axis through its midpoint. Suddenly, a compressed gas expands the rod out to a length of 160 cm. What is the rotation frequency after the expansion?

MODEL The forces push outward from the pivot and exert no torques. Thus the system's angular momentum is conserved.

VISUALIZE FIGURE 12.56 is a before-and-after pictorial representation. The angular momentum vectors $\vec{L}_i$ and $\vec{L}_f$ are perpendicular to the plane of motion.

SOLVE The particles are moving in circles, so each has angular momentum $L = mrv_t = mr^2\omega = \frac{1}{4}ml^2\omega$, where we used $r = \frac{1}{2}l$. Thus the initial angular momentum of the system is

$$L_i = \frac{1}{4}ml_i^2\omega_i + \frac{1}{4}ml_i^2\omega_i = \frac{1}{2}ml_i^2\omega_i$$

Similarly, the angular momentum after the expansion is $L_f = \frac{1}{2}ml_f^2\omega_f$. Angular momentum is conserved as the rod expands, thus

$$\frac{1}{2}ml_f^2\omega_f = \frac{1}{2}ml_i^2\omega_i$$

Solving for ω_f, we find

$$\omega_f = \left(\frac{l_i}{l_f}\right)^2\omega_i = \left(\frac{50\text{ cm}}{160\text{ cm}}\right)^2(2.0\text{ rev/s}) = 0.20\text{ rev/s}$$

ASSESS The values of the masses weren't needed. All that matters is the ratio of the lengths.

FIGURE 12.56 The system before and after the rod expands.

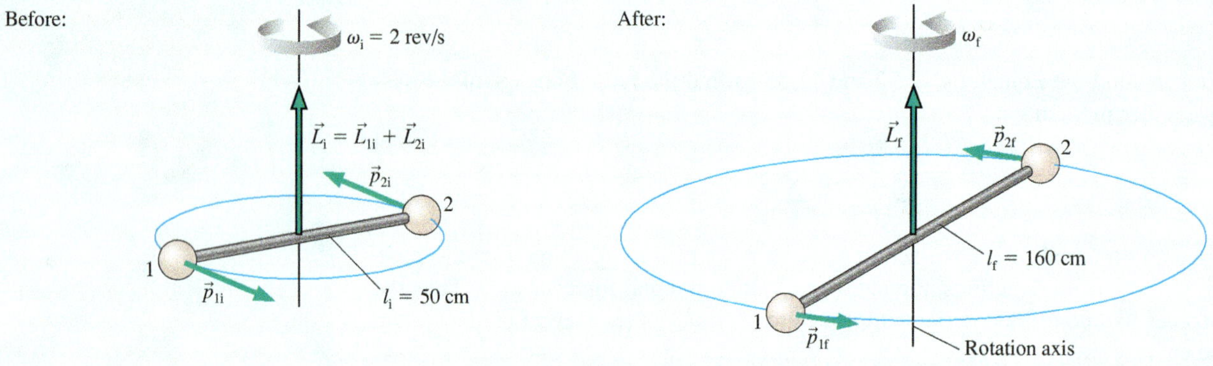

The expansion of the rod in Example 12.20 causes a dramatic slowing of the rotation. Similarly, the rotation would speed up if the weights were pulled in. This is how an ice skater controls her speed as she does a spin. Pulling in her arms decreases her moment of inertia and causes her angular velocity to increase. Similarly, extending her arms increases her moment of inertia, and her angular velocity drops until she can skate out of the spin. It's all a matter of conserving angular momentum.

Angular Momentum and Angular Velocity

The analogy between linear and rotational motion has been so consistent that you might expect one more. The Chapter 9 result $\vec{P} = M\vec{v}_{cm}$ might give us reason to anticipate that angular momentum and angular velocity are related by $\vec{L} = I\vec{\omega}$. Unfortunately, the analogy breaks down here. For an arbitrarily shaped object, the angular momentum vector and the angular velocity vector don't necessarily point in the same direction. The general relationship between $\vec{L}$ and $\vec{\omega}$ is beyond the scope of this text.

The good news is that the analogy *does* continue to hold in two important situations: the rotation of a *symmetrical* object about the symmetry axis and the rotation of any object about a fixed axle. For example, the axis of a cylinder or disk is a symmetry axis, as is any diameter through a sphere. In these two situations—which are all this textbook will consider—the angular momentum and angular velocity are related by

$$\vec{L} = I\vec{\omega} \quad \text{(rotation about a fixed axle or axis of symmetry)} \quad (12.55)$$

This relationship is shown for a spinning disk in FIGURE 12.57. Equation 12.55 is particularly important for applying the law of conservation of angular momentum.

FIGURE 12.57 The angular momentum vector of a rigid body rotating about an axis of symmetry.

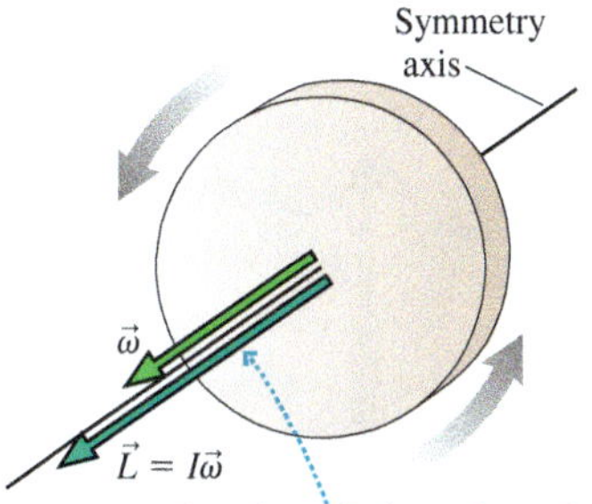

EXAMPLE 12.21 Two interacting disks

A 20-cm-diameter, 2.0 kg solid disk is rotating at 200 rpm. A 20-cm-diameter, 1.0 kg circular loop is dropped straight down onto the rotating disk. Friction causes the loop to accelerate until it is "riding" on the disk. What is the final angular velocity of the combined system?

MODEL The friction between the two objects creates torques that speed up the loop and slow down the disk. But these torques are internal to the combined disk + loop system, so $\tau_{net} = 0$ and the *total* angular momentum of the disk + loop system is conserved.

VISUALIZE FIGURE 12.58 is a before-and-after pictorial representation. Initially only the disk is rotating, at angular velocity $\vec{\omega}_i$. The rotation is about an axis of symmetry, so the angular momentum $\vec{L} = I\vec{\omega}$ is parallel to $\vec{\omega}$. At the end of the problem, $\vec{\omega}_{disk} = \vec{\omega}_{loop} = \vec{\omega}_f$.

SOLVE Both angular momentum vectors point along the rotation axis. Conservation of angular momentum tells us that the magnitude of $\vec{L}$ is unchanged. Thus

$$L_f = I_{disk}\omega_f + I_{loop}\omega_f = L_i = I_{disk}\omega_i$$

Solving for ω_f gives

$$\omega_f = \frac{I_{disk}}{I_{disk} + I_{loop}}\omega_i$$

The moments of inertia for a disk and a loop can be found in Table 12.2, leading to

$$\omega_f = \frac{\frac{1}{2}M_{disk}R^2}{\frac{1}{2}M_{disk}R^2 + M_{loop}R^2}\omega_i = 100 \text{ rpm}$$

ASSESS What appeared to be a difficult problem turns out to be fairly easy once you recognize that the total angular momentum is conserved.

FIGURE 12.58 The circular loop drops onto the rotating disk.

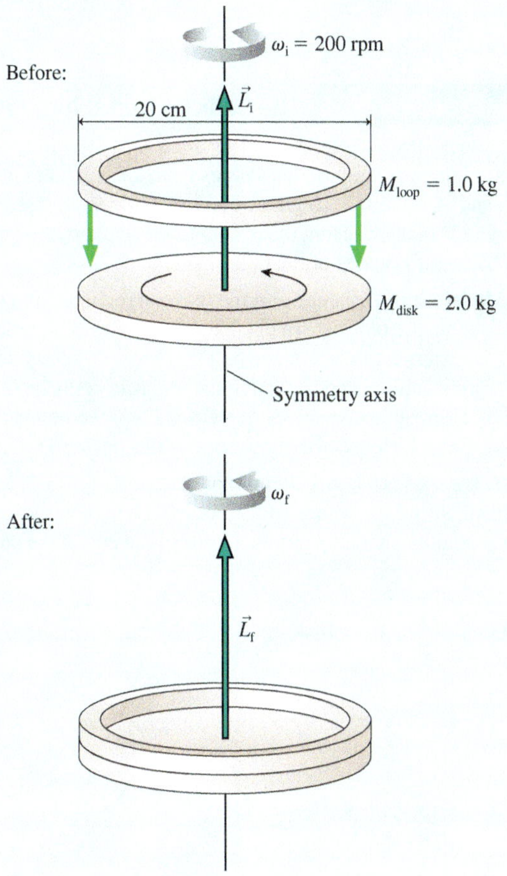

When angular momentum—a vector—is conserved, its direction—the direction of the rotation axis—must remain unchanged. This is often shown with the lecture demonstration illustrated in FIGURE 12.59 on the next page. A bicycle wheel with two handles is given a spin, then handed to an unsuspecting student. The student is asked to turn the wheel 90°. Surprisingly, this is *very hard to do.*

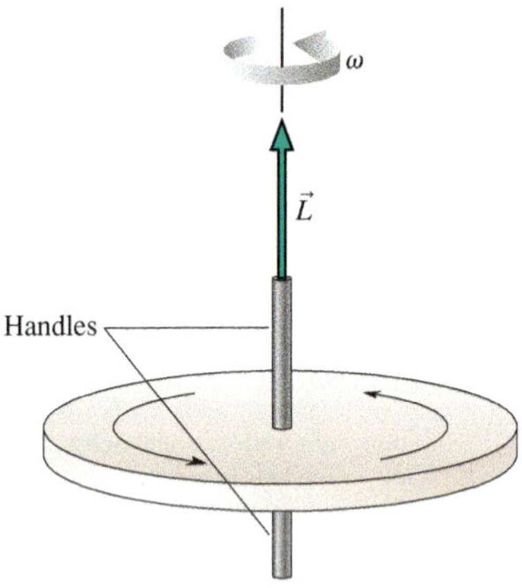

FIGURE 12.59 The vector nature of angular momentum makes it difficult to turn a rapidly spinning wheel.

The reason is that the wheel's angular momentum vector, which points straight up, is highly resistant to change. If the wheel is spinning fast, a *large* torque must be supplied to change $\vec{L}$. This directional stability of a rapidly spinning object is why gyroscopes are used as navigational devices on ships and planes. Once the axis of a spinning gyroscope is pointed north, it will maintain that direction as the ship or plane moves.

STOP TO THINK 12.5 Two buckets spin around in a horizontal circle on frictionless bearings. Suddenly, it starts to rain. As a result,

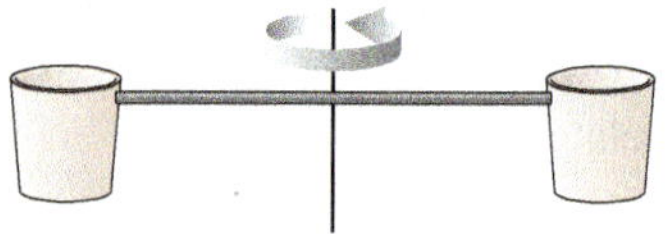

a. The buckets continue to rotate at constant angular velocity because the rain is falling vertically while the buckets move in a horizontal plane.
b. The buckets continue to rotate at constant angular velocity because the total mechanical energy of the bucket + rain system is conserved.
c. The buckets speed up because the potential energy of the rain is transformed into kinetic energy.
d. The buckets slow down because the angular momentum of the bucket + rain system is conserved.
e. Both a and b.
f. None of the above.

CHALLENGE EXAMPLE 12.22 The ballistic pendulum revisited

A 2.0 kg block hangs from the end of a 1.5 kg,1.0-m-long rod, together forming a pendulum that swings from a frictionless pivot at the top end of the rod. A 10 g bullet is fired horizontally into the block, where it sticks, causing the pendulum to swing out to a 30° angle. What was the speed of the bullet?

MODEL Model the rod as a uniform rod that can rotate around one end, and assume the block is small enough to model as a particle. There are no external torques on the bullet + block + rod system, so angular momentum is conserved in the inelastic collision. Further, the mechanical energy of the system is conserved after (but not during) the collision as the pendulum swings outward.

VISUALIZE FIGURE 12.60 is a pictorial representation. This is a two-part problem, so we've separated the collision's before-and-after from the pendulum swing's before-and-after. The end of the collision is the beginning of the swing.

FIGURE 12.60 Pictorial representation of the bullet hitting the pendulum.

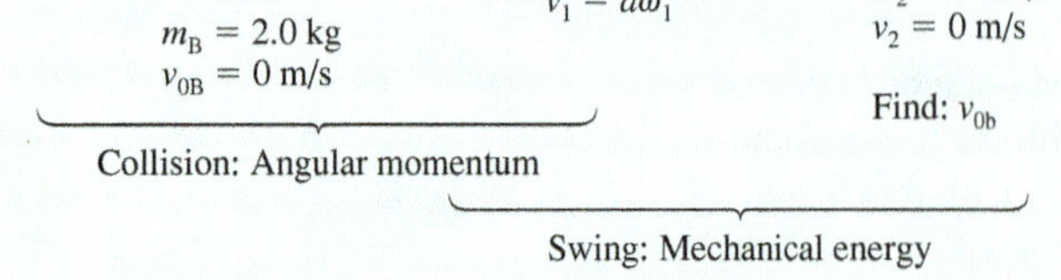

SOLVE This is a *ballistic pendulum.* Example 10.4 considered a simpler ballistic pendulum with a mass on a string, rather than on a rod, and a review of that example is highly recommended. The key to both is that a different conservation law applies to each part of the problem.

Angular momentum is conserved in the collision, thus $L_1 = L_0$. Before the collision, the angular momentum—which we'll measure about the pendulum's pivot point—is entirely that of the bullet. The angular momentum of a particle is $L = mrv\sin\beta$. An instant before the collision, just as the bullet reaches the block, $r = d$ and, because $\vec{v}$ is perpendicular to $\vec{r}$ at that instant, $\beta = 90°$. Thus $L_0 = m_b d v_{0b}$. (This is the magnitude of the angular momentum; from the right-hand rule, the angular momentum vector points out of the page.)

An instant after the collision, but before the pendulum has had time to move, the rod has angular velocity ω_1 and the block, with the embedded bullet, is moving in a circle with speed $v_1 = \omega_1 r = \omega_1 d$. The angular momentum of the block + bullet system is that of a particle, still with $\beta = 90°$, while that of the rod—an object rotating on a fixed axle—is $I_{rod}\,\omega_1$. Thus the post-collision angular momentum is

$$L_1 = (m_B + m_b)v_1 r + I_{rod}\omega_1 = (m_B + m_b)d^2\omega_1 + \frac{1}{3}m_R d^2\omega_1$$

The moment of inertia of the rod was taken from Table 12.2.

Equating the before-and-after angular momenta, then solving for v_{0b}, gives

$$m_b d v_{0b} = (m_B + m_b)d^2\omega_1 + \frac{1}{3}m_R d^2\omega_1$$

$$v_{0b} = \frac{m_B + m_b + \frac{1}{3}m_R}{m_b}d\omega_1 = 251 d\omega_1$$

Once we know ω_1, which we'll find from energy conservation in the swing, we'll be able to compute the bullet's speed.

Mechanical energy is conserved during the swing, but you must be careful to include all the energies. The kinetic energy has two components: the translational kinetic energy of the block + bullet system and the rotational kinetic energy of the rod. The gravitational potential energy also has two components: the potential energy of the block + bullet system and the potential energy of the rod. The latter changes because the center of mass moves upward as the rod swings. Thus the energy conservation statement is

$$\frac{1}{2}(m_B + m_b)v_2^2 + \frac{1}{2}I_{rod}\omega_2^2 + (m_B + m_b)gy_2 + m_R g y_{cm2} =$$

$$\frac{1}{2}(m_B + m_b)v_1^2 + \frac{1}{2}I_{rod}\omega_1^2 + (m_B + m_b)gy_1 + m_R g y_{cm1}$$

Although this looks very complicated, you should convince yourself that we've done nothing more than add up two kinetic energies and two potential energies before and after the swing.

We know that $v_2 = 0$ and $\omega_2 = 0$ at the end of the swing, and that $v_1 = d\omega_1$ at the beginning. We also know the moment of inertia of a rod pivoted at one end. Combining the potential energy terms and using $\Delta y = y_f - y_i$, we thus have

$$\frac{1}{2}\left(m_B + m_b + \frac{1}{3}m_R\right)d^2\omega_1^2 = (m_B + m_b)g\Delta y + m_R g \Delta y_{cm}$$

We see from Figure 12.60 that the block, at its highest point, is distance $d\cos\theta$ *below* the pivot. It started distance d below the pivot, so the bullet + block system *gained* height $\Delta y = d - d\cos\theta = d(1-\cos\theta)$. The rod's center of mass started distance $d/2$ below the pivot and rises only half as much as the block, so $\Delta y_{cm} = \frac{1}{2}d(1-\cos\theta)$. With these, the energy equation becomes

$$\frac{1}{2}\left(m_B + m_b + \frac{1}{3}m_R\right)d^2\omega_1^2 = (m_B + m_b + \frac{1}{2}m_R)gd(1 - \cos\theta)$$

We can now solve for ω_1:

$$\omega_1 = \sqrt{\frac{m_B + m_b + \frac{1}{2}m_R}{m_B + m_b + \frac{1}{3}m_R}\,\frac{2g(1 - \cos\theta)}{d}} = 1.70 \text{ rad/s}$$

and with that

$$v_{0b} = 251 d\omega_1 = 430 \text{ m/s}$$

ASSESS 430 m/s seems a reasonable speed for a bullet. This was a challenging problem, but one that you can solve if you focus on the problem-solving strategies—drawing a careful pictorial representation, defining the system, and thinking about which conservation laws apply—rather than hunting for the "right" equation.

SUMMARY

The goal of Chapter 12 has been to understand the physics of rotating objects.

General Principles

Rotational Dynamics

Every point on a **rigid body** rotating about a fixed axis has the same angular velocity ω and angular acceleration α.

Newton's second law for rotational motion is

$$\alpha = \frac{\tau_{net}}{I}$$

Use rotational kinematics to find angles and angular velocities.

Conservation Laws

Energy is conserved for an isolated system.

- Pure rotation $E = K_{rot} + U_g = \frac{1}{2}I\omega^2 + Mgy_{cm}$
- Rolling $E = K_{rot} + K_{cm} + U_g = \frac{1}{2}I\omega^2 + \frac{1}{2}Mv_{cm}{}^2 + Mgy_{cm}$

Angular momentum is conserved if $\vec{\tau}_{net} = \vec{0}$.

- Particle $\vec{L} = \vec{r} \times \vec{p}$
- Rotation about a symmetry axis or fixed axle $\vec{L} = I\vec{\omega}$

Important Concepts

Torque is the rotational equivalent of force:

$$\tau = rF\sin\phi = rF_t = dF$$

The vector description of torque is

$$\vec{\tau} = \vec{r} \times \vec{F}$$

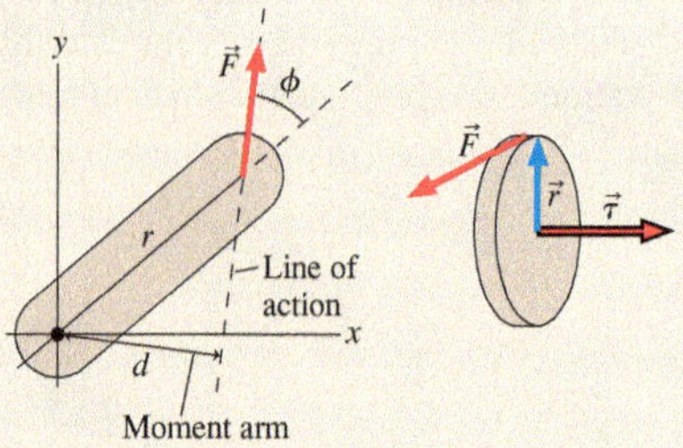

Vector description of rotation

Angular velocity $\vec{\omega}$ points along the rotation axis in the direction of the right-hand rule.

For a rigid body rotating about a fixed axle or an axis of symmetry, the angular momentum is $\vec{L} = I\vec{\omega}$.

Newton's second law is $\dfrac{d\vec{L}}{dt} = \vec{\tau}_{net}$.

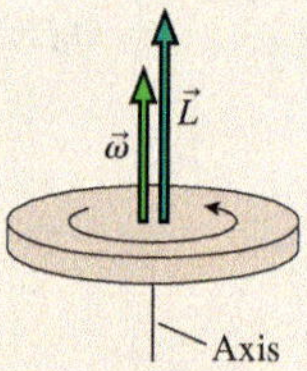

A system of particles on which there is no net force undergoes unconstrained rotation about the **center of mass**:

$$x_{cm} = \frac{1}{M}\int x\,dm \qquad y_{cm} = \frac{1}{M}\int y\,dm$$

The gravitational torque on a body can be found by treating the body as a particle with all the mass M concentrated at the center of mass.

The **moment of inertia**

$$I = \sum_i m_i r_i^2 = \int r^2\,dm$$

is the rotational equivalent of mass. The moment of inertia depends on how the mass is distributed around the axis. If I_{cm} is known, I about a parallel axis distance d away is given by the **parallel-axis theorem:** $I = I_{cm} + Md^2$.

Applications

Rotational kinematics

$\omega_f = \omega_i + \alpha\,\Delta t$

$\theta_f = \theta_i + \omega_i\,\Delta t + \frac{1}{2}\alpha(\Delta t)^2$

$v_t = r\omega \qquad a_t = r\alpha$

Rigid-body equilibrium

An object is in total equilibrium only if both $\vec{F}_{net} = \vec{0}$ and $\vec{\tau}_{net} = \vec{0}$.

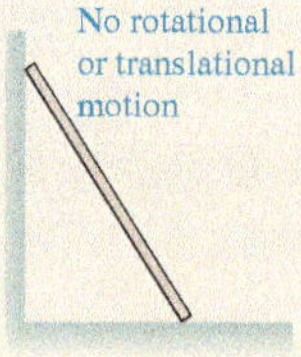

Rolling motion

For an object that rolls without slipping

$v_{cm} = R\omega$

$K = K_{rot} + K_{cm}$

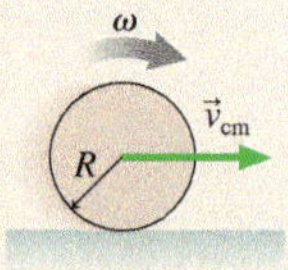

Terms and Notation

rigid body
rigid-body model
translational motion
rotational motion
combination motion
center of mass
rotational kinetic energy, K_{rot}
moment of inertia, I
parallel-axis theorem
torque, τ
line of action
moment arm, d
rolling constraint
cross product
vector product
right-hand rule
angular momentum, $\vec{L}$
law of conservation of angular momentum

CONCEPTUAL QUESTIONS

1. Is the center of mass of the dumbbell in FIGURE Q12.1 at point a, b, or c? Explain.

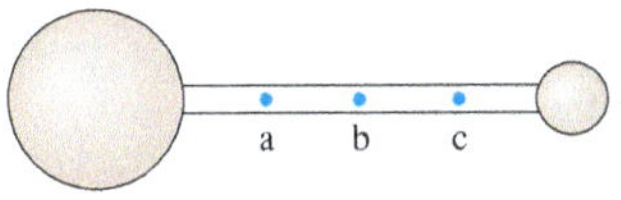

FIGURE Q12.1

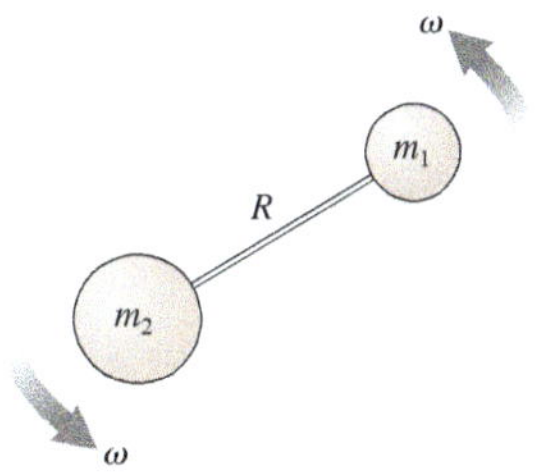

FIGURE Q12.2

2. If the angular velocity ω is held constant, by what *factor* must R change to double the rotational kinetic energy of the dumbbell in FIGURE Q12.2?
3. FIGURE Q12.3 shows three rotating disks, all of equal mass. Rank in order, from largest to smallest, their rotational kinetic energies K_a to K_c.

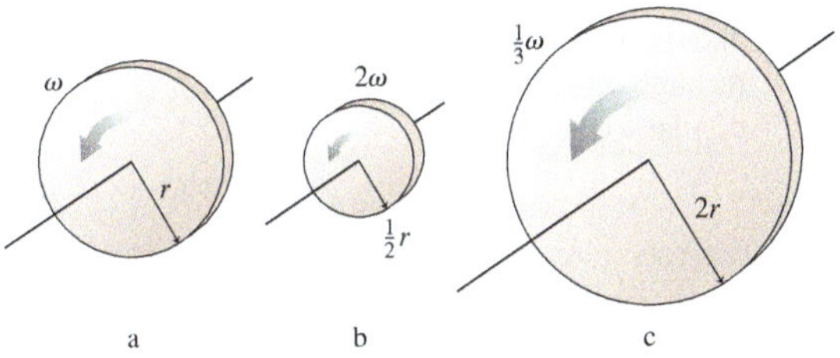

FIGURE Q12.3

4. Must an object be rotating to have a moment of inertia? Explain.
5. The moment of inertia of a uniform rod about an axis through its center is $\frac{1}{12}mL^2$. The moment of inertia about an axis at one end is $\frac{1}{3}mL^2$. Explain *why* the moment of inertia is larger about the end than about the center.
6. You have two steel spheres. Sphere 2 has twice the radius of sphere 1. By what *factor* does the moment of inertia I_2 of sphere 2 exceed the moment of inertia I_1 of sphere 1?
7. The professor hands you two spheres. They have the same mass, the same radius, and the same exterior surface. The professor claims that one is a solid sphere and the other is hollow. Can you determine which is which without cutting them open? If so, how? If not, why not?
8. Six forces are applied to the door in FIGURE Q12.8. Rank in order, from largest to smallest, the six torques τ_a to τ_f about the hinge. Explain.

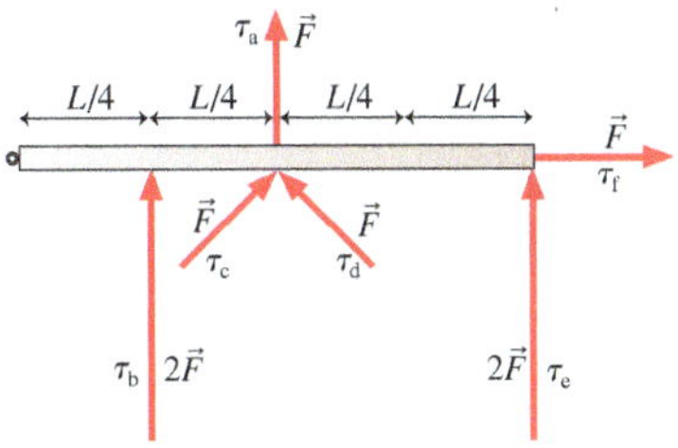

FIGURE Q12.8

9. A student gives a quick push to a ball at the end of a massless, rigid rod, as shown in FIGURE Q12.9, causing the ball to rotate clockwise in a *horizontal* circle. The rod's pivot is frictionless.
   a. As the student is pushing, is the torque about the pivot positive, negative, or zero?
   b. After the push has ended, does the ball's angular velocity (i) steadily increase; (ii) increase for awhile, then hold steady; (iii) hold steady; (iv) decrease for awhile, then hold steady; or (v) steadily decrease? Explain.
   c. Right after the push has ended, is the torque positive, negative, or zero?

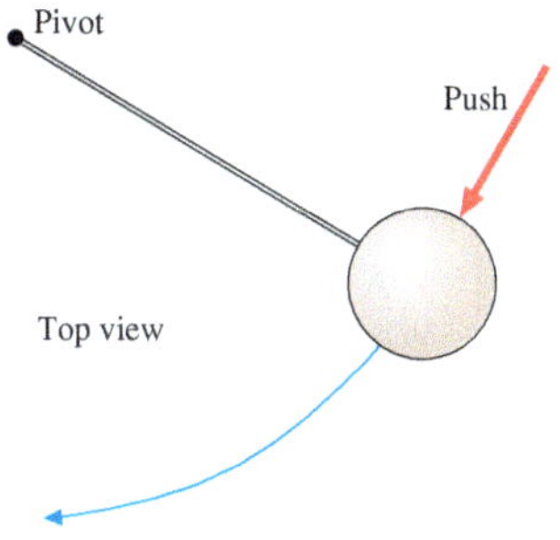

FIGURE Q12.9

10. Rank in order, from largest to smallest, the angular accelerations α_a to α_d in FIGURE Q12.10. Explain.

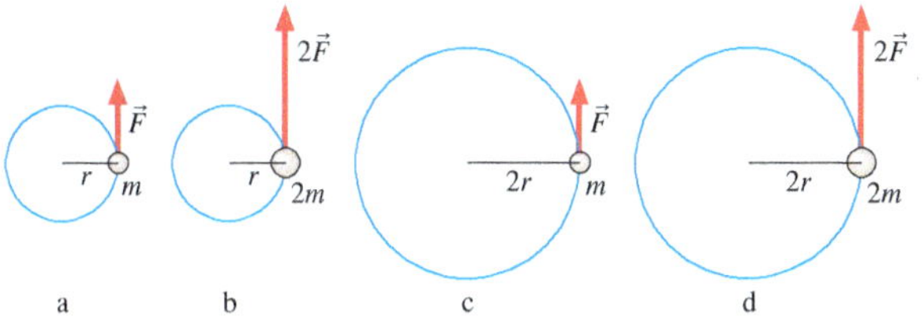

FIGURE Q12.10

11. The solid cylinder and cylindrical shell in FIGURE Q12.11 have the same mass, same radius, and turn on frictionless, horizontal axles. (The cylindrical shell has lightweight spokes connecting the shell to the axle.) A rope is wrapped around each cylinder and tied to a block. The blocks have the same mass and are held the same height above the ground. Both blocks are released simultaneously. Which hits the ground first? Or is it a tie? Explain.

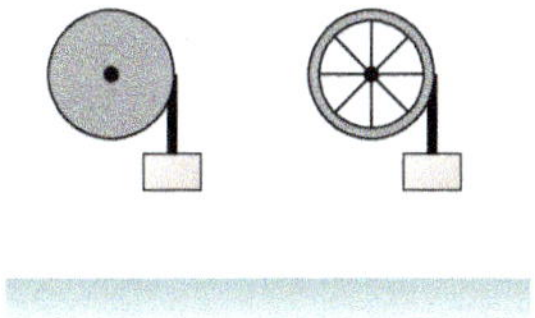

FIGURE Q12.11

12. A diver in the pike position (legs straight, hands on ankles) usually makes only one or one-and-a-half rotations. To make two or three rotations, the diver goes into a tuck position (knees bent, body curled up tight). Why?
13. Is the angular momentum of disk a in FIGURE Q12.13 larger than, smaller than, or equal to the angular momentum of disk b? Explain.

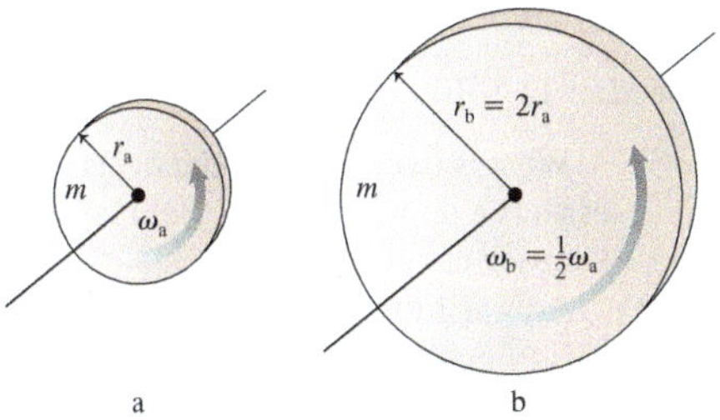

FIGURE Q12.13

EXERCISES AND PROBLEMS

Problems labeled integrate material from earlier chapters.

Exercises

Section 12.1 Rotational Motion

1. || A skater holds her arms outstretched as she spins at 180 rpm. What is the speed of her hands if they are 140 cm apart?
2. || A high-speed drill reaches 2000 rpm in 0.50 s.
   a. What is the drill's angular acceleration?
   b. Through how many revolutions does it turn during this first 0.50 s?
3. || A ceiling fan with 80-cm-diameter blades is turning at 60 rpm. Suppose the fan coasts to a stop 25 s after being turned off.
   a. What is the speed of the tip of a blade 10 s after the fan is turned off?
   b. Through how many revolutions does the fan turn while stopping?
4. ||| An 18-cm-long bicycle crank arm, with a pedal at one end, is attached to a 20-cm-diameter sprocket, the toothed disk around which the chain moves. A cyclist riding this bike increases her pedaling rate from 60 rpm to 90 rpm in 10 s.
   a. What is the tangential acceleration of the pedal?
   b. What length of chain passes over the top of the sprocket during this interval?

Section 12.2 Rotation About the Center of Mass

5. || How far from the center of the earth is the center of mass of the earth + moon system? Data for the earth and moon can be found inside the back cover of the book.
6. | The three masses shown in FIGURE EX12.6 are connected by massless, rigid rods. What are the coordinates of the center of mass?

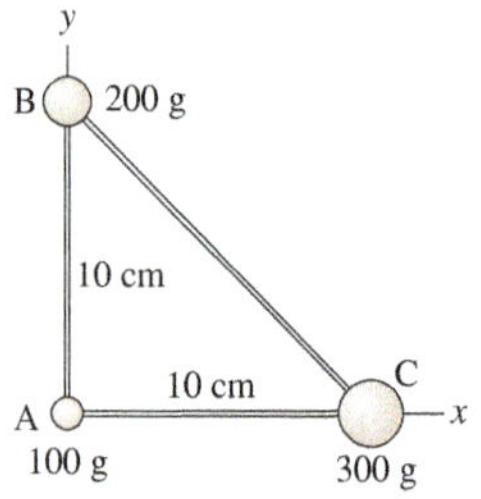

FIGURE EX12.6

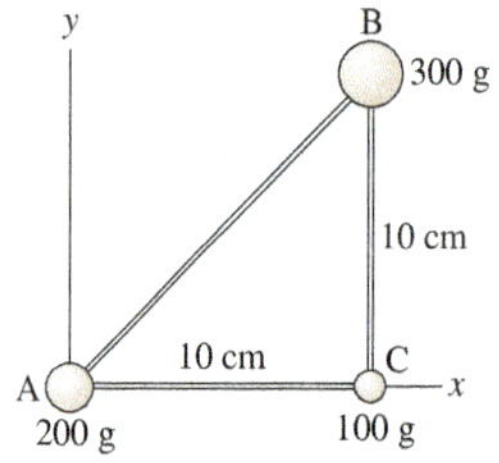

FIGURE EX12.7

7. | The three masses shown in FIGURE EX12.7 are connected by massless, rigid rods. What are the coordinates of the center of mass?
8. || A 100 g ball and a 200 g ball are connected by a 30-cm-long, massless, rigid rod. The balls rotate about their center of mass at 120 rpm. What is the speed of the 100 g ball?

Section 12.3 Rotational Energy

9. || What is the rotational kinetic energy of the earth? Assume the earth is a uniform sphere. Data for the earth can be found inside the back cover of the book.
10. || A thin, 100 g disk with a diameter of 8.0 cm rotates about an axis through its center with 0.15 J of kinetic energy. What is the speed of a point on the rim?
11. ||| The three 200 g masses in FIGURE EX12.11 are connected by massless, rigid rods.
    a. What is the triangle's moment of inertia about the axis through the center?
    b. What is the triangle's kinetic energy if it rotates about the axis at 5.0 rev/s?

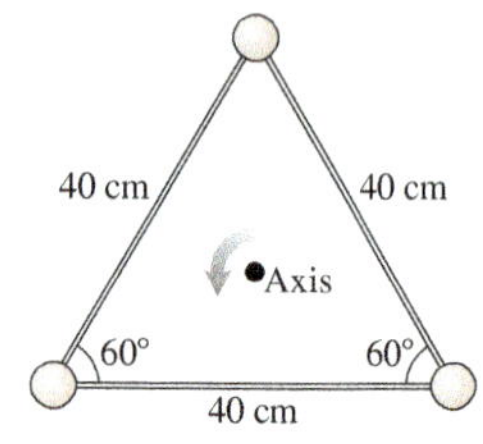

FIGURE EX12.11

12. ||| A drum major twirls a 96-cm-long, 400 g baton about its center of mass at 100 rpm. What is the baton's rotational kinetic energy?

Section 12.4 Calculating Moment of Inertia

13. || The four masses shown in FIGURE EX12.13 are connected by massless, rigid rods.
    a. Find the coordinates of the center of mass.
    b. Find the moment of inertia about an axis that passes through mass A and is perpendicular to the page.

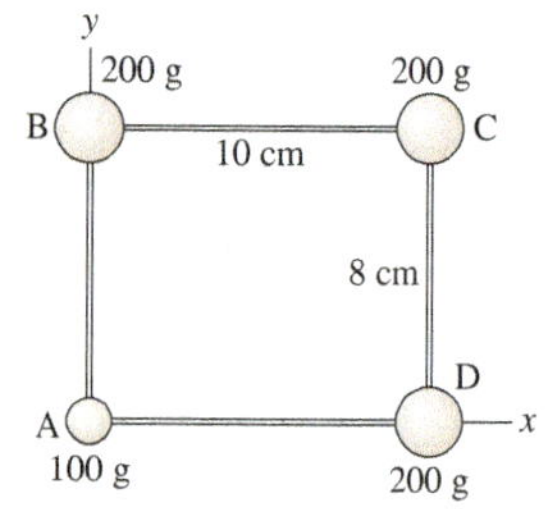

FIGURE EX12.13

14. || The four masses shown in FIGURE EX12.13 are connected by massless, rigid rods.
    a. Find the coordinates of the center of mass.
    b. Find the moment of inertia about a diagonal axis that passes through masses B and D.
15. | The three masses shown in FIGURE EX12.15 are connected by massless, rigid rods.
    a. Find the coordinates of the center of mass.
    b. Find the moment of inertia about an axis that passes through mass A and is perpendicular to the page.
    b. Find the moment of inertia about an axis that passes through masses B and C.

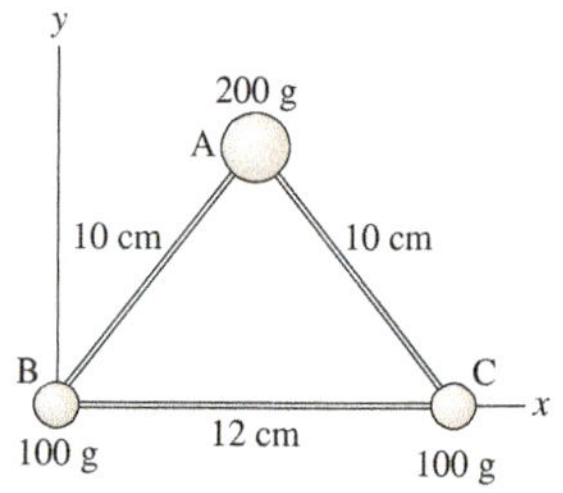

FIGURE EX12.15

16. || A 25 kg solid door is 220 cm tall, 91 cm wide. What is the door's moment of inertia for (a) rotation on its hinges and (b) rotation about a vertical axis inside the door, 15 cm from one edge?
17. || A 12-cm-diameter CD has a mass of 21 g. What is the CD's moment of inertia for rotation about a perpendicular axis (a) through its center and (b) through the edge of the disk?

Section 12.5 Torque

18. | In FIGURE EX12.18, what is the net torque about the axle?

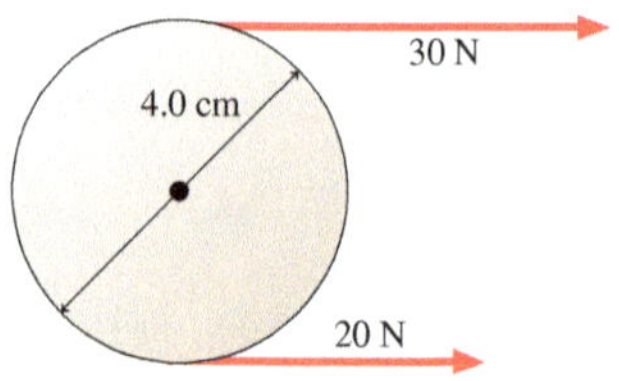

FIGURE EX12.18

19. || In FIGURE EX12.19, what is the net torque about the axle?

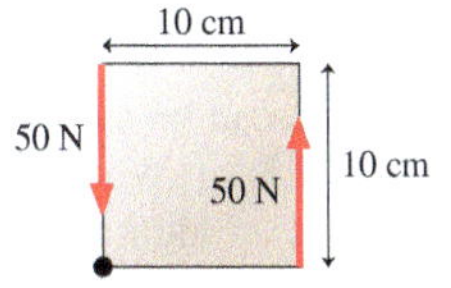

FIGURE EX12.19

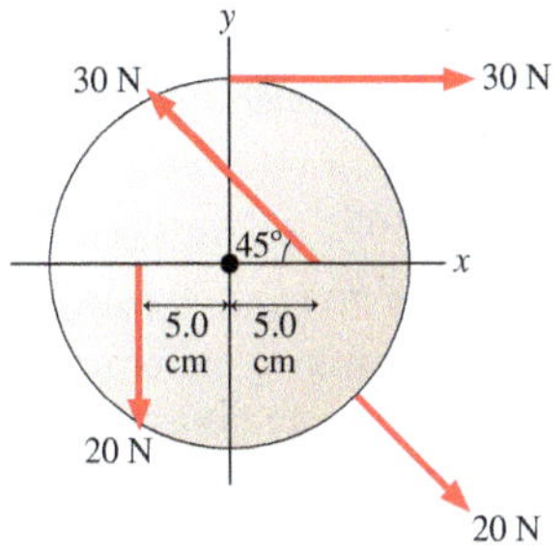

FIGURE EX12.20

20. || The 20-cm-diameter disk in FIGURE EX12.20 can rotate on an axle through its center. What is the net torque about the axle?

21. || A 4.0-m-long, 500 kg steel beam extends horizontally from the point where it has been bolted to the framework of a new building under construction. A 70 kg construction worker stands at the far end of the beam. What is the magnitude of the torque about the point where the beam is bolted into place?

22. || An athlete at the gym holds a 3.0 kg steel ball in his hand. His arm is 70 cm long and has a mass of 4.0 kg. What is the magnitude of the torque about his shoulder if he holds his arm
BIO
   a. Straight out to his side, parallel to the floor?
   b. Straight, but 45° below horizontal?

Section 12.6 Rotational Dynamics

Section 12.7 Rotation About a Fixed Axis

23. | An object's moment of inertia is 2.0 kg m^2. Its angular velocity is increasing at the rate of 4.0 rad/s per second. What is the torque on the object?

24. || An object whose moment of inertia is 4.0 kg m^2 experiences the torque shown in FIGURE EX12.24. What is the object's angular velocity at $t = 3.0$ s? Assume it starts from rest.

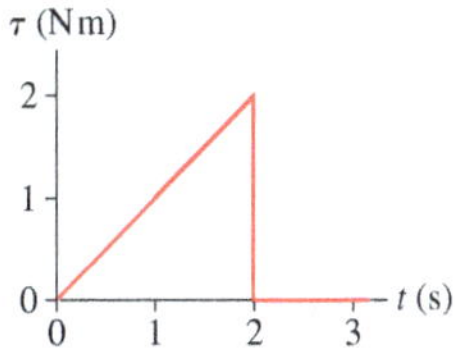

FIGURE EX12.24

25. || A 1.0 kg ball and a 2.0 kg ball are connected by a 1.0-m-long rigid, massless rod. The rod is rotating cw about its center of mass at 20 rpm. What torque will bring the balls to a halt in 5.0 s?

26. || Starting from rest, a 12-cm-diameter compact disk takes 3.0 s to reach its operating angular velocity of 2000 rpm. Assume that the angular acceleration is constant. The disk's moment of inertia is 2.5×10^{-5} kg m^2.
   a. How much torque is applied to the disk?
   b. How many revolutions does it make before reaching full speed?

27. || A 750 g, 50-cm-long metal rod is free to rotate about a frictionless axle at one end. While at rest, the rod is given a short but sharp 1000 N hammer blow at the center of the rod, aimed in a direction that causes the rod to rotate on the axle. The blow lasts a mere 2.0 ms. What is the rod's angular velocity immediately after the blow?

Section 12.8 Static Equilibrium

28. || How much torque must the pin exert to keep the rod in FIGURE EX12.28 from rotating?

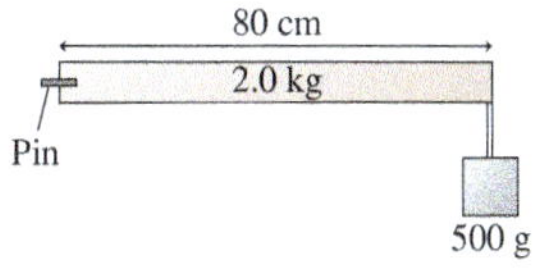

FIGURE EX12.28

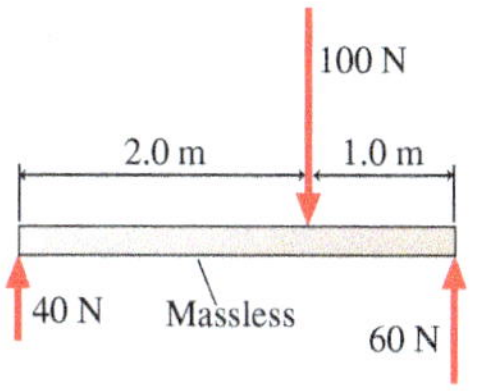

FIGURE EX12.29

29. || Is the object in FIGURE EX12.29 in equilibrium? Explain.

30. || The two objects in FIGURE EX12.30 are balanced on the pivot. What is distance d?

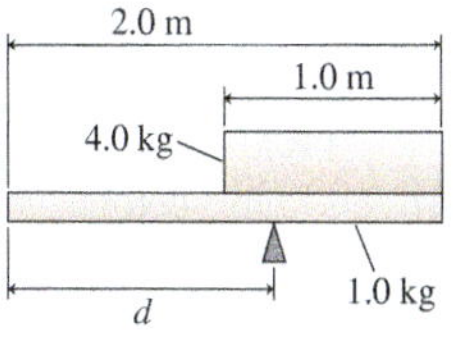

FIGURE EX12.30

31. || A 5.0 kg cat and a 2.0 kg bowl of tuna fish are at opposite ends of the 4.0-m-long seesaw of FIGURE EX12.31. How far to the left of the pivot must a 4.0 kg cat stand to keep the seesaw balanced?

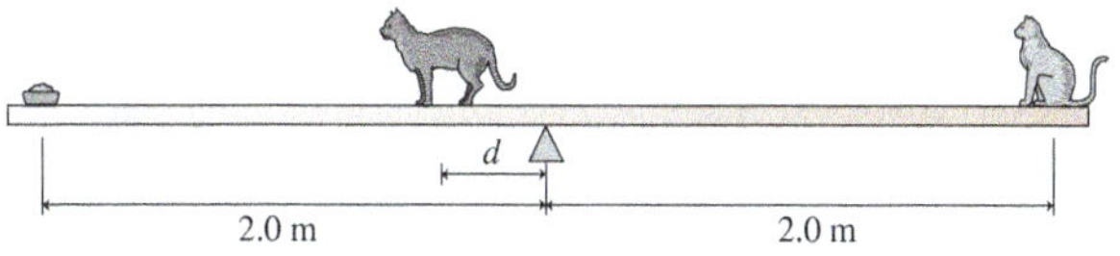

FIGURE EX12.31

Section 12.9 Rolling Motion

32. || A car tire is 60 cm in diameter. The car is traveling at a speed of 20 m/s.
   a. What is the tire's angular velocity, in rpm?
   b. What is the speed of a point at the top edge of the tire?
   c. What is the speed of a point at the bottom edge of the tire?

33. || A 500 g, 8.0-cm-diameter can is filled with uniform, dense food. It rolls across the floor at 1.0 m/s. What is the can's kinetic energy?

34. || An 8.0-cm-diameter, 400 g solid sphere is released from rest at the top of a 2.1-m-long, 25° incline. It rolls, without slipping, to the bottom.
   a. What is the sphere's angular velocity at the bottom of the incline?
   b. What fraction of its kinetic energy is rotational?

35. | A solid sphere of radius R is placed at a height of 30 cm on a 15° slope. It is released and rolls, without slipping, to the bottom. From what height should a circular hoop of radius R be released on the same slope in order to equal the sphere's speed at the bottom?

Section 12.10 The Vector Description of Rotational Motion

36. | Evaluate the cross products $\vec{A} \times \vec{B}$ and $\vec{C} \times \vec{D}$.

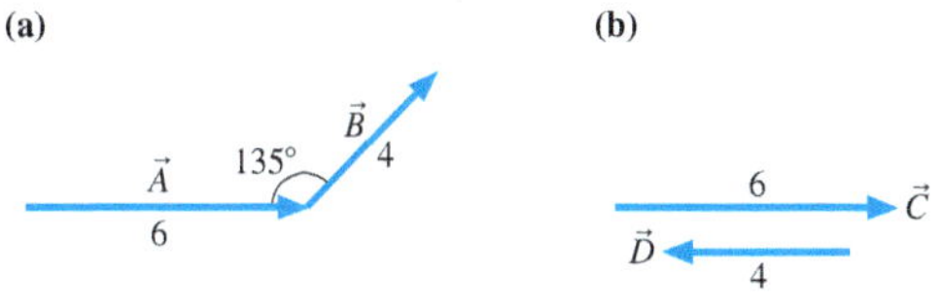

FIGURE EX12.36

37. | Evaluate the cross products $\vec{A} \times \vec{B}$ and $\vec{C} \times \vec{D}$.

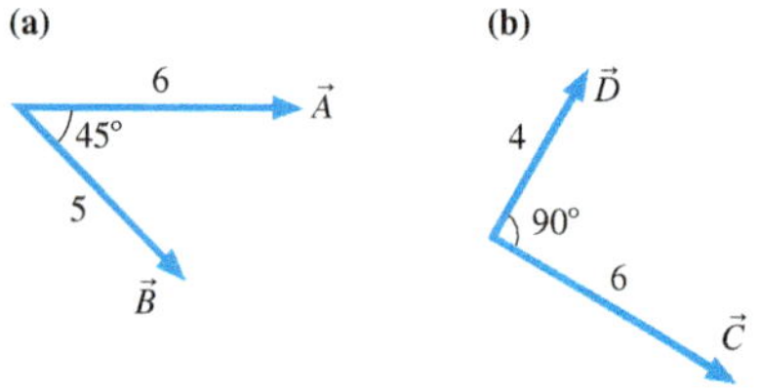

FIGURE EX12.37

38. || a. What is $(\hat{\imath} \times \hat{\jmath}) \times \hat{\imath}$?
b. What is $\hat{\imath} \times (\hat{\jmath} \times \hat{\imath})$?

39. || a. What is $\hat{\imath} \times (\hat{\imath} \times \hat{\jmath})$?
b. What is $(\hat{\imath} \times \hat{\jmath}) \times \hat{k}$?

40. || Vector $\vec{A} = 3\hat{\imath} + \hat{\jmath}$ and vector $\vec{B} = 3\hat{\imath} - 2\hat{\jmath} + 2\hat{k}$. What is the cross product $\vec{A} \times \vec{B}$?

41. | Consider the vector $\vec{C} = 3\hat{\imath}$.
a. What is a vector $\vec{D}$ such that $\vec{C} \times \vec{D} = \vec{0}$?
b. What is a vector $\vec{E}$ such that $\vec{C} \times \vec{E} = 6\hat{k}$?
c. What is a vector $\vec{F}$ such that $\vec{C} \times \vec{F} = -3\hat{\jmath}$?

42. || Force $\vec{F} = -10\hat{\jmath}$ N is exerted on a particle at $\vec{r} = (5\hat{\imath} + 5\hat{\jmath})$ m. What is the torque on the particle about the origin?

43. || What are the magnitude and direction of the angular momentum relative to the origin of the 100 g particle in FIGURE EX12.43?

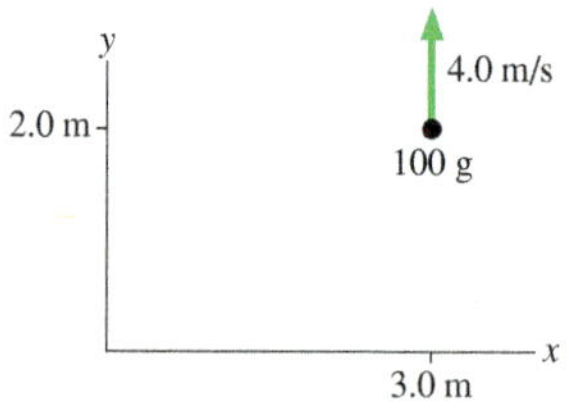

FIGURE EX12.43

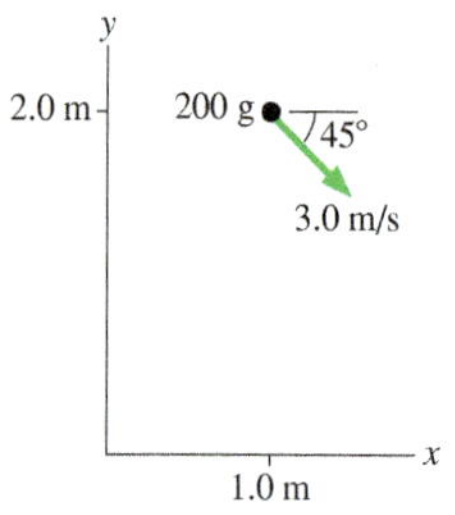

FIGURE EX12.44

44. || What are the magnitude and direction of the angular momentum relative to the origin of the 200 g particle in FIGURE EX12.44?

Section 12.11 Angular Momentum

45. || What is the angular momentum of the 500 g rotating bar in FIGURE EX12.45?

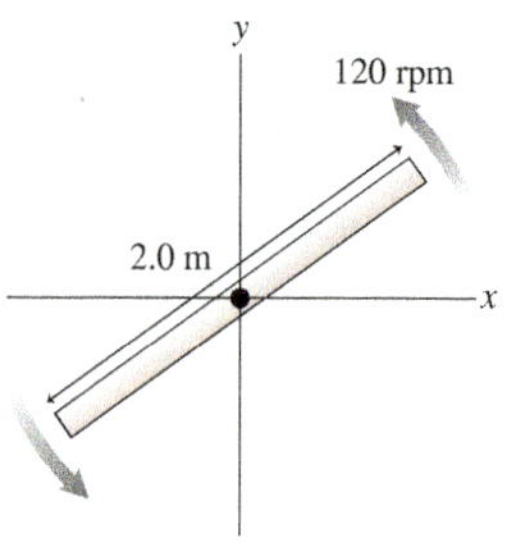

FIGURE EX12.45

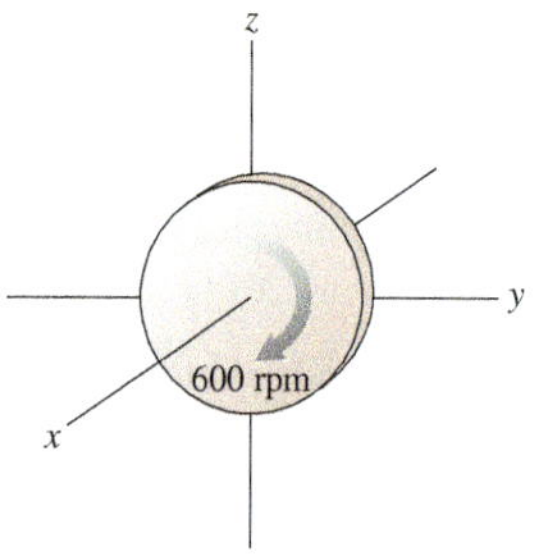

FIGURE EX12.46

46. || What is the angular momentum of the 2.0 kg, 4.0-cm-diameter rotating disk in FIGURE EX12.46?

47. || How fast, in rpm, would a 5.0 kg, 22-cm-diameter bowling ball have to spin to have an angular momentum of 0.23 kg m^2/s?

48. || A 2.0 kg, 20-cm-diameter turntable rotates at 100 rpm on frictionless bearings. Two 500 g blocks fall from above, hit the turntable simultaneously at opposite ends of a diameter, and stick. What is the turntable's angular velocity, in rpm, just after this event?

Problems

49. || A 70 kg man's arm, including the hand, can be modeled as a
BIO 75-cm-long uniform rod with a mass of 3.5 kg. When the man raises both his arms, from hanging down to straight up, by how much does he raise his center of mass?

50. ||| A 300 g ball and a 600 g ball are connected by a 40-cm-long massless, rigid rod. The structure rotates about its center of mass at 100 rpm. What is its rotational kinetic energy?

51. ||| A 60-cm-diameter wheel is rolling along at 20 m/s. What is the speed of a point at the forward edge of the wheel?

52. || An 800 g steel plate has the shape of the isosceles triangle shown in FIGURE P12.52. What are the x- and y-coordinates of the center of mass?

Hint: Divide the triangle into vertical strips of width dx, then relate the mass dm of a strip at position x to the values of x and dx.

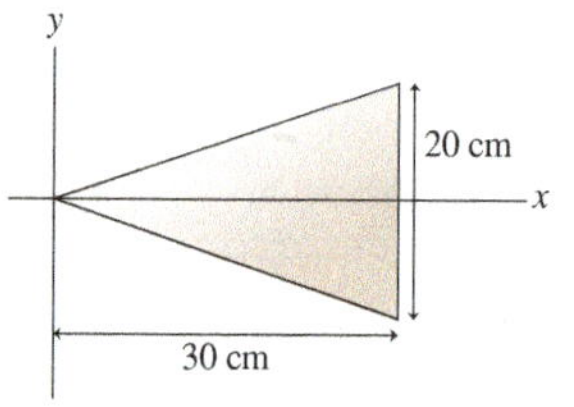

FIGURE P12.52

53. || What is the moment of inertia of a 2.0 kg, 20-cm-diameter disk for rotation about an axis (a) through the center, and (b) through the edge of the disk?

54. || Determine the moment of inertia about the axis of the object shown in FIGURE P12.54.

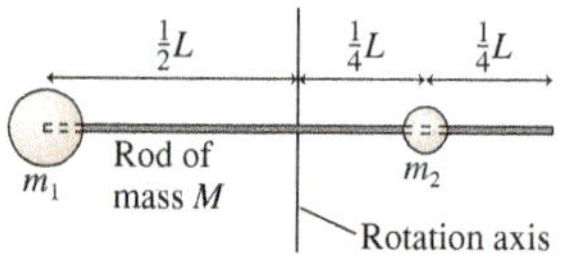

FIGURE P12.54

55. || Calculate by direct integration the moment of inertia for a thin rod of mass M and length L about an axis located distance d from one end. Confirm that your answer agrees with Table 12.2 when $d = 0$ and when $d = L/2$.

56. || a. A disk of mass M and radius R has a hole of radius r centered on the axis. Calculate the moment of inertia of the disk.
b. Confirm that your answer agrees with Table 12.2 when $r = 0$ and when $r = R$.
c. A 4.0-cm-diameter disk with a 3.0-cm-diameter hole rolls down a 50-cm-long, 20° ramp. What is its speed at the bottom? What percent is this of the speed of a particle sliding down a frictionless ramp?

57. || Calculate the moment of inertia of the rectangular plate in FIGURE P12.57 for rotation about a perpendicular axis through the center.

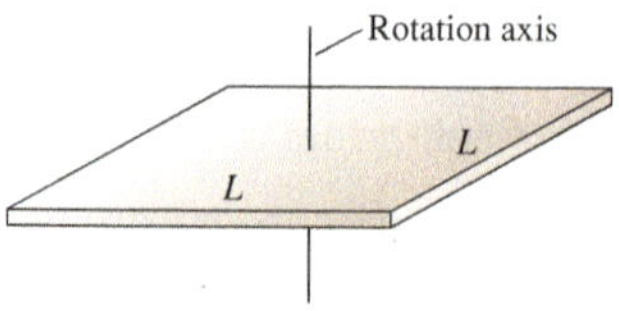

FIGURE P12.57

58. ||| Calculate the moment of inertia of the steel plate in FIGURE P12.52 for rotation about a perpendicular axis passing through the origin.

59. || A person's center of mass is easily found by having the person
BIO lie on a *reaction board.* A horizontal, 2.5-m-long, 6.1 kg reaction board is supported only at the ends, with one end resting on a scale and the other on a pivot. A 60 kg woman lies on the reaction board with her feet over the pivot. The scale reads 25 kg. What is the distance from the woman's feet to her center of mass?

60. || A 3.0-m-long ladder, as shown in Figure 12.37, leans against a frictionless wall. The coefficient of static friction between the ladder and the floor is 0.40. What is the minimum angle the ladder can make with the floor without slipping?

61. || The 3.0-m-long, 100 kg rigid beam of FIGURE P12.61 is supported at each end. An 80 kg student stands 2.0 m from support 1. How much upward force does each support exert on the beam?

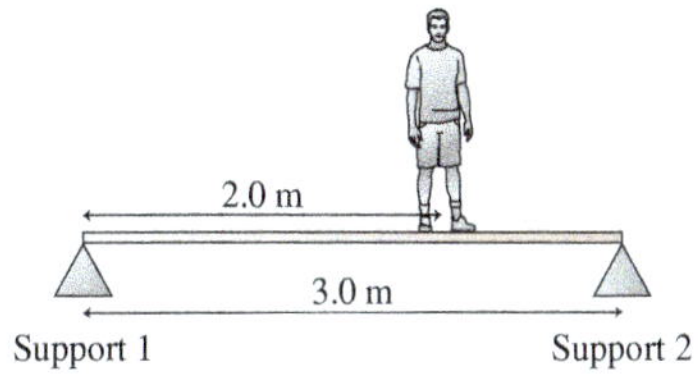

FIGURE P12.61

62. || In FIGURE P12.62, an 80 kg construction worker sits down 2.0 m from the end of a 1450 kg steel beam to eat his lunch. The cable supporting the beam is rated at 15,000 N. Should the worker be worried?

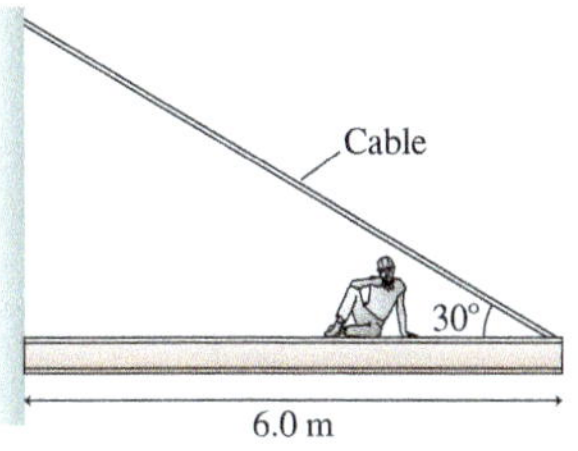

FIGURE P12.62

63. ||| A 40 kg, 5.0-m-long beam is supported by, but not attached to, the two posts in FIGURE P12.63. A 20 kg boy starts walking along the beam. How close can he get to the right end of the beam without it falling over?

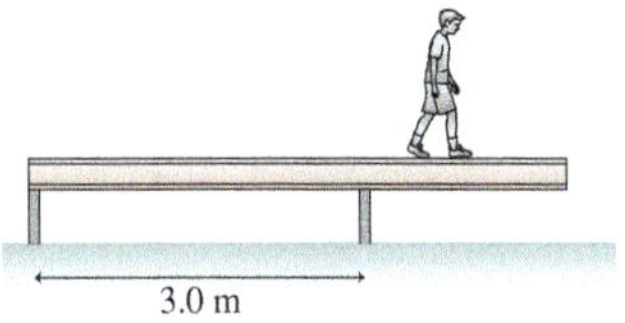

FIGURE P12.63

64. || Your task in a science contest is to stack four identical uniform bricks, each of length L, so that the top brick is as far to the right as possible without the stack falling over. Is it possible, as FIGURE P12.64 shows, to stack the bricks such that no part of the top brick is over the table? Answer this question by determining the maximum possible value of d.

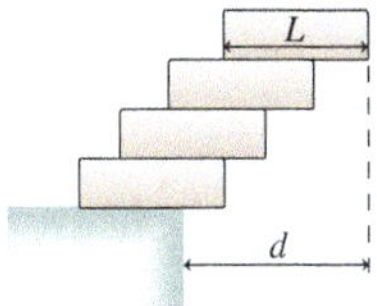

FIGURE P12.64

65. || A 120-cm-wide sign hangs from a 5.0 kg, 200-cm-long pole. A cable of negligible mass supports the end of the rod as shown in FIGURE P12.65. What is the maximum mass of the sign if the maximum tension in the cable without breaking is 300 N?

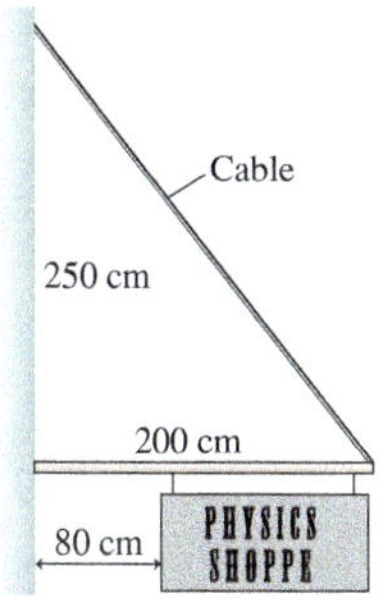

FIGURE P12.65

66. || The bunchberry flower has the fastest-moving parts ever ob-
BIO served in a plant. Initially, the stamens are held by the petals in a bent position, storing elastic energy like a coiled spring. When the petals release, the tips of the stamen act like medieval catapults, flipping through a 60° angle in just 0.30 ms to launch pollen from anther sacs at their ends. The human eye just sees a burst of pollen; only high-speed photography reveals the details. As FIGURE P12.66 shows, we can model the stamen tip as a 1.0-mm-long, 10 μg rigid rod with a 10 μg anther sac at the end. Although oversimplifying, we'll assume a constant angular acceleration.

a. How large is the "straightening torque"?
b. What is the speed of the anther sac as it releases its pollen?

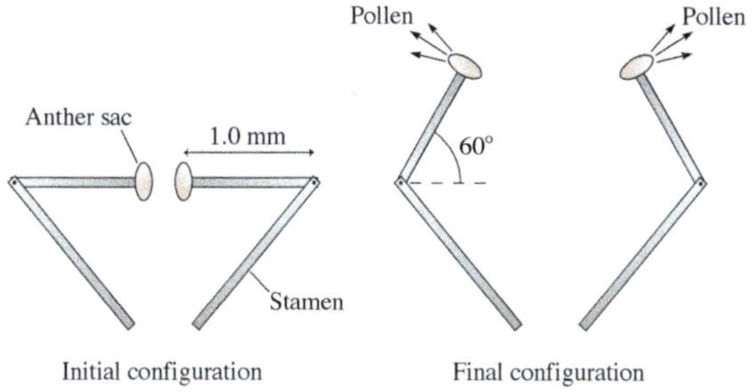

FIGURE P12.66

67. || A 60-cm-long, 500 g bar rotates in a horizontal plane on an axle that passes through the center of the bar. Compressed air is fed in through the axle, passes through a small hole down the length of the bar, and escapes as air jets from holes at the ends of the bar. The jets are perpendicular to the bar's axis. Starting from rest, the bar spins up to an angular velocity of 150 rpm at the end of 10 s.

a. How much force does each jet of escaping air exert on the bar?
b. If the axle is moved to one end of the bar while the air jets are unchanged, what will be the bar's angular velocity at the end of 10 seconds?

68. ||| Flywheels are large, massive wheels used to store energy. They can be spun up slowly, then the wheel's energy can be released quickly to accomplish a task that demands high power. An industrial flywheel has a 1.5 m diameter and a mass of 250 kg. Its maximum angular velocity is 1200 rpm.

a. A motor spins up the flywheel with a constant torque of 50 N m. How long does it take the flywheel to reach top speed?
b. How much energy is stored in the flywheel?
c. The flywheel is disconnected from the motor and connected to a machine to which it will deliver energy. Half the energy stored in the flywheel is delivered in 2.0 s. What is the average power delivered to the machine?
d. How much torque does the flywheel exert on the machine?

69. ||| The two blocks in FIGURE P12.69 are connected by a massless rope that passes over a pulley. The pulley is 12 cm in diameter and has a mass of 2.0 kg. As the pulley turns, friction at the axle exerts a torque of magnitude 0.50 N m. If the blocks are released from rest, how long does it take the 4.0 kg block to reach the floor?

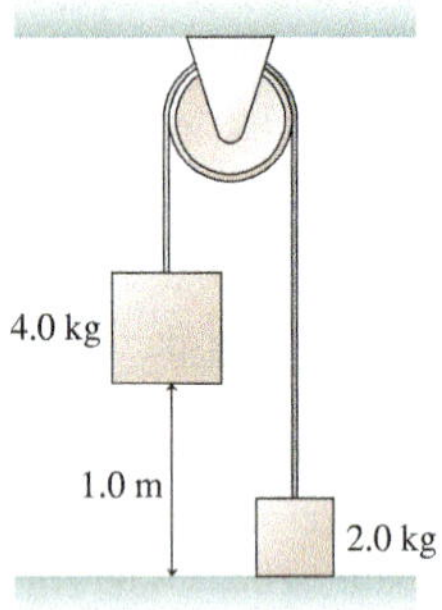

FIGURE P12.69

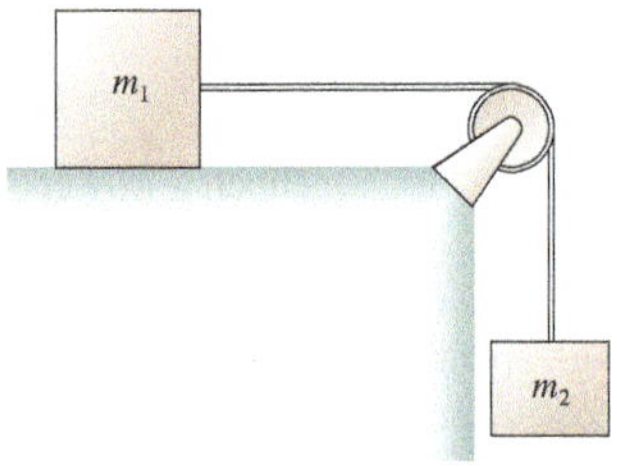

FIGURE P12.70

70. || Blocks of mass m_1 and m_2 are connected by a massless string that passes over the pulley in FIGURE P12.70. The pulley turns on frictionless bearings. Mass m_1 slides on a horizontal, frictionless surface. Mass m_2 is released while the blocks are at rest.

a. Assume the pulley is massless. Find the acceleration of m_1 and the tension in the string. This is a Chapter 7 review problem.

b. Suppose the pulley has mass m_p and radius R. Find the acceleration of m_1 and the tensions in the upper and lower portions of the string. Verify that your answers agree with part a if you set $m_p = 0$.

71. || The 2.0 kg, 30-cm-diameter disk in FIGURE P12.71 is spinning at 300 rpm. How much friction force must the brake apply to the rim to bring the disk to a halt in 3.0 s?

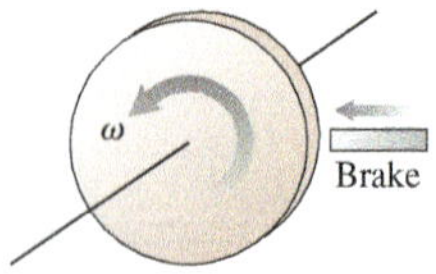

FIGURE P12.71

72. || Your engineering team has been assigned the task of measuring the properties of a new jet-engine turbine. You've previously determined that the turbine's moment of inertia is 2.6 kg m^2. The next job is to measure the frictional torque of the bearings. Your plan is to run the turbine up to a predetermined rotation speed, cut the power, and time how long it takes the turbine to reduce its rotation speed by 50%. Your data are as follows:

Rotation (rpm)	Time (s)
1500	19
1800	22
2100	25
2400	30
2700	34

Draw an appropriate graph of the data and, from the slope of the best-fit line, determine the frictional torque.

73. || A hollow sphere is rolling along a horizontal floor at 5.0 m/s when it comes to a 30° incline. How far up the incline does it roll before reversing direction?

74. || The 5.0 kg, 60-cm-diameter disk in FIGURE P12.74 rotates on an axle passing through one edge. The axle is parallel to the floor. The cylinder is held with the center of mass at the same height as the axle, then released.

a. What is the cylinder's initial angular acceleration?

b. What is the cylinder's angular velocity when it is directly below the axle?

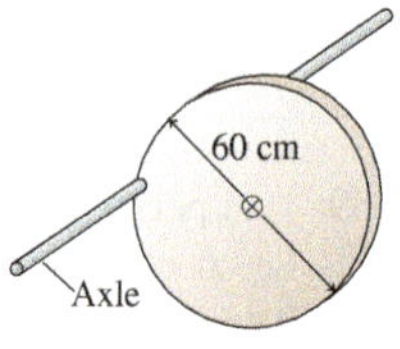

FIGURE P12.74

75. || FIGURE P12.75 shows a hoop of mass M and radius R rotating about an axle at the edge of the hoop. The hoop starts at its highest position and is given a very small push to start it rotating. At its lowest position, what are (a) the angular velocity and (b) the speed of the lowest point on the hoop?

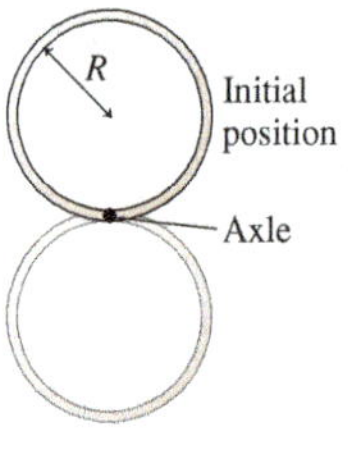

FIGURE P12.75

76. || A long, thin rod of mass M and length L is standing straight up on a table. Its lower end rotates on a frictionless pivot. A very slight push causes the rod to fall over. As it hits the table, what are (a) the angular velocity and (b) the speed of the tip of the rod?

77. || The sphere of mass M and radius R in FIGURE P12.77 is rigidly attached to a thin rod of radius r that passes through the sphere at distance $\frac{1}{2}R$ from the center. A string wrapped around the rod pulls with tension T. Find an expression for the sphere's angular acceleration. The rod's moment of inertia is negligible.

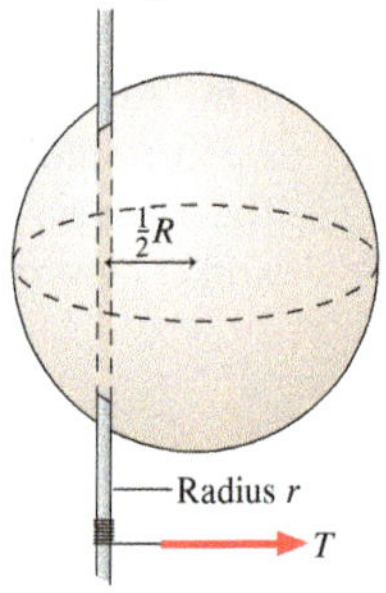

FIGURE P12.77

78. || A satellite follows the elliptical orbit shown in FIGURE P12.78. The only force on the satellite is the gravitational attraction of the planet. The satellite's speed at point a is 8000 m/s.

a. Does the satellite experience any torque about the center of the planet? Explain.

b. What is the satellite's speed at point b?

c. What is the satellite's speed at point c?

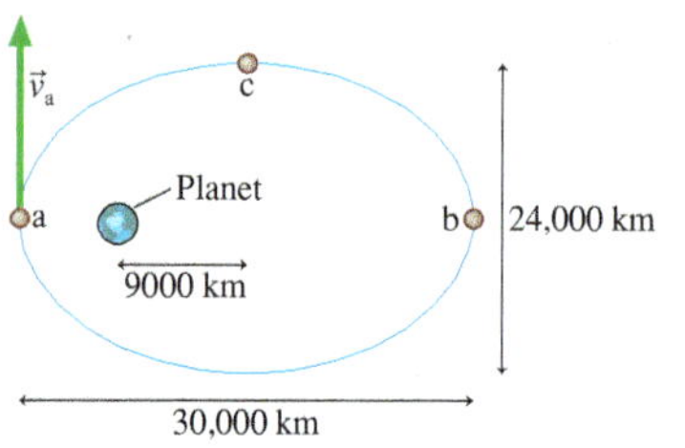

FIGURE P12.78

79. || A 10 g bullet traveling at 400 m/s strikes a 10 kg, 1.0-m-wide door at the edge opposite the hinge. The bullet embeds itself in the door, causing the door to swing open. What is the angular velocity of the door just after impact?

80. || A 200 g, 40-cm-diameter turntable rotates on frictionless bearings at 60 rpm. A 20 g block sits at the center of the turntable. A compressed spring shoots the block radially outward along a frictionless groove in the surface of the turntable. What is the turntable's rotation angular velocity when the block reaches the outer edge?

81. || A merry-go-round is a common piece of playground equipment. A 3.0-m-diameter merry-go-round with a mass of 250 kg is spinning at 20 rpm. John runs tangent to the merry-go-round at 5.0 m/s, in the same direction that it is turning, and jumps onto the outer edge. John's mass is 30 kg. What is the merry-go-round's angular velocity, in rpm, after John jumps on?
82. || A 45 kg figure skater is spinning on the toes of her skates at 1.0 rev/s. Her arms are outstretched as far as they will go. In this orientation, the skater can be modeled as a cylindrical torso (40 kg, 20 cm average diameter, 160 cm tall) plus two rod-like arms (2.5 kg each, 66 cm long) attached to the outside of the torso. The skater then raises her arms straight above her head, where she appears to be a 45 kg, 20-cm-diameter, 200-cm-tall cylinder. What is her new angular velocity, in rev/s?

Challenge Problems

83. In FIGURE CP12.83, a 200 g toy car is placed on a narrow 60-cm-diameter track with wheel grooves that keep the car going in a circle. The 1.0 kg track is free to turn on a frictionless, vertical axis. The spokes have negligible mass. After the car's switch is turned on, it soon reaches a steady speed of 0.75 m/s relative to the track. What then is the track's angular velocity, in rpm?

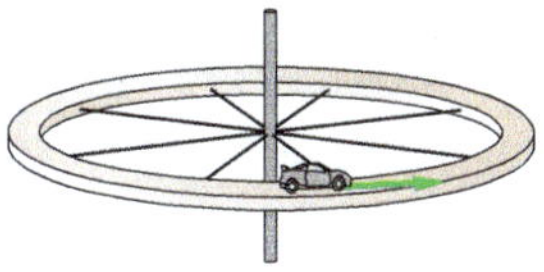

FIGURE CP12.83

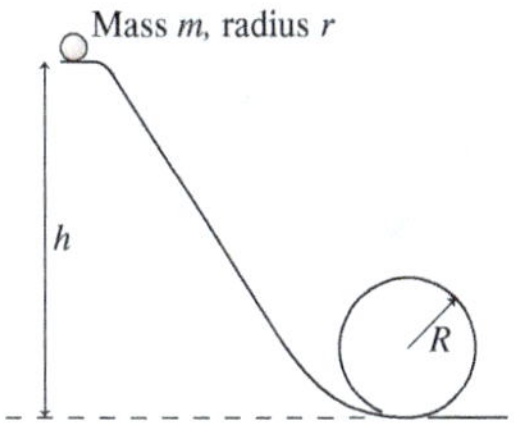

FIGURE CP12.84

84. The marble rolls down the track shown in FIGURE CP12.84 and around a loop-the-loop of radius R. The marble has mass m and radius r. What minimum height h must the track have for the marble to make it around the loop-the-loop without falling off?
85. FIGURE CP12.85 shows a triangular block of Swiss cheese sitting on a cheese board. You and your friends start to wonder what will happen if you slowly tilt the board, increasing angle θ. Emily thinks the cheese will start to slide before it topples over. Fred thinks it will topple before starting to slide. Some quick Internet research on your part reveals that the coefficient of static friction of Swiss cheese on wood is 0.90. Who is right?

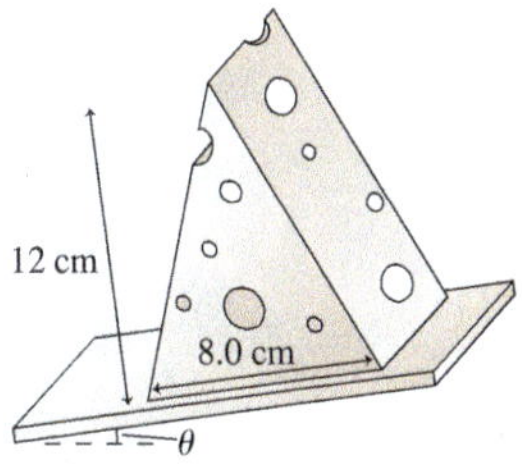

FIGURE CP12.85

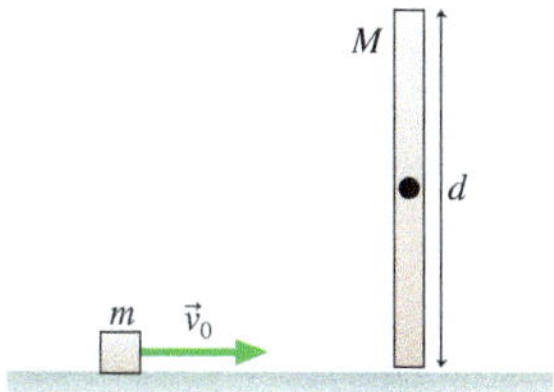

FIGURE CP12.86

86. FIGURE CP12.86 shows a cube of mass m sliding without friction at speed v_0. It undergoes a perfectly elastic collision with the bottom tip of a rod of length d and mass $M = 2m$. The rod is pivoted about a frictionless axle through its center, and initially it hangs straight down and is at rest. What is the cube's velocity—both speed and direction—after the collision?
87. A 75 g, 30-cm-long rod hangs vertically on a frictionless, horizontal axle passing through its center. A 10 g ball of clay traveling horizontally at 2.5 m/s hits and sticks to the very bottom tip of the rod. To what maximum angle, measured from vertical, does the rod (with the attached ball of clay) rotate?
88. During most of its lifetime, a star maintains an equilibrium size in which the inward force of gravity on each atom is balanced by an outward pressure force due to the heat of the nuclear reactions in the core. But after all the hydrogen "fuel" is consumed by nuclear fusion, the pressure force drops and the star undergoes a *gravitational collapse* until it becomes a *neutron star*. In a neutron star, the electrons and protons of the atoms are squeezed together by gravity until they fuse into neutrons. Neutron stars spin very rapidly and emit intense pulses of radio and light waves, one pulse per rotation. These "pulsing stars" were discovered in the 1960s and are called *pulsars*.
    a. A star with the mass ($M = 2.0 \times 10^{30}$ kg) and size ($R = 7.0 \times 10^{8}$ m) of our sun rotates once every 30 days. After undergoing gravitational collapse, the star forms a pulsar that is observed by astronomers to emit radio pulses every 0.10 s. By treating the neutron star as a solid sphere, deduce its radius.
    b. What is the speed of a point on the equator of the neutron star?

    Your answers will be somewhat too large because a star cannot be accurately modeled as a solid sphere. Even so, you will be able to show that a star, whose mass is 10^6 larger than the earth's, can be compressed by gravitational forces to a size smaller than a typical state in the United States!

STOP TO THINK ANSWERS

Stop to Think 12.1: $I_a > I_d > I_b > I_c$. The moment of inertia is smaller when the mass is more concentrated near the rotation axis.

Stop to Think 12.2: $\tau_e > \tau_a = \tau_d > \tau_b > \tau_c$. The tangential component in e is larger than 2 N.

Stop to Think 12.3: $\alpha_b > \alpha_a > \alpha_c = \alpha_d = \alpha_e$. Angular acceleration is proportional to torque and inversely proportional to the moment of inertia. The moment of inertia depends on the *square* of the radius. The tangential force component in e is the same as in d.

Stop to Think 12.4: c > d > a = b. To keep the meter stick in equilibrium, the student must supply a torque equal and opposite to the torque due to the hanging masses. Torque depends on the mass *and* on how far the mass is from the pivot point.

Stop to Think 12.5: d. There is no net torque on the bucket + rain system, so the angular momentum is conserved. The addition of mass on the outer edge of the circle increases I, so ω must decrease. Mechanical energy is not conserved because the raindrop collisions are inelastic.

13 Newton's Theory of Gravity

This beautiful galaxy consists of billions of stars orbiting the galactic center exactly as predicted by Newton's theory of gravity.

Looking Ahead The goal of Chapter 13 is to use Newton's theory of gravity to understand the motion of satellites and planets.

Copernicus and Galileo

In many ways, modern science began with Copernicus's assertion in 1543 that the planets orbit the sun rather than the sun and planets revolving around the earth.

Copernicus's ideas were confirmed a century later when Galileo, using one of the earliest telescopes, made the first detailed observations of the solar system.

In this chapter you'll learn how Newton's theory of gravity explains the motions of satellites, planets, and even the entire solar system as it revolves around the galactic center.

Kepler's Laws

Before Galileo and the telescope, Kepler used naked-eye measurements of the planets to make three major discoveries:

- The planets move in elliptical orbits.
- The planets "sweep out" equal areas in equal times.
- The square of the period is proportional to the cube of the orbit's radius.

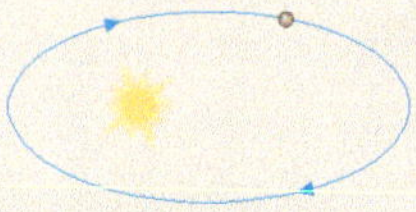

A planet's orbit is an ellipse with the sun at one focus.

Newton's Theory

Newton proposed that *any* two masses are attracted toward each other by a gravitational force

$$F_{M \text{ on } m} = F_{m \text{ on } M} = \frac{GMm}{r^2}$$

This is an *inverse-square* force law because the force depends inversely on the square of the distance between the masses.

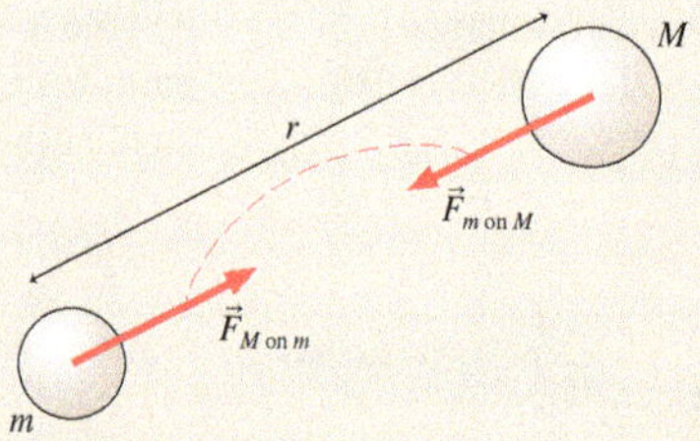

In addition, Newton proposed that his three laws of motion applied to the planets, not just to earthly objects.

We'll use Newton's theory to

- Understand the value of g, and
- Weigh the earth.

Newton's theory explains the orbits of the dust and ice particles that form Saturn's rings.

Gravitational Energy

The gravitational potential energy of Chapter 10, $U_g = mgy$, is valid only very near the surface of a planet. We'll find a more general gravitational potential energy that applies to satellites and planets.

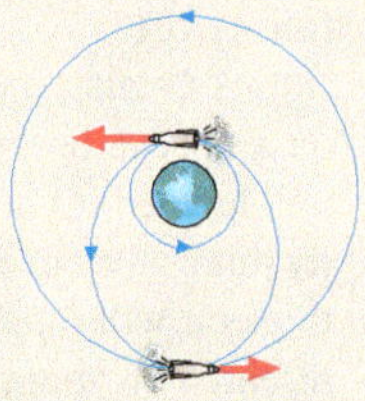

You'll learn to use energy conservation to solve orbit problems, such as how a satellite is transferred from one orbit to another.

Looking Back
Chapter 10 Potential energy and energy conservation

Orbits

We'll use Newton's theory to *derive* Kepler's three laws, providing strong evidence in favor of Newton's theory.

Communications satellites are in *geosynchronous orbits* around the earth. You'll learn how to calculate the height of these satellites.

Looking Back
Sections 8.2–8.3 Uniform circular motion and orbital motion

13.1 A Little History

The study of the structure of the universe is called **cosmology.** The ancient Greeks developed a cosmological model, with the earth at the center of the universe while the moon, the sun, the planets, and the stars were points of light turning about the earth on large "celestial spheres." This viewpoint was further expanded by the second-century Egyptian astronomer Ptolemy (the P is silent). He developed an elaborate mathematical model of the solar system that quite accurately predicted the complex planetary motions.

Then, in 1543, the medieval world was turned on its head with the publication of Nicholas Copernicus's *De Revolutionibus.* Copernicus argued that it is not the earth at rest in the center of the universe—it is the sun! Furthermore, Copernicus asserted that all of the planets revolve about the sun (hence his title) in circular orbits.

Tycho and Kepler

The greatest medieval astronomer was Tycho Brahe. For 30 years, from 1570 to 1600, Tycho compiled the most accurate astronomical observations the world had known. The invention of the telescope was still to come, but Tycho developed ingenious mechanical sighting devices that allowed him to determine the positions of stars and planets in the sky with unprecedented accuracy.

Tycho had a young mathematical assistant named Johannes Kepler. Kepler had become one of the first outspoken defenders of Copernicus, and his goal was to find evidence for circular planetary orbits in Tycho's records. To appreciate the difficulty of this task, keep in mind that Kepler was working before the development of graphs or of calculus—and certainly before calculators! His mathematical tools were algebra, geometry, and trigonometry, and he was faced with thousands upon thousands of individual observations of planetary positions measured as angles above the horizon.

Many years of work led Kepler to discover that the orbits are not circles, as Copernicus claimed, but *ellipses.* Furthermore, the speed of a planet is not constant but varies as it moves around the ellipse.

Kepler's laws, as we call them today, state that

1. Planets move in elliptical orbits, with the sun at one focus of the ellipse.
2. A line drawn between the sun and a planet sweeps out equal areas during equal intervals of time.
3. The square of a planet's orbital period is proportional to the cube of the semimajor-axis length.

FIGURE 13.1a shows that an ellipse has two *foci* (plural of *focus*), and the sun occupies one of these. The long axis of the ellipse is the *major axis,* and half the length of this axis is called the *semimajor-axis length.* As the planet moves, a line drawn from the sun to the planet "sweeps out" an area. FIGURE 13.1b shows two such areas. Kepler's discovery that the areas are equal for equal Δt implies that the planet moves faster when near the sun, slower when farther away.

All the planets except Mercury have elliptical orbits that are only very slightly distorted circles. As FIGURE 13.2 shows, a circle is an ellipse in which the two foci move to the center, effectively making one focus, and the semimajor-axis length becomes the radius. Because the mathematics of ellipses is difficult, this chapter will focus on circular orbits.

Kepler made an additional contribution that was essential to prepare the way for Newton. For Ptolemy and, later, Copernicus, the role of the sun was merely to light and warm the earth and planets. Kepler was the first to suggest that the sun was a center of force that somehow *caused* the planetary motions. Now, Kepler was working before Galileo and Newton, so he did not speak in terms of forces and centripetal accelerations. The value of his contribution was not the specific mechanism he proposed but his introduction of the idea that the sun somehow exerts forces on the planets to determine their motion.

FIGURE 13.1 The elliptical orbit of a planet about the sun.

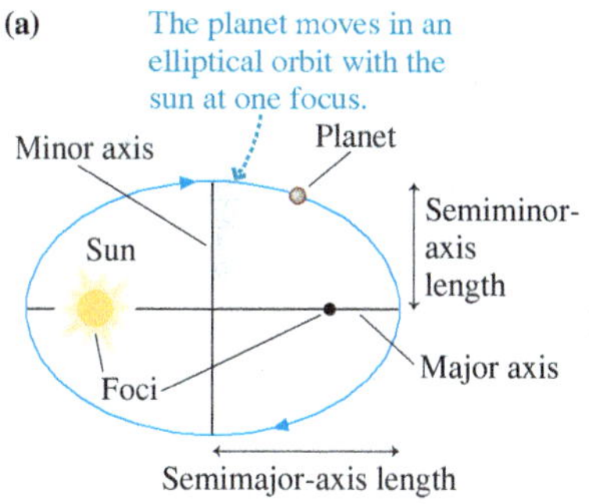

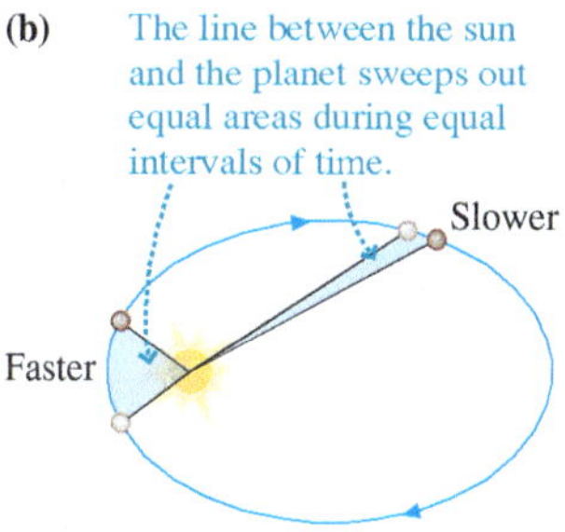

FIGURE 13.2 A circular orbit is a special case of an elliptical orbit.

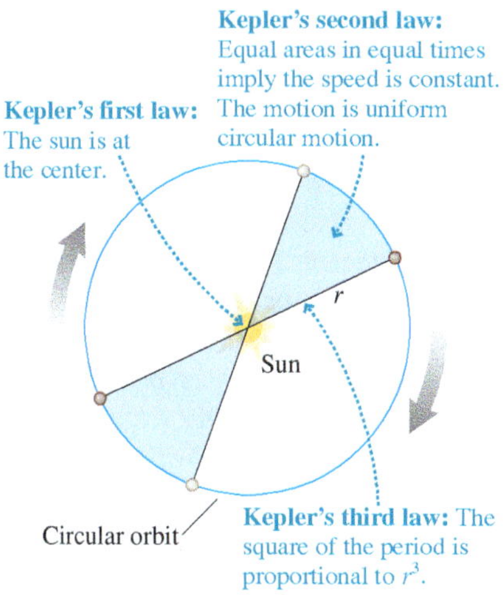

Kepler published the first two of his laws in 1609, the same year in which Galileo first turned a telescope to the heavens. Through his telescope Galileo could *see* moons orbiting Jupiter, just as Copernicus had suggested the planets orbit the sun. He could *see* that Venus has phases, like the moon, which implied its orbital motion about the sun. By the time of Galileo's death in 1642, the Copernican revolution was complete.

13.2 Isaac Newton

Isaac Newton, 1642–1727.

A popular image has Newton thinking of the idea of gravity after an apple fell on his head. This amusing story is at least close to the truth. Newton himself said that the "notion of gravitation" came to him as he "sat in a contemplative mood" and "was occasioned by the fall of an apple." It occurred to him that, perhaps, the apple was attracted to the *center* of the earth but was prevented from getting there by the earth's surface. And if the apple was so attracted, why not the moon? In other words, perhaps gravitation is a *universal* force between all objects in the universe! This is not shocking today, but no one before Newton had ever thought that the mundane motion of objects on earth had any connection at all with the stately motion of the planets through the heavens.

Newton reasoned along the following lines. Suppose the moon's circular motion around the earth is due to the pull of the earth's gravity. Then, as you learned in Chapter 8 and is shown in FIGURE 13.3, the moon must be in *free fall* with the free-fall acceleration $g_{\text{at moon}}$.

NOTE ▶ We need to be careful with notation. The symbol g_{moon} is the free-fall acceleration caused by the *moon's* gravity—that is, the acceleration of a falling object on the moon. Here we're interested in the acceleration *of* the moon by the earth's gravity, which we'll call $g_{\text{at moon}}$. ◀

FIGURE 13.3 The moon is in free fall around the earth.

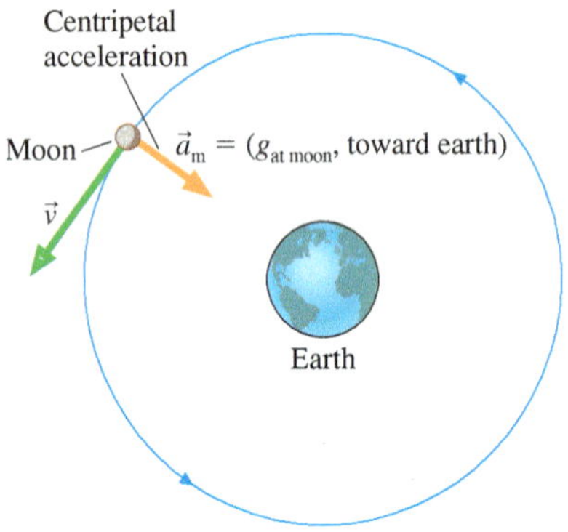

The centripetal acceleration of an object in uniform circular motion is

$$a_r = g_{\text{at moon}} = \frac{v_m^2}{r_m} \tag{13.1}$$

The moon's speed is related to the radius r_m and period T_m of its orbit by $v_m =$ circumference/period $= 2\pi r_m/T_m$. Combining these, Newton found

$$g_{\text{at moon}} = \frac{4\pi^2 r_m}{T_m^2} = \frac{4\pi^2(3.84\times10^8\text{ m})}{(2.36\times10^6\text{ s})^2} = 0.00272\text{ m/s}^2$$

Astronomical measurements had established a reasonably good value for r_m by the time of Newton, and the period $T_m = 27.3$ days was quite well known.

The moon's centripetal acceleration is significantly less than the free-fall acceleration on the earth's surface. In fact,

$$\frac{g_{\text{at moon}}}{g_{\text{on earth}}} = \frac{0.00272\text{ m/s}^2}{9.80\text{ m/s}^2} = \frac{1}{3600}$$

This is an interesting result, but it was Newton's next step that was critical. He compared the radius of the moon's orbit to the radius of the earth:

$$\frac{r_m}{R_e} = \frac{3.84\times10^8\text{ m}}{6.37\times10^6\text{ m}} = 60.2$$

NOTE ▶ We'll use a lowercase r, as in r_m, to indicate the radius of an orbit. We'll use an uppercase R, as in R_e, to indicate the radius of a star or planet. ◀

I deduced that the forces which keep the planets in their orbs must be reciprocally as the squares of their distances from the centers about which they revolve; and thereby compared the force requisite to keep the Moon in her orb with the force of gravity at the surface of the Earth; and found them answer pretty nearly.
Isaac Newton

Newton recognized that $(60.2)^2$ is almost exactly 3600. Thus, he reasoned:

- If g has the value 9.80 at the earth's surface, and
- If the force of gravity and g decrease in size depending inversely on the square of the distance from the center of the earth,
- Then g will have exactly the value it needs at the distance of the moon to cause the moon to orbit the earth with a period of 27.3 days.

His two ratios were not identical (because the earth isn't a perfect sphere and the moon's orbit isn't a perfect circle), but he found them to "answer pretty nearly" and knew that he had to be on the right track.

STOP TO THINK 13.1 A satellite orbits the earth with constant speed at a height above the surface equal to the earth's radius. The magnitude of the satellite's acceleration is

a. $4g_{\text{on earth}}$ b. $2g_{\text{on earth}}$ c. $g_{\text{on earth}}$
d. $\frac{1}{2}g_{\text{on earth}}$ e. $\frac{1}{4}g_{\text{on earth}}$ f. 0

13.3 Newton's Law of Gravity

Newton proposed that *every* object in the universe attracts *every other* object with a force that is

1. Inversely proportional to the square of the distance between the objects.
2. Directly proportional to the product of the masses of the two objects.

To make these ideas more specific, FIGURE 13.4 shows masses m_1 and m_2 separated by distance r. Each mass exerts an attractive force on the other, a force that we call the **gravitational force.** These two forces form an action/reaction pair, so $\vec{F}_{1 \text{ on } 2}$ is equal and opposite to $\vec{F}_{2 \text{ on } 1}$. The magnitude of the forces is given by Newton's law of gravity.

FIGURE 13.4 The gravitational forces on masses m_1 and m_2.

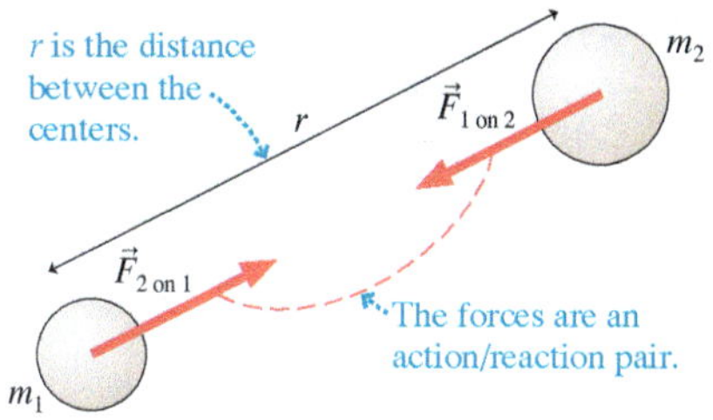

Newton's law of gravity If two objects with masses m_1 and m_2 are a distance r apart, the objects exert attractive forces on each other of magnitude

$$F_{1 \text{ on } 2} = F_{2 \text{ on } 1} = \frac{Gm_1 m_2}{r^2} \tag{13.2}$$

The forces are directed along the straight line joining the two objects.

The constant G, called the **gravitational constant,** is a proportionality constant necessary to relate the masses, measured in kilograms, to the force, measured in newtons. In the SI system of units, G has the value

$$G = 6.67 \times 10^{-11} \text{ N m}^2/\text{kg}^2$$

FIGURE 13.5 is a graph of the gravitational force as a function of the distance between the two masses. As you can see, an inverse-square force decreases rapidly.

Strictly speaking, Equation 13.2 is valid only for particles. However, Newton was able to show that this equation also applies to spherical objects, such as planets, if r is the distance between their centers. Our intuition and common sense suggest this to us, as they did to Newton. The rather difficult proof is not essential, so we will omit it.

FIGURE 13.5 The gravitational force is an inverse-square force.

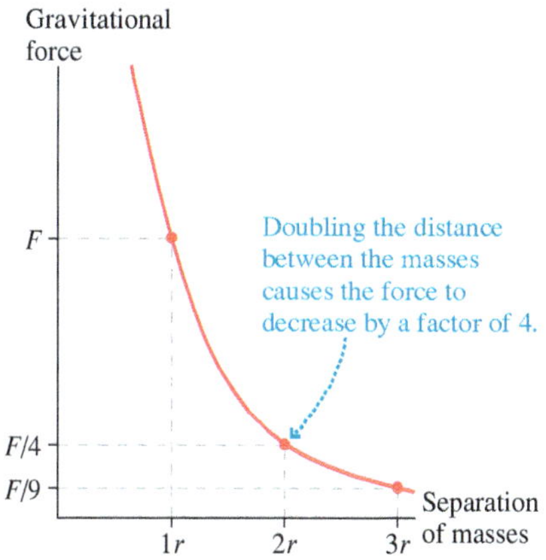

Gravitational Force and Weight

Knowing G, we can calculate the size of the gravitational force. Consider two 1.0 kg masses that are 1.0 m apart. According to Newton's law of gravity, these two masses exert an attractive gravitational force on each other of magnitude

$$F_{1 \text{ on } 2} = F_{2 \text{ on } 1} = \frac{Gm_1 m_2}{r^2}$$

$$= \frac{(6.67 \times 10^{-11} \text{ N m}^2/\text{kg}^2)(1.0 \text{ kg})(1.0 \text{ kg})}{(1.0 \text{ m})^2} = 6.67 \times 10^{-11} \text{ N}$$

This is an exceptionally tiny force, especially when compared to the gravitational force of the entire earth on each mass: $F_G = mg = 9.8$ N.

The fact that the gravitational force between two ordinary-size objects is so small is the reason we are not aware of it. As you sit there reading, you are being attracted to this book, to the person sitting next to you, and to every object around you, but the forces are so tiny in comparison to the normal forces and friction forces acting on you that they are completely undetectable. Only when one (or both) of the masses is exceptionally large—planet-size—does the force of gravity become important.

We find a more respectable result if we calculate the force *of the earth* on a 1.0 kg mass at the earth's surface:

$$F_{\text{earth on 1 kg}} = \frac{GM_e m_{1\text{ kg}}}{R_e^2}$$

$$= \frac{(6.67 \times 10^{-11}\text{ N m}^2/\text{kg}^2)(5.98 \times 10^{24}\text{ kg})(1.0\text{ kg})}{(6.37 \times 10^6\text{ m})^2} = 9.8\text{ N}$$

The dynamics of stellar motions, spanning many thousands of light years, are governed by Newton's law of gravity.

A galaxy of $\approx 10^{11}$ stars spanning a distance greater than 100,000 light years.

where the distance between the mass and the center of the earth is the earth's radius. The earth's mass M_e and radius R_e were taken from Table 13.2 in Section 13.6. This table, which is also printed inside the back cover of the book, contains astronomical data that will be used for examples and homework.

The force $F_{\text{earth on 1 kg}} = 9.8$ N is exactly the weight of a stationary 1.0 kg mass: $F_G = mg = 9.8$ N. Is this a coincidence? Of course not. Weight—the upward force of a spring scale—exactly balances the downward gravitational force, so numerically they must be equal.

Although weak, gravity is a *long-range* force. No matter how far apart two objects may be, there is a gravitational attraction between them given by Equation 13.2. Consequently, gravity is the most ubiquitous force in the universe. It not only keeps your feet on the ground, it also keeps the earth orbiting the sun, the solar system orbiting the center of the Milky Way galaxy, and the entire Milky Way galaxy performing an intricate orbital dance with other galaxies making up what is called the "local cluster" of galaxies.

The Principle of Equivalence

Newton's law of gravity depends on a rather curious assumption. The concept of *mass* was introduced in Chapter 5 by considering the relationship between force and acceleration. The *inertial mass* of an object, which is the mass that appears in Newton's second law, is found by measuring the object's acceleration a in response to force F:

$$m_{\text{inert}} = \text{inertial mass} = \frac{F}{a} \tag{13.3}$$

Gravity plays no role in this definition of mass.

The quantities m_1 and m_2 in Newton's law of gravity are being used in a very different way. Masses m_1 and m_2 govern the strength of the gravitational attraction between two objects. The mass used in Newton's law of gravity is called the **gravitational mass.** The gravitational mass of an object can be determined by measuring the attractive force exerted on it by another mass M a distance r away:

$$m_{\text{grav}} = \text{gravitational mass} = \frac{r^2 F_{M\text{ on }m}}{GM} \tag{13.4}$$

Acceleration does not enter into the definition of the gravitational mass.

These are two very different concepts of mass. Yet Newton, in his theory of gravity, asserts that the inertial mass in his second law is the very same mass that governs the strength of the gravitational attraction between two objects. The assertion that $m_{\text{grav}} = m_{\text{inert}}$ is called the **principle of equivalence.** It says that inertial mass is *equivalent to* gravitational mass.

As a hypothesis about nature, the principle of equivalence is subject to experimental verification or disproof. Many exceptionally clever experiments have looked for any difference between the gravitational mass and the inertial mass, and they have shown that any difference, if it exists at all, is less than 10 parts in a trillion! As far as we know today, the gravitational mass and the inertial mass are exactly the same thing.

But why should a quantity associated with the dynamics of motion, relating force to acceleration, have anything at all to do with the gravitational attraction? This is a question that intrigued Einstein and eventually led to his general theory of relativity, the theory about curved spacetime and black holes. General relativity is beyond the scope of this textbook, but it explains the principle of equivalence as a property of space itself.

Newton's Theory of Gravity

Newton's theory of gravity is more than just Equation 13.2. The *theory* of gravity consists of:

1. A specific force law for gravity, given by Equation 13.2, *and*
2. The principle of equivalence, *and*
3. An assertion that Newton's three laws of motion are universally applicable. These laws are as valid for heavenly bodies, the planets and stars, as for earthly objects.

Consequently, everything we have learned about forces, motion, and energy is relevant to the dynamics of satellites, planets, and galaxies.

STOP TO THINK 13.2 The figure shows a binary star system. The mass of star 2 is twice the mass of star 1. Compared to $\vec{F}_{1 \text{ on } 2}$, the magnitude of the force $\vec{F}_{2 \text{ on } 1}$ is

a. Four times as big.
b. Twice as big.
c. The same size.
d. Half as big.
e. One-quarter as big.

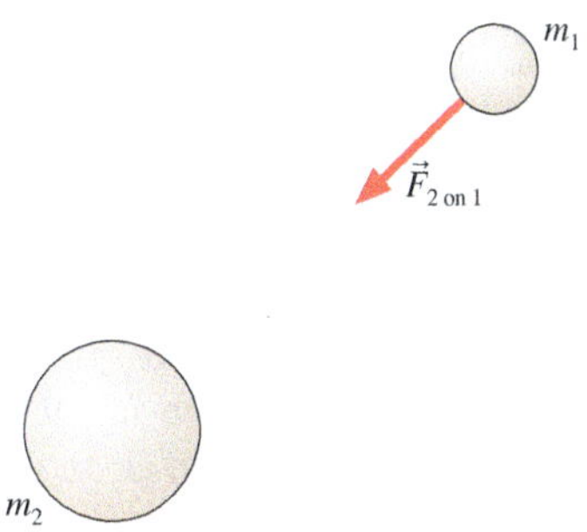

13.4 Little *g* and Big *G*

The familiar equation $F_G = mg$ works well when an object is on the surface of a planet, but mg will not help us find the force exerted on the same object if it is in orbit around the planet. Neither can we use mg to find the force of attraction between the earth and the moon. Newton's law of gravity provides a more fundamental starting point because it describes a *universal* force that exists between all objects.

To illustrate the connection between Newton's law of gravity and the familiar $F_G = mg$, FIGURE 13.6 shows an object of mass m on the surface of Planet X. Planet X inhabitant Mr. Xhzt, standing on the surface, finds that the downward gravitational force is $F_G = mg_X$, where g_X is the free-fall acceleration on Planet X.

We, taking a more cosmic perspective, reply, "Yes, that is the force *because* of a universal force of attraction between your planet and the object. The size of the force is determined by Newton's law of gravity."

We and Mr. Xhzt are both correct. Whether you think locally or globally, we and Mr. Xhzt must arrive at the *same numerical value* for the magnitude of the force. Suppose an object of mass m is on the surface of a planet of mass M and radius R. The local gravitational force is

$$F_G = mg_{\text{surface}} \tag{13.5}$$

where g_{surface} is the acceleration due to gravity at the planet's surface. The force of gravitational attraction for an object on the surface ($r = R$), as given by Newton's law of gravity, is

$$F_{M \text{ on } m} = \frac{GMm}{R^2} \tag{13.6}$$

FIGURE 13.6 Weighing an object of mass m on Planet X.

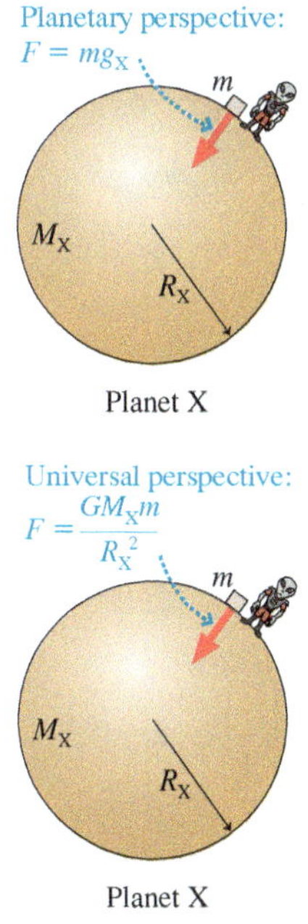

Because these are two names and two expressions for the same force, we can equate the right-hand sides to find that

$$g_{\text{surface}} = \frac{GM}{R^2} \tag{13.7}$$

We have used Newton's law of gravity to *predict* the value of g at the surface of a planet. The value depends on the mass and radius of the planet as well as on the value of G, which establishes the overall strength of the gravitational force.

The expression for g_{surface} in Equation 13.7 is valid for any planet or star. Using the mass and radius of Mars, we can predict the Martian value of g:

$$g_{\text{Mars}} = \frac{GM_{\text{Mars}}}{R_{\text{Mars}}{}^2} = \frac{(6.67 \times 10^{-11}\ \text{N m}^2/\text{kg}^2)(6.42 \times 10^{23}\ \text{kg})}{(3.37 \times 10^6\ \text{m})^2} = 3.8\ \text{m/s}^2$$

NOTE ▶ We noted in Chapter 6 that measured values of g are very slightly smaller on a rotating planet. We'll ignore rotation in this chapter. ◀

Decrease of *g* with Distance

Equation 13.7 gives g_{surface} at the surface of a planet. More generally, imagine an object of mass m at distance $r > R$ from the center of a planet. Further, suppose that gravity from the planet is the only force acting on the object. Then its acceleration, the free-fall acceleration, is given by Newton's second law:

$$g = \frac{F_{M \text{ on } m}}{m} = \frac{GM}{r^2} \tag{13.8}$$

This more general result agrees with Equation 13.7 if $r = R$, but it allows us to determine the "local" free-fall acceleration at distances $r > R$. Equation 13.8 expresses Newton's discovery, with regard to the moon, that g decreases inversely with the square of the distance.

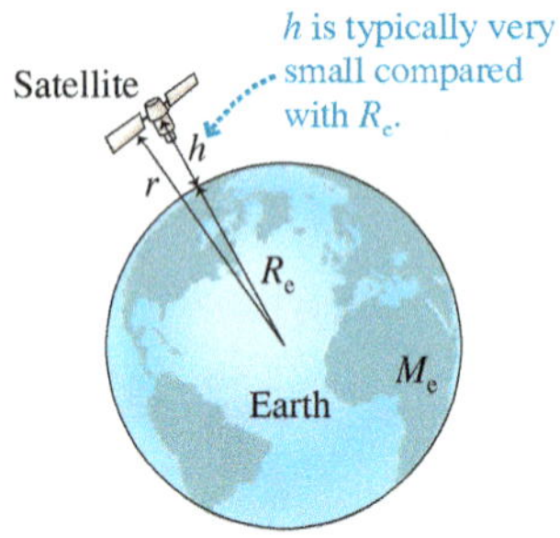

FIGURE 13.7 A satellite orbits the earth at height h.

FIGURE 13.7 shows a satellite orbiting at height h above the earth's surface. Its distance from the center of the earth is $r = R_e + h$. Most people have a mental image that satellites orbit "far" from the earth, but in reality h is typically 200 miles $\approx 3 \times 10^5$ m, while $R_e = 6.37 \times 10^6$ m. Thus the satellite is barely "skimming" the earth at a height only about 5% of the earth's radius!

The value of g at height h above the earth (i.e., above sea level) is

$$g = \frac{GM_e}{(R_e + h)^2} = \frac{GM_e}{R_e^2(1 + h/R_e)^2} = \frac{g_{\text{earth}}}{(1 + h/R_e)^2} \tag{13.9}$$

where $g_{\text{earth}} = 9.83\ \text{m/s}^2$ is the value calculated from Equation 13.7 for $h = 0$ on a nonrotating earth. Table 13.1 shows the value of g evaluated at several values of h.

TABLE 13.1 Variation of g with height above the ground

Height h	Example	g (m/s^2)
0 m	ground	9.83
4500 m	Mt. Whitney	9.82
10,000 m	jet airplane	9.80
300,000 m	space shuttle	8.90
35,900,000 m	communications satellite	0.22

NOTE ▶ The free-fall acceleration of a satellite such as the space shuttle is only slightly less than the ground-level value. An object in orbit is not "weightless" because there is no gravity in space but because it is in free fall, as you learned in Chapter 8. ◀

Weighing the Earth

We can predict g if we know the earth's mass. But how do we know the value of M_e? We cannot place the earth on a giant pan balance, so how is its mass known? Furthermore, how do we know the value of G?

Newton did not know the value of G. He could say that the gravitational force is proportional to the product $m_1 m_2$ and inversely proportional to r^2, but he had no means of knowing the value of the proportionality constant.

Determining G requires a *direct* measurement of the gravitational force between two known masses at a known separation. The small size of the gravitational force between ordinary-size objects makes this quite a feat. Yet the English scientist Henry Cavendish came up with an ingenious way of doing so with a device called a *torsion balance*. Two fairly small masses m, typically about 10 g, are placed on the ends of a lightweight rod. The rod is hung from a thin fiber, as shown in FIGURE 13.8a, and allowed to reach equilibrium.

If the rod is then rotated slightly and released, a *restoring force* will return it to equilibrium. This is analogous to displacing a spring from equilibrium, and in fact the restoring force and the angle of displacement obey a version of Hooke's law: $F_{restore} = k\Delta\theta$. The "torsion constant" k can be determined by timing the period of oscillations. Once k is known, a force that twists the rod slightly away from equilibrium can be measured by the product $k\Delta\theta$. It is possible to measure very small angular deflections, so this device can be used to determine very small forces.

Two larger masses M (typically lead spheres with $M \approx 10$ kg) are then brought close to the torsion balance, as shown in FIGURE 13.8b. The gravitational attraction that they exert on the smaller hanging masses causes a very small but measurable twisting of the balance, enough to measure $F_{M \text{ on } m}$. Because m, M, and r are all known, Cavendish was able to determine G from

$$G = \frac{F_{M \text{ on } m} r^2}{Mm} \tag{13.10}$$

FIGURE 13.8 Cavendish's experiment to measure G.

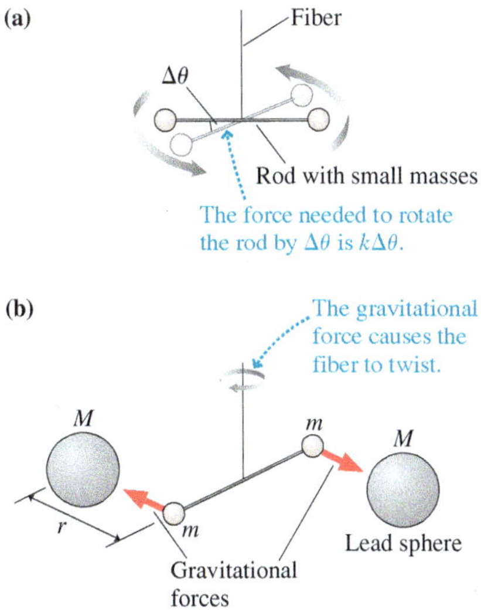

His first results were not highly accurate, but improvements over the years in this and similar experiments have produced the value of G accepted today.

With an independently determined value of G, we can return to Equation 13.7 to find

$$M_e = \frac{g_{earth} R_e^2}{G} \tag{13.11}$$

We have weighed the earth! The value of g_{earth} at the earth's surface is known with great accuracy from kinematics experiments. The earth's radius R_e is determined by surveying techniques. Combining our knowledge from these very different measurements has given us a way to determine the mass of the earth.

The gravitational constant G is what we call a *universal constant*. Its value establishes the strength of one of the fundamental forces of nature. As far as we know, the gravitational force between two masses would be the same anywhere in the universe. Universal constants tell us something about the most basic and fundamental properties of nature. You will soon meet other universal constants.

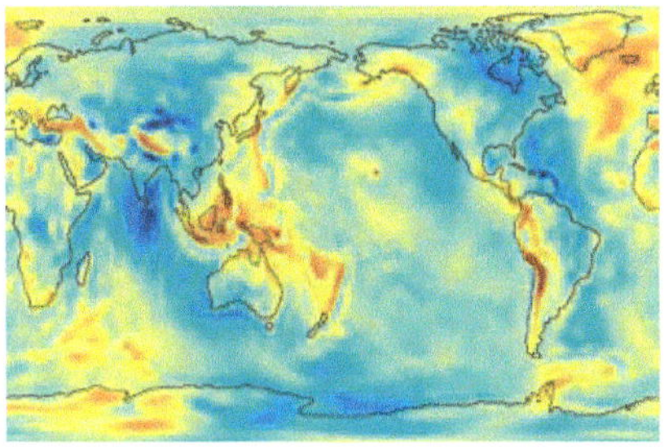

The free-fall acceleration varies slightly due to mountains and to variation in the density of the earth's crust. This map shows the *gravitational anomaly*, with red regions of slightly stronger gravity and blue regions of slightly weaker gravity. The variation is tiny, less than 0.001 m/s^2.

STOP TO THINK 13.3 A planet has four times the mass of the earth, but the acceleration due to gravity on the planet's surface is the same as on the earth's surface. The planet's radius is

a. $4R_e$ b. $2R_e$ c. R_e d. $\frac{1}{2}R_e$ e. $\frac{1}{4}R_e$

13.5 Gravitational Potential Energy

Gravitational problems are ideal for the conservation-law tools we developed in Chapters 9 through 11. Because gravity is the only force, and it is a conservative force, both the momentum and the mechanical energy of the system $m_1 + m_2$ are conserved. To employ conservation of energy, however, we need to determine an appropriate form for the gravitational potential energy for two interacting masses.

The definition of potential energy that we developed in Chapter 11 is

$$\Delta U = U_f - U_i = -W_c(\mathrm{i} \rightarrow \mathrm{f}) \tag{13.12}$$

where $W_c(\mathrm{i} \rightarrow \mathrm{f})$ is the work done by a conservative force as a particle moves from position i to position f. For a flat earth, we used $F = -mg$ and the choice that $U = 0$ at the surface ($y = 0$) to arrive at the now-familiar $U_g = mgy$. This result for U_g is valid only for $y \ll R_e$, when the earth's curvature and size are not apparent. We now need to find an expression for the gravitational potential energy of masses that interact over *large* distances.

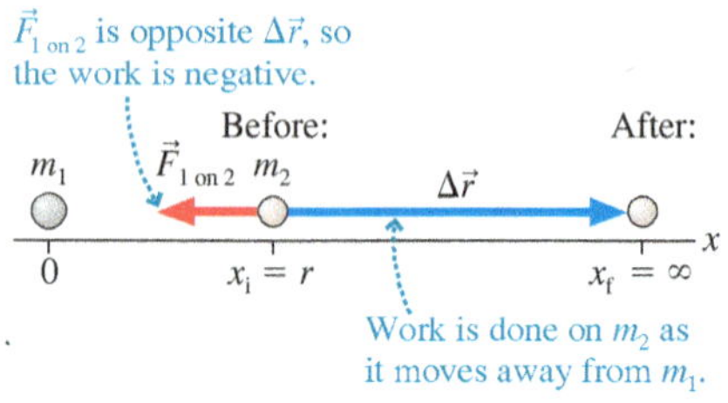

FIGURE 13.9 Calculating the work done by the gravitational force as mass m_2 moves from r to ∞.

FIGURE 13.9 shows two particles of mass m_1 and m_2. Let's calculate the work done on mass m_2 by the conservative force $\vec{F}_{1\text{ on }2}$ as m_2 moves from an initial position at distance r to a final position very far away. The force, which points to the left, is opposite the displacement; hence this force does *negative* work. Consequently, due to the minus sign in Equation 13.12, ΔU is *positive.* A pair of masses *gains* potential energy as the masses move farther apart, just as a particle near the earth's surface gains potential energy as it moves to a higher altitude.

We can establish a coordinate system with m_1 at the origin and m_2 moving along the x-axis. The gravitational force is a variable force, so we need the full definition of work:

$$W(\mathrm{i} \rightarrow \mathrm{f}) = \int_{x_i}^{x_f} F_x\, dx \tag{13.13}$$

$\vec{F}_{1\text{ on }2}$ points toward the left, so its x-component is $(F_{1\text{ on }2})_x = -Gm_1m_2/x^2$. As mass m_2 moves from $x_i = r$ to $x_f = \infty$, the potential energy changes by

$$\begin{aligned}\Delta U = U_{\text{at }\infty} - U_{\text{at }r} &= -\int_r^\infty (F_{1\text{ on }2})_x\, dx = -\int_r^\infty \left(\frac{-Gm_1m_2}{x^2}\right) dx \\ &= +Gm_1m_2 \int_r^\infty \frac{dx}{x^2} = -\frac{Gm_1m_2}{x}\bigg|_r^\infty = \frac{Gm_1m_2}{r}\end{aligned} \tag{13.14}$$

NOTE ▶ We chose to integrate along the x-axis, but the fact that gravity is a conservative force means that ΔU will have this value if m_2 moves from r to ∞ along *any* path. ◀

To proceed further, we need to choose the point where $U = 0$. We would like our choice to be valid for any star or planet, regardless of its mass and radius. This will be the case if we set $U = 0$ at the point where the interaction between the masses vanishes. According to Newton's law of gravity, the strength of the interaction is zero only when $r = \infty$. Two masses infinitely far apart will have no tendency, or potential, to move together, so we will *choose* to place the zero point of potential energy at $r = \infty$. That is, $U_{\text{at }\infty} = 0$.

This choice gives us the gravitational potential energy of masses m_1 and m_2:

FIGURE 13.10 The gravitational potential-energy curve.

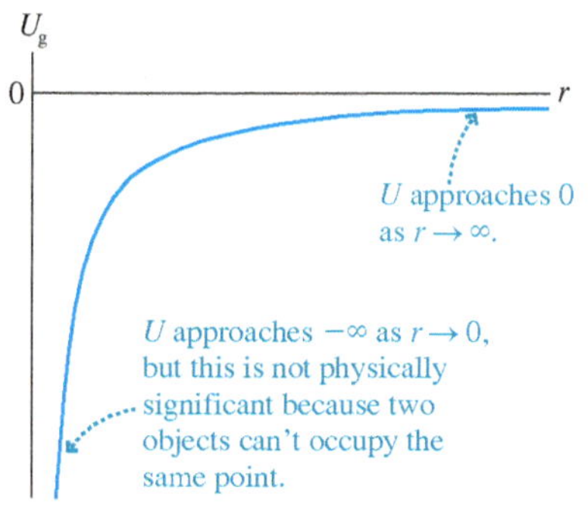

$$U_g = -\frac{Gm_1m_2}{r} \tag{13.15}$$

This is the potential energy of masses m_1 and m_2 when their *centers* are separated by distance r. FIGURE 13.10 is a graph of U_g as a function of the distance r between the masses. Notice that it asymptotically approaches 0 as $r \rightarrow \infty$.

NOTE ▶ Although Equation 13.15 looks rather similar to Newton's law of gravity, it depends only on $1/r$, *not* on $1/r^2$. ◀

It may seem disturbing that the potential energy is negative, but we encountered similar situations in Chapter 10. All a negative potential energy means is that the potential energy of the two masses at separation r is *less* than their potential energy at infinite separation. Only the *change* in U has physical significance, and the change will be the same no matter where we place the zero of potential energy.

To illustrate, suppose two masses a distance r_1 apart are released from rest. How will they move? From a force perspective, you would note that each mass experiences an attractive force and accelerates toward the other. The energy perspective of FIGURE 13.11 tells the same thing. By moving toward smaller r (that is, $r_1 \rightarrow r_2$), the system *loses* potential energy and *gains* kinetic energy while conserving E_{mech}. The system is "falling downhill," although in a more general sense than we think about on a flat earth.

FIGURE 13.11 Two masses gain kinetic energy as their separation decreases.

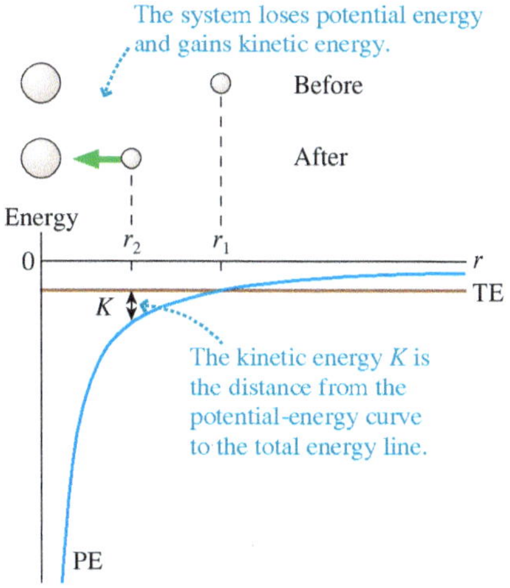

EXAMPLE 13.1 Crashing into the sun

Suppose the earth suddenly came to a halt and ceased revolving around the sun. The gravitational force would then pull it directly into the sun. What would be the earth's speed as it crashed?

MODEL Model the earth and the sun as spherical masses. This is an isolated system, so its mechanical energy is conserved.

VISUALIZE FIGURE 13.12 is a before-and-after pictorial representation for this gruesome cosmic event. The "crash" occurs as the earth touches the sun, at which point the distance between their centers is $r_2 = R_s + R_e$. The initial separation r_1 is the radius of the earth's *orbit* about the sun, not the radius of the earth.

FIGURE 13.12 Before-and-after pictorial representation of the earth crashing into the sun (not to scale).

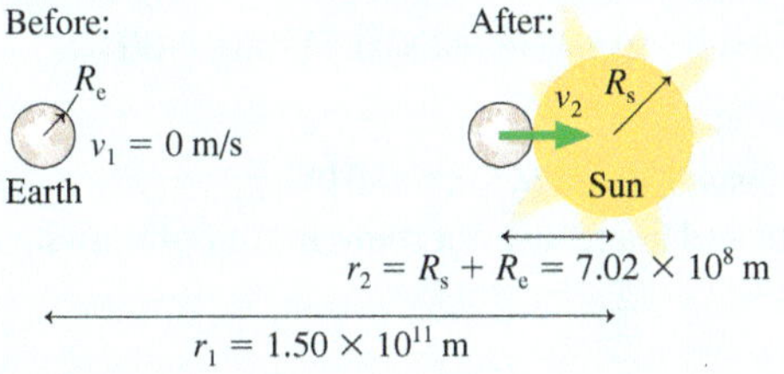

SOLVE Strictly speaking, the kinetic energy is the sum $K = K_{earth} + K_{sun}$. However, the sun is so much more massive than the earth that the lightweight earth does almost all of the moving. It is a reasonable approximation to consider the sun as remaining at rest. In that case, the energy conservation equation $K_2 + U_2 = K_1 + U_1$ is

$$\frac{1}{2}M_e v_2^2 - \frac{GM_s M_e}{R_s + R_e} = 0 - \frac{GM_s M_e}{r_1}$$

This is easily solved for the earth's speed at impact. Using data from Table 13.2, we find

$$v_2 = \sqrt{2GM_s\left(\frac{1}{R_s + R_e} - \frac{1}{r_1}\right)} = 6.13 \times 10^5 \text{ m/s}$$

ASSESS The earth would be really flying along at over 1 million miles per hour as it crashed into the sun! It is worth noting that we do not have the mathematical tools to solve this problem using Newton's second law because the acceleration is not constant. But the solution is straightforward when we use energy conservation.

EXAMPLE 13.2 Escape speed

A 1000 kg rocket is fired straight away from the surface of the earth. What speed does the rocket need to "escape" from the gravitational pull of the earth and never return? Assume a nonrotating earth.

MODEL In a simple universe, consisting of only the earth and the rocket, an insufficient launch speed will cause the rocket eventually to fall back to earth. Once the rocket finally slows to a halt, gravity will ever so slowly pull it back. The only way the rocket can escape is to never stop ($v = 0$) and thus never have a turning point! That is, the rocket must continue moving away from the earth forever. The *minimum* launch speed for escape, which is called the **escape speed,** will cause the rocket to stop ($v = 0$) only as it reaches $r = \infty$. Now ∞, of course, is not a "place," so a statement like this means that we want the rocket's speed to approach $v = 0$ asymptotically as $r \rightarrow \infty$.

VISUALIZE FIGURE 13.13 is a before-and-after pictorial representation.

FIGURE 13.13 Pictorial representation of a rocket launched with sufficient speed to escape the earth's gravity.

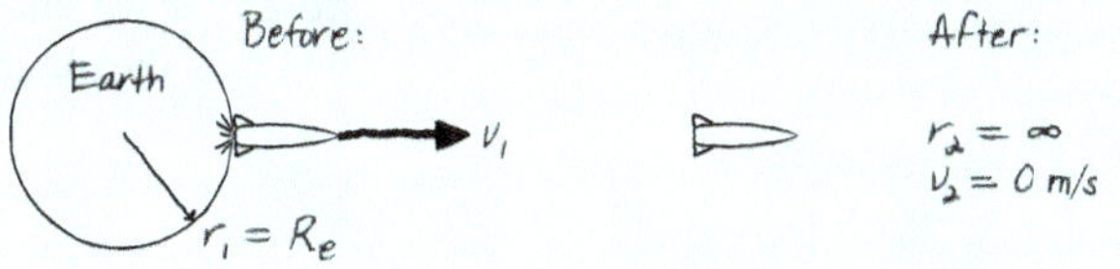

Continued

SOLVE Energy conservation $K_2 + U_2 = K_1 + U_1$ is

$$0 + 0 = \frac{1}{2}mv_1^2 - \frac{GM_e m}{R_e}$$

where we used the fact that both the kinetic and potential energy are zero at $r = \infty$. Thus the escape speed is

$$v_{escape} = v_1 = \sqrt{\frac{2GM_e}{R_e}} = 11{,}200 \text{ m/s} \approx 25{,}000 \text{ mph}$$

ASSESS The problem was mathematically easy; the difficulty was deciding how to interpret it. That is why—as you have now seen many times—the "physics" of a problem consists of thinking, interpreting, and modeling. We will see variations on this problem in the future, with both gravity and electricity, so you might want to review the *reasoning* involved. Notice that the answer does *not* depend on the rocket's mass, so this is the escape speed for any object.

The Flat-Earth Approximation

FIGURE 13.14 We can treat the earth as flat if $y \ll R_e$.

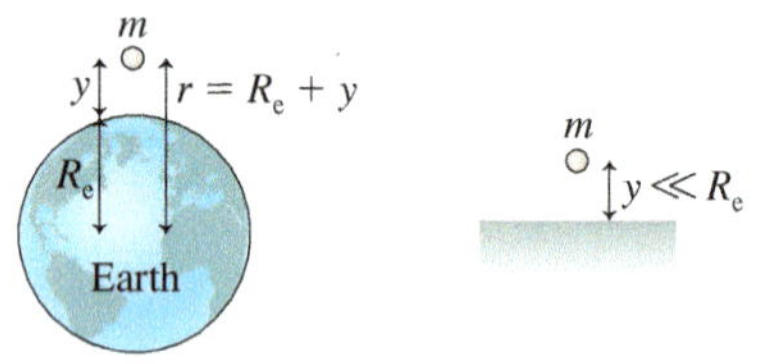

How does Equation 13.15 for the gravitational potential energy relate to our previous use of $U_g = mgy$ on a flat earth? **FIGURE 13.14** shows an object of mass m located at height y above the surface of the earth. The object's distance from the earth's center is $r = R_e + y$ and its gravitational potential energy is

$$U_g = -\frac{GM_e m}{r} = -\frac{GM_e m}{R_e + y} = -\frac{GM_e m}{R_e(1 + y/R_e)} \tag{13.16}$$

where, in the last step, we factored R_e out of the denominator.

Suppose the object is very close to the earth's surface ($y \ll R_e$). In that case, the ratio $y/R_e \ll 1$. There is an approximation you will learn about in calculus, called the *binomial approximation*, that says

$$(1 + x)^n \approx 1 + nx \qquad \text{if } x \ll 1 \tag{13.17}$$

As an illustration, you can easily use your calculator to find that $1/1.01 = 0.9901$, to four significant figures. But suppose you wrote $1.01 = 1 + 0.01$. You could then use the binomial approximation to calculate

$$\frac{1}{1.01} = \frac{1}{1 + 0.01} = (1 + 0.01)^{-1} \approx 1 + (-1)(0.01) = 0.9900$$

You can see that the approximate answer is off by only 0.01%.

If we call $y/R_e = x$ in Equation 13.16 and use the binomial approximation, with $n = -1$, we find

$$U_g(\text{if } y \ll R_e) \approx -\frac{GM_e m}{R_e}\left(1 - \frac{y}{R_e}\right) = -\frac{GM_e m}{R_e} + m\left(\frac{GM_e}{R_e^2}\right)y \tag{13.18}$$

Now the first term is just the gravitational potential energy U_0 when the object is at ground level ($y = 0$). In the second term, you can recognize $GM_e/R_e^2 = g_{earth}$ from the definition of g in Equation 13.7. Thus we can write Equation 13.18 as

$$U_g(\text{if } y \ll R_e) = U_0 + mg_{earth}\, y \tag{13.19}$$

Although we chose U_g to be zero when $r = \infty$, we are always free to change our minds. If we change the zero point of potential energy to be $U_0 = 0$ at the surface, which is the choice we made in Chapter 10, then Equation 13.19 becomes

$$U_g(\text{if } y \ll R_e) = mg_{earth}\, y \tag{13.20}$$

We can sleep easier knowing that Equation 13.15 for the gravitational potential energy is consistent with our earlier "flat-earth" expression for the potential energy.

EXAMPLE 13.3 **The speed of a satellite**

A less-than-successful inventor wants to launch small satellites into orbit by launching them straight up from the surface of the earth at very high speed.

a. With what speed should he launch the satellite if it is to have a speed of 500 m/s at a height of 400 km? Ignore air resistance.
b. By what percentage would your answer be in error if you used a flat-earth approximation?

MODEL Mechanical energy is conserved if we ignore drag.

VISUALIZE FIGURE 13.15 shows a pictorial representation.

FIGURE 13.15 Pictorial representation of a satellite launched straight up.

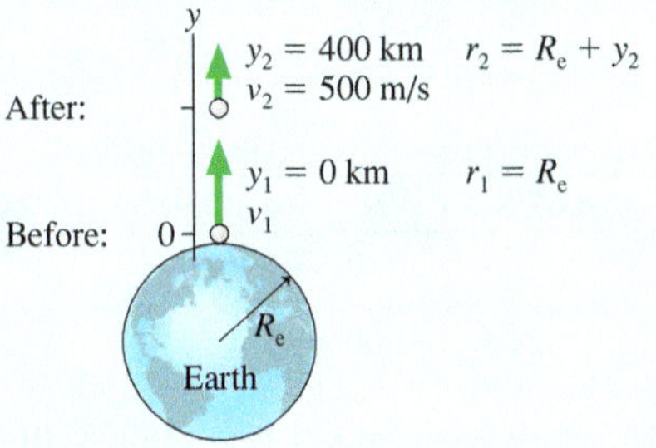

SOLVE a. Although the height is exaggerated in the figure, 400 km = 400,000 m is high enough that we cannot ignore the earth's spherical shape. The energy conservation equation $K_2 + U_2 = K_1 + U_1$ is

$$\frac{1}{2}mv_2^2 - \frac{GM_e m}{R_e + y_2} = \frac{1}{2}mv_1^2 - \frac{GM_e m}{R_e + y_1}$$

where we've written the distance between the satellite and the earth's center as $r = R_e + y$. The initial height is $y_1 = 0$. Notice that the satellite mass m cancels and is not needed. Solving for the launch speed, we have

$$v_1 = \sqrt{v_2^2 + 2GM_e\left(\frac{1}{R_e} - \frac{1}{R_e + y_2}\right)} = 2770 \text{ m/s}$$

This is about 6000 mph, much less than the escape speed.

b. The calculation is the same in the flat-earth approximation except that we use $U_g = mgy$. Thus

$$\frac{1}{2}mv_2^2 + mgy_2 = \frac{1}{2}mv_1^2 + mgy_1$$

$$v_1 = \sqrt{v_2^2 + 2gy_2} = 2840 \text{ m/s}$$

The flat-earth value of 2840 m/s is 70 m/s too big. The error, as a percentage of the correct 2770 m/s, is

$$\text{error} = \frac{70}{2770} \times 100 = 2.5\%$$

ASSESS The true speed is less than the flat-earth approximation because the force of gravity decreases with height. Launching a rocket against a decreasing force takes less effort than it would with the flat-earth force of mg at all heights.

STOP TO THINK 13.4 Rank in order, from largest to smallest, the absolute values of the gravitational potential energies of these pairs of masses. The numbers give the relative masses and distances.

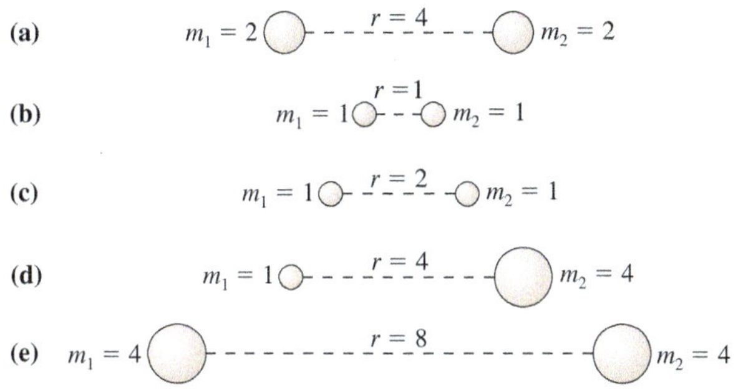

13.6 Satellite Orbits and Energies

Solving Newton's second law to find the trajectory of a mass moving under the influence of gravity is mathematically beyond this textbook. It turns out that the solution is a set of elliptical orbits, which is Kepler's first law. Kepler had no *reason* why orbits should be ellipses rather than some other shape. Newton was able to show that ellipses are a *consequence* of his theory of gravity.

The mathematics of ellipses is rather difficult, so we will restrict most of our analysis to the limiting case in which an ellipse becomes a circle. Most planetary orbits differ only very slightly from being circular. The earth's orbit, for example has a (semiminor axis/semimajor axis) ratio of 0.99986—very close to a true circle!

FIGURE 13.16 shows a massive body M, such as the earth or the sun, with a lighter body m orbiting it. The lighter body is called a **satellite,** even though it may be a planet

FIGURE 13.16 The orbital motion of a satellite due to the force of gravity.

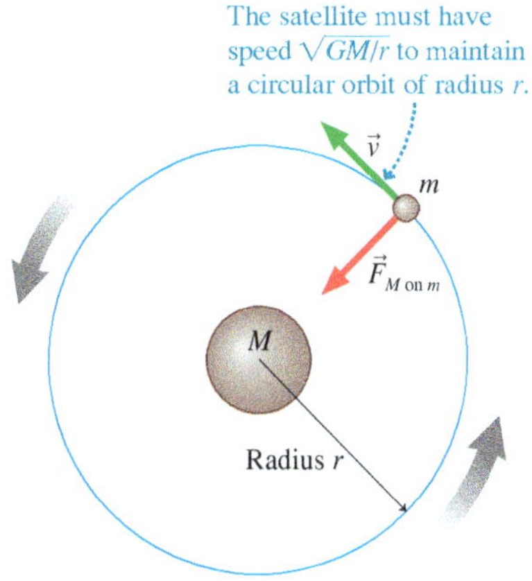

The International Space Station appears to be floating, but it's actually traveling at nearly 8000 m/s as it orbits the earth.

orbiting the sun. For circular motion, the gravitational force must provide the centripetal acceleration v^2/r. Thus Newton's second law for the satellite is

$$F_{M \text{ on } m} = \frac{GMm}{r^2} = ma_r = \frac{mv^2}{r} \tag{13.21}$$

Thus the speed of a satellite in a circular orbit is

$$v = \sqrt{\frac{GM}{r}} \tag{13.22}$$

A satellite must have this specific speed in order to have a circular orbit of radius r about the larger mass M. If the velocity differs from this value, the orbit will become elliptical rather than circular. Notice that the orbital speed does *not* depend on the satellite's mass m. This is consistent with our previous discovery, for motion on a flat earth, that motion due to gravity is independent of the mass.

EXAMPLE 13.4 The speed of the space shuttle

The space shuttle in a 300-km-high orbit ($\approx$ 180 mi) wants to capture a smaller satellite for repairs. What are the speeds of the shuttle and the satellite in this orbit?

SOLVE Despite their different masses, the shuttle, the satellite, and the astronaut working in space to make the repairs all travel side by side with the same speed. They are simply in free fall together. Using $r = R_e + h$ with $h = 300\text{ km} = 3.00 \times 10^5$ m, we find the speed

$$v = \sqrt{\frac{(6.67 \times 10^{-11}\text{ N m}^2/\text{kg}^2)(5.98 \times 10^{24}\text{ kg})}{6.67 \times 10^6\text{ m}}}$$

$$= 7730\text{ m/s} \approx 17{,}000\text{ mph}$$

ASSESS The answer depends on the mass of the earth but *not* on the mass of the satellite.

Kepler's Third Law

An important parameter of circular motion is the *period.* Recall that the period T is the time to complete one full orbit. The relationship among speed, radius, and period is

$$v = \frac{\text{circumference}}{\text{period}} = \frac{2\pi r}{T} \tag{13.23}$$

We can find a relationship between a satellite's period and the radius of its orbit by using Equation 13.22 for v:

$$v = \frac{2\pi r}{T} = \sqrt{\frac{GM}{r}} \tag{13.24}$$

Squaring both sides and solving for T give

$$T^2 = \left(\frac{4\pi^2}{GM}\right) r^3 \tag{13.25}$$

In other words, the *square* of the period is proportional to the *cube* of the radius. This is Kepler's third law. You can see that Kepler's third law is a direct consequence of Newton's law of gravity.

Table 13.2 contains astronomical information about the solar system. We can use these data to check the validity of Equation 13.25. FIGURE 13.17 is a graph of $\log T$ versus $\log r$ for all the planets in Table 13.2 except Mercury. Notice that the scales on each axis are increasing logarithmically—by *factors* of 10—rather than linearly. (Also, the vertical axis has converted T to the SI units of s.) As you can see, the graph is a straight line with a best-fit equation

$$\log T = 1.500 \log r - 9.264$$

TABLE 13.2 Useful astronomical data

Planetary body	Mean distance from sun (m)	Period (years)	Mass (kg)	Mean radius (m)
Sun	—	—	1.99×10^{30}	6.96×10^{8}
Moon	3.84×10^{8}*	27.3 days	7.36×10^{22}	1.74×10^{6}
Mercury	5.79×10^{10}	0.241	3.18×10^{23}	2.43×10^{6}
Venus	1.08×10^{11}	0.615	4.88×10^{24}	6.06×10^{6}
Earth	1.50×10^{11}	1.00	5.98×10^{24}	6.37×10^{6}
Mars	2.28×10^{11}	1.88	6.42×10^{23}	3.37×10^{6}
Jupiter	7.78×10^{11}	11.9	1.90×10^{27}	6.99×10^{7}
Saturn	1.43×10^{12}	29.5	5.68×10^{26}	5.85×10^{7}
Uranus	2.87×10^{12}	84.0	8.68×10^{25}	2.33×10^{7}
Neptune	4.50×10^{12}	165	1.03×10^{26}	2.21×10^{7}

*Distance from earth.

Taking the logarithm of both sides of Equation 13.25, and using the logarithm properties $\log a^n = n \log a$ and $\log(ab) = \log a + \log b$, we have

$$\log T = \frac{3}{2}\log r + \frac{1}{2}\log\left(\frac{4\pi^2}{GM}\right)$$

In other words, theory predicts that the slope of a $\log T$-versus-$\log r$ graph should be exactly $\frac{3}{2}$. As Figure 13.17 shows, the solar-system data agree to an impressive four significant figures. A homework problem will let you use the y-intercept of the graph to determine the mass of the sun.

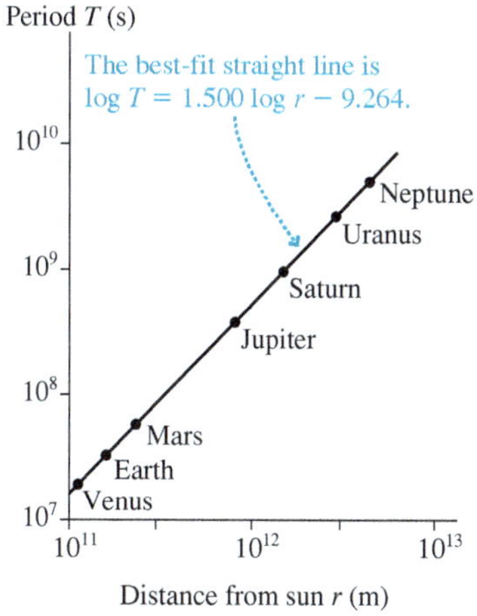

FIGURE 13.17 The graph of $\log T$ versus $\log r$ for the planetary data of Table 13.2.

A particularly interesting application of Equation 13.25 is to communications satellites that are in **geosynchronous orbits** above the earth. These satellites have a period of 24 h = 86,400 s, making their orbital motion synchronous with the earth's rotation. As a result, a satellite in such an orbit appears to remain stationary over one point on the earth's equator. Equation 13.25 allows us to compute the radius of an orbit with this period:

$$r_{geo} = R_e + h_{geo} = \left[\left(\frac{GM}{4\pi^2}\right)T^2\right]^{1/3}$$

$$= \left[\left(\frac{(6.67 \times 10^{-11}\ \text{N m}^2/\text{kg}^2)(5.98 \times 10^{24}\ \text{kg})}{4\pi^2}\right)(86{,}400\ \text{s})^2\right]^{1/3}$$

$$= 4.225 \times 10^{7}\ \text{m}$$

The height of the orbit is

$$h_{geo} = r_{geo} - R_e = 3.59 \times 10^{7}\ \text{m} = 35{,}900\ \text{km} \approx 22{,}300\ \text{mi}$$

NOTE ▶ When you use Equation 13.25, the period *must* be in SI units of s. ◀

Geosynchronous orbits are much higher than the low-earth orbits used by the space shuttle and remote-sensing satellites, where $h \approx 300$ km. Communications satellites in geosynchronous orbits were first proposed in 1948 by science fiction writer Arthur C. Clarke, 10 years before the first artificial satellite of any type!

EXAMPLE 13.5 Extrasolar planets

Astronomers have only recently seen evidence of planets orbiting nearby stars. These are called *extrasolar planets.* Suppose a planet is observed to have a 1200 day period as it orbits a star at the same distance that Jupiter is from the sun. What is the mass of the star in solar masses? (1 *solar mass* is defined to be the mass of the sun.)

SOLVE Here "day" means earth days, as used by astronomers to measure the period. Thus the planet's period in SI units is $T = 1200 \text{ days} = 1.037 \times 10^8 \text{ s}$. The orbital radius is that of Jupiter, which we can find in Table 13.2 to be $r = 7.78 \times 10^{11}$ m. Solving Equation 13.25 for the mass of the star gives

$$M = \frac{4\pi^2 r^3}{GT^2} = 2.59 \times 10^{31} \text{ kg} \times \frac{1 \text{ solar mass}}{1.99 \times 10^{30} \text{ kg}}$$
$$= 13 \text{ solar masses}$$

ASSESS This is a large, but not extraordinary, star.

STOP TO THINK 13.5 Two planets orbit a star. Planet 1 has orbital radius r_1 and planet 2 has $r_2 = 4r_1$. Planet 1 orbits with period T_1. Planet 2 orbits with period

a. $T_2 = 8T_1$ b. $T_2 = 4T_1$ c. $T_2 = 2T_1$
d. $T_2 = \frac{1}{2}T_1$ e. $T_2 = \frac{1}{4}T_1$ f. $T_2 = \frac{1}{8}T_1$

Kepler's Second Law

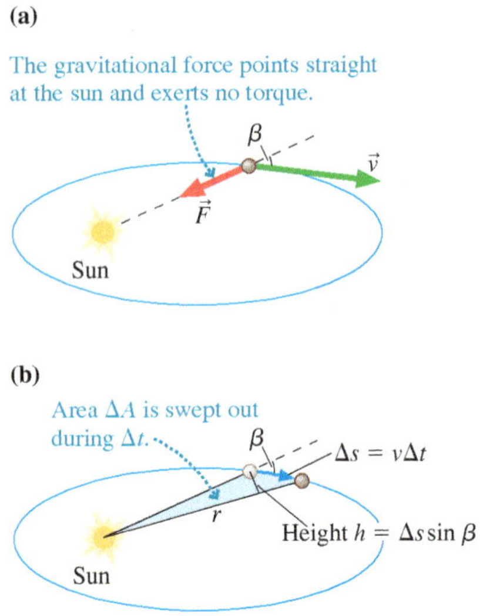

FIGURE 13.18 Angular momentum is conserved for a planet in an elliptical orbit.

FIGURE 13.18a shows a planet moving in an elliptical orbit. In Chapter 12 we defined a particle's *angular momentum* to be

$$L = mrv\sin\beta \tag{13.26}$$

where β is the angle between $\vec{r}$ and $\vec{v}$. For a circular orbit, where β is always 90°, this reduces to simply $L = mrv$.

The only force on the satellite, the gravitational force, points directly toward the star or planet that the satellite is orbiting and exerts no torque; thus **the satellite's angular momentum is conserved as it orbits.**

The satellite moves forward a small distance $\Delta s = v\Delta t$ during the small interval of time Δt. This motion defines the triangle of area ΔA shown in FIGURE 13.18b. ΔA is the area "swept out" by the satellite during Δt. You can see that the height of the triangle is $h = \Delta s \sin\beta$, so the triangle's area is

$$\Delta A = \frac{1}{2} \times \text{base} \times \text{height} = \frac{1}{2} \times r \times \Delta s \sin\beta = \frac{1}{2} rv\sin\beta\, \Delta t \tag{13.27}$$

The *rate* at which the area is swept out by the satellite as it moves is

$$\frac{\Delta A}{\Delta t} = \frac{1}{2} rv\sin\beta = \frac{mrv\sin\beta}{2m} = \frac{L}{2m} \tag{13.28}$$

The angular momentum L is conserved, so it has the same value at every point in the orbit. Consequently, the rate at which the area is swept out by the satellite is constant. This is Kepler's second law, which says that a line drawn between the sun and a planet sweeps out equal areas during equal intervals of time. We see that Kepler's second law is a consequence of the conservation of angular momentum.

Another consequence of angular momentum is that the orbital speed is constant only for a circular orbit. Consider the "ends" of an elliptical orbit, where r is a minimum or maximum. At these points, $\beta = 90°$ and thus $L = mrv$. Because L is constant, the satellite's speed at the farthest point must be less than its speed at the nearest point. In general, a satellite slows as r increases, then speeds up as r decreases, to keep its angular momentum constant.

Kepler's laws summarize observational data about the motions of the planets. They were an outstanding achievement, but they did not form a theory. Newton put forward a *theory,* a specific set of relationships between force and motion that allows *any* motion to be understood and calculated. Newton's theory of gravity has allowed us to *deduce* Kepler's laws and, thus, to understand them at a more fundamental level.

Orbital Energetics

Let us conclude this chapter by thinking about the energetics of orbital motion. We found, with Equation 13.24, that a satellite in a circular orbit must have $v^2 = GM/r$. A satellite's speed is determined entirely by the size of its orbit. The satellite's kinetic energy is thus

$$K = \frac{1}{2}mv^2 = \frac{GMm}{2r} \tag{13.29}$$

But $-GMm/r$ is the potential energy, U_g, so

$$K = -\frac{1}{2}U_g \tag{13.30}$$

This is an interesting result. In all our earlier examples, the kinetic and potential energy were two independent parameters. In contrast, a satellite can move in a circular orbit *only* if there is a very specific relationship between K and U. It is not that K and U *have* to have this relationship, but if they do not, the trajectory will be elliptical rather than circular.

Equation 13.30 gives us the mechanical energy of a satellite in a circular orbit:

$$E_{mech} = K + U_g = \frac{1}{2}U_g \tag{13.31}$$

The gravitational potential energy is negative, hence the *total* mechanical energy is also negative. Negative total energy is characteristic of a **bound system,** a system in which the satellite is bound to the central mass by the gravitational force and cannot get away. The total energy of an unbound system must be ≥ 0 because the satellite can reach infinity, where $U = 0$, while still having kinetic energy. A negative value of E_{mech} tells us that the satellite is unable to escape the central mass.

FIGURE 13.19 shows the energies of a satellite in a circular orbit as a function of the orbit's radius. Notice how $E_{mech} = \frac{1}{2}U_g$. This figure can help us understand the energetics of transferring a satellite from one orbit to another. Suppose a satellite is in an orbit of radius r_1 and we'd like it to be in a larger orbit of radius r_2. The kinetic energy at r_2 is less than at r_1 (the satellite moves more slowly in the larger orbit), but you can see that the total energy *increases* as r increases. Consequently, transferring a satellite to a larger orbit requires a net energy increase $\Delta E > 0$. Where does this increase of energy come from?

Artificial satellites are raised to higher orbits by firing their rocket motors to create a forward thrust. This force does work on the satellite, and the energy equation of Chapter 11 tells us that this work increases the satellite's energy by $\Delta E_{mech} = W_{ext}$. Thus the energy to "lift" a satellite into a higher orbit comes from the chemical energy stored in the rocket fuel.

FIGURE 13.19 The kinetic, potential, and total energy of a satellite in a circular orbit.

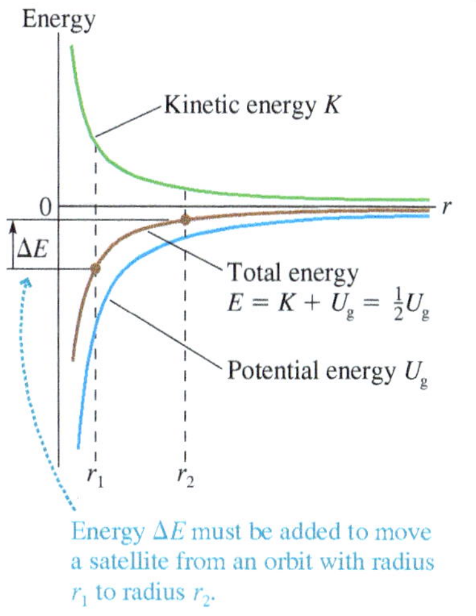

EXAMPLE 13.6 Raising a satellite

How much work must be done to boost a 1000 kg communications satellite from a low earth orbit with $h = 300$ km, where it is released by the space shuttle, to a geosynchronous orbit?

SOLVE The required work is $W_{ext} = \Delta E_{mech}$, and from Equation 13.31 we see that $\Delta E_{mech} = \frac{1}{2}\Delta U_g$. The initial orbit has radius $r_{shuttle} = R_e + h = 6.67 \times 10^6$ m. We earlier found the radius of a geosynchronous orbit to be 4.22×10^7 m. Thus

$$W_{ext} = \Delta E_{mech} = \frac{1}{2}\Delta U_g = \frac{1}{2}(-GM_e m)\left(\frac{1}{r_{geo}} - \frac{1}{r_{shuttle}}\right) = 2.52 \times 10^{10}\ \text{J}$$

ASSESS It takes a lot of energy to boost satellites to high orbits!

You might think that the way to get a satellite into a larger orbit would be to point the thrusters toward the earth and blast outward. That would work fine *if* the satellite were initially at rest and moved straight out along a linear trajectory. But an orbiting satellite is already moving and has significant inertia. A force directed straight outward would *change* the satellite's velocity vector in that direction but would not cause it to *move* along that line. (Remember all those earlier motion diagrams for motion along curved trajectories.) In addition, a force directed outward would be almost at right angles to the motion and would do essentially zero work. Navigating in space is not as easy as it appears in *Star Wars!*

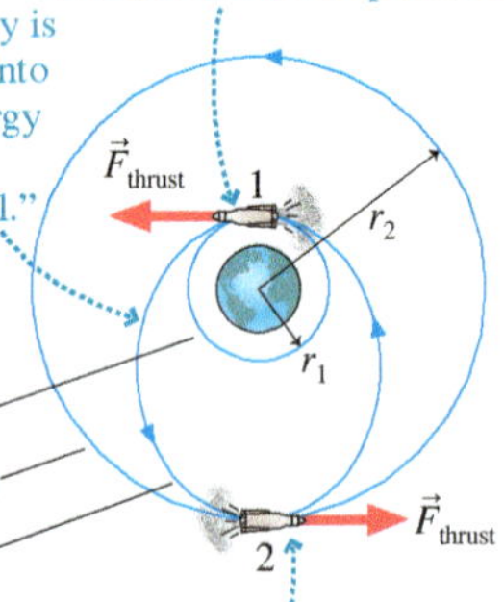

FIGURE 13.20 Transferring a satellite to a larger circular orbit.

To move the satellite in FIGURE 13.20 from the orbit with radius r_1 to the larger circular orbit of radius r_2, the thrusters are turned on at point 1 to apply a brief *forward* thrust force in the direction of motion, *tangent* to the circle. This force does a significant amount of work because the force is parallel to the displacement, so the satellite quickly gains kinetic energy ($\Delta K > 0$). But $\Delta U_g = 0$ because the satellite does not have time to change its distance from the earth during a thrust of short duration. With the kinetic energy increased, but not the potential energy, the satellite no longer meets the requirement $K = -\frac{1}{2}U_g$ for a circular orbit. Instead, it goes into an elliptical orbit.

In the elliptical orbit, the satellite moves "uphill" toward point 2 by transforming kinetic energy into potential energy. At point 2, the satellite has arrived at the desired distance from earth and has the "right" value of the potential energy, but its kinetic energy is now *less* than needed for a circular orbit. (The analysis is more complex than we want to pursue here. It will be left for a homework Challenge Problem.) If no action is taken, the satellite will continue on its elliptical orbit and "fall" back to point 1. But another *forward* thrust at point 2 increases its kinetic energy, without changing U_g, until the kinetic energy reaches the value $K = -\frac{1}{2}U_g$ required for a circular orbit. Presto! The second burn kicks the satellite into the desired circular orbit of radius r_2. The work $W_{ext} = \Delta E_{mech}$ is the *total* work done in both burns. It takes a more extended analysis to see how the work has to be divided between the two burns, but even without those details you now have enough knowledge about orbits and energy to understand the ideas that are involved.

CHALLENGE EXAMPLE 13.7 A binary star system

Astronomers discover a binary star system with a period of 90 days. Both stars have a mass twice that of the sun. How far apart are the two stars?

MODEL Model the stars as spherical masses exerting gravitational forces on each other.

VISUALIZE *An isolated system rotates around its center of mass.* FIGURE 13.21 shows the orbits and the forces. If r is the distance of each star to the center of mass—the radius of that star's orbit—then the distance between the stars is $d = 2r$.

FIGURE 13.21 The binary star system.

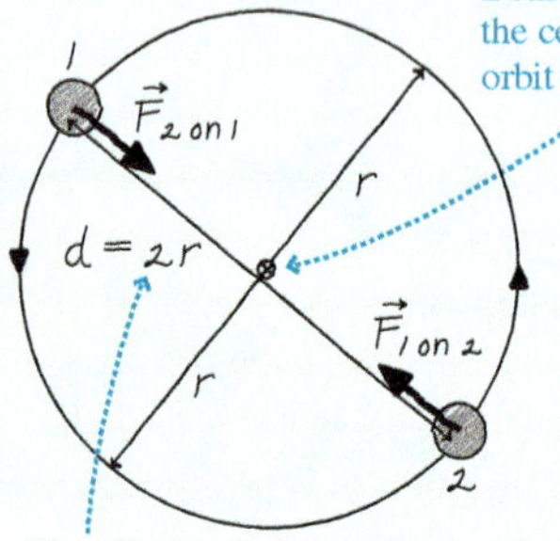

SOLVE Star 2 has only one force acting on it $\vec{F}_{1 \text{ on } 2}$, and that force has to provide the centripetal acceleration v^2/r of circular motion. Newton's second law for star 2 is

$$F_{1 \text{ on } 2} = \frac{GM_1M_2}{d^2} = \frac{GM^2}{4r^2} = Ma_r = \frac{Mv^2}{r}$$

where we used $M_1 = M_2 = M$. The equation for star 1 is identical. The star's speed is related to the period and the circumference of its orbit by $v = 2\pi r/T$. With this, the force equation becomes

$$\frac{GM^2}{4r^2} = \frac{4\pi^2 Mr}{T^2}$$

Solving for r gives

$$r = \left[\frac{GMT^2}{16\pi^2}\right]^{1/3}$$
$$= \left[\frac{(6.67 \times 10^{-11}\,\text{Nm}^2/\text{kg}^2)(2 \times 1.99 \times 10^{30}\,\text{kg})(7.78 \times 10^6\,\text{s})^2}{16\pi^2}\right]^{1/3}$$
$$= 4.67 \times 10^{10}\,\text{m}$$

The distance between the stars is $d = 2r = 9.3 \times 10^{10}$ m.

ASSESS The result is in the range of solar-system distances and thus is reasonable.

SUMMARY

The goal of Chapter 13 has been to use Newton's theory of gravity to understand the motion of satellites and planets.

General Principles

Newton's Theory of Gravity

1. Two objects with masses M and m a distance r apart exert attractive **gravitational forces** on each other of magnitude

$$F_{M\text{ on }m} = F_{m\text{ on }M} = \frac{GMm}{r^2}$$

where the **gravitational constant** is $G = 6.67 \times 10^{-11}\ \text{N m}^2/\text{kg}^2$.

2. Gravitational mass and inertial mass are equivalent.
3. Newton's three laws of motion apply to all objects in the universe.

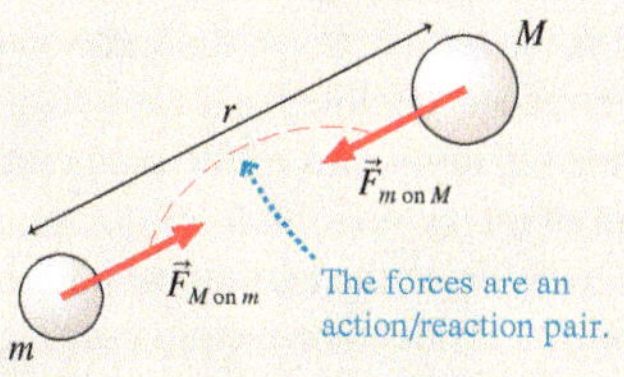

Important Concepts

Orbital motion of a planet (or satellite) is described by **Kepler's laws:**

1. Orbits are ellipses with the sun (or planet) at one focus.
2. A line between the sun and the planet sweeps out equal areas during equal intervals of time.
3. The square of the planet's period T is proportional to the cube of the orbit's semimajor axis.

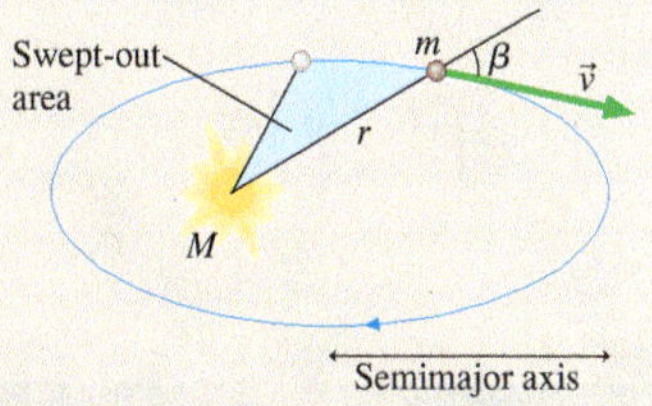

Circular orbits are a special case of an ellipse. For a circular orbit around a mass M,

$$v = \sqrt{\frac{GM}{r}} \quad \text{and} \quad T^2 = \left(\frac{4\pi^2}{GM}\right)r^3$$

Conservation of angular momentum

The angular momentum $L = mrv\sin\beta$ remains constant throughout the orbit. Kepler's second law is a consequence of this law.

Orbital energetics

A satellite's mechanical energy $E_{\text{mech}} = K + U_g$ is conserved, where the gravitational potential energy is

$$U_g = -\frac{GMm}{r}$$

For circular orbits, $K = -\frac{1}{2}U_g$ and $E_{\text{mech}} = \frac{1}{2}U_g$. Negative total energy is characteristic of a **bound system.**

Applications

For a planet of mass M and radius R,

- The free-fall acceleration on the surface is $g_{\text{surface}} = \dfrac{GM}{R^2}$
- The escape speed is $v_{\text{escape}} = \sqrt{\dfrac{2GM}{R}}$
- The radius of a geosynchronous orbit is $r_{\text{geo}} = \left(\dfrac{GM}{4\pi^2}T^2\right)^{1/3}$

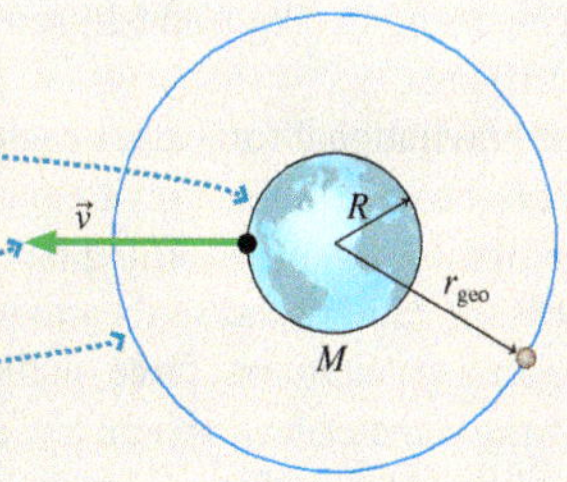

Terms and Notation

cosmology
Kepler's laws
gravitational force
Newton's law of gravity
gravitational constant, G
gravitational mass
principle of equivalence
Newton's theory of gravity
escape speed
satellite
geosynchronous orbit
bound system

CONCEPTUAL QUESTIONS

1. Is the earth's gravitational force on the sun larger than, smaller than, or equal to the sun's gravitational force on the earth? Explain.
2. The gravitational force of a star on orbiting planet 1 is F_1. Planet 2, which is twice as massive as planet 1 and orbits at twice the distance from the star, experiences gravitational force F_2. What is the ratio F_1/F_2?
3. A 1000 kg satellite and a 2000 kg satellite follow exactly the same orbit around the earth.
   a. What is the ratio F_1/F_2 of the force on the first satellite to that on the second satellite?
   b. What is the ratio a_1/a_2 of the acceleration of the first satellite to that of the second satellite?
4. How far away from the earth must an orbiting spacecraft be for the astronauts inside to be weightless? Explain.
5. A space shuttle astronaut is working outside the shuttle as it orbits the earth. If he drops a hammer, will it fall to earth? Explain why or why not.
6. The free-fall acceleration at the surface of planet 1 is 20 m/s^2. The radius and the mass of planet 2 are twice those of planet 1. What is g on planet 2?
7. *Why* is the gravitational potential energy of two masses negative? Note that saying "because that's what the equation gives" is *not* an explanation.
8. The escape speed from Planet X is 10,000 m/s. Planet Y has the same radius as Planet X but is twice as dense. What is the escape speed from Planet Y?
9. The mass of Jupiter is 300 times the mass of the earth. Jupiter orbits the sun with $T_{\text{Jupiter}} = 11.9$ yr in an orbit with $r_{\text{Jupiter}} = 5.2r_{\text{earth}}$. Suppose the earth could be moved to the distance of Jupiter and placed in a circular orbit around the sun. Which of the following describes the earth's new period? Explain.
   a. 1 yr
   b. Between 1 yr and 11.9 yr
   c. 11.9 yr
   d. More than 11.9 yr
   e. It would depend on the earth's speed.
   f. It's impossible for a planet of earth's mass to orbit at the distance of Jupiter.
10. Satellites in near-earth orbit experience a very slight drag due to the extremely thin upper atmosphere. These satellites slowly but surely spiral inward, where they finally burn up as they reach the thicker lower levels of the atmosphere. The radius decreases so slowly that you can consider the satellite to have a circular orbit at all times. As a satellite spirals inward, does it speed up, slow down, or maintain the same speed? Explain.

EXERCISES AND PROBLEMS

Problems labeled integrate material from earlier chapters.

Exercises

Section 13.3 Newton's Law of Gravity

1. || What is the ratio of the sun's gravitational force on you to the earth's gravitational force on you?
2. || The centers of a 10 kg lead ball and a 100 g lead ball are separated by 10 cm.
   a. What gravitational force does each exert on the other?
   b. What is the ratio of this gravitational force to the gravitational force of the earth on the 100 g ball?
3. || What is the ratio of the sun's gravitational force on the moon to the earth's gravitational force on the moon?
4. || A 1.0-m-diameter lead sphere has a mass of 5900 kg. A dust particle rests on the surface. What is the ratio of the gravitational force of the sphere on the dust particle to the gravitational force of the earth on the dust particle?
5. | Estimate the force of attraction between a 50 kg woman and a 70 kg man sitting 1.0 m apart.
6. || The space shuttle orbits 300 km above the surface of the earth. What is the gravitational force on a 1.0 kg sphere inside the space shuttle?

Section 13.4 Little *g* and Big *G*

7. | a. What is the free-fall acceleration at the surface of the sun?
   b. What is the sun's free-fall acceleration at the distance of the earth?
8. || What is the free-fall acceleration at the surface of (a) the moon and (b) Jupiter?
9. || A sensitive gravimeter at a mountain observatory finds that the free-fall acceleration is 0.0075 m/s^2 less than that at sea level. What is the observatory's altitude?
10. || Suppose we could shrink the earth without changing its mass. At what fraction of its current radius would the free-fall acceleration at the surface be three times its present value?
11. || Planet Z is 10,000 km in diameter. The free-fall acceleration on Planet Z is 8.0 m/s^2.
    a. What is the mass of Planet Z?
    b. What is the free-fall acceleration 10,000 km above Planet Z's north pole?

Section 13.5 Gravitational Potential Energy

12. | An astronaut on earth can throw a ball straight up to a height of 15 m. How high can he throw the ball on Mars?
13. || What is the escape speed from Jupiter?

14. || A rocket is launched straight up from the earth's surface at a speed of 15,000 m/s. What is its speed when it is very far away from the earth?

15. | A space station orbits the sun at the same distance as the earth but on the opposite side of the sun. A small probe is fired away from the station. What minimum speed does the probe need to escape the solar system?

16. || You have been visiting a distant planet. Your measurements have determined that the planet's mass is twice that of earth but the free-fall acceleration at the surface is only one-fourth as large.
   a. What is the planet's radius?
   b. To get back to earth, you need to escape the planet. What minimum speed does your rocket need?

Section 13.6 Satellite Orbits and Energies

17. | The *asteroid belt* circles the sun between the orbits of Mars and Jupiter. One asteroid has a period of 5.0 earth years. What are the asteroid's orbital radius and speed?

18. | Use information about the earth and its orbit to determine the mass of the sun.

19. || Planet X orbits the star Omega with a "year" that is 200 earth days long. Planet Y circles Omega at four times the distance of Planet X. How long is a year on Planet Y?

20. | You are the science officer on a visit to a distant solar system. Prior to landing on a planet you measure its diameter to be 1.8×10^7 m and its rotation period to be 22.3 hours. You have previously determined that the planet orbits 2.2×10^{11} m from its star with a period of 402 earth days. Once on the surface you find that the free-fall acceleration is 12.2 m/s^2. What is the mass of (a) the planet and (b) the star?

21. || Three satellites orbit a planet of radius R, as shown in FIGURE EX13.21. Satellites S_1 and S_3 have mass m. Satellite S_2 has mass $2m$. Satellite S_1 orbits in 250 minutes and the force on S_1 is 10,000 N.
   a. What are the periods of S_2 and S_3?
   b. What are the forces on S_2 and S_3?
   c. What is the kinetic-energy ratio K_1/K_3 for S_1 and S_3?

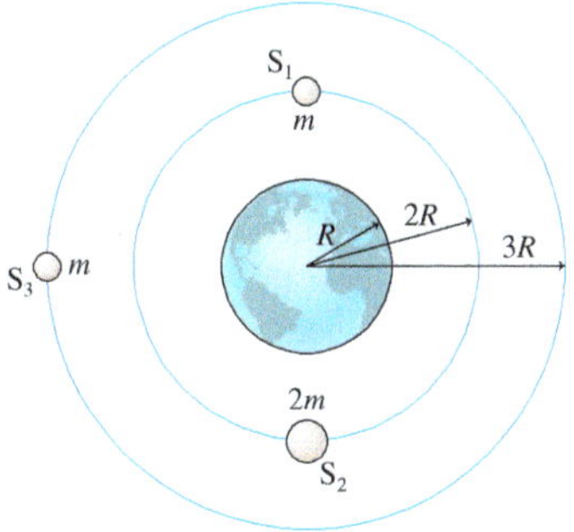

FIGURE EX13.21

22. || A satellite orbits the sun with a period of 1.0 day. What is the radius of its orbit?

23. || An earth satellite moves in a circular orbit at a speed of 5500 m/s. What is its orbital period?

24. || What are the speed and altitude of a geosynchronous satellite orbiting Mars? Mars rotates on its axis once every 24.8 hours.

Problems

25. || Two spherical objects have a combined mass of 150 kg. The gravitational attraction between them is 8.00×10^{-6} N when their centers are 20 cm apart. What is the mass of each?

26. || FIGURE P13.26 shows three masses. What are the magnitude and the direction of the net gravitational force on (a) the 20.0 kg mass and (b) the 5.0 kg mass? Give the direction as an angle cw or ccw from the y-axis.

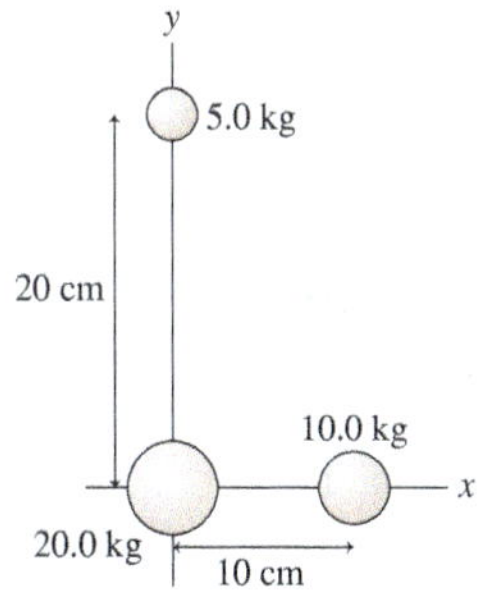

FIGURE P13.26

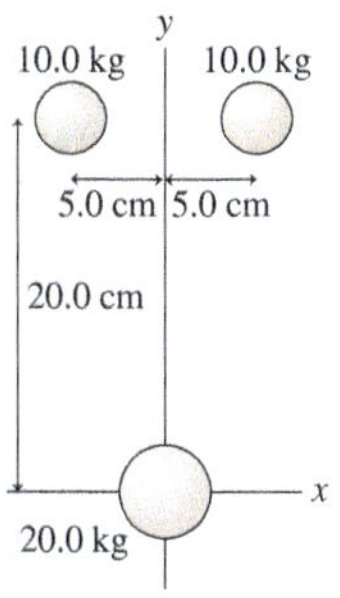

FIGURE P13.27

27. || What are the magnitude and direction of the net gravitational force on the 20.0 kg mass in FIGURE P13.27?

28. || What is the total gravitational potential energy of the three masses in FIGURE P13.26?

29. || What is the total gravitational potential energy of the three masses in FIGURE P13.27?

30. ||| Two 100 kg lead spheres are suspended from 100-m-long massless cables. The tops of the cables have been carefully anchored *exactly* 1 m apart. What is the distance between the centers of the spheres?

31. ||| A 20 kg sphere is at the origin and a 10 kg sphere is at $x = 20$ cm. At what position on the x-axis could you place a small mass such that the net gravitational force on it due to the spheres is zero?

32. || a. At what height above the earth is the acceleration due to gravity 10% of its value at the surface?
   b. What is the speed of a satellite orbiting at that height?

33. || A 1.0 kg object is released from rest 500 km ($\approx$300 miles) above the earth.
   a. What is its impact speed as it hits the ground? Ignore air resistance.
   b. What would the impact speed be if the earth were flat?
   c. By what percentage is the flat-earth calculation in error?

34. ||| An object of mass m is dropped from height h above a planet of mass M and radius R. Find an expression for the object's speed as it hits the ground.

35. ||| A projectile is shot straight up from the earth's surface at a speed of 10,000 km/h. How high does it go?

36. || Two meteoroids are heading for earth. Their speeds as they cross the moon's orbit are 2.0 km/s.
   a. The first meteoroid is heading straight for earth. What is its speed of impact?
   b. The second misses the earth by 5000 km. What is its speed at its closest point?

37. || A binary star system has two stars, each with the same mass as our sun, separated by 1.0×10^{12} m. A comet is very far away and essentially at rest. Slowly but surely, gravity pulls the comet toward the stars. Suppose the comet travels along a straight line that passes through the midpoint between the two stars. What is the comet's speed at the midpoint?

38. || Suppose that on earth you can jump straight up a distance of 50 cm. Can you escape from a 4.0-km-diameter asteroid with a mass of 1.0×10^{14} kg?

39. ||| A projectile is fired straight away from the moon from a base on the far side of the moon, away from the earth. What is the projectile's escape speed from the earth-moon system?
40. ||| Two spherical asteroids have the same radius R. Asteroid 1 has mass M and asteroid 2 has mass $2M$. The two asteroids are released from rest with distance $10R$ between their centers. What is the speed of each asteroid just before they collide?
Hint: You will need to use two conservation laws.
41. ||| Two Jupiter-size planets are released from rest 1.0×10^{11} m apart. What are their speeds as they crash together?
42. ||| A starship is circling a distant planet of radius R. The astronauts find that the free-fall acceleration at their altitude is half the value at the planet's surface. How far above the surface are they orbiting? Your answer will be a multiple of R.
43. || Three stars, each with the mass and radius of our sun, form an equilateral triangle 5.0×10^9 m on a side. If all three are simultaneously released from rest, what are their speeds as they crash together in the center?
44. ||| The two stars in a binary star system have masses 2.0×10^{30} kg and 6.0×10^{30} kg. They are separated by 2.0×10^{12} m. What are
   a. The system's rotation period, in years?
   b. The speed of each star?
45. ||| A 4000 kg lunar lander is in orbit 50 km above the surface of the moon. It needs to move out to a 300-km-high orbit in order to link up with the mother ship that will take the astronauts home. How much work must the thrusters do?
46. ||| The space shuttle is in a 250-km-high circular orbit. It needs to reach a 610-km-high circular orbit to catch the Hubble Space Telescope for repairs. The shuttle's mass is 75,000 kg. How much energy is required to boost it to the new orbit?
47. || In 2000, NASA placed a satellite in orbit around an asteroid. Consider a spherical asteroid with a mass of 1.0×10^{16} kg and a radius of 8.8 km.
   a. What is the speed of a satellite orbiting 5.0 km above the surface?
   b. What is the escape speed from the asteroid?
48. || NASA would like to place a satellite in orbit around the moon such that the satellite always remains in the same position over the lunar surface. What is the satellite's altitude?
49. || A satellite orbiting the earth is directly over a point on the equator at 12:00 midnight every two days. It is not over that point at any time in between. What is the radius of the satellite's orbit?
50. ||| FIGURE P13.50 shows two planets of mass m orbiting a star of mass M. The planets are in the same orbit, with radius r, but are always at opposite ends of a diameter. Find an exact expression for the orbital period T.
Hint: Each planet feels two forces.

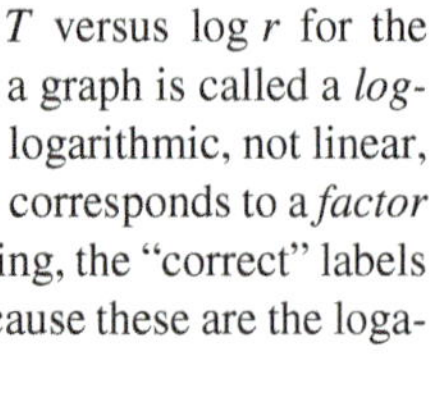

FIGURE P13.50

51. || Figure 13.17 showed a graph of $\log T$ versus $\log r$ for the planetary data given in Table 13.2. Such a graph is called a *log-log graph.* The scales in Figure 13.17 are logarithmic, not linear, meaning that each division along the axis corresponds to a *factor* of 10 increase in the value. Strictly speaking, the "correct" labels on the y-axis should be 7, 8, 9, and 10 because these are the logarithms of $10^7, \ldots, 10^{10}$.
   a. Consider two quantities u and v that are related by the expression $v^p = Cu^q$, where C is a constant. The exponents p and q are not necessarily integers. Define $x = \log u$ and $y = \log v$. Find an expression for y in terms of x.
   b. What *shape* will a graph of y versus x have? Explain.
   c. What *slope* will a graph of y versus x have? Explain.
   d. Use the experimentally determined "best-fit" line in Figure 13.17 to find the mass of the sun.
52. || Large stars can explode as they finish burning their nuclear fuel, causing a *supernova.* The explosion blows away the outer layers of the star. According to Newton's third law, the forces that push the outer layers away have *reaction forces* that are inwardly directed on the core of the star. These forces compress the core and can cause the core to undergo a *gravitational collapse.* The gravitational forces keep pulling all the matter together tighter and tighter, crushing atoms out of existence. Under these extreme conditions, a proton and an electron can be squeezed together to form a neutron. If the collapse is halted when the neutrons all come into contact with each other, the result is an object called a *neutron star,* an entire star consisting of solid nuclear matter. Many neutron stars rotate about their axis with a period of ≈ 1 s and, as they do so, send out a pulse of electromagnetic waves once a second. These stars were discovered in the 1960s and are called *pulsars.*
   a. Consider a neutron star with a mass equal to the sun, a radius of 10 km, and a rotation period of 1.0 s. What is the speed of a point on the equator of the star?
   b. What is g at the surface of this neutron star?
   c. A stationary 1.0 kg mass has a weight on earth of 9.8 N. What would be its weight on the star?
   d. How many revolutions per minute are made by a satellite orbiting 1.0 km above the surface?
   e. What is the radius of a geosynchronous orbit about the neutron star?
53. || The solar system is 25,000 light years from the center of our Milky Way galaxy. One *light year* is the distance light travels in one year at a speed of 3.0×10^8 m/s. Astronomers have determined that the solar system is orbiting the center of the galaxy at a speed of 230 km/s.
   a. Assuming the orbit is circular, what is the period of the solar system's orbit? Give your answer in years.
   b. Our solar system was formed roughly 5 billion years ago. How many orbits has it completed?
   c. The gravitational force on the solar system is the net force due to all the matter inside our orbit. Most of that matter is concentrated near the center of the galaxy. Assume that the matter has a spherical distribution, like a giant star. What is the approximate mass of the galactic center?
   d. Assume that the sun is a typical star with a typical mass. If galactic matter is made up of stars, approximately how many stars are in the center of the galaxy?

   Astronomers have spent many years trying to determine how many stars there are in the Milky Way. The number of stars seems to be only about 10% of what you found in part d. In other words, about 90% of the mass of the galaxy appears to be in some form other than stars. This is called the *dark matter* of the universe. No one knows what the dark matter is. This is one of the outstanding scientific questions of our day.
54. || Three stars, each with the mass of our sun, form an equilateral triangle with sides 1.0×10^{12} m long. (This triangle would just about fit within the orbit of Jupiter.) The triangle has to rotate, because otherwise the stars would crash together in the center. What is the period of rotation? Give your answer in years.

55. || Pluto moves in a fairly elliptical orbit around the sun. Pluto's speed at its closest approach of 4.43×10^9 km is 6.12 km/s. What is Pluto's speed at the most distant point in its orbit, where it is 7.30×10^9 km from the sun?

56. || Mercury moves in a fairly elliptical orbit around the sun. Mercury's speed is 38.8 km/s when it is at its most distant point, 6.99×10^{10} m from the sun. How far is Mercury from the sun at its closest point, where its speed is 59.0 km/s?

57. || Comets move around the sun in very elliptical orbits. At its closet approach, in 1986, Comet Halley was 8.79×10^7 km from the sun and moving with a speed of 54.6 km/s. What was the comet's speed when it crossed Neptune's orbit in 2006?

58. || A spaceship is in a circular orbit of radius r_0 about a planet of mass M. A brief but intense firing of its engine in the forward direction decreases the spaceship's speed by 50%. This causes the spaceship to move into an elliptical orbit.
   a. What is the spaceship's new speed, just after the rocket burn is complete, in terms of M, G, and r_0?
   b. In terms of r_0, what are the spaceship's maximum and minimum distances from the planet in its new orbit?

In Problems 59 through 61 you are given the equation(s) used to solve a problem. For each of these, you are to
   a. Write a realistic problem for which this is the correct equation(s).
   b. Draw a pictorial representation.
   c. Finish the solution of the problem.

59. $$\frac{(6.67 \times 10^{-11}\ \text{N m}^2/\text{kg}^2)(5.68 \times 10^{26}\ \text{kg})}{r^2} = \frac{(6.67 \times 10^{-11}\ \text{N m}^2/\text{kg}^2)(5.98 \times 10^{24}\ \text{kg})}{(6.37 \times 10^6\ \text{m})^2}$$

60. $$\frac{(6.67 \times 10^{-11}\ \text{N m}^2/\text{kg}^2)(5.98 \times 10^{24}\ \text{kg})(1000\ \text{kg})}{r^2} = \frac{(1000\ \text{kg})(1997\ \text{m/s})^2}{r}$$

61. $$\frac{1}{2}(100\ \text{kg})v_2^2 - \frac{(6.67 \times 10^{-11}\ \text{N m}^2/\text{kg}^2)(7.36 \times 10^{22}\ \text{kg})(100\ \text{kg})}{1.74 \times 10^6\ \text{m}} = 0 - \frac{(6.67 \times 10^{-11}\ \text{N m}^2/\text{kg}^2)(7.36 \times 10^{22}\ \text{kg})(100\ \text{kg})}{3.48 \times 10^6\ \text{m}}$$

Challenge Problems

62. A satellite in a circular orbit of radius r has period T. A satellite in a nearby orbit with radius $r + \Delta r$, where $\Delta r \ll r$, has the very slightly different period $T + \Delta T$.
   a. Show that
   $$\frac{\Delta T}{T} = \frac{3}{2}\frac{\Delta r}{r}$$
   b. Two earth satellites are in parallel orbits with radii 6700 km and 6701 km. One day they pass each other, 1 km apart, along a line radially outward from the earth. How long will it be until they are again 1 km apart?

63. In 1996, the Solar and Heliospheric Observatory (SOHO) was "parked" in an orbit slightly inside the earth's orbit, as shown in FIGURE CP13.63. The satellite's period in this orbit is exactly one year, so it remains fixed relative to the earth. At this point, called a *Lagrange point,* the light from the sun is never blocked by the earth, yet the satellite remains "nearby" so that data are easily transmitted to earth. What is SOHO's distance from the earth?

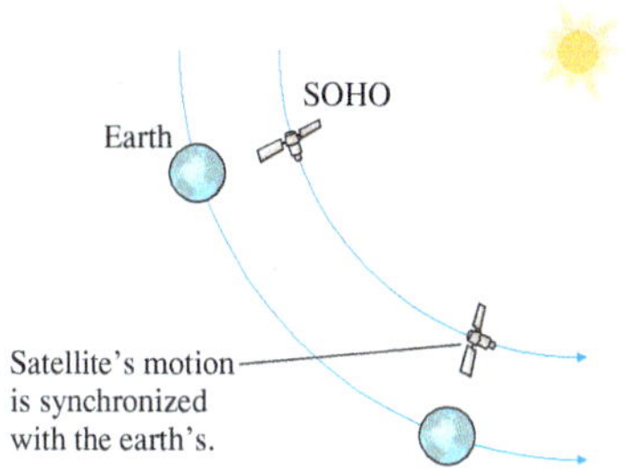

FIGURE CP13.63

Hint: Use the binomial approximation. SOHO's distance from the earth is much less than the earth's distance from the sun.

64. A projectile is fired from the earth in the direction of the earth's motion around the sun. What minimum speed must the projectile have relative to the earth to escape the solar system? Ignore the earth's rotation.
   Hint: This is a three-part problem. First find the speed a projectile at the earth's distance needs to escape the sun. Transform that speed into the earth's reference frame, then determine how fast the projectile must be launched to have this speed when far from the earth.

65. Your job with NASA is to monitor satellite orbits. One day, during a routine survey, you find that a 400 kg satellite in a 1000-km-high circular orbit is going to collide with a smaller 100 kg satellite traveling in the same orbit but in the opposite direction. Knowing the construction of the two satellites, you expect they will become enmeshed into a single piece of space debris. When you notify your boss of this impending collision, he asks you to quickly determine whether the space debris will continue to orbit or crash into the earth. What will the outcome be?

66. While visiting Planet Physics, you toss a rock straight up at 11 m/s and catch it 2.5 s later. While you visit the surface, your cruise ship orbits at an altitude equal to the planet's radius every 230 min. What are the (a) mass and (b) radius of Planet Physics?

67. A moon lander is orbiting the moon at an altitude of 1000 km. By what percentage must it decrease its speed so as to just graze the moon's surface one-half period later?

68. Let's look in more detail at how a satellite is moved from one circular orbit to another. FIGURE CP13.68 shows two circular orbits, of radii r_1 and r_2, and an elliptical orbit that connects them. Points 1 and 2 are at the ends of the semimajor axis of the ellipse.

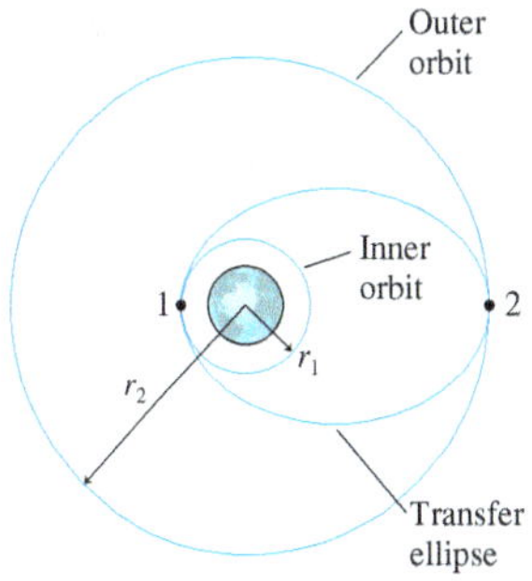

FIGURE CP13.68

   a. A satellite moving along the elliptical orbit has to satisfy two conservation laws. Use these two laws to prove that the velocities at points 1 and 2 are
   $$v_1' = \sqrt{\frac{2GM(r_2/r_1)}{r_1 + r_2}} \quad \text{and} \quad v_2' = \sqrt{\frac{2GM(r_1/r_2)}{r_1 + r_2}}$$

The prime indicates that these are the velocities on the elliptical orbit. Both reduce to Equation 13.22 if $r_1 = r_2 = r$.

b. Consider a 1000 kg communications satellite that needs to be boosted from an orbit 300 km above the earth to a geosynchronous orbit 35,900 km above the earth. Find the velocity v_1 on the inner circular orbit and the velocity v_1' at the low point on the elliptical orbit that spans the two circular orbits.

c. How much work must the rocket motor do to transfer the satellite from the circular orbit to the elliptical orbit?

d. Now find the velocity v_2' at the high point of the elliptical orbit and the velocity v_2 of the outer circular orbit.

e. How much work must the rocket motor do to transfer the satellite from the elliptical orbit to the outer circular orbit?

f. Compute the total work done and compare your answer to the result of Example 13.6.

69. FIGURE CP13.69 shows a particle of mass m at distance x from the center of a very thin cylinder of mass M and length L. The particle is outside the cylinder, so $x > L/2$.

a. Calculate the gravitational potential energy of these two masses.

b. Use what you know about the relationship between force and potential energy to find the magnitude of the gravitational force on m when it is at position x.

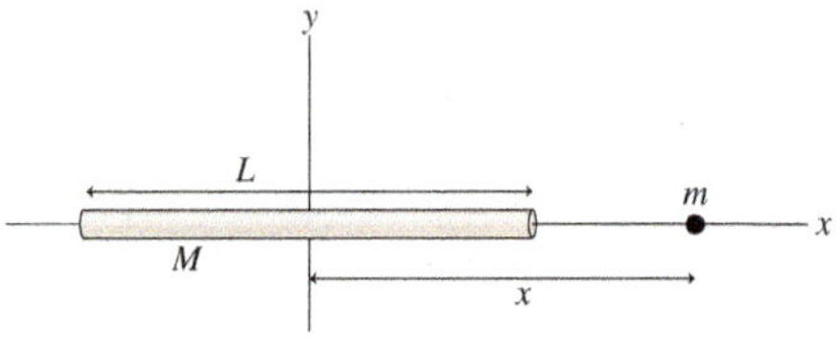

FIGURE CP13.69

70. FIGURE CP13.70 shows a particle of mass m at distance x along the axis of a very thin ring of mass M and radius R.

a. Calculate the gravitational potential energy of these two masses.

b. Use what you know about the relationship between force and potential energy to find the magnitude of the gravitational force on m when it is at position x.

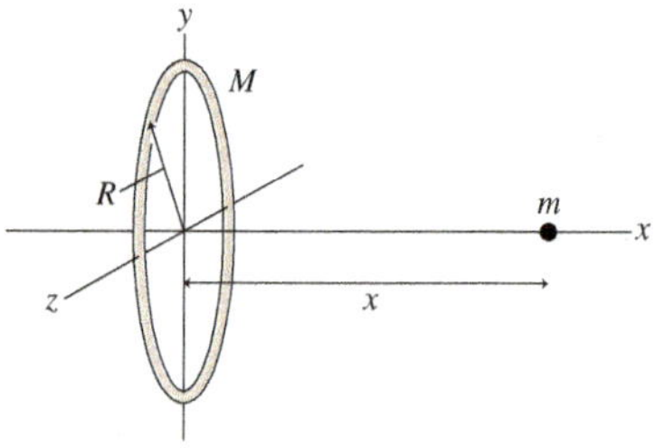

FIGURE CP13.70

STOP TO THINK ANSWERS

Stop to Think 13.1: e. The acceleration decreases inversely with the square of the distance. At height R_e, the distance from the center of the earth is $2R_e$.

Stop to Think 13.2: c. Newton's third law requires $F_{1 \text{ on } 2} = F_{2 \text{ on } 1}$.

Stop to Think 13.3: b. $g_{surface} = GM/R^2$. Because of the square, a radius twice as large balances a mass four times as large.

Stop to Think 13.4: In absolute value, $U_e > U_a = U_b = U_d > U_c$. $|U_g|$ is proportional to $m_1 m_2/r$.

Stop to Think 13.5: a. T^2 is proportional to r^3, or T is proportional to $r^{3/2}$. $4^{3/2} = 8$.

14 Oscillations

This loudspeaker cone generates sound waves by oscillating back and forth at audio frequencies.

▶ **Looking Ahead** The goal of Chapter 14 is to understand systems that oscillate with simple harmonic motion.

Simple Harmonic Motion

The most basic oscillation, with sinusoidal motion, is called **simple harmonic motion.**

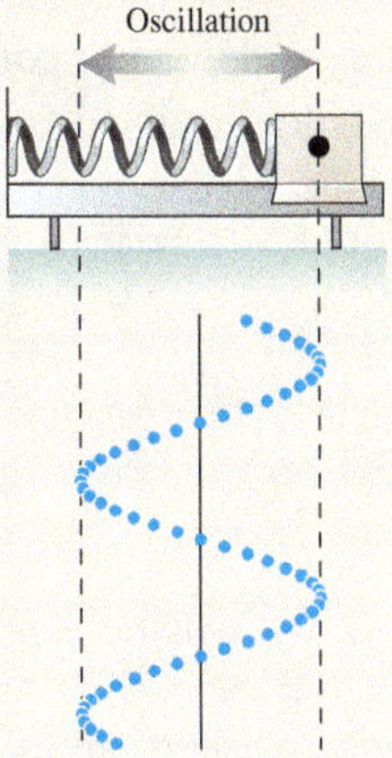

The oscillating cart is an example of simple harmonic motion. You'll learn how to use the mass and the spring constant to determine the frequency of oscillation.

In this chapter you will learn to:

- Represent simple harmonic motion both graphically and mathematically.
- Understand the dynamics of oscillating systems.
- Recognize the similarities among many types of oscillating systems.

Simple harmonic motion has a very close connection to uniform circular motion. You'll learn that an edge-on view of uniform circular motion is none other than simple harmonic motion.

◀ **Looking Back**
Section 4.5 Uniform circular motion

Springs

Simple harmonic motion occurs when there is a **linear restoring force.** The simplest example is a mass on a spring. You will learn how to determine the period of oscillation.

The "bounce" at the bottom of a bungee jump is an exhilarating example of a mass oscillating on a spring.

◀ **Looking Back**
Section 10.4 Restoring forces

Pendulums

A mass swinging at the end of a string or rod is a **pendulum.** Its motion is another example of simple harmonic motion.

The period of a pendulum is determined by the length of the string; neither the mass nor the amplitude matters. Consequently, the pendulum was the basis of time keeping for many centuries.

Energy of Oscillations

If there is no friction or other dissipation, then the mechanical energy of an oscillator is conserved. Conservation of energy will be an important tool.

The system oscillates between all kinetic energy and all potential energy

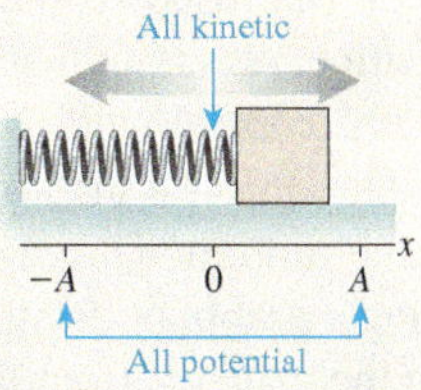

◀ **Looking Back**
Section 10.5 Elastic potential energy
Section 10.6 Energy diagrams

Damping and Resonance

If there's drag or other dissipation, then the oscillation "runs down." This is called a **damped oscillation.**

The amplitude of a damped oscillation undergoes *exponential decay.*

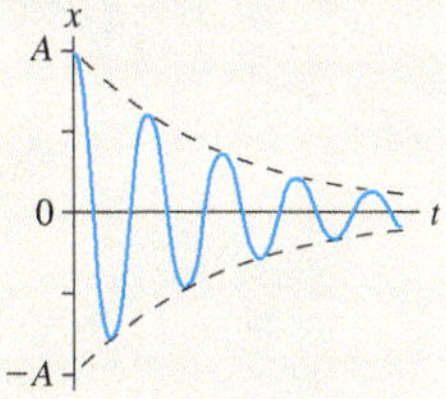

Oscillations can increase in amplitude, sometimes dramatically, when driven at their natural oscillation frequency. This is called **resonance.**

FIGURE 14.1 Examples of position-versus-time graphs for oscillating systems.

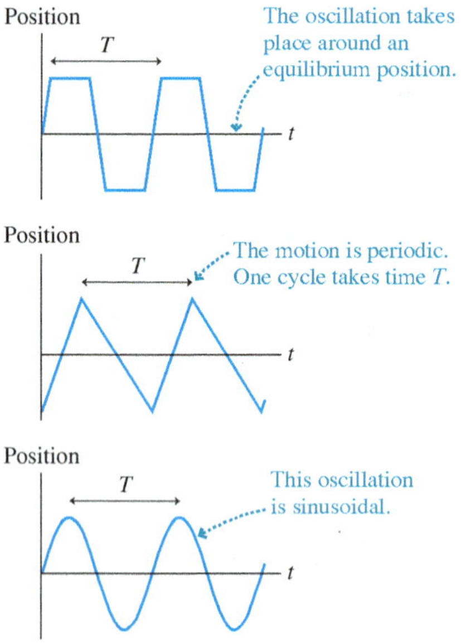

TABLE 14.1 Units of frequency

Frequency	Period
10^3 Hz = 1 kilohertz = 1 kHz	1 ms
10^6 Hz = 1 megahertz = 1 MHz	1 μs
10^9 Hz = 1 gigahertz = 1 GHz	1 ns

FIGURE 14.2 A prototype simple-harmonic-motion experiment.

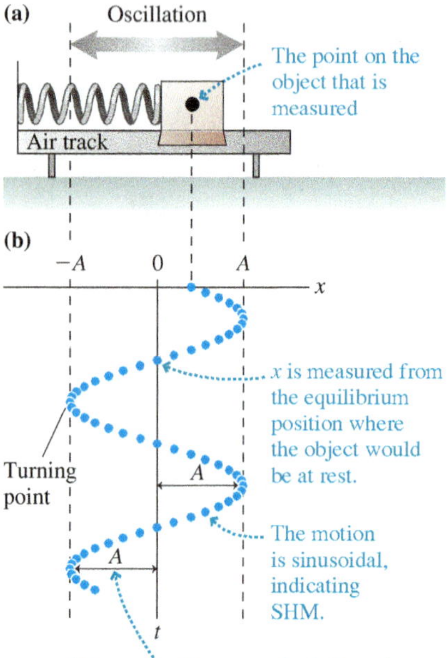

14.1 Simple Harmonic Motion

Objects or systems of objects that undergo **oscillatory motion**—a repetitive motion back and forth around an equilibrium position—are called **oscillators.** FIGURE 14.1 shows position-versus-time graphs for three different oscillating systems. Although the shapes of the graphs are different, all these oscillators have two things in common:

1. The oscillation takes place about an equilibrium position, and
2. The motion is *periodic*, repeating at regular intervals of time.

The time to complete one full cycle, or one oscillation, is called the **period** of the motion. Period is represented by the symbol T.

A closely related piece of information is the number of cycles, or oscillations, completed per second. If the period is $\frac{1}{10}$ s, then the oscillator can complete 10 cycles in one second. Conversely, an oscillation period of 10 s allows only $\frac{1}{10}$ of a cycle to be completed per second. In general, T seconds per cycle implies that $1/T$ cycles will be completed each second. The number of cycles per second is called the **frequency** f of the oscillation. The relationship between frequency and period is

$$f = \frac{1}{T} \quad \text{or} \quad T = \frac{1}{f} \tag{14.1}$$

The units of frequency are **hertz,** abbreviated Hz, named in honor of the German physicist Heinrich Hertz, who produced the first artificially generated radio waves in 1887. By definition,

$$1 \text{ Hz} \equiv 1 \text{ cycle per second} = 1 \text{ s}^{-1}$$

We will frequently deal with very rapid oscillations and make use of the units shown in Table 14.1.

NOTE ▶ Uppercase and lowercase letters *are* important. 1 MHz is 1 megahertz = 10^6 Hz, but 1 mHz is 1 millihertz = 10^{-3} Hz! ◀

EXAMPLE 14.1 Frequency and period of a loudspeaker cone

What is the oscillation period of a loudspeaker cone that vibrates back and forth 5000 times per second?

SOLVE The oscillation frequency is $f = 5000$ cycles/s = 5000 Hz = 5.0 kHz. The period is the inverse of the frequency; hence

$$T = \frac{1}{f} = \frac{1}{5000 \text{ Hz}} = 2.0 \times 10^{-4} \text{ s} = 200\ \mu\text{s}$$

A system can oscillate in many ways, but we will be especially interested in the smooth *sinusoidal* oscillation (i.e., like a sine or cosine) of the third graph in Figure 14.1. This sinusoidal oscillation, the most basic of all oscillatory motions, is called **simple harmonic motion,** often abbreviated SHM. Let's look at a graphical description before we dive into the mathematics of simple harmonic motion.

FIGURE 14.2a shows an air-track glider attached to a spring. If the glider is pulled out a few centimeters and released, it will oscillate back and forth on the nearly frictionless air track. FIGURE 14.2b shows actual results from an experiment in which a computer was used to measure the glider's position 20 times every second. This is a position-versus-time graph that has been rotated 90° from its usual orientation in order for the x-axis to match the motion of the glider.

The object's maximum displacement from equilibrium is called the **amplitude** A of the motion. The object's position oscillates between $x = -A$ and $x = +A$. When using a graph, notice that the amplitude is the distance from the *axis* to the maximum, *not* the distance from the minimum to the maximum.

FIGURE 14.3a shows the data with the graph axes in their "normal" positions. You can see that the amplitude in this experiment was $A = 0.17$ m, or 17 cm. You can also measure the period to be $T = 1.60$ s. Thus the oscillation frequency was $f = 1/T = 0.625$ Hz.

FIGURE 14.3b is a velocity-versus-time graph that the computer produced by using $\Delta x/\Delta t$ to find the slope of the position graph at each point. The velocity graph is also sinusoidal, oscillating between $-v_{max}$ (maximum speed to the left) and $+v_{max}$ (maximum speed to the right). As the figure shows,

- The instantaneous velocity is zero at the points where $x = \pm A$. These are the *turning points* in the motion.
- The maximum speed v_{max} is reached as the object passes through the equilibrium position at $x = 0$ m. The *velocity* is positive as the object moves to the right but *negative* as it moves to the left.

FIGURE 14.3 Position and velocity graphs of the experimental data.

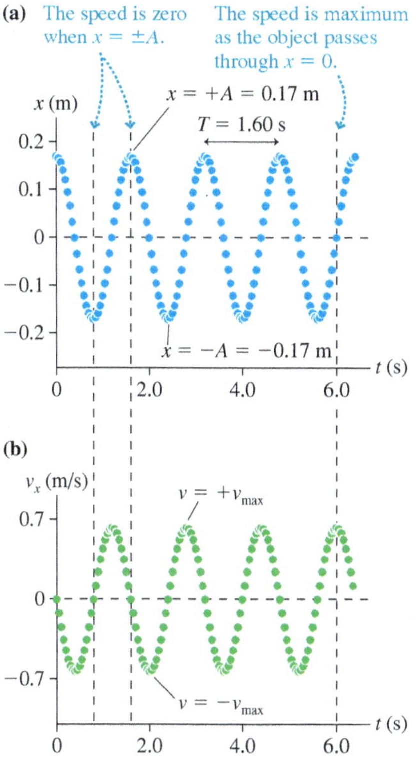

We can ask three important questions about this oscillating system:

1. How is the maximum speed v_{max} related to the amplitude A?
2. How are the period and frequency related to the object's mass m, the spring constant k, and the amplitude A?
3. Is the sinusoidal oscillation a consequence of Newton's laws?

A mass oscillating on a spring is the prototype of simple harmonic motion. Our analysis, in which we answer these questions, will be of a spring-mass system. Even so, most of what we learn will be applicable to other types of SHM.

Kinematics of Simple Harmonic Motion

FIGURE 14.4 redraws the position-versus-time graph of Figure 14.3a as a smooth curve. Although these are empirical data (we don't yet have any "theory" of oscillation) the graph for this particular motion is clearly a cosine function. The object's position is

$$x(t) = A\cos\left(\frac{2\pi t}{T}\right) \qquad (14.2)$$

where the notation $x(t)$ indicates that the position x is a *function* of time t. Because $\cos(2\pi) = \cos(0)$, it's easy to see that the position at time $t = T$ is the same as the position at $t = 0$. In other words, this is a cosine function with period T. Be sure to convince yourself that this function agrees with the five special points shown in Figure 14.4.

FIGURE 14.4 The position-versus-time graph for simple harmonic motion.

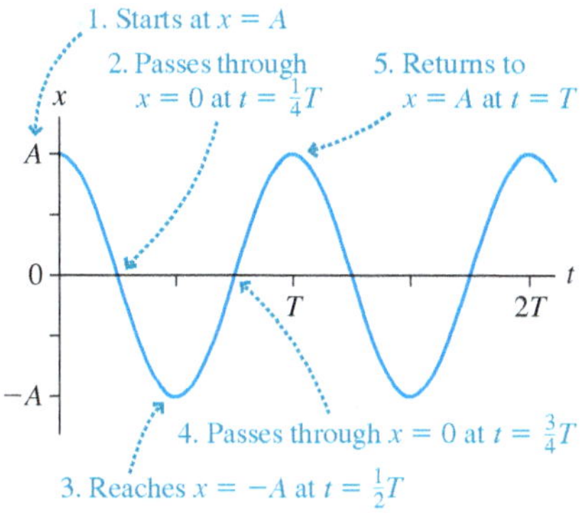

NOTE ▶ The argument of the cosine function is in *radians*. That will be true throughout this chapter. It's especially important to remember to set your calculator to radian mode before working oscillation problems. Leaving it in degree mode will lead to errors. ◀

We can write Equation 14.2 in two alternative forms. Because the oscillation frequency is $f = 1/T$, we can write

$$x(t) = A\cos(2\pi f t) \qquad (14.3)$$

Recall from Chapter 4 that a particle in circular motion has an *angular velocity* ω that is related to the period by $\omega = 2\pi/T$, where ω is in rad/s. Now that we've defined the frequency f, you can see that ω and f are related by

$$\omega\ (\text{in rad/s}) = \frac{2\pi}{T} = 2\pi f\ (\text{in Hz}) \qquad (14.4)$$

In this context, ω is called the **angular frequency.** The position can be written in terms of ω as

$$x(t) = A\cos\omega t \qquad (14.5)$$

Equations 14.2, 14.3, and 14.5 are equivalent ways to write the position of an object moving in simple harmonic motion.

TABLE 14.2 Derivatives of sine and cosine functions

$\frac{d}{dt}\big(a\sin(bt + c)\big)$	$= +ab\cos(bt + c)$
$\frac{d}{dt}\big(a\cos(bt + c)\big)$	$= -ab\sin(bt + c)$

FIGURE 14.5 Position and velocity graphs for simple harmonic motion.

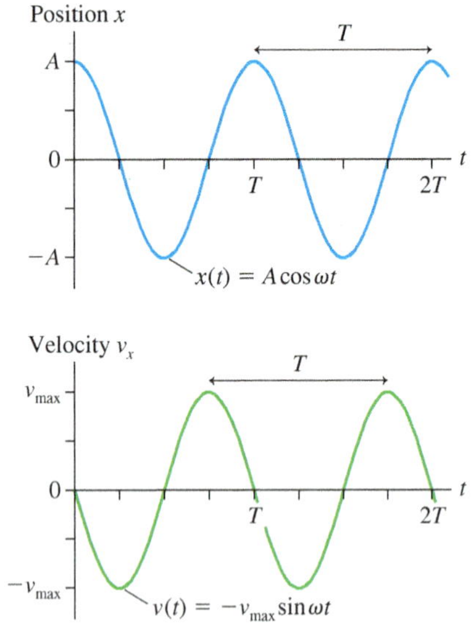

Just as the position graph was clearly a cosine function, the velocity graph shown in FIGURE 14.5 is clearly an "upside-down" sine function with the same period T. The velocity v_x, which is a function of time, can be written

$$v_x(t) = -v_{max}\sin\left(\frac{2\pi t}{T}\right) = -v_{max}\sin(2\pi f t) = -v_{max}\sin\omega t \qquad (14.6)$$

NOTE ▶ v_{max} is the maximum *speed* and thus is a *positive* number. ◀

We deduced Equation 14.6 from the experimental results, but we could equally well find it from the position function of Equation 14.2. After all, velocity is the time derivative of position. Table 14.2 on the previous page reminds you of the derivatives of the sine and cosine functions. Using the derivative of the position function, we find

$$v_x(t) = \frac{dx}{dt} = -\frac{2\pi A}{T}\sin\left(\frac{2\pi t}{T}\right) = -2\pi f A\sin(2\pi f t) = -\omega A\sin\omega t \quad (14.7)$$

Comparing Equation 14.7, the mathematical definition of velocity, to Equation 14.6, the empirical description, we see that the maximum speed of an oscillation is

$$v_{max} = \frac{2\pi A}{T} = 2\pi f A = \omega A \qquad (14.8)$$

Equation 14.8 answers the first question we posed above, which was how the maximum speed v_{max} is related to the amplitude A. Not surprisingly, the object has a greater maximum speed if you stretch the spring farther and give the oscillation a larger amplitude.

EXAMPLE 14.2 A system in simple harmonic motion

An air-track glider is attached to a spring, pulled 20.0 cm to the right, and released at $t = 0$ s. It makes 15 oscillations in 10.0 s.

a. What is the period of oscillation?
b. What is the object's maximum speed?
c. What are the position and velocity at $t = 0.800$ s?

MODEL An object oscillating on a spring is in SHM.

SOLVE a. The oscillation frequency is

$$f = \frac{15\text{ oscillations}}{10.0\text{ s}} = 1.50\text{ oscillations/s} = 1.50\text{ Hz}$$

Thus the period is $T = 1/f = 0.667$ s.

b. The oscillation amplitude is $A = 0.200$ m. Thus

$$v_{max} = \frac{2\pi A}{T} = \frac{2\pi(0.200\text{ m})}{0.667\text{ s}} = 1.88\text{ m/s}$$

c. The object starts at $x = +A$ at $t = 0$ s. This is exactly the oscillation described by Equations 14.2 and 14.6. The position at $t = 0.800$ s is

$$x = A\cos\left(\frac{2\pi t}{T}\right) = (0.200\text{ m})\cos\left(\frac{2\pi(0.800\text{ s})}{0.667\text{ s}}\right)$$
$$= (0.200\text{ m})\cos(7.54\text{ rad}) = 0.0625\text{ m} = 6.25\text{ cm}$$

The velocity at this instant of time is

$$v_x = -v_{max}\sin\left(\frac{2\pi t}{T}\right) = -(1.88\text{ m/s})\sin\left(\frac{2\pi(0.800\text{ s})}{0.667\text{ s}}\right)$$
$$= -(1.88\text{ m/s})\sin(7.54\text{ rad}) = -1.79\text{ m/s} = -179\text{ cm/s}$$

At $t = 0.800$ s, which is slightly more than one period, the object is 6.25 cm to the right of equilibrium and moving to the *left* at 179 cm/s. Notice the use of radians in the calculations.

EXAMPLE 14.3 Finding the time

A mass oscillating in simple harmonic motion starts at $x = A$ and has period T. At what time, as a fraction of T, does the object first pass through $x = \frac{1}{2}A$?

SOLVE Figure 14.4 showed that the object passes through the equilibrium position $x = 0$ at $t = \frac{1}{4}T$. This is one-quarter of the total distance in one-quarter of a period. You might expect it to take $\frac{1}{8}T$ to reach $\frac{1}{2}A$, but this is not the case because the SHM graph is not linear between $x = A$ and $x = 0$. We need to use $x(t) = A\cos(2\pi t/T)$. First, we write the equation with $x = \frac{1}{2}A$:

$$x = \frac{A}{2} = A\cos\left(\frac{2\pi t}{T}\right)$$

Then we solve for the time at which this position is reached:

$$t = \frac{T}{2\pi}\cos^{-1}\left(\frac{1}{2}\right) = \frac{T}{2\pi}\frac{\pi}{3} = \frac{1}{6}T$$

ASSESS The motion is slow at the beginning and then speeds up, so it takes longer to move from $x = A$ to $x = \frac{1}{2}A$ than it does to move from $x = \frac{1}{2}A$ to $x = 0$. Notice that the answer is independent of the amplitude A.

STOP TO THINK 14.1 An object moves with simple harmonic motion. If the amplitude and the period are both doubled, the object's maximum speed is

a. Quadrupled. b. Doubled. c. Unchanged.
d. Halved. e. Quartered.

14.2 Simple Harmonic Motion and Circular Motion

The graphs of Figure 14.5 and the position function $x(t) = A\cos\omega t$ are for an oscillation in which the object just happened to be at $x_0 = A$ at $t = 0$. But you will recall that $t = 0$ is an arbitrary choice, the instant of time when you or someone else starts a stopwatch. What if you had started the stopwatch when the object was at $x_0 = -A$, or when the object was somewhere in the middle of an oscillation? In other words, what if the oscillator had different *initial conditions.* The position graph would still show an oscillation, but neither Figure 14.5 nor $x(t) = A\cos\omega t$ would describe the motion correctly.

To learn how to describe the oscillation for other initial conditions it will help to turn to a topic you studied in Chapter 4—circular motion. There's a very close connection between simple harmonic motion and circular motion.

Imagine you have a turntable with a small ball glued to the edge. FIGURE 14.6a shows how to make a "shadow movie" of the ball by projecting a light past the ball and onto a screen. The ball's shadow oscillates back and forth as the turntable rotates. This is certainly periodic motion, with the same period as the turntable, but is it simple harmonic motion?

To find out, you could place a real object on a real spring directly below the shadow, as shown in FIGURE 14.6b. If you did so, and if you adjusted the turntable to have the same period as the spring, you would find that the shadow's motion exactly matches the simple harmonic motion of the object on the spring. **Uniform circular motion projected onto one dimension is simple harmonic motion.**

To understand this, consider the particle in FIGURE 14.7. It is in uniform circular motion, moving *counterclockwise* in a circle with radius A. As in Chapter 4, we can locate the particle by the angle ϕ measured ccw from the x-axis. Projecting the ball's shadow onto a screen in Figure 14.6 is equivalent to observing just the x-component of the particle's motion. Figure 14.7 shows that the x-component, when the particle is at angle ϕ, is

$$x = A\cos\phi \tag{14.9}$$

Recall that the particle's *angular velocity,* in rad/s, is

$$\omega = \frac{d\phi}{dt} \tag{14.10}$$

This is the rate at which the angle ϕ is increasing. If the particle starts from $\phi_0 = 0$ at $t = 0$, its angle at a later time t is simply

$$\phi = \omega t \tag{14.11}$$

As ϕ increases, the particle's x-component is

$$x(t) = A\cos\omega t \tag{14.12}$$

This is identical to Equation 14.5 for the position of a mass on a spring! Thus the x-component of a particle in uniform circular motion is simple harmonic motion.

NOTE ▶ When used to describe oscillatory motion, ω is called the *angular frequency* rather than the angular velocity. The angular frequency of an oscillator has the same numerical value, in rad/s, as the angular velocity of the corresponding particle in circular motion. ◀

FIGURE 14.6 A projection of the circular motion of a rotating ball matches the simple harmonic motion of an object on a spring.

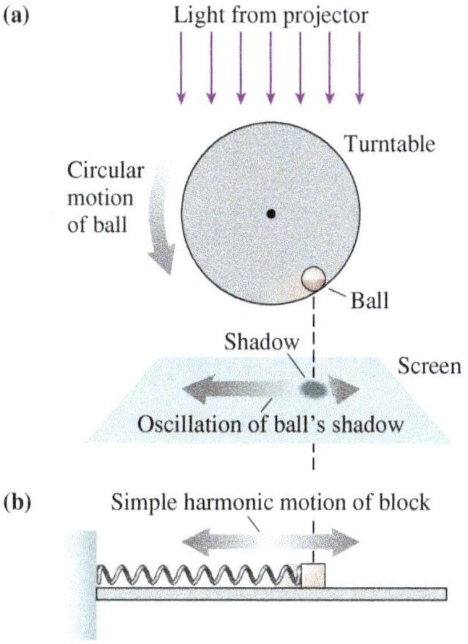

FIGURE 14.7 A particle in uniform circular motion with radius A and angular velocity ω.

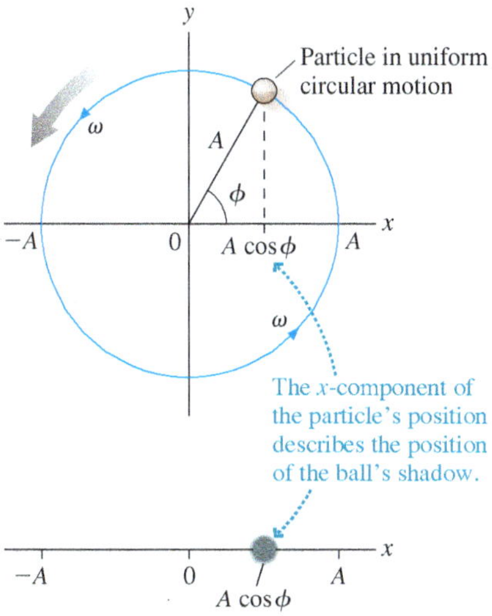

A cup on the turntable in a microwave oven moves in a circle. But from the outside, you see the cup sliding back and forth—in simple harmonic motion!

The names and units can be a bit confusing until you get used to them. It may help to notice that *cycle* and *oscillation* are not true units. Unlike the "standard meter" or the "standard kilogram," to which you could compare a length or a mass, there is no "standard cycle" to which you can compare an oscillation. Cycles and oscillations are simply counted events. Thus the frequency f has units of hertz, where $1\text{ Hz} = 1\text{ s}^{-1}$. We may *say* "cycles per second" just to be clear, but the actual units are only "per second."

The radian is the SI unit of angle. However, the radian is a *defined* unit. Further, its definition as a ratio of two lengths ($\theta = s/r$) makes it a pure number without dimensions. As we noted in Chapter 4, the unit of angle, be it radians or degrees, is really just a *name* to remind us that we're dealing with an angle. The 2π in the equation $\omega = 2\pi f$ (and in similar situations), which is stated without units, *means* 2π rad/cycle. When multiplied by the frequency f in cycles/s, it gives the frequency in rad/s. That is why, in this context, ω is called the angular *frequency.*

NOTE ▶ *Hertz* is specifically "cycles per second" or "oscillations per second." It is used for f but *not* for ω. We'll always be careful to use rad/s for ω, but you should be aware that many books give the units of ω as simply s^{-1}. ◀

The Phase Constant

FIGURE 14.8 A particle in uniform circular motion with initial angle ϕ_0.

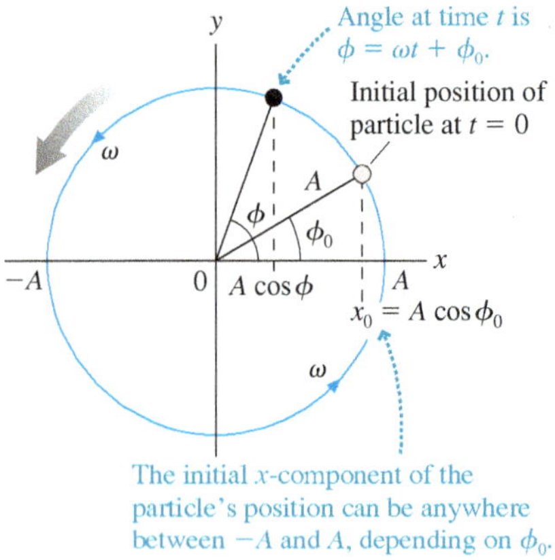

Now we're ready to consider the issue of other initial conditions. The particle in Figure 14.7 started at $\phi_0 = 0$. This was equivalent to an oscillator starting at the far right edge, $x_0 = A$. FIGURE 14.8 shows a more general situation in which the initial angle ϕ_0 can have any value. The angle at a later time t is then

$$\phi = \omega t + \phi_0 \tag{14.13}$$

In this case, the particle's projection onto the x-axis at time t is

$$x(t) = A\cos(\omega t + \phi_0) \tag{14.14}$$

If Equation 14.14 describes the particle's projection, then it must also be the position of an oscillator in simple harmonic motion. The oscillator's velocity v_x is found by taking the derivative dx/dt. The resulting equations,

$$\begin{aligned} x(t) &= A\cos(\omega t + \phi_0) \\ v_x(t) &= -\omega A\sin(\omega t + \phi_0) = -v_{max}\sin(\omega t + \phi_0) \end{aligned} \tag{14.15}$$

are the two primary kinematic equations of simple harmonic motion.

The quantity $\phi = \omega t + \phi_0$, which steadily increases with time, is called the **phase** of the oscillation. The phase is simply the *angle* of the circular-motion particle whose shadow matches the oscillator. The constant ϕ_0 is called the **phase constant.** It specifies the *initial conditions* of the oscillator.

To see what the phase constant means, set $t = 0$ in Equations 14.15:

$$\begin{aligned} x_0 &= A\cos\phi_0 \\ v_{0x} &= -\omega A\sin\phi_0 \end{aligned} \tag{14.16}$$

The position x_0 and velocity v_{0x} at $t = 0$ are the initial conditions. **Different values of the phase constant correspond to different starting points on the circle and thus to different initial conditions.**

The perfect cosine function of Figure 14.5 and the equation $x(t) = A\cos\omega t$ are for an oscillation with $\phi_0 = 0$ rad. You can see from Equations 14.16 that $\phi_0 = 0$ rad implies $x_0 = A$ and $v_0 = 0$. That is, the particle starts from rest at the point of maximum displacement.

FIGURE 14.9 illustrates these ideas by looking at three values of the phase constant: $\phi_0 = \pi/3$ rad (60°), $-\pi/3$ rad (−60°), and π rad (180°). Notice that $\phi_0 = \pi/3$ rad and $\phi_0 = -\pi/3$ rad have the same starting position, $x_0 = \frac{1}{2}A$. This is a property of the cosine function in Equation 14.16. But these are *not* the same initial conditions. In one case the oscillator starts at $\frac{1}{2}A$ while moving to the right, in the other case it starts at $\frac{1}{2}A$ while moving to the left. You can distinguish between the two by visualizing the motion.

FIGURE 14.9 Oscillations described by the phase constants $\phi_0 = \pi/3$ rad, $-\pi/3$ rad, and π rad.

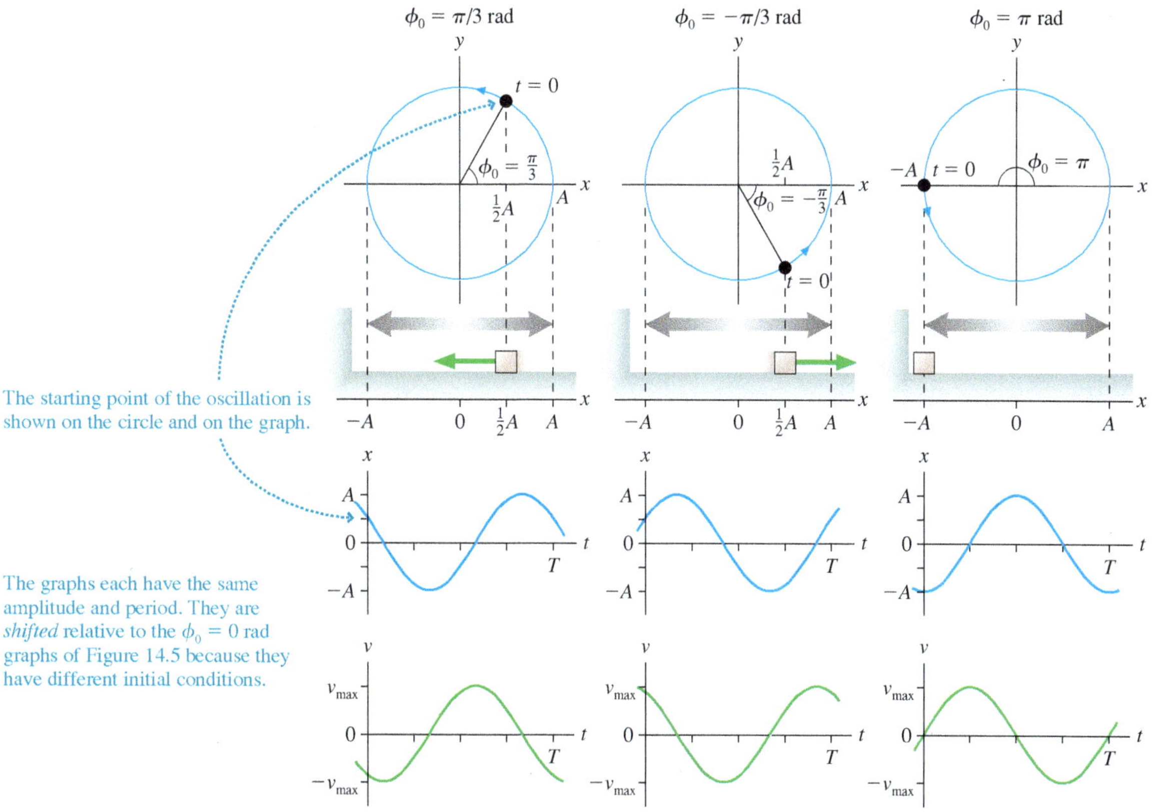

All values of the phase constant ϕ_0 between 0 and π rad correspond to a particle in the upper half of the circle and *moving to the left.* Thus v_{0x} is negative. All values of the phase constant ϕ_0 between π and 2π rad (or, as they are usually stated, between $-\pi$ and 0 rad) have the particle in the lower half of the circle and *moving to the right.* Thus v_{0x} is positive. If you're told that the oscillator is at $x = \frac{1}{2}A$ and moving to the right at $t = 0$, then the phase constant must be $\phi_0 = -\pi/3$ rad, not $+\pi/3$ rad.

EXAMPLE 14.4 Using the initial conditions

An object on a spring oscillates with a period of 0.80 s and an amplitude of 10 cm. At $t = 0$ s, it is 5.0 cm to the left of equilibrium and moving to the left. What are its position and direction of motion at $t = 2.0$ s?

MODEL An object oscillating on a spring is in simple harmonic motion.

SOLVE We can find the phase constant ϕ_0 from the initial condition $x_0 = -5.0\text{ cm} = A\cos\phi_0$. This condition gives

$$\phi_0 = \cos^{-1}\left(\frac{x_0}{A}\right) = \cos^{-1}\left(-\frac{1}{2}\right) = \pm\frac{2}{3}\pi\text{ rad} = \pm 120°$$

Because the oscillator is moving to the *left* at $t = 0$, it is in the upper half of the circular-motion diagram and must have a phase constant between 0 and π rad. Thus ϕ_0 is $\frac{2}{3}\pi$ rad. The angular frequency is

$$\omega = \frac{2\pi}{T} = \frac{2\pi}{0.80\text{ s}} = 7.85\text{ rad/s}$$

Thus the object's position at time $t = 2.0$ s is

$$x(t) = A\cos(\omega t + \phi_0)$$

$$= (10\text{ cm})\cos\left((7.85\text{ rad/s})(2.0\text{ s}) + \frac{2}{3}\pi\right)$$

$$= (10\text{ cm})\cos(17.8\text{ rad}) = 5.0\text{ cm}$$

The object is now 5.0 cm to the right of equilibrium. But which way is it moving? There are two ways to find out. The direct way is to calculate the velocity at $t = 2.0$ s:

$$v_x = -\omega A\sin(\omega t + \phi_0) = +68\text{ cm/s}$$

The velocity is positive, so the motion is to the right. Alternatively, we could note that the phase at $t = 2.0$ s is $\phi = 17.8$ rad. Dividing by π, you can see that

$$\phi = 17.8\text{ rad} = 5.67\pi\text{ rad} = (4\pi + 1.67\pi)\text{ rad}$$

The 4π rad represents two complete revolutions. The "extra" phase of 1.67π rad falls between π and 2π rad, so the particle in the circular-motion diagram is in the lower half of the circle and moving to the right.

NOTE ▶ The inverse-cosine function $\cos^{-1}$ is a *two-valued* function. Your calculator returns a single value, an angle between 0 rad and π rad. But the negative of this angle is also a solution. As Example 14.4 demonstrates, you must use additional information to choose between them. ◀

STOP TO THINK 14.2 The figure shows four oscillators at $t = 0$. Which one has the phase constant $\phi_0 = \pi/4$ rad?

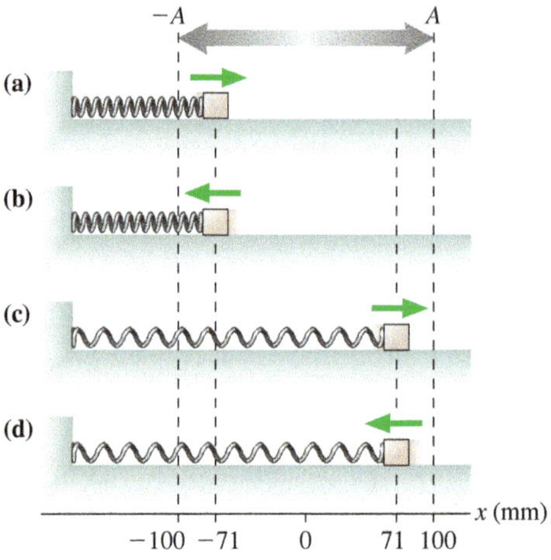

14.3 Energy in Simple Harmonic Motion

We've begun to develop the mathematical language of simple harmonic motion, but thus far we haven't included any physics. We've made no mention of the mass of the object or the spring constant of the spring. An energy analysis, using the tools of Chapters 10 and 11, is a good starting place.

FIGURE 14.10a shows an object oscillating on a spring, our prototype of simple harmonic motion. Now we'll specify that the object has mass m, the spring has spring constant k, and the motion takes place on a frictionless surface. You learned in Chapter 10 that the elastic potential energy when the object is at position x is $U_s = \frac{1}{2}k(\Delta x)^2$, where $\Delta x = x - x_e$ is the displacement from the equilibrium position x_e. In this chapter we'll always use a coordinate system in which $x_e = 0$, making $\Delta x = x$. There's no chance for confusion with gravitational potential energy, so we can omit the subscript s and write the elastic potential energy as

$$U = \frac{1}{2}kx^2 \tag{14.17}$$

Thus the mechanical energy of an object oscillating on a spring is

$$E = K + U = \frac{1}{2}mv^2 + \frac{1}{2}kx^2 \tag{14.18}$$

FIGURE 14.10b is an energy diagram, showing the potential-energy curve $U = \frac{1}{2}kx^2$ as a parabola. Recall that a particle oscillates between the *turning points* where the total energy line E crosses the potential-energy curve. The left turning point is at $x = -A$, and the right turning point is at $x = +A$. To go beyond these points would require a negative kinetic energy, which is physically impossible.

You can see that **the particle has purely potential energy at $x = \pm A$ and purely kinetic energy as it passes through the equilibrium point at $x = 0$.** At maximum displacement, with $x = \pm A$ and $v = 0$, the energy is

$$E(\text{at } x = \pm A) = U = \frac{1}{2}kA^2 \tag{14.19}$$

At $x = 0$, where $v = \pm v_{max}$, the energy is

$$E(\text{at } x = 0) = K = \frac{1}{2}m(v_{max})^2 \tag{14.20}$$

FIGURE 14.10 The energy is transformed between kinetic energy and potential energy as the object oscillates, but the mechanical energy $E = K + U$ doesn't change.

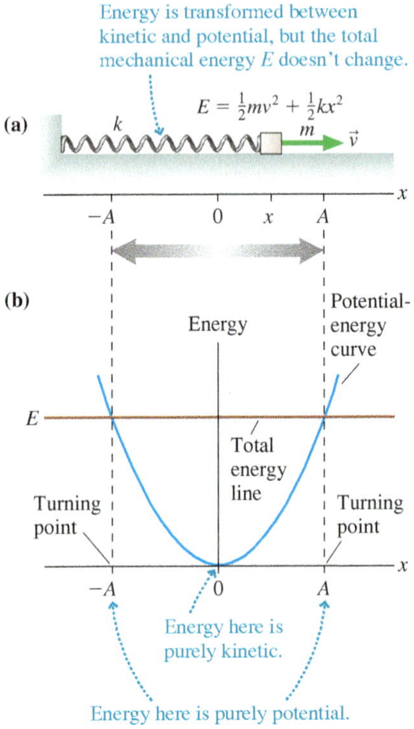

The system's mechanical energy is conserved because the surface is frictionless and there are no external forces, so the energy at maximum displacement and the energy at maximum speed, Equations 14.19 and 14.20, must be equal. That is

$$\frac{1}{2}m(v_{max})^2 = \frac{1}{2}kA^2 \tag{14.21}$$

Thus the maximum speed is related to the amplitude by

$$v_{max} = \sqrt{\frac{k}{m}}A \tag{14.22}$$

This is a relationship based on the physics of the situation.

Earlier, using kinematics, we found that

$$v_{max} = \frac{2\pi A}{T} = 2\pi fA = \omega A \tag{14.23}$$

Comparing Equations 14.22 and 14.23, we see that frequency and period of an oscillating spring are determined by the spring constant k and the object's mass m:

$$\omega = \sqrt{\frac{k}{m}} \qquad f = \frac{1}{2\pi}\sqrt{\frac{k}{m}} \qquad T = 2\pi\sqrt{\frac{m}{k}} \tag{14.24}$$

These three expressions are really only one equation. They say the same thing, but each expresses it in slightly different terms.

Equations 14.24 are the answer to the second question we posed at the beginning of the chapter, where we asked how the period and frequency are related to the object's mass m, the spring constant k, and the amplitude A. It is perhaps surprising, but **the period and frequency do not depend on the amplitude A.** A small oscillation and a large oscillation have the same period.

Because energy is conserved, we can combine Equations 14.18, 14.19, and 14.20 to write

$$E = \frac{1}{2}mv^2 + \frac{1}{2}kx^2 = \frac{1}{2}kA^2 = \frac{1}{2}m(v_{max})^2 \quad \text{(conservation of energy)} \tag{14.25}$$

Any pair of these expressions may be useful, depending on the known information. For example, you can use the amplitude A to find the speed at any point x by combining the first and second expressions for E. The speed v at position x is

$$v = \sqrt{\frac{k}{m}(A^2 - x^2)} = \omega\sqrt{A^2 - x^2} \tag{14.26}$$

FIGURE 14.11 shows graphically how the kinetic and potential energy change with time. They both oscillate but remain *positive* because x and v are squared. Energy is continuously being transformed back and forth between the kinetic energy of the moving block and the stored potential energy of the spring, but their sum remains constant. Notice that K and U both oscillate *twice* each period; make sure you understand why.

FIGURE 14.11 Kinetic energy, potential energy, and the total mechanical energy for simple harmonic motion.

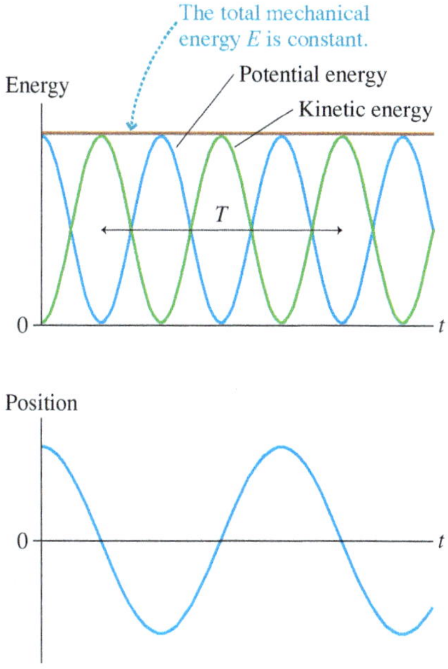

EXAMPLE 14.5 Using conservation of energy

A 500 g block on a spring is pulled a distance of 20 cm and released. The subsequent oscillations are measured to have a period of 0.80 s.

a. At what position or positions is the block's speed 1.0 m/s?
b. What is the spring constant?

MODEL The motion is SHM. Energy is conserved.

SOLVE a. The block starts from the point of maximum displacement, where $E = U = \frac{1}{2}kA^2$. At a later time, when the position is x and the speed is v, energy conservation requires

$$\frac{1}{2}mv^2 + \frac{1}{2}kx^2 = \frac{1}{2}kA^2$$

Solving for x, we find

$$x = \sqrt{A^2 - \frac{mv^2}{k}} = \sqrt{A^2 - \left(\frac{v}{\omega}\right)^2}$$

where we used $k/m = \omega^2$ from Equation 14.24. The angular frequency is easily found from the period: $\omega = 2\pi/T = 7.85$ rad/s. Thus

$$x = \sqrt{(0.20\text{ m})^2 - \left(\frac{1.0\text{ m/s}}{7.85\text{ rad/s}}\right)^2} = \pm 0.15\text{ m} = \pm 15\text{ cm}$$

There are two positions because the block has this speed on either side of equilibrium.

b. Although part a did not require that we know the spring constant, it is straightforward to find from Equation 14.24:

$$T = 2\pi\sqrt{\frac{m}{k}}$$

$$k = \frac{4\pi^2 m}{T^2} = \frac{4\pi^2(0.50\text{ kg})}{(0.80\text{ s})^2} = 31\text{ N/m}$$

STOP TO THINK 14.3

The four springs shown here have been compressed from their equilibrium position at $x = 0$ cm. When released, the attached mass will start to oscillate. Rank in order, from highest to lowest, the maximum speeds of the masses.

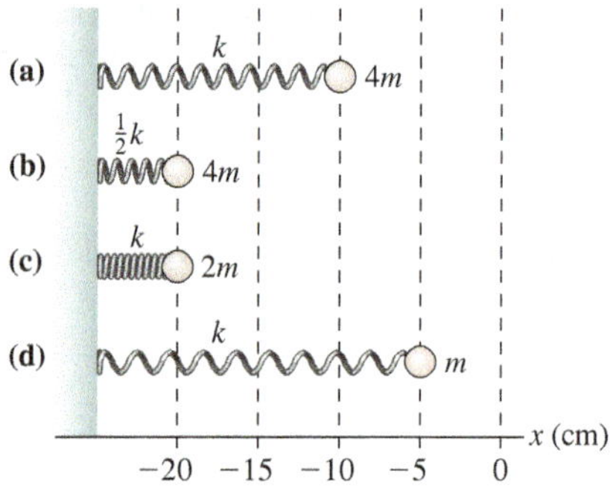

14.4 The Dynamics of Simple Harmonic Motion

Our analysis thus far has been based on the experimental observation that the oscillation of a spring "looks" sinusoidal. It's time to show that Newton's second law *predicts* sinusoidal motion.

A motion diagram will help us visualize the object's acceleration. FIGURE 14.12 shows one cycle of the motion, separating motion to the left and motion to the right to make the diagram clear. As you can see, the object's velocity is large as it passes through the equilibrium point at $x = 0$, but $\vec{v}$ is *not changing* at that point. Acceleration measures the *change* of the velocity; hence $\vec{a} = \vec{0}$ at $x = 0$.

FIGURE 14.12 Motion diagram of simple harmonic motion. The left and right motions are separated vertically for clarity but really occur along the same line.

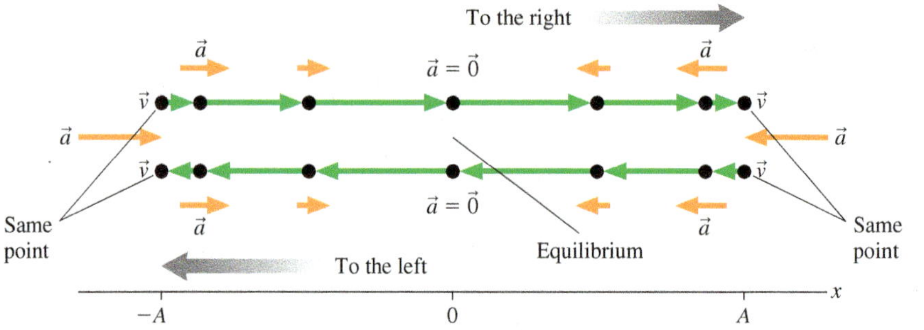

FIGURE 14.13 Position and acceleration graphs for an oscillating spring. We've chosen $\phi_0 = 0$.

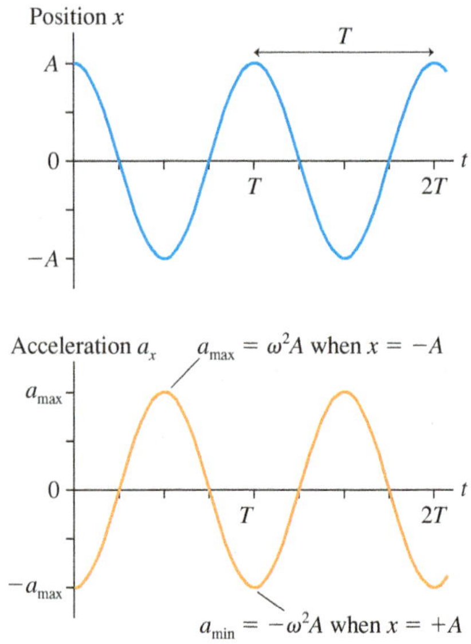

In contrast, the velocity is changing rapidly at the turning points. At the right turning point, $\vec{v}$ changes from a right-pointing vector to a left-pointing vector. Thus the acceleration $\vec{a}$ at the right turning point is large and *to the left.* In one-dimensional motion, the acceleration component a_x has a large *negative* value at the right turning point. Similarly, the acceleration $\vec{a}$ at the left turning point is large and *to the right.* Consequently, a_x has a large positive value at the left turning point.

Our motion-diagram analysis suggests that the acceleration a_x is most positive when the displacement is most negative, most negative when the displacement is a maximum, and zero when $x = 0$. This is confirmed by taking the derivative of the velocity:

$$a_x = \frac{dv_x}{dt} = \frac{d}{dt}(-\omega A \sin \omega t) = -\omega^2 A \cos \omega t \qquad (14.27)$$

then graphing it.

FIGURE 14.13 shows the position graph that we started with in Figure 14.4 and the corresponding acceleration graph. Comparing the two, you can see that the acceleration

FIGURE 14.14 The prototype of simple harmonic motion: a mass oscillating on a horizontal spring without friction.

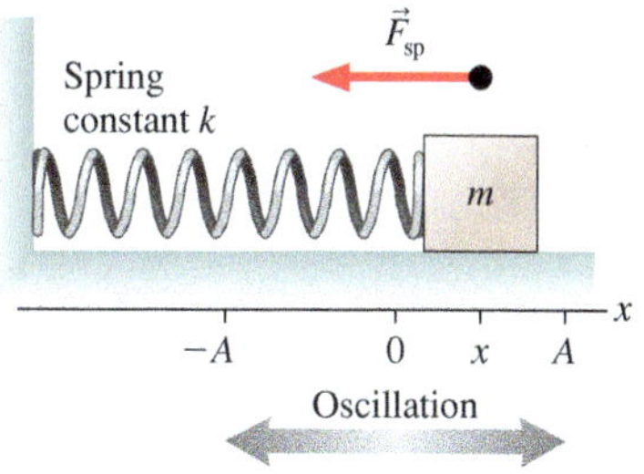

graph looks like an upside-down position graph. In fact, because $x = A\cos\omega t$, Equation 14.27 for the acceleration can be written

$$a_x = -\omega^2 x \tag{14.28}$$

That is, **the acceleration is proportional to the negative of the displacement.** The acceleration is, indeed, most positive when the displacement is most negative and is most negative when the displacement is most positive.

Recall that the acceleration is related to the net force by Newton's second law. Consider again our prototype mass on a spring, shown in FIGURE 14.14. This is the simplest possible oscillation, with no distractions due to friction or gravitational forces. We will assume the spring itself to be massless.

As you learned in Chapter 10, the spring force is given by Hooke's law:

$$(F_{sp})_x = -k\,\Delta x \tag{14.29}$$

The minus sign indicates that the spring force is a **restoring force,** a force that always points back toward the equilibrium position. If we place the origin of the coordinate system at the equilibrium position, as we've done throughout this chapter, then $\Delta x = x$ and Hooke's law is simply $(F_{sp})_x = -kx$.

The x-component of Newton's second law for the object attached to the spring is

$$(F_{net})_x = (F_{sp})_x = -kx = ma_x \tag{14.30}$$

Equation 14.30 is easily rearranged to read

$$a_x = -\frac{k}{m}x \tag{14.31}$$

You can see that Equation 14.31 is identical to Equation 14.28 if the system oscillates with angular frequency $\omega = \sqrt{k/m}$. We previously found this expression for ω from an energy analysis. Our experimental observation that the acceleration is proportional to the *negative* of the displacement is exactly what Hooke's law would lead us to expect. That's the good news.

The bad news is that a_x is not a constant. As the object's position changes, so does the acceleration. Nearly all of our kinematic tools have been based on constant acceleration. We can't use those tools to analyze oscillations, so we must go back to the very definition of acceleration:

$$a_x = \frac{dv_x}{dt} = \frac{d^2x}{dt^2}$$

Acceleration is the second derivative of position with respect to time. If we use this definition in Equation 14.31, it becomes

$$\frac{d^2x}{dt^2} = -\frac{k}{m}x \quad \text{(equation of motion for a mass on a spring)} \tag{14.32}$$

Equation 14.32, which is called the **equation of motion,** is a second-order differential equation. Unlike other equations we've dealt with, Equation 14.32 cannot be solved by direct integration. We'll need to take a different approach.

Solving the Equation of Motion

The solution to an algebraic equation such as $x^2 = 4$ is a number. The solution to a differential equation is a *function.* The x in Equation 14.32 is really $x(t)$, the position as a function of time. The solution to this equation is a function $x(t)$ whose second derivative is the function itself multiplied by $(-k/m)$.

One important property of differential equations that you will learn about in math is that the solutions are *unique.* That is, there is only *one* solution to Equation 14.32 that satisfies the initial conditions. If we were able to *guess* a solution, the uniqueness property would tell us that we had found the *only* solution. That might seem a rather

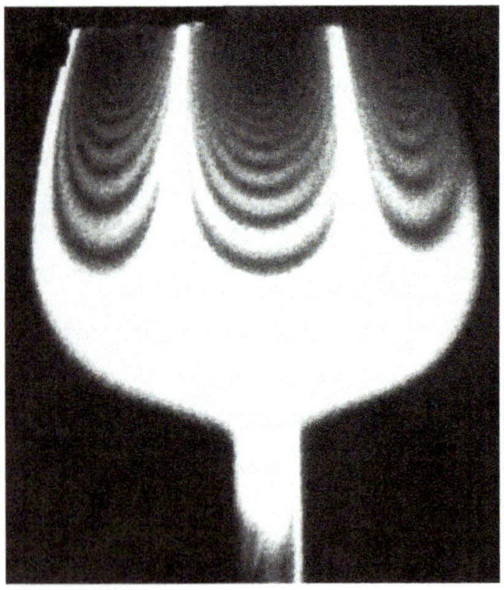

An optical technique called *interferometry* reveals the bell-like vibrations of a wine glass.

strange way to solve equations, but in fact differential equations are frequently solved by using your knowledge of what the solution needs to look like to guess an appropriate function. Let us give it a try!

We know from experimental evidence that the oscillatory motion of a spring appears to be sinusoidal. Let us *guess* that the solution to Equation 14.32 should have the functional form

$$x(t) = A\cos(\omega t + \phi_0) \tag{14.33}$$

where A, ω, and ϕ_0 are unspecified constants that we can adjust to any values that might be necessary to satisfy the differential equation.

If you were to guess that a solution to the algebraic equation $x^2 = 4$ is $x = 2$, you would verify your guess by substituting it into the original equation to see if it works. We need to do the same thing here: Substitute our guess for $x(t)$ into Equation 14.32 to see if, for an appropriate choice of the three constants, it works. To do so, we need the second derivative of $x(t)$. That is straightforward:

$$\begin{aligned} x(t) &= A\cos(\omega t + \phi_0) \\ \frac{dx}{dt} &= -\omega A\sin(\omega t + \phi_0) \\ \frac{d^2x}{dt^2} &= -\omega^2 A\cos(\omega t + \phi_0) \end{aligned} \tag{14.34}$$

If we now substitute the first and third of Equations 14.34 into Equation 14.32, we find

$$-\omega^2 A\cos(\omega t + \phi_0) = -\frac{k}{m}A\cos(\omega t + \phi_0) \tag{14.35}$$

Equation 14.35 will be true at all instants of time if and only if $\omega^2 = k/m$. There do not seem to be any restrictions on the two constants A and ϕ_0—they are determined by the initial conditions.

So we have found—by guessing!—that *the* solution to the equation of motion for a mass oscillating on a spring is

$$x(t) = A\cos(\omega t + \phi_0) \tag{14.36}$$

where the angular frequency

$$\omega = 2\pi f = \sqrt{\frac{k}{m}} \tag{14.37}$$

is determined by the mass and the spring constant.

NOTE ▶ Once again we see that the oscillation frequency is independent of the amplitude A. ◀

Equations 14.36 and 14.37 seem somewhat anticlimactic because we've been using these results for the last several pages. But keep in mind that we had been *assuming* $x = A\cos\omega t$ simply because the experimental observations "looked" like a cosine function. We've now justified that assumption by showing that Equation 14.36 really is the solution to Newton's second law for a mass on a spring. **The *theory* of oscillation, based on Hooke's law for a spring and Newton's second law, is in good agreement with the experimental observations.** This conclusion gives an affirmative answer to the last of the three questions that we asked early in the chapter, which was whether the sinusoidal oscillation of SHM is a consequence of Newton's laws.

EXAMPLE 14.6 Analyzing an oscillator

At $t = 0$ s, a 500 g block oscillating on a spring is observed moving to the right at $x = 15$ cm. It reaches a maximum displacement of 25 cm at $t = 0.30$ s.

a. Draw a position-versus-time graph for one cycle of the motion.
b. At what times during the first cycle does the mass pass through $x = 20$ cm?

MODEL The motion is simple harmonic motion.

SOLVE a. The position equation of the block is $x(t) = A\cos(\omega t + \phi_0)$. We know that the amplitude is $A = 0.25$ m and that $x_0 = 0.15$ m. From these two pieces of information we obtain the phase constant:

$$\phi_0 = \cos^{-1}\left(\frac{x_0}{A}\right) = \cos^{-1}(0.60) = \pm 0.927 \text{ rad}$$

The object is initially moving to the right, which tells us that the phase constant must be between $-\pi$ and 0 rad. Thus $\phi_0 = -0.927$ rad. The block reaches its maximum displacement $x_{max} = A$ at time $t = 0.30$ s. At that instant of time

$$x_{max} = A = A\cos(\omega t + \phi_0)$$

This can be true only if $\cos(\omega t + \phi_0) = 1$, which requires $\omega t + \phi_0 = 0$. Thus

$$\omega = \frac{-\phi_0}{t} = \frac{-(-0.927 \text{ rad})}{0.30 \text{ s}} = 3.09 \text{ rad/s}$$

Now that we know ω, it is straightforward to compute the period:

$$T = \frac{2\pi}{\omega} = 2.0 \text{ s}$$

FIGURE 14.15 graphs $x(t) = (25 \text{ cm})\cos(3.09t - 0.927)$, where t is in s, from $t = 0$ s to $t = 2.0$ s.

b. From $x = A\cos(\omega t + \phi_0)$, the time at which the mass reaches position $x = 20$ cm is

FIGURE 14.15 Position-versus-time graph for the oscillator of Example 14.6.

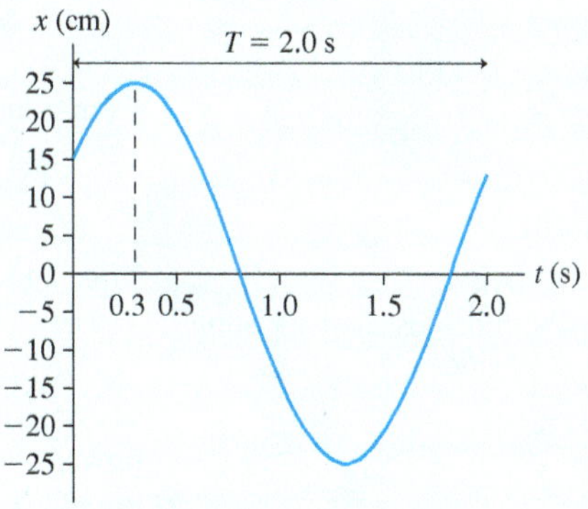

$$t = \frac{1}{\omega}\left(\cos^{-1}\left(\frac{x}{A}\right) - \phi_0\right)$$

$$= \frac{1}{3.09 \text{ rad/s}}\left(\cos^{-1}\left(\frac{20 \text{ cm}}{25 \text{ cm}}\right) + 0.927 \text{ rad}\right) = 0.51 \text{ s}$$

A calculator returns only one value of $\cos^{-1}$, in the range 0 to π rad, but we noted earlier that $\cos^{-1}$ actually has two values. Indeed, you can see in Figure 14.15 that there are two times at which the mass passes $x = 20$ cm. Because they are symmetrical on either side of $t = 0.30$ s, when $x = A$, the first point is $(0.51 \text{ s} - 0.30 \text{ s}) = 0.21$ s *before* the maximum. Thus the mass passes through $x = 20$ cm at $t = 0.09$ s and again at $t = 0.51$ s.

STOP TO THINK 14.4 This is the position graph of a mass on a spring. What can you say about the velocity and the force at the instant indicated by the dashed line?

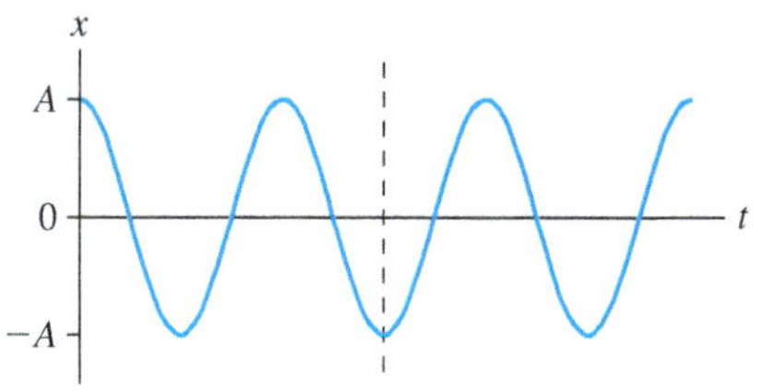

a. Velocity is positive; force is to the right.
b. Velocity is negative; force is to the right.
c. Velocity is zero; force is to the right.
d. Velocity is positive; force is to the left.
e. Velocity is negative; force is to the left.
f. Velocity is zero; force is to the left.
g. Velocity and force are both zero.

14.5 Vertical Oscillations

We have focused our analysis on a horizontally oscillating spring. But the typical demonstration you'll see in class is a mass bobbing up and down on a spring hung vertically from a support. Is it safe to assume that a vertical oscillation has the same mathematical description as a horizontal oscillation? Or does the additional force of gravity change the motion? Let us look at this more carefully.

FIGURE 14.16 shows a block of mass m hanging from a spring of spring constant k. An important fact to notice is that the equilibrium position of the block is *not* where the spring is at its unstretched length. At the equilibrium position of the block, where it hangs motionless, the spring has stretched by ΔL.

Finding ΔL is a static-equilibrium problem in which the upward spring force balances the downward gravitational force on the block. The y-component of the spring force is given by Hooke's law:

$$(F_{sp})_y = -k\,\Delta y = +k\,\Delta L \tag{14.38}$$

FIGURE 14.16 Gravity stretches the spring.

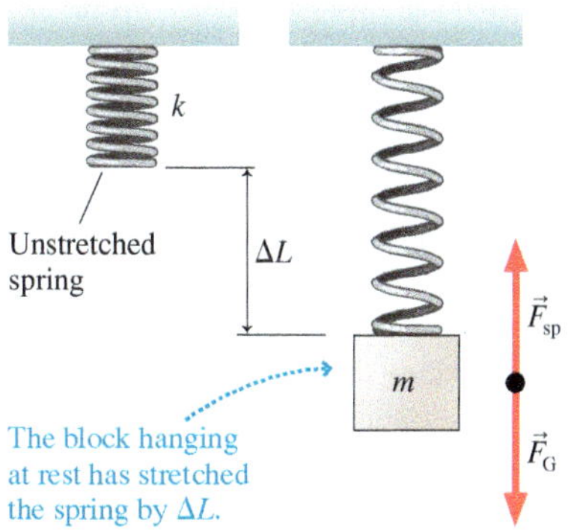

Equation 14.38 makes a distinction between ΔL, which is simply a *distance* and is a positive number, and the displacement Δy. The block is displaced downward, so $\Delta y = -\Delta L$. Newton's first law for the block in equilibrium is

$$(F_{net})_y = (F_{sp})_y + (F_G)_y = k\,\Delta L - mg = 0 \qquad (14.39)$$

from which we can find

$$\Delta L = \frac{mg}{k} \qquad (14.40)$$

This is the distance the spring stretches when the block is attached to it.

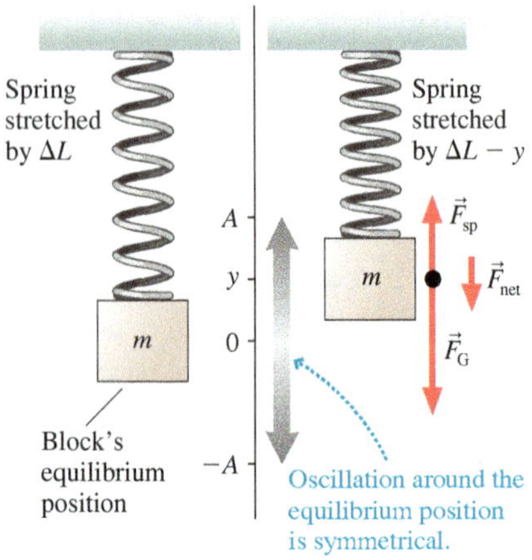

FIGURE 14.17 The block oscillates around the equilibrium position.

Let the block oscillate around this equilibrium position, as shown in FIGURE 14.17. We've now placed the origin of the y-axis at the block's equilibrium position in order to be consistent with our analyses of oscillations throughout this chapter. If the block moves upward, as the figure shows, the spring gets shorter compared to its equilibrium length, but the spring is still *stretched* compared to its unstretched length in Figure 14.16. When the block is at position y, the spring is stretched by an amount $\Delta L - y$ and hence exerts an *upward* spring force $F_{sp} = k(\Delta L - y)$. The net force on the block at this point is

$$(F_{net})_y = (F_{sp})_y + (F_G)_y = k(\Delta L - y) - mg = (k\,\Delta L - mg) - ky \qquad (14.41)$$

But $k\,\Delta L - mg$ is zero, from Equation 14.40, so the net force on the block is simply

$$(F_{net})_y = -ky \qquad (14.42)$$

Equation 14.42 for vertical oscillations is *exactly* the same as Equation 14.30 for horizontal oscillations, where we found $(F_{net})_x = -kx$. That is, the restoring force for vertical oscillations is identical to the restoring force for horizontal oscillations. The role of gravity is to determine where the equilibrium position is, but it doesn't affect the oscillatory motion around the equilibrium position.

Because the net force is the same, Newton's second law has exactly the same oscillatory solution:

$$y(t) = A\cos(\omega t + \phi_0) \qquad (14.43)$$

with, again, $\omega = \sqrt{k/m}$. **The vertical oscillations of a mass on a spring are the same simple harmonic motion as those of a block on a horizontal spring.** This is an important finding because it was not obvious that the motion would still be simple harmonic motion when gravity was included.

EXAMPLE 14.7 Bungee oscillations

An 83 kg student hangs from a bungee cord with spring constant 270 N/m. The student is pulled down to a point where the cord is 5.0 m longer than its unstretched length, then released. Where is the student, and what is his velocity 2.0 s later?

MODEL A bungee cord can be modeled as a spring. Vertical oscillations on the bungee cord are SHM.

VISUALIZE FIGURE 14.18 shows the situation.

FIGURE 14.18 A student on a bungee cord oscillates about the equilibrium position.

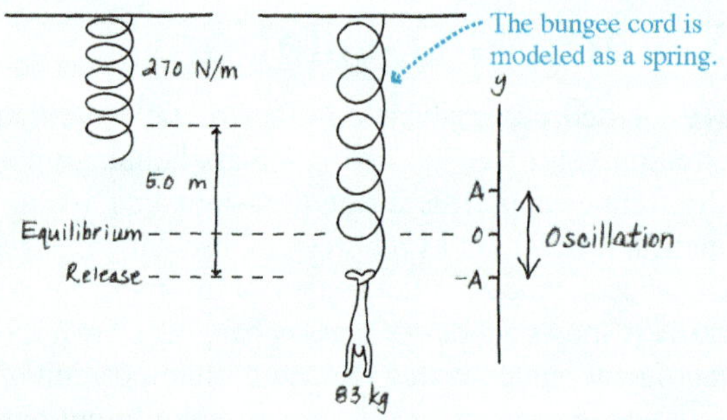

SOLVE Although the cord is stretched by 5.0 m when the student is released, this is *not* the amplitude of the oscillation. Oscillations occur around the equilibrium position, so we have to begin by finding the equilibrium point where the student hangs motionless. The cord stretch at equilibrium is given by Equation 14.40:

$$\Delta L = \frac{mg}{k} = 3.0 \text{ m}$$

Stretching the cord 5.0 m pulls the student 2.0 m below the equilibrium point, so $A = 2.0$ m. That is, the student oscillates with amplitude $A = 2.0$ m about a point 3.0 m beneath the bungee cord's original end point. The student's position as a function of time, as measured from the equilibrium position, is

$$y(t) = (2.0 \text{ m})\cos(\omega t + \phi_0)$$

where $\omega = \sqrt{k/m} = 1.80$ rad/s. The initial condition

$$y_0 = A\cos\phi_0 = -A$$

requires the phase constant to be $\phi_0 = \pi$ rad. At $t = 2.0$ s the student's position and velocity are

$$y = (2.0\text{ m})\cos\big((1.80\text{ rad/s})(2.0\text{ s}) + \pi\text{ rad}\big) = 1.8\text{ m}$$
$$v_y = -\omega A\sin(\omega t + \phi_0) = -1.6\text{ m/s}$$

The student is 1.8 m *above* the equilibrium position, or 1.2m *below* the original end of the cord. Because his velocity is negative, he's passed through the highest point and is heading down.

14.6 The Pendulum

Now let's look at another very common oscillator: a pendulum. FIGURE 14.19a shows a mass m attached to a string of length L and free to swing back and forth. The pendulum's position can be described by the arc of length s, which is zero when the pendulum hangs straight down. Because angles are measured ccw, s and θ are positive when the pendulum is to the right of center, negative when it is to the left.

Two forces are acting on the mass: the string tension $\vec{T}$ and gravity $\vec{F}_G$. It will be convenient to repeat what we did in our study of circular motion: Divide the forces into tangential components, parallel to the motion, and radial components parallel to the string. These are shown on the free-body diagram of FIGURE 14.19b.

Newton's second law for the tangential component, parallel to the motion, is

$$(F_{\text{net}})_t = \sum F_t = (F_G)_t = -mg\sin\theta = ma_t \qquad (14.44)$$

Using $a_t = d^2s/dt^2$ for acceleration "around" the circle, and noting that the mass cancels, we can write Equation 14.44 as

$$\frac{d^2s}{dt^2} = -g\sin\theta \qquad (14.45)$$

This is the equation of motion for an oscillating pendulum. The sine function makes this equation more complicated than the equation of motion for an oscillating spring.

FIGURE 14.19 The motion of a pendulum.

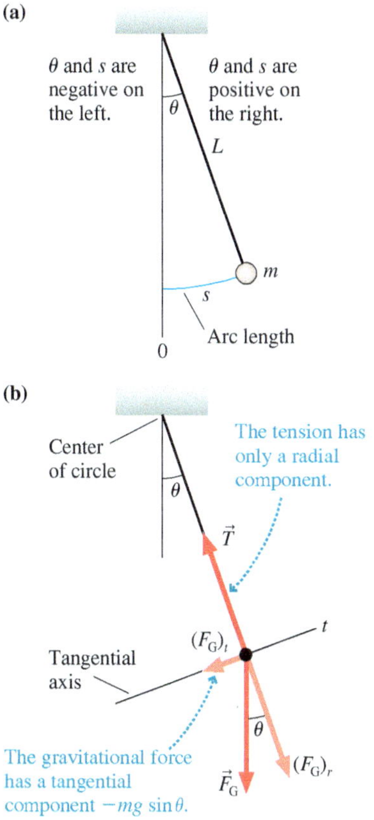

The Small-Angle Approximation

Suppose we restrict the pendulum's oscillations to *small angles* of less than about 10°. This restriction allows us to make use of an interesting and important piece of geometry.

FIGURE 14.20 shows an angle θ and a circular arc of length $s = r\theta$. A right triangle has been constructed by dropping a perpendicular from the top of the arc to the axis. The height of the triangle is $h = r\sin\theta$. Suppose that the angle θ is "small." In that case there is very little difference between h and s. If $h \approx s$, then $r\sin\theta \approx r\theta$. It follows that

$$\sin\theta \approx \theta \quad (\theta \text{ in radians})$$

FIGURE 14.20 The geometrical basis of the small-angle approximation.

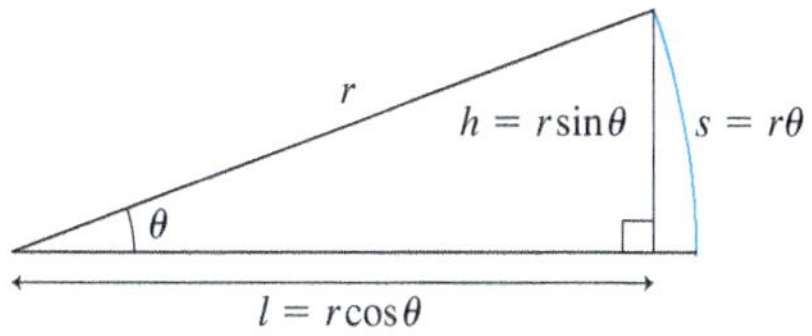

The result that $\sin\theta \approx \theta$ for small angles is called the **small-angle approximation.** We can similarly note that $l \approx r$ for small angles. Because $l = r\cos\theta$, it follows that $\cos\theta \approx 1$. Finally, we can take the ratio of sine and cosine to find $\tan\theta \approx \sin\theta \approx \theta$. Table 14.3 summarizes the small-angle approximation. We will have other occasions to use the small-angle approximation throughout the remainder of this text.

NOTE ▶ The small-angle approximation is valid *only* if angle θ is in radians! ◀

How small does θ have to be to justify using the small-angle approximation? It's easy to use your calculator to find that the small-angle approximation is good to three

TABLE 14.3 Small-angle approximations. θ must be in radians.

$\sin\theta \approx \theta$	$\tan\theta \approx \sin\theta \approx \theta$
$\cos\theta \approx 1$	

The pendulum clock has been used for hundreds of years.

significant figures, an error of $\leq 0.1\%$, up to angles of ≈ 0.10 rad ($\approx 5°$). In practice, we will use the approximation up to about 10°, but for angles any larger it rapidly loses validity and produces unacceptable results.

If we restrict the pendulum to $\theta < 10°$, we can use $\sin\theta \approx \theta$. In that case, Equation 14.44 for the net force on the mass is

$$(F_{net})_t = -mg\sin\theta \approx -mg\theta = -\frac{mg}{L}s$$

where, in the last step, we used the fact that angle θ is related to the arc length by $\theta = s/L$. Then the equation of motion becomes

$$\frac{d^2s}{dt^2} = -\frac{g}{L}s \quad (14.46)$$

This is *exactly* the same as Equation 14.32 for a mass oscillating on a spring. The names are different, with x replaced by s and k/m by g/L, but that does not make it a different equation.

Because we know the solution to the spring problem, we can immediately write the solution to the pendulum problem just by changing variables and constants:

$$s(t) = A\cos(\omega t + \phi_0) \quad \text{or} \quad \theta(t) = \theta_{max}\cos(\omega t + \phi_0) \quad (14.47)$$

The angular frequency

$$\omega = 2\pi f = \sqrt{\frac{g}{L}} \quad (14.48)$$

is determined by the length of the string. The pendulum is interesting in that **the frequency, and hence the period, is independent of the mass.** It depends only on the length of the pendulum. The amplitude A and the phase constant ϕ_0 are determined by the initial conditions, just as they were for an oscillating spring.

EXAMPLE 14.8 The maximum angle of a pendulum

A 300 g mass on a 30-cm-long string oscillates as a pendulum. It has a speed of 0.25 m/s as it passes through the lowest point. What maximum angle does the pendulum reach?

MODEL Assume that the angle remains small, in which case the motion is simple harmonic motion.

SOLVE The angular frequency of the pendulum is

$$\omega = \sqrt{\frac{g}{L}} = \sqrt{\frac{9.8\text{ m/s}^2}{0.30\text{ m}}} = 5.72\text{ rad/s}$$

The speed at the lowest point is $v_{max} = \omega A$, so the amplitude is

$$A = s_{max} = \frac{v_{max}}{\omega} = \frac{0.25\text{ m/s}}{5.72\text{ rad/s}} = 0.0437\text{ m}$$

The maximum angle, at the maximum arc length s_{max}, is

$$\theta_{max} = \frac{s_{max}}{L} = \frac{0.0437\text{ m}}{0.30\text{ m}} = 0.146\text{ rad} = 8.3°$$

ASSESS Because the maximum angle is less than 10°, our analysis based on the small-angle approximation is reasonable.

EXAMPLE 14.9 The gravimeter

Deposits of minerals and ore can alter the local value of the free-fall acceleration because they tend to be denser than surrounding rocks. Geologists use a *gravimeter*—an instrument that accurately measures the local free-fall acceleration—to search for ore deposits. One of the simplest gravimeters is a pendulum. To achieve the highest accuracy, a stopwatch is used to time 100 oscillations of a pendulum of different lengths. At one location in the field, a geologist makes the following measurements:

Length (m)	Time (s)
0.500	141.7
1.000	200.6
1.500	245.8
2.000	283.5

What is the local value of g?

MODEL Assume the oscillation angle is small, in which case the motion is simple harmonic motion with a period independent of the mass of the pendulum. Because the data are known to four significant figures (± 1 mm on the length and ± 0.1 s on the timing, both of which are easily achievable), we expect to determine g to four significant figures.

SOLVE From Equation 14.48, using $f = 1/T$, we find

$$T^2 = \left(2\pi\sqrt{\frac{L}{g}}\right)^2 = \frac{4\pi^2}{g}L$$

That is, the square of a pendulum's period is proportional to its length. Consequently, a graph of T^2 versus L should be a straight line passing through the origin with slope $4\pi^2/g$. We can use the experimentally measured slope to determine g. FIGURE 14.21 is a graph of the data, with the period found by dividing the measured time by 100.

As expected, the graph is a straight line passing through the origin. The slope of the best-fit line is 4.021 s^2/m. Consequently,

$$g = \frac{4\pi^2}{\text{slope}} = \frac{4\pi^2}{4.021\ \text{s}^2/\text{m}} = 9.818\ \text{m/s}^2$$

ASSESS The fact that the graph is linear and passes through the origin confirms our model of the situation. Had this *not* been the case, we would have had to conclude either that our model of the pendulum as a simple, small-angle pendulum was not valid or that our measurements were bad. This is an important reason for having multiple data points rather than using only one length.

FIGURE 14.21 Graph of the square of the pendulum's period versus its length.

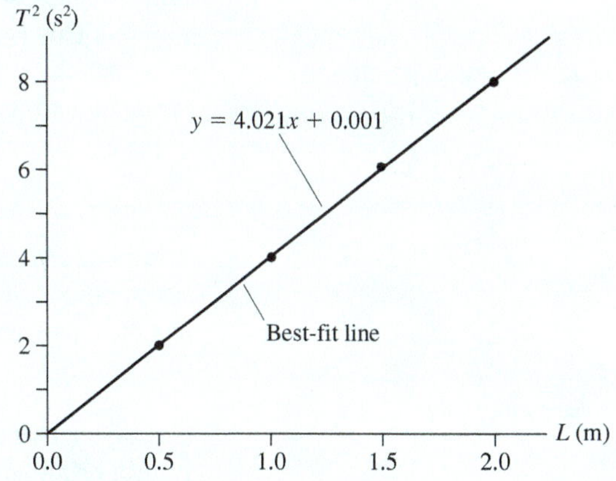

The Conditions for Simple Harmonic Motion

You can begin to see how, in a sense, we have solved *all* simple-harmonic-motion problems once we have solved the problem of the horizontal spring. The restoring force of a spring, $F_{sp} = -kx$, is directly proportional to the displacement x from equilibrium. The pendulum's restoring force, in the small-angle approximation, is directly proportional to the displacement s. A restoring force that is directly proportional to the displacement from equilibrium is called a **linear restoring force.** For *any* linear restoring force, the equation of motion is identical to the spring equation (other than perhaps using different symbols). Consequently, **any system with a linear restoring force will undergo simple harmonic motion around the equilibrium position.**

This is why an oscillating spring is the prototype of SHM. Everything that we learn about an oscillating spring can be applied to the oscillations of any other linear restoring force, ranging from the vibration of airplane wings to the motion of electrons in electric circuits. Let's summarize this information with a Tactics Box.

TACTICS BOX 14.1 Identifying and analyzing simple harmonic motion (MP)

1. If the net force acting on a particle is a linear restoring force, the motion will be simple harmonic motion around the equilibrium position.
2. The position as a function of time is $x(t) = A\cos(\omega t + \phi_0)$. The velocity as a function of time is $v_x(t) = -\omega A\sin(\omega t + \phi_0)$. The maximum speed is $v_{max} = \omega A$. The equations are given here in terms of x, but they can be written in terms of y, θ, or some other parameter if the situation calls for it.
3. The amplitude A and the phase constant ϕ_0 are determined by the initial conditions through $x_0 = A\cos\phi_0$ and $v_{0x} = -\omega A\sin\phi_0$.
4. The angular frequency ω (and hence the period $T = 2\pi/\omega$) depends on the physics of the particular situation. But ω does *not* depend on A or ϕ_0.
5. Mechanical energy is conserved. Thus $\frac{1}{2}mv_x^2 + \frac{1}{2}kx^2 = \frac{1}{2}kA^2 = \frac{1}{2}m(v_{max})^2$. Energy conservation provides a relationship between position and velocity that is independent of time.

Exercises 7–12, 15–19

The Physical Pendulum

A mass on a string is often called a *simple pendulum.* But you can also make a pendulum from any solid object that swings back and forth on a pivot under the influence of gravity. This is called a *physical pendulum.*

FIGURE 14.22 A physical pendulum.

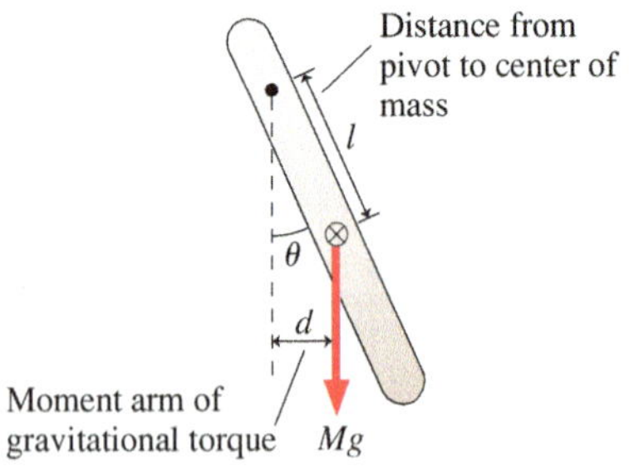

FIGURE 14.22 shows a physical pendulum of mass M for which the distance between the pivot and the center of mass is l. The moment arm of the gravitational force acting at the center of mass is $d = l\sin\theta$, so the gravitational torque is

$$\tau = -Mgd = -Mgl\sin\theta$$

The torque is negative because, for positive θ, it's causing a clockwise rotation. If we restrict the angle to being small ($\theta < 10°$), as we did for the simple pendulum, we can use the small-angle approximation to write

$$\tau = -Mgl\theta \tag{14.49}$$

Gravity causes a linear restoring torque on the pendulum—that is, the torque is directly proportional to the angular displacement θ—so we expect the physical pendulum to undergo SHM.

From Chapter 12, Newton's second law for rotational motion is

$$\alpha = \frac{d^2\theta}{dt^2} = \frac{\tau}{I}$$

where I is the object's moment of inertia about the pivot point. Using Equation 14.49 for the torque, we find

$$\frac{d^2\theta}{dt^2} = \frac{-Mgl}{I}\theta \tag{14.50}$$

Comparison with Equation 14.32 shows that this is again the SHM equation of motion, this time with angular frequency

$$\omega = 2\pi f = \sqrt{\frac{Mgl}{I}} \tag{14.51}$$

It appears that the frequency depends on the mass of the pendulum, but recall that the moment of inertia is directly proportional to M. Thus M cancels and the frequency of a physical pendulum, like that of a simple pendulum, is independent of mass.

EXAMPLE 14.10 A swinging leg as a pendulum

A student in a biomechanics lab measures the length of his leg, from hip to heel, to be 0.90 m. What is the frequency of the pendulum motion of the student's leg? What is the period?

MODEL We can model a human leg reasonably well as a rod of uniform cross section, pivoted at one end (the hip) to form a physical pendulum. The center of mass of a uniform leg is at the midpoint, so $l = L/2$.

SOLVE The moment of inertia of a rod pivoted about one end is $I = \frac{1}{3}ML^2$, so the pendulum frequency is

$$f = \frac{1}{2\pi}\sqrt{\frac{Mgl}{I}} = \frac{1}{2\pi}\sqrt{\frac{Mg(L/2)}{ML^2/3}} = \frac{1}{2\pi}\sqrt{\frac{3g}{2L}} = 0.64\text{ Hz}$$

The corresponding period is $T = 1/f = 1.6$ s. Notice that we didn't need to know the mass.

ASSESS As you walk, your legs do swing as physical pendulums as you bring them forward. The frequency is fixed by the length of your legs and their distribution of mass; it doesn't depend on amplitude. Consequently, you don't increase your walking speed by taking more rapid steps—changing the frequency is difficult. You simply take longer strides, changing the amplitude but not the frequency.

STOP TO THINK 14.5 One person swings on a swing and finds that the period is 3.0 s. A second person of equal mass joins him. With two people swinging, the period is

a. 6.0 s
b. >3.0 s but not necessarily 6.0 s
c. 3.0 s
d. <3.0 s but not necessarily 1.5 s
e. 1.5 s
f. Can't tell without knowing the length

14.7 Damped Oscillations

A pendulum left to itself gradually slows down and stops. The sound of a ringing bell gradually dies away. All real oscillators do run down—some very slowly but others quite quickly—as friction or other dissipative forces transform their mechanical energy into the thermal energy of the oscillator and its environment. An oscillation that runs down and stops is called a **damped oscillation.**

There are many possible reasons for the dissipation of energy, such as air resistance, friction, and internal forces within a metal spring as it flexes. The forces involved in dissipation are complex, but a simple *linear drag* model gives a quite accurate description of most damped oscillations. That is, we'll assume a drag force that depends linearly on the velocity as

$$\vec{D} = -b\vec{v} \quad \text{(model of the drag force)} \tag{14.52}$$

where the minus sign is the mathematical statement that the force is always opposite in direction to the velocity in order to slow the object.

The shock absorbers in cars and trucks are heavily damped springs. The vehicle's vertical motion, after hitting a rock or a pothole, is a damped oscillation.

The **damping constant** b depends in a complicated way on the shape of the object *and* on the viscosity of the air or other medium in which the particle moves. The damping constant plays the same role in our model of air resistance that the coefficient of friction does in our model of friction.

The units of b need to be such that they will give units of force when multiplied by units of velocity. As you can confirm, these units are kg/s. A value $b = 0$ kg/s corresponds to the limiting case of no resistance, in which case the mechanical energy is conserved. A typical value of b for a spring or a pendulum in air is ≤ 0.10 kg/s. Objects moving in a liquid can have significantly larger values of b.

FIGURE 14.23 shows a mass oscillating on a spring in the presence of a drag force. With the drag included, Newton's second law is

$$(F_{net})_x = (F_{sp})_x + D_x = -kx - bv_x = ma_x \tag{14.53}$$

FIGURE 14.23 An oscillating mass in the presence of a drag force.

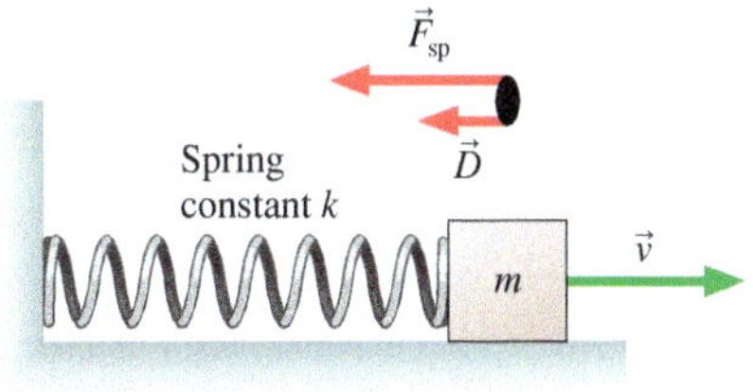

Using $v_x = dx/dt$ and $a_x = d^2x/dt^2$, we can write Equation 14.53 as

$$\frac{d^2x}{dt^2} + \frac{b}{m}\frac{dx}{dt} + \frac{k}{m}x = 0 \tag{14.54}$$

Equation 14.54 is the equation of motion of a damped oscillator. If you compare it to Equation 14.32, the equation of motion for a block on a frictionless surface, you'll see that it differs by the inclusion of the term involving dx/dt.

Equation 14.54 is another second-order differential equation. We will simply assert (and, as a homework problem, you can confirm) that the solution is

$$x(t) = Ae^{-bt/2m}\cos(\omega t + \phi_0) \quad \text{(damped oscillator)} \tag{14.55}$$

where the angular frequency is given by

$$\omega = \sqrt{\frac{k}{m} - \frac{b^2}{4m^2}} = \sqrt{\omega_0^2 - \frac{b^2}{4m^2}} \tag{14.56}$$

FIGURE 14.24 Position-versus-time graph for a damped oscillator.

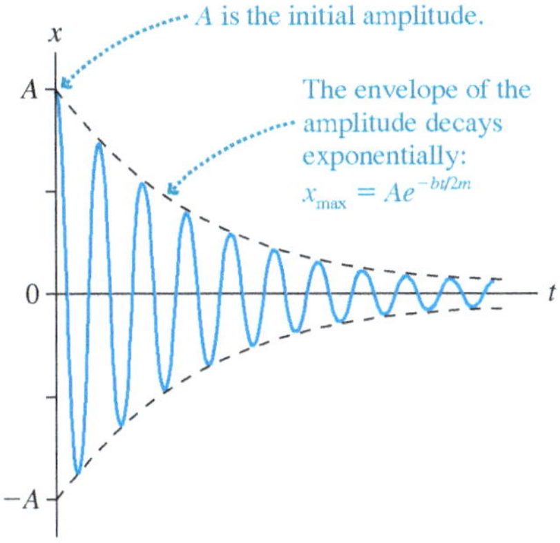

Here $\omega_0 = \sqrt{k/m}$ is the angular frequency of an undamped oscillator ($b = 0$). The constant e is the base of natural logarithms, so $e^{-bt/2m}$ is an *exponential function.*

Because $e^0 = 1$, Equation 14.55 reduces to our previous solution, $x(t) = A\cos(\omega t + \phi_0)$, when $b = 0$. This makes sense and gives us confidence in Equation 14.55. A *lightly damped* system, which oscillates many times before stopping, is one for which $b/2m \ll \omega_0$. In that case, $\omega \approx \omega_0$ is a good approximation. That is, light damping does not affect the oscillation frequency.

FIGURE 14.24 is a graph of the position $x(t)$ for a lightly damped oscillator, as given by Equation 14.55. Notice that the term $Ae^{-bt/2m}$, which is shown by the dashed line,

FIGURE 14.25 Several oscillation envelopes, corresponding to different values of the damping constant b.

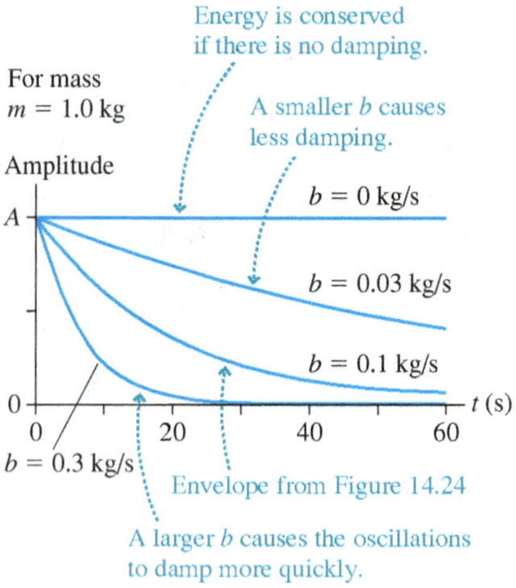

acts as a slowly varying amplitude:

$$x_{max}(t) = Ae^{-bt/2m} \tag{14.57}$$

where A is the *initial* amplitude, at $t = 0$. The oscillation keeps bumping up against this line, slowly dying out with time.

A slowly changing line that provides a border to a rapid oscillation is called the **envelope** of the oscillations. In this case, the oscillations have an *exponentially decaying envelope.* Make sure you study Figure 14.24 long enough to see how both the oscillations and the decaying amplitude are related to Equation 14.55.

Changing the amount of damping, by changing the value of b, affects how quickly the oscillations decay. FIGURE 14.25 shows just the envelope $x_{max}(t)$ for several oscillators that are identical except for the value of the damping constant b. (You need to imagine a rapid oscillation within each envelope, as in Figure 14.24.) Increasing b causes the oscillations to damp more quickly, while decreasing b makes them last longer.

MATHEMATICAL ASIDE **Exponential decay**

Exponential decay occurs in a vast number of physical systems of importance in science and engineering. Mechanical vibrations, electric circuits, and nuclear radioactivity all exhibit exponential decay.

The number $e = 2.71828\ldots$ is the base of natural logarithms in the same way that 10 is the base of ordinary logarithms. It arises naturally in calculus from the integral

$$\int \frac{du}{u} = \ln u$$

This integral—which shows up in the analysis of many physical systems—frequently leads to solutions of the form

$$u = Ae^{-v/v_0} = A\exp(-v/v_0)$$

where exp is the *exponential function.*

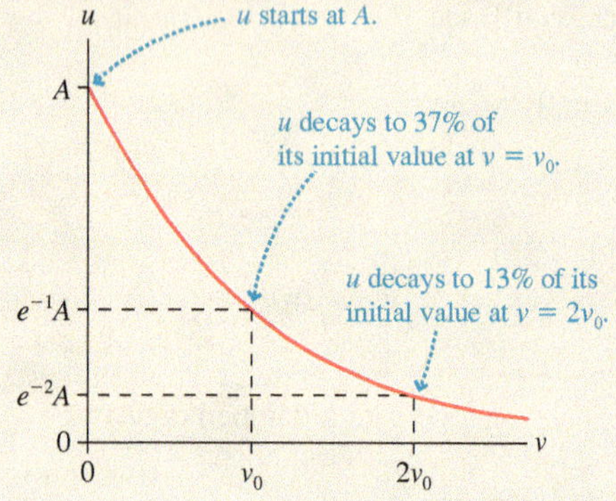

A graph of u illustrates what we mean by *exponential decay.* It starts with $u = A$ at $v = 0$ (because $e^0 = 1$) and then steadily decays, asymptotically approaching zero. The quantity v_0 is called the *decay constant.* When $v = v_0$, $u = e^{-1}A = 0.37A$. When $v = 2v_0$, $u = e^{-2}A = 0.13A$.

Arguments of functions must be pure numbers, without units. That is, we can evaluate e^{-2}, but $e^{-2\,\text{kg}}$ makes no sense. If v/v_0 is a pure number, which it must be, then the decay constant v_0 must have the same units as v. If v represents position, then v_0 is a length; if v represents time, then v_0 is a time interval. In a specific situation, v_0 is often called the *decay length* or the *decay time.* It is the length or time in which the quantity decays to 37% of its initial value.

No matter what the process is or what u represents, **a quantity that decays exponentially decays to 37% of its initial value when one decay constant has passed.** Thus exponential decay is a universal behavior. Every time you meet a new system that exhibits exponential decay, its behavior will be exactly the same as every other exponential decay. The decay curve always looks exactly like the figure shown here. Once you've learned the properties of exponential decay, you'll immediately know how to apply this knowledge to a new situation.

Energy in Damped Systems

When considering the oscillator's mechanical energy, it is useful to define the **time constant** τ (also called the *decay time*) to be

$$\tau = \frac{m}{b} \tag{14.58}$$

Because b has units of kg/s, τ has units of seconds. With this definition, we can write the oscillation amplitude as $x_{max}(t) = Ae^{-t/2\tau}$.

Because of the drag force, the mechanical energy is no longer conserved. At any particular time we can compute the mechanical energy from

$$E(t) = \frac{1}{2}k(x_{max})^2 = \frac{1}{2}k(Ae^{-t/2\tau})^2 = \left(\frac{1}{2}kA^2\right)e^{-t/\tau} = E_0 e^{-t/\tau} \quad (14.59)$$

where $E_0 = \frac{1}{2}kA^2$ is the initial energy at $t = 0$ and where we used $(z^m)^2 = z^{2m}$. In other words, **the oscillator's mechanical energy decays exponentially with time constant τ.**

As FIGURE 14.26 shows, the time constant is the amount of time needed for the energy to decay to e^{-1}, or 37%, of its initial value. We say that the time constant τ measures the "characteristic time" during which the energy of the oscillation is dissipated. Roughly two-thirds of the initial energy is gone after one time constant has elapsed, and nearly 90% has dissipated after two time constants have gone by.

For practical purposes, we can speak of the time constant as the *lifetime* of the oscillation—about how long it lasts. Mathematically, there is never a time when the oscillation is "over." The decay approaches zero asymptotically, but it never gets there in any finite time. The best we can do is define a characteristic time when the motion is "almost over," and that is what the time constant τ does.

FIGURE 14.26 Exponential decay of the mechanical energy of an oscillator.

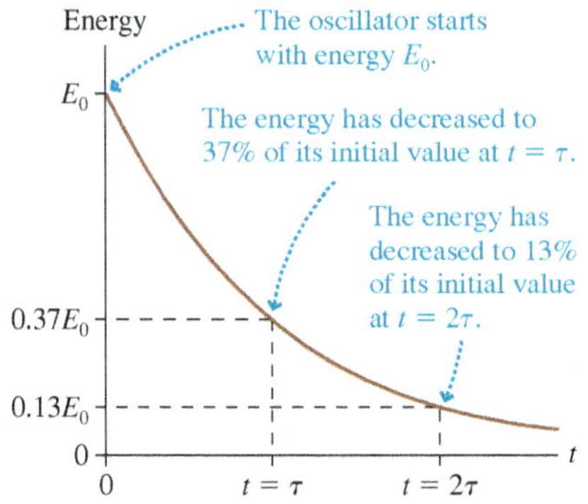

EXAMPLE 14.11 A damped pendulum

A 500 g mass swings on a 60-cm-string as a pendulum. The amplitude is observed to decay to half its initial value after 35.0 s.

a. What is the time constant for this oscillator?

b. At what time will the *energy* have decayed to half its initial value?

MODEL The motion is a damped oscillation.

SOLVE a. The initial amplitude at $t = 0$ is $x_{max} = A$. At $t = 35.0$ s the amplitude is $x_{max} = \frac{1}{2}A$. The amplitude of oscillation at time t is given by Equation 14.57:

$$x_{max}(t) = Ae^{-bt/2m} = Ae^{-t/2\tau}$$

In this case,

$$\frac{1}{2}A = Ae^{-(35.0\text{ s})/2\tau}$$

Notice that we do not need to know A itself because it cancels out. To solve for τ, we take the natural logarithm of both sides of the equation:

$$\ln\left(\frac{1}{2}\right) = -\ln 2 = \ln e^{-(35.0\text{ s})/2\tau} = -\frac{35.0\text{ s}}{2\tau}$$

This is easily rearranged to give

$$\tau = \frac{35.0\text{ s}}{2\ln 2} = 25.2\text{ s}$$

If desired, we could now determine the damping constant to be $b = m/\tau = 0.020$ kg/s.

b. The energy at time t is given by

$$E(t) = E_0 e^{-t/\tau}$$

The time at which an exponential decay is reduced to $\frac{1}{2}E_0$, half its initial value, has a special name. It is called the **half-life** and given the symbol $t_{1/2}$. The concept of the half-life is widely used in applications such as radioactive decay. To relate $t_{1/2}$ to τ, we first write

$$E(\text{at } t = t_{1/2}) = \frac{1}{2}E_0 = E_0 e^{-t_{1/2}/\tau}$$

The E_0 cancels, giving

$$\frac{1}{2} = e^{-t_{1/2}/\tau}$$

Again, we take the natural logarithm of both sides:

$$\ln\left(\frac{1}{2}\right) = -\ln 2 = \ln e^{-t_{1/2}/\tau} = -t_{1/2}/\tau$$

Finally, we solve for $t_{1/2}$:

$$t_{1/2} = \tau \ln 2 = 0.693\tau$$

This result that $t_{1/2}$ is 69% of τ is valid for any exponential decay. In this particular problem, half the energy is gone at

$$t_{1/2} = (0.693)(25.2\text{ s}) = 17.5\text{ s}$$

ASSESS The oscillator loses energy faster than it loses amplitude. This is what we should expect because the energy depends on the *square* of the amplitude.

STOP TO THINK 14.6 Rank in order, from largest to smallest, the time constants τ_a to τ_d of the decays shown in the figure. All the graphs have the same scale.

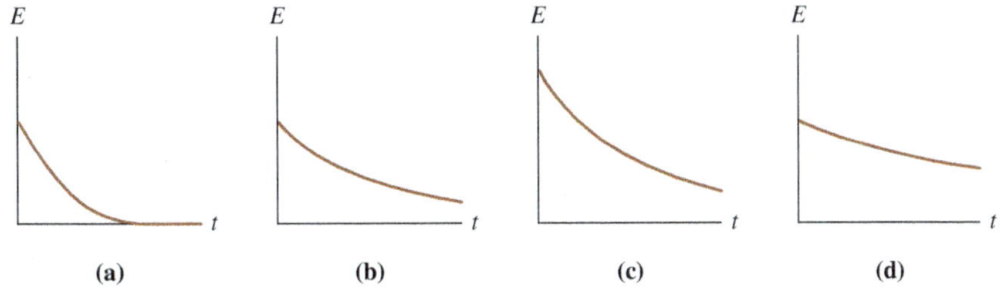

14.8 Driven Oscillations and Resonance

Thus far we have focused on the free oscillations of an isolated system. Some initial disturbance displaces the system from equilibrium, and it then oscillates freely until its energy is dissipated. These are very important situations, but they do not exhaust the possibilities. Another important situation is an oscillator that is subjected to a periodic external force. Its motion is called a **driven oscillation.**

A simple example of a driven oscillation is pushing a child on a swing, where your push is a periodic external force applied to the swing. A more complex example is a car driving over a series of equally spaced bumps. Each bump causes a periodic upward force on the car's shock absorbers, which are big, heavily damped springs. The electromagnetic coil on the back of a loudspeaker cone provides a periodic magnetic force to drive the cone back and forth, causing it to send out sound waves. Air turbulence moving across the wings of an aircraft can exert periodic forces on the wings and other aerodynamic surfaces, causing them to vibrate if they are not properly designed.

As these examples suggest, driven oscillations have many important applications. However, driven oscillations are a mathematically complex subject. We will simply hint at some of the results, saving the details for more advanced classes.

Consider an oscillating system that, when left to itself, oscillates at a frequency f_0. We will call this the **natural frequency** of the oscillator. The natural frequency for a mass on a spring is $\sqrt{k/m}/2\pi$, but it might be given by some other expression for another type of oscillator. Regardless of the expression, f_0 is simply the frequency of the system if it is displaced from equilibrium and released.

FIGURE 14.27 The response curve shows the amplitude of a driven oscillator at frequencies near its natural frequency of 2.0 Hz.

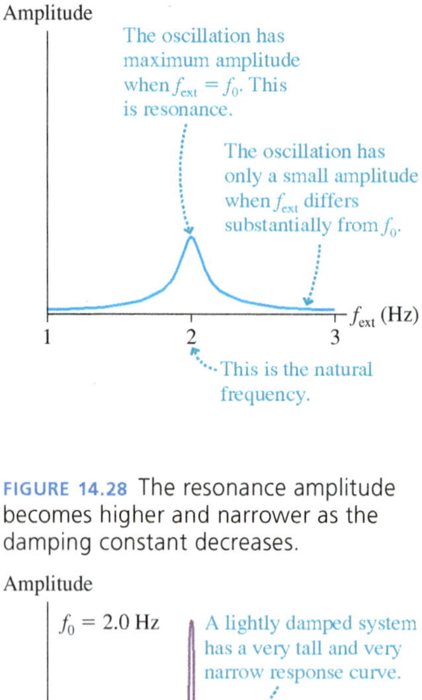

Suppose that this system is subjected to a *periodic* external force of frequency f_{ext}. This frequency, which is called the **driving frequency,** is completely independent of the oscillator's natural frequency f_0. Somebody or something in the environment selects the frequency f_{ext} of the external force, causing the force to push on the system f_{ext} times every second.

Although it is possible to solve Newton's second law with an external driving force, we will be content to look at a graphical representation of the solution. The most important result is that the oscillation amplitude depends very sensitively on the frequency f_{ext} of the driving force. The response to the driving frequency is shown in FIGURE 14.27 for a system with $m = 1.0$ kg, a natural frequency $f_0 = 2.0$ Hz, and a damping constant $b = 0.20$ kg/s. This graph of amplitude versus driving frequency, called the **response curve,** occurs in many different applications.

When the driving frequency is substantially different from the oscillator's natural frequency, at the right and left edges of Figure 14.27, the system oscillates but the amplitude is very small. The system simply does not respond well to a driving frequency that differs much from f_0. As the driving frequency gets closer and closer to the natural frequency, the amplitude of the oscillation rises dramatically. After all, f_0 is the frequency at which the system "wants" to oscillate, so it is quite happy to respond to a driving frequency near f_0. Hence the amplitude reaches a maximum when the driving frequency exactly matches the system's natural frequency: $f_{ext} = f_0$.

FIGURE 14.28 The resonance amplitude becomes higher and narrower as the damping constant decreases.

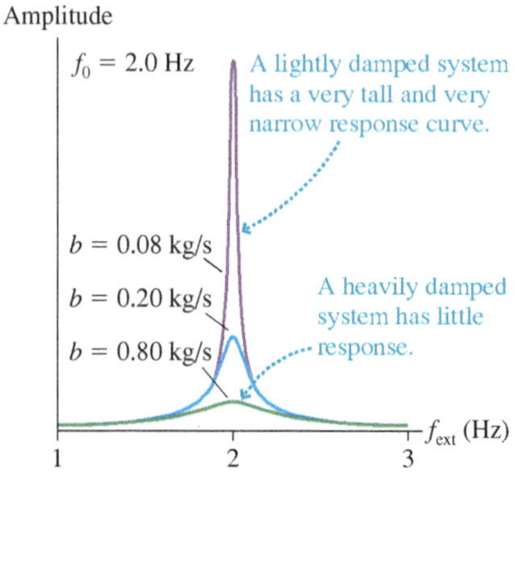

The amplitude can become exceedingly large when the frequencies match, especially if the damping constant is very small. FIGURE 14.28 shows the same oscillator with three different values of the damping constant. There's very little response if the damping constant is increased to 0.80 kg/s, but the amplitude for $f_{ext} = f_0$ becomes very large when the damping constant is reduced to 0.08 kg/s. This large-amplitude response to a driving force whose frequency matches the natural frequency of the system is a phenomenon called **resonance.** The condition for resonance is

$$f_{ext} = f_0 \quad \text{(resonance condition)} \tag{14.60}$$

Within the context of driven oscillations, the natural frequency f_0 is often called the **resonance frequency.**

An important feature of Figure 14.28 is how the amplitude and width of the resonance depend on the damping constant. A heavily damped system responds fairly little, even at resonance, but it responds to a wide range of driving frequencies. Very lightly damped systems can reach exceptionally high amplitudes, but notice that the range of frequencies to which the system responds becomes narrower and narrower as b decreases.

This allows us to understand why a few singers can break crystal goblets but not inexpensive, everyday glasses. An inexpensive glass gives a "thud" when tapped, but a fine crystal goblet "rings" for several seconds. In physics terms, the goblet has a much longer time constant than the glass. That, in turn, implies that the goblet is very lightly damped while the ordinary glass is heavily damped (because the internal forces within the glass are not those of a high-quality crystal structure).

The singer causes a sound wave to impinge on the goblet, exerting a small driving force at the frequency of the note she is singing. If the singer's frequency matches the natural frequency of the goblet—resonance! Only the lightly damped goblet, like the top curve in Figure 14.28, can reach amplitudes large enough to shatter. The restriction, though, is that its natural frequency has to be matched very precisely. The sound also has to be very loud.

A singer or musical instrument can shatter a crystal goblet by matching the goblet's natural oscillation frequency.

CHALLENGE EXAMPLE 14.12 A swinging pendulum

A pendulum consists of a massless, rigid rod with a mass at one end. The other end is pivoted on a frictionless pivot so that the rod can rotate in a complete circle. The pendulum is inverted, with the mass directly above the pivot point, then released. The speed of the mass as it passes through the lowest point is 5.0 m/s. If the pendulum later undergoes small-amplitude oscillations at the bottom of the arc, what will its frequency be?

MODEL This is a simple pendulum because the rod is massless. However, our analysis of a pendulum used the small-angle approximation. It applies only to the small-amplitude oscillations at the end, *not* to the pendulum swinging down from the inverted position. Fortunately, energy is conserved throughout, so we can analyze the big swing using conservation of mechanical energy.

VISUALIZE FIGURE 14.29 is a pictorial representation of the pendulum swinging down from the inverted position. The pendulum length is L, so the initial height is $2L$.

FIGURE 14.29 Before-and-after pictorial representation of the pendulum swinging down from an inverted position.

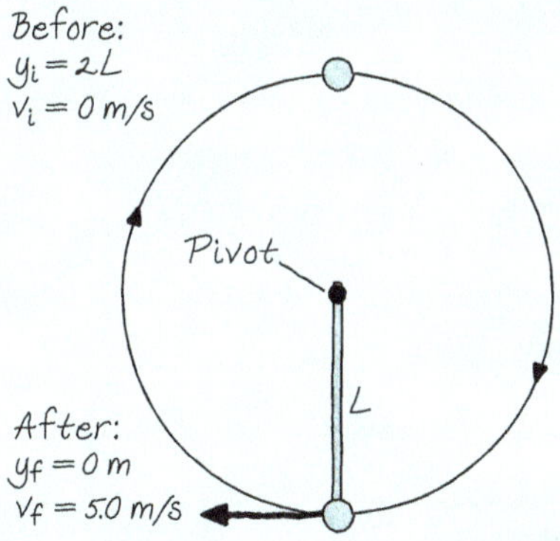

SOLVE The frequency of a simple pendulum is $f = \sqrt{g/L}/2\pi$. We're not given L, but we can find it by analyzing the pendulum's swing down from an inverted position. Mechanical energy is conserved, and the only potential energy is gravitational potential energy. Conservation of mechanical energy $K_f + U_{gf} = K_i + U_{gi}$, with $U_g = mgy$, is

$$\frac{1}{2}mv_f^2 + mgy_f = \frac{1}{2}mv_i^2 + mgy_i$$

The mass cancels, which is good since we don't know it, and two terms are zero. Thus

$$\frac{1}{2}v_f^2 = g(2L) = 2gL$$

Solving for L, we find

$$L = \frac{v_f^2}{4g} = \frac{(5.0\text{ m/s})^2}{4(9.80\text{ m/s}^2)} = 0.638\text{ m}$$

Now we can calculate the frequency:

$$f = \frac{1}{2\pi}\sqrt{\frac{g}{L}} = \frac{1}{2\pi}\sqrt{\frac{9.80\text{ m/s}^2}{0.638\text{ m}}} = 0.62\text{ Hz}$$

ASSESS The frequency corresponds to a period of about 1.5 s, which seems reasonable.

SUMMARY

The goal of Chapter 14 has been to understand systems that oscillate with simple harmonic motion.

General Principles

Dynamics

SHM occurs when a linear restoring force acts to return a system to an equilibrium position.

Horizontal spring

$(F_{net})_x = -kx$

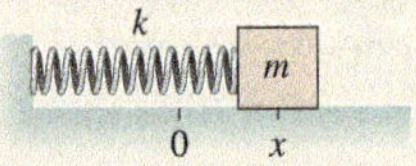

Vertical spring
The origin is at the equilibrium position $\Delta L = mg/k$.

$(F_{net})_y = -ky$

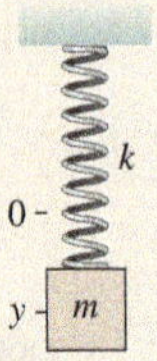

Both: $\omega = \sqrt{\dfrac{k}{m}}$ $\quad T = 2\pi\sqrt{\dfrac{m}{k}}$

Pendulum

$(F_{net})_t = -\left(\dfrac{mg}{L}\right)s$

$\omega = \sqrt{\dfrac{g}{L}}$ $\quad T = 2\pi\sqrt{\dfrac{L}{g}}$

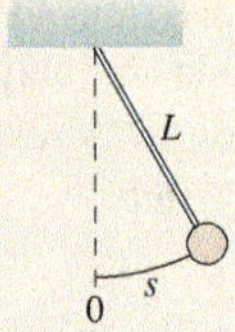

Energy

If there is **no friction** or dissipation, kinetic and potential energy are alternately transformed into each other, but the total mechanical energy $E = K + U$ is conserved.

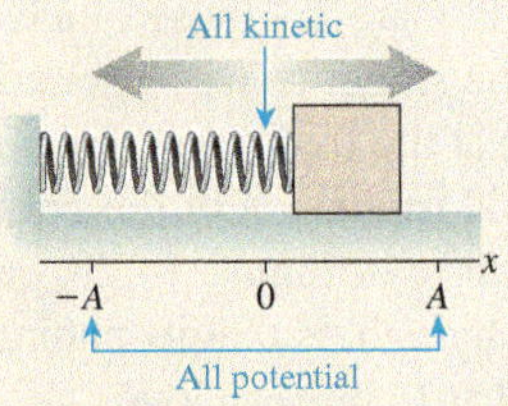

$$E = \frac{1}{2}mv^2 + \frac{1}{2}kx^2$$
$$= \frac{1}{2}m(v_{max})^2$$
$$= \frac{1}{2}kA^2$$

In a **damped system,** the energy decays exponentially

$$E = E_0 e^{-t/\tau}$$

where τ is the time constant.

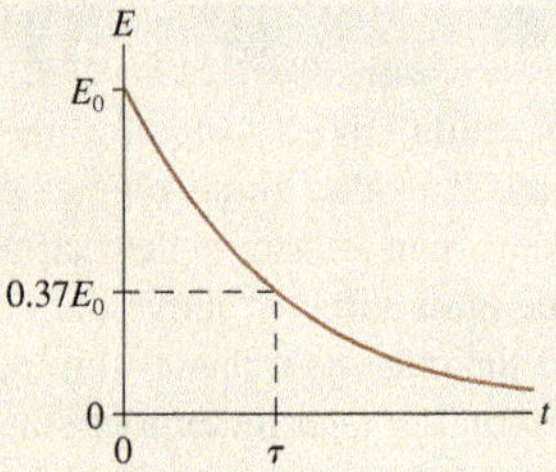

Important Concepts

Simple harmonic motion (SHM) is a sinusoidal oscillation with period T and amplitude A.

Frequency $f = \dfrac{1}{T}$

Angular frequency
$$\omega = 2\pi f = \frac{2\pi}{T}$$

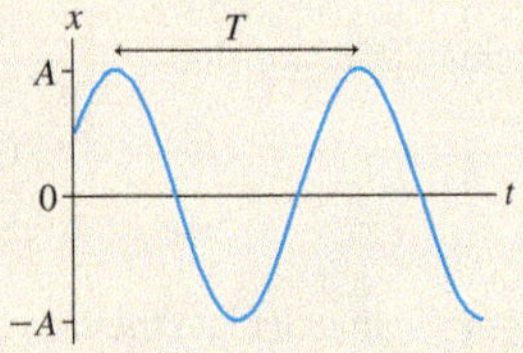

Position $x(t) = A\cos(\omega t + \phi_0)$

$$= A\cos\left(\frac{2\pi t}{T} + \phi_0\right)$$

Velocity $v_x(t) = -v_{max}\sin(\omega t + \phi_0)$ with maximum speed $v_{max} = \omega A$

Acceleration $a_x(t) = -\omega^2 x(t) = -\omega^2 A\cos(\omega t + \phi_0)$

SHM is the projection onto the x-axis of **uniform circular motion.**

$\phi = \omega t + \phi_0$ is the phase

The position at time t is

$$x(t) = A\cos\phi$$
$$= A\cos(\omega t + \phi_0)$$

The phase constant ϕ_0 determines the initial conditions:

$x_0 = A\cos\phi_0 \qquad v_{0x} = -\omega A\sin\phi_0$

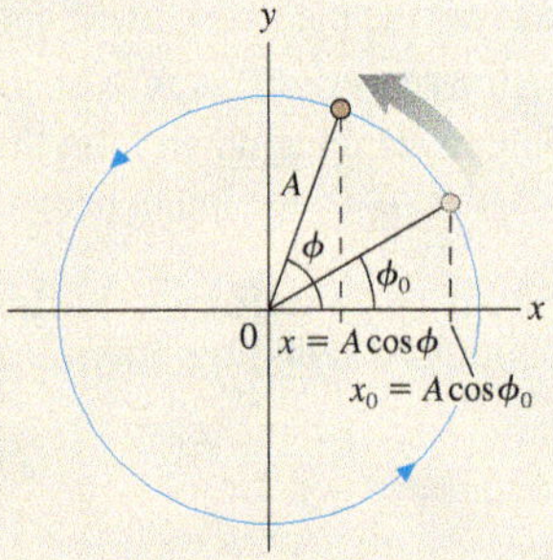

Applications

Resonance

When a system is driven by a periodic external force, it responds with a large-amplitude oscillation if $f_{ext} \approx f_0$, where f_0 is the system's natural oscillation frequency, or **resonant frequency.**

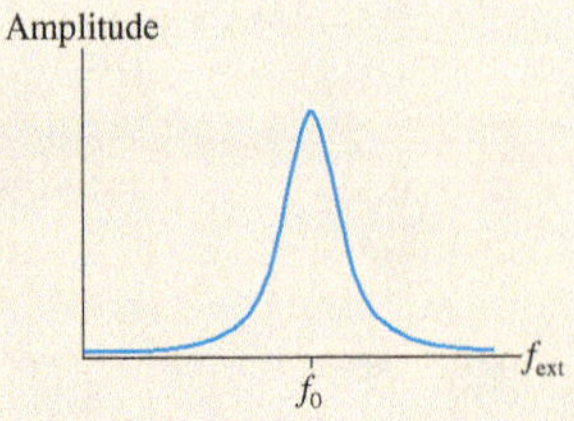

Damping

If there is a drag force $\vec{D} = -b\vec{v}$, where b is the damping constant, then (for lightly damped systems)

$$x(t) = Ae^{-bt/2m}\cos(\omega t + \phi_0)$$

The time constant for energy loss is $\tau = m/b$.

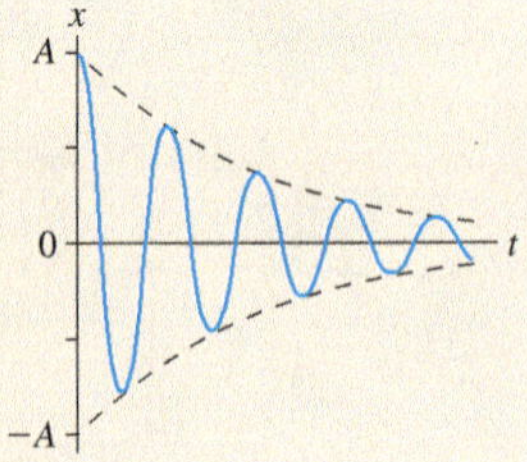

Terms and Notation

oscillatory motion	amplitude, A	linear restoring force	natural frequency, f_0
oscillator	angular frequency, ω	damped oscillation	driving frequency, f_{ext}
period, T	phase, ϕ	damping constant, b	response curve
frequency, f	phase constant, ϕ_0	envelope	resonance
hertz, Hz	restoring force	time constant, τ	resonance frequency, f_0
simple harmonic motion, SHM	equation of motion	half-life, $t_{1/2}$	
	small-angle approximation	driven oscillation	

CONCEPTUAL QUESTIONS

1. A block oscillating on a spring has period $T = 2$ s. What is the period if:
   a. The block's mass is doubled? Explain. Note that you do not know the value of either m or k, so do *not* assume any particular values for them. The required analysis involves thinking about ratios.
   b. The value of the spring constant is quadrupled?
   c. The oscillation amplitude is doubled while m and k are unchanged?
2. A pendulum on Planet X, where the value of g is unknown, oscillates with a period $T = 2$ s. What is the period of this pendulum if:
   a. Its mass is doubled? Explain. Note that you do not know the value of m, L, or g, so do not assume any specific values. The required analysis involves thinking about ratios.
   b. Its length is doubled?
   c. Its oscillation amplitude is doubled?
3. FIGURE Q14.3 shows a position-versus-time graph for a particle in SHM. What are (a) the amplitude A, (b) the angular frequency ω, and (c) the phase constant ϕ_0? Explain.

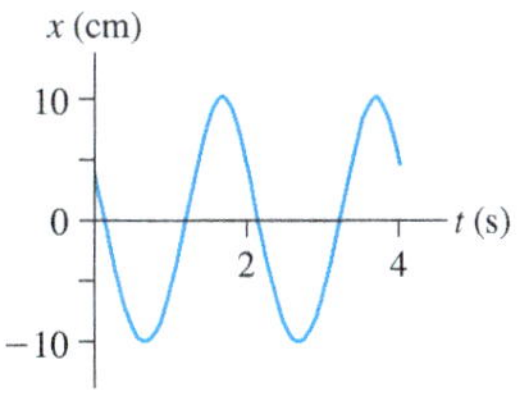

FIGURE Q14.3

4. Equation 14.25 states that $\frac{1}{2}kA^2 = \frac{1}{2}m(v_{max})^2$. What does this mean? Write a couple of sentences explaining how to interpret this equation.
5. A block oscillating on a spring has an amplitude of 20 cm. What will the amplitude be if the total energy is doubled? Explain.
6. A block oscillating on a spring has a maximum speed of 20 cm/s. What will the block's maximum speed be if the total energy is doubled? Explain.
7. FIGURE Q14.7 shows a position-versus-time graph for a particle in SHM.
   a. What is the phase constant ϕ_0? Explain.
   b. What is the phase of the particle at each of the three numbered points on the graph?

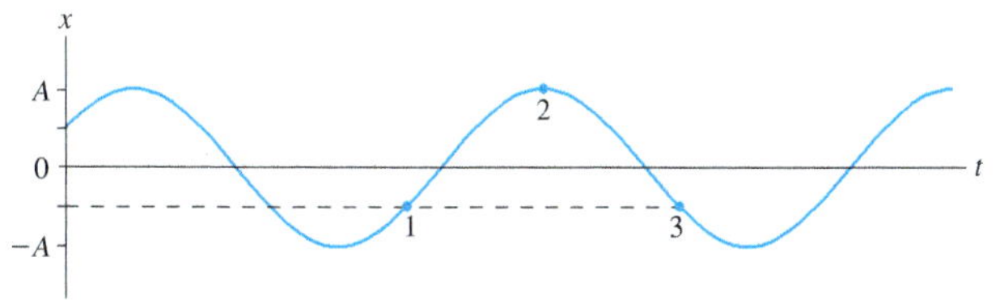

FIGURE Q14.7

8. FIGURE Q14.8 shows a velocity-versus-time graph for a particle in SHM.
   a. What is the phase constant ϕ_0? Explain.
   b. What is the phase of the particle at each of the three numbered points on the graph?

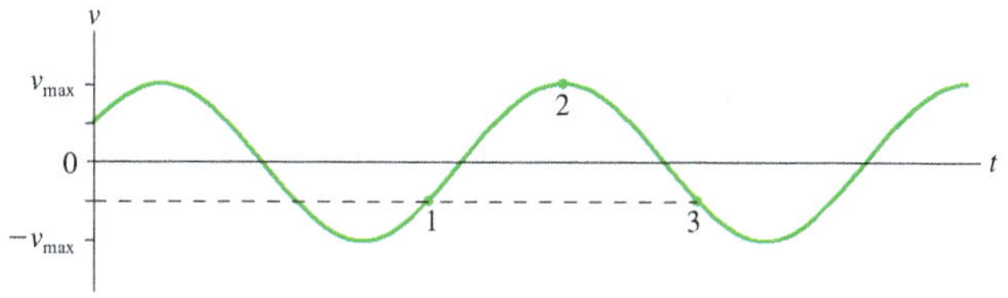

FIGURE Q14.8

9. FIGURE Q14.9 shows the potential-energy diagram and the total energy line of a particle oscillating on a spring.
   a. What is the spring's equilibrium length?
   b. Where are the turning points of the motion? Explain.
   c. What is the particle's maximum kinetic energy?
   d. What will be the turning points if the particle's total energy is doubled?

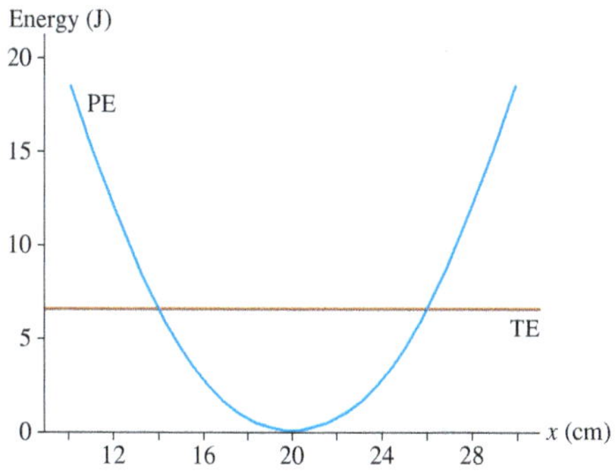

FIGURE Q14.9

10. Suppose the damping constant b of an oscillator increases.
    a. Is the medium more resistive or less resistive?
    b. Do the oscillations damp out more quickly or less quickly?
    c. Is the time constant τ increased or decreased?
11. a. Describe the difference between τ and T. Don't just *name* them; say what is different about the physical concepts they represent.
    b. Describe the difference between τ and $t_{1/2}$.
12. What is the difference between the driving frequency and the natural frequency of an oscillator?

EXERCISES AND PROBLEMS

Problems labeled integrate material from earlier chapters.

Exercises

Section 14.1 Simple Harmonic Motion

1. | When a guitar string plays the note "A," the string vibrates at 440 Hz. What is the period of the vibration?
2. | An air-track glider attached to a spring oscillates between the 10 cm mark and the 60 cm mark on the track. The glider completes 10 oscillations in 33 s. What are the (a) period, (b) frequency, (c) angular frequency, (d) amplitude, and (e) maximum speed of the glider?
3. || An air-track glider is attached to a spring. The glider is pulled to the right and released from rest at $t = 0$ s. It then oscillates with a period of 2.0 s and a maximum speed of 40 cm/s.
   a. What is the amplitude of the oscillation?
   b. What is the glider's position at $t = 0.25$ s?

Section 14.2 Simple Harmonic Motion and Circular Motion

4. | What are the (a) amplitude, (b) frequency, and (c) phase constant of the oscillation shown in FIGURE EX14.4?

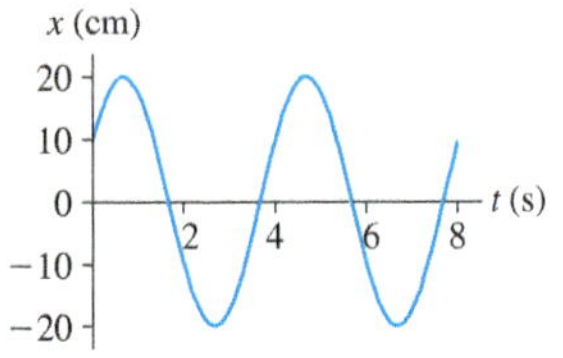

FIGURE EX14.4

5. || What are the (a) amplitude, (b) frequency, and (c) phase constant of the oscillation shown in FIGURE EX14.5?

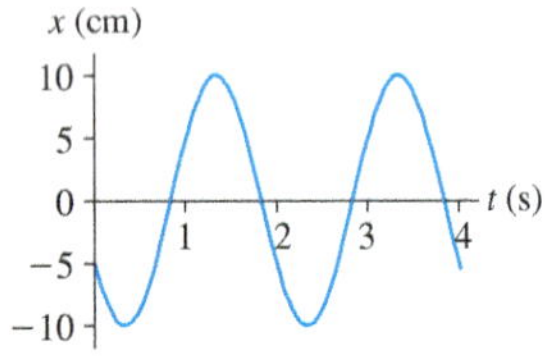

FIGURE EX14.5

6. || An object in simple harmonic motion has an amplitude of 4.0 cm, a frequency of 2.0 Hz, and a phase constant of $2\pi/3$ rad. Draw a position graph showing two cycles of the motion.
7. || An object in simple harmonic motion has an amplitude of 8.0 cm, a frequency of 0.25 Hz, and a phase constant of $-\pi/2$ rad. Draw a position graph showing two cycles of the motion.
8. | An object in simple harmonic motion has amplitude 4.0 cm and frequency 4.0 Hz, and at $t = 0$ s it passes through the equilibrium point moving to the right. Write the function $x(t)$ that describes the object's position.
9. | An object in simple harmonic motion has amplitude 8.0 cm and frequency 0.50 Hz. At $t = 0$ s it has its most negative position. Write the function $x(t)$ that describes the object's position.
10. || An air-track glider attached to a spring oscillates with a period of 1.5 s. At $t = 0$ s the glider is 5.00 cm left of the equilibrium position and moving to the right at 36.3 cm/s.
   a. What is the phase constant?
   b. What is the phase at $t = 0$ s, 0.5 s, 1.0 s, and 1.5 s?

Section 14.3 Energy in Simple Harmonic Motion

Section 14.4 The Dynamics of Simple Harmonic Motion

11. | A block attached to a spring with unknown spring constant oscillates with a period of 2.0 s. What is the period if
   a. The mass is doubled?
   b. The mass is halved?
   c. The amplitude is doubled?
   d. The spring constant is doubled?

   Parts a to d are independent questions, each referring to the initial situation.
12. || A 200 g air-track glider is attached to a spring. The glider is pushed in 10 cm and released. A student with a stopwatch finds that 10 oscillations take 12.0 s. What is the spring constant?
13. || A 200 g mass attached to a horizontal spring oscillates at a frequency of 2.0 Hz. At $t = 0$ s, the mass is at $x = 5.0$ cm and has $v_x = -30$ cm/s. Determine:
   a. The period.
   b. The angular frequency.
   c. The amplitude.
   d. The phase constant.
   e. The maximum speed.
   f. The maximum acceleration.
   g. The total energy.
   h. The position at $t = 0.40$ s.
14. | The position of a 50 g oscillating mass is given by $x(t) = (2.0\text{ cm})\cos(10t - \pi/4)$, where t is in s. Determine:
   a. The amplitude.
   b. The period.
   c. The spring constant.
   d. The phase constant.
   e. The initial conditions.
   f. The maximum speed.
   g. The total energy.
   h. The velocity at $t = 0.40$ s.
15. || A 1.0 kg block is attached to a spring with spring constant 16 N/m. While the block is sitting at rest, a student hits it with a hammer and almost instantaneously gives it a speed of 40 cm/s. What are
   a. The amplitude of the subsequent oscillations?
   b. The block's speed at the point where $x = \frac{1}{2}A$?

Section 14.5 Vertical Oscillations

16. | A spring is hanging from the ceiling. Attaching a 500 g physics book to the spring causes it to stretch 20 cm in order to come to equilibrium.
   a. What is the spring constant?
   b. From equilibrium, the book is pulled down 10 cm and released. What is the period of oscillation?
   c. What is the book's maximum speed?
17. || A spring with spring constant 15 N/m hangs from the ceiling. A ball is attached to the spring and allowed to come to rest. It is then pulled down 6.0 cm and released. If the ball makes 30 oscillations in 20 s, what are its (a) mass and (b) maximum speed?
18. || A spring is hung from the ceiling. When a block is attached to its end, it stretches 2.0 cm before reaching its new equilibrium length. The block is then pulled down slightly and released. What is the frequency of oscillation?

Section 14.6 The Pendulum

19. | A mass on a string of unknown length oscillates as a pendulum with a period of 4.0 s. What is the period if
   a. The mass is doubled?

b. The string length is doubled?
c. The string length is halved?
d. The amplitude is doubled?

Parts a to d are independent questions, each referring to the initial situation.

20. || A 200 g ball is tied to a string. It is pulled to an angle of 8.0° and released to swing as a pendulum. A student with a stopwatch finds that 10 oscillations take 12 s. How long is the string?

21. | What is the period of a 1.0-m-long pendulum on (a) the earth and (b) Venus?

22. | What is the length of a pendulum whose period on the moon matches the period of a 2.0-m-long pendulum on the earth?

23. | Astronauts on the first trip to Mars take along a pendulum that has a period on earth of 1.50 s. The period on Mars turns out to be 2.45 s. What is the free-fall acceleration on Mars?

24. || A uniform steel bar swings from a pivot at one end with a period of 1.2 s. How long is the bar?

Section 14.7 Damped Oscillations

Section 14.8 Driven Oscillations and Resonance

25. | A 2.0 g spider is dangling at the end of a silk thread. You can make the spider bounce up and down on the thread by tapping lightly on his feet with a pencil. You soon discover that you can give the spider the largest amplitude on his little bungee cord if you tap exactly once every second. What is the spring constant of the silk thread?

26. || The amplitude of an oscillator decreases to 36.8% of its initial value in 10.0 s. What is the value of the time constant?

27. || Sketch a position graph from $t = 0$ s to $t = 10$ s of a damped oscillator having a frequency of 1.0 Hz and a time constant of 4.0 s.

28. | In a science museum, a 110 kg brass pendulum bob swings at the end of a 15.0-m-long wire. The pendulum is started at exactly 8:00 A.M. every morning by pulling it 1.5 m to the side and releasing it. Because of its compact shape and smooth surface, the pendulum's damping constant is only 0.010 kg/s. At exactly 12:00 noon, how many oscillations will the pendulum have completed and what is its amplitude?

29. || Vision is blurred if the head is vibrated at 29 Hz because the vibrations are resonant with the natural frequency of the eyeball in its socket. If the mass of the eyeball is 7.5 g, a typical value, what is the effective spring constant of the musculature that holds the eyeball in the socket?
BIO

Problems

30. || FIGURE P14.30 is the velocity-versus-time graph of a particle in simple harmonic motion.
a. What is the amplitude of the oscillation?
b. What is the phase constant?
c. What is the position at $t = 0$ s?

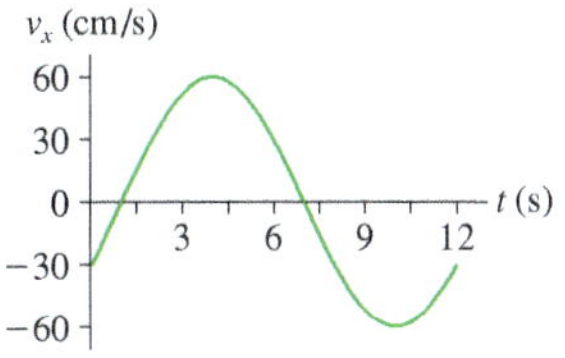

FIGURE P14.30

31. | FIGURE P14.31 is the position-versus-time graph of a particle in simple harmonic motion.
a. What is the phase constant?
b. What is the velocity at $t = 0$ s?
c. What is v_{max}?

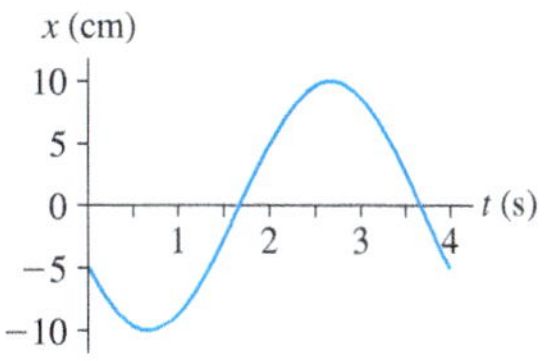

FIGURE P14.31

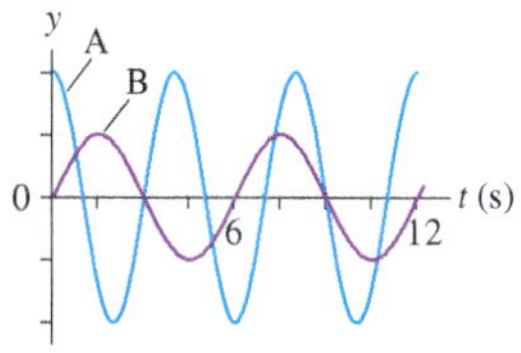

FIGURE P14.32

32. || The two graphs in FIGURE P14.32 are for two different vertical mass-spring systems. If both systems have the same mass, what is the ratio k_A/k_B of their spring constants?

33. ||| An object in SHM oscillates with a period of 4.0 s and an amplitude of 10 cm. How long does the object take to move from $x = 0.0$ cm to $x = 6.0$ cm?

34. || A 1.0 kg block oscillates on a spring with spring constant 20 N/m. At $t = 0$ s the block is 20 cm to the right of the equilibrium position and moving to the left at a speed of 100 cm/s. Determine (a) the period and (b) the amplitude.

35. || Astronauts in space cannot weigh themselves by standing on a bathroom scale. Instead, they determine their mass by oscillating on a large spring. Suppose an astronaut attaches one end of a large spring to her belt and the other end to a hook on the wall of the space capsule. A fellow astronaut then pulls her away from the wall and releases her. The spring's length as a function of time is shown in FIGURE P14.35.
BIO
a. What is her mass if the spring constant is 240 N/m?
b. What is her speed when the spring's length is 1.2 m?

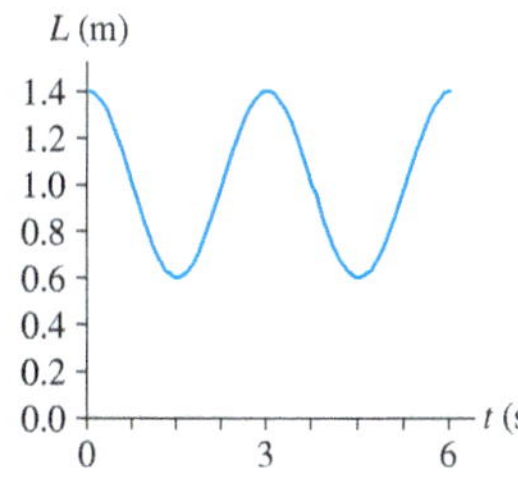

FIGURE P14.35

36. || The motion of a particle is given by $x(t) = (25 \text{ cm})\cos(10t)$, where t is in s. At what time is the kinetic energy twice the potential energy?

37. || a. When the displacement of a mass on a spring is $\frac{1}{2}A$, what fraction of the energy is kinetic energy and what fraction is potential energy?
b. At what displacement, as a fraction of A, is the energy half kinetic and half potential?

38. || For a particle in simple harmonic motion, show that $v_{max} = (\pi/2)v_{avg}$ where v_{avg} is the average speed during one cycle of the motion.

39. || A 100 g block attached to a spring with spring constant 2.5 N/m oscillates horizontally on a frictionless table. Its velocity is 20 cm/s when $x = -5.0$ cm.
a. What is the amplitude of oscillation?
b. What is the block's maximum acceleration?
c. What is the block's position when the acceleration is maximum?
d. What is the speed of the block when $x = 3.0$ cm?

40. || A block on a spring is pulled to the right and released at $t = 0$ s. It passes $x = 3.00$ cm at $t = 0.685$ s, and it passes $x = -3.00$ cm at $t = 0.886$ s.
    a. What is the angular frequency?
    b. What is the amplitude?
    Hint: $\cos(\pi - \theta) = -\cos\theta$.

41. ||| A 300 g oscillator has a speed of 95.4 cm/s when its displacement is 3.0 cm and 71.4 cm/s when its displacement is 6.0 cm. What is the oscillator's maximum speed?

42. || An ultrasonic transducer, of the type used in medical ultrasound imaging, is a very thin disk ($m = 0.10$ g) driven back and forth in SHM at 1.0 MHz by an electromagnetic coil.
BIO
    a. The maximum restoring force that can be applied to the disk without breaking it is 40,000 N. What is the maximum oscillation amplitude that won't rupture the disk?
    b. What is the disk's maximum speed at this amplitude?

43. || A 5.0 kg block hangs from a spring with spring constant 2000 N/m. The block is pulled down 5.0 cm from the equilibrium position and given an initial velocity of 1.0 m/s back toward equilibrium. What are the (a) frequency, (b) amplitude, and (c) total mechanical energy of the motion?

44. || Your lab instructor has asked you to measure a spring constant using a dynamic method—letting it oscillate—rather than a static method of stretching it. You and your lab partner suspend the spring from a hook, hang different masses on the lower end, and start them oscillating. One of you uses a meter stick to measure the amplitude, the other uses a stopwatch to time 10 oscillations. Your data are as follows:

Mass (g)	Amplitude (cm)	Time (s)
100	6.5	7.8
150	5.5	9.8
200	6.0	10.9
250	3.5	12.4

Use the best-fit line of an appropriate graph to determine the spring constant.

45. ||| A 200 g block hangs from a spring with spring constant 10 N/m. At $t = 0$ s the block is 20 cm below the equilibrium point and moving upward with a speed of 100 cm/s. What are the block's
    a. Oscillation frequency?
    b. Distance from equilibrium when the speed is 50 cm/s?
    c. Distance from equilibrium at $t = 1.0$ s?

46. || A spring with spring constant k is suspended vertically from a support and a mass m is attached. The mass is held at the point where the spring is not stretched. Then the mass is released and begins to oscillate. The lowest point in the oscillation is 20 cm below the point where the mass was released. What is the oscillation frequency?

47. || While grocery shopping, you put several apples in the spring scale in the produce department. The scale reads 20 N, and you use your ruler (which you always carry with you) to discover that the pan goes down 9.0 cm when the apples are added. If you tap the bottom of the apple-filled pan to make it bounce up and down a little, what is its oscillation frequency? Ignore the mass of the pan.

48. || A compact car has a mass of 1200 kg. Assume that the car has one spring on each wheel, that the springs are identical, and that the mass is equally distributed over the four springs.
    a. What is the spring constant of each spring if the empty car bounces up and down 2.0 times each second?
    b. What will be the car's oscillation frequency while carrying four 70 kg passengers?

49. || The two blocks in FIGURE P14.49 oscillate on a frictionless surface with a period of 1.5 s. The upper block just begins to slip when the amplitude is increased to 40 cm. What is the coefficient of static friction between the two blocks?

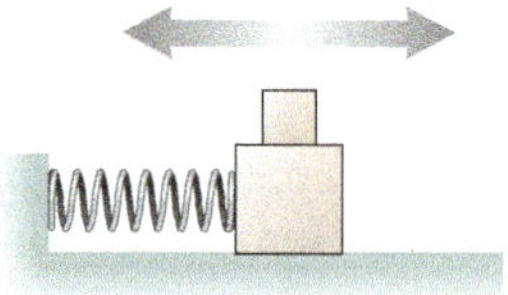

FIGURE P14.49

50. ||| It has recently become possible to "weigh" DNA molecules by measuring the influence of their mass on a nano-oscillator. FIGURE P14.50 shows a thin rectangular cantilever etched out of silicon (density 2300 kg/m^3) with a small gold dot at the end. If pulled down and released, the end of the cantilever vibrates with simple harmonic motion, moving up and down like a diving board after a jump. When bathed with DNA molecules whose ends have been modified to bind with gold, one or more molecules may attach to the gold dot. The addition of their mass causes a very slight—but measurable—decrease in the oscillation frequency.
BIO

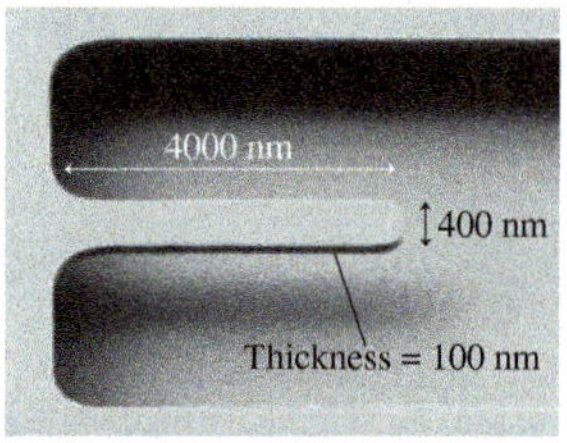

FIGURE P14.50

A vibrating cantilever of mass M can be modeled as a block of mass $\frac{1}{3}M$ attached to a spring. (The factor of $\frac{1}{3}$ arises from the moment of inertia of a bar pivoted at one end.) Neither the mass nor the spring constant can be determined very accurately— perhaps to only two significant figures—but the oscillation frequency can be measured with very high precision simply by counting the oscillations. In one experiment, the cantilever was initially vibrating at exactly 12 MHz. Attachment of a DNA molecule caused the frequency to decrease by 50 Hz. What was the mass of the DNA?

51. || It is said that Galileo discovered a basic principle of the pendulum—that the period is independent of the amplitude—by using his pulse to time the period of swinging lamps in the cathedral as they swayed in the breeze. Suppose that one oscillation of a swinging lamp takes 5.5 s.
    a. How long is the lamp chain?
    b. What maximum speed does the lamp have if its maximum angle from vertical is 3.0°?

52. || A 100 g mass on a 1.0-m-long string is pulled 8.0° to one side and released. How long does it take for the pendulum to reach 4.0° on the opposite side?

53. || Orangutans can move by brachiation, swinging like a pendulum beneath successive handholds. If an orangutan has arms that are 0.90 m long and repeatedly swings to a 20° angle, taking one swing after another, estimate its speed of forward motion in m/s. While this is somewhat beyond the range of validity of the small-angle approximation, the standard results for a pendulum are adequate for making an estimate.
BIO

54. | Show that Equation 14.51 for the angular frequency of a physical pendulum gives Equation 14.48 when applied to a simple pendulum of a mass on a string.

55. ||| A 15-cm-long, 200 g rod is pivoted at one end. A 20 g ball of clay is stuck on the other end. What is the period if the rod and clay swing as a pendulum?

56. ||| A uniform rod of mass M and length L swings as a pendulum on a pivot at distance $L/4$ from one end of the rod. Find an expression for the frequency f of small-angle oscillations.

57. ||| A solid sphere of mass M and radius R is suspended from a thin rod, as shown in FIGURE P14.57. The sphere can swing back and forth at the bottom of the rod. Find an expression for the frequency f of small-angle oscillations.

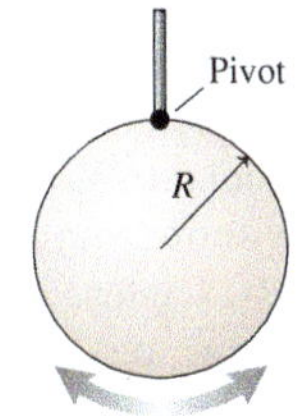

FIGURE P14.57

58. || A geologist needs to determine the local value of g. Unfortunately, his only tools are a meter stick, a saw, and a stopwatch. He starts by hanging the meter stick from one end and measuring its frequency as it swings. He then saws off 20 cm—using the centimeter markings—and measures the frequency again. After two more cuts, these are his data:

Length (cm)	Frequency (Hz)
100	0.61
80	0.67
60	0.79
40	0.96

Use the best-fit line of an appropriate graph to determine the local value of g.

59. || Interestingly, there have been several studies using cadavers
BIO to determine the moments of inertia of human body parts, information that is important in biomechanics. In one study, the center of mass of a 5.0 kg lower leg was found to be 18 cm from the knee. When the leg was allowed to pivot at the knee and swing freely as a pendulum, the oscillation frequency was 1.6 Hz. What was the moment of inertia of the lower leg about the knee joint?

60. || A 500 g air-track glider attached to a spring with spring constant 10 N/m is sitting at rest on a frictionless air track. A 250 g glider is pushed toward it from the far end of the track at a speed of 120 cm/s. It collides with and sticks to the 500 g glider. What are the amplitude and period of the subsequent oscillations?

61. || A 200 g block attached to a horizontal spring is oscillating with an amplitude of 2.0 cm and a frequency of 2.0 Hz. Just as it passes through the equilibrium point, moving to the right, a sharp blow directed to the left exerts a 20 N force for 1.0 ms. What are the new (a) frequency and (b) amplitude?

62. || FIGURE P14.62 is a top view of an object of mass m connected between two stretched rubber bands of length L. The object rests on a frictionless surface. At equilibrium, the tension in each rubber band is T. Find an expression for the frequency of oscillations *perpendicular* to the rubber bands. Assume the amplitude is sufficiently small that the magnitude of the tension in the rubber bands is essentially unchanged as the mass oscillates.

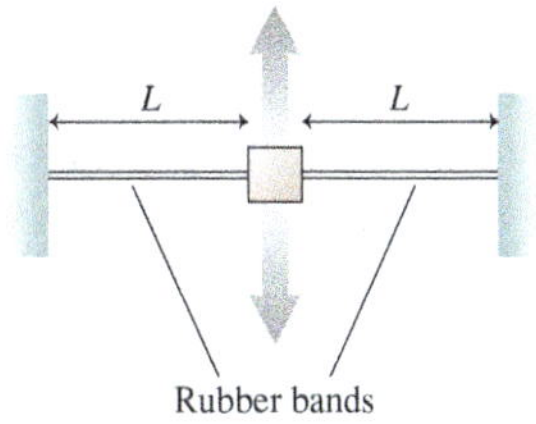

FIGURE P14.62

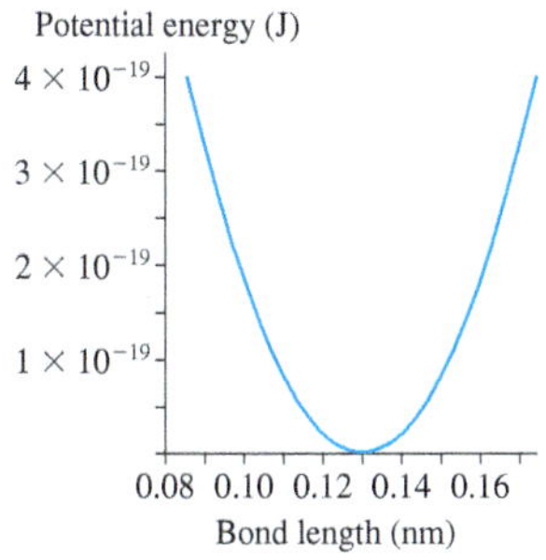

FIGURE P14.63

63. || A molecular bond can be modeled as a spring between two atoms that vibrate with simple harmonic motion. FIGURE P14.63 shows an SHM approximation for the potential energy of an HCl molecule. For $E < 4 \times 10^{-19}$ J it is a good approximation to the more accurate HCl potential-energy curve that was shown in Figure 10.31. Because the chlorine atom is so much more massive than the hydrogen atom, it is reasonable to assume that the hydrogen atom ($m = 1.67 \times 10^{-27}$ kg) vibrates back and forth while the chlorine atom remains at rest. Use the graph to estimate the vibrational frequency of the HCl molecule.

64. || An ice cube can slide around the inside of a vertical circular hoop of radius R. It undergoes small-amplitude oscillations if displaced slightly from the equilibrium position at the lowest point. Find an expression for the period of these small-amplitude oscillations.

65. || A penny rides on top of a piston as it undergoes vertical simple harmonic motion with an amplitude of 4.0 cm. If the frequency is low, the penny rides up and down without difficulty. If the frequency is steadily increased, there comes a point at which the penny leaves the surface.
   a. At what point in the cycle does the penny first lose contact with the piston?
   b. What is the maximum frequency for which the penny just barely remains in place for the full cycle?

66. || On your first trip to Planet X you happen to take along a 200 g mass, a 40-cm-long spring, a meter stick, and a stopwatch. You're curious about the free-fall acceleration on Planet X, where ordinary tasks seem easier than on earth, but you can't find this information in your Visitor's Guide. One night you suspend the spring from the ceiling in your room and hang the mass from it. You find that the mass stretches the spring by 31.2 cm. You then pull the mass down 10.0 cm and release it. With the stopwatch you find that 10 oscillations take 14.5 s. Based on this information, what is g?

67. || The 15 g head of a bobble-head doll oscillates in SHM at a frequency of 4.0 Hz.
   a. What is the spring constant of the spring on which the head is mounted?
   b. The amplitude of the head's oscillations decreases to 0.5 cm in 4.0 s. What is the head's damping constant?

68. || An oscillator with a mass of 500 g and a period of 0.50 s has an amplitude that decreases by 2.0% during each complete oscillation. If the initial amplitude is 10 cm, what will be the amplitude after 25 oscillations?

69. || A spring with spring constant 15.0 N/m hangs from the ceiling. A 500 g ball is attached to the spring and allowed to come to rest. It is then pulled down 6.0 cm and released. What is the time constant if the ball's amplitude has decreased to 3.0 cm after 30 oscillations?

70. ||| A 250 g air-track glider is attached to a spring with spring constant 4.0 N/m. The damping constant due to air resistance is 0.015 kg/s. The glider is pulled out 20 cm from equilibrium and released. How many oscillations will it make during the time in which the amplitude decays to e^{-1} of its initial value?

71. || A 200 g oscillator in a vacuum chamber has a frequency of 2.0 Hz. When air is admitted, the oscillation decreases to 60% of its initial amplitude in 50 s. How many oscillations will have been completed when the amplitude is 30% of its initial value?

72. || Prove that the expression for $x(t)$ in Equation 14.55 is a solution to the equation of motion for a damped oscillator, Equation 14.54, if and only if the angular frequency ω is given by the expression in Equation 14.56.

73. || A block on a frictionless table is connected as shown in FIGURE P14.73 to two springs having spring constants k_1 and k_2. Show that the block's oscillation frequency is given by

$$f = \sqrt{f_1^2 + f_2^2}$$

where f_1 and f_2 are the frequencies at which it would oscillate if attached to spring 1 or spring 2 alone.

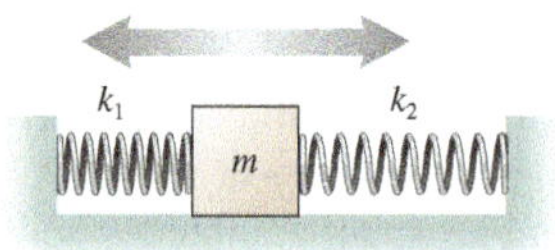

FIGURE P14.73

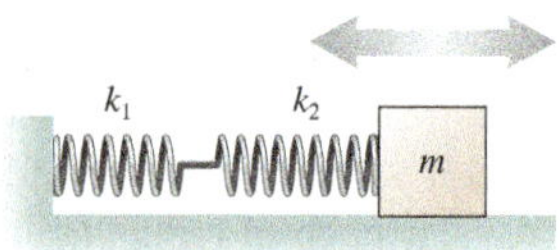

FIGURE P14.74

74. || A block on a frictionless table is connected as shown in FIGURE P14.74 to two springs having spring constants k_1 and k_2. Find an expression for the block's oscillation frequency f in terms of the frequencies f_1 and f_2 at which it would oscillate if attached to spring 1 or spring 2 alone.

Challenge Problems

75. A block hangs in equilibrium from a vertical spring. When a second identical block is added, the original block sags by 5.0 cm. What is the oscillation frequency of the two-block system?

76. A 1.00 kg block is attached to a horizontal spring with spring constant 2500 N/m. The block is at rest on a frictionless surface. A 10 g bullet is fired into the block, in the face opposite the spring, and sticks. What was the bullet's speed if the subsequent oscillations have an amplitude of 10.0 cm?

77. A spring is standing upright on a table with its bottom end fastened to the table. A block is dropped from a height 3.0 cm above the top of the spring. The block sticks to the top end of the spring and then oscillates with an amplitude of 10 cm. What is the oscillation frequency?

78. The analysis of a simple pendulum assumed that the mass was a particle, with no size. A realistic pendulum is a small, uniform sphere of mass M and radius R at the end of a massless string, with L being the distance from the pivot to the center of the sphere.
    a. Find an expression for the period of this pendulum.
    b. Suppose $M = 25$ g, $R = 1.0$ cm, and $L = 1.0$ m, typical values for a real pendulum. What is the ratio T_{real}/T_{simple}, where T_{real} is your expression from part a and T_{simple} is the expression derived in this chapter?

79. a. A mass m oscillating on a spring has period T. Suppose the mass changes very slightly from m to $m + \Delta m$, where $\Delta m \ll m$. Find an expression for ΔT, the small change in the period. Your expression should involve T, m, and Δm but not the spring constant.
    b. Suppose the period is 2.000 s and the mass increases by 0.1%. What is the new period?

80. FIGURE CP14.80 shows a 200 g uniform rod pivoted at one end. The other end is attached to a horizontal spring. The spring is neither stretched nor compressed when the rod hangs straight down. What is the rod's oscillation period? You can assume that the rod's angle from vertical is always small.

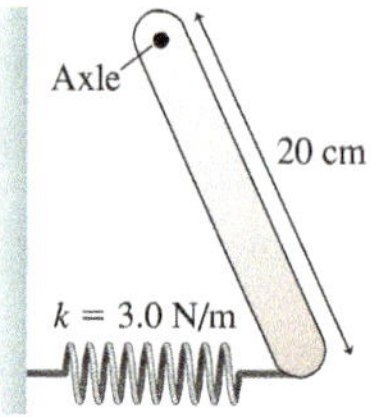

FIGURE CP14.80

STOP TO THINK ANSWERS

Stop to Think 14.1: c. $v_{max} = 2\pi A/T$. Doubling A and T leaves v_{max} unchanged.

Stop to Think 14.2: d. Think of circular motion. At 45°, the particle is in the first quadrant (positive x) and moving to the left (negative v_x).

Stop to Think 14.3: c > b > a = d. Energy conservation $\frac{1}{2}kA^2 = \frac{1}{2}m(v_{max})^2$ gives $v_{max} = \sqrt{k/m}\,A$. k or m has to be increased or decreased by a factor of 4 to have the same effect as increasing or decreasing A by a factor of 2.

Stop to Think 14.4: c. $v_x = 0$ because the slope of the position graph is zero. The negative value of x shows that the particle is left of the equilibrium position, so the restoring force is to the right.

Stop to Think 14.5: c. The period of a pendulum does not depend on its mass.

Stop to Think 14.6: $\tau_d > \tau_b = \tau_c > \tau_a$. The time constant is the time to decay to 37% of the initial height. The time constant is independent of the initial height.

PART

V

Waves and Optics

The song of a humpback whale can travel hundreds of kilometers underwater. This graph uses a procedure called wavelet analysis to study the frequency structure of a humpback whale song.

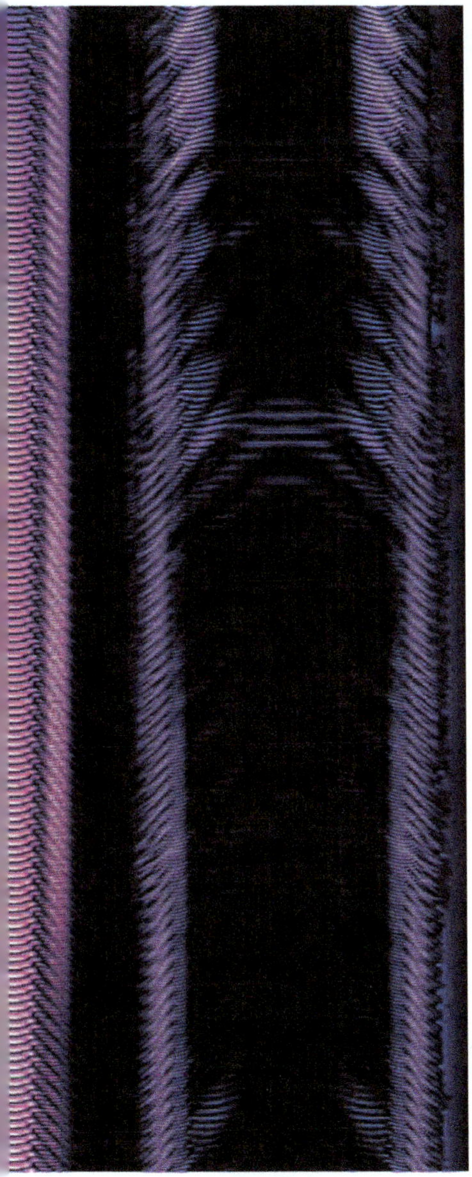

OVERVIEW

The Wave Model

Parts I–IV of this text have been primarily about the physics of particles. You've seen that macroscopic systems ranging from balls and rockets to a gas of molecules can be thought of as particles or as systems of particles. A *particle* is one of the two fundamental models of classical physics. The other, to which we now turn our attention, is a *wave.*

Waves are ubiquitous in nature. Familiar examples of waves include

- Undulating ripples on a pond.
- The swaying ground of an earthquake.
- A vibrating guitar string.
- The sweet sound of a flute.
- The colors of the rainbow.

The physics of waves is the subject of Part V, the next stage of our journey. Despite the great diversity of types and sources of waves, a single, elegant physical theory is capable of describing them all. Our exploration of wave phenomena will call upon sound waves, light waves, and vibrating strings for examples, but our goal is to emphasize the unity and coherence of the ideas that are common to *all* types of waves.

A wave, in contrast with a particle, is diffuse, spread out, not to be found at a single point in space. We will start with waves traveling outward through some medium, like the spreading ripples after a pebble hits a pool of water. These are called *traveling waves.* An investigation of what happens when waves travel through each other will lead us to *standing waves,* which are essential for understanding phenomena ranging from those as common as musical instruments and water sloshing in a tub to as complex as lasers and the electrons in atoms. We'll also study one of the most important defining characteristics of waves—their ability to exhibit *interference*.

Three chapters will be devoted to light and optics, perhaps the most important application of waves. Although light is an electromagnetic wave, your understanding of these chapters will depend on nothing more than the "waviness" of light. You can study these chapters either before or after your study of electricity and magnetism in Part VI. The electromagnetic aspects of light waves will be taken up in Chapter 34.

Our investigation of light will be aided by a second model, the *ray model,* in which light travels in straight lines, reflects from mirrors, and is focused by lenses. Many practical applications of optics, from the camera to the telescope, arc best understood with the ray model of light.

In fact, that you're able to read this book at all is due to the first optical instrument you ever used—your eyes. We will investigate the optics of the eye, learn how the cornea and lens form an image on the retina, and see how glasses or contact lenses can be used to correct the image if it is out of focus.

20 Traveling Waves

This surfer is "catching a wave." At the same time, he's seeing light waves and hearing sound waves.

▶ **Looking Ahead** The goal of Chapter 20 is to learn the basic properties of traveling waves.

The Wave Model

A **wave** is a disturbance traveling through a medium. Our goal is to develop a model —the wave model—that describes the basic properties of all waves.

The wave propagates, but the particles of the medium don't. The water molecules simply oscillate up and down as the ripples spread outward.

Two Types of Waves

You'll find that waves come in two basic types:

Transverse waves: The displacement is perpendicular to the direction of travel.

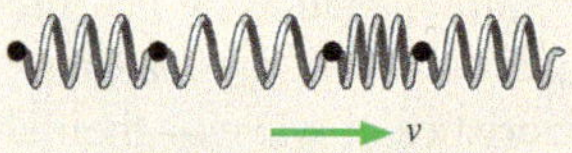

Longitudinal waves: The displacement is parallel to the direction of travel.

Sound and Light

Two types of waves are especially important: sound and light.

- Sound waves are longitudinal waves.
- Light waves are transverse waves.

You'll learn that the colors of visible light correspond to different wavelengths.

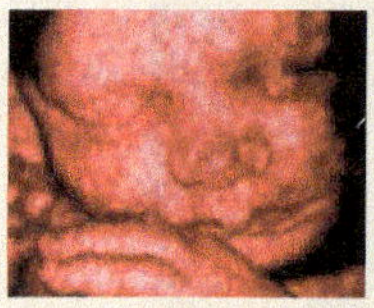

Ultrasound images are made with very-high-frequency sound waves.

Wave Properties

You'll learn that a wave is characterized by three basic properties:

- **Wave speed**: How fast it travels through the medium.
- **Wavelength**: The distance between two neighboring crests.
- **Frequency**: The number of oscillations per second.

You'll also see that wave motion is closely related to simple harmonic motion.

◀ **Looking Back**
Sections 14.1 and 14.2 Properties of simple harmonic motion

Intensity and Loudness

Waves carry energy. The rate at which a wave delivers energy to a surface is the **intensity** of the wave.

Your ears are sensitive to a remarkable range of intensities. You'll learn to use the logarithmic **decibel** scale to characterize the loudness of a sound.

Focusing the sun's light into a smaller area increases its intensity.

The Doppler Effect

The frequency and wavelength of a wave are shifted when there is relative motion between the source and the observer of the waves. This is called the **Doppler effect.**

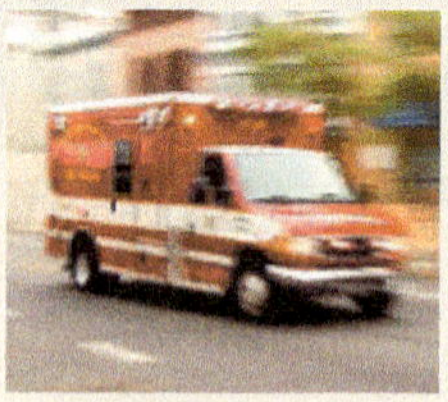

The pitch of the ambulance siren drops as it races past you. The frequency is shifted up as it approaches, then shifted down as it recedes.

20.1 The Wave Model

Balls, cars, and rockets obviously differ from one another, but the general features of their motions are well described by the *particle model* of Parts I–IV. In Part V we will explore the basic properties of waves with a **wave model,** emphasizing those aspects of wave behavior common to all waves. Although water waves, sound waves, and light waves are clearly different, the wave model will allow us to understand many of the important features they have in common.

The wave model is built around the idea of a **traveling wave,** which is an organized disturbance traveling with a well-defined wave speed. We'll begin our study of traveling waves by looking at two distinct wave motions.

Two types of traveling waves

A transverse wave

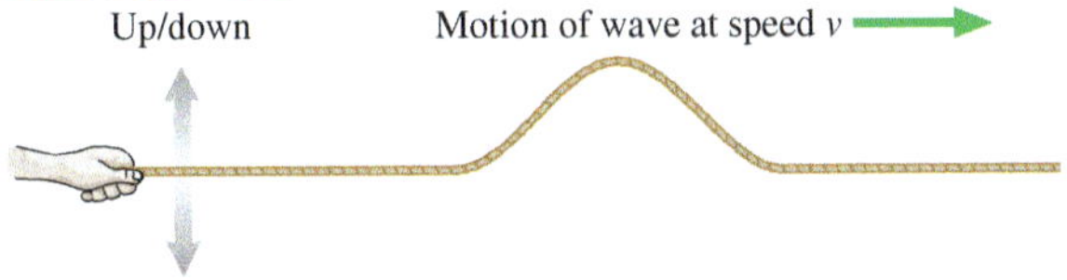

A **transverse wave** is a wave in which the displacement is *perpendicular* to the direction in which the wave travels. For example, a wave travels along a string in a horizontal direction while the particles that make up the string oscillate vertically. Electromagnetic waves are also transverse waves because the electromagnetic fields oscillate perpendicular to the direction in which the wave travels.

A longitudinal wave

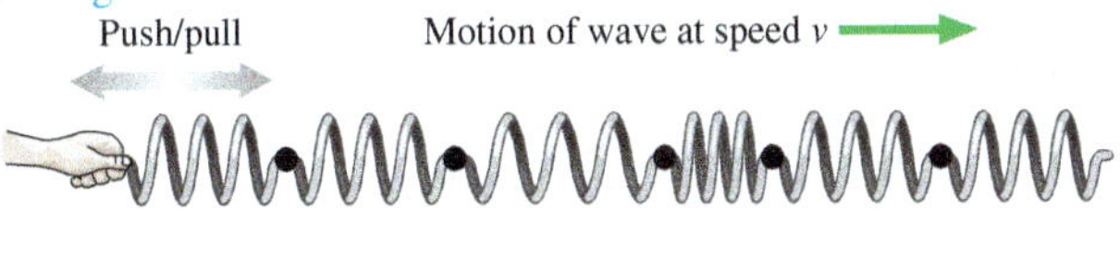

In a **longitudinal wave,** the particles in the medium move *parallel* to the direction in which the wave travels. Here we see a chain of masses connected by springs. If you give the first mass in the chain a sharp push, a disturbance travels down the chain by compressing and expanding the springs. Sound waves in gases and liquids are the most well known examples of longitudinal waves.

We can also classify waves on the basis of what is "waving":

1. **Mechanical waves** travel only within a material *medium*, such as air or water. Two familiar mechanical waves are sound waves and water waves.
2. **Electromagnetic waves,** from radio waves to visible light to x rays, are a self-sustaining oscillation of the *electromagnetic field.* Electromagnetic waves require no material medium and can travel through a vacuum.

FIGURE 20.1 Ripples on a pond are a traveling wave.

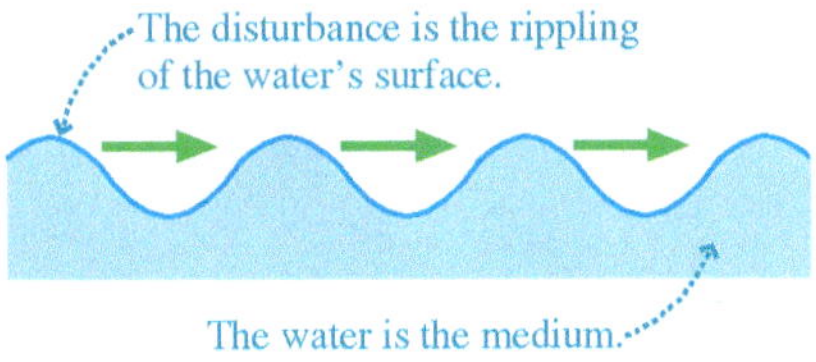

The **medium** of a mechanical wave is the substance through or along which the wave moves. For example, the medium of a water wave is the water, the medium of a sound wave is the air, and the medium of a wave on a stretched string is the string. A medium must be *elastic.* That is, a restoring force of some sort brings the medium back to equilibrium after it has been displaced or disturbed. The tension in a stretched string pulls the string back straight after you pluck it. Gravity restores the level surface of a lake after the wave generated by a boat has passed by.

As a wave passes through a medium, the atoms of the medium—we'll simply call them the particles of the medium—are displaced from equilibrium. This is a **disturbance** of the medium. The water ripples of FIGURE 20.1 are a disturbance of the water's surface. A pulse traveling down a string is a disturbance, as are the wake of a boat and the sonic boom created by a jet traveling faster than the speed of sound. **The disturbance of a wave is an *organized* motion of the particles in the medium,** in contrast to the *random* molecular motions of thermal energy.

Wave Speed

A wave disturbance is created by a *source*. The source of a wave might be a rock thrown into water, your hand plucking a stretched string, or an oscillating loudspeaker cone pushing on the air. Once created, the disturbance travels outward through the medium at the **wave speed** v. This is the speed with which a ripple moves across the water or a pulse travels down a string.

NOTE ▶ The disturbance propagates through the medium, but **the medium as a whole does not move!** The ripples on the pond (the disturbance) move outward from the splash of the rock, but there is no outward flow of water from the splash. Likewise, the particles of a string oscillate up and down but do not move in the direction of a pulse traveling along the string. **A wave transfers energy, but it does not transfer any material or substance outward from the source.** ◀

As an example, we'll prove in Section 20.3 that the wave speed on a string stretched with tension T_s is

$$v_{string} = \sqrt{\frac{T_s}{\mu}} \quad \text{(wave speed on a stretched string)} \qquad (20.1)$$

where μ is the string's **linear density,** its mass-to-length ratio:

$$\mu = \frac{m}{L} \qquad (20.2)$$

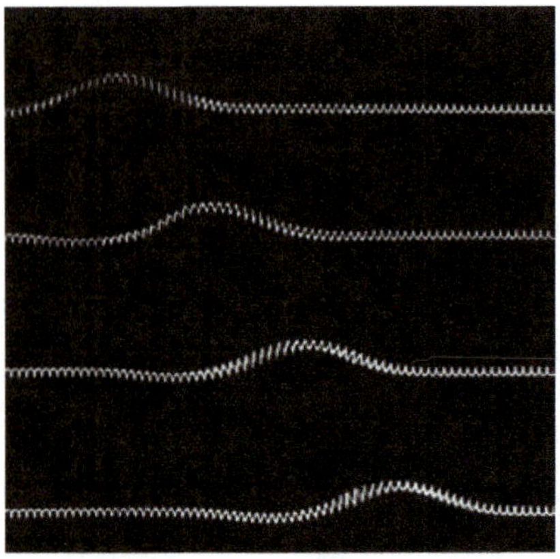

This sequence of photographs shows a wave pulse traveling along a spring.

The SI unit of linear density is kg/m. A fat string has a larger value of μ than a skinny string made of the same material. Similarly, a steel wire has a larger value of μ than a plastic string of the same diameter. We'll assume that strings are *uniform,* meaning the linear density is the same everywhere along the length of the string.

NOTE ▶ The subscript s on the symbol T_s for the string's tension distinguishes it from the symbol T for the *period* of oscillation. ◀

Equation 20.1 is the wave *speed,* not the wave velocity, so v_{string} always has a positive value. Every point on a wave travels with this speed. You can increase the wave speed either by *increasing* the string's tension (make it tighter) or by *decreasing* the string's linear density (make it skinnier). We'll examine the implications for stringed musical instruments in Chapter 21.

EXAMPLE 20.1 Measuring the linear density

In a laboratory experiment, one end of a metal wire is connected to a motion sensor. The wire is stretched horizontally to a pulley 1.50 m away, then attached to a hanging mass that provides tension. A mechanical pick plucks the horizontal segment of the wire right at the pulley, creating a small wave pulse that travels along the wire. The plucking motion starts a timer that is stopped by the motion sensor when the pulse reaches the end of the wire. Changing the hanging mass changes the time required for the pulse to travel the length of the wire. The data are as follows:

Mass (kg)	Time (ms)
0.50	31
1.00	23
1.50	18
2.00	15
2.50	14

Use the data to determine the wire's linear density.

MODEL The wave pulse is a traveling wave on a stretched string. The hanging mass is in static equilibrium.

VISUALIZE FIGURE 20.2 is a pictorial representation.

SOLVE The wave speed on the wire is determined by the wire's linear density μ and tension T_s. The hanging mass is in static equilibrium, with no net force, so we see from the free-body diagram that the tension in the wire is $T_s = F_G = Mg$. Squaring both sides of Equation 20.1 gives

FIGURE 20.2 A wave pulse on the wire.

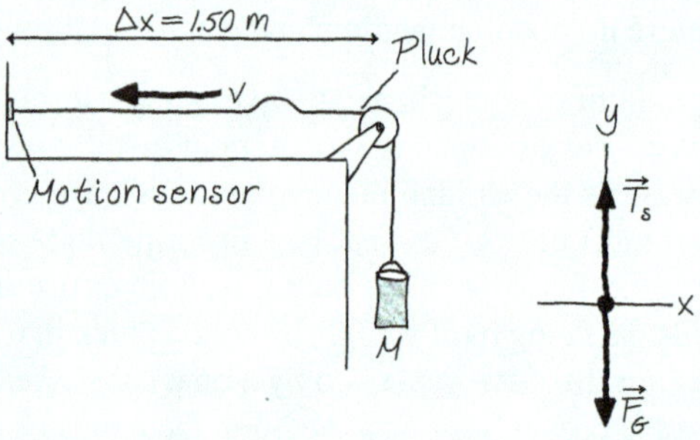

$$v^2 = \left(\frac{\Delta x}{\Delta t}\right)^2 = \frac{T_s}{\mu} = \frac{Mg}{\mu}$$

Mass M is the independent variable that we've changed, each time measuring the pulse travel time Δt, so we can rearrange the wave-speed equation as

$$(\Delta t)^2 = \frac{\mu(\Delta x)^2}{g}\frac{1}{M}$$

Theory predicts that a graph of the *square* of the travel time versus the *inverse* of the hanging mass should be a straight line passing through the origin with slope $\mu(\Delta x)^2/g$. The graph of FIGURE 20.3, with the times converted from ms to s, is indeed linear with a y-intercept of zero. The slope of the best-fit line is seen to be 4.85×10^{-4} kg s^2 (recall that spreadsheets and graphing calculators display this as 4.85E–04), from which we find the wire's linear density:

$$\mu = \frac{g \times \text{slope}}{(\Delta x)^2} = 0.0021 \text{ kg/m} = 2.1 \text{ g/m}$$

ASSESS A meter of thin wire is likely to have a mass of a few grams, so a linear density of a few g/m seems reasonable.

FIGURE 20.3 A graph of the data.

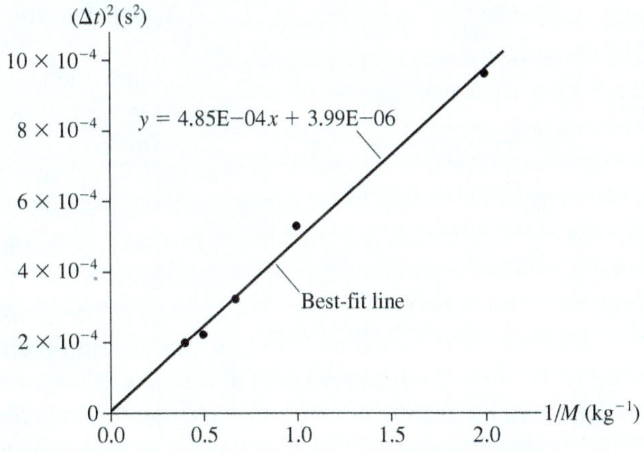

The wave speed on a string is a property of the string—its tension and linear density. In general, **the wave speed is a property *of the medium.*** The wave speed depends on the restoring forces within the medium but not at all on the shape or size of the pulse, how the pulse was generated, or how far it has traveled.

STOP TO THINK 20.1 Which of the following actions would make a pulse travel faster along a stretched string? More than one answer may be correct. If so, give all that are correct.

a. Move your hand up and down more quickly as you generate the pulse.
b. Move your hand up and down a larger distance as you generate the pulse.
c. Use a heavier string of the same length, under the same tension.
d. Use a lighter string of the same length, under the same tension.
e. Stretch the string tighter to increase the tension.
f. Loosen the string to decrease the tension.
g. Put more force into the wave.

20.2 One-Dimensional Waves

To understand waves we must deal with functions of *two* variables. Until now, we have been concerned with quantities that depend only on time, such as $x(t)$ or $v(t)$. Functions of the one variable t are all right for a particle because a particle is only in one place at a time, but a wave is not localized. It is spread out through space at each instant of time. To describe a wave mathematically requires a function that specifies not only an instant of time (when) but also a point in space (where).

Rather than leaping into mathematics, we will start by thinking about waves graphically. Consider the wave pulse shown moving along a stretched string in FIGURE 20.4. (We will consider somewhat artificial triangular and square-shaped pulses in this section to make clear where the edges of the pulse are.) The graph shows the string's displacement Δy at a particular instant of time t_1 as a function of position x along the string. This is a "snapshot" of the wave, much like what you might make with a camera whose shutter is opened briefly at t_1. A graph that shows the wave's displacement as a function of position at a single instant of time is called a **snapshot graph.** For a wave on a string, a snapshot graph is literally a picture of the wave at this instant.

FIGURE 20.5 shows a sequence of snapshot graphs as the wave of Figure 20.4 continues to move. These are like successive frames from a movie. Notice that the wave

FIGURE 20.4 A snapshot graph of a wave pulse on a string.

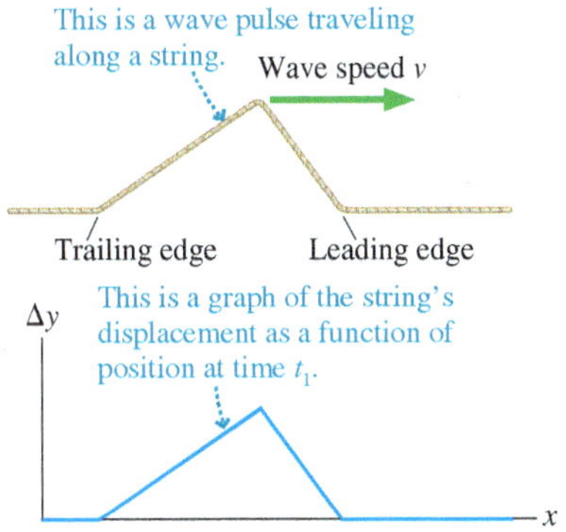

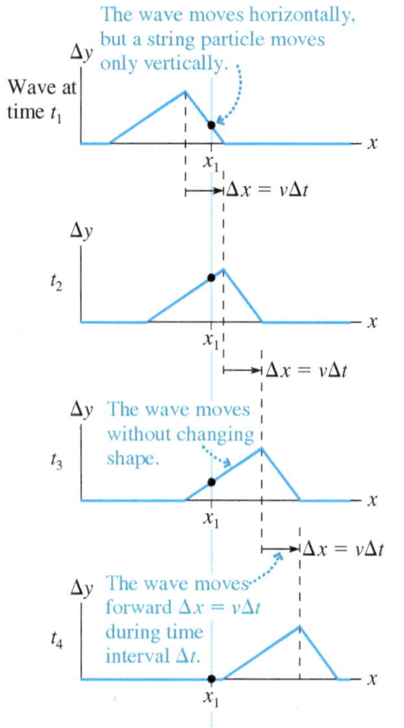

FIGURE 20.5 A sequence of snapshot graphs shows the wave in motion.

pulse moves forward distance $\Delta x = v\Delta t$ during the time interval Δt. That is, the wave moves with constant speed.

A snapshot graph tells only half the story. It tells us *where* the wave is and how it varies with position, but only at one instant of time. It gives us no information about how the wave *changes* with time. As a different way of portraying the wave, suppose we follow the dot marked on the string in Figure 20.5 and produce a graph showing how the displacement of this dot changes with time. The result, shown in FIGURE 20.6, is a displacement-versus-time graph at a single position in space. A graph that shows the wave's displacement as a function of time at a single position in space is called a **history graph.** It tells the history of that particular point in the medium.

You might think we have made a mistake; the graph of Figure 20.6 is reversed compared to Figure 20.5. It is not a mistake, but it requires careful thought to see why. As the wave moves toward the dot, the steep *leading edge* causes the dot to rise quickly. On the displacement-versus-time graph, *earlier* times (smaller values of t) are to the *left* and later times (larger t) to the right. Thus the leading edge of the wave is on the *left* side of the Figure 20.6 history graph. As you move to the right on Figure 20.6 you see the slowly falling *trailing edge* of the wave as it moves past the dot at later times.

The snapshot graph of Figure 20.4 and the history graph of Figure 20.6 portray complementary information. The snapshot graph tells us how things look throughout all of space, but at only one instant of time. The history graph tells us how things look at all times, but at only one position in space. We need them both to have the full story of the wave. An alternative representation of the wave is the series of graphs in FIGURE 20.7, where we can get a clearer sense of the wave moving forward. But graphs like these are essentially impossible to draw by hand, so it is necessary to move back and forth between snapshot graphs and history graphs.

FIGURE 20.6 A history graph for the dot on the string in Figure 20.5.

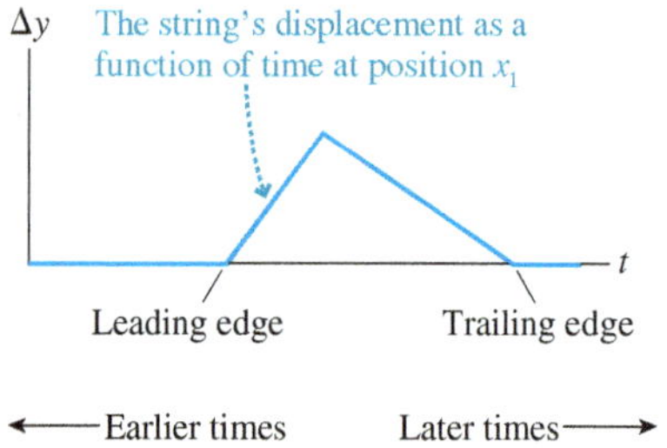

FIGURE 20.7 An alternative look at a traveling wave.

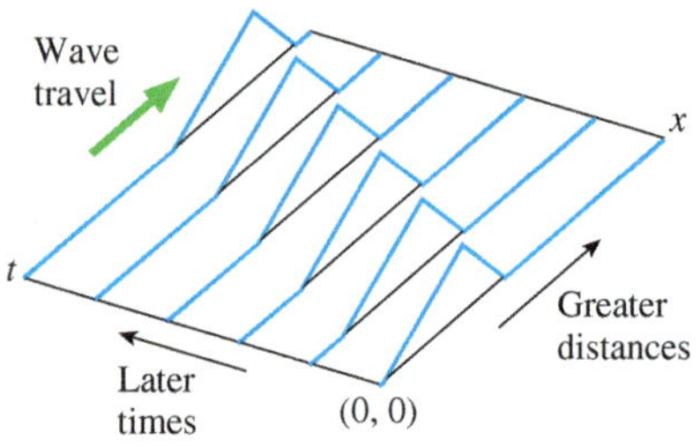

EXAMPLE 20.2 Finding a history graph from a snapshot graph

FIGURE 20.8 is a snapshot graph at $t = 0$ s of a wave moving to the right at a speed of 2.0 m/s. Draw a history graph for the position $x = 8.0$ m.

FIGURE 20.8 A snapshot graph at $t = 0$ s.

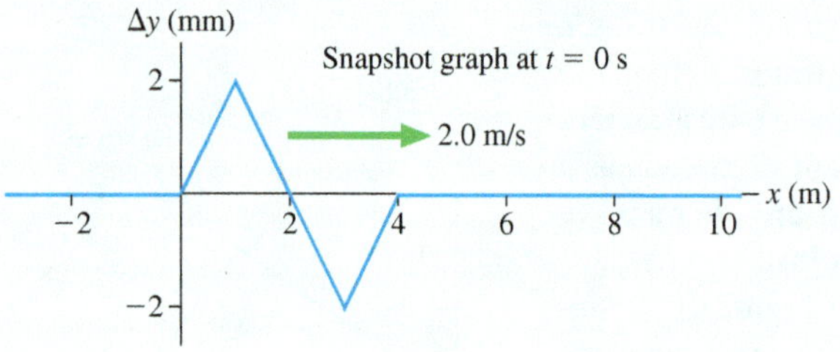

MODEL This is a wave traveling at constant speed. The pulse moves 2.0 m to the right every second.

VISUALIZE The snapshot graph of Figure 20.8 shows the wave at all points on the x-axis at $t = 0$ s. You can see that nothing is happening at $x = 8.0$ m at this instant of time because the wave has not yet reached $x = 8.0$ m. In fact, at $t = 0$ s the leading edge of the wave is still 4.0 m away from $x = 8.0$ m. Because the wave is traveling at 2.0 m/s, it will take 2.0 s for the leading edge to reach $x = 8.0$ m. Thus the history graph for $x = 8.0$ m will be zero until $t = 2.0$ s. The first part of the wave causes a *downward* displacement of the medium, so immediately after $t = 2.0$ s the displacement at $x = 8.0$ m will be negative. The negative portion of the wave pulse is 2.0 m wide and takes 1.0 s to pass $x = 8.0$ m, so the midpoint of the pulse reaches $x = 8.0$ m at $t = 3.0$ s. The positive portion takes another 1.0 s to go past, so the trailing edge of the pulse arrives at $t = 4.0$ s. You could also note that the trailing edge was initially 8.0 m away from $x = 8.0$ m and needed 4.0 s to travel that distance at 2.0 m/s. The displacement at $x = 8.0$ m returns to zero at $t = 4.0$ s and remains zero for all later times. This information is all portrayed on the history graph of FIGURE 20.9.

FIGURE 20.9 The corresponding history graph at $x = 8.0$ m.

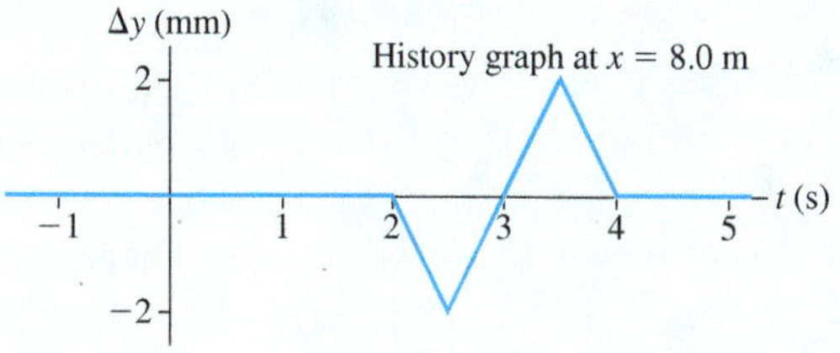

STOP TO THINK 20.2 The graph at the right is the history graph at $x = 4.0$ m of a wave traveling to the right at a speed of 2.0 m/s. Which is the history graph of this wave at $x = 0$ m?

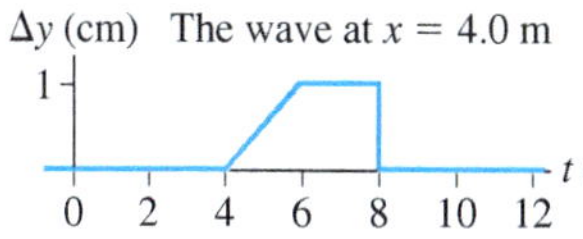

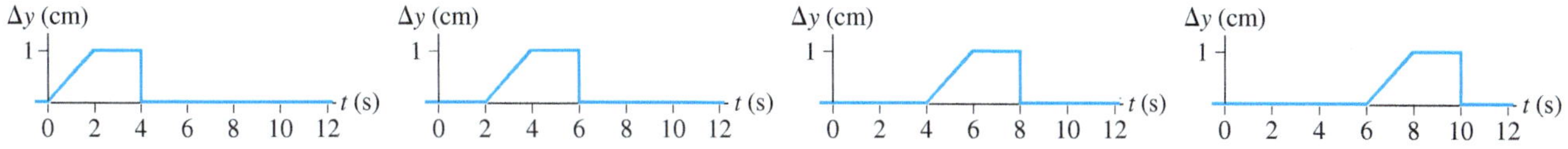

(a) (b) (c) (d)

Longitudinal Waves

For a wave on a string, a transverse wave, the snapshot graph is literally a picture of the wave. Not so for a longitudinal wave, where the particles in the medium are displaced parallel to the direction in which the wave is traveling. Thus the displacement is Δx rather than Δy, and a snapshot graph is a graph of Δx versus x.

FIGURE 20.10a is a snapshot graph of a longitudinal wave, such as a sound wave. It's purposefully drawn to have the same shape as the string wave in Example 20.2. Without practice, it's not clear what this graph tells us about the particles in the medium.

FIGURE 20.10 Visualizing a longitudinal wave.

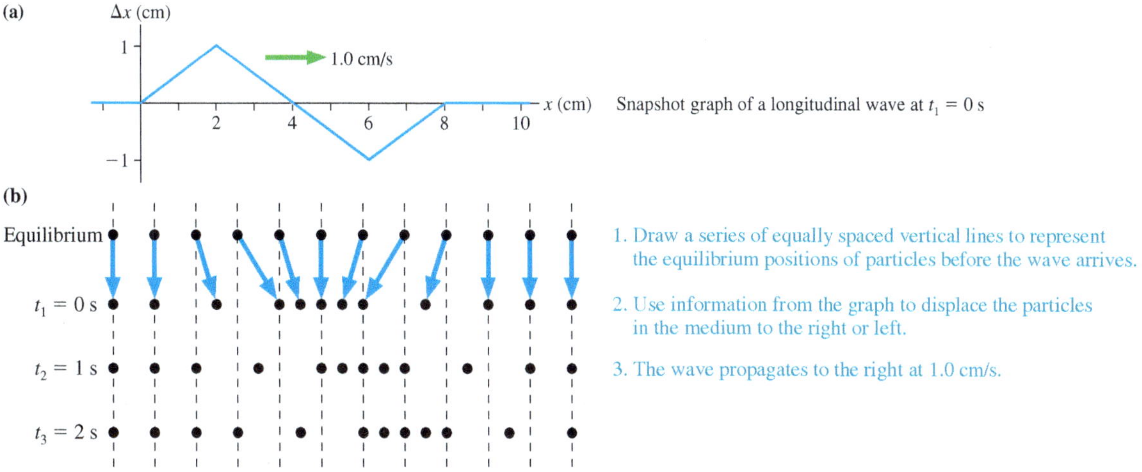

To help you find out, FIGURE 20.10b provides a tool for visualizing longitudinal waves. In the second row, we've used information from the graph to displace the particles in the medium to the right or to the left of their equilibrium positions. For example, the particle at $x = 1.0$ cm has been displaced 0.5 cm to the right because the snapshot graph shows $\Delta x = 0.5$ cm at $x = 1.0$ cm. We now have a picture of the longitudinal wave pulse at $t_1 = 0$ s. You can see that the medium is compressed to higher density at the center of the pulse and, to compensate, expanded to lower density at the leading and trailing edges. Two more lines show the medium at $t_2 = 1$ s and $t_3 = 2$ s so that you can see the wave propagating through the medium at 1.0 cm/s.

You've probably seen or participated in "the wave" at a sporting event. The wave moves around the stadium, but the people (the medium) simply undergo small displacements from their equilibrium positions.

The Displacement

A traveling wave causes the particles of the medium to be displaced from their equilibrium positions. Because one of our goals is to develop a mathematical representation to describe all types of waves, we'll use the generic symbol D to stand for the *displacement* of a wave of any type. But what do we mean by a "particle" in the medium? And what about electromagnetic waves, for which there is no medium?

For a string, where the atoms stay fixed relative to each other, you can think of either the atoms themselves or very small segments of the string as being the particles of the medium. D is then the perpendicular displacement Δy of a point on the string. For a sound wave, D is the longitudinal displacement Δx of a small volume of fluid. For any other mechanical wave, D is the appropriate displacement. Even electromagnetic waves can be described within the same mathematical representation if D is interpreted as a yet-undefined *electromagnetic field strength,* a "displacement" in a more abstract sense as an electromagnetic wave passes through a region of space.

Because the displacement of a particle in the medium depends both on *where* the particle is (position x) and on *when* you observe it (time t), D must be a function of the two variables x and t. That is,

$$D(x, t) = \text{the displacement at time } t \text{ of a particle at position } x$$

The values of *both* variables—where and when—must be specified before you can evaluate the displacement D.

20.3 Sinusoidal Waves

A wave source that oscillates with simple harmonic motion (SHM) generates a **sinusoidal wave.** For example, a loudspeaker cone that oscillates in SHM radiates a sinusoidal sound wave. The sinusoidal electromagnetic waves broadcast by television and FM radio stations are generated by electrons oscillating back and forth in the antenna wire with SHM. **The frequency f of the wave is the frequency of the oscillating source.**

FIGURE 20.11 A sinusoidal wave moving along the x-axis.

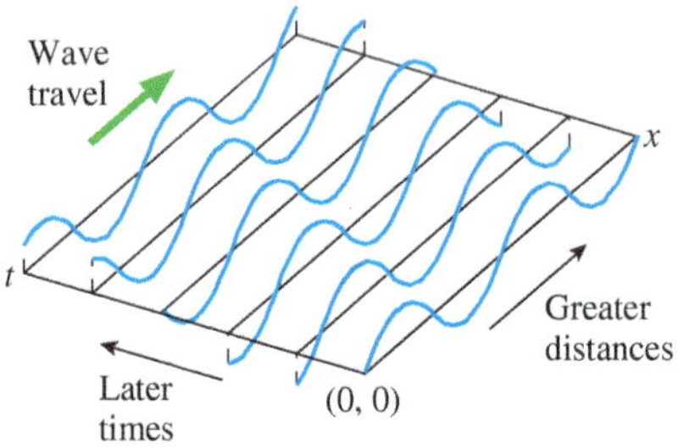

FIGURE 20.11 shows a sinusoidal wave moving through a medium. The source of the wave, which is undergoing vertical SHM, is located at $x = 0$. Notice how the wave crests move with steady speed toward larger values of x at later times t.

FIGURE 20.12a is a history graph for a sinusoidal wave, showing the displacement of the medium at one point in space. Each particle in the medium undergoes simple harmonic motion with frequency f, so this graph of SHM is identical to the graphs you learned to work with in Chapter 14. The *period* of the wave, shown on the graph, is the time interval for one cycle of the motion. The period is related to the wave frequency f by

$$T = \frac{1}{f} \tag{20.3}$$

exactly as in simple harmonic motion. The **amplitude** A of the wave is the maximum value of the displacement. The crests of the wave have displacement $D_{crest} = A$ and the troughs have displacement $D_{trough} = -A$.

FIGURE 20.12 History and snapshot graphs for a sinusoidal wave.

(a) A history graph at one point in space

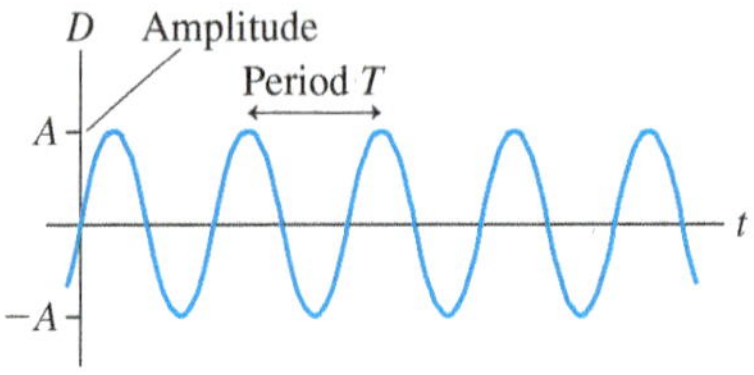

(b) A snapshot graph at one instant of time

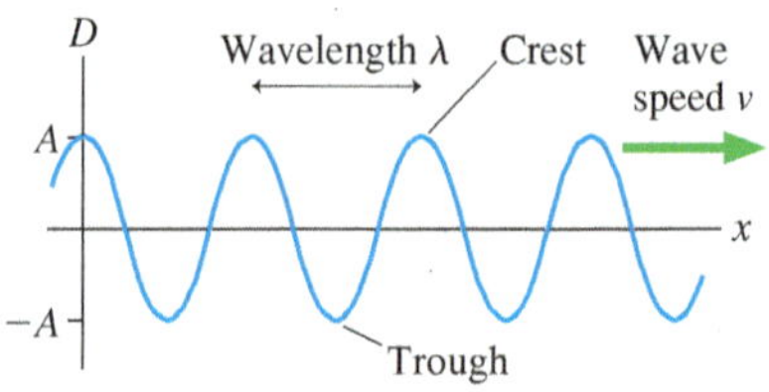

Displacement versus time is only half the story. FIGURE 20.12b shows a snapshot graph for the same wave at one instant in time. Here we see the wave stretched out in space, moving to the right with speed v. An important characteristic of a sinusoidal wave is that it is periodic *in space* as well as in time. As you move from left to right along the "frozen" wave in the snapshot graph, the disturbance repeats itself over and over. The distance spanned by one cycle of the motion is called the **wavelength** of the wave. Wavelength is symbolized by λ (lowercase Greek lambda) and, because it is a length, it is measured in units of meters. The wavelength is shown in Figure 20.12b as the distance between two crests, but it could equally well be the distance between two troughs.

NOTE ▶ Wavelength is the spatial analog of period. The period T is the *time* in which the disturbance at a single point in space repeats itself. The wavelength λ is the *distance* in which the disturbance at one instant of time repeats itself. ◀

The Fundamental Relationship for Sinusoidal Waves

There is an important relationship between the wavelength and the period of a wave. FIGURE 20.13 shows this relationship through five snapshot graphs of a sinusoidal wave at time increments of one-quarter of the period T. One full period has elapsed between the first graph and the last, which you can see by observing the motion at a fixed point on the x-axis. Each point in the medium has undergone exactly one complete oscillation.

The critical observation is that the wave crest marked by an arrow has moved one full wavelength between the first graph and the last. That is, **during a time interval of exactly one period T, each crest of a sinusoidal wave travels forward a distance of exactly one wavelength λ.** Because speed is distance divided by time, the wave speed must be

$$v = \frac{\text{distance}}{\text{time}} = \frac{\lambda}{T} \tag{20.4}$$

Because $f = 1/T$, it is customary to write Equation 20.4 in the form

$$v = \lambda f \tag{20.5}$$

Although Equation 20.5 has no special name, it is *the* fundamental relationship for periodic waves. When using it, keep in mind the *physical* meaning that **a wave moves forward a distance of one wavelength during a time interval of one period.**

NOTE ▶ Wavelength and period are defined only for *periodic* waves, so Equations 20.4 and 20.5 apply only to periodic waves. A wave pulse has a wave speed, but it doesn't have a wavelength or a period. Hence Equations 20.4 and 20.5 cannot be applied to wave pulses. ◀

Because the wave speed is a property of the medium while the wave frequency is a property of the source, it is often useful to write Equation 20.5 as

$$\lambda = \frac{v}{f} = \frac{\text{property of the medium}}{\text{property of the source}} \tag{20.6}$$

The wavelength is a *consequence* of a wave of frequency f traveling through a medium in which the wave speed is v.

FIGURE 20.13 A series of snapshot graphs at time increments of one-quarter of the period T.

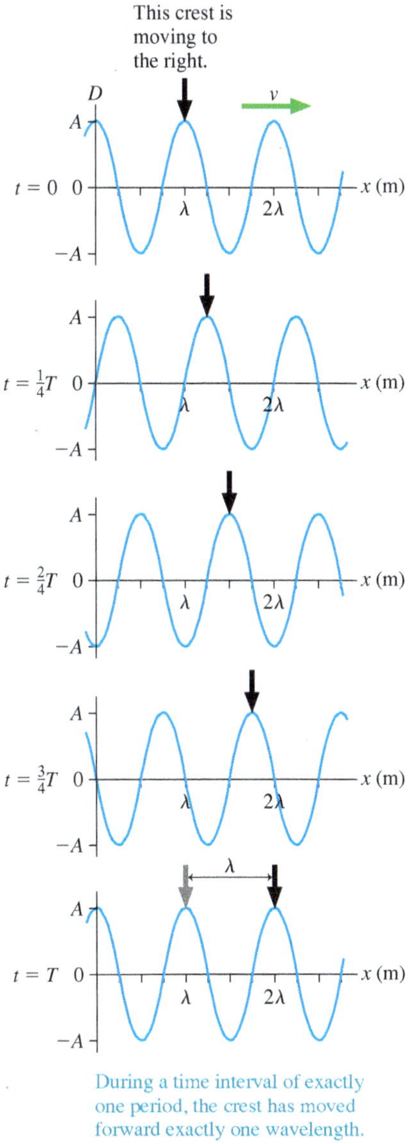

STOP TO THINK 20.3 What is the frequency of this traveling wave?

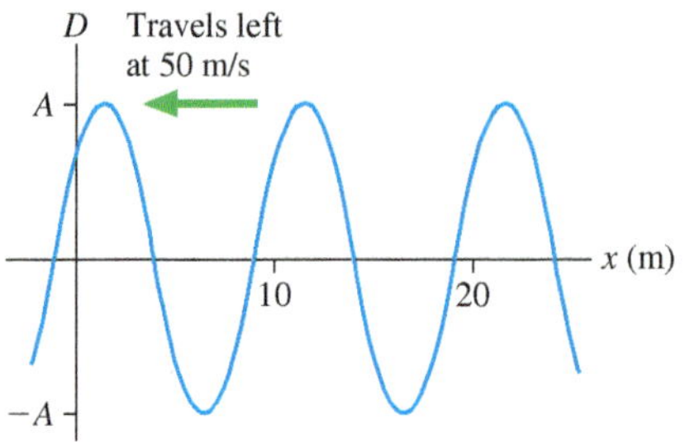

a. 0.1 Hz
b. 0.2 Hz
c. 2 Hz
d. 5 Hz
e. 10 Hz
f. 500 Hz

The Mathematics of Sinusoidal Waves

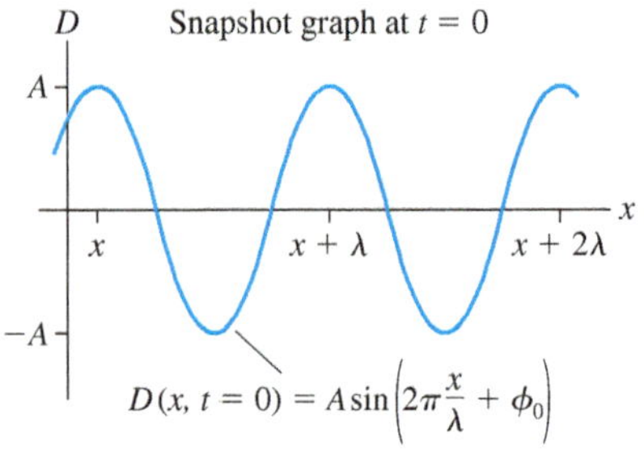

FIGURE 20.14 A sinusoidal wave is "frozen" at $t = 0$.

FIGURE 20.14 shows a snapshot graph at $t = 0$ of a sinusoidal wave. The sinusoidal function that describes the displacement of this wave is

$$D(x, t = 0) = A\sin\left(2\pi\frac{x}{\lambda} + \phi_0\right) \tag{20.7}$$

where the notation $D(x, t = 0)$ means that we've frozen the time at $t = 0$ to make the displacement a function of only x. The term ϕ_0 is a *phase constant* that characterizes the initial conditions. (We'll return to the phase constant momentarily.)

The function of Equation 20.7 is periodic with period λ. We can see this by writing

$$D(x + \lambda) = A\sin\left(2\pi\frac{(x + \lambda)}{\lambda} + \phi_0\right) = A\sin\left(2\pi\frac{x}{\lambda} + \phi_0 + 2\pi \text{ rad}\right)$$

$$= A\sin\left(2\pi\frac{x}{\lambda} + \phi_0\right) = D(x)$$

where we used the fact that $\sin(a + 2\pi \text{ rad}) = \sin a$. In other words, the disturbance created by the wave at $x + \lambda$ is exactly the same as the disturbance at x.

The next step—and it's an important step to graph—is to set the wave in motion. We can do this by replacing x in Equation 20.7 with $x - vt$. To see why this works, recall that the wave moves distance vt during time t. In other words, whatever displacement the wave has at position x at time t, the wave must have had that same displacement at position $x - vt$ at the earlier time $t = 0$. Mathematically, this idea can be captured by writing

$$D(x, t) = D(x - vt, t = 0) \tag{20.8}$$

Make sure you understand how this statement describes a wave moving in the positive x-direction at speed v.

This is what we were looking for. $D(x, t)$ is the general function describing the traveling wave. It's found by taking the function that describes the wave at $t = 0$—the function of Equation 20.7—and replacing x with $x - vt$. Thus the displacement equation of a sinusoidal wave traveling in the positive x-direction at speed v is

$$D(x, t) = A\sin\left(2\pi\frac{x - vt}{\lambda} + \phi_0\right) = A\sin\left(2\pi\left(\frac{x}{\lambda} - \frac{t}{T}\right) + \phi_0\right) \tag{20.9}$$

In the last step we used $v = \lambda f = \lambda/T$ to write $v/\lambda = 1/T$. The function of Equation 20.9 is not only periodic in space with period λ, it is also periodic in time with period T. That is, $D(x, t + T) = D(x, t)$.

It will be useful to introduce two new quantities. First, recall from simple harmonic motion the *angular frequency*

$$\omega = 2\pi f = \frac{2\pi}{T} \tag{20.10}$$

The units of ω are rad/s, although many textbooks use simply s^{-1}.

You can see that ω is 2π times the reciprocal of the period in time. This suggests that we define an analogous quantity, called the **wave number** k, that is 2π times the reciprocal of the period in space:

$$k = \frac{2\pi}{\lambda} \tag{20.11}$$

The units of k are rad/m, although many textbooks use simply m^{-1}.

NOTE ▶ The wave number k is *not* a spring constant, even though it uses the same symbol. This is a most unfortunate use of symbols, but every major textbook and professional tradition uses the same symbol k for these two very different meanings, so we have little choice but to follow along. ◀

We can use the fundamental relationship $v = \lambda f$ to find an analogous relationship between ω and k:

$$v = \lambda f = \frac{2\pi}{k}\frac{\omega}{2\pi} = \frac{\omega}{k} \tag{20.12}$$

which is usually written

$$\omega = vk \tag{20.13}$$

Equation 20.13 contains no new information. It is a variation of Equation 20.5, but one that is convenient when working with k and ω.

If we use the definitions of Equations 20.10 and 20.11, Equation 20.9 for the displacement can be written

$$D(x, t) = A\sin(kx - \omega t + \phi_0) \quad \text{(sinusoidal wave traveling in the positive } x\text{-direction)} \tag{20.14}$$

A sinusoidal wave traveling in the negative x-direction is $A\sin(kx + \omega t + \phi_0)$. Equation 20.14 is graphed versus x and t in FIGURE 20.15.

Just as it did for simple harmonic motion, the phase constant ϕ_0 characterizes the initial conditions. At $(x, t) = (0\text{ m}, 0\text{ s})$ Equation 20.14 becomes

$$D(0\text{ m}, 0\text{ s}) = A\sin\phi \tag{20.15}$$

Different values of ϕ_0 describe different initial conditions for the wave.

FIGURE 20.15 Interpreting the equation of a sinusoidal traveling wave.

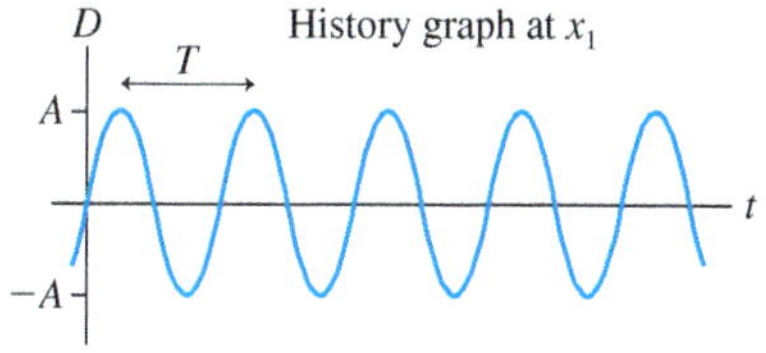

If x is fixed, $D(x_1, t) = A\sin(kx_1 - \omega t + \phi_0)$ gives a sinusoidal history graph at one point in space, x_1. It repeats every T s.

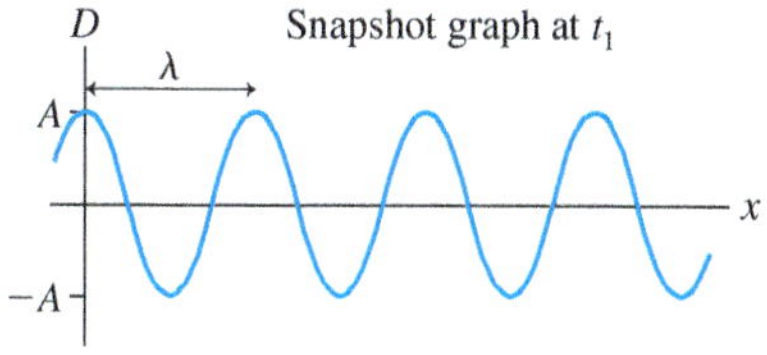

If t is fixed, $D(x, t_1) = A\sin(kx - \omega t_1 + \phi_0)$ gives a sinusoidal snapshot graph at one instant of time, t_1. It repeats every λ m.

EXAMPLE 20.3 Analyzing a sinusoidal wave

A sinusoidal wave with an amplitude of 1.00 cm and a frequency of 100 Hz travels at 200 m/s in the positive x-direction. At $t = 0$ s, the point $x = 1.00$ m is on a crest of the wave.

a. Determine the values of A, v, λ, k, f, ω, T, and ϕ_0 for this wave.
b. Write the equation for the wave's displacement as it travels.
c. Draw a snapshot graph of the wave at $t = 0$ s.

VISUALIZE The snapshot graph will be sinusoidal, but we must do some numerical analysis before we know how to draw it.

SOLVE a. There are several numerical values associated with a sinusoidal traveling wave, but they are not all independent. From the problem statement itself we learn that

$$A = 1.00\text{ cm} \qquad v = 200\text{ m/s} \qquad f = 100\text{ Hz}$$

We can then find:

$$\lambda = v/f = 2.00\text{ m}$$

$$k = 2\pi/\lambda = \pi\text{ rad/m or } 3.14\text{ rad/m}$$

Continued

$$\omega = 2\pi f = 628 \text{ rad/s}$$

$$T = 1/f = 0.0100 \text{ s} = 10.0 \text{ ms}$$

The phase constant ϕ_0 is determined by the initial conditions. We know that a wave crest, with displacement $D = A$, is passing $x_0 = 1.00$ m at $t_0 = 0$ s. Equation 20.14 at x_0 and t_0 is

$$D(x_0, t_0) = A = A\sin\big(k(1.00 \text{ m}) + \phi_0\big)$$

This equation is true only if $\sin\big(k(1.00 \text{ m}) + \phi_0\big) = 1$, which requires

$$k(1.00 \text{ m}) + \phi_0 = \frac{\pi}{2} \text{ rad}$$

Solving for the phase constant gives

$$\phi_0 = \frac{\pi}{2} \text{ rad} - (\pi \text{ rad/m})(1.00 \text{ m}) = -\frac{\pi}{2} \text{ rad}$$

b. With the information gleaned from part a, the wave's displacement is

$$D(x, t) = 1.00 \text{ cm} \times \sin\big[(3.14 \text{ rad/m})x - (628 \text{ rad/s})t - \pi/2 \text{ rad}\big]$$

Notice that we included units with A, k, ω, and ϕ_0.

c. We know that $x = 1.00$ m is a wave crest at $t = 0$ s and that the wavelength is $\lambda = 2.00$ m. Because the origin is $\lambda/2$ away from the crest at $x = 1.00$ m, we expect to find a wave trough at $x = 0$. This is confirmed by calculating $D(0 \text{ m}, 0 \text{ s}) = (1.00 \text{ cm})\sin(-\pi/2 \text{ rad}) = -1.00$ cm. FIGURE 20.16 is a snapshot graph that portrays this information.

FIGURE 20.16 A snapshot graph at $t = 0$ s of the sinusoidal wave of Example 20.3.

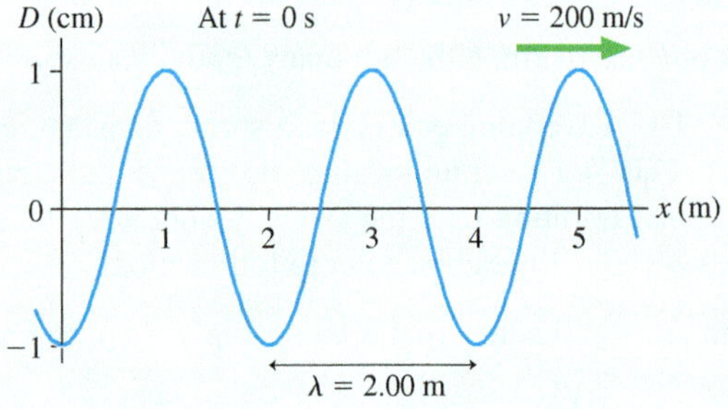

Wave Motion on a String

The displacement equation, Equation 20.14, allows us to learn more about wave motion on a string. As a wave travels along the x-axis, the points on the string oscillate back and forth in the y-direction. The displacement D of a point on the string is simply that point's y-coordinate, so Equation 20.14 for a string wave is

$$y(x, t) = A\sin(kx - \omega t + \phi_0) \tag{20.16}$$

The velocity of a particle on the string—**which is not the same as the velocity of the wave along the string**—is the time derivative of Equation 20.16:

$$v_y = \frac{dy}{dt} = -\omega A\cos(kx - \omega t + \phi_0) \tag{20.17}$$

The maximum velocity of a small segment of the string is $v_{max} = \omega A$. This is the same result we found for simple harmonic motion because the motion of the string particles is simple harmonic motion. FIGURE 20.17 shows velocity vectors *of the particles* at different points on a sinusoidal wave.

FIGURE 20.17 A snapshot graph of a wave on a string with vectors showing the velocity *of the string* at various points.

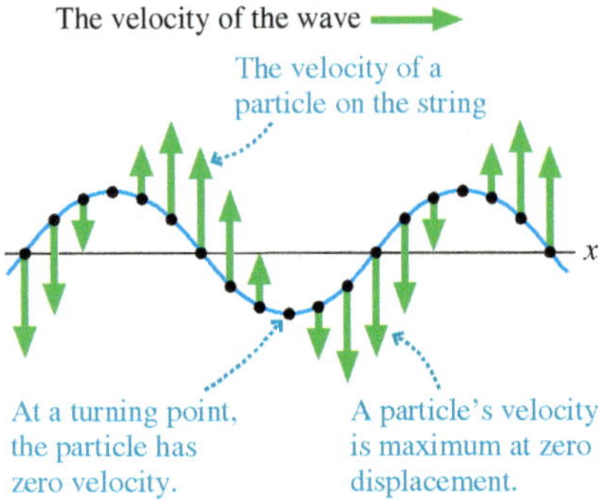

NOTE ▶ Creating a wave of larger amplitude increases the speed of particles in the medium, but it does *not* change the speed of the wave *through* the medium. ◀

Pursuing this line of thought, we can derive an expression for the wave speed along the string. FIGURE 20.18 shows a small segment of the string with length $\Delta x \ll \lambda$ right at a crest of the wave. You can see that the string's tension exerts a downward force on this piece of the string, pulling it back to equilibrium. Newton's second law for this small segment of string is

$$(F_{net})_y = ma_y = (\mu\Delta x)a_y \tag{20.18}$$

where we used the string's linear density μ to write the mass as $m = \mu\Delta x$.

From simple harmonic motion, we know that this point of maximum displacement is also the point of maximum acceleration. The acceleration of a point on the string is the time derivative of Equation 20.17:

$$a_y = \frac{dv_y}{dt} = -\omega^2 A \sin(kx - \omega t + \phi_0) \tag{20.19}$$

Thus the acceleration at the crest of the wave is $a_y = -\omega^2 A$. But the angular frequency ω with which the particles of the string oscillate is related to the wave's speed v along the string by Equation 20.13, $\omega = vk$. Thus

$$a_y = -\omega^2 A = -v^2 k^2 A \tag{20.20}$$

A large wave speed causes the particles of the string to oscillate more quickly and thus to have a larger acceleration.

You can see from Figure 20.18 that the y-component of the tension is $T_s \sin\theta$, where θ is the angle of the string at $x = \frac{1}{2}\Delta x$. θ is a *negative* angle because it is below the x-axis. This segment of string is pulled from both ends, so

$$(F_{net})_y = 2T_s \sin\theta \tag{20.21}$$

The angle θ is very small because $\Delta x \ll \lambda$, so we can use the small-angle approximation ($\sin u \approx \tan u$ if $u \ll 1$) to write

$$(F_{net})_y \approx 2T_s \tan\theta \tag{20.22}$$

where $\tan\theta$ is the slope of the string at $x = \frac{1}{2}\Delta x$.

At this specific instant, with the crest of the wave at $x = 0$, the equation of the string is

$$y = A\cos(kx)$$

The slope of the string at $x = \frac{1}{2}\Delta x$ is the derivative evaluated at that point:

$$\tan\theta = \left.\frac{dy}{dx}\right|_{\text{at } \Delta x/2} = -kA\sin(kx)\big|_{\text{at } \Delta x/2} = -kA\sin\left(\frac{k\Delta x}{2}\right)$$

Now $\Delta x \ll \lambda$, so $k\Delta x/2 = \pi\Delta x/\lambda \ll 1$. Thus the small-angle approximation ($\sin u \approx u$ if $u \ll 1$) of the slope is

$$\tan\theta \approx -kA\left(\frac{k\Delta x}{2}\right) = -\frac{k^2 A\Delta x}{2} \tag{20.23}$$

If we substitute this expression for $\tan\theta$ into Equation 20.22, we find that the net force on this little piece of string is

$$(F_{net})_y = -k^2 A T_s \Delta x \tag{20.24}$$

Now we can use Equation 20.20 for a_y and Equation 20.24 for $(F_{net})_y$ in Newton's second law. With these substitutions, Equation 20.18 becomes

$$(F_{net})_y = -k^2 A T_s \Delta x = (\mu\Delta x)a_y = -v^2 k^2 A\mu\Delta x \tag{20.25}$$

The term $-k^2 A\Delta x$ cancels, and we're left with

$$v = \sqrt{\frac{T_s}{\mu}} \tag{20.26}$$

This was the result that we stated, without proof, in Equation 20.1. Although we've derived Equation 20.26 with the assumption of a sinusoidal wave, the wave speed does not depend on the shape of the wave. Thus any wave on a stretched string will have this wave speed.

FIGURE 20.18 A small segment of string at the crest of a wave.

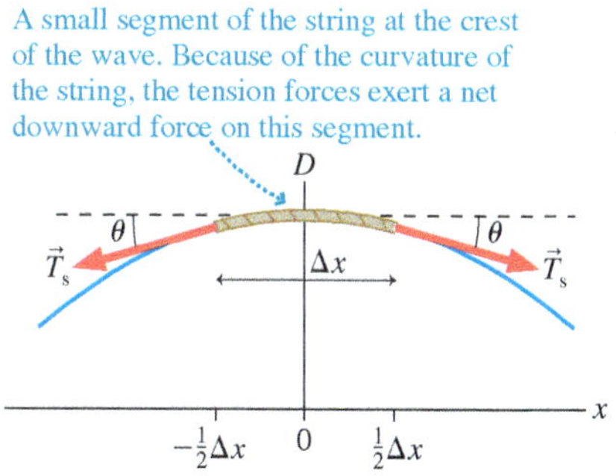

EXAMPLE 20.4 Generating a sinusoidal wave

A very long string with $\mu = 2.0$ g/m is stretched along the x-axis with a tension of 5.0 N. At $x = 0$ m it is tied to a 100 Hz simple harmonic oscillator that vibrates perpendicular to the string with an amplitude of 2.0 mm. The oscillator is at its maximum positive displacement at $t = 0$ s.

a. Write the displacement equation for the traveling wave on the string.

b. At $t = 5.0$ ms, what is the string's displacement at a point 2.7 m from the oscillator?

MODEL The oscillator generates a sinusoidal traveling wave on a string. The displacement of the wave has to match the displacement of the oscillator at $x = 0$ m.

SOLVE a. The equation for the displacement is

$$D(x,t) = A\sin(kx - \omega t + \phi_0)$$

with A, k, ω, and ϕ_0 to be determined. The wave amplitude is the same as the amplitude of the oscillator that generates the wave, so $A = 2.0$ mm. The oscillator has its maximum displacement $y_{osc} = A = 2.0$ mm at $t = 0$ s, thus

$$D(0\text{ m}, 0\text{ s}) = A\sin(\phi_0) = A$$

This requires the phase constant to be $\phi_0 = \pi/2$ rad. The wave's frequency is $f = 100$ Hz, the frequency of the source; therefore the angular frequency is $\omega = 2\pi f = 200\pi$ rad/s. We still need $k = 2\pi/\lambda$, but we do not know the wavelength. However, we have enough information to determine the wave speed, and we can then use either $\lambda = v/f$ or $k = \omega/v$. The speed is

$$v = \sqrt{\frac{T_s}{\mu}} = \sqrt{\frac{5.0\text{ N}}{0.0020\text{ kg/m}}} = 50\text{ m/s}$$

Using v, we find $\lambda = 0.50$ m and $k = 2\pi/\lambda = 4\pi$ rad/m. Thus the wave's displacement equation is

$$D(x,t) = (2.0\text{ mm}) \times \sin\left[2\pi\left((2.0\text{ m}^{-1})x - (100\text{ s}^{-1})t\right) + \pi/2\text{ rad}\right]$$

Notice that we have separated out the 2π. This step is not essential, but for some problems it makes subsequent steps easier.

b. The wave's displacement at $t = 5.0$ ms $= 0.0050$ s is

$$D(x, t = 5.0\text{ ms}) = (2.0\text{ mm})\sin(4\pi x - \pi\text{ rad} + \pi/2\text{ rad})$$
$$= (2.0\text{ mm})\sin(4\pi x - \pi/2\text{ rad})$$

At $x = 2.7$ m (calculator set to radians!), the displacement is

$$D(2.7\text{ m}, 5.0\text{ ms}) = 1.6\text{ mm}$$

20.4 Waves in Two and Three Dimensions

Suppose you were to take a photograph of ripples spreading on a pond. If you mark the location of the *crests* on the photo, your picture would look like FIGURE 20.19a. The lines that locate the crests are called **wave fronts,** and they are spaced precisely one wavelength apart. The diagram shows only a single instant of time, but you can imagine a movie in which you would see the wave fronts moving outward from the source at speed v. A wave like this is called a **circular wave.** It is a two-dimensional wave that spreads across a surface.

Although the wave fronts are circles, you would hardly notice the curvature if you observed a small section of the wave front very, very far away from the source. The wave fronts would appear to be parallel lines, still spaced one wavelength apart and traveling at speed v. A good example is an ocean wave reaching a beach. Ocean waves are generated by storms and wind far out at sea, hundreds or thousands of miles away. By the time they reach the beach where you are working on your tan, the crests appear to be straight lines. An aerial view of the ocean would show a wave diagram like FIGURE 20.19b.

Many waves of interest, such as sound waves or light waves, move in three dimensions. For example, loudspeakers and lightbulbs emit **spherical waves.** That is, the crests of the wave form a series of concentric spherical shells separated by the wavelength λ. In essence, the waves are three-dimensional ripples. It will still be useful to draw wave-front diagrams such as Figure 20.19, but now the circles are slices through the spherical shells locating the wave crests.

If you observe a spherical wave very, very far from its source, the small piece of the wave front that you can see is a little patch on the surface of a very large sphere. If the radius of the sphere is sufficiently large, you will not notice the curvature and this little patch of the wave front appears to be a plane. FIGURE 20.20 illustrates the idea of a **plane wave.**

FIGURE 20.19 The wave fronts of a circular or spherical wave.

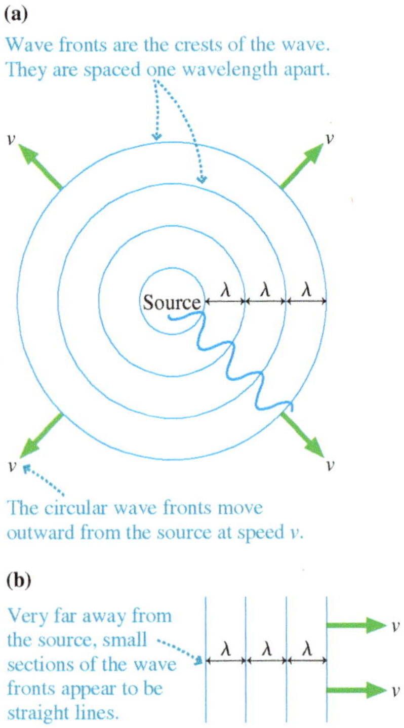

To visualize a plane wave, imagine standing on the x-axis facing a sound wave as it comes toward you from a very distant loudspeaker. Sound is a longitudinal wave, so the particles of medium oscillate toward you and away from you. If you were to locate all of the particles that, at one instant of time, were at their maximum displacement toward you, they would all be located in a plane perpendicular to the travel direction. This is one of the wave fronts in Figure 20.20, and all the particles in this plane are doing exactly the same thing at that instant of time. This plane is moving toward you at speed v. There is another plane one wavelength behind it where the molecules are also at maximum displacement, yet another two wavelengths behind the first, and so on.

Because a plane wave's displacement depends on x but not on y or z, the displacement function $D(x, t)$ describes a plane wave just as readily as it does a one-dimensional wave. Once you specify a value for x, the displacement is the same at every point in the yz-plane that slices the x-axis at that value (i.e., one of the planes shown in Figure 20.20).

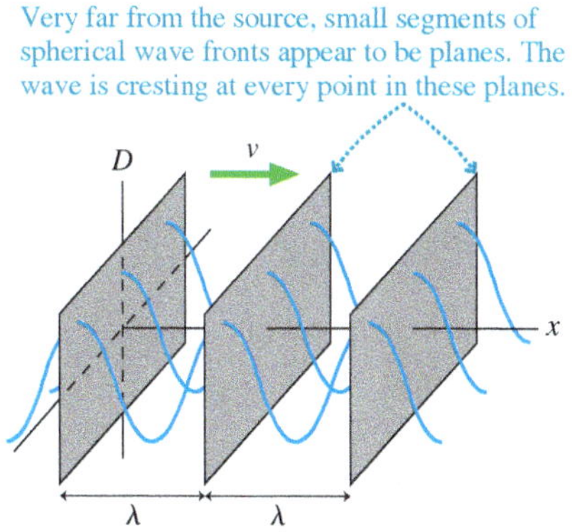

FIGURE 20.20 A plane wave.

NOTE ▶ There are no perfect plane waves in nature, but many waves of practical interest can be modeled as plane waves. ◀

We can describe a circular wave or a spherical wave by changing the mathematical description from $D(x, t)$ to $D(r, t)$, where r is the radial distance measured outward from the source. Then the displacement of the medium will be the same at every point on a spherical surface. In particular, a sinusoidal spherical wave with wave number k and angular frequency ω is written

$$D(r, t) = A(r)\sin(kr - \omega t + \phi_0) \qquad (20.27)$$

Other than the change of x to r, the only difference is that the amplitude is now a function of r. A one-dimensional wave propagates with no change in the wave amplitude. But circular and spherical waves spread out to fill larger and larger volumes of space. To conserve energy, an issue we'll look at later in the chapter, the wave's amplitude has to decrease with increasing distance r. This is why sound and light decrease in intensity as you get farther from the source. We don't need to specify exactly how the amplitude decreases with distance, but you should be aware that it does.

Phase and Phase Difference

The quantity $(kx - \omega t + \phi_0)$ is called the **phase** of the wave, denoted ϕ. The phase of a wave will be an important concept in Chapters 21 and 22, where we will explore the consequences of adding various waves together. For now, we can note that the wave fronts seen in Figures 20.19 and 20.20 are "surfaces of constant phase." To see this, use the phase to write the displacement as simply $D(x, t) = A\sin\phi$. Because each point on a wave front has the same displacement, the phase must be the same at every point.

It will be useful to know the *phase difference* $\Delta\phi$ between two different points on a sinusoidal wave. **FIGURE 20.21** shows two points on a sinusoidal wave at time t. The phase difference between these points is

$$\begin{aligned}\Delta\phi = \phi_2 - \phi_1 &= (kx_2 - \omega t + \phi_0) - (kx_1 - \omega t + \phi_0)\\ &= k(x_2 - x_1) = k\Delta x = 2\pi\frac{\Delta x}{\lambda}\end{aligned} \qquad (20.28)$$

That is, **the phase difference between two points on a wave depends on only the ratio of their separation Δx to the wavelength λ.** For example, two points on a wave separated by $\Delta x = \frac{1}{2}\lambda$ have a phase difference $\Delta\phi = \pi$ rad.

An important consequence of Equation 20.28 is that **the phase difference between two adjacent wave fronts is $\Delta\phi = 2\pi$ rad.** This follows from the fact that two adjacent wave fronts are separated by $\Delta x = \lambda$. This is an important idea. Moving from one crest of the wave to the next corresponds to changing the *distance* by λ and changing the *phase* by 2π rad.

FIGURE 20.21 The phase difference between two points on a wave.

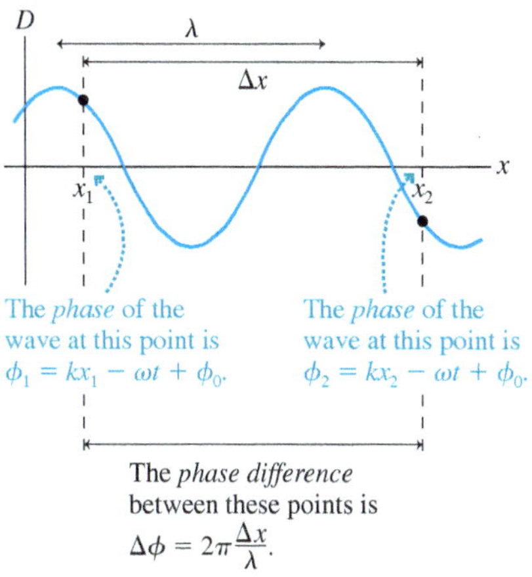

EXAMPLE 20.5 The phase difference between two points on a sound wave

A 100 Hz sound wave travels with a wave speed of 343 m/s.

a. What is the phase difference between two points 60.0 cm apart along the direction the wave is traveling?
b. How far apart are two points whose phase differs by 90°?

MODEL Treat the wave as a plane wave traveling in the positive x-direction.

SOLVE a. The phase difference between two points is

$$\Delta\phi = 2\pi\frac{\Delta x}{\lambda}$$

In this case, $\Delta x = 60.0 \text{ cm} = 0.600 \text{ m}$. The wavelength is

$$\lambda = \frac{v}{f} = \frac{343 \text{ m/s}}{100 \text{ Hz}} = 3.43 \text{ m}$$

and thus

$$\Delta\phi = 2\pi\frac{0.600 \text{ m}}{3.43 \text{ m}} = 0.350\pi \text{ rad} = 63.0°$$

b. A phase difference $\Delta\phi = 90°$ is $\pi/2$ rad. This will be the phase difference between two points when $\Delta x/\lambda = \frac{1}{4}$, or when $\Delta x = \lambda/4$. Here, with $\lambda = 3.43$ m, $\Delta x = 85.8$ cm.

ASSESS The phase difference increases as Δx increases, so we expect the answer to part b to be larger than 60 cm.

STOP TO THINK 20.4 What is the phase difference between the crest of a wave and the adjacent trough?

a. -2π rad
b. 0 rad
c. $\pi/4$ rad
d. $\pi/2$ rad
e. π rad
f. 3π rad

20.5 Sound and Light

Although there are many kinds of waves in nature, two are especially significant for us as humans. These are sound waves and light waves, the basis of hearing and seeing.

Sound Waves

We usually think of sound waves traveling in air, but sound can travel through any gas, through liquids, and even through solids. FIGURE 20.22 shows a loudspeaker cone vibrating back and forth in a fluid such as air or water. Each time the cone moves forward, it collides with the molecules and pushes them closer together. A half cycle later, as the cone moves backward, the fluid has room to expand and the density decreases a little. These regions of higher and lower density (and thus higher and lower pressure) are called **compressions** and **rarefactions.**

FIGURE 20.22 A sound wave in a fluid is a sequence of compressions and rarefactions. The variation in density and the amount of motion have been greatly exaggerated.

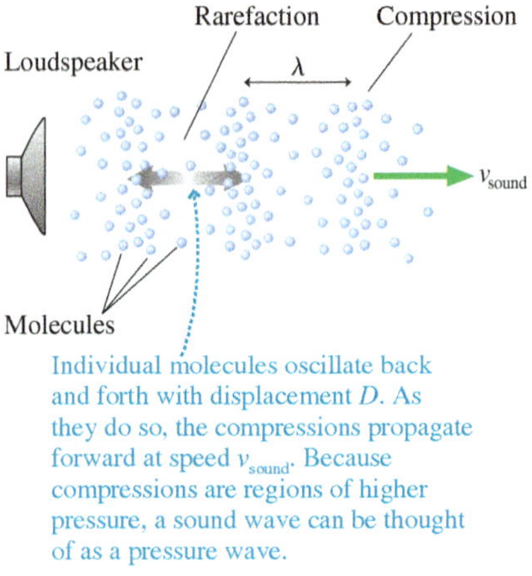

This periodic sequence of compressions and rarefactions travels outward from the loudspeaker as a longitudinal sound wave. When the wave reaches your ear, the oscillating pressure causes your eardrum to vibrate. These vibrations are transferred into your inner ear and perceived as sound.

Your ears are able to detect sinusoidal sound waves with frequencies between about 20 Hz and about 20,000 Hz, or 20 kHz. Low frequencies are perceived as "low pitch" bass notes, while high frequencies are heard as "high pitch" treble notes. Your high-frequency range of hearing can deteriorate either with age or as a result of exposure to loud sounds that damage the ear.

The speed of sound waves depends on the properties of the medium. A thermodynamic analysis of the compressions and expansions shows that the wave speed in a gas depends on the temperature and on the molecular mass of the gas. For air at room temperature (20°C),

$$v_{sound} = 343 \text{ m/s} \qquad \text{(sound speed in air at 20°C)}$$

The speed of sound is a little lower at lower temperatures and a little higher at higher temperatures. Liquids and solids are less compressible than air, and that makes the speed of sound in those media higher than in air. Table 20.1 gives the speed of sound in several substances.

TABLE 20.1 The speed of sound

Medium	Speed (m/s)
Air (0°C)	331
Air (20°C)	343
Helium (0°C)	970
Ethyl alcohol	1170
Water	1480
Granite	6000
Aluminum	6420

A speed of 343 m/s is high, but not extraordinarily so. A distance as small as 100 m is enough to notice a slight delay between when you see something, such as a person hammering a nail, and when you hear it. The time required for sound to travel 1 km is $t = (1000 \text{ m})/(343 \text{ m/s}) \approx 3$ s. You may have learned to estimate the distance to a bolt of lightning by timing the number of seconds between when you see the flash and when you hear the thunder. Because sound takes 3 s to travel 1 km, the time divided by 3 gives the distance in kilometers. Or, in English units, the time divided by 5 gives the distance in miles.

Sound waves exist at frequencies well above 20 kHz, even though humans can't hear them. These are called *ultrasonic* frequencies. Oscillators vibrating at frequencies of many MHz generate the ultrasonic waves used in ultrasound medical imaging. A 3 MHz wave traveling through water (which is basically what your body is) at a sound speed of 1480 m/s has a wavelength of about 0.5 mm. It is this very small wavelength that allows ultrasound to image very small objects. We'll see why when we study *diffraction* in Chapter 22.

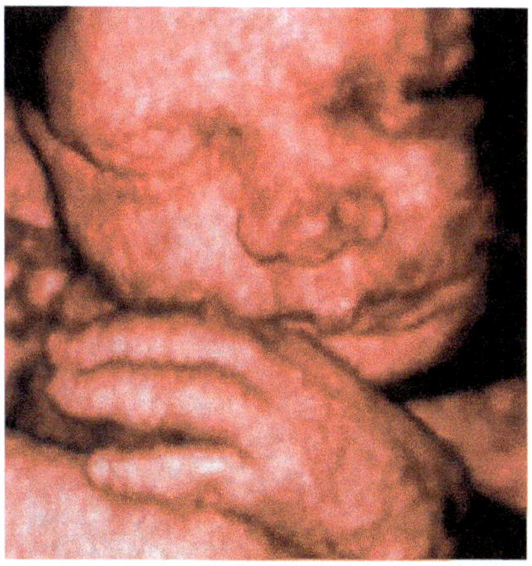

This ultrasound image is an example of using high-frequency sound waves to "see" within the human body.

EXAMPLE 20.6 Sound wavelengths

What are the wavelengths of sound waves at the limits of human hearing and at the midrange frequency of 500 Hz? Notes sung by human voices are near 500 Hz, as are notes played by striking keys near the center of a piano keyboard.

MODEL Assume a room temperature of 20°C.

SOLVE We can use the fundamental relationship $\lambda = v/f$ to find the wavelengths for sounds of various frequencies:

$$f = 20 \text{ Hz} \qquad \lambda = \frac{343 \text{ m/s}}{20 \text{ Hz}} = 17 \text{ m}$$

$$f = 500 \text{ Hz} \qquad \lambda = \frac{343 \text{ m/s}}{500 \text{ Hz}} = 0.69 \text{ m}$$

$$f = 20{,}000 \text{ Hz} \qquad \lambda = \frac{343 \text{ m/s}}{20{,}000 \text{ Hz}} = 0.017 \text{ m} = 1.7 \text{ cm}$$

ASSESS The wavelength of a 20 kHz note is a small 1.7 cm while, at the other extreme, a 20 Hz note has a huge wavelength of 17 m! This is because a wave moves forward one wavelength during a time interval of one period, and a wave traveling at 343 m/s can move 17 m during the $\frac{1}{20}$ s period of a 20 Hz note. The 69 cm wavelength of a 500 Hz note is more of a "human scale." You might note that most musical instruments are a meter or a little less in size. This is not a coincidence. You will see in the next chapter how the wavelength produced by a musical instrument is related to its size.

Electromagnetic Waves

A light wave is an *electromagnetic wave,* an oscillation of the electromagnetic field. Other electromagnetic waves, such as radio waves, microwaves, and ultraviolet light, have the same physical characteristics as light waves even though we cannot sense them with our eyes. It is easy to demonstrate that light will pass unaffected through a container from which all the air has been removed, and light reaches us from distant stars through the vacuum of interstellar space. Such observations raise interesting but difficult questions. If light can travel through a region in which there is no matter, then what is the *medium* of a light wave? What is it that is waving?

It took scientists over 50 years, most of the 19th century, to answer this question. We will examine the answers in more detail in Part IV after we introduce the ideas of electric and magnetic fields. For now we can say that light waves are a "self-sustaining oscillation of the electromagnetic field." That is, the displacement D is an electric or magnetic field. Being self-sustaining means that electromagnetic waves require *no material medium* in order to travel; hence electromagnetic waves are not mechanical waves. Fortunately, we can learn about the wave properties of light without having to understand electromagnetic fields.

It was predicted theoretically in the late 19th century, and has been subsequently confirmed, that all electromagnetic waves travel through vacuum with the same speed, called the *speed of light.* The value of the speed of light is

$$v_{\text{light}} = c = 299{,}792{,}458 \text{ m/s} \qquad \text{(electromagnetic wave speed in vacuum)}$$

where the special symbol c is used to designate the speed of light. (This is the c in Einstein's famous formula $E = mc^2$.) Now *this* is really moving—about one million times faster than the speed of sound in air!

NOTE ▶ $c = 3.00 \times 10^8$ m/s is the appropriate value to use in calculations. ◀

White light passing through a prism is spread out into a band of colors called the *visible spectrum.*

The wavelengths of light are extremely small. You will learn in Chapter 22 how these wavelengths are determined, but for now we will note that visible light is an electromagnetic wave with a wavelength (in air) in the range of roughly 400 nm (400×10^{-9} m) to 700 nm (700×10^{-9} m). Each wavelength is perceived as a different color, with the longer wavelengths seen as orange or red light and the shorter wavelengths seen as blue or violet light. A prism is able to spread the different wavelengths apart, from which we learn that "white light" is all the colors, or wavelengths, combined. The spread of colors seen with a prism, or seen in a rainbow, is called the *visible spectrum.*

If the wavelengths of light are unbelievably small, the oscillation frequencies are unbelievably large. The frequency for a 600 nm wavelength of light (orange) is

$$f = \frac{v}{\lambda} = \frac{3.00 \times 10^8 \text{ m/s}}{600 \times 10^{-9} \text{ m}} = 5.00 \times 10^{14} \text{ Hz}$$

The frequencies of light waves are roughly a factor of a trillion (10^{12}) higher than sound frequencies.

Electromagnetic waves exist at many frequencies other than the rather limited range that our eyes detect. One of the major technological advances of the 20th century was learning to generate and detect electromagnetic waves at many frequencies, ranging from low-frequency radio waves to the extraordinarily high frequencies of x rays. FIGURE 20.23 shows that the visible spectrum is a small slice of the much broader **electromagnetic spectrum.**

FIGURE 20.23 The electromagnetic spectrum from 10^6 Hz to 10^{18} Hz.

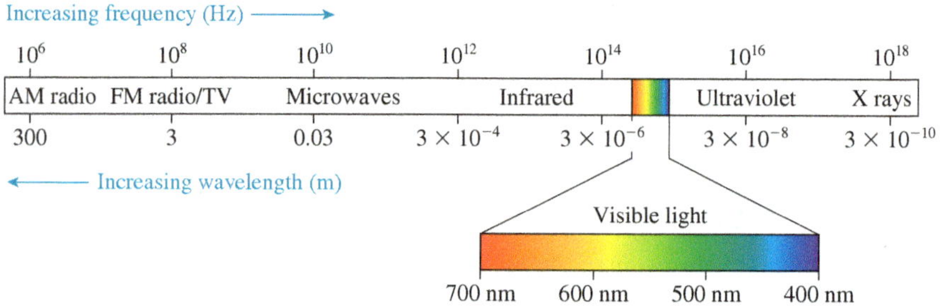

EXAMPLE 20.7 Traveling at the speed of light

A satellite exploring Jupiter transmits data to the earth as a radio wave with a frequency of 200 MHz. What is the wavelength of the electromagnetic wave, and how long does it take the signal to travel 800 million kilometers from Jupiter to the earth?

SOLVE Radio waves are sinusoidal electromagnetic waves traveling with speed c. Thus

$$\lambda = \frac{c}{f} = \frac{3.00 \times 10^8 \text{ m/s}}{2.00 \times 10^8 \text{ Hz}} = 1.5 \text{ m}$$

The time needed to travel 800×10^6 km $= 8.0 \times 10^{11}$ m is

$$\Delta t = \frac{\Delta x}{c} = \frac{8.0 \times 10^{11} \text{ m}}{3.00 \times 10^8 \text{ m/s}} = 2700 \text{ s} = 45 \text{ min}$$

The Index of Refraction

Light waves travel with speed c in a vacuum, but they slow down as they pass through transparent materials such as water or glass or even, to a very slight extent, air. The slowdown is a consequence of interactions between the electromagnetic field of the wave and the electrons in the material. The speed of light in a material is characterized by the material's **index of refraction** n, defined as

$$n = \frac{\text{speed of light in a vacuum}}{\text{speed of light in the material}} = \frac{c}{v} \tag{20.29}$$

The index of refraction of a material is always greater than 1 because $v < c$. A vacuum has $n = 1$ exactly. Table 20.2 shows the index of refraction for several materials. You can see that liquids and solids have larger indices of refraction than gases.

NOTE ▶ An accurate value for the index of refraction of air is relevant only in very precise measurements. We will assume $n_{air} = 1.00$ in this text. ◀

TABLE 20.2 Typical indices of refraction

Material	Index of refraction
Vacuum	1 exactly
Air	1.0003
Water	1.33
Glass	1.50
Diamond	2.42

If the speed of a light wave changes as it enters into a transparent material, such as glass, what happens to the light's frequency and wavelength? Because $v = \lambda f$, either λ or f or both have to change when v changes.

As an analogy, think of a sound wave in the air as it impinges on the surface of a pool of water. As the air oscillates back and forth, it periodically pushes on the surface of the water. These pushes generate the compressions of the sound wave that continues on into the water. Because each push of the air causes one compression of the water, the frequency of the sound wave in the water must be *exactly the same* as the frequency of the sound wave in the air. In other words, **the frequency of a wave is the frequency of the source. It does not change as the wave moves from one medium to another.**

The same is true for electromagnetic waves; the frequency does not change as the wave moves from one material to another.

FIGURE 20.24 shows a light wave passing through a transparent material with index of refraction n. As the wave travels through vacuum it has wavelength λ_{vac} and frequency f_{vac} such that $\lambda_{vac}f_{vac} = c$. In the material, $\lambda_{mat}f_{mat} = v = c/n$. The frequency does not change as the wave enters ($f_{mat} = f_{vac}$), so the wavelength must. The wavelength in the material is

$$\lambda_{mat} = \frac{v}{f_{mat}} = \frac{c}{nf_{mat}} = \frac{c}{nf_{vac}} = \frac{\lambda_{vac}}{n} \tag{20.30}$$

The wavelength in the transparent material is less than the wavelength in vacuum. This makes sense. Suppose a marching band is marching at one step per second at a speed of 1 m/s. Suddenly they slow their speed to $\frac{1}{2}$ m/s but maintain their march at one step per second. The only way to go slower while marching at the same pace is to take *smaller steps.* When a light wave enters a material, the only way it can go slower while oscillating at the same frequency is to have a *smaller wavelength.*

FIGURE 20.24 Light passing through a transparent material with index of refraction n.

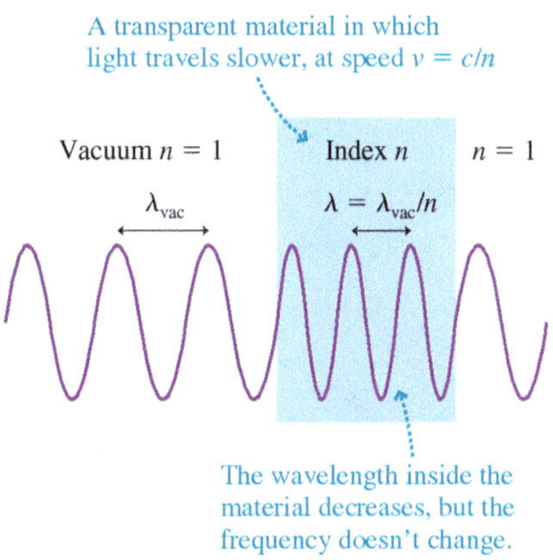

EXAMPLE 20.8 Light traveling through glass

Orange light with a wavelength of 600 nm is incident upon a 1.00-mm-thick glass microscope slide.

a. What is the light speed in the glass?
b. How many wavelengths of the light are inside the slide?

SOLVE a. From Table 20.2 we see that the index of refraction of glass is $n_{glass} = 1.50$. Thus the speed of light in glass is

$$v_{glass} = \frac{c}{n_{glass}} = \frac{3.00 \times 10^8 \text{ m/s}}{1.50} = 2.00 \times 10^8 \text{ m/s}$$

b. The wavelength inside the glass is

$$\lambda_{glass} = \frac{\lambda_{vac}}{n_{glass}} = \frac{600 \text{ nm}}{1.50} = 400 \text{ nm} = 4.00 \times 10^{-7} \text{ m}$$

N wavelengths span a distance $d = N\lambda$, so the number of wavelengths in $d = 1.00$ mm is

$$N = \frac{d}{\lambda} = \frac{1.00 \times 10^{-3} \text{ m}}{4.00 \times 10^{-7} \text{ m}} = 2500$$

ASSESS The fact that 2500 wavelengths fit within 1 mm shows how small the wavelengths of light are.

STOP TO THINK 20.5 A light wave travels through three transparent materials of equal thickness. Rank in order, from largest to smallest, the indices of refraction n_a, n_b, and n_c.

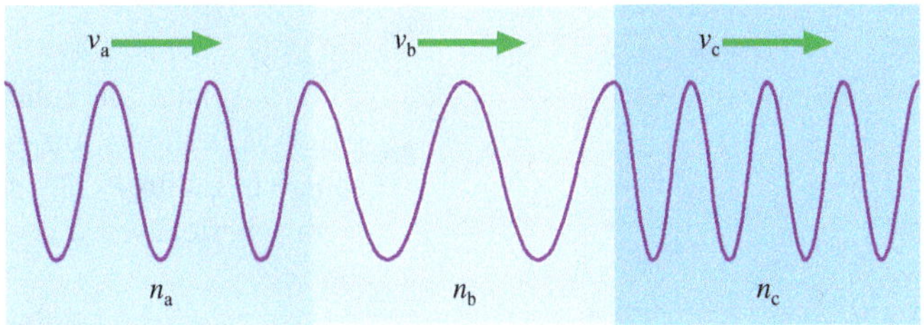

20.6 Power, Intensity, and Decibels

A traveling wave transfers energy from one point to another. The sound wave from a loudspeaker sets your eardrum into motion. Light waves from the sun warm the earth. The *power* of a wave is the rate, in joules per second, at which the wave transfers energy. As you learned in Chapter 11, power is measured in watts. A loudspeaker might emit 2 W of power, meaning that energy in the form of sound waves is radiated at the rate of 2 joules per second. A lightbulb might emit 5 W, or 5 J/s, of visible light. (In fact, this is about right for a so-called 100 watt bulb, with the other 95 W of power being emitted as heat, or infrared radiation, rather than as visible light.)

Imagine doing two experiments with a lightbulb that emits 5 W of visible light. In the first, you hang the bulb in the center of a room and allow the light to illuminate the walls. In the second experiment, you use mirrors and lenses to "capture" the bulb's light and focus it onto a small spot on one wall. This is what a computer projector does. The energy emitted by the bulb is the same in both cases, but, as you know, the light is much brighter when focused onto a small area. We would say that the focused light is more *intense* than the diffuse light that goes in all directions. Similarly, a loudspeaker that beams its sound forward into a small area produces a louder sound in that area than a speaker of equal power that radiates the sound in all directions. Quantities such as brightness and loudness depend not only on the rate of energy transfer, or power, but also on the *area* that receives that power.

FIGURE 20.25 Plane waves of power P impinge on area a with intensity $I = P/a$.

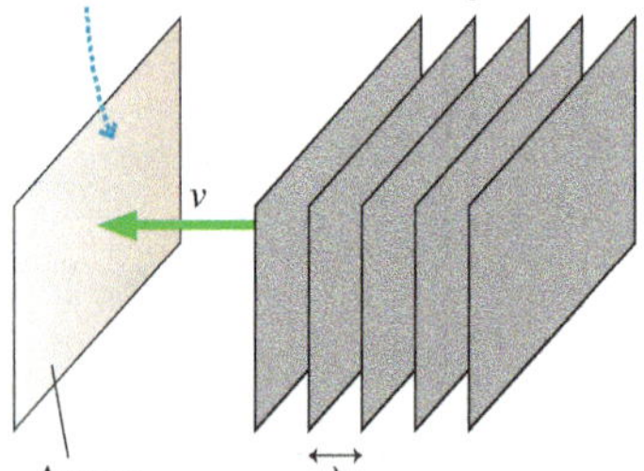

FIGURE 20.25 shows a wave impinging on a surface of area a. The surface is perpendicular to the direction in which the wave is traveling. This might be a real, physical surface, such as your eardrum or a photovoltaic cell, but it could equally well be a mathematical surface in space that the wave passes right through. If the wave has power P, we define the **intensity** I of the wave to be

$$I = \frac{P}{a} = \text{power-to-area ratio} \qquad (20.31)$$

The SI units of intensity are W/m^2. Because intensity is a power-to-area ratio, a wave focused into a small area will have a larger intensity than a wave of equal power that is spread out over a large area.

EXAMPLE 20.9 The intensity of a laser beam

A helium-neon laser, the kind that provides the familiar red light of classroom demonstrations and supermarket checkout scanners, emits 1.0 mW of light power into a 1.0-mm-diameter laser beam. What is the intensity of the laser beam?

MODEL The laser beam is a light wave.

SOLVE The light waves of the laser beam pass through a mathematical surface that is a circle of diameter 1.0 mm. The intensity of the laser beam is

$$I = \frac{P}{a} = \frac{P}{\pi r^2} = \frac{0.0010\text{ W}}{\pi(0.00050\text{ m})^2} = 1300\text{ W/m}^2$$

ASSESS This is roughly the intensity of sunlight at noon on a summer day. The difference between the sun and a small laser is not their intensities, which are about the same, but their powers. The laser has a small power of 1 mW. It can produce a very intense wave only because the area through which the wave passes is very small. The sun, by contrast, radiates a total power $P_{sun} \approx 4 \times 10^{26}$ W. This immense power is spread through *all* of space, producing an intensity of 1400 W/m^2 at a distance of 1.5×10^{11} m, the radius of the earth's orbit.

If a source of spherical waves radiates uniformly in all directions, then, as FIGURE 20.26 shows, the power at distance r is spread uniformly over the surface of a sphere of radius r. The surface area of a sphere is $a = 4\pi r^2$, so the intensity of a uniform spherical wave is

$$I = \frac{P_{source}}{4\pi r^2} \quad \text{(intensity of a uniform spherical wave)} \qquad (20.32)$$

The inverse-square dependence of r is really just a statement of energy conservation. The source emits energy at the rate P joules per second. The energy is spread over a larger and larger area as the wave moves outward. Consequently, the energy *per unit area* must decrease in proportion to the surface area of a sphere.

If the intensity at distance r_1 is $I_1 = P_{source}/4\pi r_1^2$ and the intensity at r_2 is $I_2 = P_{source}/4\pi r_2^2$, then you can see that the intensity *ratio* is

$$\frac{I_1}{I_2} = \frac{r_2^2}{r_1^2} \qquad (20.33)$$

You can use Equation 20.33 to compare the intensities at two distances from a source without needing to know the power of the source.

NOTE ▶ Wave intensities are strongly affected by reflections and absorption. Equations 20.32 and 20.33 apply to situations such as the light from a star or the sound from a firework exploding high in the air. Indoor sound does *not* obey a simple inverse-square law because of the many reflecting surfaces. ◀

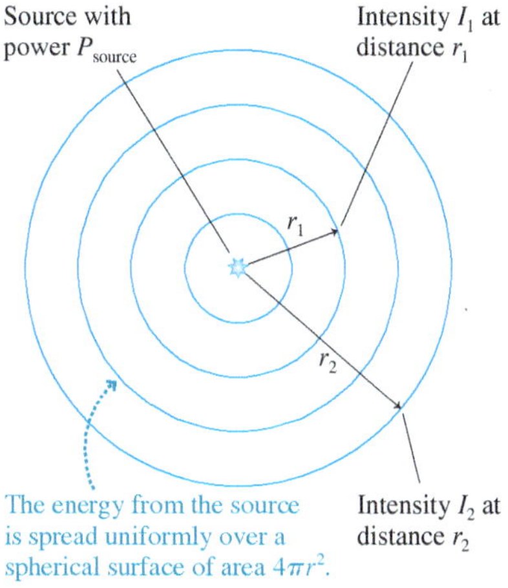

FIGURE 20.26 A source emitting uniform spherical waves.

For a sinusoidal wave, each particle in the medium oscillates back and forth in simple harmonic motion. You learned in Chapter 14 that a particle in SHM with amplitude A has energy $E = \frac{1}{2}kA^2$, where k is the spring constant of the medium, not the wave number. It is this oscillatory energy of the medium that is transferred, particle to particle, as the wave moves through the medium.

Because a wave's intensity is proportional to the rate at which energy is transferred through the medium, and because the oscillatory energy in the medium is proportional to the *square* of the amplitude, we can infer that

$$I \propto A^2 \qquad (20.34)$$

That is, **the intensity of a wave is proportional to the square of its amplitude.** If you double the amplitude of a wave, you increase its intensity by a factor of 4.

Human hearing spans an extremely wide range of intensities, from the *threshold of hearing* at $\approx 1 \times 10^{-12}$ W/m^2 (at midrange frequencies) to the *threshold of pain* at ≈ 10 W/m^2. If we want to make a scale of loudness, it's convenient and logical to place the zero of our scale at the threshold of hearing. To do so, we define the **sound intensity level,** expressed in **decibels** (dB), as

$$\beta = (10 \text{ dB}) \log_{10}\left(\frac{I}{I_0}\right) \qquad (20.35)$$

where $I_0 = 1.0 \times 10^{-12}$ W/m^2. The symbol β is the Greek letter beta. Notice that β is computed as a base-10 logarithm, not a natural logarithm.

The decibel is named after Alexander Graham Bell, inventor of the telephone. Sound intensity level is actually dimensionless because it's formed from the ratio of two intensities, so decibels are just a *name* to remind us that we're dealing with an intensity *level* rather than a true intensity.

Right at the threshold of hearing, where $I = I_0$, the sound intensity level is

$$\beta = (10 \text{ dB}) \log_{10}\left(\frac{I_0}{I_0}\right) = (10 \text{ dB}) \log_{10}(1) = 0 \text{ dB}$$

Note that 0 dB doesn't mean no sound; it means that, for most people, no sound is heard. Dogs have more sensitive hearing than humans, and most dogs can easily perceive a 0 dB sound. The sound intensity level at the pain threshold is

$$\beta = (10 \text{ dB}) \log_{10}\left(\frac{10 \text{ W/m}^2}{10^{-12} \text{ W/m}^2}\right) = (10 \text{ dB}) \log_{10}(10^{13}) = 130 \text{ dB}$$

TABLE 20.3 Sound intensity levels of common sounds

Sound	β (dB)
Threshold of hearing	0
Person breathing, at 3 m	10
A whisper, at 1 m	20
Quiet room	30
Outdoors, no traffic	40
Quiet restaurant	50
Normal conversation, at 1 m	60
Busy traffic	70
Vacuum cleaner, for user	80
Niagara Falls, at viewpoint	90
Snowblower, at 2 m	100
Stereo, at maximum volume	110
Rock concert	120
Threshold of pain	130

The major point to notice is that the sound intensity level increases by 10 dB each time the actual intensity increases by a *factor* of 10. For example, the sound intensity level increases from 70 dB to 80 dB when the sound intensity increases from 10^{-5} W/m^2 to 10^{-4} W/m^2. Perception experiments find that sound is perceived as "twice as loud" when the intensity increases by a factor of 10. In terms of decibels, we can say that the perceived loudness of a sound doubles with each increase in the sound intensity level by 10 dB.

Table 20.3 gives the sound intensity levels for a number of sounds. Although 130 dB is the threshold of pain, quieter sounds can damage your hearing. A fairly short exposure to 120 dB can cause damage to the hair cells in the ear, but lengthy exposure to sound intensity levels of over 85 dB can produce damage as well.

EXAMPLE 20.10 Blender noise

The blender making a smoothie produces a sound intensity level of 83 dB. What is the intensity of the sound? What will the sound intensity level be if a second blender is turned on?

SOLVE We can solve Equation 20.35 for the sound intensity, finding $I = I_0 \times 10^{\beta/10\,\text{dB}}$. Here we used the fact that 10 raised to a power is an "antilogarithm." In this case,

$$I = (1.0 \times 10^{-12}\ \text{W/m}^2) \times 10^{8.3} = 2.0 \times 10^{-4}\ \text{W/m}^2$$

A second blender doubles the sound power and thus raises the intensity to $I = 4.0 \times 10^{-4}$ W/m^2. The new sound intensity level is

$$\beta = (10\ \text{dB})\log_{10}\left(\frac{4.0 \times 10^{-4}\ \text{W/m}^2}{1.0 \times 10^{-12}\ \text{W/m}^2}\right) = 86\ \text{dB}$$

ASSESS In general, doubling the actual sound intensity increases the decibel level by 3 dB.

STOP TO THINK 20.6 Four trumpet players are playing the same note. If three of them suddenly stop, the sound intensity level decreases by

a. 40 dB b. 12 dB c. 6 dB d. 4 dB

20.7 The Doppler Effect

Our final topic for this chapter is an interesting effect that occurs when you are in motion relative to a wave source. It is called the *Doppler effect.* You've likely noticed that the pitch of an ambulance's siren drops as it goes past you. Why?

FIGURE 20.27a shows a source of sound waves moving away from Pablo and toward Nancy at a steady speed v_s. The subscript s indicates that this is the speed of the source, not the speed of the waves. The source is emitting sound waves of frequency f_0 as it travels. The figure is a motion diagram showing the position of the source at times $t = 0$, T, $2T$, and $3T$, where $T = 1/f_0$ is the period of the waves.

Nancy measures the frequency of the wave emitted by the *approaching source* to be f_+. At the same time, Pablo measures the frequency of the wave emitted by the *receding source* to be f_-. Our task is to relate f_+ and f_- to the source frequency f_0 and speed v_s.

After a wave crest leaves the source, its motion is governed by the properties of the medium. That is, the motion of the source cannot affect a wave that has already been emitted. Thus each circular wave front in FIGURE 20.27b is centered on the point from which it was emitted. The wave crest from point 3 was emitted just as this figure was made, but it hasn't yet had time to travel any distance.

The wave crests are bunched up in the direction the source is moving, stretched out behind it. The distance between one crest and the next is one wavelength, so the wavelength λ_+ Nancy measures is *less* than the wavelength $\lambda_0 = v/f_0$ that would be emitted if the source were at rest. Similarly, λ_- behind the source is larger than λ_0.

FIGURE 20.27 A motion diagram showing the wave fronts emitted by a source as it moves to the right at speed v_s.

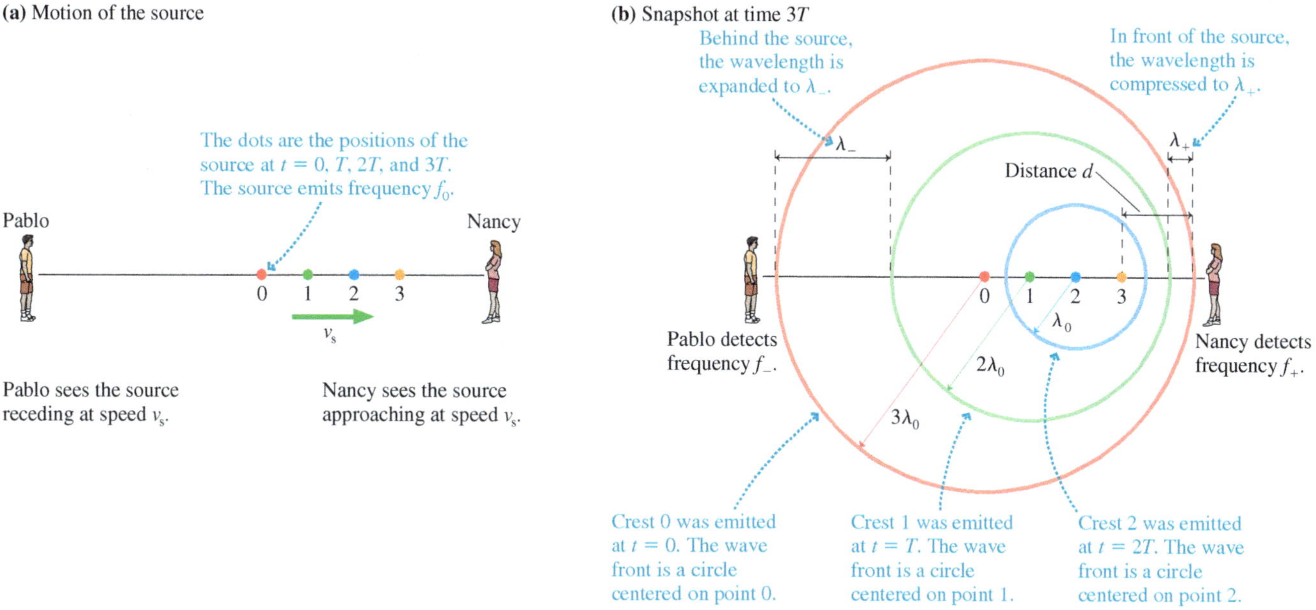

These crests move through the medium at the wave speed v. Consequently, the frequency $f_+ = v/\lambda_+$ detected by the observer whom the source is approaching is *higher* than the frequency f_0 emitted by the source. Similarly, $f_- = v/\lambda_-$ detected behind the source is *lower* than frequency f_0. This change of frequency when a source moves relative to an observer is called the **Doppler effect.**

The distance labeled d in Figure 20.27b is the difference between how far the wave has moved and how far the source has moved at time $t = 3T$. These distances are

$$\begin{aligned}\Delta x_{\text{wave}} &= vt = 3vT \\ \Delta x_{\text{source}} &= v_s t = 3v_s T\end{aligned} \tag{20.36}$$

The distance d spans three wavelengths; thus the wavelength of the wave emitted by an approaching source is

$$\lambda_+ = \frac{d}{3} = \frac{\Delta x_{\text{wave}} - \Delta x_{\text{source}}}{3} = \frac{3vT - 3v_s T}{3} = (v - v_s)T \tag{20.37}$$

You can see that our arbitrary choice of three periods was not relevant because the 3 cancels. The frequency detected in Nancy's direction is

$$f_+ = \frac{v}{\lambda_+} = \frac{v}{(v - v_s)T} = \frac{v}{(v - v_s)} f_0 \tag{20.38}$$

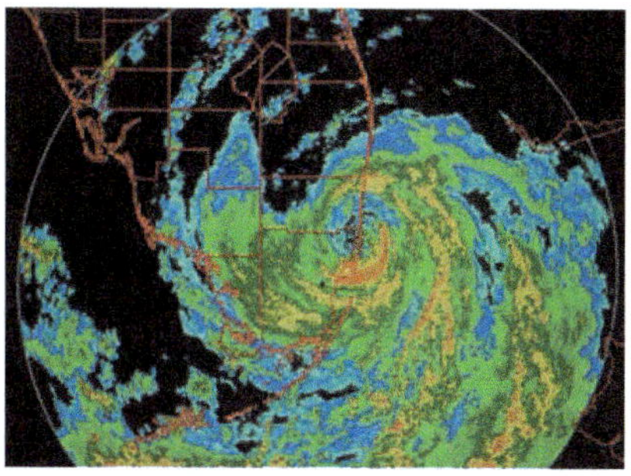

Doppler weather radar uses the Doppler shift of reflected radar signals to measure wind speeds and thus better gauge the severity of a storm.

where $f_0 = 1/T$ is the frequency of the source and is the frequency you would detect if the source were at rest. We'll find it convenient to write the detected frequency as

$$\begin{aligned}f_+ &= \frac{f_0}{1 - v_s/v} \quad \text{(Doppler effect for an approaching source)} \\ f_- &= \frac{f_0}{1 + v_s/v} \quad \text{(Doppler effect for a receding source)}\end{aligned} \tag{20.39}$$

Proof of the second version, for the frequency f_- of a receding source, is similar. You can see that $f_+ > f_0$ in front of the source, because the denominator is less than 1, and $f_- < f_0$ behind the source.

EXAMPLE 20.11 How fast are the police traveling?

A police siren has a frequency of 550 Hz as the police car approaches you, 450 Hz after it has passed you and is receding. How fast are the police traveling? The temperature is 20°C.

MODEL The siren's frequency is altered by the Doppler effect. The frequency is f_+ as the car approaches and f_- as it moves away.

SOLVE To find v_s, we rewrite Equations 20.39 as

$$f_0 = (1 + v_s/v)f_-$$
$$f_0 = (1 - v_s/v)f_+$$

We subtract the second equation from the first, giving

$$0 = f_- - f_+ + \frac{v_s}{v}(f_- + f_+)$$

This is easily solved to give

$$v_s = \frac{f_+ - f_-}{f_+ + f_-}v = \frac{100\ \text{Hz}}{1000\ \text{Hz}}343\ \text{m/s} = 34.3\ \text{m/s}$$

ASSESS If you now solve for the siren frequency when at rest, you will find $f_0 = 495$ Hz. Surprisingly, the at-rest frequency is not halfway between f_- and f_+.

A Stationary Source and a Moving Observer

Suppose the police car in Example 20.11 is at rest while you drive toward it at 34.3 m/s. You might think that this is equivalent to having the police car move toward you at 34.3 m/s, but it isn't. Mechanical waves move through a medium, and the Doppler effect depends not just on how the source and the observer move with respect to each other but also on how they move with respect to the medium. We'll omit the proof, but it's not hard to show that the frequencies heard by an observer moving at speed v_o relative to a stationary source emitting frequency f_0 are

$$\begin{aligned} f_+ &= (1 + v_o/v)f_0 \quad \text{(observer approaching a source)} \\ f_- &= (1 - v_o/v)f_0 \quad \text{(observer receding from a source)} \end{aligned} \tag{20.40}$$

A quick calculation shows that the frequency of the police siren as you approach it at 34.3 m/s is 545 Hz, not the 550 Hz you heard as it approached you at 34.3 m/s.

The Doppler Effect for Light Waves

The Doppler effect is observed for all types of waves, not just sound waves. If a source of light waves is receding from you, the wavelength λ_- that you detect is longer than the wavelength λ_0 emitted by the source.

Although the reason for the Doppler shift for light is the same as for sound waves, there is one fundamental difference. We derived Equations 20.39 for the Doppler-shifted frequencies by measuring the wave speed v relative to the medium. For electromagnetic waves in empty space, there is no medium. Consequently, we need to turn to Einstein's theory of relativity to determine the frequency of light waves from a moving source. The result, which we state without proof, is

$$\begin{aligned} \lambda_- &= \sqrt{\frac{1 + v_s/c}{1 - v_s/c}}\,\lambda_0 \quad \text{(receding source)} \\ \lambda_+ &= \sqrt{\frac{1 - v_s/c}{1 + v_s/c}}\,\lambda_0 \quad \text{(approaching source)} \end{aligned} \tag{20.41}$$

Here v_s is the speed of the source *relative to* the observer.

The light waves from a receding source are shifted to longer wavelengths ($\lambda_- > \lambda_0$). Because the longest visible wavelengths are perceived as the color red, the light from a receding source is **red shifted.** That is *not* to say that the light is red, simply that its wavelength is shifted toward the red end of the spectrum. If $\lambda_0 = 470$ nm (blue) light emitted by a rapidly receding source is detected at $\lambda_- = 520$ nm (green), we would say that the light has been red shifted. Similarly, light from an approaching source is **blue shifted,** meaning that the detected wavelengths are shorter than the emitted wavelengths ($\lambda_+ < \lambda_0$) and thus are shifted toward the blue end of the spectrum.

EXAMPLE 20.12 Measuring the velocity of a galaxy

Hydrogen atoms in the laboratory emit red light with wavelength 656 nm. In the light from a distant galaxy, this "spectral line" is observed at 691 nm. What is the speed of this galaxy relative to the earth?

MODEL The observed wavelength is longer than the wavelength emitted by atoms at rest with respect to the observer (i.e., red shifted), so we are looking at light emitted from a galaxy that is receding from us.

SOLVE Squaring the expression for λ_- in Equations 20.41 and solving for v_s give

$$v_s = \frac{(\lambda_-/\lambda_0)^2 - 1}{(\lambda_-/\lambda_0)^2 + 1}c$$

$$= \frac{(691\text{ nm}/656\text{ nm})^2 - 1}{(691\text{ nm}/656\text{ nm})^2 + 1}c$$

$$= 0.052c = 1.56 \times 10^7\text{ m/s}$$

ASSESS The galaxy is moving away from the earth at about 5% of the speed of light!

In the 1920s, an analysis of the red shifts of many galaxies led the astronomer Edwin Hubble to the conclusion that the galaxies of the universe are *all* moving apart from each other. Extrapolating backward in time must bring us to a point when all the matter of the universe—and even space itself, according to the theory of relativity—began rushing out of a primordial fireball. Many observations and measurements since have given support to the idea that the universe began in a *Big Bang* about 14 billion years ago.

As an example, FIGURE 20.28 is a Hubble Space Telescope picture of a *quasar*, short for *quasistellar object.* Quasars are extraordinarily powerful sources of light and radio waves. The light reaching us from quasars is highly red shifted, corresponding in some cases to objects that are moving away from us at greater than 90% of the speed of light. Astronomers have determined that some quasars are 10 to 12 *billion* light years away from the earth, hence the light we see was emitted when the universe was only about 25% of its present age. Today, the red shifts of distant quasars and supernovae (exploding stars) are being used to refine our understanding of the structure and evolution of the universe.

FIGURE 20.28 A Hubble Space Telescope picture of a quasar.

STOP TO THINK 20.7 Amy and Zack are both listening to the source of sound waves that is moving to the right. Compare the frequencies each hears.

a. $f_{Amy} > f_{Zack}$
b. $f_{Amy} = f_{Zack}$
c. $f_{Amy} < f_{Zack}$

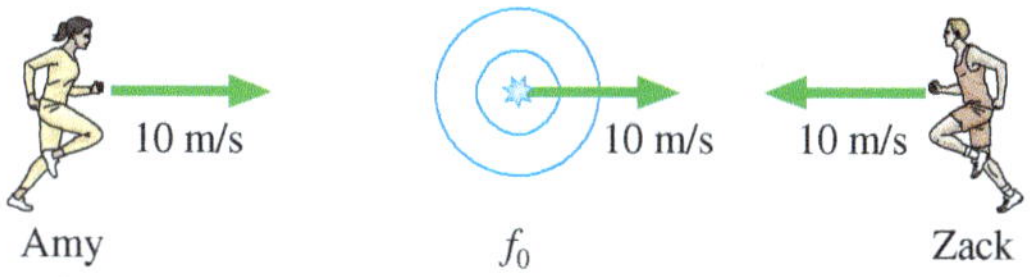

CHALLENGE EXAMPLE 20.13 Decreasing the sound

The loudspeaker on a homecoming float—mounted on a pole—is stuck playing an annoying 210 Hz tone. When the speaker is 10 m away, you measure the sound to be a loud 95 dB at 208 Hz. How long will it take for the sound intensity level to drop to a tolerable 55 dB?

MODEL The source is on a pole, so model the sound waves as uniform spherical waves. Assume a temperature of 20°C.

SOLVE The 208 Hz frequency you measure is less than the 210 Hz frequency that was emitted, so the float must be moving away from you. The Doppler effect for a receding source is

$$f_- = \frac{f_0}{1 + v_s/v}$$

We can solve this to find the speed of the float:

$$v_s = \left(\frac{f_0}{f_-} - 1\right)v = \left(\frac{210\text{ Hz}}{208\text{ Hz}} - 1\right) \times 343\text{ m/s} = 3.3\text{ m/s}$$

The sound intensity of a spherical wave decreases with the inverse square of the distance from the source. A sound intensity level β corresponds to an intensity $I = I_0 \times 10^{\beta/10\text{ dB}}$, where $I_0 = 1.0 \times 10^{-12}\text{ W/m}^2$. At the initial 95 dB, the intensity is

$$I_1 = I_0 \times 10^{9.5} = 3.2 \times 10^{-3}\text{ W/m}^2$$

At the desired 55 dB, the intensity will have dropped to

$$I_2 = I_0 \times 10^{5.5} = 3.2 \times 10^{-7}\text{ W/m}^2$$

The intensity ratio is related to the distances by

$$\frac{I_1}{I_2} = \frac{r_2^2}{r_1^2}$$

Thus the sound will have dropped to 55 dB when the distance to the speaker is

$$r_2 = \sqrt{\frac{I_1}{I_2}}r_1 = \sqrt{10^4} \times 10\text{ m} = 1000\text{ m}$$

The float has to travel $\Delta x = 990$ m, which will take

$$\Delta t = \frac{\Delta x}{v_s} = \frac{990\text{ m}}{3.3\text{ m/s}} = 300\text{ s} = 5.0\text{ min}$$

ASSESS To drop the sound intensity level by 40 dB requires decreasing the intensity by a factor of 10^4. And with the intensity depending on the inverse square of the distance, that requires increasing the distance by a factor of 100. Floats don't move very fast—3.3 m/s is about 7 mph—so needing several minutes to travel the $\approx$ 1000 m seems reasonable.

SUMMARY

The goal of Chapter 20 has been to learn the basic properties of traveling waves.

General Principles

The Wave Model

This model is based on the idea of a traveling wave, which is an organized disturbance traveling at a well-defined **wave speed** v.

- In transverse waves the displacement is perpendicular to the direction in which the wave travels.

- In longitudinal waves the particles of the medium move parallel to the direction in which the wave travels.

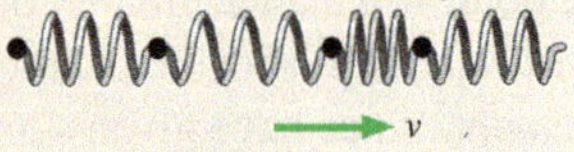

A wave transfers **energy,** but no material or substance is transferred outward from the source.

Two basic types of waves:

- Mechanical waves travel through a material medium such as water or air.
- Electromagnetic waves require no material medium and can travel through a vacuum.

For mechanical waves, such as sound waves and waves on strings, the speed of the wave is a property of the medium. Speed does not depend on the size or shape of the wave.

Important Concepts

The **displacement** D of a wave is a function of both position (where) and time (when).

- A snapshot graph shows the wave's displacement as a function of position at a single instant of time.

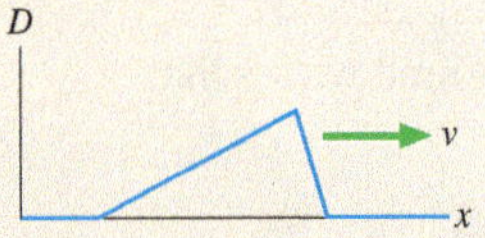

- A history graph shows the wave's displacement as a function of time at a single point in space.

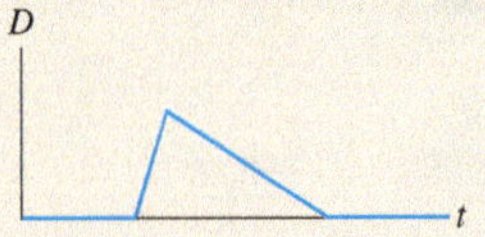

For a transverse wave on a string, the snapshot graph is a picture of the wave. The displacement of a longitudinal wave is parallel to the motion; thus the snapshot graph of a longitudinal sound wave is *not* a picture of the wave.

Sinusoidal waves are periodic in both time (period T) and space (wavelength λ):

$$D(x, t) = A\sin\left[2\pi(x/\lambda - t/T) + \phi_0\right]$$
$$= A\sin(kx - \omega t + \phi_0)$$

where A is the **amplitude,** $k = 2\pi/\lambda$ is the **wave number,** $\omega = 2\pi f = 2\pi/T$ is the **angular frequency,** and ϕ_0 is the **phase constant** that describes initial conditions.

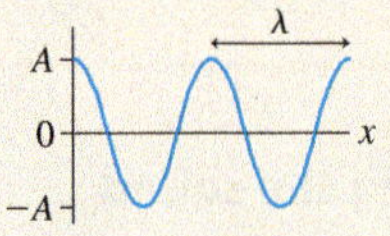

One-dimensional waves

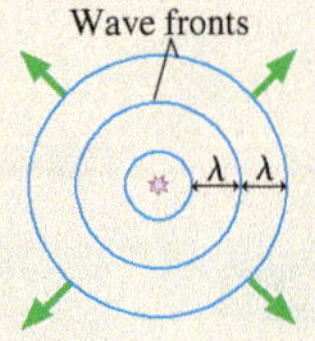

Two- and three-dimensional waves

The fundamental relationship for any sinusoidal wave is $v = \lambda f$.

Applications

- **String** (transverse): $v = \sqrt{T_s/\mu}$
- **Sound** (longitudinal): $v = 343$ m/s in 20°C air
- **Light** (transverse): $v = c/n$, where $c = 3.00 \times 10^8$ m/s is the speed of light in a vacuum and n is the material's **index of refraction**

The wave intensity is the power-to-area ratio: $I = P/a$

For a circular or spherical wave: $I = P_{source}/4\pi r^2$

The sound intensity level is

$$\beta = (10\text{ dB})\log_{10}(I/1.0 \times 10^{-12}\text{ W/m}^2)$$

The Doppler effect occurs when a wave source and detector are moving with respect to each other: the frequency detected differs from the frequency f_0 emitted.

Approaching source

$$f_+ = \frac{f_0}{1 - v_s/v}$$

Observer approaching a source

$$f_+ = (1 + v_o/v)f_0$$

Receding source

$$f_- = \frac{f_0}{1 + v_s/v}$$

Observer receding from a source

$$f_- = (1 - v_o/v)f_0$$

The Doppler effect for light uses a result derived from the theory of relativity.

Terms and Notation

wave model	wave speed, v	wave front	index of refraction, n
traveling wave	linear density, μ	circular wave	intensity, I
transverse wave	snapshot graph	spherical wave	sound intensity level, β
longitudinal wave	history graph	plane wave	decibels
mechanical waves	sinusoidal wave	phase, ϕ	Doppler effect
electromagnetic waves	amplitude, A	compression	red shifted
medium	wavelength, λ	rarefaction	blue shifted
disturbance	wave number, k	electromagnetic spectrum	

CONCEPTUAL QUESTIONS

1. The three wave pulses in FIGURE Q20.1 travel along the same stretched string. Rank in order, from largest to smallest, their wave speeds v_a, v_b, and v_c. Explain.

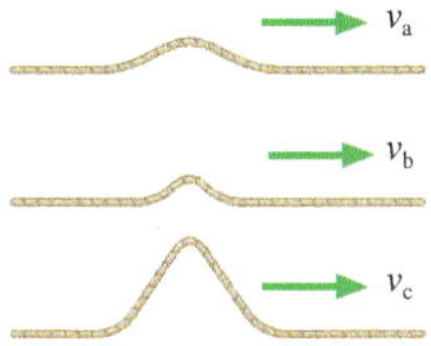

FIGURE Q20.1

2. A wave pulse travels along a stretched string at a speed of 200 cm/s. What will be the speed if:
   a. The string's tension is doubled?
   b. The string's mass is quadrupled (but its length is unchanged)?
   c. The string's length is quadrupled (but its mass is unchanged)?
   Note: Each part is independent and refers to changes made to the original string.
3. FIGURE Q20.3 is a history graph showing the displacement as a function of time at one point on a string. Did the displacement at this point reach its maximum of 2 mm *before* or *after* the interval of time when the displacement was a constant 1 mm?

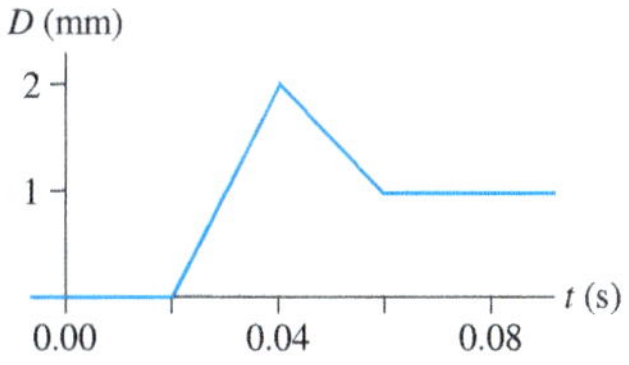

FIGURE Q20.3

4. FIGURE Q20.4 shows a snapshot graph *and* a history graph for a wave pulse on a stretched string. They describe the same wave from two perspectives.
   a. In which direction is the wave traveling? Explain.
   b. What is the speed of this wave?

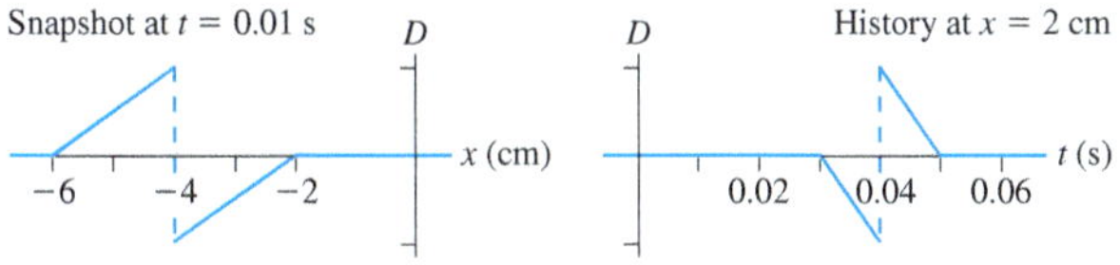

FIGURE Q20.4

5. Rank in order, from largest to smallest, the wavelengths λ_a, λ_b, and λ_c for sound waves having frequencies $f_a = 100$ Hz, $f_b = 1000$ Hz, and $f_c = 10{,}000$ Hz. Explain.
6. A sound wave with wavelength λ_0 and frequency f_0 moves into a new medium in which the speed of sound is $v_1 = 2v_0$. What are the new wavelength λ_1 and frequency f_1? Explain.
7. What are the amplitude, wavelength, frequency, and phase constant of the traveling wave in FIGURE Q20.7?

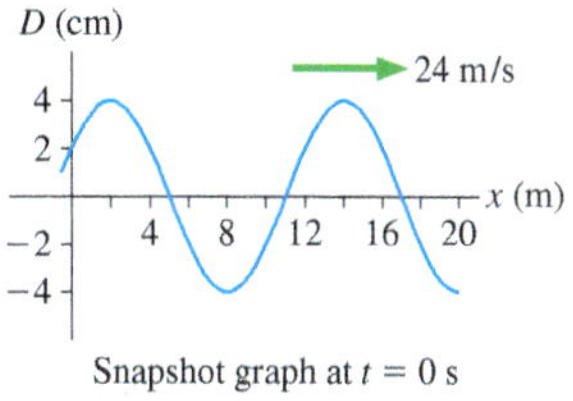

FIGURE Q20.7

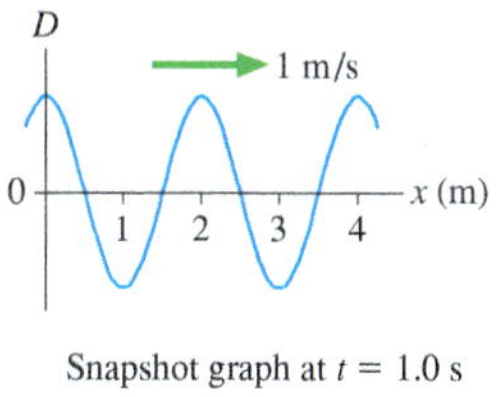

FIGURE Q20.8

8. FIGURE Q20.8 is a snapshot graph of a sinusoidal wave at $t = 1.0$ s. What is the phase constant of this wave?
9. FIGURE Q20.9 shows the wave fronts of a circular wave. What is the phase difference between (a) points A and B, (b) points C and D, and (c) points E and F?

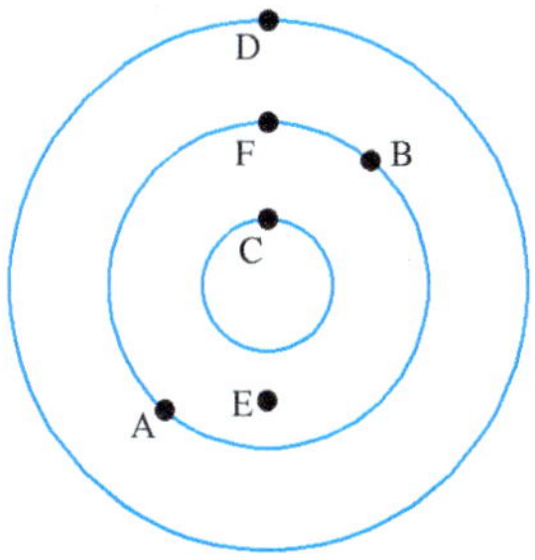

FIGURE Q20.9

10. Sound wave A delivers 2 J of energy in 2 s. Sound wave B delivers 10 J of energy in 5 s. Sound wave C delivers 2 mJ of energy in 1 ms. Rank in order, from largest to smallest, the sound powers P_A, P_B, and P_C of these three sound waves. Explain.
11. One physics professor talking produces a sound intensity level of 52 dB. It's a frightening idea, but what would be the sound intensity level of 100 physics professors talking simultaneously?
12. You are standing at $x = 0$ m, listening to a sound that is emitted at frequency f_0. The graph of FIGURE Q20.12 shows the frequency you hear during a 4-second interval. Which of the following describes the sound source? Explain your choice.

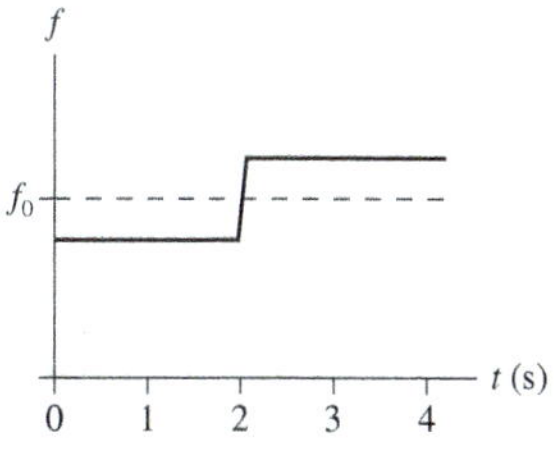

FIGURE Q20.12

   A. It moves from left to right and passes you at $t = 2$ s.
   B. It moves from right to left and passes you at $t = 2$ s.
   C. It moves toward you but doesn't reach you. It then reverses direction at $t = 2$ s.
   D. It moves away from you until $t = 2$ s. It then reverses direction and moves toward you but doesn't reach you.

EXERCISES AND PROBLEMS

Problems labeled integrate material from earlier chapters.

Exercises

Section 20.1 The Wave Model

1. | The wave speed on a string is 150 m/s when the tension is 75 N. What tension will give a speed of 180 m/s?
2. | The wave speed on a string under tension is 200 m/s. What is the speed if the tension is halved?
3. || A 25 g string is under 20 N of tension. A pulse travels the length of the string in 50 ms. How long is the string?

Section 20.2 One-Dimensional Waves

4. || Draw the history graph $D(x = 0\text{ m}, t)$ at $x = 0$ m for the wave shown in FIGURE EX20.4.

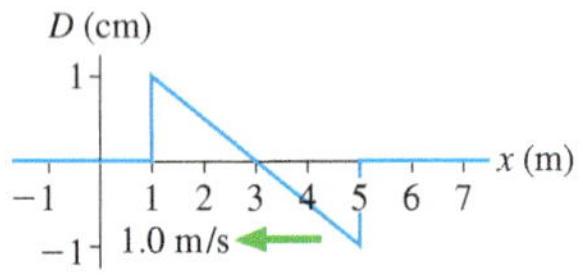

Snapshot graph of a wave at $t = 2$ s

FIGURE EX20.4

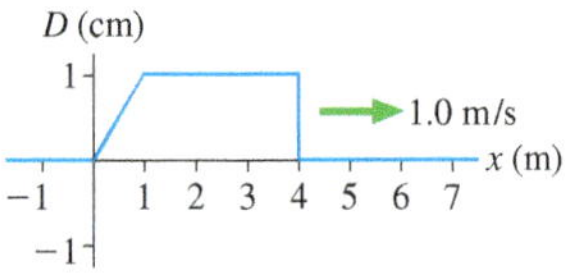

Snapshot graph of a wave at $t = 0$ s

FIGURE EX20.5

5. || Draw the history graph $D(x = 5.0\text{ m}, t)$ at $x = 5.0$ m for the wave shown in FIGURE EX20.5.
6. || Draw the snapshot graph $D(x, t = 0\text{ s})$ at $t = 0$ s for the wave shown in FIGURE EX20.6.

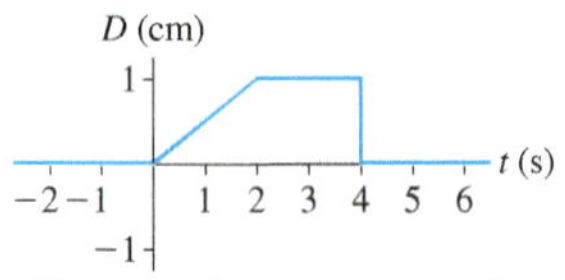

History graph of a wave at $x = 2$ m
Wave moving to the left at 1.0 m/s

FIGURE EX20.6

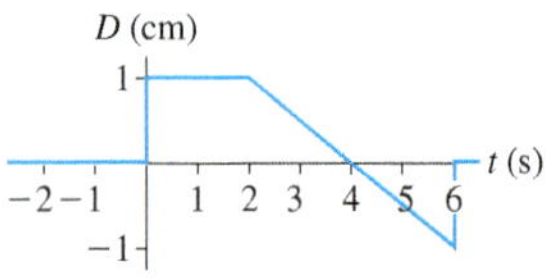

History graph of a wave at $x = 0$ m
Wave moving to the right at 1.0 m/s

FIGURE EX20.7

7. || Draw the snapshot graph $D(x, t = 1.0\text{ s})$ at $t = 1.0$ s for the wave shown in FIGURE EX20.7.
8. || FIGURE EX20.8 is the snapshot graph at $t = 0$ s of a *longitudinal* wave. Draw the corresponding picture of the particle positions, as was done in Figure 20.10b. Let the equilibrium spacing between the particles be 1.0 cm.

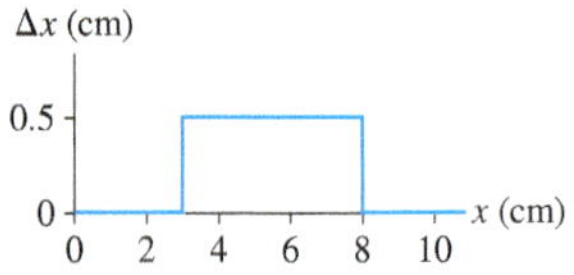

FIGURE EX20.8

9. || FIGURE EX20.9 is a picture at $t = 0$ s of the particles in a medium as a longitudinal wave is passing through. The equilibrium spacing between the particles is 1.0 cm. Draw the snapshot graph $D(x, t = 0\text{ s})$ of this wave at $t = 0$ s.

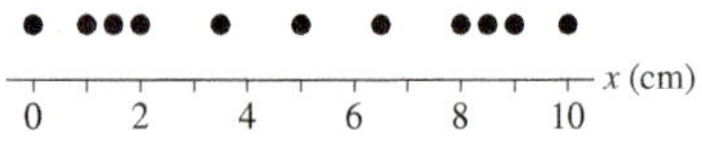

FIGURE EX20.9

Section 20.3 Sinusoidal Waves

10. | A wave travels with speed 200 m/s. Its wave number is 1.5 rad/m. What are its (a) wavelength and (b) frequency?
11. | A wave has angular frequency 30 rad/s and wavelength 2.0 m. What are its (a) wave number and (b) wave speed?
12. | The displacement of a wave traveling in the positive x-direction is $D(x, t) = (3.5\text{ cm})\sin(2.7x - 124t)$, where x is in m and t is in s. What are the (a) frequency, (b) wavelength, and (c) speed of this wave?
13. | The displacement of a wave traveling in the negative y-direction is $D(y, t) = (5.2\text{ cm})\sin(5.5y + 72t)$, where y is in m and t is in s. What are the (a) frequency, (b) wavelength, and (c) speed of this wave?
14. | What are the amplitude, frequency, and wavelength of the wave in FIGURE EX20.14?

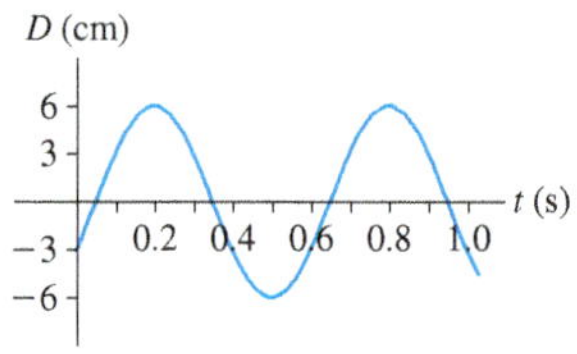

FIGURE EX20.14

History graph at $x = 0$ m
Wave traveling left at 2.0 m/s

Section 20.4 Waves in Two and Three Dimensions

15. | A spherical wave with a wavelength of 2.0 m is emitted from the origin. At one instant of time, the phase at $r = 4.0$ m is π rad. At that instant, what is the phase at $r = 3.5$ m and at $r = 4.5$ m?
16. | A circular wave travels outward from the origin. At one instant of time, the phase at $r_1 = 20$ cm is 0 rad and the phase at $r_2 = 80$ cm is 3π rad. What is the wavelength of the wave?
17. || A loudspeaker at the origin emits a 120 Hz tone on a day when the speed of sound is 340 m/s. The phase difference between two points on the x-axis is 5.5 rad. What is the distance between these two points?
18. || A sound source is located somewhere along the x-axis. Experiments show that the same wave front simultaneously reaches listeners at $x = -7.0$ m and $x = +3.0$ m.
    a. What is the x-coordinate of the source?
    b. A third listener is positioned along the positive y-axis. What is her y-coordinate if the same wave front reaches her at the same instant it does the first two listeners?

Section 20.5 Sound and Light

19. || A hammer taps on the end of a 4.00-m-long metal bar at room temperature. A microphone at the other end of the bar picks up two pulses of sound, one that travels through the metal and one that travels through the air. The pulses are separated in time by 9.00 ms. What is the speed of sound in this metal?
20. || a. What is the wavelength of a 2.0 MHz ultrasound wave traveling through aluminum?
    b. What frequency of electromagnetic wave would have the same wavelength as the ultrasound wave of part a?

21. | a. What is the frequency of an electromagnetic wave with a wavelength of 20 cm?
b. What would be the wavelength of a sound wave in water with the same frequency as the electromagnetic wave of part a?

22. | a. What is the frequency of blue light that has a wavelength of 450 nm?
b. What is the frequency of red light that has a wavelength of 650 nm?
c. What is the index of refraction of a material in which the red-light wavelength is 450 nm?

23. | a. An FM radio station broadcasts at a frequency of 101.3 MHz. What is the wavelength?
b. What is the frequency of a sound source that produces the same wavelength in 20°C air?

24. | a. Telephone signals are often transmitted over long distances by microwaves. What is the frequency of microwave radiation with a wavelength of 3.0 cm?
b. Microwave signals are beamed between two mountaintops 50 km apart. How long does it take a signal to travel from one mountaintop to the other?

25. || a. How long does it take light to travel through a 3.0-mm-thick piece of window glass?
b. Through what thickness of water could light travel in the same amount of time?

26. || Cell phone conversations are transmitted by high-frequency radio waves. Suppose the signal has wavelength 35 cm while traveling through air. What are the (a) frequency and (b) wavelength as the signal travels through 3-mm-thick window glass into your room?

27. | A light wave has a 670 nm wavelength in air. Its wavelength in a transparent solid is 420 nm.
a. What is the speed of light in this solid?
b. What is the light's frequency in the solid?

Section 20.6 Power, Intensity, and Decibels

28. || BIO A sound wave with intensity $2.0 \times 10^{-3}\ \text{W/m}^2$ is perceived to be modestly loud. Your eardrum is 6.0 mm in diameter. How much energy will be transferred to your eardrum while listening to this sound for 1.0 min?

29. || The intensity of electromagnetic waves from the sun is $1.4\ \text{kW/m}^2$ just above the earth's atmosphere. Eighty percent of this reaches the surface at noon on a clear summer day. Suppose you think of your back as a 30 cm $\times$ 50 cm rectangle. How many joules of solar energy fall on your back as you work on your tan for 1.0 h?

30. || A concert loudspeaker suspended high above the ground emits 35 W of sound power. A small microphone with a $1.0\ \text{cm}^2$ area is 50 m from the speaker.
a. What is the sound intensity at the position of the microphone?
b. How much sound energy impinges on the microphone each second?

31. || During takeoff, the sound intensity level of a jet engine is 140 dB at a distance of 30 m. What is the sound intensity level at a distance of 1.0 km?

32. | The sun emits electromagnetic waves with a power of 4.0×10^{26} W. Determine the intensity of electromagnetic waves from the sun just outside the atmospheres of Venus, the earth, and Mars.

33. | What are the sound intensity levels for sound waves of intensity (a) $3.0 \times 10^{-6}\ \text{W/m}^2$ and (b) $3.0 \times 10^{-2}\ \text{W/m}^2$?

34. | What are the intensities of sound waves with sound intensity levels (a) 46 dB and (b) 103 dB?

35. || A loudspeaker on a tall pole broadcasts sound waves equally in all directions. What is the speaker's power output if the sound intensity level is 90 dB at a distance of 20 m?

Section 20.7 The Doppler Effect

36. | A friend of yours is loudly singing a single note at 400 Hz while racing toward you at 25.0 m/s on a day when the speed of sound is 340 m/s.
a. What frequency do you hear?
b. What frequency does your friend hear if you suddenly start singing at 400 Hz?

37. | An opera singer in a convertible sings a note at 600 Hz while cruising down the highway at 90 km/h. What is the frequency heard by
a. A person standing beside the road in front of the car?
b. A person on the ground behind the car?

38. || BIO A bat locates insects by emitting ultrasonic "chirps" and then listening for echoes from the bugs. Suppose a bat chirp has a frequency of 25 kHz. How fast would the bat have to fly, and in what direction, for you to just barely be able to hear the chirp at 20 kHz?

39. | A mother hawk screeches as she dives at you. You recall from biology that female hawks screech at 800 Hz, but you hear the screech at 900 Hz. How fast is the hawk approaching?

Problems

40. || The displacement of a traveling wave is

$$D(x, t) = \begin{cases} 1\ \text{cm} & \text{if } |x - 3t| \le 1 \\ 0\ \text{cm} & \text{if } |x - 3t| > 1 \end{cases}$$

where x is in m and t in s.
a. Draw displacement-versus-position graphs from $x = -2$ m to $x = 12$ m at 1 s intervals from $t = 0$ s to $t = 3$ s.
b. Determine the wave speed from the graphs. Explain.
c. Determine the wave speed from the equation for $D(x, t)$. Does it agree with your answer to part b?

41. || FIGURE P20.41 is a history graph at $x = 0$ m of a wave traveling in the positive x-direction at 4.0 m/s.
a. What is the wavelength?
b. What is the phase constant of the wave?
c. Write the displacement equation for this wave.

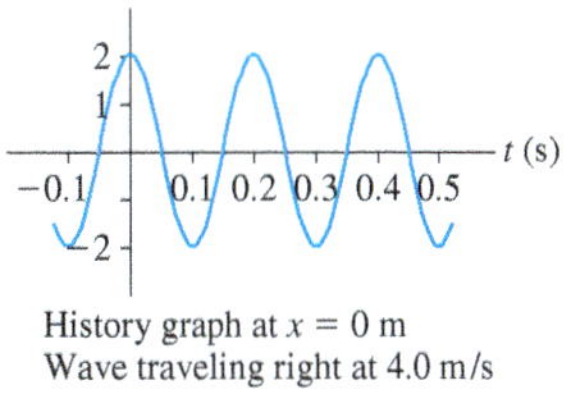

FIGURE P20.41

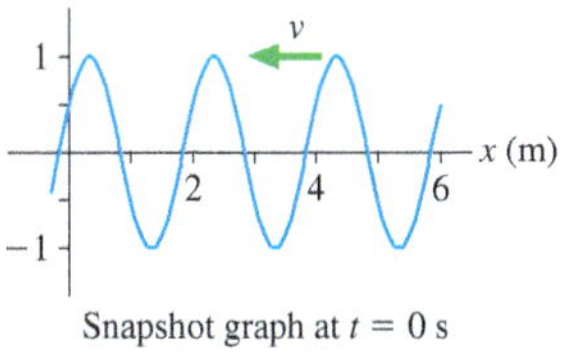

FIGURE P20.42

42. || FIGURE P20.42 is a snapshot graph at $t = 0$ s of a 5.0 Hz wave traveling to the left.
a. What is the wave speed?
b. What is the phase constant of the wave?
c. Write the displacement equation for this wave.

43. || A wave travels along a string at speed v_0. What will be the speed if the string is replaced by one made of the same material and under the same tension but having twice the radius?

44. | String 1 in FIGURE P20.44 has linear density 2.0 g/m and string 2 has linear density 4.0 g/m. A student sends pulses in both directions by quickly pulling up on the knot, then releasing it. What should the string lengths L_1 and L_2 be if the pulses are to reach the ends of the strings simultaneously?

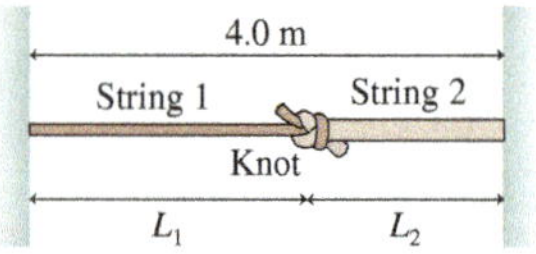

FIGURE P20.44

45. || Ships measure the distance to the ocean bottom with sonar. A pulse of sound waves is aimed at the ocean bottom, then sensitive microphones listen for the echo. FIGURE P20.45 shows the delay time as a function of the ship's position as it crosses 60 km of ocean. Draw a graph of the ocean bottom. Let the ocean surface define $y = 0$ and ocean bottom have negative values of y. This way your graph will be a picture of the ocean bottom. The speed of sound in ocean water varies slightly with temperature, but you can use 1500 m/s as an average value.

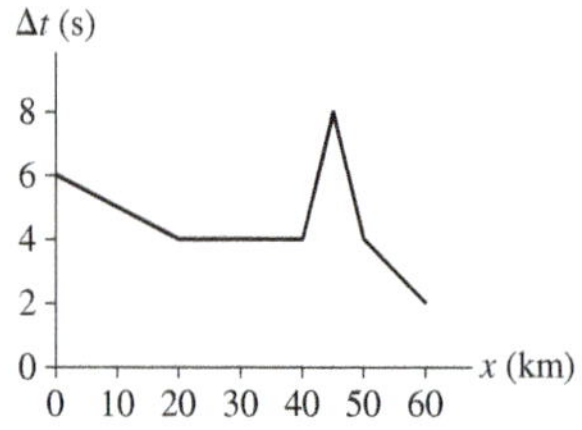

FIGURE P20.45

46. || Oil explorers set off explosives to make loud sounds, then listen for the echoes from underground oil deposits. Geologists suspect that there is oil under 500-m-deep Lake Physics. It's known that Lake Physics is carved out of a granite basin. Explorers detect a weak echo 0.94 s after exploding dynamite at the lake surface. If it's really oil, how deep will they have to drill into the granite to reach it?

47. || One cue your hearing system uses to localize a sound (i.e., to BIO tell where a sound is coming from) is the slight difference in the arrival times of the sound at your ears. Your ears are spaced approximately 20 cm apart. Consider a sound source 5.0 m from the center of your head along a line 45° to your right. What is the difference in arrival times? Give your answer in microseconds. **Hint:** You are looking for the difference between two numbers that are nearly the same. What does this near equality imply about the necessary precision during intermediate stages of the calculation?

48. || A helium-neon laser beam has a wavelength in air of 633 nm. It takes 1.38 ns for the light to travel through 30 cm of an unknown liquid. What is the wavelength of the laser beam in the liquid?

49. | A 440 Hz sound wave in 20°C air propagates into the water of a swimming pool. What are the wave's (a) frequency and (b) wavelength in the water?

50. || Earthquakes are essentially sound waves—called seismic waves—traveling through the earth. Because the earth is solid, it can support both longitudinal and transverse seismic waves. The speed of longitudinal waves, called P waves, is 8000 m/s. Transverse waves, called S waves, travel at a slower 4500 m/s. A seismograph records the two waves from a distant earthquake. If the S wave arrives 2.0 min after the P wave, how far away was the earthquake? You can assume that the waves travel in straight lines, although actual seismic waves follow more complex routes.

51. || A sound wave is described by $D(y, t) = (0.0200 \text{ mm}) \times \sin[(8.96 \text{ rad/m})y + (3140 \text{ rad/s})t + \pi/4 \text{ rad}]$, where y is in m and t is in s.
 a. In what direction is this wave traveling?
 b. Along which axis is the air oscillating?
 c. What are the wavelength, the wave speed, and the period of oscillation?

52. || A wave on a string is described by $D(x, t) = (3.0 \text{ cm}) \times \sin[2\pi(x/(2.4 \text{ m}) + t/(0.20 \text{ s}) + 1)]$, where x is in m and t is in s.
 a. In what direction is this wave traveling?
 b. What are the wave speed, the frequency, and the wave number?
 c. At $t = 0.50$ s, what is the displacement of the string at $x = 0.20$ m?

53. || A wave on a string is described by $D(x, t) = (2.00 \text{ cm}) \times \sin[(12.57 \text{ rad/m})x - (638 \text{ rad/s})t]$, where x is in m and t in s. The linear density of the string is 5.00 g/m. What are
 a. The string tension?
 b. The maximum displacement of a point on the string?
 c. The maximum speed of a point on the string?

54. | Write the displacement equation for a sinusoidal wave that is traveling in the negative y-direction with wavelength 50 cm, speed 4.0 m/s, and amplitude 5.0 cm. Assume $\phi_0 = 0$.

55. | Write the displacement equation for a sinusoidal wave that is traveling in the positive x-direction with frequency 200 Hz, speed 400 m/s, amplitude 0.010 mm, and phase constant $\pi/2$ rad.

56. | A string with linear density 2.0 g/m is stretched along the positive x-axis with tension 20 N. One end of the string, at $x = 0$ m, is tied to a hook that oscillates up and down at a frequency of 100 Hz with a maximum displacement of 1.0 mm. At $t = 0$ s, the hook is at its lowest point.
 a. What are the wave speed on the string and the wavelength?
 b. What are the amplitude and phase constant of the wave?
 c. Write the equation for the displacement $D(x, t)$ of the traveling wave.
 d. What is the string's displacement at $x = 0.50$ m and $t = 15$ ms?

57. || FIGURE P20.57 shows a snapshot graph of a wave traveling to the right along a string at 45 m/s. At this instant, what is the velocity of points 1, 2, and 3 on the string?

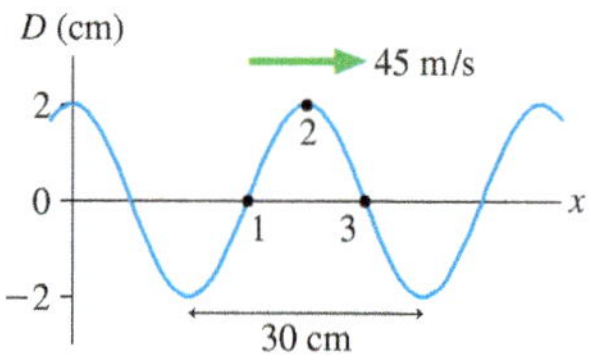

FIGURE P20.57

58. || FIGURE P20.58 shows two masses hanging from a steel wire. The mass of the wire is 60.0 g. A wave pulse travels along the wire from point 1 to point 2 in 24.0 ms. What is mass m?

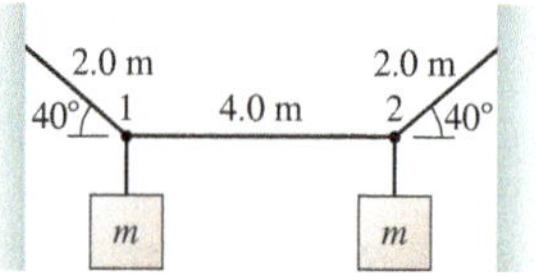

FIGURE P20.58

59. || A wire is made by welding together two metals having different densities. FIGURE P20.59 shows a 2.00-m-long section of wire centered on the junction, but the wire extends much farther in both directions. The wire is placed under 2250 N tension, then a 1500 Hz wave with an amplitude of 3.00 mm is sent down the wire. How many wavelengths (complete cycles) of the wave are in this 2.00-m-long section of the wire?

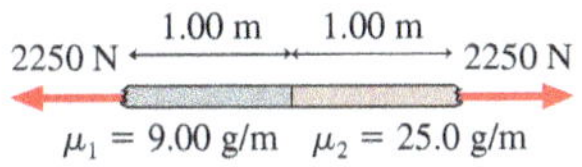

FIGURE P20.59

60. || The string in FIGURE P20.60 has linear density μ. Find an expression in terms of M, μ and θ for the speed of waves on the string.

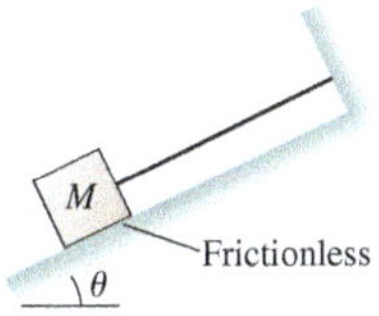

FIGURE P20.60

61. || A string that is under 50.0 N of tension has linear density 5.0 g/m. A sinusoidal wave with amplitude 3.0 cm and wavelength 2.0 m travels along the string. What is the maximum speed of a particle on the string?

62. || A sinusoidal wave travels along a stretched string. A particle on the string has a maximum speed of 2.0 m/s and a maximum acceleration of 200 m/s^2. What are the frequency and amplitude of the wave?

63. || a. A 100 W lightbulb produces 5.0 W of visible light. (The other 95 W are dissipated as heat and infrared radiation.) What is the light intensity on a wall 2.0 m away from the lightbulb?
   b. A krypton laser produces a cylindrical red laser beam 2.0 mm in diameter with 5.0 W of power. What is the light intensity on a wall 2.0 m away from the laser?

64. || An AM radio station broadcasts with a power of 25 kW at a frequency of 920 kHz. Estimate the intensity of the radio wave at a point 10 km from the broadcast antenna.

65. || LASIK eye surgery uses pulses of laser light to shave off tissue from the cornea, reshaping it. A typical LASIK laser emits a 1.0-mm-diameter laser beam with a wavelength of 193 nm. Each laser pulse lasts 15 ns and contains 1.0 mJ of light energy
BIO
   a. What is the power of one laser pulse?
   b. During the very brief time of the pulse, what is the intensity of the light wave?

66. || The sound intensity 50 m from a wailing tornado siren is 0.10 W/m^2.
   a. What is the intensity at 1000 m?
   b. The weakest intensity likely to be heard over background noise is $\approx 1\ \mu$W/m^2. Estimate the maximum distance at which the siren can be heard.

67. || The sound intensity level 5.0 m from a large power saw is 100 dB. At what distance will the sound be a more tolerable 80 dB?

68. || Two loudspeakers on elevated platforms are at opposite ends of a field. Each broadcasts equally in all directions. The sound intensity level at a point halfway between the loudspeakers is 75.0 dB. What is the sound intensity level at a point one-quarter of the way from one speaker to the other along the line joining them?

69. || Your ears are sensitive to differences in pitch, but they are not very sensitive to differences in intensity. You are not capable of detecting a difference in sound intensity level of less than 1 dB. By what factor does the sound intensity increase if the sound intensity level increases from 60 dB to 61 dB?
BIO

70. ||| The intensity of a sound source is described by an inverse-square law only if the source is very small (a point source) and only if the waves can travel unimpeded in all directions. For an extended source or in a situation where obstacles absorb or reflect the waves, the intensity at distance r can often be expressed as $I = cP_{source}/r^x$, where c is a constant and the exponent x—which would be 2 for an ideal spherical wave—depends on the situation. In one such situation, you use a sound meter to measure the sound intensity level at different distances from a source, acquiring the following data:

Distance (m)	Intensity level (dB)
1	100
3	93
10	85
30	78
100	70

Use the best-fit line of an appropriate graph to determine the exponent x that characterizes this sound source.

71. ||| A mad doctor believes that baldness can be cured by warming the scalp with sound waves. His patients sit underneath the Bald-o-Matic loudspeakers, where their heads are bathed with 93 dB of soothing 800 Hz sound waves. Suppose we model a bald head as a 16-cm-diameter hemisphere. If 0.10 J of sound energy is considered an appropriate "dose," how many minutes should each therapy session last?

72. || A physics professor demonstrates the Doppler effect by tying a 600 Hz sound generator to a 1.0-m-long rope and whirling it around her head in a horizontal circle at 100 rpm. What are the highest and lowest frequencies heard by a student in the classroom?

73. || Show that the Doppler frequency f_- of a receding source is $f_- = f_0/(1 + v_s/v)$.

74. | A starship approaches its home planet at a speed of 0.1c. When it is 54×10^6 km away, it uses its green laser beam ($\lambda = 540$ nm) to signal its approach.
   a. How long does the signal take to travel to the home planet?
   b. At what wavelength is the signal detected on the home planet?

75. | Wavelengths of light from a distant galaxy are found to be 0.5% longer than the corresponding wavelengths measured in a terrestrial laboratory. Is the galaxy approaching or receding from the earth? At what speed?

76. | You have just been pulled over for running a red light, and the police officer has informed you that the fine will be $250. In desperation, you suddenly recall an idea that your physics professor recently discussed in class. In your calmest voice, you tell the officer that the laws of physics prevented you from knowing that the light was red. In fact, as you drove toward it, the light was Doppler shifted to where it appeared green to you. "OK," says the officer, "Then I'll ticket you for speeding. The fine is $1 for every 1 km/h over the posted speed limit of 50 km/h." How big is your fine? Use 650 nm as the wavelength of red light and 540 nm as the wavelength of green light.

Challenge Problems

77. One way to monitor global warming is to measure the average temperature of the ocean. Researchers are doing this by measuring the time it takes sound pulses to travel underwater over large distances. At a depth of 1000 m, where ocean temperatures hold steady near 4°C, the average sound speed is 1480 m/s. It's known from laboratory measurements that the sound speed increases 4.0 m/s for every 1.0°C increase in temperature. In one experiment, where sounds generated near California are detected in the South Pacific, the sound waves travel 8000 km. If the smallest time change that can be reliably detected is 1.0 s, what is the smallest change in average temperature that can be measured?
78. The G string on a guitar is a 0.46-mm-diameter steel string with a linear density of 1.3 g/m. When the string is properly tuned to 196 Hz, the wave speed on the string is 250 m/s. Tuning is done by turning the tuning screw, which slowly tightens—and stretches—the string. By how many mm does a 75-cm-long G string stretch when it's first tuned?
79. A rope of mass m and length L hangs from a ceiling.
    a. Show that the wave speed on the rope a distance y above the lower end is $v = \sqrt{gy}$.
    b. Show that the time for a pulse to travel the length of the string is $\Delta t = 2\sqrt{L/g}$.
80. Some modern optical devices are made with glass whose index of refraction changes with distance from the front surface. FIGURE CP20.80 shows the index of refraction as a function of the distance into a slab of glass of thickness L. The index of refraction increases linearly from n_1 at the front surface to n_2 at the rear surface.

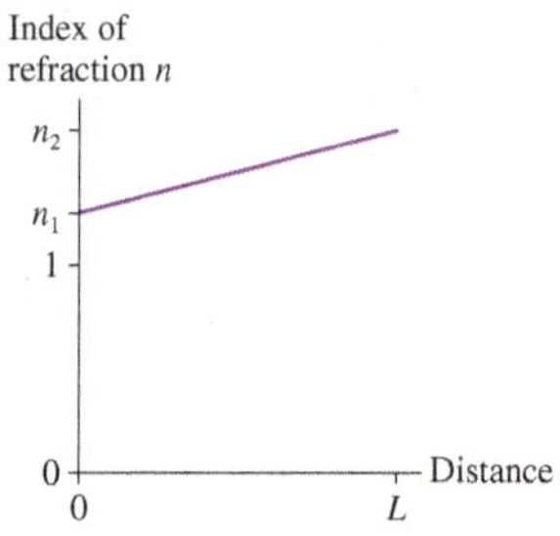

FIGURE CP20.80

    a. Find an expression for the time light takes to travel through this piece of glass.
    b. Evaluate your expression for a 1.0-cm-thick piece of glass for which $n_1 = 1.50$ and $n_2 = 1.60$.
81. A water wave is a *shallow-water wave* if the water depth d is less than $\approx \lambda/10$. It is shown in hydrodynamics that the speed of a shallow-water wave is $v = \sqrt{gd}$, so waves slow down as they move into shallower water. Ocean waves, with wavelengths of typically 100 m, are shallow-water waves when the water depth is less than $\approx$ 10 m. Consider a beach where the depth increases linearly with distance from the shore until reaching a depth of 5.0 m at a distance of 100 m. How long does it take a wave to move the last 100 m to the shore? Assume that the waves are so small that they don't break before reaching the shore.
82. BIO An important characteristic of the heart, one used to diagnose heart disease, is the *pressure difference* between the blood pressure inside the heart and the blood pressure in the aorta, the large artery leading away from the heart. The blood inside the heart is essentially at rest, but it speeds up significantly as it enters the aorta—and its speed can be measured by using the Doppler shift of reflected ultrasound.
    a. The Doppler effect enters twice in calculating the frequency of the reflection from a moving object. Suppose the object's speed v_o is very small compared to the wave speed v. Show that a good approximation for the *Doppler shift*—the difference between the reflected frequency and the incident frequency—is

$$\Delta f = 2f_0\frac{v_o}{v}$$

    b. A doctor using 2.5 MHz ultrasound measures a 6000 Hz Doppler shift as the ultrasound reflects from blood ejected from the heart into the aorta. What is the blood pressure difference, in mm of Hg, between the inside of the heart and the aorta? Assume the patient is lying down so that there is no height difference between the heart and the aorta. The density of blood is 1060 kg/m^3.

STOP TO THINK ANSWERS

Stop to Think 20.1: d and e. The wave speed depends on properties of the medium, not on how you generate the wave. For a string, $v = \sqrt{T_s/\mu}$. Increasing the tension or decreasing the linear density (lighter string) will increase the wave speed.

Stop to Think 20.2: b. The wave is traveling to the right at 2.0 m/s, so each point on the wave passes $x = 0$ m, the point of interest, 2.0 s before reaching $x = 4.0$ m. The graph has the same shape, but everything happens 2.0 s earlier.

Stop to Think 20.3: d. The wavelength—the distance between two crests—is seen to be 10 m. The frequency is $f = v/\lambda = (50 \text{ m/s})/(10 \text{ m}) = 5$ Hz.

Stop to Think 20.4: e. A crest and an adjacent trough are separated by $\lambda/2$. This is a phase difference of π rad.

Stop to Think 20.5: $n_c > n_a > n_b$. $\lambda = \lambda_{vac}/n$, so a shorter wavelength corresponds to a larger index of refraction.

Stop to Think 20.6: c. Any factor-of-2 change in intensity changes the sound intensity level by 3 dB. One trumpet is $\frac{1}{4}$ the original number, so the intensity has decreased by two factors of 2.

Stop to Think 20.7: c. Zack hears a higher frequency as he and the source approach. Amy is moving with the source, so $f_{Amy} = f_0$.

21 Superposition

This swirl of colors is due to a very thin layer of oil. Oil is clear. The colors arise from the interference of reflected light waves.

▶ **Looking Ahead** The goal of Chapter 21 is to understand and use the idea of superposition.

Standing Waves

Standing waves are created from the superposition of traveling waves bouncing back and forth between the edges of the medium.

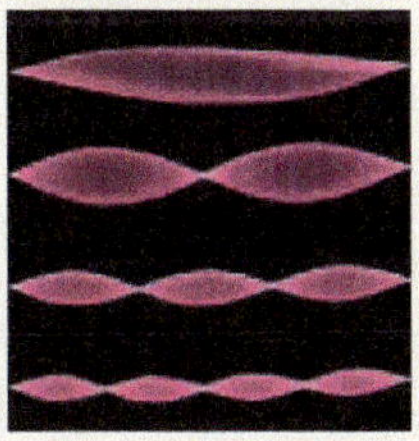

Standing waves occur in well-defined patterns called **modes,** each with its own distinct frequency. Some points on the wave, called **nodes,** do not oscillate at all.

You'll learn how to calculate the frequencies and wavelengths of standing waves on strings and in air.

Superposition

Waves can pass through each other—a characteristic that distinguishes waves from particles. As they do, their displacements add together. This is called the **principle of superposition**.

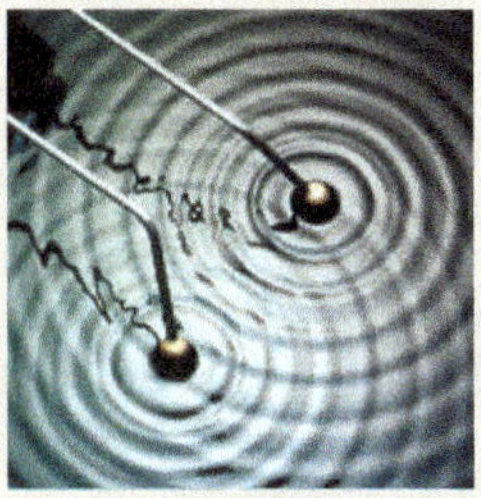

These water waves are exhibiting superposition as the ripples pass through each other.

You'll learn to analyze the response of the medium when two waves pass through each other.

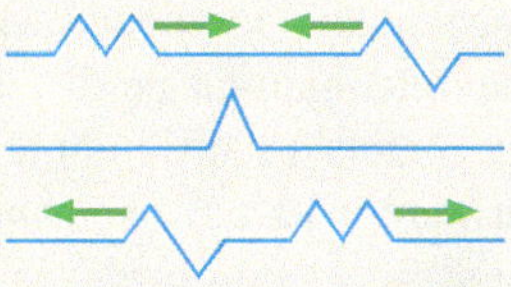

◀ **Looking Back**
Sections 20.2–20.4 Properties of traveling waves

Interference

When two sources emit waves of the same wavelength and frequency, the overlapped waves create an **interference pattern.**

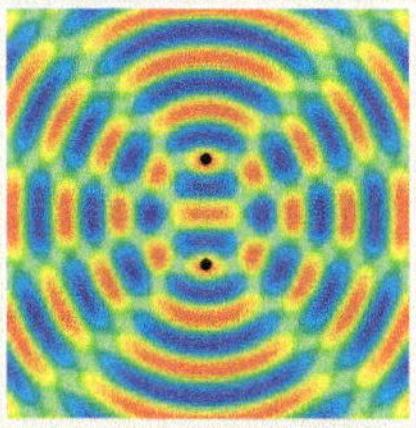

You'll learn to interpret interference patterns such as this one. The two black dots are the sources of the waves.

Constructive interference occurs where the waves add to make a larger wave. *Destructive interference* is where the waves cancel to make a smaller wave.

Applications

You'll learn how standing waves determine the notes of a guitar and other musical instruments...

...and how interference is used to design antireflection coatings for lenses.

◀ **Looking Back**
Section 20.5 Sound waves

Beats

The superposition of two waves of slightly different frequency produces a soft-loud-soft-loud-... modulation of the intensity called **beats.**

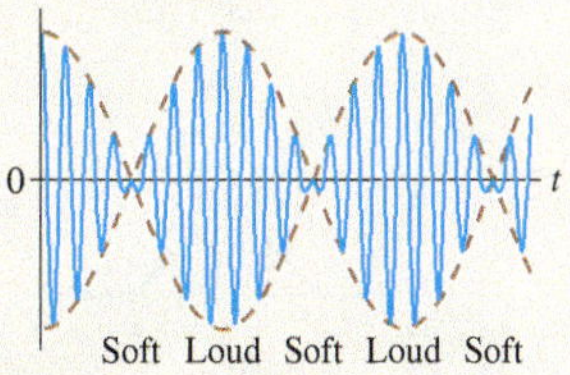

Beats are easily demonstrated with sound waves, but the concept is used in applications from ultrasonics to telecommunications.

21.1 The Principle of Superposition

FIGURE 21.1a shows two baseball players, Alan and Bill, at batting practice. Unfortunately, someone has turned the pitching machines so that pitching machine A throws baseballs toward Bill while machine B throws toward Alan. If two baseballs are launched at the same time, and with the same speed, they collide at the crossing point. Two particles cannot occupy the same point of space at the same time.

FIGURE 21.1 Unlike particles, two waves can pass directly through each other.

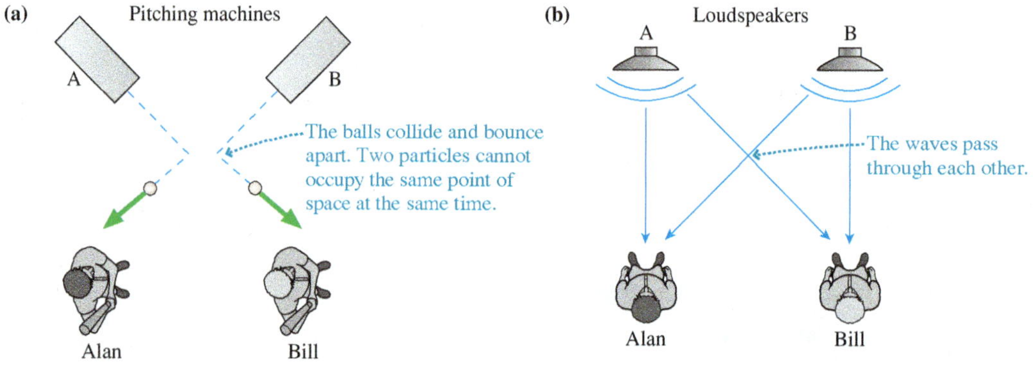

But waves, unlike particles, can pass directly through each other. In FIGURE 21.1b Alan and Bill are listening to the stereo system in the locker room after practice. Because both hear the music quite well, the sound wave that travels from loudspeaker A toward Bill must pass through the wave traveling from loudspeaker B toward Alan. What happens to the medium at a point where two waves are present simultaneously?

If wave 1 displaces a particle in the medium by D_1 and wave 2 *simultaneously* displaces it by D_2, the net displacement of the particle is simply $D_1 + D_2$. This is a very important idea because it tells us how to combine waves. It is known as the *principle of superposition.*

FIGURE 21.2 The superposition of two waves on a string as they pass through each other.

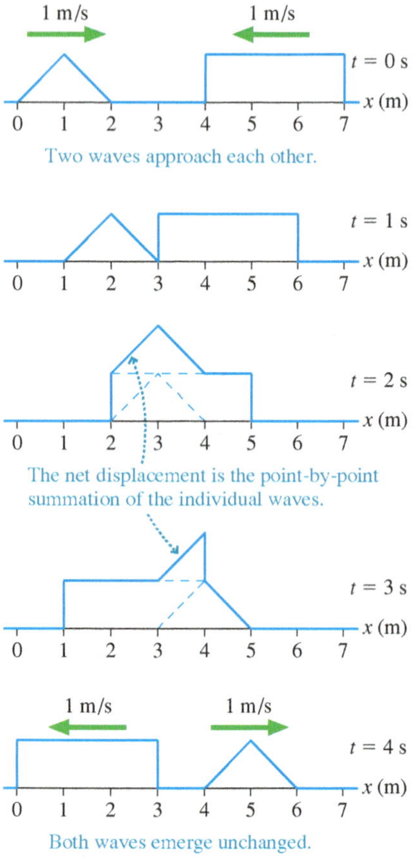

> **Principle of superposition** When two or more waves are *simultaneously* present at a single point in space, the displacement of the medium at that point is the sum of the displacements due to each individual wave.

When one object is placed on top of another, the two are said to be *superimposed.* But through some quirk in the English language, the result of superimposing objects is called a *superposition,* without the syllable "im." When one wave is "placed" on top of another wave, we have a superposition of waves.

Mathematically, the net displacement of a particle in the medium is

$$D_{\text{net}} = D_1 + D_2 + \cdots = \sum_i D_i \qquad (21.1)$$

where D_i is the displacement that would be caused by wave i alone. We will make the simplifying assumption that the displacements of the individual waves are along the same line so that we can add displacements as scalars rather than vectors.

To use the principle of superposition you must know the displacement caused by each wave if traveling alone. Then you go through the medium *point by point* and add the displacements due to each wave *at that point* to find the net displacement at that point. The outcome will be different at each and every point in the medium because the displacements are different at each point.

To illustrate, FIGURE 21.2 shows snapshot graphs taken 1 s apart of two waves traveling at the same speed (1 m/s) in opposite directions along a string. The principle of superposition comes into play wherever the waves overlap. The solid line is the sum *at each point* of the two displacements at that point. This is the displacement that you would actually observe as the two waves pass through each other.

STOP TO THINK 21.1 Two pulses on a string approach each other at speeds of 1 m/s. What is the shape of the string at $t = 6$ s?

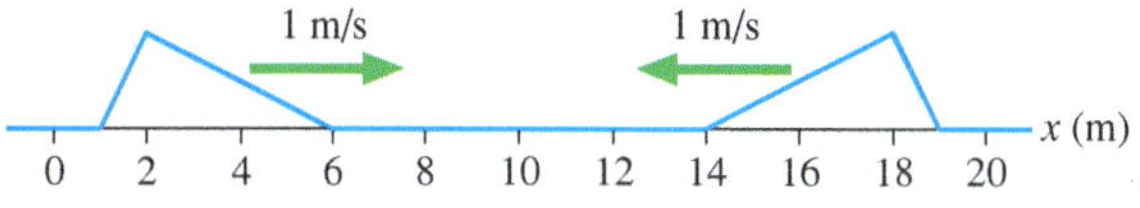

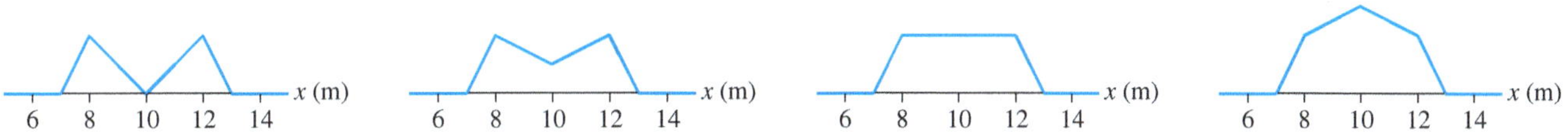

(a) (b) (c) (d)

21.2 Standing Waves

FIGURE 21.3 is a time-lapse photograph of a *standing wave* on a vibrating string. It's not obvious from the photograph, but this is actually a superposition of two waves. To understand this, consider two sinusoidal waves **with the same frequency, wavelength, and amplitude** traveling in opposite directions. For example, FIGURE 21.4a shows two waves on a string, and FIGURE 21.4b shows nine snapshot graphs, at intervals of $\frac{1}{8}T$. The dots identify two of the crests to help you visualize the wave movement.

At *each point,* the net displacement—the superposition—is found by adding the red displacement and the green displacement. FIGURE 21.4c shows the result. It is the wave you would actually observe. The blue dot shows that the blue wave is moving neither right nor left. The wave of Figure 21.4c is called a **standing wave** because the crests and troughs "stand in place" as the wave oscillates.

FIGURE 21.3 A vibrating string is an example of a standing wave.

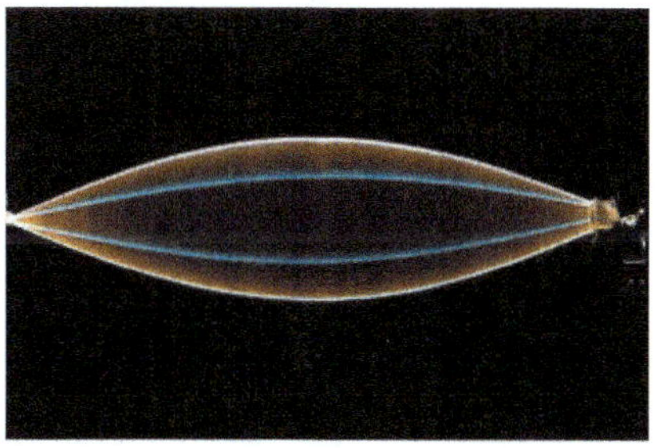

FIGURE 21.4 The superposition of two sinusoidal waves traveling in opposite directions.

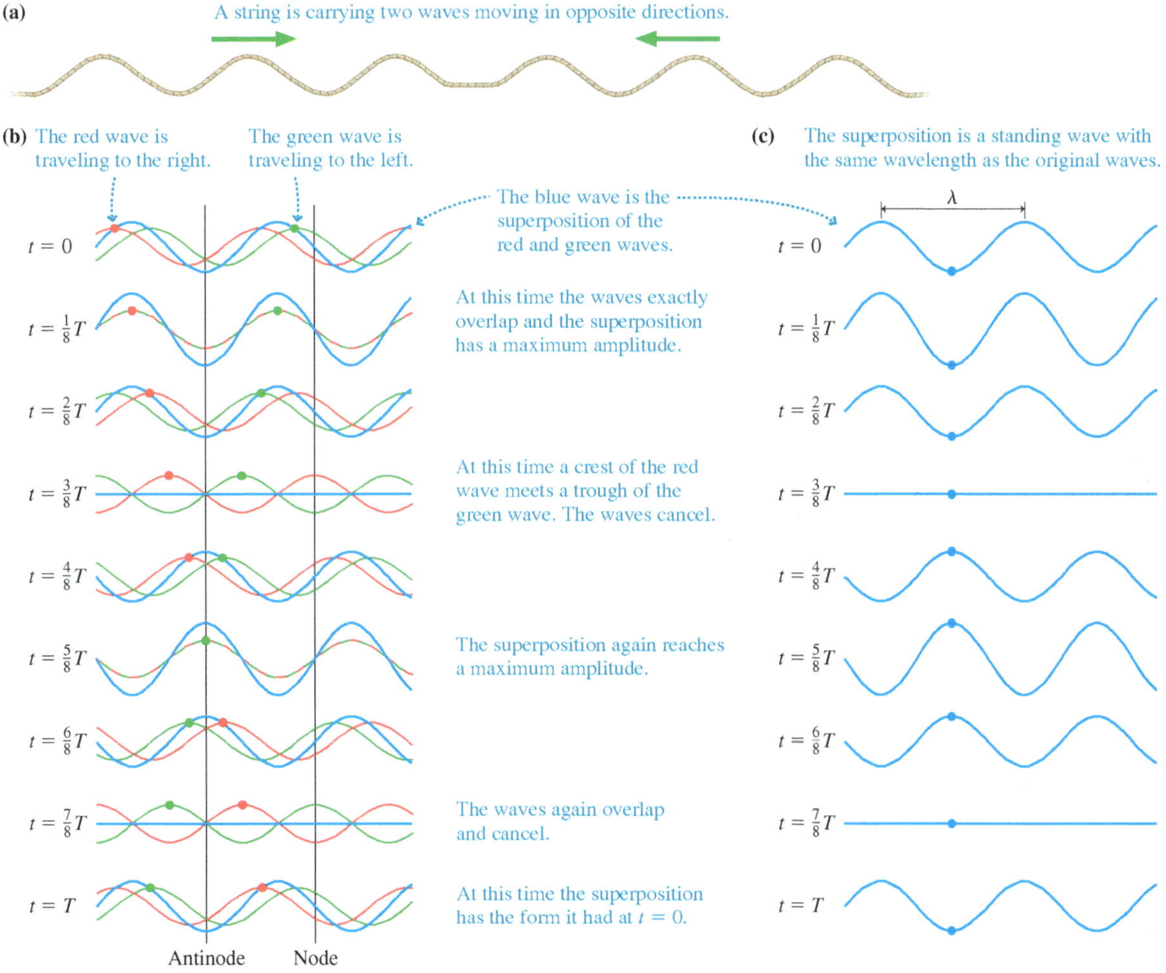

FIGURE 21.5 Standing waves are often represented as they would be seen in a time-lapse photograph.

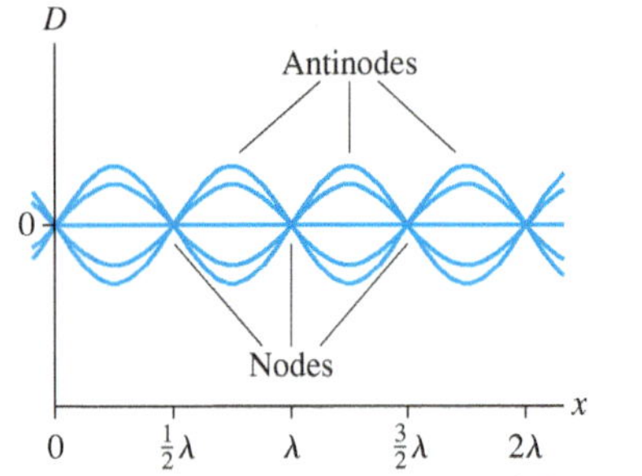

FIGURE 21.6 The intensity of a standing wave is maximum at the antinodes, zero at the nodes.

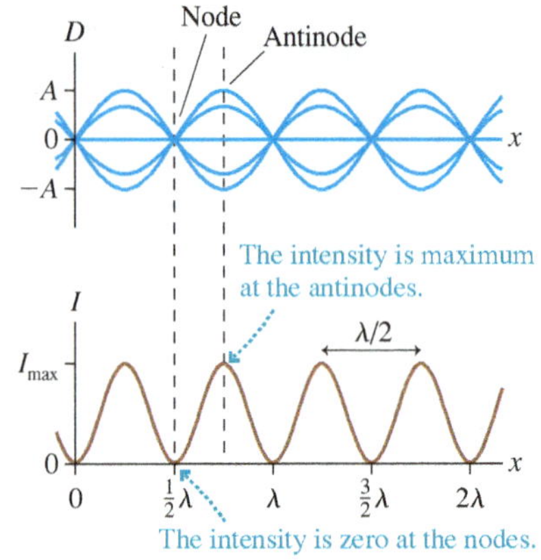

This photograph shows the Tacoma Narrows suspension bridge on the day in 1940 when it experienced a catastrophic standing-wave oscillation that led to its collapse. Aerodynamic forces caused the amplitude of a particular resonant mode of the bridge to increase dramatically until the bridge failed. In this photo, the red line shows the original line of the deck of the bridge. You can clearly see the large amplitude of the oscillation and the node at the center of the span.

Nodes and Antinodes

FIGURE 21.5 has collapsed the nine graphs of Figure 21.4b into a single graphical representation of a standing wave. Compare this to the Figure 21.3 photograph of a vibrating string. A striking feature of a standing-wave pattern is the existence of **nodes,** points that *never move!* **The nodes are spaced $\lambda/2$ apart.** Halfway between the nodes are the points where the particles in the medium oscillate with maximum displacement. These points of maximum amplitude are called **antinodes,** and you can see that they are also spaced $\lambda/2$ apart.

It seems surprising and counterintuitive that some particles in the medium have no motion at all. To understand how this happens, look carefully at the two traveling waves in Figure 21.4a. You will see that the nodes occur at points where at *every instant* of time the displacements of the two traveling waves have equal magnitudes but *opposite signs.* Thus the superposition of the displacements at these points is always zero. The antinodes correspond to points where the two displacements have equal magnitudes and the *same sign* at all times.

Two waves 1 and 2 are said to be *in phase* at a point where D_1 is *always* equal to D_2. The superposition at that point yields a wave whose amplitude is twice that of the individual waves. This is called a point of *constructive interference.* The antinodes of a standing wave are points of constructive interference between the two traveling waves.

In contrast, two waves are said to be *out of phase* at points where D_1 is *always* equal to $-D_2$. Their superposition gives a wave with zero amplitude—no wave at all! This is a point of *destructive interference.* The nodes of a standing wave are points of destructive interference. We will defer the main discussion of constructive and destructive interference until later in this chapter, but you'll then recognize that you're seeing constructive and destructive interference at the antinodes and nodes of a standing wave.

In Chapter 20 you learned that the *intensity* of a wave is proportional to the square of the amplitude: $I \propto A^2$. You can see in FIGURE 21.6 that maximum intensity occurs at the antinodes and that the intensity is zero at the nodes. If this is a sound wave, the loudness is maximum at the antinodes and zero at the nodes. A standing light wave is bright at the antinodes, dark at the nodes. The key idea is that **the intensity is maximum at points of constructive interference and zero at points of destructive interference.**

The Mathematics of Standing Waves

A sinusoidal wave traveling to the right along the x-axis with angular frequency $\omega = 2\pi f$, wave number $k = 2\pi/\lambda$, and amplitude a is

$$D_R = a\sin(kx - \omega t) \tag{21.2}$$

An equivalent wave traveling to the left is

$$D_L = a\sin(kx + \omega t) \tag{21.3}$$

We previously used the symbol A for the wave amplitude, but here we will use a lowercase a to represent the amplitude of each individual wave and reserve A for the amplitude of the net wave.

According to the principle of superposition, the net displacement of the medium when both waves are present is the sum of D_R and D_L:

$$D(x, t) = D_R + D_L = a\sin(kx - \omega t) + a\sin(kx + \omega t) \tag{21.4}$$

We can simplify Equation 21.4 by using the trigonometric identity

$$\sin(\alpha \pm \beta) = \sin\alpha\cos\beta \pm \cos\alpha\sin\beta$$

Doing so gives

$$\begin{aligned} D(x, t) &= a(\sin kx\cos\omega t - \cos kx\sin\omega t) + a(\sin kx\cos\omega t + \cos kx\sin\omega t) \\ &= (2a\sin kx)\cos\omega t \end{aligned} \tag{21.5}$$

It is useful to write Equation 21.5 as

$$D(x, t) = A(x)\cos\omega t \qquad (21.6)$$

where the **amplitude function** $A(x)$ is defined as

$$A(x) = 2a\sin kx \qquad (21.7)$$

The amplitude reaches a maximum value $A_{max} = 2a$ at points where $\sin kx = 1$.

The displacement $D(x, t)$ given by Equation 21.6 is neither a function of $x - vt$ nor a function of $x + vt$; hence it is *not* a traveling wave. Instead, the $\cos\omega t$ term in Equation 21.6 describes a medium in which each point oscillates in simple harmonic motion with frequency $f = \omega/2\pi$. The function $A(x) = 2a\sin kx$ gives the amplitude of the oscillation for a particle at position x.

FIGURE 21.7 graphs Equation 21.6 at several different instants of time. Notice that the graphs are identical to those of Figure 21.5, showing us that Equation 21.6 is the mathematical description of a standing wave.

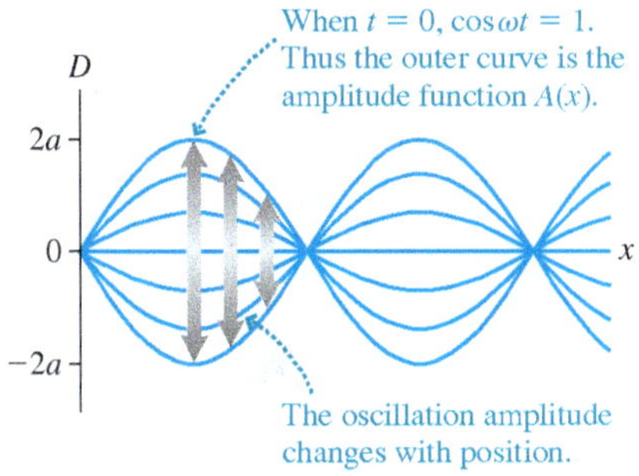

FIGURE 21.7 The net displacement resulting from two counter-propagating sinusoidal waves.

The nodes of the standing wave are the points at which the amplitude is zero. They are located at positions x for which

$$A(x) = 2a\sin kx = 0 \qquad (21.8)$$

The sine function is zero if the angle is an integer multiple of π rad, so Equation 21.8 is satisfied if

$$kx_m = \frac{2\pi x_m}{\lambda} = m\pi \qquad m = 0, 1, 2, 3, \ldots \qquad (21.9)$$

Thus the position x_m of the mth node is

$$x_m = m\frac{\lambda}{2} \qquad m = 0, 1, 2, 3, \ldots \qquad (21.10)$$

You can see that the spacing between two adjacent nodes is $\lambda/2$, in agreement with Figure 21.6. The nodes are *not* spaced by λ, as you might have expected.

EXAMPLE 21.1 Node spacing on a string

A very long string has a linear density of 5.0 g/m and is stretched with a tension of 8.0 N. 100 Hz waves with amplitudes of 2.0 mm are generated at the ends of the string.

a. What is the node spacing along the resulting standing wave?
b. What is the maximum displacement of the string?

MODEL Two counter-propagating waves of equal frequency create a standing wave.

VISUALIZE The standing wave will look like Figure 21.5.

SOLVE a. The speed of the waves on the string is

$$v = \sqrt{\frac{T_s}{\mu}} = \sqrt{\frac{8.0\text{ N}}{0.0050\text{ kg/m}}} = 40\text{ m/s}$$

and the wavelength is

$$\lambda = \frac{v}{f} = \frac{40\text{ m/s}}{100\text{ Hz}} = 0.40\text{ m} = 40\text{ cm}$$

Thus the spacing between adjacent nodes is $\lambda/2 = 20$ cm.

b. The maximum displacement is $A_{max} = 2a = 4.0$ mm.

21.3 Standing Waves on a String

Wiggling both ends of a very long string is not a practical way to generate standing waves. Instead, as in the photograph in Figure 21.3, standing waves are usually seen on a string that is fixed at both ends. To understand why this condition causes standing waves, we need to examine what happens when a traveling wave encounters a discontinuity.

FIGURE 21.8a on the next page shows a *discontinuity* between a string with a larger linear density and one with a smaller linear density. The tension is the same in both strings, so the wave speed is slower on the left, faster on the right. Whenever a wave encounters a discontinuity, some of the wave's energy is *transmitted* forward and some is *reflected.*

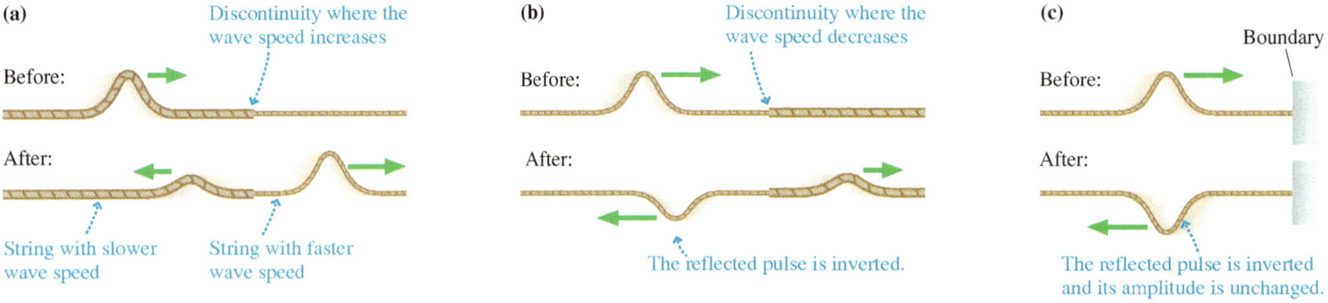

FIGURE 21.8 A wave reflects when it encounters a discontinuity or a boundary.

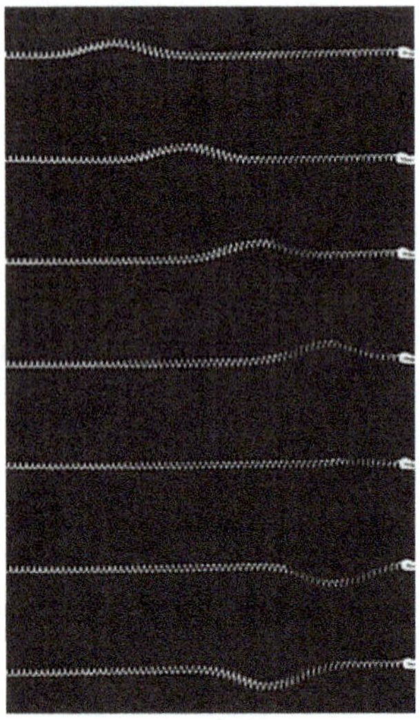

FIGURE 21.9 A strobe photo of a pulse traveling along a rope-like spring.

Light waves exhibit an analogous behavior when they encounter a piece of glass. Most of the light wave's energy is transmitted through the glass, which is why glass is transparent, but a small amount of energy is reflected. That is how you see your reflection dimly in a storefront window.

In FIGURE 21.8b, an incident wave encounters a discontinuity at which the wave speed decreases. In this case, the reflected pulse is *inverted.* A positive displacement of the incident wave becomes a negative displacement of the reflected wave. Because $\sin(\phi + \pi) = -\sin\phi$, we say that the reflected wave has a *phase change of* π *upon reflection.* This aspect of reflection will be important later in the chapter when we look at the interference of light waves.

The wave in FIGURE 21.8c reflects from a *boundary.* You can think of this as Figure 21.8b in the limit that the string on the right becomes infinitely massive. Thus the reflection in Figure 21.8c looks like that of Figure 21.8b with one exception: Because there is no transmitted wave, *all* the wave's energy is reflected. Hence **the amplitude of a wave reflected from a boundary is unchanged.** FIGURE 21.9 is a sequence of strobe photos in which you see a pulse on a rope-like spring reflecting from a boundary at the right of the photo. The reflected pulse is inverted but otherwise unchanged.

Creating Standing Waves

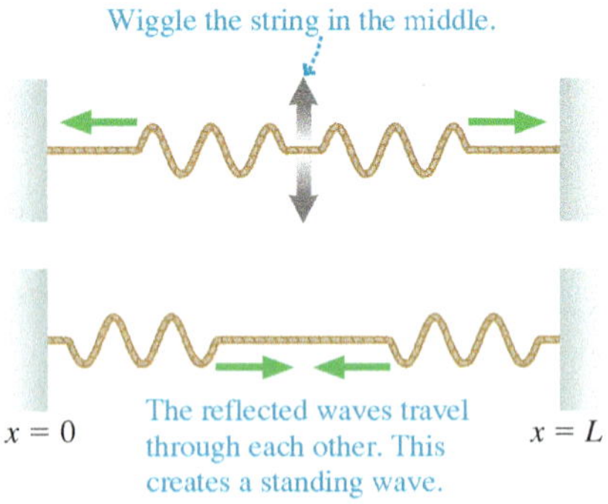

FIGURE 21.10 Reflections at the two boundaries cause a standing wave on the string.

FIGURE 21.10 shows a string of length L tied at $x = 0$ and $x = L$. If you wiggle the string in the middle, sinusoidal waves travel outward in both directions and soon reach the boundaries. Because the speed of a reflected wave does not change, **the wavelength and frequency of a reflected sinusoidal wave are unchanged.** Consequently, reflections at the ends of the string cause two waves of *equal amplitude and wavelength* to travel in opposite directions along the string. As we've just seen, these are the conditions that cause a standing wave!

To connect the mathematical analysis of standing waves in Section 21.2 with the physical reality of a string tied down at the ends, we need to impose *boundary conditions.* A **boundary condition** is a mathematical statement of any constraint that *must* be obeyed at the boundary or edge of a medium. Because the string is tied down at the ends, the displacements at $x = 0$ and $x = L$ must be zero at all times. Thus the standing-wave boundary conditions are $D(x = 0, t) = 0$ and $D(x = L, t) = 0$. Stated another way, we require nodes at both ends of the string.

We found that the displacement of a standing wave is $D(x, t) = (2a\sin kx)\cos\omega t$. This equation already satisfies the boundary condition $D(x = 0, t) = 0$. That is, the origin has already been located at a node. The second boundary condition, at $x = L$, requires $D(x = L, t) = 0$. This condition will be met at all times if

$$2a\sin kL = 0 \qquad \text{(boundary condition at } x = L\text{)} \tag{21.11}$$

Equation 21.11 will be true if $\sin kL = 0$, which in turn requires

$$kL = \frac{2\pi L}{\lambda} = m\pi \qquad m = 1, 2, 3, 4, \ldots \tag{21.12}$$

kL must be a multiple of $m\pi$, but $m = 0$ is excluded because L can't be zero.

For a string of fixed length L, the only quantity in Equation 21.12 that can vary is λ. That is, the boundary condition is satisfied only if the wavelength has one of the values

$$\lambda_m = \frac{2L}{m} \qquad m = 1, 2, 3, 4, \ldots \qquad (21.13)$$

A standing wave can exist on the string *only* if its wavelength is one of the values given by Equation 21.13. The mth possible wavelength $\lambda_m = 2L/m$ is just the right size so that its mth node is located at the end of the string (at $x = L$).

NOTE ▶ Other wavelengths, which would be perfectly acceptable wavelengths for a traveling wave, cannot exist as a *standing* wave of length L because they cannot meet the boundary conditions requiring a node at each end of the string. ◀

If standing waves are possible only for certain wavelengths, then only a few specific oscillation frequencies are allowed. Because $\lambda f = v$ for a sinusoidal wave, the oscillation frequency corresponding to wavelength λ_m is

$$f_m = \frac{v}{\lambda_m} = \frac{v}{2L/m} = m\frac{v}{2L} \qquad m = 1, 2, 3, 4, \ldots \qquad (21.14)$$

The lowest allowed frequency

$$f_1 = \frac{v}{2L} \qquad \text{(fundamental frequency)} \qquad (21.15)$$

which corresponds to wavelength $\lambda_1 = 2L$, is called the **fundamental frequency** of the string. The allowed frequencies can be written in terms of the fundamental frequency as

$$f_m = mf_1 \qquad m = 1, 2, 3, 4, \ldots \qquad (21.16)$$

The allowed standing-wave frequencies are all integer multiples of the fundamental frequency. The higher-frequency standing waves are called **harmonics,** with the $m = 2$ wave at frequency f_2 called the *second harmonic,* the $m = 3$ wave called the *third harmonic,* and so on.

FIGURE 21.11 graphs the first four possible standing waves on a string of fixed length L. These possible standing waves are called the **modes** of the string, or sometimes the *normal modes.* Each mode, numbered by the integer m, has a unique wavelength and frequency. Keep in mind that these drawings simply show the *envelope,* or outer edge, of the oscillations. The string is continuously oscillating at all positions between these edges, as we showed in more detail in Figure 21.5.

There are three things to note about the modes of a string.

1. m is the number of *antinodes* on the standing wave, not the number of nodes. You can tell a string's mode of oscillation by counting the number of antinodes.
2. The *fundamental mode,* with $m = 1$, has $\lambda_1 = 2L$, not $\lambda_1 = L$. Only half of a wavelength is contained between the boundaries, a direct consequence of the fact that the spacing between nodes is $\lambda/2$.
3. The frequencies of the normal modes form a series: $f_1, 2f_1, 3f_1, 4f_1, \ldots$. The fundamental frequency f_1 can be found as the *difference* between the frequencies of any two adjacent modes. That is, $f_1 = \Delta f = f_{m+1} - f_m$.

FIGURE 21.12 is a time-exposure photograph of the $m = 3$ standing wave on a string. The nodes and antinodes are quite distinct. The string vibrates three times faster for the $m = 3$ mode than for the fundamental $m = 1$ mode.

FIGURE 21.11 The first four modes for standing waves on a string of length L.

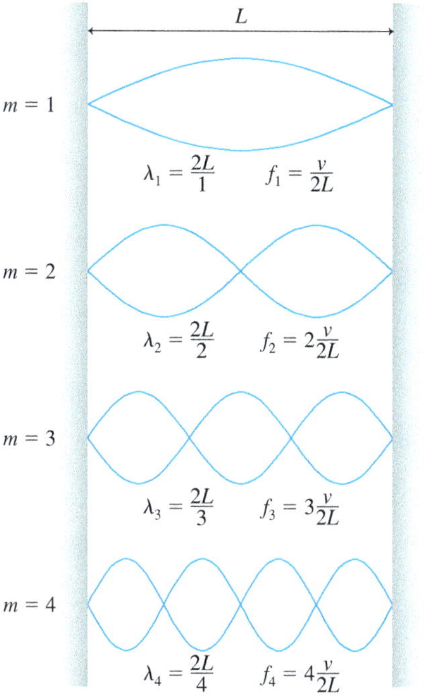

FIGURE 21.12 Time-exposure photograph of the $m = 3$ standing-wave mode on a stretched string.

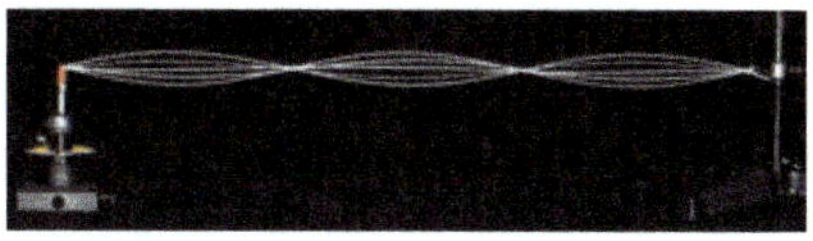

EXAMPLE 21.2 Measuring g

Standing-wave frequencies can be measured very accurately. Consequently, standing waves are often used in experiments to make accurate measurements of other quantities. One such experiment, shown in FIGURE 21.13, uses standing waves to measure the free-fall acceleration g. A heavy mass is suspended from a 1.65-m-long, 5.85 g steel wire; then an oscillating magnetic field (because steel is magnetic) is used to excite the $m = 3$ standing wave on the wire. Measuring the frequency for different masses yields the following data:

FIGURE 21.13 An experiment to measure g with standing waves.

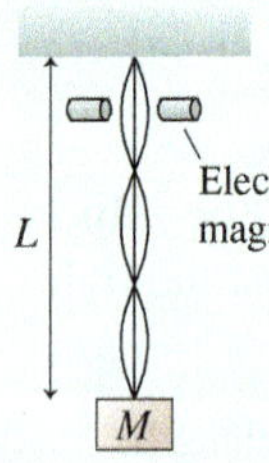

Mass (kg)	f_3 (Hz)
2.00	68
4.00	97
6.00	117
8.00	135
10.00	152

Analyze these data to determine the local value of g.

MODEL The hanging mass creates tension in the wire. This establishes the wave speed along the wire and thus the frequencies of standing waves. Masses of a few kg might stretch the wire a mm or so, but that doesn't change the length L until the third decimal place. The mass of the wire itself is insignificant in comparison to that of the hanging mass. We'll be justified in determining g to three significant figures.

SOLVE The frequency of the third harmonic is

$$f_3 = \frac{3}{2}\frac{v}{L}$$

The wave speed on the wire is

$$v = \sqrt{\frac{T_s}{\mu}} = \sqrt{\frac{Mg}{m/L}} = \sqrt{\frac{MgL}{m}}$$

where Mg is the weight of the hanging mass, and thus the tension in the wire, while m is the mass of the wire. Combining these two equations, we have

$$f_3 = \frac{3}{2}\sqrt{\frac{Mg}{mL}} = \frac{3}{2}\sqrt{\frac{g}{mL}}\sqrt{M}$$

Squaring both sides gives

$$f_3^2 = \frac{9g}{4mL}M$$

A graph of the square of the standing-wave frequency versus mass M should be a straight line passing through the origin with slope $9g/4mL$. We can use the experimental slope to determine g.

FIGURE 21.14 is a graph of f_3^2 versus M. The slope of the best-fit line is 2289 $\text{kg}^{-1}\text{s}^{-2}$, from which we find

$$g = \text{slope} \times \frac{4mL}{9}$$
$$= 2289\ \text{kg}^{-1}\text{s}^{-2} \times \frac{4(0.00585\ \text{kg})(1.65\ \text{m})}{9} = 9.82\ \text{m/s}^2$$

FIGURE 21.14 Graph of the data.

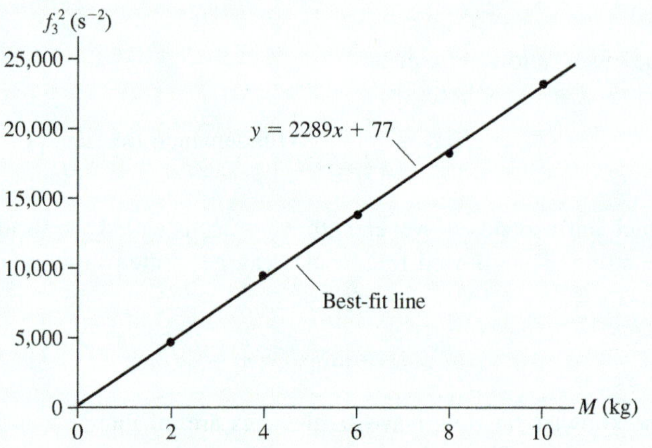

ASSESS The fact that the graph is linear and passes through the origin confirms our model. This is an important reason for having multiple data points rather than using only one mass.

STOP TO THINK 21.2 A standing wave on a string vibrates as shown at the right. Suppose the string tension is quadrupled while the frequency and the length of the string are held constant. Which standing-wave pattern is produced?

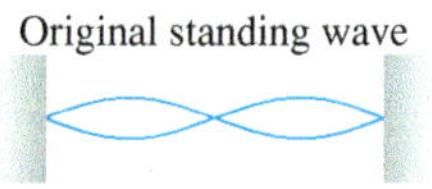

Standing Electromagnetic Waves

Because electromagnetic waves are transverse waves, a standing electromagnetic wave is very much like a standing wave on a string. Standing electromagnetic waves can be established between two parallel mirrors that reflect light back and forth. The mirrors are boundaries, analogous to the boundaries at the ends of a string. In fact, this is exactly how a laser operates. The two facing mirrors in FIGURE 21.15 form what is called a *laser cavity*.

Because the mirrors act like the points to which a string is tied, the light wave must have a node at the surface of each mirror. One of the mirrors is only partially reflective, to allow some light to escape and form the laser beam, but this doesn't affect the boundary condition.

Because the boundary conditions are the same, Equations 21.13 and 21.14 for λ_m and f_m apply to a laser just as they do to a vibrating string. The primary difference is the size of the wavelength. A typical laser cavity has a length $L \approx 30$ cm, and visible light has a wavelength $\lambda \approx 600$ nm. The standing light wave in a laser cavity has a mode number m that is approximately

$$m = \frac{2L}{\lambda} \approx \frac{2 \times 0.30\text{ m}}{6.00 \times 10^{-7}\text{ m}} = 1{,}000{,}000$$

In other words, the standing light wave inside a laser cavity has approximately one million antinodes! This is a consequence of the very short wavelength of light.

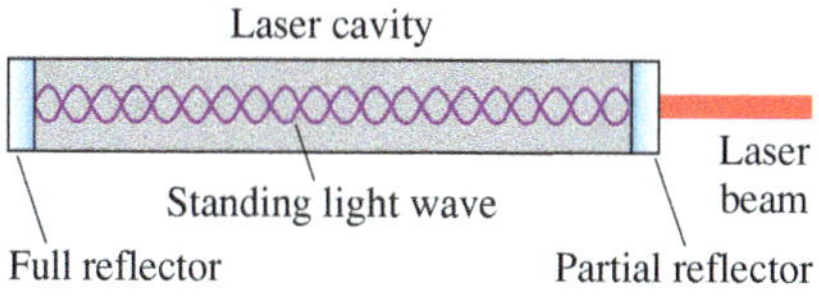

FIGURE 21.15 A laser contains a standing light wave between two parallel mirrors.

EXAMPLE 21.3 The standing light wave inside a laser

Helium-neon lasers emit the red laser light commonly used in classroom demonstrations and supermarket checkout scanners. A helium-neon laser operates at a wavelength of precisely 632.9924 nm when the spacing between the mirrors is 310.372 mm.

a. In which mode does this laser operate?
b. What is the next longest wavelength that could form a standing wave in this laser cavity?

MODEL The light wave forms a standing wave between the two mirrors.

VISUALIZE The standing wave looks like Figure 21.15.

SOLVE a. We can use $\lambda_m = 2L/m$ to find that m (the mode) is

$$m = \frac{2L}{\lambda_m} = \frac{2(0.310372\text{ m})}{6.329924 \times 10^{-7}\text{ m}} = 980{,}650$$

There are 980,650 antinodes in the standing light wave.

b. The next longest wavelength that can fit in this laser cavity will have one fewer node. It will be the $m = 980{,}649$ mode and its wavelength will be

$$\lambda = \frac{2L}{m} = \frac{2(0.310372\text{ m})}{980{,}649} = 632.9930\text{ nm}$$

ASSESS The wavelength increases by a mere 0.0006 nm when the mode number is decreased by 1.

Microwaves, with a wavelength of a few centimeters, can also set up standing waves. This is not always good. If the microwaves in a microwave oven form a standing wave, there are nodes where the electromagnetic field intensity is always zero. These nodes cause cold spots where the food does not heat. Although designers of microwave ovens try to prevent standing waves, ovens usually do have cold spots spaced $\lambda/2$ apart at nodes in the microwave field. A turntable in a microwave oven keeps the food moving so that no part of your dinner remains at a node.

21.4 Standing Sound Waves and Musical Acoustics

A long, narrow column of air, such as the air in a tube or pipe, can support a *longitudinal* standing sound wave. Longitudinal waves are somewhat trickier than string waves because a graph—showing displacement *parallel* to the tube—is not a picture of the wave.

To illustrate the ideas, FIGURE 21.16 on the next page is a series of three graphs and pictures that show the $m = 2$ standing wave inside a column of air closed at both ends. We call this a *closed-closed tube.* The air at the closed ends cannot oscillate because the air molecules are pressed up against the wall, unable to move; hence **a closed end of a column of air must be a displacement node.** Thus the boundary conditions—nodes at the ends—are the same as for a standing wave on a string.

Although the graph looks familiar, it is now a graph of *longitudinal* displacement. At $t = 0$, positive displacements in the left half and negative displacements in the right half cause all the air molecules to converge at the center of the tube. The density and pressure rise at the center and fall at the ends—a *compression* and *rarefaction* in the terminology of Chapter 20. A half cycle later, the molecules have rushed to the ends

FIGURE 21.16 This time sequence of graphs and pictures illustrates the $m = 2$ standing sound wave in a closed-closed tube of air.

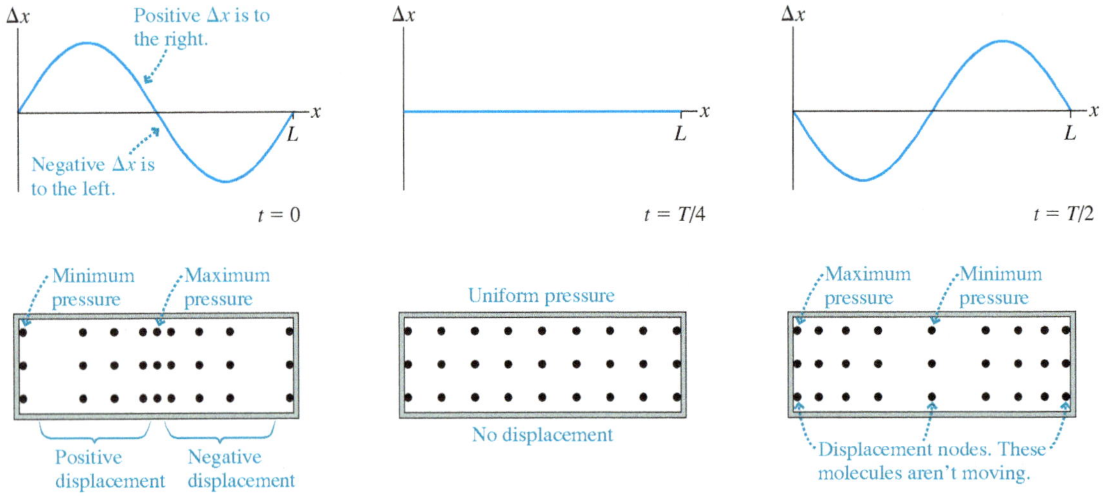

FIGURE 21.17 The $m = 2$ longitudinal standing wave can be represented as a displacement wave or as a pressure wave.

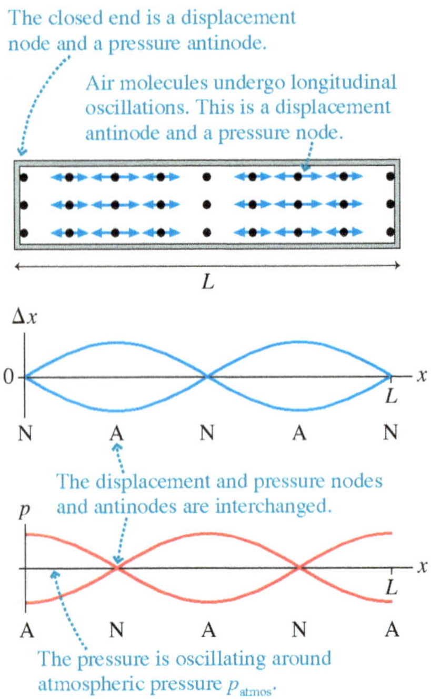

of the tube. Now the pressure is maximum at the ends, minimum in the center. Try to visualize the air molecules sloshing back and forth this way.

FIGURE 21.17 combines these illustrations into a single picture showing where the molecules are oscillating (antinodes) and where they're not (nodes). A graph of the displacement Δx looks just like the $m = 2$ graph of a standing wave on a string. Because the boundary conditions are the same, the possible wavelengths and frequencies of standing waves in a closed-closed tube are the same as for a string of the same length.

It is often useful to think of sound as a *pressure wave* rather than a displacement wave, and the bottom graph in Figure 21.17 shows the $m = 2$ pressure standing wave in a closed-closed tube. Notice that the pressure is oscillating around p_{atmos}, its equilibrium value. **The nodes and antinodes of the pressure wave are interchanged with those of the displacement wave,** and a careful study of Figure 21.16 reveals why. The gas molecules are alternately pushed up against the ends of the tube, then pulled away, causing the pressure at the closed ends to oscillate with maximum amplitude—an antinode.

EXAMPLE 21.4 Singing in the shower

A shower stall is 2.45 m (8 ft) tall. For what frequencies less than 500 Hz are there standing sound waves in the shower stall?

MODEL The shower stall, to a first approximation, is a column of air 2.45 m long. It is closed at the ends by the ceiling and floor. Assume a 20°C speed of sound.

VISUALIZE A standing sound wave will have nodes at the ceiling and the floor. The $m = 2$ mode will look like Figure 21.17 rotated 90°.

SOLVE The fundamental frequency for a standing sound wave in this air column is

$$f_1 = \frac{v}{2L} = \frac{343\text{ m/s}}{2(2.45\text{ m})} = 70\text{ Hz}$$

The possible standing-wave frequencies are integer multiples of the fundamental frequency. These are 70 Hz, 140 Hz, 210 Hz, 280 Hz, 350 Hz, 420 Hz, and 490 Hz.

ASSESS The many possible standing waves in a shower cause the sound to *resonate,* which helps explain why some people like to sing in the shower. Our approximation of the shower stall as a one-dimensional tube is actually a bit too simplistic. A three-dimensional analysis would find additional modes, making the "sound spectrum" even richer.

Air columns closed at both ends are of limited interest unless, as in Example 21.4, you are inside the column. Columns of air that *emit* sound are open at one or both ends. Many musical instruments fit this description. For example, a flute is a tube of air open at both ends. The flutist blows across one end to create a standing wave inside the tube, and a note of that frequency is emitted from both ends of the flute. (The blown end of a flute is open on the side, rather than across the tube. That is necessary for practical reasons of how flutes are played, but from a physics perspective this is the "end" of the tube because it opens the tube to the atmosphere.) A trumpet, however, is open at the bell end but is *closed* by the player's lips at the other end.

You saw earlier that a wave is partially transmitted and partially reflected at a discontinuity. When a sound wave traveling through a tube of air reaches an open end, some of the wave's energy is transmitted out of the tube to become the sound that you hear and some portion of the wave is reflected back into the tube. These reflections, analogous to the reflection of a string wave from a boundary, allow standing sound waves to exist in a tube of air that is open at one or both ends.

Not surprisingly, the *boundary condition* at the open end of a column of air is not the same as the boundary condition at a closed end. The air pressure at the open end of a tube is constrained to match the atmospheric pressure of the surrounding air. Consequently, the open end of a tube must be a pressure node. Because pressure nodes and antinodes are interchanged with those of the displacement wave, **an open end of an air column is required to be a displacement antinode.** (A careful analysis shows that the antinode is actually just outside the open end, but for our purposes we'll assume the antinode is exactly at the open end.)

FIGURE 21.18 shows displacement and pressure graphs of the first three standing-wave modes of a tube closed at both ends (a *closed-closed tube*), a tube open at both ends (an *open-open tube*), and a tube open at one end but closed at the other (an *open-closed tube*), all with the same length L. Notice the pressure and displacement boundary conditions. The standing wave in the open-open tube looks like the closed-closed tube except that the positions of the nodes and antinodes are interchanged. In both cases there are m half-wavelength segments between the ends; thus the wavelengths and frequencies of an open-open tube and a closed-closed tube are the same as those of a string tied at both ends:

FIGURE 21.18 The first three standing sound wave modes in columns of air with different boundary conditions.

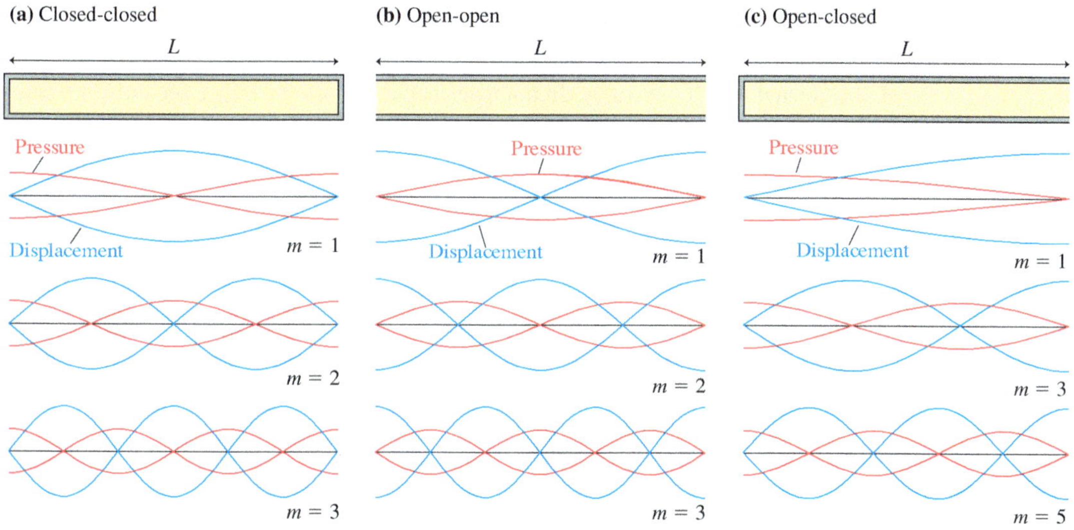

$$\begin{cases} \lambda_m = \dfrac{2L}{m} \\ f_m = m\dfrac{v}{2L} = mf_1 \end{cases} \qquad \begin{matrix} m = 1, 2, 3, 4, \ldots \\ \text{(open-open or closed-closed tube)} \end{matrix} \qquad (21.17)$$

The open-closed tube is different. The fundamental mode has only one-quarter of a wavelength in a tube of length L; hence the $m = 1$ wavelength is $\lambda_1 = 4L$. This is twice the λ_1 wavelength of an open-open or a closed-closed tube. Consequently, **the fundamental frequency of an open-closed tube is half that of an open-open or a closed-closed tube of the same length.** It will be left as a homework problem for you to show that the possible wavelengths and frequencies of an open-closed tube of length L are

$$\begin{cases} \lambda_m = \dfrac{4L}{m} \\ f_m = m\dfrac{v}{4L} = mf_1 \end{cases} \qquad \begin{array}{l} m = 1, 3, 5, 7, \ldots \\ \text{(open-closed tube)} \end{array} \qquad (21.18)$$

Notice that m in Equation 21.18 takes on only *odd* values.

EXAMPLE 21.5 Resonances of the ear canal

The eardrum, which transmits sounds vibrations to the sensory organs of the inner ear, lies at the end of the ear canal. For adults, the ear canal is about 2.5 cm in length. What frequency standing waves can occur in the ear canal that are within the range of human hearing? The speed of sound in the warm air of the ear canal is 350 m/s.

MODEL The ear canal is open to the air at one end, closed by the eardrum at the other. We can model it as an open-closed tube. The standing waves will be those of Figure 21.18c.

SOLVE The lowest standing-wave frequency is the fundamental frequency for a 2.5-cm-long open-closed tube:

$$f_1 = \frac{v}{4L} = \frac{350 \text{ m/s}}{4(0.025 \text{ m})} = 3500 \text{ Hz}$$

Standing waves also occur at the harmonics, but an open-closed tube has only odd harmonics. These are

$$f_3 = 3f_1 = 10{,}500 \text{ Hz}$$
$$f_5 = 5f_1 = 17{,}500 \text{ Hz}$$

Higher harmonics are beyond the range of human hearing, as discussed in Section 20.5.

ASSESS The ear canal is short, so we expected the standing-wave frequencies to be relatively high. The air in your ear canal responds readily to sounds at these frequencies—what we call a *resonance* of the ear canal—and transmits theses sounds to the eardrum. Consequently, your ear actually is slightly more sensitive to sounds with frequencies around 3500 Hz and 10,500 Hz than to sounds at nearby frequencies.

STOP TO THINK 21.3 An open-open tube of air supports standing waves at frequencies of 300 Hz and 400 Hz and at no frequencies between these two. The second harmonic of this tube has frequency

a. 100 Hz b. 200 Hz c. 400 Hz d. 600 Hz e. 800 Hz

Musical Instruments

An important application of standing waves is to musical instruments. Instruments such as the guitar, the piano, and the violin have strings fixed at the ends and tightened to create tension. A disturbance generated on the string by plucking, striking, or bowing it creates a standing wave on the string.

The fundamental frequency of a vibrating string is

$$f_1 = \frac{v}{2L} = \frac{1}{2L}\sqrt{\frac{T_s}{\mu}}$$

where T_s is the tension in the string and μ is its linear density. The fundamental frequency is the musical note you hear when the string is sounded. Increasing the tension in the string raises the fundamental frequency, which is how stringed instruments are tuned.

NOTE ▶ v is the wave speed *on the string,* not the speed of sound in air. ◀

For the guitar or the violin, the strings are all the same length and under approximately the same tension. Were that not the case, the neck of the instrument would tend to twist toward the side of higher tension. The strings have different frequencies because they differ in linear density: The lower-pitched strings are "fat" while the higher-pitched strings are "skinny." This difference changes the frequency by changing the wave speed. *Small* adjustments are then made in the tension to bring each string to the exact desired frequency. Once the instrument is tuned, you play it by using your fingertips to alter the effective length of the string. As you shorten the string's length, the frequency and pitch go up.

The strings on a harp vibrate as standing waves. Their frequencies determine the notes that you hear.

A piano covers a much wider range of frequencies than a guitar or violin. This range cannot be produced by changing only the linear densities of the strings. The high end would have strings too thin to use without breaking, and the low end would have solid rods rather than flexible wires! So a piano is tuned through a combination of changing the linear density *and* the length of the strings. The bass note strings are not only fatter, they are also longer.

With a wind instrument, blowing into the mouthpiece creates a standing sound wave inside a tube of air. The player changes the notes by using her fingers to cover holes or open valves, changing the length of the tube and thus its frequency. The fact that the holes are on the side makes very little difference; the first open hole becomes an antinode because the air is free to oscillate in and out of the opening.

A wind instrument's frequency depends on the speed of sound *inside* the instrument. But the speed of sound depends on the temperature of the air. When a wind player first blows into the instrument, the air inside starts to rise in temperature. This increases the sound speed, which in turn raises the instrument's frequency for each note until the air temperature reaches a steady state. Consequently, wind players must "warm up" before tuning their instrument.

Many wind instruments have a "buzzer" at one end of the tube, such as a vibrating reed on a saxophone or vibrating lips on a trombone. Buzzers generate a continuous range of frequencies rather than single notes, which is why they sound like a "squawk" if you play on just the mouthpiece without the rest of the instrument. When a buzzer is connected to the body of the instrument, most of those frequencies cause no response of the air molecules. But the frequency from the buzzer that matches the fundamental frequency of the instrument causes the buildup of a large-amplitude response at just that frequency—a standing-wave resonance. This is the energy input that generates and sustains the musical note.

EXAMPLE 21.6 Flutes and clarinets

A clarinet is 66.0 cm long. A flute is nearly the same length, with 63.5 cm between the hole the player blows across and the end of the flute. What are the frequencies of the lowest note and the next higher harmonic on a flute and on a clarinet? The speed of sound in warm air is 350 m/s.

MODEL The flute is an open-open tube, open at the end as well as at the hole the player blows across. A clarinet is an open-closed tube because the player's lips and the reed seal the tube at the upper end.

SOLVE The lowest frequency is the fundamental frequency. For the flute, an open-open tube, this is

$$f_1 = \frac{v}{2L} = \frac{350\text{ m/s}}{2(0.635\text{ m})} = 275\text{ Hz}$$

The clarinet, an open-closed tube, has

$$f_1 = \frac{v}{4L} = \frac{350\text{ m/s}}{4(0.660\text{ m})} = 133\text{ Hz}$$

The next higher harmonic on the flute's open-open tube is $m = 2$ with frequency $f_2 = 2f_1 = 550$ Hz. An open-closed tube has only odd harmonics, so the next higher harmonic of the clarinet is $f_3 = 3f_1 = 399$ Hz.

ASSESS The clarinet plays a much lower note than the flute—musically, about an octave lower—because it is an open-closed tube. It's worth noting that neither of our fundamental frequencies is exactly correct because our open-open and open-closed tube models are a bit too simplified to adequately describe a real instrument. However, both calculated frequencies are close because our models do capture the essence of the physics.

A vibrating string plays the musical note corresponding to the fundamental frequency f_1, so stringed instruments must use several strings to obtain a reasonable range of notes. In contrast, wind instruments can sound at the second or third harmonic of the tube of air (f_2 or f_3). These higher frequencies are sounded by *overblowing* (flutes, brass instruments) or with keys that open small holes to encourage the formation of an antinode at that point (clarinets, saxophones). The controlled use of these higher harmonics gives wind instruments a wide range of notes.

21.5 Interference in One Dimension

One of the most basic characteristics of waves is the ability of two waves to combine into a single wave whose displacement is given by the principle of superposition. The pattern resulting from the superposition of two waves is often called **interference.** A standing wave is the interference pattern produced when two waves of equal frequency travel in opposite directions. In this section we will look at the interference of two waves traveling in the *same* direction.

FIGURE 21.19 Two overlapped waves travel along the x-axis.

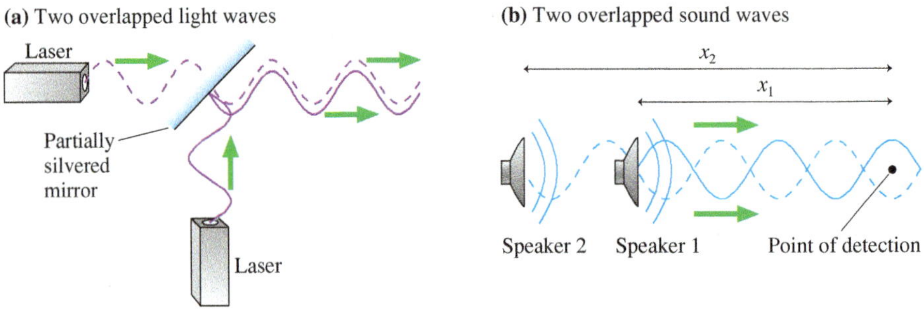

FIGURE 21.19a shows two light waves impinging on a partially silvered mirror. Such a mirror partially transmits and partially reflects each wave, causing two *overlapped* light waves to travel along the x-axis to the right of the mirror. Or consider the two loudspeakers in FIGURE 21.19b. The sound wave from loudspeaker 2 passes just to the side of loudspeaker 1; hence two overlapped sound waves travel to the right along the x-axis. We want to find out what happens when two overlapped waves travel in the same direction along the same axis.

Figure 21.19b shows a point on the x-axis where the overlapped waves are detected, either by your ear or by a microphone. This point is distance x_1 from loudspeaker 1 and distance x_2 from loudspeaker 2. (We will use loudspeakers and sound waves for most of our examples, but our analysis is valid for any wave.) What is the amplitude of the combined waves at this point?

FIGURE 21.20 Waves from three sources having phase constants $\phi_0 = 0$ rad, $\phi_0 = \pi/2$ rad, and $\phi_0 = \pi$ rad.

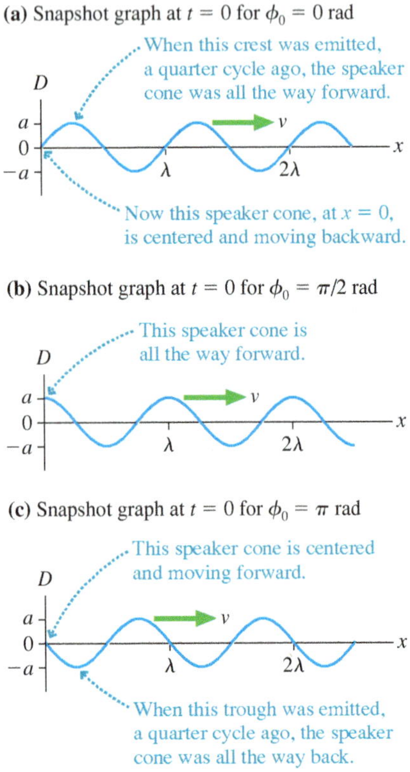

Throughout this section, **we will assume that the waves are sinusoidal, have the same frequency and amplitude, and travel to the right along the x-axis.** Thus we can write the displacements of the two waves as

$$D_1(x_1, t) = a\sin(kx_1 - \omega t + \phi_{10}) = a\sin\phi_1$$
$$D_2(x_2, t) = a\sin(kx_2 - \omega t + \phi_{20}) = a\sin\phi_2 \qquad (21.19)$$

where ϕ_1 and ϕ_2 are the *phases* of the waves. Both waves have the same wave number $k = 2\pi/\lambda$ and the same angular frequency $\omega = 2\pi f$.

The phase constants ϕ_{10} and ϕ_{20} are characteristics of *the sources,* not the medium. FIGURE 21.20 shows snapshot graphs at $t = 0$ of waves emitted by three sources with phase constants $\phi_0 = 0$ rad, $\phi_0 = \pi/2$ rad, and $\phi_0 = \pi$ rad. You can see that **the phase constant tells us what the source is doing at $t = 0$.** For example, a loudspeaker at its center position and moving backward at $t = 0$ has $\phi_0 = 0$ rad. Looking back at Figure 21.19b, you can see that loudspeaker 1 has phase constant $\phi_{10} = 0$ rad and loudspeaker 2 has $\phi_{20} = \pi$ rad.

NOTE ▶ We will often consider *identical sources,* by which we mean that $\phi_{20} = \phi_{10}$. That is, the sources oscillate in phase. ◀

Let's examine overlapped waves graphically before diving into the mathematics. FIGURE 21.21 shows two important situations. In part a, the crests of the two waves are aligned as they travel along the x-axis. In part b, the crests of one wave align with the troughs of the other wave. The graphs and the wave fronts are slightly displaced from each other so that you can see what each wave is doing, but the *physical situation* is one in which the waves are traveling *on top of* each other. Recall, from Chapter 20, that the wave fronts shown in the middle panel locate the crests of the waves.

The two waves of FIGURE 21.21a have the same displacement at every point: $D_1(x) = D_2(x)$. Two waves that are aligned crest to crest and trough to trough are said to be **in phase.** Waves that are in phase march along "in step" with each other.

When we combine two in-phase waves, using the principle of superposition, the net displacement at each point is twice the displacement of each individual wave. The superposition of two waves to create a traveling wave with an amplitude *larger* than either individual wave is called **constructive interference.** When the waves are exactly in phase, giving $A = 2a$, we have *maximum constructive interference.*

In FIGURE 21.21b, where the crests of one wave align with the troughs of the other, the waves march along "out of step" with $D_1(x) = -D_2(x)$ at every point. Two waves that are aligned crest to trough are said to be *180° out of phase* or, more generally, just **out of phase.** A superposition of two waves to create a wave with an amplitude smaller than either individual wave is called **destructive interference.** In this case, because $D_1 = -D_2$, the net displacement is *zero* at *every point* along the axis. The combination of two waves that cancel each other to give no wave is called *perfect destructive interference.*

NOTE ▶ Perfect destructive interference occurs only if the two waves have equal wavelengths and amplitudes, as we're assuming. Two waves of unequal amplitudes can interfere destructively, but the cancellation won't be perfect. ◀

FIGURE 21.21 Constructive and destructive interference of two waves traveling along the x-axis.

(a) Constructive interference

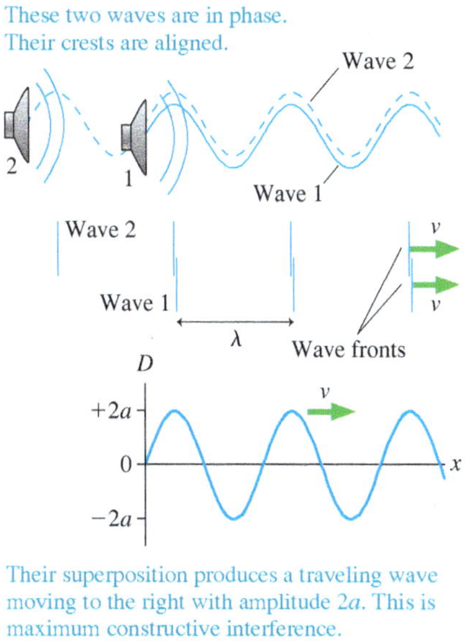

(b) Destructive interference

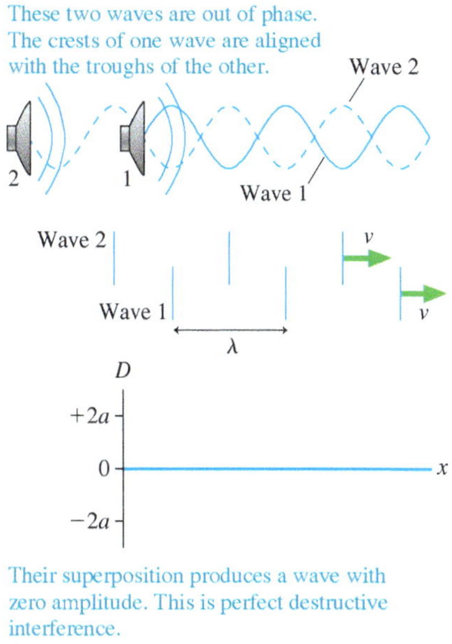

The Phase Difference

To understand interference, we need to focus on the *phases* of the two waves, which are

$$\begin{aligned} \phi_1 &= kx_1 - \omega t + \phi_{10} \\ \phi_2 &= kx_2 - \omega t + \phi_{20} \end{aligned} \tag{21.20}$$

The difference between the two phases ϕ_1 and ϕ_2, called the **phase difference** $\Delta\phi$, is

$$\begin{aligned} \Delta\phi = \phi_2 - \phi_1 &= (kx_2 - \omega t + \phi_{20}) - (kx_1 - \omega t + \phi_{10}) \\ &= k(x_2 - x_1) + (\phi_{20} - \phi_{10}) \\ &= 2\pi\frac{\Delta x}{\lambda} + \Delta\phi_0 \end{aligned} \tag{21.21}$$

You can see that there are two contributions to the phase difference. $\Delta x = x_2 - x_1$, the distance between the two sources, is called **path-length difference.** It is the extra distance traveled by wave 2 on the way to the point where the two waves are combined. $\Delta\phi_0 = \phi_{20} - \phi_{10}$ is the *inherent phase difference* between the sources.

The condition of being in phase, where crests are aligned with crests and troughs with troughs, is $\Delta\phi = 0$, 2π, 4π, or any integer multiple of 2π rad. Thus the condition for maximum constructive interference is

Maximum constructive interference:

$$\Delta\phi = 2\pi\frac{\Delta x}{\lambda} + \Delta\phi_0 = m \cdot 2\pi \text{ rad} \qquad m = 0, 1, 2, 3, \ldots \tag{21.22}$$

For identical sources, which have $\Delta\phi_0 = 0$ rad, maximum constructive interference occurs when $\Delta x = m\lambda$. That is, **two identical sources produce maximum constructive interference when the path-length difference is an integer number of wavelengths.**

FIGURE 21.22 Two identical sources one wavelength apart.

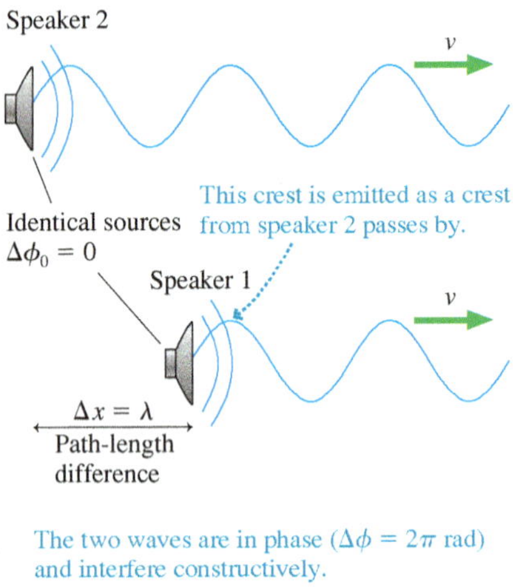

FIGURE 21.22 shows two identical sources (i.e., the two loudspeakers are doing the same thing at the same time), so $\Delta\phi_0 = 0$ rad. The path-length difference Δx is the extra distance traveled by the wave from loudspeaker 2 before it combines with loudspeaker 1. In this case, $\Delta x = \lambda$. Because a wave moves forward exactly one wavelength during one period, loudspeaker 1 emits a crest exactly as a crest of wave 2 passes by. The two waves are "in step," with $\Delta\phi = 2\pi$ rad, so the two waves interfere constructively to produce a wave of amplitude $2a$.

Perfect destructive interference, where the crests of one wave are aligned with the troughs of the other, occurs when two waves are *out of phase,* meaning that $\Delta\phi = \pi$, 3π, 5π, or any odd multiple of π rad. Thus the condition for perfect destructive interference is

Perfect destructive interference:

$$\Delta\phi = 2\pi\frac{\Delta x}{\lambda} + \Delta\phi_0 = \left(m + \frac{1}{2}\right) \cdot 2\pi \text{ rad} \qquad m = 0, 1, 2, 3, \ldots \qquad (21.23)$$

For identical sources, which have $\Delta\phi_0 = 0$ rad, perfect destructive interference occurs when $\Delta x = (m + \frac{1}{2})\lambda$. **That is, two identical sources produce perfect destructive interference when the path-length difference is a half-integer number of wavelengths.**

Two waves can be out of phase because the sources are located at different positions, because the sources themselves are out of phase, or because of a combination of these two. FIGURE 21.23 illustrates these ideas by showing three different ways in which two waves interfere destructively. Each of these three arrangements creates waves with $\Delta\phi = \pi$ rad.

FIGURE 21.23 Destructive interference three ways.

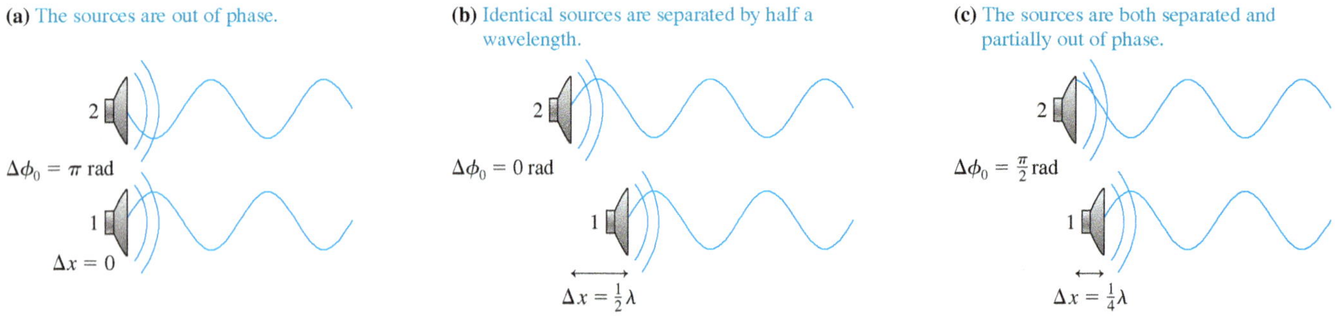

NOTE ▶ Don't confuse the phase difference of the waves ($\Delta\phi$) with the phase difference of the sources ($\Delta\phi_0$). It is $\Delta\phi$, the phase difference of the waves, that governs interference. ◀

EXAMPLE 21.7 Interference between two sound waves

You are standing in front of two side-by-side loudspeakers playing sounds of the same frequency. Initially there is almost no sound at all. Then one of the speakers is moved slowly away from you. The sound intensity increases as the separation between the speakers increases, reaching a maximum when the speakers are 0.75 m apart. Then, as the speaker continues to move, the intensity starts to decrease. What is the distance between the speakers when the sound intensity is again a minimum?

MODEL The changing sound intensity is due to the interference of two overlapped sound waves.

VISUALIZE Moving one speaker relative to the other changes the phase difference between the waves.

SOLVE A minimum sound intensity implies that the two sound waves are interfering destructively. Initially the loudspeakers are side by side, so the situation is as shown in Figure 21.23a with $\Delta x = 0$ and $\Delta\phi_0 = \pi$ rad. That is, the speakers themselves are out of phase. Moving one of the speakers does not change $\Delta\phi_0$, but it does change the path-length difference Δx and thus increases the overall phase difference $\Delta\phi$. Constructive interference, causing maximum intensity, is reached when

$$\Delta\phi = 2\pi\frac{\Delta x}{\lambda} + \Delta\phi_0 = 2\pi\frac{\Delta x}{\lambda} + \pi = 2\pi \text{ rad}$$

where we used $m = 1$ because this is the first separation giving constructive interference. The speaker separation at which this occurs is $\Delta x = \lambda/2$. This is the situation shown in FIGURE 21.24.

Because $\Delta x = 0.75$ m is $\lambda/2$, the sound's wavelength is $\lambda = 1.50$ m. The next point of destructive interference, with $m = 1$, occurs when

$$\Delta\phi = 2\pi\frac{\Delta x}{\lambda} + \Delta\phi_0 = 2\pi\frac{\Delta x}{\lambda} + \pi = 3\pi \text{ rad}$$

Thus the distance between the speakers when the sound intensity is again a minimum is

$$\Delta x = \lambda = 1.50 \text{ m}$$

ASSESS A separation of λ gives constructive interference for two *identical* speakers ($\Delta\phi_0 = 0$). Here the phase difference of π rad between the speakers (one is pushing forward as the other pulls back) gives destructive interference at this separation.

FIGURE 21.24 Two out-of-phase sources generate waves that are in phase if the sources are one half-wavelength apart.

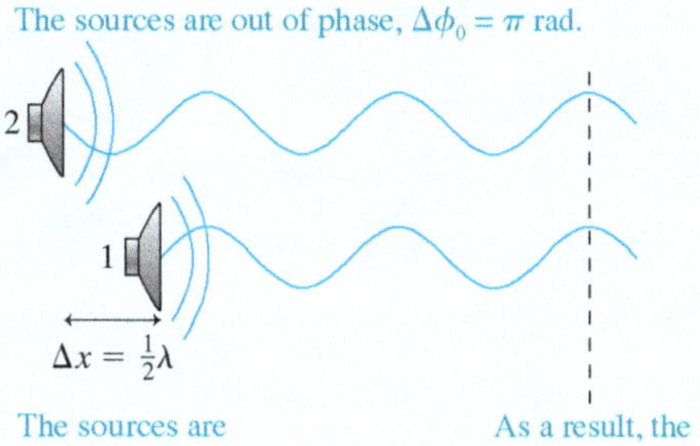

STOP TO THINK 21.4 Two loudspeakers emit waves with $\lambda = 2.0$ m. Speaker 2 is 1.0 m in front of speaker 1. What, if anything, can be done to cause constructive interference between the two waves?

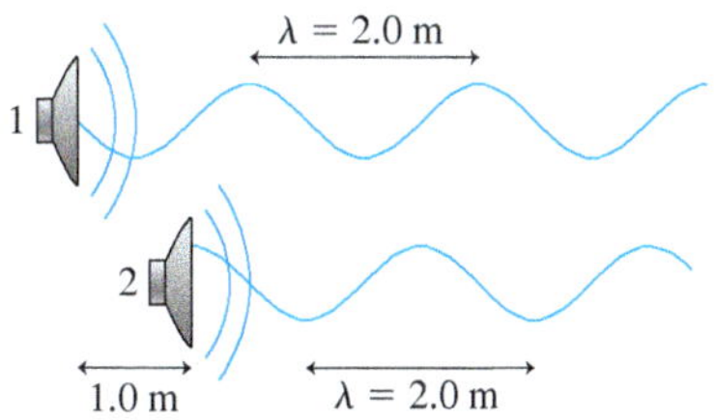

a. Move speaker 1 forward (to the right) 1.0 m.
b. Move speaker 1 forward (to the right) 0.5 m.
c. Move speaker 1 backward (to the left) 0.5 m.
d. Move speaker 1 backward (to the left) 1.0 m.
e. Nothing. The situation shown already causes constructive interference.
f. Constructive interference is not possible for any placement of the speakers.

21.6 The Mathematics of Interference

Let's look more closely at the superposition of two waves. As two waves of equal amplitude and frequency travel together along the x-axis, the net displacement of the medium is

$$\begin{aligned} D = D_1 + D_2 &= a\sin(kx_1 - \omega t + \phi_{10}) + a\sin(kx_2 - \omega t + \phi_{20}) \\ &= a\sin\phi_1 + a\sin\phi_2 \end{aligned} \tag{21.24}$$

where the phases ϕ_1 and ϕ_2 were defined in Equation 21.20.

A useful trigonometric identity is

$$\sin\alpha + \sin\beta = 2\cos\left[\tfrac{1}{2}(\alpha - \beta)\right]\sin\left[\tfrac{1}{2}(\alpha + \beta)\right] \tag{21.25}$$

This identity is certainly not obvious, although it is easily proven by working backward from the right side. We can use this identity to write the net displacement of Equation 21.24 as

$$D = \left[2a\cos\left(\frac{\Delta\phi}{2}\right)\right]\sin(kx_{avg} - \omega t + (\phi_0)_{avg}) \tag{21.26}$$

where $\Delta\phi = \phi_2 - \phi_1$ is the phase difference between the two waves, exactly as in Equation 21.21. $x_{avg} = (x_1 + x_2)/2$ is the average distance to the two sources and $(\phi_0)_{avg} = (\phi_{10} + \phi_{20})/2$ is the average phase constant of the sources.

The sine term shows that the superposition of the two waves is still a traveling wave. An observer would see a sinusoidal wave moving along the x-axis with the *same* wavelength and frequency as the original waves.

FIGURE 21.25 The interference of two waves for three different values of the phase difference.

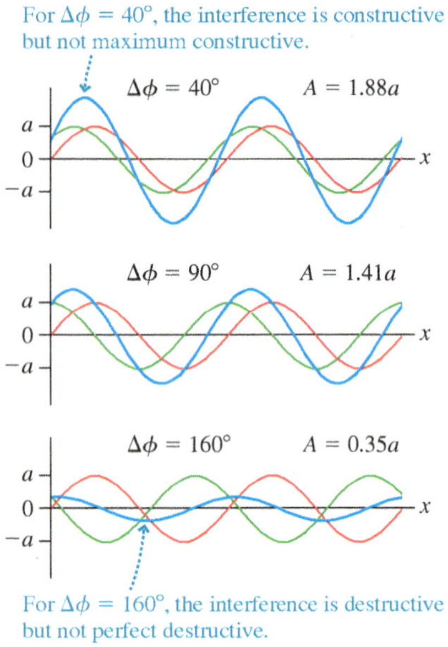

But how *big* is this wave compared to the two original waves? They each had amplitude a, but the amplitude of their superposition is

$$A = \left|2a\cos\left(\frac{\Delta\phi}{2}\right)\right| \tag{21.27}$$

where we have used an absolute value sign because amplitudes must be positive. Depending upon the phase difference of the two waves, the amplitude of their superposition can be anywhere from zero (perfect destructive interference) to $2a$ (maximum constructive interference).

The amplitude has its maximum value $A = 2a$ if $\cos(\Delta\phi/2) = \pm 1$. This occurs when

$$\Delta\phi = m \cdot 2\pi \qquad \text{(maximum amplitude } A = 2a\text{)} \tag{21.28}$$

where m is an integer. Similarly, the amplitude is zero if $\cos(\Delta\phi/2) = 0$, which occurs when

$$\Delta\phi = \left(m + \tfrac{1}{2}\right) \cdot 2\pi \qquad \text{(minimum amplitude } A = 0\text{)} \tag{21.29}$$

Equations 21.28 and 21.29 are identical to the conditions of Equations 21.22 and 21.23 for constructive and destructive interference. We initially found these conditions by considering the alignment of the crests and troughs. Now we have confirmed them with an algebraic addition of the waves.

It is entirely possible, of course, that the two waves are neither exactly in phase nor exactly out of phase. Equation 21.27 allows us to calculate the amplitude of the superposition for any value of the phase difference. As an example, **FIGURE 21.25** shows the calculated interference of two waves that differ in phase by 40°, by 90°, and by 160°.

EXAMPLE 21.8 More interference of sound waves

Two loudspeakers emit 500 Hz sound waves with an amplitude of 0.10 mm. Speaker 2 is 1.00 m behind speaker 1, and the phase difference between the speakers is 90°. What is the amplitude of the sound wave at a point 2.00 m in front of speaker 1?

MODEL The amplitude is determined by the interference of the two waves. Assume that the speed of sound has a room-temperature (20°C) value of 343 m/s.

SOLVE The amplitude of the sound wave is

$$A = |2a\cos(\Delta\phi/2)|$$

where $a = 0.10$ mm and the phase difference between the waves is

$$\Delta\phi = \phi_2 - \phi_1 = 2\pi\frac{\Delta x}{\lambda} + \Delta\phi_0$$

The sound's wavelength is

$$\lambda = \frac{v}{f} = \frac{343 \text{ m/s}}{500 \text{ Hz}} = 0.686 \text{ m}$$

Distances $x_1 = 2.00$ m and $x_2 = 3.00$ m are measured from the speakers, so the path-length difference is $\Delta x = 1.00$ m. We're given that the inherent phase difference between the speakers is $\Delta\phi_0 = \pi/2$ rad. Thus the phase difference at the observation point is

$$\Delta\phi = 2\pi\frac{\Delta x}{\lambda} + \Delta\phi_0 = 2\pi\frac{1.00 \text{ m}}{0.686 \text{ m}} + \frac{\pi}{2} \text{ rad} = 10.73 \text{ rad}$$

and the amplitude of the wave at this point is

$$A = \left|2a\cos\left(\frac{\Delta\phi}{2}\right)\right| = \left|(0.200 \text{ mm})\cos\left(\frac{10.73}{2}\right)\right| = 0.121 \text{ mm}$$

ASSESS The interference is constructive because $A > a$, but less than maximum constructive interference.

Application: Thin-Film Optical Coatings

The shimmering colors of soap bubbles and oil slicks, as seen in the photo at the beginning of the chapter, are due to the interference of light waves. In fact, the idea of light-wave interference in one dimension has an important application in the optics industry, namely the use of **thin-film optical coatings.** These films, less than 1 μm (10^{-6} m) thick, are placed on glass surfaces, such as lenses, to control reflections from the glass. Antireflection coatings on the lenses in cameras, microscopes, and other optical equipment are examples of thin-film coatings.

FIGURE 21.26 shows a light wave of wavelength λ approaching a piece of glass that has been coated with a transparent film of thickness d whose index of refraction is n. The air-film boundary is a discontinuity at which the wave speed suddenly decreases, and you saw earlier, in Figure 21.8, that a discontinuity causes a reflection. Most of the light is transmitted into the film, but a little bit is reflected.

Furthermore, you saw in Figure 21.8 that the wave reflected from a discontinuity at which the speed decreases is *inverted* with respect to the incident wave. For a sinusoidal wave, which we're now assuming, the inversion is represented mathematically as a phase shift of π rad. The speed of a light wave decreases when it enters a material with a *larger* index of refraction. Thus **a light wave that reflects from a boundary at which the index of refraction increases has a phase shift of π rad.** There is no phase shift for the reflection from a boundary at which the index of refraction decreases. The reflection in Figure 21.26 is from a boundary between air ($n_{air} = 1.00$) and a transparent film with $n_{film} > n_{air}$, so the reflected wave is inverted due to the phase shift of π rad.

FIGURE 21.26 The two reflections, one from the coating and one from the glass, interfere.

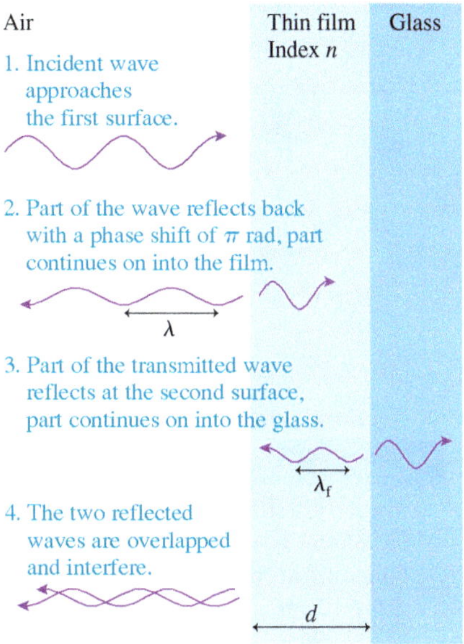

When the transmitted wave reaches the glass, most of it continues on into the glass but a portion is reflected back to the left. We'll assume that the index of refraction of the glass is larger than that of the film, $n_{glass} > n_{film}$, so this reflection also has a phase shift of π rad. This second reflection, after traveling back through the film, passes back into the air. There are now *two* equal-frequency waves traveling to the left, and these waves will interfere. If the two reflected waves are *in phase,* they will interfere constructively to cause a *strong reflection.* If the two reflected waves are *out of phase,* they will interfere destructively to cause a *weak reflection* or, if their amplitudes are equal, *no reflection* at all.

This suggests practical uses for thin-film optical coatings. The reflections from glass surfaces, even if weak, are often undesirable. For example, reflections degrade the performance of optical equipment. These reflections can be eliminated by coating the glass with a film whose thickness is chosen to cause *destructive* interference of the two reflected waves. This is an *antireflection coating.*

Antireflection coatings use the interference of light waves to nearly eliminate reflections from glass surfaces.

The amplitude of the reflected light depends on the phase difference between the two reflected waves. This phase difference is

$$\Delta\phi = \phi_2 - \phi_1 = (kx_2 + \phi_{20} + \pi \text{ rad}) - (kx_1 + \phi_{10} + \pi \text{ rad}) \\ = 2\pi\frac{\Delta x}{\lambda_f} + \Delta\phi_0 \qquad (21.30)$$

where we explicitly included the reflection phase shift of each wave. In this case, because *both* waves had a phase shift of π rad, the reflection phase shifts cancel.

The wavelength λ_f is the wavelength *in the film* because that's where the path-length difference Δx occurs. You learned in Chapter 20 that the wavelength in a transparent material with index of refraction n is $\lambda_f = \lambda/n$, where the unsubscripted λ is the wavelength in vacuum or air. That is, λ is the wavelength that we measure on "our" side of the air-film boundary.

The path-length difference between the two waves is $\Delta x = 2d$ because wave 2 travels through the film *twice* before rejoining wave 1. The two waves have a common origin—the initial division of the incident wave at the front surface of the film—so the inherent phase difference is $\Delta\phi_0 = 0$. Thus the phase difference of the two reflected waves is

$$\Delta\phi = 2\pi\frac{2d}{\lambda/n} = 2\pi\frac{2nd}{\lambda} \qquad (21.31)$$

The interference is constructive, causing a strong reflection, when $\Delta\phi = m\cdot 2\pi$ rad. So when both reflected waves have a phase of π rad, constructive interference occurs for wavelengths

$$\lambda_C = \frac{2nd}{m} \qquad m = 1, 2, 3, \ldots \qquad \text{(constructive interference)} \qquad (21.32)$$

You will notice that m starts with 1, rather than 0, in order to give meaningful results. Destructive interference, with minimum reflection, requires $\Delta\phi = \left(m - \frac{1}{2}\right) \cdot 2\pi$ rad. This—again, when both waves have a phase shift of π rad—occurs for wavelengths

$$\lambda_D = \frac{2nd}{m - \frac{1}{2}} \qquad m = 1, 2, 3, \ldots \qquad \text{(destructive interference)} \qquad (21.33)$$

We've used $m - \frac{1}{2}$, rather than $m + \frac{1}{2}$, so that m can start with 1 to match the condition for constructive interference.

NOTE ▶ The exact condition for constructive or destructive interference is satisfied for only a few discrete wavelengths λ. Nonetheless, reflections are strongly enhanced (nearly constructive interference) for a range of wavelengths near λ_C. Likewise, there is a range of wavelengths near λ_D for which the reflection is nearly canceled. ◀

EXAMPLE 21.9 Designing an antireflection coating

Magnesium fluoride (MgF_2) is used as an antireflection coating on lenses. The index of refraction of MgF_2 is 1.39. What is the thinnest film of MgF_2 that works as an antireflection coating at $\lambda = 510$ nm, near the center of the visible spectrum?

MODEL Reflection is minimized if the two reflected waves interfere destructively.

SOLVE The film thicknesses that cause destructive interference at wavelength λ are

$$d = \left(m - \frac{1}{2}\right)\frac{\lambda}{2n}$$

The thinnest film has $m = 1$. Its thickness is

$$d = \frac{\lambda}{4n} = \frac{510 \text{ nm}}{4(1.39)} = 92 \text{ nm}$$

The film thickness is significantly less than the wavelength of visible light!

ASSESS The reflected light is completely eliminated (perfect destructive interference) only if the two reflected waves have equal amplitudes. In practice, they don't. Nonetheless, the reflection is reduced from $\approx 4\%$ of the incident intensity for "bare glass" to well under 1%. Furthermore, the intensity of reflected light is much reduced across most of the visible spectrum (400–700 nm), even though the phase difference deviates more and more from π rad as the wavelength moves away from 510 nm. It is the increasing reflection at the ends of the visible spectrum ($\lambda \approx 400$ nm and $\lambda \approx 700$ nm), where $\Delta\phi$ deviates significantly from π rad, that gives a reddish-purple tinge to the lenses on cameras and binoculars. Homework problems will let you explore situations where only one of the two reflections has a reflection phase shift of π rad.

21.7 Interference in Two and Three Dimensions

Ripples on a lake move in two dimensions. The glow from a lightbulb spreads outward as a spherical wave. A circular or spherical wave can be written

$$D(r, t) = a\sin(kr - \omega t + \phi_0) \qquad (21.34)$$

where r is the distance measured outward from the source. Equation 21.34 is our familiar wave equation with the one-dimensional coordinate x replaced by a more general radial coordinate r. Strictly speaking, the amplitude a of a circular or spherical wave diminishes as r increases. However, we will assume that a remains essentially constant over the region in which we study the wave. FIGURE 21.27 shows the wave-front diagram for a circular or spherical wave. Recall that the wave fronts represent the *crests* of the wave and are spaced by the wavelength λ.

FIGURE 21.27 A circular or spherical wave.

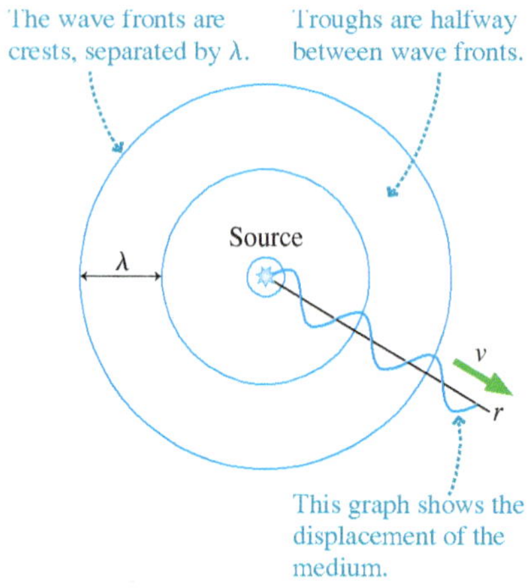

What happens when two circular or spherical waves overlap? For example, imagine two paddles oscillating up and down on the surface of a pond. We will assume that the two paddles oscillate with the same frequency and amplitude and that they are in phase. FIGURE 21.28 shows the wave fronts of the two waves. The ripples overlap as they travel, and, as was the case in one dimension, this causes interference.

Constructive interference with $A = 2a$ occurs where two crests align or two troughs align. Several locations of constructive interference are marked in Figure 21.28. Intersecting wave fronts are points where two crests are aligned. It's a bit harder to visualize, but two troughs are aligned when a midpoint between two wave fronts is overlapped with another midpoint between two wave fronts. Destructive interference with $A = 0$ occurs where the crest of one wave aligns with a trough of the other wave. Several points of destructive interference are also indicated in Figure 21.28.

A picture on a page is static, but **the wave fronts are in motion.** Try to imagine the wave fronts of Figure 21.28 expanding outward as new circular rings are born at the sources. The waves will move forward half a wavelength during half a period, causing the crests in Figure 21.28 to be replaced by troughs while the troughs become crests.

The important point to recognize is that **the motion of the waves does not affect the points of constructive and destructive interference.** Points in the figure where two crests overlap will become points where two troughs overlap, but this overlap is still constructive interference. Similarly, points in the figure where a crest and a trough overlap will become a point where a trough and a crest overlap—still destructive interference.

The mathematical description of interference in two or three dimensions is very similar to that of one-dimensional interference. The net displacement of a particle in the medium is

$$D = D_1 + D_2 = a\sin(kr_1 - \omega t + \phi_{10}) + a\sin(kr_2 - \omega t + \phi_{20}) \quad (21.35)$$

The only difference between Equation 21.35 and the earlier one-dimensional Equation 21.24 is that the linear coordinates x_1 and x_2 have been changed to radial coordinates r_1 and r_2. Thus our conclusions are unchanged. The superposition of the two waves yields a wave traveling outward with amplitude

$$A = \left|2a\cos\left(\frac{\Delta\phi}{2}\right)\right| \quad (21.36)$$

where the phase difference, with x replaced by r, is now

$$\Delta\phi = 2\pi\frac{\Delta r}{\lambda} + \Delta\phi_0 \quad (21.37)$$

The term $2\pi(\Delta r/\lambda)$ is the phase difference that arises when the waves travel different distances from the sources to the point at which they combine. Δr itself is the *path-length difference*. As before, $\Delta\phi_0$ is any inherent phase difference of the sources themselves.

Maximum constructive interference with $A = 2a$ occurs, just as in one dimension, at those points where $\cos(\Delta\phi/2) = \pm 1$. Similarly, perfect destructive interference occurs at points where $\cos(\Delta\phi/2) = 0$. The conditions for constructive and destructive interference are

Maximum constructive interference:

$$\Delta\phi = 2\pi\frac{\Delta r}{\lambda} + \Delta\phi_0 = m \cdot 2\pi$$

$$m = 0, 1, 2, \ldots \quad (21.38)$$

Perfect destructive interference:

$$\Delta\phi = 2\pi\frac{\Delta r}{\lambda} + \Delta\phi_0 = \left(m + \frac{1}{2}\right) \cdot 2\pi$$

For two identical sources (i.e., sources that oscillate in phase with $\Delta\phi_0 = 0$), the conditions for constructive and destructive interference are simple:

$$\text{Constructive:} \quad \Delta r = m\lambda$$

$$\text{Destructive:} \quad \Delta r = \left(m + \frac{1}{2}\right)\lambda \qquad \text{(identical sources)} \quad (21.39)$$

FIGURE 21.28 The overlapping ripple patterns of two sources. Several points of constructive and destructive interference are noted.

Two in-phase sources emit circular or spherical waves.

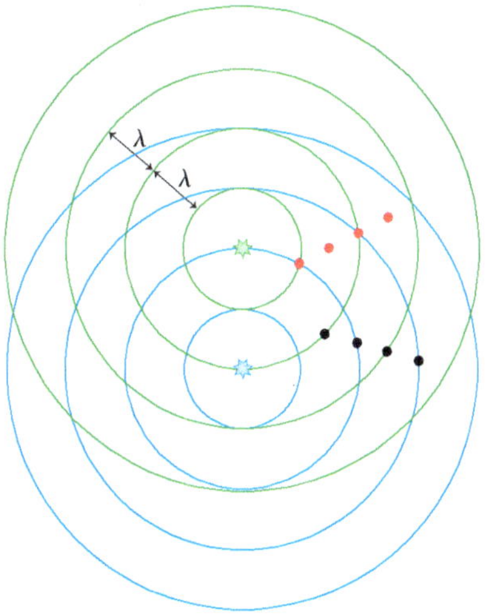

• Points of constructive interference. A crest is aligned with a crest, or a trough with a trough.

• Points of destructive interference. A crest is aligned with a trough of another wave.

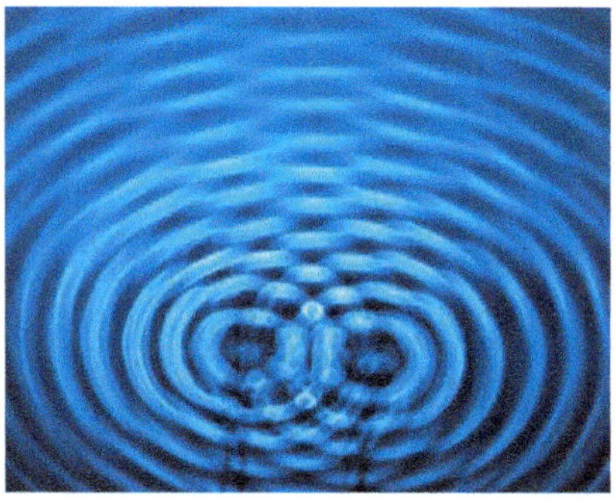

Two overlapping water waves create an interference pattern.

FIGURE 21.29 The path-length difference Δr determines whether the interference at a particular point is constructive or destructive.

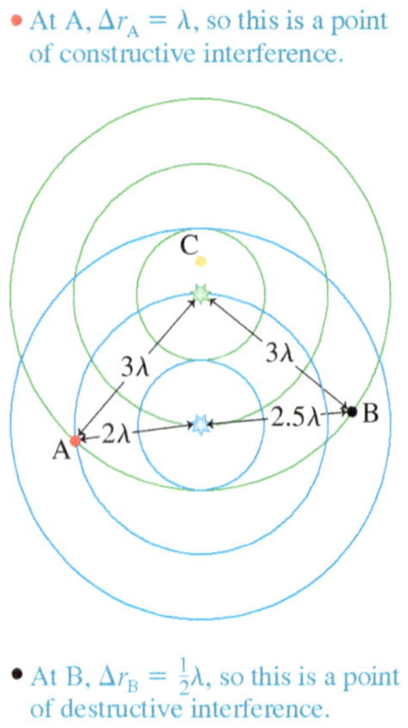

The waves from two identical sources interfere constructively at points where the path-length difference is an integer number of wavelengths because, for these values of Δr, crests are aligned with crests and troughs with troughs. **The waves interfere destructively at points where the path-length difference is a half-integer number of wavelengths** because, for these values of Δr, crests are aligned with troughs. These two statements are the essence of interference.

NOTE ▶ Equation 21.39 applies only if the sources are in phase. If the sources are not in phase, you must use the more general Equation 21.38 to locate the points of constructive and destructive interference. ◀

Wave fronts are spaced exactly one wavelength apart; hence we can measure the distances r_1 and r_2 simply by counting the rings in the wave-front pattern. In FIGURE 21.29, which is based on Figure 21.28, point A is distance $r_1 = 3\lambda$ from the first source and $r_2 = 2\lambda$ from the second. The path-length difference is $\Delta r_A = 1\lambda$, the condition for the maximum constructive interference of identical sources. Point B has $\Delta r_B = \frac{1}{2}\lambda$, so it is a point of perfect destructive interference.

NOTE ▶ Interference is determined by Δr, the path-length *difference,* rather than by r_1 or r_2. ◀

STOP TO THINK 21.5 The interference at point C in Figure 21.29 is

a. Maximum constructive.
b. Constructive, but less than maximum.
c. Perfect destructive.
d. Destructive, but not perfect.
e. There is no interference at point C.

We can now locate the points of maximum constructive interference, for which $\Delta r = m\lambda$, by drawing a line through *all* the points at which $\Delta r = 0$, another line through all the points at which $\Delta r = \lambda$, and so on. These lines, shown in red in FIGURE 21.30, are called **antinodal lines.** They are analogous to the antinodes of a standing wave, hence the name. An antinode is a *point* of maximum constructive interference; for circular waves, oscillation at maximum amplitude occurs along a continuous *line.* Similarly, destructive interference occurs along lines called **nodal lines.** The displacement is *always zero* along these lines, just as it is at a node in a standing-wave pattern.

FIGURE 21.30 The points of constructive and destructive interference fall along antinodal and nodal lines.

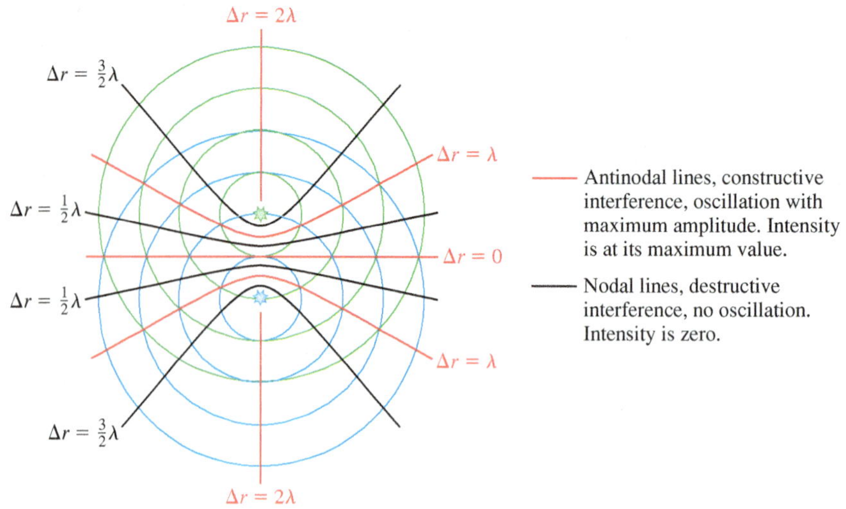

A Problem-Solving Strategy for Interference Problems

The information in this section is the basis of a strategy for solving interference problems. This strategy applies equally well to interference in one dimension if you use Δx instead of Δr.

PROBLEM-SOLVING STRATEGY 21.1 Interference of two waves

MODEL Make simplifying assumptions, such as assuming waves are circular and of equal amplitude.

VISUALIZE Draw a picture showing the sources of the waves and the point where the waves interfere. Give relevant dimensions. Identify the distances r_1 and r_2 from the sources to the point. Note any phase difference $\Delta\phi_0$ between the two sources.

SOLVE The interference depends on the path-length difference $\Delta r = r_2 - r_1$ and the source phase difference $\Delta\phi_0$.

$$\text{Constructive:} \quad \Delta\phi = 2\pi\frac{\Delta r}{\lambda} + \Delta\phi_0 = m \cdot 2\pi$$

$$\text{Destructive:} \quad \Delta\phi = 2\pi\frac{\Delta r}{\lambda} + \Delta\phi_0 = \left(m + \frac{1}{2}\right) \cdot 2\pi \qquad m = 0, 1, 2, \ldots$$

For identical sources ($\Delta\phi_0 = 0$), the interference is maximum constructive if $\Delta r = m\lambda$, perfect destructive if $\Delta r = \left(m + \frac{1}{2}\right)\lambda$.

ASSESS Check that your result has the correct units, is reasonable, and answers the question.

Exercise 18

EXAMPLE 21.10 Two-dimensional interference between two loudspeakers

Two loudspeakers in a plane are 2.0 m apart and in phase with each other. Both emit 700 Hz sound waves into a room where the speed of sound is 341 m/s. A listener stands 5.0 m in front of the loudspeakers and 2.0 m to one side of the center. Is the interference at this point maximum constructive, perfect destructive, or in between? How will the situation differ if the loudspeakers are out of phase?

MODEL The two speakers are sources of in-phase, spherical waves. The overlap of these waves causes interference.

VISUALIZE FIGURE 21.31 shows the loudspeakers and defines the distances r_1 and r_2 to the point of observation. The figure includes dimensions and notes that $\Delta\phi_0 = 0$ rad.

FIGURE 21.31 Pictorial representation of the interference between two loudspeakers.

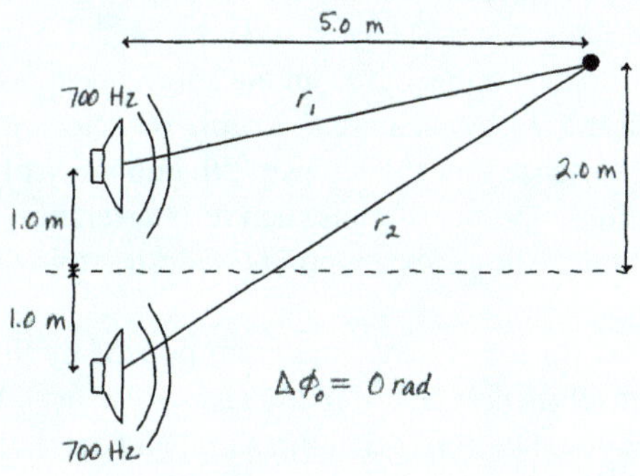

SOLVE It's not r_1 and r_2 that matter, but the *difference* Δr between them. From the geometry of the figure we can calculate that

$$r_1 = \sqrt{(5.0\text{ m})^2 + (1.0\text{ m})^2} = 5.10\text{ m}$$

$$r_2 = \sqrt{(5.0\text{ m})^2 + (3.0\text{ m})^2} = 5.83\text{ m}$$

Thus the path-length difference is $\Delta r = r_2 - r_1 = 0.73$ m. The wavelength of the sound waves is

$$\lambda = \frac{v}{f} = \frac{341\text{ m/s}}{700\text{ Hz}} = 0.487\text{ m}$$

In terms of wavelengths, the path-length difference is $\Delta r/\lambda = 1.50$, or

$$\Delta r = \frac{3}{2}\lambda$$

Because the sources are in phase ($\Delta\phi_0 = 0$), this is the condition for *destructive* interference. If the sources were out of phase ($\Delta\phi_0 = \pi$ rad), then the phase difference of the waves at the listener would be

$$\Delta\phi = 2\pi\frac{\Delta r}{\lambda} + \Delta\phi_0 = 2\pi\left(\frac{3}{2}\right) + \pi\text{ rad} = 4\pi\text{ rad}$$

This is an integer multiple of 2π rad, so in this case the interference would be *constructive.*

ASSESS Both the path-length difference *and* any inherent phase difference of the sources must be considered when evaluating interference.

Picturing Interference

A *contour map* is a useful way to visualize an interference pattern. FIGURE 21.32a shows the superposition of the waves from two identical sources ($\Delta\phi_0 = 0$) emitting waves with $\lambda = 1$ m. The sources, indicated with black dots, are located two wavelengths apart at $y = \pm 1$ m. Positive displacements are shown in red, with the deepest red representing the maximum displacement of the wave at this instant in time. These are the points where the crests of the individual waves interfere constructively to give $D = 2a$. Negative displacements are blue, with the darkest blue being the most negative displacement of the wave. These are also points of constructive interference, with two troughs overlapping to give $D = -2a$.

FIGURE 21.32 A contour map of the interference pattern of two sources. The graph on the right side of each figure shows the wave intensity along a vertical line at $x = 4$ m.

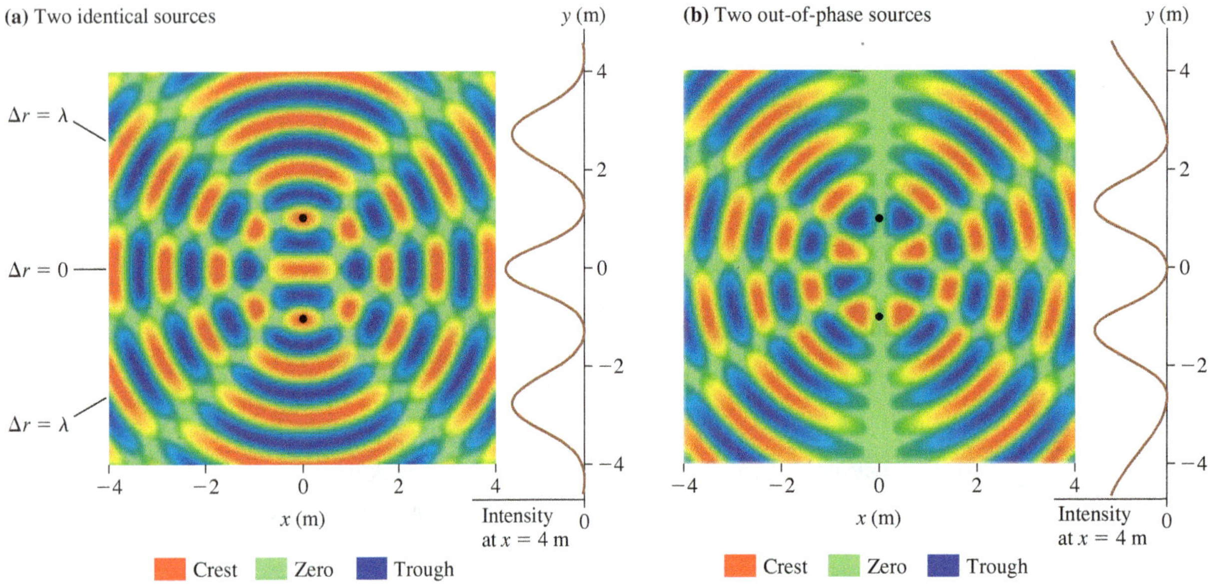

To understand this figure, try to visualize the waves expanding outward from the center. The red-blue-red-blue-red-· · · pattern of crests and troughs moves outward along the antinodal lines as a *traveling wave* of amplitude $A = 2a$. Nothing ever happens along the nodal lines, where the amplitude is always zero.

Suppose you were to observe the *intensity* of the wave as it crosses the vertical line at $x = 4$ m on the right edge of the figure. If, for example, these are sound waves, you could listen to (or measure with a microphone) the sound intensity as you walk from $(x, y) = (4\text{ m}, -4\text{ m})$ at the bottom of the figure to $(x, y) = (4\text{ m}, 4\text{ m})$ at the top. The intensity is zero as you cross the nodal lines at $y \approx \pm 1$ m $\left(\Delta r = \frac{1}{2}\lambda\right)$. The intensity is maximum at the antinodal lines at $y = 0$ ($\Delta r = 0$) and $y \approx \pm 2.5$ m ($\Delta r = \lambda$), where a wave of maximum amplitude streams out from the sources.

The intensity is shown in the rather unusual graph on the right side of Figure 21.32a. It is unusual in the sense that the intensity, the quantity of interest, is graphed to the left. The peaks are the points of constructive interference, where you would measure maximum amplitude. The zeros are points of destructive interference, where the intensity is zero.

FIGURE 21.32b is a contour map of the interference pattern produced by the same two sources but with the sources themselves now out of phase ($\Delta\phi_0 = \pi$ rad). We'll leave the investigation of this figure to you, but notice that the nodal and antinodal lines are reversed from those of Figure 21.32a.

EXAMPLE 21.11 The intensity of two interfering loudspeakers

Two loudspeakers in a plane are 6.0 m apart and in phase. They emit equal-amplitude sound waves with a wavelength of 1.0 m. Each speaker alone creates sound with intensity I_0. An observer at point A is 10 m in front of the plane containing the two loudspeakers and centered between them. A second observer at point B is 10 m directly in front of one of the speakers. In terms of I_0, what are the intensity I_A at point A and the intensity I_B at point B?

MODEL The two speakers are sources of in-phase waves. The overlap of these waves causes interference.

VISUALIZE FIGURE 21.33 shows the two loudspeakers and the two points of observation. Distances r_1 and r_2 are defined for point B.

FIGURE 21.33 Pictorial representation of the interference between two loudspeakers.

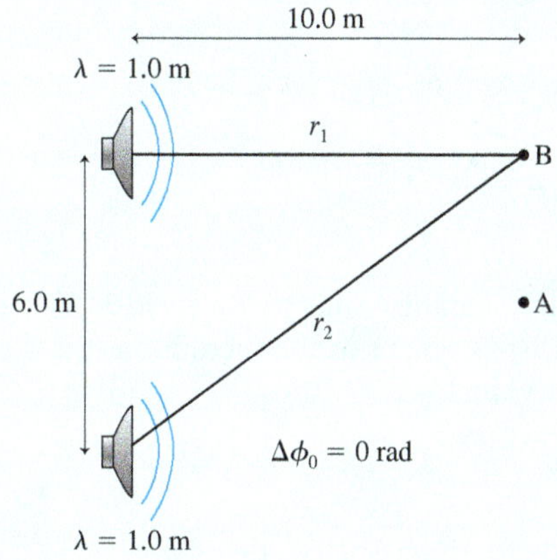

SOLVE Let the amplitude of the wave from each speaker be a. The intensity of a wave is proportional to the square of the amplitude, so the intensity of each speaker alone is $I_0 = ca^2$, where c is an unknown proportionality constant. Point A is a point of constructive interference because the speakers are in phase ($\Delta\phi_0 = 0$) and the path-length difference is $\Delta r = 0$. The amplitude at this point is given by Equation 21.36:

$$A_A = \left|2a\cos\left(\frac{\Delta\phi}{2}\right)\right| = 2a\cos(0) = 2a$$

Consequently, the intensity at this point is

$$I_A = cA_A^2 = c(2a)^2 = 4ca^2 = 4I_0$$

The intensity at A is four times that of either speaker played alone.

At point B, the path-length difference is

$$\Delta r = \sqrt{(10.0 \text{ m})^2 + (6.0 \text{ m})^2} - 10.0 \text{ m} = 1.662 \text{ m}$$

The phase difference of the waves at this point is

$$\Delta\phi = 2\pi\frac{\Delta r}{\lambda} = 2\pi\frac{1.662 \text{ m}}{1.0 \text{ m}} = 10.44 \text{ rad}$$

Consequently, the amplitude at B is

$$A_B = \left|2a\cos\left(\frac{\Delta\phi}{2}\right)\right| = |2a\cos(5.22 \text{ rad})| = 0.972a$$

Thus the intensity at this point is

$$I_B = cA_B^2 = c(0.972a)^2 = 0.95ca^2 = 0.95I_0$$

ASSESS Although B is directly in front of one of the speakers, superposition of the two waves results in an intensity that is less than it would be if this speaker played alone.

STOP TO THINK 21.6 These two loudspeakers are in phase. They emit equal-amplitude sound waves with a wavelength of 1.0 m. At the point indicated, is the interference maximum constructive, perfect destructive, or something in between?

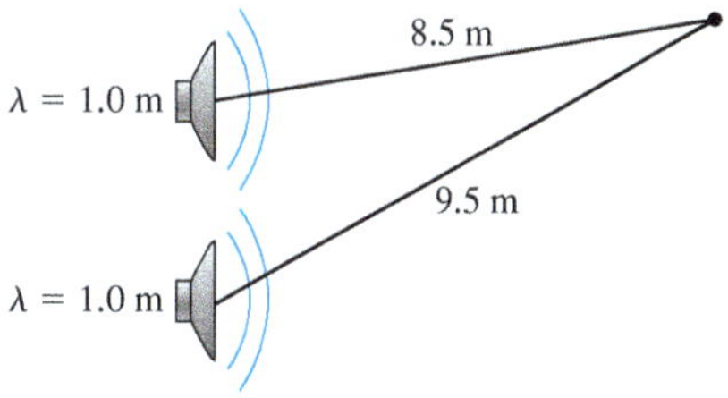

21.8 Beats

Thus far we have looked at the superposition of sources having the same wavelength and frequency. We can also use the principle of superposition to investigate a phenomenon that is easily demonstrated with two sources of slightly different frequency.

If you listen to two sounds with very different frequencies, such as a high note and a low note, you hear two distinct tones. But if the frequency difference is very small, just one or two hertz, then you hear a single tone whose intensity is *modulated* once or twice every second. That is, the sound goes up and down in volume, loud, soft, loud, soft, . . . , making a distinctive sound pattern called **beats.**

Consider two sinusoidal waves traveling along the x-axis with angular frequencies $\omega_1 = 2\pi f_1$ and $\omega_2 = 2\pi f_2$. The two waves are

$$\begin{aligned} D_1 &= a\sin(k_1x - \omega_1t + \phi_{10}) \\ D_2 &= a\sin(k_2x - \omega_2t + \phi_{20}) \end{aligned} \tag{21.40}$$

where the subscripts 1 and 2 indicate that the frequencies, wave numbers, and phase constants of the two waves may be different.

To simplify the analysis, let's make several assumptions:

1. The two waves have the same amplitude a,
2. A detector, such as your ear, is located at the origin ($x = 0$),
3. The two sources are in phase ($\phi_{10} = \phi_{20}$), and
4. The source phases happen to be $\phi_{10} = \phi_{20} = \pi$ rad.

None of these assumptions is essential to the outcome. All could be otherwise and we would still come to basically the same conclusion, but the mathematics would be far messier. Making these assumptions allows us to emphasize the physics with the least amount of mathematics.

With these assumptions, the two waves as they reach the detector at $x = 0$ are

$$\begin{aligned} D_1 &= a\sin(-\omega_1t + \pi) = a\sin\omega_1t \\ D_2 &= a\sin(-\omega_2t + \pi) = a\sin\omega_2t \end{aligned} \tag{21.41}$$

where we've used the trigonometric identity $\sin(\pi - \theta) = \sin\theta$. The principle of superposition tells us that the *net* displacement of the medium at the detector is the sum of the displacements of the individual waves. Thus

$$D = D_1 + D_2 = a(\sin\omega_1t + \sin\omega_2t) \tag{21.42}$$

Earlier, for interference, we used the trigonometric identity

$$\sin\alpha + \sin\beta = 2\cos\left[\tfrac{1}{2}(\alpha - \beta)\right]\sin\left[\tfrac{1}{2}(\alpha + \beta)\right]$$

We can use this identity again to write Equation 21.42 as

$$\begin{aligned} D &= 2a\cos\left[\tfrac{1}{2}(\omega_1 - \omega_2)t\right]\sin\left[\tfrac{1}{2}(\omega_1 + \omega_2)t\right] \\ &= \left[2a\cos(\omega_{mod}t)\right]\sin(\omega_{avg}t) \end{aligned} \tag{21.43}$$

where $\omega_{avg} = \frac{1}{2}(\omega_1 + \omega_2)$ is the *average* angular frequency and $\omega_{mod} = \frac{1}{2}(\omega_1 - \omega_2)$ is called the *modulation frequency.*

We are interested in the situation when the two frequencies are very nearly equal: $\omega_1 \approx \omega_2$. In that case, ω_{avg} hardly differs from either ω_1 or ω_2 while ω_{mod} is very near to—but not exactly—zero. When ω_{mod} is very small, the term $\cos(\omega_{mod}t)$ oscillates *very* slowly. We have grouped it with the $2a$ term because, together, they provide a slowly changing "amplitude" for the rapid oscillation at frequency ω_{avg}.

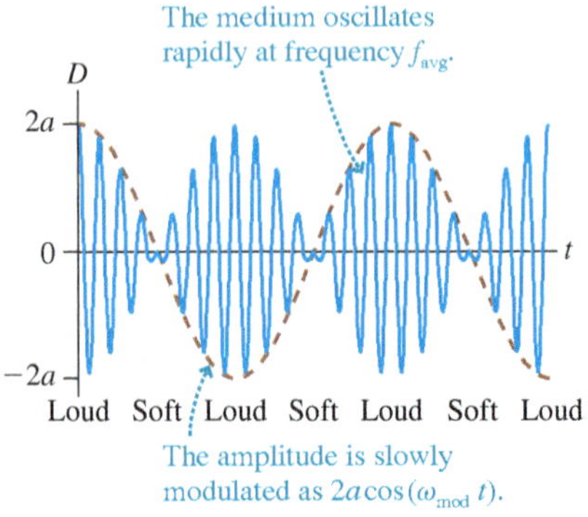

FIGURE 21.34 Beats are caused by the superposition of two waves of nearly identical frequency.

FIGURE 21.34 is a history graph of the wave at the detector ($x = 0$). It shows the oscillation of the air against your eardrum at frequency $f_{avg} = \omega_{avg}/2\pi = \frac{1}{2}(f_1 + f_2)$. This oscillation determines the note you hear; it differs little from the two notes at frequencies f_1 and f_2. We are especially interested in the time-dependent amplitude, shown as a dashed line, that is given by the term $2a\cos(\omega_{mod}t)$. This periodically varying amplitude is called a **modulation** of the wave, which is where ω_{mod} gets its name.

As the amplitude rises and falls, the sound alternates as loud, soft, loud, soft, and so on. But that is exactly what you hear when you listen to beats! The alternating loud and soft sounds arise from the two waves being alternately in phase and out of phase, causing constructive and then destructive interference.

Imagine two people walking side by side at just slightly different paces. Initially both of their right feet hit the ground together, but after a while they get out of step. A little bit later they are back in step and the process alternates. The sound waves are doing the same. Initially the crests of each wave, of amplitude a, arrive together at your ear and the net displacement is doubled to $2a$. But after a while the two waves, being of slightly different frequency, get out of step and a crest of one arrives with a trough of the other. When this happens, the two waves cancel each other to give a net displacement of zero. This process alternates over and over, loud and soft.

Notice, from the figure, that the sound intensity rises and falls *twice* during one cycle of the modulation envelope. Each "loud-soft-loud" is one beat, so the **beat frequency** f_{beat}, which is the number of beats per second, is *twice* the modulation frequency $f_{mod} = \omega_{mod}/2\pi$. From the above definition of ω_{mod}, the beat frequency is

$$f_{beat} = 2f_{mod} = 2\frac{\omega_{mod}}{2\pi} = 2\cdot\frac{1}{2}\left(\frac{\omega_1}{2\pi} - \frac{\omega_2}{2\pi}\right) = f_1 - f_2 \qquad (21.44)$$

where, to keep f_{beat} from being negative, we will always let f_1 be the larger of the two frequencies. The beat frequency is simply the *difference* between the two individual frequencies.

EXAMPLE 21.12 Detecting bats with beats

The little brown bat is a common species in North America. It emits echolocation pulses at a frequency of 40 kHz, well above the range of human hearing. To allow researchers to "hear" these bats, the bat detector shown in FIGURE 21.35 combines the bat's sound wave at frequency f_1 with a wave of frequency f_2 from a tunable oscillator. The resulting beat frequency is then amplified and sent to a loudspeaker. To what frequency should the tunable oscillator be set to produce an audible beat frequency of 3 kHz?

SOLVE Combining two waves with different frequencies gives a beat frequency

$$f_{beat} = f_1 - f_2$$

A beat frequency will be generated at 3 kHz if the oscillator frequency and the bat frequency *differ* by 3 kHz. An oscillator frequency of either 37 kHz or 43 kHz will work nicely.

ASSESS The electronic circuitry of radios, televisions, and cell phones makes extensive use of *mixers* to generate difference frequencies.

FIGURE 21.35 The operation of a bat detector.

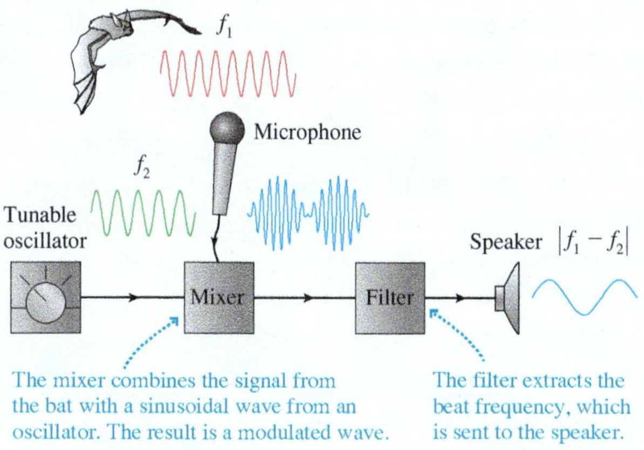

Beats aren't limited to sound waves. FIGURE 21.36 shows a graphical example of beats. Two "fences" of slightly different frequencies are superimposed on each other. The difference in the two frequencies is two lines per inch. You can confirm, with a ruler, that the figure has two "beats" per inch, in agreement with Equation 21.44.

Beats are important in many other situations. For example, you have probably seen movies where rotating wheels seem to turn slowly backward. Why is this? Suppose the movie camera is shooting at 30 frames per second but the wheel is rotating 32 times per second. The combination of the two produces a "beat" of 2 Hz, meaning that the wheel appears to rotate only twice per second. The same is true if the wheel is rotating 28 times per second, but in this case, where the wheel frequency slightly lags the camera frequency, it appears to rotate *backward* twice per second!

FIGURE 21.36 A graphical example of beats.

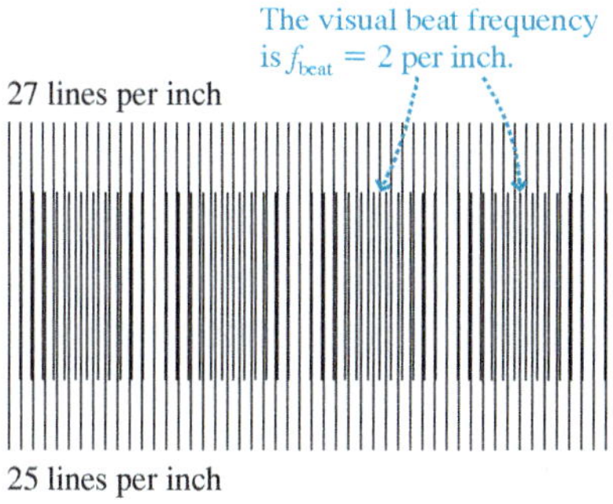

STOP TO THINK 21.7 You hear three beats per second when two sound tones are generated. The frequency of one tone is 610 Hz. The frequency of the other is

a. 604 Hz
b. 607 Hz
c. 613 Hz
d. 616 Hz
e. Either a or d.
f. Either b or c.

CHALLENGE EXAMPLE 21.13 An airplane landing system

Your firm has been hired to design a system that allows airplane pilots to make instrument landings in rain or fog. You've decided to place two radio transmitters 50 m apart on either side of the runway. These two transmitters will broadcast the same frequency, but out of phase with each other. This will cause a nodal line to extend straight off the end of the runway. As long as the airplane's receiver is silent, the pilot knows she's directly in line with the runway. If she drifts to one side or the other, the radio will pick up a signal and sound a warning beep. To have sufficient accuracy, the first intensity maxima need to be 60 m on either side of the nodal line at a distance of 3.0 km. What frequency should you specify for the transmitters?

MODEL The two transmitters are sources of out-of-phase, circular waves. The overlap of these waves produces an interference pattern.

VISUALIZE For out-of-phase sources, the center line—with zero path-length difference—is a nodal line of perfect destructive interference because the two signals always arrive out of phase. **FIGURE 21.37** shows the nodal line, extending straight off the runway, and the first antinodal line—the points of maximum constructive interference—on either side. Comparing this to Figure 21.30, where the two sources were in phase, you can see that the nodal and antinodal lines have been reversed.

FIGURE 21.37 Pictorial representation of the landing system.

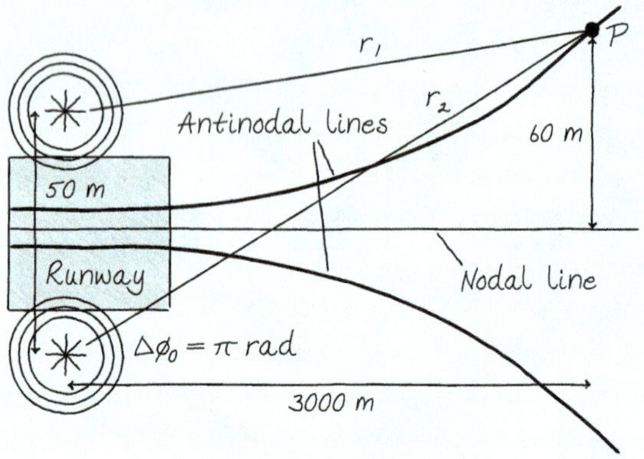

SOLVE Point P, 60 m to the side at a distance of 3000 m, needs to be a point of maximum constructive interference. The distances are

$$r_1 = \sqrt{(3000\text{ m})^2 + (60\text{ m} - 25\text{ m})^2} = 3000.204\ m$$

$$r_2 = \sqrt{(3000\text{ m})^2 + (60\text{ m} + 25\text{ m})^2} = 3001.204\ m$$

We needed to keep several extra significant figures because we're looking for the difference between two numbers that are almost the same. The path-length difference at P is

$$\Delta r = r_2 - r_1 = 1.000\text{ m}$$

We know, for out-of-phase transmitters, that the phase difference of the sources is $\Delta\phi_0 = \pi$ rad. The first maximum will occur where the phase difference between the waves is $\Delta\phi = 1 \cdot 2\pi$ rad. Thus the condition that we must satisfy at P is

$$\Delta\phi = 2\pi\text{ rad} = 2\pi\frac{\Delta r}{\lambda} + \pi\text{ rad}$$

Solving for λ, we find

$$\lambda = 2\,\Delta r = 2.00\text{ m}$$

Consequently, the required frequency is

$$f = \frac{c}{\lambda} = \frac{3.00 \times 10^8\text{ m/s}}{2.00\text{ m}} = 1.50 \times 10^8\text{ Hz} = 150\text{ MHz}$$

ASSESS 150 MHz is slightly higher than the frequencies of FM radio ($\approx$ 100 MHz) but is well within the radio frequency range. Notice that the condition to be satisfied at P is that the path-length difference must be $\frac{1}{2}\lambda$. This makes sense. A path-length difference of $\frac{1}{2}\lambda$ contributes π rad to the phase difference. When combined with the π rad from the out-of-phase sources, the total phase difference of 2π rad creates constructive interference.

SUMMARY

The goal of Chapter 21 has been to understand and use the idea of superposition.

General Principles

Principle of Superposition

The displacement of a medium when more than one wave is present is the sum of the displacements due to each individual wave.

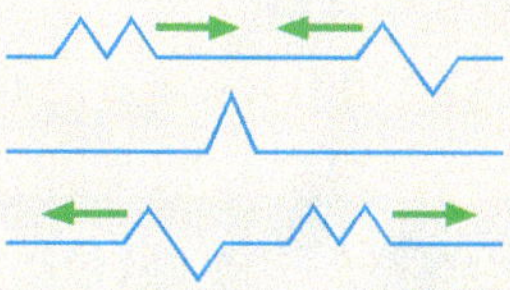

Important Concepts

Standing waves are due to the superposition of two traveling waves moving in opposite directions.

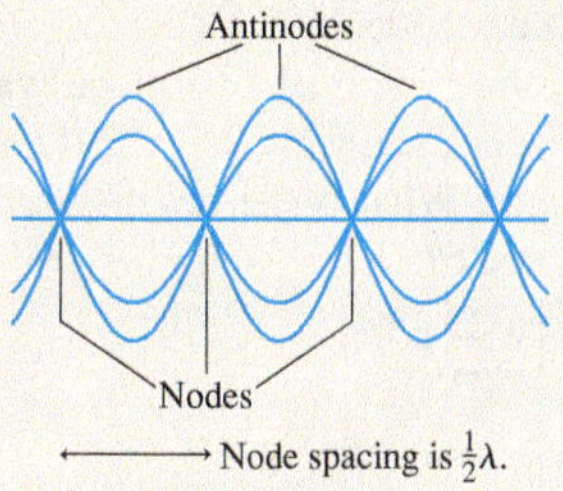

The amplitude at position x is

$$A(x) = 2a \sin kx$$

where a is the amplitude of each wave.

The boundary conditions determine which standing-wave frequencies and wavelengths are allowed. The allowed standing waves are **modes** of the system.

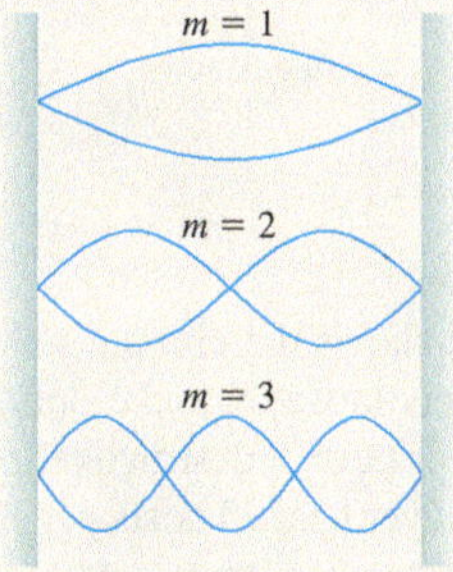

Standing waves on a string

Interference

In general, the superposition of two or more waves into a single wave is called interference.

Maximum constructive interference occurs where crests are aligned with crests and troughs with troughs. These waves are in phase. The maximum displacement is $A = 2a$.

Perfect destructive interference occurs where crests are aligned with troughs. These waves are out of phase. The amplitude is $A = 0$.

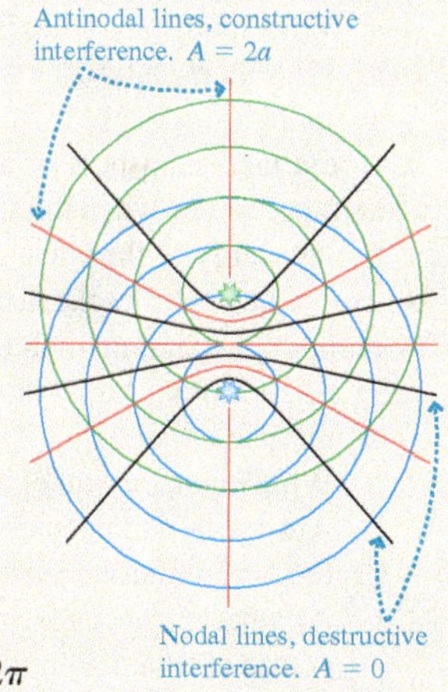

Interference depends on the phase difference $\Delta\phi$ between the two waves.

$$\text{Constructive: } \Delta\phi = 2\pi\frac{\Delta r}{\lambda} + \Delta\phi_0 = m \cdot 2\pi$$

$$\text{Destructive: } \Delta\phi = 2\pi\frac{\Delta r}{\lambda} + \Delta\phi_0 = \left(m + \frac{1}{2}\right) \cdot 2\pi$$

Δr is the path-length difference of the two waves, and $\Delta\phi_0$ is any phase difference between the sources. For identical sources (in phase, $\Delta\phi_0 = 0$):

Interference is constructive if the path-length difference $\Delta r = m\lambda$.

Interference is destructive if the path-length difference $\Delta r = \left(m + \frac{1}{2}\right)\lambda$.

The amplitude at a point where the phase difference is $\Delta\phi$ is $A = \left|2a\cos\left(\frac{\Delta\phi}{2}\right)\right|$.

Applications

Boundary conditions

Strings, electromagnetic waves, and sound waves in closed-closed tubes must have nodes at both ends:

$$\lambda_m = \frac{2L}{m} \qquad f_m = m\frac{v}{2L} = mf_1$$

where $m = 1, 2, 3, \ldots$.

The frequencies and wavelengths are the same for a sound wave in an open-open tube, which has antinodes at both ends.

A sound wave in an open-closed tube must have a node at the closed end but an antinode at the open end. This leads to

$$\lambda_m = \frac{4L}{m} \qquad f_m = m\frac{v}{4L} = mf_1$$

where $m = 1, 3, 5, 7, \ldots$.

Beats (loud-soft-loud-soft modulations of intensity) occur when two waves of slightly different frequency are superimposed.

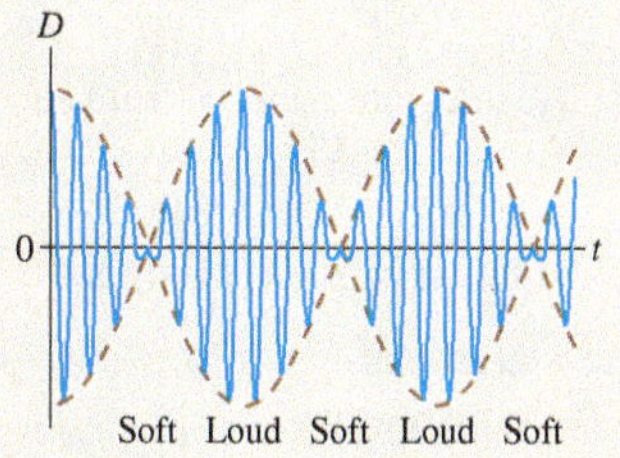

The beat frequency between waves of frequencies f_1 and f_2 is

$$f_{beat} = f_1 - f_2$$

Terms and Notation

principle of superposition
standing wave
node
antinode
amplitude function, $A(x)$
boundary condition
fundamental frequency, f_1
harmonic
mode
interference
in phase
constructive interference
out of phase
destructive interference
phase difference, $\Delta\phi$
path-length difference, Δx or Δr
thin-film optical coating
antinodal line
nodal line
beats
modulation
beat frequency, f_{beat}

CONCEPTUAL QUESTIONS

1. FIGURE Q21.1 shows a standing wave oscillating on a string at frequency f_0.
   a. What mode (m-value) is this?
   b. How many antinodes will there be if the frequency is doubled to $2f_0$?

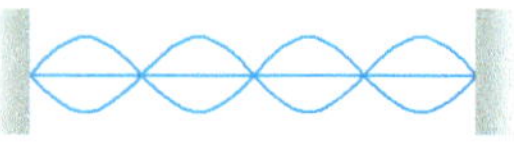

FIGURE Q21.1

2. If you take snapshots of a standing wave on a string, there are certain instants when the string is totally flat. What has happened to the energy of the wave at those instants?
3. FIGURE Q21.3 shows the displacement of a standing sound wave in a 32-cm-long horizontal tube of air open at both ends.
   a. What mode (m-value) is this?
   b. Are the air molecules moving horizontally or vertically? Explain.
   c. At what distances from the left end of the tube do the molecules oscillate with maximum amplitude?
   d. At what distances from the left end of the tube does the air pressure oscillate with maximum amplitude?

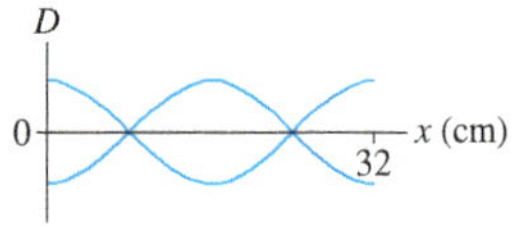

FIGURE Q21.3

4. An organ pipe is tuned to exactly 384 Hz when the room temperature is 20°C. If the room temperature later increases to 22°C, does the pipe's frequency increase, decrease, or stay the same? Explain.
5. If you pour liquid into a tall, narrow glass, you may hear sound with a steadily rising pitch. What is the source of the sound? And why does the pitch rise as the glass fills?
6. A flute filled with helium will, until the helium escapes, play notes at a much higher pitch than normal. Why?
7. In music, two notes are said to be an *octave* apart when one note is exactly twice the frequency of the other. Suppose you have a guitar string playing frequency f_0. To increase the frequency by an octave, to $2f_0$, by what factor would you have to (a) increase the tension or (b) decrease the length?
8. FIGURE Q21.8 is a snapshot graph of two plane waves passing through a region of space. Each wave has a 2.0 mm amplitude and the same wavelength. What is the net displacement of the medium at points a, b, and c?

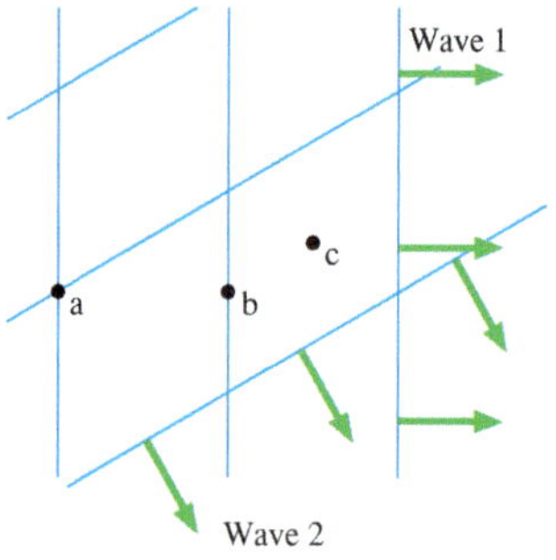

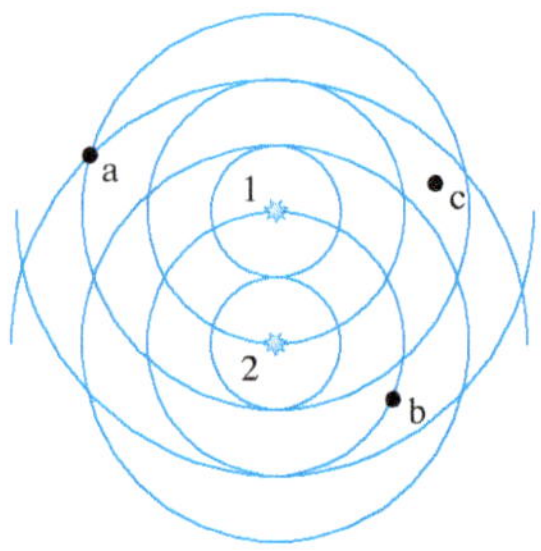

FIGURE Q21.8

FIGURE Q21.9

9. FIGURE Q21.9 shows the circular waves emitted by two in-phase sources. Are points a, b, and c points of maximum constructive interference or perfect destructive interference? Explain.
10. A trumpet player hears 5 beats per second when she plays a note and simultaneously sounds a 440 Hz tuning fork. After pulling her tuning valve out to slightly increase the length of her trumpet, she hears 3 beats per second against the tuning fork. Was her initial frequency 435 Hz or 445 Hz? Explain.

EXERCISES AND PROBLEMS

Problems labeled integrate material from earlier chapters.

Exercises

Section 21.1 The Principle of Superposition

1. | FIGURE EX21.1 is a snapshot graph at $t = 0$ s of two waves approaching each other at 1.0 m/s. Draw six snapshot graphs, stacked vertically, showing the string at 1 s intervals from $t = 1$ s to $t = 6$ s.

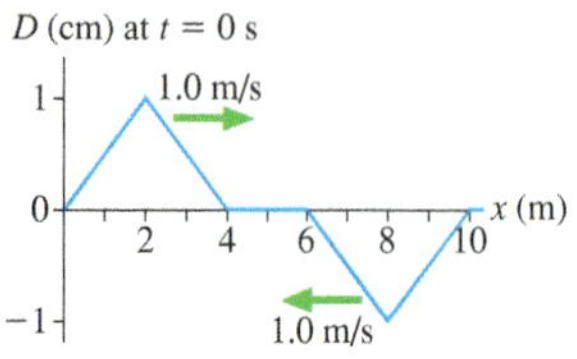

FIGURE EX21.1

2. | FIGURE EX21.2 is a snapshot graph at $t = 0$ s of two waves approaching each other at 1.0 m/s. Draw six snapshot graphs, stacked vertically, showing the string at 1 s intervals from $t = 1$ s to $t = 6$ s.

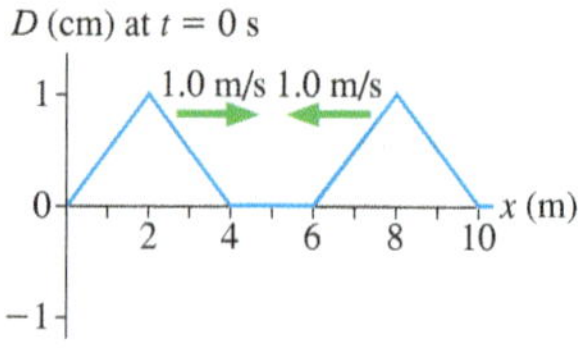

FIGURE EX21.2

3. || FIGURE EX21.3 is a snapshot graph at $t = 0$ s of two waves approaching each other at 1.0 m/s. Draw four snapshot graphs, stacked vertically, showing the string at $t = 2$, 4, 6, and 8 s.

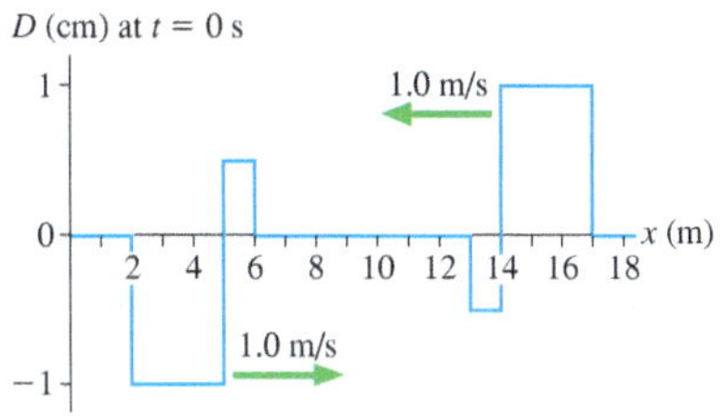

FIGURE EX21.3

4. || FIGURE EX21.4a is a snapshot graph at $t = 0$ s of two waves approaching each other at 1.0 m/s.
   a. At what time was the snapshot graph in FIGURE EX21.4b taken?
   b. Draw a history graph of the string at $x = 5.0$ m from $t = 0$ s to $t = 6$ s.

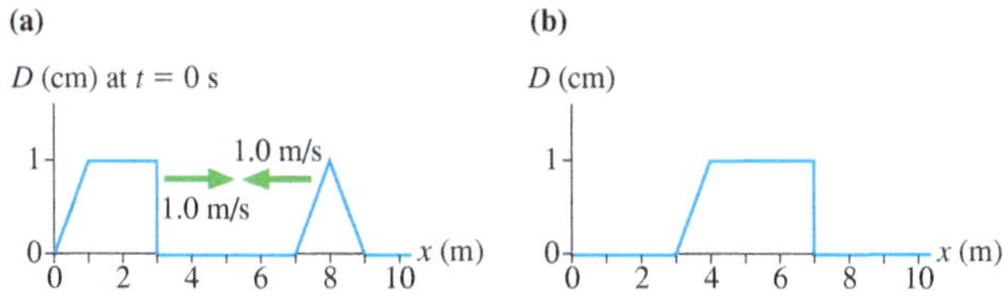

FIGURE EX21.4

Section 21.2 Standing Waves

Section 21.3 Standing Waves on a String

5. || FIGURE EX21.5 is a snapshot graph at $t = 0$ s of two waves moving to the right at 1.0 m/s. The string is fixed at $x = 8.0$ m. Draw four snapshot graphs, stacked vertically, showing the string at $t = 2$, 4, 6, and 8 s.

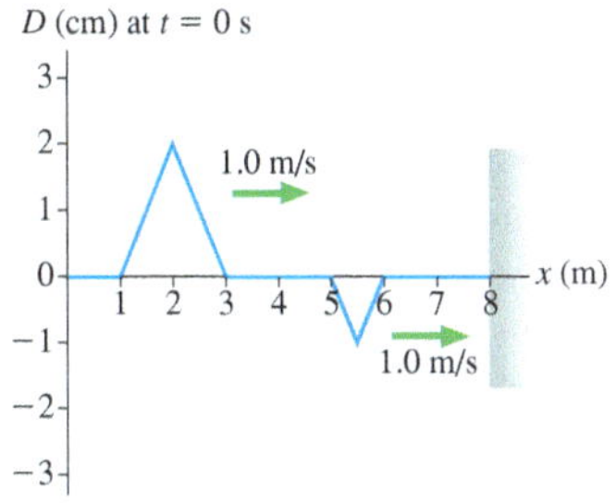

FIGURE EX21.5

6. | FIGURE EX21.6 shows a standing wave oscillating at 100 Hz on a string. What is the wave speed?

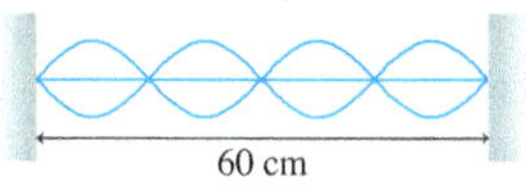

FIGURE EX21.6

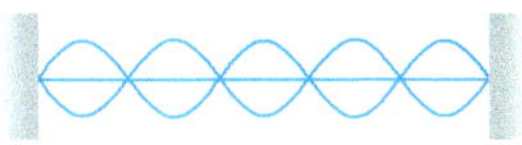

FIGURE EX21.7

7. || FIGURE EX21.7 shows a standing wave on a 2.0-m-long string that has been fixed at both ends and tightened until the wave speed is 40 m/s. What is the frequency?

8. || FIGURE EX21.8 shows a standing wave that is oscillating at frequency f_0.
   a. How many antinodes will there be if the frequency is doubled to $2f_0$? Explain.

FIGURE EX21.8

   b. If the tension in the string is increased by a factor of four, for what frequency, in terms of f_0, will the string continue to oscillate as a standing wave with four antinodes?

9. | a. What are the three longest wavelengths for standing waves on a 240-cm-long string that is fixed at both ends?
   b. If the frequency of the second-longest wavelength is 50 Hz, what is the frequency of the third-longest wavelength?

10. | Standing waves on a 1.0-m-long string that is fixed at both ends are seen at successive frequencies of 36 Hz and 48 Hz.
   a. What are the fundamental frequency and the wave speed?
   b. Draw the standing-wave pattern when the string oscillates at 48 Hz.

11. || A heavy piece of hanging sculpture is suspended by a 90-cm-long, 5.0 g steel wire. When the wind blows hard, the wire hums at its fundamental frequency of 80 Hz. What is the mass of the sculpture?

12. | A carbon dioxide laser is an infrared laser. A CO_2 laser with a cavity length of 53.00 cm oscillates in the $m = 100{,}000$ mode. What are the wavelength and frequency of the laser beam?

Section 21.4 Standing Sound Waves and Musical Acoustics

13. | What are the three longest wavelengths for standing sound waves in a 121-cm-long tube that is (a) open at both ends and (b) open at one end, closed at the other?

14. | FIGURE EX21.14 shows a standing sound wave in an 80-cm-long tube. The tube is filled with an unknown gas. What is the speed of sound in this gas?

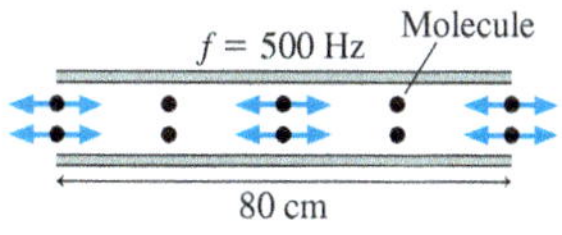

FIGURE EX21.14

15. || The fundamental frequency of an open-open tube is 1500 Hz when the tube is filled with 0°C helium. What is its frequency when filled with 0°C air?

16. | BIO We can make a simple model of the human vocal tract as an open-closed tube extending from the opening of the mouth to the diaphragm. What is the length of this tube if its fundamental frequency equals a typical speech frequency of 250 Hz? The speed of sound in the warm air is 350 m/s.

17. || The lowest note on a grand piano has a frequency of 27.5 Hz. The entire string is 2.00 m long and has a mass of 400 g. The vibrating section of the string is 1.90 m long. What tension is needed to tune this string properly?

18. || A violin string is 30 cm long. It sounds the musical note A (440 Hz) when played without fingering. How far from the end of the string should you place your finger to play the note C (523 Hz)?

Section 21.5 Interference in One Dimension

Section 21.6 The Mathematics of Interference

19. || Two loudspeakers emit sound waves along the x-axis. The sound has maximum intensity when the speakers are 20 cm apart. The sound intensity decreases as the distance between the speakers is increased, reaching zero at a separation of 60 cm.
   a. What is the wavelength of the sound?
   b. If the distance between the speakers continues to increase, at what separation will the sound intensity again be a maximum?

20. || Two loudspeakers in a 20°C room emit 686 Hz sound waves along the x-axis.
   a. If the speakers are in phase, what is the smallest distance between the speakers for which the interference of the sound waves is perfectly destructive?
   b. If the speakers are out of phase, what is the smallest distance between the speakers for which the interference of the sound waves is maximum constructive?

21. | What is the thinnest film of MgF_2 ($n = 1.39$) on glass that produces a strong reflection for orange light with a wavelength of 600 nm?

22. || A very thin oil film ($n = 1.25$) floats on water ($n = 1.33$). What is the thinnest film that produces a strong reflection for green light with a wavelength of 500 nm?

Section 21.7 Interference in Two and Three Dimensions

23. || FIGURE EX21.23 shows the circular wave fronts emitted by two wave sources.
   a. Are these sources in phase or out of phase? Explain.
   b. Make a table with rows labeled P, Q, and R and columns labeled r_1, r_2, Δr, and C/D. Fill in the table for points P, Q, and R, giving the distances as multiples of λ and indicating, with a C or a D, whether the interference at that point is constructive or destructive.

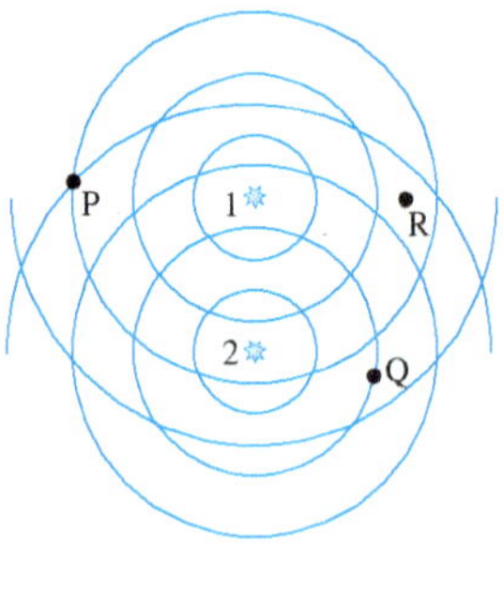

FIGURE EX21.23

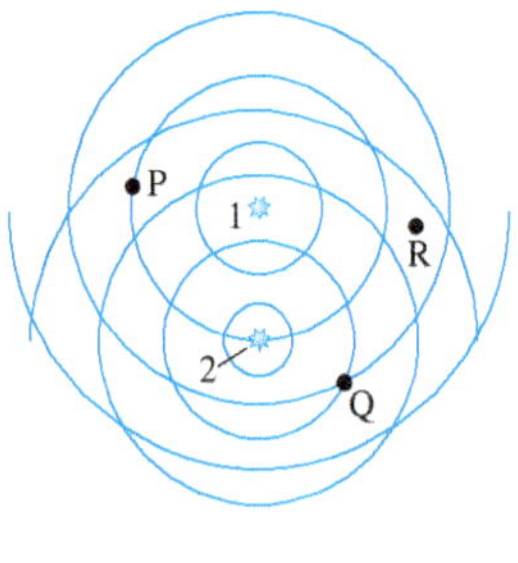

FIGURE EX21.24

24. || FIGURE EX21.24 shows the circular wave fronts emitted by two wave sources.
   a. Are these sources in phase or out of phase? Explain.
   b. Make a table with rows labeled P, Q, and R and columns labeled r_1, r_2, Δr, and C/D. Fill in the table for points P, Q, and R, giving the distances as multiples of λ and indicating, with a C or a D, whether the interference at that point is constructive or destructive.

25. || Two in-phase speakers 2.0 m apart in a plane are emitting 1800 Hz sound waves into a room where the speed of sound is 340 m/s. Is the point 4.0 m in front of one of the speakers, perpendicular to the plane of the speakers, a point of maximum constructive interference, perfect destructive interference, or something in between?

26. || Two out-of-phase radio antennas at $x = \pm 300$ m on the x-axis are emitting 3.0 MHz radio waves. Is the point $(x, y) = (300 \text{ m}, 800 \text{ m})$ a point of maximum constructive interference, perfect destructive interference, or something in between?

Section 21.8 Beats

27. | Two strings are adjusted to vibrate at exactly 200 Hz. Then the tension in one string is increased slightly. Afterward, three beats per second are heard when the strings vibrate at the same time. What is the new frequency of the string that was tightened?

28. | A flute player hears four beats per second when she compares her note to a 523 Hz tuning fork (the note C). She can match the frequency of the tuning fork by pulling out the "tuning joint" to lengthen her flute slightly. What was her initial frequency?

29. | Two microwave signals of nearly equal wavelengths can generate a beat frequency if both are directed onto the same microwave detector. In an experiment, the beat frequency is 100 MHz. One microwave generator is set to emit microwaves with a wavelength of 1.250 cm. If the second generator emits the longer wavelength, what is that wavelength?

Problems

30. || Two waves on a string travel in opposite directions at 100 m/s. FIGURE P21.30 shows a snapshot graph of the string at $t = 0$ s, when the two waves are overlapped, and a snapshot graph of the left-traveling wave at $t = 0.050$ s. Draw a snapshot graph of the right-traveling wave at $t = 0.050$ s.

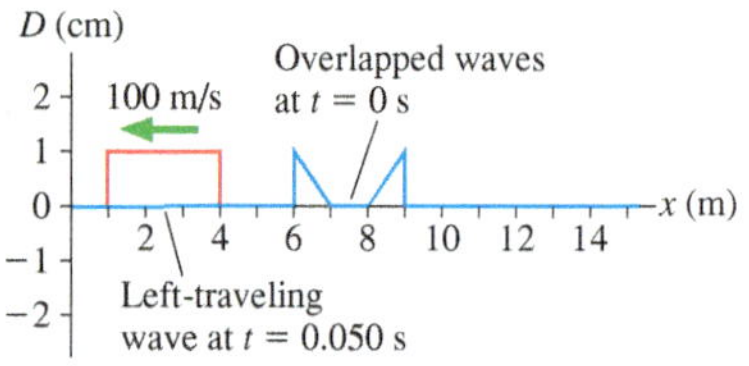

FIGURE P21.30

31. | A 2.0-m-long string vibrates at its second-harmonic frequency with a maximum amplitude of 2.0 cm. One end of the string is at $x = 0$ cm. Find the oscillation amplitude at $x = 10$, 20, 30, 40, and 50 cm.

32. || A string vibrates at its third-harmonic frequency. The amplitude at a point 30 cm from one end is half the maximum amplitude. How long is the string?

33. || A string of length L vibrates at its fundamental frequency. The amplitude at a point $\frac{1}{4}L$ from one end is 2.0 cm. What is the amplitude of each of the traveling waves that form this standing wave?

34. || Two sinusoidal waves with equal wavelengths travel along a string in opposite directions at 3.0 m/s. The time between two successive instants when the antinodes are at maximum height is 0.25 s. What is the wavelength?

35. || BIO Tendons are, essentially, elastic cords stretched between two fixed ends. As such, they can support standing waves. A woman has a 20-cm-long Achilles tendon—connecting the heel to a muscle in the calf—with a cross-section area of 90 mm^2. The density of tendon tissue is 1100 kg/m^3. For a reasonable tension of 500 N, what will be the fundamental frequency of her Achilles tendon?

36. || BIO Biologists think that some spiders "tune" strands of their web to give enhanced response at frequencies corresponding to those at which desirable prey might struggle. Orb spider web silk has a typical diameter of 20 μm, and spider silk has a density of 1300 kg/m^3. To have a fundamental frequency at 100 Hz, to what tension must a spider adjust a 12-cm-long strand of silk?

37. || A particularly beautiful note reaching your ear from a rare Stradivarius violin has a wavelength of 39.1 cm. The room is slightly warm, so the speed of sound is 344 m/s. If the string's linear density is 0.600 g/m and the tension is 150 N, how long is the vibrating section of the violin string?

38. || A violinist places her finger so that the vibrating section of a 1.0 g/m string has a length of 30 cm, then she draws her bow across it. A listener nearby in a 20°C room hears a note with a wavelength of 40 cm. What is the tension in the string?

39. || A steel wire is used to stretch the spring of FIGURE P21.39. An oscillating magnetic field drives the steel wire back and forth. A standing wave with three antinodes is created when the spring is stretched 8.0 cm. What stretch of the spring produces a standing wave with two antinodes?

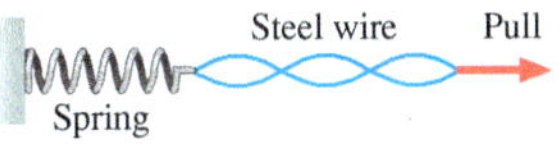

FIGURE P21.39

40. || Astronauts visiting Planet X have a 250-cm-long string whose mass is 5.00 g. They tie the string to a support, stretch it horizontally over a pulley 2.00 m away, and hang a 4.00 kg mass on the free end. Then the astronauts begin to excite standing waves on the horizontal portion of the string. Their data are as follows:

m	Frequency (Hz)
1	31
2	66
3	95
4	130
5	162

Use the best-fit line of an appropriate graph to determine the value of g, the free-fall acceleration on Planet X.

41. || A 75 g bungee cord has an equilibrium length of 1.20 m. The cord is stretched to a length of 1.80 m, then vibrated at 20 Hz. This produces a standing wave with two antinodes. What is the spring constant of the bungee cord?

42. || A metal wire under tension T_0 vibrates at its fundamental frequency. For what tension will the second-harmonic frequency be the same as the fundamental frequency at tension T_0?

43. ||| In a laboratory experiment, one end of a horizontal string is tied to a support while the other end passes over a frictionless pulley and is tied to a 1.5 kg sphere. Students determine the frequencies of standing waves on the horizontal segment of the string, then they raise a beaker of water until the hanging 1.5 kg sphere is completely submerged. The frequency of the fifth harmonic with the sphere submerged exactly matches the frequency of the third harmonic before the sphere was submerged. What is the diameter of the sphere?

44. ||| What is the fundamental frequency of the steel wire in FIGURE P21.44?

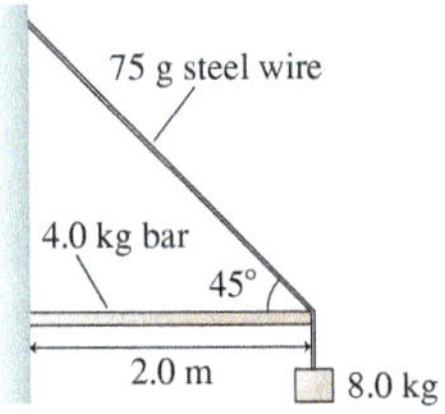

FIGURE P21.44

45. || The two strings in FIGURE P21.45 are of equal length and are being driven at equal frequencies. The linear density of the left string is μ_0. What is the linear density of the right string?

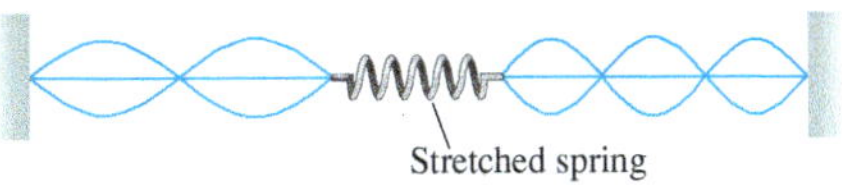

FIGURE P21.45

46. | Microwaves pass through a small hole into the "microwave cavity" of FIGURE P21.46. What frequencies between 10 GHz and 20 GHz will create standing waves in the cavity?

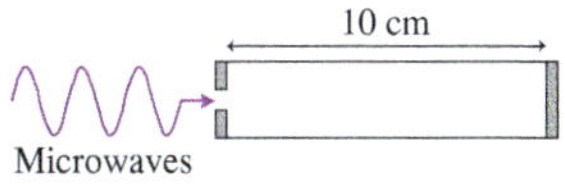

FIGURE P21.46

47. || An open-open organ pipe is 78.0 cm long. An open-closed pipe has a fundamental frequency equal to the third harmonic of the open-open pipe. How long is the open-closed pipe?

48. | A narrow column of 20°C air is found to have standing waves at frequencies of 390 Hz, 520 Hz, and 650 Hz and at no frequencies in between these. The behavior of the tube at frequencies less than 390 Hz or greater than 650 Hz is not known.

a. Is this an open-open tube or an open-closed tube? Explain.
b. How long is the tube?

49. || BIO Deep-sea divers often breathe a mixture of helium and oxygen to avoid getting the "bends" from breathing high-pressure nitrogen. The helium has the side effect of making the divers' voices sound odd. Although your vocal tract can be roughly described as an open-closed tube, the way you hold your mouth and position your lips greatly affects the standing-wave frequencies of the vocal tract. This is what allows different vowels to sound different. The "ee" sound is made by shaping your vocal tract to have standing-wave frequencies at, normally, 270 Hz and 2300 Hz. What will these frequencies be for a helium-oxygen mixture in which the speed of sound at body temperature is 750 m/s? The speed of sound in air at body temperature is 350 m/s.

50. || In 1866, the German scientist Adolph Kundt developed a technique for accurately measuring the speed of sound in various gases. A long glass tube, known today as a Kundt's tube, has a vibrating piston at one end and is closed at the other. Very finely ground particles of cork are sprinkled in the bottom of the tube before the piston is inserted. As the vibrating piston is slowly moved forward, there are a few positions that cause the cork particles to collect in small, regularly spaced piles along the bottom. FIGURE P21.50 shows an experiment in which the tube is filled with pure oxygen and the piston is driven at 400 Hz. What is the speed of sound in oxygen?

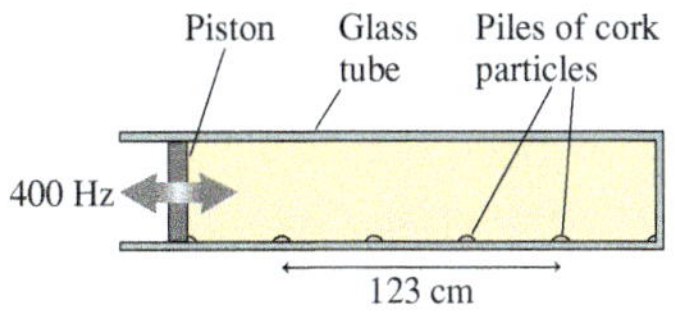

FIGURE P21.50

51. || The 40-cm-long tube of FIGURE P21.51 has a 40-cm-long insert that can be pulled in and out. A vibrating tuning fork is held next to the tube. As the insert is slowly pulled out, the sound from the tuning fork creates standing waves in the tube when the total length L is 42.5 cm, 56.7 cm, and 70.9 cm. What is the frequency of the tuning fork? Assume $v_{sound} = 343$ m/s.

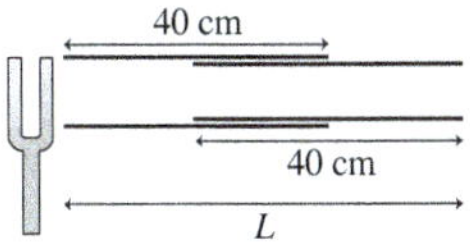

FIGURE P21.51

52. || A 1.0-m-tall vertical tube is filled with 20°C water. A tuning fork vibrating at 580 Hz is held just over the top of the tube as the water is slowly drained from the bottom. At what water heights, measured from the bottom of the tube, will there be a standing wave in the tube above the water?

53. || A 25-cm-long wire with a linear density of 20 g/m passes across the open end of an 85-cm-long open-closed tube of air. If the wire, which is fixed at both ends, vibrates at its fundamental frequency, the sound wave it generates excites the second vibrational mode of the tube of air. What is the tension in the wire? Assume $v_{sound} = 340$ m/s.

54. || A longitudinal standing wave can be created in a long, thin aluminum rod by stroking the rod with very dry fingers. This is often done as a physics demonstration, creating a high-pitched, very annoying whine. From a wave perspective, the standing wave is equivalent to a sound standing wave in an open-open tube. As FIGURE P21.54 shows, both ends of the rod are anti-nodes. What is the fundamental frequency of a 2.0-m-long aluminum rod?

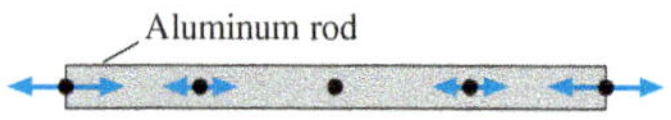

FIGURE P21.54

55. || An old mining tunnel disappears into a hillside. You would like to know how long the tunnel is, but it's too dangerous to go inside. Recalling your recent physics class, you decide to try setting up standing-wave resonances inside the tunnel. Using your subsonic amplifier and loudspeaker, you find resonances at 4.5 Hz and 6.3 Hz, and at no frequencies between these. It's rather chilly inside the tunnel, so you estimate the sound speed to be 335 m/s. Based on your measurements, how far is it to the end of the tunnel?

56. || Analyze the standing sound waves in an open-closed tube to show that the possible wavelengths and frequencies are given by Equation 21.18.

57. ||| Two in-phase loudspeakers emit identical 1000 Hz sound waves along the x-axis. What distance should one speaker be placed behind the other for the sound to have an amplitude 1.5 times that of each speaker alone?

58. || Two loudspeakers emit sound waves of the same frequency along the x-axis. The amplitude of each wave is a. The sound intensity is minimum when speaker 2 is 10 cm behind speaker 1. The intensity increases as speaker 2 is moved forward and first reaches maximum, with amplitude $2a$, when it is 30 cm in front of speaker 1. What is
   a. The wavelength of the sound?
   b. The phase difference between the two loudspeakers?
   c. The amplitude of the sound (as a multiple of a) if the speakers are placed side by side?

59. ||| Two loudspeakers emit sound waves along the x-axis. A listener in front of both speakers hears a maximum sound intensity when speaker 2 is at the origin and speaker 1 is at $x = 0.50$ m. If speaker 1 is slowly moved forward, the sound intensity decreases and then increases, reaching another maximum when speaker 1 is at $x = 0.90$ m.
   a. What is the frequency of the sound? Assume $v_{sound} = 340$ m/s.
   b. What is the phase difference between the speakers?

60. | A sheet of glass is coated with a 500-nm-thick layer of oil ($n = 1.42$).
   a. For what *visible* wavelengths of light do the reflected waves interfere constructively?
   b. For what *visible* wavelengths of light do the reflected waves interfere destructively?
   c. What is the color of reflected light? What is the color of transmitted light?

61. || A manufacturing firm has hired your company, Acoustical Consulting, to help with a problem. Their employees are complaining about the annoying hum from a piece of machinery. Using a frequency meter, you quickly determine that the machine emits a rather loud sound at 1200 Hz. After investigating, you tell the owner that you cannot solve the problem entirely, but you can at least improve the situation by eliminating reflections of this sound from the walls. You propose to do this by installing mesh screens in front of the walls. A portion of the sound will reflect from the mesh; the rest will pass through the mesh and reflect from the wall. How far should the mesh be placed in front of the wall for this scheme to work?

62. || A soap bubble is essentially a very thin film of water ($n = 1.33$) surrounded by air. The colors that you see in soap bubbles are produced by interference.
   a. Derive an expression for the wavelengths λ_C for which constructive interference causes a strong reflection from a soap bubble of thickness d.

   Hint: Think about the reflection phase shifts at both boundaries.
   b. What visible wavelengths of light are strongly reflected from a 390-nm-thick soap bubble? What color would such a soap bubble appear to be?

63. || Two radio antennas are separated by 2.0 m. Both broadcast identical 750 MHz waves. If you walk around the antennas in a circle of radius 10 m, how many maxima will you detect?

64. || You are standing 2.5 m directly in front of one of the two loudspeakers shown in FIGURE P21.64. They are 3.0 m apart and both are playing a 686 Hz tone in phase. As you begin to walk directly away from the speaker, at what distances from the speaker do you hear a *minimum* sound intensity? The room temperature is 20°C.

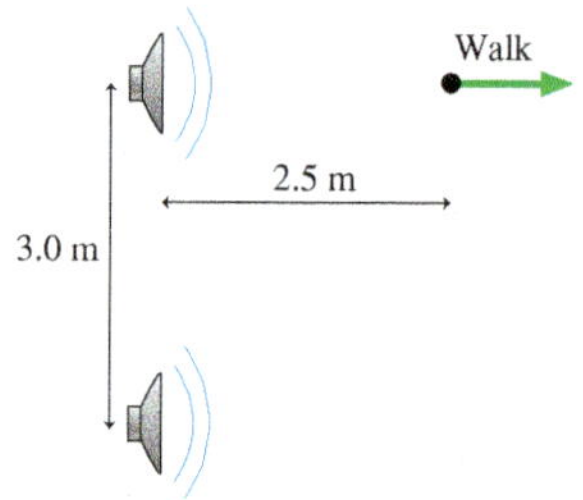

FIGURE P21.64

65. || Two loudspeakers in a plane, 5.0 apart, are playing the same frequency. If you stand 12.0 m in front of the plane of the speakers, centered between them, you hear a sound of maximum intensity. As you walk parallel to the plane of the speakers, staying 12.0 m in front of them, you first hear a minimum of sound intensity when you are directly in front of one of the speakers. What is the frequency of the sound? Assume a sound speed of 340 m/s.

66. || Two in-phase loudspeakers are located at (x, y) coordinates $(-3.0$ m, $+2.0$ m$)$ and $(-3.0$ m, -2.0 m$)$. They emit identical sound waves with a 2.0 m wavelength and amplitude a. Determine the amplitude of the sound at the five positions on the y-axis ($x = 0$) with $y = 0.0$ m, 0.5 m, 1.0 m, 1.5 m, and 2.0 m.

67. || Two identical loudspeakers separated by distance Δx each emit sound waves of wavelength λ and amplitude a along the x-axis. What is the minimum value of the ratio $\Delta x/\lambda$ for which the amplitude of their superposition is also a?

68. || Two radio antennas are 100 m apart along a north-south line. They broadcast identical radio waves at a frequency of 3.0 MHz. Your job is to monitor the signal strength with a handheld receiver. To get to your first measuring point, you walk 800 m east from the midpoint between the antennas, then 600 m north.
   a. What is the phase difference between the waves at this point?
   b. Is the interference at this point maximum constructive, perfect destructive, or somewhere in between? Explain.
   c. If you now begin to walk farther north, does the signal strength increase, decrease, or stay the same? Explain.

69. || The three identical loudspeakers in FIGURE P21.69 play a 170 Hz tone in a room where the speed of sound is 340 m/s. You are standing 4.0 m in front of the middle speaker. At this point, the amplitude of the wave from each speaker is a.
   a. What is the amplitude at this point?
   b. How far must speaker 2 be moved to the left to produce a maximum amplitude at the point where you are standing?
   c. When the amplitude is maximum, by what factor is the sound intensity greater than the sound intensity from a single speaker?

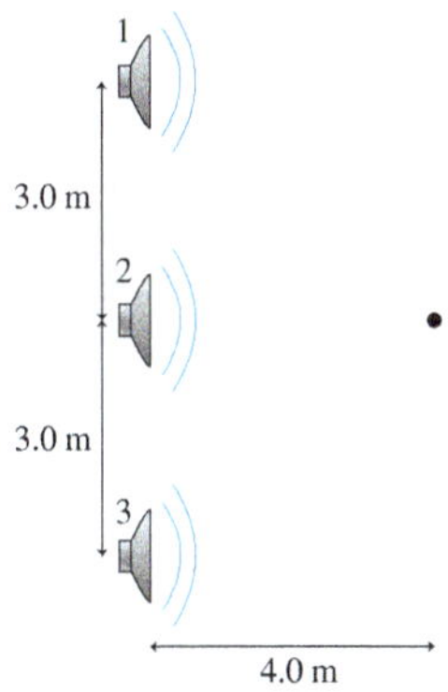

FIGURE P21.69

70. | Piano tuners tune pianos by listening to the beats between the *harmonics* of two different strings. When properly tuned, the note A should have a frequency of 440 Hz and the note E should be at 659 Hz.
   a. What is the frequency difference between the third harmonic of the A and the second harmonic of the E?
   b. A tuner first tunes the A string very precisely by matching it to a 440 Hz tuning fork. She then strikes the A and E strings simultaneously and listens for beats between the harmonics. What beat frequency indicates that the E string is properly tuned?
   c. The tuner starts with the tension in the E string a little low, then tightens it. What is the frequency of the E string when she hears four beats per second?

71. || A flutist assembles her flute in a room where the speed of sound is 342 m/s. When she plays the note A, it is in perfect tune with a 440 Hz tuning fork. After a few minutes, the air inside her flute has warmed to where the speed of sound is 346 m/s.
   a. How many beats per second will she hear if she now plays the note A as the tuning fork is sounded?
   b. How far does she need to extend the "tuning joint" of her flute to be in tune with the tuning fork?

72. || Two loudspeakers face each other from opposite walls of a room. Both are playing exactly the same frequency, thus setting up a standing wave with distance $\lambda/2$ between antinodes. Assume that λ is much less than the room width, so there are many antinodes.
   a. Yvette starts at one speaker and runs toward the other at speed v_Y. As the does so, she hears a loud-soft-loud modulation of the sound intensity. From your perspective, as you sit at rest in the room, Yvette is running through the nodes and antinodes of the standing wave. Find an expression for the number of sound maxima she hears per second.
   b. From Yvette's perspective, the two sound waves are Doppler shifted. They're not the same frequency, so they don't create a standing wave. Instead, she hears a loud-soft-loud modulation of the sound intensity because of beats. Find an expression for the beat frequency that Yvette hears.
   c. Are your answers to parts a and b the same or different? *Should* they be the same or different?

73. || Two loudspeakers emit 400 Hz notes. One speaker sits on the ground. The other speaker is in the back of a pickup truck. You hear eight beats per second as the truck drives away from you. What is the truck's speed?

Challenge Problems

74. a. The frequency of a standing wave on a string is f when the string's tension is T_s. If the tension is changed by the *small* amount ΔT_s, without changing the length, show that the frequency changes by an amount Δf such that

$$\frac{\Delta f}{f} = \frac{1}{2}\frac{\Delta T_s}{T_s}$$

   b. Two identical strings vibrate at 500 Hz when stretched with the same tension. What percentage increase in the tension of one of the strings will cause five beats per second when both strings vibrate simultaneously?

75. A 280 Hz sound wave is directed into one end of the trombone slide seen in FIGURE CP21.75. A microphone is placed at the other end to record the intensity of sound waves that are transmitted through the tube. The straight sides of the slide are 80 cm in length and 10 cm apart with a semicircular bend at the end. For what slide extensions s will the microphone detect a maximum of sound intensity?

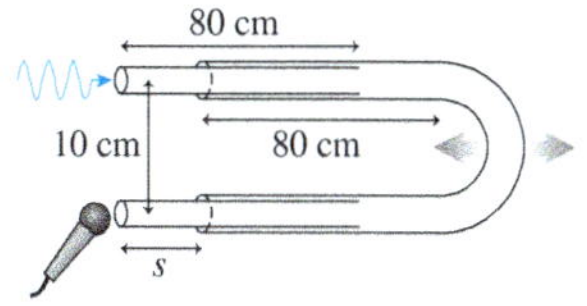

FIGURE CP21.75

76. As the captain of the scientific team sent to Planet Physics, one of your tasks is to measure g. You have a long, thin wire labeled 1.00 g/m and a 1.25 kg weight. You have your accurate space cadet chronometer but, unfortunately, you seem to have forgotten a meter stick. Undeterred, you first find the midpoint of the wire by folding it in half. You then attach one end of the wire to the wall of your laboratory, stretch it horizontally to pass over a pulley at the midpoint of the wire, then tie the 1.25 kg weight to the end hanging over the pulley. By vibrating the wire, and measuring time with your chronometer, you find that the wire's second-harmonic frequency is 100 Hz. Next, with the 1.25 kg weight still tied to one end of the wire, you attach the other end to the ceiling to make a pendulum. You find that the pendulum requires 314 s to complete 100 oscillations. Pulling out your trusty calculator, you get to work. What value of g will you report back to headquarters?

77. When mass M is tied to the bottom of a long, thin wire suspended from the ceiling, the wire's second-harmonic frequency is 200 Hz. Adding an additional 1.0 kg to the hanging mass increases the second-harmonic frequency to 245 Hz. What is M?

78. Ultrasound has many medical applications, one of which is to monitor fetal heartbeats by reflecting ultrasound off a fetus in the womb.
BIO

a. Consider an object moving at speed v_o toward an at-rest source that is emitting sound waves of frequency f_0. Show that the reflected wave (i.e., the echo) that returns to the source has a Doppler-shifted frequency

$$f_{echo} = \left(\frac{v + v_o}{v - v_o}\right) f_0$$

where v is the speed of sound in the medium.

b. Suppose the object's speed is much less than the wave speed: $v_o \ll v$. Then $f_{echo} \approx f_0$, and a microphone that is sensitive to these frequencies will detect a beat frequency if it listens to f_0 and f_{echo} simultaneously. Use the binomial approximation and other appropriate approximations to show that the beat frequency is $f_{beat} \approx (2v_o/v)f_0$.

c. The reflection of 2.40 MHz ultrasound waves from the surface of a fetus's beating heart is combined with the 2.40 MHz wave to produce a beat frequency that reaches a maximum of 65 Hz. What is the maximum speed of the surface of the heart? The speed of ultrasound waves within the body is 1540 m/s.

d. Suppose the surface of the heart moves in simple harmonic motion at 90 beats/min. What is the amplitude in mm of the heartbeat?

79. A water wave is called a *deep-water wave* if the water's depth is more than one-quarter of the wavelength. Unlike the waves we've considered in this chapter, the speed of a deep-water wave depends on its wavelength:

$$v = \sqrt{\frac{g\lambda}{2\pi}}$$

Longer wavelengths travel faster. Let's apply this to standing waves. Consider a diving pool that is 5.0 m deep and 10.0 m wide. Standing water waves can set up across the width of the pool. Because water sloshes up and down at the sides of the pool, the boundary conditions require antinodes at $x = 0$ and $x = L$. Thus a standing water wave resembles a standing sound wave in an open-open tube.

a. What are the wavelengths of the first three standing-wave modes for water in the pool? Do they satisfy the condition for being deep-water waves? Draw a graph of each.

b. What are the wave speeds for each of these waves?

c. Derive a general expression for the frequencies f_m of the possible standing waves. Your expression should be in terms of m, g, and L.

d. What are the oscillation *periods* of the first three standing-wave modes?

80. The broadcast antenna of an AM radio station is located at the edge of town. The station owners would like to beam all of the energy into town and none into the countryside, but a single antenna radiates energy equally in all directions. FIGURE CP21.80 shows two parallel antennas separated by distance L. Both antennas broadcast a signal at wavelength λ, but antenna 2 can delay its broadcast relative to antenna 1 by a time interval Δt in order to create a phase difference $\Delta\phi_0$ between the sources. Your task is to find values of L and Δt such that the waves interfere constructively on the town side and destructively on the country side.

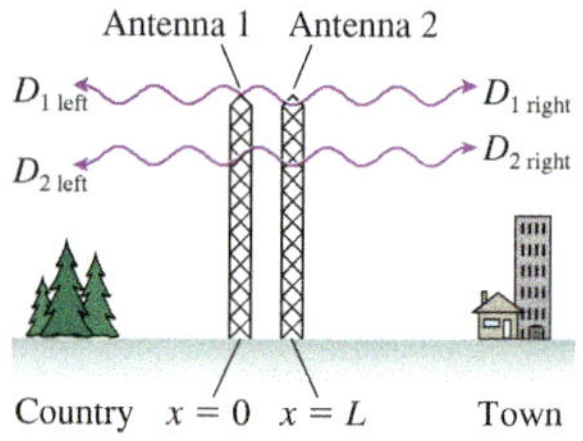

FIGURE CP21.80

Let antenna 1 be at $x = 0$. The wave that travels to the right is $a\sin[2\pi(x/\lambda - t/T)]$. The left wave is $a\sin[2\pi(-x/\lambda - t/T)]$. (It must be this, rather than $a\sin[2\pi(x/\lambda + t/T)]$, so that the two waves match at $x = 0$.) Antenna 2 is at $x = L$. It broadcasts wave $a\sin[2\pi((x - L)/\lambda - t/T) + \phi_{20}]$ to the right and wave $a\sin[2\pi(-(x - L)/\lambda - t/T) + \phi_{20}]$ to the left.

a. What is the smallest value of L for which you can create perfect constructive interference on the town side and perfect destructive interference on the country side? Your answer will be a multiple or fraction of the wavelength λ.

b. What phase constant ϕ_{20} of antenna 2 is needed?

c. What fraction of the oscillation period T must Δt be to produce the proper value of ϕ_{20}?

d. Evaluate both L and Δt for the realistic AM radio frequency of 1000 KHz.

Comment: This is a simple example of what is called a *phased array,* where phase differences between identical emitters are used to "steer" the radiation in a particular direction. Phased arrays are widely used in radar technology.

STOP TO THINK ANSWERS

Stop to Think 21.1: c. The figure shows the two waves at $t = 6$ s and their superposition. The superposition is the *point-by-point* addition of the displacements of the two individual waves.

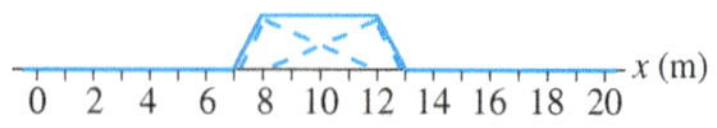

Stop to Think 21.2: a. The allowed standing-wave frequencies are $f_m = m(v/2L)$, so the mode number of a standing wave of frequency f is $m = 2Lf/v$. Quadrupling T_s increases the wave speed v by a factor of 2. The initial mode number was 2, so the new mode number is 1.

Stop to Think 21.3: b. 300 Hz and 400 Hz are allowed standing waves, but they are not f_1 and f_2 because $400\text{ Hz} \neq 2 \times 300\text{ Hz}$. Because there's a 100 Hz difference between them, these must be $f_3 = 3 \times 100$ Hz and $f_4 = 4 \times 100$ Hz, with a fundamental frequency $f_1 = 100$ Hz. Thus the second harmonic is $f_2 = 2 \times 100\text{ Hz} = 200$ Hz.

Stop to Think 21.4: c. Shifting the top wave 0.5 m to the left aligns crest with crest and trough with trough.

Stop to Think 21.5: a. $r_1 = 0.5\lambda$ and $r_2 = 2.5\lambda$, so $\Delta r = 2.0\lambda$. This is the condition for maximum constructive interference.

Stop to Think 21.6: Maximum constructive. The path-length difference is $\Delta r = 1.0\text{ m} = \lambda$. For identical sources, interference is constructive when Δr is an integer multiple of λ.

Stop to Think 21.7: f. The beat frequency is the difference between the two frequencies.

PART V

SUMMARY

Waves and Optics

We end our study of waves a long distance from where we started. Who would have guessed, as we examined our first pulse on a string, that we would end up discussing the resolution of microscopes? But despite the wide disparity between string waves, sound waves, and light waves, a few key ideas have stayed with us throughout Part V: the principle of superposition, interference and diffraction, and standing waves. As part of your final study of waves, you should trace the influence of these ideas through the chapters of Part V.

One point we have tried to emphasize is the *unity* of wave physics. We did not need separate theories of string waves and sound waves and light waves. Instead, a few basic ideas enabled us to understand waves of all types. By focusing on similarities, we have been able to analyze vibrating guitar strings and antireflection coatings on lenses in a single part of this book.

Unfortunately, the physics of waves is not as easily summarized as the physics of particles. Newton's laws and the conservation laws are two very general sets of principles about particles, principles that allowed us to develop the powerful problem-solving strategies of Parts I and II. You probably noticed that we have not found any general problem-solving strategies for wave problems.

This is not to say that wave physics has no structure. Rather, the knowledge structure of waves and optics rests more heavily on *phenomena* than on general principles. Unlike the knowledge structure of Newtonian mechanics, which was a "pyramid of ideas," the knowledge structure of waves is a logical grouping of the major topics you studied. This is a different way of structuring knowledge, but it still provides you with a mental framework for analyzing and thinking about wave problems.

KNOWLEDGE STRUCTURE V Waves and Optics

ESSENTIAL CONCEPTS	Wave speed, wavelength, frequency, phase, wave front, and ray
BASIC GOALS	What are the distinguishing features of waves? How does a wave travel through a medium? How does a medium respond to the presence of more than one wave? What is light and what are its properties?
GENERAL PRINCIPLES	Principle of superposition $v = \lambda f$ for periodic waves

Traveling Waves

- The wave speed v is a property of the medium.
- The motion of particles in the medium is distinct from the motion of the wave.
- Snapshot graphs and history graphs show the same wave from different perspectives.
- The Doppler effect of shifted frequencies is observed whenever the wave source or the detector is moving.

Standing Waves

- Standing waves are the superposition of waves moving in opposite directions.
- Nodes and antinodes are spaced by $\lambda/2$.
- Only certain discrete frequencies are allowed, depending on the boundary conditions.

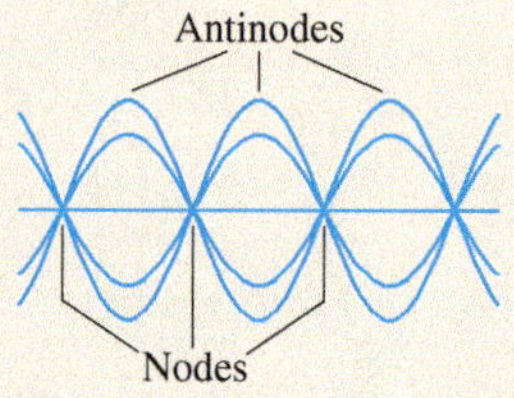

Interference

- Interference is constructive—crests align with crests—if two waves are in phase: $\Delta\phi = 0, 2\pi, 4\pi, \ldots$. The wave is enhanced.
- Interference is destructive—crests align with troughs—if two waves are out of phase: $\Delta\phi = \pi, 3\pi, 5\pi, \ldots$. The wave is reduced.
- The phase difference depends on the path-length difference Δr and on any phase difference between the sources.
- Beats occur when $f_1 \neq f_2$.

Light and Optics

- The wave model, used for interference and diffraction, is appropriate when apertures are comparable in size to the wavelength.

Single-slit diffraction: 

Double-slit interference: 

- The ray model, used for mirrors and lenses, is appropriate when apertures are much larger than the wavelength.
- Diffraction, a wave effect, limits the best possible resolution of a lens.

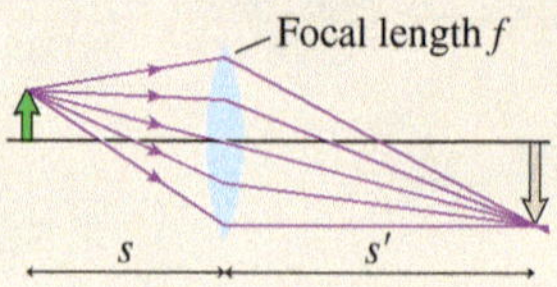

ONE STEP BEYOND

Tsunami!

In December 2004, an earthquake off the Indonesian coast produced a devastating water wave, a *tsunami,* that caused tremendous destruction and loss of life around the edges of the Indian Ocean, often thousands of miles from the earthquake's epicenter. The tsunami was a dramatic reminder of the power of the earth's forces and an impressive illustration of the energy carried by waves.

The Indian Ocean tsunami of 2004 was caused when a very large earthquake disrupted the seafloor along a fault line, pushing one side of the fault up several meters. This dramatic shift in the seafloor produced an almost instantaneous rise in the surface of the ocean above, much like giving a quick shake to one end of a rope. This was the disturbance that produced the tsunami. And just as shaking one end of rope causes a pulse to travel along it, the resulting water wave propagated throughout the Indian Ocean, as we see in the figure, carrying energy from the earthquake.

This computer simulation of the tsunami looks much like the ripples that spread out when you drop a pebble into a pond, but on an immensely larger scale. The individual wave pulses are up to 100 km wide, and the leading wave front spans more than 5000 km.

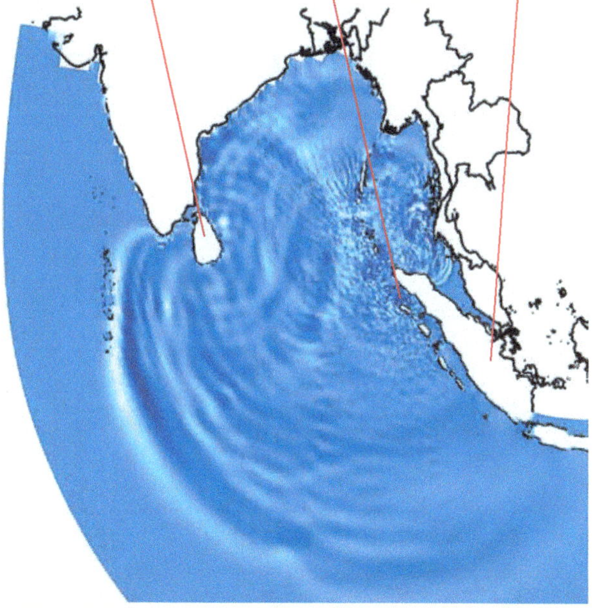

A frame from a computer simulation of the tsunami, showing the Indian Ocean about three hours after the earthquake. Notice the interference pattern to the east of Sri Lanka, where incoming waves and reflected waves are superimposed.

Technically, a tsunami is a "shallow-water wave," even in the deep ocean, because the scale of the wave (roughly 100 km) is much larger than the depth of the ocean (typically 4 km). Consequently, a tsunami travels differently than normal ocean waves. Unlike normal waves on the surface, whose speed is independent of depth, the speed of a shallow-water wave is determined by the depth of the ocean: The greater the depth, the greater the speed. In the deep ocean, a tsunami travels at hundreds of kilometers per hour, about the speed of a jet plane. This great speed allows a tsunami to cross oceans in only a few hours.

The height of the tsunami as it raced across the open ocean was about half a meter. Why should such a small wave—one that ships didn't even notice as it passed—be so fearsome? It's the width of the wave that matters. The wave pulse may have been only half a meter high, but it was about 100 km wide. In other words, the tsunami far from land was a half-meter-high, 100-km-wide wall of water. This is a tremendous amount of water displaced upward, and thus the tsunami was carrying a tremendous amount of energy.

As a tsunami nears shore, the ocean depth decreases and—because its speed is determined by depth—the tsunami begins to slow. This is when the awesome power of a tsunami begins to become apparent. As the leading edge of the wave slows, the trailing edge, still 100 km away and traveling much faster in deeper water, quickly begins to catch up. Water is nearly incompressible. As the width of the wave pulse decreases, the water begins to pile up higher and higher and the wave increases dramatically in height. The Indian Ocean tsunami had a height of up to 15 m (50 ft) as it came ashore.

Despite its height, a tsunami doesn't break and crash on the beach like a normal wave. The wave pulse may have narrowed dramatically from its 100 km width in the open ocean, but it is still several kilometers wide. Thus a tsunami reaching shore is more like a huge water surge than a typical wave—a wall of water that moves onto the shore and just keeps on coming. In many places, the Indian Ocean tsunami reached 2 km inland.

The impact of the Indian Ocean tsunami was devastating, but it was the first tsunami for which scientists were able to use satellites and ocean sensors to make planet-wide measurements. An analysis of the data, including computer simulations like the one seen here, has helped us better understand the physics of these ocean waves. We won't be able to stop future tsunamis, but with a better knowledge of how they are formed and how they travel, we will be better able to warn people to get out of their way.

Mathematics Review

Algebra

Using exponents: $a^{-x} = \dfrac{1}{a^x}$ $\quad a^x a^y = a^{(x+y)}$ $\quad \dfrac{a^x}{a^y} = a^{(x-y)}$ $\quad (a^x)^y = a^{xy}$

$a^0 = 1$ $\quad a^1 = a$ $\quad a^{1/n} = \sqrt[n]{a}$

Fractions: $\left(\dfrac{a}{b}\right)\left(\dfrac{c}{d}\right) = \dfrac{ac}{bd}$ $\quad \dfrac{a/b}{c/d} = \dfrac{ad}{bc}$ $\quad \dfrac{1}{1/a} = a$

Logarithms: If $a = e^x$, then $\ln(a) = x$ $\quad \ln(e^x) = x$ $\quad e^{\ln(x)} = x$

$\ln(ab) = \ln(a) + \ln(b)$ $\quad \ln\left(\dfrac{a}{b}\right) = \ln(a) - \ln(b)$ $\quad \ln(a^n) = n\ln(a)$

The expression $\ln(a + b)$ cannot be simplified.

Linear equations: The graph of the equation $y = ax + b$ is a straight line. a is the slope of the graph. b is the y-intercept.

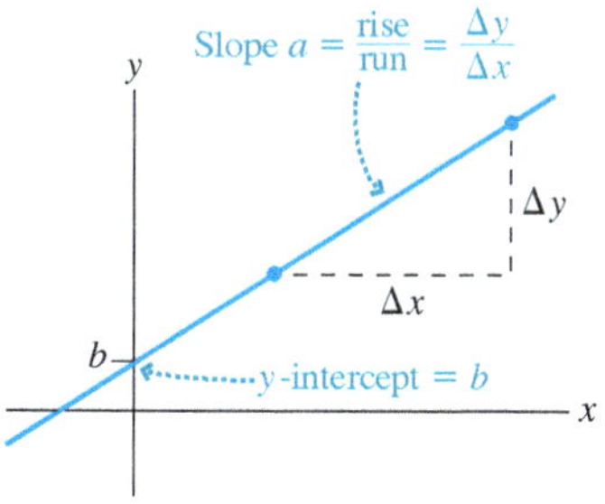

Proportionality: To say that y is proportional to x, written $y \propto x$, means that $y = ax$, where a is a constant. Proportionality is a special case of linearity. A graph of a proportional relationship is a straight line that passes through the origin. If $y \propto x$, then

$$\frac{y_1}{y_2} = \frac{x_1}{x_2}$$

Quadratic equation: The quadratic equation $ax^2 + bx + c = 0$ has the two solutions $x = \dfrac{-b \pm \sqrt{b^2 - 4ac}}{2a}$.

Geometry and Trigonometry

Area and volume:

Rectangle

$A = ab$

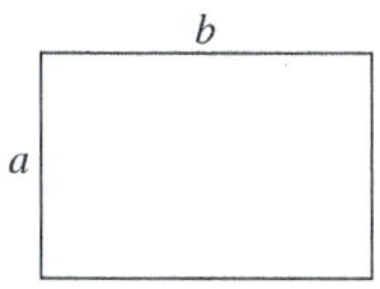

Rectangular box

$V = abc$

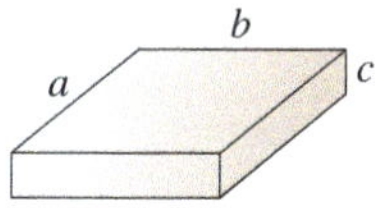

Triangle

$A = \frac{1}{2}ab$

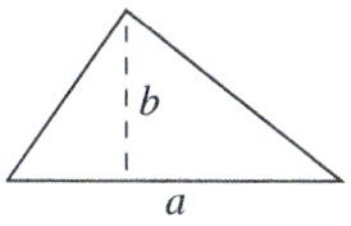

Right circular cylinder

$V = \pi r^2 l$

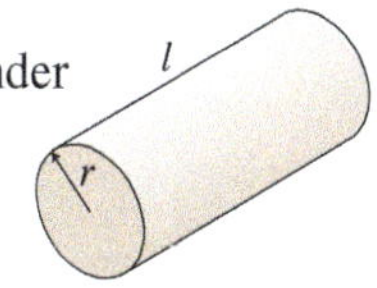

Circle

$C = 2\pi r$

$A = \pi r^2$

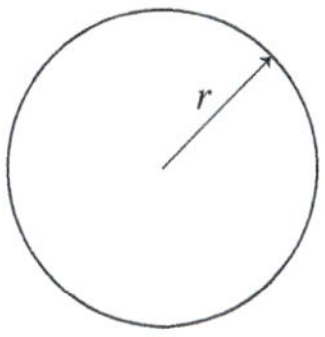

Sphere

$A = 4\pi r^2$

$V = \frac{4}{3}\pi r^3$

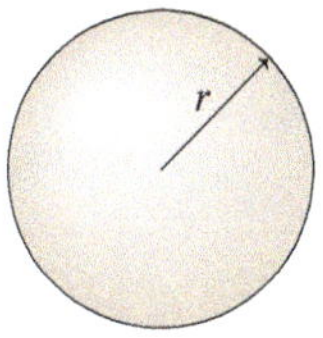

Arc length and angle: The angle θ in radians is defined as $\theta = s/r$.

The arc length that spans angle θ is $s = r\theta$.

$2\pi \text{ rad} = 360°$

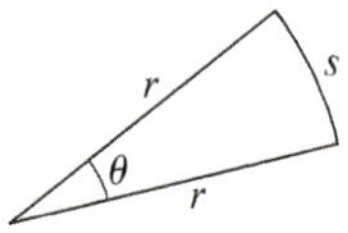

Right triangle: Pythagorean theorem $c = \sqrt{a^2 + b^2}$ or $a^2 + b^2 = c^2$

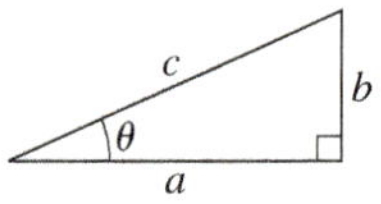

$$\sin\theta = \frac{b}{c} = \frac{\text{far side}}{\text{hypotenuse}} \qquad \theta = \sin^{-1}\left(\frac{b}{c}\right)$$

$$\cos\theta = \frac{a}{c} = \frac{\text{adjacent side}}{\text{hypotenuse}} \qquad \theta = \cos^{-1}\left(\frac{a}{c}\right)$$

$$\tan\theta = \frac{b}{a} = \frac{\text{far side}}{\text{adjacent side}} \qquad \theta = \tan^{-1}\left(\frac{b}{a}\right)$$

General triangle: $\alpha + \beta + \gamma = 180° = \pi \text{ rad}$

Law of cosines $c^2 = a^2 + b^2 - 2ab\cos\gamma$

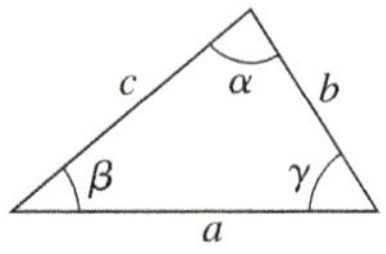

Identities:

$$\tan\alpha = \frac{\sin\alpha}{\cos\alpha} \qquad \sin^2\alpha + \cos^2\alpha = 1$$

$$\sin(-\alpha) = -\sin\alpha \qquad \cos(-\alpha) = \cos\alpha$$

$$\sin(\alpha \pm \beta) = \sin\alpha\cos\beta \pm \cos\alpha\sin\beta \qquad \cos(\alpha \pm \beta) = \cos\alpha\cos\beta \mp \sin\alpha\sin\beta$$

$$\sin(2\alpha) = 2\sin\alpha\cos\alpha \qquad \cos(2\alpha) = \cos^2\alpha - \sin^2\alpha$$

$$\sin(\alpha \pm \pi/2) = \pm\cos\alpha \qquad \cos(\alpha \pm \pi/2) = \mp\sin\alpha$$

$$\sin(\alpha \pm \pi) = -\sin\alpha \qquad \cos(\alpha \pm \pi) = -\cos\alpha$$

Expansions and Approximations

Binomial expansion: $(1 + x)^n = 1 + nx + \frac{n(n-1)}{2}x^2 + \cdots$

Binomial approximation: $(1 + x)^n \approx 1 + nx \quad \text{if} \quad x \ll 1$

Trigonometric expansions:

$$\sin\alpha = \alpha - \frac{\alpha^3}{3!} + \frac{\alpha^5}{5!} - \frac{\alpha^7}{7!} + \cdots \quad \text{for } \alpha \text{ in rad}$$

$$\cos\alpha = 1 - \frac{\alpha^2}{2!} + \frac{\alpha^4}{4!} - \frac{\alpha^6}{6!} + \cdots \quad \text{for } \alpha \text{ in rad}$$

Small-angle approximation: If $\alpha \ll 1$ rad, then $\sin\alpha \approx \tan\alpha \approx \alpha$ and $\cos\alpha \approx 1$.

The small-angle approximation is excellent for $\alpha < 5°$ (≈ 0.1 rad) and generally acceptable up to $\alpha \approx 10°$.

Calculus

The letters a and n represent constants in the following derivatives and integrals.

Derivatives

$$\frac{d}{dx}(a) = 0$$

$$\frac{d}{dx}(ax) = a$$

$$\frac{d}{dx}\left(\frac{a}{x}\right) = -\frac{a}{x^2}$$

$$\frac{d}{dx}(ax^n) = anx^{n-1}$$

$$\frac{d}{dx}\big(\ln(ax)\big) = \frac{1}{x}$$

$$\frac{d}{dx}(e^{ax}) = ae^{ax}$$

$$\frac{d}{dx}\big(\sin(ax)\big) = a\cos(ax)$$

$$\frac{d}{dx}\big(\cos(ax)\big) = -a\sin(ax)$$

Integrals

$$\int x\,dx = \frac{1}{2}x^2$$

$$\int x^2\,dx = \frac{1}{3}x^3$$

$$\int \frac{1}{x^2}\,dx = -\frac{1}{x}$$

$$\int x^n\,dx = \frac{x^{n+1}}{n+1} \qquad n \neq -1$$

$$\int \frac{dx}{x} = \ln x$$

$$\int \frac{dx}{a+x} = \ln(a+x)$$

$$\int \frac{x\,dx}{a+x} = x - a\ln(a+x)$$

$$\int \frac{dx}{\sqrt{x^2 \pm a^2}} = \ln\left(x + \sqrt{x^2 \pm a^2}\right)$$

$$\int \frac{x\,dx}{\sqrt{x^2 \pm a^2}} = \sqrt{x^2 \pm a^2}$$

$$\int \frac{dx}{x^2 + a^2} = \frac{1}{a}\tan^{-1}\left(\frac{x}{a}\right)$$

$$\int \frac{dx}{(x^2 + a^2)^2} = \frac{1}{2a^3}\tan^{-1}\left(\frac{x}{a}\right) + \frac{x}{2a^2(x^2 + a^2)}$$

$$\int \frac{dx}{(x^2 \pm a^2)^{3/2}} = \frac{\pm x}{a^2\sqrt{x^2 \pm a^2}}$$

$$\int \frac{x\,dx}{(x^2 \pm a^2)^{3/2}} = -\frac{1}{\sqrt{x^2 \pm a^2}}$$

$$\int e^{ax}\,dx = \frac{1}{a}e^{ax}$$

$$\int xe^{-x}\,dx = -(x+1)e^{-x}$$

$$\int x^2e^{-x}\,dx = -(x^2 + 2x + 2)e^{-x}$$

$$\int \sin(ax)\,dx = -\frac{1}{a}\cos(ax)$$

$$\int \cos(ax)\,dx = \frac{1}{a}\sin(ax)$$

$$\int \sin^2(ax)\,dx = \frac{x}{2} - \frac{\sin(2ax)}{4a}$$

$$\int \cos^2(ax)\,dx = \frac{x}{2} + \frac{\sin(2ax)}{4a}$$

$$\int_0^\infty x^n e^{-ax}\,dx = \frac{n!}{a^{n+1}}$$

$$\int_0^\infty e^{-ax^2}\,dx = \frac{1}{2}\sqrt{\frac{\pi}{a}}$$

Periodic Table of Elements

Key: Atomic number — 27; Symbol — Co; Atomic mass — 58.9

Period																		
1	1 H 1.0																	2 He 4.0
2	3 Li 6.9	4 Be 9.0											5 B 10.8	6 C 12.0	7 N 14.0	8 O 16.0	9 F 19.0	10 Ne 20.2
3	11 Na 23.0	12 Mg 24.3	Transition elements										13 Al 27.0	14 Si 28.1	15 P 31.0	16 S 32.1	17 Cl 35.5	18 Ar 39.9
4	19 K 39.1	20 Ca 40.1	21 Sc 45.0	22 Ti 47.9	23 V 50.9	24 Cr 52.0	25 Mn 54.9	26 Fe 55.8	27 Co 58.9	28 Ni 58.7	29 Cu 63.5	30 Zn 65.4	31 Ga 69.7	32 Ge 72.6	33 As 74.9	34 Se 79.0	35 Br 79.9	36 Kr 83.8
5	37 Rb 85.5	38 Sr 87.6	39 Y 88.9	40 Zr 91.2	41 Nb 92.9	42 Mo 95.9	43 Tc [98]	44 Ru 101.1	45 Rh 102.9	46 Pd 106.4	47 Ag 107.9	48 Cd 112.4	49 In 114.8	50 Sn 118.7	51 Sb 121.8	52 Te 127.6	53 I 126.9	54 Xe 131.3
6	55 Cs 132.9	56 Ba 137.3	71 Lu 175.0	72 Hf 178.5	73 Ta 180.9	74 W 183.9	75 Re 186.2	76 Os 190.2	77 Ir 192.2	78 Pt 195.1	79 Au 197.0	80 Hg 200.6	81 Tl 204.4	82 Pb 207.2	83 Bi 209.0	84 Po [209]	85 At [210]	86 Rn [222]
7	87 Fr [223]	88 Ra [226]	103 Lr [262]	104 Rf [265]	105 Db [268]	106 Sg [271]	107 Bh [272]	108 Hs [270]	109 Mt [276]	110 Ds [281]	111 Rg [280]	112 Cn [285]	113	114	115	116	117	118

Lanthanides 6	57 La 138.9	58 Ce 140.1	59 Pr 140.9	60 Nd 144.2	61 Pm 144.9	62 Sm 150.4	63 Eu 152.0	64 Gd 157.3	65 Tb 158.9	66 Dy 162.5	67 Ho 164.9	68 Er 167.3	69 Tm 168.9	70 Yb 173.0
Actinides 7	89 Ac [227]	90 Th 232.0	91 Pa 231.0	92 U 238.0	93 Np [237]	94 Pu [244]	95 Am [243]	96 Cm [247]	97 Bk [247]	98 Cf [251]	99 Es [252]	100 Fm [257]	101 Md [258]	102 No [259]

Inner transition elements

An atomic mass in brackets is that of the longest-lived isotope of an element with no stable isotopes.

C APPENDIX

Atomic and Nuclear Data

Atomic Number (Z)	Element	Symbol	Mass Number (A)	Atomic Mass (u)	Percent Abundance	Decay Mode	Half-Life $t_{1/2}$
0	(Neutron)	n	1	1.008 665		β^-	10.4 min
1	Hydrogen	H	1	1.007 825	99.985	stable	
	Deuterium	D	2	2.014 102	0.015	stable	
	Tritium	T	3	3.016 049		β^-	12.33 yr
2	Helium	He	3	3.016 029	0.000 1	stable	
			4	4.002 602	99.999 9	stable	
			6	6.018 886		β^-	0.81 s
3	Lithium	Li	6	6.015 121	7.50	stable	
			7	7.016 003	92.50	stable	
			8	8.022 486		β^-	0.84 s
4	Beryllium	Be	7	7.016 928		EC	53.3 days
			9	9.012 174	100	stable	
			10	10.013 534		β^-	1.5×10^6 yr
5	Boron	B	10	10.012 936	19.90	stable	
			11	11.009 305	80.10	stable	
			12	12.014 352		β^-	0.020 2 s
6	Carbon	C	10	10.016 854		β^+	19.3 s
			11	11.011 433		β^+	20.4 min
			12	12.000 000	98.90	stable	
			13	13.003 355	1.10	stable	
			14	14.003 242		β^-	5 730 yr
			15	15.010 599		β^-	2.45 s
7	Nitrogen	N	12	12.018 613		β^+	0.011 0 s
			13	13.005 738		β^+	9.96 min
			14	14.003 074	99.63	stable	
			15	15.000 108	0.37	stable	
			16	16.006 100		β^-	7.13 s
			17	17.008 450		β^-	4.17 s
8	Oxygen	O	14	14.008 595		EC	70.6 s
			15	15.003 065		β^+	122 s
			16	15.994 915	99.76	stable	
			17	16.999 132	0.04	stable	
			18	17.999 160	0.20	stable	
			19	19.003 577		β^-	26.9 s
9	Fluorine	F	17	17.002 094		EC	64.5 s
			18	18.000 937		β^+	109.8 min
			19	18.998 404	100	stable	
			20	19.999 982		β^-	11.0 s
10	Neon	Ne	19	19.001 880		β^+	17.2 s
			20	19.992 435	90.48	stable	
			21	20.993 841	0.27	stable	
			22	21.991 383	9.25	stable	

Atomic Number (Z)	Element	Symbol	Mass Number (A)	Atomic Mass (u)	Percent Abundance	Decay Mode	Half-Life $t_{1/2}$
11	Sodium	Na	22	21.994 434		β^+	2.61 yr
			23	22.989 770	100	stable	
			24	23.990 961		β^-	14.96 hr
12	Magnesium	Mg	24	23.985 042	78.99	stable	
			25	24.985 838	10.00	stable	
			26	25.982 594	11.01	stable	
13	Aluminum	Al	27	26.981 538	100	stable	
			28	27.981 910		β^-	2.24 min
14	Silicon	Si	28	27.976 927	92.23	stable	
			29	28.976 495	4.67	stable	
			30	29.973 770	3.10	stable	
			31	30.975 362		β^-	2.62 hr
15	Phosphorus	P	30	29.978 307		β^+	2.50 min
			31	30.973 762	100	stable	
			32	31.973 908		β^-	14.26 days
16	Sulfur	S	32	31.972 071	95.02	stable	
			33	32.971 459	0.75	stable	
			34	33.967 867	4.21	stable	
			35	34.969 033		β^-	87.5 days
			36	35.967 081	0.02	stable	
17	Chlorine	Cl	35	34.968 853	75.77	stable	
			36	35.968 307		β^-	3.0×10^5 yr
			37	36.965 903	24.23	stable	
18	Argon	Ar	36	35.967 547	0.34	stable	
			38	37.962 732	0.06	stable	
			39	38.964 314		β^-	269 yr
			40	39.962 384	99.60	stable	
			42	41.963 049		β^-	33 yr
19	Potassium	K	39	38.963 708	93.26	stable	
			40	39.964 000	0.01	β^+	1.28×10^9 yr
			41	40.961 827	6.73	stable	
20	Calcium	Ca	40	39.962 591	96.94	stable	
			42	41.958 618	0.64	stable	
			43	42.958 767	0.13	stable	
			44	43.955 481	2.08	stable	
			47	46.954 547		β^-	4.5 days
			48	47.952 534	0.18	stable	
24	Chromium	Cr	50	49.946 047	4.34	stable	
			52	51.940 511	83.79	stable	
			53	52.940 652	9.50	stable	
			54	53.938 883	2.36	stable	
26	Iron	Fe	54	53.939 613	5.9	stable	
			55	54.938 297		EC	2.7 yr
			56	55.934 940	91.72	stable	
			57	56.935 396	2.1	stable	
			58	57.933 278	0.28	stable	

Atomic Number (Z)	Element	Symbol	Mass Number (A)	Atomic Mass (u)	Percent Abundance	Decay Mode	Half-Life $t_{1/2}$
27	Cobalt	Co	59	58.933 198	100	stable	
			60	59.933 820		β^-	5.27 yr
28	Nickel	Ni	58	57.935 346	68.08	stable	
			60	59.930 789	26.22	stable	
			61	60.931 058	1.14	stable	
			62	61.928 346	3.63	stable	
			64	63.927 967	0.92	stable	
29	Copper	Cu	63	62.929 599	69.17	stable	
			65	64.927 791	30.83	stable	
47	Silver	Ag	107	106.905 091	51.84	stable	
			109	108.904 754	48.16	stable	
48	Cadmium	Cd	106	105.906 457	1.25	stable	
			109	108.904 984		EC	462 days
			110	109.903 004	12.49	stable	
			111	110.904 182	12.80	stable	
			112	111.902 760	24.13	stable	
			113	112.904 401	12.22	stable	
			114	113.903 359	28.73	stable	
			116	115.904 755	7.49	stable	
53	Iodine	I	127	126.904 474	100	stable	
			129	128.904 984		β^-	1.6×10^7 yr
			131	130.906 124		β^-	8 days
54	Xenon	Xe	128	127.903 531	1.9	stable	
			129	128.904 779	26.4	stable	
			130	129.903 509	4.1	stable	
			131	130.905 069	21.2	stable	
			132	131.904 141	26.9	stable	
			133	132.905 906		β^-	5.4 days
			134	133.905 394	10.4	stable	
			136	135.907 215	8.9	stable	
55	Cesium	Cs	133	132.905 436	100	stable	
			137	136.907 078		β^-	30 yr
56	Barium	Ba	131	130.906 931		EC	12 days
			133	132.905 990		EC	10.5 yr
			134	133.904 492	2.42	stable	
			135	134.905 671	6.59	stable	
			136	135.904 559	7.85	stable	
			137	136.905 816	11.23	stable	
			138	137.905 236	71.70	stable	
79	Gold	Au	197	196.966 543	100	stable	
81	Thallium	Tl	203	202.972 320	29.524	stable	
			205	204.974 400	70.476	stable	
			207	206.977 403		β^-	4.77 min
82	Lead	Pb	204	203.973 020	1.4	stable	
			205	204.974 457		EC	1.5×10^7 yr

Atomic Number (Z)	Element	Symbol	Mass Number (A)	Atomic Mass (u)	Percent Abundance	Decay Mode	Half-Life $t_{1/2}$
			206	205.974 440	24.1	stable	
			207	206.975 871	22.1	stable	
			208	207.976 627	52.4	stable	
			210	209.984 163		α, β^-	22.3 yr
			211	210.988 734		β^-	36.1 min
83	Bismuth	Bi	208	207.979 717		EC	3.7×10^5 yr
			209	208.980 374	100	stable	
			211	210.987 254		α	2.14 min
			215	215.001 836		β^-	7.4 min
84	Polonium	Po	209	208.982 405		α	102 yr
			210	209.982 848		α	138.38 days
			215	214.999 418		α	0.001 8 s
			218	218.008 965		α, β^-	3.10 min
85	Astatine	At	218	218.008 685		α, β^-	1.6 s
			219	219.011 294		α, β^-	0.9 min
86	Radon	Rn	219	219.009 477		α	3.96 s
			220	220.011 369		α	55.6 s
			222	222.017 571		α, β^-	3.823 days
87	Francium	Fr	223	223.019 733		α, β^-	22 min
88	Radium	Ra	223	223.018 499		α	11.43 days
			224	224.020 187		α	3.66 days
			226	226.025 402		α	1 600 yr
			228	228.031 064		β^-	5.75 yr
89	Actinium	Ac	227	227.027 749		α, β^-	21.77 yr
			228	228.031 015		β^-	6.15 hr
90	Thorium	Th	227	227.027 701		α	18.72 days
			228	228.028 716		α	1.913 yr
			229	229.031 757		α	7 300 yr
			230	230.033 127		α	75.000 yr
			231	231.036 299		α, β^-	25.52 hr
			232	232.038 051	100	α	1.40×10^{10} yr
			234	234.043 593		β^-	24.1 days
91	Protactinium	Pa	231	231.035 880		α	32.760 yr
			234	234.043 300		β^-	6.7 hr
92	Uranium	U	233	233.039 630		α	1.59×10^5 yr
			234	234.040 946		α	2.45×10^5 yr
			235	235.043 924	0.72	α	7.04×10^8 yr
			236	236.045 562		α	2.34×10^7 yr
			238	238.050 784	99.28	α	4.47×10^9 yr
93	Neptunium	Np	236	236.046 560		EC	1.15×10^5 yr
			237	237.048 168		α	2.14×10^6 yr
94	Plutonium	Pu	238	238.049 555		α	87.7 yr
			239	239.052 157		α	2.412×10^4 yr
			240	240.053 808		α	6 560 yr
			242	242.058 737		α	3.73×10^6 yr

ActivPhysics OnLine™ Activities www.masteringphysics.com

D

APPENDIX

The following list gives the activity numbers and titles of the ActivPhysics activities available in the Pearson eText and the Study Area of MasteringPhysics, followed by the corresponding textbook page references.

APPENDIX D

PhET Simulations

www.masteringphysics.com

The following list gives the titles of the PhET simulations available in the Pearson eText and in the Study Area of MasteringPhysics, with the corresponding textbook section and page references.

*Indicates an associated tutorial is available in the MasteringPhysics Item Library.

Answers

Answers to Odd-Numbered Exercises and Problems

Chapter 1

1.

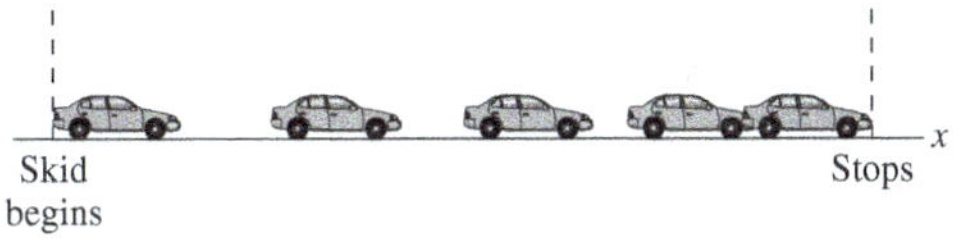

3.

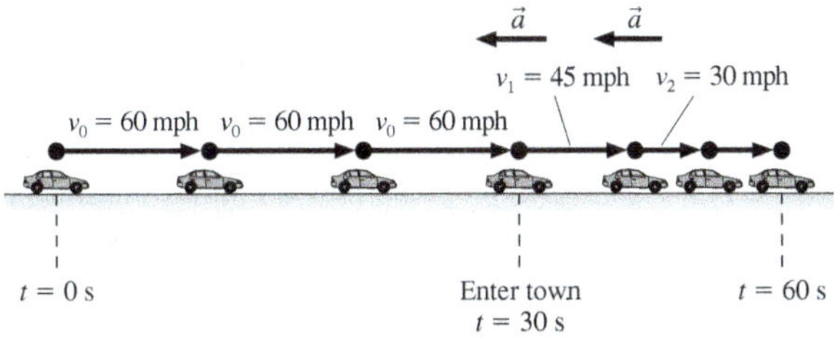

5.

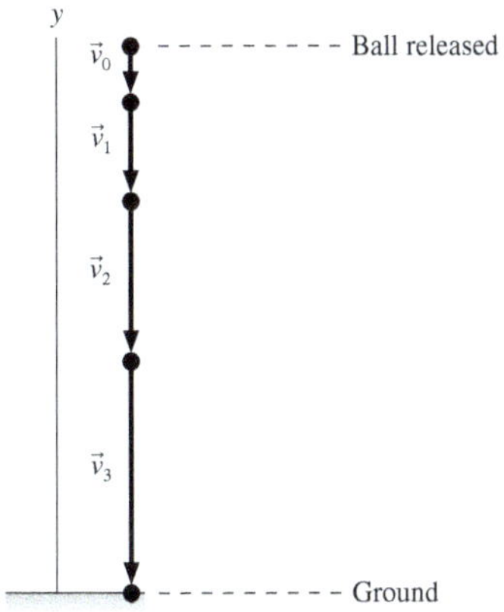

7.

9. a. Greater
b.

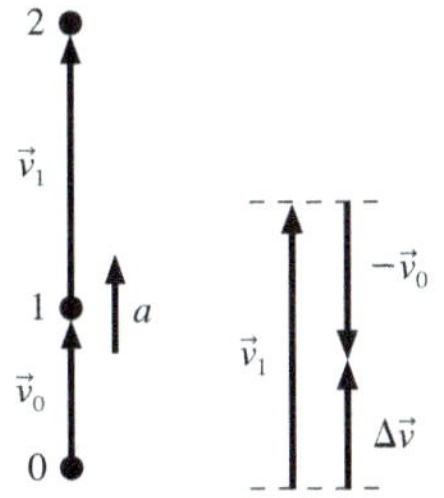

11. a.

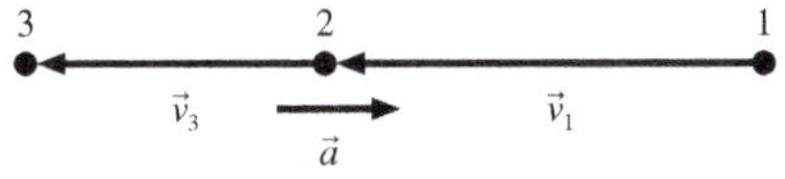

b.

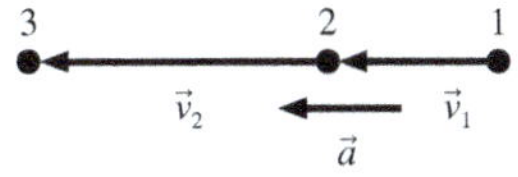

13.

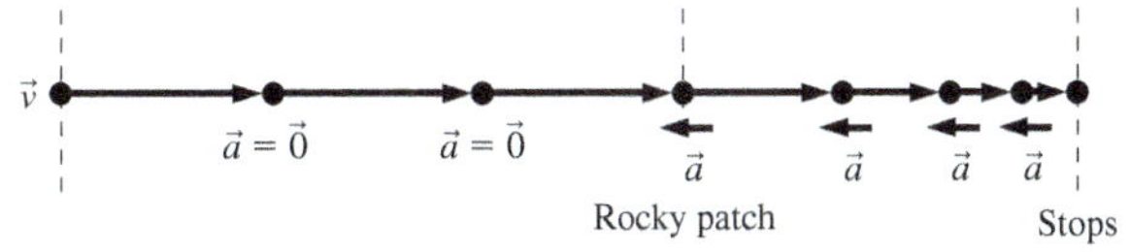

15.

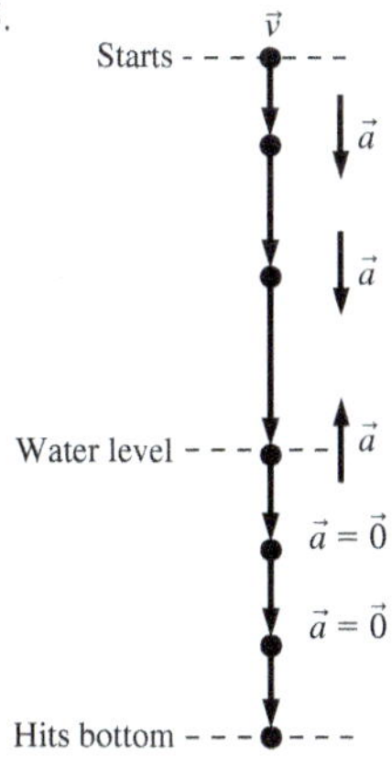

17.

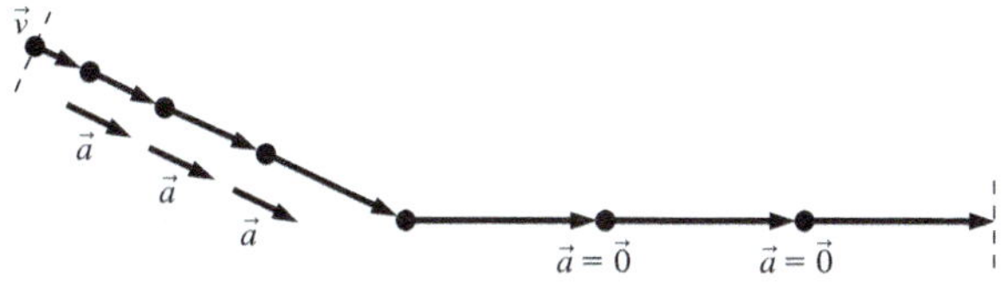

21.

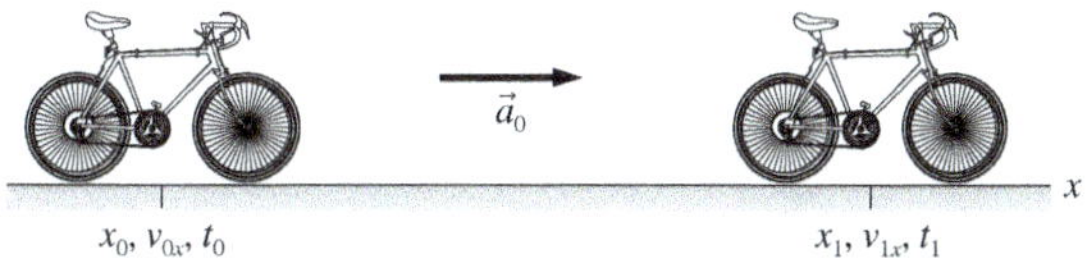

Known

$v_{0x} = 0$ m/s $t_0 = 0$ s $x_0 = 0$ m

$a_0 = 1.5$ m/s^2

$v_1 = 7.5$ m/s

Find

x_1

23. a. 6.15×10^{-3} s b. 27.2×10^3 m c. 31.1 m/s d. 7.2×10^{-2} m/s
25. a. 4×10^3 s b. 9×10^4 s c. 3×10^7 s d. 65.5 m/s
27. a. 12 in b. 50 mph c. 3 mi d. 0.3 in
29. a. 1.95×10^3 b. 65.1 c. 2.3 d. 0.0800
31. 50 ft or 15 m, about 8 times my height
33. 9.8×10^{-9} m/s = 40 μm/h
35. **Pictorial representation**

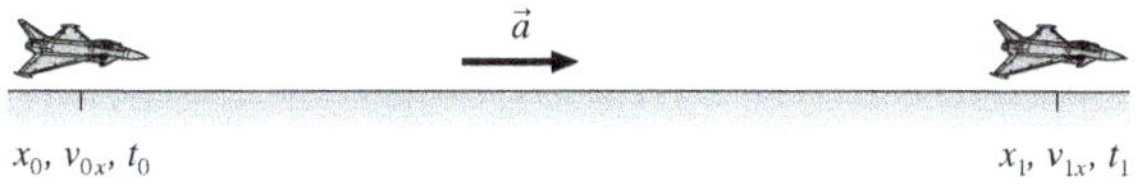

Motion diagram

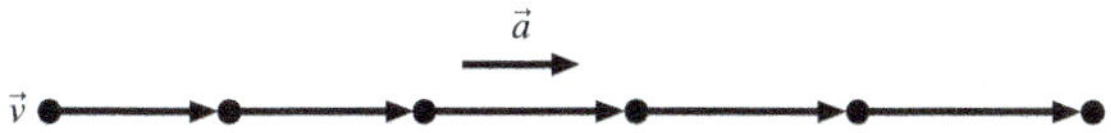

Known

$x_0 = 0$ m $x_1 = 4$ km

$v_{0x} = 300$ m/s $t_0 = 0$ s

Find

a_x

37. **Pictorial representation** **Motion diagram**

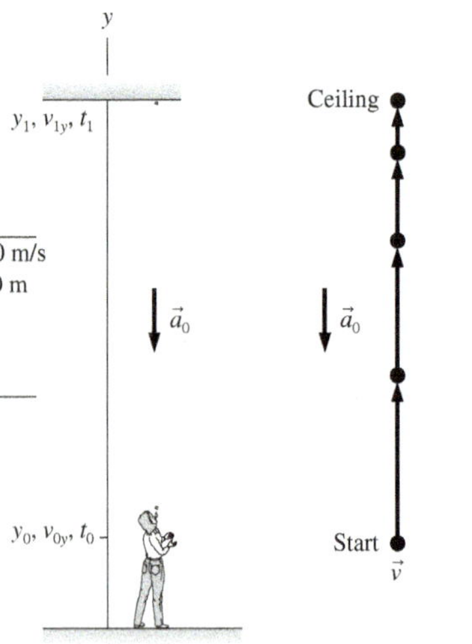

Known
$y_0 = 0$ m $v_{0y} = 10$ m/s
$t_0 = 0$ s $y_1 = 3.0$ m
$a_{0y} < 0$

Find
t_1

39. **Pictorial representation**

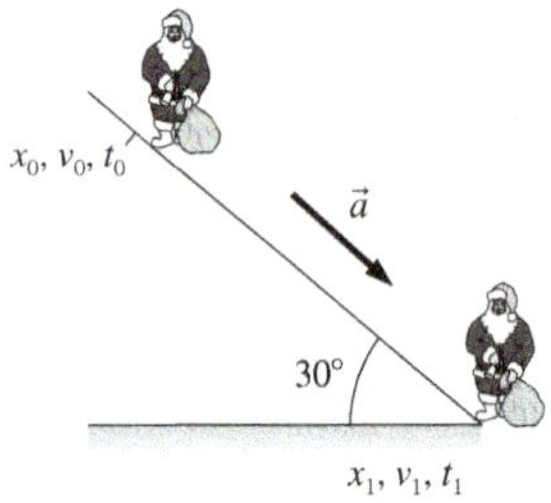

Known
$x_0 = 0$ m $x_1 = 10$ m/s
$t_0 = 0$ s $a_x = (9.8 \text{ m/s}^2) \sin(30°)$
$v_{0x} = 0$ m/s

Find
v_{1x}

Motion diagram

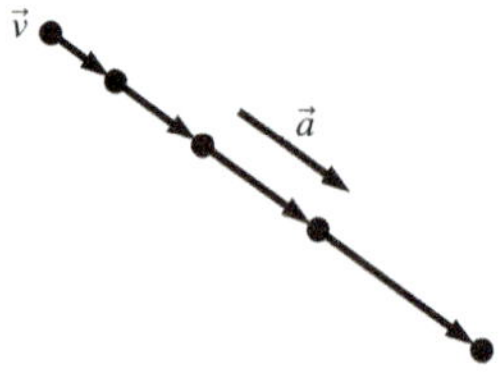

41. **Pictorial representation**

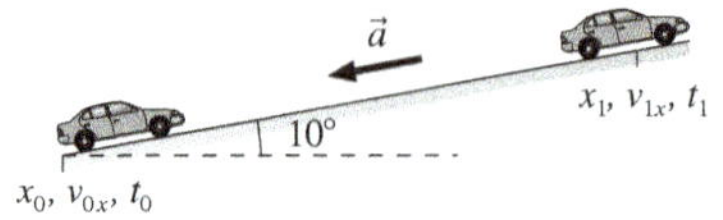

Motion diagram

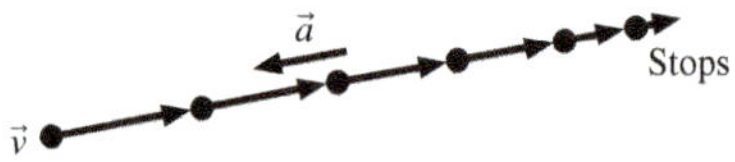

Known
$x_0 = 0$ m $v_{1x} = 0$ m/s
$t_0 = 0$ s $a_x = -(9.8 \text{ m/s}^2) \sin(10°)$
$v_{0x} = 30$ m/s

Find
x_1

43. **Pictorial representation**

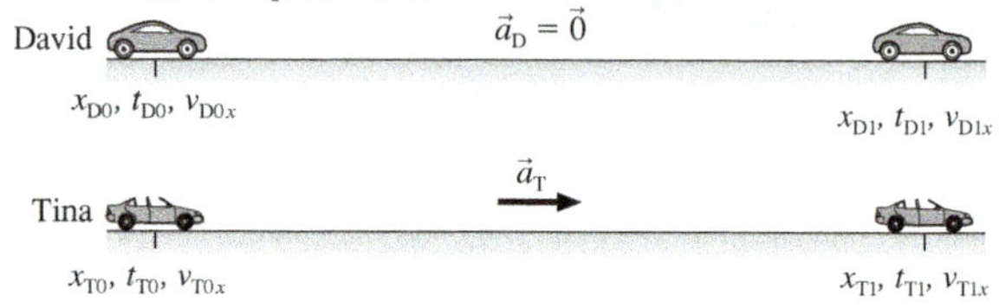

Motion diagram

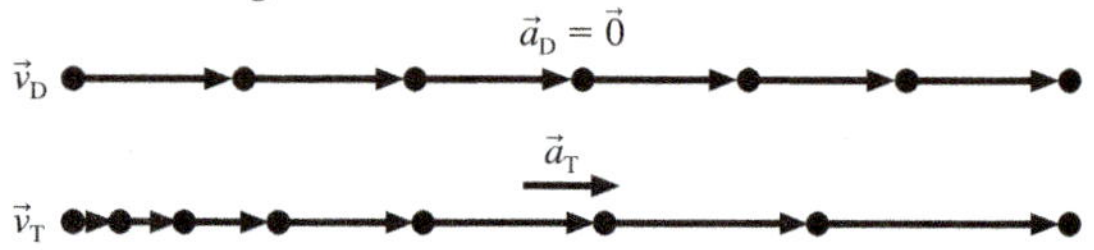

Known
$x_{D0} = 0$ m $x_{T0} = 0$ m
$t_{D0} = 0$ s $t_{T0} = 0$ s
$v_{D0x} = 30$ m/s $v_{T0x} = 0$ m/s
$a_{D0x} = 0 \text{ m/s}^2$

Find
a_{Tx}

49.

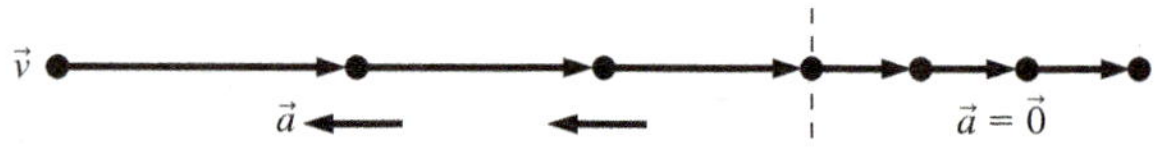

51.

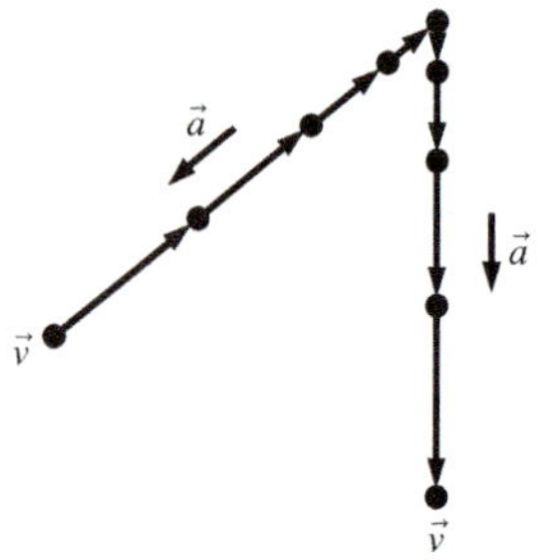

53. Smallest: $6.4 \times 10^3 \text{ m}^2$, largest: $8.3 \times 10^3 \text{ m}^2$

55.

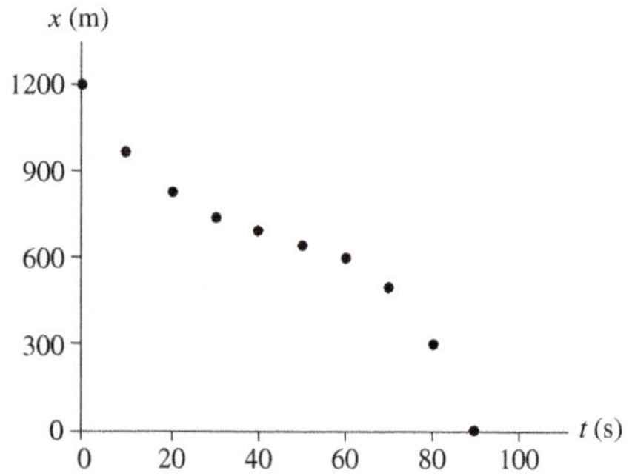

Chapter 2

1. a. Beth b. 20 min
3. a. 48 mph b. 50 mph
5. a.

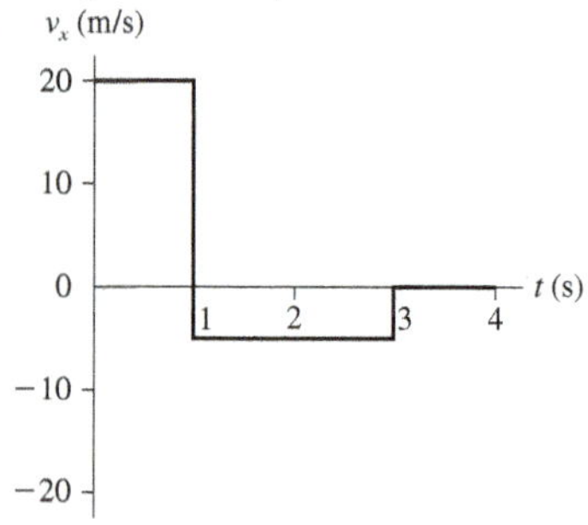

b. 1 s

7. 8.0 cm
9.

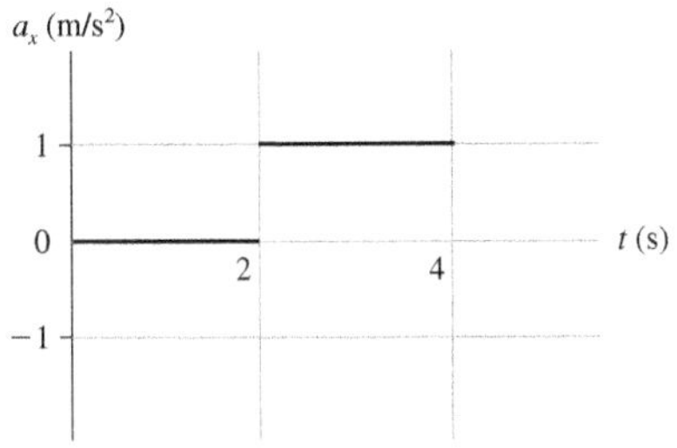

11. a. 6.0 m b. 4.0 m/s c. 2.0 m/s^2
13. 8.8 m/s^2; this is reasonable for a jet
15. −2.8 m/s^2
17. a. 78.4 m b. −39.2 m/s
19. 3.2 s
21. a. 64 m b. 7.1 s
23. a. 7 m b. 7 m/s c. 4 m/s^2
25. 16 m/s
27. −10 m/s, −20 m/s, 75 m/s
29. a. Zero at t = 0 s, 1 s, 2 s, 3 s, . . . ; most positive at t = 0.5 s, 2.5 s, most negative at t = 1.5 s, 3.5 s, . . .
   b.

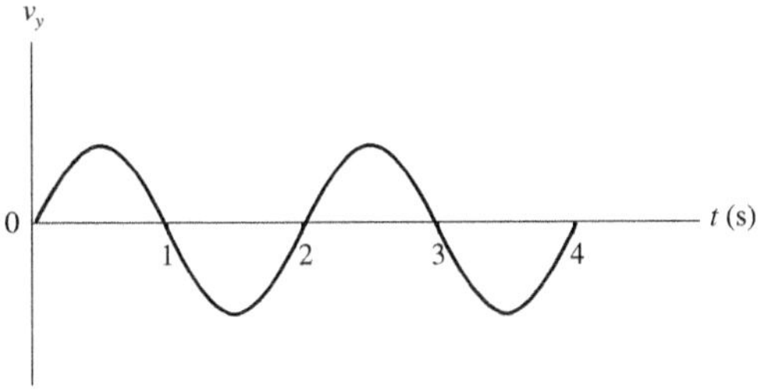

31. a. 0 s and 3 s b. 12 m and −18 m/s^2; −15 m and 18 m/s^2
33. 2.0 m/s^3
35.

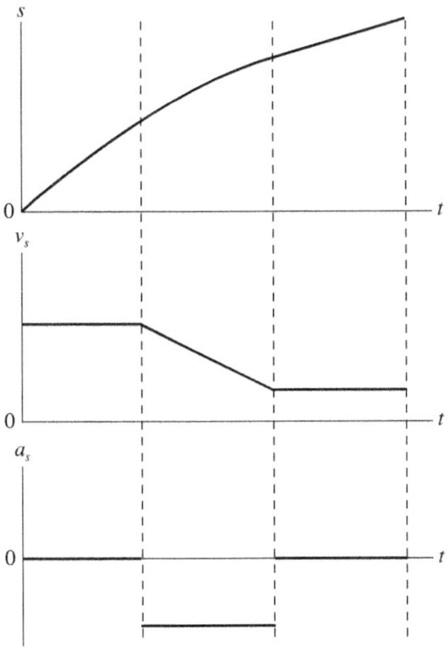

37.

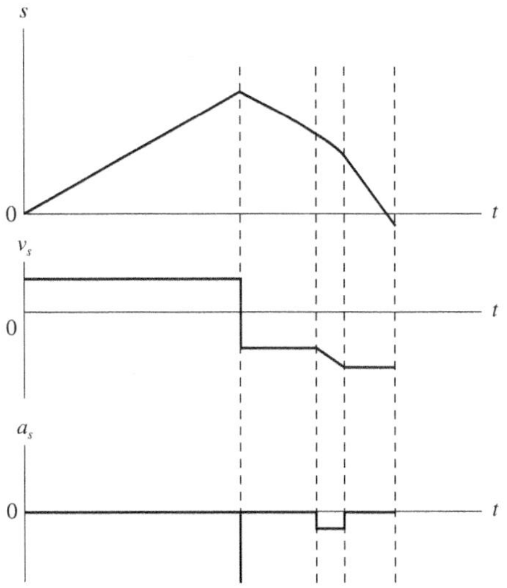

39.

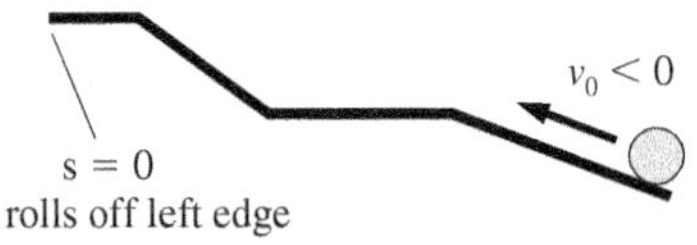

Ball rolls off left edge

41. a. 2.7 m/s^2 b. 28% c. 1.3×10^2 m = 4.3×10^2 ft
43. a. 100 m b. −2 m/s^2 c. 11.5 s
45. a. 5 m b. 22 m/s
47. 5.2 cm
49. a. 54.8 km b. 228 s
   c.

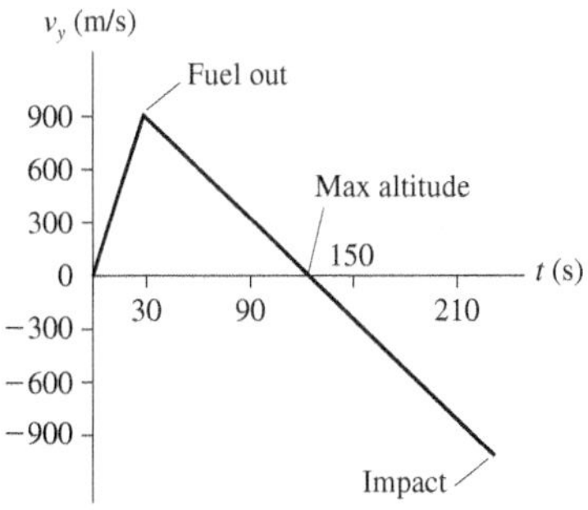

51. 19.7 m
53. 216 m
55. 9.9 m/s
57. a. $v_f = \sqrt{2gh}$ b. 2.4 m/s
59. Yes
61. a. gh/d b. 70 m/s^2
63. 5.7 m/s
65. a. 900 m b. 60 m/s
67. 17.2 m
69. 4.4 m/s^2
71. a. Yes b. 7.9 m/s^2
73. c. 17.2 m/s
75. c. 750 m
77. 5.5 m/s^2
79. a. 10 s b. 3.8 m/s^2 c. 5.6%
81. 12.5 m/s
83. 4500 m/s^2

Chapter 3

1. a.

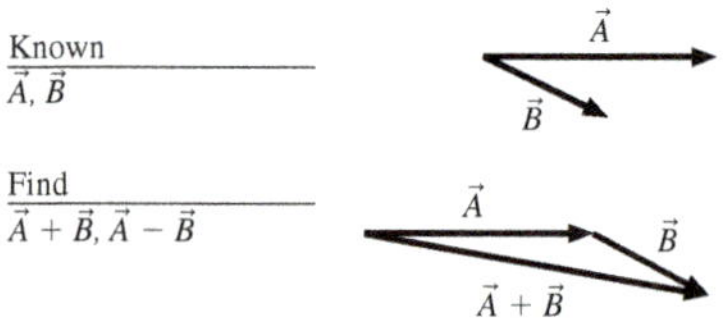

b.

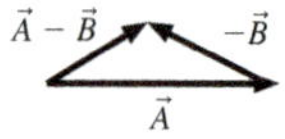

3. a. $-E\cos\theta$, $E\sin\theta$ b. $E_x = -E\sin\phi$, $E_y = E\cos\phi$
5. 6 m
7. a. 0 m/s, −10 m/s b. 17 m/s², −10 m/s² c. −60 N, 80 N
9. 2.2 T, 27° below the $+x$-axis
11. a. 7.2, 56° below the $+x$-axis
    b. 94 m, 58° above the $+x$-axis
    c. 45 m/s, 63° above the $-x$-axis
    d. 6.3 m/s², 18° right of the $-y$-axis
13. a. $1\hat{\imath} + 3\hat{\jmath}$
    b.

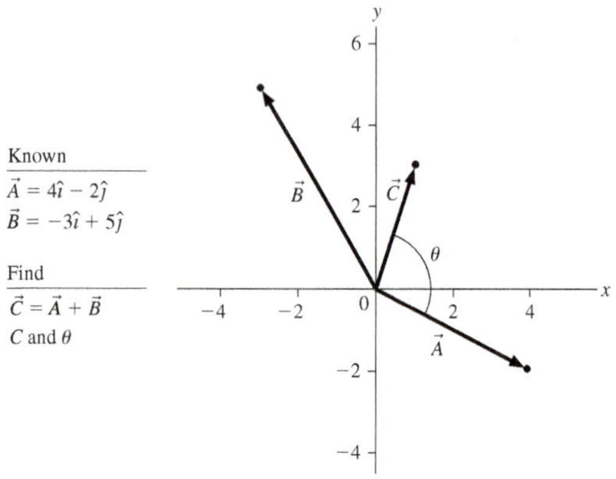

    c. 3.2, 72° above the $+x$-axis
15. a. $10\hat{\imath} + 2\hat{\jmath}$
    b.

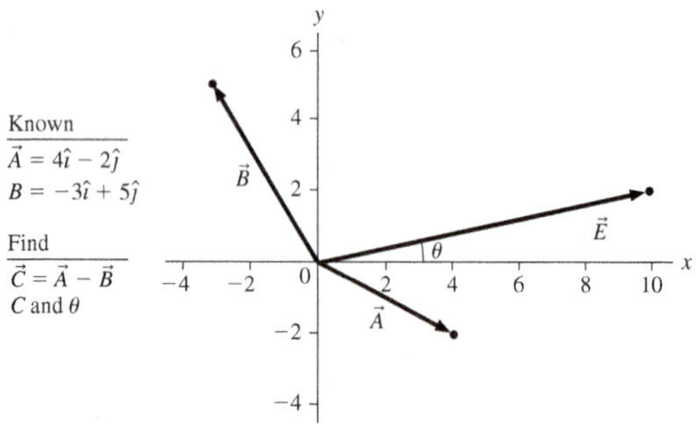

    c. 10, 11° above the $+x$-axis
17. a. $B_x = -4.3$ m, $B_y = 2.5$ m b. $B_x = -2.5$ m, $B_y = 4.3$ m
19. a.

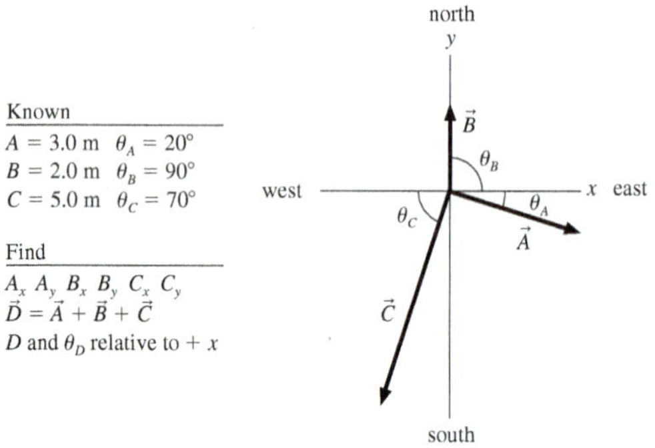

    b. $\vec{A} = (2.8\hat{\imath} + 1.0\hat{\jmath})$ m, $\vec{B} = (2.0\hat{\jmath})$ m, $\vec{C} = (-1.7\hat{\imath} - 4.7\hat{\jmath})$ m
    c. 3.9 m, 74° below the $+x$-axis
21. a. 0 m, 26 m, 160 m b. $(10\hat{\imath} + 8.0\hat{\jmath})t$ m/s
    c. 0 m/s, 26 m/s, 64 m/s
23. a. $-4\hat{\imath} + 3\hat{\jmath}$, b. 5.0, 37° above the $-x$-axis
25. $-1\hat{\imath} - 3\hat{\jmath}$
27. $\vec{B} = \dfrac{1}{\sqrt{2}}\hat{\imath} + \dfrac{1}{\sqrt{2}}\hat{\jmath}$
29. a. 100 m lower b. 5.0 km
31. a.

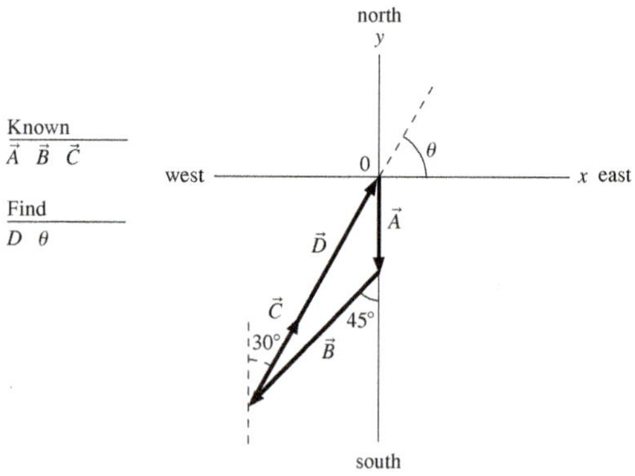

    b. 360 m, 59° north of east c. Yes
33. 7.5 m
35. a. 34° b. 1.7 m/s
37. a. 50 m
    b.

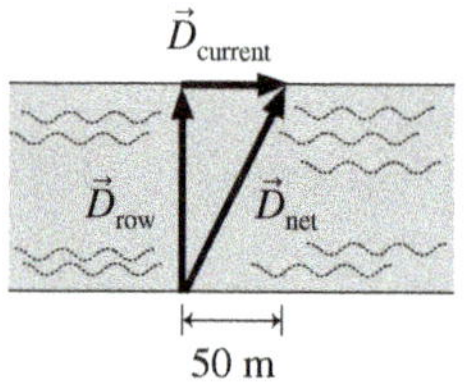

39. −15 m/s
41. 49° west of south
43. a. 0.50 N up-slope b. 1.7 N
    c. 1.7 N, 73° clockwise from the $-x$-axis

Chapter 4

1. a.

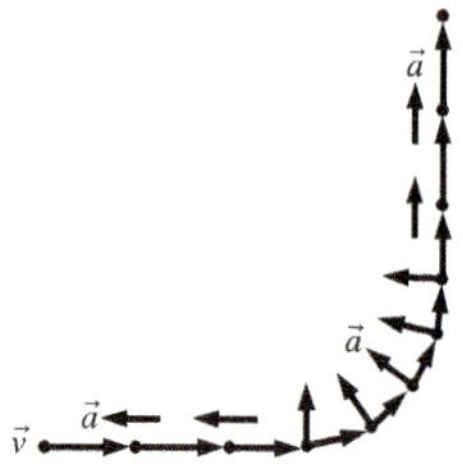

3. H
5. E
7. 111 m
9. a. 62° above the $+x$-axis b. 180 cm
11. a. $\vec{v}(0) = (2.0\hat{\imath} + 4.0\hat{\jmath})$ m/s, $\vec{v}(2) = (2.0\hat{\imath} + 0.0\hat{\jmath})$ m/s,
    $\vec{v}(3) = (2.0\hat{\imath} - 2.0\hat{\jmath})$ m/s b. 2.0 m/s² c. 63° above the $+x$-axis
13. a. 0.064 s b. 780 m/s
15. 2.0 km/h
17. a. 300 m b. 3.2 m/s
19. a. 0 rad/s b. $-\dfrac{\pi}{2}$ rad/s c. 3π rad/s

21.

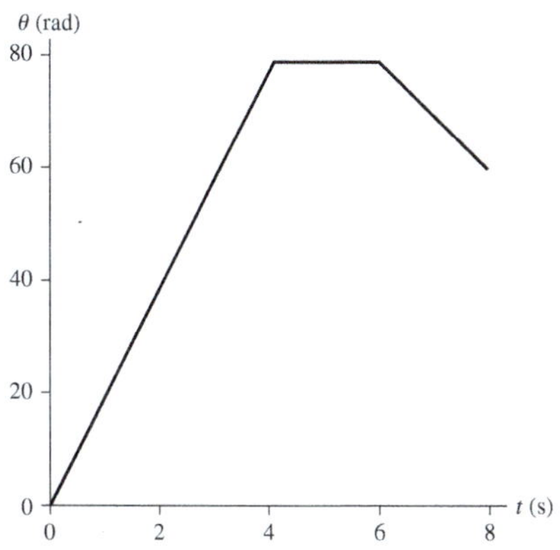

23. 8.0°/h
25. 1680 km/h, 1040 mph
27. a. 3.0×10^4 m/s b. 2.0×10^{-7} rad/s c. 6.0×10^{-3} m/s^2
29. a. -100 rad/s^2 b. 0 rad/s^2 c. 50 rad/s^2
31. 9.5 rev
33. 47 rad/s^2
35. a. -2.6 m/s^2 b. 31 rev
37. $\vec{r} = (710\hat{\imath} - 400\hat{\jmath} + 160\hat{k}) \times 10^3$ km
39. a. $\dfrac{v_0^2 \sin^2\theta}{2g}$

b. $h = 14.4$ m, 28.8 m, 43.2 m; $d = 99.8$ m, 115.2 m, 99.8 m
41. a. 81 m higher b. 34 m
43. a. 6 times farther, or 276 m farther
b. 6 times longer, or 12.8 s longer
45. Clears by 1.0 m
47. No
49. 3.7 m
51. 470 m/s^2
53. a. 42° west of north b. 45 s
55. 69 m/s at 21° from the vertical
57. a. 40 m/s^2 b. 80 m/s^2
59. a. 0.97 m/s^2 b. $14g$
61. a. 420 m/s b. 200 m/s
63. a. 12 m/s b. 36 rev
65. a. -100 rad/s^2 b. 50 rev
67. 0.75 rad/s^2
69. a. $v = \sqrt{2\alpha\Delta\theta R}$ b. $a = 2\alpha\Delta\theta R$
71. 26 rad/s^2
73. 5500 rpm
75. b. 30 m west
77. 6.6×10^{12} m/s^2
79. 110 m/s
81. 4.8 m/s
83. 59 m away, 11 m lower
85. 10°

Chapter 5

1. Tension, gravity
3. Thrust, normal force, gravity, air resistance
5. Gravity, air resistance
7. a. 2.4 m/s^2 b. 0.60 m/s^2
9. 12/25
11. 3.7 s
13. a. 1 N b. 2 N
15. 3.6 N
17. a. ≈0.05 N b. ≈100 N
19.

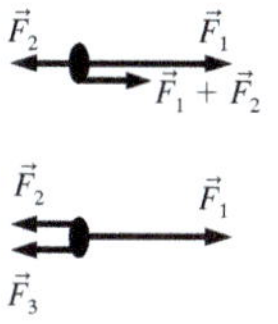

25.

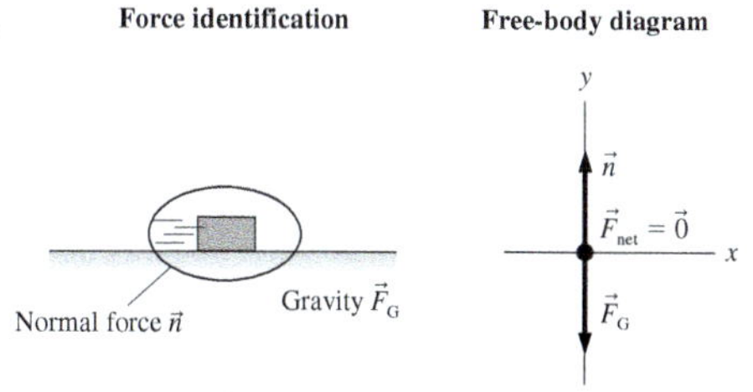

27.

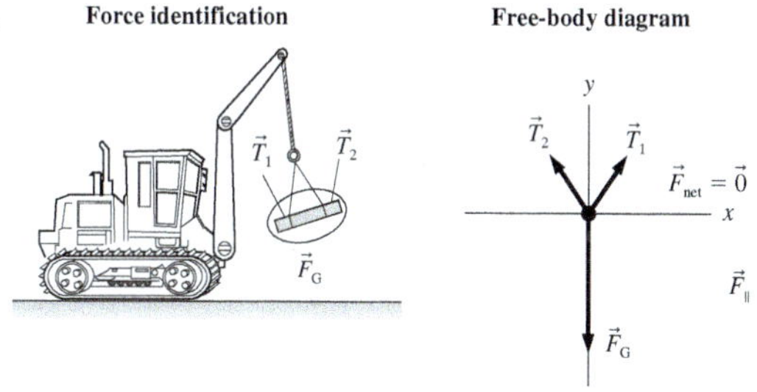

29.

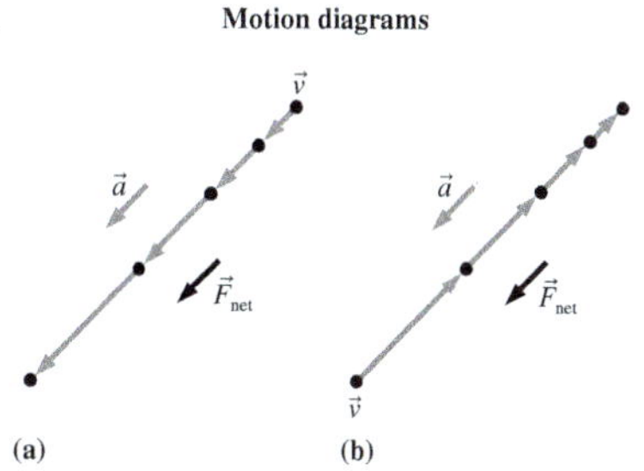

31.

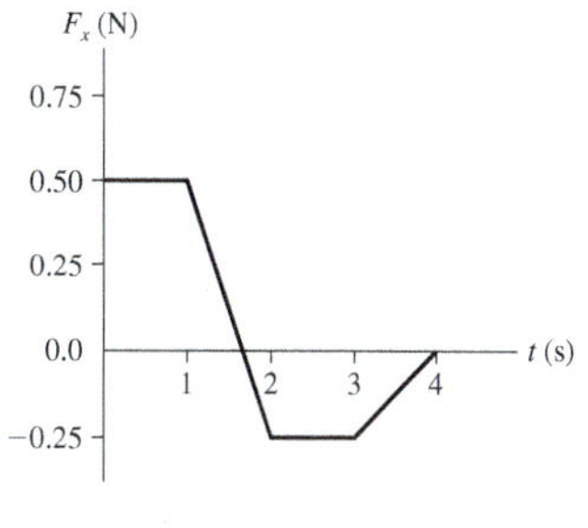

33.

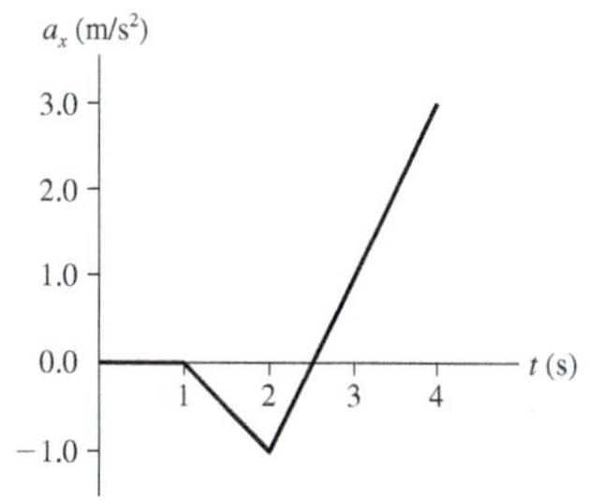

35. a. 5.0 m/s^2 b. 30 m/s^2
c. 10 m/s^2 d. 2.5 m/s^2

37.

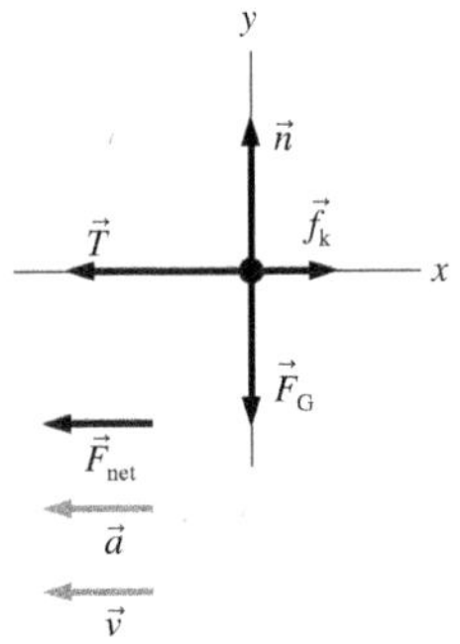

39.

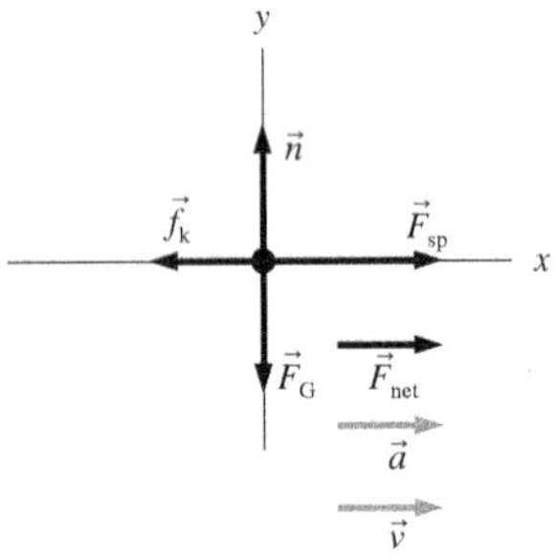

41.

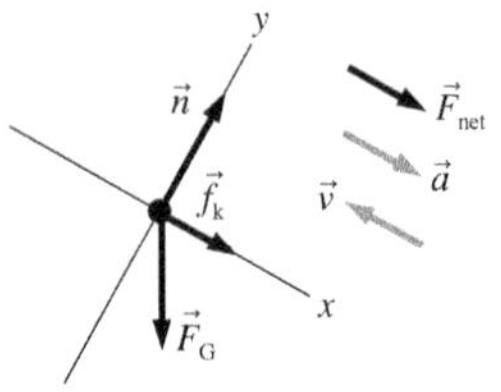

43.

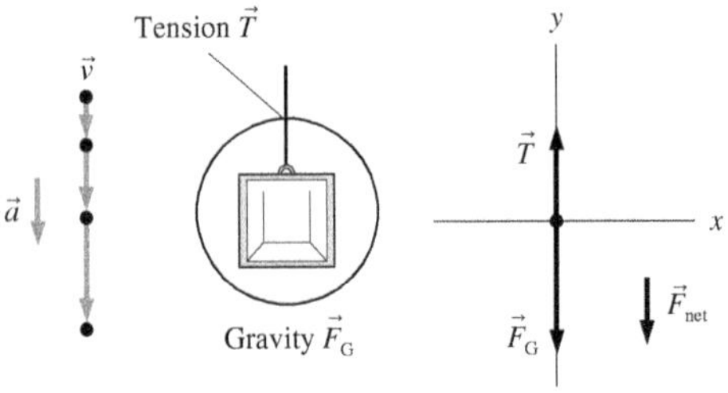

45.

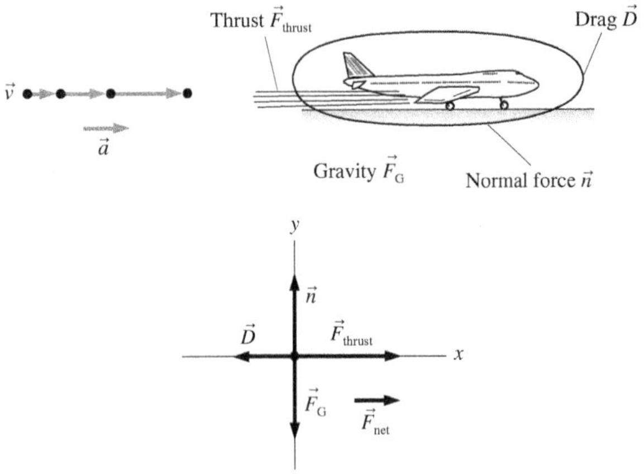

47.

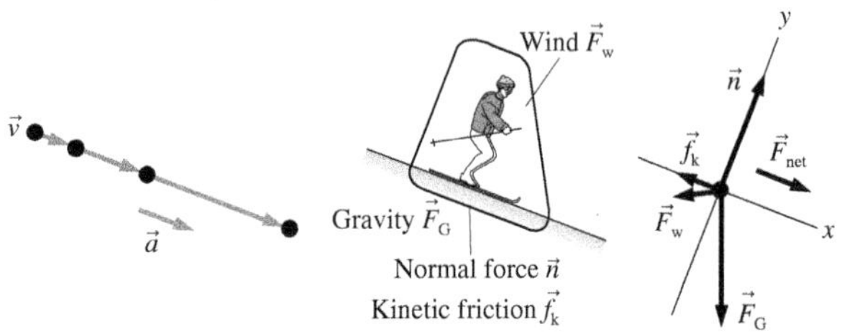

49.

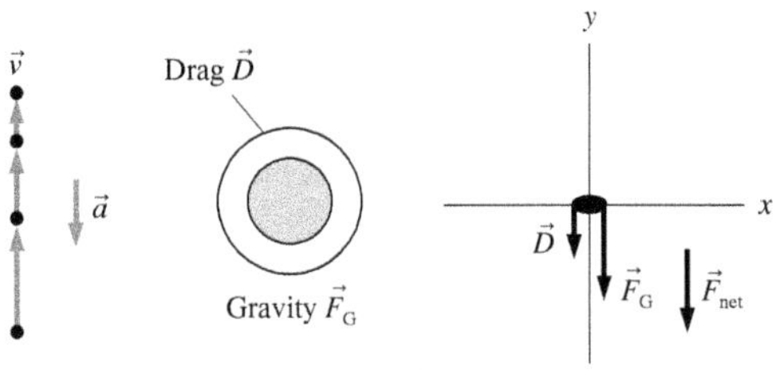

51.

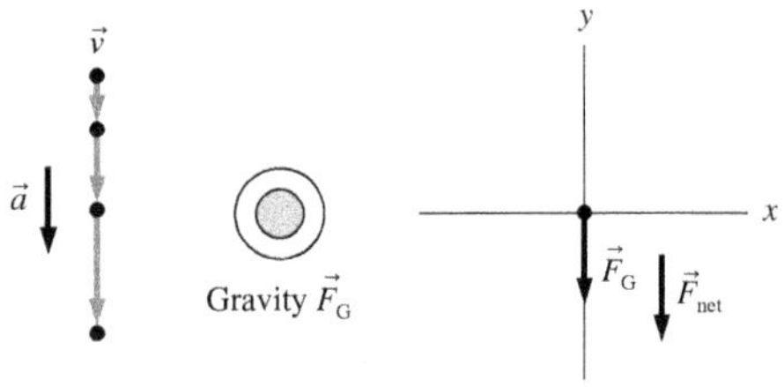

53.

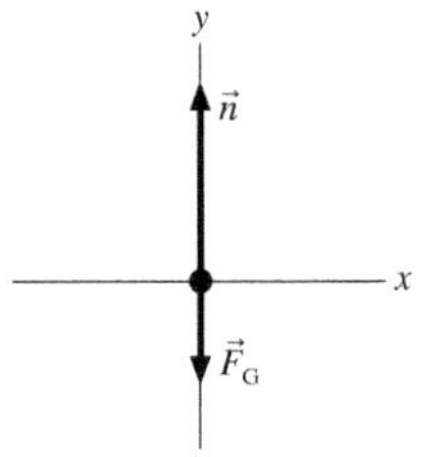

55.

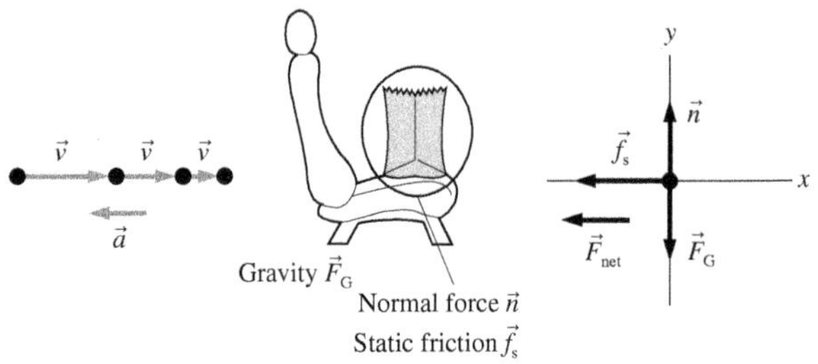

57. a.

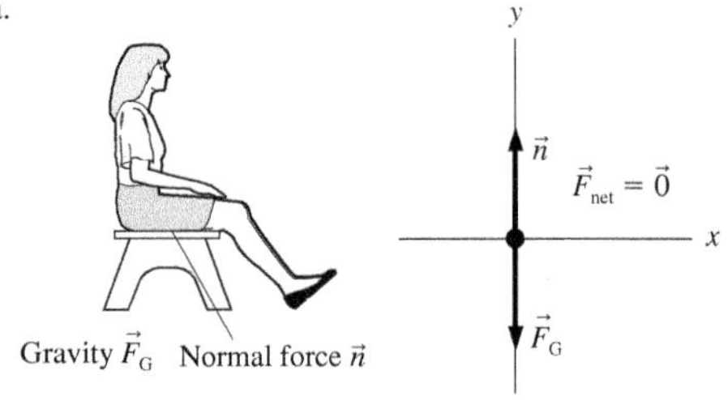

f.

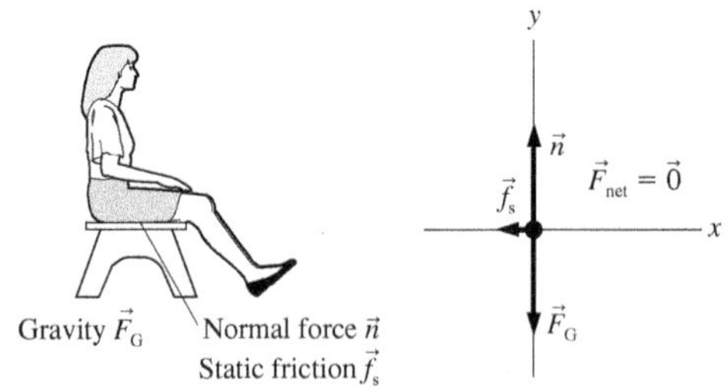

Chapter 6

1. $T_1 = 86.7$ N, $T_2 = 50.0$ N
3. 147 N
5. 800 N
7. a. $a_x = 1.5$ m/s^2, $a_y = 0$ m/s^2 b. $a_x = -0.28$ m/s^2, $a_y = 0$ m/s^2
9. 0 m/s^2, 4 m/s
11. a. 490 N b. 490 N c. 740 N d. 240 N
13. a. 540 N b. $m = 55$ kg; $mg = 210$ N
15. 1035 N, 740 N, 590 N
17. 0.250
19. a.–b.

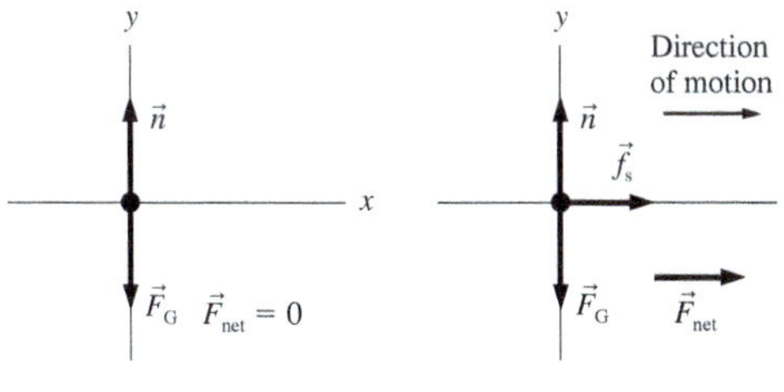

(a) (b)

c. 4.9 m/s^2

21. 10,000 N
23. 2.6 km
25. 69 g
27. 9 m/s
29. a. 0.0036 N b. 0.010 N
31. a. Down b. 77 kg c. 39 m
33. a. 59 N b. 67.8° c. 79 N
35. a. −12 N b. 12 N
37. 3.1 m
39. a. 6700 N b. 600 μs

c.

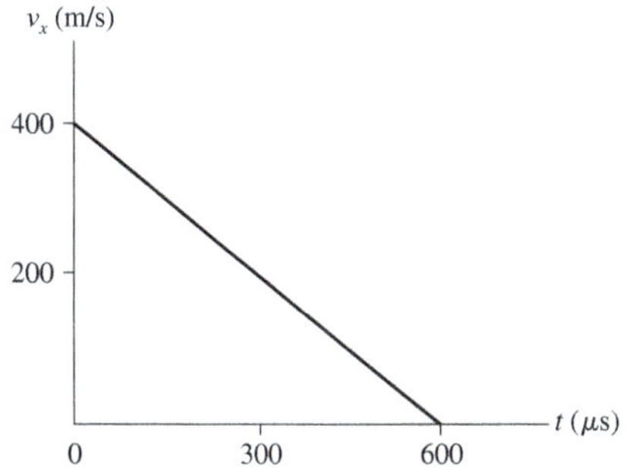

41. a. $\sqrt{2g(h-\mu_k\sqrt{L^2-h^2})}$ b. 9.9 m/s
43. 0.165
45. 2.7 m/s
47. 190 kN
49. 0.12
51. a. $\dfrac{v_0^2}{2\mu_s g}$ b. 14 m
53. a. Yes b. Yes
55. Stay at rest
57. 37 m/s
59. 3/8 in
61. a. 0 N b. 220 N
63. a. $\dfrac{F_0}{m}\dfrac{T}{2}$ b. $\dfrac{F_0}{m}\dfrac{T^2}{3}$
65. a. 9.4×10^{-10} N, 5.7×10^{-13} N b. 1.8 m/s^2, 130 m/s^2
67. b. 0.36 s
69. c. 102 m
71. c. 2.8 m/s^2
73. 13 m/s^2
75. a. $v_0 e^{-\frac{6\pi\eta R t}{m}}$ b. 61 s
77. b. 160 s, 480 s b. No

Chapter 7

1. a. **Interaction diagram**

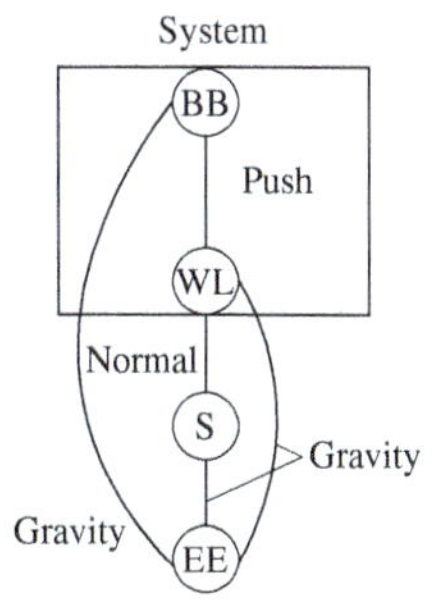

BB = Barbells
WL = Weight lifter
S = Surface EE = Entire Earth

b. The system is the weightlifter and barbell.

c. **Free-body diagrams**

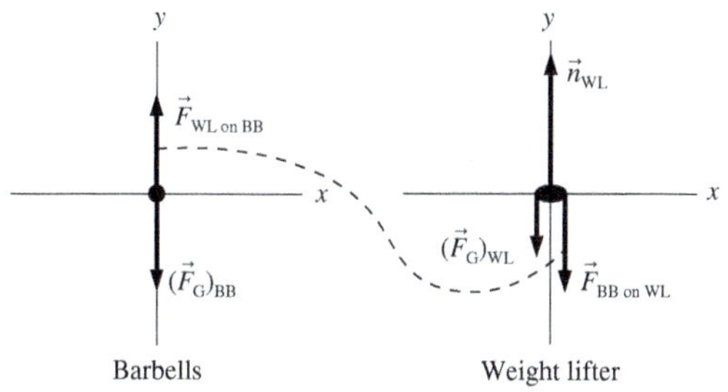

3. a. **Interaction diagram**

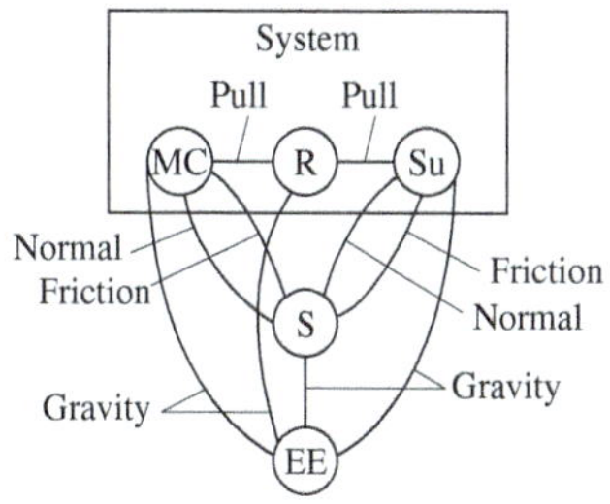

MC = Mountain climber
R = Rope Su = Supply bag
S = Surface
EE = Entire Earth

b. The system consists of the mountain climber, rope, and bag of supplies.

c. **Free-body diagrams**

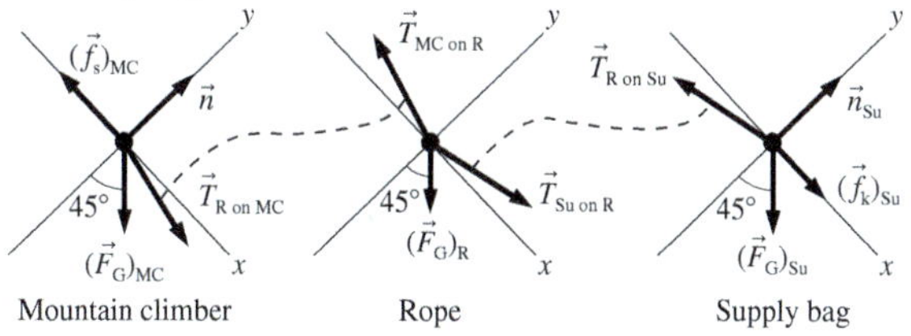

5. a. **Interaction diagram**

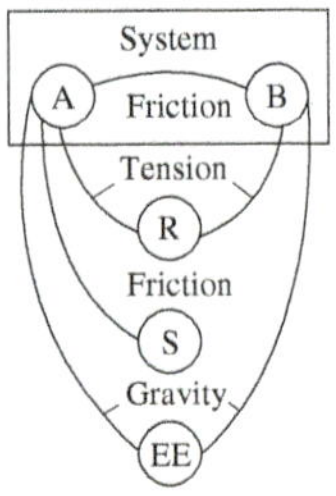

b. The system consists of the two blocks.

c. **Free-body diagrams**

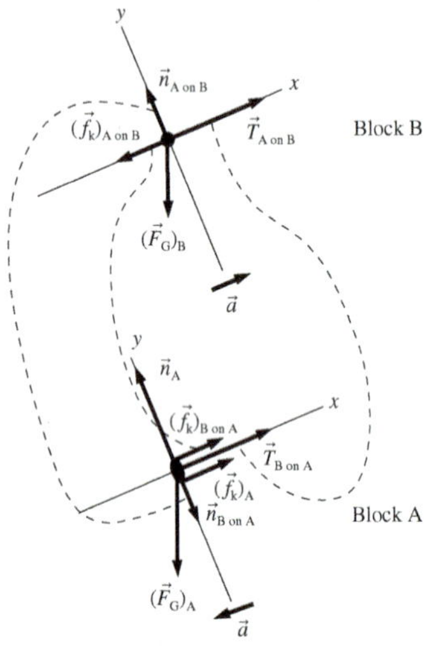

7. a. 7.8×10^2 N b. 1.6×10^3 N
9. a. 3000 N b. 3000 N
11. 5.0 kg
13. a. 32 N b. 19 N c. 16 N d. 3.2 N
15. a. 20 N b. 21 N
17. a. ~ 0.05 N b. ~ 100 N
19. a. **Interaction diagram**

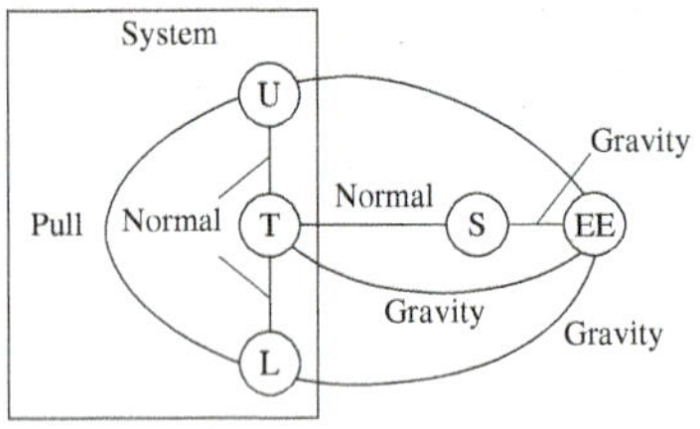

U = Upper magnet
L = Lower magnet
T = Table
S = Surface
EE = Entire Earth

Known
$(F_G)_T = 20.0$ N
$(F_G)_U = 2.0$ N
$(F_G)_L = 2.0$ N
$F_{U\ on\ L} = 3(F_G)_L$

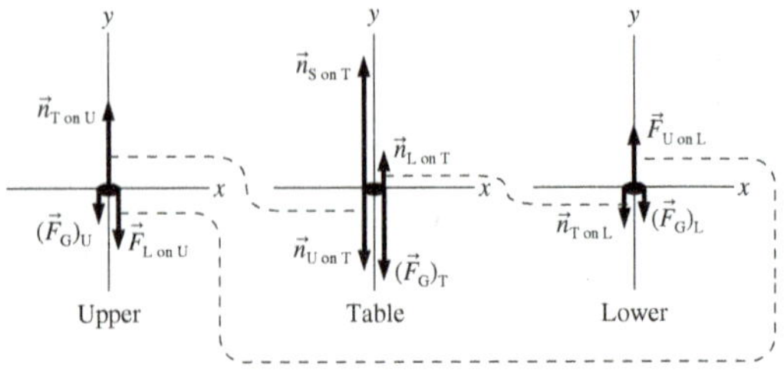

b. $(F_G)_U = 2.0$ N, $n_{T\ on\ U} = 8.0$ N, $F_{L\ on\ U} = 6.0$ N, $(F_G)_T = 20$ N, $n_{U\ on\ T} = 8.0$ N, $n_{L\ on\ T} = 4.0$ N, $n_{S\ on\ T} = 24$ N, $n_{S\ on\ T} = 24$ N, $n_{T\ on\ L} = 4.0$ N, $F_{U\ on\ L} = 6.0$ N

21. 60 N
23. 2.7×10^2 N
25. No
27. 99 m
29. a. 2.3×10^2 N b. 0.20 m/s
31. 1.5 s
33. a. 3.9 N b. 2.2 m/s^2
35. 1.8 m/s^2
37. 160 N
39. $T_1 = 100$ N, $T_2 = T_3 = T_5 = F = 50$ N, $T_4 = 150$ N
41. a. 1.8 kg b. 1.3 m/s^2
43. a. 0.67 m b. Slides back down
45. a. 8.2×10^3 N b. 4.8×10^2 N
47. $a = 2T/m - \mu_r g$
49. $F = (m_1 + m_2)g \tan\theta$
53. a. 1.0 m/s b. 90 N c. 90 N
55. 3.3 m/s^2
57. a. $2a_{3y} + a_{2y} + a_{1y} = 0$ m/s^2
c. $a_{1y} - 2.2$ m/s^2, $a_{2y} = 2.9$ m/s^2, $a_{3y} = -0.32$ m/s^2

Chapter 8

1. 39 m
3. a. 56 h b. 0.092° c. Yes
5. 6.8 kN
7. 6.6×10^{15} rev/s
9. 2.01×10^{20} N
11. 1.58 m/s^2
13. 22 m/s
15. 3
17. 30 rpm
19. 8.2 s
21. 4.5 m/s
23. 105 m
25. a. $y = \dfrac{1}{2}x$ b. Straight line c. 1090 m
27. a. 165 m b. Straight line
29. a. 24.0 h b. 0.223 m/s^2 c. 0 N
31. 5.5 m/s
33. a. $\sqrt{\dfrac{g}{L\sin\theta}}$ b. 72 rpm
35. 4 m/s
37. a. 2.9 m/s b. 14 N
39. Horizontal circle
41. $a_A = -9.8$ m/s^2, $a_B = -12.9$ m/s^2, $a_C = -6.7$ m/s^2
43. a. 320 N, 1400 N b. 5.7 s
45. a. $\sqrt{\dfrac{g}{L}}$ b. 30 rpm
47. 0.38 N
49. a. $\sqrt{gL}$ b. 2.6 m/s, 5.9 mph
51. 1.4 m
53. 13 N
55. a. 6.6 rad/s b. 44 N
57. b. 20 rad/s = 190 rpm
59. a. $\theta = \frac{1}{2}\tan^{-1}(mg/F)$ b. 11.5%
61. $\sqrt{gy}$
63. $T_1 = 14.2$ N, $T_2 = 8.3$ N
65. 37 km
67. $\dfrac{rv_0}{r + v_0\mu_k t}$

Chapter 9

1. a. 4.5×10^4 kg m/s b. 8.0 kg m/s
3. 4 N s
5. 1.5×10^3 N
7. 800 N in the $-x$-direction
9. 1.0 m/s to the left
11. a. 1.5×10^4 N s b. 30 s, 110 m/s
13. 0.2 s
15. 5.0×10^2 kg
17. 4.8 m/s
19. 3.0×10^2 m/s
21. 3.6 m/s
23. $(2\hat{\imath} + 4\hat{\jmath})$ kg m/s
25. 45° north of east at 1.7 m/s
27. a. 6.4 m/s b. $F_{avg} = 3.6 \times 10^2\ \text{N} = 612(F_G)_B$
29. 0.497 m
31. 7.5×10^{-10} kg m/s
33. 8.0×10^2 N
35. $v_f = \dfrac{m_1}{(m_1 + m_2)}\sqrt{\dfrac{2dF}{m_1}}$
37. $2.2 \times 10^{-10}\%$
39. 32 m/s
41. $v_0/\sqrt{2}$, 45° east of north
43. a. $v_{bullet} = \dfrac{m + M}{m}\sqrt{2\mu_k g d}$ b. 4.4×10^2 m/s
45. 28 m/s
47. 20 m/s downward
49. 2.0 m/s
51. 8.6 m/s
53. $3v_0$
55. 16 m/s
57. 4.5 km
59. 14.0 u
61. a. -1.4×10^{-22} kg m/s
    b. and c. 1.4×10^{-22} kg m/s in the direction of the electron
63. 0.85 m/s, 72° below the $+x$-axis
65. 2.0×10^3 m/s
67. b. $(v_{ix})_2 = 6.0$ m/s
69. c. $(v_{fx})_1 = -12$ m/s
71. 1.5×10^7 m/s in the forward direction
73. 1.2 km/s

Chapter 10

1. The bullet
3. 112 km/h
5. a. 25.1 m b. 10 m/s c. 22 m/s
7. 2
9. 4.5 m/s
11. 4.2 m/s
13. 98 N/m
15. 31 cm
17. a. 49 N b. 1450 N/m c. 3.4 cm
19. 18 J
21. 4.0 cm
23. Go hungry
25. a. Right b. 17.3 m/s c. $x = 1.0$ m and 6.0 m
27. a. 7.7 m/s b. 10 m/s
29. 0.86 m/s and 2.9 m/s
31. a. -5.0 m/s and 5.0 m/s b. 2.5 m/s
33. a. 0.34 m
35. 81 m/s
37. a. 2.3 m/s b. $v(\theta) = \sqrt{9 - 1.96(1 - \cos\theta)}$ m/s
39. 9.7 J
41. $v_0/\sqrt{2}$
43. a. 2.6 m/s b. 33 cm
45. 2.0×10^5 N/m
47. a. 2.2×10^4 N/m b. 18 m/s
49. a. 3.3 m/s b. 11.8 cm c. 0.83 m/s, 6.5 cm
51. a. $\sqrt{\left[2\left(\dfrac{m+M}{m}\right)^2 gd - \left(\dfrac{m+M}{m^2}\right)\dfrac{M^2g^2}{k} + k\left(\dfrac{m+M}{m^2}\right)(d - Mg/k)^2\right]}$
    b. 450 m/s
53. $\frac{5}{2}R$
55. 7.9 m/s
57. 100 g ball: 0.80 m/s to the left; 400 g ball: 2.2 m/s to the right
59. a. $x_1 = \dfrac{\pi}{3}$ and $x_2 = \dfrac{2\pi}{3}$ b. $\dfrac{\pi}{3}$: unstable; $\dfrac{2\pi}{3}$: stable
61. a. Yes b. 1.0 mJ/m
63. c. 15 m/s
65. c. 32 cm
67. 93 cm
69. 20 N/m
71. a. 1.5 m b. 20 cm
73. a. 4.6 cm b. $v_{2\,kg} = 1.33$ m/s and $v_{1\,kg} = 5.3$ m/s
75. 100 g ball rebounds to 79°; 200 g ball rebounds to 14.7°

Chapter 11

1. a. -18 b. 10
3. a. 125° b. 67°
5. a. 11 b. -4.6 c. 0
7. a. -6.0 J b. 12 J
9. 0 J
11. 1.25×10^4 J, -7.92×10^3 J, -4.58×10^3 J
13. AB = 0 J, BC = 0 J, CD = -4 J, DE = 4 J
15. 3.7 m/s, 6.6 m/s, 9.7 m/s
17. 8 N
19. -60 N at $x = 1$ m, 15 N at $x = 4$ m
21. 2.5 N, 0.40 N, 0.16 N
23. 1360 m/s
25.

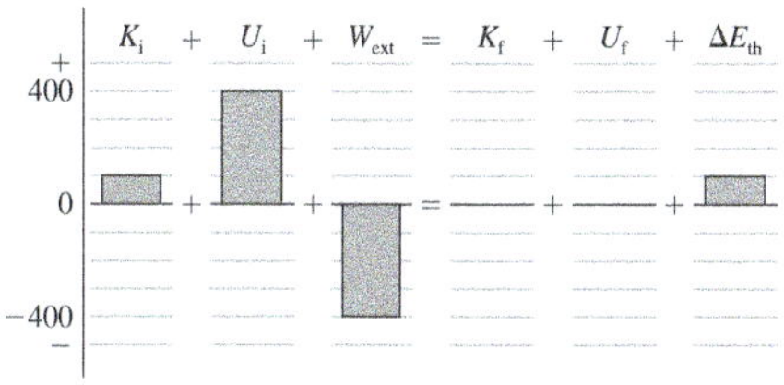

27. -1 J of work is done to the environment
29. a. 9.80×10^5 J b. 1.96×10^4 W
31. 42 m²
33. 5.5×10^4 liters
37. a. 50 J b. 50 J c. 50 J, yes
39. a.

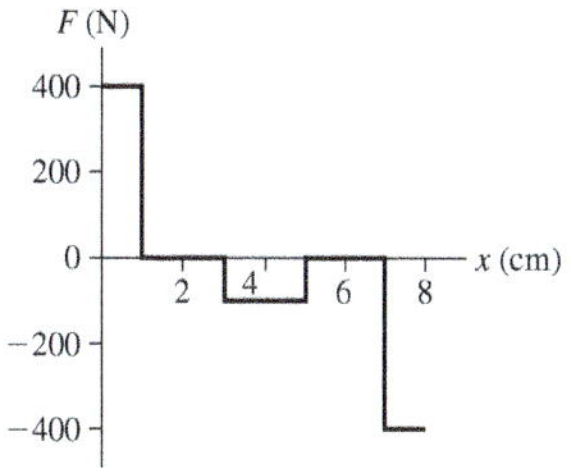

b. -2 J c. 22 m/s

41. a.

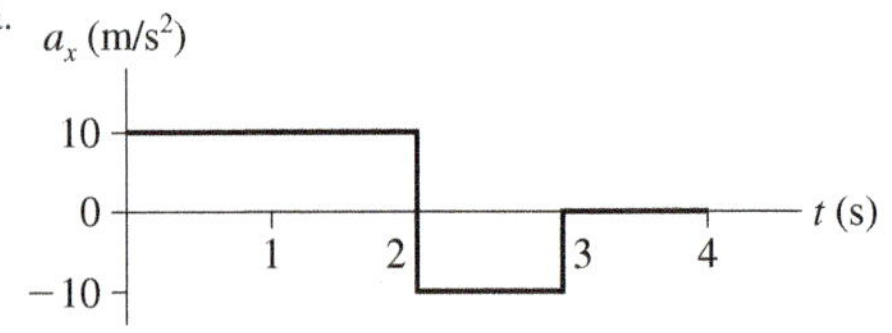

b.

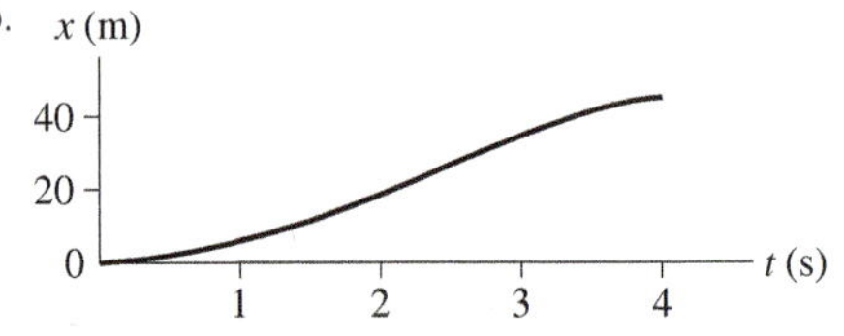

c.

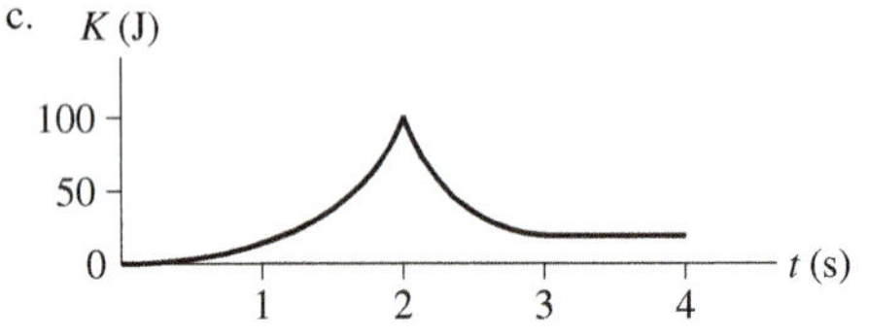

d.

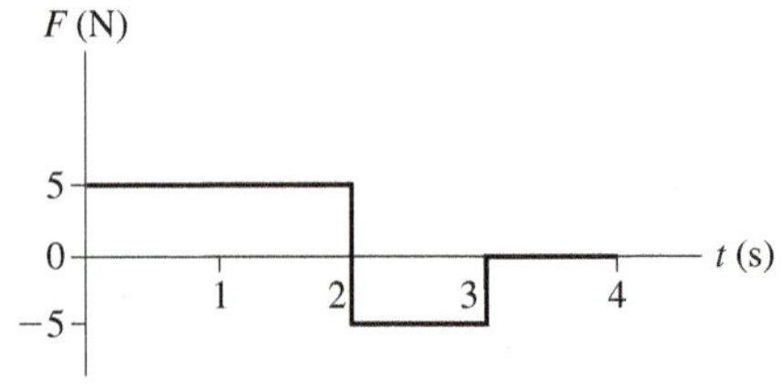

e. 10 N s, −5 N s   f. 20 m/s, 10 m/s

g.

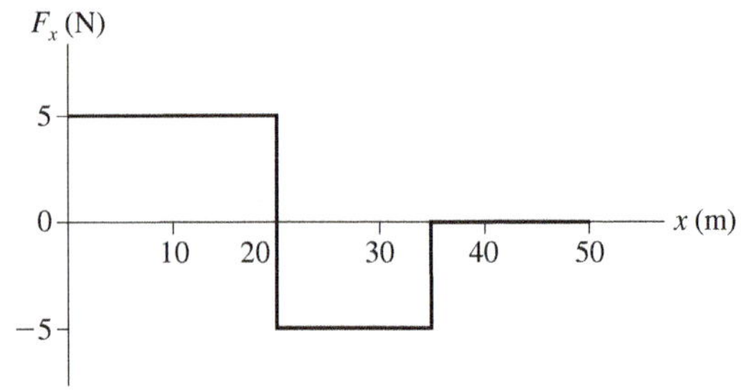

h. 100 J, −75 J   i. 20 m/s, 10 m/s

43. a. 2.3×10^2 J   b. 2.3×10^2 N   c. 6.8 kW
45. 16 m/s
47. 0.54 m
49. 0.12 km
51. a. $v_f = \sqrt{2gh(m - \mu_k M)/(M + m)}$ b. $v_f = \sqrt{2gmh/(M + m)}$
53. 10 m/s
55. a. 0.51 m   b. 0.38 m
57. a. 14 m/s   b. 32 m
61. a. N   b. m^{-1} c. $\pi/(2c)$

d.

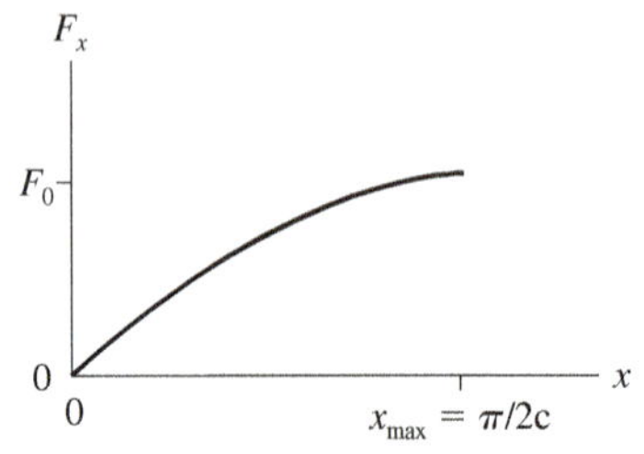

e. $\sqrt{v_0^2 + 2F_0/(mc)}$

63. a. 78 J   b. 1.3 W
65. a. −0.25 kJ   b. 2.6×10^5 kg
67. a. 6.53 m/s²   b. 11/7 m/s   c. 1.79 s
69 c. 6.3 m/s
71. c. 3.2 kW
73. 6.7 m
75. 24 W

Chapter 12

1. 13.2 m/s
3. a. 1.5 m/s   b. 13 rev
5. 4.7×10^6 m
7. $x_{cm} = 6.7$, $y_{cm} = 5.0$
9. 2.57×10^{29} J
11. a. 0.032 kg m²   b. 16 J
13. a. (5.7 cm, 4.6 cm)   b. 0.0066 kg m²
15. a. (0.060 m, 0.040 m)   b. 0.0020 kg m²   c. 0.0013 kg m²
17. a. 3.8×10^{-5} kg m²   b. 1.14×10^{-4} kg m²
19. 4.3 N m
21. 12.5 kN m
23. 8.0 N m
25. 0.28 N m in the ccw direction
27. 8.0 rad/s
29. No
31. 1.5 m
33. 0.38 J
35. 43 cm
37. a. (21, into the page)   b. (24, out of the page)
39. a. $-\hat{j}$ b. $\vec{0}$
41. a. $n\hat{\imath}$ b. $2\hat{j}$ c. $1\hat{k}$
43. $1.20\hat{k}$ kg m²/s
45. (2.1 kg m²/s, out of the page)
47. 91 rpm
49. 7.5 cm
51. 28 m/s
53. a. 0.010 kg m²   b. 0.030 kg m²
55. $\dfrac{M}{3L}[(L - d)^3 + d^3]$
57. $\dfrac{1}{6}ML^2$
59. 0.91 m
61. $F_1 = 750$ N, $F_2 = 1000$ N
63. 1.0 m
65. 31 kg
67. a. 39 mN   b. 38 rpm
69. 1.1 s
71. 1.6 N
73. 4.3 m
75. a. $\sqrt{2g/R}$ b. $\sqrt{8gR}$
77. $\dfrac{20Tr}{13MR^2}$
79. 1.2 rad/s
81. 22 rpm
83. 4.0 rpm
85. Emily
87. 67°

Chapter 13

1. 6.00×10^{-4}
3. 2.18
5. 2.3×10^{-7} N

7. a. $274\ m/s^2$ b. $5.90 \times 10^{-3}\ m/s^2$
9. 2.4 km
11. a. 3.0×10^{24} kg b. $0.89\ m/s^2$
13. 60.2 km/s
15. 4.21×10^4 m/s
17. 4.4×10^{11} m, 1.7×10^4 m/s
19. 1600 earth days
21. a. $T_2 = 250$ min, $T_3 = 459$ min b. $F_2 = 20{,}000$ N, $F_3 = 4{,}440$ N
  c. 1.50
23. 4.2 h
25. 46 kg and 104 kg
27. $3.0 \times 10^{-7}\hat{j}$ N
29. -1.96×10^{-7} J
31. 12 cm
33. a. 3.02 km/s b. 3.13 km/s c. 3.6%
35. 4.2×10^5 m
37. 33 km/s
39. 2.78 km/s
41. 3.0×10^4 m/s
43. 3.7×10^5 m/s
45. 6.7×10^8 J
47. a. 7.0 m/s b. 12 m/s
49. 6.71×10^7 m
51. a. $y = \left(\dfrac{q}{p}\right)x + \dfrac{\log C}{p}$ b. Straight line
  c. q/p d. 1.996×10^{30} kg
53. a. 2.1×10^8 y b. 24 c. 1.9×10^{41} kg d. 9.4×10^{10}
55. 3.71 km/s
57. 4.49 km/s
59. c. 6.21×10^7 m
61. c. 1680 m/s
63. 1.50×10^9 m
65. Crash
67. 11.8%
69. a. $-\dfrac{GMm}{L}\ln\left(\dfrac{x + L/2}{x - L/2}\right)$ b. $-GMm\left(\dfrac{4}{4x^2 - L^2}\right), x \geq \dfrac{L}{2}$

Chapter 14

1. 2.27 ms
3. a. 13 cm b. 9.0 cm
5. a. 10 cm b. 0.50 Hz c. +120°
9. $x(t) = (8.0\ \text{cm})\cos[(\pi\ \text{rad/s})t - \pi\ \text{rad}]$
11. a. 2.8 s b. 1.4 s c. 2.0 s d. 1.4 s
13. a. 0.50 s b. 4π rad/s c. 5.5 cm d. 0.45 rad
  e. 70 cm/s f. $8.8\ m/s^2$ g. 0.049 J h. 3.8 cm
15. a. 10 cm b. 35 cm/s
17. a. 0.17 kg b. 0.57 m/s
19. a. 4.0 s b. 5.7 s c. 2.8 s d. 4.0 s
21. a. 2.0 s b. 2.1 s
23. $3.67\ m/s^2$
25. 0.079 N/m
27.

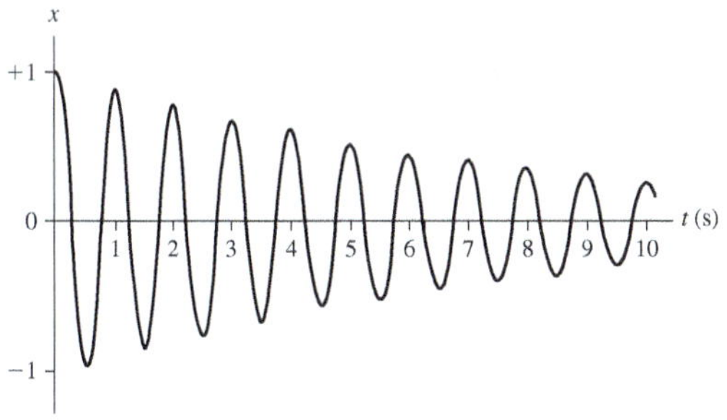

29. 250 N/m
31. a. $\frac{2}{3}\pi$ rad b. −13.6 cm/s c. 15.7 cm/s
33. 0.41 s
35. a. 55 kg b. 0.73 m/s
37. a. $\dfrac{3}{4}$ b. $\dfrac{A}{\sqrt{2}}$
39. a. 6.4 cm b. $160\ cm/s^2$ c. −6.4 cm d. 28 cm/s
41. 1.02 m/s
43. a. 3.2 Hz b. 7.1 cm 5.0 J
45. a. 1.1 Hz b. 23 cm c. 4.1 cm below equilibrium point
47. 1.7 Hz
49. 0.72
51. a. 7.5 m b. 0.45 m/s
53. 0.65 m/s
55. 0.66 s
57. $\dfrac{1}{2\pi}\sqrt{\dfrac{5}{7}\dfrac{g}{R}}$
59. $8.7 \times 10^{-2}\ kg\,m^2$
61. a. 2.0 Hz b. 1.2 cm
63. 7.9×10^{13} Hz
65. a. Highest point b. 2.5 Hz
67. a. 9.5 N/m b. 0.010 kg/s
69. 25 s
71. 236 oscillations
75. 1.6 Hz
77. 1.8 Hz
79. a. $\Delta T = \dfrac{T}{2}\dfrac{\Delta m}{m}$ b. 2.001s

Chapter 20

1. 110 N
3. 2.0 m
5.

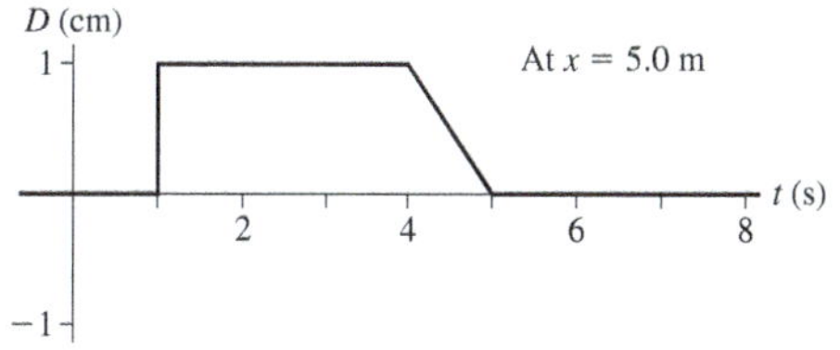

7.

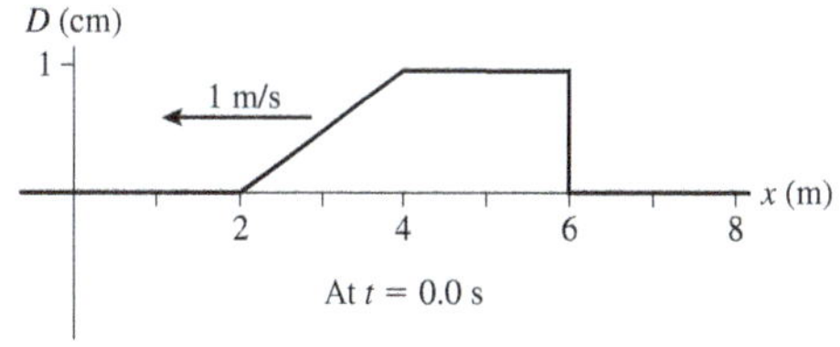

9.

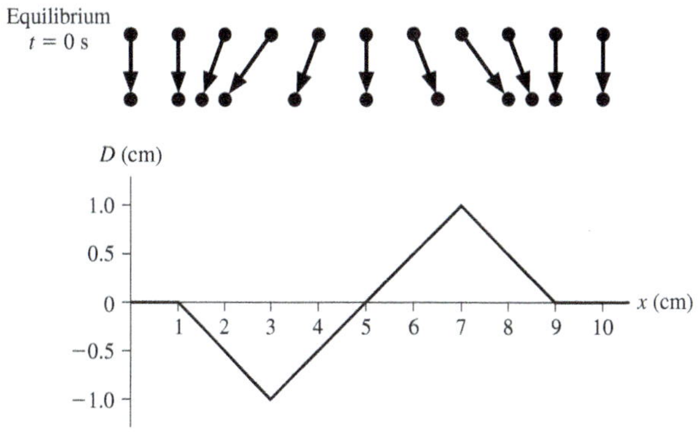

11. a. 3.1 rad/m b. 9.5 m/s
13. a. 11 Hz b. 1.1 m c. 13 m/s
15. $\frac{\pi}{2}$ rad, $\frac{3}{2}\pi$ rad
17. 2.5 m
19. 1500 m/s
21. a. 1.5 GHz b. 990 nm
23. a. 2.96 m b. 116 Hz
25. a. 1.5×10^{-11} s b. 3.4 mm
27. a. 1.88×10^{8} m/s b. 4.48×10^{14} Hz
29. 6.0×10^{5} J
31. 110 dB
33. a. 65 dB b. 105 dB
35. 5.0 W
37. a. 650 Hz b. 560 Hz
39. 38.1 m/s
41. a. 0.80 m b. $\frac{1}{2}\pi$ rad
c. $D(x, t) = (2.0 \text{ mm})\sin(2.5\pi x - 10\pi t + \frac{1}{2}\pi)$
43. $\frac{v_0}{2}$
45.

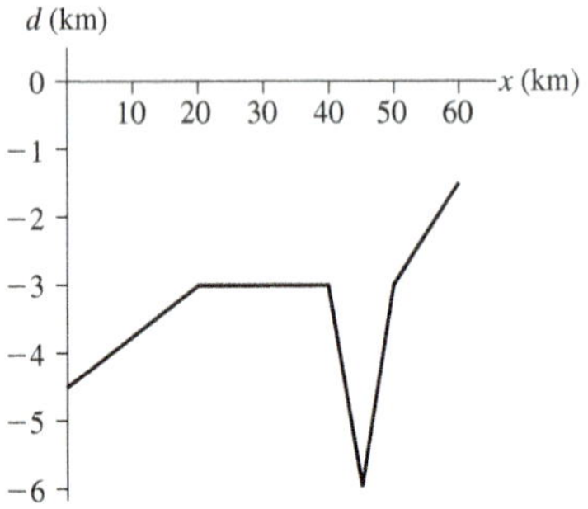

47. 410 ms
49. a. 440 Hz b. 3.4 m
51. a. $-y$-direction b. y-axis c. 0.701 m, 350 m/s, 2.00 ms
53. a. 12.6 N b. 2.00 cm c. 12.8 m/s
55. $D(x, t) = (0.010 \text{ mm})\sin[(\pi \text{ rad/m})x - (400\pi \text{ rad/s})t + \frac{1}{2}\pi \text{ rad}]$
57. −19 m/s, 0 m/s, 19 m/s
59. 8
61. 9.4 m/s
63. a. 0.095 W/m^2 b. 1.6 MW/m^2
65. a. 6.67×10^{4} W b. 8.5×10^{10} W/m^2
67. 50 m
69. 1.3
71. 21 min
75. Receding at 1.5×10^{6} m/s
77. 0.07°C
81. 29 s

Chapter 21

1.

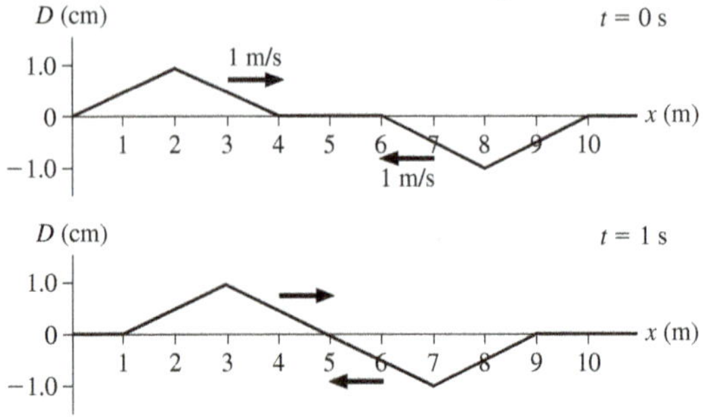

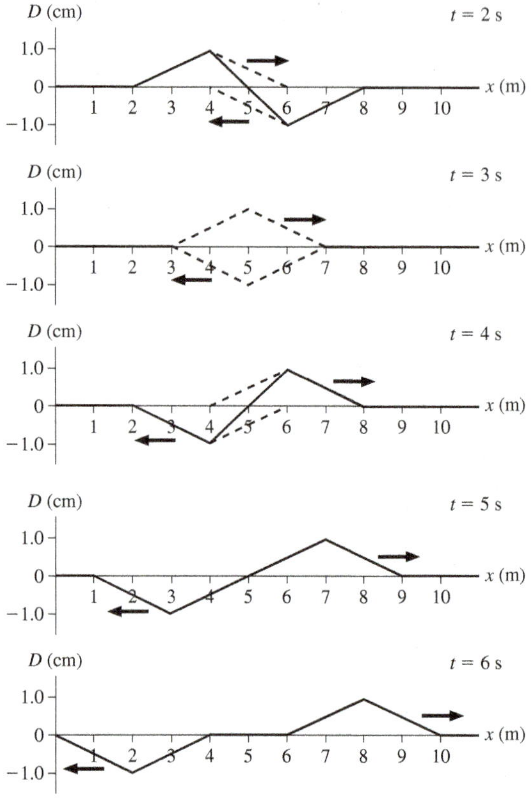

3.

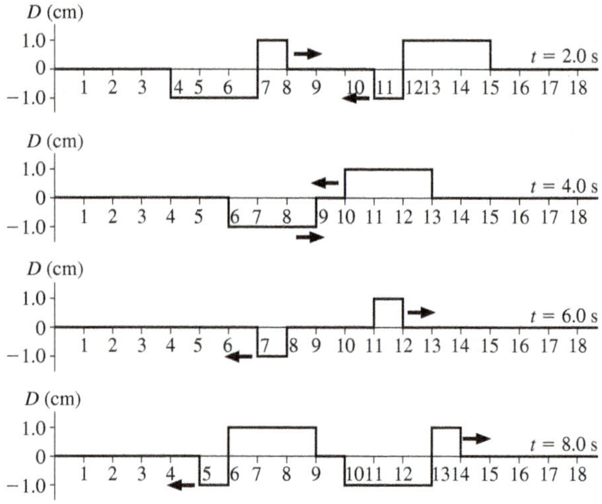

5.

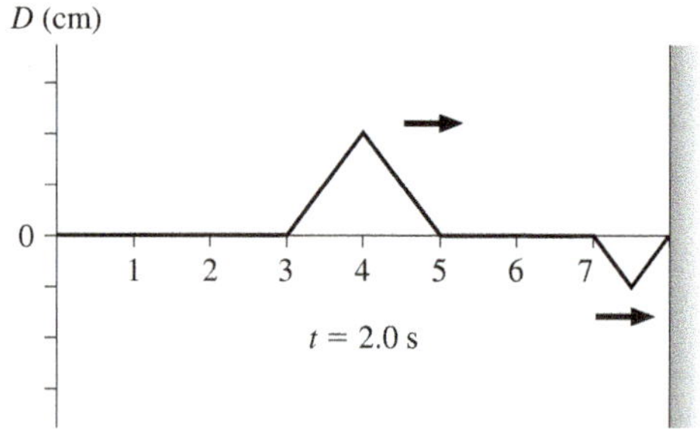

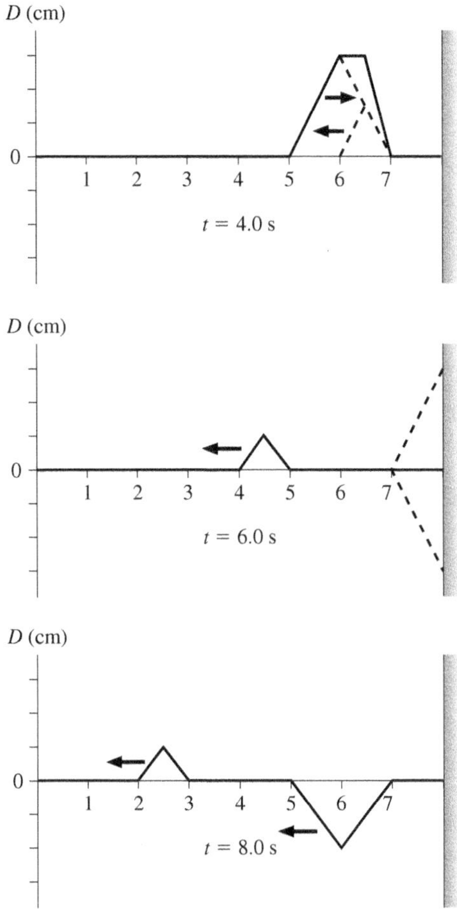

7. 50 Hz
9. a. 4.8 m, 2.4 m, 1.6 m b. 75 Hz
11. 12 kg
13. a. 2.42 m, 1.21 m, 0.807 m b. 4.84 m, 1.61 m, 0.968 m
15. 512 Hz
17. 2180 N
19. a. 80 cm b. 100 cm
21. 216 nm
23. a. In phase
b.

	r_1	r_2	Δr	C/D
P	3λ	4λ	λ	C
Q	$\frac{7}{2}\lambda$	2λ	$\frac{3}{2}\lambda$	D
R	$\frac{5}{2}\lambda$	$\frac{7}{2}\lambda$	λ	C

25. Perfect destructive
27. 203 Hz
29. 1.26 cm
31. $A(x = 10\text{ cm}) = 0.62\text{ cm}$, $A(x = 20\text{ cm}) = 1.18\text{ cm}$, $A(x = 30\text{ cm}) = 1.62\text{ cm}$, $A(x = 40\text{ cm}) = 1.90\text{ cm}$, $A(x = 50\text{ cm}) = 2.00\text{ cm}$
33. 1.4 cm
35. 180 Hz
37. 28.4 cm
39. 18 cm
41. 140 N/m
43. 6.1 cm
45. $9\mu_0/4$
47. 13.0 cm
49. 580 Hz, 4.9 kHz
51. 12.1 kHz
53. 450 N
55. 93 m
57. 7.9 cm
59. a. 850 Hz b. $-\pi/2$ rad
61. 7.2 cm
63. 20
65. 170 Hz
67. 1/3
69. a. a b. 1.0 m c. 9
71. a. 5 b. 4.6 mm
73. 7.0 m/s
75. 4.0 cm, 35 cm, 65 cm
77. 2.0 kg
79. a. $\lambda_1 = 20.0\text{ m}$, $\lambda_2 = 10.0\text{ m}$, $\lambda_3 = 6.67\text{ m}$
b. $v_1 = 5.59\text{ m/s}$, $v_2 = 3.95\text{ m/s}$, $v_3 = 3.22\text{ m/s}$
d. $T_1 = 3.58\text{ s}$, $T_2 = 2.53\text{ s}$, $T_3 = 2.07\text{ s}$

Credits

Page **xvi:** Educational Development Center, Inc.
Page **xxix:** IBM Research, Almaden Research Center.

PART I OPENER

Page **1:** Sharon Green/Pacific Stock/Photolibrary.

CHAPTER 1

Page **2** Top: iStockphoto. Page **2** Bottom: Bureau International des Poids et Mesures. Page **3** Top left: Jason Edwards/National Geographic/Alamy. Page **3** Top right: Shutterstock. Page **3** Bottom left: Richard Megna/Fundamental Photographs. Page **3** Bottom right: Brand X Pictures/Alamy. Page **10** Top: Shutterstock. Page **10** Bottom: Shutterstock. Page **13:** Transtock/Alamy. Page **21:** Shutterstock. Page **23:** U.S. Department of Commerce. Page **24:** Bureau International des Poids et Mesures.

CHAPTER 2

Page **33** Top: Steve Allen/Photo Researchers, Inc. Page **33** Bottom: Dominic Ebenbichler/Reuters. Page **34:** Nicosan/Alamy. Page **40:** NASA. Page **52:** Jim Sugar/Corbis. Page **53:** Michel Bureau/BIOSphoto/Photolibrary.

CHAPTER 3

Page **69:** Simon Hathaway/Alamy. Page **69** Bottom: Paul Chesley/National Geographic Stock. Page **74:** Patrick LaCroix/Alamy.

CHAPTER 4

Page **85** Top: Danita Delimont/Alamy. Page **85** Middle: Richard Bartz. Page **85** Bottom: Shutterstock. Page **91:** Tony Freeman/PhotoEdit, Inc. Page **93:** Richard Megna/Fundamental Photographs. Page **98:** Photolibrary RF. Page **102:** Guy Croft SciTech/Alamy.

CHAPTER 5

Page **116** Top: p17/ZUMA Press/Newscom. Page **116** 2nd from Top: Tony Freeman/PhotoEdit, Inc. Page **116** Middle left: Pearson Science/Creative Digital Vision. Page **116** Middle 2nd from left: Dave Pattison/Alamy. Page**116** Middle 3rd from left: Randy Allbritton/Getty Images - PhotoDisc. Page **116** Middle right: 81a/Alamy. Page **116** Bottom: Shutterstock. Page **117** Top: Tony Freeman/PhotoEdit, Inc. Page **117** 2nd from Top: Brian Drake/Photolibrary. Page **117** 3rd from Top: Getty Images/Comstock Images. Page **117** 4th from Top: Dorling Kindersley. Page **117** 5th from Top: Shutterstock. Page **117** Bottom: Pearson Science/Creative Digital Vision. Page **129:** fStop/Alamy. Page **130:** Photodisc/AGE Fotostock.

CHAPTER 6

Page **138** Top: Tom Tracy Photography/Alamy. Page **138** Bottom left: NASA. Page **138** Bottom right: NASA. Page **139:** Akira Kaede/Photodisc/Getty Images. Page **147:** NASA. Page **152** Top: Pearson Science/Eric Schrader. Page **152** Bottom: Patrick Behar/Agence Vandystadt/Photo Researchers, Inc. Page **164:** Richard Megna/Fundamental Photographs.

CHAPTER 7

Page **167** Top: Shutterstock. Page **167** Bottom: iStockphoto. Page **169:** Shutterstock. Page **171:** Koji Aoki/Alfo Foto Agency RF/AGE Fotostock. Page **189:** Peter Langer/DanitaDelimont.com/Newscom.

CHAPTER 8

Page **191** Top: imagebroker.net/SuperStock. Page **191** Left: Antony Nettle/Alamy. Page **191** Right: Tony Freeman/PhotoEdit, Inc. Page **193:** Getty Images/Digital Vision. Page **195:** Shutterstock. Page **200:** Getty Images/Digital Vision. Page **201:** NASA.

PART II OPENER

Page **218:** Hank Morgan/Photo Researchers, Inc.

CHAPTER 9

Page **220** Top: Alexey Stiop/Alamy. Page **220** Middle left: Shutterstock. Page **220** Middle right: iStockphoto. Page **220** Bottom: Shutterstock. Page **221:** Russ Kinne/Jupiter Images RF. Page **228:** bl1/ZUMA Press/Newscom. Page **236:** Richard Megna/Fundamental Photographs.

CHAPTER 10

Page **245** Top: Oleksiy Maksymenko Photography/Alamy. Page **245** Bottom: Dorling Kindersley. Page **246** Left: Kyodo/Newscom. Page **246** Middle: Thinkstock/Stockbyte/John Foxx. Page **246** Right: Shutterstock. Page **257:** Paul Harris/Paul Harris BWP Media USA/Newscom. Page **258:** iStockphoto. Page **265:** Dorling Kindersley.

CHAPTER 11

Page **278** Top: REUTERS/Bobby Yip. Page **278** Middle: x99/ZUMA Press/Newscom. Page **278** Bottom: Shutterstock. Page **281:** Max Faulkner/MCT/Newscom. Page **289:** PETER COSGROVE/AFP/Newscom. Page **298:** JOEL SAGET/AFP/Getty Images/Newscom.

PART III OPENER

Page **310:** NASA-GSFC, data from NOAA GOES.

CHAPTER 12

Page **312** Top: Gouhier-Hahn-Nebinger/MCT/Newscom. Page **312** Middle: Fotolia. Page **312** Bottom: Visions of America, LLC/Alamy. Page **322:** iStockphoto. Page **330:** Shutterstock. Page **333:** Chris Nash/Getty Images. Page **334:** Richard Megna/Fundamental Photographs. Page **341:** PCN Photography/Alamy.

CHAPTER 13

Page **354** Top: NASA. Page **354** Left: NASA. Page **354** Right: PNT.GOV. Page **356:** American Institute of Physics/Emilio Segre Visual Archives. Page **358:** Corbis RF. Page **361:** NASA/Ralph White. Page **366:** NASA.

CHAPTER 14

Page **377** Top: iStockphoto. Page **377** Right: iStockphoto. Page **377** Left: Superstock/Art Life Images. Page **382:** Serge Kozak/AGE Fotostock. Page **388:** Thomas D. Rossing. Page **392:** Richard Megna/Fundamental Photographs. Page **395:** Hornbil Images/Alamy. Page **399:** Martin Bough/Fundamental Photographs. Page **404:** Harold G. Craighead.

PART V OPENER

Page **558:** AguaSonic Acoustics/Photo Researchers, Inc.

CHAPTER 20

Page **560** Top: EpicStockMedia/Shutterstock. Page **560** Middle left: iStockphoto. Page **560** Middle: B. Benoit/Photo Researchers, Inc. Page **560** Middle right: David Parker/Photo Researchers, Inc. Page **560** Bottom left: Oleksiy Maksymenko/Alamy. Page **560** Bottom right: Aaron Kohr/Shutterstock. Page **561:** Dudarev Mikhail/Shutterstock. Page **562:**

Uri Haber-Schaim - From PSSC Physics 7th Edition by Haber-Schaim, Dodge, Gardner, Shore. Kendall/Hunt Publishing Company, Dubuque, Iowa (C)1991. Page **566:** Aflo Foto Agency/Alamy. Page **575:** B. Benoit/Photo Researchers, Inc. Page **576:** David Parker/Photo Researchers, Inc. Page **581:** NOAA/AP Images. Page **583:** Space Telescope Science Institute.

CHAPTER 21

Page **591** Top: Peter Aprahamian/SPL/Photo Researchers, Inc. Page **591** Middle left: Pearson Science/Creative Digital Vision. Page **591** Middle: Richard Megna/Fundamental Photographs. Page **591** Middle right: Randy Knight. Page **591** Bottom left: Olga Miltsova/Shutterstock. Page **591** Bottom right: Pearson Science/Eric Schrader. Page **593:** Richard Megna/Fundamental Photographs. Page **594:** Prelinger Archives. Page **596:** Education Development Center, Inc. Page **597:** Pearson Science/Creative Digital Vision. Page **603:** Brian Atkinson/Alamy. Page **609:** Pearson Science/Eric Schrader. Page **611:** Richard Megna/Fundamental Photographs. Page **614** Both: Randy Knight.

Index

N

Q

R

S

T

U

Astronomical Data

Planetary body	Mean distance from sun (m)	Period (years)	Mass (kg)	Mean radius (m)
Sun	—	—	1.99×10^{30}	6.96×10^{8}
Moon	3.84×10^{8}*	27.3 days	7.36×10^{22}	1.74×10^{6}
Mercury	5.79×10^{10}	0.241	3.18×10^{23}	2.43×10^{6}
Venus	1.08×10^{11}	0.615	4.88×10^{24}	6.06×10^{6}
Earth	1.50×10^{11}	1.00	5.98×10^{24}	6.37×10^{6}
Mars	2.28×10^{11}	1.88	6.42×10^{23}	3.37×10^{6}
Jupiter	7.78×10^{11}	11.9	1.90×10^{27}	6.99×10^{7}
Saturn	1.43×10^{12}	29.5	5.68×10^{26}	5.85×10^{7}
Uranus	2.87×10^{12}	84.0	8.68×10^{25}	2.33×10^{7}
Neptune	4.50×10^{12}	165	1.03×10^{26}	2.21×10^{7}

*Distance from earth

Typical Coefficients of Friction

Material	Static μ_s	Kinetic μ_k	Rolling μ_r
Rubber on concrete	1.00	0.80	0.02
Steel on steel (dry)	0.80	0.60	0.002
Steel on steel (lubricated)	0.10	0.05	
Wood on wood	0.50	0.20	
Wood on snow	0.12	0.06	
Ice on ice	0.10	0.03	

Melting/Boiling Temperatures and Heats of Transformation

Substance	T_m (°C)	L_f (J/kg)	T_b (°C)	L_v (J/kg)
Water	0	3.33×10^{5}	100	22.6×10^{5}
Nitrogen (N_2)	−210	0.26×10^{5}	−196	1.99×10^{5}
Ethyl alcohol	−114	1.09×10^{5}	78	8.79×10^{5}
Mercury	−39	0.11×10^{5}	357	2.96×10^{5}
Lead	328	0.25×10^{5}	1750	8.58×10^{5}

Properties of Materials

Substance	ρ (kg/m^3)	c (J/kg K)
Air at STP*	1.28	
Ethyl alcohol	790	2400
Gasoline	680	
Glycerin	1260	
Mercury	13,600	140
Oil (typical)	900	
Seawater	1030	
Water	1000	4190
Aluminum	2700	900
Copper	8920	385
Gold	19,300	129
Ice	920	2090
Iron	7870	449
Lead	11,300	128
Silicon	2330	703

*Standard temperature (0°C) and pressure (1 atm)

Molar Specific Heats of Gases

Gas	C_P (J/mol K)	C_V (J/mol K)
Monatomic Gases		
He	20.8	12.5
Ne	20.8	12.5
Ar	20.8	12.5
Diatomic Gases		
H_2	28.7	20.4
N_2	29.1	20.8
O_2	29.2	20.9

Indices of Refraction

Material	Index of refraction
Vacuum	1 exactly
Air	1.0003
Water	1.33
Glass	1.50
Diamond	2.42

Resistivity and Conductivity of Conductors

Metals	Resistivity (Ω m)	Conductivity (Ω^{-1} m^{-1})
Aluminum	2.8×10^{-8}	3.5×10^{7}
Copper	1.7×10^{-8}	6.0×10^{7}
Gold	2.4×10^{-8}	4.1×10^{7}
Iron	9.7×10^{-8}	1.0×10^{7}
Silver	1.6×10^{-8}	6.2×10^{7}
Tungsten	5.6×10^{-8}	1.8×10^{7}
Nichrome	1.5×10^{-6}	6.7×10^{5}
Carbon	3.5×10^{-5}	2.9×10^{4}

Atomic and Nuclear Data

Atom	Z	Mass (u)	Mass (MeV/c^2)
Electron	—	0.00055	0.51
Proton	—	1.00728	938.28
Neutron	—	1.00866	939.57
^{1}H	1	1.00783	938.79
^{2}H	1	2.01410	
^{4}He	2	4.00260	
^{12}C	6	12.00000	
^{14}C	6	14.00324	
^{14}N	7	14.00307	
^{16}O	8	15.99492	
^{20}Ne	10	19.99244	
^{27}Al	13	26.98154	
^{40}Ar	18	39.96238	
^{207}Pb	82	206.97444	
^{238}U	92	238.05078	

Hydrogen Atom Energies and Radii

n	E_n (eV)	r_n (nm)
1	−13.60	0.053
2	−3.40	0.212
3	−1.51	0.476
4	−0.85	0.848
5	−0.54	1.322

Work Functions of Metals

Metal	E_0 (eV)
Potassium	2.30
Sodium	2.75
Aluminum	4.28
Tungsten	4.55
Iron	4.65
Copper	4.70
Gold	5.10